Applied Calc Business, Management, Life and the Social Sciences

Waner | Costenoble | Crum

CENGAGE Learning™

Australia • Brazil • Japan • Korea • Mexico • Singapore • Spain • United Kingdom • United States

CENGAGE Learning™

Applied Calculus for Business, Management, Life and the Social Sciences

APPLIED CALCULUS, Fifth Edition
Stefan Waner | Steven R. Costenoble

Executive Editors:
Maureen Staudt
Michael Stranz

Senior Project Development Manager:
Linda deStefano

Marketing Specialist:
Courtney Sheldon

Senior Production/Manufacturing Manager:
Donna M. Brown

PreMedia Manager:
Joel Brennecke

Sr. Rights Acquisition Account Manager:
Todd Osborne

Cover Image:
Getty Images*

*Unless otherwise noted, all cover images used by Custom Solutions, a part of Cengage Learning, have been supplied courtesy of Getty Images with the exception of the Earthview cover image, which has been supplied by the National Aeronautics and Space Administration (NASA).

ISBN-13: 978-1-111-47088-3

ISBN-10: 1-111-47088-X

Cengage Learning
5191 Natorp Boulevard
Mason, Ohio 45040
USA

Cengage Learning is a leading provider of customized learning solutions with office locations around the globe, including Singapore, the United Kingdom, Australia, Mexico, Brazil, and Japan. Locate your local office at: **international.cengage.com/region.**

Cengage Learning products are represented in Canada by Nelson Education, Ltd.
For your lifelong learning solutions, visit **www.cengage.com /custom.**
Visit our corporate website at **www.cengage.com.**

Printed in the United States of America

TABLE OF CONTENTS

PREFACE

Applied Calculus, Fifth Edition, is intended as a one- or two-term course for students majoring in business, the social sciences, or the liberal arts. Like earlier editions, the Fifth Edition of *Applied Calculus* is designed to address the challenge of generating enthusiasm and mathematical sophistication in an audience that is too often under-prepared for and uninspired by traditional mathematics courses. We meet this challenge by focusing on real-life applications that students can relate to, by presenting mathematical concepts intuitively and thoroughly, and by employing a writing style that is informal, engaging, and occasionally humorous.

In renewing our commitment to these goals, we have further improved the text by revising the Table of Contents—combining some sections and rearranging others—to promote better flow and organization, enhancing the technology notes to include step-by-step instructions at point of use, and adding notes to the student to provide further clarification and explanation where necessary. Please see the "New to This Edition" section that follows for additional information about these changes.

The Fifth Edition continues to implement support for a wide range of instructional paradigms: from settings using little or no technology to courses taught in computerized classrooms, and from classes in which a single form of technology is used exclusively to those incorporating several technologies. We feature three forms of technology in this text—TI-83/84 Plus graphing calculators, Excel spreadsheets, and powerful online utilities created specifically for this book—in a way that allows them to be integrated or omitted. In particular, our comprehensive support for Excel, both in the text and online, is highly relevant for students who are studying business and economics, where skill with spreadsheets may be vital to their future careers.

Our Approach to Pedagogy

Real World Orientation We are particularly proud of the diversity, breadth, and abundance of examples and exercises included in this edition. A large number are based on real, referenced data from business, economics, the life sciences, and the social sciences. Examples and exercises based on dated information have generally been replaced by more current versions; applications based on unique or historically interesting data have been kept.

Adapting real data for pedagogical use can be tricky; available data can be numerically complex, intimidating for students, or incomplete. We have modified and streamlined many of the real world applications, rendering them as tractable as any "made-up" application. At the same time, we have been careful to strike a pedagogically sound balance between applications based on real data and more traditional "generic" applications. Thus, the density and selection of real data-based applications have been tailored to the pedagogical goals and appropriate difficulty level for each section.

Readability We would like students to read this book. We would also like students to *enjoy* reading this book. Thus, we have written the book in a conversational and student-oriented style, and have made frequent use of question-and-answer dialogues to encourage the development of the student's mathematical curiosity and intuition. We hope that this text will give the student insight into how a mathematician develops and thinks about mathematical ideas and their applications.

Rigor We feel that mathematical rigor can work easily in conjunction with the kind of applied focus and conceptual approach that are earmarks of this book. We have, especially in the Fifth Edition, worked hard to ensure that we are always mathematically honest without being unnecessarily formal. Sometimes we do this through the question-and-answer dialogues and sometimes through the "Before we go on . . ." discussions that follow examples, but always in a manner designed to provoke the interest of the student.

Five Elements of Mathematical Pedagogy to Address Different Learning Styles The "Rule of Four" is a common theme in many texts. Implementing this approach, we discuss many of the central concepts **numerically**, **graphically** and **algebraically**, and clearly delineate these distinctions. The fourth element, **verbal communication** of mathematical concepts, is emphasized through our discussions on translating English sentences into mathematical statements, and our Communication and Reasoning exercises at the end of each section. A fifth element, **interactivity**, is integrated through expanded use of question-and-answer dialogues, but is seen most dramatically within the student Web site. Using this resource, students can use interactive tutorials specific to concepts and examples covered in sections and online utilities that automate a variety of tasks, from graphing to regression and matrix algebra. Added recently to the site are more challenging "game" tutorials, with randomized questions, scoring systems, and even "health points."

Exercise Sets The substantial collection of exercises provides a wealth of material that can be used to challenge students at almost every level of preparation, and includes everything from straightforward drill exercises to interesting and rather challenging applications. The exercise sets have been carefully graded to increase in complexity from basic exercises and exercises that are similar to examples in the text to more interesting and advanced ones, marked in this edition as "more advanced" for easy reference. There are also several much more difficult exercises, designated as "challenging." The advanced and challenging exercises encourage students to think beyond the straightforward situations and calculations in the earlier exercises. We have also included, in virtually every section, interesting applications based on real data, Communication and Reasoning exercises that help students articulate mathematical concepts, and exercises ideal for the use of technology.

Many of the scenarios used in application examples and exercises are revisited several times throughout the book. Thus, students will find themselves using a variety of techniques, from graphing through the use of derivatives and elasticity, to analyze the same application. Reusing scenarios and important functions provides unifying threads and shows students the complex texture of real-life problems.

New to This Edition

Content

- Chapter 1 (page 39): Chapter 1 now includes a new section, *Functions and Models*, in which we bring together a variety of applied topics such as revenue, profit, demand, supply, and change over time that occur throughout the book. We have also streamlined the rest of the chapter; functions are now introduced in a single section rather than two, and linear functions and models are similarly discussed in a single section.

- Chapter 4 (page 285): Derivatives of sums and constant multiples are now moved to the beginning of Chapter 4 on techniques of differentiation (where they properly belong).

- Chapter 5 (page 369): Second and higher order derivatives are now discussed in their own section with a great deal of added material on concavity and acceleration.

- Chapter 8 (page 583): We have changed the exposition of constrained maxima and minima to give more emphasis to Lagrange multipliers over substitution methods. Functions of several variables are now discussed in a single section rather than two.
- We now include the discussion of derivatives and integrals of functions involving absolute values in simple closed form, thus expanding the variety of functions available for modeling.
- We have expanded the list of additional optional sections, available to include in custom-published versions of the text, and now offer sections on: Taylor polynomials, the chain rule for multivariate calculus, calculus applied to probability, the extreme value theorem and optimization with boundary constraints for functions of several variables, and determinants and Cramer's rule.

Current Topics in the Applications

Case Study **Reducing Sulfur Emissions**

Norbert Schaefer/CORBIS

The Environmental Protection Agency (EPA) wishes to formulate a policy that will encourage utilities to reduce sulfur emissions. Its goal is to reduce annual emissions of sulfur dioxide by a total of 10 million tons from the current level of 25 million tons by imposing a fixed charge for every ton of sulfur released into the environment per year. As a consultant to the EPA, you must determine the amount to be charged per ton of sulfur emissions.

You would like first to know the cost to the utility industry of reducing sulfur emissions. In other words, you would like to have a cost function of the form

$$C(q) = \text{Cost of removing } q \text{ tons of sulfur dioxide.}$$

Unfortunately, you do not have such a function handy. You do, however, have the following data, which show the *marginal* cost (that is, the *rate of change* of cost) to the utility industry of reducing sulfur emissions at several levels of reduction.[68]

Reduction (tons) q	8,000,000	10,000,000	12,000,000
Marginal Cost ($ per ton) $C'(q)$	270	360	779

The table tells you that $C'(8{,}000{,}000) = \$270$ per ton, $C'(10{,}000{,}000) = \$360$ per ton, and $C'(12{,}000{,}000) = \$779$ per ton. Recalling that $C'(q)$ is the slope of the tangent to the graph of the cost function, you can see from the table that this slope is positive and increasing as q increases, so the graph of the cost function has the general shape shown in Figure 39.

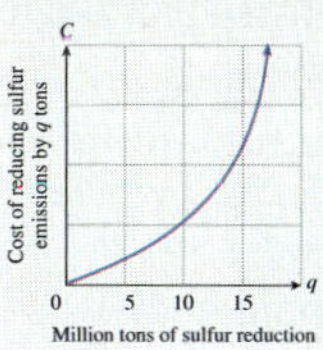

Figure 39

- **Case Studies** Each chapter ends with a section entitled "Case Study," an extended application that uses and illustrates the central ideas of the chapter, focusing on the development of mathematical models appropriate to the topics. The Case Studies have been extensively revised, and in many cases completely replaced by new ones that reflect topics of current interest.
- We have added numerous real data exercises and examples based on topics that are either of intense current interest or of general interest to contemporary students, including *Facebook, XBoxes, iPhones, eBay,* real estate foreclosures and home construction, subprime mortgages, stock market gyrations, travel to Cancun, and the oil industry in the United States and Mexico. (Also see the Index of Companies, in the inside back cover, of the corporations we reference in the applications.)

Exercises

76. ▼ ***Information Highway*** The amount of information transmitted each month in the early years of the Internet (1988 to 1994) can be modeled by the equation

$$q(t) = \frac{2e^{0.69t}}{3 + 1.5e^{-0.4t}} \qquad (0 \le t \le 6)$$

where q is the amount of information transmitted each month in billions of data packets and t is the number of years since the start of 1988.[33]

a. Use technology to estimate $q'(2)$.
b. Assume that it costs $5 to transmit a million packets of data. What is the marginal cost $C'(q)$?
c. How fast was the cost increasing at the start of 1990?

Money Stock *Exercises 77–80 are based on the following demand function for money (taken from a question on the GRE Economics Test):*

$$M_d = 2 \times y^{0.6} \times r^{-0.3} \times p$$

where

M_d = *demand for nominal money balances (money stock)*
y = *real income*
r = *an index of interest rates*
p = *an index of prices*

These exercises also use the idea of ***percentage rate of growth****:*

$$\text{Percentage Rate of Growth of } M = \frac{\text{Rate of Growth of } M}{M} = \frac{dM/dt}{M}$$

- Exercises that are not based entirely on examples in the text are designated as "more advanced" (and indicated by an icon in the exercise set) as a guide for students and instructors.
- We have expanded the exercise sets themselves and carefully reorganized them to gradually increase in level and to include more basic skills exercises that carefully follow the examples.

87. ▼ Formulate a simple procedure for deciding whether to apply first the chain rule, the product rule, or the quotient rule when finding the derivative of a function.

88. ▼ Give an example of a function f with the property that calculating $f'(x)$ requires use of the following rules in the given order: (1) the chain rule, (2) the quotient rule, and (3) the chain rule.

89. ◆ Give an example of a function f with the property that calculating $f'(x)$ requires use of the chain rule five times in succession.

90. ◆ What can you say about composites of linear functions?

- Many more of the exercises now have "hints" that either refer to an example in the text where a similar problem is solved, or offer some advice to the student.
- We have added numerous new "communication and reasoning" exercises—many dealing with common student errors and misconceptions—and further expanded the chapter review exercise sections.

Pedagogy

* **NOTE** See the section on exponents in the algebra review to brush up on negative and fractional exponents.

using Technology
See the Technology Guide at the end of the chapter for detailed instructions on how to calculate the average rate of change of the function in Example 3 using a TI-83/84 Plus or Excel. Here is an outline:

TI-83/84 Plus
`Y1=-8X2+144X+150`
Home screen: `(Y1(9.5)-Y1(8))/(9.5-8)`
[More details on page 280.]

Excel
Enter the headings t, $G(t)$, Rate of Change in A1–C1
t-values 8, 9.5 in A2–A3
`=-8*A2^2+144*A2+150`
in B2 and copy down to B3
`=(B3-B2)/(A3-A2)` in C2.
[More details on page 282.]

- **Supplementary Notes** We have added new Notes to the student, located in the side column of the text. These Notes include a wide variety of additional information for the student—further explanation or clarification of a concept, reminders of previously learned material or references for further study, and additional tips for using technology.
- **Technology Notes** have been enhanced to include step-by-step instructions and keystrokes at point of use, to enable better integration of graphing calculators and spreadsheets. As always, these notes can be omitted without loss of continuity.

Hallmark Features

- **Question-and-Answer Dialogue** We frequently use informal question-and-answer dialogues that anticipate the kind of questions that may occur to the student and also guide the student through the development of new concepts. This feature has been streamlined, as has the "Frequently Asked Questions" feature at the end of each section.

Q: *Do we always have to calculate the limit of the difference quotient to find a formula for the derivative function?*

A: As it turns out, no. In Section 4.1 we will start to look at shortcuts for finding derivatives that allow us to bypass the definition of the derivative in many cases.

- **Before we go on . . . feature** Most examples are followed by supplementary discussions, which may include a check on the answer, a discussion of the feasibility and significance of a solution, or an in-depth look at what the solution means.
- **Quick Examples** Most definition boxes include quick, straightforward examples that a student can use to solidify each new concept.

Functions with Equal Limits

If $f(x) = g(x)$ for all x except possibly $x = a$, then

$$\lim_{x \to a} f(x) = \lim_{x \to a} g(x).$$

Quick Example

$\dfrac{x^2 - 1}{x - 1} = x + 1$ for all x except $x = 1$. Write $\dfrac{x^2-1}{x-1}$ as $\dfrac{(x+1)(x-1)}{x-1}$ and cancel the $(x-1)$

Therefore,

$$\lim_{x \to 1} \frac{x^2 - 1}{x - 1} = \lim_{x \to 1} (x + 1) = 1 + 1 = 2.$$

- **Communication and Reasoning Exercises for Writing and Discussion** These are exercises designed to broaden the student's grasp of the mathematical concepts and develop modeling skills. They include exercises in which the student is asked to provide his or her own examples to illustrate a point or design an application with a given solution. They also include fill-in-the-blank type exercises and exercises that invite discussion and debate. These exercises often have no single correct answer.

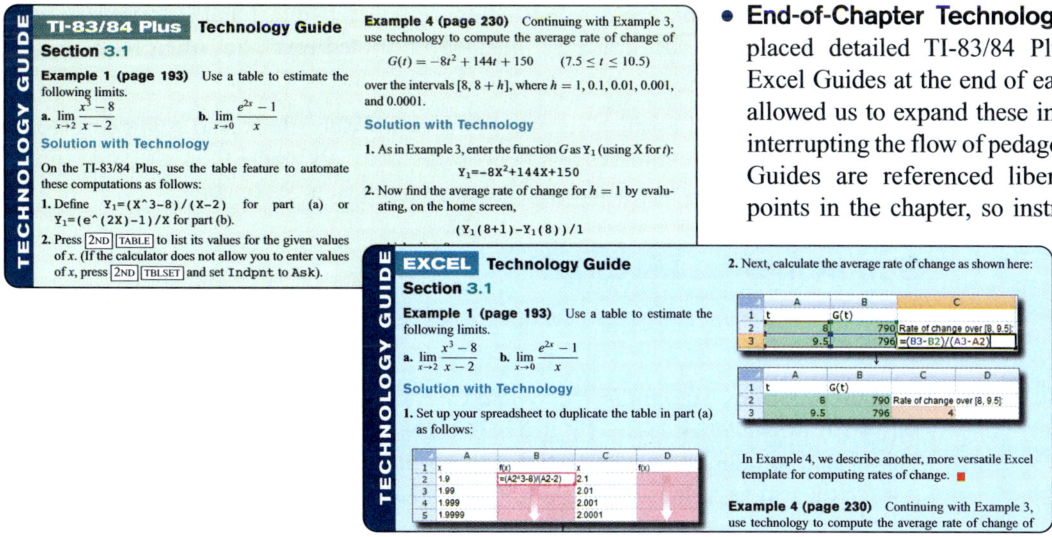

TECHNOLOGY GUIDE

TI-83/84 Plus Technology Guide

Section 3.1

Example 1 (page 193) Use a table to estimate the following limits.

a. $\lim_{x\to 2} \frac{x^3-8}{x-2}$ **b.** $\lim_{x\to 0} \frac{e^{2x}-1}{x}$

Solution with Technology

On the TI-83/84 Plus, use the table feature to automate these computations as follows:

1. Define Y1=(X^3-8)/(X-2) for part (a) or Y1=(e^(2X)-1)/X for part (b).
2. Press 2ND TABLE to list its values for the given values of x. (If the calculator does not allow you to enter values of x, press 2ND TBLSET and set Indpnt to Ask).

Example 4 (page 230) Continuing with Example 3, use technology to compute the average rate of change of

$G(t) = -8t^2 + 144t + 150 \quad (7.5 \le t \le 10.5)$

over the intervals $[8, 8+h]$, where $h = 1, 0.1, 0.01, 0.001,$ and 0.0001.

Solution with Technology

1. As in Example 3, enter the function G as Y1 (using X for t):

Y1=-8X²+144X+150

2. Now find the average rate of change for $h = 1$ by evaluating, on the home screen,

(Y1(8+1)-Y1(8))/1

TECHNOLOGY GUIDE

EXCEL Technology Guide

Section 3.1

Example 1 (page 193) Use a table to estimate the following limits.

a. $\lim_{x\to 2} \frac{x^3-8}{x-2}$ **b.** $\lim_{x\to 0} \frac{e^{2x}-1}{x}$

Solution with Technology

1. Set up your spreadsheet to duplicate the table in part (a) as follows:

	A	B	C	D
1	x	f(x)	x	f(x)
2	1.9	=(A2^3-8)/(A2-2)	2.1	
3	1.99		2.01	
4	1.999		2.001	
5	1.9999		2.0001	

2. Next, calculate the average rate of change as shown here:

	A	B	C
1	t	G(t)	
2	8	790	Rate of change over [8, 9.5]:
3	9.5	796	=(B3-B2)/(A3-A2)

	A	B	C	D
1	t	G(t)		
2	8	790	Rate of change over [8, 9.5]:	
3	9.5	796		4

In Example 4, we describe another, more versatile Excel template for computing rates of change. ■

Example 4 (page 230) Continuing with Example 3, use technology to compute the average rate of change of

- **End-of-Chapter Technology Guides** We have placed detailed TI-83/84 Plus and Microsoft® Excel Guides at the end of each chapter. This has allowed us to expand these instructions while not interrupting the flow of pedagogy in the text. These Guides are referenced liberally at appropriate points in the chapter, so instructors and students can easily use this material or not, as they prefer. Groups of exercises for which the use of technology is suggested or required appear throughout the exercise sets.

The Web Site

The authors' Web site, accessible through www.AppliedCalc.com, has been evolving for several years, with growing recognition. Students, raised in an environment in which computers suffuse both work and play, can use their Web browsers to engage with the material in an active way. The following features of the authors' Web site are fully integrated with the text and can be used as a personalized study resource:

- **Interactive Tutorials** Highly interactive tutorials, with guided exercises that parallel the text and a great deal of help and feedback to assist the student.
- **Game Versions of Tutorials** More challenging tutorials with randomized questions that work as games (complete with "health" scores, "health vials," and an assessment of one's performance at the end of the game) are offered alongside the traditional tutorials. These game tutorials, which mirror the traditional "more gentle" tutorials, do not give the student the answers, but instead offer hints in exchange for health points, so that just staying alive (not running out of health) can be quite challenging. Because the questions are randomized and scores are automatically calculated, these tutorials can be used for in-class quizzes, as the authors themselves have done.
- **Detailed Chapter Summaries** Comprehensive summaries with interactive elements review all the basic definitions and problem solving techniques discussed in each chapter. These are a terrific pre-test study tool for students.
- **Downloadable Excel Tutorials** Detailed Excel tutorials are available for almost every section of the book. These interactive tutorials expand on the examples given in the text.
- **Online Utilities** Our collection of easy-to-use online utilities, written in JavaScript, allows students to solve many of the technology-based application exercises directly on the Web page. The utilities available include a function grapher and evaluator that also does derivatives, regression tools, and a numerical integration tool. These utilities require nothing more than a standard Web browser.
- **Chapter True-False Quizzes** Short quizzes based on the key concepts in each chapter assist the student in further mastery of the material.

- **Supplemental Topics** We include complete interactive text and exercise sets for a selection of topics not ordinarily included in printed texts, but often requested by instructors.
- **Spanish** A parallel Spanish version of the entire Web site is also being developed. All of the chapter summaries and many of the tutorials, game tutorials, and utilities are already available in Spanish, with many more resources to come.

Supplemental Material

For Students

Student Solutions Manual *by Waner and Costenoble*
ISBN-10: **1439049904**, ISBN-13: **9781439049907**
The student solutions manual provides worked-out solutions to the odd-numbered exercises in the text as well as complete solutions to all the chapter review tests.

Microsoft Excel Computer Laboratory Manual *by Anne D. Henriksen*
ISBN-10: **0538733209**, ISBN-13: **9780538733205**
This laboratory manual illustrates how Microsoft Excel can be used to solve real-world problems in a variety of scientific, technical, and business disciplines. It provides hands-on experience to demonstrate for students that calculus is a valuable tool for solving practical, real-world problems, while helping students increase their knowledge of Microsoft Excel. The manual is a set of self-contained computer exercises that are meant to be used over the course of a 15-week semester in a separate, 75-minute computer laboratory period. The weekly labs parallel the material in the text.

NetTutor™
Cengage Learning is pleased to provide students with online tutoring and chatting capabilities through NetTutor. NetTutor utilizes a WorldWideWhiteboard application, a Web-based application that allows students and tutors to interact with one another through text and images, and also offers audio and video over Internet Protocol or VOIP, where voice is transmitted through the Internet. The WorldWide Whiteboard software offers the ability to easily create custom groups, which contain course-specific functions, symbol palettes and buttons.

For Instructors

Instructor's Solution Manual *by Waner and Costenoble*
ISBN-10: **1439049831**, ISBN-13: **9781439049839**
The instructor's solutions manual provides worked-out solutions to all of the exercises in the text.

PowerLecture CD with ExamView
ISBN-10: **1439049890**, ISBN-13: **9781439049891**
This comprehensive CD-ROM contains the Instructor's Solutions Manual, PowerPoint lecture notes, and ExamView computerized testing to create, deliver, and customize tests. The PowerLecture CD also includes a multimedia library containing all of the art from the book in MS PowerPoint as well as individual JPEG files.

Test Bank
The test bank—available on the companion Web site—contains numerous multiple choice and free response questions for those instructors who prefer a more traditional method of test preparation.

Enhanced WebAssign

Instant feedback and ease of use are just two reasons why WebAssign is the most widely used homework system in higher education. WebAssign allows you to assign, collect, grade, and record homework assignments via the Web. Now this proven homework system has been enhanced to include links to textbook sections, video examples, and problem-specific tutorials. Enhanced WebAssign is more than a homework system—it is a complete learning system for math students.

Acknowledgments

This project would not have been possible without the contributions and suggestions of numerous colleagues, students, and friends. We are particularly grateful to our colleagues at Hofstra and elsewhere who used and gave us useful feedback on previous editions. We are also grateful to everyone at Cengage Learning for their encouragement and guidance throughout the project. Specifically, we would like to thank Carolyn Crockett and Liz Covello for their unflagging enthusiasm and Jeannine Lawless for whipping the book into shape.

We would also like to thank Jerrold Grossman for his meticulous critique of the manuscript, Lynn Lustberg and the production team for their patience in dealing with our often shifting demands, and the numerous reviewers and accuracy checkers who provided many helpful suggestions that have shaped the development of this book.

Doug Burkholder, *Lenoir-Rhyne University*
Leslie Cohn, *The Citadel*
Melanie Fulton, *High Point University*
Jerrold W. Grossman, *Oakland University*
Celeste Hernandez, *Richland College*
Dean Leoni, *Edmonds Community College*
Michael J. Kallaher, *Washington State University*
Karla Karstans, *University of Vermont*
Vincent Koehler, *University of Vermont*
Karl Kruppstadt, *University of Minnesota, Duluth*
Robert H. Lewis, *Fordham University*
David Miller, *William Paterson University*
Phillip Miller, *Indiana University Southeast*
Jack Y. Narayan, *SUNY Oswego*
Sergei Ovchinnikov, *San Francisco State University*
Lauri Papay, *Santa Clara University*
Jean Nicholas Pesticau, *Suffolk County Community College*
Leela Rakesh, *Central Michigan University*
Nelson de la Rosa, *Miami Dade College, Kendall*
Arthur Rosenthal, *Salem State College*
Carol H. Serotta, *Cabrini College*
Mary Ann Teel, *University of North Texas*
Tzvetalin S. Vassilev, *North Carolina Central University*
Marie A. Vitulli, *University of Oregon*
Richard West, *Francis Marion University*
Donna Wilson, *University of Texas at El Paso*

Stefan Waner
Steven R. Costenoble

0 Precalculus Review

DreamPictures/Taxi/Getty Images

Web Site

www.AppliedCalc.org

- At the Web site you will find section-by-section interactive tutorials for further study and practice.

Introduction

In this appendix we review some topics from algebra that you need to know to get the most out of this book. This appendix can be used either as a refresher course or as a reference.

There is one crucial fact you must always keep in mind: The letters used in algebraic expressions stand for numbers. All the rules of algebra are just facts about the arithmetic of numbers. If you are not sure whether some algebraic manipulation you are about to do is legitimate, try it first with numbers. If it doesn't work with numbers, it doesn't work.

0.1 Real Numbers

The **real numbers** are the numbers that can be written in decimal notation, including those that require an infinite decimal expansion. The set of real numbers includes all integers, positive and negative; all fractions; and the irrational numbers, those with decimal expansions that never repeat. Examples of irrational numbers are

$$\sqrt{2} = 1.414213562373\ldots$$

and

$$\pi = 3.141592653589\ldots$$

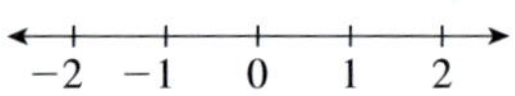

Figure 1

It is very useful to picture the real numbers as points on a line. As shown in Figure 1, larger numbers appear to the right, in the sense that if $a < b$ then the point corresponding to b is to the right of the one corresponding to a.

Intervals

Some subsets of the set of real numbers, called **intervals**, show up quite often and so we have a compact notation for them.

Interval Notation

Here is a list of types of intervals along with examples.

	Interval	*Description*	*Picture*	*Example*
Closed	$[a, b]$	Set of numbers x with $a \le x \le b$	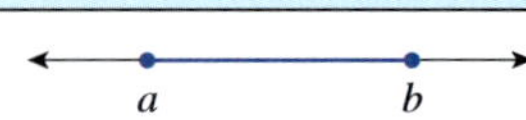 (includes end points)	$[0, 10]$
Open	(a, b)	Set of numbers x with $a < x < b$	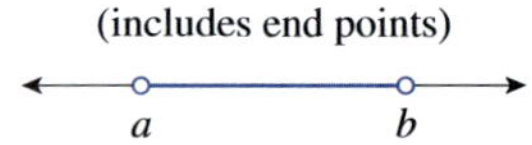 (excludes end points)	$(-1, 5)$
Half-Open	$(a, b]$	Set of numbers x with $a < x \le b$		$(-3, 1]$
	$[a, b)$	Set of numbers x with $a \le x < b$		$[0, 5)$

Infinite	$[a, +\infty)$	Set of numbers x with $a \le x$	a	$[10, +\infty)$
	$(a, +\infty)$	Set of numbers x with $a < x$	a	$(-3, +\infty)$
	$(-\infty, b]$	Set of numbers x with $x \le b$	b	$(-\infty, -3]$
	$(-\infty, b)$	Set of numbers x with $x < b$	b	$(-\infty, 10)$
	$(-\infty, +\infty)$	Set of all real numbers		$(-\infty, +\infty)$

Operations

There are five important operations on real numbers: addition, subtraction, multiplication, division, and exponentiation. "Exponentiation" means raising a real number to a power; for instance, $3^2 = 3 \cdot 3 = 9$; $2^3 = 2 \cdot 2 \cdot 2 = 8$.

A note on technology: Most graphing calculators and spreadsheets use an asterisk * for multiplication and a caret sign ^ for exponentiation. Thus, for instance, 3×5 is entered as `3*5`, $3x$ as `3*x`, and 3^2 as `3^2`.

When we write an expression involving two or more operations, like

$$2 \cdot 3 + 4$$

or

$$\frac{2 \cdot 3^2 - 5}{4 - (-1)}$$

we need to agree on the order in which to do the operations. Does $2 \cdot 3 + 4$ mean $(2 \cdot 3) + 4 = 10$ or $2 \cdot (3 + 4) = 14$? We all agree to use the following rules for the order in which we do the operations.

Standard Order of Operations

Parentheses and Fraction Bars First, calculate the values of all expressions inside parentheses or brackets, working from the innermost parentheses out, before using them in other operations. In a fraction, calculate the numerator and denominator separately before doing the division.

Quick Examples

1. $6(2 + [3 - 5] - 4) = 6(2 + (-2) - 4) = 6(-4) = -24$
2. $\dfrac{(4-2)}{3(-2+1)} = \dfrac{2}{3(-1)} = \dfrac{2}{-3} = -\dfrac{2}{3}$
3. $3/(2+4) = \dfrac{3}{2+4} = \dfrac{3}{6} = \dfrac{1}{2}$
4. $(x + 4x)/(y + 3y) = 5x/(4y)$

Exponents Next, perform exponentiation.

Quick Examples

1. $2 + 4^2 = 2 + 16 = 18$
2. $(2 + 4)^2 = 6^2 = 36$
Note the difference.

3. $2\left(\frac{3}{4-5}\right)^2 = 2\left(\frac{3}{-1}\right)^2 = 2(-3)^2 = 2 \times 9 = 18$

4. $2(1 + 1/10)^2 = 2(1.1)^2 = 2 \times 1.21 = 2.42$

Multiplication and Division Next, do all multiplications and divisions, from left to right.

Quick Examples

1. $2(3 - 5)/4 \cdot 2 = 2(-2)/4 \cdot 2$ — Parentheses first
$= -4/4 \cdot 2$ — Left-most product
$= -1 \cdot 2 = -2$ — Multiplications and divisions, left to right

2. $2(1 + 1/10)^2 \times 2/10 = 2(1.1)^2 \times 2/10$ — Parentheses first
$= 2 \times 1.21 \times 2/10$ — Exponent
$= 4.84/10 = 0.484$ — Multiplications and divisions, left to right

3. $4\frac{2(4-2)}{3(-2 \cdot 5)} = 4\frac{2(2)}{3(-10)} = 4\frac{4}{-30} = \frac{16}{-30} = -\frac{8}{15}$

Addition and Subtraction Last, do all additions and subtractions, from left to right.

Quick Examples

1. $2(3 - 5)^2 + 6 - 1 = 2(-2)^2 + 6 - 1 = 2(4) + 6 - 1 = 8 + 6 - 1 = 13$

2. $\left(\frac{1}{2}\right)^2 - (-1)^2 + 4 = \frac{1}{4} - 1 + 4 = -\frac{3}{4} + 4 = \frac{13}{4}$

3. $3/2 + 4 = 1.5 + 4 = 5.5$
4. $3/(2 + 4) = 3/6 = 1/2 = 0.5$
Note the difference.

5. $4/2^2 + (4/2)^2 = 4/2^2 + 2^2 = 4/4 + 4 = 1 + 4 = 5$

Entering Formulas

Any good calculator or spreadsheet will respect the standard order of operations. However, we must be careful with division and exponentiation and use parentheses as necessary. The following table gives some examples of simple mathematical expressions and their equivalents in the functional format used in most graphing calculators, spreadsheets, and computer programs.

Mathematical Expression	*Formula*	*Comments*
$\frac{2}{3-x}$	`2/(3-x)`	Note the use of parentheses instead of the fraction bar. If we omit the parentheses, we get the expression shown next.
$\frac{2}{3}-x$	`2/3-x`	The calculator follows the usual order of operations.
$\frac{2}{3\times 5}$	`2/(3*5)`	Putting the denominator in parentheses ensures that the multiplication is carried out first. The asterisk is usually used for multiplication in graphing calculators and computers.
$\frac{2}{x}\times 5$	`(2/x)*5`	Putting the fraction in parentheses ensures that it is calculated first. Some calculators will interpret `2/3*5` as $\frac{2}{3\times 5}$, but `2/3(5)` as $\frac{2}{3}\times 5$.
$\frac{2-3}{4+5}$	`(2-3)/(4+5)`	Note once again the use of parentheses in place of the fraction bar.
2^3	`2^3`	The caret ^ is commonly used to denote exponentiation.
2^{3-x}	`2^(3-x)`	Be careful to use parentheses to tell the calculator where the exponent ends. Enclose the *entire exponent* in parentheses.
2^3-x	`2^3-x`	Without parentheses, the calculator will follow the usual order of operations: exponentiation and then subtraction.
3×2^{-4}	`3*2^(-4)`	On some calculators, the negation key is separate from the minus key.
$2^{-4\times 3}\times 5$	`2^(-4*3)*5`	Note once again how parentheses enclose the entire exponent.
$100\left(1+\frac{0.05}{12}\right)^{60}$	`100*(1+0.05/12)^60`	This is a typical calculation for compound interest.
$PV\left(1+\frac{r}{m}\right)^{mt}$	`PV*(1+r/m)^(m*t)`	This is the compound interest formula. PV is understood to be a single number (present value) and not the product of P and V (or else we would have used `P*V`).

$\frac{2^{3-2}\times 5}{y-x}$	`2^(3-2)*5/(y-x)`
	or
	`(2^(3-2)*5)/(y-x)`
		Notice again the use of parentheses to hold the denominator together. We could also have enclosed the numerator in parentheses, although this is optional. (Why?)

$\frac{2^y+1}{2-4^{3x}}$	`(2^y+1)/(2-4^(3*x))`	Here, it is necessary to enclose both the numerator and the denominator in parentheses.
$2^y+\frac{1}{2}-4^{3x}$	`2^y+1/2-4^(3*x)`	This is the effect of leaving out the parentheses around the numerator and denominator in the previous expression.

Accuracy and Rounding

When we use a calculator or computer, the results of our calculations are often given to far more decimal places than are useful. For example, suppose we are told that a square has an area of 2.0 square feet and we are asked how long its sides are. Each side is the square root of the area, which the calculator tells us is

$$\sqrt{2} \approx 1.414213562$$

However, the measurement of 2.0 square feet is probably accurate to only two digits, so our estimate of the lengths of the sides can be no more accurate than that. Therefore, we round the answer to two digits:

$$\text{Length of one side} \approx 1.4 \text{ feet}$$

The digits that follow 1.4 are meaningless. The following guide makes these ideas more precise.

Significant Digits, Decimal Places, and Rounding

The number of **significant digits** in a decimal representation of a number is the number of digits that are not leading zeros after the decimal point (as in .0005) or trailing zeros before the decimal point (as in 5,400,000). We say that a value is **accurate to *n* significant digits** if only the first *n* significant digits are meaningful.

When to Round

After doing a computation in which all the quantities are accurate to no more than *n* significant digits, round the final result to *n* significant digits.

Quick Examples

1. 0.00067 has two significant digits. *The 000 before 67 are leading zeros.*
2. 0.000670 has three significant digits. *The 0 after 67 is significant.*
3. 5,400,000 has two or more significant digits. *We can't say how many of the zeros are trailing.**
4. 5,400,001 has 7 significant digits. *The string of zeros is not trailing.*
5. Rounding 63,918 to three significant digits gives 63,900.
6. Rounding 63,958 to three significant digits gives 64,000.
7. $\pi = 3.141592653\ldots$ $\frac{22}{7} = 3.142857142\ldots$ Therefore, $\frac{22}{7}$ is an approximation of π that is accurate to only three significant digits (3.14).
8. $4.02(1 + 0.02)^{1.4} \approx 4.13$ *We rounded to three significant digits.*

*If we obtained 5,400,000 by rounding 5,401,011, then it has three significant digits because the zero after the 4 is significant. On the other hand, if we obtained it by rounding 5,411,234, then it has only two significant digits. The use of scientific notation avoids this ambiguity: 5.40×10^6 (or 5.40 E6 on a calculator or computer) is accurate to three digits and 5.4×10^6 is accurate to two.

One more point, though: If, in a long calculation, you round the intermediate results, your final answer may be even less accurate than you think. As a general rule,

When calculating, don't round intermediate results. Rather, use the most accurate results obtainable or have your calculator or computer store them for you.

When you are done with the calculation, *then* round your answer to the appropriate number of digits of accuracy.

0.1 EXERCISES

Calculate each expression in Exercises 1–24, giving the answer as a whole number or a fraction in lowest terms.

1. $2(4+(-1))(2\cdot -4)$

2. $3+([4-2]\cdot 9)$

3. `20/(3*4)-1`

4. `2-(3*4)/10`

5. $\dfrac{3+([3+(-5)])}{3-2\times 2}$

6. $\dfrac{12-(1-4)}{2(5-1)\cdot 2-1}$

7. `(2-5*(-1))/1-2*(-1)`

8. `2-5*(-1)/(1-2*(-1))`

9. $2\cdot(-1)^2/2$

10. $2+4\cdot 3^2$

11. $2\cdot 4^2+1$

12. $1-3\cdot(-2)^2\times 2$

13. `3^2+2^2+1`

14. `2^(2^2-2)`

15. $\dfrac{3-2(-3)^2}{-6(4-1)^2}$

16. $\dfrac{1-2(1-4)^2}{2(5-1)^2\cdot 2}$

17. `10*(1+1/10)^3`

18. `121/(1+1/10)^2`

19. $3\left(\dfrac{-2\cdot 3^2}{-(4-1)^2}\right)$

20. $-\left(\dfrac{8(1-4)^2}{-9(5-1)^2}\right)$

21. $3\left(1-\left(-\dfrac{1}{2}\right)^2\right)^2+1$

22. $3\left(\dfrac{1}{9}-\left(\dfrac{2}{3}\right)^2\right)^2+1$

23. `(1/2)^2-1/2^2`

24. `2/(1^2)-(2/1)^2`

Convert each expression in Exercises 25–50 into its technology formula equivalent as in the table in the text.

25. $3\times(2-5)$

26. $4+\dfrac{5}{9}$

27. $\dfrac{3}{2-5}$

28. $\dfrac{4-1}{3}$

29. $\dfrac{3-1}{8+6}$

30. $3+\dfrac{3}{2-9}$

31. $3-\dfrac{4+7}{8}$

32. $\dfrac{4\times 2}{\left(\frac{2}{3}\right)}$

33. $\dfrac{2}{3+x}-xy^2$

34. $3+\dfrac{3+x}{xy}$

35. $3.1x^3-4x^{-2}-\dfrac{60}{x^2-1}$

36. $2.1x^{-3}-x^{-1}+\dfrac{x^2-3}{2}$

37. $\dfrac{\left(\frac{2}{3}\right)}{5}$

38. $\dfrac{2}{\left(\frac{3}{5}\right)}$

39. $3^{4-5}\times 6$

40. $\dfrac{2}{3+5^{7-9}}$

41. $3\left(1+\dfrac{4}{100}\right)^{-3}$

42. $3\left(\dfrac{1+4}{100}\right)^{-3}$

43. $3^{2x-1}+4^x-1$

44. $2^{x^2}-(2^{2x})^2$

45. 2^{2x^2-x+1}

46. $2^{2x^2-x}+1$

47. $\dfrac{4e^{-2x}}{2-3e^{-2x}}$

48. $\dfrac{e^{2x}+e^{-2x}}{e^{2x}-e^{-2x}}$

49. $3\left(1-\left(-\dfrac{1}{2}\right)^2\right)^2+1$

50. $3\left(\dfrac{1}{9}-\left(\dfrac{2}{3}\right)^2\right)^2+1$

0.2 Exponents and Radicals

In Section 1 we discussed exponentiation, or "raising to a power"; for example, $2^3=2\cdot 2\cdot 2$. In this section we discuss the algebra of exponentials more fully. First, we look at *integer* exponents: cases in which the powers are positive or negative whole numbers.

Integer Exponents

Positive Integer Exponents

If a is any real number and n is any positive integer, then by a^n we mean the quantity $a \cdot a \cdot \cdots \cdot a$ (n times); thus, $a^1 = a, a^2 = a \cdot a, a^5 = a \cdot a \cdot a \cdot a \cdot a$. In the expression a^n the number n is called the **exponent**, and the number a is called the **base**.

Quick Examples

$3^2 = 9$ $\qquad$ $2^3 = 8$

$0^{34} = 0$ $\qquad$ $(-1)^5 = -1$

$10^3 = 1{,}000$ $\qquad$ $10^5 = 100{,}000$

Negative Integer Exponents

If a is any real number *other than zero* and n is any positive integer, then we define

$$a^{-n} = \frac{1}{a^n} = \frac{1}{a \cdot a \cdot \cdots \cdot a} \quad (n \text{ times})$$

Quick Examples

$2^{-3} = \frac{1}{2^3} = \frac{1}{8}$ $\qquad$ $1^{-27} = \frac{1}{1^{27}} = 1$

$x^{-1} = \frac{1}{x^1} = \frac{1}{x}$ $\qquad$ $(-3)^{-2} = \frac{1}{(-3)^2} = \frac{1}{9}$

$y^7y^{-2} = y^7\frac{1}{y^2} = y^5$ $\qquad$ 0^{-2} is not defined

Zero Exponent

If a is any real number other than zero, then we define

$$a^0 = 1$$

Quick Examples

$3^0 = 1$ $\qquad$ $1{,}000{,}000^0 = 1$

0^0 is not defined

When combining exponential expressions, we use the following identities.

Exponent Identity	Quick Examples
1. $a^m a^n = a^{m+n}$	$2^3 2^2 = 2^{3+2} = 2^5 = 32$ $x^3 x^{-4} = x^{3-4} = x^{-1} = \frac{1}{x}$ $\frac{x^3}{x^{-2}} = x^3 \frac{1}{x^{-2}} = x^3 x^2 = x^5$
2. $\frac{a^m}{a^n} = a^{m-n}$ if $a \neq 0$	$\frac{4^3}{4^2} = 4^{3-2} = 4^1 = 4$ $\frac{x^3}{x^{-2}} = x^{3-(-2)} = x^5$ $\frac{3^2}{3^4} = 3^{2-4} = 3^{-2} = \frac{1}{9}$
3. $(a^n)^m = a^{nm}$	$(3^2)^2 = 3^4 = 81$ $(2^x)^2 = 2^{2x}$
4. $(ab)^n = a^n b^n$	$(4 \cdot 2)^2 = 4^2 2^2 = 64$ $(-2y)^4 = (-2)^4 y^4 = 16y^4$
5. $\left(\frac{a}{b}\right)^n = \frac{a^n}{b^n}$ if $b \neq 0$	$\left(\frac{4}{3}\right)^2 = \frac{4^2}{3^2} = \frac{16}{9}$ $\left(\frac{x}{-y}\right)^3 = \frac{x^3}{(-y)^3} = -\frac{x^3}{y^3}$

Caution

- In the first two identities, the bases of the expressions must be the same. For example, the first gives $3^2 3^4 = 3^6$, but does *not* apply to $3^2 4^2$.
- People sometimes invent their own identities, such as $a^m + a^n = a^{m+n}$, which is wrong! (Try it with $a = m = n = 1$.) If you wind up with something like $2^3 + 2^4$, you are stuck with it; there are no identities around to simplify it further. (You can factor out 2^3, but whether or not that is a simplification depends on what you are going to do with the expression next.)

EXAMPLE 1 Combining the Identities

$$\frac{(x^2)^3}{x^3} = \frac{x^6}{x^3} \quad \text{By (3)}$$
$$= x^{6-3} \quad \text{By (2)}$$
$$= x^3$$

$$\frac{(x^4 y)^3}{y} = \frac{(x^4)^3 y^3}{y} \quad \text{By (4)}$$
$$= \frac{x^{12} y^3}{y} \quad \text{By (3)}$$
$$= x^{12} y^{3-1} \quad \text{By (2)}$$
$$= x^{12} y^2$$

EXAMPLE 2 Eliminating Negative Exponents

Simplify the following and express the answer using no negative exponents.

a. $\dfrac{x^4y^{-3}}{x^5y^2}$ **b.** $\left(\dfrac{x^{-1}}{x^2y}\right)^5$

Solution

a. $\dfrac{x^4y^{-3}}{x^5y^2} = x^{4-5}y^{-3-2} = x^{-1}y^{-5} = \dfrac{1}{xy^5}$

b. $\left(\dfrac{x^{-1}}{x^2y}\right)^5 = \dfrac{(x^{-1})^5}{(x^2y)^5} = \dfrac{x^{-5}}{x^{10}y^5} = \dfrac{1}{x^{15}y^5}$

Radicals

If a is any non-negative real number, then its **square root** is the non-negative number whose square is a. For example, the square root of 16 is 4, because $4^2 = 16$. We write the square root of n as $\sqrt{n}$. (Roots are also referred to as **radicals.**) It is important to remember that $\sqrt{n}$ is never negative. Thus, for instance, $\sqrt{9}$ is 3, and not -3, even though $(-3)^2 = 9$. If we want to speak of the "negative square root" of 9, we write it as $-\sqrt{9} = -3$. If we want to write both square roots at once, we write $\pm\sqrt{9} = \pm 3$.

The **cube root** of a real number a is the number whose cube is a. The cube root of a is written as $\sqrt[3]{a}$ so that, for example, $\sqrt[3]{8} = 2$ (because $2^3 = 8$). Note that we can take the cube root of any number, positive, negative, or zero. For instance, the cube root of -8 is $\sqrt[3]{-8} = -2$ because $(-2)^3 = -8$. Unlike square roots, the cube root of a number may be negative. In fact, the cube root of a always has the same sign as a.

Higher roots are defined similarly. The **fourth root** of the *non-negative* number a is defined as the non-negative number whose fourth power is a, and written $\sqrt[4]{a}$. The **fifth root** of any number a is the number whose fifth power is a, and so on.

Note We cannot take an even-numbered root of a negative number, but we can take an odd-numbered root of any number. Even roots are always positive, whereas odd roots have the same sign as the number we start with. ■

EXAMPLE 3 *n*th Roots

$\sqrt{4} = 2$	Because $2^2 = 4$
$\sqrt{16} = 4$	Because $4^2 = 16$
$\sqrt{1} = 1$	Because $1^2 = 1$
If $x \geq 0$, then $\sqrt{x^2} = x$	Because $x^2 = x^2$
$\sqrt{2} \approx 1.414213562$	$\sqrt{2}$ is not a whole number.
$\sqrt{1+1} = \sqrt{2} \approx 1.414213562$	First add, then take the square root.*
$\sqrt{9+16} = \sqrt{25} = 5$	Contrast with $\sqrt{9} + \sqrt{16} = 3 + 4 = 7$.

*In general, $\sqrt{a+b}$ means the square root of the *quantity* $(a + b)$. The radical sign acts as a pair of parentheses or a fraction bar, telling us to evaluate what is inside before taking the root. (See the Caution on next page.)

$\dfrac{1}{\sqrt{2}} = \dfrac{\sqrt{2}}{2}$ Multiply top and bottom by $\sqrt{2}$.

$\sqrt[3]{27} = 3$ Because $3^3 = 27$

$\sqrt[3]{-64} = -4$ Because $(-4)^3 = -64$

$\sqrt[4]{16} = 2$ Because $2^4 = 16$

$\sqrt[4]{-16}$ is not defined Even-numbered root of a negative number

$\sqrt[5]{-1} = -1$, since $(-1)^5 = -1$ Odd-numbered root of a negative number

$\sqrt[n]{-1} = -1$ if n is any odd number

Q: *In the example we saw that $\sqrt{x^2} = x$ if x is non-negative. What happens if x is negative?*

A: If x is negative, then x^2 is positive, and so $\sqrt{x^2}$ is still defined as the non-negative number whose square is x^2. This number must be $|x|$, the **absolute value of x**, which is the non-negative number with the same size as x. For instance, $|-3| = 3$, while $|3| = 3$, and $|0| = 0$. It follows that

$$\sqrt{x^2} = |x|$$

for every real number x, positive or negative. For instance,

$$\sqrt{(-3)^2} = \sqrt{9} = 3 = |-3|$$

and $$\sqrt{3^2} = \sqrt{9} = 3 = |3|.$$

In general, we find that

$$\sqrt[n]{x^n} = x \text{ if } n \text{ is odd, and } \sqrt[n]{x^n} = |x| \text{ if } n \text{ is even.}$$

We use the following identities to evaluate radicals of products and quotients.

Radicals of Products and Quotients

If a and b are any real numbers (non-negative in the case of even-numbered roots), then

$\sqrt[n]{ab} = \sqrt[n]{a}\sqrt[n]{b}$ Radical of a product = Product of radicals

$\sqrt[n]{\dfrac{a}{b}} = \dfrac{\sqrt[n]{a}}{\sqrt[n]{b}}$ if $b \neq 0$ Radical of a quotient = Quotient of radicals

Notes

- The first rule is similar to the rule $(a \cdot b)^2 = a^2b^2$ for the square of a product, and the second rule is similar to the rule $\left(\dfrac{a}{b}\right)^2 = \dfrac{a^2}{b^2}$ for the square of a quotient.
- ***Caution*** There is no corresponding identity for addition:

$$\sqrt{a+b} \text{ is } \textit{not} \text{ equal to } \sqrt{a} + \sqrt{b}$$

(Consider $a = b = 1$, for example.) Equating these expressions is a common error, so be careful!

Quick Examples

1. $\sqrt{9\cdot 4} = \sqrt{9}\sqrt{4} = 3\times 2 = 6$ — Alternatively, $\sqrt{9\cdot 4} = \sqrt{36} = 6$
2. $\sqrt{\dfrac{9}{4}} = \dfrac{\sqrt{9}}{\sqrt{4}} = \dfrac{3}{2}$
3. $\dfrac{\sqrt{2}}{\sqrt{5}} = \dfrac{\sqrt{2}\sqrt{5}}{\sqrt{5}\sqrt{5}} = \dfrac{\sqrt{10}}{5}$
4. $\sqrt{4(3+13)} = \sqrt{4(16)} = \sqrt{4}\sqrt{16} = 2\times 4 = 8$
5. $\sqrt[3]{-216} = \sqrt[3]{(-27)8} = \sqrt[3]{-27}\sqrt[3]{8} = (-3)2 = -6$
6. $\sqrt{x^3} = \sqrt{x^2\cdot x} = \sqrt{x^2}\sqrt{x} = x\sqrt{x} \quad \text{if } x \geq 0$
7. $\sqrt{\dfrac{x^2+y^2}{z^2}} = \dfrac{\sqrt{x^2+y^2}}{\sqrt{z^2}} = \dfrac{\sqrt{x^2+y^2}}{|z|}$ — We can't simplify the numerator any further.

Rational Exponents

We already know what we mean by expressions such as x^4 and a^{-6}. The next step is to make sense of *rational* exponents: exponents of the form p/q with p and q integers as in $a^{1/2}$ and $3^{-2/3}$.

Q: *What should we mean by $a^{1/2}$?*

A: The overriding concern here is that all the exponent identities should remain true. In this case the identity to look at is the one that says that $(a^m)^n = a^{mn}$. This identity tells us that

$$(a^{1/2})^2 = a^1 = a.$$

That is, $a^{1/2}$, when squared, gives us a. But that must mean that $a^{1/2}$ is the *square root* of a, or

$$a^{1/2} = \sqrt{a}.$$

A similar argument tells us that, if q is any positive whole number, then

$$a^{1/q} = \sqrt[q]{a}, \text{ the } q\text{th root of } a.$$

Notice that if a is negative, this makes sense only for q odd. To avoid this problem, we usually stick to positive a.

Q: *If p and q are integers (q positive), what should we mean by $a^{p/q}$?*

A: By the exponent identities, $a^{p/q}$ should equal both $(a^p)^{1/q}$ and $(a^{1/q})^p$. The first is the qth root of a^p, and the second is the pth power of $a^{1/q}$, which gives us the following.

Conversion Between Rational Exponents and Radicals

If a is any non-negative number, then

$$a^{p/q} = \sqrt[q]{a^p} = \left(\sqrt[q]{a}\right)^p.$$

↑ Using exponents ↑ ↑ Using radicals

In particular,

$$a^{1/q} = \sqrt[q]{a}, \text{ the } q\text{th root of } a.$$

Notes

- If a is negative, all of this makes sense only if q is odd.
- All of the exponent identities continue to work when we allow rational exponents p/q. In other words, we are free to use all the exponent identities even though the exponents are not integers. ■

Quick Examples

1. $4^{3/2} = (\sqrt{4})^3 = 2^3 = 8$
2. $8^{2/3} = (\sqrt[3]{8})^2 = 2^2 = 4$
3. $9^{-3/2} = \dfrac{1}{9^{3/2}} = \dfrac{1}{(\sqrt{9})^3} = \dfrac{1}{3^3} = \dfrac{1}{27}$
4. $\dfrac{\sqrt{3}}{\sqrt[3]{3}} = \dfrac{3^{1/2}}{3^{1/3}} = 3^{1/2-1/3} = 3^{1/6} = \sqrt[6]{3}$
5. $2^2 2^{7/2} = 2^2 2^{3+1/2} = 2^2 2^3 2^{1/2} = 2^5 2^{1/2} = 2^5\sqrt{2}$

EXAMPLE 4 Simplifying Algebraic Expressions

Simplify the following.

a. $\dfrac{(x^3)^{5/3}}{x^3}$ **b.** $\sqrt[4]{a^6}$ **c.** $\dfrac{(xy)^{-3}y^{-3/2}}{x^{-2}\sqrt{y}}$

Solution

a. $\dfrac{(x^3)^{5/3}}{x^3} = \dfrac{x^5}{x^3} = x^2$

b. $\sqrt[4]{a^6} = a^{6/4} = a^{3/2} = a \cdot a^{1/2} = a\sqrt{a}$

c. $\dfrac{(xy)^{-3}y^{-3/2}}{x^{-2}\sqrt{y}} = \dfrac{x^{-3}y^{-3}y^{-3/2}}{x^{-2}y^{1/2}} = \dfrac{1}{x^{-2+3}y^{1/2+3+3/2}} = \dfrac{1}{xy^5}$

Converting Between Rational, Radical, and Exponent Form

In calculus we must often convert algebraic expressions involving powers of x, such as $\dfrac{3}{2x^2}$, into expressions in which x does not appear in the denominator, such as $\dfrac{3}{2}x^{-2}$. Also, we must often convert expressions with radicals, such as $\dfrac{1}{\sqrt{1+x^2}}$, into expressions with no radicals and all powers in the numerator, such as $(1+x^2)^{-1/2}$. In these cases, we are converting from **rational form** or **radical form** to **exponent form.**

Rational Form

An expression is in **rational form** if it is written with positive exponents only.

Quick Examples

1. $\frac{2}{3x^2}$ is in rational form.
2. $\frac{2x^{-1}}{3}$ is not in rational form because the exponent of x is negative.
3. $\frac{x}{6}+\frac{6}{x}$ is in rational form.

Radical Form

An expression is in **radical form** if it is written with integer powers and roots only.

Quick Examples

1. $\frac{2}{5\sqrt[3]{x}}+\frac{2}{x}$ is in radical form.
2. $\frac{2x^{-1/3}}{5}+2x^{-1}$ is not in radical form because $x^{-1/3}$ appears.
3. $\frac{1}{\sqrt{1+x^2}}$ is in radical form, but $(1+x^2)^{-1/2}$ is not.

Exponent Form

An expression is in **exponent form** if there are no radicals and all powers of unknowns occur in the numerator. We write such expressions as sums or differences of terms of the form

Constant × (Expression with $x)^p$ As in $\frac{1}{3}x^{-3/2}$

Quick Examples

1. $\frac{2}{3}x^4-3x^{-1/3}$ is in exponent form.
2. $\frac{x}{6}+\frac{6}{x}$ is not in exponent form because the second expression has x in the denominator.
3. $\sqrt[3]{x}$ is not in exponent form because it has a radical.
4. $(1+x^2)^{-1/2}$ is in exponent form, but $\frac{1}{\sqrt{1+x^2}}$ is not.

EXAMPLE 5 Converting from One Form to Another

Convert the following to rational form:

a. $\frac{1}{2}x^{-2} + \frac{4}{3}x^{-5}$ **b.** $\frac{2}{\sqrt{x}} - \frac{2}{x^{-4}}$

Convert the following to radical form:

c. $\frac{1}{2}x^{-1/2} + \frac{4}{3}x^{-5/4}$ **d.** $\frac{(3+x)^{-1/3}}{5}$

Convert the following to exponent form:

e. $\frac{3}{4x^2} - \frac{x}{6} + \frac{6}{x} + \frac{4}{3\sqrt{x}}$ **f.** $\frac{2}{(x+1)^2} - \frac{3}{4\sqrt[5]{2x-1}}$

Solution For (a) and (b), we eliminate negative exponents as we did in Example 2:

a. $\frac{1}{2}x^{-2} + \frac{4}{3}x^{-5} = \frac{1}{2} \cdot \frac{1}{x^2} + \frac{4}{3} \cdot \frac{1}{x^5} = \frac{1}{2x^2} + \frac{4}{3x^5}$

b. $\frac{2}{\sqrt{x}} - \frac{2}{x^{-4}} = \frac{2}{\sqrt{x}} - 2x^4$

For (c) and (d), we rewrite all terms with fractional exponents as radicals:

c.
$$\frac{1}{2}x^{-1/2} + \frac{4}{3}x^{-5/4} = \frac{1}{2} \cdot \frac{1}{x^{1/2}} + \frac{4}{3} \cdot \frac{1}{x^{5/4}}$$
$$= \frac{1}{2} \cdot \frac{1}{\sqrt{x}} + \frac{4}{3} \cdot \frac{1}{\sqrt[4]{x^5}} = \frac{1}{2\sqrt{x}} + \frac{4}{3\sqrt[4]{x^5}}$$

d. $\frac{(3+x)^{-1/3}}{5} = \frac{1}{5(3+x)^{1/3}} = \frac{1}{5\sqrt[3]{3+x}}$

For (e) and (f), we eliminate any radicals and move all expressions involving x to the numerator:

e.
$$\frac{3}{4x^2} - \frac{x}{6} + \frac{6}{x} + \frac{4}{3\sqrt{x}} = \frac{3}{4}x^{-2} - \frac{1}{6}x + 6x^{-1} + \frac{4}{3x^{1/2}}$$
$$= \frac{3}{4}x^{-2} - \frac{1}{6}x + 6x^{-1} + \frac{4}{3}x^{-1/2}$$

f.
$$\frac{2}{(x+1)^2} - \frac{3}{4\sqrt[5]{2x-1}} = 2(x+1)^{-2} - \frac{3}{4(2x-1)^{1/5}}$$
$$= 2(x+1)^{-2} - \frac{3}{4}(2x-1)^{-1/5}$$

Solving Equations with Exponents

EXAMPLE 6 Solving Equations

Solve the following equations:

a. $x^3 + 8 = 0$ **b.** $x^2 - \frac{1}{2} = 0$ **c.** $x^{3/2} - 64 = 0$

Solution

a. Subtracting 8 from both sides gives $x^3 = -8$. Taking the cube root of both sides gives $x = -2$.

b. Adding $\frac{1}{2}$ to both sides gives $x^2 = \frac{1}{2}$. Thus, $x = \pm\sqrt{\frac{1}{2}} = \pm\frac{1}{\sqrt{2}}$.

c. Adding 64 to both sides gives $x^{3/2} = 64$. Taking the reciprocal (2/3) power of both sides gives

$$(x^{3/2})^{2/3} = 64^{2/3}$$

$$x^1 = \left(\sqrt[3]{64}\right)^2 = 4^2 = 16$$

so $$x = 16.$$

0.2 EXERCISES

Evaluate the expressions in Exercises 1–16.

1. 3^3 **2.** $(-2)^3$ **3.** $-(2 \cdot 3)^2$ **4.** $(4 \cdot 2)^2$

5. $\left(\frac{-2}{3}\right)^2$ **6.** $\left(\frac{3}{2}\right)^3$ **7.** $(-2)^{-3}$ **8.** -2^{-3}

9. $\left(\frac{1}{4}\right)^{-2}$ **10.** $\left(\frac{-2}{3}\right)^{-2}$ **11.** $2 \cdot 3^0$ **12.** $3 \cdot (-2)^0$

13. $2^3\,2^2$ **14.** $3^2 3$ **15.** $2^2 2^{-1} 2^4 2^{-4}$ **16.** $5^2 5^{-3} 5^2 5^{-2}$

Simplify each expression in Exercises 17–30, expressing your answer in rational form.

17. $x^3 x^2$ **18.** $x^4 x^{-1}$ **19.** $-x^2 x^{-3} y$ **20.** $-xy^{-1} x^{-1}$

21. $\frac{x^3}{x^4}$ **22.** $\frac{y^5}{y^3}$ **23.** $\frac{x^2y^2}{x^{-1}y}$ **24.** $\frac{x^{-1}y}{x^2y^2}$

25. $\frac{(xy^{-1}z^3)^2}{x^2yz^2}$ **26.** $\frac{x^2yz^2}{(xyz^{-1})^{-1}}$ **27.** $\left(\frac{xy^{-2}z}{x^{-1}z}\right)^3$

28. $\left(\frac{x^2y^{-1}z^0}{xyz}\right)^2$ **29.** $\left(\frac{x^{-1}y^{-2}z^2}{xy}\right)^{-2}$ **30.** $\left(\frac{xy^{-2}}{x^2y^{-1}z}\right)^{-3}$

Convert the expressions in Exercises 31–36 to rational form.

31. $3x^{-4}$ **32.** $\frac{1}{2}x^{-4}$ **33.** $\frac{3}{4}x^{-2/3}$

34. $\frac{4}{5}y^{-3/4}$ **35.** $1 - \frac{0.3}{x^{-2}} - \frac{6}{5}x^{-1}$ **36.** $\frac{1}{3x^{-4}} + \frac{0.1x^{-2}}{3}$

Evaluate the expressions in Exercises 37–56, rounding your answer to four significant digits where necessary.

37. $\sqrt{4}$ **38.** $\sqrt{5}$ **39.** $\sqrt{\frac{1}{4}}$

40. $\sqrt{\frac{1}{9}}$ **41.** $\sqrt{\frac{16}{9}}$ **42.** $\sqrt{\frac{9}{4}}$

43. $\frac{\sqrt{4}}{5}$ **44.** $\frac{6}{\sqrt{25}}$ **45.** $\sqrt{9} + \sqrt{16}$

46. $\sqrt{25} - \sqrt{16}$ **47.** $\sqrt{9 + 16}$ **48.** $\sqrt{25 - 16}$

49. $\sqrt[3]{8 - 27}$ **50.** $\sqrt[4]{81 - 16}$ **51.** $\sqrt[3]{27/8}$

52. $\sqrt[3]{8 \times 64}$ **53.** $\sqrt{(-2)^2}$ **54.** $\sqrt{(-1)^2}$

55. $\sqrt{\frac{1}{4}(1 + 15)}$ **56.** $\sqrt{\frac{1}{9}(3 + 33)}$

Simplify the expressions in Exercises 57–64, given that x, y, z, a, b, and c are positive real numbers.

57. $\sqrt{a^2b^2}$ **58.** $\sqrt{\frac{a^2}{b^2}}$ **59.** $\sqrt{(x + 9)^2}$

60. $(\sqrt{x + 9})^2$ **61.** $\sqrt[3]{x^3(a^3 + b^3)}$ **62.** $\sqrt[4]{\frac{x^4}{a^4b^4}}$

63. $\sqrt{\frac{4xy^3}{x^2y}}$ **64.** $\sqrt{\frac{4(x^2 + y^2)}{c^2}}$

Convert the expressions in Exercises 65–84 to exponent form.

65. $\sqrt{3}$ **66.** $\sqrt{8}$ **67.** $\sqrt{x^3}$

68. $\sqrt[3]{x^2}$ **69.** $\sqrt[3]{xy^2}$ **70.** $\sqrt{x^2y}$

71. $\frac{x^2}{\sqrt{x}}$ **72.** $\frac{x}{\sqrt{x}}$ **73.** $\frac{3}{5x^2}$

74. $\frac{2}{5x^{-3}}$ **75.** $\frac{3x^{-1.2}}{2} - \frac{1}{3x^{2.1}}$ **76.** $\frac{2}{3x^{-1.2}} - \frac{x^{2.1}}{3}$

77. $\frac{2x}{3} - \frac{x^{0.1}}{2} + \frac{4}{3x^{1.1}}$ **78.** $\frac{4x^2}{3} + \frac{x^{3/2}}{6} - \frac{2}{3x^2}$

79. $\frac{3\sqrt{x}}{4} - \frac{5}{3\sqrt{x}} + \frac{4}{3x\sqrt{x}}$ **80.** $\frac{3}{5\sqrt{x}} - \frac{5\sqrt{x}}{8} + \frac{7}{2\sqrt[3]{x}}$

81. $\frac{3\sqrt[5]{x^2}}{4} - \frac{7}{2\sqrt{x^3}}$ **82.** $\frac{1}{8x\sqrt{x}} - \frac{2}{3\sqrt[5]{x^3}}$

83. $\frac{1}{(x^2 + 1)^3} - \frac{3}{4\sqrt[3]{(x^2 + 1)}}$ **84.** $\frac{2}{3(x^2 + 1)^{-3}} - \frac{3\sqrt[3]{(x^2 + 1)^7}}{4}$

Convert the expressions in Exercises 85–96 to radical form.

85. $2^{2/3}$ **86.** $3^{4/5}$ **87.** $x^{4/3}$ **88.** $y^{7/4}$

89. $(x^{1/2}y^{1/3})^{1/5}$ **90.** $x^{-1/3}y^{3/2}$ **91.** $-\frac{3}{2}x^{-1/4}$ **92.** $\frac{4}{5}x^{3/2}$

93. $0.2x^{-2/3} + \dfrac{3}{7x^{-1/2}}$ **94.** $\dfrac{3.1}{x^{-4/3}} - \dfrac{11}{7}x^{-1/7}$

95. $\dfrac{3}{4(1-x)^{5/2}}$ **96.** $\dfrac{9}{4(1-x)^{-7/3}}$

Simplify the expressions in Exercises 97–106.

97. $4^{-1/2}4^{7/2}$ **98.** $2^{1/a}/2^{2/a}$ **99.** $3^{2/3}3^{-1/6}$

100. $2^{1/3}2^{-1}2^{2/3}2^{-1/3}$ **101.** $\dfrac{x^{3/2}}{x^{5/2}}$ **102.** $\dfrac{y^{5/4}}{y^{3/4}}$

103. $\dfrac{x^{1/2}y^2}{x^{-1/2}y}$ **104.** $\dfrac{x^{-1/2}y}{x^2y^{3/2}}$

105. $\left(\dfrac{x}{y}\right)^{1/3}\left(\dfrac{y}{x}\right)^{2/3}$ **106.** $\left(\dfrac{x}{y}\right)^{-1/3}\left(\dfrac{y}{x}\right)^{1/3}$

Solve each equation in Exercises 107–120 for x, rounding your answer to four significant digits where necessary.

107. $x^2 - 16 = 0$ **108.** $x^2 - 1 = 0$

109. $x^2 - \dfrac{4}{9} = 0$ **110.** $x^2 - \dfrac{1}{10} = 0$

111. $x^2 - (1+2x)^2 = 0$ **112.** $x^2 - (2-3x)^2 = 0$

113. $x^5 + 32 = 0$ **114.** $x^4 - 81 = 0$

115. $x^{1/2} - 4 = 0$ **116.** $x^{1/3} - 2 = 0$

117. $1 - \dfrac{1}{x^2} = 0$ **118.** $\dfrac{2}{x^3} - \dfrac{6}{x^4} = 0$

119. $(x-4)^{-1/3} = 2$ **120.** $(x-4)^{2/3} + 1 = 5$

0.3 Multiplying and Factoring Algebraic Expressions

Multiplying Algebraic Expressions

Distributive Law

The **distributive law** for real numbers states that

$$a(b \pm c) = ab \pm ac$$
$$(a \pm b)c = ac \pm bc$$

for any real numbers a, b, and c.

Quick Examples

1. $2(x-3)$ is *not* equal to $2x - 3$ but is equal to $2x - 2(3) = 2x - 6$.
2. $x(x+1) = x^2 + x$
3. $2x(3x-4) = 6x^2 - 8x$
4. $(x-4)x^2 = x^3 - 4x^2$
5. $(x+2)(x+3) = (x+2)x + (x+2)3$
   $= (x^2+2x) + (3x+6) = x^2 + 5x + 6$
6. $(x+2)(x-3) = (x+2)x - (x+2)3$
   $= (x^2+2x) - (3x+6) = x^2 - x - 6$

There is a quicker way of expanding expressions like the last two, called the "FOIL" method (First, Outer, Inner, Last). Consider, for instance, the expression $(x+1)(x-2)$. The FOIL method says: Take the product of the first terms: $x \cdot x = x^2$, the product of the outer terms: $x \cdot (-2) = -2x$, the product of the inner terms: $1 \cdot x = x$, and the product of the last terms: $1 \cdot (-2) = -2$, and then add them all up, getting $x^2 - 2x + x - 2 = x^2 - x - 2$.

EXAMPLE 1 FOIL

a. $(x-2)(2x+5) = 2x^2 + 5x - 4x - 10 = 2x^2 + x - 10$

$\uparrow$ First $\uparrow$ Outer $\uparrow$ Inner $\uparrow$ Last

b. $(x^2+1)(x-4) = x^3 - 4x^2 + x - 4$

c. $(a-b)(a+b) = a^2 + ab - ab - b^2 = a^2 - b^2$

d. $(a+b)^2 = (a+b)(a+b) = a^2 + ab + ab + b^2 = a^2 + 2ab + b^2$

e. $(a-b)^2 = (a-b)(a-b) = a^2 - ab - ab + b^2 = a^2 - 2ab + b^2$

The last three are particularly important and are worth memorizing.

Special Formulas

$(a-b)(a+b) = a^2 - b^2$	Difference of two squares
$(a+b)^2 = a^2 + 2ab + b^2$	Square of a sum
$(a-b)^2 = a^2 - 2ab + b^2$	Square of a difference

Quick Examples

1. $(2-x)(2+x) = 4 - x^2$
2. $(1+a)(1-a) = 1 - a^2$
3. $(x+3)^2 = x^2 + 6x + 9$
4. $(4-x)^2 = 16 - 8x + x^2$

Here are some longer examples that require the distributive law.

EXAMPLE 2 Multiplying Algebraic Expressions

a. $$\begin{aligned}(x+1)(x^2+3x-4) &= (x+1)x^2 + (x+1)3x - (x+1)4 \\ &= (x^3+x^2) + (3x^2+3x) - (4x+4) \\ &= x^3 + 4x^2 - x - 4\end{aligned}$$

b. $$\begin{aligned}\left(x^2 - \frac{1}{x} + 1\right)(2x+5) &= \left(x^2 - \frac{1}{x} + 1\right)2x + \left(x^2 - \frac{1}{x} + 1\right)5 \\ &= (2x^3 - 2 + 2x) + \left(5x^2 - \frac{5}{x} + 5\right) \\ &= 2x^3 + 5x^2 + 2x + 3 - \frac{5}{x}\end{aligned}$$

c. $$\begin{aligned}(x-y)(x-y)(x-y) &= (x^2 - 2xy + y^2)(x-y) \\ &= (x^2 - 2xy + y^2)x - (x^2 - 2xy + y^2)y \\ &= (x^3 - 2x^2y + xy^2) - (x^2y - 2xy^2 + y^3) \\ &= x^3 - 3x^2y + 3xy^2 - y^3\end{aligned}$$

Factoring Algebraic Expressions

We can think of factoring as applying the distributive law in reverse—for example,

$$2x^2 + x = x(2x + 1),$$

which can be checked by using the distributive law. Factoring is an art that you will learn with experience and the help of a few useful techniques.

Factoring Using a Common Factor

To use this technique, locate a **common factor**—a term that occurs as a factor in each of the expressions being added or subtracted (for example, x is a common factor in $2x^2 + x$, because it is a factor of both $2x^2$ and x). Once you have located a common factor, "factor it out" by applying the distributive law.

Quick Examples

1. $2x^3 - x^2 + x$ has x as a common factor, so

$$2x^3 - x^2 + x = x(2x^2 - x + 1)$$

2. $2x^2 + 4x$ has $2x$ as a common factor, so

$$2x^2 + 4x = 2x(x + 2)$$

3. $2x^2y + xy^2 - x^2y^2$ has xy as a common factor, so

$$2x^2y + xy^2 - x^2y^2 = xy(2x + y - xy)$$

4. $(x^2 + 1)(x + 2) - (x^2 + 1)(x + 3)$ has $x^2 + 1$ as a common factor, so

$$\begin{aligned}(x^2 + 1)(x + 2) - (x^2 + 1)(x + 3) &= (x^2 + 1)[(x + 2) - (x + 3)]\\ &= (x^2 + 1)(x + 2 - x - 3)\\ &= (x^2 + 1)(-1) = -(x^2 + 1)\end{aligned}$$

5. $12x(x^2 - 1)^5(x^3 + 1)^6 + 18x^2(x^2 - 1)^6(x^3 + 1)^5$ has $6x(x^2 - 1)^5(x^3 + 1)^5$ as a common factor, so

$$\begin{aligned}&12x(x^2 - 1)^5(x^3 + 1)^6 + 18x^2(x^2 - 1)^6(x^3 + 1)^5\\ &= 6x(x^2 - 1)^5(x^3 + 1)^5[2(x^3 + 1) + 3x(x^2 - 1)]\\ &= 6x(x^2 - 1)^5(x^3 + 1)^5(2x^3 + 2 + 3x^3 - 3x)\\ &= 6x(x^2 - 1)^5(x^3 + 1)^5(5x^3 - 3x + 2)\end{aligned}$$

We would also like to be able to reverse calculations such as $(x + 2)(2x - 5) = 2x^2 - x - 10$. That is, starting with the expression $2x^2 - x - 10$, we would like to **factor** it to get the expression $(x + 2)(2x - 5)$. An expression of the form $ax^2 + bx + c$, where a, b, and c are real numbers, is called a **quadratic** expression in x. Thus, given a quadratic expression $ax^2 + bx + c$, we would like to write it in the form $(dx + e)(fx + g)$ for some real numbers d, e, f, and g. There are some quadratics, such as $x^2 + x + 1$, that cannot be factored in this form at all. Here, we consider only quadratics that do factor, and in such a way that the numbers d, e, f, and g are integers (whole numbers; other cases are discussed in Section 5). The usual technique of factoring such quadratics is a "trial and error" approach.

Factoring Quadratics by Trial and Error

To factor the quadratic $ax^2 + bx + c$, factor ax^2 as $(a_1x)(a_2x)$ (with a_1 positive) and c as c_1c_2, and then check whether or not $ax^2 + bx + c = (a_1x \pm c_1)(a_2x \pm c_2)$. If not, try other factorizations of ax^2 and c.

Quick Examples

1. To factor $x^2 - 6x + 5$, first factor x^2 as $(x)(x)$, and 5 as (5)(1):

$(x + 5)(x + 1) = x^2 + 6x + 5.$ No good

$(x - 5)(x - 1) = x^2 - 6x + 5.$ Desired factorization

2. To factor $x^2 - 4x - 12$, first factor x^2 as $(x)(x)$, and -12 as $(1)(-12)$, $(2)(-6)$, or $(3)(-4)$. Trying them one by one gives

$(x + 1)(x - 12) = x^2 - 11x - 12.$ No good

$(x - 1)(x + 12) = x^2 + 11x - 12.$ No good

$(x + 2)(x - 6) = x^2 - 4x - 12.$ Desired factorization

3. To factor $4x^2 - 25$, we can follow the above procedure, or recognize $4x^2 - 25$ as the difference of two squares:

$$4x^2 - 25 = (2x)^2 - 5^2 = (2x - 5)(2x + 5).$$

Note: Not all quadratic expressions factor. In Section 5 we look at a test that tells us whether or not a given quadratic factors.

Here are examples requiring either a little more work or a little more thought.

EXAMPLE 3 Factoring Quadratics

Factor the following: **a.** $4x^2 - 5x - 6$ **b.** $x^4 - 5x^2 + 6$

Solution

a. Possible factorizations of $4x^2$ are $(2x)(2x)$ or $(x)(4x)$. Possible factorizations of -6 are $(1)(-6)$, $(2)(-3)$. We now systematically try out all the possibilities until we come up with the correct one.

$(2x)(2x)$ and $(1)(-6)$:	$(2x + 1)(2x - 6) = 4x^2 - 10x - 6$	No good
$(2x)(2x)$ and $(2)(-3)$:	$(2x + 2)(2x - 3) = 4x^2 - 2x - 6$	No good
$(x)(4x)$ and $(1)(-6)$:	$(x + 1)(4x - 6) = 4x^2 - 2x - 6$	No good
$(x)(4x)$ and $(2)(-3)$:	$(x + 2)(4x - 3) = 4x^2 + 5x - 6$	Almost!
Change signs:	$(x - 2)(4x + 3) = 4x^2 - 5x - 6$	Correct

b. The expression $x^4 - 5x^2 + 6$ is not a quadratic, you say? Correct. It's a quartic (a fourth degree expression). However, it looks rather like a quadratic. In fact, it is quadratic *in* x^2, meaning that it is

$$(x^2)^2 - 5(x^2) + 6 = y^2 - 5y + 6$$

where $y = x^2$. The quadratic $y^2 - 5y + 6$ factors as

$$y^2 - 5y + 6 = (y - 3)(y - 2)$$

so

$$x^4 - 5x^2 + 6 = (x^2 - 3)(x^2 - 2)$$

This is a sometimes useful technique.

Our last example is here to remind you why we should want to factor polynomials in the first place. We shall return to this in Section 5.

EXAMPLE 4 Solving a Quadratic Equation by Factoring

Solve the equation $3x^2 + 4x - 4 = 0$.

Solution We first factor the left-hand side to get

$$(3x - 2)(x + 2) = 0.$$

Thus, the product of the two quantities $(3x - 2)$ and $(x + 2)$ is zero. Now, if a product of two numbers is zero, one of the two must be zero. In other words, either $3x - 2 = 0$, giving $x = \frac{2}{3}$, or $x + 2 = 0$, giving $x = -2$. Thus, there are two solutions: $x = \frac{2}{3}$ and $x = -2$.

0.3 EXERCISES

Expand each expression in Exercises 1–22.

1. $x(4x + 6)$
2. $(4y - 2)y$
3. $(2x - y)y$
4. $x(3x + y)$
5. $(x + 1)(x - 3)$
6. $(y + 3)(y + 4)$
7. $(2y + 3)(y + 5)$
8. $(2x - 2)(3x - 4)$
9. $(2x - 3)^2$
10. $(3x + 1)^2$
11. $\left(x + \frac{1}{x}\right)^2$
12. $\left(y - \frac{1}{y}\right)^2$
13. $(2x - 3)(2x + 3)$
14. $(4 + 2x)(4 - 2x)$
15. $\left(y - \frac{1}{y}\right)\left(y + \frac{1}{y}\right)$
16. $(x - x^2)(x + x^2)$
17. $(x^2 + x - 1)(2x + 4)$
18. $(3x + 1)(2x^2 - x + 1)$
19. $(x^2 - 2x + 1)^2$
20. $(x + y - xy)^2$
21. $(y^3 + 2y^2 + y)(y^2 + 2y - 1)$
22. $(x^3 - 2x^2 + 4)(3x^2 - x + 2)$

In Exercises 23–30, factor each expression and simplify as much as possible.

23. $(x + 1)(x + 2) + (x + 1)(x + 3)$
24. $(x + 1)(x + 2)^2 + (x + 1)^2(x + 2)$
25. $(x^2 + 1)^5(x + 3)^4 + (x^2 + 1)^6(x + 3)^3$
26. $10x(x^2 + 1)^4(x^3 + 1)^5 + 15x^2(x^2 + 1)^5(x^3 + 1)^4$
27. $(x^3 + 1)\sqrt{x + 1} - (x^3 + 1)^2\sqrt{x + 1}$
28. $(x^2 + 1)\sqrt{x + 1} - \sqrt{(x + 1)^3}$
29. $\sqrt{(x + 1)^3} + \sqrt{(x + 1)^5}$
30. $(x^2 + 1)\sqrt[3]{(x + 1)^4} - \sqrt[3]{(x + 1)^7}$

*In Exercises 31–48, **(a)** factor the given expression; **(b)** set the expression equal to zero and solve for the unknown (x in the odd-numbered exercises and y in the even-numbered exercises).*

31. $2x + 3x^2$
32. $y^2 - 4y$
33. $6x^3 - 2x^2$
34. $3y^3 - 9y^2$
35. $x^2 - 8x + 7$
36. $y^2 + 6y + 8$
37. $x^2 + x - 12$
38. $y^2 + y - 6$
39. $2x^2 - 3x - 2$
40. $3y^2 - 8y - 3$
41. $6x^2 + 13x + 6$
42. $6y^2 + 17y + 12$
43. $12x^2 + x - 6$
44. $20y^2 + 7y - 3$
45. $x^2 + 4xy + 4y^2$
46. $4y^2 - 4xy + x^2$
47. $x^4 - 5x^2 + 4$
48. $y^4 + 2y^2 - 3$

0.4 Rational Expressions

Rational Expression

A **rational expression** is an algebraic expression of the form $\dfrac{P}{Q}$, where P and Q are simpler expressions (usually polynomials) and the denominator Q is not zero.

Quick Examples

1. $\dfrac{x^2 - 3x}{x}$ $\quad P = x^2 - 3x,\ Q = x$
2. $\dfrac{x + \frac{1}{x} + 1}{2x^2y + 1}$ $\quad P = x + \dfrac{1}{x} + 1,\ Q = 2x^2y + 1$
3. $3xy - x^2$ $\quad P = 3xy - x^2,\ Q = 1$

Algebra of Rational Expressions

We manipulate rational expressions in the same way that we manipulate fractions, using the following rules:

	Algebraic Rule	Quick Example
Product:	$\dfrac{P}{Q} \cdot \dfrac{R}{S} = \dfrac{PR}{QS}$	$\dfrac{x+1}{x} \cdot \dfrac{x-1}{2x+1} = \dfrac{(x+1)(x-1)}{x(2x+1)} = \dfrac{x^2-1}{2x^2+x}$
Sum:	$\dfrac{P}{Q} + \dfrac{R}{S} = \dfrac{PS+RQ}{QS}$	$\dfrac{2x-1}{3x+2} + \dfrac{1}{x} = \dfrac{(2x-1)x + 1(3x+2)}{x(3x+2)} = \dfrac{2x^2+2x+2}{3x^2+2x}$
Difference:	$\dfrac{P}{Q} - \dfrac{R}{S} = \dfrac{PS-RQ}{QS}$	$\dfrac{x}{3x+2} - \dfrac{x-4}{x} = \dfrac{x^2-(x-4)(3x+2)}{x(3x+2)} = \dfrac{-2x^2+10x+8}{3x^2+2x}$
Reciprocal:	$\dfrac{1}{\left(\frac{P}{Q}\right)} = \dfrac{Q}{P}$	$\dfrac{1}{\left(\frac{2xy}{3x-1}\right)} = \dfrac{3x-1}{2xy}$
Quotient:	$\dfrac{\left(\frac{P}{Q}\right)}{\left(\frac{R}{S}\right)} = \dfrac{P}{Q} \cdot \dfrac{S}{R} = \dfrac{PS}{QR}$	$\dfrac{\left(\frac{x}{x-1}\right)}{\left(\frac{y-1}{y}\right)} = \dfrac{xy}{(x-1)(y-1)} = \dfrac{xy}{xy-x-y+1}$
Cancellation:	$\dfrac{PR}{QR} = \dfrac{P}{Q}$	$\dfrac{(x-1)(xy+4)}{(x^2y-8)(x-1)} = \dfrac{xy+4}{x^2y-8}$

Caution Cancellation of summands is *invalid*. For instance,

$$\frac{\not{x} + (2xy^2 - y)}{\not{x} + 4y} = \frac{(2xy^2 - y)}{4y} \quad ✗ \textit{ WRONG!} \quad \text{Do } \textit{not} \text{ cancel a summand.}$$

$$\frac{\not{x}(2xy^2 - y)}{4\not{x}y} = \frac{(2xy^2 - y)}{4y} \quad ✔ \textit{ CORRECT} \quad \text{Do cancel a factor.}$$

Here are some examples that require several algebraic operations.

EXAMPLE 1 Simplifying Rational Expressions

a. $$\frac{\left(\frac{1}{x+y} - \frac{1}{x}\right)}{y} = \frac{\left(\frac{x-(x+y)}{x(x+y)}\right)}{y} = \frac{\left(\frac{-y}{x(x+y)}\right)}{y} = \frac{-y}{xy(x+y)} = -\frac{1}{x(x+y)}$$

b. $$\frac{(x+1)(x+2)^2 - (x+1)^2(x+2)}{(x+2)^4} = \frac{(x+1)(x+2)[(x+2)-(x+1)]}{(x+2)^4}$$

$$= \frac{(x+1)(x+2)(x+2-x-1)}{(x+2)^4} = \frac{(x+1)(x+2)}{(x+2)^4} = \frac{x+1}{(x+2)^3}$$

c. $$\frac{2x\sqrt{x+1} - \frac{x^2}{\sqrt{x+1}}}{x+1} = \frac{\left(\frac{2x\left(\sqrt{x+1}\right)^2 - x^2}{\sqrt{x+1}}\right)}{x+1} = \frac{2x(x+1) - x^2}{(x+1)\sqrt{x+1}}$$

$$= \frac{2x^2 + 2x - x^2}{(x+1)\sqrt{x+1}} = \frac{x^2 + 2x}{\sqrt{(x+1)^3}} = \frac{x(x+2)}{\sqrt{(x+1)^3}}$$

0.4 EXERCISES

Rewrite each expression in Exercises 1–16 as a single rational expression, simplified as much as possible.

1. $\frac{x-4}{x+1} \cdot \frac{2x+1}{x-1}$

2. $\frac{2x-3}{x-2} \cdot \frac{x+3}{x+1}$

3. $\frac{x-4}{x+1} + \frac{2x+1}{x-1}$

4. $\frac{2x-3}{x-2} + \frac{x+3}{x+1}$

5. $\frac{x^2}{x+1} - \frac{x-1}{x+1}$

6. $\frac{x^2-1}{x-2} - \frac{1}{x-1}$

7. $\frac{1}{\left(\frac{x}{x-1}\right)} + x - 1$

8. $\frac{2}{\left(\frac{x-2}{x^2}\right)} - \frac{1}{x-2}$

9. $\frac{1}{x}\left[\frac{x-3}{xy} + \frac{1}{y}\right]$

10. $\frac{y^2}{x}\left[\frac{2x-3}{y} + \frac{x}{y}\right]$

11. $\frac{(x+1)^2(x+2)^3 - (x+1)^3(x+2)^2}{(x+2)^6}$

12. $\frac{6x(x^2+1)^2(x^3+2)^3 - 9x^2(x^2+1)^3(x^3+2)^2}{(x^3+2)^6}$

13. $\frac{(x^2-1)\sqrt{x^2+1} - \frac{x^4}{\sqrt{x^2+1}}}{x^2+1}$

14. $\frac{x\sqrt{x^3-1} - \frac{3x^4}{\sqrt{x^3-1}}}{x^3-1}$

15. $\frac{\frac{1}{(x+y)^2} - \frac{1}{x^2}}{y}$

16. $\frac{\frac{1}{(x+y)^3} - \frac{1}{x^3}}{y}$

0.5 Solving Polynomial Equations

Polynomial Equation

A **polynomial equation** in one unknown is an equation that can be written in the form

$$ax^n + bx^{n-1} + \cdots + rx + s = 0$$

where $a, b, \ldots, r$, and s are constants.

We call the largest exponent of x appearing in a nonzero term of a polynomial the **degree** of that polynomial.

Quick Examples

1. $3x + 1 = 0$ has degree 1 because the largest power of x that occurs is $x = x^1$. Degree 1 equations are called **linear** equations.
2. $x^2 - x - 1 = 0$ has degree 2 because the largest power of x that occurs is x^2. Degree 2 equations are also called **quadratic equations**, or just **quadratics**.
3. $x^3 = 2x^2 + 1$ is a degree 3 polynomial (or **cubic**) in disguise. It can be rewritten as $x^3 - 2x^2 - 1 = 0$, which is in the standard form for a degree 3 equation.
4. $x^4 - x = 0$ has degree 4. It is called a **quartic**.

Now comes the question: How do we solve these equations for x? This question was asked by mathematicians as early as 1600 BC. Let's look at these equations one degree at a time.

Solution of Linear Equations

By definition, a linear equation can be written in the form

$$ax + b = 0. \qquad \text{a and b are fixed numbers with $a \neq 0$.}$$

Solving this is a nice mental exercise: Subtract b from both sides and then divide by a, getting $x = -b/a$. Don't bother memorizing this formula; just go ahead and solve linear equations as they arise. If you feel you need practice, see the exercises at the end of the section.

Solution of Quadratic Equations

By definition, a quadratic equation has the form

$$ax^2 + bx + c = 0. \qquad \text{a, b, and c are fixed numbers and $a \neq 0$.[1]}$$

[1] What happens if $a = 0$?

The solutions of this equation are also called the **roots** of $ax^2 + bx + c$. We're assuming that you saw quadratic equations somewhere in high school but may be a little hazy about the details of their solution. There are two ways of solving these equations—one works sometimes, and the other works every time.

Solving Quadratic Equations by Factoring (works sometimes)

If we can factor* a quadratic equation $ax^2 + bx + c = 0$, we can solve the equation by setting each factor equal to zero.

Quick Examples

1. $x^2 + 7x + 10 = 0$

   $(x + 5)(x + 2) = 0$ — Factor the left-hand side.

   $x + 5 = 0$ or $x + 2 = 0$ — If a product is zero, one or both factors is zero.

   Solutions: $x = -5$ and $x = -2$
2. $2x^2 - 5x - 12 = 0$

   $(2x + 3)(x - 4) = 0$ — Factor the left-hand side.

   $2x + 3 = 0$ or $x - 4 = 0$

   Solutions: $x = -3/2$ and $x = 4$

*See the section on factoring for a review of how to factor quadratics.

Test for Factoring

The quadratic $ax^2 + bx + c$, with a, b, and c being integers (whole numbers), factors into an expression of the form $(rx + s)(tx + u)$ with r, s, t, and u integers precisely when the quantity $b^2 - 4ac$ is a perfect square. (That is, it is the square of an integer.) If this happens, we say that the quadratic **factors over the integers**.

Quick Examples

1. $x^2 + x + 1$ has $a = 1$, $b = 1$, and $c = 1$, so $b^2 - 4ac = -3$, which is not a perfect square. Therefore, this quadratic does not factor over the integers.
2. $2x^2 - 5x - 12$ has $a = 2$, $b = -5$, and $c = -12$, so $b^2 - 4ac = 121$. Because $121 = 11^2$, this quadratic does factor over the integers. (We factored it above.)

Solving Quadratic Equations with the Quadratic Formula (works every time)

The solutions of the general quadratic $ax^2 + bx + c = 0$ $(a \neq 0)$ are given by

$$x = \frac{-b \pm \sqrt{b^2 - 4ac}}{2a}.$$

We call the quantity $\Delta = b^2 - 4ac$ the **discriminant** of the quadratic (Δ is the Greek letter delta), and we have the following general rules:

- If Δ is positive, there are two distinct real solutions.
- If Δ is zero, there is only one real solution: $x = -\frac{b}{2a}$. (Why?)
- If Δ is negative, there are no real solutions.

Quick Examples

1. $2x^2 - 5x - 12 = 0$ has $a = 2$, $b = -5$, and $c = -12$.
$$x = \frac{-b \pm \sqrt{b^2 - 4ac}}{2a} = \frac{5 \pm \sqrt{25 + 96}}{4} = \frac{5 \pm \sqrt{121}}{4} = \frac{5 \pm 11}{4}$$
$$= \frac{16}{4} \text{ or } -\frac{6}{4} = 4 \text{ or } -3/2$$
Δ is positive in this example.

2. $4x^2 = 12x - 9$ can be rewritten as $4x^2 - 12x + 9 = 0$, which has $a = 4$, $b = -12$, and $c = 9$.
$$x = \frac{-b \pm \sqrt{b^2 - 4ac}}{2a} = \frac{12 \pm \sqrt{144 - 144}}{8} = \frac{12 \pm 0}{8} = \frac{12}{8} = \frac{3}{2}$$
Δ is zero in this example.

3. $x^2 + 2x - 1 = 0$ has $a = 1$, $b = 2$, and $c = -1$.
$$x = \frac{-b \pm \sqrt{b^2 - 4ac}}{2a} = \frac{-2 \pm \sqrt{8}}{2} = \frac{-2 \pm 2\sqrt{2}}{2} = -1 \pm \sqrt{2}$$
The two solutions are $x = -1 + \sqrt{2} = 0.414\ldots$ and $x = -1 - \sqrt{2} = -2.414\ldots$
Δ is positive in this example.

4. $x^2 + x + 1 = 0$ has $a = 1$, $b = 1$, and $c = 1$. Because $\Delta = -3$ is negative, there are no real solutions.
Δ is negative in this example.

Q: *This is all very useful, but where does the quadratic formula come from?*

A: To see where it comes from, we will solve a general quadratic equation using "brute force." Start with the general quadratic equation.

$$ax^2 + bx + c = 0.$$

First, divide out the nonzero number a to get

$$x^2 + \frac{bx}{a} + \frac{c}{a} = 0.$$

Now we **complete the square:** Add and subtract the quantity $\frac{b^2}{4a^2}$ to get

$$x^2 + \frac{bx}{a} + \frac{b^2}{4a^2} - \frac{b^2}{4a^2} + \frac{c}{a} = 0.$$

We do this to get the first three terms to factor as a perfect square:

$$\left(x + \frac{b}{2a}\right)^2 - \frac{b^2}{4a^2} + \frac{c}{a} = 0.$$

(Check this by multiplying out.) Adding $\frac{b^2}{4a^2} - \frac{c}{a}$ to both sides gives:

$$\left(x + \frac{b}{2a}\right)^2 = \frac{b^2}{4a^2} - \frac{c}{a} = \frac{b^2 - 4ac}{4a^2}.$$

Taking square roots gives

$$x + \frac{b}{2a} = \frac{\pm\sqrt{b^2 - 4ac}}{2a}.$$

Finally, adding $-\frac{b}{2a}$ to both sides yields the result:

$$x = -\frac{b}{2a} + \frac{\pm\sqrt{b^2 - 4ac}}{2a}$$

or

$$x = \frac{-b \pm \sqrt{b^2 - 4ac}}{2a}.$$

Solution of Cubic Equations

By definition, a cubic equation can be written in the form

$$ax^3 + bx^2 + cx + d = 0.$$ a, b, c, and d are fixed numbers and $a \neq 0$.

Now we get into something of a bind. Although there is a perfectly respectable formula for the solutions, it is very complicated and involves the use of complex numbers rather heavily.[2] So we discuss instead a much simpler method that *sometimes* works nicely. Here is the method in a nutshell.

Solving Cubics by Finding One Factor

Start with a given cubic equation $ax^3 + bx^2 + cx + d = 0$.

Step 1 By trial and error, find one solution $x = s$. If a, b, c, and d are integers, the only possible *rational* solutions* are those of the form $s = \pm$(factor of d)/(factor of a).

Step 2 It will now be possible to factor the cubic as

$$ax^3 + bx^2 + cx + d = (x - s)(ax^2 + ex + f) = 0$$

To find $ax^2 + ex + f$, divide the cubic by $x - s$, using long division.†

Step 3 The factored equation says that either $x - s = 0$ or $ax^2 + ex + f = 0$. We already know that s is a solution, and now we see that the other solutions are the roots of the quadratic. Note that this quadratic may or may not have any real solutions, as usual.

* There may be *irrational* solutions, however; for example, $x^3 - 2 = 0$ has the single solution $x = \sqrt[3]{2}$.

†Alternatively, use "synthetic division," a shortcut that would take us too far afield to describe.

[2] It was when this formula was discovered in the 16th century that complex numbers were first taken seriously. Although we would like to show you the formula, it is too large to fit in this footnote.

Quick Example

To solve the cubic $x^3 - x^2 + x - 1 = 0$, we first find a single solution. Here, $a = 1$ and $d = -1$. Because the only factors of ± 1 are ± 1, the only possible rational solutions are $x = \pm 1$. By substitution, we see that $x = 1$ is a solution. Thus, $(x - 1)$ is a factor. Dividing by $(x - 1)$ yields the quotient $(x^2 + 1)$. Thus,

$$x^3 - x^2 + x - 1 = (x - 1)(x^2 + 1) = 0$$

so that either $x - 1 = 0$ or $x^2 + 1 = 0$.

Because the discriminant of the quadratic $x^2 + 1$ is negative, we don't get any real solutions from $x^2 + 1 = 0$, so the only real solution is $x = 1$.

Possible Outcomes When Solving a Cubic Equation

If you consider all the cases, there are three possible outcomes when solving a cubic equation:

1. One real solution (as in the Quick Example above)
2. Two real solutions (try, for example, $x^3 + x^2 - x - 1 = 0$)
3. Three real solutions (see the next example)

EXAMPLE 1 Solving a Cubic

Solve the cubic $2x^3 - 3x^2 - 17x + 30 = 0$.

Solution First we look for a single solution. Here, $a = 2$ and $d = 30$. The factors of a are ± 1 and ± 2, and the factors of d are $\pm 1, \pm 2, \pm 3, \pm 5, \pm 6, \pm 10, \pm 15$, and ± 30. This gives us a large number of possible ratios: $\pm 1, \pm 2, \pm 3, \pm 5, \pm 6, \pm 10, \pm 15, \pm 30, \pm 1/2, \pm 3/2, \pm 5/2, \pm 15/2$. Undaunted, we first try $x = 1$ and $x = -1$, getting nowhere. So we move on to $x = 2$, and we hit the jackpot, because substituting $x = 2$ gives $16 - 12 - 34 + 30 = 0$. Thus, $(x - 2)$ is a factor. Dividing yields the quotient $2x^2 + x - 15$. Here is the calculation:

$$\begin{array}{r l}
 & 2x^2 + x - 15 \\
x - 2 & \overline{)2x^3 - 3x^2 - 17x + 30} \\
 & \underline{2x^3 - 4x^2} \\
 & \quad\quad x^2 - 17x \\
 & \quad\quad \underline{x^2 - \ \ 2x} \\
 & \quad\quad\quad -15x + 30 \\
 & \quad\quad\quad \underline{-15x + 30} \\
 & \quad\quad\quad\quad\quad 0.
\end{array}$$

Thus,

$$2x^3 - 3x^2 - 17x + 30 = (x - 2)(2x^2 + x - 15) = 0.$$

Setting the factors equal to zero gives either $x - 2 = 0$ or $2x^2 + x - 15 = 0$. We could solve the quadratic using the quadratic formula, but, luckily, we notice that it factors as

$$2x^2 + x - 15 = (x + 3)(2x - 5).$$

Thus, the solutions are $x = 2$, $x = -3$ and $x = 5/2$.

Solution of Higher-Order Polynomial Equations

Logically speaking, our next step should be a discussion of quartics, then quintics (fifth degree equations), and so on forever. Well, we've got to stop somewhere, and cubics may be as good a place as any. On the other hand, since we've gotten so far, we ought to at least tell you what is known about higher order polynomials.

Quartics Just as in the case of cubics, there is a formula to find the solutions of quartics.[3]

Quintics and Beyond All good things must come to an end, we're afraid. It turns out that there is no "quintic formula." In other words, there is no single algebraic formula or collection of algebraic formulas that gives the solutions to all quintics. This question was settled by the Norwegian mathematician Niels Henrik Abel in 1824 after almost 300 years of controversy about this question. (In fact, several notable mathematicians had previously claimed to have devised formulas for solving the quintic, but these were all shot down by other mathematicians—this being one of the favorite pastimes of practitioners of our art.) The same negative answer applies to polynomial equations of degree 6 and higher. It's not that these equations don't have solutions; it's just that they can't be found using algebraic formulas.[4] However, there are certain special classes of polynomial equations that can be solved with algebraic methods. The way of identifying such equations was discovered around 1829 by the French mathematician Évariste Galois.[5]

[3] See, for example, *First Course in the Theory of Equations* by L. E. Dickson (New York: Wiley, 1922), or *Modern Algebra* by B. L. van der Waerden (New York: Frederick Ungar, 1953).

[4] What we mean by an "algebraic formula" is a formula in the coefficients using the operations of addition, subtraction, multiplication, division, and the taking of radicals. Mathematicians call the use of such formulas in solving polynomial equations "solution by radicals." If you were a math major, you would eventually go on to study this under the heading of Galois theory.

[5] Both Abel (1802–1829) and Galois (1811–1832) died young. Abel died of tuberculosis at the age of 26, while Galois was killed in a duel at the age of 20.

0.5 EXERCISES

Solve the equations in Exercises 1–12 for x (mentally, if possible).

1. $x + 1 = 0$ **2.** $x - 3 = 1$ **3.** $-x + 5 = 0$

4. $2x + 4 = 1$ **5.** $4x - 5 = 8$ **6.** $\frac{3}{4}x + 1 = 0$

7. $7x + 55 = 98$ **8.** $3x + 1 = x$ **9.** $x + 1 = 2x + 2$

10. $x + 1 = 3x + 1$ **11.** $ax + b = c \quad (a \neq 0)$

12. $x - 1 = cx + d \quad (c \neq 1)$

By any method, determine all possible real solutions of each equation in Exercises 13–30. Check your answers by substitution.

13. $2x^2 + 7x - 4 = 0$

14. $x^2 + x + 1 = 0$

15. $x^2 - x + 1 = 0$

16. $2x^2 - 4x + 3 = 0$

17. $2x^2 - 5 = 0$

18. $3x^2 - 1 = 0$

19. $-x^2 - 2x - 1 = 0$

20. $2x^2 - x - 3 = 0$

21. $\frac{1}{2}x^2 - x - \frac{3}{2} = 0$

22. $-\frac{1}{2}x^2 - \frac{1}{2}x + 1 = 0$

23. $x^2 - x = 1$

24. $16x^2 = -24x - 9$

25. $x = 2 - \frac{1}{x}$

26. $x + 4 = \frac{1}{x - 2}$

27. $x^4 - 10x^2 + 9 = 0$

28. $x^4 - 2x^2 + 1 = 0$

29. $x^4 + x^2 - 1 = 0$

30. $x^3 + 2x^2 + x = 0$

Find all possible real solutions of each equation in Exercises 31–44.

31. $x^3 + 6x^2 + 11x + 6 = 0$

32. $x^3 - 6x^2 + 12x - 8 = 0$

33. $x^3 + 4x^2 + 4x + 3 = 0$

34. $y^3 + 64 = 0$

35. $x^3 - 1 = 0$

36. $x^3 - 27 = 0$

37. $y^3 + 3y^2 + 3y + 2 = 0$

38. $y^3 - 2y^2 - 2y - 3 = 0$

39. $x^3 - x^2 - 5x + 5 = 0$

40. $x^3 - x^2 - 3x + 3 = 0$

41. $2x^6 - x^4 - 2x^2 + 1 = 0$

42. $3x^6 - x^4 - 12x^2 + 4 = 0$

43. $(x^2 + 3x + 2)(x^2 - 5x + 6) = 0$

44. $(x^2 - 4x + 4)^2(x^2 + 6x + 5)^3 = 0$

0.6 Solving Miscellaneous Equations

Equations often arise in calculus that are not polynomial equations of low degree. Many of these complicated-looking equations can be solved easily if you remember the following, which we used in the previous section:

Solving an Equation of the Form $P \cdot Q = 0$

If a product is equal to 0, then at least one of the factors must be 0. That is, if $P \cdot Q = 0$, then either $P = 0$ or $Q = 0$.

Quick Examples

1. $x^5 - 4x^3 = 0$

$x^3(x^2 - 4) = 0$ — Factor the left-hand side.

Either $x^3 = 0$ or $x^2 - 4 = 0$ — Either $P = 0$ or $Q = 0$.

$x = 0, 2$ or -2. — Solve the individual equations.

2. $(x^2 - 1)(x + 2) + (x^2 - 1)(x + 4) = 0$

$(x^2 - 1)[(x + 2) + (x + 4)] = 0$ — Factor the left-hand side.

$(x^2 - 1)(2x + 6) = 0$

Either $x^2 - 1 = 0$ or $2x + 6 = 0$ — Either $P = 0$ or $Q = 0$.

$x = -3, -1$, or 1. — Solve the individual equations.

EXAMPLE 1 Solving by Factoring

Solve $12x(x^2-4)^5(x^2+2)^6+12x(x^2-4)^6(x^2+2)^5=0$.

Solution

Again, we start by factoring the left-hand side:

$$\begin{aligned}&12x(x^2-4)^5(x^2+2)^6+12x(x^2-4)^6(x^2+2)^5\\&\quad=12x(x^2-4)^5(x^2+2)^5[(x^2+2)+(x^2-4)]\\&\quad=12x(x^2-4)^5(x^2+2)^5(2x^2-2)\\&\quad=24x(x^2-4)^5(x^2+2)^5(x^2-1).\end{aligned}$$

Setting this equal to 0, we get:

$$24x(x^2-4)^5(x^2+2)^5(x^2-1)=0,$$

which means that at least one of the factors of this product must be zero. Now it certainly cannot be the 24, but it could be the x: $x=0$ is one solution. It could also be that

$$(x^2-4)^5=0$$

or

$$x^2-4=0,$$

which has solutions $x=\pm 2$. Could it be that $(x^2+2)^5=0$? If so, then $x^2+2=0$, but this is impossible because $x^2+2\geq 2$, no matter what x is. Finally, it could be that $x^2-1=0$, which has solutions $x=\pm 1$. This gives us five solutions to the original equation:

$$x=-2,-1,0,1,\text{ or }2.$$

EXAMPLE 2 Solving by Factoring

Solve $(x^2-1)(x^2-4)=10$.

Solution Watch out! You may be tempted to say that $x^2-1=10$ or $x^2-4=10$, but this does not follow. If two numbers multiply to give you 10, what must they be? There are lots of possibilities: 2 and 5, 1 and 10, $-500{,}000$ and -0.00002 are just a few. The fact that the left-hand side is factored is nearly useless to us if we want to solve this equation. What we will have to do is multiply out, bring the 10 over to the left, and hope that we can factor what we get. Here goes:

$$\begin{aligned}x^4-5x^2+4&=10\\x^4-5x^2-6&=0\\(x^2-6)(x^2+1)&=0\end{aligned}$$

(Here we used a sometimes useful trick that we mentioned in Section 3: We treated x^2 like x and x^4 like x^2, so factoring x^4-5x^2-6 is essentially the same as factoring x^2-5x-6.) *Now* we are allowed to say that one of the factors must be 0: $x^2-6=0$ has solutions $x=\pm\sqrt{6}=\pm 2.449\ldots$ and $x^2+1=0$ has no real solutions. Therefore, we get exactly two solutions, $x=\pm\sqrt{6}=\pm 2.449\ldots$.

To solve equations involving rational expressions, the following rule is very useful.

Solving an Equation of the Form $P/Q = 0$

$$\text{If } \frac{P}{Q} = 0, \text{ then } P = 0.$$

How else could a fraction equal 0? If that is not convincing, multiply both sides by Q (which cannot be 0 if the quotient is defined).

Quick Example

$$\frac{(x+1)(x+2)^2 - (x+1)^2(x+2)}{(x+2)^4} = 0$$

$(x+1)(x+2)^2 - (x+1)^2(x+2) = 0$ — If $\frac{P}{Q} = 0$, then $P = 0$.

$(x+1)(x+2)[(x+2) - (x+1)] = 0$ — Factor.

$(x+1)(x+2)(1) = 0$

Either $x + 1 = 0$ or $x + 2 = 0$,

$x = -1$ or $x = -2$

$x = -1$ — $x = -2$ does not make sense in the original equation: it makes the denominator 0. So it is not a solution and $x = -1$ is the only solution.

EXAMPLE 3 Solving a Rational Equation

Solve $1 - \dfrac{1}{x^2} = 0$.

Solution Write 1 as $\frac{1}{1}$, so that we now have a difference of two rational expressions:

$$\frac{1}{1} - \frac{1}{x^2} = 0.$$

To combine these we can put both over a common denominator of x^2, which gives

$$\frac{x^2 - 1}{x^2} = 0.$$

Now we can set the numerator, $x^2 - 1$, equal to zero. Thus,

$$x^2 - 1 = 0$$

so

$$(x - 1)(x + 1) = 0,$$

giving $x = \pm 1$.

Before we go on... This equation could also have been solved by writing

$$1 = \frac{1}{x^2}$$

and then multiplying both sides by x^2. ■

EXAMPLE 4 Another Rational Equation

Solve $\frac{2x-1}{x} + \frac{3}{x-2} = 0$.

Solution We *could* first perform the addition on the left and then set the top equal to 0, but here is another approach. Subtracting the second expression from both sides gives

$$\frac{2x-1}{x} = \frac{-3}{x-2}$$

Cross-multiplying [multiplying both sides by both denominators—that is, by $x(x-2)$] now gives

$$(2x-1)(x-2) = -3x$$

so

$$2x^2 - 5x + 2 = -3x.$$

Adding $3x$ to both sides gives the quadratic equation

$$2x^2 - 2x + 1 = 0.$$

The discriminant is $(-2)^2 - 4 \cdot 2 \cdot 1 = -4 < 0$, so we conclude that there is no real solution.

Before we go on... Notice that when we said that $(2x-1)(x-2) = -3x$, we were *not* allowed to conclude that $2x-1 = -3x$ or $x-2 = -3x$. ■

EXAMPLE 5 A Rational Equation with Radicals

Solve $\frac{\left(2x\sqrt{x+1} - \frac{x^2}{\sqrt{x+1}}\right)}{x+1} = 0$.

Solution Setting the top equal to 0 gives

$$2x\sqrt{x+1} - \frac{x^2}{\sqrt{x+1}} = 0.$$

This still involves fractions. To get rid of the fractions, we could put everything over a common denominator ($\sqrt{x+1}$) and then set the top equal to 0, or we could multiply the whole equation by that common denominator in the first place to clear fractions. If we do the second, we get

$$2x(x+1) - x^2 = 0$$
$$2x^2 + 2x - x^2 = 0$$
$$x^2 + 2x = 0.$$

Factoring,

$$x(x+2) = 0$$

so either $x = 0$ or $x + 2 = 0$, giving us $x = 0$ or $x = -2$. Again, one of these is not really a solution. The problem is that $x = -2$ cannot be substituted into $\sqrt{x+1}$, because we would then have to take the square root of -1, and we are not allowing ourselves to do that. Therefore, $x = 0$ is the only solution.

0.6 EXERCISES

Solve the following equations:

1. $x^4 - 3x^3 = 0$

2. $x^6 - 9x^4 = 0$

3. $x^4 - 4x^2 = -4$

4. $x^4 - x^2 = 6$

5. $(x+1)(x+2) + (x+1)(x+3) = 0$

6. $(x+1)(x+2)^2 + (x+1)^2(x+2) = 0$

7. $(x^2+1)^5(x+3)^4 + (x^2+1)^6(x+3)^3 = 0$

8. $10x(x^2+1)^4(x^3+1)^5 - 10x^2(x^2+1)^5(x^3+1)^4 = 0$

9. $(x^3+1)\sqrt{x+1} - (x^3+1)^2\sqrt{x+1} = 0$

10. $(x^2+1)\sqrt{x+1} - \sqrt{(x+1)^3} = 0$

11. $\sqrt{(x+1)^3} + \sqrt{(x+1)^5} = 0$

12. $(x^2+1)\sqrt[3]{(x+1)^4} - \sqrt[3]{(x+1)^7} = 0$

13. $(x+1)^2(2x+3) - (x+1)(2x+3)^2 = 0$

14. $(x^2-1)^2(x+2)^3 - (x^2-1)^3(x+2)^2 = 0$

15. $\dfrac{(x+1)^2(x+2)^3 - (x+1)^3(x+2)^2}{(x+2)^6} = 0$

16. $\dfrac{6x(x^2+1)^2(x^2+2)^4 - 8x(x^2+1)^3(x^2+2)^3}{(x^2+2)^8} = 0$

17. $\dfrac{2(x^2-1)\sqrt{x^2+1} - \frac{x^4}{\sqrt{x^2+1}}}{x^2+1} = 0$

18. $\dfrac{4x\sqrt{x^3-1} - \frac{3x^4}{\sqrt{x^3-1}}}{x^3-1} = 0$

19. $x - \dfrac{1}{x} = 0$

20. $1 - \dfrac{4}{x^2} = 0$

21. $\dfrac{1}{x} - \dfrac{9}{x^3} = 0$

22. $\dfrac{1}{x^2} - \dfrac{1}{x+1} = 0$

23. $\dfrac{x-4}{x+1} - \dfrac{x}{x-1} = 0$

24. $\dfrac{2x-3}{x-1} - \dfrac{2x+3}{x+1} = 0$

25. $\dfrac{x+4}{x+1} + \dfrac{x+4}{3x} = 0$

26. $\dfrac{2x-3}{x} - \dfrac{2x-3}{x+1} = 0$

0.7 The Coordinate Plane

Q: *Just what is the xy-plane?*

A: The *xy*-plane is an infinite flat surface with two perpendicular lines, usually labeled the ***x*-axis** and ***y*-axis**. These axes are calibrated as shown in Figure 2. (Notice also how the plane is divided into four **quadrants**.)

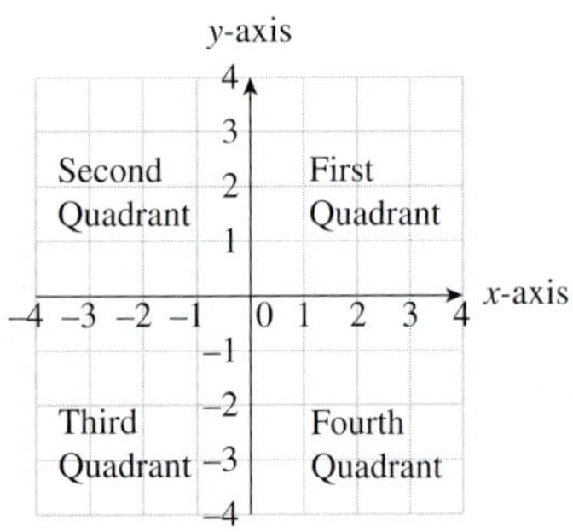

The *xy*-plane

Figure 2

Thus, the *xy*-plane is nothing more than a very large—in fact, infinitely large—flat surface. The purpose of the axes is to allow us to locate specific positions, or **points**, on the plane, with the use of **coordinates**. (If Captain Picard wants to have himself beamed to a specific location, he must supply its coordinates, or he's in trouble.)

Q: *So how do we use coordinates to locate points?*

A: The rule is simple. Each point in the plane has two coordinates, an ***x*-coordinate** and a ***y*-coordinate**. These can be determined in two ways:

1. The *x*-coordinate measures a point's distance to the right or left of the *y*-axis. It is positive if the point is to the right of the axis, negative if it is to the left of the axis, and 0 if it is on the axis. The *y*-coordinate measures a point's distance above or below the *x*-axis. It is positive if the point is above the axis, negative if it is below the axis, and 0 if it is on the axis. Briefly, the *x*-coordinate tells us the *horizontal* position (distance left or right), and the *y*-coordinate tells us the *vertical* position (height).

2. Given a point P, we get its x-coordinate by drawing a vertical line from P and seeing where it intersects the x-axis. Similarly, we get the y-coordinate by extending a horizontal line from P and seeing where it intersects the y-axis.

This way of assigning coordinates to points in the plane is often called the system of **Cartesian** coordinates, in honor of the mathematician and philosopher René Descartes (1596–1650), who was the first to use them extensively.

Here are a few examples to help you review coordinates.

EXAMPLE 1 Coordinates of Points

a. Find the coordinates of the indicated points. (See Figure 3. The grid lines are placed at intervals of one unit.)

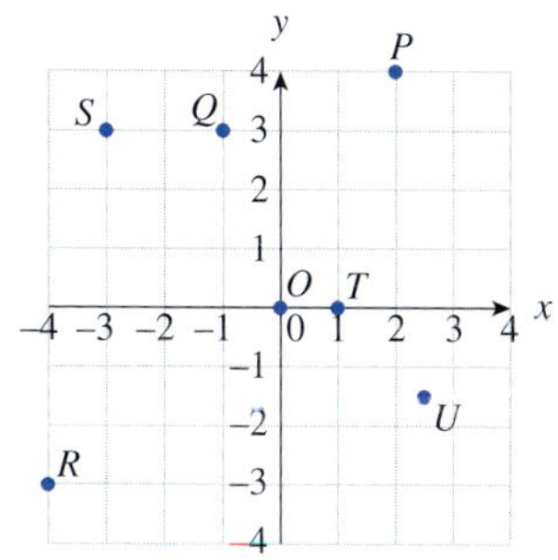

Figure 3

b. Locate the following points in the xy-plane.

$$A(2, 3),\ B(-4, 2),\ C(3, -2.5),\ D(0, -3),\ E(3.5, 0),\ F(-2.5, -1.5)$$

Solution

a. Taking them in alphabetical order, we start with the origin O. This point has height zero and is also zero units to the right of the y-axis, so its coordinates are (0, 0). Turning to P, dropping a vertical line gives $x = 2$ and extending a horizontal line gives $y = 4$. Thus, P has coordinates (2, 4). For practice, determine the coordinates of the remaining points, and check your work against the list that follows:

$$Q(-1, 3),\ R(-4, -3),\ S(-3, 3),\ T(1, 0),\ U(2.5, -1.5)$$

b. In order to locate the given points, we start at the origin (0, 0), and proceed as follows. (See Figure 4.)

To locate A, we move 2 units to the right and 3 up, as shown.

To locate B, we move -4 units to the right (that is, 4 to the *left*) and 2 up, as shown.

To locate C, we move 3 units right and 2.5 down.

We locate the remaining points in a similar way.

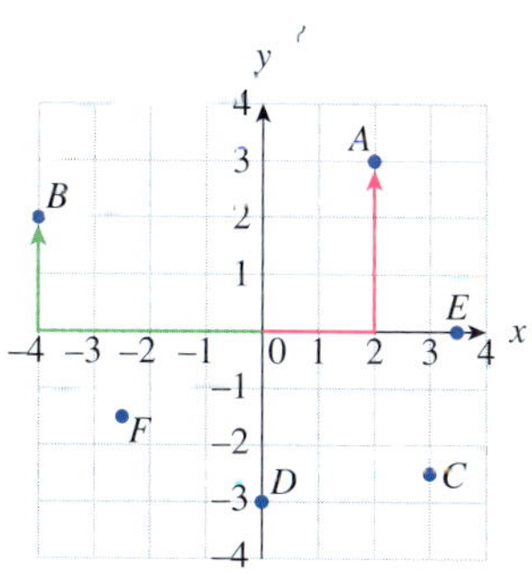

Figure 4

The Graph of an Equation

One of the more surprising developments of mathematics was the realization that equations, which are algebraic objects, can be represented by graphs, which are geometric objects. The kinds of equations that we have in mind are equations in x and y, such as

$$y = 4x - 1, \quad 2x^2 - y = 0, \quad y = 3x^2 + 1, \quad y = \sqrt{x-1}.$$

The **graph** of an equation in the two variables x and y consists of all points (x, y) in the plane whose coordinates are solutions of the equation.

EXAMPLE 2 Graph of an Equation

Obtain the graph of the equation $y - x^2 = 0$.

Solution We can solve the equation for y to obtain $y = x^2$. Solutions can then be obtained by choosing values for x and then computing y by squaring the value of x, as shown in the following table:

x	-3	-2	-1	0	1	2	3
$y = x^2$	9	4	1	0	1	4	9

Plotting these points (x, y) gives the following picture (left side of Figure 5), suggesting the graph on the right in Figure 5.

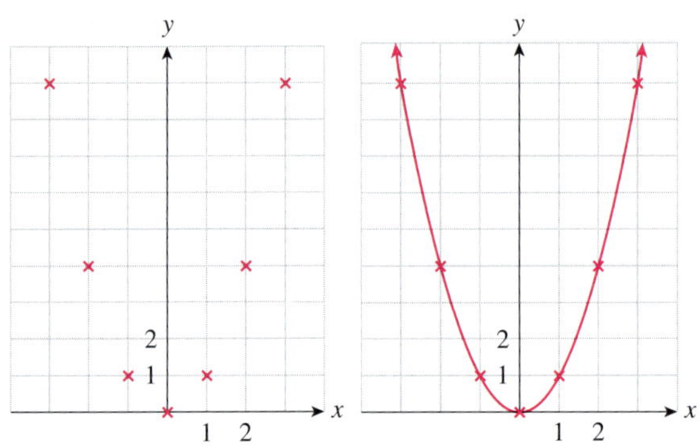

Figure 5

Distance

The distance between two points in the xy-plane can be expressed as a function of their coordinates, as follows:

Distance Formula

The distance between the points $P(x_1, y_1)$ and $Q(x_2, y_2)$ is

$$d = \sqrt{(x_2 - x_1)^2 + (y_2 - y_1)^2} = \sqrt{(\Delta x)^2 + (\Delta y)^2}.$$

Derivation

The distance d is shown in the figure below.

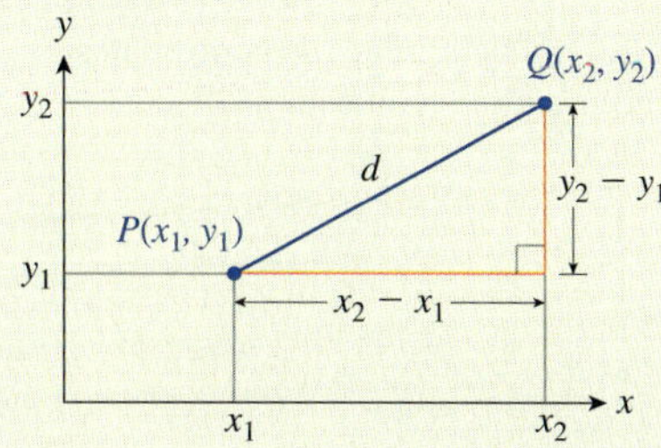

By the Pythagorean theorem applied to the right triangle shown, we get

$$d^2 = (x_2 - x_1)^2 + (y_2 - y_1)^2.$$

Taking square roots (d is a distance, so we take the positive square root), we get the distance formula. Notice that if we switch x_1 with x_2 or y_1 with y_2, we get the same result.

Quick Examples

1. The distance between the points $(3, -2)$ and $(-1, 1)$ is
$$d = \sqrt{(-1-3)^2 + (1+2)^2} = \sqrt{25} = 5.$$
2. The distance from (x, y) to the origin $(0, 0)$ is
$$d = \sqrt{(x-0)^2 + (y-0)^2} = \sqrt{x^2 + y^2}.$$ Distance to the origin

The set of all points (x, y) whose distance from the origin $(0, 0)$ is a fixed quantity r is a circle centered at the origin with radius r. From the second Quick Example, we get the following equation for the circle centered at the origin with radius r:

$$\sqrt{x^2 + y^2} = r.$$ Distance from the origin $= r$.

Squaring both sides gives the following equation:

Equation of the Circle of Radius *r* Centered at the Origin

$$x^2 + y^2 = r^2$$

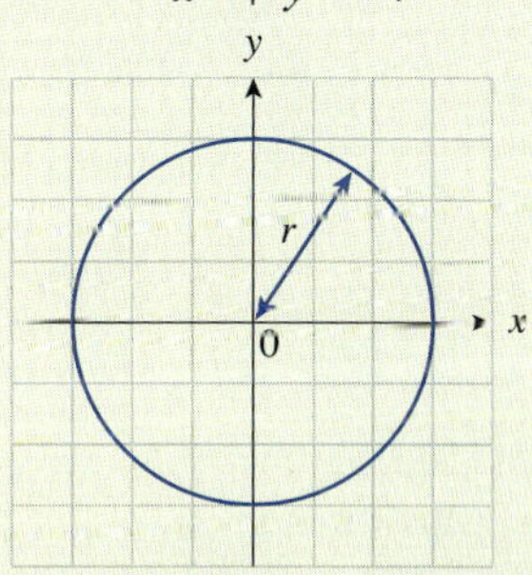

Quick Examples

1. The circle of radius 1 centered at the origin has equation $x^2 + y^2 = 1$.
2. The circle of radius 2 centered at the origin has equation $x^2 + y^2 = 4$.

0.7 EXERCISES

1. Referring to the following figure, determine the coordinates of the indicated points as accurately as you can.

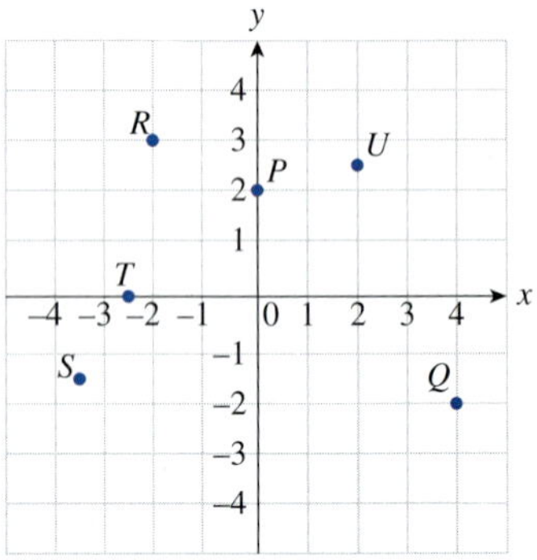

2. Referring to the following figure, determine the coordinates of the indicated points as accurately as you can.

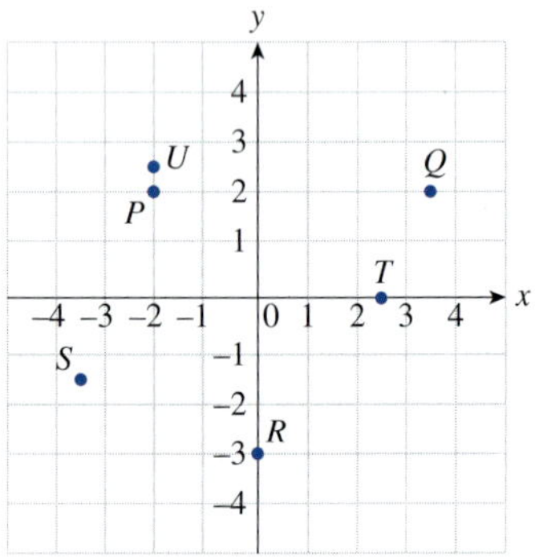

3. Graph the following points.

$P(4, 4)$, $Q(4, -4)$, $R(3, 0)$, $S(4, 0.5)$, $T(0.5, 2.5)$, $U(-2, 0)$, $V(-4, 4)$

4. Graph the following points.

$P(4, -2)$, $Q(2, -4)$, $R(1, -3)$, $S(-4, 2)$, $T(2, -1)$, $U(-2, 0)$, $V(-4, -4)$

Sketch the graphs of the equations in Exercises 5–12.

5. $x + y = 1$

6. $y - x = -1$

7. $2y - x^2 = 1$

8. $2y + \sqrt{x} = 1$

9. $xy = 4$

10. $x^2y = -1$

11. $xy = x^2 + 1$

12. $xy = 2x^3 + 1$

In Exercises 13–16, find the distance between the given pairs of points.

13. $(1, -1)$ and $(2, -2)$

14. $(1, 0)$ and $(6, 1)$

15. $(a, 0)$ and $(0, b)$

16. (a, a) and (b, b)

17. Find the value of k such that $(1, k)$ is equidistant from $(0, 0)$ and $(2, 1)$.

18. Find the value of k such that (k, k) is equidistant from $(-1, 0)$ and $(0, 2)$.

19. Describe the set of points (x, y) such that $x^2 + y^2 = 9$.

20. Describe the set of points (x, y) such that $x^2 + y^2 = 0$.

1

Functions and Linear Models

Case Study Modeling Spending on Internet Advertising

You are the new director of *Impact Advertising Inc.'s* Internet division, which has enjoyed a steady 0.25% of the Internet advertising market. You have drawn up an ambitious proposal to expand your division in light of your anticipation that Internet advertising will continue to skyrocket. The VP in charge of Financial Affairs feels that current projections (based on a linear model) do not warrant the level of expansion you propose. **How can you persuade the VP that those projections do not fit the data convincingly?**

Jeff Titcomb/Photographer's Choice / Getty Images

Introduction

To analyze recent trends in spending on Internet advertising and to make reasonable projections, we need a mathematical model of this spending. Where do we start? To apply mathematics to real-world situations like this, we need a good understanding of basic mathematical concepts. Perhaps the most fundamental of these concepts is that of a function: a relationship that shows how one quantity depends on another. Functions may be described numerically and, often, algebraically. They can also be described graphically—a viewpoint that is extremely useful.

The simplest functions—the ones with the simplest formulas and the simplest graphs—are linear functions. Because of their simplicity, they are also among the most useful functions and can often be used to model real-world situations, at least over short periods of time. In discussing linear functions, we will meet the concepts of slope and rate of change, which are the starting point of the mathematics of change.

In the last section of this chapter, we discuss *simple linear regression*: construction of linear functions that best fit given collections of data. Regression is used extensively in applied mathematics, statistics, and quantitative methods in business. The inclusion of regression utilities in computer spreadsheets like Excel® makes this powerful mathematical tool readily available for anyone to use.

algebra Review
For this chapter, you should be familiar with real numbers and intervals. To review this material, see **Chapter 0.**

1.1 Functions from the Numerical and Algebraic Viewpoints

The following table gives the approximate number of Facebook users at various times since its establishment early in 2004.[1]

Year t (Since start of 2004)	0	0.5	1	1.5	2	2.5	3	3.5	4	4.5
Facebook Members n (Millions)	0	0.5	1	2	5.5	7	12	30	58	80

Let's write $n(0)$ for the number of members (in millions) at time $t = 0$, $n(0.5)$ for the number at time $t = 0.5$, and so on. (We read $n(0)$ as "n of 0.") Thus, $n(0) = 0$, $n(0.5) = 0.5$, $n(1) = 1$, $n(1.5) = 2, \ldots, n(4.5) = 80$. In general, we write $n(t)$ for the number of members (in millions) at time t. We call n a **function** of the variable t, meaning that for each value of t between 0 and 4.5, n gives us a single corresponding number $n(t)$ (the number of members at that time).

In general, we think of a function as a way of producing new objects from old ones. The functions we deal with in this text produce new numbers from old numbers. The numbers we have in mind are the *real* numbers, including not only positive and negative integers and fractions but also numbers like $\sqrt{2}$ or π. (See Appendix A for more on real numbers.) For this reason, the functions we use are called **real-valued functions of a real variable**. For example, the function n takes the year since the start of 2004 as input and returns the number of Facebook members as output (Figure 1).

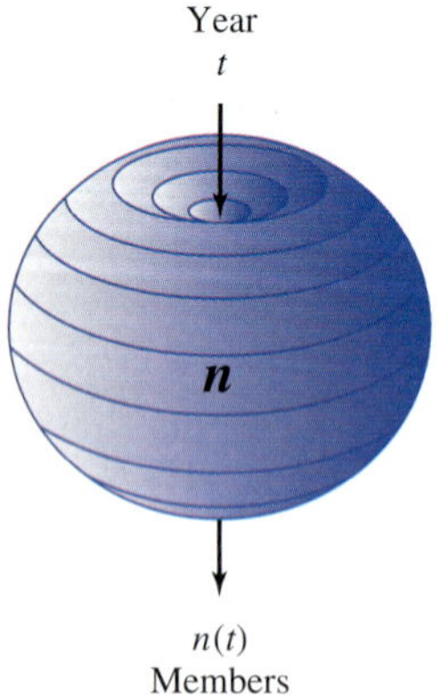

Figure 1

[1] Sources: www.facebook.com, www.insidehighered.com (Some data are interpolated.)

The variable t is called the **independent** variable, while n is called the **dependent variable** as its value depends on t. A function may be specified in several different ways. Here, we have specified the function n **numerically** by giving the values of the function for a number of values of the independent variable, as in the preceding table.

Q: *For which values of t does it make sense to ask for n(t)? In other words, for which years t is the function n defined?*

A: Because $n(t)$ refers to the number of members from the start of 2004 to the middle of 2008, $n(t)$ is defined when t is any number between 0 and 4.5, that is, when $0 \le t \le 4.5$. Using interval notation (see Chapter 0), we can say that $n(t)$ is defined when t is in the interval [0, 4.5].

The set of values of the independent variable for which a function is defined is called its **domain** and is a necessary part of the definition of the function. Notice that the preceding table gives the value of $n(t)$ at only some of the infinitely many possible values in the domain [0, 4.5]. The domain of a function is not always specified explicitly; if no domain is specified for the function f, we take the domain to be the largest set of numbers x for which $f(x)$ makes sense. This "largest possible domain" is sometimes called the **natural domain**.

The previous Facebook data can also be represented on a graph by plotting the given pairs of numbers $(t, n(t))$ in the xy-plane. (See Figure 2.) We have connected successive points by line segments. In general, the **graph** of a function f consists of all points $(x, f(x))$ in the plane with x in the domain of f.

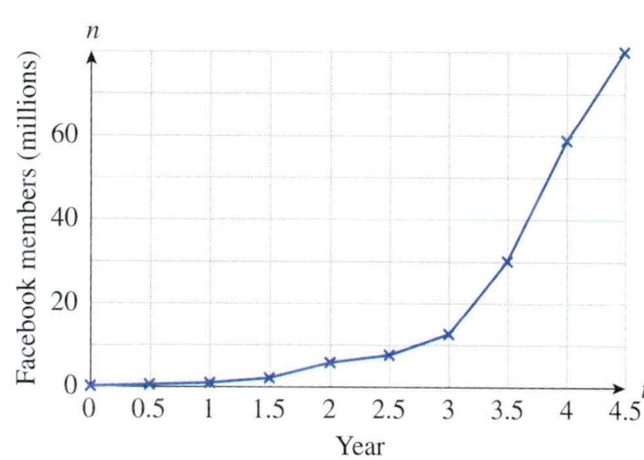

Figure 2

*** NOTE** In a graphically defined function, we can never know the y coordinates of points exactly; no matter how accurately a graph is drawn, we can obtain only *approximate* values of the coordinates of points. That is why we have been using the word *estimate* rather than *calculate* and why we say "$n(3) \approx 12$" rather than "$n(3) = 12$."

In Figure 2 we specified the function n **graphically** by using a graph to display its values. Suppose now that we had only the graph without the table of data. We could use the graph to find approximate values of n. For instance, to find $n(3)$ from the graph, we do the following:

1. Find the desired value of t at the bottom of the graph ($t = 3$ in this case).
2. Estimate the height (n-coordinate) of the corresponding point on the graph (around 12 in this case).

Thus, $n(3) \approx 12$ million members.*

In some cases we may be able to use an algebraic formula to calculate the function, and we say that the function is specified **algebraically**. These are not the only ways in which a function can be specified; for instance, it could also be specified **verbally**, as in "Let $n(t)$ be the number of Facebook members, in millions, t years since the start of 2004."* Notice that any function can be represented graphically by plotting the points $(x, f(x))$ for a number of values of x in its domain.

* **NOTE** Specifying a function verbally in this way is useful for understanding what the function is doing, but it gives no numerical information.

Here is a summary of the terms we have just introduced.

Functions

A **real-valued function f of a real-valued variable x** assigns to each real number x in a specified set of numbers, called the **domain** of f, a unique real number $f(x)$, read "f of x." The variable x is called the **independent variable**, and f is called the **dependent variable**. A function is usually specified **numerically** using a table of values, **graphically** using a graph, or **algebraically** using a formula. The **graph of a function** consists of all points $(x, f(x))$ in the plane with x in the domain of f.

Quick Examples

1. **A function specified numerically:** Take $f(t)$ to be the amount of freon (in tons) produced in developing countries in year t since 2000, represented by the following table:

t (Year Since 2000)	$f(t)$ (Tons of Freon 22)
0	100
2	140
4	200
6	270
8	400
10	590

Source: *New York Times*, February 23, 2007, p. C1.

The domain of f is [0, 10], the independent variable is t, the number of years since 2000, and the dependent variable is f, the number of tons of freon produced in a year in developing countries. Some values of f are:

$f(0) = 100$ 100 tons of freon were produced in developing countries in 2000.

$f(6) = 270$ 270 tons of freon were produced in developing countries in 2006.

$f(10) = 590$ 590 tons of freon were produced in developing countries in 2010.

Graph of f: Plotting the pairs $(t, f(t))$ gives the following graph:

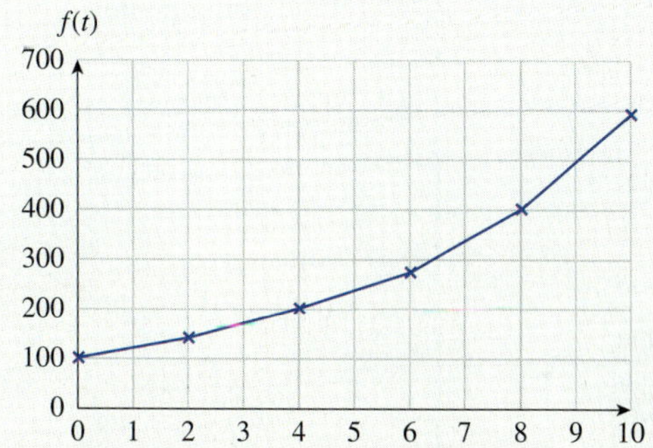

2. **A function specified graphically:** Take $p(t)$ to be the home price index as a percentage change from 2003 in year t, represented by the following graph:

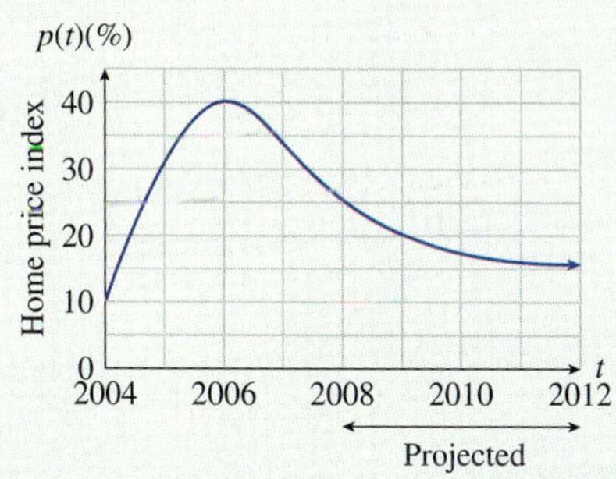

S&P/Case-Shiller Home Price Index. Source: Standard & Poors/Bloomberg Financial Markets/*New York Times*, September 29, 2007, p. C3. Projection is the authors'.*

* **NOTE** *added in July 2009:* Our projection turned out to be wrong: The index fell further than we anticipated and wound up negative! This error illustrates the pitfalls of *extrapolation*—a point we will discuss after Example 1.

The domain of p is [2004, 2012], the independent variable is the the year t, and the dependent variable is the percentage p above the 2003 value. Some values of p are:

$p(2004) \approx 10$ In 2004 the index was about 10% above the 2003 value.

$p(2006) \approx 40$ In 2006 the index was about 40% above the 2003 value.

$p(2009) \approx 20$ In 2009 the index was about 20% above the 2003 value.

3. **A function specified algebraically:** Let $f(x) = \frac{1}{x}$. The function f is specified algebraically. The independent variable is x and the dependent variable is f. The natural domain of f consists of all real numbers except zero because $f(x)$ makes sense for all values of x other than $x = 0$. Some specific values of f are

$$f(2) = \frac{1}{2} \qquad f(3) = \frac{1}{3} \qquad f(-1) = \frac{1}{-1} = -1$$

$f(0)$ is not defined because 0 is not in the domain of f.

4. The graph of a function: Let $f(x) = x^2$, with domain the set of all real numbers. To draw the graph of f, first choose some convenient values of x in the domain and compute the corresponding y-coordinates $f(x)$:

x	−3	−2	−1	0	1	2	3
$f(x) = x^2$	9	4	1	0	1	4	9

Plotting these points $(x, f(x))$ gives the picture on the left, suggesting the graph on the right.*

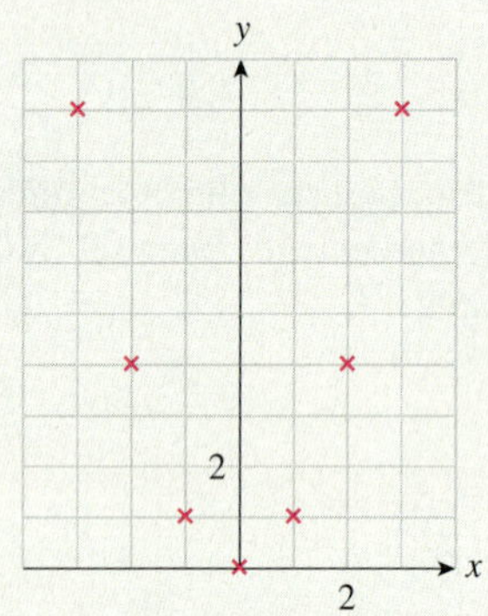

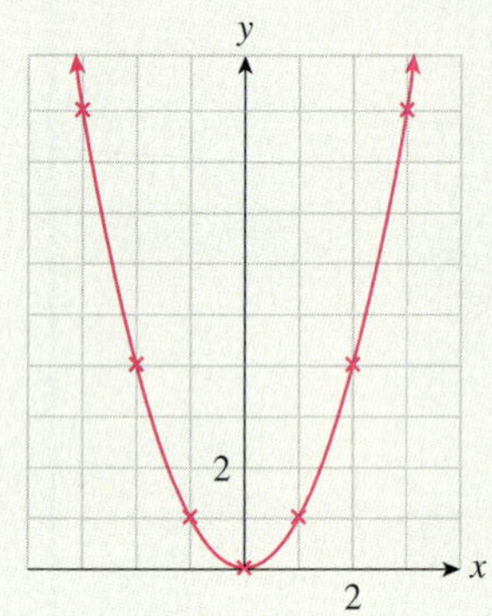

(This particular curve happens to be called a **parabola**, and its lowest point, at the origin, is called its **vertex**.)

✱ NOTE If you plot more points, you will find that they lie on a smooth curve as shown. That is why we did not use line segments to connect the points.

Bartomeu Amengual/Index Stock Imagery / PhotoLibrary

EXAMPLE 1 iPod Sales

The number of iPods sold by Apple Inc. each year from 2004 through 2007 can be approximated by

$$f(x) = -x^2 + 20x + 3 \text{ million iPods} \qquad (0 \le x \le 3)$$

where x is the number of years since 2004.*

a. What is the domain of f? Compute $f(0)$, $f(1)$, $f(2)$, and $f(3)$. What do these answers tell you about iPod sales? Is $f(-1)$ defined?

b. Compute $f(a)$, $f(-b)$, and $f(a + h)$ assuming that the quantities a, $-b$, and $a + h$ are in the domain of f.

c. Sketch the graph of f. Does the shape of the curve suggest that iPod sales were accelerating or decelerating?

Solution

a. The domain of f is the set of numbers x with $0 \le x \le 3$—that is, the interval [0, 3]. If we substitute 0 for x in the formula for $f(x)$, we get

$$f(0) = -(0)^2 + 20(0) + 3 = 3$$ In 2004 approximately 3 million iPods were sold.

* Source for data: Apple quarterly earnings reports at www.apple.com/investor/.

so $f(0) = 3$. Similarly,

$$f(1) = -(1)^2 + 20(1) + 3 = 22 \qquad \text{In 2005 approximately 22 million iPods were sold.}$$
$$f(2) = -(2)^2 + 20(2) + 3 = 39 \qquad \text{In 2006 approximately 39 million iPods were sold.}$$
$$f(2) = -(3)^2 + 20(3) + 3 = 54 \qquad \text{In 2007 approximately 54 million iPods were sold.}$$

Because -1 is not in the domain of f, $f(-1)$ is not defined.

b. To find $f(a)$ we substitute a for x in the formula for $f(x)$ to get

$$f(a) = -a^2 + 20a + 3. \qquad \text{Substitute } a \text{ for } x.$$

Similarly,

$$f(-b) = -(-b)^2 + 20(-b) + 3 \qquad \text{Substitute } -b \text{ for } x.$$
$$= -b^2 - 20b + 3 \qquad (-b)^2 = b^2$$
$$f(a+h) = -(a+h)^2 + 20(a+h) + 3 \qquad \text{Substitute } (a+h) \text{ for } x.$$
$$= -(a^2 + 2ah + h^2) + 20a + 20h + 3 \qquad \text{Expand.}$$
$$= -a^2 - 2ah - h^2 + 20a + 20h + 3$$

Note how we placed parentheses around the quantities at which we evaluated the function. If we omitted any of these parentheses, we would likely get errors.

$$f(-b) = -(-b)^2 + 20(-b) + 3 \ ✓ \qquad \textbf{NOT} \ --b^2 + 20(-b) + 3 \ ×$$
$$f(a+h) = -(a+h)^2 + 20(a+h) + 3 \ ✓ \qquad \textbf{NOT} \ -a + h^2 + 20a + h + 3 \ ×$$

c. To draw the graph of f, we plot points of the form $(x, f(x))$ for several values of x in the domain of f. Let us use the values we computed in part (a):

x	0	1	2	3
$f(x) = -x^2 + 20x + 3$	3	22	39	54

Graphing these points gives the graph shown in Figure 3(a), suggesting the curve shown in Figure 3(b).

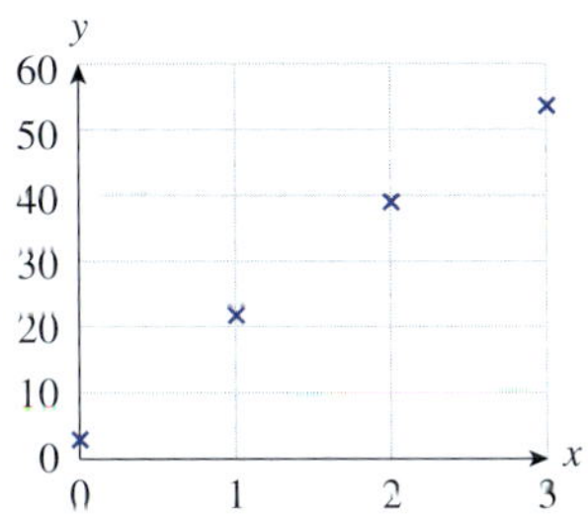

Figure 3(a)

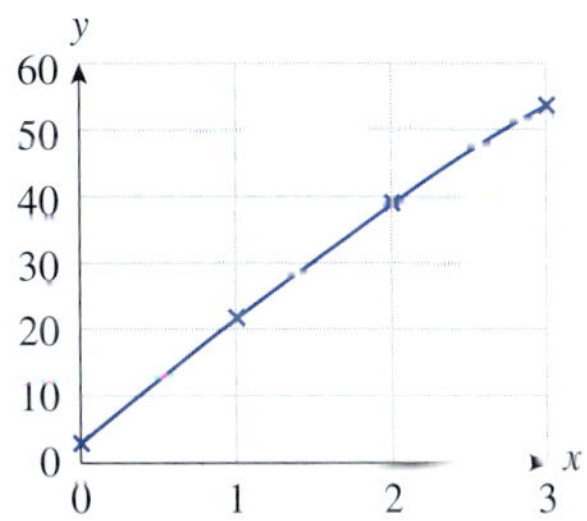

Figure 3(b)

The graph is becoming less steep as we move from left to right, suggesting that iPod sales were decelerating slightly. (This concept will be made more precise in Chapter 4.)

using Technology

See the Technology Guide at the end of the chapter for detailed instructions on how to obtain the table of values and graph in Example 1 using a TI-83/84 Plus or Excel. Here is an outline:

TI-83/84 Plus

Table of values:
`Y1=-X^2+20X+3`
[2ND] [TABLE].
Graph: [WINDOW];
Xmin = 0, Xmax = 3
[ZOOM] [0].
[More details on page 110.]

Excel

Table of values: Headings x and $f(x)$ in A1–B1; t-values 0, 1, 2, 3 in A2–A5.
`=-1*A2^2+20*A2+3`
in B2; copy down through B5.
Graph: Highlight A1 through B5 and insert a Scatter chart.
[More details on page 116.]

Web Site

www.AppliedCalc.org
Go to the Function Evaluator and Grapher under Online Utilities and enter
`-x^2+20x+3`
for y_1. To obtain a table of values, enter the x-values 0, 1, 2, 3 in the Evaluator box and press "Evaluate."
Graph: Set Xmin = 0 and Xmax = 3, and press "Plot Graphs."

➡ Before we go on... The following table compares the value of f in Example 1 with the actual sales figures:

Year x	0	1	2	3
$f(x) = -x^2 + 20x + 3$	3	22	39	54
Actual iPod Sales (millions)	4	22	39	52

The actual figures are stated here only for integer values of x; for instance, $x = 2$ gives the sales for the year ending December 2006. But what were, for instance, the sales for the year ending June 2007 ($x = 2.5$)? This is where our formula comes in handy: We can use the formula for f to **interpolate**; that is, to find sales at values of x between those that are stated:

$$f(2.5) = -(2.5)^2 + 20(2.5) + 3 = 46.75 \approx 47 \text{ million iPods}$$

We can also use the formula to **extrapolate**; that is, to predict sales at values of x *outside* the domain—say, for $x = 3.5$ (the year ending June 2008):

$$f(3.5) = -(3.5)^2 + 20(3.5) + 3 = 60.75 \approx 61 \text{ million iPods}$$

As a general rule, extrapolation is far less reliable than interpolation: Predicting the future from current data is difficult, especially given the vagaries of the marketplace.

We call the algebraic function f an **algebraic model** of iPod sales because it uses an algebraic formula to model—or mathematically represent (approximately)—the annual sales. The particular kind of algebraic model we used is called a **quadratic model**. (See the end of this section for the names of some commonly used models.) ■

Note Equation and Function Notation

Instead of using *function notation*

$$f(x) = -x^2 + 20x + 3 \qquad \text{Function notation}$$

we could use *equation notation*

$$y = -x^2 + 20x + 3 \qquad \text{Equation notation}$$

(the choice of the letter y is a convention) and we say that "y is a function of x." When we write a function in this form, the variable x is the independent variable and y is the dependent variable.

We could also write the above function as $f = -x^2 + 20x + 3$, in which case the dependent variable would be f. ■

Look again at the graph of the number of Facebook users in Figure 2. From year 0 through year 3, the membership appears to curve gently upward, but then increases quite dramatically from year 3 to year 4. This behavior can be modeled by using two different functions: one for the interval [0, 3] and another for the interval [3, 4]. (See Figure 4.)

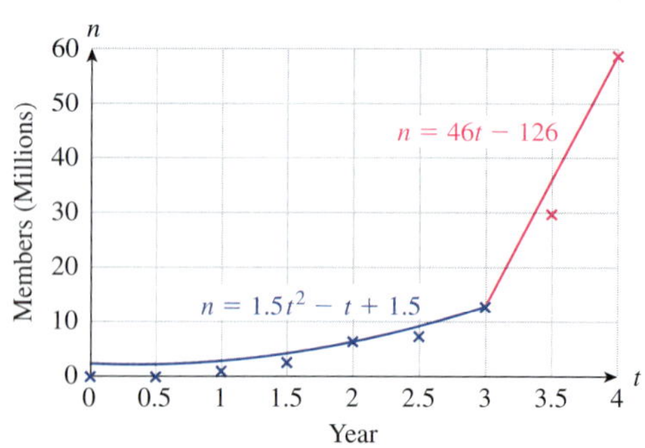

Figure 4

A function specified by two or more different formulas like this is called a **piecewise-defined function**.

using Technology

See the Technology Guide at the end of the chapter for detailed instructions on how to obtain the values and graph of n in Example 2 using a TI-83/84 Plus or Excel. Here is an outline:

TI-83/84 Plus

Table of values:
`Y1=(X≤3)(1.5X^2-X+1.5)+(X>3)(46X-126)`
[2ND] [TABLE].

Graph: [WINDOW]; Xmin = 0, Xmax = 5; [ZOOM] [0].
[More details on page 110.]

Excel

Table of values: Headings t and $n(t)$ in A1 and B1; t-values 0, 0.5, 1, 1.5, . . . , 5 in A2–A12.
`=(A2<=3)*(1.5*A2^2-A2+1.5)+(A2>3)*(46*A2-126)`
in B2; copy down through B12.

Graph: Highlight A1–B10; insert Scatter chart. [More details on page 117.]

Web Site

www.AppliedCalc.org
Go to the Function Evaluator and Grapher under Online Utilities, and enter
`(x<=3)(1.5x^2-x+1.5)+(x>3)(46x-126)`
for y_1. To obtain a table of values, enter the x-values 0, 0.5, 1, 1.5, . . . , 5 in the Evaluator box, and press "Evaluate."
Graph: Set Xmin = 0 and Xmax = 5, and press "Plot Graphs."

EXAMPLE 2 A Piecewise-Defined Function: Facebook Membership

The number $n(t)$ of Facebook members can be approximated by the following function of time t in years ($t = 0$ represents January 2004):*

$$n(t) = \begin{cases} 1.5t^2 - t + 1.5 & \text{if } 0 \le t \le 3 \\ 46t - 126 & \text{if } 3 < t \le 5 \end{cases} \quad \text{million members}$$

(Its graph is shown in Figure 4.) What was the approximate membership of Facebook in January 2005, January 2007, and June 2008? Sketch the graph of n by plotting several points.

Solution We evaluate the given function at the corresponding values of t:

Jan. 2005 ($t = 1$): $n(1) = 1.5(1)^2 - 1 + 1.5 = 2$ — Use the first formula because $0 \le t \le 3$.

Jan. 2007 ($t = 3$): $n(3) = 1.5(3)^2 - 3 + 1.5 = 12$ — Use the first formula because $0 \le t \le 3$.

June 2008 ($t = 4.5$): $n(4.5) = 46(4.5) - 126 = 81$ — Use the second formula because $3 < t \le 5$.

Thus, the number of Facebook members was approximately 2 million in January 2005, 12 million in January 2007, and 81 million in June 2008.

To sketch the graph of n, we use a table of values of $n(t)$ (some of which we have already calculated above), plot the points, and connect them to sketch the graph:

t	0	1	2	3	3.5	4	4.5	5
$n(t)$	1.5	2	5.5	12	35	58	81	104

First formula (t = 0 to 3) — Second formula (t = 3.5 to 5)

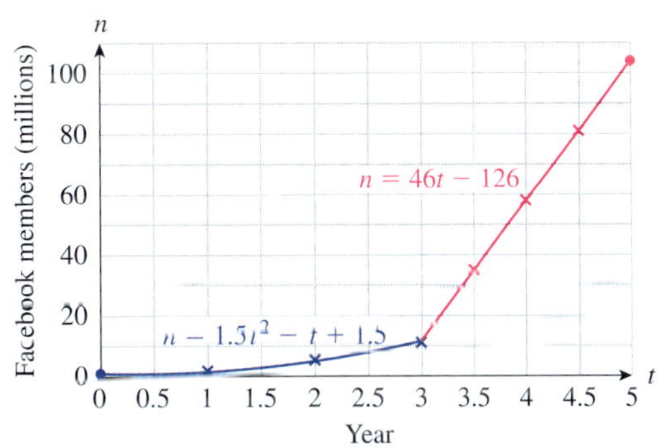

Figure 5

The graph (Figure 5) has the following features:

1. The first formula (the curve) is used for $0 \le t \le 3$.
2. The second formula (ascending line) is used for $3 < t \le 5$.

---

* Source for data: www.facebook.com/press.php.

3. The domain is [0, 5], so the graph is cut off at $t = 0$ and $t = 5$.
4. The solid dots at the ends indicate the endpoints of the domain.

using Technology

See the Technology Guide at the end of the chapter for detailed instructions on how to obtain the values and graph of f in Example 3 using a TI-83/84 Plus or Excel. Here is an outline:

TI-83/84 Plus

```
Y1=(X<-1)*(-1)+
(-1≤X)*(X≤1)*X+
(1<X)*(X^2-1)
```

Table of values: 2ND TABLE.
Graph: WINDOW; Xmin = −4, Xmax = 2; ZOOM 0.
[More details on page 111.]

Excel

Headings x and $f(x)$ in A1 and B1; x-values −4, −3.9, . . . , 2 in A2–A62.

```
=(A2<-1)*(-1)+
(-1<=A2)*(A2<=1)*A2
+(1<A2)*(A2^2-1)
```

in B2; copy down through B62.
Graph: Highlight A1–B62; insert Scatter chart. [More details on page 117.]

Web Site

www.AppliedCalc.org
Go to the Function Evaluator and Grapher under Online Utilities and enter

```
=(x<-1)*(-1)+(-1<=x)*
(x<=1)*x+(1<x)*(x^2-1)
```

for y_1. To obtain a table of values, enter the x-values 0, 0.5, 1, 1.5, . . . , 3 in the Evaluator box and press "Evaluate."
Graph: Set Xmin = −4 and Xmax = 2, and press "Plot Graphs."

EXAMPLE 3 More Complicated Piecewise-Defined Functions

Let f be the function specified by

$$f(x) = \begin{cases} -1 & \text{if } -4 \le x < -1 \\ x & \text{if } -1 \le x \le 1 \\ x^2 - 1 & \text{if } 1 < x \le 2 \end{cases}.$$

a. What is the domain of f? Find $f(-2)$, $f(-1)$, $f(0)$, $f(1)$, and $f(2)$.
b. Sketch the graph of f.

Solution

a. The domain of f is $[-4, 2]$, because $f(x)$ is specified only when $-4 \le x \le 2$.

$f(-2) = -1$ — We used the first formula because $-4 \le x < -1$.
$f(-1) = -1$ — We used the second formula because $-1 \le x \le 1$.
$f(0) = 0$ — We used the second formula because $-1 \le x \le 1$.
$f(1) = 1$ — We used the second formula because $-1 \le x \le 1$.
$f(2) = 2^2 - 1 = 3$ — We used the third formula because $1 < x \le 2$.

b. To sketch the graph by hand, we first sketch the three graphs $y = -1$, $y = x$, and $y = x^2 - 1$, and then use the appropriate portion of each (Figure 6).

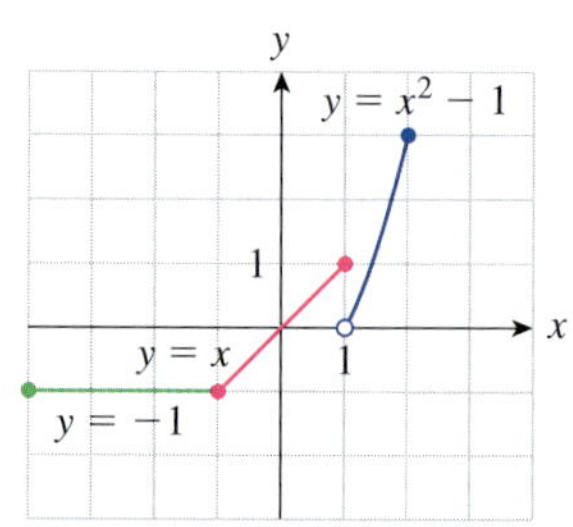

Figure 6

Note that solid dots indicate points on the graph, whereas the open dots indicate points *not* on the graph. For example, when $x = 1$, the inequalities in the formula tell us that we are to use the middle formula (x) rather than the bottom one ($x^2 - 1$). Thus, $f(1) = 1$, not 0, so we place a solid dot at (1, 1) and an open dot at (1, 0).

Vertical Line Test

Every point in the graph of a function has the form $(x, f(x))$ for some x in the domain of f. Because f assigns a *single* value $f(x)$ to each value of x in the domain, it follows that, in the graph of f, there should be only one y corresponding to any such value of x—namely, $y = f(x)$. In other words, *the graph of a function cannot contain two or more points with the same x-coordinate—that is, two or more points on the same vertical line.* On the other hand, a vertical line at a value of x not in the domain will not contain any points in the graph. This gives us the following rule.

Vertical-Line Test

For a graph to be the graph of a function, every vertical line must intersect the graph in *at most* one point.

Quick Examples

As illustrated below, only graph B passes the vertical line test, so only graph B is the graph of a function.

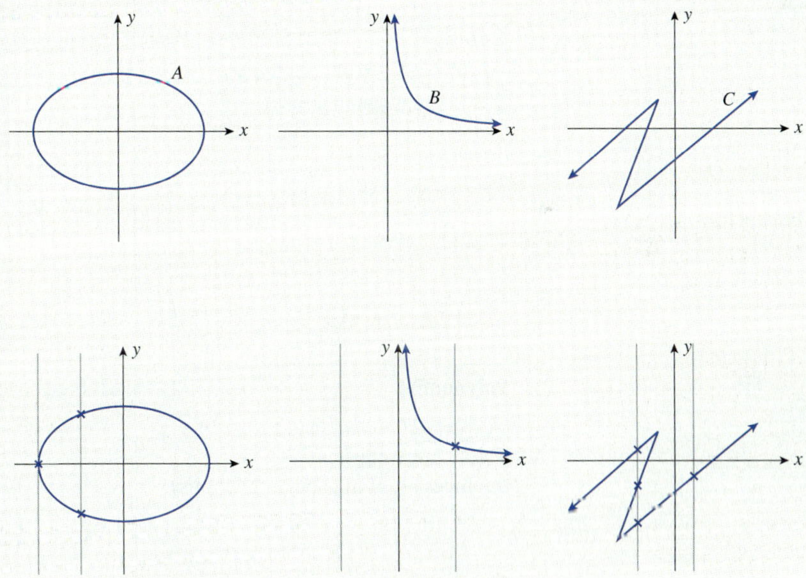

Table 1 lists some common types of functions that are often used to model real world situations.

Table 1 A Compendium of Functions and Their Graphs

Type of Function	***Examples***	
Linear $f(x) = mx + b$ m, b constant Graphs of linear functions are straight lines. The quantity m is the **slope** of the line; the quantity b is the ***y*-intercept** of the line. [See Section 1.3.]	$y = x$ 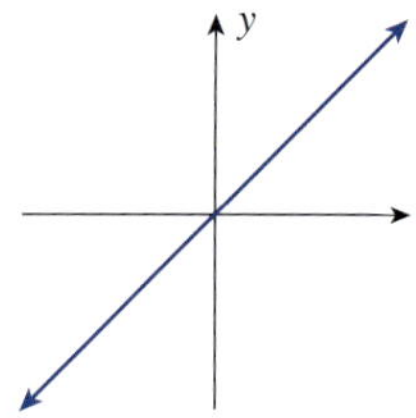	$y = -2x + 2$ 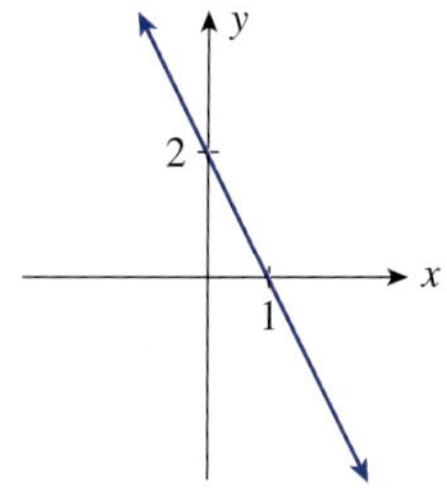
Technology formulas:	`x`	`-2*x+2`
Quadratic $f(x) = ax^2 + bx + c$ a, b, c constant ($a \neq 0$) Graphs of quadratic functions are called **parabolas**. [See Section 2.1.]	$y = x^2$ 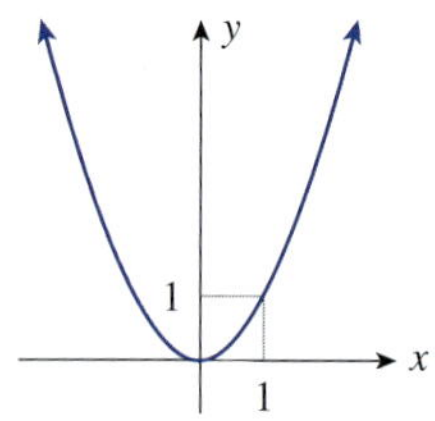	$y = -2x^2 + 2x + 4$ 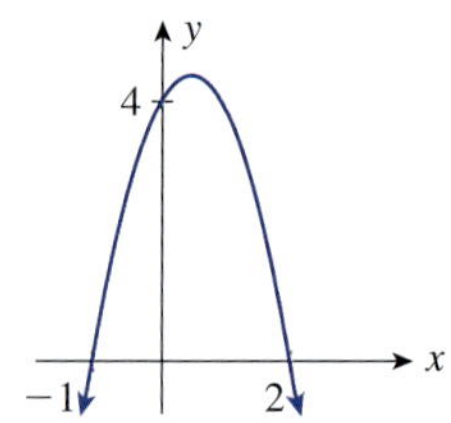
Technology formulas:	`x^2`	`-2*x^2+2*x+4`
Cubic $f(x) = ax^3 + bx^2 + cx + d$ a, b, c, d constant ($a \neq 0$)	$y = x^3$ 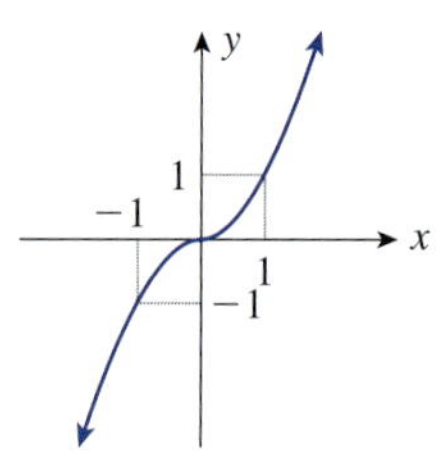	$y = -x^3 + 3x^2 + 1$ 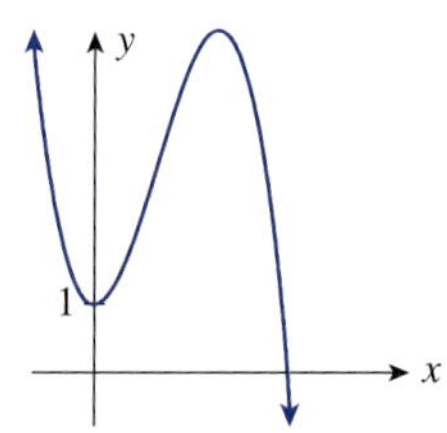
Technology formulas:	`x^3`	`-x^3+3*x^2+1`
Polynomial $f(x) = ax^n + bx^{n-1} + \ldots + rx + s$ $a, b, \ldots, r, s$ constant (includes all of the above functions)	All the above, and $f(x) = x^6 - 2x^5 - 2x^4 + 4x^2$ 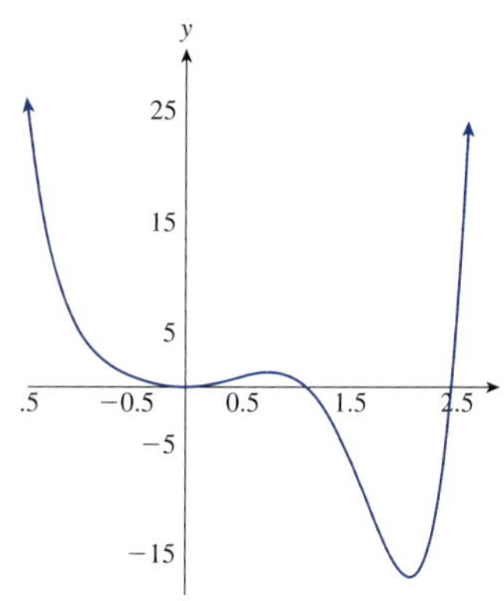	
Technology formula:	`x^6-2x^5-2x^4+4x^2`	

Table 1 (*Continued*)

Type of Function	*Examples*	
Exponential $f(x) = Ab^x$ A, b constant ($b > 0$ and $b \neq 1$) The y-coordinate is multiplied by b every time x increases by 1.	$y = 2^x$ 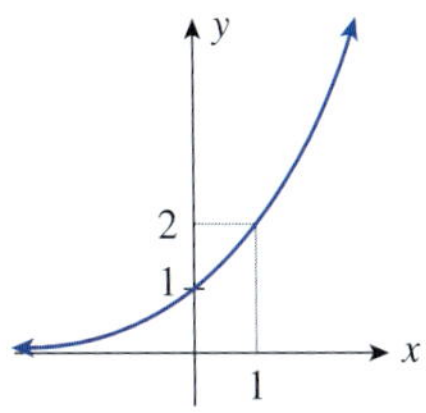 y is doubled every time x increases by 1.	$y = 4(0.5)^x$ 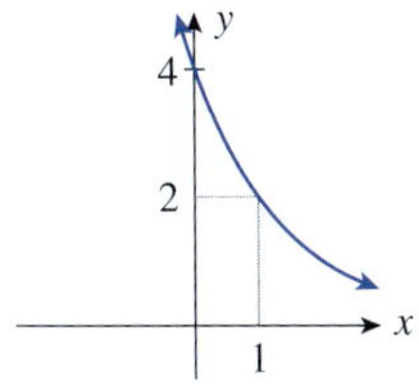 y is halved every time x increases by 1.
Technology formulas:	`2^x`	`4*0.5^x`
Rational $f(x) = \dfrac{P(x)}{Q(x)}$; $P(x)$ and $Q(x)$ polynomials The graph of $y = 1/x$ is a **hyperbola**. The domain excludes zero because $1/0$ is not defined.	$y = \dfrac{1}{x}$ 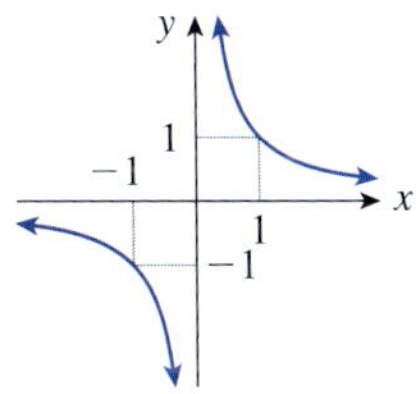	$y = \dfrac{x}{x-1}$ 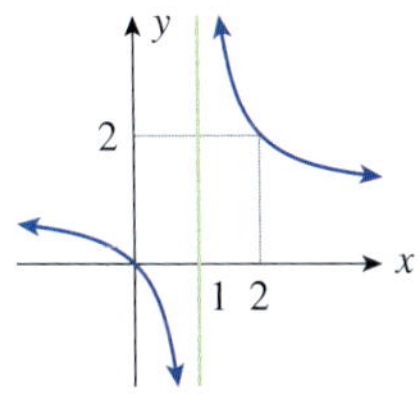
Technology formulas:	`1/x`	`x/(x-1)`
Absolute value For x positive or zero, the graph of $y = \lvert x \rvert$ is the same as that of $y = x$. For x negative or zero, it is the same as that of $y = -x$.	$y = \lvert x \rvert$ 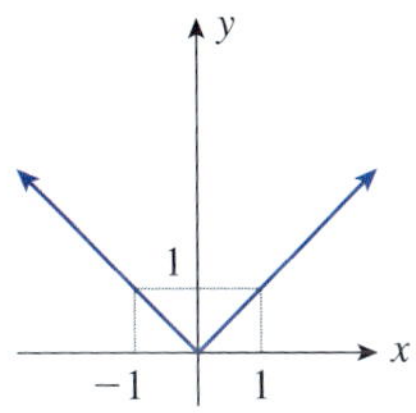	$y = \lvert 2x + 2 \rvert$ 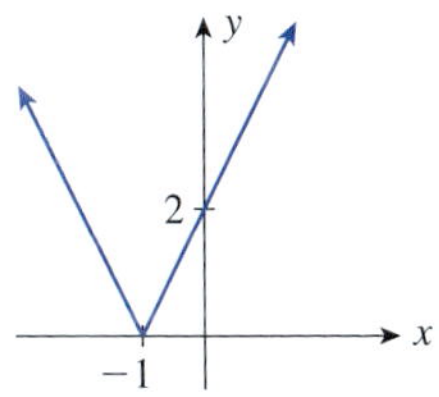
Technology formulas:	`abs(x)`	`abs(2*x+2)`
Square Root The domain of $y = \sqrt{x}$ must be restricted to the nonnegative numbers, because the square root of a negative number is not real. Its graph is the top half of a horizontally oriented parabola.	$y = \sqrt{x}$ 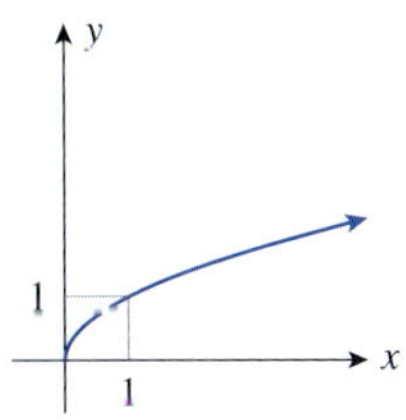	$y = \sqrt{4x - 2}$ 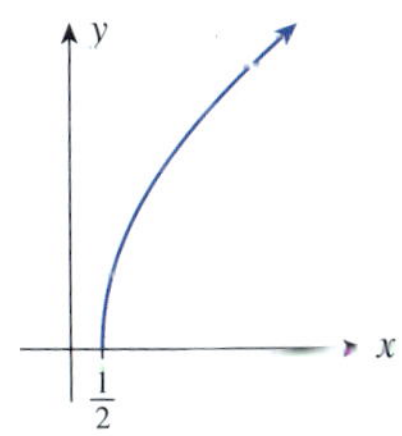
Technology formulas:	`x^0.5 or √(x)`	`(4*x-2)^0.5 or √(4*x-2)`

Go to the Web site and follow the path

Online Text
→ **New Functions from Old: Scaled and Shifted Functions**

where you will find complete online interactive text, examples, and exercises on scaling and translating the graph of a function by changing the formula.

Functions and models other than linear ones are called **nonlinear**.

1.1 EXERCISES

▼ more advanced ◆ challenging

T indicates exercises that should be solved using technology

In Exercises 1–4, evaluate or estimate each expression based on the following table. HINT [See Quick Example 1 on page 42.]

x	−3	−2	−1	0	1	2	3
$f(x)$	1	2	4	2	1	0.5	0.25

1. a. $f(0)$ **b.** $f(2)$ **2. a.** $f(-1)$ **b.** $f(1)$

3. a. $f(2) - f(-2)$ **b.** $f(-1)f(-2)$ **c.** $-2f(-1)$

4. a. $f(1) - f(-1)$ **b.** $f(1)f(-2)$ **c.** $3f(-2)$

In Exercises 5–8, use the graph of the function f to find approximations of the given values. HINT [See Example 1.]

5. ◆

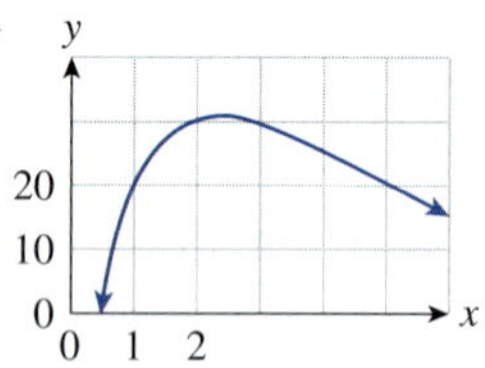

a. $f(1)$ **b.** $f(2)$ **c.** $f(3)$ **d.** $f(5)$ **e.** $f(3) - f(2)$

6.

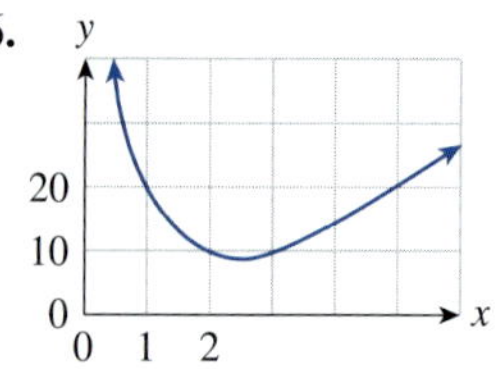

a. $f(1)$ **b.** $f(2)$ **c.** $f(3)$ **d.** $f(5)$ **e.** $f(3) - f(2)$

7.

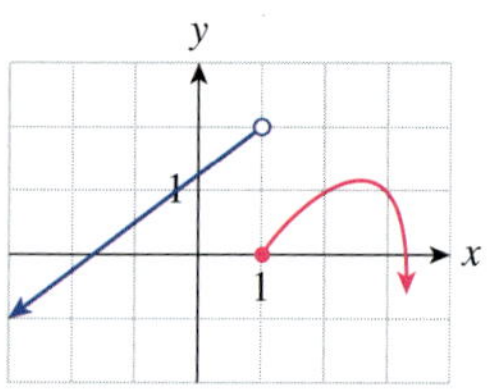

a. $f(-3)$ **b.** $f(0)$ **c.** $f(1)$ **d.** $f(2)$ **e.** $\dfrac{f(3) - f(2)}{3 - 2}$

8.

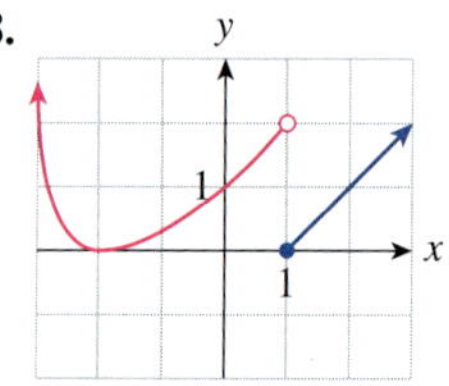

a. $f(-2)$ **b.** $f(0)$ **c.** $f(1)$ **d.** $f(3)$ **e.** $\dfrac{f(3) - f(1)}{3 - 1}$

In Exercises 9–12, say whether or not $f(x)$ is defined for the given values of x. If it is defined, give its value. HINT [See Quick Example 3 page 43.]

9. $f(x) = x - \dfrac{1}{x^2}$, with domain $(0, +\infty)$

a. $x = 4$ **b.** $x = 0$ **c.** $x = -1$

10. $f(x) = \dfrac{2}{x} - x^2$, with domain $[2, +\infty)$

a. $x = 4$ **b.** $x = 0$ **c.** $x = 1$

11. $f(x) = \sqrt{x + 10}$, with domain $[-10, 0)$

a. $x = 0$ **b.** $x = 9$ **c.** $x = -10$

12. $f(x) = \sqrt{9 - x^2}$, with domain $(-3, 3)$

a. $x = 0$ **b.** $x = 3$ **c.** $x = -3$

13. Given $f(x) = 4x - 3$, find **a.** $f(-1)$ **b.** $f(0)$ **c.** $f(1)$ **d.** $f(y)$ **e.** $f(a + b)$ HINT [See Example 1.]

14. Given $f(x) = -3x + 4$, find **a.** $f(-1)$ **b.** $f(0)$ **c.** $f(1)$ **d.** $f(y)$ **e.** $f(a + b)$

15. Given $f(x) = x^2 + 2x + 3$, find **a.** $f(0)$ **b.** $f(1)$ **c.** $f(-1)$ **d.** $f(-3)$ **e.** $f(a)$ **f.** $f(x + h)$ HINT [See Example 1.]

16. Given $g(x) = 2x^2 - x + 1$, find **a.** $g(0)$ **b.** $g(-1)$ **c.** $g(r)$ **d.** $g(x + h)$

17. Given $g(s) = s^2 + \dfrac{1}{s}$, find **a.** $g(1)$ **b.** $g(-1)$ **c.** $g(4)$ **d.** $g(x)$ **e.** $g(s + h)$ **f.** $g(s + h) - g(s)$

18. Given $h(r) = \dfrac{1}{r + 4}$, find **a.** $h(0)$ **b.** $h(-3)$ **c.** $h(-5)$ **d.** $h(x^2)$ **e.** $h(x^2 + 1)$ **f.** $h(x^2) + 1$

In Exercises 19–24, graph the given functions. Give the technology formula and use technology to check your graph. We suggest that you become familiar with these graphs in addition to those in Table 2. HINT [See Quick Example 4 on page 44.]

19. $f(x) = -x^3$ (domain $(-\infty, +\infty)$)

20. $f(x) = x^3$ (domain $[0, +\infty)$)

21. $f(x) = x^4$ (domain $(-\infty, +\infty)$)

22. $f(x) = \sqrt[3]{x}$ (domain $(-\infty, +\infty)$)

23. $f(x) = \dfrac{1}{x^2}$ $(x \neq 0)$ **24.** $f(x) = x + \dfrac{1}{x}$ $(x \neq 0)$

In Exercises 25 and 26, match the functions to the graphs. Using technology to draw the graphs is suggested, but not required.

25. T **a.** $f(x) = x$ $(-1 \le x \le 1)$

b. $f(x) = -x$ $(-1 \le x \le 1)$

c. $f(x) = \sqrt{x}$ $(0 < x < 4)$

d. $f(x) = x + \dfrac{1}{x} - 2$ $(0 < x < 4)$

e. $f(x) = |x|$ $(-1 \le x \le 1)$

f. $f(x) = x - 1$ $(-1 \le x \le 1)$

(I)

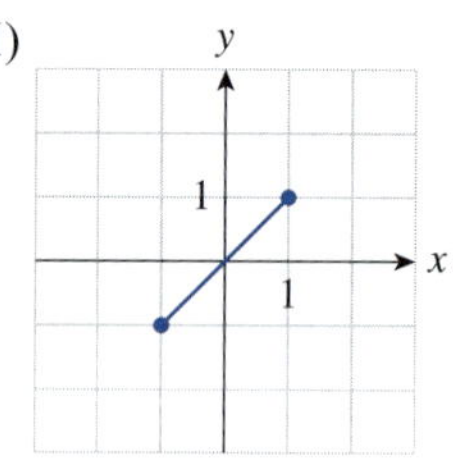

(II)

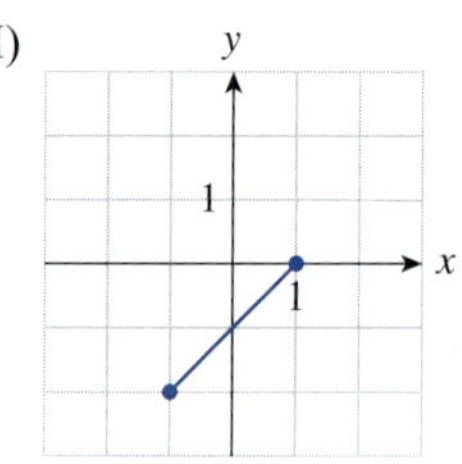

(III)

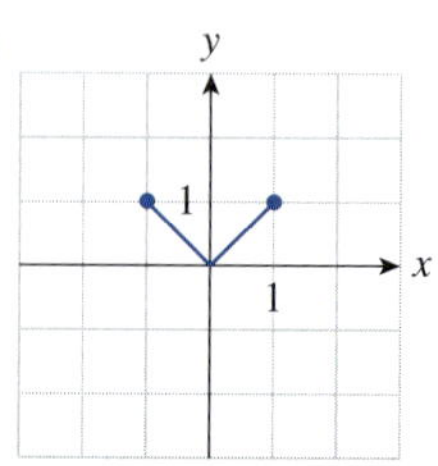

(IV)

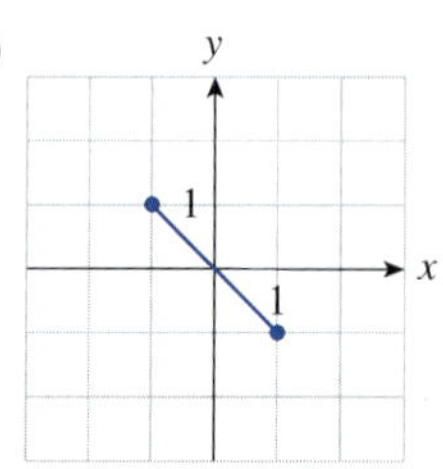

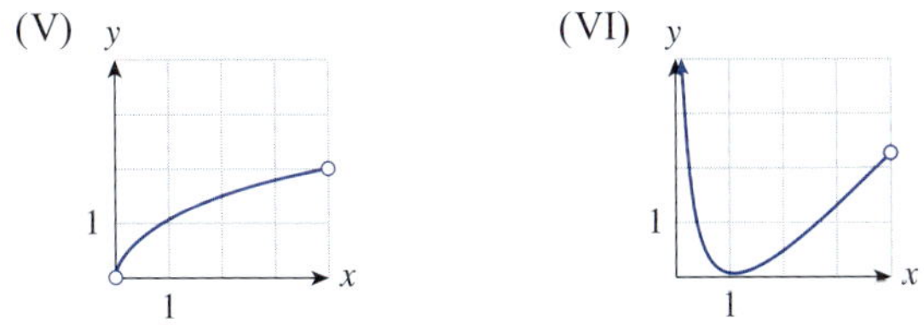

26. T **a.** $f(x) = -x + 4 \quad (0 < x \le 4)$

b. $f(x) = 2 - |x| \quad (-2 < x \le 2)$

c. $f(x) = \sqrt{x+2} \quad (-2 < x \le 2)$

d. $f(x) = -x^2 + 2 \quad (-2 < x \le 2)$

e. $f(x) = \dfrac{1}{x} - 1 \quad (0 < x \le 4)$

f. $f(x) = x^2 - 1 \quad (-2 < x \le 2)$

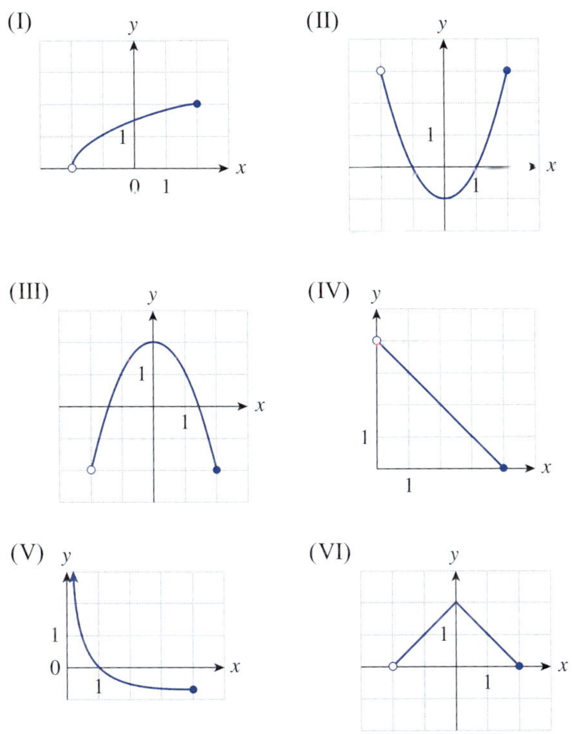

T *In Exercises 27–30, first give the technology formula for the given function and then use technology to evaluate the function for the given values of x (when defined there).*

27. T $f(x) = 0.1x^2 - 4x + 5$; $x = 0, 1, \ldots, 10$

28. T $g(x) = 0.4x^2 - 6x - 0.1$; $x = -5, -4, \ldots, 4, 5$

29. T $h(x) = \dfrac{x^2 - 1}{x^2 + 1}$; $x = 0.5, 1.5, 2.5, \ldots, 10.5$ (Round all answers to four decimal places.)

30. T $r(x) = \dfrac{2x^2 + 1}{2x^2 - 1}$; $x = -1, 0, 1, \ldots, 9$ (Round all answers to four decimal places.)

In Exercises 31–36, sketch the graph of the given function, evaluate the given expressions, and then use technology to duplicate the graphs. Give the technology formula. HINT [See Example 2.]

31. $f(x) = \begin{cases} x & \text{if } -4 \le x < 0 \\ 2 & \text{if } 0 \le x \le 4 \end{cases}$

a. $f(-1)$ **b.** $f(0)$ **c.** $f(1)$

32. $f(x) = \begin{cases} -1 & \text{if } -4 \le x \le 0 \\ x & \text{if } 0 < x \le 4 \end{cases}$

a. $f(-1)$ **b.** $f(0)$ **c.** $f(1)$

33. $f(x) = \begin{cases} x^2 & \text{if } -2 < x \le 0 \\ 1/x & \text{if } 0 < x \le 4 \end{cases}$

a. $f(-1)$ **b.** $f(0)$ **c.** $f(1)$

34. $f(x) = \begin{cases} -x^2 & \text{if } -2 < x \le 0 \\ \sqrt{x} & \text{if } 0 < x < 4 \end{cases}$

a. $f(-1)$ **b.** $f(0)$ **c.** $f(1)$

35. $f(x) = \begin{cases} x & \text{if } -1 < x \le 0 \\ x+1 & \text{if } 0 < x \le 2 \\ x & \text{if } 2 < x \le 4 \end{cases}$

a. $f(0)$ **b.** $f(1)$ **c.** $f(2)$ **d.** $f(3)$ HINT [See Example 3.]

36. $f(x) = \begin{cases} -x & \text{if } -1 < x \le 0 \\ x-2 & \text{if } 0 \le x \le 2 \\ -x & \text{if } 2 < x \le 4 \end{cases}$

a. $f(0)$ **b.** $f(1)$ **c.** $f(2)$ **d.** $f(3)$

In Exercises 37–40, find and simplify ***(a)*** $f(x+h) - f(x)$ ***(b)*** $\dfrac{f(x+h) - f(x)}{h}$

37. ▼ $f(x) = x^2$

38. ▼ $f(x) = 3x - 1$

39. ▼ $f(x) = 2 - x^2$

40. ▼ $f(x) = x^2 + x$

APPLICATIONS

41. ***Oil Imports from Mexico*** The following table shows U.S. oil imports from Mexico, for 2001–2006 ($t = 1$ represents 2001):[2]

***t* (year since 2000)**	1	2	3	4	5	6
***I* (million gallons/day)**	1.35	1.5	1.55	1.6	1.5	1.5

a. Find $I(3)$, $I(5)$, and $I(6)$. Interpret your answers.

b. What is the domain of I?

c. Represent I graphically, and use your graph to estimate $I(4.5)$. Interpret your answer. HINT [See Quick Example 1 on page 42.]

[2]Figures are approximate. Source: Energy Information Administration: Pemex/*New York Times*, March 9, 2007, p. C4.

42. *Oil Production in Mexico* The following table shows oil production by **Pemex**, Mexico's national oil company, for 2001–2006 ($t = 1$ represents 2001):[3]

***t* (year since 2000)**	1	2	3	4	5	6
***P* (million gallons/day)**	3.1	3.2	3.4	3.4	3.4	3.3

a. Find $P(3)$, $P(4)$, and $P(6)$. Interpret your answers.
b. What is the domain of P?
c. Represent P graphically, and use your graph to estimate $P(1.5)$. Interpret your answer.

Housing Starts *Exercises 43–46 refer to the following graph, which shows the number f(t) of housing starts in the U.S. each year from 2000 through 2007 (t = 0 represents 2000, and f(t) is the number of housing starts in year t in thousands of units).*[4]

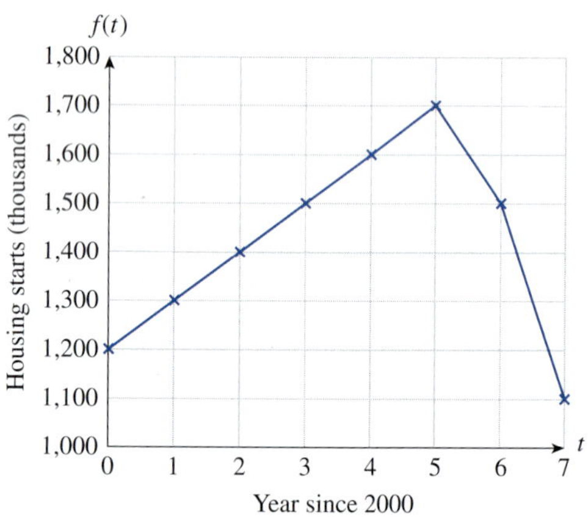

43. Estimate $f(4)$, $f(5)$, and $f(6.5)$. Interpret your answers.

44. Estimate $f(3)$, $f(6)$, and $f(5.5)$. Interpret your answers.

45. Which has the larger magnitude: $f(5) - f(0)$ or $f(7) - f(5)$? Interpret the answer.

46. Which has the larger magnitude: $f(7) - f(6)$ or $f(5) - f(0)$? Interpret the answer.

47. *Trade with China* The value of U.S. trade with China from 1994 through 2004 can be approximated by

$$C(t) = 3t^2 - 7t + 50 \text{ billion dollars}$$

(t is time in years since 1994).[5]

a. Find an appropriate domain of C. Is $t \geq 0$ an appropriate domain? Why or why not?
b. Compute $C(10)$. What does the answer say about trade with China?

48. *Scientific Research* The number of research articles in *Physical Review* that were written by researchers in the United States from 1983 through 2003 can be approximated by

$$A(t) = -0.01t^2 + 0.24t + 3.4 \text{ hundred articles}$$

(t is time in years since 1983).[6]

a. Find an appropriate domain of A. Is $t \leq 20$ an appropriate domain? Why or why not?
b. Compute $A(10)$. What does the answer say about the number of research articles?

49. *Satellite Radio Losses* The following graph shows the approximate annual loss $L(t)$ of **Sirius Satellite Radio** for the perod 2001–2006 ($t = 1$ represents the start of 2001).[7]

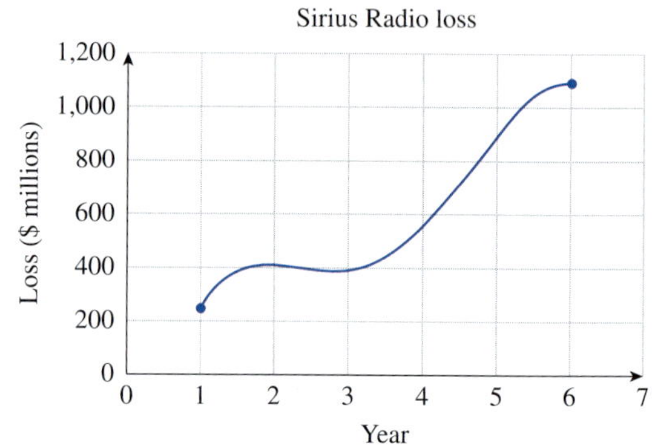

a. What is the domain of L?
b. Estimate $L(2)$, $L(5)$, and $L(6)$. Interpret your answers.
c. At approximately which value of t is $L(t)$ increasing most rapidly? Interpret the result.

50. *Satellite Radio Losses* The following graph shows the approximate annual loss $L(t)$ of **XM Satellite Radio** for the period 2001–2006 ($t = 1$ represents the start of 2001).

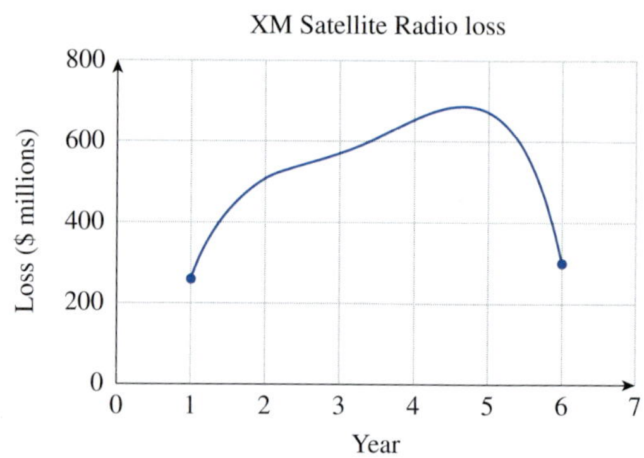

a. What is the domain of L?
b. Estimate $L(2)$, $L(3.5)$, and $L(6)$. Interpret your answers.
c. At approximately which value of t is $L(t)$ increasing most rapidly? Interpret the result.

[3] Figures are approximate. Source: Energy Information Administration: Pemex/*New York Times*, March 9, 2007, p. C4.

[4] Figures are rounded. Source: www.census.gov (2007).

[5] Based on a regression by the authors. Source for data: U.S. Census Bureau/*New York Times*, September 23, 2004, p. C1.

[6] Based on a regression by the authors. Source for data: The American Physical Society/*New York Times*, May 3, 2003, p. A1.

[7] Source: Bloomberg Financial Markets, Sirius Satellite Radio/*New York Times*, February 20, 2008, p. C2.

51. ***Processor Speeds*** The processor speed, in megahertz, of Intel processors could be approximated by the following function of time t in years since the start of 1995:[8]

$$P(t) = \begin{cases} 75t + 200 & \text{if } 0 \le t \le 4 \\ 600t - 1900 & \text{if } 4 < t \le 9 \end{cases}$$

a. Evaluate $P(0)$, $P(4)$, and $P(5)$ and interpret the results.
b. Sketch the graph of P and use your graph to estimate when processor speeds first reached 2.0 gigahertz (1 gigahertz = 1,000 megahertz).
c. T Use technology to generate a table of values for $P(t)$ with $t = 0, 1, \ldots, 9$.

52. ***Leading Economic Indicators*** The value of the Conference Board Index of 10 economic indicators in the United States could be approximated by the following function of time t in months since the end of December 2002:[9]

$$E(t) = \begin{cases} 0.4t + 110 & \text{if } 6 \le t \le 15 \\ -0.2t + 119 & \text{if } 15 < t \le 20 \end{cases}$$

a. Estimate $E(10)$, $E(15)$, and $E(20)$ and interpret the results.
b. Sketch the graph of E and use your graph to estimate when the index first reached 115.
c. T Use technology to generate a table of values for $E(t)$ with $t = 6, 7, \ldots, 20$.

53. ▼ ***Income Taxes*** The U.S. Federal income tax is a function of taxable income. Write T for the tax owed on a taxable income of I dollars. For tax year 2007, the function T for a single taxpayer was specified as follows:

If your taxable income was			*of the*
	But not		*amount*
Over—	*over—*	*Your tax is*	*over—*
$0	7,825	.... 10%	$0
7,825	31,850	$782.50 + 15%	$7,825
31,850	77,100	4,386.25 + 25%	$31,850
77,100	160,850	15,698.75 + 28%	$77,100
160,850	349,700	39,148.75 + 33%	$160,850
349,700	....	101,469.25 + 35%	$349,700

What was the tax owed by a single taxpayer on a taxable income of \$26,000? On a taxable income of \$65,000?

54. ▼ ***Income Taxes*** The income tax function T in Exercise 53 can also be written in the following form:

$$T(I) = \begin{cases} 0.10I & \text{if } 0 < I \le 7{,}825 \\ 782.50 + 0.15(I - 7{,}825) & \text{if } 7{,}825 < I \le 31{,}850 \\ 4{,}386.25 + 0.25(I - 31{,}850) & \text{if } 31{,}850 < I \le 77{,}100 \\ 15{,}698.75 + 0.28(I - 77{,}100) & \text{if } 77{,}100 < I \le 160{,}850 \\ 39{,}148.75 + 0.33(I - 160{,}850) & \text{if } 160{,}850 < I \le 349{,}700 \\ 101{,}469.25 + 0.35(I - 349{,}700) & \text{if } I > 349{,}700 \end{cases}$$

What was the tax owed by a single taxpayer on a taxable income of \$25,000? On a taxable income of \$125,000?

55. T ▼ ***Acquisition of Language*** The percentage $p(t)$ of children who can speak in at least single words by the age of t months can be approximated by the equation[10]

$$p(t) = 100\left(1 - \frac{12{,}200}{t^{4.48}}\right) \quad (t \ge 8.5)$$

a. Give a technology formula for p.
b. Graph p for $8.5 \le t \le 20$ and $0 \le p \le 100$.
c. Create a table of values of p for $t = 9, 10, \ldots, 20$ (rounding answers to one decimal place).
d. What percentage of children can speak in at least single words by the age of 12 months?
e. By what age are 90% or more children speaking in at least single words?

56. T ▼ ***Acquisition of Language*** The percentage $p(t)$ of children who can speak in sentences of five or more words by the age of t months can be approximated by the equation[11]

$$p(t) = 100\left(1 - \frac{5.27 \times 10^{17}}{t^{12}}\right) \quad (t \ge 30)$$

a. Give a technology formula for p.
b. Graph p for $30 \le t \le 45$ and $0 \le p \le 100$.
c. Create a table of values of p for $t = 30, 31, \ldots, 40$ (rounding answers to one decimal place).
d. What percentage of children can speak in sentences of five or more words by the age of 36 months?
e. By what age are 75% or more children speaking in sentences of five or more words?

COMMUNICATION AND REASONING EXERCISES

57. If the market price m of gold varies with time t, then the independent variable is ___ and the dependent variable is ___.

58. Complete the following sentence: If weekly profit P is specified as a function of selling price s, then the independent variable is ___ and the dependent variable is ___.

59. Complete the following: The function notation for the equation $y = 4x^2 - 2$ is ___ .

[8] Source: Sandpile.org/*New York Times*, May 17, 2004, p. C1.

[9] Source: The Conference Board/*New York Times*, November 19, 2004, p. C7.

[10] The model is the authors' and is based on data presented in the article *The Emergence of Intelligence* by William H. Calvin, *Scientific American*, October, 1994, pp. 101–107.

[11] Ibid.

60. Complete the following: The equation notation for $C(t) = -0.34t^2 + 0.1t$ is ____.

61. True or false? Every graphically specified function can also be specified numerically. Explain.

62. True or false? Every algebraically specified function can also be specified graphically. Explain.

63. True or false? Every numerically specified function with domain [0, 10] can also be specified algebraically. Explain.

64. True or false? Every graphically specified function can also be specified algebraically. Explain.

65. ▼ True or false? Every function can be specified numerically.

66. ▼ Which supplies more information about a situation: a numerical model or an algebraic model?

67. ▼ Why is the following assertion false? "If $f(x) = x^2 - 1$, then $f(x+h) = x^2 + h - 1$."

68. ▼ Why is the following assertion false? "If $f(2) = 2$ and $f(4) = 4$, then $f(3) = 3$."

69. How do the graphs of two functions differ if they are specified by the same formula but have different domains?

70. How do the graphs of two functions $f(x)$ and $g(x)$ differ if $g(x) = f(x) + 10$? (Try an example.)

71. ▼ How do the graphs of two functions $f(x)$ and $g(x)$ differ if $g(x) = f(x - 5)$? (Try an example.)

72. ▼ How do the graphs of two functions $f(x)$ and $g(x)$ differ if $g(x) = f(-x)$? (Try an example.)

1.2 Functions and Models

The functions we used in Examples 1 and 2 in Section 1.1 are **mathematical models** of real-life situations, because they model, or represent, situations in mathematical terms.

Mathematical Modeling

To mathematically model a situation means to represent it in mathematical terms. The particular representation used is called a **mathematical model** of the situation. Mathematical models do not always represent a situation perfectly or completely. Some (like Example 1 of Section 1.1) represent a situation only approximately, whereas others represent only some aspects of the situation.

Quick Examples

Situation	Model
1. The temperature is now 10°F and increasing by 20° per hour.	$T(t) = 10 + 20t$ (t = time in hours, T = temperature)
2. I invest \$1,000 at 5% interest compounded quarterly. Find the value of the investment after t years.	$A(t) = 1{,}000\left(1 + \frac{0.05}{4}\right)^{4t}$ This is the compound interest formula we will study in Example 6.
3. I am fencing a rectangular area whose perimeter is 100 ft. Find the area as a function of the width x.	Take y to be the length, so the perimeter is $100 = x + y + x + y = 2(x + y)$ so $x + y = 50$. Thus the length is $y = 50 - x$. Area $A = xy = x(50 - x)$.
4. iPod sales	The function $f(x) = -x^2 + 20x + 3$ in Example 1 of Section 1.1 is an **algebraic model** of iPod sales.

5. Facebook membership

The function

$$n(t) = \begin{cases} 1.5t^2 - t + 1.5 & \text{if } 0 \le t \le 3 \\ 46t - 126 & \text{if } 3 < t \le 4 \end{cases}$$

in Example 2 of Section 1.1 is a **piecewise algebraic model** of Facebook membership.

Analytical and Curve-Fitting Models

Quick Examples 1–3 are **analytical models**, obtained by analyzing the situation being modeled, whereas Quick Examples 4 and 5 are **curve-fitting models**, obtained by finding mathematical formulas that approximate observed data.

Cost, Revenue, and Profit Models

Photodisc/Getty Images

EXAMPLE 1 Modeling Cost: Cost Function

As of October 2007, Yellow Cab Chicago's rates were \$1.90 on entering the cab plus \$1.60 for each mile.*

a. Find the cost C of an x-mile trip.

b. Use your answer to calculate the cost of a 40-mile trip.

c. What is the cost of the second mile? What is the cost of the tenth mile?

d. Graph C as a function of x.

Solution

a. We are being asked to find how the cost C depends on the length x of the trip, or to find C as a function of x. Here is the cost in a few cases:

Cost of a 1-mile trip: $C = 1.60(1) + 1.90 = 3.50$ 1 mile @ \$1.60 per mile plus \$1.90

Cost of a 2-mile trip: $C = 1.60(2) + 1.90 = 5.10$ 2 miles @ \$1.60 per mile plus \$1.90

Cost of a 3-mile trip: $C = 1.60(3) + 1.90 = 6.70$ 3 miles @ \$1.60 per mile plus \$1.90

Do you see the pattern? The cost of an x-mile trip is given by the linear function

$$C(x) = 1.60x + 1.90.$$

Notice that the cost function is a sum of two terms: the **variable cost** $1.60x$, which depends on x, and the **fixed cost** 1.90, which is independent of x:

$$\text{Cost} = \text{Variable Cost} + \text{Fixed Cost}.$$

The quantity 1.60 by itself is the incremental cost per mile; you might recognize it as the *slope* of the given linear function. In this context we call 1.60 the **marginal cost**. You might recognize the fixed cost 1.90 as the *C-intercept* of the given linear function.

* According to their Web site at www.yellowcabchicago.com/.

b. We can use the formula for the cost function to calculate the cost of a 40-mile trip as

$$C(40) = 1.60(40) + 1.90 = \$65.90.$$

c. To calculate the cost of the second mile, we *could* proceed as follows:

Find the cost of a 1-mile trip: $C(1) = 1.60(1) + 1.90 = \$3.50$.
Find the cost of a 2-mile trip: $C(2) = 1.60(2) + 1.90 = \$5.10$.
Therefore, the cost of the second mile is $\$5.10 - \$3.50 = \$1.60$.

But notice that this is just the marginal cost. In fact, the marginal cost is the cost of each additional mile, so we could have done this more simply:

$$\text{Cost of second mile} = \text{Cost of tenth mile} = \text{Marginal cost} = \$1.60$$

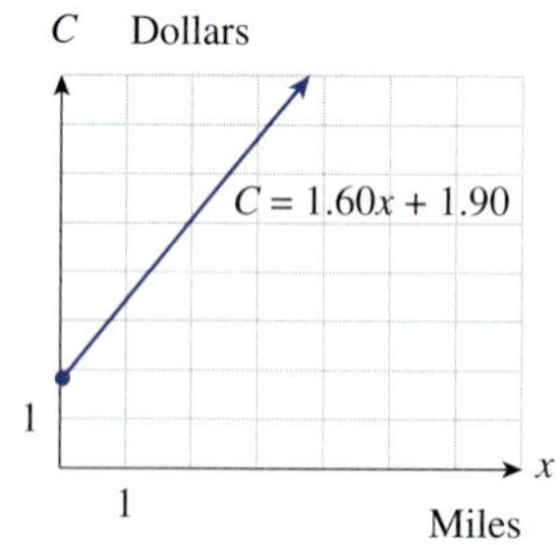

Figure 7

d. Figure 7 shows the graph of the cost function, which we can interpret as a *cost vs. miles* graph. The fixed cost is the starting height on the left, while the marginal cost is the slope of the line: It rises 1.60 units per unit of x. (See Section 1.3 for a discussion of properties of straight lines.)

➡ **Before we go on...** The cost function in Example 1 is an example of an *analytical model:* We derived the form of the cost function from a knowledge of the cost per mile and the fixed cost.

As we discussed on page 46 in Section 1.1, we can use equation notation to specify a function. In equation notation, the function C in Example 1 is

$$C = 1.60x + 1.90. \quad \text{Equation notation}$$

The independent variable is x and the dependent variable is C. Function notation and equation notation, using the same letter for the function name and the dependent variable, are often used interchangeably. It is important to be able to switch back and forth between function notation and equation notation easily. ■

Here is a summary of some terms we used in Example 1, along with an introduction to some new terms:

Cost, Revenue, and Profit Functions

A **cost function** specifies the cost C as a function of the number of items x. Thus, $C(x)$ is the cost of x items, and has the form

$$\text{Cost} = \text{Variable cost} + \text{Fixed cost}$$

where the variable cost is a function of x and the fixed cost is a constant. A cost function of the form

$$C(x) = mx + b$$

is called a **linear cost function**; the variable cost is mx and the fixed cost is b. The slope m, the **marginal cost**, measures the incremental cost per item.

The **revenue** resulting from one or more business transactions is the total payment received, sometimes called the gross proceeds. If $R(x)$ is the revenue from selling x items at a price of m each, then R is the linear function $R(x) = mx$ and the selling price m can also be called the **marginal revenue**.

The **profit**, on the other hand, is the *net* proceeds, or what remains of the revenue when costs are subtracted. If the profit depends linearly on the number of items, the slope m is called the **marginal profit**. Profit, revenue, and cost are related by the following formula.

$$\text{Profit} = \text{Revenue} - \text{Cost}$$
$$P = R - C$$

If the profit is negative, say $-\$500$, we refer to a **loss** (of \$500 in this case). To **break even** means to make neither a profit nor a loss. Thus, break even occurs when $P = 0$, or

$$R = C. \qquad \text{Break even}$$

The **break-even point** is the number of items x at which break even occurs.

Quick Example

If the daily cost (including operating costs) of manufacturing x T-shirts is $C(x) = 8x + 100$, and the revenue obtained by selling x T-shirts is $R(x) = 10x$, then the daily profit resulting from the manufacture and sale of x T-shirts is

$$P(x) = R(x) - C(x) = 10x - (8x + 100) = 2x - 100.$$

Break even occurs when $P(x) = 0$, or $x = 50$.

EXAMPLE 2 Cost, Revenue, and Profit

The annual operating cost of *YSport Fitness* gym is estimated to be

$$C(x) = 100{,}000 + 160x - 0.2x^2 \qquad (0 \le x \le 400)$$

where x is the number of members. Annual revenue from membership averages \$800 per member. What is the variable cost? What is the fixed cost? What is the profit function? How many members must *YSport* have to make a profit? What will happen if it has fewer members? If it has more?

Solution The variable cost is the part of the cost function that depends on x:

$$\text{Variable cost} = 160x - 0.2x^2.$$

The fixed cost is the constant term:

$$\text{Fixed cost} = 100{,}000.$$

The annual revenue *YSport* obtains from a single member is \$800. So, if it has x members, it earns an annual revenue of

$$R(x) = 800x.$$

For the profit, we use the formula

$$\begin{aligned} P(x) &= R(x) - C(x) && \text{Formula for profit} \\ &= 800x - (100{,}000 + 160x - 0.2x^2) && \text{Substitute } R(x) \text{ and } C(x) \\ &= -100{,}000 + 640x + 0.2x^2. \end{aligned}$$

using Technology

Excel has a feature called "Goal Seek" which can be used to find the point of intersection of the cost and revenue graphs numerically rather than graphically. See the downloadable Excel tutorial for this section at the Web site.

To make a profit, *YSport* needs to do better than break even, so let us find the break-even point: the value of x such that $P(x) = 0$. All we have to do is set $P(x) = 0$ and solve for x:

$$-100{,}000 + 640x + 0.2x^2 = 0.$$

Notice that we have a quadratic equation $ax^2 + bx + c = 0$ with $a = 0.2$, $b = 640$, and $c = -100{,}000$. Its solution is given by the quadratic formula:

$$x = \frac{-b \pm \sqrt{b^2 - 4ac}}{2a} = \frac{-640 \pm \sqrt{640^2 + 4(0.2)(100{,}000)}}{2(0.2)}$$
$$\approx \frac{-640 \pm 699.71}{2(0.2)}$$
$$\approx 149.3 \text{ or } -3{,}349.3.$$

We reject the negative solution (as the domain is [0, 400]) and conclude that $x \approx 149.3$ members. To make a profit, should *YSport* have 149 members or 150 members? To decide, take a look at Figure 8, which shows two graphs: On the top we see the graph of revenue and cost, and on the bottom we see the graph of the profit function.

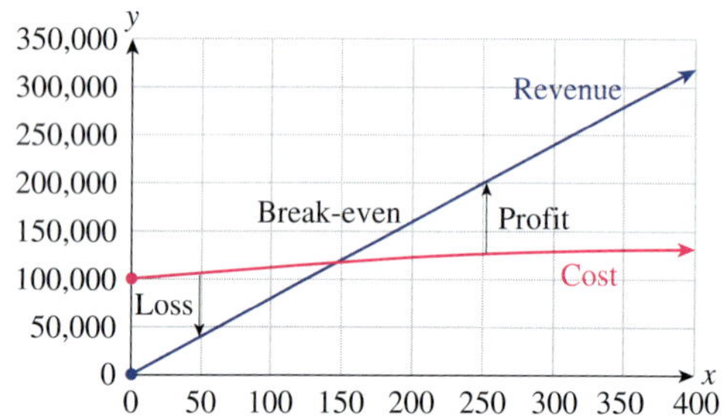

Cost: $C(x) = 100{,}000 + 160x - 0.2x^2$
Revenue: $R(x) = 800x$
Break even occurs at the point of intersection of the graphs of revenue and cost.

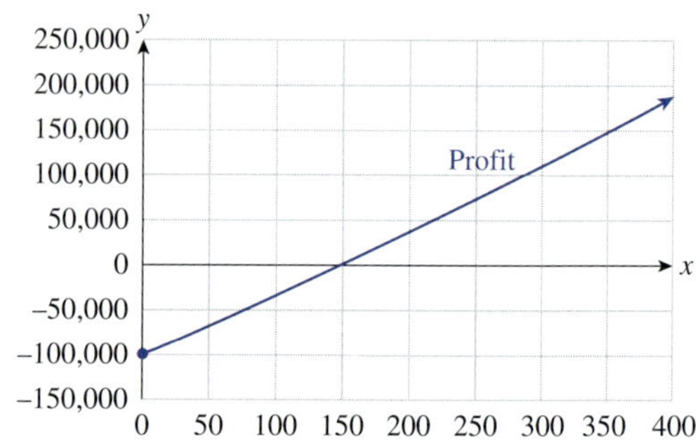

Profit: $P(x) = -100{,}000 + 640x + 0.2x^2$
Break even occurs when $P(x) = 0$

Figure 8

For values of x less than the break-even point of 149.3, $P(x)$ is negative, so the company will have a loss. For values of x greater than the break-even point, $P(x)$ is positive, so the company will make a profit. Thus, *YSport Fitness* needs at least 150 members to make a profit. (Note that we rounded 149.3 up to 150 in this case.)

Demand and Supply Models

The demand for a commodity usually goes down as its price goes up. It is traditional to use the letter q for the (quantity of) demand, as measured, for example, in sales. Consider the following example.

EXAMPLE 3 Demand: Private Schools

The demand for private schools in Michigan depends on the tuition cost and can be approximated by

$$q = 77.8p^{-0.11} \text{ thousand students} \qquad (200 \le p \le 2{,}200) \qquad \text{Demand curve}$$

where p is the net tuition cost in dollars.*

a. Use technology to plot the demand function.

b. What is the effect on demand if the tuition cost is increased from \$1,000 to \$1,500?

Solution

a. The demand function is given by $q(p) = 77.8p^{-0.11}$. Its graph is known as a **demand curve**. (Figure 9)

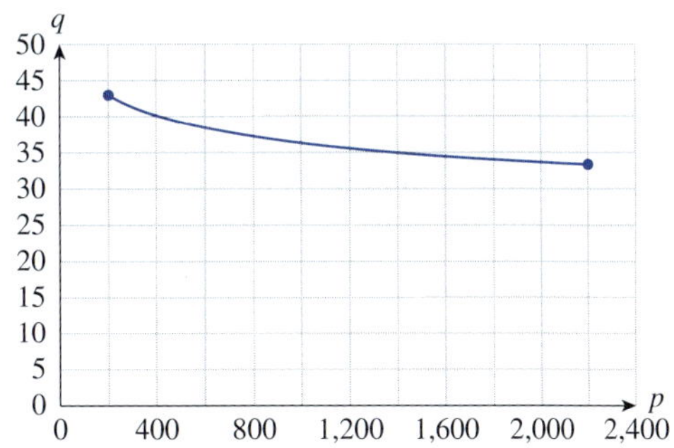

Technology Formula: `y = 77.8*x^(-0.11)`

Figure 9

b. The demand at tuition costs of \$1,000 and \$1,500 are

$$q(1{,}000) = 77.8(1{,}000)^{-0.11} \approx 36.4 \text{ thousand students}$$
$$q(1{,}500) = 77.8(1{,}500)^{-0.11} \approx 34.8 \text{ thousand students}$$

The change in demand is therefore

$$q(1{,}500) - q(1{,}000) \approx 34.8 - 36.4 = -1.6 \text{ thousand students}$$

* The tuition cost is net cost: tuition minus tax credit. The model is based on data in "The Universal Tuition Tax Credit: A Proposal to Advance Personal Choice in Education," Patrick L. Anderson, Richard McLellan, J.D., Joseph P. Overton, J.D., Gary Wolfram, Ph.D., Mackinac Center for Public Policy, www.mackinac.org/

We have seen that a demand function gives the number of items consumers are willing to buy at a given price, and a higher price generally results in a lower demand. However, as the price rises, suppliers will be more inclined to produce these items (as opposed to spending their time and money on other products), so supply will generally rise. A **supply function** gives q, the number of items suppliers are willing to make available for sale,* as a function of p, the price per item.

* **NOTE** Although a bit confusing at first, it is traditional to use the same letter q for the quantity of supply and the quantity of demand, particularly when we want to compare them, as in the next example.

Demand, Supply, and Equilibrium Price

A **demand equation** or **demand function** expresses demand q (the number of items demanded) as a function of the unit price p (the price per item). A **supply equation** or **supply function** expresses supply q (the number of items a supplier is willing to make available) as a function of the unit price p (the price per item). It is usually the case that demand decreases and supply increases as the unit price increases.

Demand and supply are said to be in **equilibrium** when demand equals supply. The corresponding values of p and q are called the **equilibrium price** and **equilibrium demand**. To find the equilibrium price, determine the unit price p where the demand and supply curves cross (sometimes we can determine this value analytically by setting demand equal to supply and solving for p). To find the equilibrium demand, evaluate the demand (or supply) function at the equilibrium price.

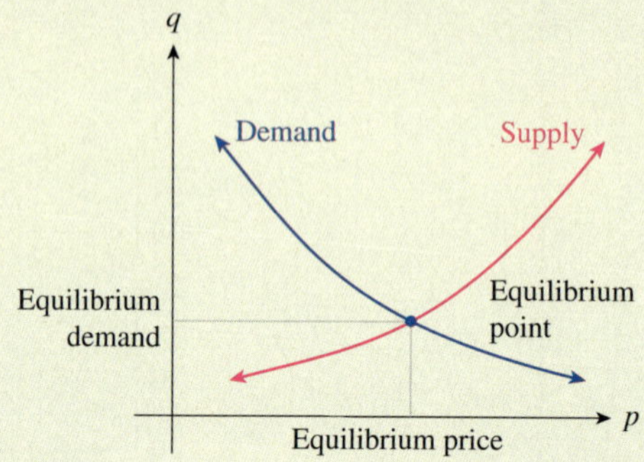

Quick Example

If the demand for your exclusive T-shirts is $q = -20p + 800$ shirts sold per day and the supply is $q = 10p - 100$ shirts per day, then the equilibrium point is obtained when demand = supply:

$$-20p + 800 = 10p - 100$$
$$30p = 900, \text{ giving } p = \$30$$

The equilibrium price is therefore \$30 and the equilibrium demand is $q = -20(30) + 800 = 200$ shirts per day. What happens at prices other than the equilibrium price is discussed in Example 4.

EXAMPLE 4 Demand, Supply, and Equilibrium Price

Continuing with Example 3, suppose that private school institutions are willing to create private schools to accommodate

$$q = 30.4 + 0.006p \text{ thousand students} \qquad (200 \le p \le 2{,}200) \qquad \text{Supply curve}$$

who pay a net tuition of p dollars.

a. Graph the demand curve of Example 3 and the supply curve given here on the same set of axes. Use your graph to estimate, to the nearest \$100, the tuition at which the demand equals the supply. Approximately how many students will be accommodated at that price, known as the **equilibrium price**?

b. What happens if the tuition is higher than the equilibrium price? What happens if it is lower?

c. Estimate the shortage or surplus of openings at private schools if tuition is set at \$1,200.

Solution

a. Figure 10 shows the graphs of demand $q = 77.8p^{-0.11}$ and supply $q = 30.4 + 0.006p$. (See the margin note for a brief description of how to plot them.)

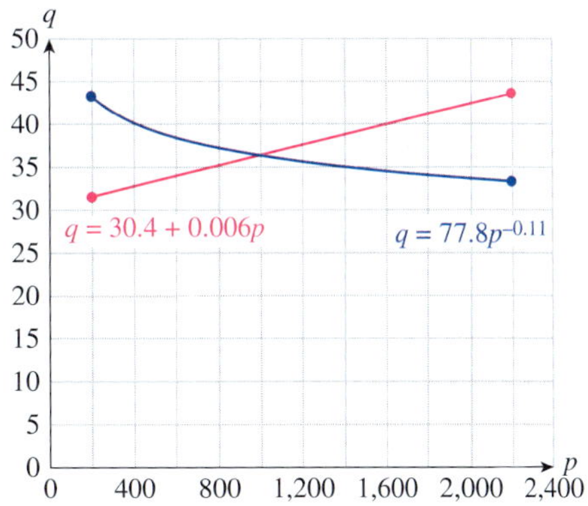

Demand: $q = 77.8p^{-0.11}$
Supply: $q = 30.4 + 0.006p$

Figure 10

The lines cross close to $p = \$1{,}000$, so we conclude that demand = supply when $p \approx \$1{,}000$ (to the nearest \$100). This is the (approximate) equilibrium tuition price. At that price, we can estimate the demand or supply at around

$$\text{Demand: } q = 77.8(1000)^{-0.11} \approx 36.4$$

$$\text{Supply: } q = 30.4 + 0.006(1000) = 36.4 \qquad \text{Demand = Supply at equilibrium}$$

or 36,400 students.

b. Take a look at Figure 11, which shows what happens if schools charge more or less than the equilibrium price. If tuition is, say, \$1,800, then the supply will be larger than demand and there will be a surplus of available openings at private schools. Similarly, if tuition is less—say \$400—then the supply will be less than the demand, and there will be a shortage of available openings.

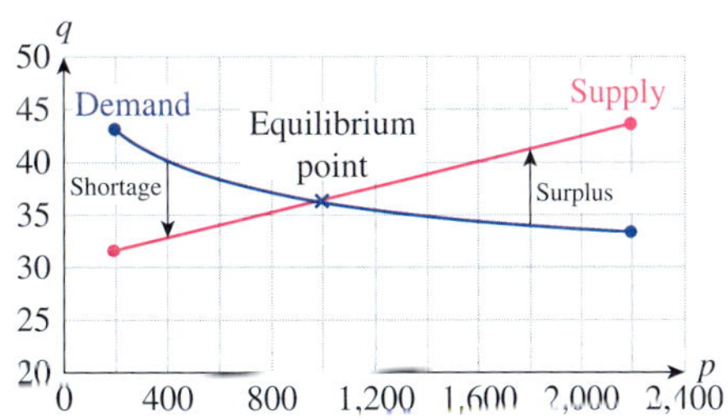

Figure 11

c. The discussion in part (b) shows that if tuition is set at \$1,200 there will be a surplus of available openings. To estimate that number, we calculate the projected demand and supply when $p = \$1{,}200$:

$$\text{Demand: } q = 77.8(1{,}200)^{-0.11} \approx 35.7 \text{ thousand seats}$$

$$\text{Supply: } q = 30.4 + 0.006(1{,}200) = 37.6 \text{ thousand seats}$$

$$\text{Surplus} = \text{Supply} - \text{Demand} \approx 37.6 - 35.7 = 1.9 \text{ thousand seats}$$

So, there would be a surplus of around 1,900 available seats.

using Technology

See the Technology Guide at the end of the chapter for detailed instructions on how to obtain the graph in Example 4 using a TI-83/84 Plus or Excel. Here is an outline:

TI-83/84 Plus

Graphs:
`Y1=77.8*X^(-0.11)`
`Y2=30.4+0.006*X`
Xmin = 200, Xmax = 2200; ZOOM 0. [More details on page 111.]

Excel

Graphs: Headings *p*, Demand, Supply in A1–C1; *p*-values 200, 300, . . . , 2200 in A2–A22.
`=77.8*A2^(-0.11)` in B2
`=30.4+0.006*A2` in C2
copy down through C22. Highlight A1–C22; insert Scatter chart. [More details on page 118.]

Web Site

www.AppliedCalc.org
Go to the Function Evaluator and Grapher under Online Utilities, enter
`77.8*x^(-0.11)` for y_1 and `30.4+0.006*x` for y_2.
Graph: Set Xmin = 200 and Xmax = 2200, and press "Plot Graphs."

➡ **Before we go on...** We just saw in Example 4 that if tuition is less than the equilibrium price there will be a shortage. If schools were to raise their tuition toward the equilibrium, they could create and fill more openings and increase revenue, because it is the supply equation—and not the demand equation—that determines what one can sell below the equilibrium price. On the other hand, if they were to charge more than the equilibrium price, they will be left with a possibly costly surplus of unused openings (and will want to lower tuition to reduce the surplus). Prices tend to move toward the equilibrium, so supply tends to equal demand. When supply equals demand, we say that the market **clears**. ■

Modeling Change Over Time

Things around us change with time. Thus, there are many quantities, such as your income or the temperature in Honolulu, that are natural to think of as functions of time. Example 1 on page 44 (on iPod sales) and Example 2 on page 47 (on Facebook membership) in Section 1.1 are models of change over time. Both of those models are curve-fitting models: We used algebraic functions to approximate observed data.

In the next example we are asked to select from among several curve-fitting models for given data.

EXAMPLE 5 Model Selection: Casino Revenue

The following table shows annual revenues in 2000–2006 from slot machines and video poker machines at the *Mohegan Sun* casino in Connecticut ($t = 0$ represents 2000).*

t	0	1	2	3	4	5	6
Revenue ($ million)	550	680	750	820	850	890	910

Consider the following four models:

(1) $r(t) = 57t + 607$ Linear model

(2) $r(t) = -9t^2 + 110t + 560$ Quadratic model

(3) $r(t) = 608(1.08)^t$ Exponential model

(4) $r(t) = \dfrac{930}{1 + 0.67(1.7)^{-t}}$ Logistic model

a. Which models fit the data significantly better than the rest?

b. Of the models you selected in part (a), which gives the most reasonable prediction for 2010?

* Source: Connecticut Division of Special Revenue/*New York Times*, Sept. 23, 2007, p. C1.

using Technology

See the Technology Guide at the end of the chapter for detailed instructions on how to obtain the table and graphs in Example 5 using a TI-83/84 Plus or Excel. Here is an outline:

TI-83/84 Plus

[STAT] EDIT; enter the values of t in L_1, and $r(t)$ in L_2. Turn on Plot1 in Y= screen. [ZOOM] [9]

Adding a curve:

`Y1=-9*X^2+110X+560`

[GRAPH]

[More details on page 112.]

Excel

Table of values: Headings t, Revenue, $r(t)$ in A1–C1; t-values in A2–A8. Revenues in B2–B8. Formula in C2:

`=-9*A2^2+110*A2+560`

Copy down through C8.

Graph: Highlight A1–C8; insert Scatter chart. [More details on page 118.]

Web Site

www.AppliedCalc.org

In the Function Evaluator and Grapher utility, enter the data and model(s) as shown below. Set xMin = 0, xMax = 6 and press "Plot Graphs."

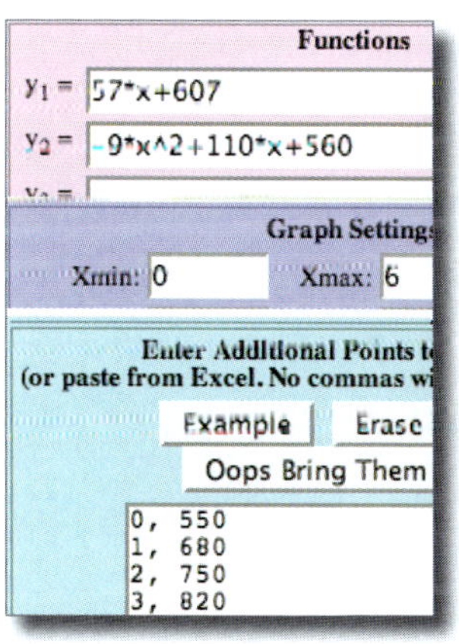

Solution

a. The following table shows the original data together with the values for all four models:

t	0	1	2	3	4	5	6
Revenue ($ million)	550	680	750	820	850	890	910
Linear: $r(t) = 57t + 607$ Technology: `57*x+607`	607	664	721	778	835	892	949
Quadratic: $r(t) = -9t^2 + 110t + 560$ Technology: `-9*x^2+110*x+560`	560	661	744	809	856	885	896
Exponential: $r(t) = 608(1.08)^t$ Technology: `608*(1.08)^x`	608	657	709	766	827	893	965
Logistic: $r(t) = \dfrac{930}{1 + 0.67(1.7)^{-t}}$ Technology: `930/(1+0.67*1.7^(-x))`	557	667	755	818	861	888	905

Notice first that all the values for the quadratic and logistic models are close to the actual revenue values. This cannot be said for the linear and exponential models. Figure 12 shows the original data together with the graph of each model. Notice

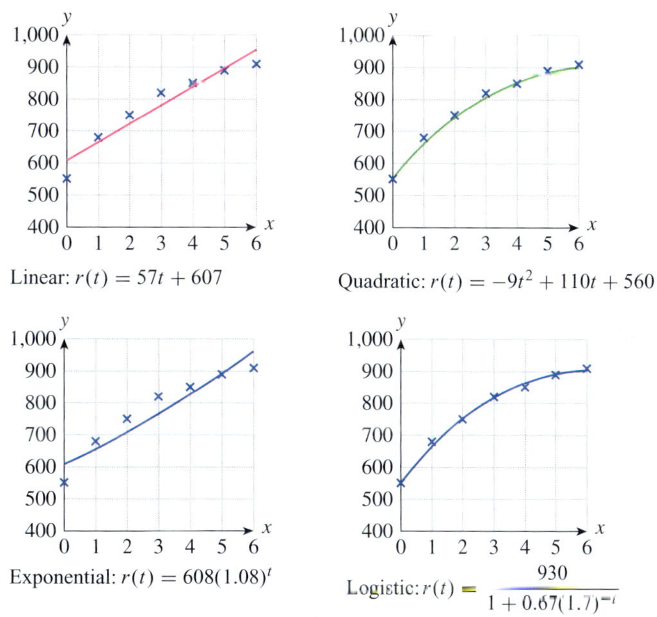

Linear: $r(t) = 57t + 607$ — Quadratic: $r(t) = -9t^2 + 110t + 560$ — Exponential: $r(t) = 608(1.08)^t$ — Logistic: $r(t) = \dfrac{930}{1 + 0.67(1.7)^{-t}}$

Figure 12

again that the quadratic and logistic curves appear to follow the data more closely than the other two. We therefore conclude that the quadratic and logistic models fit the data significantly better than the others.

b. Although the quadratic and logistic models both appear to fit the data well, they do not both extrapolate to give reasonable predictions for 2010:

$$\text{Quadratic Model: } r(10) = -9(10)^2 + 110(10) + 560 \approx 760$$

$$\text{Logistic Model: } r(10) = \frac{930}{1 + 0.67(1.7)^{-10}} \approx 927$$

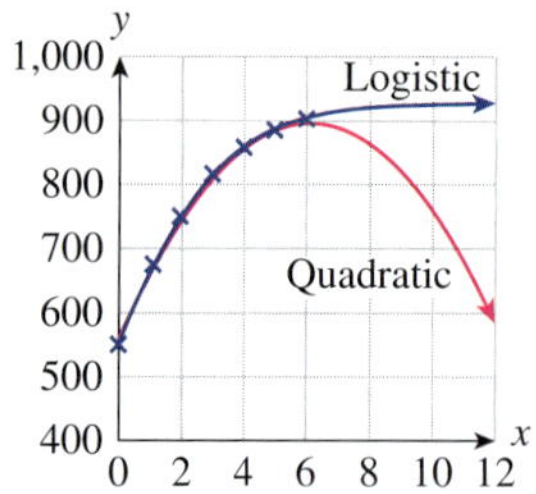

Figure 13

Notice that the quadratic model predicts a significant *decline* in casino revenues whereas the logistic model predicts a more reasonable modest increase. This discrepancy can be seen quite dramatically in Figure 13.

We now derive an analytical model of change over time based on the idea of **compound interest**. Suppose you invest \$500 (the **present value**) in an investment account with an annual yield of 15%, and the interest is reinvested at the end of every year (we say that the interest is **compounded** or **reinvested** once a year). Let t represent the number of years since you made the initial \$500 investment. Each year, the investment is worth 115% (or 1.15 times) of its value the previous year. The **future value** A of your investment changes over time t, so we think of A as a function of t. The following table illustrates how we can calculate the future value for several values of t:

t	0	1	2	3
Future Value $A(t)$	500	575	661.25	760.44
	A	$500(1.15)$	$500(1.15)^2$	$500(1.15)^3$

×1.15 ×1.15 ×1.15

Thus, $A(t) = 500(1.15)^t$. A traditional way to write this formula is

$$A(t) = P(1+r)^t$$

where P is the present value ($P = 500$) and r is the annual interest rate ($r = 0.15$).

If, instead of compounding the interest once a year, we compound it every three months (four times a year), we would earn one quarter of the interest ($r/4$ of the current investment) every three months. Because this would happen $4t$ times in t years, the formula for the future value becomes

$$A(t) = P\left(1+\frac{r}{4}\right)^{4t}.$$

Compound Interest

If an amount **(present value)** P is invested for t years at an annual rate of r, and if the interest is compounded (reinvested) m times per year, then the **future value** A is

$$A(t) = P\left(1+\frac{r}{m}\right)^{mt}.$$

A special case is **interest compounded once a year:**

$$A(t) = P(1+r)^t.$$

Quick Example

If \$2,000 is invested for two and a half years in a mutual fund with an annual yield of 12.6% and the earnings are reinvested each month, then $P = 2{,}000$, $r = 0.126$, $m = 12$, and $t = 2.5$, which gives

$$A(2.5) = 2{,}000\left(1+\frac{0.126}{12}\right)^{12\times 2.5}$$

```
2000*(1+0.126/12)^(12*2.5)
```

$$= 2{,}000(1.0105)^{30} = \$2{,}736.02.$$

EXAMPLE 6 Compound Interest: Investments

Consider the scenario in the preceding Quick Example: You invest \$2,000 in a mutual fund with an annual yield of 12.6% and the interest is reinvested each month.

a. Find the associated exponential model.

b. T Use a table of values to estimate the year during which the value of your investment reaches \$5,000.

c. Use a graph to confirm your answer in part (b).

Solution

a. Apply the formula

$$A(t) = P\left(1 + \frac{r}{m}\right)^{mt}$$

with $P = 2{,}000$, $r = 0.126$, and $m = 12$. We get

$$A(t) = 2{,}000\left(1 + \frac{0.126}{12}\right)^{12t}$$

$$A(t) = 2{,}000(1.0105)^{12t}$$ `2000*(1+0.126/12)^(12*t)`

This is the exponential model. (What would happen if we left out the last set of parentheses in the technology formula?)

b. We need to find the value of t for which $A(t) = \$5{,}000$, so we need to solve the equation

$$5{,}000 = 2{,}000(1.0105)^{12t}$$

In Section 2.3 we will learn how to use logarithms to do this algebraically, but we can answer the question now using a graphing calculator, a spreadsheet, or the Function Evaluator and Grapher utility at the Web site. Just enter the model and compute the balance at the end of several years. Here are examples of tables obtained using three forms of technology:

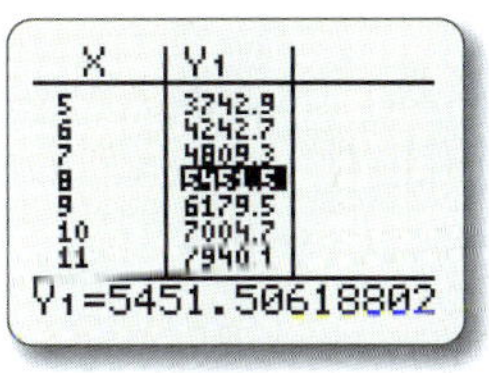

TI-83/84 Plus

	A	B
1	t	A
2	0	\$ 2,000.00
3	1	\$ 2,267.07
4	2	\$ 2,569.81
5	3	\$ 2,912.98
6	4	\$ 3,301.97
7	5	\$ 3,742.91
8	6	\$ 4,242.72
9	7	\$ 4,809.29
10	8	\$ 5,451.51
11	9	\$ 6,179.49

Excel

x-Values	y_1-Values
3	2912.98
4	3301.97
5	3742.91
6	4242.72
7	4809.29
8	5451.51

Web site

Because the balance first exceeds \$5,000 at $t = 8$ (the end of year 8) your investment has reached \$5,000 during year 8.

using Technology

TI-83/84 Plus

`Y1=2000(1+0.126/12)^(12X)`

[2ND] [TABLE]

Excel

Headings *t* and *A* in A1–B1; *t*-values 0–11 in A2–A13.

`=2000*(1+0.126/12)^(12*A2)`

in B2; copy down through B13.

Web Site

www.AppliedCalc.org

Go to the Function Evaluator and Grapher under Online Utilities, and enter

`2000(1+0.126/12)^(12x)`

for y_1. Scroll down to the Evaluator, enter the values 0–11 under *x*-values and press "Evaluate."

c. Figure 14 shows the graph of $A(t) = 2{,}000(1.0105)^{12t}$ together with the horizontal line $y = 5{,}000$.

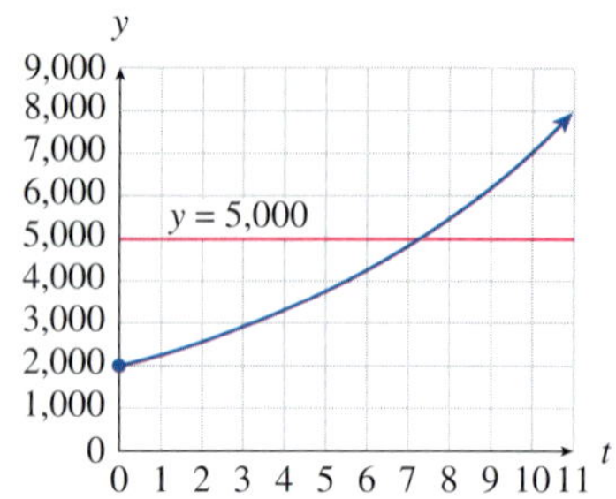

Technology format: `2000*1.0105^(12*x)`

Figure 14

The graphs cross between $t = 7$ and $t = 8$, confirming that year 8 is the first year during which the value of the investment reaches $5,000.

The compound interest examples we saw above are instances of **exponential growth:** a quantity whose magnitude is an increasing exponential function of time. The decay of unstable radioactive isotopes provides instances of **exponential decay:** a quantity whose magnitude is a *decreasing* exponential function of time. For example, carbon 14, an unstable isotope of carbon, decays exponentially to nitrogen. Because carbon 14 decay is extremely slow, it has important applications in the dating of fossils.

EXAMPLE 7 Exponential Decay: Carbon Dating

The amount of carbon 14 remaining in a sample that originally contained A grams is approximately

$$C(t) = A(0.999879)^t$$

where t is time in years.

a. What percentage of the original amount remains after one year? After two years?

b. Graph the function C for a sample originally containing 50g of carbon 14, and use your graph to estimate how long, to the nearest 1,000 years, it takes for half the original carbon 14 to decay.

c. A fossilized plant unearthed in an archaeological dig contains 0.50 gram of carbon 14 and is known to be 50,000 years old. How much carbon 14 did the plant originally contain?

Solution Notice that the given model is exponential as it has the form $f(t) = Ab^t$. (See page 51.)

a. At the start of the first year, $t = 0$, so there are

$$C(0) = A(0.999879)^0 = A \text{ grams.}$$

At the end of the first year, $t = 1$, so there are

$$C(1) = A(0.999879)^1 = 0.999879A \text{ grams;}$$

that is, 99.9879% of the original amount remains. After the second year, the amount remaining is

$$C(2) = A(0.999879)^2 \approx 0.999758A \text{ grams,}$$

or about 99.9758% of the original sample.

b. For a sample originally containing 50g of carbon 14, $A = 50$, so $C(t) = 50(0.999879)^t$. Its graph is shown in Figure 15.

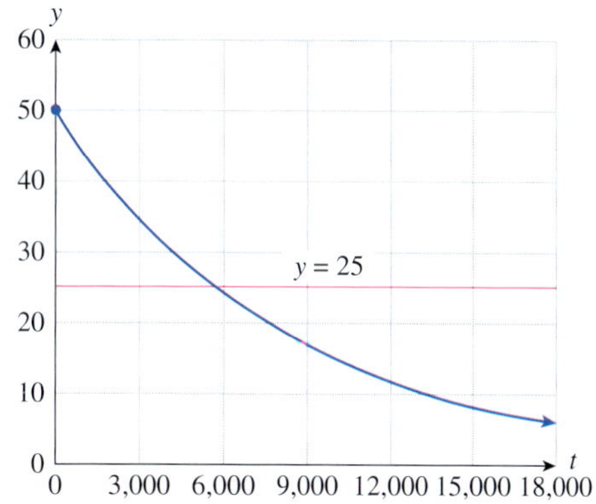

Technology format: `50*0.999879^x`

Figure 15

We have also plotted the line $y = 25$ on the same graph. The graphs intersect at the point where the original sample has decayed to 25g: about $t = 6{,}000$ years.

c. We are given the following information: $C = 0.50$, $A =$ the unknown, and $t = 50{,}000$. Substituting gives

$$0.50 = A(0.999879)^{50{,}000}.$$

Solving for A gives

$$A = \frac{0.5}{0.999879^{50{,}000}} \approx 212 \text{ grams.}$$

Thus, the plant originally contained 212 grams of carbon 14.

Before we go on... The formula we used for A in Example 7(c) has the form

$$A(t) = \frac{C}{0.999879^t},$$

which gives the original amount of carbon 14 t years ago in terms of the amount C that is left now. A similar formula can be used in finance to find the present value, given the future value. ■

1.2 EXERCISES

▼ more advanced ◆ challenging

T indicates exercises that should be solved using technology

1. ***Resources*** You now have 200 sound files on your hard drive, and this number is increasing by 10 sound files each day. Find a mathematical model for this situation. HINT [See Quick Example 1 on page 56.]

2. ***Resources*** The amount of free space left on your hard drive is now 50 gigabytes (GB) and is decreasing by 5 GB/month. Find a mathematical model for this situation.

3. ***Soccer*** My rectangular soccer field site has a length equal to twice its width. Find its area in terms of its length x. HINT [See Quick Example 3 on page 56.]

4. ***Cabbage*** My rectangular cabbage patch has a total area of 100 sq. ft. Find its perimeter in terms of the width x.

5. ***Vegetables*** I want to fence in a square vegetable patch. The fencing for the east and west sides costs \$4 per foot, and the fencing for the north and south sides costs only \$2 per foot. Find the total cost of the fencing as a function of the length of a side x.

6. ***Orchids*** My square orchid garden abuts my house so that the house itself forms the northern boundary. The fencing for the southern boundary costs \$4 per foot, and the fencing for the east and west sides costs \$2 per foot. Find the total cost of the fencing as a function of the length of a side x.

7. ***Cost*** A piano manufacturer has a daily fixed cost of \$1,200 and a marginal cost of \$1,500 per piano. Find the cost $C(x)$ of manufacturing x pianos in one day. Use your function to answer the following questions:

   a. On a given day, what is the cost of manufacturing 3 pianos?
   b. What is the cost of manufacturing the 3rd piano that day?

c. What is the cost of manufacturing the 11th piano that day?
d. What is the variable cost? What is the fixed cost? What is the marginal cost?
e. Graph C as a function of x. HINT [See Example 1.]

8. ***Cost*** The cost of renting tuxes for the Choral Society's formal is \$20 down, plus \$88 per tux. Express the cost C as a function of x, the number of tuxedos rented. Use your function to answer the following questions.

a. What is the cost of renting 2 tuxes?
b. What is the cost of the 2nd tux?
c. What is the cost of the 4,098th tux?
d. What is the variable cost? What is the marginal cost?
e. Graph C as a function of x.

9. ***Break-Even Analysis*** Your college newspaper, *The Collegiate Investigator*, has fixed production costs of \$70 per edition and marginal printing and distribution costs of 40¢ per copy. *The Collegiate Investigator* sells for 50¢ per copy.

a. Write down the associated cost, revenue, and profit functions. HINT [See Examples 1 and 2.]
b. What profit (or loss) results from the sale of 500 copies of *The Collegiate Investigator*?
c. How many copies should be sold in order to break even?

10. ***Break-Even Analysis*** The Audubon Society at Enormous State University (ESU) is planning its annual fund-raising "Eat-a-thon." The society will charge students 50¢ per serving of pasta. The only expenses the society will incur are the cost of the pasta, estimated at 15¢ per serving, and the \$350 cost of renting the facility for the evening.

a. Write down the associated cost, revenue, and profit functions.
b. How many servings of pasta must the Audubon Society sell in order to break even?
c. What profit (or loss) results from the sale of 1,500 servings of pasta?

11. ***Break-Even Analysis*** Gymnast Clothing manufactures expensive hockey jerseys for sale to college bookstores in runs of up to 200. Its cost (in dollars) for a run of x hockey jerseys is

$$C(x) = 2{,}000 + 10x + 0.2x^2 \quad (0 \le x \le 200)$$

Gymnast Clothing sells the jerseys at \$100 each. Find the revenue and profit functions. How many should Gymnast Clothing manufacture to make a profit? HINT [See Example 2.]

12. ***Break-Even Analysis*** Gymnast Clothing also manufactures expensive soccer cleats for sale to college bookstores in runs of up to 500. Its cost (in dollars) for a run of x pairs of cleats is

$$C(x) = 3{,}000 + 8x + 0.1x^2 \quad (0 \le x \le 500)$$

Gymnast Clothing sells the cleats at \$120 per pair. Find the revenue and profit functions. How many should Gymnast Clothing manufacture to make a profit?

13. ***Break-Even Analysis: School Construction Costs*** The cost, in millions of dollars, of building a two-story high school in New York State was estimated to be

$$C(x) = 1.7 + 0.12x - 0.0001x^2 \quad (20 \le x \le 400)$$

where x is the number of thousands of square feet.[12] Suppose that you are contemplating building a for-profit two-story high school and estimate that your total revenue will be \$0.1 million dollars per thousand square feet. What is the profit function? What size school should you build in order to break even?

14. ***Break-Even Analysis: School Construction Costs*** The cost, in millions of dollars, of building a three-story high school in New York State was estimated to be

$$C(x) = 1.7 + 0.14x - 0.0001x^2 \quad (20 \le x \le 400)$$

where x is the number of thousands of square feet.[13] Suppose that you are contemplating building a for-profit three-story high school and estimate that your total revenue will be \$0.2 million dollars per thousand square feet. What is the profit function? What size school should you build in order to break even?

15. ▼ ***Profit Analysis—Aviation*** The hourly operating cost of a Boeing 747-100, which seats up to 405 passengers, is estimated to be \$5,132.[14] If an airline charges each passenger a fare of \$100 per hour of flight, find the hourly profit P it earns operating a 747-100 as a function of the number of passengers x. (Be sure to specify the domain.) What is the least number of passengers it must carry in order to make a profit? HINT [The cost function is constant (Variable cost = 0).]

16. ▼ ***Profit Analysis—Aviation*** The hourly operating cost of a McDonnell Douglas DC 10-10, which seats up to 295 passengers, is estimated to be \$3,885.[15] If an airline charges each passenger a fare of \$100 per hour of flight, find the hourly profit P it earns operating a DC 10-10 as a function of the number of passengers x. (Be sure to specify the domain.) What is the least number of passengers it must carry in order to make a profit? HINT [The cost function is constant (Variable cost = 0).]

17. ▼ ***Break-Even Analysis*** *(based on a question from a CPA exam)* The Oliver Company plans to market a new product. Based on its market studies, Oliver estimates that it can sell up to 5,500 units in 2005. The selling price will be \$2 per unit. Variable costs are estimated to be 40% of total revenue. Fixed costs are estimated to be \$6,000 for 2005. How many units should the company sell to break even?

18. ▼ ***Break-Even Analysis*** *(based on a question from a CPA exam)* The Metropolitan Company sells its latest product at a unit price of \$5. Variable costs are estimated to be 30% of the total revenue, while fixed costs amount to \$7,000 per month. How many units should the company sell per month in order to break even, assuming that it can sell up to 5,000 units per month at the planned price?

---

[12] The model is the authors'. Source for data: *Project Labor Agreements and Public Construction Cost in New York State*, Paul Bachman and David Tuerck, Beacon Hill Institute at Suffolk University, April 2006, www.beaconhill.org.

[13] Ibid.

[14] In 1992. Source: Air Transportation Association of America.

[15] Ibid.

19. ◆ ***Break-Even Analysis*** *(from a CPA exam)* Given the following notations, write a formula for the break-even sales level.

SP = Selling price per unit
FC = Total fixed cost
VC = Variable cost per unit

20. ◆ ***Break-Even Analysis*** *(based on a question from a CPA exam)* Given the following notation, give a formula for the total fixed cost.

SP = Selling price per unit
VC = Variable cost per unit
BE = Break even sales level in units

21. ◆ ***Break-Even Analysis—Organized Crime*** The organized crime boss and perfume king Butch (Stinky) Rose has daily overheads (bribes to corrupt officials, motel photographers, wages for hit men, explosives, and so on) amounting to \$20,000 per day. On the other hand, he has a substantial income from his counterfeit perfume racket: He buys imitation French perfume (Chanel № 22.5) at \$20 per gram, pays an additional \$30 per 100 grams for transportation, and sells it via his street thugs for \$600 per gram. Specify Stinky's profit function, $P(x)$, where x is the quantity (in grams) of perfume he buys and sells, and use your answer to calculate how much perfume should pass through his hands per day in order that he break even.

22. ◆ ***Break-Even Analysis—Disorganized Crime*** Butch (Stinky) Rose's counterfeit Chanel № 22.5 racket has run into difficulties; it seems that the *authentic* Chanel № 22.5 perfume is selling for less than his counterfeit perfume. However, he has managed to reduce his fixed costs to zero, and his overall costs are now \$400 per gram plus \$30 per gram transportation costs and commission. (The perfume's smell is easily detected by specially trained Chanel Hounds, and this necessitates elaborate packaging measures.) He therefore decides to sell it for \$420 per gram in order to undercut the competition. Specify Stinky's profit function, $P(x)$, where x is the quantity (in grams) of perfume he buys and sells, and use your answer to calculate how much perfume should pass through his hands per day in order that he break even. Interpret your answer.

23. ***Demand for Monorail Service, Las Vegas*** The demand for monorail service in Las Vegas can be approximated by

$$q(p) = 64p^{-0.76} \text{ thousand rides per day} \quad (3 \le p \le 5)$$

where p is the cost per ride in dollars.[16]

a. Graph the demand function.
b. What is the result on demand if the cost per ride is increased from \$3.00 to \$3.50? HINT [See Example 3.]

24. ***Demand for Monorail Service, Mars*** The demand for monorail service on the Utarek monorail, which links the three urbynes (or districts) of Utarek, Mars, can be approximated by

$$q(p) = 30p^{-0.49} \text{ million rides per day} \quad (3 \le p \le 5)$$

where p is the cost per ride in zonars ($\overline{\overline{\text{Z}}}$).[17]

a. Graph the demand function.
b. What is the result on demand if the cost per ride is decreased from $\overline{\overline{\text{Z}}}$5.00 to $\overline{\overline{\text{Z}}}$3.50?

25. ▼ ***Demand*** The demand for Sigma Mu Fraternity plastic brownie dishes is

$$q(p) = 361{,}201 - (p + 1)^2$$

where q represents the number of brownie dishes Sigma Mu can sell each month at a price of p¢. Use this function to determine:

a. The number of brownie dishes Sigma Mu can sell each month if the price is set at 50¢.
b. The number of brownie dishes they can unload each month if they give them away.
c. The lowest price at which Sigma Mu will be unable to sell any dishes.

26. ▼ ***Revenue*** The total weekly revenue earned at Royal Ruby Retailers is given by

$$R(p) = -\frac{4}{3}p^2 + 80p$$

where p is the price (in dollars) RRR charges per ruby. Use this function to determine:

a. The weekly revenue, to the nearest dollar, when the price is set at \$20/ruby.
b. The weekly revenue, to the nearest dollar, when the price is set at \$200/ruby. (Interpret your result.)
c. The price RRR should charge in order to obtain a weekly revenue of \$1,200.

27. ***Equilibrium Price*** The demand for your hand-made skateboards, in weekly sales, is $q = -3p + 700$ if the selling price is \$$p$. You are prepared to supply $q = 2p - 500$ per week at the price \$$p$. At what price should you sell your skateboards so that there is neither a shortage nor a surplus? HINT [See Quick Example on page 62.]

28. ***Equilibrium Price*** The demand for your factory-made skateboards, in weekly sales, is $q = -5p + 50$ if the selling price is \$$p$. If you are selling them at that price, you can obtain $q = 3p - 30$ per week from the factory. At what price should you sell your skateboards so that there is neither a shortage nor a surplus?

29. ***Equilibrium Price: Cell Phones*** Worldwide quarterly sales of Nokia® cell phones was approximately $q = -p + 156$ million phones when the wholesale price[18] was \$$p$.

a. If Nokia was prepared to supply $q = 4p - 394$ million phones per quarter at a wholesale price of \$$p$, what would be the equilibrium price?
b. The actual wholesale price was \$105 in the fourth quarter of 2004. Estimate the projected shortage or surplus at that price. HINT [See Quick Example on page 62 and also Example 4.]

---

[16] Source: *New York Times*, Februrary 10, 2007, p. A9.

[17] $\overline{\overline{\text{Z}}}$ designates Zonars, the official currency of Mars. See www.marsnext.com for details of the Mars colony, its commerce, and its culture.

[18] Source: Embedded.com/Company reports December, 2004.

30. ***Equilibrium Price: Cell Phones*** Worldwide annual sales of all cell phones is approximately $-10p + 1,600$ million phones when the wholesale price[19] is $\$p$.

a. If manufacturers are prepared to supply $q = 14p - 800$ million phones per year at a wholesale price of $\$p$, what would be the equilibrium price?

b. The actual wholesale price was projected to be $80 in the fourth quarter of 2008. Estimate the projected shortage or surplus at that price.

31. T ***Demand for Monorail Service, Las Vegas*** The demand for monorail service in Las Vegas can be modeled by

$$q = 64p^{-0.76} \text{ thousand rides per day}$$

where p is the fare the Las Vegas Monorail Company charges in dollars.[20] Assume the company is prepared to provide service for

$$q = 2.5p + 15.5 \text{ thousand rides per day}$$

at a fare of $\$p$.

a. Graph the demand and supply equations, and use your graph to estimate the equilibrium price (to the nearest 50¢).

b. Estimate, to the nearest 10 rides, the shortage or surplus of monorail service at the December 2005 fare of $5 per ride.

32. T ***Demand for Monorail Service, Mars*** The demand for monorail service in the three urbynes (or districts) of Utarek, Mars can be modeled by

$$q = 31p^{-0.49} \text{ million rides per day}$$

where p is the fare the Utarek Monorail Cooperative charges in zonars (Z̿).[21] Assume the cooperative is prepared to provide service for

$$q = 2.5p + 17.5 \text{ thousand rides per day}$$

at a fare of Z̿p.

a. Graph the demand and supply equations, and use your graph to estimate the equilibrium price (to the nearest 0.50 zonars).

b. Estimate the shortage or surplus of monorail service at the December 2085 fare of Z̿1 per ride.

33. ▼ ***Toxic Waste Treatment*** The cost of treating waste by removing PCPs goes up rapidly as the quantity of PCPs removed goes up. Here is a possible model:

$$C(q) = 2,000 + 100q^2$$

where q is the reduction in toxicity (in pounds of PCPs removed per day) and $C(q)$ is the daily cost (in dollars) of this reduction.

a. Find the cost of removing 10 pounds of PCPs per day.

b. Government subsidies for toxic waste cleanup amount to

$$S(q) = 500q$$

where q is as above and $S(q)$ is the daily dollar subsidy. Calculate the net cost function $N(q)$ (the cost of removing q pounds of PCPs per day after the subsidy is taken into account), given the cost function and subsidy above, and find the net cost of removing 20 pounds of PCPs per day.

34. ▼ ***Dental Plans*** A company pays for its employees' dental coverage at an annual cost C given by

$$C(q) = 1,000 + 100\sqrt{q}$$

where q is the number of employees covered and $C(q)$ is the annual cost in dollars.

a. If the company has 100 employees, find its annual outlay for dental coverage.

b. Assuming that the government subsidizes coverage by an annual dollar amount of

$$S(q) = 200q$$

calculate the net cost function $N(q)$ to the company, and calculate the net cost of subsidizing its 100 employees. Comment on your answer.

35. ***Spending on Corrections in the 1990s*** The following table shows the annual spending by all states in the United States on corrections ($t = 0$ represents the year 1990):[22]

Year (t)	0	2	4	6	7
Spending ($ billion)	16	18	22	28	30

a. Which of the following functions best fits the given data? (*Warning*: None of them fits exactly, but one fits more closely than the others.) HINT [See Example 5.]

(A) $S(t) = -0.2t^2 + t + 16$
(B) $S(t) = 0.2t^2 + t + 16$
(C) $S(t) = t + 16$

b. Use your answer to part (a) to "predict" spending on corrections in 1998, assuming that the trend continued.

36. ***Spending on Corrections in the 1990s*** Repeat Exercise 35, this time choosing from the following functions:

a. $S(t) = 16 + 2t$
b. $S(t) = 16 + t + 0.5t^2$
c. $S(t) = 16 + t - 0.5t^2$

Freon Production *Exercises 37 and 38 are based on the following data in Quick Example 1 on page 42 showing the amount of ozone-damaging Freon (in tons) produced in developing countries in year t since 2000:*

[19] Wholesale price projections are the authors'. Source for sales prediction: I-Stat/NDR December, 2004.

[20] The model is the authors'. Source for data: *New York Times*, Februrary 10, 2007, p. A9.

[21] Z̿ designates Zonars, the official currency of Mars. See www. marsnext. com for details of the Mars colony, its commerce, and its culture.

[22] Data are rounded. Source: National Association of State Budget Officers/*The New York Times*, February 28, 1999, p. A1.

t (Year Since 2000)	F (Tons of Freon 22)
0	100
2	140
4	200
6	270
8	400
10	590

Source: *New York Times*, February 23, 2007, p. C1

37. a. Which two of the following models best fit the given data?

(A) $f(t) = 98(1.2^t)$
(B) $f(t) = 4.6t^2 + 1.2t + 109$
(C) $f(t) = 47t + 48$
(D) $f(t) = 98(1.2^{-t})$

b. Of the two models you chose in part (a), which predicts the larger amount of freon in 2020? How much freon does that model predict?

38. Repeat Exercise 37 using the following models:

a. $f(t) = -4.6t^2 + 1.2t + 109$
b. $f(t) = \dfrac{2500}{1 + 22(1.2^{-t})}$
c. $f(t) = 65t - 60$
d. $f(t) = 4.6t^2 + 1.2t + 109$

39. ***Soccer Gear*** The East Coast College soccer team is planning to buy new gear for its road trip to California. The cost per shirt depends on the number of shirts the team orders as shown in the following table:

x (Shirts ordered)	5	25	40	100	125
$A(x)$ (Cost/shirt, \$)	22.91	21.81	21.25	21.25	22.31

a. Which of the following functions best models the data?

(A) $A(x) = 0.005x + 20.75$
(B) $A(x) = 0.01x + 20 + \frac{25}{x}$
(C) $A(x) = 0.0005x^2 - 0.07x + 23.25$
(D) $A(x) = 25.5(1.08)^{(x-5)}$

b. T Graph the model you chose in part (a) for $10 \le x \le 100$. Use your graph to estimate the lowest cost per shirt and the number of shirts the team should order to obtain the lowest price per shirt.

40. ***Hockey Gear*** The South Coast College hockey team wants to purchase wool hats for its road trip to Alaska. The cost per hat depends on the number of hats the team orders as shown in the following table:

x (Hats ordered)	5	25	40	100	125
$A(x)$ (Cost/hat, \$)	25.50	23.50	24.63	30.25	32.70

a. Which of the following functions best models the data?

(A) $A(x) = 0.05x + 20.75$
(B) $A(x) = 0.1x + 20 + \frac{25}{x}$
(C) $A(x) = 0.0008x^2 - 0.07x + 23.25$
(D) $A(x) = 25.5(1.08)^{(x-5)}$

b. T Graph the model you chose in part (a) with $5 \le x \le 30$. Use your graph to estimate the lowest cost per hat and the number of hats the team should order to obtain the lowest price per hat.

41. ***Value of Euro*** The following table shows the approximate value V of one Euro in U.S. dollars from its introduction in January 2000 to January 2008. ($t = 0$ represents January 2000.)[23]

t (Year)	0	2	8
V (Value in \$)	1.00	0.90	1.40

Which of the following kinds of models would best fit the given data? Explain your choice of model. (A, a, b, c, and m are constants.)

(A) Linear: $V(t) = mt + b$
(B) Quadratic: $V(t) = at^2 + bt + c$
(C) Exponential: $V(t) = Ab^t$

42. ***Household Income*** The following table shows the approximate average household income in the United States in 1990, 1995, and 2003. ($t = 0$ represents 1990.)[24]

t (Year)	0	5	13
H (Household Income in \$1,000)	30	35	43

Which of the following kinds of models would best fit the given data? Explain your choice of model. (A, a, b, c, and m are constants.)

(A) Linear: $H(t) = mt + b$
(B) Quadratic: $H(t) = at^2 + bt + c$
(C) Exponential: $H(t) = Ab^t$

43. ***Investments*** In 2007, **E*TRADE Financial** was offering 4.94% interest on its online savings accounts, with interest reinvested monthly.[25] Find the associated exponential model for the value of a \$5,000 deposit after t years. Assuming this rate of return continued for 7 years, how much would a deposit of \$5,000 at the beginning of 2007 be worth at the start of 2014? (Answer to the nearest \$1.)

[23] Source: Bloomberg Financial Markets.

[24] In current dollars, unadjusted for inflation. Source: U.S. Census Bureau; "Table H-5. Race and Hispanic Origin of Householder—Households by Median and Mean Income: 1967 to 2003"; published August 27, 2004; www.census.gov/hhes/income/histinc/h05.html.

[25] Interest rate based on annual percentage yield. Source: www.us.etrade.com, December 2007.

44. ***Investments*** In 2007, **ING Direct** was offering 4.14% interest on its online Orange Savings Account, with interest reinvested quarterly.[26] Find the associated exponential model for the value of a \$4,000 deposit after t years. Assuming this rate of return continued for eight years, how much would a deposit of \$4,000 at the beginning of 2007 be worth at the start of 2015? (Answer to the nearest \$1.)

45. **T** ***Investments*** Refer to Exercise 43. At the start of which year will an investment of \$5,000 made at the beginning of 2007 first exceed \$7,500?

46. **T** ***Investments*** Refer to Exercise 44. In which year will an investment of \$4,000 made at the beginning of 2007 first exceed \$6,000?

47. ***Carbon Dating*** A fossil originally contained 104 grams of carbon 14. Refer to the formula for $C(t)$ in Example 7 and estimate the amount of carbon 14 left in the sample after 10,000 years, 20,000 years, and 30,000 years. HINT [See Example 7.]

48. ***Carbon Dating*** A fossil contains 4.06 grams of carbon 14. Refer to the formula for $C(t)$ in Example 7 and estimate the amount of carbon 14 in the sample 10,000 years, 20,000 years, and 30,000 years ago.

49. ***Carbon Dating*** A fossil contains 4.06 grams of carbon 14. It is estimated that the fossil originally contained 46 grams of carbon 14. By calculating the amount left after 5,000 years, 10,000 years, . . . , 35,000 years, estimate the age of the sample to the nearest 5,000 years. (Refer to the formula for $C(t)$ in Example 7.)

50. ***Carbon Dating*** A fossil contains 2.8 grams of carbon 14. It is estimated that the fossil originally contained 104 grams of carbon 14. By calculating the amount 5,000 years, 10,000 years, . . . , 35,000 years ago, estimate the age of the sample to the nearest 5,000 years. (Refer to the formula for $C(t)$ in Example 7.)

51. ***Radium Decay*** The amount of radium 226 remaining in a sample that originally contained A grams is approximately

$$C(t) = A(0.999567)^t$$

where t is time in years.

   a. Find, to the nearest whole number, the percentage of iodine-131 left in an originally pure sample after 1,000 years, 2,000 years, and 3,000 years.

   b. Use a graph to estimate, to the nearest 100 years, when one half of a sample of 100 grams will have decayed.

52. ***Iodine Decay*** The amount of iodine 131 remaining in a sample that originally contained A grams is approximately

$$C(t) = A(0.9175)^t$$

where t is time in days.

   a. Find, to the nearest whole number, the percentage of iodine 131 left in an originally pure sample after 2 days, 4 days, and 6 days.

   b. Use a graph to estimate, to the nearest day, when one half of a sample of 100 grams will have decayed.

COMMUNICATION AND REASONING EXERCISES

53. If the population of the lunar station at Clavius has a population of $P = 200 + 30t$ where t is time in years since the station was established, then the population is increasing by ____ per year.

54. My bank balance can be modeled by $B(t) = 5{,}000 - 200t$ dollars, where t is time in days since I opened the account. The balance on my account is __________ by \$200 per day.

55. Classify the following model as analytical or curve-fitting, and give a reason for your choice: The price of gold was \$700 on Monday, \$710 on Tuesday, and \$700 on Wednesday. Therefore, the price can be modeled by $p(t) = -10t^2 + 20t + 700$ where t is the day since Monday.

56. Classify the following model as analytical or curve-fitting, and give a reason for your choice: The width of a small animated square on my computer screen is currently 10 mm and is growing by 2 mm per second. Therefore, its area can be modeled by $a(t) = (10 + 2t)^2$ square mm where t is time in seconds.

57. Fill in the missing information for the following *analytical model* (answers may vary): ____________. Therefore, the cost of downloading a movie can be modeled by $c(t) = 4 - 0.2t$, where t is time in months since January.

58. Repeat Exercise 57, but this time regard the given model as a *curve-fitting model.*

59. Fill in the blanks: In a linear cost function, the _______ cost is x times the _______ cost.

60. Complete the following sentence: In a linear cost function, the marginal cost is the ______________.

61. ▼ We said on page 61 that the demand for a commodity generally goes down as the price goes up. Assume that the demand for a certain commodity goes up as the price goes up. Is it still possible for there to be an equilibrium price? Explain with the aid of a demand and supply graph.

62. ▼ What would happen to the price of a certain commodity if the demand was always greater than the supply? Illustrate with a demand and supply graph.

63. You have a set of data points showing the sales of videos on your Web site versus time that are closely approximated by two different mathematical models. Give one criterion that would lead you to choose one over the other. (Answers may vary.)

64. Would it ever be reasonable to use a quadratic model $s(t) = at^2 + bt + c$ to predict long-term sales if a is negative? Explain.

[26] Interest rate based on annual percentage yield. Source: www.home.ingdirect.com, December 2007.

1.3 Linear Functions and Models

Linear functions are among the simplest functions and are perhaps the most useful of all mathematical functions.

Linear Function

A **linear function** is one that can be written in the form

$f(x) = mx + b$ Function form

or

$y = mx + b$ Equation form

Quick Example

$f(x) = 3x - 1$

$y = 3x - 1$

where m and b are fixed numbers. (The names m and b are traditional.*)

* **NOTE** Actually, c is sometimes used instead of b. As for m, there has even been some research lately into the question of its origin, but no one knows exactly why the letter m is used.

Linear Functions from the Numerical and Graphical Point of View

The following table shows values of $y = 3x - 1$ ($m = 3$, $b = -1$) for some values of x:

x	−4	−3	−2	−1	0	1	2	3	4
y	−13	−10	−7	−4	−1	2	5	8	11

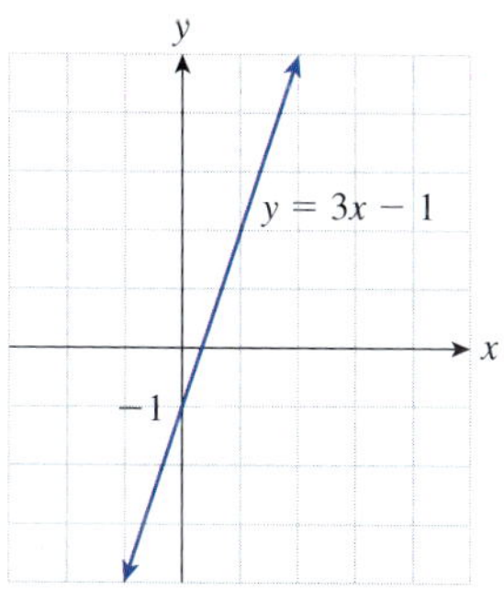

Figure 16

Its graph is shown in Figure 16.

Looking first at the table, notice that that setting $x = 0$ gives $y = -1$, the value of b.

Numerically, b is the value of y when $x = 0$.

On the graph, the corresponding point $(0, -1)$ is the point where the graph crosses the y-axis, and we say that $b = -1$ is the ***y*-intercept** of the graph (Figure 17).

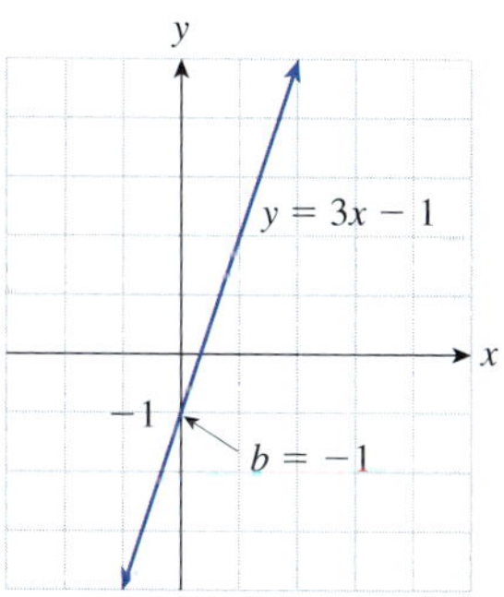

y-intercept $= b = -1$
Graphically, b is the y-intercept of the graph.

Figure 17

What about m? Looking once again at the table, notice that y increases by $m = 3$ units for every increase of 1 unit in x. This is caused by the term $3x$ in the formula: For every increase of 1 in x, we get an increase of $3 \times 1 = 3$ in y.

Numerically, y increases by m units for every 1-unit increase of x.

Likewise, for every increase of 2 in x we get an increase of $3 \times 2 = 6$ in y. In general, if x increases by some amount, y will increase by three times that amount. We write:

Change in $y = 3 \times$ Change in x.

The Change in a Quantity: Delta Notation

If a quantity q changes from q_1 to q_2, the **change in q** is just the difference:

$$\begin{aligned}\text{Change in } q &= \text{Second value} - \text{First value}\\ &= q_2 - q_1\end{aligned}$$

Mathematicians traditionally use Δ (delta, the Greek equivalent of the Roman letter D) to stand for change, and write the change in q as Δq.

$$\Delta q = \text{Change in } q = q_2 - q_1$$

Quick Examples

1. If x is changed from 1 to 3, we write

$$\Delta x = \text{Second value} - \text{First value} = 3 - 1 = 2.$$

2. Looking at our linear function, we see that when x changes from 1 to 3, y changes from 2 to 8. So,

$$\Delta y = \text{Second value} - \text{First value} = 8 - 2 = 6.$$

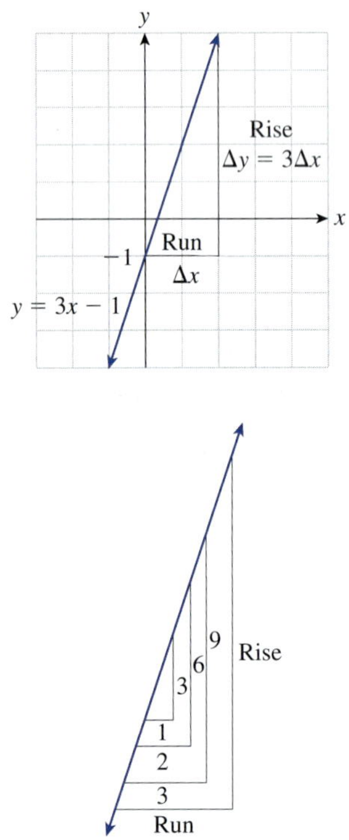

Slope = m = 3
Graphically, m is the slope of the graph.

Figure 18

Using delta notation, we can now write, for our linear function,

$$\Delta y = 3\Delta x \qquad \text{Change in } y = 3 \times \text{Change in } x.$$

or

$$\frac{\Delta y}{\Delta x} = 3.$$

Because the value of y increases by exactly 3 units for every increase of 1 unit in x, the graph is a straight line rising by 3 units for every 1 unit we go to the right. We say that we have a **rise** of 3 units for each **run** of 1 unit. Because the value of y changes by $\Delta y = 3\Delta x$ units for every change of Δx units in x, in general we have a rise of $\Delta y = 3\Delta x$ units for each run of Δx units (Figure 18). Thus, we have a rise of 6 for a run of 2, a rise of 9 for a run of 3, and so on. So, $m = 3$ is a measure of the steepness of the line; we call m the **slope of the line:**

$$\text{Slope} = m = \frac{\Delta y}{\Delta x} = \frac{\text{Rise}}{\text{Run}}$$

In general (replace the number 3 by a general number m), we can say the following.

The Roles of *m* and *b* in the Linear Function *f(x)* = *mx* + *b*

Role of *m*

Numerically If $y = mx + b$, then y changes by m units for every 1-unit change in x. A change of Δx units in x results in a change of $\Delta y = m\Delta x$ units in y. Thus,

$$m = \frac{\Delta y}{\Delta x} = \frac{\text{Change in } y}{\text{Change in } x}.$$

Graphically m is the slope of the line $y = mx + b$:

$$m = \frac{\Delta y}{\Delta x} = \frac{\text{Rise}}{\text{Run}} = \text{Slope}.$$

For positive m, the graph rises m units for every 1-unit move to the right, and rises $\Delta y = m\Delta x$ units for every Δx units moved to the right. For negative m, the graph drops $|m|$ units for every 1-unit move to the right, and drops $|m|\Delta x$ units for every Δx units moved to the right.

Graph of $y = mx + b$

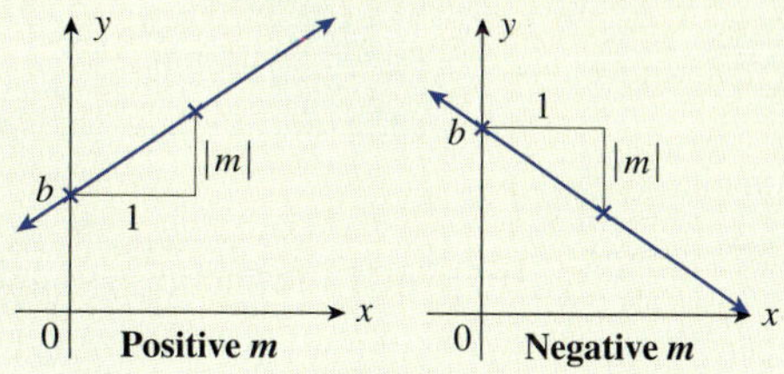

Role of *b*

Numerically When $x = 0$, $y = b$.

Graphically b is the y-intercept of the line $y = mx + b$.

Quick Examples

1. $f(x) = 2x + 1$ has slope $m = 2$ and y-intercept $b = 1$. To sketch the graph, we start at the y-intercept $b = 1$ on the y-axis, and then move 1 unit to the right and up $m = 2$ units to arrive at a second point on the graph. Now connect the two points to obtain the graph on the left.

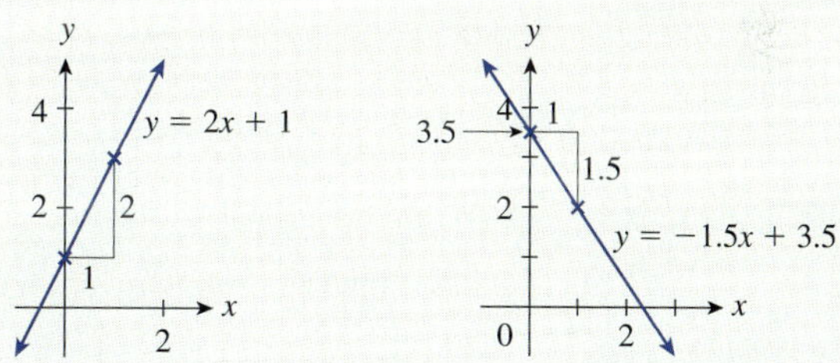

2. The line $y = -1.5x + 3.5$ has slope $m = -1.5$ and y-intercept $b = 3.5$. Because the slope is negative, the graph (above right) goes *down* 1.5 units for every 1 unit it moves to the right.

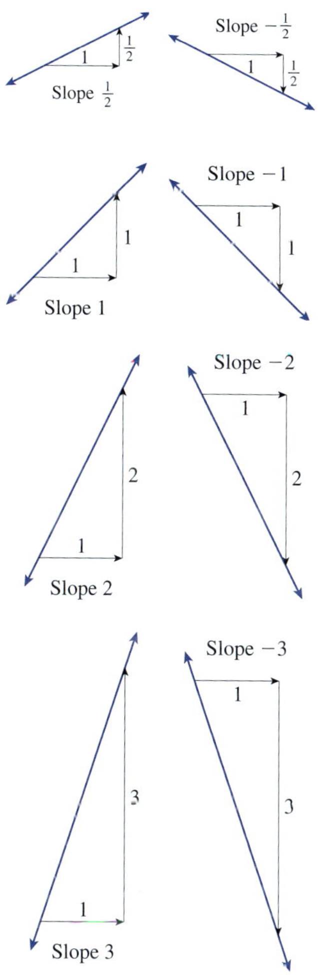

Figure 19

It helps to be able to picture what different slopes look like, as in Figure 19. Notice that the larger the absolute value of the slope, the steeper is the line.

using Technology

See the Technology Guide at the end of the chapter for detailed instructions on how to obtain a table with the successive quotients $m = \Delta y/\Delta x$ for the functions f and g in Example 1 using a TI-83/84 Plus or Excel. These tables show at a glance that f is not linear. Here is an outline:

TI-83/84 Plus

STAT EDIT; Enter values of x and $f(x)$ in lists L_1 and L_2. Highlight the heading L_3 and enter the following formula (including the quotes):

`"ΔList(L2)/ΔList(L1)"`

[More details on page 110.]

Excel

Enter headings x, $f(x)$, Df/Dx in cells A1–C1, and the corresponding values from one of the tables in cells A2–B8. Enter `=(B3-B2)/(A3-A2)` in cell C2, and copy down through C8. [More details on page 116.]

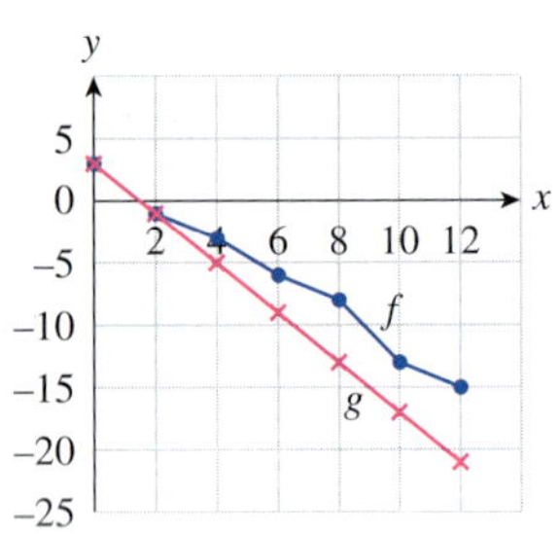

Figure 20

EXAMPLE 1 Recognizing Linear Data Numerically and Graphically

Which of the following two tables gives the values of a linear function? What is the formula for that function?

x	0	2	4	6	8	10	12
$f(x)$	3	−1	−3	−6	−8	−13	−15

x	0	2	4	6	8	10	12
$g(x)$	3	−1	−5	−9	−13	−17	−21

Solution The function f cannot be linear: If it were, we would have $\Delta f = m\Delta x$ for some fixed number m. However, although the change in x between successive entries in the table is $\Delta x = 2$ each time, the change in f is not the same each time. Thus, the ratio $\Delta f/\Delta x$ is not the same for every successive pair of points.

On the other hand, the ratio $\Delta g/\Delta x$ is the same each time, namely,

$$\frac{\Delta g}{\Delta x} = \frac{-4}{2} = -2$$

Δx: $2 - 0 = 2$, $4 - 2 = 2$, $6 - 4 = 2$, $8 - 6 = 2$, $10 - 8 = 2$, $12 - 10 = 2$

x	0	2	4	6	8	10	12
$g(x)$	3	−1	−5	−9	−13	−17	−21

Δg: $(-1) - 3 = -4$, $-5 - (-1) = -4$, $-9 - (-5) = -4$, $-13 - (-9) = -4$, $-17 - (-13) = -4$, $-21 - (-17) = -4$

Thus, g is linear with slope $m = -2$. By the table, $g(0) = 3$, hence $b = 3$. Thus,

$$g(x) = -2x + 3. \qquad \text{Check that this formula gives the values in the table.}$$

If you graph the points in the tables defining f and g above, it becomes easy to see that g is linear and f is not; the points of g lie on a straight line (with slope -2), whereas the points of f do not lie on a straight line (Figure 20).

Finding a Linear Equation from Data

If we happen to know the slope and y-intercept of a line, writing down its equation is straightforward. For example, if we know that the slope is 3 and the y-intercept is -1, then the equation is $y = 3x - 1$. Sadly, the information we are given is seldom so convenient. For instance, we may know the slope and a point other than the y intercept, two points on the line, or other information. We therefore need to know how to use the information we are given to obtain the slope and the intercept.

Computing the Slope

We can always determine the slope of a line if we are given two (or more) points on the line, because any two points—say (x_1, y_1) and (x_2, y_2)—determine the line, and hence its slope. To compute the slope when given two points, recall the formula

$$\text{Slope} = m = \frac{\text{Rise}}{\text{Run}} = \frac{\Delta y}{\Delta x}.$$

To find its slope, we need a run Δx and corresponding rise Δy. In Figure 21, we see that we can use $\Delta x = x_2 - x_1$, the change in the x-coordinate from the first point to the second, as our run, and $\Delta y = y_2 - y_1$, the change in the y-coordinate, as our rise. The resulting formula for computing the slope is given in the box.

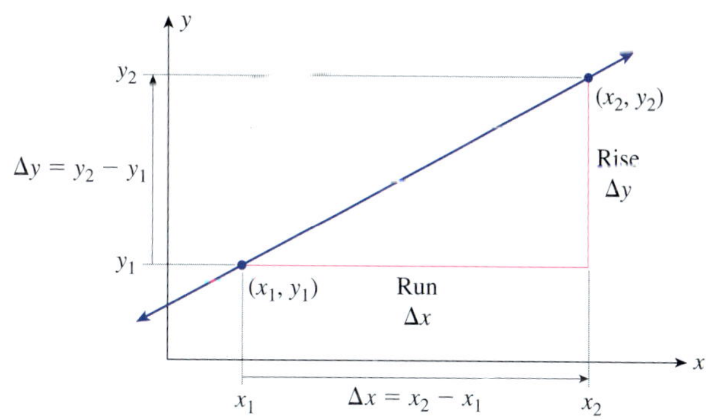

Figure 21

Computing the Slope of a Line

We can compute the slope m of the line through the points (x_1, y_1) and (x_2, y_2) by using

$$m = \frac{\Delta y}{\Delta x} = \frac{y_2 - y_1}{x_2 - x_1}.$$

Quick Examples

1. The slope of the line through $(x_1, y_1) = (1, 3)$ and $(x_2, y_2) = (5, 11)$ is

$$m = \frac{\Delta y}{\Delta x} = \frac{y_2 - y_1}{x_2 - x_1} = \frac{11 - 3}{5 - 1} = \frac{8}{4} = 2.$$

Notice that we can use the points in the reverse order: If we take $(x_1, y_1) = (5, 11)$ and $(x_2, y_2) = (1, 3)$, we obtain the same answer:

$$m = \frac{\Delta y}{\Delta x} = \frac{y_2 - y_1}{x_2 - x_1} = \frac{3 - 11}{1 - 5} = \frac{-8}{-4} = 2.$$

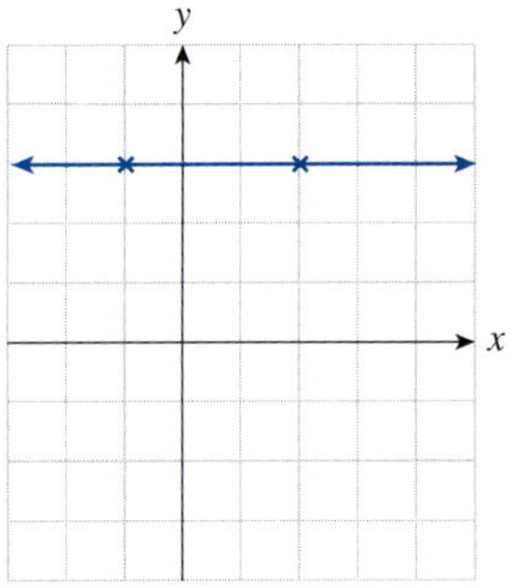

Figure 22

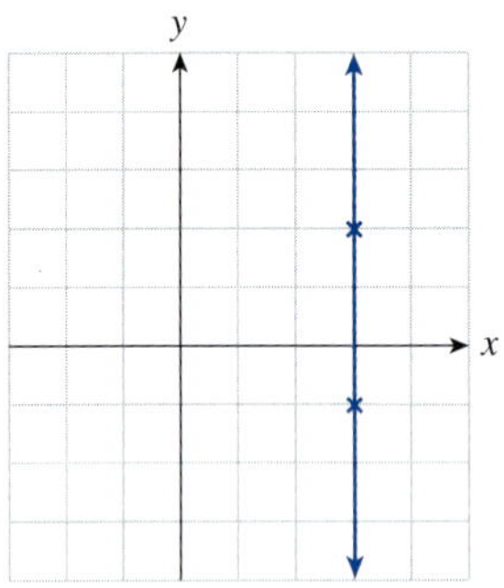

Vertical lines have undefined slope.

Figure 23

2. The slope of the line through $(x_1, y_1) = (1, 2)$ and $(x_2, y_2) = (2, 1)$ is

$$m = \frac{\Delta y}{\Delta x} = \frac{y_2 - y_1}{x_2 - x_1} = \frac{1-2}{2-1} = \frac{-1}{1} = -1.$$

3. The slope of the line through (2, 3) and (−1, 3) is

$$m = \frac{\Delta y}{\Delta x} = \frac{y_2 - y_1}{x_2 - x_1} = \frac{3-3}{-1-2} = \frac{0}{-3} = 0.$$

A line of slope 0 has 0 rise, so is a *horizontal* line, as shown in Figure 22.

4. The line through (3, 2) and (3, −1) has slope

$$m = \frac{\Delta y}{\Delta x} = \frac{y_2 - y_1}{x_2 - x_1} = \frac{-1-2}{3-3} = \frac{-3}{0},$$

which is undefined. The line passing through these points is *vertical*, as shown in Figure 23.

Computing the *y*-Intercept

Once we know the slope m of a line, and also the coordinates of a point (x_1, y_1), then we can calculate its intercept b as follows: The equation of the line must be

$$y = mx + b,$$

where b is as yet unknown. To determine b we use the fact that the line must pass through the point (x_1, y_1), so that (x_1, y_1) satisfies the equation $y = mx + b$. In other words,

$$y_1 = mx_1 + b.$$

Solving for b gives

$$b = y_1 - mx_1.$$

In summary:

Computing the *y*-Intercept of a Line

The y-intercept of the line passing through (x_1, y_1) with slope m is

$$b = y_1 - mx_1.$$

Quick Example

The line through (2, 3) with slope 4 has

$$b = y_1 - mx_1 = 3 - (4)(2) = -5.$$

Its equation is therefore

$$y = mx + b = 4x - 5.$$

EXAMPLE 2 Finding Linear Equations

Find equations for the following straight lines.

a. Through the points $(1, 2)$ and $(3, -1)$

b. Through $(2, -2)$ and parallel to the line $3x + 4y = 5$

c. Horizontal and through $(-9, 5)$

d. Vertical and through $(-9, 5)$

Solution

a. To write down the equation of the line, we need the slope m and the y-intercept b.

- ***Slope*** Because we are given two points on the line, we can use the slope formula:

$$m = \frac{y_2 - y_1}{x_2 - x_1} = \frac{-1 - 2}{3 - 1} = -\frac{3}{2}$$

- ***Intercept*** We now have the slope of the line, $m = -3/2$, and also a point—we have two to choose from, so let us choose $(x_1, y_1) = (1, 2)$. We can now use the formula for the y-intercept:

$$b = y_1 - mx_1 = 2 - \left(-\frac{3}{2}\right)(1) = \frac{7}{2}$$

Thus, the equation of the line is

$$y = -\frac{3}{2}x + \frac{7}{2}. \qquad y = mx + b$$

b. Proceeding as before,

- ***Slope*** We are not given two points on the line, but we are given a parallel line. We use the fact that *parallel lines have the same slope*. (Why?) We can find the slope of $3x + 4y = 5$ by solving for y and then looking at the coefficient of x:

$$y = -\frac{3}{4}x + \frac{5}{4} \qquad \text{To find the slope, solve for } y.$$

so the slope is $-3/4$.

- ***Intercept*** We now have the slope of the line, $m = -3/4$, and also a point $(x_1, y_1) = (2, -2)$. We can now use the formula for the y-intercept:

$$b = y_1 - mx_1 = -2 - \left(-\frac{3}{4}\right)(2) = -\frac{1}{2}$$

Thus, the equation of the line is

$$y = -\frac{3}{4}x - \frac{1}{2}. \qquad y = mx + b$$

c. We are given a point: $(-9, 5)$. Furthermore, we are told that the line is horizontal, which tells us that the slope is $m = 0$. Therefore, all that remains is the calculation of the y-intercept:

$$b = y_1 - mx_1 = 5 - (0)(-9) = 5$$

using Technology

See the Technology Guide at the end of the chapter for detailed instructions on how to obtain the slope and intercept in Example 2(a) using a TI-83/84 Plus or Excel. Here is an outline:

TI-83/84 Plus

STAT EDIT; Enter values of x and y in lists L_1 and L_2.
Slope: Highlight the heading L_3 and enter
"ΔList(L_2)/ΔList(L_1)"
Intercept: Highlight the heading L_4 and enter
"L_2-sum(L_3)*L_1" [More details on page 110.]

Excel

Enter headings x, y, m, b in cells A1–D1, and the values (x, y) in cells A2–B3. Enter
`=(B3-B2)/(A3-A2)`
in cell C2, and
`=B2-C2*A2`
in cell D2. [More details on page 116.]

so the equation of the line is

$$y = 5. \qquad y = mx + b$$

d. We are given a point: $(-9, 5)$. This time, we are told that the line is vertical, which means that the slope is undefined. Thus, we can't express the equation of the line in the form $y = mx + b$. (This formula makes sense only when the slope m of the line is defined.) What can we do? Well, here are some points on the desired line:

$$(-9, 1), (-9, 2), (-9, 3), \ldots,$$

so $x = -9$ and $y = \textit{anything}$. If we simply say that $x = -9$, then these points are all solutions, so the equation is $x = -9$.

Applications: Linear Models

Using linear functions to describe or approximate relationships in the real world is called **linear modeling**.

Recall from Section 1.2 that a **cost function** specifies the cost C as a function of the number of items x.

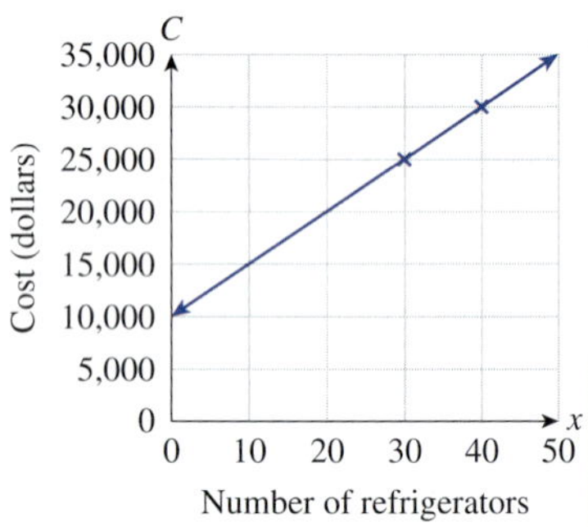

Figure 24

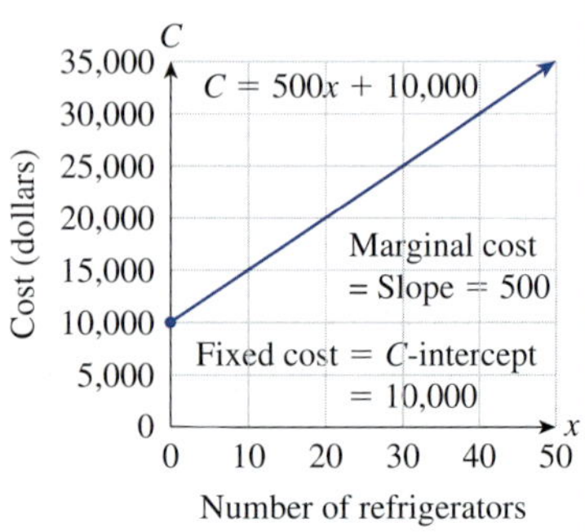

Figure 25

using Technology

To obtain the cost equation for Example 3 with technology, apply the Technology note for Example 2(a) to the given points (30, 25,000) and (40, 30,000) on the graph of the cost equation.

EXAMPLE 3 Linear Cost Function from Data

The manager of the FrozenAir Refrigerator factory notices that on Monday it cost the company a total of \$25,000 to build 30 refrigerators and on Tuesday it cost \$30,000 to build 40 refrigerators. Find a linear cost function based on this information. What is the daily fixed cost, and what is the marginal cost?

Solution We are seeking the cost C as a linear function of x, the number of refrigerators sold:

$$C = mx + b.$$

We are told that $C = 25{,}000$ when $x = 30$, and this amounts to being told that $(30, 25{,}000)$ is a point on the graph of the cost function. Similarly, $(40, 30{,}000)$ is another point on the line (Figure 24).

We can use the two points on the line to construct the linear cost equation:

- ***Slope*** $m = \dfrac{C_2 - C_1}{x_2 - x_1} = \dfrac{30{,}000 - 25{,}000}{40 - 30} = 500$ — C plays the role of y.
- ***Intercept*** $b = C_1 - mx_1 = 25{,}000 - (500)(30) = 10{,}000$ — We used the point $(x_1, C_1) = (30, 25{,}000)$.

The linear cost function is therefore

$$C(x) = 500x + 10{,}000.$$

Because $m = 500$ and $b = 10{,}000$ the factory's fixed cost is \$10,000 each day, and its marginal cost is \$500 per refrigerator. (See page 58 in Section 1.2.) These are illustrated in Figure 25.

Before we go on... Recall that, in general, the slope m measures the number of units of change in y per 1-unit change in x, so it is measured in units of y per unit of x:

Units of Slope = Units of y per unit of x

In Example 3, y is the cost C, measured in dollars, and x is the number of items, measured in refrigerators. Hence,

Units of Slope = Units of y per Unit of x = Dollars per refrigerator

The y-intercept b, being a value of y, is measured in the same units as y. In Example 3, b is measured in dollars. ■

In Section 1.2 we saw that a **demand function** specifies the demand q as a function of the price p per item, whereas a **supply function** specifies the supply q as a function of unit price p.

EXAMPLE 4 Linear Demand and Supply Functions from Data

You run a small supermarket, and must determine how much to charge for Hot'n'Spicy brand baked beans. The following chart shows weekly sales figures (the demand) for Hot'n'Spicy at two different prices, as well as the number of cans per week that you are prepared to place on sale (the supply) at these prices.

Price/Can	\$0.50	\$0.75
Demand (cans sold/week)	400	350
Supply (cans placed on sale/week)	300	500

a. Model these data with linear demand and supply functions. (See Example 4 in Section 1.2.)

b. How do we interpret the slope and q intercept of the demand equation? How do we interpret the slope of the supply equation?

c. Find the equilibrium price and graph demand and supply on the same set of axes. What happens if you charge more than the equilibrium price? What happens if you charge less?

Solution

a. Recall that a demand equation or demand function expresses demand q (in this case, the number of cans of beans sold per week) as a function of the unit price p (in this case, price per can). We model the demand using the two points we are given: (0.50, 400) and (0.75, 350).

$$\textbf{\textit{Slope:}}\ m = \frac{q_2 - q_1}{p_2 - p_1} = \frac{350 - 400}{0.75 - 0.50} = \frac{-50}{0.25} = -200$$

$$\textbf{\textit{Intercept:}}\ b = q_1 - mp_1 = 400 - (-200)(0.50) = 500$$

So, the demand equation is

$$q = -200p + 500. \qquad q = mp + b$$

To model the supply, we use the first and third rows of the table. We are again given two points: (0.50, 300) and (0.75, 500).

$$\textbf{\textit{Slope:}}\ m = \frac{q_2 - q_1}{p_2 - p_1} = \frac{500 - 300}{0.75 - 0.50} = \frac{200}{0.25} = 800$$

$$\textbf{\textit{Intercept:}}\ b = q_1 - mp_1 = 300 - (800)(0.50) = -100$$

So, the supply equation is

$$q = 800p - 100.$$

b. The key to interpreting the slope in the demand and supply equations is to recall (see the "Before we go on" note at the end of Example 3) that we measure the slope in *units of y per unit of x*. Let us consider the demand and supply equations separately:

Demand equation: Here, $m = -200$, and the units of m are units of q per unit of p, or the number of cans sold per dollar change in the price. Since m is negative, we see that the number of cans sold decreases as the price increases. We conclude that the weekly sales will drop by 200 cans per \$1-increase in the price.

To interpret the q intercept for the demand equation, recall that it gives the q-coordinate when $p = 0$. Hence, it is the number of cans the supermarket can "sell" every week if it were to give them away.*

* **NOTE** Does this seem realistic? Demand is not always unlimited if items were given away. For instance, campus newspapers are sometimes given away, and yet piles of them are often left untaken. Also see the "Before we go on..." discussion at the end of this example.

Supply equation: Here, $m = 800$, and the units of m are the number of cans you are prepared to supply per dollar change in the price. We conclude that the weekly supply will increase by 800 cans per \$1-increase in the price. (We do not interpret the q-intercept in the case of the supply equation; one cannot have a negative supply. See the "Before we go on" discussion at the end of the example.)

c. To find where the demand equals the supply, we equate the two functions:

$$\begin{aligned} \text{Demand} &= \text{Supply} \\ -200p + 500 &= 800p - 100 \\ -1000p &= -600, \end{aligned}$$

so

$$p = \frac{-600}{-1000} = \$0.60$$

This is the equilibrium price, as discussed in Example 4 of Section 1.2. We can find the corresponding demand by substituting 0.60 for p in the demand (or supply) equation.

$$\text{Equilibrium demand} = -200(0.60) + 500 = 380 \text{ cans per week}$$

So, to balance supply and demand, you should charge \$0.60 per can of Hot'n'Spicy beans and you should place 380 cans on sale each week.

using Technology

To obtain the demand and supply equations for Example 4 with technology, apply the Technology note for Example 2(a) to the points (0.50, 400) and (0.75, 350) on the graph of the demand equation, and (0.50, 300) and (0.75, 500) on the graph of the supply equation.

➡ Before we go on... As we saw in Example 4 in Section 1.2, charging more than the equilibrium price will result in a surplus of Hot'n'Spicy beans, and charging less will result in a shortage. (See Figure 26.)

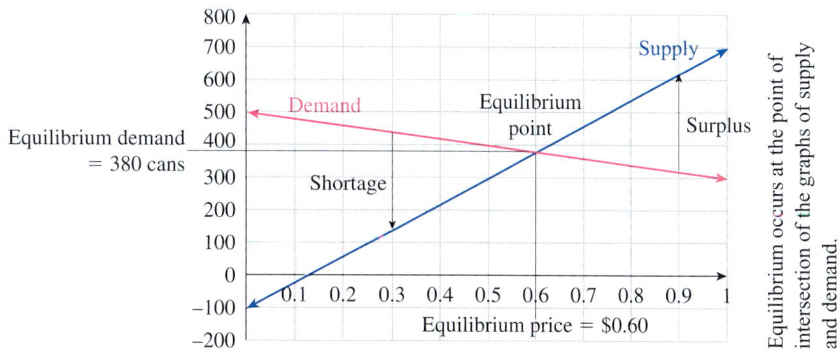

Figure 26

Q: *Just how reliable are the linear models used in Example 4?*

A: The *actual* demand and supply graphs could in principle be obtained by tabulating demand and supply figures for a large number of different prices. If the resulting points were plotted on the *pq* plane, they would probably suggest curves and not a straight line. However, if you looked at a small enough portion of any curve, you could closely *approximate* it by a straight line. In other words, *over a small range of values of p, a linear model is accurate.* Linear models of real-world situations are generally reliable only for small ranges of the variables. (This point will come up again in some of the exercises.)

■

The next example illustrates modeling change over time t with a linear function of t.

EXAMPLE 5 Modeling Change Over Time: Growth of Sales

The worldwide market for portable navigation devices was expected to grow from 50 million units in 2007 to around 530 million units in 2015.*

a. Use this information to model annual worldwide sales of portable navigation devices as a linear function of time t in years since 2007. What is the significance of the slope?

b. Use the model to predict when annual sales of mobile navigation devices will reach 440 million units.

Solution

a. Since we are interested in worldwide sales s of portable navigation devices as a function of time, we take time t to be the independent coordinate (playing the role of x) and the annual sales s, in million of units, to be the dependent coordinate (in the role of y). Notice that 2007 corresponds to $t = 0$ and 2015 corresponds to $t = 8$, so we are given

* Sales were expected to grow to more than 500 million in 2015 according to a January 2008 press release by Telematics Research Group. Source: www.telematicsresearch.com.

the coordinates of two points on the graph of sales s as a function of time t: (0, 50) and (8, 530). We model the sales using these two points:

$$m = \frac{s_2 - s_1}{t_2 - t_1} = \frac{530 - 50}{8 - 0} = \frac{480}{8} = 60$$

$$b = s_1 - mt_1 = 50 - (60)(0) = 50$$

So, $s = 60t + 50$ million units. $s = mt + b$

The slope m is measured in units of s per unit of t; that is, millions of devices per year, and is thus the *rate of change of annual sales*. To say that $m = 60$ is to say that annual sales are increasing at a rate of 60 million devices per year.

b. Our model of annual sales as a function of time is

$$s = 60t + 50 \text{ million units.}$$

Annual sales of mobile portable devices will reach 440 million when $s = 440$, or

$$440 = 60t + 50$$

Solving for t, $60t = 440 - 50 = 390$

$$t = \frac{390}{60} = 6.5 \text{ years},$$

which is midway through 2013. Thus annual sales are expected to reach 440 million midway through 2013.

using Technology

To use technology to obtain s as a function of t in Example 5, apply the Technology note for Example 2(a) to the points (0, 50) and (8, 530) on its graph.

EXAMPLE 6 Velocity

You are driving down the Ohio Turnpike, watching the mileage markers to stay awake. Measuring time in hours after you see the 20-mile marker, you see the following markers each half hour:

Time (h)	0	0.5	1	1.5	2
Marker (mi)	20	47	74	101	128

Find your location s as a function of t, the number of hours you have been driving. (The number s is also called your **position** or **displacement**.)

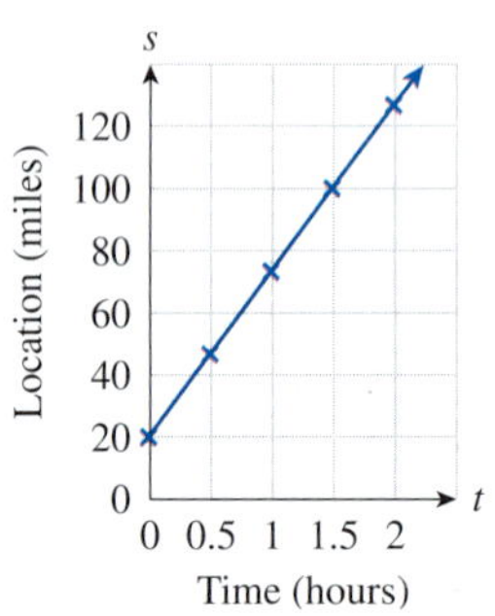

Figure 27

Solution If we plot the location s versus the time t, the five markers listed give us the graph in Figure 27.

These points appear to lie along a straight line. We can verify this by calculating how far you traveled in each half hour. In the first half hour, you traveled $47 - 20 = 27$ miles. In the second half hour you traveled $74 - 47 = 27$ miles also. In fact, you traveled exactly 27 miles each half hour. The points we plotted lie on a straight line that rises 27 units for every 0.5 unit we go to the right, for a slope of $27/0.5 = 54$.

To get the equation of that line, notice that we have the s-intercept, which is the starting marker of 20. Thus, the equation of s as a function of time t is

$$s(t) = 54t + 20.$$ We used s in place of y and t in place of x.

using Technology

To use technology to obtain s as a function of t in Example 6, apply the Technology note for Example 2(a) to the points (0, 20) and (1, 74) on its graph.

Notice the significance of the slope: For every hour you travel, you drive a distance of 54 miles. In other words, you are traveling at a constant velocity of 54 mph. We have uncovered a very important principle:

In the graph of displacement versus time, velocity is given by the slope.

Linear Change Over Time

If a quantity q is a linear function of time t,

$$q = mt + b,$$

then the slope m measures the **rate of change** of q, and b is the quantity at time $t = 0$, the **initial quantity**. If q represents the position of a moving object, then the rate of change is also called the **velocity**.

Units of *m* and *b*

The units of measurement of m are units of q per unit of time; for instance, if q is income in dollars and t is time in years, then the rate of change m is measured in dollars per year.

The units of b are units of q; for instance, if q is income in dollars and t is time in years, then b is measured in dollars.

Quick Example

If the accumulated revenue from sales of your video game software is given by $R = 2{,}000t + 500$ dollars, where t is time in years from now, then you have earned \$500 in revenue so far, and the accumulated revenue is increasing at a rate of \$2,000 per year.

Examples 3–6 share the following common theme.

General Linear Models

If $y = mx + b$ is a linear model of changing quantities x and y, then the slope m is the rate at which y is increasing per unit increase in x, and the y-intercept b is the value of y that corresponds to $x = 0$.

Units of *m* and *b*

The slope m is measured in units of y per unit of x, and the intercept b is measured in units of y.

Quick Example

If the number n of spectators at a soccer game is related to the number g of goals your team has scored so far by the equation $n = 20g + 4$, then you can expect 4 spectators if no goals have been scored and 20 additional spectators per additional goal scored.

FAQs

What to Use as x and y, and How to Interpret a Linear Model

Q: *In a problem where I must find a linear relationship between two quantities, which quantity do I use as x and which do I use as y?*

A: The key is to decide which of the two quantities is the independent variable, and which is the dependent variable. Then use the independent variable as x and the dependent variable as y. In other words, *y depends on x.*

Here are examples of phrases that convey this information, usually of the form *Find y [dependent variable] in terms of x [independent variable]:*

- Find the cost in terms of the number of items. $y = \text{Cost}, x = \#\text{ Items}$
- How does color depend on wavelength? $y = \text{Color}, x = \text{Wavelength}$

If no information is conveyed about which variable is intended to be independent, then you can use whichever is convenient.

Q: *How do I interpret a general linear model $y = mx + b$?*

A: The key to interpreting a linear model is to remember the units we use to measure m and b:

The slope m is measured in units of y per unit of x; the intercept b is measured in units of y.

For instance, if $y = 4.3x + 8.1$ and you know that x is measured in feet and y in kilograms, then you can already say, "y is 8.1 kilograms when $x = 0$ feet, and increases at a rate of 4.3 kilograms per foot" without even knowing anything more about the situation!

1.3 EXERCISES

▼ more advanced ◆ challenging

T indicates exercises that should be solved using technology

In Exercises 1–6, a table of values for a linear function is given. Fill in the missing value and calculate m in each case.

1.

x	−1	0	1
y	5	8	

2.

x	−1	0	1
y	−1	−3	

3.

x	2	3	5
$f(x)$	−1	−2	

4.

x	2	4	5
$f(x)$	−1	−2	

5.

x	−2	0	2
$f(x)$	4		10

6.

x	0	3	6
$f(x)$	−1		−5

In Exercises 7–10, first find f(0), if not supplied, and then find the equation of the given linear function.

7.

x	−2	0	2	4
$f(x)$	−1	−2	−3	−4

8.

x	−6	−3	0	3
$f(x)$	1	2	3	4

9.

x	-4	-3	-2	-1
$f(x)$	-1	-2	-3	-4

10.

x	1	2	3	4
$f(x)$	4	6	8	10

In each of Exercises 11–14, decide which of the two given functions is linear and find its equation. HINT [See Example 1.]

11.

x	0	1	2	3	4
$f(x)$	6	10	14	18	22
$g(x)$	8	10	12	16	22

12.

x	-10	0	10	20	30
$f(x)$	-1.5	0	1.5	2.5	3.5
$g(x)$	-9	-4	1	6	11

13.

x	0	3	6	10	15
$f(x)$	0	3	5	7	9
$g(x)$	-1	5	11	19	29

14.

x	0	3	5	6	9
$f(x)$	2	6	9	12	15
$g(x)$	-1	8	14	17	26

In Exercises 15–24, find the slope of the given line, if it is defined.

15. $y = -\frac{3}{2}x - 4$

16. $y = \frac{2x}{3} + 4$

17. $y = \frac{x+1}{6}$

18. $y = -\frac{2x-1}{3}$

19. $3x + 1 = 0$

20. $8x - 2y = 1$

21. $3y + 1 = 0$

22. $2x + 3 = 0$

23. $4x + 3y = 7$

24. $2y + 3 = 0$

In Exercises 25–38, graph the given equation. HINT [See Quick Examples on page 77.]

25. $y = 2x - 1$

26. $y = x - 3$

27. $y = -\frac{2}{3}x + 2$

28. $y = -\frac{1}{2}x + 3$

29. $y + \frac{1}{4}x = -4$

30. $y - \frac{1}{4}x = -2$

31. $7x - 2y = 7$

32. $2x - 3y = 1$

33. $3x = 8$

34. $2x = -7$

35. $6y = 9$

36. $3y = 4$

37. $2x = 3y$

38. $3x = -2y$

In Exercises 39–54, calculate the slope, if defined, of the straight line through the given pair of points. Try to do as many as you can without writing anything down except the answer. HINT [See Quick Examples on page 79.]

39. $(0, 0)$ and $(1, 2)$

40. $(0, 0)$ and $(-1, 2)$

41. $(-1, -2)$ and $(0, 0)$

42. $(2, 1)$ and $(0, 0)$

43. $(4, 3)$ and $(5, 1)$

44. $(4, 3)$ and $(4, 1)$

45. $(1, -1)$ and $(1, -2)$

46. $(-2, 2)$ and $(-1, -1)$

47. $(2, 3.5)$ and $(4, 6.5)$

48. $(10, -3.5)$ and $(0, -1.5)$

49. $(300, 20.2)$ and $(400, 11.2)$

50. $(1, -20.2)$ and $(2, 3.2)$

51. $(0, 1)$ and $\left(-\frac{1}{2}, \frac{3}{4}\right)$

52. $\left(\frac{1}{2}, 1\right)$ and $\left(-\frac{1}{2}, \frac{3}{4}\right)$

53. (a, b) and (c, d) $(a \neq c)$

54. (a, b) and (c, b) $(a \neq c)$

55. In the following figure, estimate the slopes of all line segments.

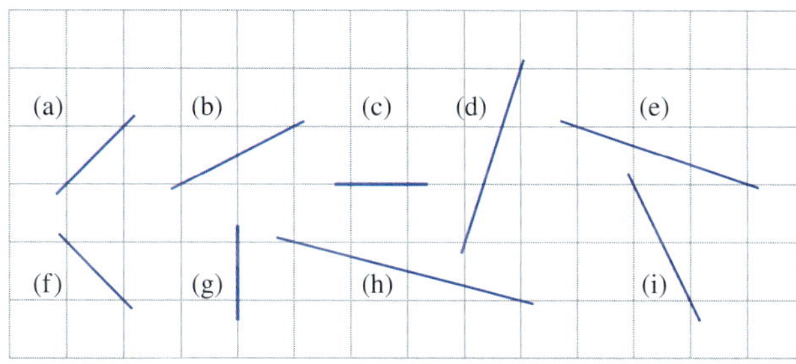

56. In the following figure, estimate the slopes of all line segments.

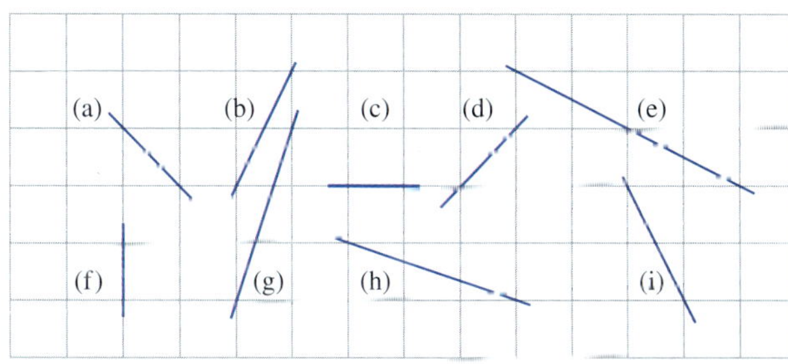

In Exercises 57–74, find a linear equation whose graph is the straight line with the given properties. HINT [See Example 2.]

57. Through $(1, 3)$ with slope 3

58. Through $(2, 1)$ with slope 2

59. Through $(1, -\frac{3}{4})$ with slope $\frac{1}{4}$

60. Through $(0, -\frac{1}{3})$ with slope $\frac{1}{3}$

61. Through $(20, -3.5)$ and increasing at a rate of 10 units of y per unit of x.

62. Through $(3.5, -10)$ and increasing at a rate of 1 unit of y per 2 units of x.

63. Through $(2, -4)$ and $(1, 1)$

64. Through $(1, -4)$ and $(-1, -1)$

65. Through $(1, -0.75)$ and $(0.5, 0.75)$

66. Through $(0.5, -0.75)$ and $(1, -3.75)$

67. Through $(6, 6)$ and parallel to the line $x + y = 4$

68. Through $(1/3, -1)$ and parallel to the line $3x - 4y = 8$

69. Through $(0.5, 5)$ and parallel to the line $4x - 2y = 11$

70. Through $(1/3, 0)$ and parallel to the line $6x - 2y = 11$

71. ▼ Through $(0, 0)$ and (p, q)

72. ▼ Through (p, q) and parallel to $y = rx + s$

73. ▼ Through (p, q) and (r, q) $(p \neq r)$

74. ▼ Through (p, q) and (r, s) $(p \neq r)$

APPLICATIONS

75. ***Cost*** The RideEm Bicycles factory can produce 100 bicycles in a day at a total cost of \$10,500 and it can produce 120 bicycles in a day at a total cost of \$11,000. What are the company's daily fixed costs, and what is the marginal cost per bicycle? HINT [See Example 3.]

76. ***Cost*** A soft-drink manufacturer can produce 1,000 cases of soda in a week at a total cost of \$6,000, and 1,500 cases of soda at a total cost of \$8,500. Find the manufacturer's weekly fixed costs and marginal cost per case of soda.

77. ***Cost: iPods*** It cost Apple approximately \$800 to manufacture 5 30-gigabyte video iPods and \$3,700 to manufacture 25.* Obtain the corresponding linear cost function. What was the cost to manufacture each additional iPod? Use the cost function to estimate the cost of manufacturing 100 iPods.

78. ***Cost: Xboxes*** If it costs Microsoft \$4,500 to manufacture 8 Xbox 360s and \$8,900 to manufacture 16,† obtain the corresponding linear cost function. What was the cost to manufacture each additional Xbox? Use the cost function to estimate the cost of manufacturing 50 Xboxes.

79. ***Demand*** Sales figures show that your company sold 1,960 pen sets each week when they were priced at \$1/pen set and 1,800 pen sets each week when they were priced at \$5/pen set. What is the linear demand function for your pen sets? HINT [See Example 4.]

80. ***Demand*** A large department store is prepared to buy 3,950 of your neon-colored shower curtains per month for \$5 each, but only 3,700 shower curtains per month for \$10 each. What is the linear demand function for your neon-colored shower curtains?

81. ***Demand for Cell Phones*** The following table shows worldwide sales of Nokia cell phones and their average wholesale prices in 2004:[27]

Quarter	***Second***	***Fourth***
Wholesale Price (\$)	111	105
Sales (millions)	45.4	51.4

a. Use the data to obtain a linear demand function for (Nokia) cell phones, and use your demand equation to predict sales if Nokia lowered the price further to \$103.

b. Fill in the blanks: For every ____ increase in price, sales of cell phones decrease by ___ units.

82. ***Demand for Cell Phones*** The following table shows projected worldwide sales of (all) cell phones and wholesale prices:[28]

Year	***2004***	***2008***
Wholesale Price (\$)	100	80
Sales (millions)	600	800

a. Use the data to obtain a linear demand function for cell phones, and use your demand equation to predict sales if the price was set at \$85.

b. Fill in the blanks: For every ____ increase in price, sales of cell phones decrease by ___ units.

83. ***Demand for Monorail Service, Las Vegas*** In 2005, the Las Vegas monorail charged \$3 per ride and had an average ridership of about 28,000 per day. In December, 2005 the Las Vegas Monorail Company raised the fare to \$5 per ride, and average ridership in 2006 plunged to around 19,000 per day.[29]

a. Use the given information to find a linear demand equation.

b. Give the units of measurement and interpretation of the slope.

c. What would be the effect on ridership of raising the fare to \$6 per ride?

*Source for cost data: Manufacturing & Technology News, July 31, 2007, Volume 14, No. 14, www.manufacturingnews.com.

†Source for estimate of marginal cost: www.isuppli.com.

[27] Source: Embedded.com/Companyreports December 2004.

[28] Wholesale price projections are the authors'. Source for sales prediction: I-Stat/NDR December, 2004.

[29] Source: *New York Times,* February 10, 2007, p. A9.

84. ***Demand for Monorail Service, Mars*** The Utarek monorail, which links the three urbynes (or districts) of Utarek, Mars, charged Z̿5 per ride[30] and sold about 14 million rides per day. When the Utarek City Council lowered the fare to Z̿3 per ride, the number of rides increased to 18 million per day.

a. Use the given information to find a linear demand equation.
b. Give the units of measurement and interpretation of the slope.
c. What would be the effect on ridership of raising the fare to Z̿10 per ride?

85. ***Equilibrium Price*** You can sell 90 pet chias per week if they are marked at \$1 each, but only 30 each week if they are marked at \$2/chia. Your chia supplier is prepared to sell you 20 chias each week if they are marked at \$1/chia, and 100 each week if they are marked at \$2 per chia.

a. Write down the associated linear demand and supply functions.
b. At what price should the chias be marked so that there is neither a surplus nor a shortage of chias? HINT [See Example 4.]

86. ***Equilibrium Price*** The demand for your college newspaper is 2,000 copies each week if the paper is given away free of charge, and drops to 1,000 each week if the charge is 10¢/copy. However, the university is prepared to supply only 600 copies per week free of charge, but will supply 1,400 each week at 20¢ per copy.

a. Write down the associated linear demand and supply functions.
b. At what price should the college newspapers be sold so that there is neither a surplus nor a shortage of papers?

87. ***Pasta Imports*** During the period 1990–2001, U.S. imports of pasta increased from 290 million pounds in 1990 ($t = 0$) by an average of 40 million pounds/year.[31]

a. Use these data to express q, the annual U.S. imports of pasta (in millions of pounds), as a linear function of t, the number of years since 1990.
b. Use your model to estimate U.S. pasta imports in 2005, assuming the import trend continued.

88. ***Mercury Imports*** During the period 2210 2220, Martian imports of mercury (from the planet of that name) increased from 550 million kg in 2210 ($t = 0$) by an average of 60 million kg/year.

a. Use these data to express h, the annual Martian imports of mercury (in millions of kilograms), as a linear function of t, the number of years since 2010.
b. Use your model to estimate Martian mercury imports in 2230, assuming the import trend continued.

89. ***Satellite Radio Subscriptions*** The number of Sirius Satellite Radio subscribers grew from 0.3 million in 2003 to 3.2 million in 2005.[32]

a. Use this information to find a linear model for the number N of subscribers (in millions) as a function of time t in years since 2000.
b. Give the units of measurement and interpretation of the slope.
c. Use the model from part (a) to predict the 2006 figure. (The actual 2006 figure was approximately 6 million subscribers.)

90. ***Freon Production*** The production of ozone-layer damaging Freon 22 (chlorodifluoromethane) in developing countries rose from 200 tons in 2004 to a projected 590 tons in 2010.[33]

a. Use this information to find a linear model for the amount F of Freon 22 (in tons) as a function of time t in years since 2000.
b. Give the units of measurement and interpretation of the slope.
c. Use the model from part (a) to estimate the 2008 figure and compare it with the actual projection of 400 tons.

91. ***Velocity*** The position of a model train, in feet along a railroad track, is given by

$$s(t) = 2.5t + 10$$

after t seconds.

a. How fast is the train moving?
b. Where is the train after 4 seconds?
c. When will it be 25 feet along the track?

92. ***Velocity*** The height of a falling sheet of paper, in feet from the ground, is given by

$$s(t) = -1.8t + 9$$

after t seconds.

a. What is the velocity of the sheet of paper?
b. How high is it after 4 seconds?
c. When will it reach the ground?

93. ▼ ***Fast Cars*** A police car was traveling down Ocean Parkway in a high speed chase from Jones Beach. It was at Jones Beach at exactly 10 pm ($t = 10$) and was at Oak Beach, 13 miles from Jones Beach, at exactly 10:06 pm.

a. How fast was the police car traveling? HINT [See Example 6.]
b. How far was the police car from Jones Beach at time t?

[30] Z̿ designates Zonars, the official currency in Mars. See www.marsnext.com for details of the Mars colony, its commerce, and its culture.

[31] Data are rounded. Sources: Department of Commerce/*New York Times*, September 5, 1995, p. D4; International Trade Administration, March 31, 2002, www.ita.doc.gov/.

[32] Figures are approximate. Source: Sirius Satellite Radio/*New York Times,* February 20, 2008, p. A1.

[33] Figures are approximate. Source: Lampert Kuijpers (Panel of the Montreal Protocol), National Bureau of Statistics in China, via CEIC DSata/*New York Times,* February 23, 2007, p. C1.

94. ▼ ***Fast Cars*** The car that was being pursued by the police in Exercise 93 was at Jones Beach at exactly 9:54 pm ($t = 9.9$) and passed Oak Beach (13 miles from Jones Beach) at exactly 10:06 pm, where it was overtaken by the police.

a. How fast was the car traveling? HINT [See Example 6.]
b. How far was the car from Jones Beach at time t?

95. ▼ ***Fahrenheit and Celsius*** In the Fahrenheit temperature scale, water freezes at 32°F and boils at 212°F. In the Celsius scale, water freezes at 0°C and boils at 100°C. Assuming that the Fahrenheit temperature F and the Celsius temperature C are related by a linear equation, find F in terms of C. Use your equation to find the Fahrenheit temperatures corresponding to 30°C, 22°C, −10°C, and −14°C, to the nearest degree.

96. ▼ ***Fahrenheit and Celsius*** Use the information about Celsius and Fahrenheit given in Exercise 95 to obtain a linear equation for C in terms of F, and use your equation to find the Celsius temperatures corresponding to 104°F, 77°F, 14°F, and −40°F, to the nearest degree.

97. ▼ ***Income*** The well-known romance novelist Celestine A. Lafleur (a.k.a. Bertha Snodgrass) has decided to sell the screen rights to her latest book, *Henrietta's Heaving Heart*, to Boxoffice Success Productions for \$50,000. In addition, the contract ensures Ms. Lafleur royalties of 5% of the net profits.[34] Express her income I as a function of the net profit N, and determine the net profit necessary to bring her an income of \$100,000. What is her marginal income (share of each dollar of net profit)?

98. ▼ ***Income*** Due to the enormous success of the movie *Henrietta's Heaving Heart* based on a novel by Celestine A. Lafleur (see the Exercise 97), Boxoffice Success Productions decides to film the sequel, *Henrietta, Oh Henrietta*. At this point, Bertha Snodgrass (whose novels now top the best seller lists) feels she is in a position to demand \$100,000 for the screen rights and royalties of 8% of the net profits. Express her income I as a function of the net profit N and determine the net profit necessary to bring her an income of \$1,000,000. What is her marginal income (share of each dollar of net profit)?

99. ▼ ***Biology*** The Snowtree cricket behaves in a rather interesting way: The rate at which it chirps depends linearly on the temperature. One summer evening you hear a cricket chirping at a rate of 140 chirps/minute, and you notice that the temperature is 80°F. Later in the evening the cricket has slowed down to 120 chirps/minute, and you notice that the temperature has dropped to 75°F. Express the temperature T as a function of the cricket's rate of chirping r. What is the temperature if the cricket is chirping at a rate of 100 chirps/minute?

[34] Percentages of net profit are commonly called "monkey points." Few movies ever make a net profit on paper, and anyone with any clout in the business gets a share of the *gross*, not the net.

100. ▼ ***Muscle Recovery Time*** Most workout enthusiasts will tell you that muscle recovery time is about 48 hours. But it is not quite as simple as that; the recovery time ought to depend on the number of sets you do involving the muscle group in question. For example, if you do no sets of biceps exercises, then the recovery time for your biceps is (of course) zero. Let's assume that if you do three sets of exercises on a muscle group, then its recovery time is 48 hours. Use these data to write a linear function that gives the recovery time (in hours) in terms of the number of sets affecting a particular muscle. Use this model to calculate how long it would take your biceps to recover if you did 15 sets of curls. Comment on your answer with reference to the usefulness of a linear model.

101. ***Television Advertising*** The cost, in millions of dollars, of a 30-second television ad during the Super Bowl in the years 1990 to 2007 can be approximated by the following piecewise linear function ($t = 0$ represents 1990):[35]

$$C(t) = \begin{cases} 0.08t + 0.6 & \text{if } 0 \le t < 8 \\ 0.13t + 0.20 & \text{if } 8 \le t \le 17 \end{cases}$$

How fast and in what direction was the cost of an ad during the Super Bowl changing in 2006?

102. ***Processor Speeds*** The processor speed, in megahertz (MHz), of Intel processors could be approximated by the following function of time t in years since the start of 1995:[36]

$$P(t) = \begin{cases} 180t + 200 & \text{if } 0 \le t \le 5 \\ 3000t - 13{,}900 & \text{if } 5 < t \le 12 \end{cases}$$

How fast and in what direction was processor speed changing in 2005?

103. ▼ ***Investment in Gold*** Following are some approximate values of the Amex Gold BUGS Index.[37]

Year	1995	2000	2007
Index	200	50	470

Take t to be the year since 1995 and y to be the BUGS index.

a. Model the 1995 and 2000 data with a linear equation.
b. Model the 2000 and 2007 data with a linear equation.
c. Use the results of parts (a) and (b) to obtain a piecewise linear model of the gold BUGS index for 1995–2007.
d. Use your model to estimate the index in 2002.

104. ▼ ***Unemployment*** The following table shows the number of unemployed persons in the U.S. in 1994, 2000, and 2008.[38]

[35] Sources for data: *New York Times*, January 26, 2001, p. C1, http://money.cnn.com.

[36] Sources for data: Sandpile.org/*New York Times*, May 17, 2004, p. C1, www.Intel.com.

[37] BUGS stands for "basket of unhedged gold stocks." Figures are approximate. Sources: www.321gold.com, Bloomberg Financial Markets/*New York Times*, Sept 7, 2003, p. BU8, www.amex.com.

[38] Figures are seasonally adjusted and rounded. Source: U.S. Department of Labor, December, 2004, www.data.bls.gov.

Year	1994	2000	2008
Unemployment (Millions)	9	6	7

Take t to be the year since 1994 and y to be the number (in millions) of unemployed persons.

a. Model the 1994 and 2000 data with a linear equation.
b. Model the 2000 and 2008 data with a linear equation.
c. Use the results of parts (a) and (b) to obtain a piecewise linear model of the number (in millions) of unemployed persons for 1994–2008.
d. Use your model to estimate the number of unemployed persons in 2002.

105. ▼ ***Employment in Mexico*** The number of workers employed in manufacturing jobs in Mexico was 3 million in 1995, rose to 4.1 million in 2000, and then dropped to 3.5 million in 2004.[39] Model this number N as a piecewise-linear function of the time t in years since 1995, and use your model to estimate the number of manufacturing jobs in Mexico in 2002. (Take the units of N to be millions.)

106. ▼ ***Mortgage Delinquencies*** The percentage of borrowers in the highest risk category who were delinquent on their payments decreased from 9.7% in 2001 to 4.3% in 2004 and then shot up to 10.3% in 2007.[40] Model this percentage P as a piecewise-linear function of the time t in years since 2001, and use your model to estimate the percentage of delinquent borrowers in 2006.

COMMUNICATION AND REASONING EXERCISES

107. How would you test a table of values of x and y to see if it comes from a linear function?

108. You have ascertained that a table of values of x and y corresponds to a linear function. How do you find an equation for that linear function?

109. To what linear function of x does the linear equation $ax + by = c$ $(b \neq 0)$ correspond? Why did we specify $b \neq 0$?

110. Complete the following. The slope of the line with equation $y = mx + b$ is the number of units that ____ increases per unit increase in ____.

111. Complete the following. If, in a straight line, y is increasing three times as fast as x, then its ____ is ____.

112. Suppose that y is decreasing at a rate of 4 units per 3-unit increase of x. What can we say about the slope of the linear relationship between x and y? What can we say about the intercept?

113. If y and x are related by the linear expression $y = mx + b$, how will y change as x changes if m is positive? negative? zero?

114. Your friend April tells you that $y = f(x)$ has the property that, whenever x is changed by Δx, the corresponding change in y is $\Delta y = -\Delta x$. What can you tell her about f?

115. T ▼ Consider the following worksheet:

◇	A	B	C	D
1	x	y	m	b
2	1	2	=(B3-B2)/(A3-A2)	=B2-C2*A2
3	3	-1	Slope	Intercept

What is the effect on the slope of increasing the y-coordinate of the second point (the point whose coordinates are in Row 3)? Explain.

116. T ▼ Referring to the worksheet in Exercise 115, what is the effect on the slope of increasing the x-coordinate of the second point (the point whose coordinates are in row 3)? Explain.

117. If y is measured in bootlags[41] and x is measured in $\overline{\overline{\mathrm{Z}}}$ (zonars, the designated currency in Utarek, Mars)[42] and $y = mx + b$, then m is measured in ______ and b is measured in ______.

118. If the slope in a linear relationship is measured in miles per dollar, then the independent variable is measured in ______ and the dependent variable is measured in ______.

119. If a quantity is changing linearly with time, and it increases by 10 units in the first day, what can you say about its behavior in the third day?

120. The quantities Q and T are related by a linear equation of the form

$$Q = mT + b.$$

When $T = 0$, Q is positive, but decreases to a negative quantity when T is 10. What are the signs of m and b. Explain your answers.

121. ▼ The velocity of an object is given by $v = 0.1t + 20$ m/sec, where t is time in seconds. The object is

(A) moving with fixed speed **(B)** accelerating
(C) decelerating **(D)** impossible to say from the given information

122. ▼ The position of an object is given by $x = 0.2t - 4$, where t is time in seconds. The object is

(A) moving with fixed speed **(B)** accelerating
(C) decelerating **(D)** impossible to say from the given information

[39] Source: *New York Times*, February 18, 2007, p. WK4.

[40] The 2007 figure is projected from data through October 2006. Source: *New York Times*, February 18, 2007, p. BU9.

[41] An ancient Martian unit of length; one bootlag is the mean distance from a Martian's foreleg to its rearleg.

[42] Source: www.marsnext.com/comm/zonars.html.

123. ▼ Suppose the cost function is $C(x) = mx + b$ (with m and b positive), the revenue function is $R(x) = kx$ $(k > m)$ and the number of items is increased from the break-even quantity. Does this result in a loss, a profit, or is it impossible to say? Explain your answer.

124. ▼ You have been constructing a demand equation, and you obtained a (correct) expression of the form $p = mq + b$, whereas you would have preferred one of the form $q = mp + b$. Should you simply switch p and q in the answer, should you start again from scratch, using p in the role of x and q in the role of y, or should you solve your demand equation for q? Give reasons for your decision.

1.4 Linear Regression

We have seen how to find a linear model given two data points: We find the equation of the line that passes through them. However, we often have more than two data points, and they will rarely all lie on a single straight line, but may often come close to doing so. The problem is to find the line coming *closest* to passing through all of the points.

Suppose, for example, that we are conducting research for a company interested in expanding into Mexico. Of interest to us would be current and projected growth in that country's economy. The following table shows past and projected per capita gross domestic product (GDP)[43] of Mexico for 2000–2012.[44]

Year t ($t = 0$ represents 2000)	0	2	4	6	8	10	12	14
Per Capita GDP y (\$1,000)	9	9	10	11	11	12	13	13

A plot of these data suggests a roughly linear growth of the GDP (Figure 28(a)).

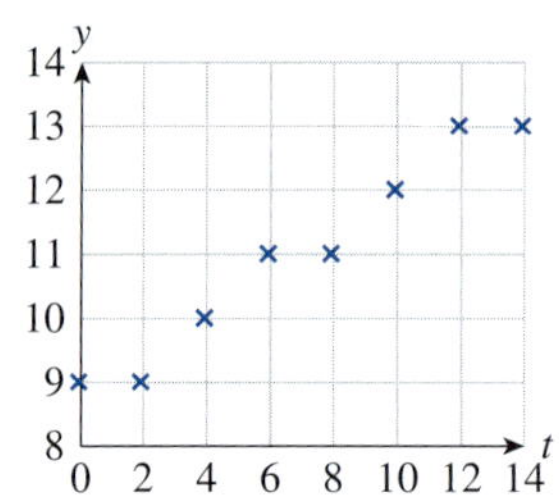

Figure 28(a)

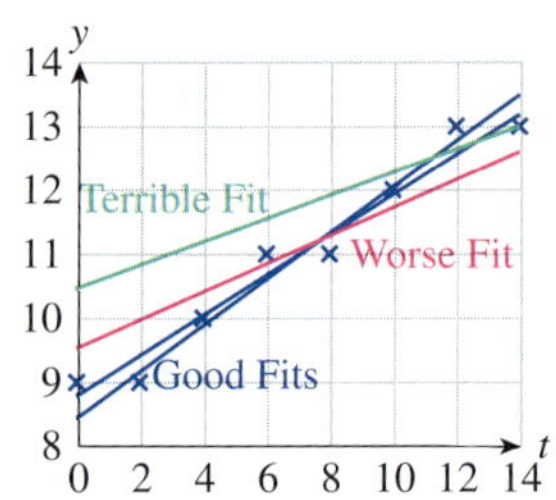

Figure 28(b)

These points suggest a roughly linear relationship between t and y, although they clearly do not all lie on a single straight line. Figure 28(b) shows the points together with several lines, some fitting better than others. Can we precisely measure which lines fit better than others? For instance, which of the two lines labeled as "good" fits in Figure 28(b) models the data more accurately? We begin by considering, for each value of t, the difference between the actual GDP (the **observed value**) and the GDP predicted by a linear equation (the **predicted value**). The difference between the predicted value and the observed value is called the **residual**.

$$\text{Residual} = \text{Observed Value} - \text{Predicted Value}$$

[43] The GDP is a measure of the total market value of all goods and services produced within a country.

[44] Data are approximate. Sources: CIA World Factbook/www.indexmundi.com, www.economist.com.

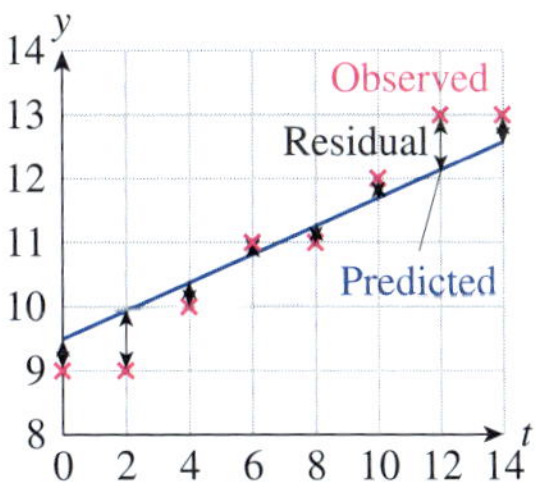

Figure 29

On the graph, the residuals measure the vertical distances between the (observed) data points and the line (Figure 29) and they tell us how far the linear model is from predicting the actual GDP.

The more accurate our model, the smaller the residuals should be. We can combine all the residuals into a single measure of accuracy by adding their *squares*. (We square the residuals in part to make them all positive.*) The sum of the squares of the residuals is called the **sum-of-squares error, SSE**. Smaller values of SSE indicate more accurate models.

Here are some definitions and formulas for what we have been discussing.

* **NOTE** Why not add the absolute values of the residuals instead? Mathematically, using the squares rather than the absolute values results in a simpler and more elegant solution. Further, using the squares always results in a *single* best-fit line in cases where the x-coordinates are all different, whereas this is not the case if we use absolute values.

Observed and Predicted Values

Suppose we are given a collection of data points $(x_1, y_1), \ldots, (x_n, y_n)$. The n quantities $y_1, y_2, \ldots, y_n$ are called the **observed y-values**. If we model these data with a linear equation

$$\hat{y} = mx + b, \qquad \hat{y} \text{ stands for "estimated } y\text{" or "predicted } y\text{."}$$

then the y-values we get by substituting the given x-values into the equation are called the **predicted y-values**:

$$\hat{y}_1 = mx_1 + b \qquad \text{Substitute } x_1 \text{ for } x.$$
$$\hat{y}_2 = mx_2 + b \qquad \text{Substitute } x_2 \text{ for } x.$$
$$\ldots$$
$$\hat{y}_n = mx_n + b \qquad \text{Substitute } x_n \text{ for } x.$$

Quick Example

Consider the three data points (0, 2), (2, 5), and (3, 6). The observed y-values are $y_1 = 2$, $y_2 = 5$, and $y_3 = 6$. If we model these data with the equation $\hat{y} = x + 2.5$, then the predicted values are:

$$\hat{y}_1 = x_1 + 2.5 = 0 + 2.5 = 2.5$$
$$\hat{y}_2 = x_2 + 2.5 = 2 + 2.5 = 4.5$$
$$\hat{y}_3 = x_3 + 2.5 = 3 + 2.5 = 5.5$$

Residuals and Sum-of-Squares Error (SSE)

If we model a collection of data $(x_1, y_1), \ldots, (x_n, y_n)$ with a linear equation $\hat{y} = mx + b$, then the **residuals** are the n quantities (Observed Value – Predicted Value):

$$(y_1 - \hat{y}_1), (y_2 - \hat{y}_2), \ldots, (y_n - \hat{y}_n).$$

The **sum-of-squares error (SSE)** is the sum of the squares of the residuals:

$$\text{SSE} = (y_1 - \hat{y}_1)^2 + (y_2 - \hat{y}_2)^2 + \cdots + (y_n - \hat{y}_n)^2.$$

Quick Example

For the data and linear approximation given above, the residuals are:

$$y_1 - \hat{y}_1 = 2 - 2.5 = -0.5$$
$$y_2 - \hat{y}_2 = 5 - 4.5 = 0.5$$
$$y_3 - \hat{y}_3 = 6 - 5.5 = 0.5$$

and $\quad \text{SSE} = (-0.5)^2 + (0.5)^2 + (0.5)^2 = 0.75.$

using Technology

See the Technology Guide at the end of the chapter for detailed instructions on how to obtain the table and graphs in Example 1 using a TI-83/84 Plus or Excel. Here is an outline:

TI-83/84 Plus

STAT EDIT Enter the values of t in L_1, and y (observed) in L_2.
Predicted y: Highlight the heading L_3 and enter `0.5*L1+8`
Squares of residuals: Highlight the heading L_4 and enter `"(L2-L3)^2"`
SSE: Enter `sum(L4)` on the home screen.
Graph: `Y1=0.5x+8` Then turn on Plot1 in Y = screen ZOOM (STAT) [More details on page 110.]

Excel

Enter the headings t, y (observed), y (predicted), Residual^2, m and b in A1–F1.
Enter the values of t in A2–A9 and y (observed) in B2–B9.
Enter the values 0.25 for m and 9 for b in E2–F2
Predicted y: In C2 enter `=$E$2*A2+$F$2` and copy down to C9.
Squares of residuals: In D2 enter `=(B2-C2)^2` and copy down to D9.
SSE: In any vacant cell to the right of Column D, enter `=SUM(D2:D9)`
Graph: Highlight A1–C9 and insert a Scatter Chart.
[More details on page 116.]

EXAMPLE 1 Computing the Sum-of-Squares Error

Using the data above on the GDP in Mexico, compute SSE, the sum-of-squares error, for the linear models $y = 0.5t + 8$ and $y = 0.25t + 9$. Which model is the better fit?

Solution We begin by creating a table showing the values of t, the observed (given) values of y, and the values predicted by the first model.

Year t	Observed y	Predicted $\hat{y} = 0.5t + 8$
0	9	8
2	9	9
4	10	10
6	11	11
8	11	12
10	12	13
12	13	14
14	13	15

We now add two new columns for the residuals and their squares.

Year t	Observed y	Predicted $\hat{y} = 0.5t + 8$	Residual $y - \hat{y}$	Residual² $(y - \hat{y})^2$
0	9	8	$9 - 8 = 1$	$1^2 = 1$
2	9	9	$9 - 9 = 0$	$0^2 = 0$
4	10	10	$10 - 10 = 0$	$0^2 = 0$
6	11	11	$11 - 11 = 0$	$0^2 = 0$
8	11	12	$11 - 12 = -1$	$(-1)^2 = 1$
10	12	13	$12 - 13 = -1$	$(-1)^2 = 1$
12	13	14	$13 - 14 = -1$	$(-1)^2 = 1$
14	13	15	$13 - 15 = -2$	$(-2)^2 = 4$

SSE, the sum of the squares of the residuals, is then the sum of the entries in the last column,

$$\text{SSE} = 8.$$

Repeating the process using the second model, $0.25t + 9$, yields the following table.

Year t	Observed y	Predicted $\hat{y} = 0.25t + 9$	Residual $y - \hat{y}$	Residual² $(y - \hat{y})^2$
0	9	9	$9 - 9 = 0$	$0^2 = 0$
2	9	9.5	$9 - 9.5 = -0.5$	$(-0.5)^2 = 0.25$
4	10	10	$10 - 10 = 0$	$0^2 = 0$
6	11	10.5	$11 - 10.5 = 0.5$	$0.5^2 = 0.25$
8	11	11	$11 - 11 = 0$	$0^2 = 0$
10	12	11.5	$12 - 11.5 = 0.5$	$0.5^2 = 0.25$
12	13	12	$13 - 12 = 1$	$1^2 = 1$
14	13	12.5	$13 - 12.5 = 0.5$	$0.5^2 = 0.25$

This time, SSE = 2 and so the second model is a better fit.

Figure 30 shows the data points and the two linear models in question.

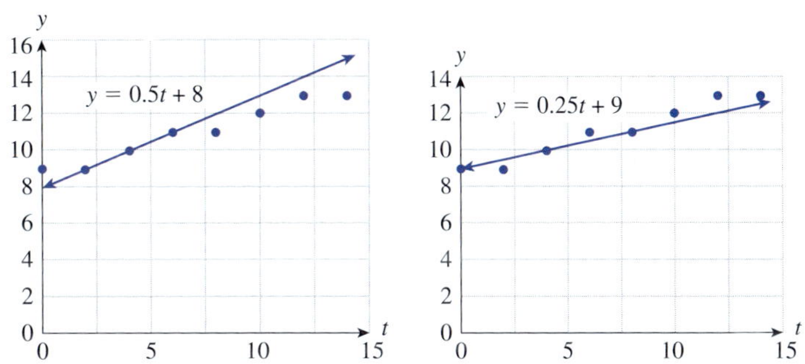

Figure 30

➡ Before we go on...

Q: *It seems clear from the figure that the second model in Example 1 gives a better fit. Why bother to compute SSE to tell me this?*

A: The difference between the two models we chose is so great that it is clear from the graphs which is the better fit. However, if we used a third model with $m = .25$ and $b = 9.1$, then its graph would be almost indistinguishable from that of the second, but a slightly better fit as measured by SSE = 1.68.

■

Among all possible lines, there ought to be one with the least possible value of SSE—that is, the greatest possible accuracy as a model. The line (and there is only one such line) that minimizes the sum of the squares of the residuals is called the **regression line**, the **least-squares line**, or the **best-fit line**.

To find the regression line, we need a way to find values of m and b that give the smallest possible value of SSE. As an example, let us take the second linear model in the example above. We said in the "Before we go on" discussion that increasing b from 9 to 9.1 had the desirable effect of decreasing SSE from 2 to 1.68. We could then increase m to 0.26, further reducing SSE to 1.328. Imagine this as a kind of game: Alternately alter the values of m and b by small amounts until SSE is as small as you can make it. This works, but is extremely tedious and time-consuming.

Fortunately, there is an algebraic way to find the regression line. Here is the calculation. To justify it rigorously requires calculus of several variables or linear algebra.

Regression Line

The **regression line (least squares line, best-fit line)** associated with the points $(x_1, y_1), (x_2, y_2), \ldots, (x_n, y_n)$ is the line that gives the minimum sum-of-squares error (SSE). The regression line is

$$y = mx + b,$$

where m and b are computed as follows:

$$m = \frac{n(\sum xy) - (\sum x)(\sum y)}{n(\sum x^2) - (\sum x)^2}$$

$$b = \frac{\sum y - m(\sum x)}{n}$$

$$n = \text{number of data points.}$$

The quantities m and b are called the **regression coefficients**.

Here, "$\sum$" means "the sum of." Thus, for example,

$$\sum x = \text{Sum of the } x\text{-values} = x_1 + x_2 + \cdots + x_n$$

$$\sum xy = \text{Sum of products} = x_1y_1 + x_2y_2 + \cdots + x_ny_n$$

$$\sum x^2 = \text{Sum of the squares of the } x\text{-values} = x_1{}^2 + x_2{}^2 + \cdots + x_n{}^2.$$

On the other hand,

$$\left(\sum x\right)^2 = \text{Square of } \sum x = \text{Square of the sum of the } x\text{-values.}$$

EXAMPLE 2 Per Capita Gross Domestic Product in Mexico

In Example 1 we considered the following data on the per capita gross domestic product (GDP) of Mexico:

Year x ($x = 0$ represents 2000)	0	2	4	6	8	10	12	14
Per Capita GDP y (\$1,000)	9	9	10	11	11	12	13	13

Find the best-fit linear model for these data and use the model to predict the per capita GDP in Mexico in 2016.

using Technology

See the Technology Guide at the end of the chapter for detailed instructions on how to obtain the regression line and graph in Example 2 using a TI-83/84 Plus or Excel. Here is an outline:

TI-83/84 Plus

STAT EDIT Enter the values of x in L_1, and y in L_2.
Regression equation: STAT CALC option #4:
`LinReg(ax+b)`
Graph: Y= VARS 5 EQ 1, then ZOOM 9 [More details on page 110.]

Excel

Enter the values of x in A2–A9 and y in B2–B9.
Graph: Highlight A2–B9 and create a Scatter Plot.
Regression Line: Add a linear trendline with the option "Display equation on chart." OR, after entering x and y,
Regression equation: Enter
`=LINEST(B2:B9, A2:A9)`
in cells C2–D2 then press Control-Shift-Enter to get m and b. [More details on page 116.]

Web Site

Follow
Chapter 1
→ Math Tools for Chapter 1
→ Simple Regression Utility

Solution Let's organize our work in the form of a table, where the original data are entered in the first two columns and the bottom row contains the column sums.

	x	y	xy	x^2
	0	9	0	0
	2	9	18	4
	4	10	40	16
	6	11	66	36
	8	11	88	64
	10	12	120	100
	12	13	156	144
	14	13	182	196
$\sum$ (Sum)	56	88	670	560

Because there are $n = 8$ data points, we get

$$m = \frac{n(\sum xy) - (\sum x)(\sum y)}{n(\sum x^2) - (\sum x)^2} = \frac{8(670) - (56)(88)}{8(560) - (56)^2} \approx 0.321$$

and $$b = \frac{\sum y - m(\sum x)}{n} \approx \frac{88 - (0.321)(56)}{8} \approx 8.75.$$

So, the regression line is

$$y = 0.321x + 8.75.$$

To predict the per capita GDP in Mexico in 2016 we substitute $x = 16$ and get $y \approx 14$, or \$14,000 per capita.

Figure 31 shows the data points and the regression line (which has SSE ≈ 0.643; a lot lower than in Example 1.).

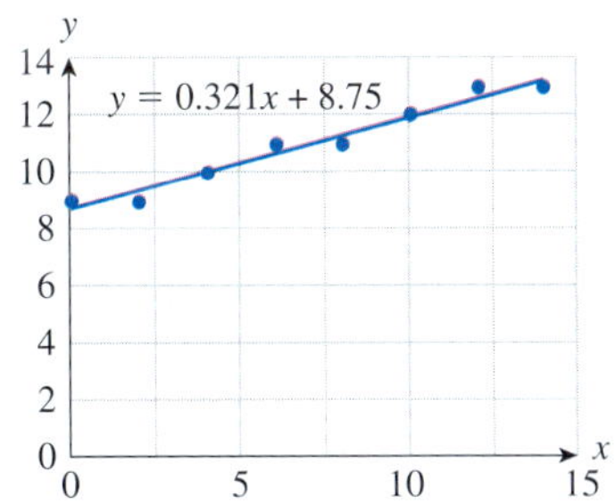

Figure 31

Coefficient of Correlation

If all the data points do not lie on one straight line, we would like to be able to measure how closely they can be approximated by a straight line. Recall that SSE measures the sum of the squares of the deviations from the regression line; therefore it constitutes a measurement of goodness of fit. (For instance, if SSE = 0, then all the points lie on a straight line.) However, SSE depends on the units we use to measure y, and also on the number of data points (the more data points we use, the larger SSE tends to be). Thus, while we can (and do) use SSE to compare the goodness of fit of two lines to the same data, we cannot use it to compare the goodness of fit of one line to one set of data with that of another to a different set of data.

To remove this dependency, statisticians have found a related quantity that can be used to compare the goodness of fit of lines to different sets of data. This quantity, called the **coefficient of correlation** or **correlation coefficient**, and usually denoted r, is between -1 and 1. The closer r is to -1 or 1, the better the fit. For an *exact* fit, we would have $r = -1$ (for a line with negative slope) or $r = 1$ (for a line with positive slope). For a bad fit, we would have r close to 0. Figure 32 shows several collections of data points with least squares lines and the corresponding values of r.

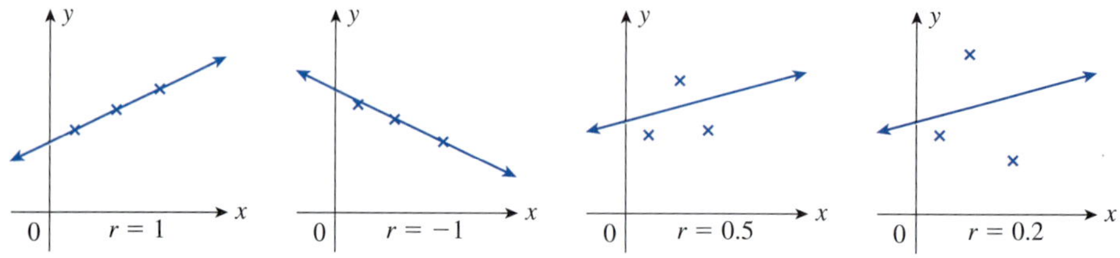

Figure 32

using Technology

See the Technology Guide at the end of the chapter for detailed instructions on how to obtain the correlation coefficient in Example 2 using a TI-83/84 Plus or Excel. Here is an outline:

TI-83/84 Plus

[2ND] [CATALOG] `DiagnosticOn` Then [STAT] CALC option #4: `LinReg(ax+b)` [More details on page 110.]

Excel

Graph the data.
Add a trendline and select the option to "Display R-squared value on chart."
OR, after entering the values of x and y,
Enter `=LINEST(B2:B9,A2:A9,TRUE,TRUE)` in cells C2–D6 then press Control-Shift-Enter. [More details on page 116.]

Web Site

Chapter 1
→ Math Tools for Chapter 1
→ Simple Regression Utility

Correlation Coefficient

The coefficient of correlation of the n data points $(x_1, y_1), (x_2, y_2), \ldots, (x_n, y_n)$ is

$$r = \frac{n(\sum xy) - (\sum x)(\sum y)}{\sqrt{n(\sum x^2) - (\sum x)^2} \cdot \sqrt{n(\sum y^2) - (\sum y)^2}}.$$

It measures how closely the data points $(x_1, y_1), (x_2, y_2), \ldots, (x_n, y_n)$ fit the regression line. (The value r^2 is sometimes called the **coefficient of determination**.)

Interpretation

If r is positive, the regression line has positive slope; if r is negative, the regression line has negative slope.

If $r = 1$ or -1, then all the data points lie exactly on the regression line; if it is close to ± 1, then all the data points are close to the regression line.

If r is close to 0, then y does not depend linearly on x.

EXAMPLE 3 Computing the Coefficient of Correlation

Find the correlation coefficient for the data in Example 2. Is the regression line a good fit?

Solution The formula for r requires $\sum x, \sum x^2, \sum xy, \sum y$, and $\sum y^2$. We have all of these except for $\sum y^2$, which we find in a new column as shown.

	x	y	xy	x^2	y^2
	0	9	0	0	81
	2	9	18	4	81
	4	10	40	16	100
	6	11	66	36	121
	8	11	88	64	121
	10	12	120	100	144
	12	13	156	144	169
	14	13	182	196	169
$\sum$ (Sum)	**56**	**88**	**670**	**560**	**986**

Substituting these values into the formula we get

$$r = \frac{n(\sum xy) - (\sum x)(\sum y)}{\sqrt{n(\sum x^2) - (\sum x)^2} \cdot \sqrt{n(\sum y^2) - (\sum y)^2}}$$

$$= \frac{8(670) - (56)(88)}{\sqrt{8(560) - 56^2} \cdot \sqrt{8(986) - 88^2}}$$

$$\approx 0.982.$$

Thus, the fit is a fairly good one, that is, the original points lie nearly along a straight line, as can be confirmed from the graph in Example 2.

1.4 EXERCISES

▼ more advanced ◆ challenging

T indicates exercises that should be solved using technology

In Exercises 1–4, compute the sum-of-squares error (SSE) by hand for the given set of data and linear model. HINT [See Example 1.]

1. (1, 1), (2, 2), (3, 4); $y = x - 1$

2. (0, 1), (1, 1), (2, 2); $y = x + 1$

3. (0, −1), (1, 3), (4, 6), (5, 0); $y = -x + 2$

4. (2, 4), (6, 8), (8, 12), (10, 0); $y = 2x - 8$

T *In Exercises 5–8, use technology to compute the sum-of-squares error (SSE) for the given set of data and linear models. Indicate which linear model gives the better fit.*

5. (1, 1), (2, 2), (3, 4); **a.** $y = 1.5x - 1$ **b.** $y = 2x - 1.5$

6. (0, 1), (1, 1), (2, 2); **a.** $y = 0.4x + 1.1$ **b.** $y = 0.5x + 0.9$

7. (0, −1), (1, 3), (4, 6), (5, 0); **a.** $y = 0.3x + 1.1$ **b.** $y = 0.4x + 0.9$

8. (2, 4), (6, 8), (8, 12), (10, 0); **a.** $y = -0.1x + 7$ **b.** $y = -0.2x + 6$

Find the regression line associated with each set of points in Exercises 9–12. Graph the data and the best-fit line. (Round all coefficients to 4 decimal places.) HINT [See Example 2.]

9. (1, 1), (2, 2), (3, 4)

10. (0, 1), (1, 1), (2, 2)

11. (0, −1), (1, 3), (4, 6), (5, 0)

12. (2, 4), (6, 8), (8, 12), (10, 0)

In the next two exercises, use correlation coefficients to determine which of the given sets of data is best fit by its associated regression line and which is fit worst. Is it a perfect fit for any of the data sets? HINT [See Example 4.]

13. a. {(1, 3), (2, 4), (5, 6)}
b. {(0, −1), (2, 1), (3, 4)}
c. {(4, −3), (5, 5), (0, 0)}

14. a. {(1, 3), (−2, 9), (2, 1)}
b. {(0, 1), (1, 0), (2, 1)}
c. {(0, 0), (5, −5), (2, −2.1)}

APPLICATIONS

15. *Worldwide Cell Phone Sales* Following are forecasts of worldwide annual cell phone handset sales:[45]

Year x	3	5	7
Sales y (millions)	500	600	800

($x = 3$ represents 2003). Complete the following table and obtain the associated regression line. (Round coefficients to 2 decimal places.) HINT [See Example 3.]

	x	y	xy	x^2
	3	500		
	5	600		
	7	800		
Totals				

Use your regression equation to project the 2008 sales.

16. *Investment in Gold* Following are approximate values of the Amex Gold BUGS Index:[46]

Year x	0	4	7
Index y	50	250	470

[45]Source: In-StatMDR, www.in-stat.com/, July, 2004.

[46]BUGS stands for "basket of unhedged gold stocks." Figures are approximate. Sources: www.321gold.com, Bloomberg Financial Markets/ *New York Times*, Sept 7, 2003, p. BU8, www.amex.com.

($x = 0$ represents 2000). Complete the following table and obtain the associated regression line. (Round coefficients to 2 decimal places.)

	x	y	xy	x^2
	0	50		
	4	250		
	7	470		
Totals				

Use your regression equation to estimate the 2005 index to the nearest whole number.

17. ***E-Commerce*** The following chart shows second quarter total retail e-commerce sales in the United States in 2000, 2004, and 2007 ($t = 0$ represents 2000):[47]

Year t	0	4	7
Sales ($ billion)	6	16	32

Find the regression line (round coefficients to two decimal places) and use it to estimate second quarter retail e-commerce sales in 2006.

18. ***Retail Inventories*** The following chart shows total January retail inventories in the United States in 2000, 2005, and 2007 ($t = 0$ represents 2000):[48]

Year t	0	5	7
Inventory ($ billion)	380	450	480

Find the regression line (round coefficients to two decimal places) and use it to estimate January retail inventories in 2004.

19. ***Oil Recovery*** The Texas Bureau of Economic Geology published a study on the economic impact of using carbon dioxide enhanced oil recovery (EOR) technology to extract additional oil from fields that have reached the end of their conventional economic life. The following table gives the approximate number of jobs for the citizens of Texas that would be created at various levels of recovery.[49]

Percent Recovery (%)	20	40	80	100
Jobs Created (millions)	3	6	9	15

Find the regression line and use it to estimate the number of jobs that would be created at a recovery level of 50%.

20. ***Oil Recovery*** (Refer to Exercise 19.) The following table gives the approximate economic value associated with various levels of oil recovery in Texas.[50]

Percent Recovery (%)	10	40	50	80
Economic Value ($ billions)	200	900	1000	2000

Find the regression line and use it to estimate the economic value associated with a recovery level of 70%.

21. T ***Soybean Production*** The following table shows soybean production, in millions of tons, in Brazil's *Cerrados* region, as a function of the cultivated area, in millions of acres.[51]

Area (millions of acres)	25	30	32	40	52
Production (millions of tons)	15	25	30	40	60

a. Use technology to obtain the regression line, and to show a plot of the points together with the regression line. (Round coefficients to two decimal places.)

b. Interpret the slope of the regression line.

22. T ***Trade with Taiwan*** The following table shows U.S. exports to Taiwan as a function of U.S. imports from Taiwan, based on trade figures in the period 1990–2003.[52]

Imports ($ billions)	22	24	27	35	25
Exports ($ billions)	12	15	20	25	17

a. Use technology to obtain the regression line, and to show a plot of the points together with the regression line. (Round coefficients to two decimal places.)

b. Interpret the slope of the regression line.

T *Exercises 23 and 24 are based on the following table comparing the number of registered automobiles, trucks, and motorcycles in Mexico for various years from 1980 to 2005.*[53]

Year	Automobiles (millions)	Trucks (millions)	Motorcycles (millions)
1980	4.0	1.5	0.28
1985	5.3	2.1	0.25
1990	6.6	3.0	0.25
1995	7.5	3.6	0.13
2000	10.2	4.9	0.29
2005	14.7	7.1	0.61

[47] Figures are rounded. Source: US Census Bureau, www.census.gov, November 2007.

[48] Ibid.

[49] Source: "CO2–Enhanced Oil Recovery Resource Potential in Texas: Potential Positive Economic Impacts," Texas Bureau of Economic Geology, April 2004, www.rrc.state.tx.us/tepc/CO2-EOR_white_paper.pdf.

[50] Ibid.

[51] Source: Brazil Agriculture Ministry/*New York Times*, December 12, 2004, p. N32.

[52] Source: Taiwan Directorate General of Customs/*New York Times*, December 13, 2004, p. C7.

[53] Source: Instituto Nacional de Estadística y Geografía (INEGI), www.inegi.org.mx.

23. **T** ▼ ***Automobiles and Motorcycles in Mexico***

a. Use $x =$ Number of automobiles (in millions) and $y =$ Number of motorcycles (in millions), and use technology to obtain the regression equation and graph the associated points and regression line. Round coefficients to two significant digits.

b. What does the slope tell you about the relationship between the number of automobiles and the number of motorcycles?

c. Use technology to obtain the coefficient of correlation r. Does the value of r suggest a strong correlation between x and y?

24. **T** ▼ ***Automobiles and Trucks in Mexico***

a. Use $x =$ Number of automobiles and $y =$ Number of trucks, and use technology to obtain the regression equation and graph the associated points and regression line. Round coefficients to two significant digits.

b. What does the slope tell you about the relationship between the number of automobiles and the number of trucks?

c. Use technology to obtain the coefficient of correlation r Does the value of r suggest a strong correlation between x and y?

25. **T** ▼ ***NY City Housing Costs: Downtown*** The following table shows the average price of a two-bedroom apartment in downtown New York City from 1994 to 2004. ($t = 0$ represents 1994.[54])

Year t	0	2	4	6	8	10
Price p ($ million)	0.38	0.40	0.60	0.95	1.20	1.60

a. Use technology to obtain the linear regression line, with regression coefficients rounded to two decimal places, and plot the regression line and the given points.

b. Does the graph suggest that a non-linear relationship between t and p would be more appropriate than a linear one? Why?

c. Use technology to obtain the residuals. What can you say about the residuals in support of the claim in part (b)?

26. **T** ▼ ***Fiber-Optic Connections*** The following table shows the number of fiber-optic cable connections to homes in the U.S. from 2000–2004 ($t = 0$ represents 2000):[55]

Year t	0	1	2	3	4
Connections c (Thousands)	0	10	25	65	150

a. Use technology to obtain the linear regression line, with regression coefficients rounded to two decimal places, and plot the regression line and the given points.

b. Does the graph suggest that a non-linear relationship between t and p would be more appropriate than a linear one? Why?

c. Use technology to obtain the residuals. What can you say about the residuals in support of the claim in part (b)?

COMMUNICATION AND REASONING EXERCISES

27. Why is the regression line associated with the two points (a, b) and (c, d) the same as the line that passes through both? (Assume that $a \neq c$.)

28. What is the smallest possible sum-of-squares error if the given points happen to lie on a straight line? Why?

29. If the points $(x_1, y_1), (x_2, y_2), \ldots, (x_n, y_n)$ lie on a straight line, what can you say about the regression line associated with these points?

30. If all but one of the points $(x_1, y_1), (x_2, y_2), \ldots, (x_n, y_n)$ lie on a straight line, must the regression line pass through all but one of these points?

31. ▼ Verify that the regression line for the points $(0, 0)$, $(-a, a)$, and (a, a) has slope 0. What is the value of r? (Assume that $a \neq 0$.)

32. ▼ Verify that the regression line for the points $(0, a)$, $(0, -a)$, and $(a, 0)$ has slope 0. What is the value of r? (Assume that $a \neq 0$.)

33. ▼ Must the regression line pass through at least one of the data points? Illustrate your answer with an example.

34. ▼ Why must care be taken when using mathematical models to extrapolate?

---

[54] Data are rounded and 2004 figure is an estimate. Source: Miller Samuel/*New York Times*, March 28, 2004, p. RE 11.

[55] Source: Render, Vanderslice & Associates/*New York Times*, October 11, 2004, p. C1.

CHAPTER 1 REVIEW

KEY CONCEPTS

Web Site www.AppliedCalc.org

Go to the student Web site at www.AppliedCalc.org to find a comprehensive and interactive Web-based summary of Chapter 1.

1.1 Functions from the Numerical and Algebraic Viewpoints

Real-valued function f of a real-valued variable x, domain *p. 42*

Independent and dependent variables *p. 42*

Graph of the function f *p. 42*

Numerically specified function *p. 42*

Graphically specified function *p. 43*

Algebraically defined function *p. 43*

Piecewise-defined function *p. 47*

Vertical line test *p. 49*

Common types of algebraic functions and their graphs *p. 50*

1.2 Functions and Models

Mathematical model *p. 56*

Analytical model *p. 57*

Curve-fitting model *p. 57*

Cost, revenue, and profit; marginal cost, revenue, and profit; break-even point *p. 58–59*

Demand, supply, and equilibrium price *p. 61–62*

Selecting a model *p. 64*

Compound interest *p. 66*

Exponential growth and decay *p. 68*

1.3 Linear Functions and Models

Linear function $f(x) = mx + b$ *p. 75*

Change in q: $\Delta q = q_2 - q_1$ *p. 76*

Slope of a line:

$$m = \frac{\Delta y}{\Delta x} = \frac{\text{Change in } y}{\text{Change in } x} \quad p.\ 76$$

Interpretations of m *p. 76*

Interpretation of b: y-intercept *p. 77*

Recognizing linear data *p. 78*

Computing the slope of a line *p. 79*

Slopes of horizontal and vertical lines *p. 80*

Computing the y-intercept *p. 80*

Linear modeling *p. 82*

Linear cost *p. 82*

Linear demand and supply *p. 83*

Linear change over time; rate of change; velocity *p. 85*

General linear models *p. 87*

1.4 Linear Regression

Observed and predicted values *p. 95*

Residuals and sum-of-squares error (SSE) *p. 95*

Regression line (least squares line, best-fit line) *p. 98*

Correlation coefficient; coefficient of determination *p. 100*

REVIEW EXERCISES

In each of Exercises 1–4, use the graph of the function f to find approximations of the given values.

1.

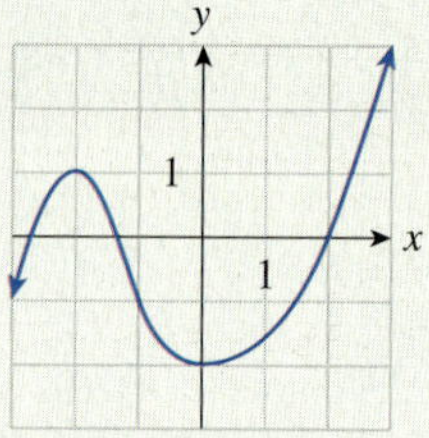

a. $f(-2)$ **b.** $f(0)$ **c.** $f(2)$ **d.** $f(2) - f(-2)$

2.

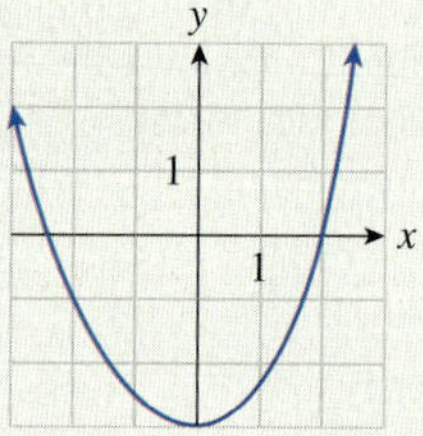

a. $f(-2)$ **b.** $f(0)$ **c.** $f(2)$ **d.** $f(2) - f(-2)$

3.

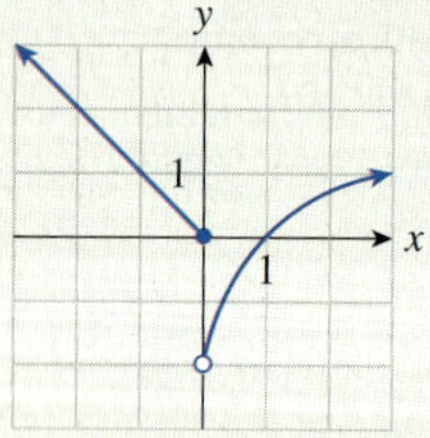

a. $f(-1)$ **b.** $f(0)$ **c.** $f(1)$ **d.** $f(1) - f(-1)$

4.

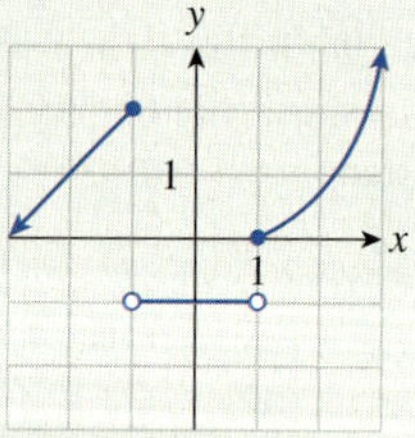

a. $f(-1)$ **b.** $f(0)$ **c.** $f(1)$ **d.** $f(1) - f(-1)$

In each of Exercises 5–8, graph the given function or equation.

5. $y = -2x + 5$

6. $2x - 3y = 12$

7. $y = \begin{cases} \frac{1}{2}x & \text{if } -1 \le x \le 1 \\ x - 1 & \text{if } 1 < x \le 3 \end{cases}$

8. $(x) = 4x - x^2$ with domain $[0, 4]$

In each of Exercises 9–14, decide whether the specified values come from a linear, quadratic, exponential, or absolute value function.

	x	−2	0	1	2	4
9.	$f(x)$	4	2	1	0	2
10.	$g(x)$	−5	−3	−2	−1	1
11.	$h(x)$	1.5	1	0.75	0.5	0
12.	$k(x)$	0.25	1	2	4	16
13.	$u(x)$	0	4	3	0	−12
14.	$w(x)$	32	8	4	2	0.5

In each of Exercises 15–18, find the equation of the specified line.

15. Through (3, 2) with slope −3

16. Through (−1, 2) and (1, 0)

17. Through (1, 2) parallel to $x - 2y = 2$

18. With slope 1/2 crossing $3x + y = 6$ at its x-intercept

In Exercises 19 and 20, determine which of the given lines better fits the given points.

19. (−1, 1), (1, 2), (2, 0); $y = -x/2 + 1$ or $y = -x/4 + 1$

20. (−2, −1), (−1, 1), (0, 1), (1, 2), (2, 4), (3, 3); $y = x + 1$ or $y = x/2 + 1$

In Exercises 21 and 22, find the line that best fits the given points and compute the correlation coefficient.

21. (−1, 1), (1, 2), (2, 0)

22. (−2, −1), (−1, 1), (0, 1), (1, 2), (2, 4), (3, 3)

APPLICATIONS

23. John Sean O'Hagan is CEO of OHaganBooks.com, an online bookstore. Since the establishment of the company Web site six years ago, the number of visitors to the site has grown quite dramatically, as indicated by the following table:

Year t	0	1	2	3	4	5	6
Website Traffic $V(t)$ (visits/day)	100	300	1,000	3,300	10,500	33,600	107,400

a. Graph the function V as a function of time t. Which of the following types of function seem to fit the curve best: linear, quadratic, or exponential?

b. Compute the ratios $\frac{V(1)}{V(0)}, \frac{V(2)}{V(1)}, \ldots,$ and $\frac{V(6)}{V(5)}$. What do you notice?

c. Use the result of part (b) to predict Web site traffic next year (to the nearest 100).

24. As the online bookstore, OHaganBooks.com, has grown in popularity, the sales manager has been monitoring book sales as a function of the traffic at your site (measured in "hits" per day) and has obtained the following model.

$$n(x) = \begin{cases} 0.02x & \text{if } 0 \le x \le 1,000 \\ 0.025x - 5 & \text{if } 1,000 < x \le 2,000 \end{cases}$$

where $n(x)$ is the average number of books sold in a day in which there are x hits at the site.

a. On average, how many books per day does the model predict the bookstore will sell when it has 500 hits in a day? 1,000 hits in a day? 1,500 hits in a day?

b. What does the coefficient 0.025 tell you about book sales?

c. According to the model, how many hits per day will be needed in order to sell an average of 30 books per day?

25. Monthly sales of books at OHaganBooks.com have increased quite dramatically over the past few months, but now appear to be leveling off. Here are the sales figures for the past 6 months.

Month t	1	2	3	4	5	6
Daily Book Sales	12.5	37.5	62.5	72.0	74.5	75.0

a. Which of the following models best approximates the data?

(A) $S(t) = \frac{300}{4 + 100(5^{-t})}$ **(B)** $S(t) = 13.3t + 8.0$

(C) $S(t) = -2.3t^2 + 30.0t - 3.3$ **(D)** $S(t) = 7(3^{0.5t})$

b. What do each of the above models predict for the sales in the next few months: rising, falling, or leveling off?

26. To increase business at OHaganBooks.com, John O'Hagan plans to place more banner ads at well-known Internet portals. So far, he has the following data on the average number of hits per day at OHaganBooks.com versus monthly advertising expenditures.

Advertising Expenditure ($/Month)	$2,000	$5,000
Website Traffic (Hits/Day)	1,900	2,050

He decides to construct a linear model giving the average number of hits h per day as a function of the advertising expenditure c.

a. What is the model he obtains?

b. Based on your model, how much traffic can he anticipate if he budgets $6,000 per month for banner ads?

c. The goal is to eventually increase traffic at the site to an average of 2,500 hits per day. Based on the model, how much should be spent on banner ads in order to accomplish this?

27. A month ago John O'Hagan increased expenditure on banner ads to $6,000 per month, and noticed that the traffic at OHaganBooks.com has not increased to the level predicted by the linear model in Exercise 26. Fitting a quadratic function to the available data gives the model

$$h = -0.000005c^2 + 0.085c + 1750,$$

where h is the daily traffic (hits) at the Web site, and c is the monthly advertising expenditure.

a. According to this model, what is the current traffic at the site?

b. Does this model give a reasonable prediction of traffic at expenditures larger than $8,500 per month? Why?

28. Besides selling books, OHaganBooks.com is generating revenue through its new online publishing service. Readers pay a fee to download the entire text of a novel. Author royalties and copyright fees cost you an average of $4 per novel, and the monthly cost of operating and maintaining the service amounts to $900 per month. The company is currently charging readers $5.50 per novel.

a. What are the associated cost, revenue, and profit functions?

b. How many novels must be sold per month in order to break even?

c. If the charge is lowered to $5.00 per novel, how many books must be sold in order to break even?

29. In order to generate a profit from its online publishing service, OHaganBooks.com needs to know how the demand for novels depends on the price it charges. During the first month of the service, it was charging $10 per novel, and sold 350. Lowering the price to $5.50 per novel had the effect of increasing demand to 620 novels per month.

a. Use the given data to construct a linear demand equation.

b. Use the demand equation you constructed in part (a) to estimate the demand if you raised the price to $15 per novel.

c. Using the information on cost given in Exercise 28, determine which of the three prices ($5.50, $10, and $15) would result in the largest profit, and the size of that profit.

30. It is now several months later, and OHaganBooks.com has tried selling its online novels at a variety of prices, with the following results.

Price	$5.50	$10	$12	$15
Demand (Monthly sales)	620	350	300	100

a. Use the given data to obtain a linear regression model of demand. (Round coefficients to four decimal places.)

b. Use the demand model you constructed in part (a) to estimate the demand if the company charged $8 per novel. (Round the answer to the nearest novel.)

Case Study: Modeling Spending on Internet Advertising

Jeff Titcomb/Photographer's Choice/Getty Images

It is 2008 and you are the new director of Impact Advertising Inc.'s Internet division, which has enjoyed a steady 0.25% of the Internet advertising market. You have drawn up an ambitious proposal to expand your division in light of your anticipation that Internet advertising will continue to skyrocket. However, upper management sees things differently and, based on the following email, does not seem likely to approve the budget for your proposal.

TO: JCheddar@impact.com (J. R. Cheddar)
CC: CVODoylePres@impact.com (C. V. O'Doyle, CEO)
FROM: SGLombardoVP@impact.com (S. G. Lombardo, VP Financial Affairs)
SUBJECT: Your Expansion Proposal
DATE: May 30, 2008

Hi John:

Your proposal reflects exactly the kind of ambitious planning and optimism we like to see in our new upper management personnel. Your presentation last week was most impressive, and obviously reflected a great deal of hard work and preparation.

I am in full agreement with you that Internet advertising is on the increase. Indeed, our Market Research department informs me that, based on a regression of the most recently available data, Internet advertising revenue in the United States will continue to grow at a rate of approximately $3 billion per year. This translates into approximately $7.5 million in increased revenues per year for Impact, given our 0.25% market share. This rate of expansion is exactly what our planned 2008 budget anticipates. Your proposal, on the other hand, would require a budget of approximately *twice* the 2008 budget allocation, even though your proposal provides no hard evidence to justify this degree of financial backing.

At this stage, therefore, I am sorry to say that I am inclined not to approve the funding for your project, although I would be happy to discuss this further with you. I plan to present my final decision on the 2008 budget at next week's divisional meeting.

Regards, Sylvia

Refusing to admit defeat, you contact the Market Research department and request the details of their projections on Internet advertising. They fax you the following information:[56]

Year	2002	2003	2004	2005	2006	2007
Internet Advertising Revenue ($ billion)	6.0	7.3	9.6	12.5	16.9	21.2

Regression Model: $y = 3.0771x + 4.5571$ (x = years since 2002)

Correlation Coefficient: $r = 0.9805$

Now you see where the VP got that $3 billion figure: The slope of the regression equation is close to 3, indicating a rate of increase of just over $3 billion per year. Also, the correlation coefficient is very high—an indication that the linear model fits the data well. In view of this strong evidence, it seems difficult to argue that revenues will increase by significantly more than the projected $3 billion per year. To get a better picture of what's going on, you decide to graph the data together with the regression line in your spreadsheet program. What you get is shown in Figure 33. You immediately notice that the data points seem to suggest a curve, and not a straight line. Then again, perhaps the suggestion of a curve is an illusion. Thus there are, you surmise, two possible interpretations of the data:

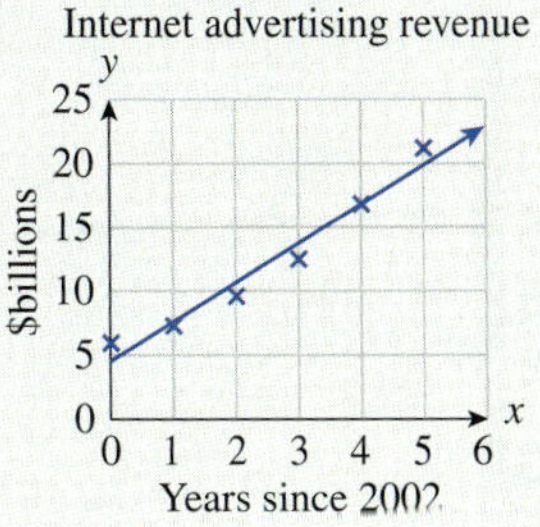

Figure 33

1. (Your first impression) As a function of time, Internet advertising revenue is nonlinear, and is in fact accelerating (the rate of change is increasing), so a linear model is inappropriate.
2. (Devil's advocate) Internet advertising revenue *is* a linear function of time; the fact that the points do not lie on the regression line is simply a consequence of random factors, such as the state of the economy, the stock market performance, and so on.

You suspect that the VP will probably opt for the second interpretation and discount the graphical evidence of accelerating growth by claiming that it is an illusion: a "statistical fluctuation." That is, of course, a possibility, but you wonder how likely it really is.

For the sake of comparison, you decide to try a regression based on the simplest non-linear model you can think of—a quadratic function.

$$y = ax^2 + bx + c$$

Your spreadsheet allows you to fit such a function with a click of the mouse. The result is the following.

$$y = 0.4179x^2 + 0.9879x + 5.95 \quad (x = \text{number of years since 2002})$$

$$r = 0.9996$$

See Note.*

*** NOTE** Note that this r is *not* the linear correlation coefficient we defined on p. 100; what this r measures is how closely the *quadratic* regression model fits the data.

Figure 34 shows the graph of the regression function together with the original data.

Aha! The fit is visually far better, and the correlation coefficient is even higher! Further, the quadratic model predicts 2008 revenue as

$$y = 0.4179(6)^2 + 0.9879(6) + 5.95 \approx \$26.9 \text{ billion},$$

which is $5.7 billion above the 2007 spending figure in the table above. Given Impact Advertising's 0.25% market share, this translates into an increase in revenues of $14.2 million, which is almost double the estimate predicted by the linear model!

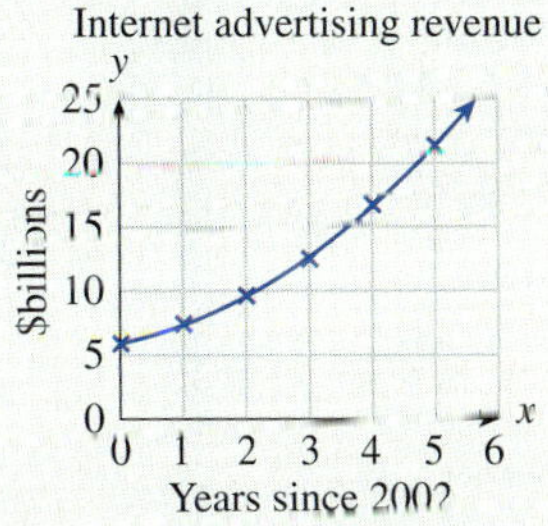

Figure 34

[56] Figures are rounded. Source: IAB Internet Advertising Revenue Report: An Industry Survey Conducted by PricewaterhouseCoopers and Sponsored by the Interactive Advertising Bureau (IAB); 2007 Full-Year Results, May 2008 (www.iab.net/insights_research/1357).

You quickly draft an email to Lombardo, and are about to click "Send" when you decide, as a precaution, to check with a statistician. He tells you to be cautious: The value of r will always tend to increase if you pass from a linear model to a quadratic one due to an increase in "degrees of freedom."* A good way to test whether a quadratic model is more appropriate than a linear one is to compute a statistic called the "p-value" associated with the coefficient of x^2. A low value of p indicates a high degree of confidence that the coefficient of x^2 cannot be zero (see below). Notice that if the coefficient of x^2 *is* zero, then you have a linear model.

* **NOTE** The number of degrees of freedom in a regression model is 1 less than the number of coefficients. For a linear model, it is 1 (there are two coefficients: the slope m and the intercept b), and for a quadratic model it is 2. For a detailed discussion, consult a text on regression analysis.

You can, your friend explains, obtain the p-value using Excel as follows. First, set up the data in columns, with an extra column for the values of x^2.

	A	B	C
1	x	x^2	y
2	0	0	6
3	1	1	7.3
4	2	4	9.6
5	3	9	12.5
6	4	16	16.9
7	5	25	21.2

Next, in the "Analysis" section of the Data tab, choose "Data analysis." (If this command is not available, you will need to load the Analysis ToolPak add-in.) Choose "Regression" from the list that appears, and in the resulting dialogue box enter the location of the data and where you want to put the results as shown in Figure 35.

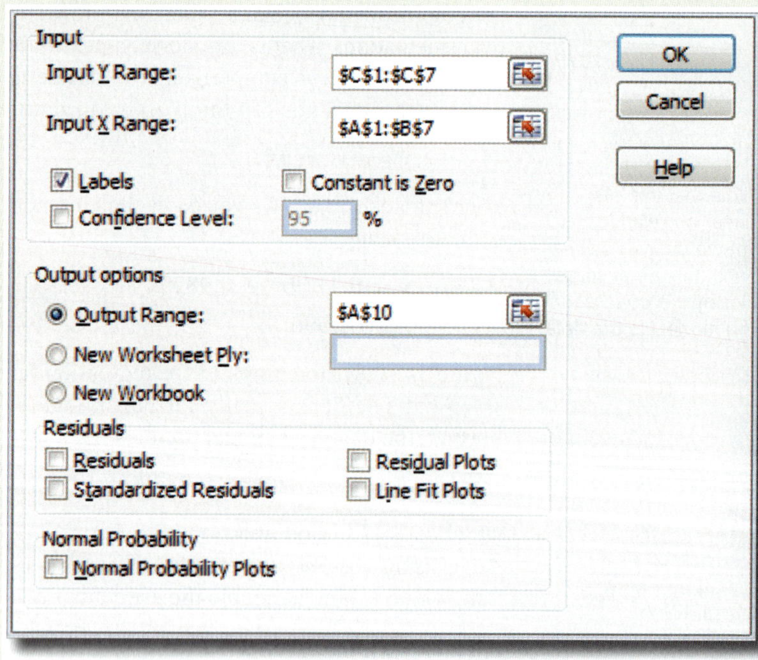

Figure 35

Clicking "OK" then gives you a large chart of statistics. The p-value you want is in the very last row of the data: $p = 0.001474$.

Q: What does p actually measure?

A: *Roughly speaking,* $1 - p = .9985$ gives the degree of confidence you can have (99.85%) in asserting that the coefficient of x^2 is not zero. (Technically, p is the probability—allowing for random fluctuation in the data—that, if the coefficient of x^2 were in fact zero, the "t-statistic" (11.332, right next to the p-value on the spreadsheet) could be as large as it is.

In short, you can go ahead and send your email with 99% confidence!

EXERCISES

Suppose you are given the following data for the spending on Internet advertising in a hypothetical country in which Impact Advertising also has a 0.25% share of the market.

Year	1999	2000	2001	2002	2003	2004	2005
Spending on Advertising ($ billion)	0	0.3	1.5	2.6	3.4	4.3	5.0

1. Obtain a linear regression model and the correlation coefficient r. According to the model, at what rate is spending on Internet Advertising increasing in this country? How does this translate to annual revenues for Impact Advertising?
2. Use a spreadsheet or other technology to graph the data together with the best-fit line. Does the graph suggest a quadratic model (parabola)?
3. Test your impression in the preceding exercise by using technology to fit a quadratic function and graphing the resulting curve together with the data. Does the graph suggest that the quadratic model is appropriate?
4. Perform a regression analysis and find the associated p-value. What does it tell you about the appropriateness of a quadratic model?

TI-83/84 Plus Technology Guide

Section 1.1

Example 1(a) and (c) (page 44) The number of iPods sold by Apple Inc. each year from 2004 through 2007 can be approximated by $f(x) = -x^2 + 20x + 3$ million iPods $(0 \le x \le 3)$ where x is the number of years since 2004. Compute $f(0)$, $f(1)$, $f(2)$, and $f(3)$ to obtain the graph of f.

Solution with Technology

There are several ways to evaluate an algebraically defined function on a graphing calculator such as the TI-83/84 Plus. First, enter the function in the Y = screen, as

`Y1 = -X^2+20*X+3` Negative (-) and minus (−) are different keys on the TI-83/84 Plus.

or `Y1 = -X²+20X+3`

(See Chapter 0 for a discussion of technology formulas.) Then, to evaluate $f(0)$, for example, enter the following in the home screen:

$Y_1(0)$ This evaluates the function Y_1 at 0.

Alternatively, you can use the table feature: After entering the function under Y_1, press [2ND] [TBLSET], and set `Indpnt` to Ask. (You do this once and for all; it will permit you to specify values for x in the table screen.) Then, press [2ND] [TABLE], and you will be able to evaluate the function at several values of x. Here is a table showing the values requested:

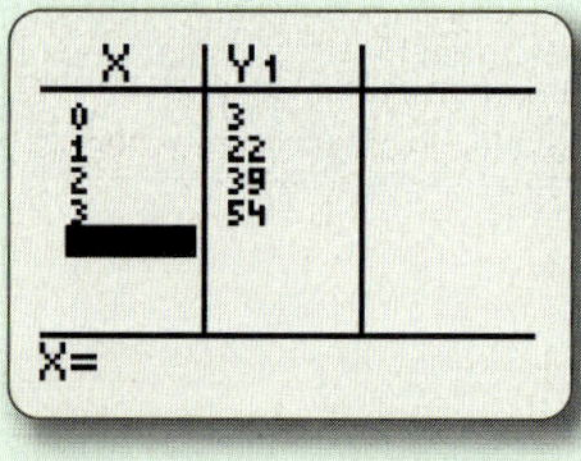

To obtain the graph, press [WINDOW], set Xmin = 0, Xmax = 3 (the range of x-values we are interested in), Ymin = 0, Ymax = 60 (we estimated Ymin and Ymax from the corresponding set of y-values in the table) and press [GRAPH] to obtain the curve:

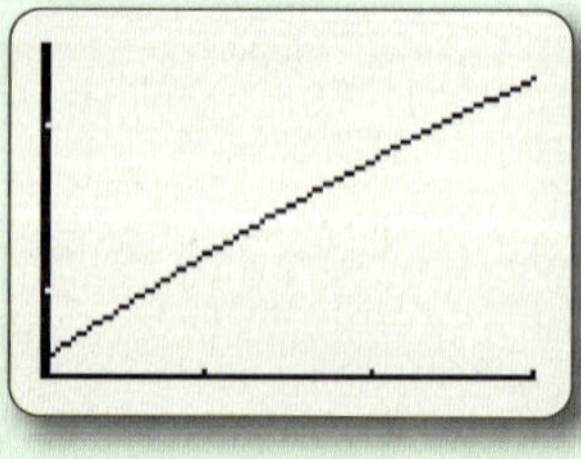

Alternatively, you can avoid having to estimate Ymin and Ymax by pressing ZoomFit ([ZOOM] [0]), which automatically sets Ymin and Ymax to the smallest and greatest values of y in the specified range for x.

Example 2 (page 47) The number $n(t)$ of Facebook members can be approximated by the following function of time t in years ($t = 0$ represents January 2004):

$$n(t) = \begin{cases} 1.5t^2 - t + 1.5 & \text{if } 0 \le t \le 3 \\ 46t - 126 & \text{if } 3 < t \le 5 \end{cases} \quad \text{million members}$$

(Its graph is shown in Figure 4.) Obtain a table showing the values $n(t)$ for $t = 0, 1, 2, 3, 4, 5$, and also obtain the graph of n.

Solution with Technology

The following formula defines the function n on the TI-83/84 Plus:

`Y1=(X≤3)(1.5X²-X+1.5)+(X>3)(46X-126)`

The logical operators (≤ and >, for example) can be found by pressing [2ND] [TEST].

When `X` is less than or equal to 3, the logical expression (`X≤3`) evaluates to 1 because it is true, and the expression (`X>3`) evaluates to 0 because it is false. The value of the function is therefore given by the expression (`1.5X²-X+1.5`). When `X` is greater than 3, the expression (`X≤3`) evaluates to 0 while the expression (`X>3`) evaluates to 1, so the value of the function is given by the expression (`46X-126`).

As in Example 1, you can use the Table feature to compute several values of the function at once by pressing [2ND] [TABLE].

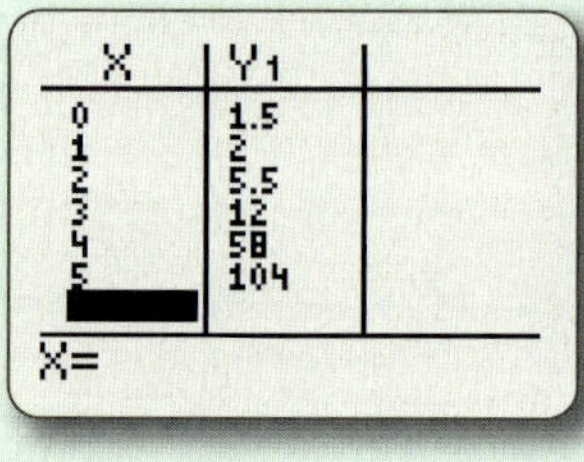

For the graph, you can proceed as Example 1: Press [WINDOW], set Xmin = 0, Xmax = 5, Ymin = 0, Ymax = 110 (see the y-values in the table) and press [GRAPH]:

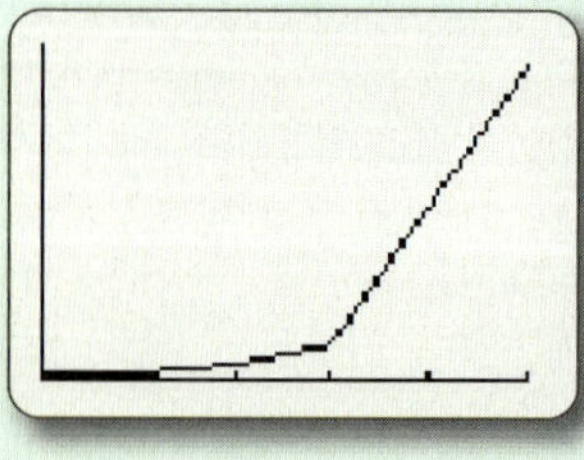

Example 3 (page 48) Obtain a table of values and the graph of the function f specified by

$$f(x) = \begin{cases} -1 & \text{if } -4 \le x < -1 \\ x & \text{if } -1 \le x \le 1 \\ x^2 - 1 & \text{if } 1 < x \le 2 \end{cases} .$$

Solution with Technology

Enter this function as

`(X<-1)*(-1)` + `(-1≤X and X≤1)*X` + `(1<X)*(X^2-1)`

First part — Second part — Third part

The logical operator `and` is found in the `TEST LOGIC` menu. The following alternative formula will also work:

`(X<-1)*(-1)+(-1≤X)*(X≤1)*X+(1<X)*(X^2-1)`

As in Example 1, you can use the Table feature to compute several values of the function at once by pressing [2ND] [TABLE] and obtain the graph by setting Xmin $= -4$, Xmax $= 2$, Ymin $= -1$, and Ymax $= 5$:

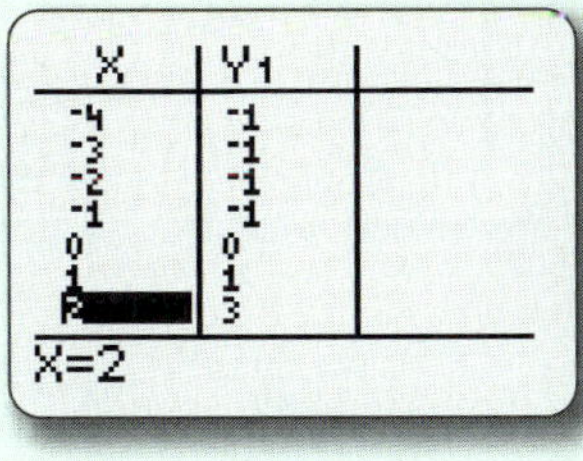

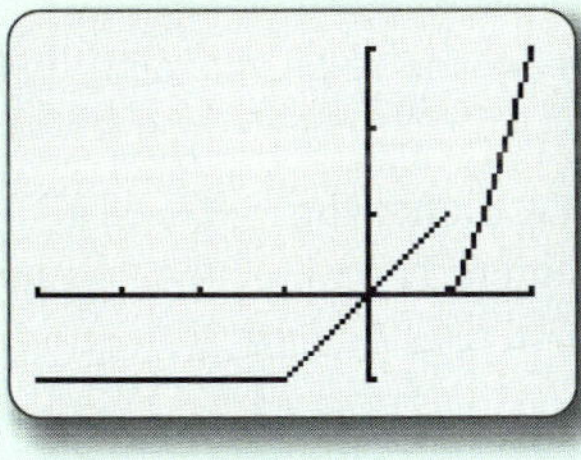

Section 1.2

Example 4(a) (page 62) The demand and supply curves for private schools in Michigan are $q = 77.8p^{-0.11}$ and $q = 30.4 + 0.006p$ thousand students, respectively $(200 \le p \le 2{,}200)$, where p is the net tuition cost in dollars. Graph the demand and supply curves on the same set of axes. Use your graph to estimate, to the nearest \$100, the tuition at which the demand equals the supply (equilibrium price). Approximately how many students will be accommodated at that price?

Solution with Technology

To obtain the graphs of demand and supply:

1. Enter `Y1=77.8*X^(-0.11)` and `Y2=30.4+0.006*X` in the "Y=" screen.
2. Press [WINDOW], enter Xmin $= 200$, Xmax $= 2200$, Ymin $= 0$, Ymax $= 50$; then ([ZOOM] [0]) for the graph:

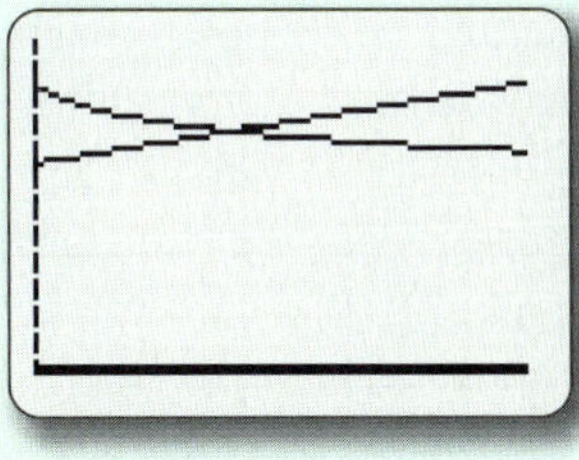

3. To estimate the equilibrium price, press [TRACE] and use the arrow keys to follow the curve to the approximate point of intersection (around X $= 1008$).

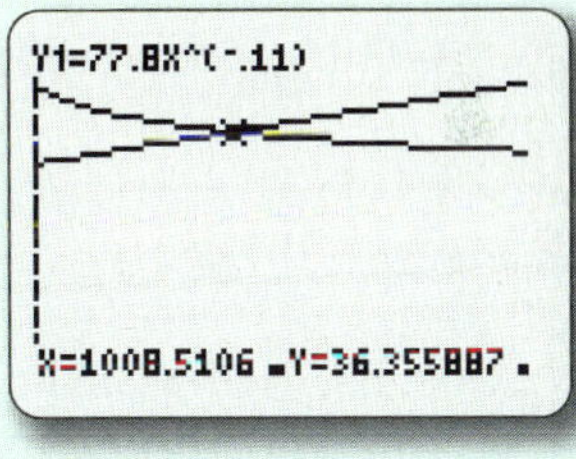

4. For a more accurate estimate, zoom in by pressing [ZOOM] and selecting Option 1 `ZBox`.
5. Move the curser to a point slightly above and to the left of the intersection, press [ENTER], and then move the curser to a point slightly below and to the right and press [ENTER] again to obtain a box.

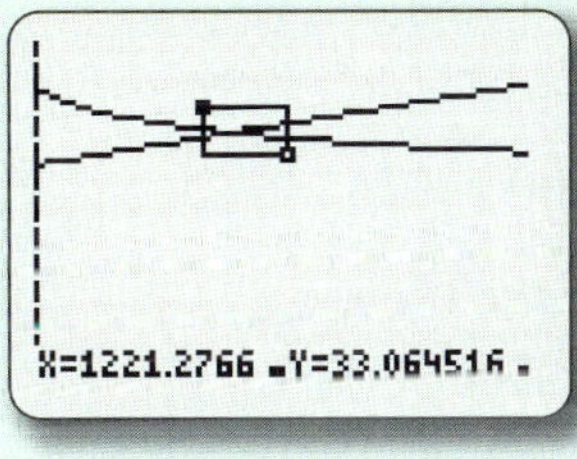

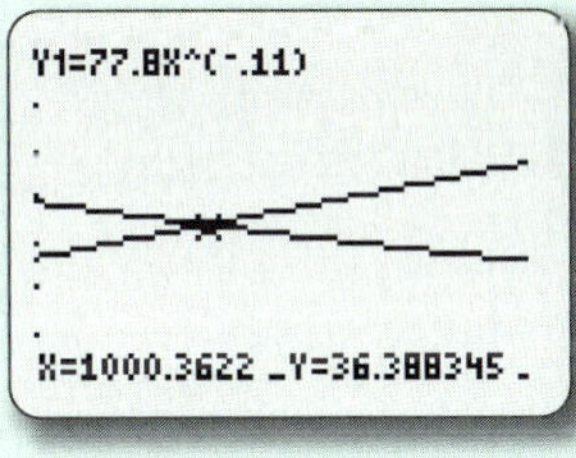

6. Now press ENTER again for a zoomed-in view of the intersection.
7. You can now use TRACE to obtain the intersection coordinates more accurately: X ≈ 1,000, representing a tuition cost of $1,000. The associated demand is the Y-coordinate: around 36.4 thousand students.

Example 5 (page 64) The following table shows annual slot machine and video poker machine revenues in 1996–2006 at the *Mohegan Sun* casino ($t = 0$ represents 2000).

t	0	1	2	3	4	5	6
Revenue ($ million)	550	680	750	820	850	890	910

Consider the following four models:

(1) $r(t) = 57t + 607$ Linear model

(2) $r(t) = -9t^2 + 110t + 560$ Quadratic model

(3) $r(t) = 608(1.08)^t$ Exponential model

(4) $r(t) = \dfrac{930}{1 + 0.67(1.7)^{-t}}$ Logistic model

a. Which models fit the data significantly better than the rest?

b. Of the models you selected in part (a), which gives the most reasonable prediction for 2010?

Solution with Technology

1. First enter the actual revenue data in the stat list editor (STAT EDIT) with the values of t in L_1, and the values of $r(t)$ in L_2.

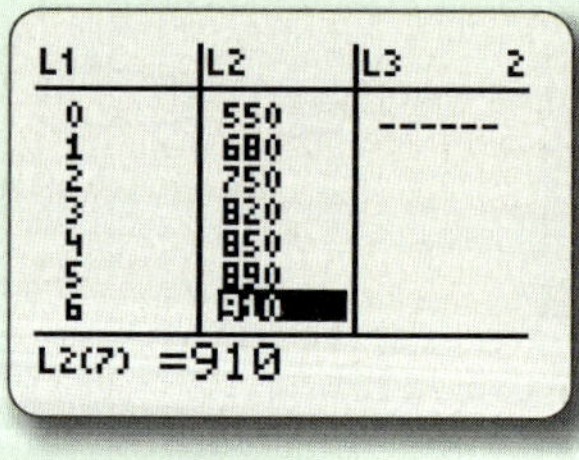

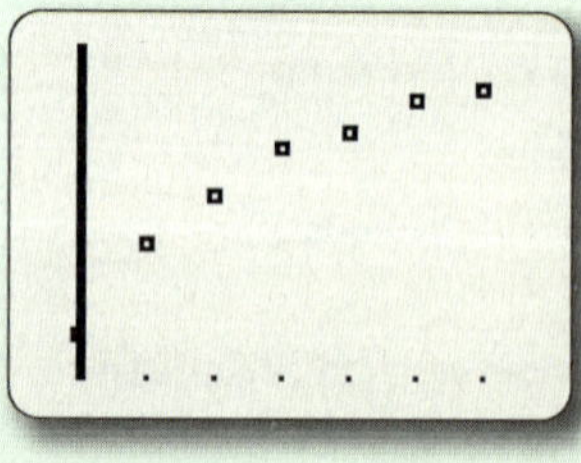

2. Now go to the Y= window and turn `Plot1` on by selecting it and pressing ENTER. (You can also turn it on in the 2ND STAT PLOT screen.) Then press ZoomStat (ZOOM 9) to obtain a plot of the points (above).
3. To see any of the four curves plotted along with the points, enter its formula in the Y= screen (for instance, `Y1=-9*X^2+110X+560` for the second model) and press GRAPH (figure on top below).

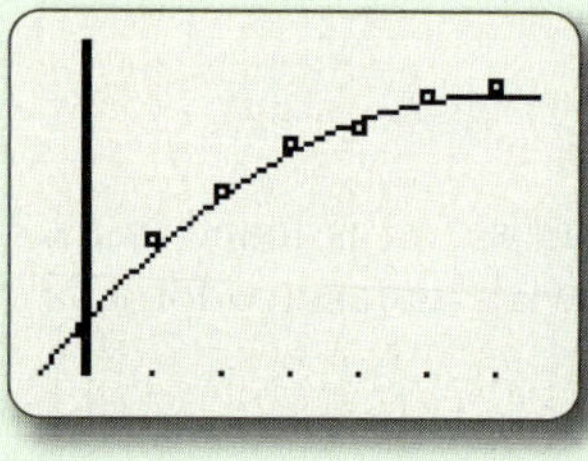

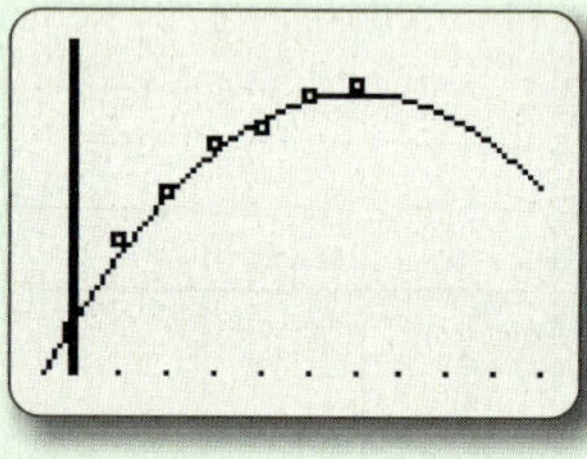

4. To see the extrapolation of the curve to 2010, just change Xmax to 10 (in the WINDOW screen) and press GRAPH again (figure above bottom).
5. Now change Y_1 to see similar graphs for the remaining curves.
6. When you are done, turn `Plot1` off again so that the points you entered do not show up in other graphs.

Section 1.3

Example 1 (page 78) Which of the following two tables gives the values of a linear function? What is the formula for that function?

x	0	2	4	6	8	10	12
$f(x)$	3	−1	−3	−6	−8	−13	−15

x	0	2	4	6	8	10	12
$g(x)$	3	−1	−5	−9	−13	−17	−21

Solution with Technology

We can use the "List" feature in the TI-83/84 Plus to automatically compute the successive quotients $m = \Delta y/\Delta x$ for either f or g as follows:

1. Use the stat list editor ([STAT] EDIT) to enter the values of x and $f(x)$ in the first two columns, called L_1 and L_2. (If there is already data in a column you want to use, you can clear it by highlighting the column heading (for example, L_1) using the arrow key, and pressing [CLEAR] [ENTER].)

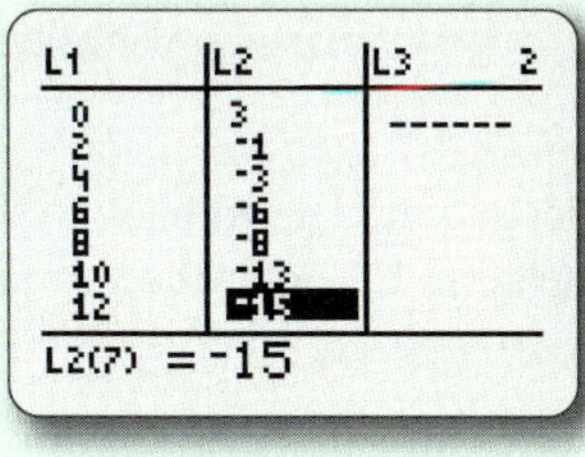

2. Highlight the heading L_3 by using the arrow keys, and enter the following formula (with the quotes, as explained below):

   "ΔList(L₂)/ΔList(L₁)" ΔList is found under [LIST] OPS; L_1 is [2ND] [1]

   The "ΔList" function computes the differences between successive elements of a list, returning a list with one less element. The formula above then computes the quotients $\Delta y/\Delta x$ in the list L_3.

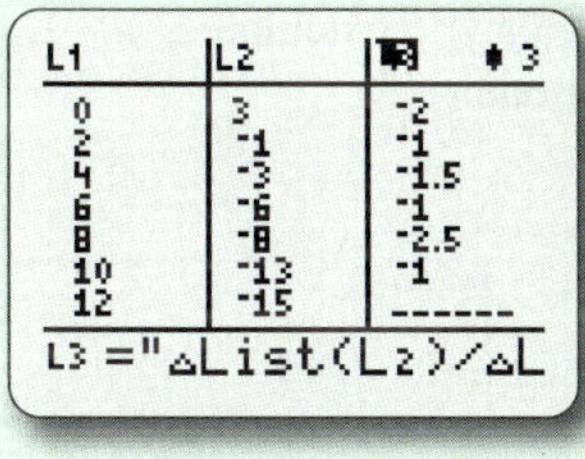

As you can see in the third column, $f(x)$ is not linear.

3. To redo the computation for $g(x)$, all you need to do is edit the values of L_2 in the stat list editor. By putting quotes around the formula we used for L_3, we told the calculator to remember the formula, so it automatically recalculates the values.

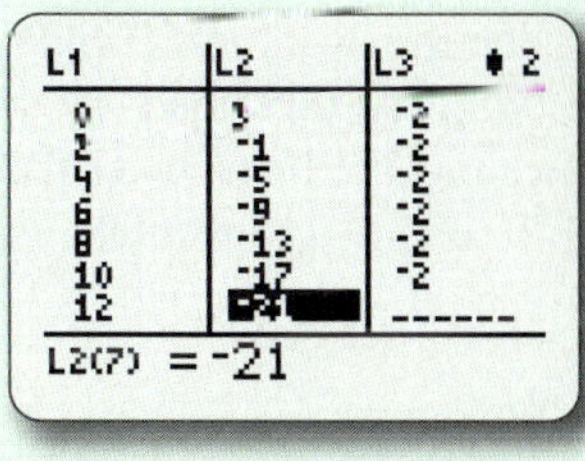

Example 2(a) (page 81) Find the equation of the line through the points (1, 2) and (3, −1).

Solution with Technology

1. Enter the coordinates of the given points in the stat list editor ([STAT] EDIT) with the values of x in L_1, and the values of y in L_2.

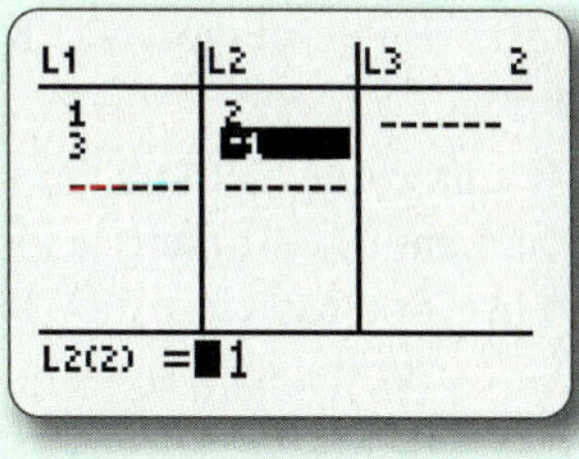

2. To compute the slope $m = \Delta y/\Delta x$, highlight the heading L_3 and enter the following formula (with the quotes—see the discussion of Example 1 above):

   "ΔList(L₂)/ΔList(L₁)" ΔList is found under [LIST] OPS

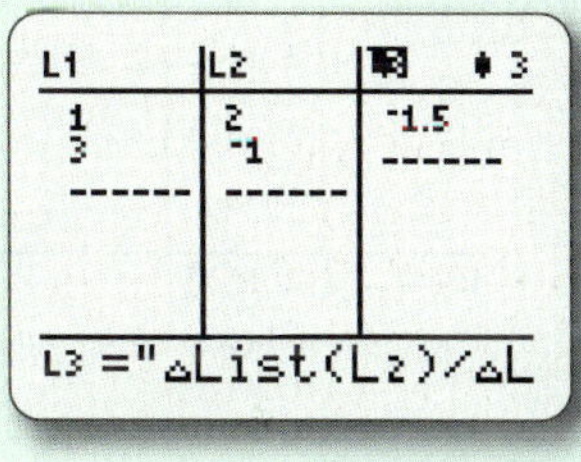

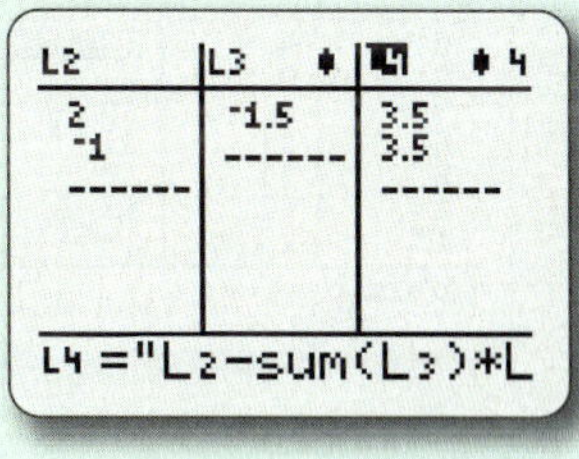

3. To compute the intercept $b = y_1 - mx_1$, highlight the heading L_4 and enter the following formula (again with the quotes):

   "L₂-sum(L₃)*(L₁)" sum() is found under [LIST] MATH

   The intercept will then appear twice in column L_4. (The calculator computes the intercept twice: using $(x_1, y_1) = (1, 2)$ and $(x_1, y_1) = (3, -1)$: If you used "L₃" instead of "sum(L₃)" in the formula, the TI would return an error, as it cannot multiply columns of different dimensions.)

Section 1.4

Example 1 (page 96) Using the data on the GDP in Mexico given on page 96, compute SSE, the sum-of-squares error, for the linear models $y = 0.5t + 8$ and $y = 0.25t + 9$, and graph the data with the given models.

Solution with Technology

We can use the "List" feature in the TI-83/84 Plus to automate the computation of SSE.

1. Use the stat list editor (STAT EDIT) to enter the given data in the lists L_1 and L_2. (If there is already data in a list you want to use, you can clear it by highlighting the column heading (for example, L_1) using the arrow key, and pressing CLEAR ENTER.)
2. To compute the predicted values, highlight the heading L_3 using the arrow keys, and enter the formula for the predicted values (figure on the top below):

   `0.5*L1+8` L_1 is 2ND 1

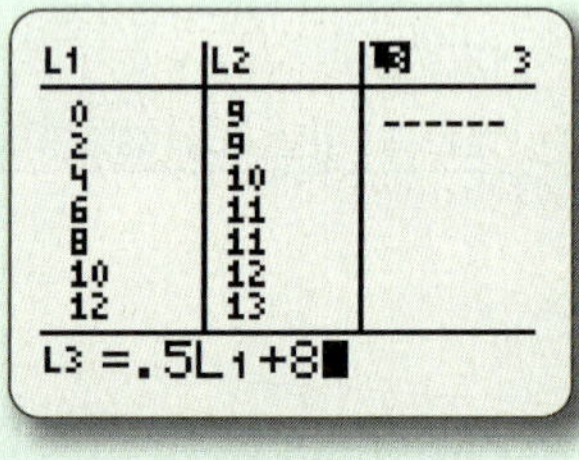

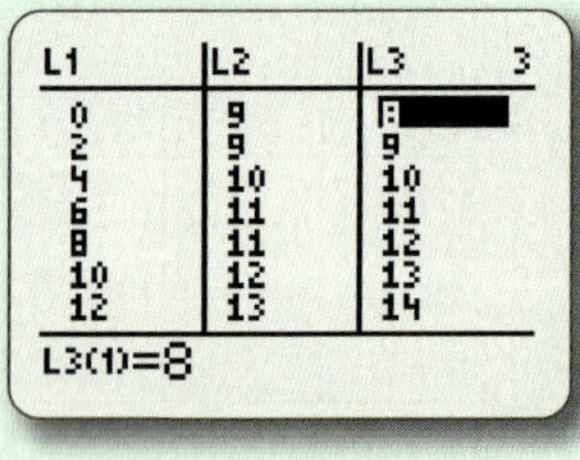

   Pressing ENTER again will fill column 3 with the predicted values (above bottom). Note that only seven of the eight data points can be seen on the screen at one time.
3. Highlight the heading L_4 and enter the following formula (including the quotes):

   `"(L2-L3)^2"`   Squaring the residuals
4. Pressing ENTER will fill column 4 with the squares of the residuals. (Putting quotes around the formula will allow us to easily check the second model, as we shall see.)
5. To compute SSE, the sum of the entries in L_4, go to the home screen and enter `sum(L4)`. The sum function is found by pressing 2ND LIST and selecting MATH.

6. To check the second model, go back to the List screen, highlight the heading L_3, enter the formula for the second model, `0.25*L1+9`, and press ENTER.

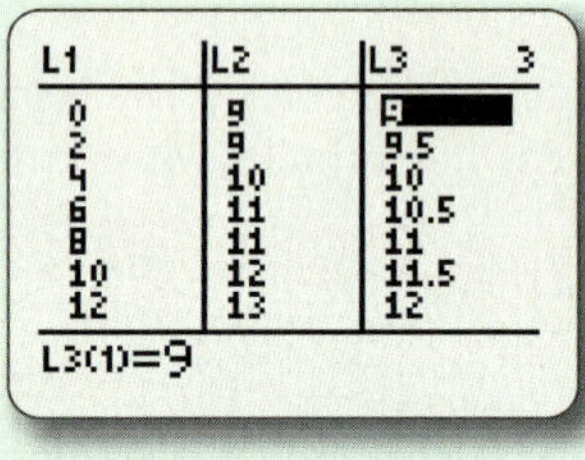

7. Because we put quotes around the formula for the residuals in L_4, the TI-83/84 Plus remembered the formula and automatically recalculated the values. On the home screen we can again calculate `sum(L4)` to get SSE for the second model.

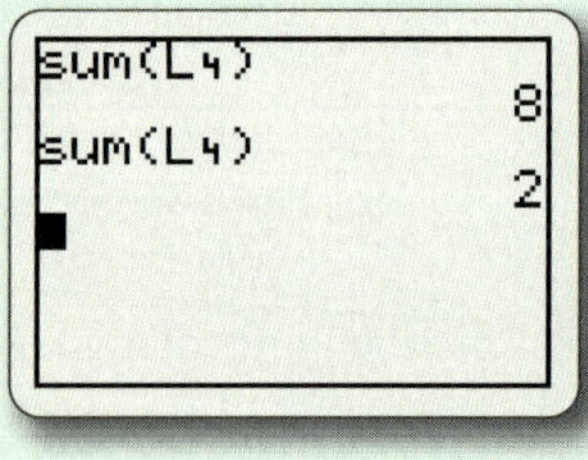

   The second model gives a much smaller SSE, and so is the better fit.
8. You can also use the TI-83/84 Plus to plot both the original data points and the two lines. Turn PLOT1 on in the STAT PLOTS window, obtained by pressing 2ND STAT PLOTS. To show the lines, enter them in the "Y=" screen as usual. To obtain a convenient window showing

all the points and the lines, press [ZOOM] and choose option #9: `ZoomStat`.

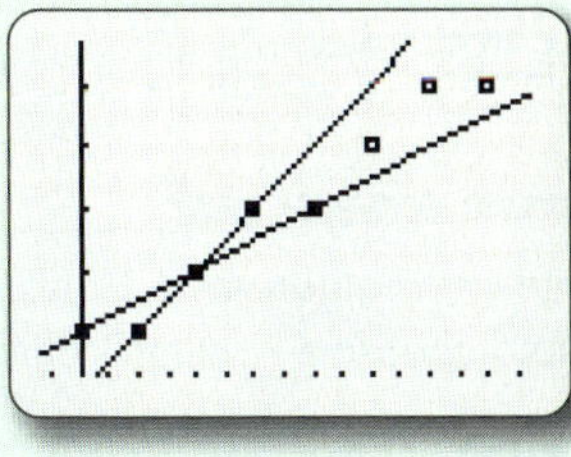

Example 2 (page 98) Consider the following data on the per capita gross domestic product (GDP) of Mexico:

Year x ($x = 0$ represents 2000)	0	2	4	6	8	10	12	14
Per Capita GDP y ($1,000)	9	9	10	11	11	12	13	13

Find the best-fit linear model for these data.

Solution with Technology

1. Enter the data in the TI-83/84 Plus using the List feature, putting the x-coordinates in L_1 and the y-coordinates in L_2, just as in Example 1.
2. Press [STAT], select CALC, and choose option #4: `LinReg(ax+b)`. Pressing [ENTER] will cause the equation of the regression line to be displayed in the home screen:

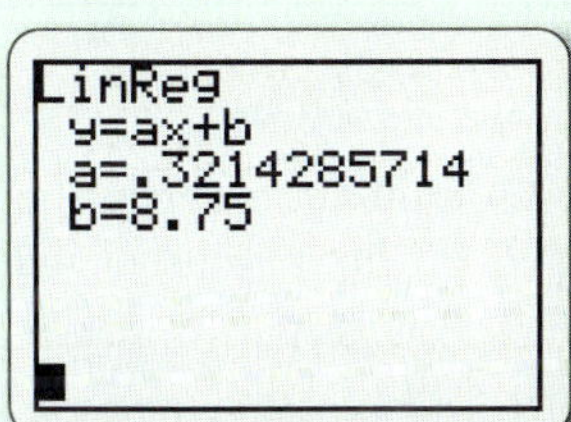

So, the regression line is $y \approx 0.321x + 8.75$.

3. To graph the regression line without having to enter it by hand in the "`Y=`" screen, press [Y=], clear the contents of Y_1, press [VARS], choose option #5: `Statistics`, select `EQ, and then choose #1:RegEQ`. The regression equation will then be entered under Y_1.
4. To simultaneously show the data points, press [2ND] [STATPLOTS] and turn PLOT1 on as in Example 1. To obtain a convenient window showing all the points and the line, press [ZOOM] and choose option #9: `ZoomStat`.

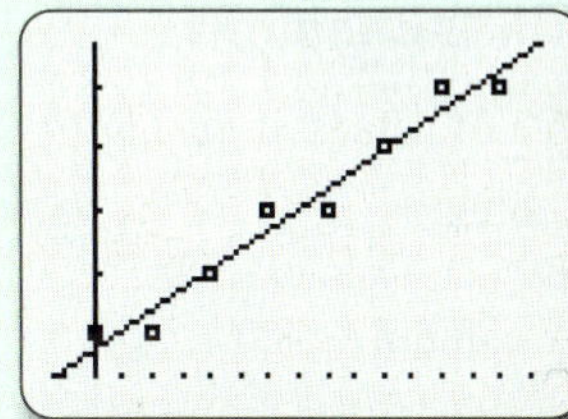

Example 3 (page 100) Find the correlation coefficient for the data in Example 2.

Solution with Technology

To find the correlation coefficient using a TI-83/84 Plus you need to tell the calculator to show you the coefficient at the same time that it shows you the regression line.

1. Press [2ND] [CATALOG] and select "`DiagnosticOn`" from the list. The command will be pasted to the home screen, and you should then press [ENTER] to execute the command.
2. Once you have done this, the "`LinReg(ax+b)`" command (see the previous discussion for Example 2). The example will show you not only a and b, but r and r^2 as well:

EXCEL Technology Guide

Section 1.1

Example 1(a) and (c) (page 44) The number of iPods sold by Apple Inc. each year from 2004 through 2007 can be approximated by $f(x) = -x^2 + 20x + 3$ million iPods ($0 \le x \le 3$) where x is the number of years since 2004. Compute $f(0)$, $f(1)$, $f(2)$, and $f(3)$ and obtain the graph of f.

Solution with Technology

To create a table of values of *f* using Excel:

1. Set up two columns—one for the values of *x* and one for the values of *f*(*x*). Then enter the sequence of values 0, 1, 2, 3 in the *x* column as shown.

	A	B
1	x	f(x)
2	0	
3	1	
4	2	
5	3	

2. Now enter the formula for *f* in cell B2. Below are the technology formula for *f* and the version we use in Excel:

Technology formula	Excel formula
`-1*x^2+20*x+3`	`=-1*A2^2+20*A2+3`

Notice that we have preceded the Excel formula by an equal sign (=) and replaced each occurrence of *x* by the name of the cell holding the value of *x* (cell A2 in this case).

	A	B	C
1	x	f(x)	
2	0	=-1*A2^2+20*A2+3	
3	1		
4	2		
5	3		
6			

Note Instead of typing in the name of the cell "A2" each time, you can simply click on the cell A2, and "A2" will be inserted automatically. ■

3. Now highlight cell B2 and drag the **fill handle** (the little dot at the lower right-hand corner of the selection) down until you reach row 5 as shown below on the left, to obtain the result shown to the right.

	A	B	C
1	x	f(x)	
2	0	=-1*A2^2+20*A2+3	
3	1		
4	2		
5	3		

	A	B
1	x	f(x)
2	0	3
3	1	22
4	2	39
5	3	54

4. To graph the data, highlight A1 through B5, select the Insert tab, and choose a "Scatter" chart:

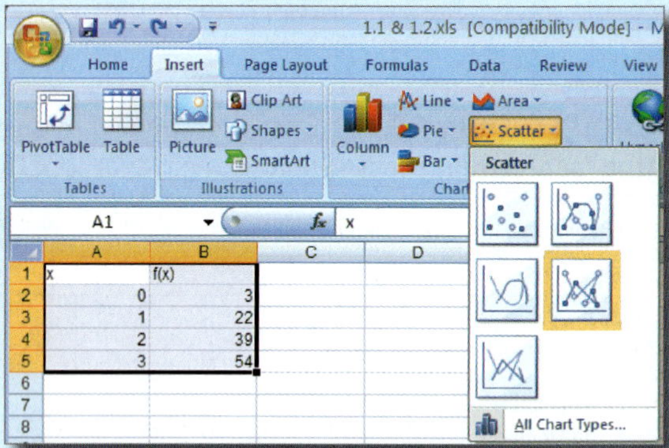

For a graph with data points connected by line segments, select the style shown on the right to obtain the graph shown below:

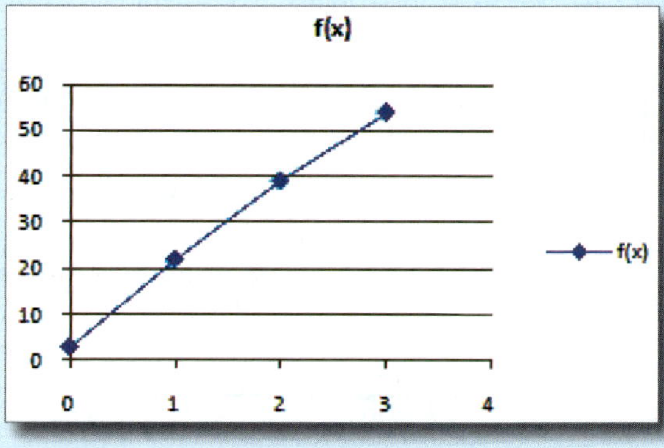

You can then adjust the appearance of the graph by right-clicking on its various elements. (Experiment!)

Q: *The formula* `-x^2+20*x+3` *will give the same result as* `-1*x^2+20*x+3` *on a graphing calculator (and the Web site). So why did we use* `-1*x^2` *in the Excel formula?*

A: Oddly enough, Excel calculates `-x^2` as though it were `(-x)^2`, which is the same as x^2, whereas calculators and computer programs all use the conventional mathematical interpretation: $-x^2$. To avoid having to worry about this, just remember to always use a coefficient in front of powers of *x*, even if the coefficient is −1.

Example 2 (page 47) The number $n(t)$ of Facebook members can be approximated by the following function of time t in years ($t = 0$ represents January 2004):

$$n(t) = \begin{cases} 1.5t^2 - t + 1.5 & \text{if } 0 \le t \le 3 \\ 46t - 126 & \text{if } 3 < t \le 5 \end{cases} \quad \text{million members}$$

(Its graph is shown in Figure 4.) Obtain a table showing the values $n(t)$ for $t = 0, 0.5, 1, \ldots, 5$ and also obtain the graph of n.

Solution with Technology

We can generate a table of values of $n(t)$ for $t = 0$, 0.5, 1, . . . , 5 as follows:

1. First set up two columns—one for the values of t and one for the values of $n(t)$. To enter the sequence of values 0, 0.5, 1, . . . , 5 in the t column, start by entering the first two values, 0 and 0.5, highlight both of them, and drag the **fill handle** (the little dot at the lower right-hand corner of the selection) down until you reach Row 12. (Why 12?)

	A	B
1	t	n(t)
2	0	
3	0.5	
4		
5		
6		
7		
8		
9		
10		
11		
12		

2. We must now enter the formula for n in cell B2. The following formula defines the function n in Excel:

```
=(x<=3)*(1.5*x^2-x+1.5)+
(x>3)*(46*x-126)
```

When x is less than or equal to 3, the logical expression (x<=3) evaluates to 1 because it is true, and the expression (x>3) evaluates to 0 because it is false. The value of the function is therefore given by the expression (1.5*x^2-x+1.5). When x is greater than 3, the expression (x<=3) evaluates to 0 while the expression (x>3) evaluates to 1, so the value of the function is given by the expression (46*x-126). We therefore enter the formula

```
=(A2<=3)*(1.5*A2^2-A2+1.5)
+(A2>3)*(46*A2-126)
```

in cell B2 and then copy down to cell B12:

	A	B	C	D	E	F
1	t	n(t)				
2	0	=(A2<=3)*(1.5*A2^2-A2+1.5)+(A2>3)*(46*A2-126)				
3	0.5					
4	1					
5	1.5					
6	2					
7	2.5					
8	3					
9	3.5					
10	4					
11	4.5					
12	5					

→

	A	B
1	t	n(t)
2	0	1.5
3	0.5	1.375
4	1	2
5	1.5	3.375
6	2	5.5
7	2.5	8.375
8	3	12
9	3.5	35
10	4	58
11	4.5	81
12	5	104

We can now read off the values asked for in the original example: Jan. 2005 ($t = 1$): $n(1) = 2$, Jan. 2007 ($t = 3$): $n(3) = 12$, June 2008 ($t = 4.5$): $n(4.5) = 81$.

3. To graph the data, highlight A1 through B10, select the Insert tab, and choose "Scatter" as in Example 1(c), obtaining the following graph:

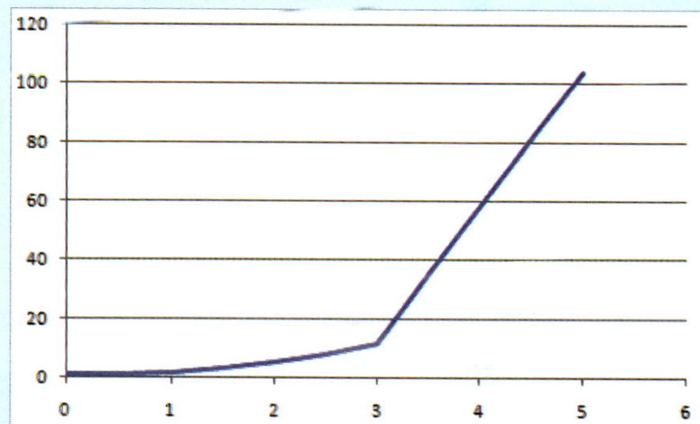

Example 3 (page 48) Obtain a table of values and the graph of the function f specified by

$$f(x) = \begin{cases} -1 & \text{if } -4 \le x < -1 \\ x & \text{if } -1 \le x \le 1 \\ x^2 - 1 & \text{if } 1 < x \le 2 \end{cases}.$$

Solution with Technology

The setup in Excel is almost identical to that in Example 2, but because the third part of the formula specifying f is not linear, we need to plot more points in Excel to get a smooth graph.

1. For the values of x in column A, use the values −4, −3.9, −3.8, . . . , 2. For a smoother curve, plot more points. You can find a general purpose Excel Graphing Worksheet that automates some of this on the Web site by following

Chapter 1 → Excel Tutorials → Section 1.1

TECHNOLOGY GUIDE

2. For $f(x)$, use the formula:

`=(A2<-1)*(-1)+(-1<=A2)*(A2<=1)*A2+(1<A2)*(A2^2-1)`

First part — Second part — Third part

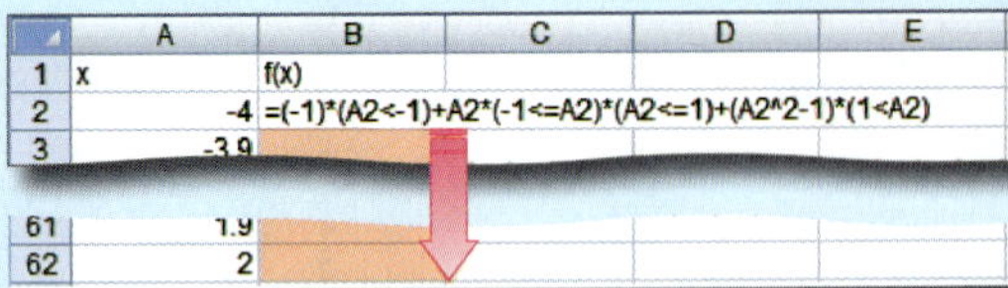

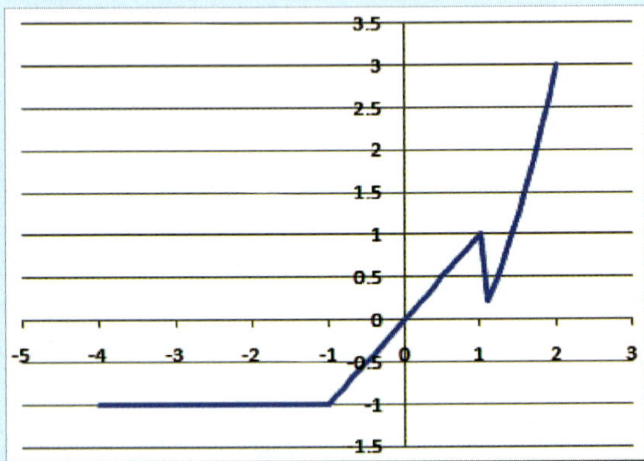

Notice that Excel does not handle the transition at $x = 1$ correctly and connects the two parts of the graph with a spurious line segment.

Section 1.2

Example 4(a) (page 62) The demand and supply curves for private schools in Michigan are $q = 77.8p^{-0.11}$ and $q = 30.4 + 0.006p$ thousand students respectively $(200 \leq p \leq 2{,}200)$ where p is the net tuition cost in dollars. Graph the demand and supply curves on the same set of axes. Use your graph to estimate, to the nearest \$100, the tuition at which the demand equals the supply (equilibrium price). Approximately how many students will be accommodated at that price?

Solution with Technology

To obtain the graphs of demand and supply:

1. Enter the headings p, Demand, and Supply in cells A1–C1 and the p-values 200, 300, . . . , 2,200 in A2–A22.
2. Now enter the formulas for the demand and supply functions in cells B2 and C2:

Demand: `=77.8*A2^(-0.11)` in cell B2

Supply: `=30.4+0.006*A2` in cell C2

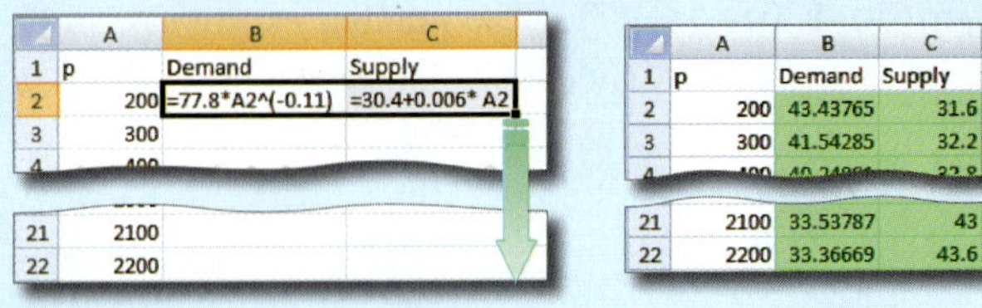

3. Now use the fill handle to copy down to C22 as shown above.
4. To graph the data, highlight A1 through C22, select the Insert tab, and choose a "Scatter" chart:

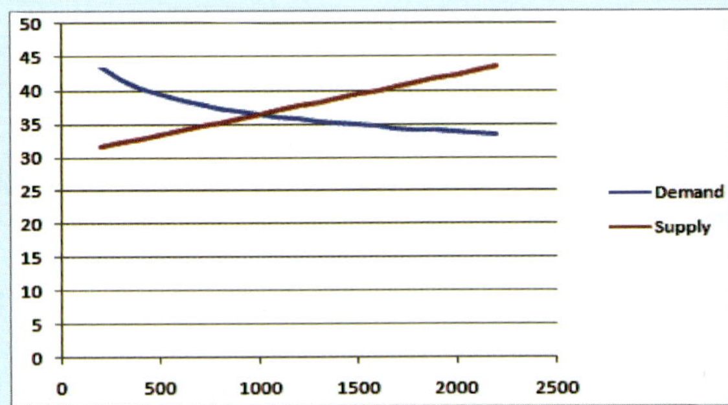

5. If you place the cursor as close as you can get to the intersection point (or just look at the table of values) you will see that the curves cross close to $p = \$1{,}000$ (to the nearest \$100).
6. To more accurately determine where the curves cross, you can narrow down the range of values shown on the x-axis by changing the p-values to 990, 991, …, 1010.

Example 5 (page 64) The following table shows annual slot machine and video poker machine revenues in 1996–2006 at the *Mohegan Sun* casino ($t = 0$ represents 2000).

t	0	1	2	3	4	5	6
Reveneue ($ million)	550	680	750	820	850	890	910

Consider the following four models:

(1) $r(t) = 57t + 607$ — Linear model

(2) $r(t) = -9t^2 + 110t + 560$ — Quadratic model

(3) $r(t) = 608(1.08)^t$ — Exponential model

(4) $r(t) = \dfrac{930}{1 + 0.67(1.7)^{-t}}$ — Logistic model

a. Which models fit the data significantly better than the rest?

b. Of the models you selected in part (a), which gives the most reasonable prediction for 2010?

Solution with Technology

1. First create a scatter plot of the given data by tabulating the data as shown below, selecting the Insert tab, and choosing a "Scatter" chart:

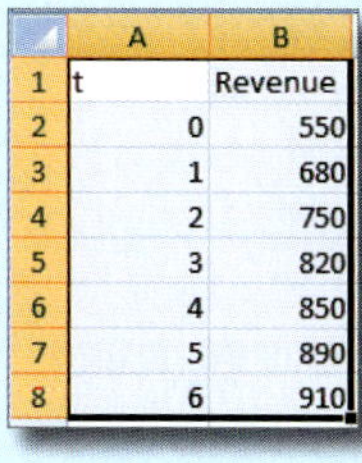

	A	B
1	t	Revenue
2	0	550
3	1	680
4	2	750
5	3	820
6	4	850
7	5	890
8	6	910

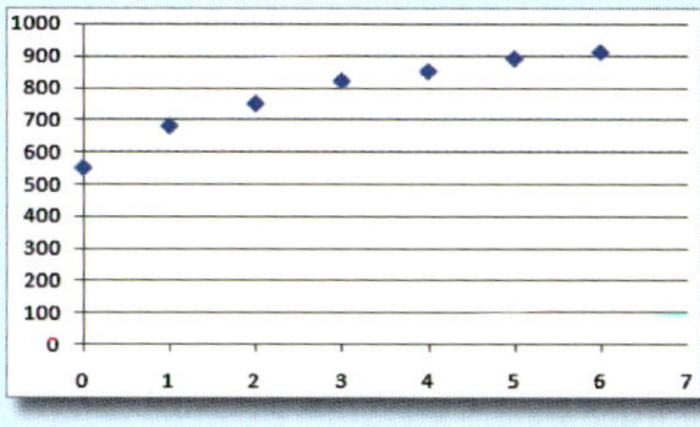

2. In column C use the formula for the model you are interested in seeing; for example, model (2): `=-9*A2^2+110*A2+560`.

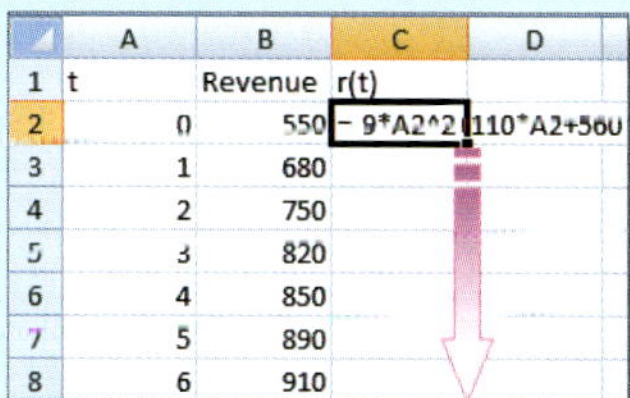

	A	B	C	D
1	t	Revenue	r(t)	
2	0	550	=-9*A2^2+110*A2+560	
3	1	680		
4	2	750		
5	3	820		
6	4	850		
7	5	890		
8	6	910		

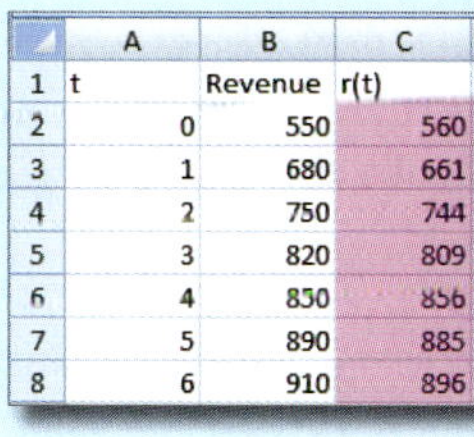

	A	B	C
1	t	Revenue	r(t)
2	0	550	560
3	1	680	661
4	2	750	744
5	3	820	809
6	4	850	856
7	5	890	885
8	6	910	896

3. To adjust the graph to include the graph of the model you have added, click once on the graph—the effect will be to outline the data you have graphed in columns A and B—and then use the fill handle at the bottom of column B to extend the selection to column C as shown:

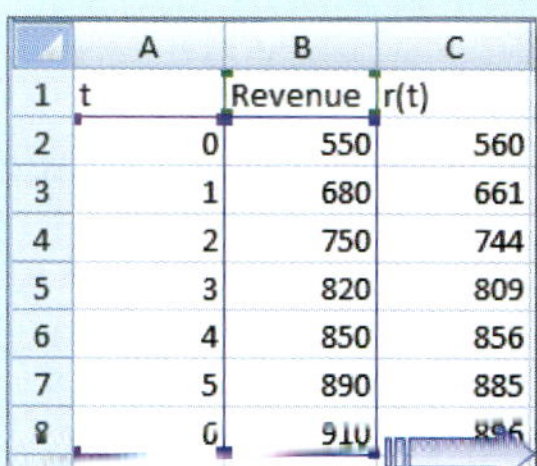

	A	B	C
1	t	Revenue	r(t)
2	0	550	560
3	1	680	661
4	2	750	744
5	3	820	809
6	4	850	856
7	5	890	885
8	6	910	896

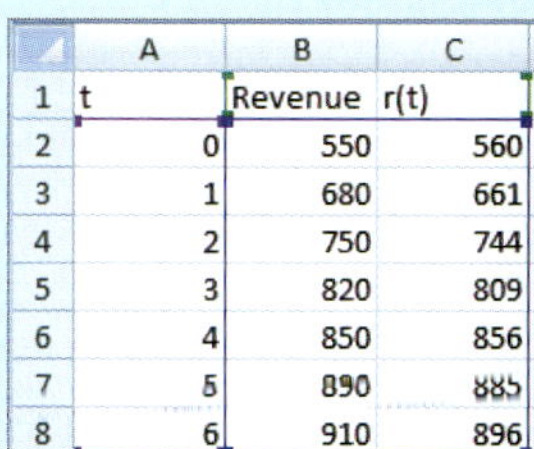

	A	B	C
1	t	Revenue	r(t)
2	0	550	560
3	1	680	661
4	2	750	744
5	3	820	809
6	4	850	856
7	5	890	885
8	6	910	896

The graph will now include markers showing the values of both the actual revenue and the model you inserted in column C.

4. Right-click on any of the markers corresponding to column B in the graph, select "Format data series" and, under Line Color select "Solid line," and under Marker Options, select "None." The effect will be as shown below, with the model represented by a curve, and the actual data points represented by dots:

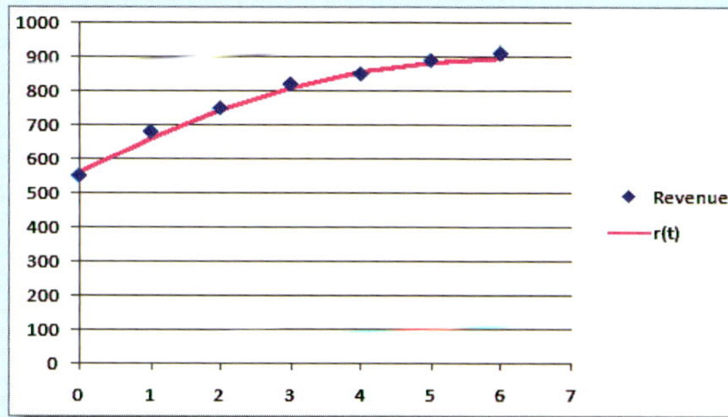

5. To see the extrapolation of the curve to 2010, add the values 7, 8, 9, 10 to Column A (notice that the values of $r(t)$ will automatically be computed in column C as you type), click on the graph, and use the fill handle at the base of column C to include the new data in the graph:

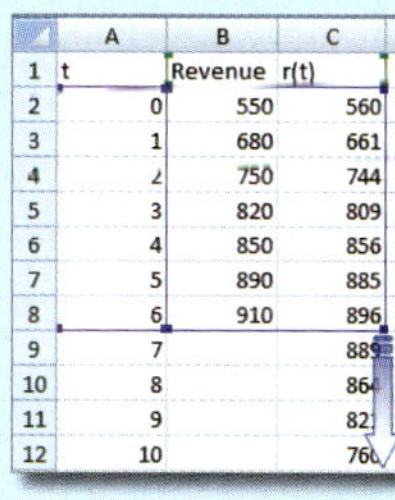

	A	B	C
1	t	Revenue	r(t)
2	0	550	560
3	1	680	661
4	2	750	744
5	3	820	809
6	4	850	856
7	5	890	885
8	6	910	896
9	7		889
10	8		864
11	9		82
12	10		760

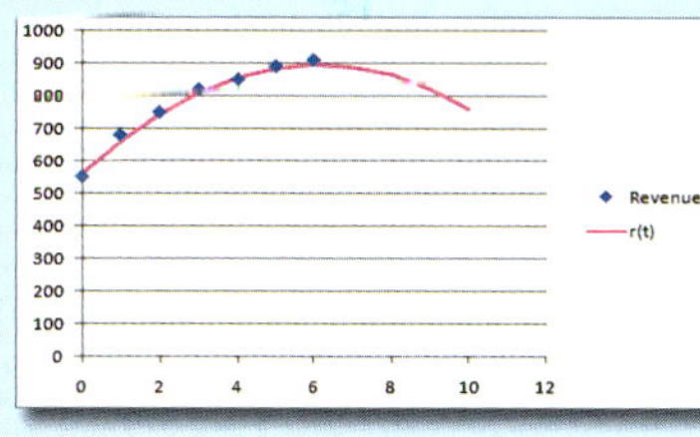

6. To see the plots for the remaining curves, change the formula in column B (and don't forget to copy the new formula down to cell C12 when you do so).

Section 1.3

Example 1 (page 78) Which of the following two tables gives the values of a linear function? What is the formula for that function?

x	0	2	4	6	8	10	12
$f(x)$	3	−1	−3	−6	−8	−13	−15

x	0	2	4	6	8	10	12
$g(x)$	3	−1	−5	−9	−13	−17	−21

Solution with Technology

The following worksheet shows how you can compute the successive quotients $m = \Delta y/\Delta x$, and hence check whether a given set of data shows a linear relationship, in which case all the quotients will be the same. (The shading indicates that the formula is to be copied down only as far as cell C7. Why not cell C8?)

	A	B	C
1	x	f(x)	m
2	0	3	=(B3-B2)/(A3-A2)
3	2	-1	
4	4	-3	
5	6	-6	
6	8	-8	
7	10	-13	
8	12	-15	

Here are the results for both $f(x)$ and $g(x)$:

	A	B	C
1	x	f(x)	m
2	0	3	-2
3	2	-1	-1
4	4	-3	-1.5
5	6	-6	-1
6	8	-8	-2.5
7	10	-13	-1
8	12	-15	

	A	B	C
1	x	g(x)	m
2	0	3	-2
3	2	-1	-2
4	4	-5	-2
5	6	-9	-2
6	8	-13	-2
7	10	-17	-2
8	12	-21	

Example 2(a) (page 81) Find the equation of the line through the points (1, 2) and (3, −1).

Solution with Technology

1. Enter the x- and y-coordinates in columns A and B as shown below on the left.

	A	B
1	x	y
2	1	2
3	3	-1

	A	B	C	D
1	x	y	m	b
2	1	2	=(B3-B2)/(A3-A2)	=B2-C2*A2
3	3	-1	Slope	Intercept

2. Add the headings m and b in C1–D1, and then the formulas for the slope and intercept in C2–D2 as shown above on the right. The result will be as shown below:

	A	B	C	D
1	x	y	m	b
2	1	2	-1.5	3.5
3	3	-1	Slope	Intercept

Section 1.4

Example 1 (page 96) Using the data on the GDP in Mexico given on page 96, compute SSE, the sum-of-squares error, for the linear models $y = 0.5t + 8$ and $y = 0.25t + 9$, and graph the data with the given models.

Solution with Technology

1. Begin by setting up your worksheet with the observed data in two columns, t and y, and the predicted data for the first model in the third.

	A	B	C	D	E	F
1	t	y (Observed)	y (Predicted)		m	b
2	0	9	=$E$2*A2+$F$2		0.5	8
3	2	9				
4	4	10				
5	6	11				
6	8	11				
7	10	12				
8	12	13				
9	14	13				

Notice that, instead of using the numerical equation for the first model in column C, we used absolute references to the cells containing the slope m and the intercept b. This way, we can switch from one linear model to the next by changing only m and b in cells E2 and F2. (We have deliberately left column D empty in anticipation of the next step.)

2. In column D we compute the squares of the residuals using the Excel formula:

```
=(B2-C2)^2.
```

	A	B	C	D	E	F
1	t	y (Observed)	y (Predicted)	Residual^2	m	b
2	0	9	8	=(B2-C2)^2	0.5	8
3	2	9	9			
4	4	10	10			
5	6	11	11			
6	8	11	12			
7	10	12	13			
8	12	13	14			
9	14	13	15			

3. We now compute SSE in cell F4 by summing the entries in column D:

	A	B	C	D	E	F
1	t	y (Observed)	y (Predicted)	Residual^2	m	b
2	0	9	8	1	0.5	8
3	2	9	9	0		
4	4	10	10	0	SSE:	=SUM(D2:D9)
5	6	11	11	0		
6	8	11	12	1		
7	10	12	13	1		
8	12	13	14	1		
9	14	13	15	4		

Here is the completed worksheet:

	A	B	C	D	E	F
1	t	y (Observed)	y (Predicted)	Residual^2	m	b
2	0	9	8	1	0.5	8
3	2	9	9	0		
4	4	10	10	0	SSE:	8
5	6	11	11	0		
6	8	11	12	1		
7	10	12	13	1		
8	12	13	14	1		
9	14	13	15	4		

4. Changing m to 0.25 and b to 9 gives the sum of squares error for the second model, SSE = 2.

	A	B	C	D	E	F
1	t	y (Observed)	y (Predicted)	Residual^2	m	b
2	0	9	9	0	0.25	9
3	2	9	9.5	0.25		
4	4	10	10	0	SSE:	2
5	6	11	10.5	0.25		
6	8	11	11	0		
7	10	12	11.5	0.25		
8	12	13	12	1		
9	14	13	12.5	0.25		

Thus, the second model is a better fit.

5. To plot both the original data points and each of the two lines, use a scatterplot to graph the data in columns A through C in each of the previous last two worksheets.

$y = 0.5t + 8$ $\qquad\qquad$ $y = 0.25t + 9$

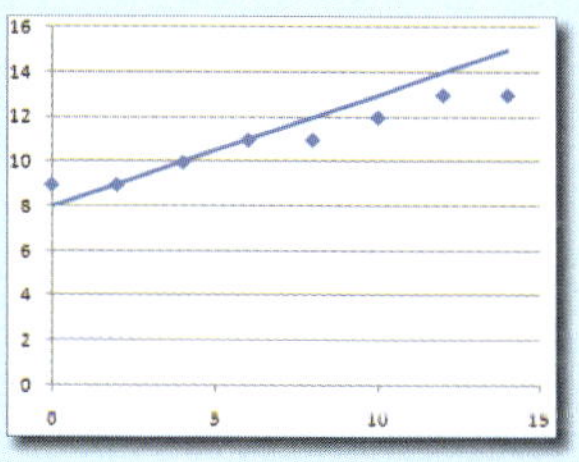

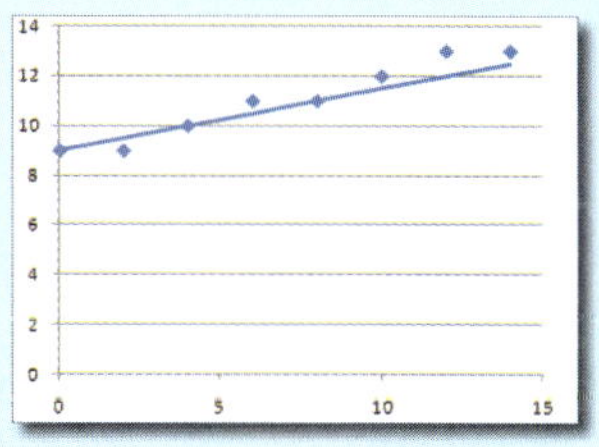

Example 2 (page 98) Consider the following data on the per capita gross domestic product (GDP) of Mexico:

Year x ($x = 0$ represents 2000)	0	2	4	6	8	10	12	14
Per Capita GDP y ($1,000)	9	9	10	11	11	12	13	13

Find the best-fit linear model for these data.

Solution with Technology

Here are two Excel shortcuts for linear regression; one graphical and one based on an Excel formula.

Using the Trendline

1. Start with the original data and a "scatter plot."

	A	B
1	x	y
2	0	9
3	2	9
4	4	10
5	6	11
6	8	11
7	10	12
8	12	13
9	14	13

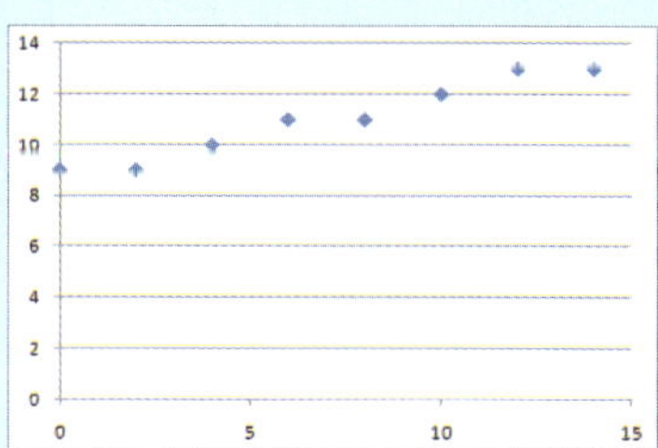

2. Click on the chart, select the Layout tab, click the Trendline button, and choose "More Trendline Options." In the Format Trendline dialogue box, select a Linear trendline (the default) and check the option to "Display Equation on chart."

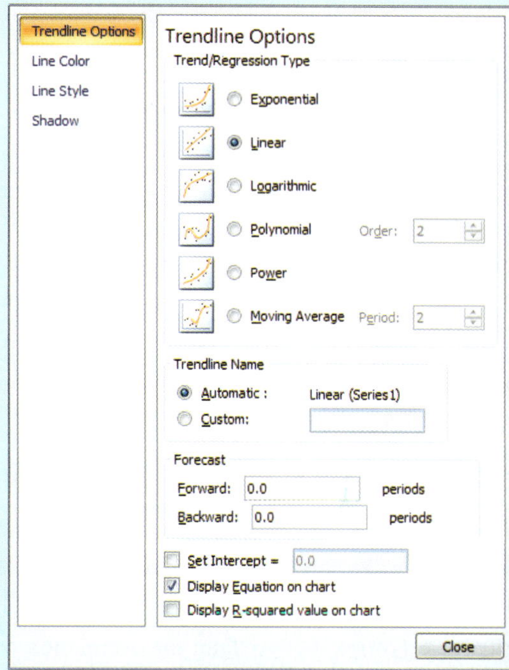

You may have to move the equation to make it readable.

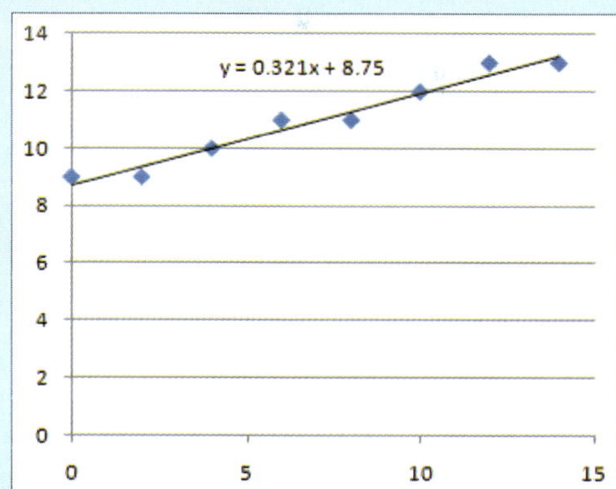

Using a Formula

Alternatively, you can use the "LINEST" function (for "linear estimate").

1. Enter your data as previously, and select a block of unused cells two wide and one tall; for example C2:D2. Then enter the formula

```
=LINEST(B2:B9,A2:A9)
```

as shown.

	A	B	C	D
1	x	y	m	b
2	0	9	=LINEST(B2:B9,A2:A9)	
3	2	9		
4	4	10		
5	6	11		
6	8	11		
7	10	12		
8	12	13		
9	14	13		

2. Press Control-Shift-Enter. The result should look like this.

	A	B	C	D
1	x	y	m	b
2	0	9	0.321429	8.75
3	2	9		
4	4	10		
5	6	11		
6	8	11		
7	10	12		
8	12	13		
9	14	13		

The values of m and b appear in cells C2 and D2 as shown.

Example 3 (page 100) Find the correlation coefficient for the data in Example 2.

Solution with Technology

1. In Excel, when you add a trendline to a chart you can select the option "Display R-squared value on chart" to show the value of r^2 on the chart (it is common to examine r^2, which takes on values between 0 and 1, instead of r).

2. Alternatively, the LINEST function we used in Example 2 can be used to display quite a few statistics about a best fit line, including r^2. Instead of selecting a block of cells two wide and one tall as we did in Example 2, we select one two wide and *five* tall. We now enter the requisite LINEST formula with two additional arguments set to "TRUE" as shown.

	A	B	C	D	E
1	x	y	m	b	
2	0	9	=LINEST(B2:B9,A2:A9,TRUE,TRUE)		
3	2	9			
4	4	10			
5	6	11			
6	8	11			
7	10	12			
8	12	13			
9	14	13			
10					

3. Press Control-Shift-Enter. The result should look something like this.

	A	B	C	D
1	x	y	m	b
2	0	9	0.321428571	8.75
3	2	9	0.025253814	0.211288564
4	4	10	0.964285714	0.327326835
5	6	11	162	6
6	8	11	17.35714286	0.642857143
7	10	12		
8	12	13		
9	14	13		

The values of m and b appear in cells C2 and D2 as before, and the value of r^2 in cell C4. (Among the other numbers shown is SSE in cell D6. For the meanings of the remaining numbers shown, see the online help for LINEST in Excel; a good course in statistics wouldn't hurt, either.)

2 Nonlinear Functions and Models

Web Site
www.AppliedCalc.org

- Inverse functions
- Using and deriving algebraic properties of logarithms

Case Study Checking up on Malthus

In 1798 Thomas R. Malthus (1766–1834) published an influential pamphlet, later expanded into a book, titled *An Essay on the Principle of Population as It Affects the Future Improvement of Society*. One of his main contentions was that population grows geometrically (exponentially), while the supply of resources such as food grows only arithmetically (linearly). Some 200 years later, you have been asked to check the validity of Malthus's contention. **How do you go about doing so?**

Robert Nickelsberg/Getty Images

Introduction

To see if Malthus was right, we need to see if the data fit the models (linear and exponential) that he suggested or if other models would be better. We saw in Chapter 1 how to fit a linear model. In this chapter we discuss how to construct models that use various *nonlinear* functions.

The nonlinear functions we consider in this chapter are the *quadratic* functions, the simplest nonlinear functions; the *exponential* functions, essential for discussing many kinds of growth and decay, including the growth (and decay) of money in finance and the initial growth of an epidemic; the *logarithmic* functions, needed to fully understand the exponential functions; and the *logistic* functions, used to model growth with an upper limit, such as the spread of an epidemic.

algebra Review
For this chapter, you should be familiar with the algebra reviewed in **Chapter 0, Section 2.**

2.1 Quadratic Functions and Models

In Chapter 1 we studied linear functions. Linear functions are useful, but in real-life applications, they are often accurate for only a limited range of values of the variables. The relationship between two quantities is often best modeled by a curved line rather than a straight line. The simplest function with a graph that is not a straight line is a *quadratic* function.

Quadratic Function

A **quadratic function** of the variable x is a function that can be written in the form

$$f(x) = ax^2 + bx + c \qquad \text{Function form}$$

or

$$y = ax^2 + bx + c \qquad \text{Equation form}$$

where a, b, and c are fixed numbers (with $a \neq 0$).

Quick Examples

1. $f(x) = 3x^2 - 2x + 1$ — $a = 3, b = -2, c = 1$
2. $g(x) = -x^2$ — $a = -1, b = 0, c = 0$
3. $R(p) = -5{,}600p^2 + 14{,}000p$ — $a = -5{,}600, b = 14{,}000, c = 0$

Every quadratic function $f(x) = ax^2 + bx + c$ $(a \neq 0)$ has a **parabola** as its graph. Following is a summary of some features of parabolas that we can use to sketch the graph of any quadratic function.*

* **NOTE** We shall not fully justify the formula for the vertex and the axis of symmetry until we have studied some calculus, although it is possible to do so with just algebra.

Features of a Parabola

The graph of $f(x) = ax^2 + bx + c$ $(a \neq 0)$ is a **parabola**. If $a > 0$ the parabola opens upward (concave up) and if $a < 0$ it opens downward (concave down):

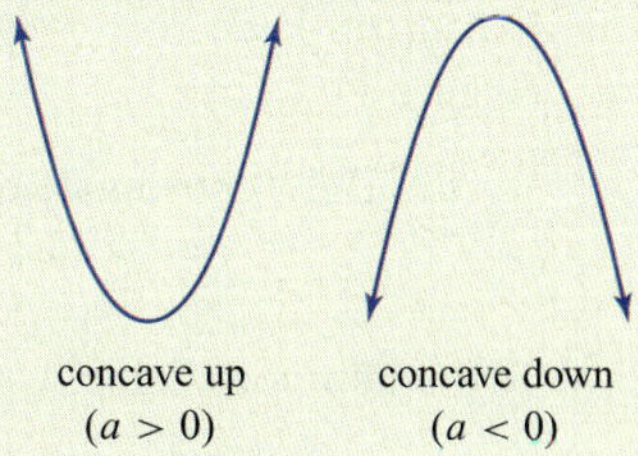

Vertex, Intercepts, and Symmetry

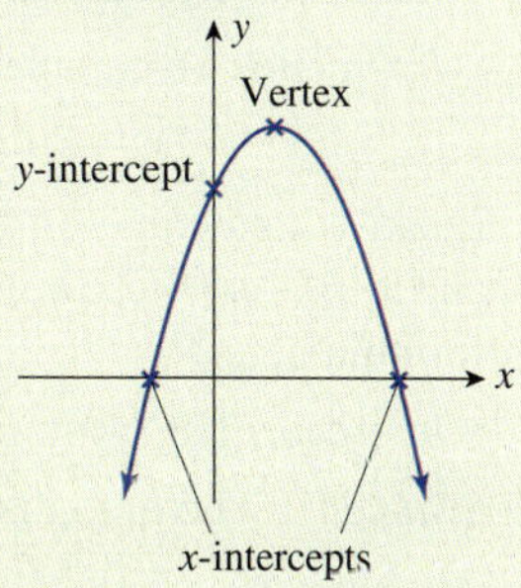

Vertex The vertex is the turning point of the parabola (see the above figure). Its x-coordinate is $-\dfrac{b}{2a}$. Its y-coordinate is $f\left(-\dfrac{b}{2a}\right)$.

x-Intercepts (if any) These occur when $f(x) = 0$; that is, when

$$ax^2 + bx + c = 0.$$

Solve this equation for x by either factoring or using the quadratic formula. The x-intercepts are

$$x = \frac{-b \pm \sqrt{b^2 - 4ac}}{2a}.$$

If the **discriminant** $b^2 - 4ac$ is positive, there are two x-intercepts. If it is zero, there is a single x-intercept (at the vertex). If it is negative, there are no x-intercepts (so the parabola doesn't touch the x-axis at all).

y-Intercept This occurs when $x = 0$, so

$$y = a(0)^2 + b(0) + c = c.$$

Symmetry The parabola is symmetric with respect to the vertical line through the vertex, which is the line $x = -\dfrac{b}{2a}$.

Note that the x-intercepts can also be written as

$$x = -\frac{b}{2a} \pm \frac{\sqrt{b^2 - 4ac}}{2a},$$

making it clear that they are located symmetrically on either side of the line $x = -b/(2a)$. This partially justifies the claim that the whole parabola is symmetric with respect to this line.

Figure 1

EXAMPLE 1 Sketching the Graph of a Quadratic Function

Sketch the graph of $f(x) = x^2 + 2x - 8$ by hand.

Solution Here, $a = 1$, $b = 2$, and $c = -8$. Because $a > 0$, the parabola is concave up (Figure 1).

Vertex: The x coordinate of the vertex is

$$x = -\frac{b}{2a} = -\frac{2}{2} = -1.$$

To get its y coordinate, we substitute the value of x back into $f(x)$ to get

$$y = f(-1) = (-1)^2 + 2(-1) - 8 = 1 - 2 - 8 = -9.$$

Thus, the coordinates of the vertex are $(-1, -9)$.

x-Intercepts: To calculate the x-intercepts (if any), we solve the equation

$$x^2 + 2x - 8 = 0.$$

Luckily, this equation factors as $(x + 4)(x - 2) = 0$. Thus, the solutions are $x = -4$ and $x = 2$, so these values are the x-intercepts. (We could also have used the quadratic formula here.)

y-Intercept: The y-intercept is given by $c = -8$.

Symmetry: The graph is symmetric around the vertical line $x = -1$.

Now we can sketch the curve as in Figure 2. (As we see in the figure, it is helpful to plot additional points by using the equation $y = x^2 + 2x - 8$, and to use symmetry to obtain others.)

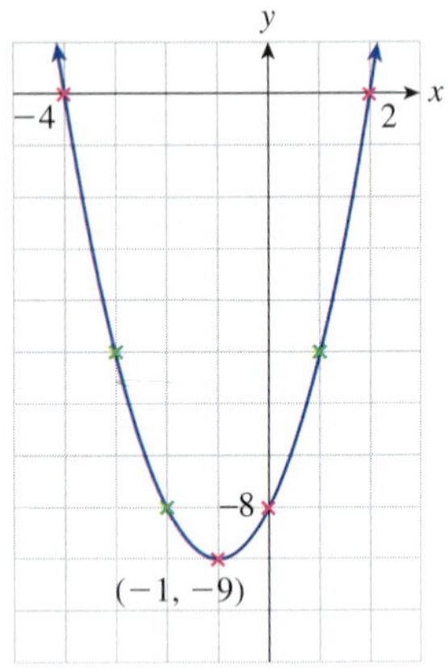

$y = x^2 + 2x - 8$

Figure 2

EXAMPLE 2 One x-Intercept and No x-Intercepts

Sketch the graph of each quadratic function, showing the location of the vertex and intercepts.

a. $f(x) = 4x^2 - 12x + 9$

b. $g(x) = -\frac{1}{2}x^2 + 4x - 12$

Solution

a. We have $a = 4$, $b = -12$, and $c = 9$. Because $a > 0$, this parabola is concave up.

Vertex: $x = -\frac{b}{2a} = \frac{12}{8} = \frac{3}{2}$ — *x coordinate of vertex*

$$y = f\left(\frac{3}{2}\right) = 4\left(\frac{3}{2}\right)^2 - 12\left(\frac{3}{2}\right) + 9 = 0$$ — *y coordinate of vertex*

Thus, the vertex is at the point $(3/2, 0)$.

x-Intercepts:

$$4x^2 - 12x + 9 = 0$$
$$(2x - 3)^2 = 0$$

The only solution is $2x - 3 = 0$, or $x = 3/2$. Note that this coincides with the vertex, which lies on the x-axis.

y-Intercept: $c = 9$

Symmetry: The graph is symmetric around the vertical line $x = 3/2$.

The graph is the narrow parabola shown in Figure 3. (As we remarked in Example 1, plotting additional points and using symmetry helps us obtain an accurate sketch.)

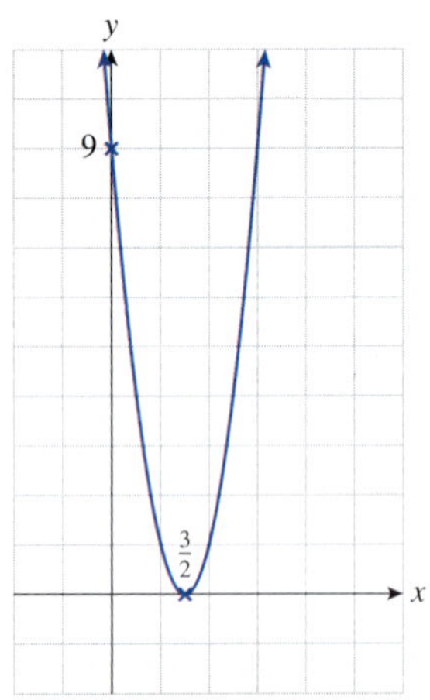

$y = 4x^2 - 12x + 9$

Figure 3

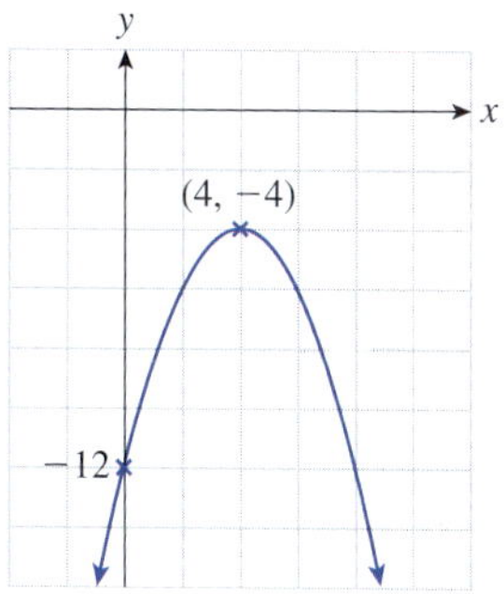

$y = -\frac{1}{2}x^2 + 4x - 12$

Figure 4

b. Here, $a = -1/2$, $b = 4$, and $c = -12$. Because $a < 0$, the parabola is concave down. The vertex has x coordinate $-b/(2a) = 4$, with corresponding y coordinate $f(4) = -\frac{1}{2}(4)^2 + 4(4) - 12 = -4$. Thus, the vertex is at $(4, -4)$.

For the x-intercepts, we must solve $-\frac{1}{2}x^2 + 4x - 12 = 0$. If we try to use the quadratic formula, we discover that the discriminant is $b^2 - 4ac = 16 - 24 = -8$. Because the discriminant is negative, there are no solutions of the equation, so there are no x-intercepts.

The y-intercept is given by $c = -12$, and the graph is symmetric around the vertical line $x = 4$.

Because there are no x-intercepts, the graph lies entirely below the x-axis, as shown in Figure 4. (Again, you should plot additional points and use symmetry to ensure that your sketch is accurate.)

using Technology

See the Technology Guide at the end of the chapter for detailed instructions on how to do the calculations and the graph in part (a) of Example 2 using a TI-83/84 Plus or Excel.

APPLICATIONS

Recall that the **revenue** resulting from one or more business transactions is the total payment received. Thus, if q units of some item are sold at p dollars per unit, the revenue resulting from the sale is

$$\text{revenue} = \text{price} \times \text{quantity}$$
$$R = pq.$$

EXAMPLE 3 Demand and Revenue

Alien Publications, Inc. predicts that the demand equation for the sale of its latest illustrated sci-fi novel *Episode 93: Yoda vs. Alien* is

$$q = -2{,}000p + 150{,}000$$

where q is the number of books it can sell each year at a price of $\$p$ per book. What price should Alien Publications, Inc., charge to obtain the maximum annual revenue?

Solution The total revenue depends on the price, as follows:

$$\begin{aligned} R &= pq && \text{Formula for revenue.} \\ &= p(-2{,}000p + 150{,}000) && \text{Substitute for } q \text{ from demand equation.} \\ &= -2{,}000p^2 + 150{,}000p. && \text{Simplify.} \end{aligned}$$

We are after the price p that gives the maximum possible revenue. Notice that what we have is a quadratic function of the form $R(p) = ap^2 + bp + c$, where $a = -2{,}000$, $b = 150{,}000$, and $c = 0$. Because a is negative, the graph of the function is a parabola, concave down, so its vertex is its highest point (Figure 5). The p coordinate of the vertex is

$$p = -\frac{b}{2a} = -\frac{150{,}000}{-4{,}000} = 37.5.$$

This value of p gives the highest point on the graph and thus gives the largest value of $R(p)$. We may conclude that Alien Publications, Inc., should charge \$37.50 per book to maximize its annual revenue.

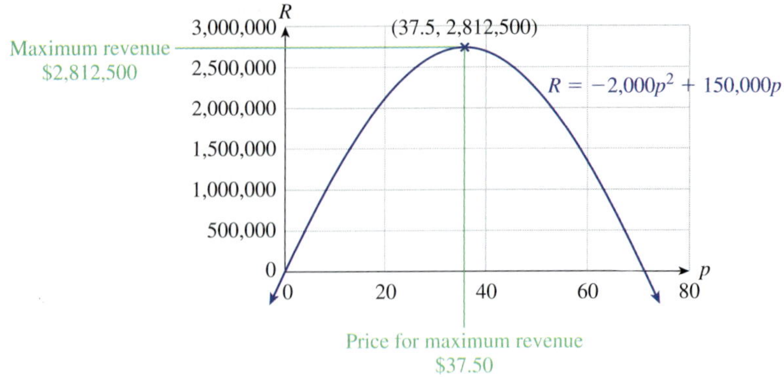

Figure 5

➡ **Before we go on...** You might ask what the maximum annual revenue is for the publisher in Example 3. Because $R(p)$ gives us the revenue at a price of $\$p$, the answer is $R(37.5) = -2{,}000\,(37.5)^2 + 150{,}000(37.5) = 2{,}812{,}500$. In other words, the company will earn total annual revenues from this book amounting to \$2,812,500. ■

EXAMPLE 4 Demand, Revenue, and Profit

As the operator of *YSport Fitness* gym, you calculate your demand equation to be

$$q = -0.06p + 84$$

where q is the number of members in the club and p is the annual membership fee you charge.

a. Your annual operating costs are a fixed cost of \$20,000 per year plus a variable cost of \$20 per member. Find the annual revenue and profit as functions of the membership price p.

b. At what price should you set the annual membership fee to obtain the maximum revenue? What is the maximum possible revenue?

c. At what price should you set the annual membership fee to obtain the maximum profit? What is the maximum possible profit? What is the corresponding revenue?

Solution

a. The annual revenue is given by

$$\begin{aligned} R &= pq && \text{Formula for revenue.} \\ &= p(-0.06p + 84) && \text{Substitute for } q \text{ from demand equation.} \\ &= -0.06p^2 + 84p. && \text{Simplify.} \end{aligned}$$

The annual cost C is given by

$$C = 20{,}000 + 20q. \qquad \text{\$20,000 plus \$20 per member}$$

However, this is a function of q, and not p. To express C as a function of p we substitute for q using the demand equation $q = -0.06p + 84$:

$$\begin{aligned} C &= 20{,}000 + 20(-0.06p + 84) \\ &= 20{,}000 - 1.2p + 1{,}680 \\ &= -1.2p + 21{,}680. \end{aligned}$$

Thus, the profit function is:

$$\begin{aligned} P &= R - C && \text{Formula for profit.} \\ &= -0.06p^2 + 84p - (-1.2p + 21{,}680) && \text{Substitute for revenue and cost.} \\ &= -0.06p^2 + 85.2p - 21{,}680. \end{aligned}$$

b. From part (a) the revenue function is given by

$$R = -0.06p^2 + 84p.$$

This is a quadratic function ($a = -0.06$, $b = 84$, $c = 0$) whose graph is a concave-down parabola (Figure 6).

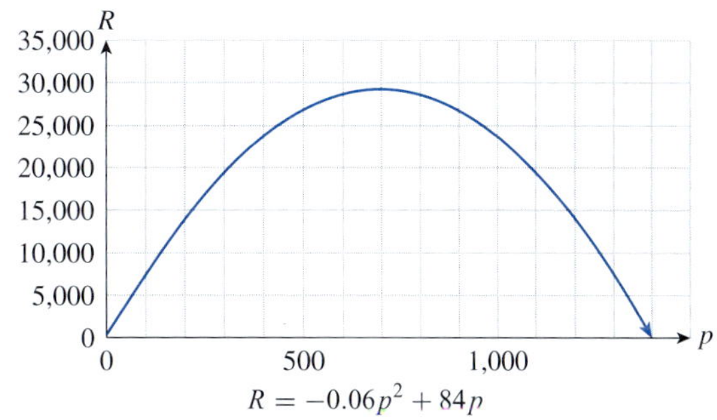

Figure 6

The maximum revenue corresponds to the highest point of the graph: the vertex, of which the p coordinate is

$$p = -\frac{b}{2a} = -\frac{84}{2(-0.06)} \approx \$700.$$

This is the membership fee you should charge for the maximum revenue. The corresponding maximum revenue is given by the y coordinate of the vertex in Figure 6:

$$R(700) = -0.06(700)^2 + 84(700) = \$29{,}400.$$

c. From part (a), the profit function is given by

$$P = -0.06p^2 + 85.2p - 21{,}680.$$

Like the revenue function, the profit function is quadratic ($a = -0.06$, $b = 85.2$, $c = -21{,}680$). Figure 7 shows both the revenue and profit functions

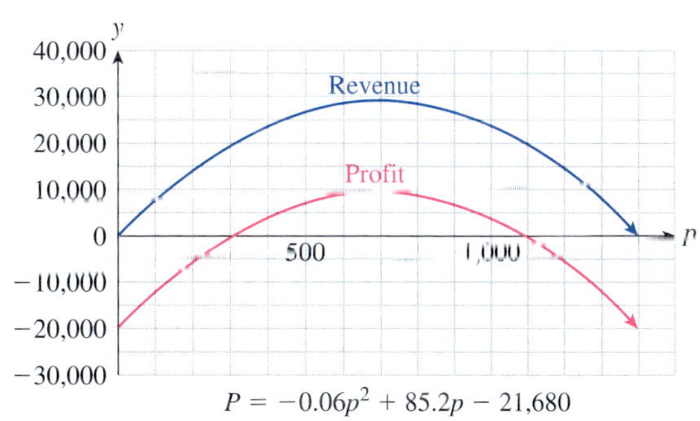

Figure 7

The maximum profit corresponds to the vertex, whose p coordinate is

$$p = -\frac{b}{2a} = -\frac{85.2}{2(-0.06)} \approx \$710.$$

This is the membership fee you should charge for the maximum profit. The corresponding maximum profit is given by the y coordinate of the vertex of the profit curve in Figure 7:

$$P(710) = -0.06(710)^2 + 85.2(710) - 21{,}680 = \$8{,}566.$$

The corresponding revenue is

$$R(710) = -0.06(710)^2 + 84(710) = \$29{,}394,$$

slightly less than the maximum possible revenue of $29,400.

➡ **Before we go on...** The result of part (c) of Example 4 tells us that the vertex of the profit curve in Figure 7 is slightly to the right of the vertex in the revenue curve. However, the difference is tiny compared to the scale of the graphs, so the graphs appear to be parallel. ■

Q: *Charging $710 membership brings in less revenue than charging $700. So why charge $710?*

A: A membership fee of $700 does bring in slightly larger revenue than a fee of $710, but it also brings in a slightly larger membership, which in turn raises the operating expense and has the effect of *lowering* the profit slightly (to $8,560). In other words, the slightly higher fee, while bringing in less revenue, also lowers the cost, and the net result is a larger profit.

Fitting a Quadratic Function to Data: Quadratic Regression

In Section 1.4 we saw how to fit a regression line to a collection of data points. Here, we see how to use technology to obtain the **quadratic regression curve** associated with a set of points. The quadratic regression curve is the quadratic curve $y = ax^2 + bx + c$ that best fits the data points in the sense that the associated sum-of-squares error (SSE—see Section 1.4) is a minimum. Although there are algebraic methods for obtaining the quadratic regression curve, it is normal to use technology to do this.

EXAMPLE 5 T Freon Production

The following table shows total and projected production of ozone-layer damaging Freon 22 (chlorodifluoromethane) in developing countries ($t = 0$ represents 2000).*

Year t	0	2	4	6	8	10
Tons of Freon F	100	140	200	270	400	590

a. Is a linear model appropriate for these data?

b. Find the quadratic model

$$F(t) = at^2 + bt + c$$

that best fits the data.

* Figures are approximate. Source: Lampert Kuijpers (Panel of the Montreal Protocol), National Bureau of Statistics in China, via CEIC DSata/*New York Times*, February 23, 2007, p. C1.

using Technology

See the Technology Guide at the end of the chapter for detailed instructions on how to obtain the regression curve and graph for Example 5 using a TI-83/84 Plus or Excel. Here is an outline:

TI-83/84 Plus

STAT EDIT values of t in L_1 and values of F in L_2

Regression curve: STAT CALC option #5 QuadReg ENTER

Graph: Y= VARS 5 EQ 1, then ZOOM 9

[More details on page 185.]

Excel

t- and F-values in Columns A and B

Graph: Highlight data and insert a Scatter chart.

Regression curve: Right-click a datapoint and add polynomial order 2 trendline with option to show equation on chart.

[More details on page 188.]

Web Site

www.AppliedCalc.org

Student Home

→ Online Utilities

→ Simple Regression

Enter the data in the x and y columns and press "y=ax^2+bx+c".

Solution

a. To see whether a linear model is appropriate, we plot the data points and the regression line using one of the methods of Example 2 in Section 1.4 (Figure 8).

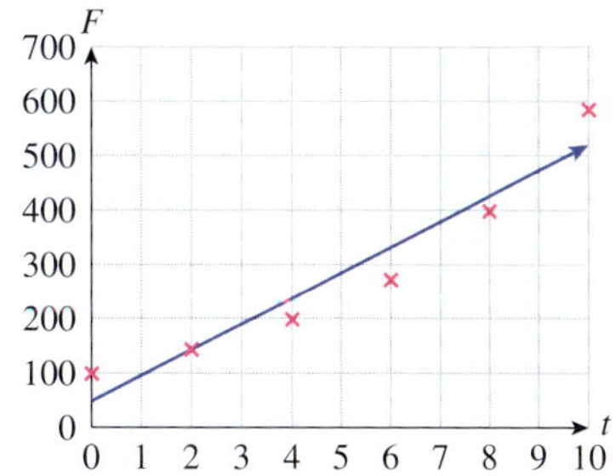

Figure 8

From the graph, we can see that the given data suggest a curve and not a straight line: The observed points are above the regression line at the ends but below in the middle. (We would expect the data points from a linear relation to fall randomly above and below the regression line.)

b. The quadratic model that best fits the data is the quadratic regression model. As with linear regression, there are algebraic formulas to compute a, b, and c, but they are rather involved. However, we exploit the fact that these formulas are built into graphing calculators, spreadsheets, and other technology, and obtain the regression curve using technology (see Figure 9):

$$F(t) = 4.6t^2 + 1.2t + 110 \quad \text{Coefficients rounded to two significant digits}$$

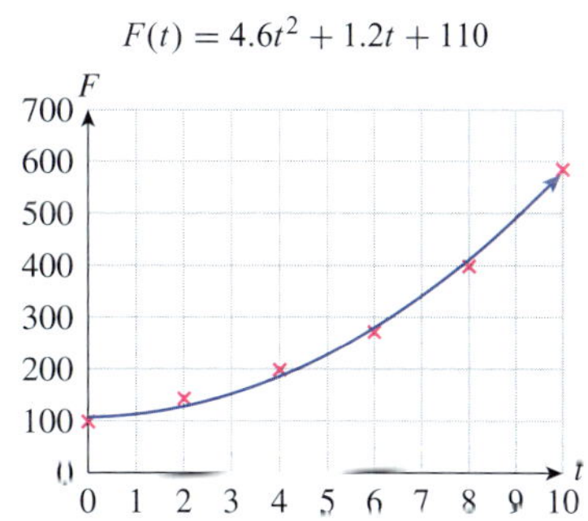

Figure 9

Notice from the previous graphs that the quadratic regression model appears to give a far better fit than the linear regression model. This impression is supported by the values of SSE: For the linear regression model SSE ≈ 13,400; for the quadratic regression model SSE is much smaller, approximately 730, indicating a much better fit.

2.1 EXERCISES

▼ more advanced ◆ challenging

T indicates exercises that should be solved using technology

In Exercises 1–10, sketch the graphs of the quadratic functions, indicating the coordinates of the vertex, the y-intercept, and the x-intercepts (if any). HINT [See Example 1.]

1. $f(x) = x^2 + 3x + 2$ **2.** $f(x) = -x^2 - x$

3. $f(x) = -x^2 + 4x - 4$ **4.** $f(x) = x^2 + 2x + 1$

5. $f(x) = -x^2 - 40x + 500$ **6.** $f(x) = x^2 - 10x - 600$

7. $f(x) = x^2 + x - 1$ **8.** $f(x) = x^2 + \sqrt{2}x + 1$

9. $f(x) = x^2 + 1$ **10.** $f(x) = -x^2 + 5$

In Exercises 11–14, for each demand equation, express the total revenue R as a function of the price p per item, sketch the graph of the resulting function, and determine the price p that maximizes total revenue in each case. HINT [See Example 3.]

11. $q = -4p + 100$ **12.** $q = -3p + 300$

13. $q = -2p + 400$ **14.** $q = -5p + 1{,}200$

T *In Exercises 15–18, use technology to find the quadratic regression curve through the given points. (Round all coefficients to four decimal places.)* HINT [See Example 5.]

15. $\{(1, 2), (3, 5), (4, 3), (5, 1)\}$

16. $\{(-1, 2), (-3, 5), (-4, 3), (-5, 1)\}$

17. $\{(-1, 2), (-3, 5), (-4, 3)\}$

18. $\{(2, 5), (3, 5), (5, 3)\}$

APPLICATIONS

19. ***World Military Expenditure*** The following chart shows total military and arms trade expenditure from 1990 to 2008 ($t = 0$ represents 1990).[1]

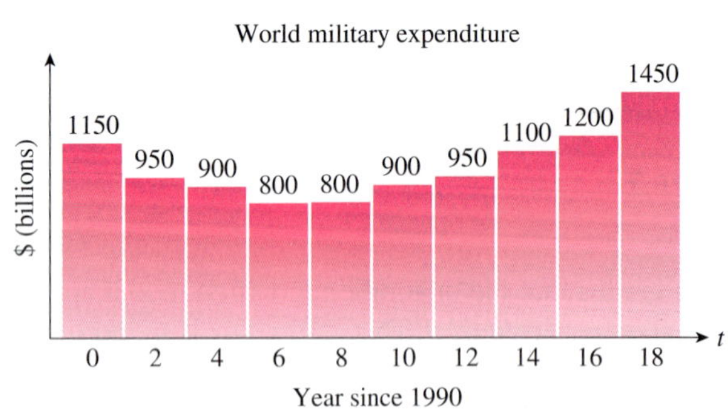

Source: www.globalissues.org/Geopolitics/ArmsTrade/Spending.asp

a. If you want to model the expenditure figures with a function of the form

$$f(t) = at^2 + bt + c,$$

would you expect the coefficient a to be positive or negative? Why? HINT [See "Features of a Parabola," page 125.]

b. Which of the following models best approximates the data given? (Try to answer this without actually computing values.)

(A) $f(t) = 5t^2 - 80t - 1{,}150$
(B) $f(t) = -5t^2 - 80t + 1{,}150$
(C) $f(t) = 5t^2 - 80t + 1{,}150$
(D) $f(t) = -5t^2 - 80t - 1{,}150$

c. What is the nearest year that would correspond to the vertex of the graph of the correct model from part (b)? What is the danger of extrapolating the data in either direction?

20. ***Education Expenditure*** The following chart shows the percentage of the U.S. Discretionary Budget allocated to education from 2003 to 2009 ($t = 3$ represents the start of 2003).

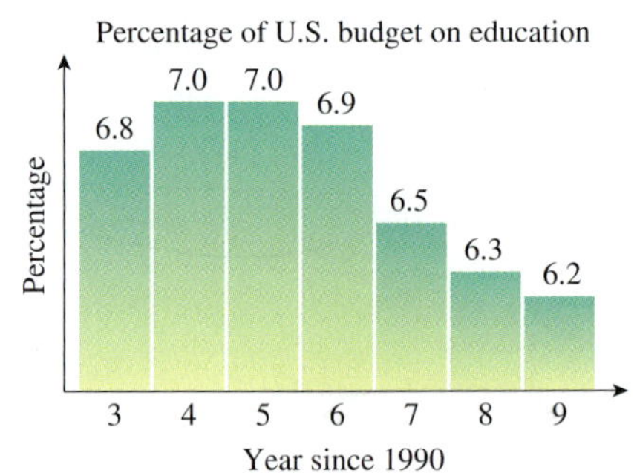

Source: www.globalissues.org/Geopolitics/ArmsTrade/Spending.asp

a. If you want to model the percentage figures with a function of the form

$$f(t) = at^2 + bt + c,$$

would you expect the coefficient a to be positive or negative? Why? HINT [See "Features of a Parabola," page 125.]

b. Which of the following models best approximates the data given? (Try to answer this without actually computing values)

(A) $f(t) = 0.04t^2 + 0.3t - 6$
(B) $f(t) = -0.04t^2 + 0.3t + 6$
(C) $f(t) = 0.04t^2 + 0.3t + 6$
(D) $f(t) = -0.04t^2 + 0.3t - 6$

c. What is the nearest year that would correspond to the vertex of the graph of the correct model from part (b)? What is the danger of extrapolating the data in either direction?

[1]Approximate figures in constant 2005 dollars. The 2008 figure is an estimate, based on the increase in U.S. military expenditure.

21. ***Oil Imports from Mexico*** Daily oil imports to the United States from Mexico can be approximated by

$$I(t) = -0.015t^2 + 0.1t + 1.4 \quad \text{million barrels/day} \quad (0 \le t \le 8)$$

where t is time in years since the start of 2000.[2] According to the model, in what year were oil imports to the United States greatest? How many barrels per day were imported that year? HINT [See Example 1.]

22. ***Oil Production in Mexico*** Daily oil production by **Pemex**, Mexico's national oil company, for 2001–2009 can be approximated by

$$P(t) = -0.022t^2 + 0.2t + 2.9 \quad \text{million barrels/day} \quad (1 \le t \le 9)$$

where t is time in years since the start of 2000.[3] According to the model, in what year was oil production by Pemex greatest? How many barrels per day were produced that year?

23. ***Fuel Efficiency*** The fuel efficiency (in miles per gallon) of an SUV depends on its weight according to the formula[4]

$$E = 0.000\,001\,6x^2 - 0.016x + 54 \quad (1{,}800 \le x \le 5{,}400)$$

where x is the weight of an SUV in pounds. According to the model, what is the weight of the least fuel-efficient SUV? Would you trust the model for weights greater than the answer you obtained? Explain.

24. ***Global Warming*** The amount of carbon dioxide (in pounds per 15,000 cubic miles) released by a typical SUV depends on its fuel efficiency according to the formula[5]

$$W = 32x^2 - 2{,}080x + 44{,}000 \quad (12 \le x \le 33)$$

where x is the fuel efficiency of an SUV in miles per gallon. According to the model, what is the fuel efficiency of the SUV with the least carbon dioxide pollution? Comment on the reliability of the model for fuel efficiencies that exceed your answer.

25. ***Revenue*** The market research department of the Better Baby Buggy Co. predicts that the demand equation for its buggies is given by $q = -0.5p + 140$, where q is the number of buggies it can sell in a month if the price is $\$p$ per buggy. At what price should it sell the buggies to get the largest revenue? What is the largest monthly revenue? HINT [See Example 3.]

26. ***Revenue*** The Better Baby Buggy Co. has just come out with a new model, the Turbo. The market research department predicts that the demand equation for Turbos is given by $q = -2p + 320$, where q is the number of buggies it can sell in a month if the price is $\$p$ per buggy. At what price should it sell the buggies to get the largest revenue? What is the largest monthly revenue?

27. ***Revenue*** Pack-Em-In Real Estate is building a new housing development. The more houses it builds, the less people will be willing to pay, due to the crowding and smaller lot sizes. In fact, if it builds 40 houses in this particular development, it can sell them for \$200,000 each, but if it builds 60 houses, it will only be able to get \$160,000 each. Obtain a linear demand equation and hence determine how many houses Pack-Em-In should build to get the largest revenue. What is the largest possible revenue? HINT [See Example 3.]

28. ***Revenue*** Pack-Em-In has another development in the works. If it builds 50 houses in this development, it will be able to sell them at \$190,000 each, but if it builds 70 houses, it will get only \$170,000 each. Obtain a linear demand equation and hence determine how many houses it should build to get the largest revenue. What is the largest possible revenue?

29. ▼ ***Revenue from Monorail Service, Las Vegas*** In 2005, the Las Vegas monorail charged \$3 per ride and had an average ridership of about 28,000 per day. In December 2005 the **Las Vegas Monorail Company** raised the fare to \$5 per ride, and average ridership in 2006 plunged to around 19,000 per day.[6]
    a. Use the given information to find a linear demand equation.
    b. Find the price the company should have charged to maximize revenue from ridership. What is the corresponding daily revenue?
    c. The Las Vegas Monorail Company would have needed \$44.9 million in revenues from ridership to break even in 2006. Would it have been possible to break even in 2006 by charging a suitable price?

30. ▼ ***Revenue from Monorail Service, Mars*** The Utarek monorail, which links the three urbynes (or districts) of Utarek, Mars, charged Ƶ5 per ride[7] and sold about 14 million rides per day. When the Utarek City Council lowered the fare to Ƶ3 per ride, the number of rides increased to 18 million per day.
    a. Use the given information to find a linear demand equation.
    b. Find the price the City Council should have charged to maximize revenue from ridership. What is the corresponding daily revenue?
    c. The City Council would have needed to raise Ƶ48 billion in revenues from ridership each Martian year (670 days[8]) to finance the new Mars organism research lab. Would this have been possible by charging a suitable price?

---

[2] Source: Energy Information Administration/Pemex/http://www.eia.doe.gov.

[3] Figures are approximate, and 2008–2009 figures are projections by the Department of Energy. Source: Energy Information Administration: Pemex.

[4] Fuel efficiency assumes 50% city driving and 50% highway driving. The model is based on a quadratic regression using data from 18 models of SUV. Source for data: Environmental Protection Agency, National Highway Traffic Safety Administration, American Automobile Manufacturers' Association, Ford Motor Company/*The New York Times*, November 30, 1997, p. 43.

[5] Ibid.

[6] Source: *New York Times*, Februrary 10, 2007, p. A9.

[7] Ƶ designates Zonars, the official currency in Mars. See www.marsnext.com for details of the Mars colony, its commerce, and its culture.

[8] As measured in Mars days. The actual length of a Mars year is about 670.55 Mars days, so frequent leap years are designated by the Mars Planetary Authority to adjust.

31. ***Web Site Profit*** You operate a gaming Web site, www.mudbeast.net, where users must pay a small fee to log on. When you charged \$2 the demand was 280 log-ons per month. When you lowered the price to \$1.50, the demand increased to 560 log-ons per month.

a. Construct a linear demand function for your Web site and hence obtain the monthly revenue R as a function of the log-on fee x.

b. Your Internet provider charges you a monthly fee of \$30 to maintain your site. Express your monthly profit P as a function of the log-on fee x, and hence determine the log-on fee you should charge to obtain the largest possible monthly profit. What is the largest possible monthly profit? HINT [See Example 4.]

32. ***T-Shirt Profit*** Two fraternities, Sig Ep and Ep Sig, plan to raise money jointly to benefit homeless people on Long Island. They will sell Yoda vs. Alien T-shirts in the student center, but are not sure how much to charge. Sig Ep treasurer Augustus recalls that they once sold 400 shirts in a week at \$8 per shirt, but Ep Sig treasurer Julius has solid research indicating that it is possible to sell 600 per week at \$4 per shirt.

a. Based on this information, construct a linear demand equation for Yoda vs. Alien T-shirts, and hence obtain the weekly revenue R as a function of the unit price x.

b. The university administration charges the fraternities a weekly fee of \$500 for use of the Student Center. Write down the monthly profit P as a function of the unit price x, and hence determine how much the fraternities should charge to obtain the largest possible weekly profit. What is the largest possible weekly profit? HINT [See Example 4.]

33. ***Web Site Profit*** The latest demand equation for your gaming Web site, www.mudbeast.net, is given by

$$q = -400x + 1{,}200$$

where q is the number of users who log on per month and x is the log-on fee you charge. Your Internet provider bills you as follows:

Site maintenance fee:	\$20 per month
High-volume access fee:	50¢ per log-on

Find the monthly cost as a function of the log-on fee x. Hence, find the monthly profit as a function of x and determine the log-on fee you should charge to obtain the largest possible monthly profit. What is the largest possible monthly profit?

34. ***T-Shirt Profit*** The latest demand equation for your Yoda vs. Alien T-shirts is given by

$$q = -40x + 600$$

where q is the number of shirts you can sell in one week if you charge \$$x$ per shirt. The Student Council charges you \$400 per week for use of their facilities, and the T-shirts cost you \$5 each. Find the weekly cost as a function of the unit price x. Hence, find the weekly profit as a function of x and determine the unit price you should charge to obtain the largest possible weekly profit. What is the largest possible weekly profit?

35. ▼ ***Nightclub Management*** You have just opened a new nightclub, Russ' Techno Pitstop, but are unsure of how high to set the cover charge (entrance fee). One week you charged \$10 per guest and averaged 300 guests per night. The next week you charged \$15 per guest and averaged 250 guests per night.

a. Find a linear demand equation showing the number of guests q per night as a function of the cover charge p.

b. Find the nightly revenue R as a function of the cover charge p.

c. The club will provide two free non-alcoholic drinks for each guest, costing the club \$3 per head. In addition, the nightly overheads (rent, salaries, dancers, DJ, etc.) amount to \$3,000. Find the cost C as a function of the cover charge p.

d. Now find the profit in terms of the cover charge p, and hence determine the entrance fee you should charge for a maximum profit.

36. ▼ ***Television Advertising*** As sales manager for Montevideo Productions, Inc., you are planning to review the prices you charge clients for television advertisement development. You currently charge each client an hourly development fee of \$2,500. With this pricing structure, the demand, measured by the number of contracts Montevideo signs per month, is 15 contracts. This is down 5 contracts from the figure last year, when your company charged only \$2,000.

a. Construct a linear demand equation giving the number of contracts q as a function of the hourly fee p Montevideo charges for development.

b. On average, Montevideo bills for 50 hours of production time on each contract. Give a formula for the total revenue obtained by charging \$$p$ per hour.

c. The costs to Montevideo Productions are estimated as follows:

Fixed costs:	\$120,000 per month
Variable costs:	\$80,000 per contract

Express Montevideo Productions' monthly cost **(i)** as a function of the number q of contracts and **(ii)** as a function of the hourly production charge p.

d. Express Montevideo Productions' monthly profit as a function of the hourly development fee p and hence the price it should charge to maximize the profit.

37. T ***World Military Expenditure*** The following table shows total military and arms trade expenditure in 1994, 1998, and 2006. (See Exercise 19; $t = 4$ represents 1994.)[9]

Year t	4	8	16
Military Expenditure (\$ billion)	900	800	1,200

[9] Approximate figures in constant 2005 dollars. The 2008 figure is an estimate, based on the increase in U.S. military expenditure. Source: www.globalissues.org/Geopolitics/ArmsTrade/Spending.asp.

Find a quadratic model for these data, and use your model to estimate world military expenditure in 2008. Compare your answer with the actual figure shown in Exercise 19. HINT [See Example 5.]

38. ***Education Expenditure*** The following table shows the percentage of the U.S. Discretionary Budget allocated to education in 2003, 2005, and 2009. (See Exercise 20; $t = 3$ represents the start of 2003.)[10]

Year t	3	5	9
Percentage	6.8	7	6.2

Find a quadratic model for these data, and use your model to estimate the percentage of the U.S. Discretionary Budget allocated to education in 2008. Compare your answer with the actual figure shown in Exercise 20.

39. ***iPhone Sales*** The following table shows Apple iPhone sales from the 2nd quarter in 2007 through the second quarter in 2008 ($t = 2$ represents the second quarter of 2007):[11]

Quarter t	2	3	4	5	6
iPhone Sales (thousands)	270	1,119	2,315	1,703	717

a. Find a quadratic regression model for these data. (Round coefficients to the nearest whole number.) Graph the model together with the data.

b. What does the model predict for iPhone sales in the third quarter of 2008 ($t = 7$) to the nearest 1,000 units? Comment on the answer, and ascertain the actual third quarter sales in 2008 (Apple's fiscal fourth quarter).

40. ***Facebook Membersip*** The following table gives the approximate number of Facebook users at various times since its establishment early in 2004.[12]

Year t (since start of 2004)	0	1	2	2.5	3	3.5	4	4.5
Facebook Members n (millions)	0	1	5.5	7	12	30	58	80

a. Find a quadratic regression model for these data. (Round coefficients to the nearest whole number.) Graph the model, together with the data.

b. Assuming the trend had continued, estimate the number of members at the start of 2010 to the nearest 10 million members.

c. Is the quadratic model appropriate for long-term prediction of the number of members? Why?

COMMUNICATION AND REASONING EXERCISES

41. What can you say about the graph of $f(x) = ax^2 + bx + c$ if $a = 0$?

42. What can you say about the graph of $f(x) = ax^2 + bx + c$ if $c = 0$?

43. Multiple choice: Following is the graph of $f(x) = ax^2 + bx + c$:

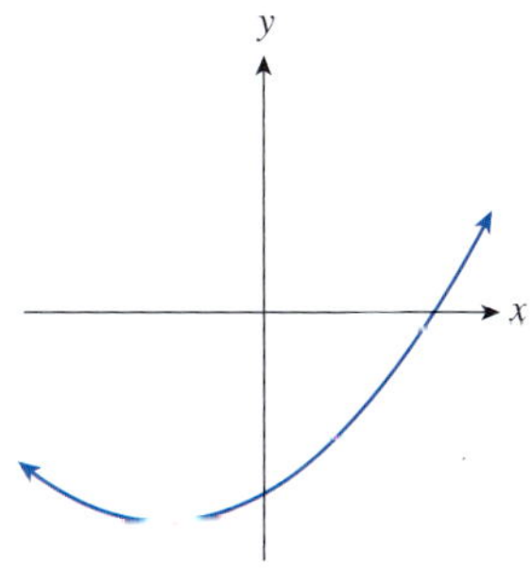

(A) a is positive and c is positive.
(B) a is negative and c is positive.
(C) a is positive and c is negative.
(D) a is negative and c is negative.

44. Multiple choice: Following is the graph of $f(x) = ax^2 + bx + c$:

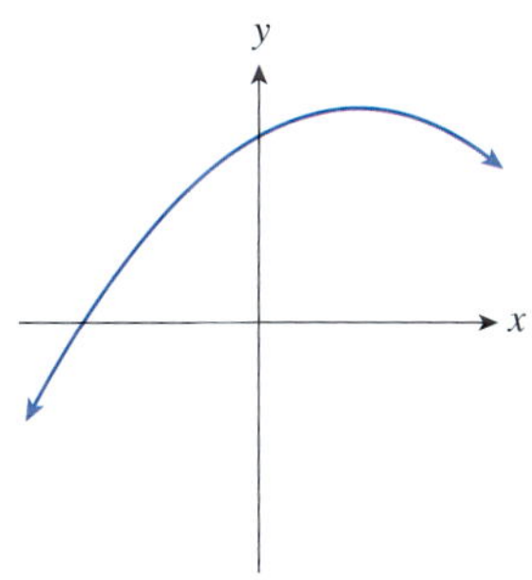

(A) a is positive and c is positive.
(B) a is negative and c is positive.
(C) a is positive and c is negative.
(D) a is negative and c is negative.

45. ▼ Refer to the graph of $f(x) = ax^2 + bx + c$ in Exercise 43. Is b positive or negative? Why?

46. ▼ Refer to the graph of $f(x) = ax^2 + bx + c$ in Exercise 44. Is b positive or negative? Why?

47. Suppose the graph of revenue as a function of unit price is a parabola that is concave down. What is the significance of the coordinates of the vertex, the x-intercepts, and the y-intercept?

48. Suppose the height of a stone thrown vertically upward is given by a quadratic function of time. What is the significance of the coordinates of the vertex, the (possible) x-intercepts, and the y-intercept?

49. How might you tell, roughly, whether a set of data should be modeled by a quadratic rather than by a linear equation?

[10] Source: www.globalissues.org/Geopolitics/ArmsTrade/Spending.asp.

[11] Source: Apple financial statements, www.apple.com.

[12] Sources: Some data are interpolated. (www.facebook.com/, www.insidehighered.com).

50. A member of your study group tells you that, because the following set of data does not suggest a straight line, the data are best modeled by a quadratic.

x	0	2	4	6	8
y	1	2	1	0	1

Comment on her suggestion.

51. Is a quadratic model useful for long-term prediction of sales of an item? Why?

52. Of what use is a quadratic model, if not for long-term prediction?

53. ▼ Explain why, if demand is a linear function of unit price p (with negative slope), then there must be a *single value of* p that results in the maximum revenue.

54. ▼ Explain why, if the average cost of a commodity is given by $y = 0.1x^2 - 4x - 2$, where x is the number of units sold, there is a single choice of x that results in the lowest possible average cost.

55. ▼ If the revenue function for a particular commodity is $R(p) = -50p^2 + 60p$, what is the (linear) demand function? Give a reason for your answer.

56. ▼ If the revenue function for a particular commodity is $R(p) = -50p^2 + 60p + 50$, can the demand function be linear? What is the associated demand function?

2.2 Exponential Functions and Models

The quadratic functions we discussed in Section 2.1 can be used to model many nonlinear situations. However, exponential functions give better models in some applications, including population growth, radioactive decay, the growth or depreciation of financial investments, and many other phenomena. (We already saw some of these applications in Section 1.2.)

To work effectively with exponential functions, we need to know the laws of exponents. The following list, taken from the algebra review in Chapter 0, gives the laws of exponents we will be using.

The Laws of Exponents

If b and c are positive and x and y are any real numbers, then the following laws hold:

Law	Quick Examples	
1. $b^x b^y = b^{x+y}$	$2^3 2^2 = 2^5 = 32$	$2^{3-x} = 2^3 2^{-x}$
2. $\dfrac{b^x}{b^y} = b^{x-y}$	$\dfrac{4^3}{4^2} = 4^{3-2} = 4^1 = 4$	$3^{x-2} = \dfrac{3^x}{3^2} = \dfrac{3^x}{9}$
3. $\dfrac{1}{b^x} = b^{-x}$	$9^{-0.5} = \dfrac{1}{9^{0.5}} = \dfrac{1}{3}$	$2^{-x} = \dfrac{1}{2^x}$
4. $b^0 = 1$	$(3.3)^0 = 1$	$x^0 = 1$ if $x \neq 0$
5. $(b^x)^y = b^{xy}$	$(2^3)^2 = 2^6 = 64$	$\left(\dfrac{1}{2}\right)^x = (2^{-1})^x = 2^{-x}$
6. $(bc)^x = b^x c^x$	$(4 \cdot 2)^2 = 4^2 2^2 = 64$	$10^x = 5^x 2^x$
7. $\left(\dfrac{b}{c}\right)^x = \dfrac{b^x}{c^x}$	$\left(\dfrac{4}{3}\right)^2 = \dfrac{4^2}{3^2} = \dfrac{16}{9}$	$\left(\dfrac{1}{2}\right)^x = \dfrac{1^x}{2^x} = \dfrac{1}{2^x}$

Here are the functions we will study in this section.

Exponential Function

An **exponential function** has the form

$$f(x) = Ab^x, \qquad \text{Technology: } \texttt{A*b^x}$$

where A and b are constants with $A \neq 0$ and b positive and not equal to 1. We call b the **base** of the exponential function.

Quick Examples

1. $f(x) = 2^x$ — $A = 1, b = 2$; Technology: `2^x`
   $f(1) = 2^1 = 2$ — `2^1`
   $f(-3) = 2^{-3} = \frac{1}{8}$ — `2^(-3)`
   $f(0) = 2^0 = 1$ — `2^0`
2. $g(x) = 20(3^x)$ — $A = 20, b = 3$; Technology: `20*3^x`
   $g(2) = 20(3^2) = 20(9) = 180$ — `20*3^2`
   $g(-1) = 20(3^{-1}) = 20\left(\frac{1}{3}\right) = 6\frac{2}{3}$ — `20*3^(-1)`
3. $h(x) = 2^{-x} = \left(\frac{1}{2}\right)^x$ — $A = 1, b = \frac{1}{2}$; Technology: `2^(-x)` or `(1/2)^x`
   $h(1) = 2^{-1} = \frac{1}{2}$ — `2^(-1)` or `(1/2)^1`
   $h(2) = 2^{-2} = \frac{1}{4}$ — `2^(-2)` or `(1/2)^2`
4. $k(x) = 3 \cdot 2^{-4x} = 3(2^{-4})^x$ — $A = 3, b = 2^{-4}$; Technology: `3*2^(-4*x)`
   $k(-2) = 3 \cdot 2^{-4(-2)}$ — `3*2^(-4*(-2))`
   $= 3 \cdot 2^8 = 3 \cdot 256 = 768$

Exponential Functions from the Numerical and Graphical Points of View

The following table shows values of $f(x) = 3(2^x)$ for some values of x ($A = 3, b = 2$):

x	-3	-2	-1	0	1	2	3
$f(x)$	$\frac{3}{8}$	$\frac{3}{4}$	$\frac{3}{2}$	3	6	12	24

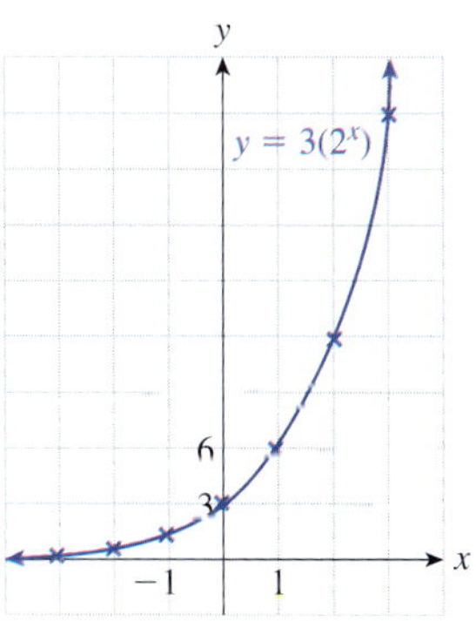

`3*2^x`

Figure 10

Its graph is shown in Figure 10.

Notice that the y-intercept is $A = 3$ (obtained by setting $x = 0$). In general:

In the graph of $f(x) = Ab^x$, A is the y-intercept, or the value of y when $x = 0$.

What about b? Notice from the table that the value of y is multiplied by $b = 2$ for every increase of 1 in x. If we decrease x by 1, the y coordinate gets *divided* by $b = 2$.

The value of y is multiplied by b for every one-unit increase of x.

x	-3	-2	-1	0	1	2	3
$f(x)$	$\frac{3}{8}$	$\frac{3}{4}$	$\frac{3}{2}$	3	6	12	24

Multiply by 2

On the graph, if we move one unit to the right from any point on the curve, the y coordinate doubles. Thus, the curve becomes dramatically steeper as the value of x increases. This phenomenon is called **exponential growth**. (See Section 1.2.)

Exponential Function Numerically and Graphically

For the exponential function $f(x) = Ab^x$:

Role of *A*

$f(0) = A$, so A is the y-intercept of the graph of f.

Role of *b*

If x increases by 1, $f(x)$ is multiplied by b.
If x increases by 2, $f(x)$ is multiplied by b^2.
⋮
If x increases by Δx, $f(x)$ is multiplied by $b^{\Delta x}$.

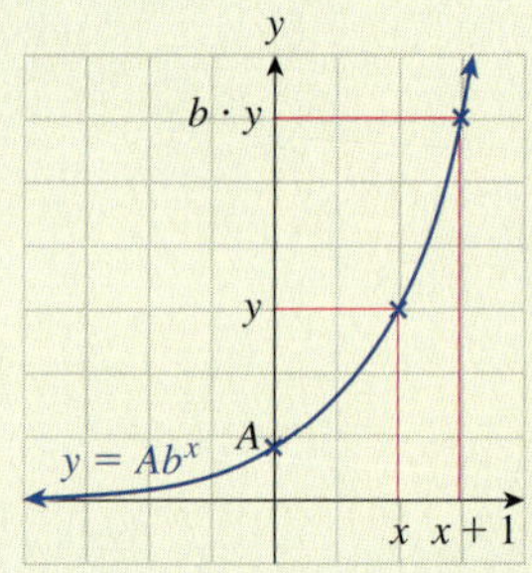

If x increases by 1, y is multiplied by b.

Quick Examples

1. $f_1(x) = 2^x$, $f_2(x) = \left(\frac{1}{2}\right)^x = 2^{-x}$

	A	B	C
1	x	2^x	2^(-x)
2	-3	1/8	8
3	-2	1/4	4
4	-1	1/2	2
5	0	1	1
6	1	2	1/2
7	2	4	1/4
8	3	8	1/8

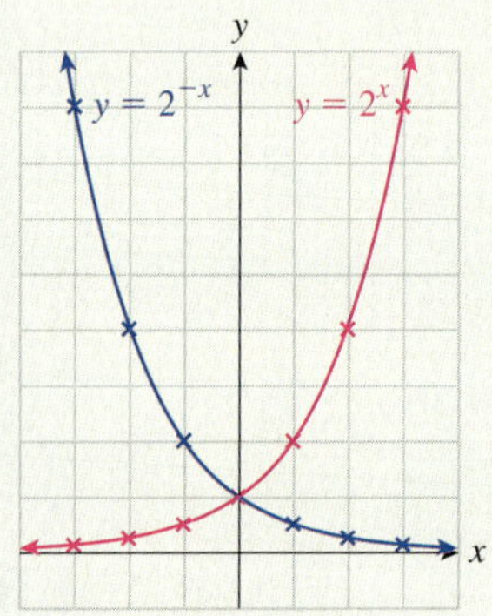

Technology: `2^x; 2^(-x)`

When x increases by 1, $f_2(x)$ is multiplied by $\frac{1}{2}$. The function $f_1(x) = 2^x$ illustrates exponential growth, while $f_2(x) = \left(\frac{1}{2}\right)^x$ illustrates the opposite phenomenon: **exponential decay**.

2. $f_1(x) = 2^x$, $f_2(x) = 3^x$, $f_3(x) = 1^x$ (Can you see why f_3 is not an exponential function?)

	A	B	C	D
1	x	2^x	3^x	1^x
2	-3	1/8	1/27	1
3	-2	1/4	1/9	1
4	-1	1/2	1/3	1
5	0	1	1	1
6	1	2	3	1
7	2	4	9	1
8	3	8	27	1

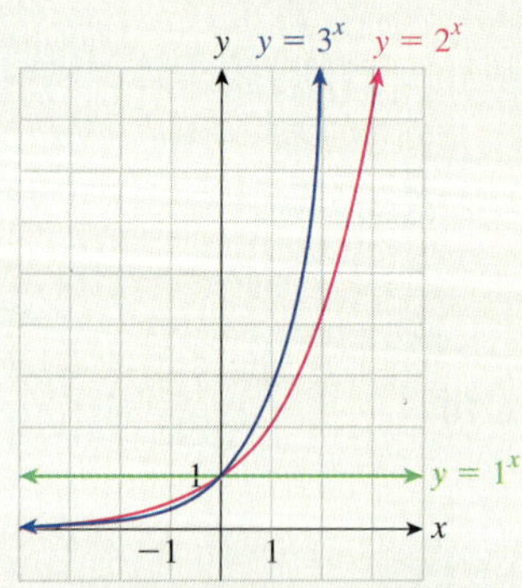

If x increases by 1, 3^x is multiplied by 3. Note also that all the graphs pass through (0, 1). (Why?)

EXAMPLE 1 Recognizing Exponential Data Numerically and Graphically

Some of the values of two functions, f and g, are given in the following table:

x	-2	-1	0	1	2
$f(x)$	-7	-3	1	5	9
$g(x)$	$\frac{2}{9}$	$\frac{2}{3}$	2	6	18

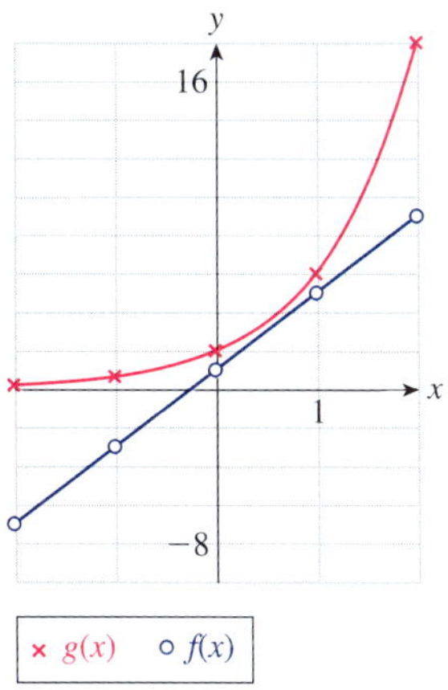

Figure 11

One of these functions is linear, and the other is exponential. Which is which?

Solution Remember that a linear function increases (or decreases) by the same amount every time x increases by 1. The values of f behave this way: Every time x increases by 1, the value of $f(x)$ increases by 4. Therefore, f is a linear function with a *slope* of 4. Because $f(0) = 1$, we see that

$$f(x) = 4x + 1$$

is a linear formula that fits the data.

On the other hand, every time x increases by 1, the value of $g(x)$ is *multiplied* by 3. Because $g(0) = 2$, we find that

$$g(x) = 2(3^x)$$

is an exponential function fitting the data.

We can visualize the two functions f and g by plotting the data points (Figure 11). The data points for $f(x)$ clearly lie along a straight line, whereas the points for $g(x)$ lie along a curve. The y coordinate of each point for $g(x)$ is 3 times the y coordinate of the preceding point, demonstrating that the curve is an exponential one.

In Section 1.3, we discussed a method for calculating the equation of the line that passes through two given points. In the following example, we show the method for calculating the equation of the exponential curve through two given points.

EXAMPLE 2 Finding the Exponential Curve through Two Points

Find an equation of the exponential curve through (1, 6.3) and (4, 170.1).

Solution We want an equation of the form

$$y = Ab^x \quad (b > 0).$$

Substituting the coordinates of the given points, we get

$$6.3 = Ab^1 \qquad \text{Substitute (1, 6.3).}$$
$$170.1 = Ab^4. \qquad \text{Substitute (4, 170.1).}$$

If we now divide the second equation by the first, we get

$$\frac{170.1}{6.3} = \frac{Ab^4}{Ab} = b^3$$
$$b^3 = 27$$
$$b = 27^{1/3} \qquad \text{Take reciprocal power of both sides.}$$
$$b = 3.$$

Now that we have b, we can substitute its value into the first equation to obtain

$$6.3 = 3A \qquad \text{Substitute } b = 3 \text{ into the equation } 6.3 = Ab^1.$$

$$A = \frac{6.3}{3} = 2.1.$$

We have both constants, $A = 2.1$ and $b = 3$, so the model is

$$y = 2.1(3^x).$$

Example 6 will show how to use technology to fit an exponential function to two or more data points.

APPLICATIONS

Recall some terminology we mentioned earlier: A quantity y experiences **exponential growth** if $y = Ab^t$ with $b > 1$. (Here we use t for the independent variable, thinking of time.) It experiences **exponential decay** if $y = Ab^t$ with $0 < b < 1$. We already saw several examples of exponential growth and decay in Section 1.2.

EXAMPLE 3 Exponential Growth and Decay

a. Compound Interest (See Section 1.2 Example 6.) If \$2,000 is invested in a mutual fund with an annual yield of 12.6% and the earnings are reinvested each month, then the future value after t years is

$$A(t) = P\left(1 + \frac{r}{m}\right)^{mt} = 2{,}000\left(1 + \frac{0.126}{12}\right)^{12t} = 2{,}000(1.0105)^{12t},$$

which can be written as $2{,}000(1.0105^{12})^t$, so $A = 2{,}000$ and $b = 1.0105^{12}$. This is an example of exponential growth, because $b > 1$.

b. Carbon Decay (See Section 1.2 Example 7.) The amount of carbon 14 remaining in a sample that originally contained A grams is approximately

$$C(t) = A(0.999879)^t.$$

This is an instance of exponential decay, because $b < 1$.

Before we go on... Refer again to part (a). In Example 6(b) of Section 1.2 we showed how to use technology to answer questions such as the following: "When, to the nearest year, will the value of your investment reach \$5,000?" ■

The next example shows an application to public health.

EXAMPLE 4 Exponential Growth: Epidemics

In the early stages of the AIDS epidemic during the 1980s, the number of cases in the United States was increasing by about 50% every 6 months. By the start of 1983, there were approximately 1,600 AIDS cases in the United States.*

a. Assuming an exponential growth model, find a function that predicts the number of people infected t years after the start of 1983.

b. Use the model to estimate the number of people infected by October 1, 1986, and also by the end of that year.

* Data based on regression of the 1982–1986 figures. Source for data: Centers for Disease Control and Prevention. HIV/AIDS Surveillance Report, 2000;12 (No. 2).

Solution

a. One way of finding the desired exponential function is to reason as follows: At time $t = 0$ (January 1, 1983), the number of people infected was 1,600, so $A = 1{,}600$. Every 6 months, the number of cases increased to 150% of the number 6 months earlier—that is, to 1.50 times that number. Each year, it therefore increased to $(1.50)^2 = 2.25$ times the number one year earlier. Hence, after t years, we need to multiply the original 1,600 by 2.25^t, so the model is

$$y = 1{,}600(2.25^t) \text{ cases.}$$

Alternatively, if we wish to use the method of Example 2, we need two data points. We are given one point: (0, 1,600). Because y increased by 50% every 6 months, 6 months later it reached $1{,}600 + 800 = 2{,}400$ $(t = 0.5)$. This information gives a second point: (0.5, 2,400). We can now apply the method in Example 2 to find the model above.

b. October 1, 1986, corresponds to $t = 3.75$ (because October 1 is 9 months, or $9/12 = 0.75$ of a year after January 1). Substituting this value of t in the model gives

$$y = 1{,}600(2.25^{3.75}) \approx 33{,}481 \text{ cases}$$ 1600*2.25^3.75

By the end of 1986, the model predicts that

$$y = 1{,}600(2.25^4) = 41{,}006 \text{ cases.}$$

(The actual number of cases was around 41,700.)

Before we go on... Increasing the number of cases by 50% every 6 months couldn't continue for very long and this is borne out by observations. If increasing by 50% every 6 months did continue, then by January 2003 $(t = 20)$, the number of infected people would have been

$$1{,}600(2.25^{20}) \approx 17{,}700{,}000{,}000$$

a number that is more than 50 times the size of the U.S. population! Thus, although the exponential model is fairly reliable in the early stages of the epidemic, it is unreliable for predicting long-term trends. ■

Epidemiologists use more sophisticated models to measure the spread of epidemics, and these models predict a leveling-off phenomenon as the number of cases becomes a significant part of the total population. We discuss such a model, the **logistic function,** in Section 2.4.

The Number e and More Applications

In nature we find examples of growth that occurs *continuously*, as though "interest" is being added more often than every second or fraction of a second. To model this, we need to see what happens to the compound interest formula of Section 1.2 as we let m (the number of times interest is added per year) become extremely large. Something very interesting does happen: We end up with a more compact and elegant formula than we began with. To see why, let's look at a very simple situation.

Suppose we invest \$1 in the bank for 1 year at 100% interest, compounded m times per year. If $m = 1$, then 100% interest is added every year, and so our money doubles at the end of the year. In general, the accumulated capital at the end of the year is

$$A = 1\left(1 + \frac{1}{m}\right)^m = \left(1 + \frac{1}{m}\right)^m.$$ (1+1/m)^m

	A	B
1	m	(1+1/m)^m
2	1	2
3	10	2.59374246
4	100	2.704813829
5	1000	2.716923932
6	10000	2.718145927
7	100000	2.718268237
8	1000000	2.718280469
9	10000000	2.718281694
10	100000000	2.718281786
11	1000000000	2.718282031

Now, we are interested in what A becomes for large values of m. On the left is an Excel sheet showing the quantity $\left(1+\frac{1}{m}\right)^m$ for larger and larger values of m.

Something interesting *does* seem to be happening! The numbers appear to be getting closer and closer to a specific value. In mathematical terminology, we say that the numbers **converge** to a fixed number, 2.71828..., called the **limiting value*** of the quantities $\left(1+\frac{1}{m}\right)^m$. This number, called e, is one of the most important in mathematics. The number e is irrational, just as the more familiar number π is, so we cannot write down its exact numerical value. To 20 decimal places,

$$e = 2.71828182845904523536\ldots.$$

We now say that, if \$1 is invested for 1 year at 100% interest **compounded continuously**, the accumulated money at the end of that year will amount to $\$e = \$2.72$ (to the nearest cent). But what about the following more general question?

*** NOTE** See Chapter 3 for more on limits.

Q: *What about a more general scenario: If we invest an amount \$P for t years at an interest rate of r, compounded continuously, what will be the accumulated amount A at the end of that period?*

A: In the special case above (P, t, and r all equal 1), we took the compound interest formula and let m get larger and larger. We do the same more generally, after a little preliminary work with the algebra of exponentials.

$$A = P\left(1+\frac{r}{m}\right)^{mt}$$

$$= P\left(1+\frac{1}{(m/r)}\right)^{mt} \qquad \text{Substituting } \frac{r}{m} = \frac{1}{(m/r)}$$

$$= P\left(1+\frac{1}{(m/r)}\right)^{(m/r)rt} \qquad \text{Substituting } m = \left(\frac{m}{r}\right)r$$

$$= P\left[\left(1+\frac{1}{(m/r)}\right)^{(m/r)}\right]^{rt} \qquad \text{Using the rule } a^{bc} = (a^b)^c$$

For continuous compounding of interest, we let m, and hence m/r, get very large. This affects only the term in brackets, which converges to e, and we get the formula

$$A = Pe^{rt}.$$

Q: *How do I obtain powers of e or e itself on a TI-83/84 Plus or in Excel?*

A: On the TI-83/84 Plus, enter e^x as `e^(x)`, where `e^(` can be obtained by pressing [2ND] [LN]. To obtain the number e on the TI-83/84 Plus, enter `e^(1)`. Excel has a built-in function called `EXP`; `EXP(x)` gives the value of e^x. To obtain the number e in Excel, enter `=EXP(1)`.

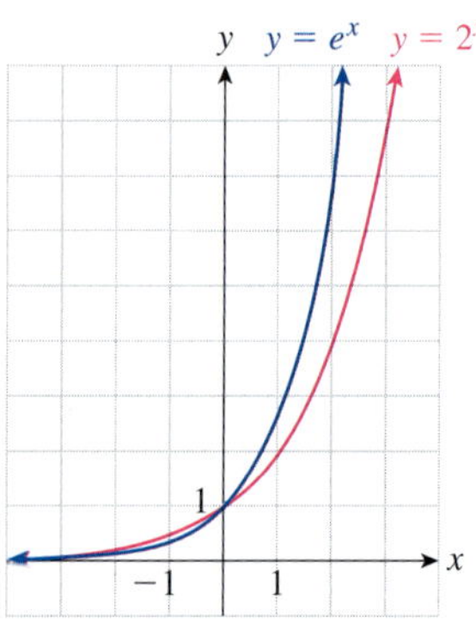

Technology formula: `e^(x)` or `EXP(x)`

Figure 12

Figure 12 shows the graph of $y = e^x$ with that of $y = 2^x$ for comparison.

The Number e and Continuous Compounding

The number e is the limiting value of the quantities $\left(1+\frac{1}{m}\right)^m$ as m gets larger and larger, and has the value 2.71828182845904523536...

If \$$P$ is invested at an annual interest rate r compounded continuously, the accumulated amount after t years is

$$A(t) = Pe^{rt}. \qquad \texttt{P*e^(r*t) or P*EXP(r*t)}$$

Quick Examples

1. If \$100 is invested in an account that bears 15% interest compounded continuously, at the end of 10 years the investment will be worth

$$A(10) = 100e^{(0.15)(10)} = \$448.17.$$

`100*e^(0.15*10)` or `100*EXP(0.15*10)`

2. If \$1 is invested in an account that bears 100% interest compounded continuously, at the end of x years, the investment will be worth

$$A(x) = e^x \text{ dollars.}$$

EXAMPLE 5 Continuous Compounding

a. You invest \$10,000 at Fastrack Savings & Loan, which pays 6% compounded continuously. Express the balance in your account as a function of the number of years t and calculate the amount of money you will have after 5 years.

b. Your friend has just invested \$20,000 in Constant Growth Funds, whose stocks are continuously *declining* at a rate of 6% per year. How much will her investment be worth in 5 years?

c. T During which year will the value of your investment first exceed that of your friend?

Solution

a. We use the continuous growth formula with $P = 10{,}000$, $r = 0.06$, and t variable, getting

$$A(t) = Pe^{rt} = 10{,}000e^{0.06t}.$$

In five years,

$$\begin{aligned} A(5) &= 10{,}000e^{0.06(5)} \\ &= 10{,}000e^{0.3} \\ &\approx \$13{,}498.59. \end{aligned}$$

b. Because the investment is depreciating, we use a negative value for r and take $P = 20{,}000$, $r = -0.06$, and $t = 5$, getting

$$\begin{aligned} A(t) &= Pe^{rt} = 20{,}000e^{-0.06t} \\ A(5) &= 20{,}000e^{-0.06(5)} \\ &= 20{,}000e^{-0.3} \\ &\approx \$14{,}816.36. \end{aligned}$$

c. We can answer the question now using a graphing calculator, a spreadsheet, or the Function Evaluator and Grapher tool at the Web site. Just enter the exponential models of parts (a) and (b) and create tables to compute the values at the end of several years:

TI-83/84 Plus

Y1=10000e^(0.06X)
Y2=20000e^(-0.06X)
[2ND] [TABLE]

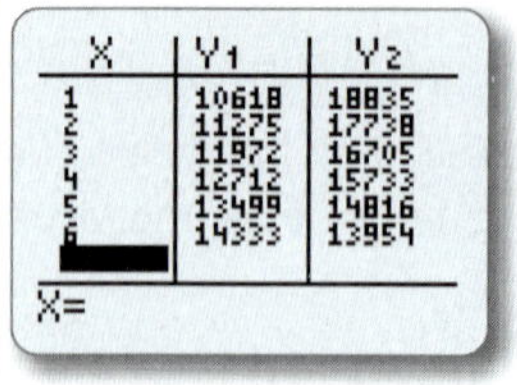

Excel

Headings *t*, Me, Friend in A1–C1; *t*-values 1, . . . , 6 in A2–A7.
=10000*exp(0.06*A2)
=20000*exp(-0.06*A2)
in B2 and C2 and copy down to C7.

	A	B	C
1	t	Me	Friend
2	1	10618.365	18835.291
3	2	11274.969	17738.409
4	3	11972.174	16705.404
5	4	12712.492	15732.557
6	5	13498.588	14816.364
7	6	14333.294	13953.527

Web Site www.AppliedCalc.org

10000e^(0.06X) under y_1
20000e^(-0.06X) under y_2
Enter 1, . . . , 6 as the *x*-Values in the Evaluator and press "Evaluate."

x-Values	y_1-Values	y_2-Values
1	10618	18835
2	11275	17738
3	11972	16705
4	12712	15733
5	13499	14816
6	14333	13954

From the table, we see that the value of your investment overtakes that of your friend after $t = 5$ (the end of year 5) and before $t = 6$ (the end of year 6). Thus your investment first exceeds that of your friend sometime during year 6.

➡ Before we go on...

Q: *How does continuous compounding compare with monthly compounding?*

A: To repeat the calculation in part (a) of Example 5 using monthly compounding instead of continuous compounding, we use the compound interest formula with $P = 10{,}000$, $r = 0.06$, $m = 12$, and $t = 5$ and find

$$A(5) = 10{,}000(1 + 0.06/12)^{60} \approx \$13{,}488.50.$$

Thus, continuous compounding earns you approximately \$10 more than monthly compounding on a 5-year, \$10,000 investment. This is little to get excited about. ■

If we write the continuous compounding formula $A(t) = Pe^{rt}$ as $A(t) = P(e^r)^t$, we see that $A(t)$ is an exponential function of t, where the base is $b = e^r$, so we have really not introduced a new kind of function. In fact, exponential functions are often written in this way:

Exponential Functions: Alternative Form

We can write any exponential function in the following alternative form:

$$f(x) = Ae^{rx}$$

where A and r are constants. If r is positive, f models exponential growth; if r is negative, f models exponential decay.

Quick Examples

1. $f(x) = 100e^{0.15x}$ — Exponential growth $A = 100, r = 0.15$
2. $f(t) = Ae^{-0.000\,121\,01t}$ — Exponential decay of carbon 14; $r = -0.000\,121\,01$
3. $f(t) = 100e^{0.15t} = 100\left(e^{0.15}\right)^t$
$= 100(1.1618)^t$ — Converting Ae^{rt} to the form Ab^t

We will see in Chapter 4 that the exponential function with base e exhibits some interesting properties when we measure its rate of change, and this is the real mathematical importance of e.

Exponential Regression

Starting with a set of data that suggests an exponential curve, we can use technology to compute the exponential regression curve in much the same way as we did for the quadratic regression curve in Example 5 of Section 2.1.

EXAMPLE 6 Exponential Regression: Health Expenditures

The following table shows annual expenditure on health in the United States from 1980 through 2010 ($t = 0$ represents 1980).*

Year t	0	5	10	15	20	25	30
Expenditure ($ billion)	246	427	696	990	1,310	1,920	2,750

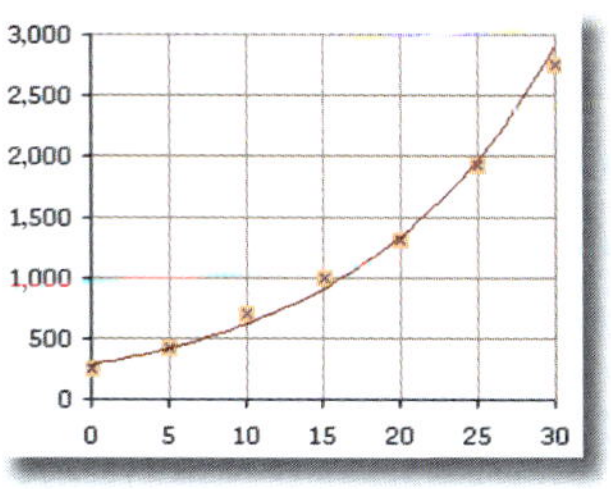

Figure 13

using Technology

See the Technology Guide at the end of the chapter for detailed instructions on how to obtain the regression curve and graph for Example 6 using a TI-83/84 Plus or Excel. Here is an outline:

TI-83/84 Plus

[STAT] EDIT values of t in L_1 and values of C in L_2

Regression curve: [STAT] CALC option #0 ExpReg [ENTER]

Graph: [Y=] [VARS] [5] EQ [1], then [ZOOM] [9]

[More details on page 186.]

a. Find the exponential regression model

$$C(t) = Ab^t$$

for the annual expenditure.

b. Use the regression model to estimate the expenditure in 2002 ($t = 22$; the actual expenditure was approximately $1,550 billion).

Solution

a. We use technology to obtain the exponential regression curve (see Figure 13):

$$C(t) \approx 282.33(1.0808)^t \quad \text{Coefficients rounded}$$

b. Using the model $C(t) \approx 282.33(1.0808)^t$ we find that

$$C(22) \approx 282.33(1.0808)^{22} \approx \$1,560 \text{ billion}$$

which is close to the actual number of around $1,550 billion.

* Data are rounded. 2005 and 2010 figures are projections. Source: Centers for Medicare and Medicaid Services, "National Health Expenditures," 2002 version, released January 2004; www.cms.hhs.gov/statistics/nhe/.

Excel
t- and C-values in columns A and B
Graph: Highlight data and insert a Scatter chart.
Regression curve: Right-click a datapoint and add exponential Trendline with option to show equation on chart.
[More details on page 189.]

Web Site
www.AppliedCalc.org
Student Home
→ Online Utilities
→ Simple Regression
Enter the data in the x and y columns and press "y=a(b^x)".

Before we go on... We said in the preceding section that the regression curve gives the smallest value of the sum-of-squares error, SSE (the sum of the squares of the residuals). However, exponential regression as computed via technology generally minimizes the sum of the squares of the residuals of the *logarithms* (logarithms are discussed in the next section). Using logarithms allows one easily to convert an exponential function into a linear one and then use linear regression formulas. However, in Section 2.4, we will discuss a way of using Excel's Solver to minimize SSE directly, which allows us to find the best-fit exponential curve directly without the need for devices to simplify the mathematics. If we do this, we obtain a very different equation:

$$C(t) \approx 316.79(1.0747^t).$$

If you plot this function, you will notice that it seems to fit the data more closely than the regression curve. ■

FAQs

When to Use an Exponential Model for Data Points, and When to Use e in Your Model

Q: *Given a set of data points that appear to be curving upward, how can I tell whether to use a quadratic model or an exponential model?*

A: Here are some things to look for:

- Do the data values appear to double at regular intervals? (For example, do the values approximately double every 5 units?) If so, then an exponential model is appropriate. If it takes longer and longer to double, then a quadratic model may be more appropriate.
- Do the values first decrease to a low point and then increase? If so, then a quadratic model is more appropriate.

It is also helpful to use technology to graph both the regression quadratic and exponential curves and to visually inspect the graphs to determine which gives the closest fit to the data.

Q: *We have two ways of writing exponential functions: $f(x) = Ab^x$ and $f(x) = Ae^{rx}$. How do we know which one to use?*

A: The two forms are equivalent, and it is always possible to convert from one form to the other.* So, use whichever form seems to be convenient for a particular situation. For instance, $f(t) = A(3^t)$ conveniently models exponential growth that is tripling every unit of time, whereas $f(t) = Ae^{0.06t}$ conveniently models an investment with continuous compounding at 6%.

* **NOTE** Quick Example 3 on page 145 shows how to convert Ae^{rx} to Ab^x. Conversion from Ab^x to Ae^{rx} involves logarithms: $r = \ln b$.

2.2 EXERCISES

▼ more advanced ◆ challenging

T indicates exercises that should be solved using technology

For each function in Exercises 1–12, compute the missing values in the following table and supply a valid technology formula for the given function: HINT [See Quick Examples on page 137.]

x	−3	−2	−1	0	1	2	3
$f(x)$							

1. $f(x) = 4^x$

2. $f(x) = 3^x$

3. $f(x) = 3^{-x}$

4. $f(x) = 4^{-x}$

5. $g(x) = 2(2^x)$

6. $g(x) = 2(3^x)$

7. $h(x) = -3(2^{-x})$

8. $h(x) = -2(3^{-x})$

9. $r(x) = 2^x - 1$

10. $r(x) = 2^{-x} + 1$

11. $s(x) = 2^{x-1}$

12. $s(x) = 2^{1-x}$

Using a chart of values, graph each of the functions in Exercises 13–18. (Use $-3 \le x \le 3$.)

13. $f(x) = 3^{-x}$

14. $f(x) = 4^{-x}$

15. $g(x) = 2(2^x)$

16. $g(x) = 2(3^x)$

17. $h(x) = -3(2^{-x})$

18. $h(x) = -2(3^{-x})$

In Exercises 19–24, the values of two functions, f and g, are given in a table. One, both, or neither of them may be exponential. Decide which, if any, are exponential, and give the exponential models for those that are. HINT [See Example 1.]

19.

x	−2	−1	0	1	2
f(x)	0.5	1.5	4.5	13.5	40.5
g(x)	8	4	2	1	$\frac{1}{2}$

20.

x	−2	−1	0	1	2
f(x)	$\frac{1}{2}$	1	2	4	8
g(x)	3	0	−1	0	3

21.

x	−2	−1	0	1	2
f(x)	22.5	7.5	2.5	7.5	22.5
g(x)	0.3	0.9	2.7	8.1	16.2

22.

x	−2	−1	0	1	2
f(x)	0.3	0.9	2.7	8.1	24.3
g(x)	3	1.5	0.75	0.375	0.1875

23.

x	−2	−1	0	1	2
f(x)	100	200	400	600	800
g(x)	100	20	4	0.8	0.16

24.

x	−2	−1	0	1	2
f(x)	0.8	0.2	0.1	0.05	0.025
g(x)	80	40	20	10	2

T *For each function in Exercises 25–30, supply a valid technology formula and then use technology to compute the missing values in the following table:* HINT [See Quick Examples on page 137.]

x	−3	−2	−1	0	1	2	3
f(x)							

25. $f(x) = e^{-2x}$

26. $g(x) = e^{x/5}$

27. $h(x) = 1.01(2.02^{-4x})$

28. $h(x) = 3.42(3^{-x/5})$

29. $r(x) = 50\left(1+\frac{1}{3.2}\right)^{2x}$

30. $r(x) = 0.043\left(4.5 - \frac{5}{1.2}\right)^{-x}$

In Exercises 31–38, supply a valid technology formula for the given function.

31. 2^{x-1}

32. 2^{-4x}

33. $\dfrac{2}{1-2^{-4x}}$

34. $\dfrac{2^{3-x}}{1-2^x}$

35. $\dfrac{(3+x)^{3x}}{x+1}$

36. $\dfrac{20.3^{3x}}{1+20.3^{2x}}$

37. $2e^{(1+x)/x}$

38. $\dfrac{2e^{2/x}}{x}$

T *On the same set of axes, use technology to graph the pairs of functions in Exercises 39–46 with $-3 \le x \le 3$. Identify which graph corresponds to which function.* HINT [See Quick Examples on page 138.]

39. $f_1(x) = 1.6^x,\ f_2(x) = 1.8^x$

40. $f_1(x) = 2.2^x,\ f_2(x) = 2.5^x$

41. $f_1(x) = 300(1.1^x),\ f_2(x) = 300(1.1^{2x})$

42. $f_1(x) = 100(1.01^{2x}),\ f_2(x) = 100(1.01^{3x})$

43. $f_1(x) = 2.5^{1.02x},\ f_2(x) = e^{1.02x}$

44. $f_1(x) = 2.5^{-1.02x},\ f_2(x) = e^{-1.02x}$

45. $f_1(x) = 1{,}000(1.045^{-3x}),\ f_2(x) = 1{,}000(1.045^{3x})$

46. $f_1(x) = 1{,}202(1.034^{-3x}),\ f_2(x) = 1{,}202(1.034^{3x})$

For Exercises 47–54, model the data using an exponential function $f(x) = Ab^x$. HINT [See Example 1.]

47.

x	0	1	2
f(x)	500	250	125

48.

x	0	1	2
f(x)	500	1,000	2,000

49.

x	0	1	2
f(x)	10	30	90

50.

x	0	1	2
f(x)	90	30	10

51.

x	0	1	2
f(x)	500	225	101.25

52.

x	0	1	2
f(x)	5	3	1.8

53.

x	1	2
f(x)	−110	−121

54.

x	1	2
f(x)	−41	−42.025

Find equations for exponential functions that pass through the pairs of points given in Exercises 55–62. (Round all coefficients to 4 decimal places when necessary.) HINT [See Example 2.]

55. Through (2, 36) and (4, 324)

56. Through (2, −4) and (4, −16)

57. Through (−2, −25) and (1, −0.2)

58. Through (1, 1.2) and (3, 0.108)

59. Through (1, 3) and (3, 6)

60. Through (1, 2) and (4, 6)

61. Through (2, 3) and (6, 2)

62. Through (−1, 2) and (3, 1)

Obtain exponential functions in the form $f(t) = Ae^{rt}$ in Exercises 63–66. HINT [See Example 5.]

63. $f(t)$ is the value after t years of a \$5,000 investment earning 10% interest compounded continuously.

64. $f(t)$ is the value after t years of a \$2,000 investment earning 5.3% interest compounded continuously.

65. $f(t)$ is the value after t years of a \$1,000 investment depreciating continuously at an annual rate of 6.3%.

66. $f(t)$ is the value after t years of a \$10,000 investment depreciating continuously at an annual rate of 60%.

T *In Exercises 67–70, use technology to find the exponential regression function through the given points. (Round all coefficients to 4 decimal places.)* HINT [See Example 6.]

67. $\{(1, 2), (3, 5), (4, 9), (5, 20)\}$

68. $\{(-1, 2), (-3, 5), (-4, 9), (-5, 20)\}$

69. $\{(-1, 10), (-3, 5), (-4, 3)\}$

70. $\{(3, 3), (4, 5), (5, 10)\}$

APPLICATIONS

71. ***Aspirin*** Soon after taking an aspirin, a patient has absorbed 300 mg of the drug. After two hours, only 75 mg remain, find an exponential model for the amount of aspirin in the bloodstream after t hours, and use your model to find the amount of aspirin in the bloodstream after 5 hours. HINT [See Example 2.]

72. ***Alcohol*** After a large number of drinks, a person has a blood alcohol level of 200 mg/dL (milligrams per deciliter). If the amount of alcohol in the blood decays exponentially, and after 2 hours, 112.5 mg/dL remain, find an exponential model for the person's blood alcohol level, and use your model to estimate the person's blood alcohol level after 4 hours. HINT [See Example 2.]

73. ***Freon Production*** The production of ozone-layer damaging Freon 22 (chlorodifluoromethane) in developing countries rose from 200 tons in 2004 to a projected 590 tons in 2010.[13]

a. Use this information to find both a linear model and an exponential model for the amount F of Freon 22 (in tons) as a function of time t in years since 2000. (Round all coefficients to three significant digits.) HINT [See Example 2.] Which of these models would you judge to be more appropriate to the data shown below:

t (year since 2000)	0	2	4	6	8	10
F (tons of Freon 22)	100	140	200	270	400	590

b. Use the better of the two models from part (a) to predict the 2008 figure and compare it with the projected figure above.

74. ***Satellite Radio Subscriptions*** The number of Sirius Satellite Radio subscribers grew from 0.3 million in 2003 to 3.2 million in 2005.[14]

a. Use this information to find both a linear model and an exponential model for the number N of subscribers (in millions) as a function of time t in years since 2000. (Round all coefficients to three significant digits.) HINT [See Example 2.] Which of these models would you judge to be more appropriate to the data shown below:

t (year since 2000)	3	4	5
N (millions of subscribers)	0.3	1.0	3.2

b. Use the better of the two models from part (a) to predict the 2006 figure. (The actual 2006 figure was approximately 6 million subscribers.)

75. ▼ ***U.S. Population*** The U.S. population was 180 million in 1960 and 303 million in 2008.[15]

a. Use these data to give an exponential growth model showing the U.S. population P as a function of time t in years since 1960. Round coefficients to 6 significant digits. HINT [See Example 2.]

b. By experimenting, determine the smallest number of significant digits to which you should round the coefficients in part (a) in order to obtain the correct 2008 population figure accurate to three significant digits.

c. Using the model in part (a), predict the population in 2020.

76. ▼ ***World Population*** World population was estimated at 2.56 billion people in 1950 and 6.72 billion people in 2008.[16]

a. Use these data to give an exponential growth model showing the world population P as a function of time t in years since 1950. Round coefficients to 6 significant digits. HINT [See Example 2.]

b. By experimenting, determine the smallest number of significant digits to which you should round the coefficients in part (a) in order to obtain the correct 2008 population figure to three significant digits.

c. Assuming the exponential growth model from part (a), estimate the world population in the year 1000. Comment on your answer.

77. ▼ ***Frogs*** Frogs have been breeding like flies at the Enormous State University (ESU) campus! Each year, the pledge class of the Epsilon Delta fraternity is instructed to tag all the frogs residing on the ESU campus. Two years ago they managed to tag all 50,000 of them (with little Epsilon Delta Fraternity tags).

[13] Figures are approximate. Source: Lampert Kuijpers (Panel of the Montreal Protocol), National Bureau of Statistics in China, via CEIC DSata/*New York Times*, February 23, 2007, p. C1.

[14] Figures are approximate. Source: Sirius Satellite Radio/*New York Times*, February 20, 2008, p. A1.

[15] Figures are rounded to three significant digits. Source: U.S. Census Bureau, www.census.gov/.

[16] Ibid.

This year's pledge class discovered that last year's tags had all fallen off, and they wound up tagging a total of 75,000 frogs.

a. Find an exponential model for the frog population.

b. Assuming exponential population growth, and that all this year's tags have fallen off, how many tags should Epsilon Delta order for next year's pledge class?

78. ▼ ***Flies*** Flies in Suffolk County have been breeding like frogs! Three years ago the Health Commission caught 4,000 flies in a trap in one hour. This year it caught 7,000 flies in one hour.

a. Find an exponential model for the fly population.

b. Assuming exponential population growth, how many flies should the commission expect to catch next year?

79. ***Bacteria*** A bacteria culture starts with 1,000 bacteria and doubles in size every 3 hours. Find an exponential model for the size of the culture as a function of time t in hours and use the model to predict how many bacteria there will be after 2 days. HINT [See Example 4.]

80. ***Bacteria*** A bacteria culture starts with 1,000 bacteria. Two hours later there are 1,500 bacteria. Find an exponential model for the size of the culture as a function of time t in hours, and use the model to predict how many bacteria there will be after 2 days. HINT [See Example 4.]

81. ***SARS*** In the early stages of the deadly SARS (Severe Acute Respiratory Syndrome) epidemic in 2003, the number of cases was increasing by about 18% each day.[17] On March 17, 2003 (the first day for which statistics were reported by the World Health Organization) there were 167 cases. Find an exponential model that predicts the number of cases t days after March 17, 2003, and use it to estimate the number of cases on March 31, 2003. (The actual reported number of cases was 1,662.)

82. ***SARS*** A few weeks into the deadly SARS (Severe Acute Respiratory Syndrome) epidemic in 2003, the number of cases was increasing by about 4% each day.[18] On April 1, 2003 there were 1,804 cases. Find an exponential model that predicts the number of cases t days after April 1, 2003, and use it to estimate the number of cases on April 30, 2003. (The actual reported number of cases was 5,663.)

83. ***Investments*** In 2007, E*TRADE Financial was offering 4.94% interest on its online savings accounts, with interest reinvested monthly.[19] Find the associated exponential model for the value of a \$5,000 deposit after t years. Assuming this rate of return continued for seven years, how much would a deposit of \$5,000 at the beginning of 2007 be worth at the end of 2014? (Answer to the nearest \$1.) HINT [See Example 9; you saw this exercise before in Section 1.2.]

84. ***Investments*** In 2007, ING Direct was offering 4.14% interest on its online Orange Savings Account, with interest reinvested quarterly.[20] Find the associated exponential model for the value of a \$4,000 deposit after t years. Assuming this rate of return continued for eight years, how much would a deposit of \$4,000 at the beginning of 2007 be worth at the end of 2015? (Answer to the nearest \$1.) HINT [See Example 9; you saw this exercise before in Section 1.2.]

85. T ***Investments*** Refer to Exercise 83. At the start of which year will an investment of \$5,000 made at the beginning of 2007 first exceed \$7,500?

86. T ***Investments*** Refer to Exercise 84. At the start of which year will an investment of \$4,000 made at the beginning of 2007 first exceed \$6,000?

87. ***Investments*** Rock Solid Bank & Trust is offering a CD (certificate of deposit) that pays 4% compounded continuously. How much interest would a \$1,000 deposit earn over 10 years? HINT [See Example 5.]

88. ***Savings*** FlybynightSavings.com is offering a savings account that pays 31% interest compounded continuously. How much interest would a deposit of \$2,000 earn over 10 years?

89. ***Home Sales*** Sales of new houses in the United States declined continuously over the period 2005–2008 at a rate of 30% per year from 1.3 million in 2005.[21] Write down a formula that predicts sales of new houses t years after 2005. Use your model to estimate sales of new houses in 2008 and 2009.

90. ***Home Prices*** The median price of a home in the United States declined continuously over the period 2005–2008 at a rate of 5.5% per year from around \$230 thousand in 2005.[22] Write down a formula that predicts the median price of a home t years after 2005. Use your model to estimate the median home price in 2007 and 2010. HINT [See Example 5.]

91. ***Global Warming*** The most abundant greenhouse gas is carbon dioxide. According to a United Nations "worst-case scenario" prediction, the amount of carbon dioxide in the atmosphere (in parts of volume per million) can be approximated by

$$C(t) \approx 277e^{0.00353t} \text{ parts per million} \quad (0 \le t \le 350)$$

where t is time in years since 1750.[23]

a. Use the model to estimate the amount of carbon dioxide in the atmosphere in 1950, 2000, 2050, and 2100.

b. According to the model, when, to the nearest decade, will the level surpass 700 parts per million?

[17] Source: World Health Organization, www.who.int/.

[18] Ibid.

[19] Interest rate based on annual percentage yield. Source: http://us.etrade.com, December, 2007.

[20] Interest rate based on annual percentage yield. Source: www.home.ingdirect.com, December, 2007.

[21] Based on figures released by the U.S. Census, www.census.gov/.

[22] Based on figures released by the National Association of Realtors, www.realtor.org/.

[23] Exponential regression based on the 1750 figure and the 2100 UN prediction. Sources: Tom Boden/Oak Ridge National Laboratory, Scripps Institute of Oceanography/University of California, International Panel on Climate Change/*New York Times*, December 1, 1997, p. F1.

92. ***Global Warming*** Repeat Exercise 91 using the United Nations "midrange scenario" prediction:

$$C(t) \approx 277e^{0.00267t} \text{ parts per million} \quad (0 \le t \le 350)$$

where t is time in years since 1750.

93. ***New York City Housing Costs: Downtown*** The following table shows the average price of a two-bedroom apartment in downtown New York City during the real estate boom from 1994 to 2004.[24]

t	0 (1994)	2	4	6	8	10 (2004)
Price ($ million)	0.38	0.40	0.60	0.95	1.20	1.60

a. Use exponential regression to model the price $P(t)$ as a function of time t since 1994. Include a sketch of the points and the regression curve. (Round the coefficients to 3 decimal places.) HINT [See Example 6.]

b. Extrapolate your model to estimate the cost of a two-bedroom downtown apartment in 2005.

94. ***New York City Housing Costs: Uptown*** The following table shows the average price of a two-bedroom apartment in uptown New York City during the real estate boom from 1994 to 2004.[25]

t	0 (1994)	2	4	6	8	10 (2004)
Price ($ million)	0.18	0.18	0.19	0.2	0.35	0.4

a. Use exponential regression to model the price $P(t)$ as a function of time t since 1994. Include a sketch of the points and the regression curve. (Round the coefficients to 3 decimal places.)

b. Extrapolate your model to estimate the cost of a two-bedroom uptown apartment in 2005.

95. ***Facebook*** The following table gives the approximate number of **Facebook** memberships at various times since early in 2005.[26]

Year t (since start of 2005)	0	0.5	1	1.5	2	2.5	3	3.5
Facebook Members n (millions)	1	2	5.5	7	12	30	58	80

a. Use exponential regression to model **Facebook** membership as a function of time in years since the start of 2005, and graph the data points and regression curve. (Round coefficients to 3 decimal places.)

b. Fill in the missing quantity: According to the model, **Facebook** membership each year was ____ times that of the year before.

c. Use your model to estimate **Facebook** membership in early 2009 to the nearest million.

96. ***Freon Production*** (Refer to Exercise 73.) The following table shows estimated total Freon 22 production in various years since 2000.[27]

***t* (year since 2000)**	0	2	4	6	8	10
***F* (tons of Freon 22)**	100	140	200	270	400	590

a. Use exponential regression to model Freon production as a function of time in years since 2000, and graph the data points and regression curve. (Round coefficients to 3 decimal places.)

b. Fill in the missing quantity: According to the model, Freon production each year was ____ times that of the year before.

c. Use your model to estimate Freon production in 2009 to the nearest ton.

COMMUNICATION AND REASONING EXERCISES

97. Which of the following three functions will be largest for large values of x?

(A) $f(x) = x^2$ **(B)** $r(x) = 2^x$ **(C)** $h(x) = x^{10}$

98. Which of the following three functions will be smallest for large values of x?

(A) $f(x) = x^{-2}$ **(B)** $r(x) = 2^{-x}$ **(C)** $h(x) = x^{-10}$

99. What limitations apply to using an exponential function to model growth in real-life situations? Illustrate your answer with an example.

100. Explain in words why 5% per year compounded continuously yields more interest than 5% per year compounded monthly.

101. The following commentary and graph appeared in www.politicalcalculations.blogspot.com on August 30 2005:[28]

> One of the neater blogs I've recently encountered is The Real Returns, which offers a wealth of investing, market and economic data. Earlier this month, The Real Returns posted data related to the recent history of U.S. median house prices over the period from 1963 to 2004. The original source of the housing data is the U.S. Census Bureau.

---

[24] Data are rounded and 2004 figure is an estimate. Source: Miller Samuel/*New York Times*, March 28, 2004, p. RE 11.

[25] Ibid.

[26] Sources: www.facebook.com/, www.insidehighered.com (Some data are interpolated.)

[27] Figures are approximate. Source: Lampert Kuijpers (Panel of the Montreal Protocol), National Bureau of Statistics in China, via CEIC DSata/*New York Times*, February 23, 2007, p. C1.

[28] The graph was recreated by the authors using the blog author's data source.
Source for article: www.politicalcalculations.blogspot.com/2005/08/projecting-us-median-housing-prices.html.
Source for data: http://therealreturns.blogspot.com/2005_08_01_archive.html.

Well, that kind of data deserves some curve-fitting and a calculator to estimate what the future U.S. median house price might be, so Political Calculations has extracted the data from 1973 onward to create the following chart:

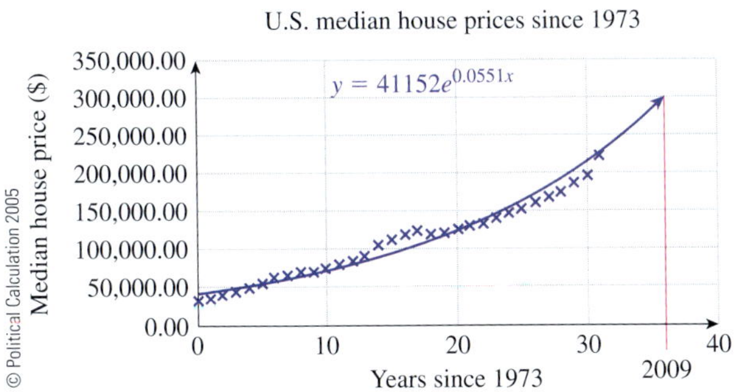

Comment on the article and graph. HINT [See Exercise 90.]

102. ▼ Refer to Exercise 101. Of what possible predictive use, then, is the kind of exponential model given by the blogger in the article referred to?

103. ▼ Describe two real-life situations in which a linear model would be more appropriate than an exponential model, and two situations in which an exponential model would be more appropriate than a linear model.

104. ▼ Describe a real-life situation in which a quadratic model would be more appropriate than an exponential model and one in which an exponential model would be more appropriate than a quadratic model.

105. How would you check whether data points of the form $(1, y_1)$, $(2, y_2)$, $(3, y_3)$ lie on an exponential curve?

106. ▼ You are told that the points $(1, y_1)$, $(2, y_2)$, $(3, y_3)$ lie on an exponential curve. Express y_3 in terms of y_1 and y_2.

107. ▼ Your local banker tells you that the reason his bank doesn't compound interest continuously is that it would be too demanding of computer resources because the computer would need to spend a great deal of time keeping all accounts updated. Comment on his reasoning.

108. ▼ Your other local banker tells you that the reason *her* bank doesn't offer continuously compounded interest is that it is equivalent to offering a fractionally higher interest rate compounded daily. Comment on her reasoning.

2.3 Logarithmic Functions and Models

Logarithms were invented by John Napier (1550–1617) in the late 16th century as a means of aiding calculation. His invention made possible the prodigious hand calculations of astronomer Johannes Kepler (1571–1630), who was the first to describe accurately the orbits and the motions of the planets. Today, computers and calculators have done away with that use of logarithms, but many other uses remain. In particular, the logarithm is used to model real-world phenomena in numerous fields, including physics, finance, and economics.

From the equation

$$2^3 = 8$$

we can see that the power to which we need to raise 2 in order to get 8 is 3. We abbreviate the phrase "the power to which we need to raise 2 in order to get 8" as "$\log_2 8$." Thus, another way of writing the equation $2^3 = 8$ is

$$\log_2 8 = 3 \qquad \text{The power to which we need to raise 2 in order to get 8 is 3.}$$

This is read "the base 2 logarithm of 8 is 3" or "the log, base 2, of 8 is 3."

Here is the general definition.

Base *b* Logarithm

The **base *b* logarithm of *x***, $\log_b x$, is the power to which we need to raise b in order to get x. Symbolically,

$\log_b x = y$	means	$b^y = x$.
Logarithmic form		*Exponential form*

Quick Examples

1. The following table lists some exponential equations and their equivalent logarithmic forms:

Exponential Form	$10^3 = 1000$	$4^2 = 16$	$3^3 = 27$	$5^1 = 5$	$7^0 = 1$	$4^{-2} = \frac{1}{16}$	$25^{1/2} = 5$
Logarithmic Form	$\log_{10} 1000 = 3$	$\log_4 16 = 2$	$\log_3 27 = 3$	$\log_5 5 = 1$	$\log_7 1 = 0$	$\log_4 \frac{1}{16} = -2$	$\log_{25} 5 = \frac{1}{2}$

2. $\log_3 9$ = the power to which we need to raise 3 in order to get 9. Because $3^2 = 9$, this power is 2, so $\log_3 9 = 2$.
3. $\log_{10} 10{,}000$ = the power to which we need to raise 10 in order to get 10,000. Because $10^4 = 10{,}000$, this power is 4, so $\log_{10} 10{,}000 = 4$.
4. $\log_3 \frac{1}{27}$ is the power to which we need to raise 3 in order to get $\frac{1}{27}$. Because $3^{-3} = \frac{1}{27}$ this power is -3, so $\log_3 \frac{1}{27} = -3$.
5. $\log_b 1 = 0$ for every positive number b other than 1 because $b^0 = 1$.

Note The number $\log_b x$ is defined only if b and x are both positive and $b \neq 1$. Thus, it is impossible to compute, say, $\log_3(-9)$ (because there is no power of 3 that equals -9), or $\log_1(2)$ (because there is no power of 1 that equals 2). ■

Logarithms with base 10 and base e are frequently used, so they have special names and notations.

Common Logarithm, Natural Logarithm

The following are standard abbreviations.

		TI-83/84 Plus & Excel Formula
Base 10: $\log_{10} x = \log x$	*Common Logarithm*	`log(x)`
Base e: $\log_e x = \ln x$	*Natural Logarithm*	`ln(x)`

Quick Examples

Logarithmic Form	Exponential Form
1. $\log 10{,}000 = 4$	$10^4 = 10{,}000$
2. $\log 10 = 1$	$10^1 = 10$
3. $\log \frac{1}{10{,}000} = -4$	$10^{-4} = \frac{1}{10{,}000}$
4. $\ln e = 1$	$e^1 = e$
5. $\ln 1 = 0$	$e^0 = 1$
6. $\ln 2 = 0.69314718\ldots$	$e^{0.69314718\ldots} = 2$

Some technologies (such as calculators) do not permit direct calculation of logarithms other than common and natural logarithms. To compute logarithms with other bases with these technologies, we can use the following formula:

Change-of-Base Formula

$$\log_b a = \frac{\log a}{\log b} = \frac{\ln a}{\ln b}$$ Change-of-base formula*

Quick Examples

1. $\log_{11} 9 = \dfrac{\log 9}{\log 11} \approx 0.91631$ `log(9)/log(11)`

2. $\log_{11} 9 = \dfrac{\ln 9}{\ln 11} \approx 0.91631$ `ln(9)/ln(11)`

3. $\log_{3.2}\left(\dfrac{1.42}{3.4}\right) \approx -0.75065$ `log(1.42/3.4)/log(3.2)`

Using Technology to Compute Logarithms

To compute $\log_b x$ using technology, use the following formulas:

TI-83/84 Plus	`log(x)/log(b)`	Example: $\log_2(16)$ is `log(16)/log(2)`
Excel:	`=LOG(x,b)`	Example: $\log_2(16)$ is `=LOG(16,2)`

* **NOTE** Here is a quick explanation of why this formula works: To calculate $\log_b a$, we ask, "to what power must we raise b to get a?" To check the formula, we try using $\log a/\log b$ as the exponent.

$$b^{\frac{\log a}{\log b}} = (10^{\log b})^{\frac{\log a}{\log b}} \quad (\text{because } b = 10^{\log b})$$
$$= 10^{\log a} = a$$

so this exponent works!

One important use of logarithms is to solve equations in which the unknown is in the exponent.

EXAMPLE 1 Solving Equations with Unknowns in the Exponent

Solve the following equations

a. $5^{-x} = 125$ **b.** $3^{2x-1} = 6$ **c.** $100(1.005)^{3x} = 200$

Solution

a. Write the given equation $5^{-x} = 125$ in logarithmic form:

$$-x = \log_5 125$$

This gives $x = -\log_5 125 = -3$.

b. In logarithmic form, $3^{2x-1} = 6$ becomes

$$2x - 1 = \log_3 6$$
$$2x = 1 + \log_3 6$$

giving
$$x = \frac{1 + \log_3 6}{2} \approx \frac{1 + 1.6309}{2} \approx 1.3155.$$

c. We cannot write the given equation, $100(1.005)^{3x} = 200$, directly in exponential form. We must first divide both sides by 100:

$$1.005^{3x} = \frac{200}{100} = 2$$
$$3x = \log_{1.005} 2$$
$$x = \frac{\log_{1.005} 2}{3} \approx \frac{138.9757}{3} \approx 46.3252.$$

Now that we know what logarithms are, we can talk about functions based on logarithms:

Logarithmic Function

A **logarithmic function** has the form

$$f(x) = \log_b x + C \qquad (b \text{ and } C \text{ are constants with } b > 0,\ b \neq 1)$$

or, alternatively,

$$f(x) = A \ln x + C. \qquad (A, C \text{ constants with } A \neq 0)$$

Quick Examples

1. $f(x) = \log x$
2. $g(x) = \ln x - 5$
3. $h(x) = \log_2 x + 1$
4. $k(x) = 3.2 \ln x + 7.2$

Q: *What is the difference between the two forms of the logarithmic function?*

A: None, really, they're equivalent: We can start with an equation in the first form and use the change-of-base formula to rewrite it:

$$\begin{aligned} f(x) &= \log_b x + C \\ &= \frac{\ln x}{\ln b} + C \qquad \text{Change-of-base formula} \\ &= \left(\frac{1}{\ln b}\right) \ln x + C. \end{aligned}$$

Our function now has the form $f(x) = A \ln x + C$, where $A = 1/\ln b$. We can go the other way as well, to rewrite $A \ln x + C$ in the form $\log_b x + C$.

EXAMPLE 2 Graphs of Logarithmic Functions

a. Sketch the graph of $f(x) = \log_2 x$ by hand.

b. Use technology to compare the graph in part (a) with the graphs of $\log_b x$ for $b = 1/4, 1/2$, and 4.

Solution

a. To sketch the graph of $f(x) = \log_2 x$ by hand, we begin with a table of values. Because $\log_2 x$ is not defined when $x = 0$, we choose several values of x close to zero and also some larger values, all chosen so that their logarithms are easy to compute:

x	$\frac{1}{8}$	$\frac{1}{4}$	$\frac{1}{2}$	1	2	4	8
$f(x) = \log_2 x$	-3	-2	-1	0	1	2	3

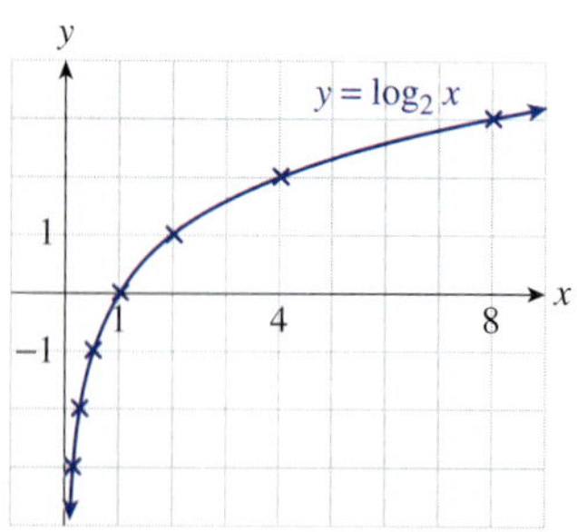

Figure 14

Graphing these points and joining them by a smooth curve gives us Figure 14.

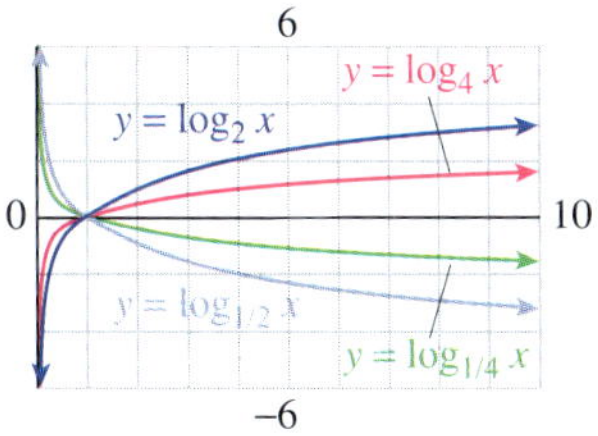

Figure 15

b. We enter the logarithmic functions in graphing utilities as follows (note the use of the change-of-base formula in the TI-83/84 Plus version):

TI-83/84 Plus	**Excel**
`Y1=log(X)/log(0.25)`	`=LOG(x,0.25)`
`Y2=log(X)/log(0.5)`	`=LOG(x,0.5)`
`Y3=log(X)/log(2)`	`=LOG(x,2)`
`Y4=log(X)/log(4)`	`=LOG(x,4)`

Figure 15 shows the resulting graphs.

➡ **Before we go on...** Notice that the graphs of the logarithmic functions in Example 2 all pass through the point (1, 0). (Why?) Notice further that the graphs of the logarithmic functions with bases less than 1 are upside-down versions of the others. Finally, how are these graphs related to the graphs of exponential functions? ■

Below are some important algebraic properties of logarithms we shall use throughout the rest of this section.

Web Site
Go to the student Web site at **www.AppliedCalc.org** and follow the path

Chapter 2

→ Using and Deriving Algebraic Properties of Logarithms

to find a list of logarithmic identities and a discussion on where they come from.

Logarithm Identities

The following identities hold for all positive bases $a \neq 1$ and $b \neq 1$, all positive numbers x and y, and every real number r. These identities follow from the laws of exponents.

Identity	**Quick Examples**
1. $\log_b(xy) = \log_b x + \log_b y$	$\log_2 16 = \log_2 8 + \log_2 2$
2. $\log_b\left(\frac{x}{y}\right) = \log_b x - \log_b y$	$\log_2\left(\frac{5}{3}\right) = \log_2 5 - \log_2 3$
3. $\log_b(x^r) = r\log_b x$	$\log_2(6^5) = 5\log_2 6$
4. $\log_b b = 1;\ \log_b 1 = 0$	$\log_2 2 = 1;\ \ln e = 1;\ \log_{11} 1 = 0$
5. $\log_b\left(\frac{1}{x}\right) = -\log_b x$	$\log_2\left(\frac{1}{3}\right) = -\log_2 3$
6. $\log_b x = \frac{\log_a x}{\log_a b}$	$\log_2 5 = \frac{\log_{10} 5}{\log_{10} 2} = \frac{\log 5}{\log 2}$

Relationship with Exponential Functions

The following two identities demonstrate that the operations of taking the base b logarithm and raising b to a power are *inverse* to each other.*

Identity	**Quick Example**
1. $\log_b(b^x) = x$ *In words:* The power to which you raise b in order to get b^x is x (!)	$\log_2(2^7) = 7$
2. $b^{\log_b x} = x$ *In words:* Raising b to the power to which it must be raised to get x, yields x (!)	$5^{\log_5 8} = 8$

* **NOTE**
Go to the student Web site at **www.AppliedCalc.org** and follow the path

Chapter 2

→ Inverse Functions

for a general discussion of inverse functions, including further discussion of the relationship between logarithmic and exponential functions.

APPLICATIONS

EXAMPLE 3 Investments: How Long?

Global bonds sold by Mexico are yielding an average of 5.2% per year.* At that interest rate, how long will it take a $1,000 investment to be worth $1,500 if the interest is compounded monthly?

Solution Substituting $A = 1{,}500$, $P = 1{,}000$, $r = 0.052$, and $m = 12$ in the compound interest equation gives

$$\begin{aligned} A(t) &= P\left(1 + \frac{r}{m}\right)^{mt} \\ 1{,}500 &= 1{,}000\left(1 + \frac{0.052}{12}\right)^{12t} \\ &\approx 1{,}000(1.004333)^{12t} \end{aligned}$$

and we must solve for t. We first divide both sides by 1,000, getting an equation in exponential form:

$$1.5 = 1.004333^{12t}$$

In logarithmic form, this becomes

$$12t = \log_{1.004333}(1.5).$$

We can now solve for t:

$$t = \frac{\log_{1.004333}(1.5)}{12} \approx 7.8 \text{ years}$$

```
log(1.5)/(log(1.004333)*12)
```

Thus, it will take approximately 7.8 years for a $1,000 investment to be worth $1,500.

* In 2008 (bonds maturing 03/03/2015). Source: www.bloomberg.com.

Before we go on... We can use the logarithm identities to solve the equation

$$1.5 = 1.004333^{12t}$$

that arose in Example 3 (and also more general equations with unknowns in the exponent) by taking the natural logarithm of both sides:

$$\begin{aligned} \ln 1.5 &= \ln(1.004333^{12t}) \\ &= 12t \ln 1.004333 \qquad \text{By Identity 3} \end{aligned}$$

We can now solve this for t to get

$$t = \frac{\ln 1.5}{12 \ln 1.004333}$$

which, by the change-of-base formula, is equivalent to the answer we got in Example 3. ■

EXAMPLE 4 Half-Life

a. The weight of carbon 14 that remains in a sample that originally contained A grams is given by

$$C(t) = A(0.999879)^t$$

where t is time in years. Find the **half-life**, the time it takes half of the carbon 14 in a sample to decay.

b. Repeat part (a) using the following alternative form of the exponential model in part (a):

$$C(t) = Ae^{-0.000\,121\,01t}$$ See page 145.

c. Another radioactive material has a half-life of 7,000 years. Find an exponential decay model in the form

$$R(t) = Ae^{-kt}$$

for the amount of undecayed material remaining. (The constant k is called the **decay constant**.)

d. How long will it take for 99.95% of the substance in a sample of the material in part (c) to decay?

Solution

a. We want to find the value of t for which $C(t) =$ the weight of undecayed carbon 14 left $=$ half the original weight $= 0.5A$. Substituting, we get

$$0.5A = A(0.999879)^t$$

Dividing both sides by A gives

$$0.5 = 0.999879^t \quad \text{Exponential form}$$

$$t = \log_{0.999879} 0.5 \approx 5{,}728 \text{ years} \quad \text{Logarithmic form}$$

b. This is similar to part (a): We want to solve the equation

$$0.5A = Ae^{-0.000\,121\,01t}$$

for t. Dividing both sides by A gives

$$0.5 = e^{-0.000\,121\,01t}$$

Taking the natural logarithm of both sides gives

$$\ln(0.5) = \ln(e^{-0.000\,121\,01t}) = -0.000\,121\,01t \quad \text{Identity 3: } \ln(e^a) = a \ln e = a$$

$$t = \frac{\ln(0.5)}{-0.000\,121\,01} \approx 5{,}728 \text{ years}$$

as we obtained in part (a).

c. This time we are given the half-life, which we can use to find the exponential model $R(t) = Ae^{-kt}$. At time $t = 0$, the amount of radioactive material is

$$R(0) = Ae^0 = A$$

Because half of the sample decays in 7,000 years, this sample will decay to $0.5A$ grams in 7,000 years ($t = 7{,}000$). Substituting this information gives

$$0.5A = Ae^{-k(7{,}000)}.$$

Canceling A and taking natural logarithms (again using Identity 3) gives

$$\ln(0.5) = -7{,}000k$$

so the decay constant k is

$$k = -\tfrac{\ln(0.5)}{7{,}000} \approx 0.000\,099\,021.$$

Therefore, the model is

$$R(t) = Ae^{-0.000\,099\,021t}.$$

d. If 99.95% of the substance in a sample has decayed, then the amount of undecayed material left is 0.05% of the original amount, or $0.0005A$. We have

$$0.0005A = Ae^{-0.000\,099\,021t}$$
$$0.0005 = e^{-0.000\,099\,021t}$$
$$\ln(0.0005) = -0.000\,099\,021t$$
$$t = \frac{\ln(0.0005)}{-0.000\,099\,021} \approx 76{,}760 \text{ years.}$$

Before we go on...

Q: *In parts (a) and (b) of Example 4 we were given two different forms of the model for carbon 14 decay. How do we convert an exponential function in one form to the other?*

A: We have already seen (See Quick Example 3 on page 145) how to convert from the form $f(t) = Ae^{rt}$ in part (b) to the form $f(t) = Ab^t$ in part (a). To go the other way, start with the model in part (a), and equate it to the desired form:

$$C(t) = A(0.999\,879)^t = Ae^{rt}.$$

To solve for r, cancel the As and take the natural logarithm of both sides:

$$t\ln(0.999\,879) = rt\ln e = rt$$

so $r = \ln(0.999\,879) \approx -0.000\,121\,007$,

giving

$$C(t) = Ae^{-0.000\,121\,007t}$$

as in part (b).

We can use the work we did in parts (b) and (c) of the above example to obtain a formula for the decay constant in an exponential decay model for any radioactive substance when we know its half-life. Write the half-life as t_h. Then the calculation in part (b) gives

$$k = -\frac{\ln(0.5)}{t_h} = \frac{\ln 2}{t_h}. \qquad -\ln(0.5) = -\ln\left(\frac{1}{2}\right) = \ln 2$$

Multiplying both sides by t_h gives us the relationship $t_h k = \ln 2$.

Exponential Decay Model and Half-Life

An **exponential decay function** has the form

$$Q(t) = Q_0 e^{-kt}. \qquad Q_0,\ k \text{ both positive}$$

Q_0 represents the value of Q at time $t = 0$, and k is the **decay constant.** The decay constant k and half-life t_h for Q are related by

$$t_h k = \ln 2.$$

Quick Examples

1. $Q(t) = Q_0 e^{-0.000\,121\,01t}$ is the decay function for carbon 14 (see Example 4b).
2. If $t_h = 10$ years, then $10k = \ln 2$, so $k = \dfrac{\ln 2}{10} \approx 0.06931$ so the decay model is

$$Q(t) = Q_0 e^{-0.06931t}.$$

3. If $k = 0.0123$, then $t_h(0.0123) = \ln 2$, so the half-life is

$$t_h = \frac{\ln 2}{0.0123} \approx 56.35 \text{ years.}$$

We can repeat the analysis above for exponential growth models:

Exponential Growth Model and Doubling Time

An **exponential growth function** has the form

$$Q(t) = Q_0 e^{kt}. \qquad Q_0,\ k \text{ both positive}$$

Q_0 represents the value of Q at time $t = 0$, and k is the **growth constant**. The growth constant k and doubling time t_d for Q are related by

$$t_d k = \ln 2.$$

Quick Examples

1. $P(t) = 1{,}000e^{0.05t}$ $1,000 invested at 5% annually with interest compounded continuously
2. If $t_d = 10$ years, then $10k = \ln 2$, so $k = \dfrac{\ln 2}{10} \approx 0.06931$ and the growth model is

$$Q(t) = Q_0 e^{0.06931t}$$

3. If $k = 0.0123$, then $t_d(0.0123) = \ln 2$, so the doubling time is

$$t_d = \frac{\ln 2}{0.0123} \approx 56.35 \text{ years.}$$

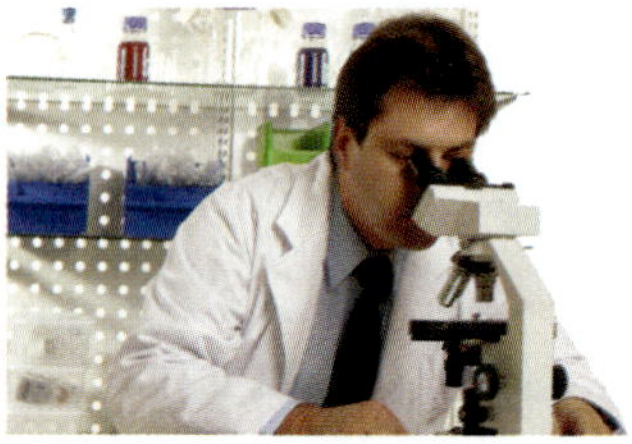

Showface/Dreamstime.com

Logarithmic Regression

If we start with a set of data that suggests a logarithmic curve we can, by repeating the methods from previous sections, use technology to find the logarithmic regression curve $y = \log_b x + C$ approximating the data.

using Technology

See the Technology Guide at the end of the chapter for detailed instructions on how to obtain the regression curve and graph for Example 5 using a TI-83/84 Plus or Excel. Here is an outline:

TI-83/84 Plus

STAT EDIT values of t in L_1 and values of S in L_2

Regression curve: STAT CALC option #9 LnReg ENTER

Graph: Y= VARS 5 EQ 1, then ZOOM 9

[More details on page 186.]

Excel

t- and S-values in Columns A and B

Graph: Highlight data and insert a Scatter chart.

Regression curve: Right-click a datapoint and add logarithmic Trendline with option to show equation on chart.

[More details on page 189.]

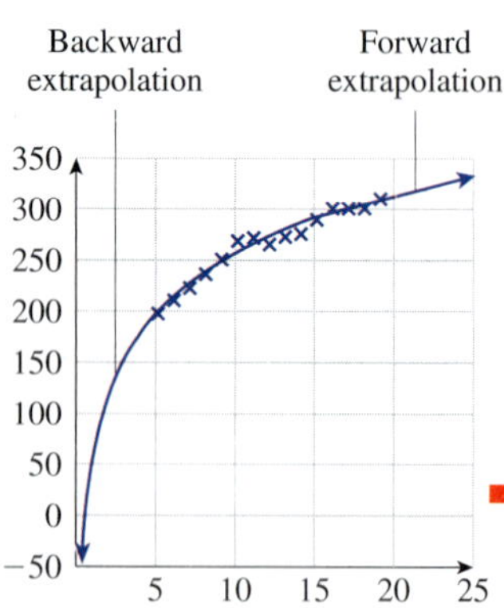

Figure 16

EXAMPLE 5 Research & Development

The following table shows the total spent on research and development in the United States, in billions of dollars, for the period 1995–2009 (t is the year since 1990).*

Year t	5	6	7	8	9	10	11	12
Spending ($ billions)	199	210	222	235	250	268	271	265
Year t	13	14	15	16	17	18	19	
Spending ($ billions)	272	275	289	299	300	304	309	

Find the best-fit logarithmic model of the form

$$S(t) = A \ln t + C$$

and use the model to project total spending on research in 2012, assuming the trend continues.

Solution We use technology to get the following regression model:

$$S(t) = 83.01 \ln t + 64.42. \qquad \text{Coefficients rounded}$$

Because 2012 is represented by $t = 22$, we have

$$S(22) = 83.01 \ln(22) + 64.42 \approx 320. \qquad \text{Why did we round the result to two significant digits?}$$

So, research and development spending is projected to be around $321 billion in 2012.

*Data are approximate and are given in constant 2000 dollars. 2007–2009 figures are projections.
Source: National Science Foundation, Division of Science Resources Statistics, National Patterns of R&D Resources. www.nsf.gov/statistics/, August 2008.

Before we go on... The model in Example 5 seems to give reasonable estimates when we extrapolate forward, but extrapolating backward is quite another matter: The logarithm curve drops sharply to the left of the given range and becomes negative for small values of t (Figure 16). ■

2.3 EXERCISES

▼ more advanced ◆ challenging

T indicates exercises that should be solved using technology

In Exercises 1–4, complete the given tables. HINT [See Quick Examples on page 152.]

1.

Exponential Form	$10^4 = 10{,}000$	$4^2 = 16$	$3^3 = 27$	$5^1 = 5$	$7^0 = 1$	$4^{-2} = \frac{1}{16}$
Logarithmic Form						

2.

Exponential Form	$4^3 = 64$	$10^{-1} = 0.1$	$2^8 = 256$	$5^0 = 1$	$(0.5)^2 = 0.25$	$6^{-2} = \frac{1}{36}$
Logarithmic Form						

3.

Exponential Form						
Logarithmic Form	$\log_{0.5} 0.25 = 2$	$\log_5 1 = 0$	$\log_{10} 0.1 = -1$	$\log_4 64 = 3$	$\log_2 256 = 8$	$\log_2 \frac{1}{4} = -2$

4.

Exponential Form						
Logarithmic Form	$\log_5 5 = 1$	$\log_4 \frac{1}{16} = -2$	$\log_4 16 = 2$	$\log_{10} 10{,}000 = 4$	$\log_3 27 = 3$	$\log_7 1 = 0$

In Exercises 5–12, use logarithms to solve the given equation. (Round answers to four decimal places.) HINT [See Example 1.]

5. $3^x = 5$

6. $4^x = 3$

7. $5^{-2x} = 40$

8. $6^{3x+1} = 30$

9. $4.16e^x = 2$

10. $5.3(10^x) = 2$

11. $5(1.06^{2x+1}) = 11$

12. $4(1.5^{2x-1}) = 8$

In Exercises 13–18, graph the given function. HINT [See Example 2.]

13. $f(x) = \log_4 x$

14. $f(x) = \log_5 x$

15. $f(x) = \log_4(x - 1)$

16. $f(x) = \log_5(x + 1)$

17. $f(x) = \log_{1/4} x$

18. $f(x) = \log_{1/5} x$

In Exercises 19–22 find the associated exponential decay or growth model. HINT [See Quick Examples on page 159.]

19. $Q = 1{,}000$ when $t = 0$; half-life $= 1$

20. $Q = 2{,}000$ when $t = 0$; half-life $= 5$

21. $Q = 1{,}000$ when $t = 0$; doubling time $= 2$

22. $Q = 2{,}000$ when $t = 0$; doubling time $= 5$

In Exercises 23–26 find the associated half-life or doubling time. HINT [See Quick Examples on page 159.]

23. $Q = 1{,}000e^{0.5t}$

24. $Q = 1{,}000e^{-0.025t}$

25. $Q = Q_0e^{-4t}$

26. $Q = Q_0e^{t}$

In Exercises 27–32 convert the given exponential function to the form indicated. Round all coefficients to four significant digits. HINT [See Example 4 "Before we go on."]

27. $f(x) = 4e^{2x}$; $f(x) = Ab^x$

28. $f(x) = 2.1e^{-0.1x}$; $f(x) = Ab^x$

29. $f(t) = 2.1(1.001)^t$; $f(t) = Q_0e^{kt}$

30. $f(t) = 23.4(0.991)^t$; $f(t) = Q_0e^{-kt}$

31. $f(t) = 10(0.987)^t$; $f(t) = Q_0e^{-kt}$

32. $f(t) = 2.3(2.2)^t$; $f(t) = Q_0e^{kt}$

APPLICATIONS

33. ***Investments*** How long will it take a \$500 investment to be worth \$700 if it is continuously compounded at 10% per year? (Give the answer to two decimal places.) HINT [See Example 3.]

34. ***Investments*** How long will it take a \$500 investment to be worth \$700 if it is continuously compounded at 15% per year? (Give the answer to two decimal places.) HINT [See Example 3.]

35. ***Investments*** How long, to the nearest year, will it take an investment to triple if it is continuously compounded at 10% per year? HINT [See Example 3.]

36. ***Investments*** How long, to the nearest year, will it take me to become a millionaire if I invest \$1,000 at 10% interest compounded continuously? HINT [See Example 3.]

37. ***Investments*** I would like my investment to double in value every 3 years. At what rate of interest would I need to invest it, assuming the interest is compounded continuously? HINT [See Quick Examples page 159.]

38. ***Depreciation*** My investment in OHaganBooks.com stocks is losing half its value every 2 years. Find and interpret the associated decay rate. HINT [See Quick Examples page 159.]

39. ***Carbon Dating*** The amount of carbon 14 remaining in a sample that originally contained A grams is given by

$$C(t) = A(0.999879)^t$$

where t is time in years. If tests on a fossilized skull reveal that 99.95% of the carbon 14 has decayed, how old, to the nearest 1,000 years, is the skull? HINT [See Example 4.]

40. ***Carbon Dating*** Refer back to Exercise 39. How old, to the nearest 1,000 years, is a fossil in which only 30% of the carbon 14 has decayed? HINT [See Example 4.]

Long-Term Investments *Exercises 41–48 are based on the following table, which lists interest rates on long-term investments (based on 10-year government bonds) in several countries in 2008.*[29] HINT [See Example 4.]

Country	U.S.	Japan	Canada	Germany	Australia
Yield	3.9%	1.5%	3.8%	4.3%	5.9%

41. Assuming that you invest \$10,000 in the United States, how long (to the nearest year) must you wait before your investment is worth \$15,000 if the interest is compounded annually?

42. Assuming that you invest \$10,000 in Japan, how long (to the nearest year) must you wait before your investment is worth \$15,000 if the interest is compounded annually?

43. If you invest \$10,400 in Canada and the interest is compounded monthly, how many months will it take for your investment to grow to \$20,000?

44. If you invest \$10,400 in the United States, and the interest is compounded monthly, how many months will it take for your investment to grow to \$20,000?

45. How long, to the nearest year, will it take an investment in Australia to double its value if the interest is compounded every six months?

46. How long, to the nearest year, will it take an investment in Germany to double its value if the interest is compounded every six months?

47. If the interest on a long-term U.S. investment is compounded continuously, how long will it take the value of an investment to double? (Give the answer correct to two decimal places.)

48. If the interest on a long-term Australia investment is compounded continuously, how long will it take the value of an investment to double? (Give an answer correct to two decimal places.)

[29] Approximate interest rates based on 10-year government bonds and similar investments. Source: www.bloomberg.com.

49. ***Half-life*** The amount of radium 226 remaining in a sample that originally contained A grams is approximately

$$C(t) = A(0.999\,567)^t$$

where t is time in years. Find the half-life to the nearest 100 years. HINT [See Example 4a.]

50. ***Half-life*** The amount of iodine 131 remaining in a sample that originally contained A grams is approximately

$$C(t) = A(0.9175)^t$$

where t is time in days. Find the half-life to two decimal places. HINT [See Example 4a.]

51. ***Automobiles*** The rate of auto thefts triples every 6 months.

a. Determine, to two decimal places, the base b for an exponential model $y = Ab^t$ of the rate of auto thefts as a function of time in months.

b. Find the doubling time to the nearest tenth of a month. HINT [(a) See Section 2.2 Example 2. (b) See Quick Examples page 159.]

52. ***Televisions*** The rate of television thefts is doubling every 4 months.

a. Determine, to two decimal places, the base b for an exponential model $y = Ab^t$ of the rate of television thefts as a function of time in months.

b. Find the tripling time to the nearest tenth of a month. HINT [(a) See Section 2.2 Example 2. (b) See Quick Examples page 159.]

53. ***Half-life*** The half-life of cobalt 60 is 5 years.

a. Obtain an exponential decay model for cobalt 60 in the form $Q(t) = Q_0e^{-kt}$. (Round coefficients to three significant digits.)

b. Use your model to predict, to the nearest year, the time it takes one third of a sample of cobalt 60 to decay. HINT [See Quick Examples page 159.]

54. ***Half-life*** The half-life of strontium 90 is 28 years.

a. Obtain an exponential decay model for strontium 90 in the form $Q(t) = Q_0e^{-kt}$. (Round coefficients to three significant digits.)

b. Use your model to predict, to the nearest year, the time it takes three-fifths of a sample of strontium 90 to decay. HINT [See Quick Examples page 159.]

55. ***Radioactive Decay*** Uranium 235 is used as fuel for some nuclear reactors. It has a half-life of 710 million years. How long will it take 10 grams of uranium 235 to decay to 1 gram? (Round your answer to three significant digits.)

56. ***Radioactive Decay*** Plutonium 239 is used as fuel for some nuclear reactors, and also as the fissionable material in atomic bombs. It has a half-life of 24,400 years. How long would it take 10 grams of plutonium 239 to decay to 1 gram? (Round your answer to three significant digits.)

57. ▼***Aspirin*** Soon after taking an aspirin, a patient has absorbed 300 mg of the drug. If the amount of aspirin in the bloodstream decays exponentially, with half being removed every 2 hours, find, to the nearest 0.1 hour, the time it will take for the amount of aspirin in the bloodstream to decrease to 100 mg.

58. ▼ ***Alcohol*** After a large number of drinks, a person has a blood alcohol level of 200 mg/dL (milligrams per deciliter). If the amount of alcohol in the blood decays exponentially, with one fourth being removed every hour, find the time it will take for the person's blood alcohol level to decrease to 80 mg/dL. HINT [See Example 4.]

59. ▼ ***Radioactive Decay*** You are trying to determine the half-life of a new radioactive element you have isolated. You start with 1 gram, and 2 days later you determine that it has decayed down to 0.7 grams. What is its half-life? (Round your answer to three significant digits.) HINT [First find an exponential model, then see Example 4.]

60. ▼ ***Radioactive Decay*** You have just isolated a new radioactive element. If you can determine its half-life, you will win the Nobel Prize in physics. You purify a sample of 2 grams. One of your colleagues steals half of it, and three days later you find that 0.1 gram of the radioactive material is still left. What is the half-life? (Round your answer to three significant digits.) HINT [First find an exponential model, then see Example 4.]

61. T ***Population Aging*** The following table shows the percentage of U.S. residents over the age of 65 in 1950, 1960, . . . , 2010 (t is time in years since 1900):[30]

t (Year since 1900)	50	60	70	80	90	100	110
P (% over 65)	8.2	9.2	9.9	11.3	12.6	12.6	13

a. Find the logarithmic regression model of the form $P(t) = A \ln t + C$. (Round the coefficients to four significant digits). HINT [See Example 5.]

b. In 1940, 6.9% of the population was over 65. To how many significant digits does the model reflect this figure?

c. Which of the following is correct? The model, if extrapolated into the indefinite future, predicts that

(A) The percentage of U.S. residents over the age of 65 will increase without bound.

(B) The percentage of U.S. residents over the age of 65 will level off at around 14.2%.

(C) The percentage of U.S. residents over the age of 65 will eventually decrease.

62. T ***Population Aging*** The following table shows the percentage of U.S. residents over the age of 85 in 1950, 1960, . . . , 2010 (t is time in years since 1900).[31]

t (Year since 1900)	50	60	70	80	90	100	110
P (% over 85)	0.4	0.5	0.7	1	1.2	1.6	1.9

a. Find the logarithmic regression model of the form $P(t) = A \ln t + C$. (Round the coefficients to four significant digits). HINT [See Example 5.]

b. In 2020, 2.1% of the population is projected to be over 85. To how many significant digits does the model reflect this figure?

c. Which of the following is correct? If you increase A by 0.1 and decrease C by 0.1 in the logarithmic model, then

(A) The new model predicts eventually lower percentages.

(B) The long-term prediction is essentially the same.

(C) The new model predicts eventually higher percentages.

63. T ***Research & Development: Industry*** The following table shows the total spent on research and development by industry in the United States, billions of dollars, for the period 1995–2009 (t is the year since 1990).[32]

Year t	5	6	7	8	9	10	11	12
Spending ($ billions)	118	129	140	150	165	183	181	170
Year t	13	14	15	16	17	18	19	
Spending ($ billions)	172	172	182	191	194	197	200	

Find the logarithmic regression model of the form $S(t) = A \ln t + C$ with coefficients A and C rounded to 2 decimal places. Also obtain a graph showing the data points and the regression curve. In which direction is it more reasonable to extrapolate the model? Why?

64. T ***Research & Development: Federal*** The following table shows the total spent on research and development by the Federal government in the United States, in billions of dollars, for the period 1995–2009 (t is the year since 1990).[33]

Year t	5	6	7	8	9	10	11	12
Spending ($ billions)	17	17	17	17	18	18	20	21
Year t	13	14	15	16	17	18	19	
Spending ($ billions)	23	23	25	24	24	23	23	

Find the logarithmic regression model of the form $S(t) = A \ln t + C$ with coefficients A and C rounded to 2 decimal

[30] Source: U.S. Census Bureau.

[31] Ibid.

[32] Non-federal funding by industry. Data are approximate and are given in constant 2000 dollars. 2007–2009 figures are projections. Source: National Science Foundation, Division of Science Resources Statistics, National Patterns of R&D Resources. www.nsf.gov/statistics/, August 2008.

[33] Federal funding excluding grants to industry and nonprofit organizations. Data are approximate and are given in constant 2000 dollars. 2007–2009 figures are projections. Source: National Science Foundation, Division of Science Resources Statistics, National Patterns of R&D Resources. www.nsf.gov/statistics/, August 2008.

places. Also obtain a graph showing the data points and the regression curve. In which direction is it more reasonable to extrapolate the model? Why? HINT [See Example 5.]

65. ▼ ***Richter Scale*** The **Richter scale** is used to measure the intensity of earthquakes. The Richter scale rating of an earthquake is given by the formula

$$R = \frac{2}{3}(\log E - 11.8)$$

where E is the energy released by the earthquake (measured in ergs[34]).

a. The San Francisco earthquake of 1906 registered $R = 8.2$ on the Richter scale. How many ergs of energy were released?

b. In 1989 another San Francisco earthquake registered 7.1 on the Richter scale. Compare the two: The energy released in the 1989 earthquake was what percentage of the energy released in the 1906 quake?

c. Solve the equation given above for E in terms of R.

d. Use the result of part (c) to show that if two earthquakes registering R_1 and R_2 on the Richter scale release E_1 and E_2 ergs of energy, respectively, then

$$\frac{E_2}{E_1} = 10^{1.5(R_2 - R_1)}.$$

e. Fill in the blank: If one earthquake registers 2 points more on the Richter scale than another, then it releases ___ times the amount of energy.

66. ▼ ***Sound Intensity*** The loudness of a sound is measured in **decibels**. The decibel level of a sound is given by the formula

$$D = 10 \log \frac{I}{I_0},$$

where D is the decibel level (dB), I is its intensity in watts per square meter (W/m^2), and $I_0 = 10^{-12}$ W/m^2 is the intensity of a barely audible "threshold" sound. A sound intensity of 90 dB or greater causes damage to the average human ear.

a. Find the decibel levels of each of the following, rounding to the nearest decibel:

Whisper:	115×10^{-12} W/m^2
TV (average volume from 10 feet):	320×10^{-7} W/m^2
Loud music:	900×10^{-3} W/m^2
Jet aircraft (from 500 feet):	100 W/m^2

b. Which of the sounds above damages the average human ear?

c. Solve the given equation to express I in terms of D.

d. Use the answer to part (c) to show that if two sounds of intensity I_1 and I_2 register decibel levels of D_1 and D_2 respectively, then

$$\frac{I_2}{I_1} = 10^{0.1(D_2 - D_1)}.$$

e. Fill in the blank: If one sound registers one decibel more than another, then it is ___ times as intense.

[34] An erg is a unit of energy. One erg is the amount of energy it takes to move a mass of one gram one centimeter in one second.

67. ▼ ***Sound Intensity*** The decibel level of a TV set decreases with the distance from the set according to the formula

$$D = 10 \log\left(\frac{320 \times 10^7}{r^2}\right)$$

where D is the decibel level and r is the distance from the TV set in feet.

a. Find the decibel level (to the nearest decibel) at distances of 10, 20, and 50 feet.

b. Express D in the form $D = A + B \log r$ for suitable constants A and B. (Round A and D to two significant digits.)

c. How far must a listener be from a TV so that the decibel level drops to 0? (Round the answer to two significant digits.)

68. ▼ ***Acidity*** The acidity of a solution is measured by its pH, which is given by the formula

$$pH = -\log(H^+)$$

where H^+ measures the concentration of hydrogen ions in moles per liter.[35] The pH of pure water is 7. A solution is referred to as *acidic* if its pH is below 7 and as *basic* if its pH is above 7.

a. Calculate the pH of each of the following substances.

Blood:	3.9×10^{-8} moles/liter
Milk:	4.0×10^{-7} moles/liter
Soap solution:	1.0×10^{-11} moles/liter
Black coffee:	1.2×10^{-7} moles/liter

b. How many moles of hydrogen ions are contained in a liter of acid rain that has a pH of 5.0?

c. Complete the following sentence: If the pH of a solution increases by 1.0, then the concentration of hydrogen ions _________.

COMMUNICATION AND REASONING EXERCISES

69. On the same set of axes, graph $y = \ln x$, $y = A \ln x$, and $y = A \ln x + C$ for various choices of *positive* A and C. What is the effect on the graph of $y = \ln x$ of multiplying by A? What is the effect of then adding C?

70. On the same set of axes, graph $y = -\ln x$, $y = A \ln x$, and $y = A \ln x + C$ for various choices of *negative* A and C. What is the effect on the graph of $y = \ln x$ of multiplying by A? What is the effect of then adding C?

71. Why is the logarithm of a negative number not defined?

72. Of what use are logarithms, now that they are no longer needed to perform complex calculations?

73. Your company's market share is undergoing steady growth. Explain why a logarithmic function is *not* appropriate for long-term future prediction of your market share.

[35] A mole corresponds to about 6.0×10^{23} hydrogen ions. (This number is known as Avogadro's number.)

74. Your company's market share is undergoing steady growth. Explain why a logarithmic function is *not* appropriate for long-term backward extrapolation of your market share.

75. If $y = 4^x$, then $x =$ _______.

76. If $y = \log_6 x$, then $x =$ _______.

77. Simplify: $2^{\log_2 8}$.

78. Simplify: $e^{\ln x}$.

79. Simplify: $\ln(e^x)$.

80. Simplify: $\ln\sqrt{a}$.

81. ▼ If a town's population is increasing exponentially with time, how is time increasing with population? Explain.

82. ▼ If a town's population is increasing logarithmically with time, how is time increasing with population? Explain.

83. ▼ If two quantities Q_1 and Q_2 are logarithmic functions of time t, show that their sum, $Q_1 + Q_2$ is also a logarithmic function of time t.

84. T ▼ In Exercise 83 we saw that the sum of two logarithmic functions is a logarithmic function. In Exercises 63 and 64 you modeled research and development expenditure by industry and government. Now do a logarithmic regression on the sum of the two sets of figures. Does the result coincide with the sum of the two individual regression models? What does your answer tell you about the sum of logarithmic regression models?

2.4 Logistic Functions and Models

Figure 17 shows the percentage of Internet-connected U.S. households that have broadband connections as a function of time t in years ($t = 0$ represents 2000).[36]

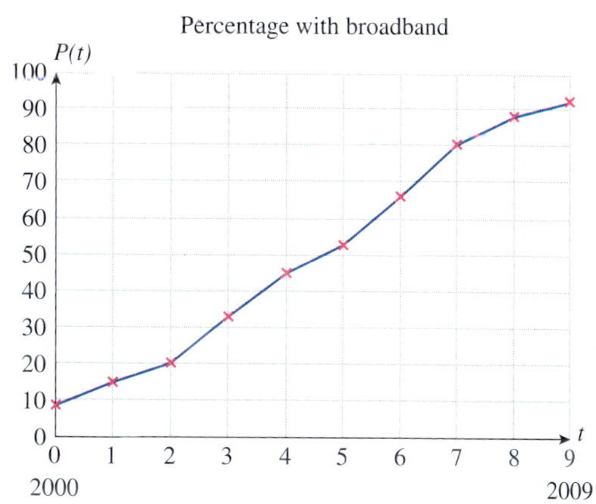

Figure 17

The left-hand part of the curve in Figure 17, from $t = 0$ to, say, $t = 4$, looks roughly like exponential growth: P behaves (roughly) like an exponential function, with the y-coordinates growing by a factor of around 1.5 per year. Then, as the market starts to become saturated, the growth of P slows and its value approaches a "saturation" point near 100%. **Logistic** functions have just this kind of behavior, growing exponentially at first and then leveling off. In addition to modeling the demand for a new technology or product, logistic functions are often used in epidemic and population modeling. In an epidemic, the number of infected people often grows exponentially at first and then slows when a significant proportion of the entire susceptible population is infected and the epidemic has "run its course." Similarly, populations may grow exponentially at first and then slow as they approach the capacity of the available resources.

[36] 2009 figure is estimated. Source: www.Nielsen.com.

Logistic Function

A **logistic function** has the form

$$f(x) = \frac{N}{1 + Ab^{-x}}$$

for nonzero constants N, A, and b (A and b positive and $b \neq 1$).

Quick Example

$N = 6, A = 2, b = 1.1$ gives $f(x) = \dfrac{6}{1 + 2(1.1^{-x})}$ `6/(1+2*1.1^-x)`

$f(0) = \dfrac{6}{1+2} = 2$ The y-intercept is $N/(1 + A)$.

$f(1{,}000) = \dfrac{6}{1 + 2(1.1^{-1{,}000})} \approx \dfrac{6}{1+0} = 6 = N$ When x is large, $f(x) \approx N$.

Graph of a Logistic Function

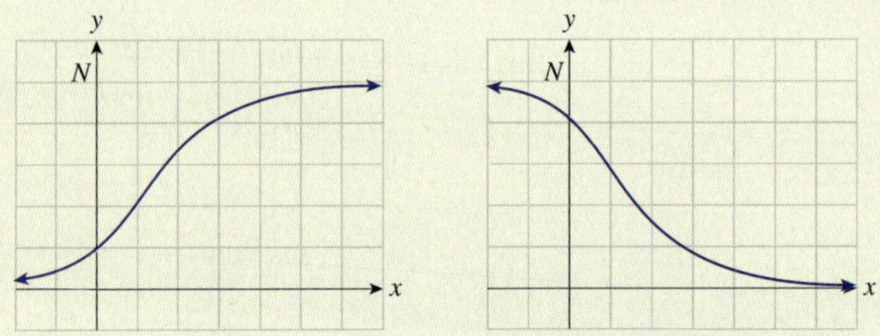

$b > 1$ $0 < b < 1$

$$y = \frac{N}{1 + Ab^{-x}}$$

Properties of the Logistic Curve $y = \dfrac{N}{1 + Ab^{-x}}$

- The graph is an S-shaped curve sandwiched between the horizontal lines $y = 0$ and $y = N$. N is called the **limiting value** of the logistic curve.
- If $b > 1$ the graph rises; if $b < 1$, the graph falls.
- The y-intercept is $\dfrac{N}{1 + A}$.
- The curve is steepest when $t = \dfrac{\ln A}{\ln b}$. We will see why in Chapter 5.

Note If we write b^{-x} as e^{-kx} (where $k = \ln b$), we get the following alternative form of the logistic function:

$$f(x) = \frac{N}{1 + Ae^{-kx}}.$$

Q: *How does the constant b affect the graph?*

A: To understand the role of b, we first rewrite the logistic function by multiplying top and bottom by b^x:

$$\begin{aligned} f(x) &= \frac{N}{1 + Ab^{-x}} \\ &= \frac{Nb^x}{(1 + Ab^{-x})b^x} \\ &= \frac{Nb^x}{b^x + A} \qquad \text{Because } b^{-x}b^x = 1 \end{aligned}$$

For values of x close to 0, the quantity b^x is close to 1, so the denominator is approximately $1 + A$, giving

$$f(x) \approx \frac{Nb^x}{1 + A} = \left(\frac{N}{1 + A}\right) b^x.$$

In other words, $f(x)$ is approximately exponential with base b for values of x close to 0. Put another way, if x represents time, then initially the logistic function behaves like an exponential function.

To summarize:

Logistic Function for Small *x* and the Role of *b*

For small values of x, we have

$$\frac{N}{1 + Ab^{-x}} \approx \left(\frac{N}{1 + A}\right) b^x.$$

Thus, for small x, the logistic function grows approximately exponentially with base b.

Quick Examples

1. Let

$$f(x) = \frac{50}{1 + 24(3^{-x})}. \qquad N = 50, A = 24, b = 3$$

Then

$$f(x) \approx \left(\frac{50}{1 + 24}\right)(3^x) = 2(3^x)$$

for small values of x. The following figure compares their graphs:

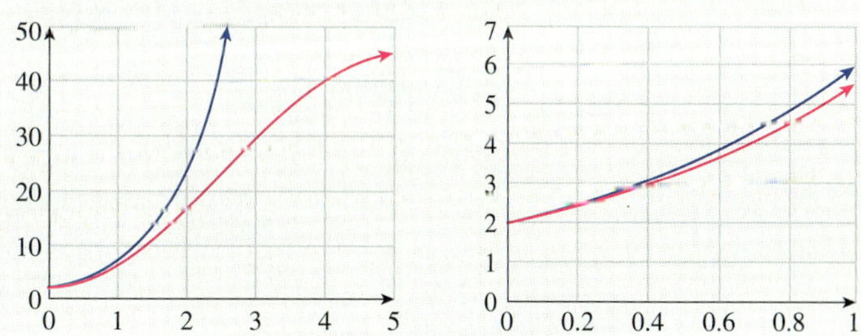

The upper curve is the exponential curve.

Modeling with the Logistic Function

EXAMPLE 1 Epidemics

A flu epidemic is spreading through the U.S. population. An estimated 150 million people are susceptible to this particular strain, and it is predicted that all susceptible people will eventually become infected. There are 10,000 people already infected, and the number is doubling every 2 weeks. Use a logistic function to model the number of people infected. Hence predict when, to the nearest week, 1 million people will be infected.

Solution Let t be time in weeks, and let $P(t)$ be the total number of people infected at time t. We want to express P as a logistic function of t, so that

$$P(t) = \frac{N}{1 + Ab^{-t}}.$$

We are told that, in the long run, 150 million people will be infected, so that

$$N = 150{,}000{,}000. \qquad \text{Limiting value of } P$$

At the current time ($t = 0$), 10,000 people are infected, so

$$10{,}000 = \frac{N}{1 + A} = \frac{150{,}000{,}000}{1 + A} \qquad \text{Value of } P \text{ when } t = 0$$

Solving for A gives

$$10{,}000(1 + A) = 150{,}000{,}000$$
$$1 + A = 15{,}000$$
$$A = 14{,}999.$$

What about b? At the beginning of the epidemic (t near 0), P is growing approximately exponentially, doubling every 2 weeks. Using the technique of Section 2.2, we find that the exponential curve passing through the points (0, 10,000) and (2, 20,000) is

$$y = 10{,}000(\sqrt{2})^t$$

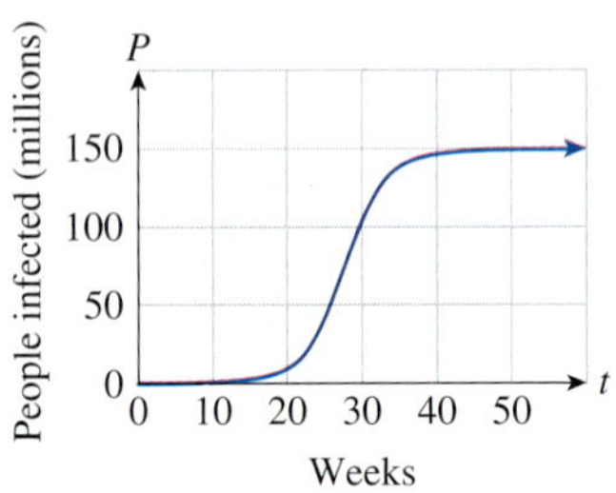

Figure 18

giving us $b = \sqrt{2}$. Now that we have the constants N, A, and b, we can write down the logistic model:

$$P(t) = \frac{150{,}000{,}000}{1 + 14{,}999(\sqrt{2})^{-t}}$$

The graph of this function is shown in Figure 18.

Now we tackle the question of prediction: When will 1 million people be infected? In other words: When is $P(t) = 1{,}000{,}000$?

$$1{,}000{,}000 = \frac{150{,}000{,}000}{1 + 14{,}999(\sqrt{2})^{-t}}$$
$$1{,}000{,}000[1 + 14{,}999(\sqrt{2})^{-t}] = 150{,}000{,}000$$
$$1 + 14{,}999(\sqrt{2})^{-t} = 150$$
$$14{,}999(\sqrt{2})^{-t} = 149$$
$$(\sqrt{2})^{-t} = \frac{149}{14{,}999}$$
$$-t = \log_{\sqrt{2}}\left(\frac{149}{14{,}999}\right) \approx -13.31 \qquad \text{Logarithmic form}$$

Thus, 1 million people will be infected by about the 13th week.

Before we go on... We said earlier that the logistic curve is steepest when $t = \frac{\ln A}{\ln b}$. In Example 1, this occurs when $t = \frac{\ln 14{,}999}{\ln\sqrt{2}} \approx 28$ weeks into the epidemic. At that time, the number of cases is growing most rapidly (look at the apparent slope of the graph at the corresponding point). ■

Logistic Regression

Let us go back to the data on the percentage of Internet-connected households with broadband and try to estimate the percentage of households that will have broadband in the long term. In order to be able to make predictions such as this, we require a model for the data, so we will need to do some form of regression.

EXAMPLE 2 Internet-Connected Households with Broadband

Here are the data graphed in Figure 17:

Year (t)	0	1	2	3	4	5	6	7	8	9
Percentage with Broadband (%) (P)	8	14	20	33	45	53	66	80	88	92

Find a logistic regression curve of the form

$$P(t) = \frac{N}{1 + Ab^{-t}}.$$

In the long term, what percentage of Internet-connected households with broadband does the model predict?

Solution We can use technology to obtain the following regression model:

$$P(t) \approx \frac{104.047}{1 + 10.453(1.642)^{-t}}$$ Coefficients rounded to 3 decimal places

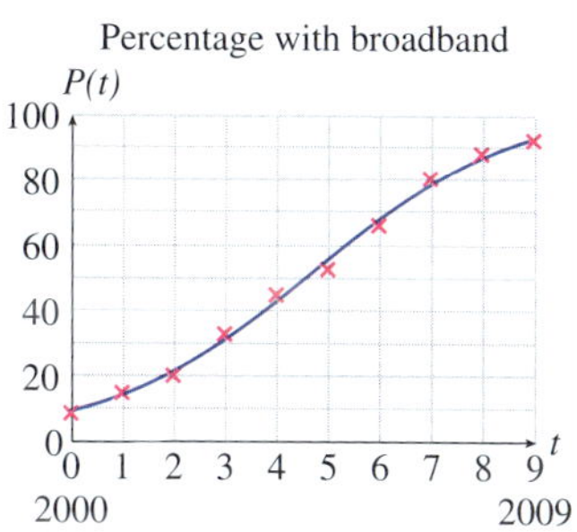

Figure 19

Its graph and the original data are shown in Figure 19.

Because $N \approx 104$, the model "predicts" that 104% of all Internet-connected households will have broadband in the long term. Clearly, this makes no sense; a reasonable value of N is 100, as it seems reasonable that, in the long term, all Internet connections will be broadband. There are various curve-fitting procedures that permit one to fix one of the parameters (N in this case) beforehand and adjust the others (A and b here) to obtain the best possible fit. In the Excel Technology Guide at the end of the chapter we show how to do this using Excel's "Solver" Add-in. Here is the best-fit model with N fixed at 100:

$$P(t) \approx \frac{100}{1 + 10.783(1.691)^{-t}}$$ Coefficients rounded to 3 decimal places

Although this second model is a slightly poorer fit to the data (SSE ≈ 31) than the first one (SSE ≈ 26), it has the advantage that it better predicts the long-term behavior we expect.

using Technology

See the Technology Guide at the end of the chapter for detailed instructions on how to obtain the regression curve and graph for Example 5 using a TI-83/84 Plus or Excel.

Here is an outline:

TI-83/84 Plus

STAT EDIT values of t in L_1 and values of P in L_2

Regression curve: STAT CALC option #B Logistic ENTER

Graph: Y= VARS 5 EQ 1, then ZOOM 9 [More details on page 187.]

Excel

Use the Solver Add-in to obtain the best-fit logistic curves; one allowing N to vary, and another fixing $N = 100$. [More details on page 189.]

Before we go on... As we noted in Example 2, logistic regression programs generally estimate all three constants N, A, and b for a model $y = \frac{N}{1 + Ab^{-x}}$. However, there are times, as in both Examples 1 and 2, when we already know the limiting value N and require estimates of only A and b. In such cases, we can use technology like Excel Solver to find A and b for the best-fit curve with N fixed as described in Example 2. Alternatively, we can use exponential regression to compute estimates of A and b as follows: First rewrite the logistic equation as

$$\frac{N}{y} = 1 + Ab^{-x},$$

so that

$$\frac{N}{y} - 1 = Ab^{-x} = A(b^{-1})^x.$$

This equation gives $N/y - 1$ as an exponential function of x. Thus, if we do exponential regression using the data points $(x, N/y - 1)$, we can obtain estimates for A and b^{-1} (and hence b). This is done in Exercises 35 and 36.

It is important to note that the resulting curve is not the best-fit curve (in the sense of minimizing SSE; See the "Before we go on" discussion on page 146 after Example 6 in Section 2.2) and will be thus be different from that obtained using the method in Example 2. ■

2.4 EXERCISES

▼ more advanced ◆ challenging

T indicates exercises that should be solved using technology

In Exercises 1–6, find N, A, and b, give a technology formula for the given function, and use technology to sketch its graph for the given range of values of x. HINT [See Quick Examples on page 166.]

1. $f(x) = \frac{7}{1 + 6(2^{-x})}$; $[0, 10]$

2. $g(x) = \frac{4}{1 + 0.333(4^{-x})}$; $[0, 2]$

3. $f(x) = \frac{10}{1 + 4(0.3^{-x})}$; $[-5, 5]$

4. $g(x) = \frac{100}{1 + 5(0.5^{-x})}$; $[-5, 5]$

5. $h(x) = \frac{2}{0.5 + 3.5(1.5^{-x})}$; $[0, 15]$
(First divide top and bottom by 0.5.)

6. $k(x) = \frac{17}{2 + 6.5(1.05^{-x})}$; $[0, 100]$
(First divide top and bottom by 2.)

In Exercises 7–10, find the logistic function f with the given properties. HINT [See Example 1.]

7. $f(0) = 10$, f has limiting value 200, and for small values of x, f is approximately exponential and doubles with every increase of 1 in x.

8. $f(0) = 1$, f has limiting value 10, and for small values of x, f is approximately exponential and grows by 50% with every increase of 1 in x.

9. f has limiting value 6 and passes through $(0, 3)$ and $(1, 4)$. HINT [First find A, then substitute.]

10. f has limiting value 4 and passes through $(0, 1)$ and $(1, 2)$. HINT [First find A, then substitute.]

In Exercises 11–16, choose the logistic function that best approximates the given curve.

11.

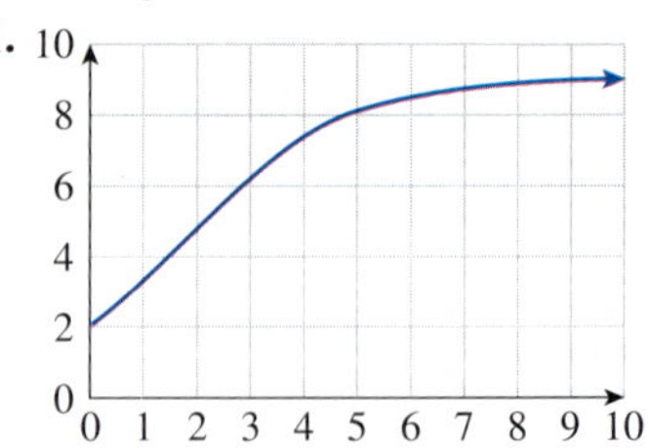

(A) $f(x) = \frac{6}{1 + 0.5(3^{-x})}$

(B) $f(x) = \frac{9}{1 + 3.5(2^{-x})}$

(C) $f(x) = \frac{9}{1 + 0.5(1.01)^{-x}}$

12.

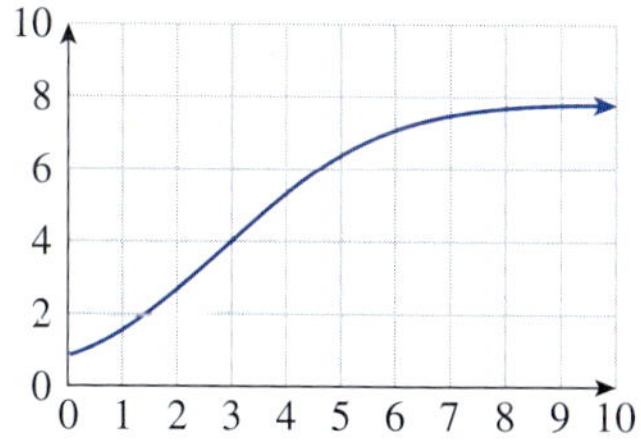

(A) $f(x) = \dfrac{8}{1 + 7(2)^{-x}}$ (B) $f(x) = \dfrac{8}{1 + 3(2)^{-x}}$

(C) $f(x) = \dfrac{6}{1 + 11(5)^{-x}}$

13.

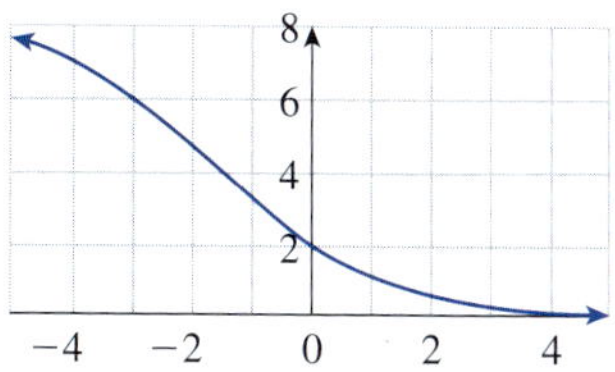

(A) $f(x) = \dfrac{8}{1 + 7(0.5)^{-x}}$ (B) $f(x) = \dfrac{8}{1 + 3(0.5)^{-x}}$

(C) $f(x) = \dfrac{8}{1 + 3(2)^{-x}}$

14.

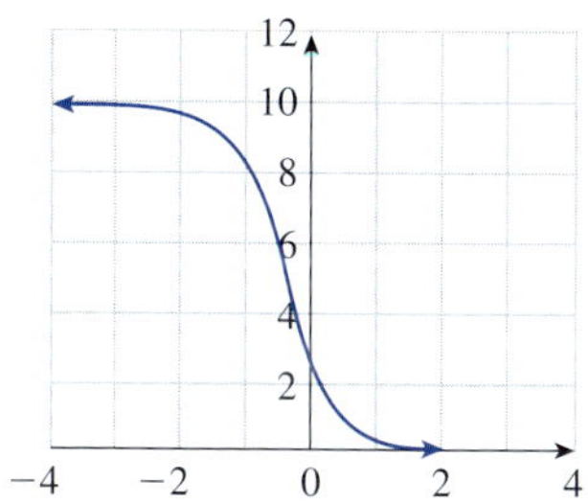

(A) $f(x) = \dfrac{10}{1 + 3(1.01)^{-x}}$ (B) $f(x) = \dfrac{8}{1 + 7(0.1)^{-x}}$

(C) $f(x) = \dfrac{10}{1 + 3(0.1)^{-x}}$

15.

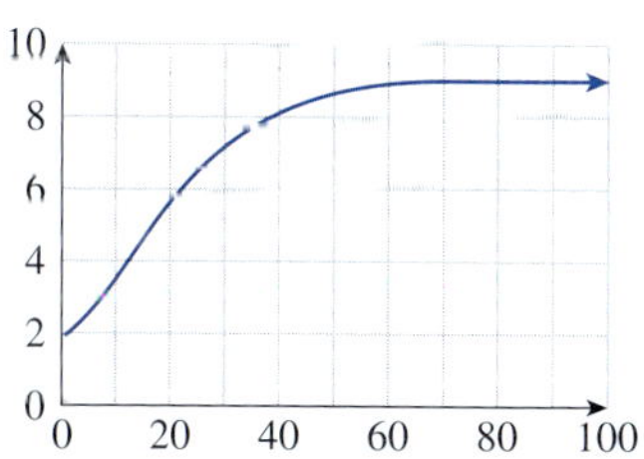

(A) $f(x) = \dfrac{18}{2 + 7(5)^{-x}}$ (B) $f(x) = \dfrac{18}{2 + 3(1.1)^{-x}}$

(C) $f(x) = \dfrac{18}{2 + 7(1.1)^{-x}}$

16.

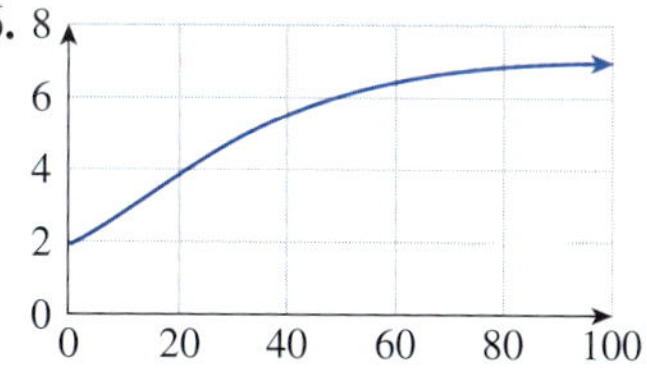

(A) $f(x) = \dfrac{14}{2 + 5(15)^{-x}}$ (B) $f(x) = \dfrac{14}{1 + 13(1.05)^{-x}}$

(C) $f(x) = \dfrac{14}{2 + 5(1.05)^{-x}}$

T *In Exercises 17–20, use technology to find a logistic regression curve* $y = \dfrac{N}{1 + Ab^{-x}}$ *approximating the given data. Draw a graph showing the data points and regression curve. (Round b to three significant digits and A and N to two significant digits.)* HINT [See Example 2.]

17.

x	0	20	40	60	80	100
y	2.1	3.6	5.0	6.1	6.8	6.9

18.

x	0	30	60	90	120	150
y	2.8	5.8	7.9	9.4	9.7	9.9

19.

x	0	20	40	60	80	100
y	30.1	11.6	3.8	1.2	0.4	0.1

20.

x	0	30	60	90	120	150
y	30.1	20	12	7.2	3.8	2.4

APPLICATIONS

21. ***Subprime Mortgages*** The following graph shows the approximate percentage of mortgages issued in the United States that are subprime (normally classified as risky) as well as the logistic regression curve:[37]

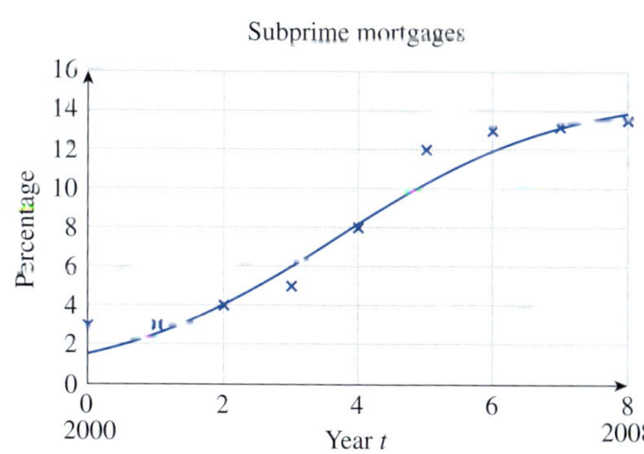

[37]2008 figure is an estimate. Sources: Mortgage Bankers Association, UBS.

a. Which of the following logistic functions best approximates the curve? (t is the number of years since the start of 2000.) Try to determine the correct model without actually computing data points. HINT [See Properties of Logistic Curves on page 166.]

(A) $A(t) = \dfrac{15.0}{1 + 8.6(1.8)^{-t}}$

(B) $A(t) = \dfrac{2.0}{1 + 6.8(0.8)^{-t}}$

(C) $A(t) = \dfrac{2.0}{1 + 6.8(1.8)^{-t}}$

(D) $A(t) = \dfrac{15.0}{1 + 8.6(0.8)^{-t}}$

b. According to the model you selected, during which year was the percentage growing fastest? HINT [See the "Before we go on" discussion after Example 1.]

22. ***Subprime Mortgage Debt*** The following graph shows the approximate value of subprime (normally classified as risky) mortgage debt outstanding in the United States as well as the logistic regression curve:[38]

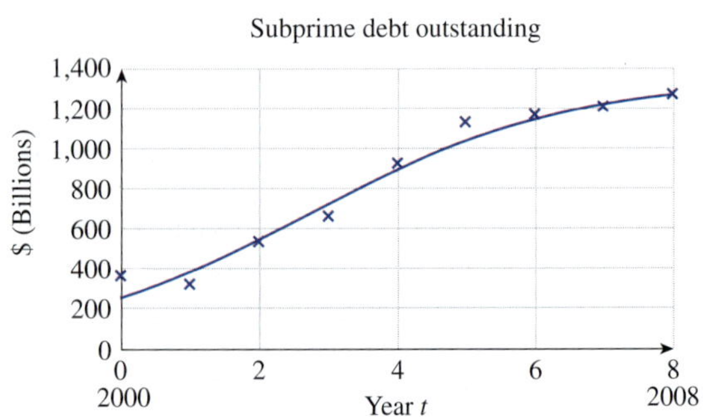

a. Which of the following logistic functions best approximates the curve? (t is the number of years since the start of 2000.) Try to determine the correct model without actually computing data points. HINT [See Properties of Logistic Curves on page 166.]

(A) $A(t) = \dfrac{1850}{1 + 5.36(1.8)^{-t}}$

(B) $A(t) = \dfrac{1350}{1 + 4.2(1.7)^{-t}}$

(C) $A(t) = \dfrac{1020}{1 + 5.3(1.8)^{-t}}$

(D) $A(t) = \dfrac{1300}{1 + 4.2(0.9)^{-t}}$

b. According to the model you selected, during which year was outstanding debt growing fastest? HINT [See the "Before we go on" discussion after Example 1.]

23. ***Scientific Research*** The following graph shows the number of research articles in the prominent journal *Physical Review* that were written by researchers in Europe during 1983–2003 ($t = 0$ represents 1983).[39]

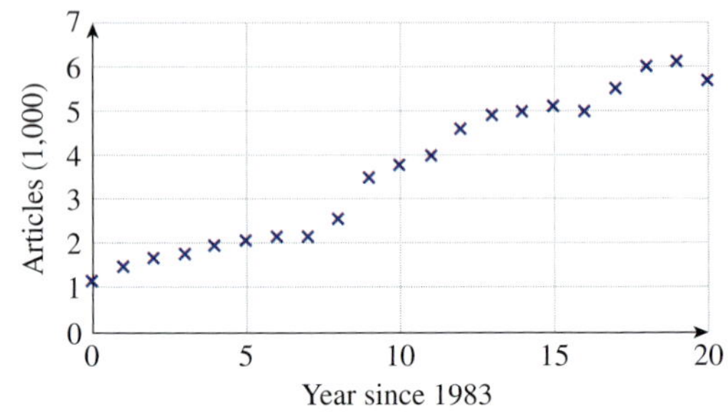

a. Which of the following logistic functions best models the data? (t is the number of years since 1983.) Try to determine the correct model without actually computing data points.

(A) $A(t) = \dfrac{7.0}{1 + 5.4(1.2)^{-t}}$

(B) $A(t) = \dfrac{4.0}{1 + 3.4(1.2)^{-t}}$

(C) $A(t) = \dfrac{4.0}{1 + 3.4(0.8)^{-t}}$

(D) $A(t) = \dfrac{7.0}{1 + 5.4(6.2)^{-t}}$

b. According to the model you selected, at what percentage was the number of articles growing around 1985?

24. ***Scientific Research*** The following graph shows the percentage, above 25%, of research articles in the prominent journal *Physical Review* that were written by researchers in the United States during 1983–2003 ($t = 0$ represents 1983).[40]

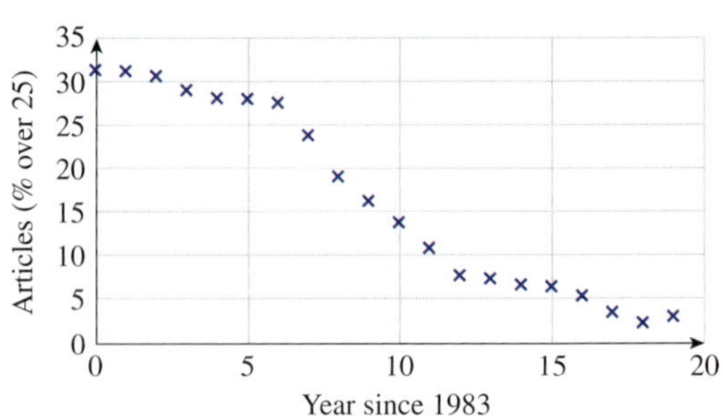

[38] 2008–2009 figures are estimates. Source: www.data360.org/dataset.aspx?Data_Set_Id=9549.

[39] Source: The American Physical Society/*New York Times* May 3, 2003, p. A1.

[40] Ibid.

a. Which of the following logistic functions best models the data? (t is the number of years since 1983 and P is the actual percentage.) Try to determine the correct model without actually computing data points.

(A) $P(t) = \dfrac{36}{1 + 0.06(0.02)^{-t}}$

(B) $P(t) = \dfrac{12}{1 + 0.06(1.7)^{-t}}$

(C) $P(t) = \dfrac{12}{1 + 0.06(0.7)^{-t}}$

(D) $P(t) = \dfrac{36}{1 + 0.06(0.7)^{-t}}$

b. According to the model you selected, how fast was the value of P declining around 1985?

25. ***Computer Use*** The following graph shows the actual percentage of U.S. households with a computer as a function of household income (the data points) and a logistic model of these data (the curve).[41]

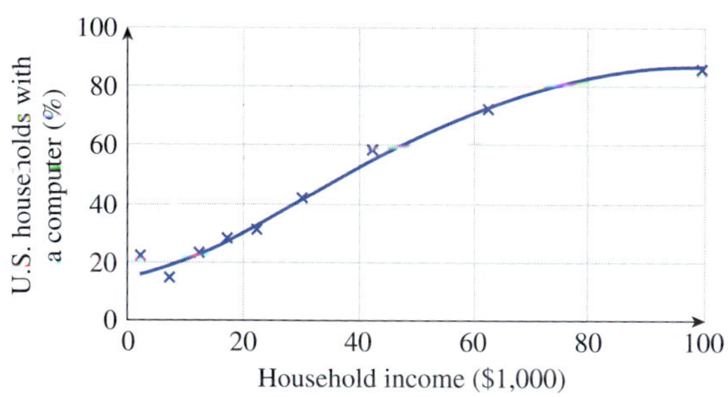

The logistic model is

$$P(x) = \frac{91}{1 + 5.35(1.05)^{-x}} \text{ percent}$$

where x is the household income in thousands of dollars.

a. According to the model, what percentage of extremely wealthy households had computers?

b. For low incomes, the logistic model is approximately exponential. Which exponential model best approximates $P(x)$ for small x?

c. According to the model, 50% of households of what income had computers in 2000? (Round the answer to the nearest $1,000).

26. ***Internet Use*** The following graph shows the actual percentage of U.S. residents who used the Internet at home as a function of income (the data points) and a logistic model of these data (the curve).[42]

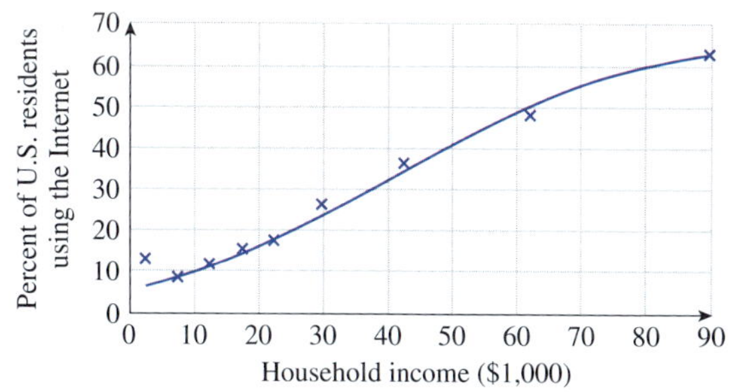

The logistic model is given by

$$P(x) = \frac{64.2}{1 + 9.6(1.06)^{-x}} \text{ percent}$$

where x is an individual's income in thousands of dollars.

a. According to the model, what percentage of extremely wealthy people used the Internet at home?

b. For low incomes, the logistic model is approximately exponential. Which exponential model best approximates $P(x)$ for small x?

c. According to the model, 50% of individuals with what income used the Internet at home in 2000? (Round the answer to the nearest $1,000).

27. ***Epidemics*** There are currently 1,000 cases of Venusian flu in a total susceptible population of 10,000 and the number of cases is increasing by 25% each day. Find a logistic model for the number of cases of Venusian flu and use your model to predict the number of flu cases a week from now. HINT [Example 1.]

28. ***Epidemics*** Last year's epidemic of Martian flu began with a single case in a total susceptible population of 10,000. The number of cases was increasing initially by 40% per day. Find a logistic model for the number of cases of Martian flu and use your model to predict the number of flu cases 3 weeks into the epidemic. HINT [Example 1.]

29. ***Sales*** You have sold 100 "I ♥ Calculus" T-shirts and sales appear to be doubling every five days. You estimate the total market for "I ♥ Calculus" T-shirts to be 3,000. Give a logistic model for your sales and use it to predict, to the nearest day, when you will have sold 700 T-shirts.

30. ***Sales*** In Russia the average consumer drank two servings of Coca-Cola® in 1993. This amount appeared to be increasing exponentially with a doubling time of 2 years.[43] Given a long-range market saturation estimate of 100 servings per year, find a logistic model for the consumption of Coca-Cola in Russia and use your model to predict when, to the nearest year, the average consumption will be 50 servings per year.

[41] Income levels are midpoints of income brackets. (The top income level is an estimate.) Source: NTIA and ESA, U.S. Department of Commerce, using U.S. Bureau of the Census Current Population, 2000.

[42] Ibid.

[43] The doubling time is based on retail sales of Coca-Cola products in Russia. Sales in 1993 were double those in 1991, and were expected to double again by 1995. Source: *New York Times*, September 26, 1994, p. D2.

31. **T** ***Scientific Research*** The following chart shows some the data shown in the graph in Exercise 23 ($t = 0$ represents 1983).[44]

Year, t	0	5	10	15	20
Research Articles, A (1,000)	1.2	2.1	3.8	5.1	5.7

a. What is the logistic regression model for the data? (Round all coefficients to two significant digits.) At what value does the model predict that the number of research articles will level off? HINT [See Example 2.]

b. According to the model, how many *Physical Review* articles were published by U.S. researchers in 2000 ($t = 17$)? (The actual number was about 5,500 articles.)

32. **T** ***Scientific Research*** The following chart shows some the data shown in the graph in Exercise 24 ($t = 0$ represents 1983).[45]

Year, t	0	5	10	15	20
Percentage, P (over 25)	36	28	16	7	3

a. What is the logistic regression model for the data? (Round all coefficients to two significant digits.) HINT [See Example 2.]

b. According to the model, what percentage of *Physical Review* articles were published by researchers in the United States in 2000 ($t = 17$)? (The actual figure was about 30.1%.)

33. **T** ***College Basketball: Men*** The following table shows the number of NCAA men's college basketball teams in the United States for various years since 1990.[46]

t (year since 1990)	0	5	10	11	12	13	14	15	16	17	18
Teams	767	868	932	937	936	967	981	983	984	994	1,000

a. What is the logistic regression model for the data? (Round all parameters to three significant digits.) At what value does the model predict that the number of basketball teams will level off?

b. According to the model, for what value of t is the regression curve steepest? Interpret the answer.

c. Interpret the coefficient b in the context of the number of men's basketball teams.

34. **T** ***College Basketball: Women*** The following table shows the number of NCAA women's college basketball teams in the United States for various years since 1990.[47]

t (year since 1990)	0	5	10	11	12	13
Teams	1,549	1,732	1,888	1,895	1,911	1,976
t (year since 1990)	14	15	16	17	18	
Teams	1,989	2,019	2,002	2,040	2,060	

a. What is the logistic regression model for the data? (Round all parameters to three significant digits.) At what value does the model predict that the number of basketball teams will level off?

b. According to the model, for what value of t is the regression curve steepest? Interpret the answer.

c. Interpret the coefficient b in the context of the number of women's basketball teams.

T *Exercises 35 and 36 are based on the discussion following Example 2. If the limiting value N is known, then*

$$\frac{N}{y} - 1 = A(b^{-1})^x$$

and so $N/y - 1$ is an exponential function of x. In Exercises 35 and 36, use the given value of N and the data points $(x, N/y - 1)$ to obtain A and b, and hence a logistic model.

35. **T** ◆ ***Population: Puerto Rico*** The following table and graph show the population of Puerto Rico in thousands from 1950 to 2025.[48]

t (year since 1950)	0	10	20	30	40	50
Population (thousands)	2,220	2,360	2,720	3,210	3,540	3,820
t (year since 1950)	55	60	65	70	75	
Population (thousands)	3,910	3,990	4,050	4,080	4,100	

[44] Source: The American Physical Society/*New York Times* May 3, 2003, p. A1.

[45] Ibid.

[46] 2007 and 2008 figures are estimates. Source: The 2008 Statistical Abstract, www.census.gov/.

[47] 2007 and 2008 figures are estimates. Source: The 2008 Statistical Abstract, www.census.gov/.

[48] Figures from 2010 on are U.S. census projections. Source: The 2008 Statistical Abstract, www.census.gov/.

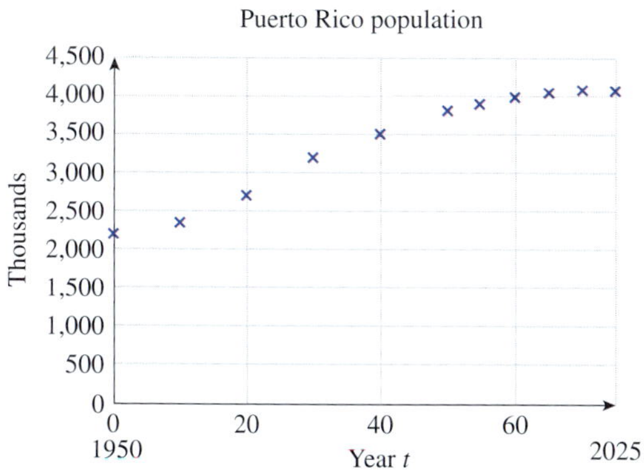

Take t to be the number of years since 1950, and find a logistic model based on the assumption that, eventually, the population of Puerto Rico will grow to 4.5 million. (Round coefficients to 4 decimal places.) In what year does your model predict the population of Puerto Rico will first reach 4.0 million?

36. T ◆ ***Population: Virgin Islands*** The following table and graph show the population of the Virgin Islands in thousands from 1950 to 2025.[49]

t (year since 1950)	0	10	20	30	40	50
Population (thousands)	27	33	63	98	104	106
t (year since 1950)	55	60	65	70	75	
Population (thousands)	108	108	107	107	108	

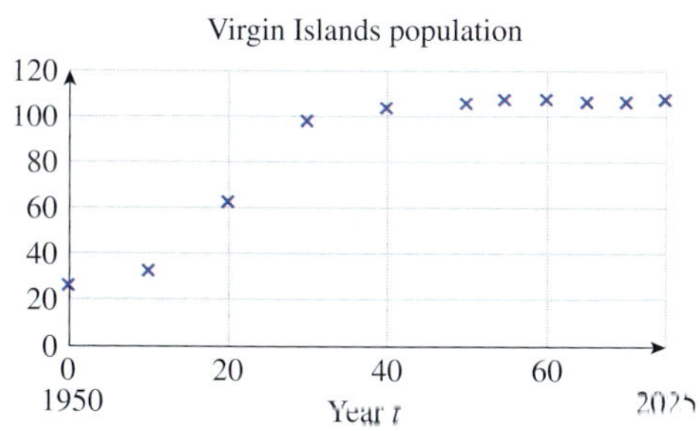

[49]Figures from 2010 on are U.S. census projections. Source: The 2008 Statistical Abstract, www.census.gov/.

Take t to be the number of years since 1950, and find a logistic model based on the assumption that, eventually, the population of the Virgin Islands will grow to 110,000. (Round coefficients to 4 decimal places.) In what year does your model predict the population of the Virgin Islands first reached 80,000?

COMMUNICATION AND REASONING EXERCISES

37. Logistic functions are commonly used to model the spread of epidemics. Given this fact, explain why a logistic function is also useful to model the spread of a new technology.

38. Why is a logistic function more appropriate than an exponential function for modeling the spread of an epidemic?

39. Give one practical use for logistic regression.

40. Refer to an exercise or example in this section to find a scenario in which a logistic model may not be a good predictor of long-term behavior.

41. What happens to the function $P(t) = \dfrac{N}{1 + Ab^{-t}}$ if we replace b^{-t} by b^t when $b > 1$? If $b < 1$?

42. What happens to the function $P(t) = \dfrac{N}{1 + Ab^{-t}}$ if $A = 0$? If $A < 0$?

43. ▼ We said that the logistic curve $y = \dfrac{N}{1 + Ab^{-t}}$ is steepest when $t = \dfrac{\ln A}{\ln b}$. Show that the corresponding value of y is $N/2$.

HINT [Use the fact that $\dfrac{\ln A}{\ln b} = \ln_b A$.]

44. ▼ We said that the logistic curve $y = \dfrac{N}{1 + Ab^{-t}}$ is steepest when $t = \dfrac{\ln A}{\ln b}$. For which values of A and b is this value of t positive, zero, and negative?

CHAPTER 2 REVIEW

KEY CONCEPTS

Web Site www.AppliedCalc.org
Go to the student Web site at www.AppliedCalc.org to find a comprehensive and interactive Web-based summary of Chapter 2.

2.1 Quadratic Functions and Models

A **quadratic function** has the form $f(x) = ax^2 + bx + c$ *p. 124*
The graph of $f(x) = ax^2 + bx + c$ $(a \neq 0)$ is a **parabola** *p. 125*
The x coordinate of the **vertex** is $-\frac{b}{2a}$. The y coordinate is $f\left(-\frac{b}{2a}\right)$ *p. 125*
***x*-intercepts** (if any) occur at
$$x = \frac{-b \pm \sqrt{b^2 - 4ac}}{2a} \quad p.\ 125$$
The ***y*-intercept** occurs at $y = c$ *p. 125*
The parabola is **symmetric** with respect to the vertical line through the vertex *p. 125*
Sketching the graph of a quadratic function *p. 126*
Application to maximizing revenue *p. 127*
Application to maximizing profit *p. 128*
Finding the quadratic regression curve *p. 130*

2.2 Exponential Functions and Models

An **exponential function** has the form $f(x) = Ab^x$ *p. 137*
Roles of the constants A and b in an exponential function $f(x) = Ab^x$ *p. 138*
Recognizing exponential data *p. 139*
Finding the exponential curve through two points *p. 139*
Application to compound interest *p. 140*
Application to exponential decay (carbon dating) *p. 140*
Application to exponential growth (epidemics) *p. 140*
The number e and continuous compounding *p. 142*
Alternative form of an exponential function: $f(x) = Ae^{rx}$ *p. 144*
Finding the exponential regression curve *p. 145*

2.3 Logarithmic Functions and Models

The **base *b* logarithm of *x***: $y = \log_b x$ means $b^y = x$ *p. 151*
Common logarithm, $\log x = \log_{10} x$, and **natural logarithm**, $\ln x = \log_e x$ *p. 152*
Change of base formula *p. 153*
Solving equations with unknowns in the exponent *p. 153*
A **logarithmic function** has the form $f(x) = Ab^x$ *p. 154*
Graphs of logarithmic functions *p. 154*
Logarithm identities *p. 155*
Application to investments. (How long?) *p. 156*
Application to half-life *p. 157*
Exponential decay models and half-life *p. 159*
Exponential growth models and doubling time *p. 159*
Finding the logarithmic regression curve *p. 160*

2.4 Logistic Functions and Models

A **logistic function** has the form
$$f(x) = \frac{N}{1 + Ab^{-x}} \quad p.\ 166$$
Properties of the logistic curve, point where curve is steepest *p. 166*
Logistic function for small x, the role of b *p. 167*
Application to epidemics *p. 168*
Finding the logistic regression curve *p. 169*

REVIEW EXERCISES

Sketch the graph of the quadratic functions in Exercises 1 and 2, indicating the coordinates of the vertex, the y-intercept, and the x-intercepts (if any).

1. $f(x) = x^2 + 2x - 3$

2. $f(x) = -x^2 - x - 1$

In Exercises 3 and 4, the values of two functions, f and g, are given in a table. One, both, or neither of them may be exponential. Decide which, if any, are exponential, and give the exponential models for those that are.

3.

x	−2	−1	0	1	2
$f(x)$	20	10	5	2.5	1.25
$g(x)$	8	4	2	1	0

4.

x	−2	−1	0	1	2
$f(x)$	8	6	4	2	1
$g(x)$	$\frac{3}{4}$	$\frac{3}{2}$	3	6	12

In Exercises 5 and 6 graph given the pairs of functions on the same set of axes with $-3 \le x \le 3$.

5. $f(x) = \frac{1}{2}(3^x)$; $g(x) = \frac{1}{2}(3^{-x})$

6. $f(x) = 2(4^x)$; $g(x) = 2(4^{-x})$

T *On the same set of axes, use technology to graph the pairs of functions in Exercises 7 and 8 for the given range of x. Identify which graph corresponds to which function.*

7. $f(x) = e^x$; $g(x) = e^{0.8x}$; $-3 \le x \le 3$

8. $f(x) = 2(1.01)^x$; $g(x) = 2(0.99)^x$; $-100 \le x \le 100$

In Exercises 9–14, compute the indicated quantity.

9. A \$3,000 investment earns 3% interest, compounded monthly. Find its value after 5 years.

10. A \$10,000 investment earns 2.5% interest, compounded quarterly. Find its value after 10 years.

11. An investment earns 3% interest, compounded monthly and is worth \$5,000 after 10 years. Find its initial value.

12. An investment earns 2.5% interest, compounded quarterly and is worth \$10,000 after 10 years. Find its initial value.

13. A \$3,000 investment earns 3% interest, compounded continuously. Find its value after 5 years.

14. A \$10,000 investment earns 2.5% interest, compounded continuously. Find its value after 10 years.

In Exercises 15–18, find a formula of the form $f(x) = Ab^x$ using the given information.

15. $f(0) = 4.5$; the value of f triples for every half-unit increase in x.

16. $f(0) = 5$; the value of f decreases by 75% for every one-unit increase in x.

17. $f(1) = 2,\ f(3) = 18.$

18. $f(1) = 10,\ f(3) = 5.$

In Exercises 19–22, use logarithms to solve the given equation for x.

19. $3^{-2x} = 4$ **20.** $2^{2x^2-1} = 2$

21. $300(10^{3x}) = 315$ **22.** $P(1+i)^{mx} = A$

On the same set of axes, graph the pairs of functions in Exercises 23 and 24.

23. $f(x) = \log_3 x;\ g(x) = \log_{(1/3)} x$

24. $f(x) = \log x;\ g(x) = \log_{(1/10)} x$

In Exercises 25–28, use the given information to find an exponential model of the form $Q = Q_0e^{-kt}$ or $Q = Q_0e^{kt}$, as appropriate. Round all coefficients to three significant digits when rounding is necessary.

25. Q is the amount of radioactive substance with a half-life of 100 years in a sample originally containing 5g (t is time in years).

26. Q is the number of cats on an island whose cat population was originally 10,000 but is being cut in half every 5 years (t is time in years).

27. Q is the diameter (in cm) of a circular patch of mold on your roommate's damp towel you have been monitoring with morbid fascination. You measured the patch at 2.5 cm across four days ago, and have observed that it is doubling in diameter every two days (t is time in days).

28. Q is the population of cats on another island whose cat population was originally 10,000 but is doubling every 15 months (t is time in months).

In Exercises 29–32, find the time required, to the nearest 0.1 year, for the investment to reach the desired goal.

29. \$2,000 invested at 4%, compounded monthly; goal: \$3,000.

30. \$2,000 invested at 6.75%, compounded daily; goal: \$3,000.

31. \$2,000 invested at 3.75%, compounded continuously; goal: \$3,000.

32. \$1,000 invested at 100%, compounded quarterly; goal: \$1,200.

In Exercises 33–36, find equations for the logistic functions of x with the stated properties.

33. Through (0, 100), initially increasing by 50% per unit of x, and limiting value 900.

34. Initially exponential of the form $y = 5(1.1)^x$ with limiting value 25.

35. Passing through (0, 5) and decreasing from a limiting value of 20 to 0 at a rate of 20% per unit of x when x is near 0.

36. Initially exponential of the form $y = 2(0.8)^x$ with a value close to 10 when $x = -60$.

APPLICATIONS

37. ***Web Site Traffic*** The daily traffic ("hits per day") at OHaganBooks.com seems to depend on the monthly expenditure on advertising through banner ads on well-known Internet portals. The following model, based on information collected over the past few months, shows the approximate relationship:

$$h = -0.000005c^2 + 0.085c + 1{,}750$$

where h is the average number of hits per day at OHaganBooks.com, and c is the monthly advertising expenditure.

a. According to the model, what monthly advertising expenditure will result in the largest volume of traffic at OHaganBooks.com? What is that volume?

b. In addition to predicting a maximum volume of traffic, the model predicts that the traffic will eventually drop to zero if the advertising expenditure is increased too far. What expenditure (to the nearest dollar) results in no Web site traffic?

c. What feature of the formula for this quadratic model indicates that it will predict an eventual decline in traffic as advertising expenditure increases?

38. ***Broadband Access*** Pablo Pelogrande, a new summer intern at OHaganBooks.com in 2010, was asked by John O'Hagan to research the extent of broadband access in the United States Pelogrande found some old data online on broadband access from the start of 2001 to the end of 2003 and used it to construct the following quadratic model of the growth rate of broadband access:

$n(t) = 2t^2 - 6t + 12$ million new American adults with broadband per year

(t is time in years; $t = 0$ represents the start 2000).[50]

a. What is an appropriate domain of n?

b. According to the model, when was the growth rate at a minimum?

c. Does the model predict a zero growth rate at any particular time? If so, when?

d. What feature of the formula for this quadratic model indicates that the growth rate eventually increases?

e. Does the fact that $n(t)$ decreases for $t \le 1.5$ suggest that the number of broadband users actually declined before June 2001? Explain.

[50] Based on data for 2001–2003. Source for data: Pew Internet and American Life Project data memos dated May 18, 2003 and April 19, 2004, available at www.pewinternet.org.

f. Pelogande extrapolated the model in order to estimate the growth rate at the beginning of 2010 and 2011. What did he find? Comment on the answer.

39. ***Revenue and Profit*** Some time ago, a consultant formulated the following linear model of demand for online novels:

$$q = -60p + 950$$

where q is the monthly demand for OHaganBooks.com's online novels at a price of p dollars per novel.

a. Use this model to express the monthly revenue as a function of the unit price p, and hence determine the price you should charge for a maximum monthly revenue.

b. Author royalties and copyright fees cost the company an average of $4 per novel, and the monthly cost of operating and maintaining the online publishing service amounts to $900 per month. Express the monthly profit P as a function of the unit price p, and hence determine the unit price you should charge for a maximum monthly profit. What is the resulting profit (or loss)?

40. ***Revenue and Profit*** Billy-Sean O'Hagan is John O'Hagan's son and a freshman in college. He notices that the demand for the college newspaper was 2,000 copies each week when the paper was given away free of charge, but dropped to 1,000 each week when the college started charging 10¢/copy.

a. Write down the associated linear demand function.

b. Use your demand function to express the revenue as a function of the unit price p, and hence determine the price the college should charge for a maximum revenue. At that price, what is the revenue from sales of one edition of the newspaper?

c. It costs the college 4¢ to produce each copy of the paper, plus an additional fixed cost of $200. Express the profit P as a function of the unit price p, and hence determine the unit price the college should charge for a maximum monthly profit (or minimum loss). What is the resulting profit (or loss)?

41. ***Lobsters*** Marjory Duffin, CEO of Duffin Press, is particularly fond of having steamed lobster at working lunches with executives from OHaganBooks.com, and is therefore alarmed by the news that the yearly lobster harvest from New York's Long Island Sound has been decreasing dramatically since 1997. Indeed, the size of the annual harvest can be approximated by

$$n(t) = 10(0.66^t) \text{ million pounds}$$

where t is time in years since June, 1997.[51]

a. The model tells us that the harvest was ____ million pounds in 1997 and decreasing by ___% each year.

b. What does the model predict for the 2005 harvest?

42. ***Stock Prices*** In the period immediately following its initial public offering (IPO), OHaganBooks.com's stock was doubling in value every three hours. If you bought $10,000 worth of the stock when it was first offered, how much was your stock worth after eight hours?

[51] Based on a regression model. Source for data: NY State Department of Environmental Conservation/*Newsday*, October 6, 2004, p. A4.

43. ***Lobsters*** (See Exercise 41.) Marjory Duffin has just left John O'Hagan, CEO of OHaganBooks.com, a frantic phone message to the effect that yearly lobster harvest from New York's Long Island Sound has just dipped below 100,000 pounds, making that planned lobster working lunch more urgent than ever. What year is it?

44. ***Stock Prices*** We saw in Exercise 42 that OHaganBooks.com's stock was doubling in value every three hours, following its IPO. If you bought $10,000 worth of the stock when it was first offered, how long from the initial offering did it take your investment to reach $50,000?

45. ***Lobsters*** We saw in Exercise 41 that the Long Island Sound lobster harvest was given by $n(t) = 10(0.66^t)$ million pounds t years after 1997. However, in 2007, thanks to the efforts of Duffin Press, Inc. it turned around and started increasing by 24% each year.[52] What, to the nearest 1,000 pounds, was the actual size of the harvest in 2010?

46. ***Stock Prices*** We saw in Exercise 42 that OHaganBooks.com's stock was doubling in value every three hours, following its IPO. After 10 hours of trading, the stock turns around and starts losing one third of its value every four hours. How long (from the initial offering) will it be before your stock is once again worth $10,000?

47. T ***Lobsters*** The model in Exercise 41 was based on the data shown in the following chart:

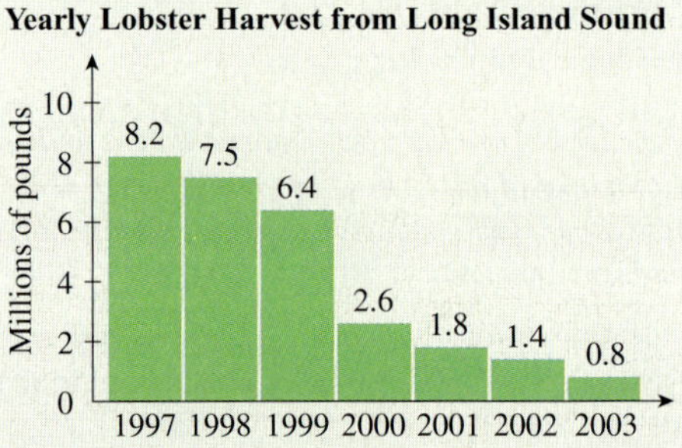

Use the data to obtain an exponential regression curve of the form $n(t) = Ab^t$, with $t = 0$ corresponding to 1997 and coefficients rounded to two significant digits.

48. T ***Stock Prices*** The actual stock price of OHaganBooks.com in the hours following its IPO is shown in the following chart:

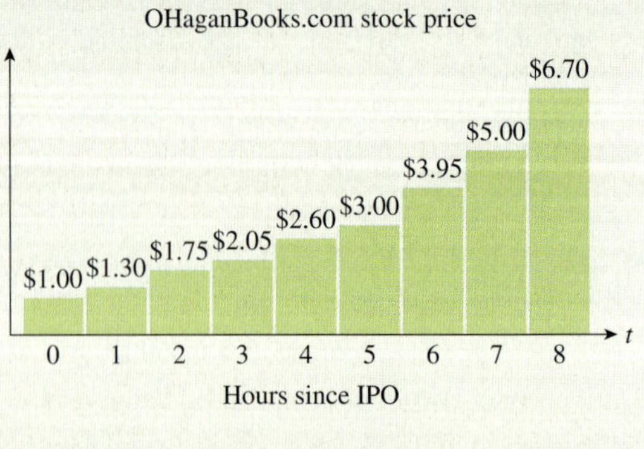

[52] This claim, like Duffin Press, is fiction.

Use the data to obtain an exponential regression curve of the form $P(t) = Ab^t$, with $t = 0$ the time in hours since the IPO and coefficients rounded to 3 significant digits. At the end of which hour will the stock price first be above \$10?

49. ***Hardware Life*** *(Based on a question from the GRE economics exam)* To estimate the rate at which new computer hard drives will have to be retired, OHaganBooks.com uses the "survivor curve":

$$L_x = L_0 e^{-x/t}$$

where

L_x = number of surviving hard drives at age x
L_0 = number of hard drives initially
t = average life in years.

All of the following are implied by the curve *except:*

(A) Some of the equipment is retired during the first year of service.
(B) Some equipment survives three average lives.
(C) More than half the equipment survives the average life.
(D) Increasing the average life of equipment by using more durable materials would increase the number surviving at every age.
(E) The number of survivors never reaches zero.

50. ***Sales*** OHaganBooks.com modeled its weekly sales over a period of time with the function

$$s(t) = 6{,}050 + \frac{4{,}470}{1 + 14(1.73^{-t})}$$

as shown in the following graph (t is measured in weeks):

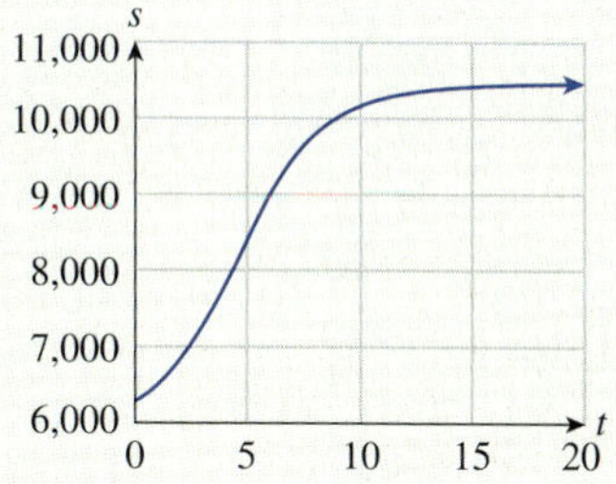

a. As time goes on, it appears that weekly sales are leveling off. At what value are they leveling off?
b. When did weekly sales rise above 10,000?
c. When, to the nearest week, were sales rising most rapidly?

Case Study Checking up on Malthus

Robert Nickelsberg/Getty Images

In 1798 Thomas R. Malthus (1766–1834) published an influential pamphlet, later expanded into a book, titled *An Essay on the Principle of Population As It Affects the Future Improvement of Society.* One of his main contentions was that population grows geometrically (exponentially) while the supply of resources such as food grows only arithmetically (linearly). This led him to the pessimistic conclusion that population would always reach the limits of subsistence and precipitate famine, war, and ill-health, unless population could be checked by other means. He advocated "moral restraint," which includes the pattern of late marriage common in Western Europe at the time and is now common in most developed countries, and which leads to a lower reproduction rate.

Two hundred years later, you have been asked to check the validity of Malthus's contention. That population grows geometrically, at least over short periods of time, is commonly assumed. That resources grow linearly is more questionable. You decide to check the actual production of a common crop, wheat, in the United States. Agricultural statistics like these are available from the U.S. government on the Internet, through the U.S. Department of Agriculture's National Agricultural Statistics Service (NASS). As of 2008, this service was available at www.nass.usda.gov/. Looking through this site, you locate data on the annual production of all wheat in the United States from 1900 through 2008.

Web Site
www.AppliedCalc.org
To download an Excel sheet with the data used in the case study, go to Everything for Calculus, scroll down to the case study for Chapter 2, and click on "Wheat Production Data (Excel)."

Year	1900	1901	. . .	2007	2008
Thousands of Bushels	599,315	762,546	. . .	2,066,722	2,462,418

Graphing these data (using Excel, for example), you obtain the graph in Figure 20.

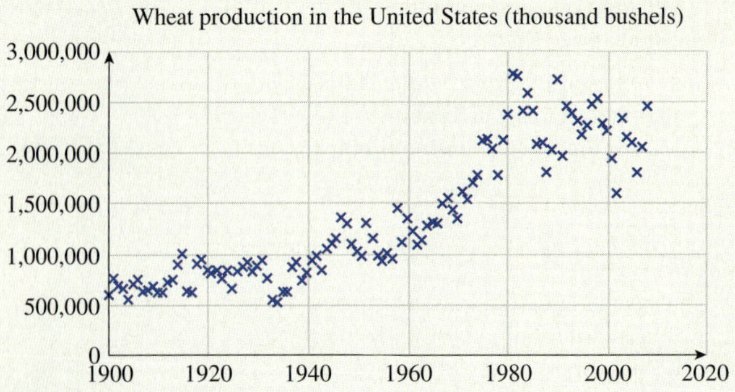

Figure 20

This does not look very linear, particularly in the last half of the 20th century, but you continue checking the mathematics. Using Excel's built-in linear regression capabilities, you find that the line that best fits these data, shown in Figure 21, has $r^2 = 0.7997$. (Recall the discussion of the correlation coefficient r in Section 1.4. A similar statistic is available for other types of regression as well.)

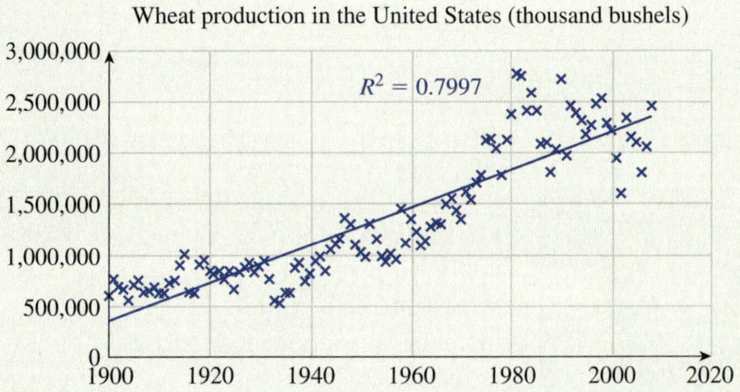

Figure 21

*** NOTE** Recall that the residuals are defined as $y_{\text{Observed}} - y_{\text{Predicted}}$ (see Section 1.4) and are the vertical distances between the observed data points and the regression line.

Although that is a fairly high correlation, you notice that the residuals* are not distributed randomly: The actual wheat production starts out higher than the line, dips below the line from about 1920 to about 1970, then rises above the line, and finally appears to dip below the line around 2002. This behavior seems to suggest a logistic curve or perhaps a cubic curve. On the other hand, it is also possible that the apparent dip at the end of the data is not statistically significant—it could be nothing more than a transitory fluctuation in the wheat production industry—so perhaps we should also consider models that do not bend downward, like exponential and quadratic models.

Following is a comparison of the four proposed models (with coefficients rounded to three significant digits). For the independent variable, we used t = time in years since 1900. SSE is the sum-of-squares error.

Quadratic

$P(t) \approx 110t^2 - 6{,}740t + 572{,}000$

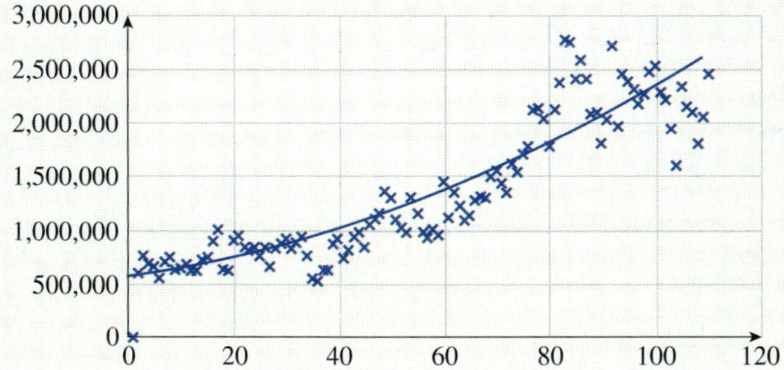

$\text{SSE} \approx 8.367 \times 10^{12}$

Cubic

$P(t) \approx -5.25t^3 + 960t^2 - 29{,}800t + 893{,}000$

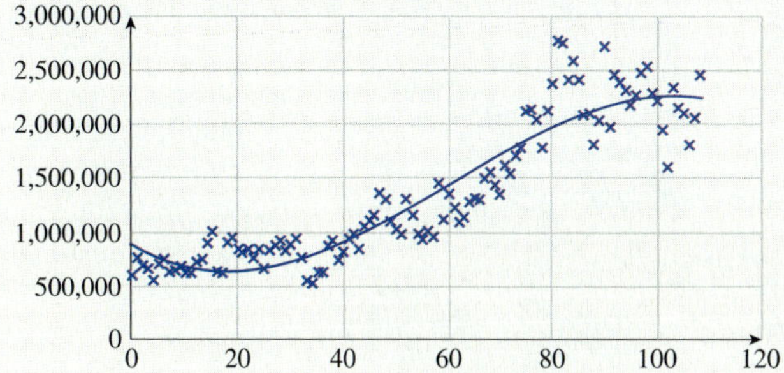

$\text{SSE} \approx 6.573 \times 10^{12}$

Exponential

$P(t) \approx 570{,}000e^{0.0141t}$

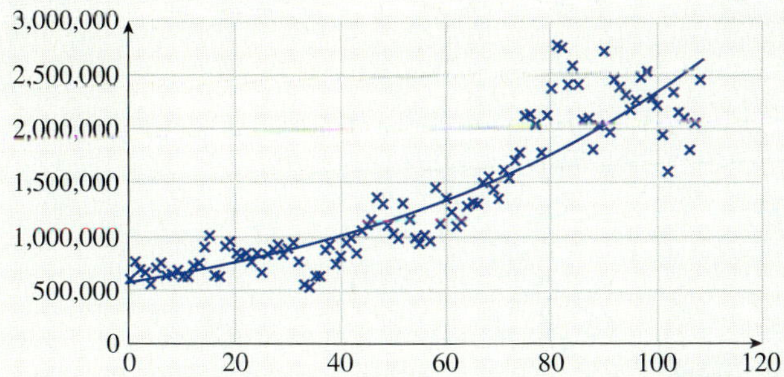

$\text{SSE} \approx 8.944 \times 10^{12}$

Logistic

$$P(t) \approx \frac{3{,}690{,}000}{1 + 6.91(1.02^{-t})}$$

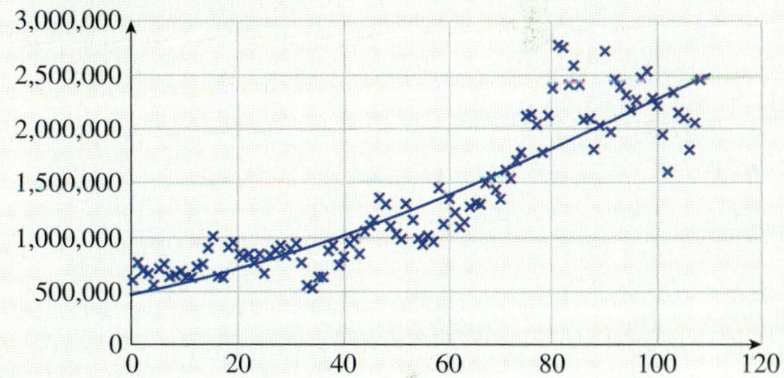

$\text{SSE} \approx 8.104 \times 10^{12}$

The model that appears to best fit the data seems to be the cubic model; both visually and by virtue of SSE. Notice also that the cubic model predicts a *decrease* in the production of wheat in the near term (see Figure 22).

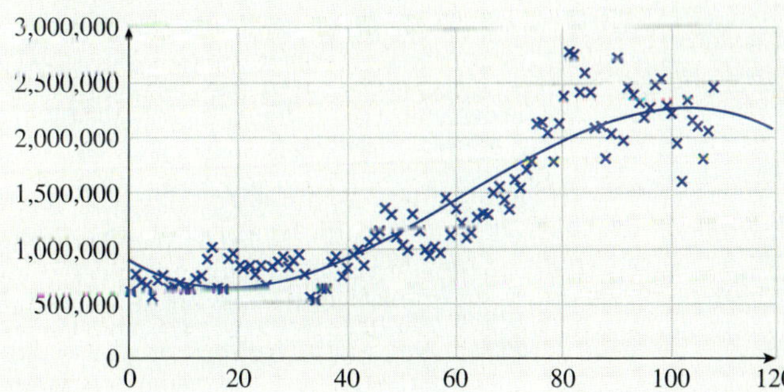

Figure 22

So you prepare a report that documents your findings and concludes that things are even worse than Malthus predicted, at least as far as wheat production in the United States is concerned: The supply is deceasing while the population is still increasing more-or-less exponentially. (See Exercise 75 in Section 2.2.)

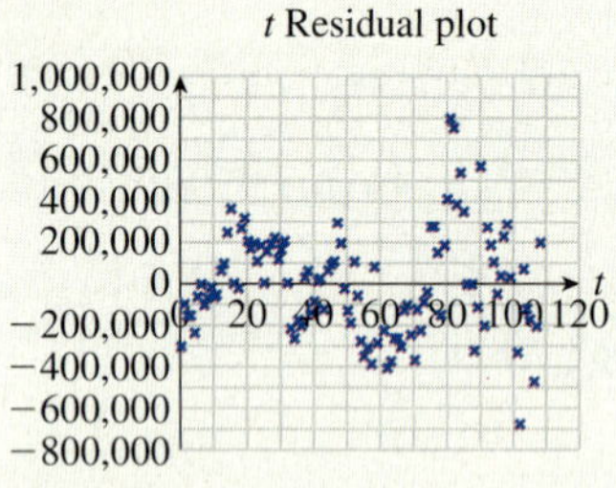

Figure 23

You are about to hit "Send," which will dispatch copies of your report to a significant number of people on whom the success of your career depends, when you notice something strange about the pattern of data in Figure 22: The observed data points appear to hug the regression curve quite closely for small values of t, but appear to become more and more scattered as the value of t increases. In the language of residuals, the residuals are small for small values of t but then tend to get larger with increasing t. Figure 23 shows a plot of the residuals that shows this trend even more clearly.

This reminds you vaguely of something that came up in your college business statistics course, so you consult the textbook from that class that (fortunately) you still own and discover that a pattern of residuals with increasing magnitude suggests that, instead of modeling y versus t directly, you instead model $\ln y$ versus t. (The residuals for large values of t will then be scaled down by the logarithm.)

Figure 24 shows the resulting plot together with the regression line (what we call the "linear transformed model").

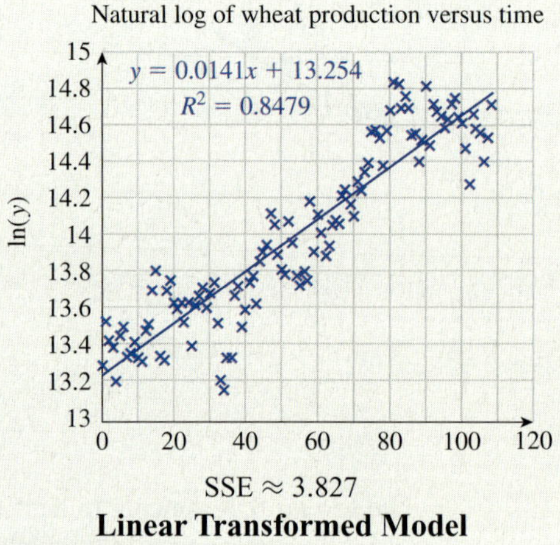

Figure 24

Notice that this time, the regression patterns no longer suggest an obvious curve. Further, they no longer appear to grow with increasing t. Although SSE is dramatically lower than the values for the earlier models, the contrast is a false one; the units of y are now different, and comparing SSE with that or the earlier models is like comparing apples and oranges. While SSE depends on the units of measurement used, the coefficient of determination r^2 discussed in Section 1.4 is independent of the units used. A similar statistic is available for other types of regression as well, as well as something called "adjusted r^2."

The value of r^2 for the transformed model is approximately 0.8497, while r^2 for the cubic model* is about 0.8219, which is fairly close.

* **NOTE** The "adjusted r^2" from statistics that corrects for model size.

Q: *If the cubic model and the linear transformed model have similar values of r^2, how do I decide which is more appropriate?*

A: The cubic model, if extrapolated, predicts unrealistically that the production of wheat will plunge in the near future, but the linear transformed model sees the recent drop-off as just one of several market fluctuations that show up in the residuals. You should therefore favor the more reasonable linear transformed model.

Q: *The linear transformed model gives us ln y versus t. What does it say about y versus t?*

A: Accurately write down the equation of the transformed linear model, being careful to replace y by $\ln y$:

$$\ln y = 0.0140613t + 13.2537$$

Rewriting this is exponential form gives

$$\begin{aligned} y &= e^{0.0140613t+13.2537} \\ &= e^{13.2537}e^{0.0140613t} \\ &\approx 570,000e^{0.0141t}, \end{aligned}$$

Coefficients rounded to 3 digits

which is exactly the exponential model we found earlier! (In fact, using the natural logarithm transformation is the standard method of computing the regression exponential curve.)

Q: *What of the logistic model; should that not be the most realistic?*

A: The logistic model seems as though it *ought* to be the most appropriate, because wheat production cannot reasonably be expected to continue increasing exponentially forever; eventually resource limitations must lead to a leveling-off of wheat production. Such a leveling off, if it occurred before the population started to level off, would seem to vindicate Malthus' pessimistic predictions. However, the logistic regression model looks suspect: It seems to fit the curve more poorly than any of the other models considered, suggesting that, as yet, we do not have significant evidence that any leveling-off is occurring. Wheat production—even if it is logistic—appears still in the early (exponential) stage of growth. In general, for a logistic model to be reliable in its prediction of the leveling-off value N, we would need to see significant evidence of leveling-off in the data. (See, however, Exercise 2 following.)

You now conclude that wheat production for the past 100 years is better described as increasing exponentially than linearly, contradicting Malthus, and moreover that it shows no sign of leveling off as yet.

EXERCISES

1. Use the wheat production data starting at 1950 to construct the exponential regression model in two ways: directly, and using a linear transformed model as above. (Round coefficients to three digits.) Comparing the growth constant k of your model with that of the exponential model based on the data from 1900 on. How would you interpret the difference?
2. Compute the least squares logistic model for the data in the preceding exercise. (Round coefficients to 3 significant digits.) At what level does it predict that wheat production will level off? [Note on using Excel Solver for logistic regression: Before running Solver, press Options in the Solver window and turn "Automatic Scaling" on. This adjusts the algorithm for the fact

that the constants A, N, and b have vastly different orders of magnitude.] Give two graphs: One showing the data with the exponential regression model, and the other showing the data with the logistic regression model. Which model gives a better fit visually? Justify your observation by computing SSE directly for both models. Comment on your answer in terms of Malthus' assertions.

3. Find the production figures for another common crop grown in the United States. Compare the linear, quadratic, cubic, exponential, and logistic models. What can you conclude?

4. Below are the census figures for the population of the United States (in thousands) from 1820 to 2010.[53] Compare the linear, quadratic, and exponential models. What can you conclude?

Population of the United States (1,000)

1820	**1830**	**1840**	**1850**	**1860**	**1870**	**1880**	**1890**	**1900**	**1910**
9,638	12,861	17,063	23,192	31,443	38,558	50,189	62,980	76,212	92,228
1920	**1930**	**1940**	**1950**	**1960**	**1970**	**1980**	**1990**	**2000**	**2010**
106,022	123,203	132,165	151,326	179,323	203,302	226,542	248,710	281,422	308,936

[53] Source: Bureau of the Census, U.S. Department of Commerce.

TI-83/84 Plus Technology Guide

Section 2.1

Example 2 (page 126) Sketch the graph of each quadratic function, showing the location of the vertex and intercepts.

a. $f(x) = 4x^2 - 12x + 9$ **b.** $g(x) = -\frac{1}{2}x^2 + 4x - 12$

Solution with Technology

We will do part (a).

1. Start by storing the coefficients a, b, c using

   `4→A:-12→B:9→C`

   [STO>] gives the arrow [ALPHA] [.] gives the colon

2. Save your quadratic as Y_1 using the Y= screen:

   Y_1=AX^2+BX+C

3. To obtain the x-coordinate of the vertex, enter its formula as shown:

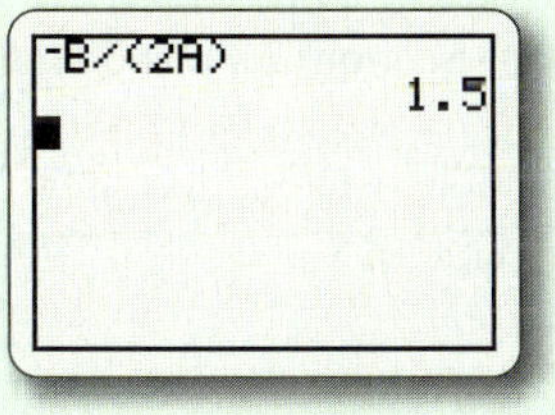

4. The y-coordinate of the vertex can be obtained from the Table screen by entering $x = 1.5$ as shown. (If you can't enter values of x, press [2ND] [TBLSET], and set Indpnt to Ask.)

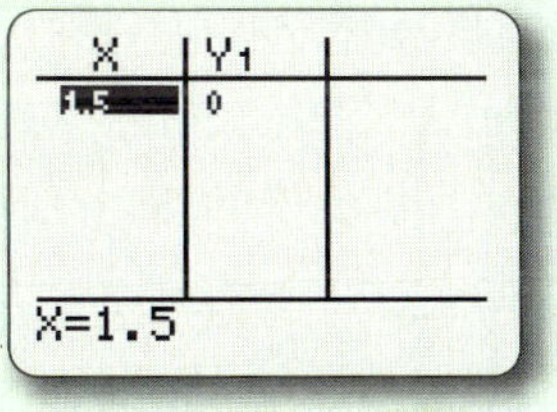

From the table, we see that the vertex is at the point (1.5, 0).

5. To obtain the x-intercepts, enter the quadratic formula on the home screen as shown:

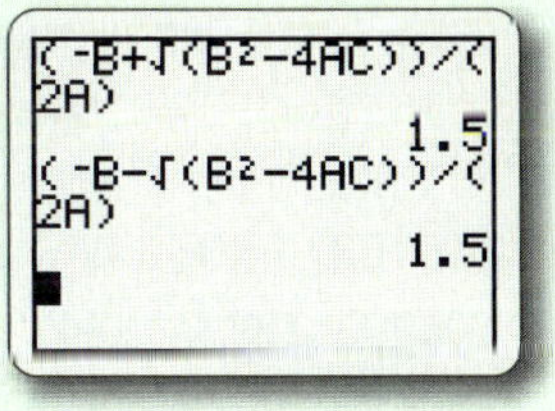

Because both intercepts agree, we conclude that the graph intersects the x-axis on a single point (at the vertex).

6. To graph the function, we need to select good values for Xmin and Xmax. In general, we would like our graph to show the vertex as well as all the intercepts. To see the vertex, make sure that its x coordinate (1.5) is between Xmin and Xmax. To see the x-intercepts, make sure that they are also between Xmin and Xmax. To see the y-intercept, make sure that $x = 0$ is between Xmin and Xmax. Thus, to see everything, choose Xmin and Xmax so that the interval [xMin, xMax] contains the x coordinate of the vertex, the x-intercepts, and 0. For this example, we can choose an interval like $[-1, 3]$.

7. Once xMin and xMax are chosen, you can obtain convenient values of yMin and yMax by pressing [ZOOM] and selecting the option ZoomFit. (Make sure that your quadratic equation is entered in the Y= screen before doing this!)

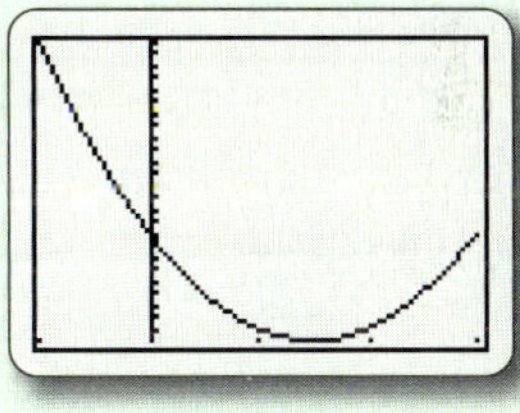

Example 5(b) (page 130) The following table shows total and projected production of ozone-layer damaging Freon 22 (chlorodifluoromethane) in developing countries ($t = 0$ represents 2000).

Year t	0	2	4	6	8	10
Tons of Freon F	100	140	200	270	400	590

Find the quadratic regression model.

Solution with Technology

1. Using [STAT] EDIT enter the data with the x-coordinates (values of t) in L_1 and the y-coordinates (values of F) in L_2, just as in Section 1.4:

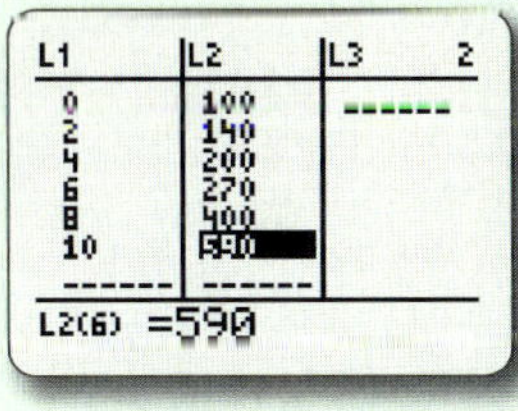

2. Press `STAT`, select `CALC`, and choose option #5 `QuadReg`. Pressing `ENTER` gives the quadratic regression curve in the home screen:

$y \approx 4.598x^2 + 1.161x + 108.9$ Coefficients rounded to four decimal places

3. Now go to the `Y=` window and turn `Plot1` on by selecting it and pressing `ENTER`. (You can also turn it on in the `2ND` STAT PLOT screen.)
4. Next, enter the regression equation in the `Y=` screen by pressing `Y=` clearing out whatever function is there, and pressing `VARS` `5` and selecting `EQ` (Option #1: `RegEq`):

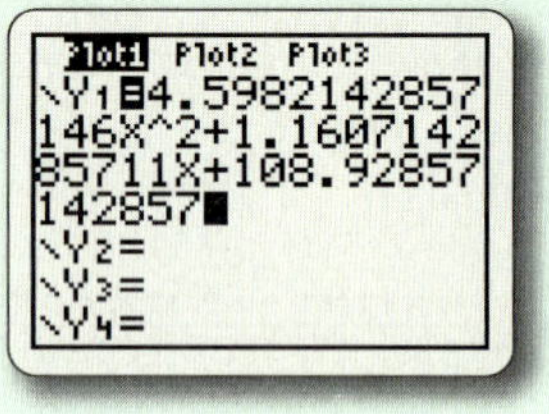

5. To obtain a convenient window showing all the points and the lines, press `ZOOM` and choose option #9: `ZoomStat`:

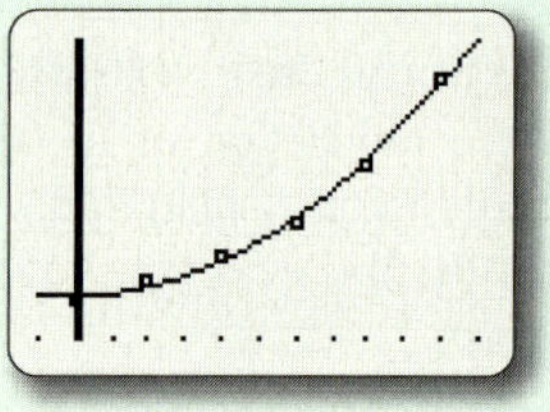

Note When you are done viewing the graph, it is a good idea to turn PLOT1 off again to avoid errors in graphing or data points showing up in your other graphs. ■

Section 2.2

Example 6(a) (page 145) The following table shows annual expenditure on health in the United States from 1980 through 2010 ($t = 0$ represents 1980).

Year t	0	5	10	15	20	25	30
Expenditure ($ billion)	246	427	696	990	1,310	1,920	2,750

Find the exponential regression model $C(t) = Ab^t$.

Solution with Technology

This is very similar to Example 5 in Section 2.1 (see the Technology Guide for Section 2.1):

1. Use `STAT` `EDIT` to enter the above table of values.

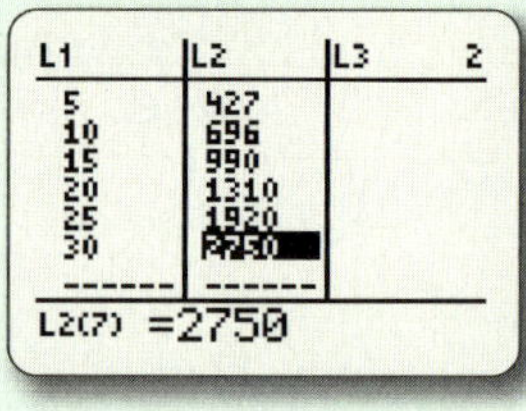

2. Press `STAT`, select `CALC`, and choose option #0 `ExpReg`. Pressing `ENTER` gives the exponential regression curve in the home screen:

$C(t) \approx 282.33(1.0808)^t$ Coefficients rounded

3. To graph the points and regression line in the same window, turn Stat Plot on (see the Technology Guide for Example 5 in Section 2.1) and enter the regression equation in the `Y=` screen by pressing `Y=`, clearing out whatever function is there, and pressing `VARS` `5` and selecting `EQ` (Option 1: `RegEq`). Then press `ZOOM` and choose option #9: `ZoomStat` to see the graph.

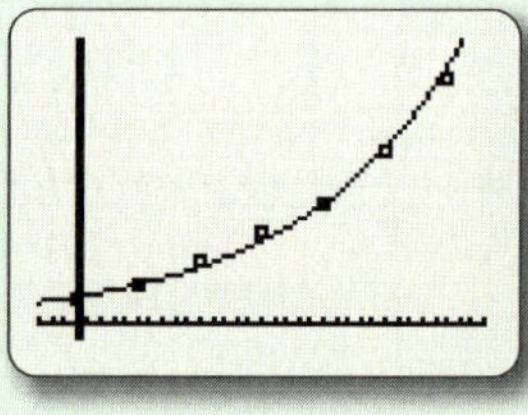

Note When you are done viewing the graph, it is a good idea to turn PLOT1 off again to avoid errors in graphing or data points showing up in your other graphs. ■

Section 2.3

Example 5 (page 160) The following table shows the total spent in on research and development in the United States, in billions of dollars, for the period 1995–2009 (t is the year since 1990).

Year t	5	6	7	8	9	10	11	12
Spending (\$ billions)	199	210	222	235	250	268	271	265
Year t	13	14	15	16	17	18	19	
Spending (\$ billions)	272	275	289	299	300	304	309	

Find the best-fit logarithmic model of the form:

$$S(t) = A \ln t + C.$$

Solution with Technology

This is very similar to Example 5 in Section 2.1 (see the Technology Guide for Section 2.1):

1. Use `STAT` `EDIT` to enter the above table of values.
2. Press `STAT`, select `CALC`, and choose the #9 `LnReg`. Pressing `ENTER` gives the logarithmic regression curve in the home screen:

$$S(t) = 83.01 \ln t + 64.42 \qquad \text{Coefficients rounded}$$

3. To graph the points and regression line in the same window, turn Stat Plot on (see the Technology Guide for Example 5 in Section 2.1) and enter the regression equation in the `Y=` screen by pressing `Y=`, clearing out whatever function is there, and pressing `VARS` `5` and selecting `EQ` (Option 1: `RegEq`). Then press `ZOOM` and choose option #9: `ZoomStat` to see the graph.

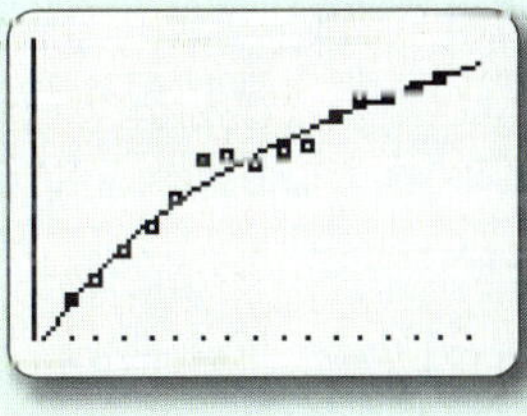

Section 2.4

Example 2 (page 169) The following table shows the percentage of U.S. Internet-connected households that have broadband connections as a function of time t in years ($t = 0$ represents 2000).

Year (t)	0	1	2	3	4	5	6	7	8	9
Percentage with Broadband (%) (P)	8	14	20	33	45	53	66	80	88	92

Find a logistic regression curve of the form

$$P(t) = \frac{N}{1 + Ab^{-t}}.$$

Solution with Technology

This is very similar to Example 5 in Section 2.1 (see the Technology Guide for Section 2.1):

1. Use `STAT` `EDIT` to enter the above table of values.
2. Press `STAT`, select `CALC`, and choose the #B `Logistic`. Pressing `ENTER` gives the logistic regression curve in the home screen:

$$P(t) \approx \frac{104.046}{1 + 10.453e^{-0.49578t}} \qquad \text{Coefficients rounded}$$

This is not exactly the form we are seeking, but we can convert it to that form by writing

$$e^{-0.49578t} = (e^{0.49578})^{-t} \approx 1.642^{-t}$$

so,

$$P(t) \approx \frac{104.046}{1 + 10.453(1.642)^{-t}}.$$

3. To graph the points and regression line in the same window, turn Stat Plot on (see the Technology Guide for Example 5 in Section 2.1) and enter the regression equation in the `Y=` screen by pressing `Y=`, clearing out whatever function is there, and pressing `VARS` `5` and selecting `EQ` (Option 1: `RegEq`). Then press `ZOOM` and choose option #9: `ZoomStat` to see the graph.

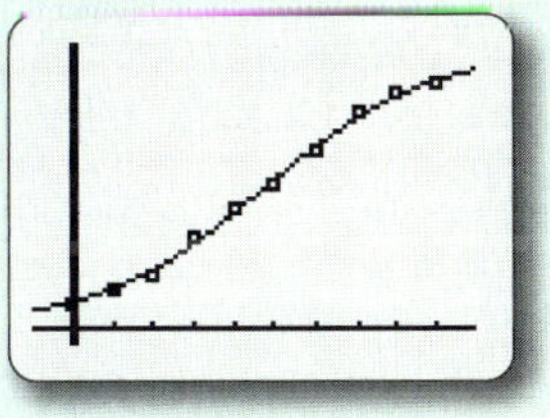

TECHNOLOGY GUIDE

EXCEL Technology Guide

Section 2.1

Example 2 (page 126) Sketch the graph of each quadratic function, showing the location of the vertex and intercepts.

a. $f(x) = 4x^2 - 12x + 9$ **b.** $g(x) = -\frac{1}{2}x^2 + 4x - 12$

Solution with Technology

We can set up a worksheet so that all we have to enter are the coefficients a, b, and c, and a range of x-values for the graph. Here is a possible layout that will plot 101 points using the coefficients for part (a) (similar to the Excel Graphing Worksheet we mentioned in Example 3 of Section 1.1, which can be found on the Web site by following Chapter 1 → Excel Tutorials → Section 1.1.)

1. First, we compute the x coordinates:

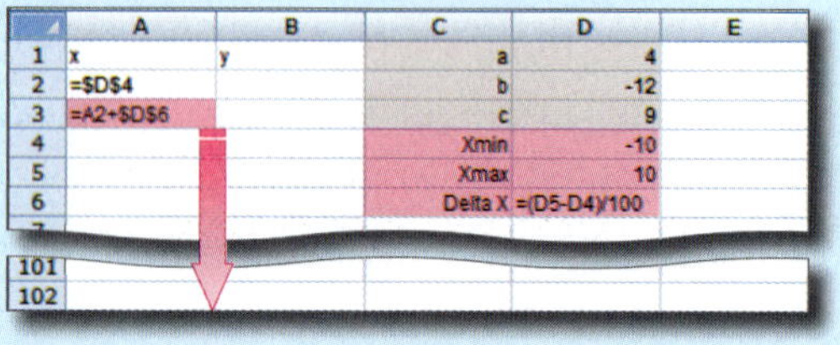

2. To add the y coordinates, we use the technology formula

```
a*x^2+b*x+c
```

replacing a, b, and c with (absolute) references to the cells containing their values.

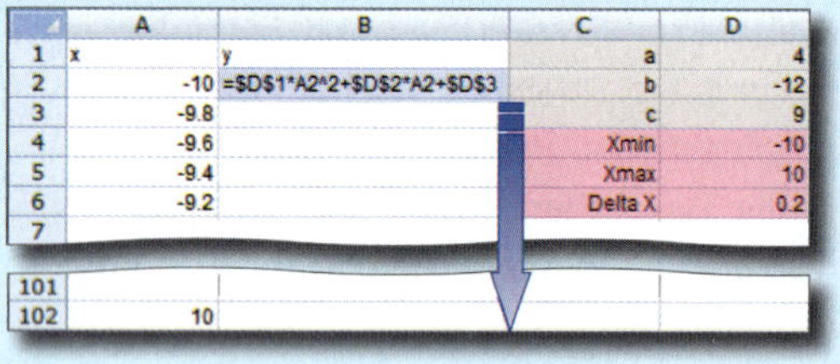

3. Graphing the data in columns A and B gives the graph shown here:

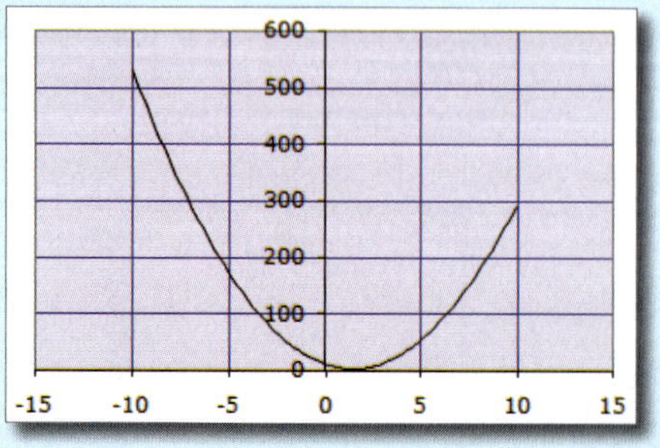

$y = 4x^2 - 12x + 9$

4. We can go further and compute the exact coordinates of the vertex and intercepts:

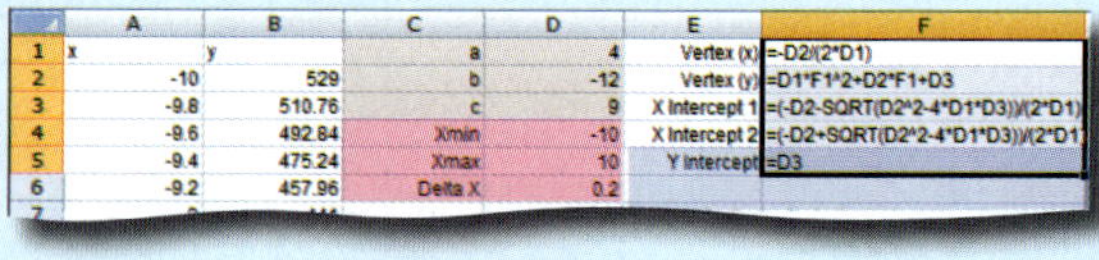

The completed sheet should look like this:

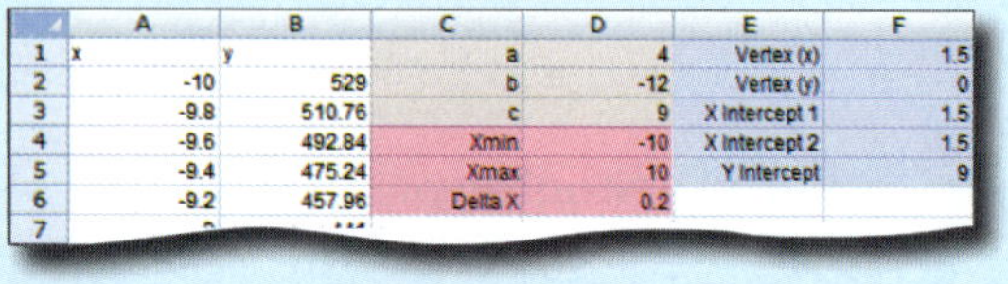

We can now save this sheet as a template to handle all quadratic functions. For instance, to do part (b), we just change the values of a, b, and c in column D to $a = -1/2$, $b = 4$, and $c = -12$.

Example 5(b) (page 130) The following table shows total and projected production of ozone-layer damaging Freon 22 (chlorodifluoromethane) in developing countries ($t = 0$ represents 2000).

Year t	0	2	4	6	8	10
Tons of Freon F	100	140	200	270	400	590

Find the quadratic regression model.

Solution with Technology

As in Section 1.4, Example 3, we start with a scatter plot of the original data, and add a trendline:

1. Start with the original data and a "Scatter plot." (See Section 1.2 Example 5.)

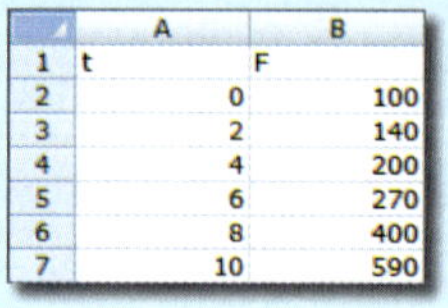

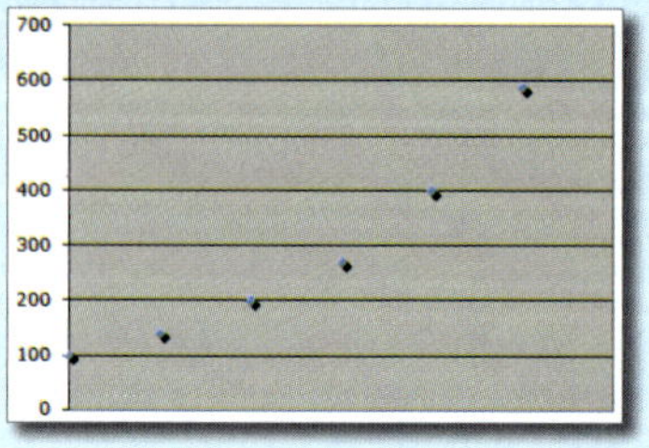

2. Click on the chart, select the Layout tab, click the Trendline button, and choose "More Trendline Options." (Alternatively, right-click on any data point in the chart and select "Add Trendline.") Then select a "Polynomial"

type of order 2 and check the option to "Display Equation on chart."

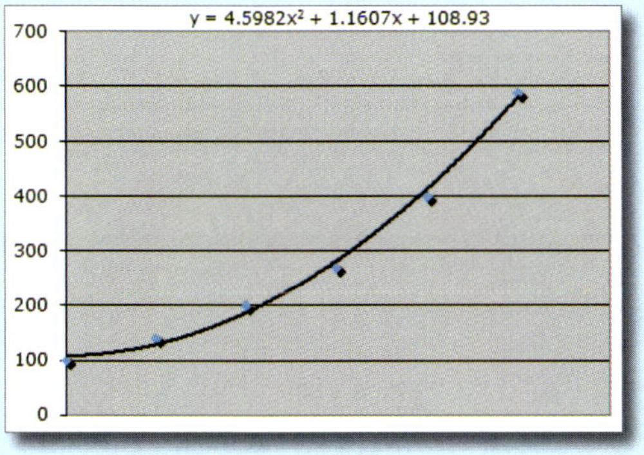

Section 2.2

Example 6(a) (page 145) The following table shows annual expenditure on health in the United States from 1980 through 2010 ($t = 0$ represents 1980).

Year t	0	5	10	15	20	25	30
Expenditure ($ billion)	246	427	696	990	1,310	1,920	2,750

Find the exponential regression model $C(t) = Ab^t$.

Solution with Technology

This is very similar to Example 5 in Section 2.1 (see the Technology Guide for Section 2.1):

1. Start with a "Scatter plot" of the observed data.
2. Click on the chart, select the Layout tab, click the Trendline button, and choose "More Trendline Options." (Alternatively, right-click on any data point in the chart and select "Add Trendline.") Then select an "Exponential" type and check the option to "Display Equation on chart."

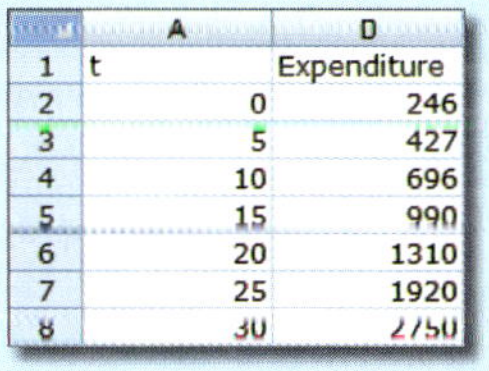

	A	B
1	t	Expenditure
2	0	246
3	5	427
4	10	696
5	15	990
6	20	1310
7	25	1920
8	30	2750

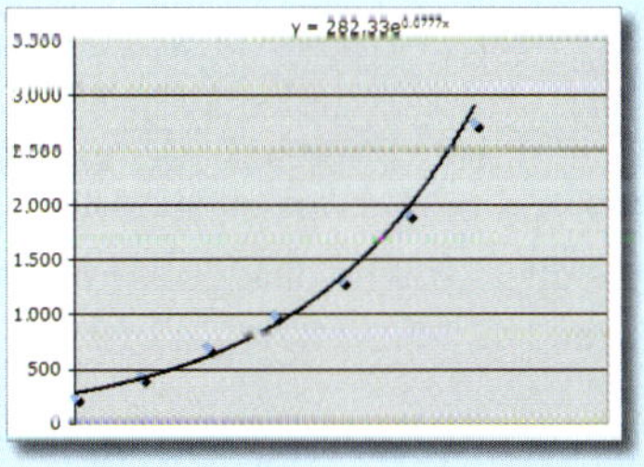

Notice that the regression curve is given in the form Ae^{kt} rather than Ab^t. To transform it, write

$$\begin{aligned} 282.33e^{0.0777t} &= 282.33(e^{0.0777})^t \\ &\approx 282.33(1.0808)^t. \end{aligned}$$

$e^{0.0777} \approx 1.0808$

Section 2.3

Example 5 (page 160) The following table shows the total spent in on research and development in the United States, in billions of dollars, for the period 1995–2009 (t is the year since 1990).

Year t	5	6	7	8	9	10	11	12
Spending ($ billions)	199	210	222	235	250	268	271	265
Year t	13	14	15	16	17	18	19	
Spending ($ billions)	272	275	289	299	300	304	309	

Find the best-fit logarithmic model of the form

$$S(t) = A \ln t + C.$$

Solution with Technology

This is very similar to Example 5 in Section 2.1 (see the Technology Guide for Section 2.1): We start, as usual, with a "Scatter plot" of the observed data and add a `Logarithmic` trendline. Here is the result:

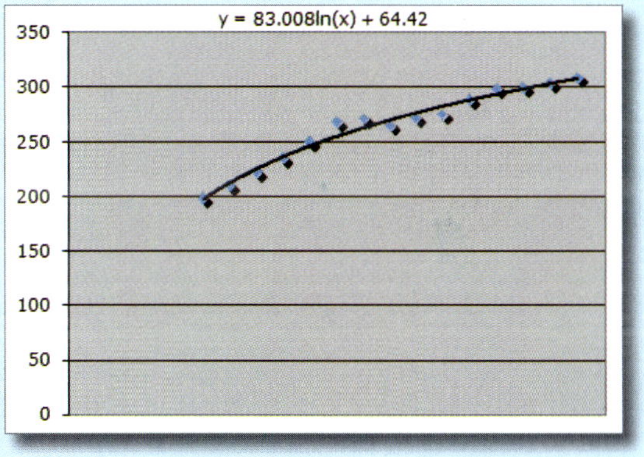

Section 2.4

Example 2 (page 169) The following table shows the percentage of U.S. Internet-connected households that have broadband connections as a function of time t in years ($t = 0$ represents 2000).

Year (t)	0	1	2	3	4	5	6	7	8	9
Percentage with Broadband (%) (P)	8	14	20	33	45	53	66	80	88	92

Find a logistic regression curve of the form

$$P(t) = \frac{N}{1 + Ab^{-t}}.$$

Solution with Technology

Excel does not have a built-in logistic regression calculation, so we use an alternative method that works for any type of regression curve.

1. First use rough estimates for N, A, and b, and compute the sum-of-squares error (SSE; see Section 1.5) directly:

	A	B	C	D	E	F	G
1	t	P (Observed)	P (Predicted)	Residual^2	N	A	b
2	0	8	=$E$2/(1+$F$2*$G$2^(-A2))	=(B2-C2)^2	100	10	1.5
3	1	14					
4	2	20					
5	3	33				SSE	
6	4	45				=SUM(D2:D11)	
7	5	53					
8	6	66					
9	7	80					
10	8	88					
11	9	92					

Cells E2:G2 contain our initial rough estimates of N, A, and b. For N, we used 100 (notice that the y coordinates do appear to level off around 100). For A, we used the fact that the y-intercept is $N/(1 + A)$. In other words,

$$8 = \frac{100}{1 + A}.$$

Because a very rough estimate is all we are after, using $A = 10$ will do just fine. For b, we chose 1.5 as the values of P appear to be increasing by around 50% per year initially (again, this is rough).

2. Cell C2 contains the formula for $P(t)$, and the square of the resulting residual is computed in D2.

3. Cell F6 will contain SSE. The completed spreadsheet should look like this:

	A	B	C	D	E	F	G
1	t	P (Observed)	P (Predicted)	Residual^2	N	A	b
2	0	8	9.090909091	1.19008264	100	10	1.5
3	1	14	13.04347826	0.91493384			
4	2	20	18.36734694	2.66555602			
5	3	33	25.23364486	60.3162722		SSE	
6	4	45	33.60995851	129.733045		1158.4	
7	5	53	43.1616341	96.7934435			
8	6	66	53.25054785	162.54853			
9	7	80	63.08047303	286.270393			
10	8	88	71.93290209	258.151635			
11	9	92	79.35733581	159.836958			

The best-fit curve will result from values of N, A, and b that give a minimum value for SSE. We shall use Excel's "Solver," found in the "Analysis" group on the "Data" tab. (If "Solver" does not appear in the Analysis group, you will have to install the Solver Add-in using the Excel Options dialogue.) Figure 25 shows the Solver

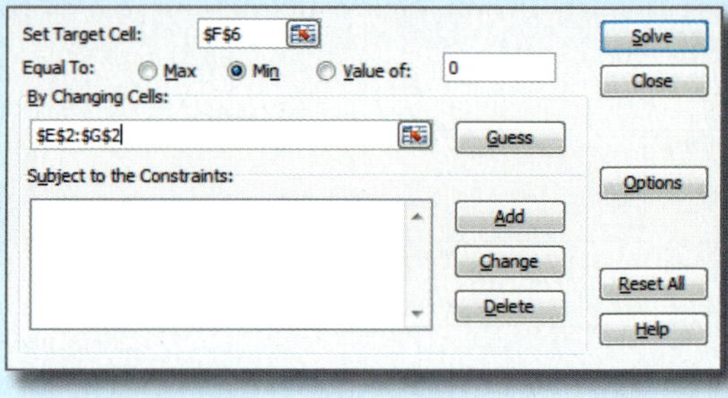

Figure 25

dialogue box with the necessary fields completed to solve the problem.

- The Target Cell refers to the cell that contains SSE.
- "Min" is selected because we are minimizing SSE.
- "Changing Cells" are obtained by selecting the cells that contain the current values of N, A, and b.

4. When you have filled in the values for the three items above, press "Solve" and tell Solver to Keep Solver Solution when done. You will find $N \approx 104.047$, $A \approx 10.453$, and $b \approx 1.642$ so

$$P(t) \approx \frac{104.047}{1 + 10.453(1.642)^{-t}}.$$

If you use a scatter plot to graph the data in columns A, B and C, you will obtain the following graph:

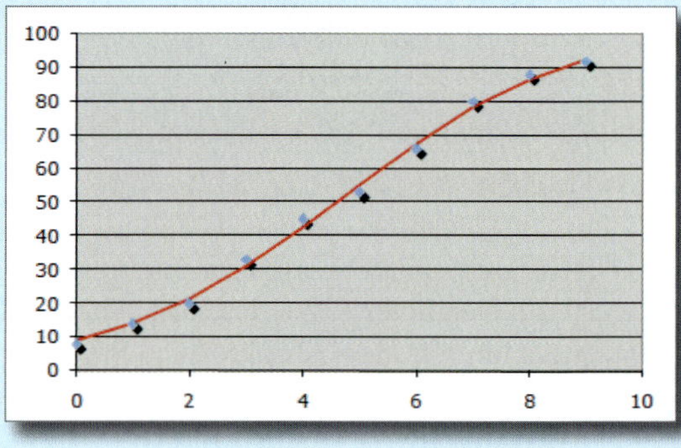

In order to find a model in which $N = 100$, all we need to do is enter 100 in cell E2 and tell Solver that the only cells it should change are F2 and G2, corresponding to A and b, by entering $F$2:$G$2 in the "By Changing Cells" box.

3

Introduction to the Derivative

Case Study Reducing Sulfur Emissions

The Environmental Protection Agency (EPA) wants to formulate a policy that will encourage utilities to reduce sulfur emissions. Its goal is to reduce annual emissions of sulfur dioxide by a total of 10 million tons from the current level of 25 million tons by imposing a fixed charge for every ton of sulfur released into the environment per year. The EPA has some data showing the marginal cost to utilities of reducing sulfur emissions. As a consultant to the EPA, you must determine the amount to be charged per ton of sulfur emissions in light of these data.

Norbert Schaefer/CORBIS

Web Site

At the Web site you will find:

- Section by section tutorials, including game tutorials with randomized quizzes
- A detailed chapter summary
- A true/false quiz
- Additional review exercises
- Graphers, Excel tutorials, and other resources
- The following extra topics:

  Sketching the graph of the derivative

  Continuity and differentiability

Introduction

In the world around us, everything is changing. The mathematics of change is largely about the rate of change: how fast and in which direction the change is occurring. Is the Dow Jones average going up, and if so, how fast? If I raise my prices, how many customers will I lose? If I launch this missile, how fast will it be traveling after two seconds, how high will it go, and where will it come down?

We have already discussed the concept of rate of change for linear functions (straight lines), where the slope measures the rate of change. But this works only because a straight line maintains a constant rate of change along its whole length. Other functions rise faster here than there—or rise in one place and fall in another—so that the rate of change varies along the graph. The first achievement of calculus is to provide a systematic and straightforward way of calculating (hence the name) these rates of change. To describe a changing world, we need a language of change, and that is what calculus is.

The history of calculus is an interesting story of personalities, intellectual movements, and controversy. Credit for its invention is given to two mathematicians: Isaac Newton (1642–1727) and Gottfried Leibniz (1646–1716). Newton, an English mathematician and scientist, developed calculus first, probably in the 1660s. We say "probably" because, for various reasons, he did not publish his ideas until much later. This allowed Leibniz, a German mathematician and philosopher, to publish his own version of calculus first, in 1684. Fifteen years later, stirred up by nationalist fervor in England and on the continent, controversy erupted over who should get the credit for the invention of calculus. The debate got so heated that the Royal Society (of which Newton and Leibniz were both members) set up a commission to investigate the question. The commission decided in favor of Newton, who happened to be president of the society at the time. The consensus today is that both mathematicians deserve credit because they came to the same conclusions working independently. This is not really surprising: Both built on well-known work of other people, and it was almost inevitable that someone would put it all together at about that time.

algebra Review

For this chapter, you should be familiar with the algebra reviewed in **Chapter 0, Section 2.**

3.1 Limits: Numerical and Graphical Approaches

Rates of change are calculated by derivatives, but an important part of the definition of the derivative is something called a **limit**. Arguably, much of mathematics since the 18th century has revolved around understanding, refining, and exploiting the idea of the limit. The basic idea is easy, but getting the technicalities right is not.

Evaluating Limits Numerically

Start with a very simple example: Look at the function $f(x) = 2 + x$ and ask: What happens to $f(x)$ as x approaches 3? The following table shows the value of $f(x)$ for values of x close to and on either side of 3:

x approaching 3 from the left → ← x approaching 3 from the right

x	2.9	2.99	2.999	2.9999	3	3.0001	3.001	3.01	3.1
$f(x) = 2 + x$	4.9	4.99	4.999	4.9999		5.0001	5.001	5.01	5.1

We have left the entry under 3 blank to emphasize that when calculating the limit of $f(x)$ as x *approaches* 3, we are not interested in its value when x *equals* 3.

Notice from the table that the closer x gets to 3 from either side, the closer $f(x)$ gets to 5. We write this as

$$\lim_{x \to 3} f(x) = 5. \qquad \text{The limit of } f(x) \text{, as } x \text{ approaches 3, equals 5.}$$

Q: *Why all the fuss? Can't we simply substitute $x = 3$ and avoid having to use a table?*

A: This happens to work for *some* functions, but not for *all* functions. The following example illustrates this point.

EXAMPLE 1 Estimating a Limit Numerically

Use a table to estimate the following limits:

a. $\lim_{x \to 2} \frac{x^3 - 8}{x - 2}$ **b.** $\lim_{x \to 0} \frac{e^{2x} - 1}{x}$

Solution

a. We cannot simply substitute $x = 2$, because the function $f(x) = \frac{x^3 - 8}{x - 2}$ is not defined at $x = 2$. (Why?)* Instead, we use a table of values as we did above, with x approaching 2 from both sides.

x approaching 2 from the left → ← x approaching 2 from the right

x	1.9	1.99	1.999	1.9999	2	2.0001	2.001	2.01	2.1
$f(x) = \frac{x^3 - 8}{x - 2}$	11.41	11.9401	11.9940	11.9994		12.0006	12.0060	12.0601	12.61

We notice that as x approaches 2 from either side, $f(x)$ appears to be approaching 12. This suggests that the limit is 12, and we write

$$\lim_{x \to 2} \frac{x^3 - 8}{x - 2} = 12.$$

b. The function $g(x) = \frac{e^{2x} - 1}{x}$ is not defined at $x = 0$ (nor can it even be simplified to one which *is* defined at $x = 0$). In the following table, we allow x to approach 0 from both sides:

x approaching 0 from the left → ← x approaching 0 from the right

x	−0.1	−0.01	−0.001	−0.0001	0	0.0001	0.001	0.01	0.1
$g(x) = \frac{e^{2x} - 1}{x}$	1.8127	1.9801	1.9980	1.9998		2.0002	2.0020	2.0201	2.2140

The table suggests that $\lim_{x \to 0} \frac{e^{2x} - 1}{x} = 2$.

*** NOTE** However, if you factor $x^3 - 8$, you will find that $f(x)$ can be simplified to a function which *is* defined at $x = 2$. This point will be discussed (and this example redone) in Section 3.3. The function in part (b) cannot be simplified by factoring.

using Technology

We can automate the computations in Example 1 using a graphing calculator or Excel. See the Technology Guides at the end of the section to find out how to create tables like these using a TI-83/84 Plus or Excel. Here is an outline for part (a):

TI-83/84 Plus

Home screen: `Y1=(X^3-8)/(X-2)`

[2ND] [TBLSET] Indpnt set to Ask

[2ND] [TABLE] Enter some values of x from the example: 1.9, 1.99, 1.999 . . .

[More details on page 280.]

Excel
Enter the headings x, $f(x)$ in A1–B1 and again in C1–D1. In A2–A5 enter 1.9, 1.99, 1.999, 1.9999. In C1–C5 enter 2.1, 2.01, 2.001, 2.0001. Enter `=(A2^3-8)/(A2-2)` in B2 and copy down to B5. Copy and paste the same formula in D2–D5. [More details on page 282.]

Web Site
www.AppliedCalc.org
Student Home → Online Utilities → Function Evaluator and Grapher Enter `(x^3-8)/(x-2)` for y_1. To obtain a table of values, enter the various x-values in the Evaluator box, and press "Evaluate."

➡ **Before we go on...** Although the table *suggests* that the limit in Example 1 part (b) is 2, it by no means establishes that fact conclusively. It is *conceivable* (though not in fact the case here) that putting $x = 0.000000087$ could result in $g(x) = 426$. Using a table can only suggest a value for the limit. In the next two sections we shall discuss algebraic techniques for finding limits. ■

Before we continue, let us make a more formal definition.

Definition of a Limit

If $f(x)$ approaches the number L as x approaches (but is not equal to) a from both sides, then we say that $f(x)$ **approaches L as $x \to a$** ("x approaches a") or that the **limit** of $f(x)$ as $x \to a$ is L. More precisely, *we can make $f(x)$ be as close to L as we like by choosing any x sufficiently close to (but not equal to) a on either side*. We write

$$\lim_{x\to a} f(x) = L$$

or

$$f(x) \to L \text{ as } x \to a.$$

If $f(x)$ *fails* to approach *a single fixed number* as x approaches a from both sides, then we say that $f(x)$ **has no limit** as $x \to a$, or

$$\lim_{x\to a} f(x) \textbf{ does not exist.}$$

Quick Examples

1. $\lim_{x\to 3}(2 + x) = 5$ — See discussion before Example 1.
2. $\lim_{x\to -2}(3x) = -6$ — As x approaches -2, $3x$ approaches -6.
3. $\lim_{x\to 0}(x^2 - 2x + 1)$ exists. — In fact, the limit is 1.
4. $\lim_{x\to 5}\frac{1}{x} = \frac{1}{5}$ — As x approaches 5, $\frac{1}{x}$ approaches $\frac{1}{5}$.
5. $\lim_{x\to 2}\frac{x^3 - 8}{x - 2} = 12$ — See Example 1. (We cannot just put $x = 2$ here.)

(For examples where the limit does not exist, see Example 2.)

Notes

1. It is important that $f(x)$ approach the same number as x approaches a from either side. For instance, if $f(x)$ approaches 5 for $x = 1.9, 1.99, 1.999, \ldots$, but approaches 4 for $x = 2.1, 2.01, 2.001, \ldots$, then the limit as $x \to 2$ does not exist. (See Example 2 for such a situation.)
2. It may happen that $f(x)$ does not approach any fixed number at all as $x \to a$ from either side. In this case, we also say that the limit does not exist.
3. If a happens to be an endpoint of the domain of f, then x can only approach a from one side. In this case, we relax the definition of the limit and require only that x approach a from the side it can. For example, $f(x) = \sqrt{x-1}$ has natural domain $[1, +\infty)$, and we say that

$$\lim_{x\to 1}\sqrt{x - 1} = 0$$

even though x can only approach 1 from the right. ■

The following example gives instances in which a stated limit does not exist.

EXAMPLE 2 Limits Do Not Always Exist

Do the following limits exist?

a. $\lim_{x\to 0} \frac{1}{x^2}$ **b.** $\lim_{x\to 0} \frac{|x|}{x}$ **c.** $\lim_{x\to 2} \frac{1}{x-2}$

Solution

a. Here is a table of values for $f(x) = \frac{1}{x^2}$, with x approaching 0 from both sides.

x approaching 0 from the left → ← x approaching 0 from the right

x	−0.1	−0.01	−0.001	−0.0001	0	0.0001	0.001	0.01	0.1
$f(x) = \frac{1}{x^2}$	100	10,000	1,000,000	100,000,000		100,000,000	1,000,000	10,000	100

The table shows that as x gets closer to zero on either side, $f(x)$ gets larger and larger **without bound**—that is, if you name any number, no matter how large, $f(x)$ will be even larger than that if x is sufficiently close to 0. Because $f(x)$ is not approaching any real number, we conclude that $\lim_{x\to 0} \frac{1}{x^2}$ does not exist. Because $f(x)$ is becoming arbitrarily large, we also say that $\lim_{x\to 0} \frac{1}{x^2}$ **diverges to** $+\infty$, or just

$$\lim_{x\to 0} \frac{1}{x^2} = +\infty$$

Note This is not meant to imply that the limit exists; the symbol $+\infty$ does not represent any real number. We write $\lim_{x\to a} f(x) = +\infty$ to indicate two things: (1) the limit does not exist and (2) the function gets large without bound as x approaches a. ■

b. Here is a table of values for $f(x) = \frac{|x|}{x}$, with x approaching 0 from both sides.

x approaching 0 from the left → ← x approaching 0 from the right

x	−0.1	−0.01	−0.001	−0.0001	0	0.0001	0.001	0.01	0.1
$f(x) = \frac{\lvert x\rvert}{x}$	−1	−1	−1	−1		1	1	1	1

The table shows that $f(x)$ does not approach the same limit as x approaches 0 from both sides. There appear to be two *different* limits: the limit as we approach 0 from the left and the limit as we approach from the right. We write

$$\lim_{x\to 0^-} f(x) = -1$$

read as "the limit as x approaches 0 from the left (or from below) is −1" and

$$\lim_{x\to 0^+} f(x) = 1$$

read as "the limit as x approaches 0 from the right (or from above) is 1." These are called the **one-sided limits** of $f(x)$. In order for f to have a **two-sided limit**, the two one-sided limits must be equal. Because they are not, we conclude that $\lim_{x\to 0} f(x)$ does not exist.

c. Near $x = 2$, we have the following table of values for $f(x) = \dfrac{1}{x-2}$:

x approaching 2 from the left → ← x approaching 2 from the right

x	1.9	1.99	1.999	1.9999	2	2.0001	2.001	2.01	2.1
$f(x) = \dfrac{1}{x-2}$	−10	−100	−1000	−10,000		10,000	1000	100	10

Because $\dfrac{1}{x-2}$ is approaching no (single) real number as $x \to 2$, we see that $\lim_{x\to 2} \dfrac{1}{x-2}$ does not exist. Notice also that $\dfrac{1}{x-2}$ diverges to $+\infty$ as $x \to 2$ from the positive side (right half of the table) and to $-\infty$ as $x \to 2$ from the left (left half of the table). In other words,

$$\lim_{x\to 2^-} \frac{1}{x-2} = -\infty$$

$$\lim_{x\to 2^+} \frac{1}{x-2} = +\infty$$

$$\lim_{x\to 2} \frac{1}{x-2} \text{ does not exist.}$$

In another useful kind of limit, we let x approach either $+\infty$ or $-\infty$, by which we mean that we let x get arbitrarily large or let x become an arbitrarily large negative number. The next example illustrates this.

EXAMPLE 3 Limits at Infinity

Use a table to estimate: **a.** $\lim_{x\to +\infty} \dfrac{2x^2-4x}{x^2-1}$ and **b.** $\lim_{x\to -\infty} \dfrac{2x^2-4x}{x^2-1}$.

Solution

a. By saying that x is "approaching $+\infty$," we mean that x is getting larger and larger without bound, so we make the following table:

x approaching $+\infty$ →

x	10	100	1,000	10,000	100,000
$f(x) = \dfrac{2x^2-4x}{x^2-1}$	1.6162	1.9602	1.9960	1.9996	2.0000

(Note that we are only approaching $+\infty$ from the left because we can hardly approach it from the right!) What seems to be happening is that $f(x)$ is approaching 2. Thus we write

$$\lim_{x\to +\infty} f(x) = 2$$

b. Here, x is approaching $-\infty$, so we make a similar table, this time with x assuming negative values of greater and greater magnitude (read this table from right to left):

$\leftarrow x$ approaching $-\infty$

x	$-100{,}000$	$-10{,}000$	$-1{,}000$	-100	-10
$f(x) = \dfrac{2x^2 - 4x}{x^2 - 1}$	2.0000	2.0004	2.0040	2.0402	2.4242

Once again, $f(x)$ is approaching 2. Thus, $\lim_{x \to -\infty} f(x) = 2$.

Estimating Limits Graphically

We can often estimate a limit from a graph, as the next example shows.

EXAMPLE 4 Estimating Limits Graphically

The graph of a function f is shown in Figure 1. (Recall that the solid dots indicate points on the graph, and the hollow dots indicate points not on the graph.)

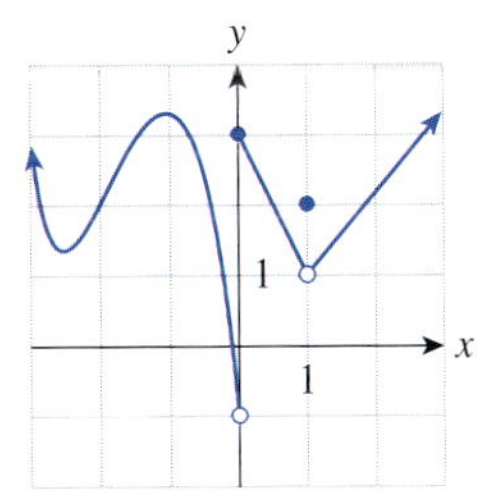

Figure 1

From the graph, analyze the following limits.

a. $\lim\limits_{x \to -2} f(x)$ **b.** $\lim\limits_{x \to 0} f(x)$ **c.** $\lim\limits_{x \to 1} f(x)$ **d.** $\lim\limits_{x \to +\infty} f(x)$

Solution Since we are given only a graph of f, we must analyze these limits graphically.

a. Imagine that Figure 1 was drawn on a graphing calculator equipped with a trace feature that allows us to move a cursor along the graph and see the coordinates as we go. To simulate this, place a pencil point on the graph to the left of $x = -2$, and move it along the curve so that the x-coordinate approaches -2. (See Figure 2.) We evaluate the limit numerically by noting the behavior of the y-coordinates.*

However, we can see directly from the graph that the y-coordinate approaches 2. Similarly, if we place our pencil point to the right of $x = -2$ and move it to the left, the y coordinate will approach 2 from that side as well (Figure 3). Therefore, as x approaches -2 from either side, $f(x)$ approaches 2, so

$$\lim_{x \to -2} f(x) = 2.$$

* **NOTE** For a visual animation of this process, look at the online tutorial for this section at the Web site.

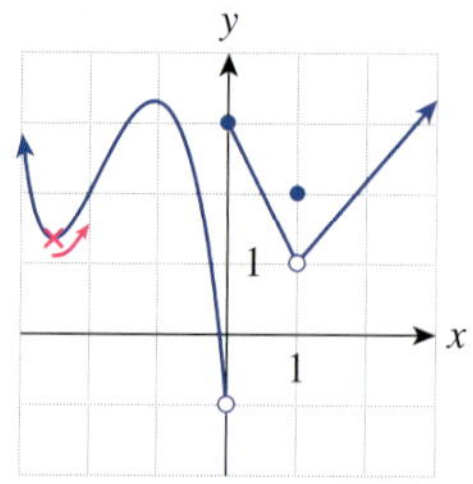

Figure 2

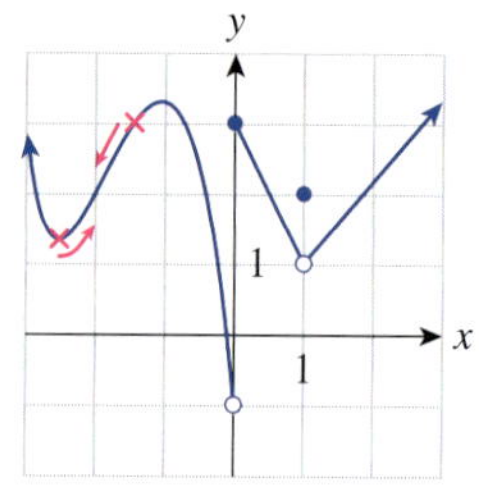

Figure 3

b. This time we move our pencil point toward $x = 0$. Referring to Figure 4, if we start from the left of $x = 0$ and approach 0 (by moving right), the y-coordinate approaches -1. However, if we start from the right of $x = 0$ and approach 0 (by moving left), the y-coordinate approaches 3. Thus (see Example 2),

$$\lim_{x \to 0^-} f(x) = -1$$

and

$$\lim_{x \to 0^+} f(x) = 3.$$

Because these limits are not equal, we conclude that

$$\lim_{x \to 0} f(x) \text{ does not exist.}$$

In this case there is a "break" in the graph at $x = 0$, and we say that the function is **discontinuous** at $x = 0$. (See Section 3.2.)

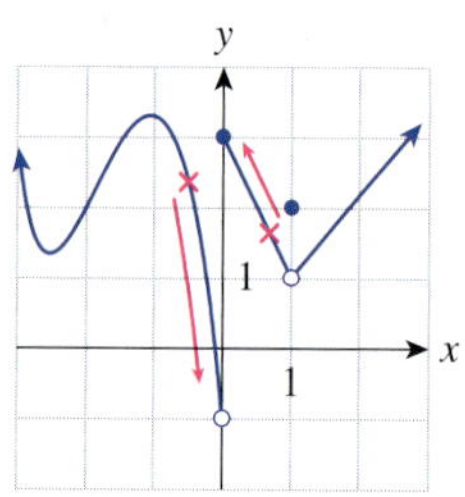

Figure 4

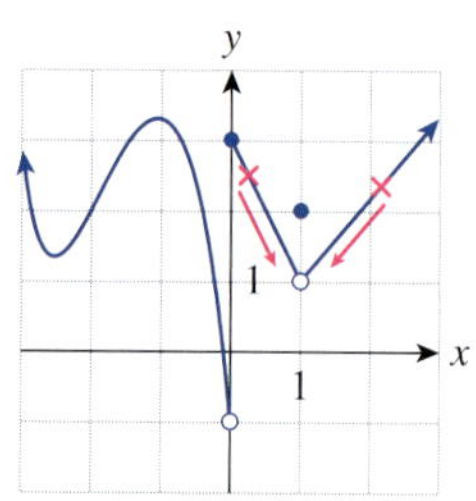

Figure 5

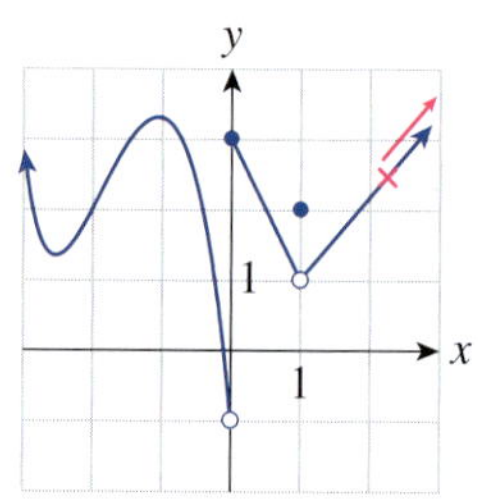

Figure 6

c. Once more we think about a pencil point moving along the graph with the x-coordinate this time approaching $x = 1$ from the left and from the right (Figure 5). As the x-coordinate of the point approaches 1 from either side, the y-coordinate approaches 1 also. Therefore,

$$\lim_{x \to 1} f(x) = 1.$$

d. For this limit, x is supposed to approach infinity. We think about a pencil point moving along the graph further and further to the right as shown in Figure 6.

As the x-coordinate gets larger, the y-coordinate also gets larger and larger without bound. Thus, $f(x)$ diverges to $+\infty$:

$$\lim_{x \to +\infty} f(x) = +\infty.$$

Similarly,

$$\lim_{x \to -\infty} f(x) = +\infty.$$

➡ **Before we go on...** In Example 4(c) $\lim_{x\to 1} f(x) = 1$ but $f(1) = 2$ (why?). Thus, $\lim_{x\to 1} f(x) \neq f(1)$. In other words, the limit of $f(x)$ as x *approaches* 1 is not the same as the value of f at $x = 1$. Always keep in mind that when we evaluate a limit as $x \to a$, *we do not care about the value of the function at* $x = a$. We only care about the value of $f(x)$ as x *approaches* a. In other words, $f(a)$ may or may not equal $\lim_{x\to a} f(x)$. ■

Here is a summary of the graphical method we used in Example 4, together with some additional information:

Evaluating Limits Graphically

To decide whether $\lim_{x\to a} f(x)$ exists and to find its value if it does:

1. Draw the graph of $f(x)$ by hand or with graphing technology.
2. Position your pencil point (or the Trace cursor) on a point of the graph to the right of $x = a$.
3. Move the point *along the graph* toward $x = a$ from the right and read the y-coordinate as you go. The value the y-coordinate approaches (if any) is the limit $\lim_{x\to a^+} f(x)$.
4. Repeat Steps 2 and 3, this time starting from a point on the graph to the left of $x = a$, and approaching $x = a$ along the graph from the left. The value the y-coordinate approaches (if any) is $\lim_{x\to a^-} f(x)$.
5. If the left and right limits both exist and have the same value L, then $\lim_{x\to a} f(x) = L$. Otherwise, the limit does not exist. The value $f(a)$ has no relevance whatsoever.
6. To evaluate $\lim_{x\to +\infty} f(x)$, move the pencil point toward the far right of the graph and estimate the value the y-coordinate approaches (if any). For $\lim_{x\to -\infty} f(x)$, move the pencil point toward the far left.
7. If $x = a$ happens to be an endpoint of the domain of f, then only a one-sided limit is possible at $x = a$. For instance, if the domain is $(-\infty, 4]$, then $\lim_{x\to 4^-} f(x)$ can be computed, but not $\lim_{x\to 4^+} f(x)$. In this case, we have said that $\lim_{x\to 4} f(x)$ is just the one-sided limit $\lim_{x\to 4^-} f(x)$.

In the next example we use both the numerical and graphical approaches.

EXAMPLE 5 Infinite Limit

Does $\lim\limits_{x\to 0^+} \dfrac{1}{x}$ exist?

Solution

Numerical Method Because we are asked for only the right hand limit, we need only list values of x approaching 0 from the right.

← x approaching 0 from the right

x	0	0.0001	0.001	0.01	0.1
$f(x) = \dfrac{1}{x}$		10,000	1,000	100	10

What seems to be happening as x approaches 0 from the right is that $f(x)$ is increasing without bound, as in Example 4(d). That is, if you name any number, no matter how large, $f(x)$ will be even larger than that if x is sufficiently close to zero. Thus, the limit diverges to $+\infty$, so

$$\lim_{x\to 0^+} \frac{1}{x} = +\infty$$

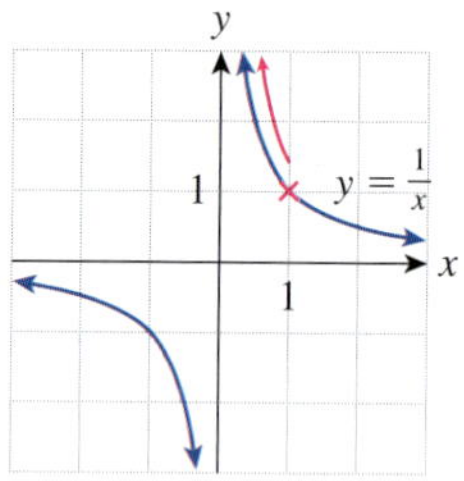

Figure 7

Graphical Method Recall that the graph of $f(x) = \frac{1}{x}$ is the standard hyperbola shown in Figure 7. The figure also shows the pencil point moving so that its x-coordinate approaches 0 from the right. Because the point moves along the graph, it is forced to go higher and higher. In other words, its y-coordinate becomes larger and larger, approaching $+\infty$. Thus, we conclude that

$$\lim_{x\to 0^+} \frac{1}{x} = +\infty$$

➡ **Before we go on...** In Example 5(a) you should also check that

$$\lim_{x\to 0^-} \frac{1}{x} = -\infty.$$

We say that as x approaches 0 from the left, $\frac{1}{x}$ diverges to $-\infty$. Also, check that

$$\lim_{x\to +\infty} \frac{1}{x} = \lim_{x\to -\infty} \frac{1}{x} = 0. ■$$

APPLICATION

EXAMPLE 6 Broadband

The percentage of U.S. Internet-connected households that have broadband connections can be modeled by

$$P(t) = \frac{100}{1 + 10.8(1.7)^{-t}} \quad (t \geq 0)$$

where t is time in years since 2000.*

a. Estimate $\lim_{t\to +\infty} P(t)$ and interpret the answer.

b. Estimate $\lim_{t\to 0^+} P(t)$ and interpret the answer.

Solution

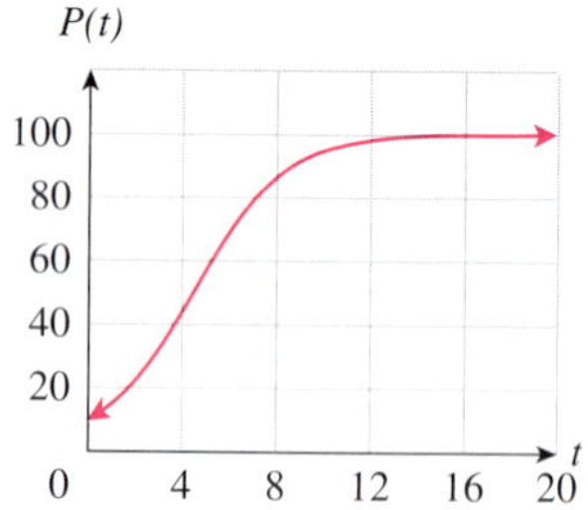

Tech Formula:
`100/(1+10.8*1.7^(-x))`

Figure 8

a. Figure 8 shows a plot of $P(t)$ for $0 \leq t \leq 20$.
Using either the numerical or the graphical approach, we find

$$\lim_{t\to +\infty} P(t) = \lim_{t\to +\infty} \frac{100}{1 + 10.8(1.7)^{-t}} = 100.$$

Thus, in the long term (as t gets larger and larger), the percentage of U.S. Internet-connected households that have broadband is expected to approach 100%.

* See Example 2 in Section 2.4. Source for data: www.Nielsen.com.

b. The limit here is

$$\lim_{t\to 0^+} P(t) = \lim_{t\to 0^+} \frac{100}{1+10.8(1.7)^{-t}} \approx 8.475.$$

(Notice that in this case, we can simply put $t = 0$ to evaluate this limit.) Thus, the closer t gets to 0 (representing 2000) the closer $P(t)$ gets to 8.475%, meaning that, in 2000, about 8.5% of Internet-connected households had broadband.

FAQs

Determining When a Limit Does or Does Not Exist

Q: *If I substitute $x = a$ in the formula for a function and find that the function is not defined there, it means that $\lim_{x\to a} f(x)$ does not exist, right?*

A: Wrong. The limit may still exist, as in Example 1, or may not exist, as in Example 2. In general, whether or not $\lim_{x\to a} f(x)$ exists has nothing to do with $f(a)$, but rather the value of f when x is *very close to, but not equal to a.*

Q: *Is there a quick and easy way of telling from a graph whether $\lim_{x\to a} f(x)$ exists?*

A: Yes. If you cover up the portion of the graph corresponding to $x = a$, and it appears as though the visible part of the graph could be made into a continuous curve by filling in a suitable point at $x = a$, then the limit exists. (The "suitable point" need not be $(a, f(a))$.) Otherwise, it does not. Try this method with the curves in Example 4.

3.1 EXERCISES

▼ more advanced ◆ challenging

T indicates exercises that should be solved using technology

Estimate the limits in Exercises 1–18 numerically. HINT [See Example 1.]

1. $\lim_{x\to 0} \frac{x^2}{x+1}$

2. $\lim_{x\to 0} \frac{x-3}{x-1}$

3. $\lim_{x\to 2} \frac{x^2-4}{x-2}$

4. $\lim_{x\to 2} \frac{x^2-1}{x-2}$

5. $\lim_{x\to -1} \frac{x^2+1}{x+1}$

6. $\lim_{x\to -1} \frac{x^2+2x+1}{x+1}$

7. $\lim_{x\to +\infty} \frac{3x^2+10x-1}{2x^2-5x}$ HINT [See Example 3.]

8. $\lim_{x\to +\infty} \frac{6x^2+5x+100}{3x^2-9}$ HINT [See Example 3.]

9. $\lim_{x\to -\infty} \frac{x^5-1{,}000x^4}{2x^5+10{,}000}$

10. $\lim_{x\to -\infty} \frac{x^6+3{,}000x^3+1{,}000{,}000}{2x^6+1{,}000x^3}$

11. $\lim_{x\to +\infty} \frac{10x^2+300x+1}{5x+2}$

12. $\lim_{x\to +\infty} \frac{2x^4+20x^3}{1{,}000x^6+6}$

13. $\lim_{x\to +\infty} \frac{10x^2+300x+1}{5x^3+2}$

14. $\lim_{x\to +\infty} \frac{2x^4+20x^3}{1{,}000x^3+6}$

15. $\lim_{x\to 2} e^{x-2}$

16. $\lim_{x\to +\infty} e^{-x}$

17. $\lim_{x\to +\infty} xe^{-x}$

18. $\lim_{x\to -\infty} xe^{x}$

In each of Exercises 19–30, the graph of f is given. Use the graph to compute the quantities asked for. HINT [See Example 4.]

19. a. $\lim_{x\to 1} f(x)$ **b.** $\lim_{x\to -1} f(x)$

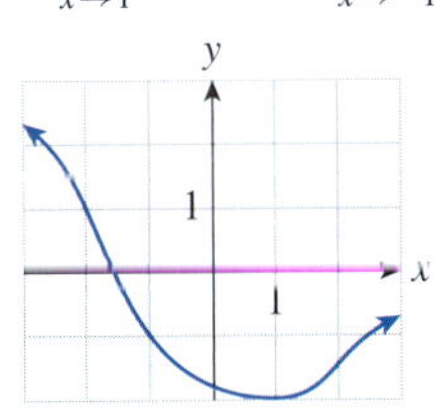

20. a. $\lim_{x\to -1} f(x)$ **b.** $\lim_{x\to 1} f(x)$

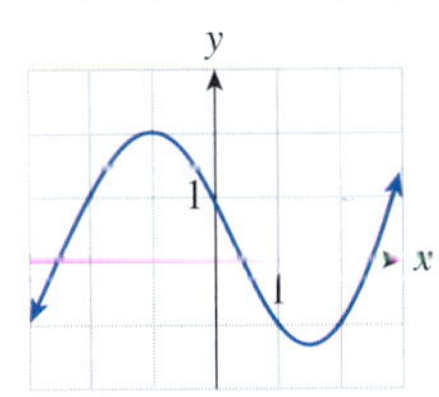

21. **a.** $\lim_{x\to 0} f(x)$ **b.** $\lim_{x\to 2} f(x)$ **c.** $\lim_{x\to -\infty} f(x)$ **d.** $\lim_{x\to +\infty} f(x)$

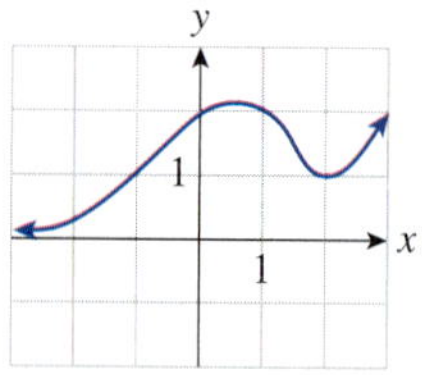

22. **a.** $\lim_{x\to -1} f(x)$ **b.** $\lim_{x\to 1} f(x)$ **c.** $\lim_{x\to +\infty} f(x)$ **d.** $\lim_{x\to -\infty} f(x)$

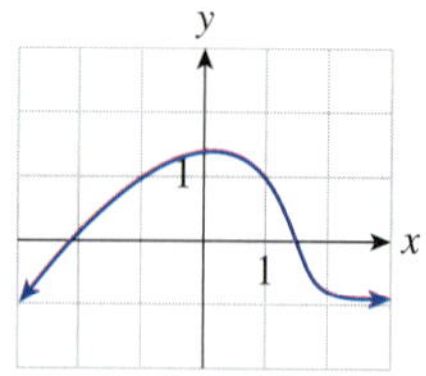

23. **a.** $\lim_{x\to 2} f(x)$ **b.** $\lim_{x\to 0^+} f(x)$ **c.** $\lim_{x\to 0^-} f(x)$ **d.** $\lim_{x\to 0} f(x)$ **e.** $f(0)$ **f.** $\lim_{x\to -\infty} f(x)$

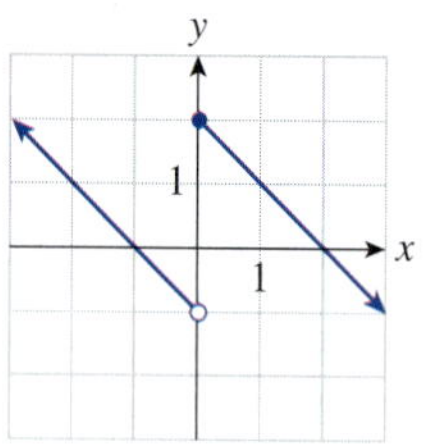

24. **a.** $\lim_{x\to 3} f(x)$ **b.** $\lim_{x\to 1^+} f(x)$ **c.** $\lim_{x\to 1^-} f(x)$ **d.** $\lim_{x\to 1} f(x)$ **e.** $f(1)$ **f.** $\lim_{x\to +\infty} f(x)$

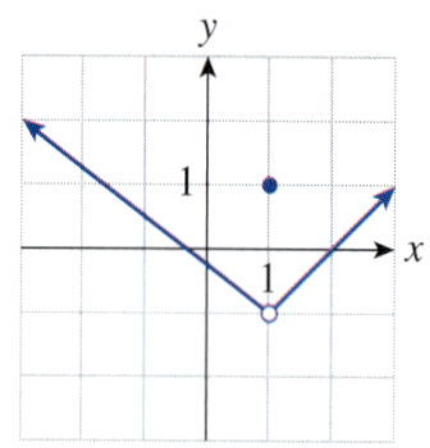

25. **a.** $\lim_{x\to -2} f(x)$ **b.** $\lim_{x\to -1^+} f(x)$ **c.** $\lim_{x\to -1^-} f(x)$ **d.** $\lim_{x\to -1} f(x)$ **e.** $f(-1)$ **f.** $\lim_{x\to +\infty} f(x)$

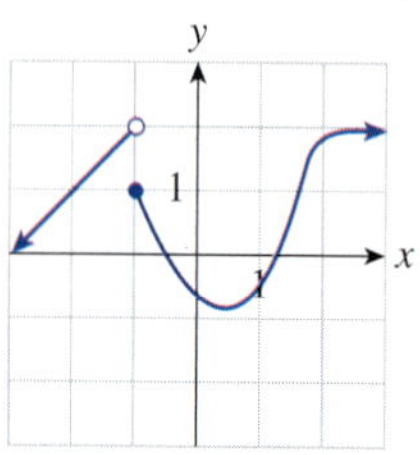

26. **a.** $\lim_{x\to -1} f(x)$ **b.** $\lim_{x\to 0^+} f(x)$ **c.** $\lim_{x\to 0^-} f(x)$ **d.** $\lim_{x\to 0} f(x)$ **e.** $f(0)$ **f.** $\lim_{x\to -\infty} f(x)$

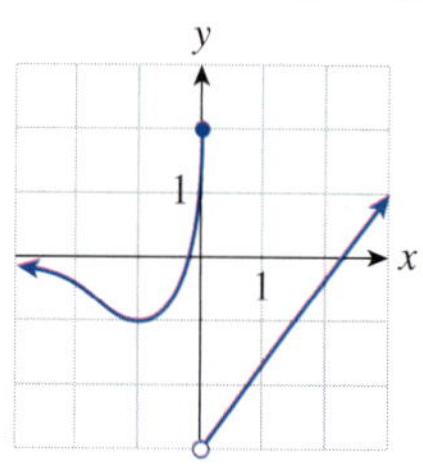

27. **a.** $\lim_{x\to -1} f(x)$ **b.** $\lim_{x\to 0^+} f(x)$ **c.** $\lim_{x\to 0^-} f(x)$ **d.** $\lim_{x\to 0} f(x)$ **e.** $f(0)$ **f.** $\lim_{x\to +\infty} f(x)$

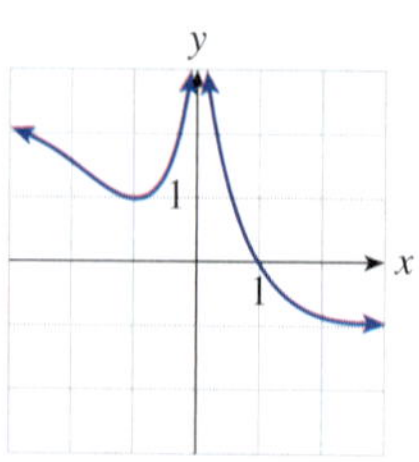

28. **a.** $\lim_{x\to 1} f(x)$ **b.** $\lim_{x\to 0^+} f(x)$ **c.** $\lim_{x\to 0^-} f(x)$ **d.** $\lim_{x\to 0} f(x)$ **e.** $f(0)$ **f.** $\lim_{x\to -\infty} f(x)$

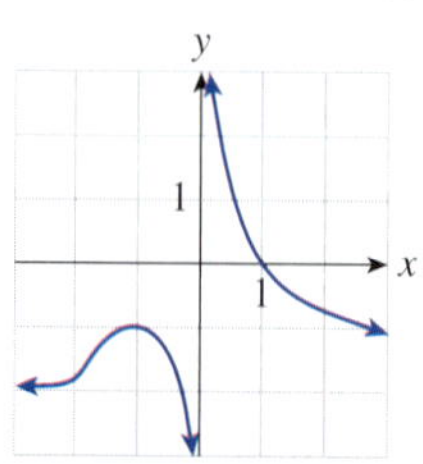

29. **a.** $\lim_{x\to -1} f(x)$ **b.** $\lim_{x\to 0^+} f(x)$ **c.** $\lim_{x\to 0^-} f(x)$ **d.** $\lim_{x\to 0} f(x)$ **e.** $f(0)$ **f.** $f(-1)$

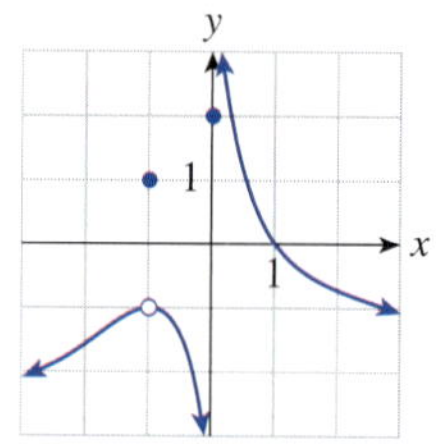

30. **a.** $\lim_{x\to 0^-} f(x)$ **b.** $\lim_{x\to 1^+} f(x)$ **c.** $\lim_{x\to 0} f(x)$ **d.** $\lim_{x\to 1} f(x)$ **e.** $f(0)$ **f.** $f(1)$

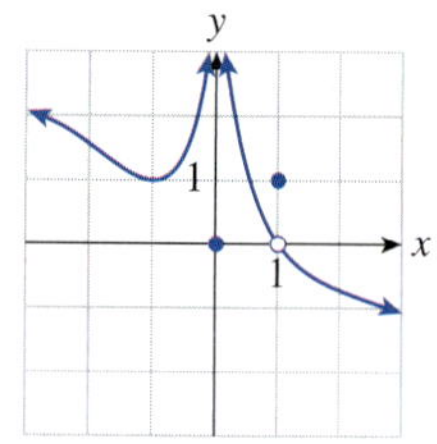

APPLICATIONS

31. ***Economic Growth*** The value of sold goods in Mexico can be approximated by

$$v(t) = 210 - 62e^{-0.05t} \text{ trillion pesos per month} \quad (t \geq 0)$$

where t is time in months since January 2005.[1] Numerically estimate $\lim_{t\to +\infty} v(t)$ and interpret the answer. HINT [See Example 6.]

32. ***Housing Starts*** Housing starts in the United States can be approximated by

$$n(t) = \frac{1}{12}(1.1 + 1.2e^{-0.08t}) \text{ million homes per month} \quad (t \geq 0)$$

where t is time in months since January 2006.[2] Numerically estimate $\lim_{t\to +\infty} n(t)$ and interpret the answer. HINT [See Example 6.]

33. ***Scientific Research*** The number of research articles per year, in thousands, in the prominent journal *Physical Review* written by researchers in Europe can be modeled by

$$A(t) = \frac{7.0}{1 + 5.4(1.2)^{-t}}$$

where t is time in years ($t = 0$ represents 1983).[3] Numerically estimate $\lim_{t\to +\infty} A(t)$ and interpret the answer. HINT [See Example 6.]

34. ***Scientific Research*** The percentage of research articles in the prominent journal *Physical Review* written by researchers in the United States can be modeled by

$$A(t) = 25 + \frac{36}{1 + 0.6(0.7)^{-t}},$$

where t is time in years ($t = 0$ represents 1983).[4] Numerically estimate $\lim_{t\to +\infty} A(t)$ and interpret the answer. HINT [See Example 6.]

---

[1] Source: Instituto Nacional de Estadística y Geografía (INEGI), www.inegi.org.mx.

[2] Source for data: *New York Times*, February 17, 2007, p. C3.

[3] Based on data from 1983 to 2003. Source: The American Physical Society/*New York Times*, May 3, 2003, p. A1.

[4] Ibid.

35. ***SAT Scores by Income*** The following bar graph shows U.S. verbal SAT scores as a function of parents' income level:[5]

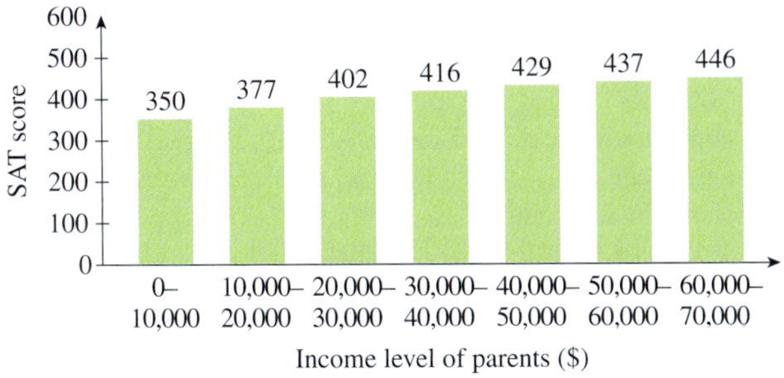

These data can be modeled by

$$S(x) = 470 - 136(0.974)^x.$$

where $S(x)$ is the average SAT verbal score of a student whose parents' income is x thousand dollars per year. Numerically estimate $\lim_{x\to+\infty} S(x)$ and interpret the result.

36. ***SAT Scores by Income*** The following bar graph shows U.S. math SAT scores as a function of parents' income level:[6]

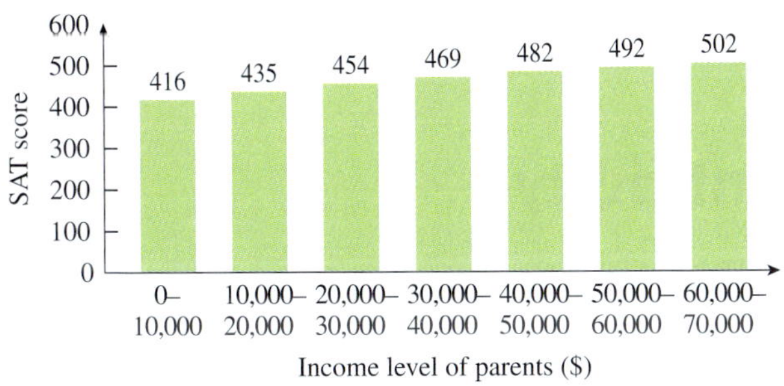

These data can be modeled by

$$S(x) = 535 - 136(0.979)^x,$$

where $S(x)$ is the average math SAT score of a student whose parents' income is x thousand dollars per year. Numerically estimate $\lim_{x\to+\infty} S(x)$ and interpret the result.

37. ***Home Prices*** The following graph shows the approximate value of the home price index as a percentage change from 2003.[7]

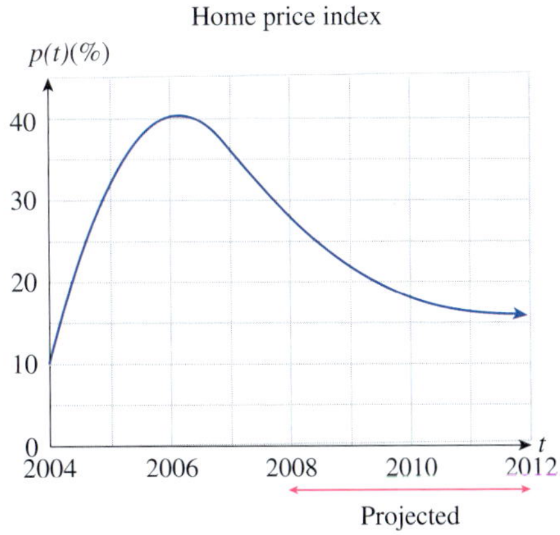

Estimate $\lim_{t\to+\infty} p(t)$ and interpret your answer.

38. ***Existing Home Sales*** The following graph shows the approximate value of existing home sales as a percentage change from 2003.[8]

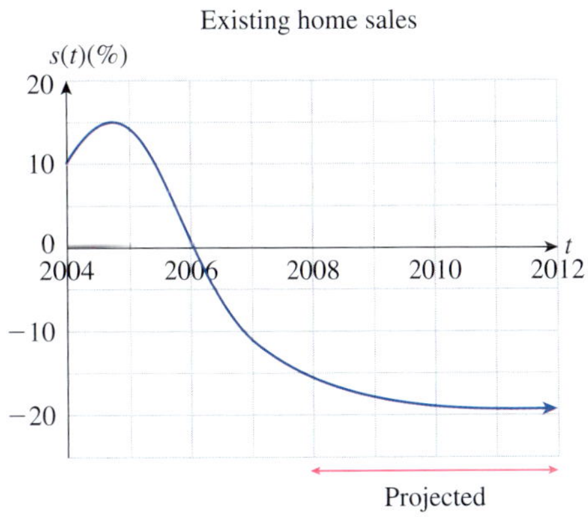

Estimate $\lim_{t\to+\infty} s(t)$ and interpret your answer.

39. ***Electric Rates*** The cost of electricity in Portland, Oregon, for residential customers increased suddenly on October 1, 2001, from around \$0.06 to around \$0.08 per kilowatt hour.[9] Let $C(t)$ be this cost at time t, and take $t = 1$ to represent October 1, 2001. What does the given information tell you about $\lim_{t\to1} C(t)$? HINT [See Example 4(b).]

40. ***Airline Stocks*** Prior to the September 11, 2001 attacks, United Airlines stock was trading at around \$35 per share. Immediately following the attacks, the share price dropped by \$15.[10] Let $U(t)$ be this cost at time t, and take $t = 11$ to represent September 11, 2001. What does the given information tell you about $\lim_{t\to11} U(t)$? HINT [See Example 4(b).]

[5] Based on 1994 data. Source: The College Board/*New York Times*, March 5, 1995, p. E16.

[6] Ibid.

[7] S&P/Case-Shiller Home Price Index. Source: Standard & Poors/*New York Times*, September 29, 2007, p. C3. Projection is the authors'.

[8] Source: Bloomberg Finiancial Markets/*New York Times*, September 29, 2007, p. C3. Projection is the authors'.

[9] Source: Portland General Electric/*New York Times*, February 2, 2002, p. C1.

[10] Stock prices are approximate.

Foreign Trade *Annual U.S. imports from China in the years 1996 through 2003 can be approximated by*

$$I(t) = t^2 + 3.5t + 50 \qquad (1 \le t \le 9)$$

billion dollars, where t represents time in years since 1995. Annual U.S. exports to China in the same years can be approximated by

$$E(t) = 0.4t^2 - 1.6t + 14$$

billion dollars.[11] *Exercises 41 and 42 are based on these models.*

41. ▼ Assuming the trends shown in the above models continued indefinitely, numerically estimate

$$\lim_{t\to+\infty} I(t) \text{ and } \lim_{t\to+\infty} \frac{I(t)}{E(t)},$$

interpret your answers, and comment on the results.

42. ▼ Repeat Exercise 41, this time calculating

$$\lim_{t\to+\infty} E(t) \text{ and } \lim_{t\to+\infty} \frac{E(t)}{I(t)}.$$

COMMUNICATION AND REASONING EXERCISES

43. Describe the method of evaluating limits numerically. Give at least one disadvantage of this method.

[11] Based on quadratic regression using data from the U.S. Census Bureau Foreign Trade Division Web site www.census.gov/foreign-trade/sitc1/ as of December 2004.

44. Describe the method of evaluating limits graphically. Give at least one disadvantage of this method.

45. Your friend Dion, a business student, claims that the study of limits that do not exist is completely unrealistic and has nothing to do with the world of business. Give two examples from the world of business that might convince him that he is wrong.

46. Your other friend Fiona claims that the study of limits is a complete farce; all you ever need to do to find the limit as x approaches a is substitute $x = a$. Give two examples that show she is wrong.

47. ▼ What is wrong with the following statement? "Because $f(a)$ is not defined, $\lim_{x\to a} f(x)$ does not exist."

48. ▼ What is wrong with the following statement? "Because $f(a)$ is defined, $\lim_{x\to a} f(x)$ exists."

49. ▼ What is wrong with the following statement? "If $f(a)$ is defined, then $\lim_{x\to a} f(x)$ exists and equals $f(a)$."

50. ▼ If $D(t)$ is the Dow Jones Average at time t and $\lim_{t\to+\infty} D(t) = +\infty$, is it possible that the Dow will fluctuate indefinitely into the future?

51. ◆ Give an example of a function f with $\lim_{x\to 1} f(x) = f(2)$.

52. ◆ If $S(t)$ represents the size of the universe in billions of light years at time t years since the big bang and $\lim_{t\to+\infty} S(t) = 130{,}000$, is it possible that the universe will continue to expand forever?

3.2 Limits and Continuity

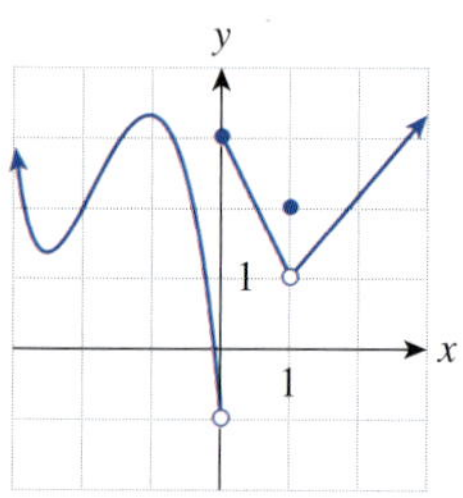

Figure 9

In Section 2.1 we saw examples of graphs that had various kinds of "breaks" or "jumps." For instance, in Example 4 we looked at the graph in Figure 9. This graph appears to have breaks, or **discontinuities**, at $x = 0$ and at $x = 1$. At $x = 0$ we saw that $\lim_{x\to 0} f(x)$ does not exist because the left- and right-hand limits are not the same. Thus, the discontinuity at $x = 0$ seems to be due to the fact that the limit does not exist there. On the other hand, at $x = 1$, $\lim_{x\to 1} f(x)$ *does* exist (it is equal to 1), but is not equal to $f(1) = 2$.

Thus, we have identified two kinds of discontinuity:

1. Points where the limit of the function does not exist.

$x = 0$ in Figure 9 because $\lim_{x\to 0} f(x)$ does not exist.

2. Points where the limit exists but does not equal the value of the function.

$x = 1$ in Figure 9 because $\lim_{x\to 1} f(x) = 1 \neq f(1)$

On the other hand, there is no discontinuity at, say, $x = -2$, where we find that $\lim_{x\to -2} f(x)$ exists and equals 2 and $f(-2)$ is also equal to 2. In other words,

$$\lim_{x\to -2} f(x) = 2 = f(-2).$$

The point $x = -2$ is an example of a point where f is **continuous**. (Notice that you can draw the portion of the graph near $x = -2$ without lifting your pencil from the paper.) Similarly, f is continuous at *every* point other than $x = 0$ and $x = 1$. Here is the mathematical definition.

Continuous Function

Let f be a function and let a be a number in the domain of f. Then f is **continuous at** $\boldsymbol{a}$ if

a. $\lim_{x \to a} f(x)$ exists, and

b. $\lim_{x \to a} f(x) = f(a)$.

The function f is said to be **continuous on its domain** if it is continuous at each point in its domain.

If f is not continuous at a particular a in its domain, we say that f is **discontinuous** at a or that f has a **discontinuity** at a. Thus, a discontinuity can occur at $x = a$ if either

a. $\lim_{x \to a} f(x)$ does not exist, or

b. $\lim_{x \to a} f(x)$ exists but is not equal to $f(a)$.

Quick Examples

1. The function shown in Figure 9 is continuous at $x = -1$ and $x = 2$. It is discontinuous at $x = 0$ and $x = 1$, and so is not continuous on its domain.
2. The function $f(x) = x^2$ is continuous on its domain. (Think of its graph, which contains no breaks).
3. The function f whose graph is shown on the left in the following figure is continuous on its domain. (Although the graph breaks at $x = 2$, that is not a point of its domain.) The function g whose graph is shown on the right is not continuous on its domain because it has a discontinuity at $x = 2$. (Here, $x = 2$ is a point of the domain of g.)

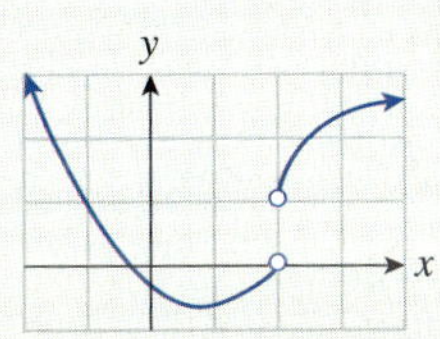

$y = f(x)$: Continuous on its domain

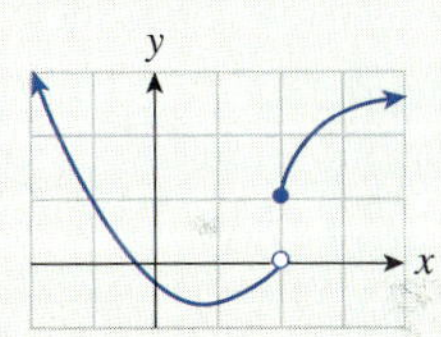

$y = g(x)$:0 Not continuous on its domain

Note If the number a is not in the domain of f—that is, if $f(a)$ is not defined—we will not consider the question of continuity at a. A function cannot be continuous at a point not in its domain, and it cannot be discontinuous there either. ■

EXAMPLE 1 Continuous and Discontinuous Functions

Which of the following functions are continuous on their domains?

a. $h(x) = \begin{cases} x + 3 & \text{if } x \le 1 \\ 5 - x & \text{if } x > 1 \end{cases}$

b. $k(x) = \begin{cases} x + 3 & \text{if } x \le 1 \\ 1 - x & \text{if } x > 1 \end{cases}$

c. $f(x) = \dfrac{1}{x}$

d. $g(x) = \begin{cases} \dfrac{1}{x} & \text{if } x \ne 0 \\ 0 & \text{if } x = 0 \end{cases}$

Solution

a and **b.** The graphs of h and k are shown in Figure 10.

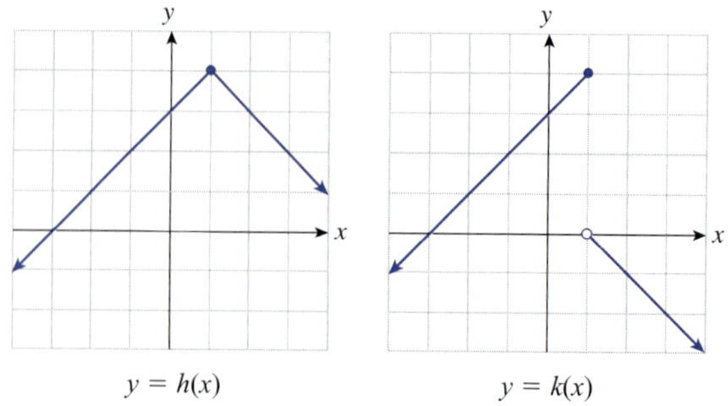

Figure 10

Even though the graph of h is made up of two different line segments, it is continuous at every point of its domain, including $x = 1$ because

$$\lim_{x\to 1} h(x) = 4 = h(1).$$

On the other hand, $x = 1$ is also in the domain of k, but $\lim_{x\to 1} k(x)$ does not exist. Thus, k is discontinuous at $x = 1$ and thus not continuous on its domain.

c and **d.** The graphs of f and g are shown in Figure 11.

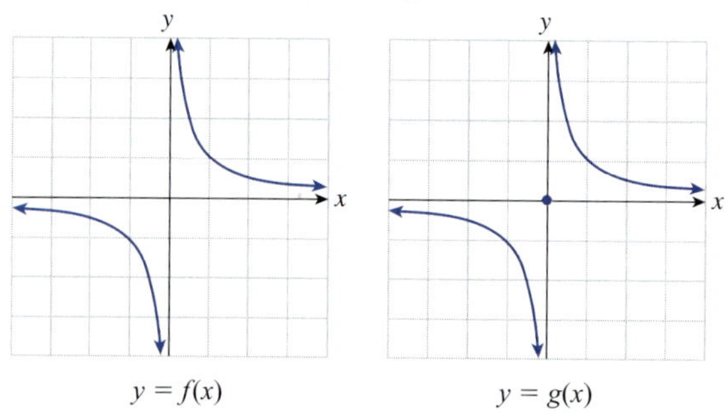

Figure 11

The domain of f consists of all real numbers except 0 and f is continuous at all such numbers. (Notice that 0 is not in the domain of f, so the question of continuity at 0 does not arise.) Thus, f is continuous on its domain.

The function g, on the other hand, has its domain expanded to include 0, so we now need to check whether g is continuous at 0. From the graph, it is easy to see that g is discontinuous there because $\lim_{x\to 0} g(x)$ does not exist. Thus, g is not continuous on its domain because it is discontinuous at 0.

using Technology

We can use technology to draw (approximate) graphs of the functions in Example 1(a), (b), and (c). Here are the technology formulas that will work for the TI-83/84 Plus, Excel, and Web site function evaluator and grapher. (In the TI-83/84 Plus, replace <= by ≤. In Excel, replace *x* by a cell reference and insert an equals sign in front of the formula.)

a. `(x+3)*(x<=1)+(5-x)*(x>1)`

b. `(x+3)*(x<=1)+(1-x)*(x>1)`

c. `(1/x)`

Observe in each case how technology handles the breaks in the curves.

➡ Before we go on...

Q: *Wait a minute! How can a function like $f(x) = 1/x$ be continuous when its graph has a break in it?*

A: We are not claiming that f is continuous *at every real number.* What we are saying is that f is continuous *on its domain;* the break in the graph occurs at a point not in the domain of f. In other words, f is continuous on the set of all nonzero real numbers; it is not continuous on the set of *all* real numbers because it is not even defined on that set.

■

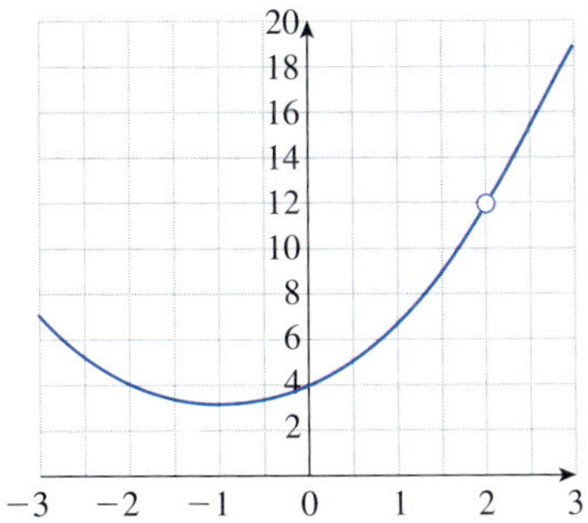

Figure 12

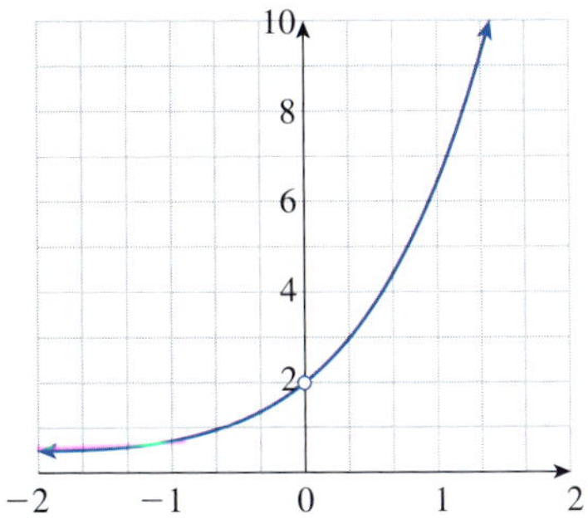

Figure 13

EXAMPLE 2 Continuous Except at a Point

In each case, say what, if any, value of $f(a)$ would make f continuous at a.

a. $f(x) = \dfrac{x^3 - 8}{x - 2}$; $a = 2$ **b.** $f(x) = \dfrac{e^{2x} - 1}{x}$; $a = 0$ **c.** $f(x) = \dfrac{|x|}{x}$; $a = 0$

Solution

a. In Figure 12 we see the graph of $f(x) = \dfrac{x^3 - 8}{x - 2}$. The point corresponding to $x = 2$ is missing because f is not (yet) defined there. (Your graphing utility will probably miss this subtlety and render a continuous curve. See the technology note in the margin.) To turn f into a function that is continuous at $x = 2$, we need to "fill in the gap" so as to obtain a continuous curve. Since the graph suggests that the missing point is (2, 12), let us define $f(2) = 12$.

Does f now become continuous if we take $f(2) = 12$? From the graph, or Example 1(a) of Section 3.1,

$$\lim_{x \to 2} f(x) = \lim_{x \to 2} \frac{x^3 - 8}{x - 2} = 12,$$

which is now equal to $f(2)$. Thus, $\lim_{x \to 2} f(x) = f(2)$, showing that f is now continuous at $x = 2$.

b. In Example 1(b) of the preceding section, we saw that

$$\lim_{x \to 0} f(x) = \lim_{x \to 0} \frac{e^{2x} - 1}{x} = 2$$

and so, as in part (a), we must define $f(0) = 2$. This is confirmed by the graph, shown in Figure 13.

c. We considered the function $f(x) = |x|/x$ in Example 2 in Section 3.1. Its graph is shown in Figure 14.

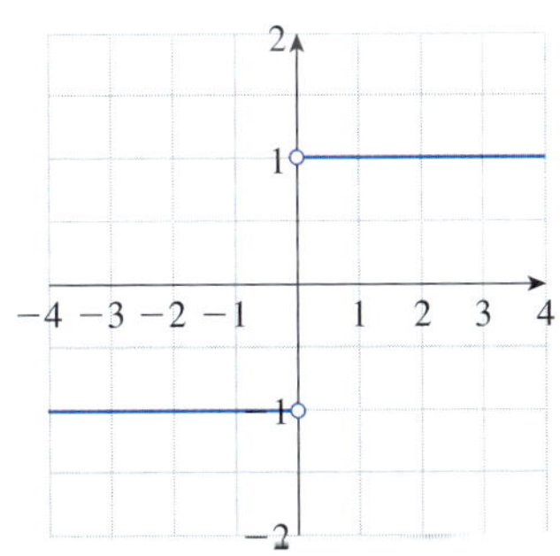

Figure 14

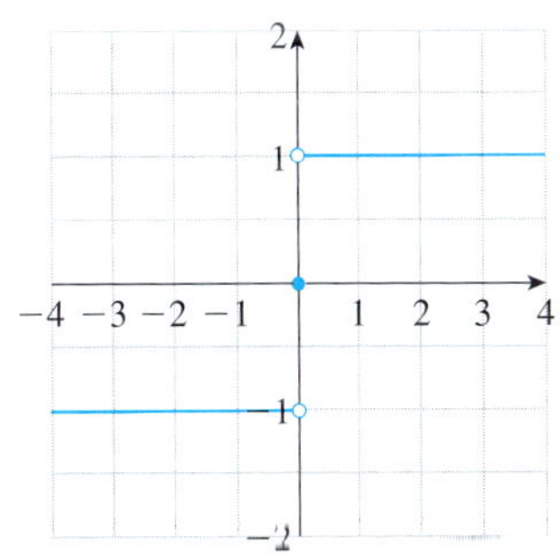

Figure 15

Now we encounter a problem: No matter how we try to fill in the gap at $x = 0$, the result will be a discontinuous function. For example, setting $f(0) = 0$ will result in the discontinuous function shown in Figure 15. We conclude that it is impossible to assign any value to $f(0)$ to turn f into a function that is continuous at $x = 0$.

We can also see this result algebraically. In Example 2 of Section 3.1, we saw that $\lim_{x \to 0} \dfrac{|x|}{x}$ does not exist. Thus, the resulting function will fail to be continuous at 0, no matter how we define $f(0)$.

using Technology

It is instructive to see how technology handles the functions in Example 2. Here are the technology formulas that will work for the TI-83/84 Plus, Excel, and Web site function evaluator and grapher. (In Excel, replace `x` by a cell reference and insert an equal sign in front of the formula.)

a. `(x^3-8)/(x-2)`

b. `(e^(2x)-1)/x`

Excel:

`-(exp(2*A2)-1)/A2`

c. `abs(x)/x`

In each case, compare the graph rendered by technology with the corresponding figure in Example 2.

A function not defined at an isolated point is said to have a **singularity** at that point. The function in part (a) of Example 2 has a singularity at $x = 2$, and the functions in parts (b) and (c) have singularities at $x = 0$. The functions in parts (a) and (b) have *removable* singularities because we can make these functions continuous at $x = a$ by properly defining $f(a)$. The function in part (c) has an **essential singularity** because we cannot make f continuous at $x = a$ just by defining $f(a)$ properly.

3.2 EXERCISES

▼ more advanced ◆ challenging

T indicates exercises that should be solved using technology

In Exercises 1–12, the graph of a function f is given. Determine whether f is continuous on its domain. If it is not continuous on its domain, say why. HINT [See Quick Examples page 205.]

1.
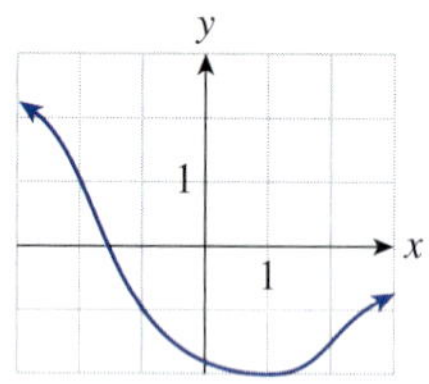

2.
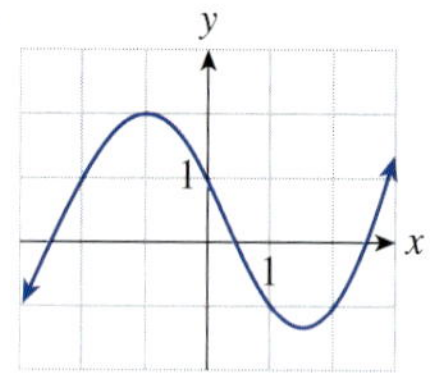

3.
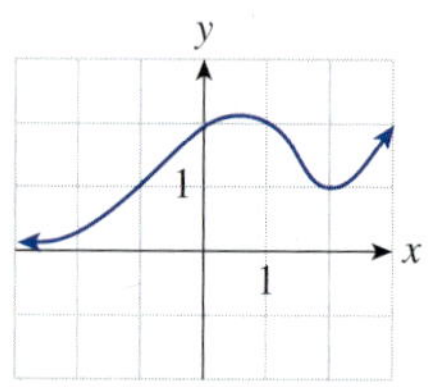

4.
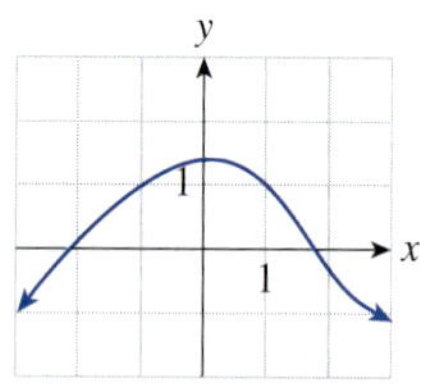

5.
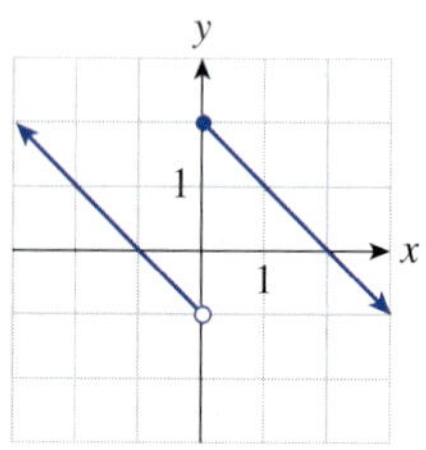

6.
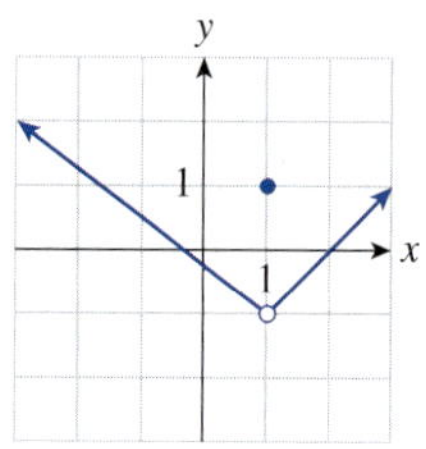

7.
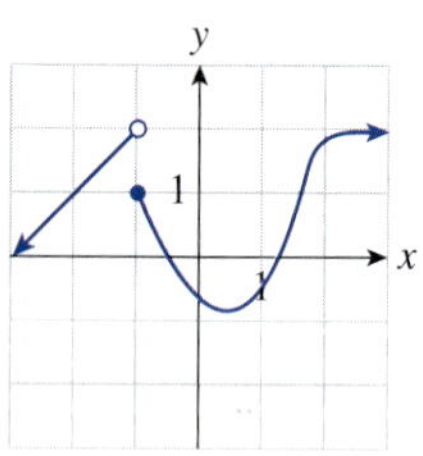

8.
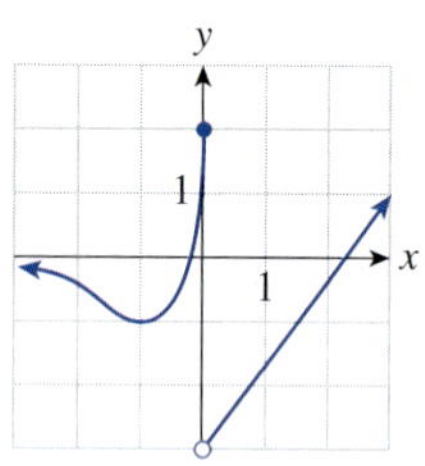

9.
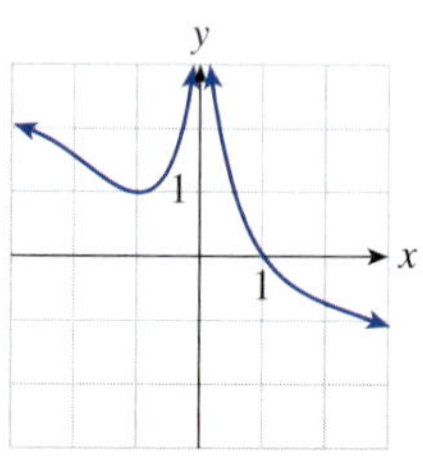

10.
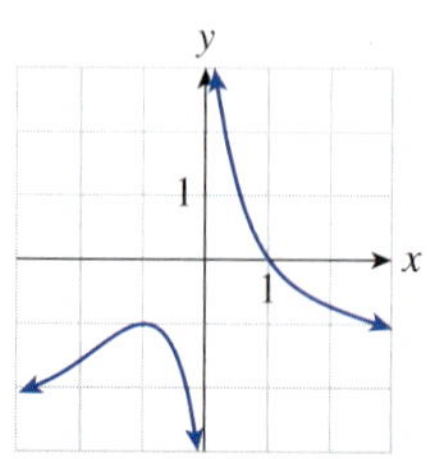

11.
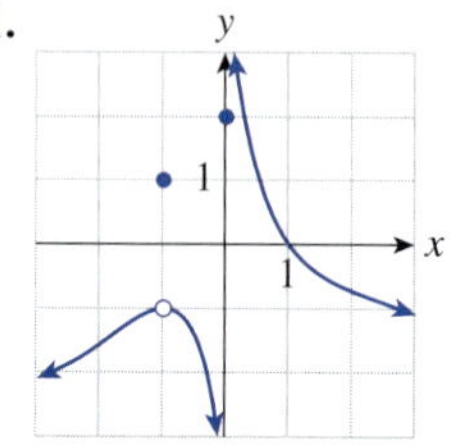

12.
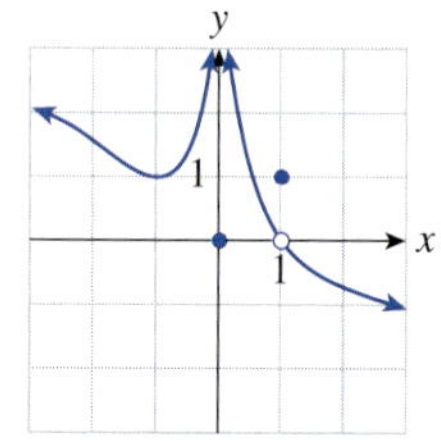

In Exercises 13 and 14, identify which (if any) of the given graphs represent functions continuous on their domains. HINT [See Quick Examples page 205.]

13. (A)
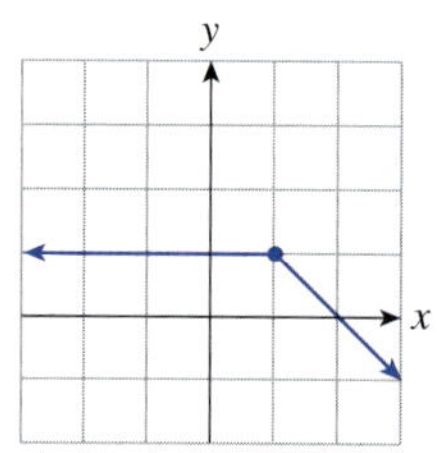

(B)
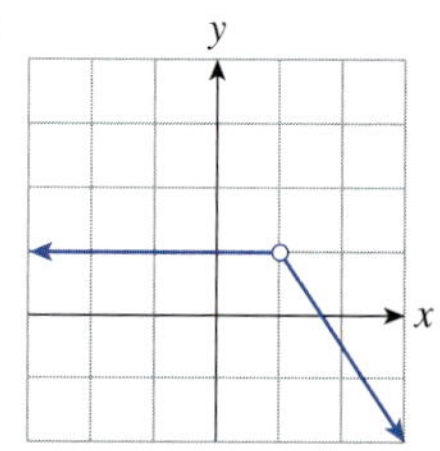

(C)
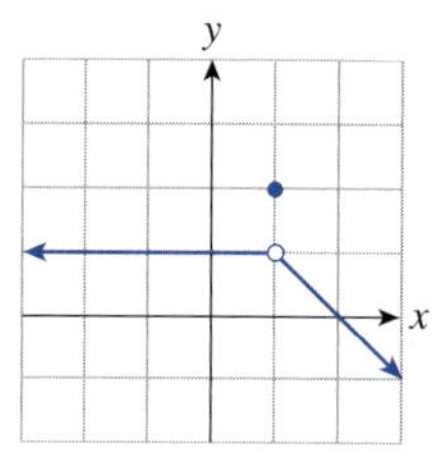

(D)
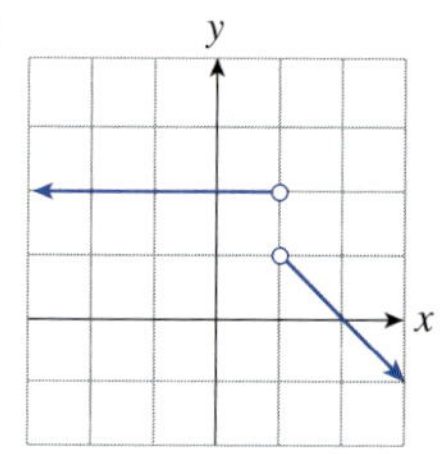

(E)
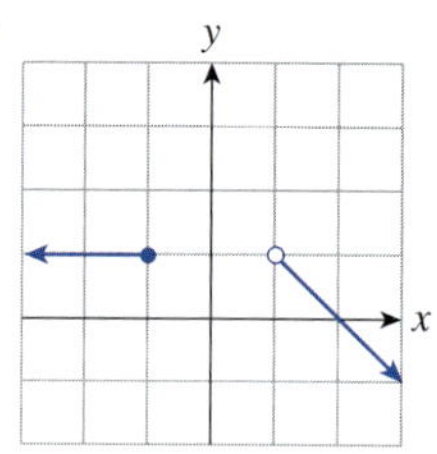

14. (A)

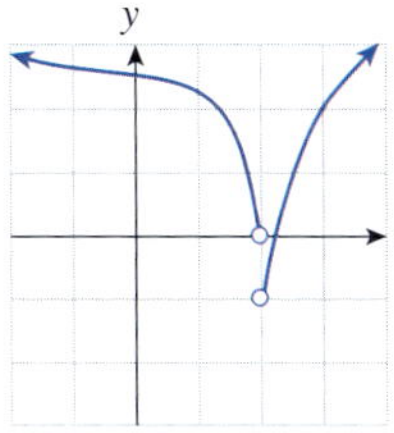

(B)

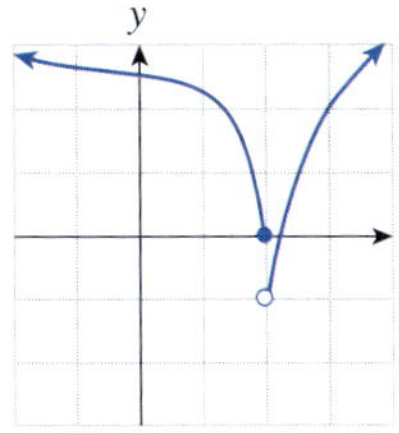

(C)

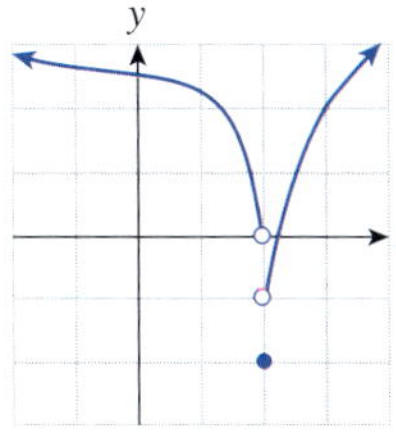

(D)

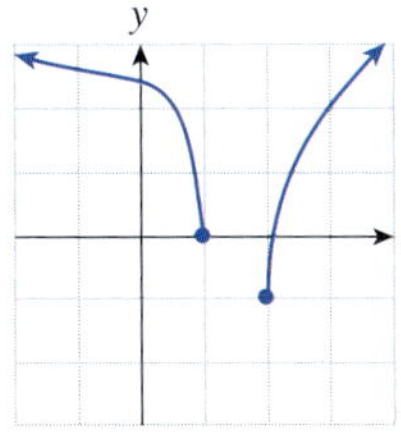

(E)

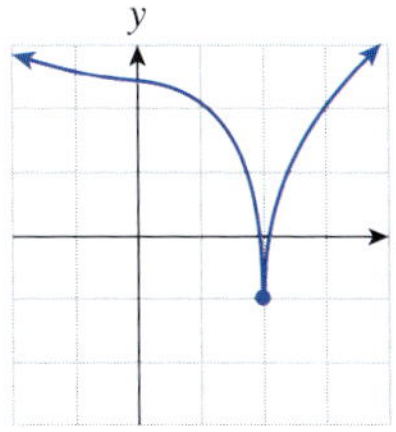

In Exercises 15–22, use a graph of f or some other method to determine what, if any, value to assign to f(a) to make f continuous at x = a. HINT [See Example 2.]

15. $f(x) = \dfrac{x^2 - 2x + 1}{x - 1}$; $a = 1$

16. $f(x) = \dfrac{x^2 + 3x + 2}{x + 1}$; $a = -1$

17. $f(x) = \dfrac{x}{3x^2 - x}$; $a = 0$

18. $f(x) = \dfrac{x^2 - 3x}{x + 4}$; $a = -4$

19. $f(x) = \dfrac{3}{3x^2 - x}$; $a = 0$

20. $f(x) = \dfrac{x - 1}{x^3 - 1}$; $a = 1$

21. $f(x) = \dfrac{1 - e^x}{x}$; $a = 0$

22. $f(x) = \dfrac{1 + e^x}{1 - e^x}$; $a = 0$

In Exercises 23–32, use a graph to determine whether the given function is continuous on its domain. If it is not continuous on its domain, list the points of discontinuity. HINT [See Example 1.]

23. $f(x) = |x|$

24. $f(x) = \dfrac{|x|}{x}$

25. $g(x) = \dfrac{1}{x^2 - 1}$

26. $g(x) = \dfrac{x - 1}{x + 2}$

27. $f(x) = \begin{cases} x + 2 & \text{if } x < 0 \\ 2x - 1 & \text{if } x \geq 0 \end{cases}$

28. $f(x) = \begin{cases} 1 - x & \text{if } x \leq 1 \\ x - 1 & \text{if } x > 1 \end{cases}$

29. $h(x) = \begin{cases} \dfrac{|x|}{x} & \text{if } x \neq 0 \\ 0 & \text{if } x = 0 \end{cases}$

30. $h(x) = \begin{cases} \dfrac{1}{x^2} & \text{if } x \neq 0 \\ 2 & \text{if } x = 0 \end{cases}$

31. $g(x) = \begin{cases} x + 2 & \text{if } x < 0 \\ 2x + 2 & \text{if } x \geq 0 \end{cases}$

32. $g(x) = \begin{cases} 1 - x & \text{if } x \leq 1 \\ x + 1 & \text{if } x > 1 \end{cases}$

COMMUNICATION AND REASONING EXERCISES

33. If a function is continuous on its domain, is it continuous at every real number? Explain.

34. True or false? The graph of a function that is continuous on its domain is a continuous curve with no breaks in it. Explain your answer.

35. True or false? The graph of a function that is continuous at every real number is a continuous curve with no breaks in it. Explain your answer.

36. True or false? If the graph of a function is a continuous curve with no breaks in it, then the function is continuous on its domain. Explain your answer.

37. ▼ Give a formula for a function that is continuous on its domain but whose graph consists of three distinct curves.

38. ▼ Give a formula for a function that is not continuous at $x = -1$ but is not discontinuous there either.

39. ▼ Draw the graph of a function that is discontinuous at every integer.

40. ▼ Draw the graph of a function that is continuous on its domain but whose graph has a break at every integer.

41. ▼ Describe a real-life scenario in the stock market that can be modeled by a discontinuous function.

42. ▼ Describe a real-life scenario in your room that can be modeled by a discontinuous function.

3.3 Limits and Continuity: Algebraic Approach

Although numerical and graphical estimation of limits is effective, the estimates these methods yield may not be perfectly accurate. The algebraic method, when it can be used, will always yield an exact answer. Moreover, algebraic analysis of a function often enables us to take a function apart and see "what makes it tick."

Let's start with the function $f(x) = 2 + x$ and ask: What happens to $f(x)$ as x approaches 3? To answer this algebraically, notice that as x gets closer and closer to 3, the quantity $2 + x$ must get closer and closer to $2 + 3 = 5$. Hence,

$$\lim_{x\to 3} f(x) = \lim_{x\to 3}(2 + x) = 2 + 3 = 5$$

Q: *Is that all there is to the algebraic method? Just substitute $x = a$?*

A: Under certain circumstances: Notice that by substituting $x = 3$ we *evaluated the function at $x = 3$*. In other words, we relied on the fact that

$$\lim_{x\to 3} f(x) = f(3).$$

In Section 3.2 we said that a function satisfying this equation is *continuous* at $x = 3$.

Thus,

If we know that the function f is continuous at a point a, we can compute $\lim_{x\to a} f(x)$ by simply substituting $x = a$ into $f(x)$.

To use this fact, we need to know how to recognize continuous functions when we see them. Geometrically, they are easy to spot: A function is continuous at $x = a$ if its graph has no break at $x = a$. Algebraically, a large class of functions are known to be continuous on their domains—those, roughly speaking, that are *specified by a single formula*.

We can be more precise: A **closed-form function** is any function that can be obtained by combining constants, powers of x, exponential functions, radicals, logarithms, absolute values, trigonometric functions (and some other functions we do not encounter in this text) into a *single* mathematical formula by means of the usual arithmetic operations and composition of functions. (They can be as complicated as we like.)

Closed-Form Functions

A function is **written in closed form** if it is specified by combining constants, powers of x, exponential functions, radicals, logarithms, absolute values, trigonometric functions (and some other functions we do not encounter in this text) into a *single* mathematical formula by means of the usual arithmetic operations and composition of functions. A **closed-form function** is any function that can be written in closed form.

Quick Examples

1. $3x^2 - |x| + 1$, $\dfrac{\sqrt{x^2-1}}{6x-1}$, $e^{-\frac{4x^2-1}{x}}$, and $\sqrt{\log_3(x^2-1)}$ are written in closed form, so they are all closed-form functions.

2. $f(x) = \begin{cases} -1 & \text{if } x \le -1 \\ x^2 + x & \text{if } -1 < x \le 1 \\ 2 - x & \text{if } 1 < x \le 2 \end{cases}$ is not written in closed-form because $f(x)$ is not expressed by a *single* mathematical formula.*

* **NOTE** It is possible to rewrite some piecewise defined functions in closed form (using a single formula), but not this particular function, so *f*(*x*) is not a closed-form function.

What is so special about closed-form functions is the following theorem.

Theorem 3.1 Continuity of Closed-Form Functions

Every closed-form function is continuous on its domain. Thus, if f is a closed-form function and $f(a)$ is defined, we have $\lim_{x \to a} f(x) = f(a)$.

Quick Example

$f(x) = 1/x$ is a closed-form function, and its natural domain consists of all real numbers except 0. Thus, f is continuous at every nonzero real number. That is,

$$\lim_{x \to a} \frac{1}{x} = \frac{1}{a}$$

provided $a \neq 0$.

Mathematics majors spend a great deal of time studying the proof of this theorem. We ask you to accept it without proof.

EXAMPLE 1 Limit of a Closed-Form Function at a Point in Its Domain

Evaluate $\lim_{x \to 1} \dfrac{x^3 - 8}{x - 2}$ algebraically.

Solution First, notice that $(x^3 - 8)/(x - 2)$ is a closed-form function because it is specified by a single algebraic formula. Also, $x = 1$ is in the domain of this function. Therefore,

$$\lim_{x \to 1} \frac{x^3 - 8}{x - 2} = \frac{1^3 - 8}{1 - 2} = 7.$$

Before we go on... In Example 1, the point $x = 2$ is not in the domain of the function $(x^3 - 8)/(x - 2)$, so we cannot evaluate $\lim_{x \to 2} \dfrac{x^3 - 8}{x - 2}$ by substituting $x = 2$. However—and this is the key to finding such limits—some preliminary algebraic simplification will allow us to obtain a closed-form function with $x = 2$ in its domain, as we shall see in Example 2. ■

EXAMPLE 2 Limit of a Closed-Form Function at a Point Not in Its Domain: Simplifying to Obtain the Limit

Evaluate $\lim_{x \to 2} \dfrac{x^3 - 8}{x - 2}$ algebraically.

Solution Again, although $(x^3 - 8)/(x - 2)$ is a closed-form function, $x = 2$ is not in its domain. Thus, we cannot obtain the limit by substitution. Instead, we first simplify $f(x)$ to obtain a new function with $x = 2$ in its domain. To do this, notice first that the numerator can be factored as

$$x^3 - 8 = (x - 2)(x^2 + 2x + 4).$$

Thus,

$$\frac{x^3 - 8}{x - 2} = \frac{(x - 2)(x^2 + 2x + 4)}{x - 2} = x^2 + 2x + 4.$$

Once we have canceled the offending $(x - 2)$ in the denominator, we are left with a closed-form function *with 2 in its domain*. Thus,

$$\lim_{x \to 2} \frac{x^3 - 8}{x - 2} = \lim_{x \to 2}(x^2 + 2x + 4)$$
$$= 2^2 + 2(2) + 4 = 12. \quad \text{Substitute } x = 2.$$

This confirms the answer we found numerically in Example 1 in Section 3.1.

➡ Before we go on... Notice that in Example 2, before simplification, the substitution $x = 2$ yields

$$\frac{x^3 - 8}{x - 2} = \frac{8 - 8}{2 - 2} = \frac{0}{0}.$$

Worse than the fact that 0/0 is undefined, it also conveys absolutely no information as to what the limit might be. (The limit turned out to be 12!) We therefore call the expression 0/0 an **indeterminate form**. Once simplified, the function became $x^2 + 2x + 4$, which, upon the substitution $x = 2$, yielded 12—no longer an indeterminate form. In general, we have the following rule of thumb:

> *If the substitution $x = a$ yields the indeterminate form 0/0, try simplifying by the method in Example 2.*

We will say more about indeterminate forms in Example 3. ■

Q: *There is something suspicious about Example 2. If 2 was not in the domain before simplifying but was in the domain after simplifying, we must have changed the function, right?*

A: Correct. In fact, when we said that

$$\frac{x^3 - 8}{x - 2} = x^2 + 2x + 4$$

Domain excludes 2 Domain includes 2

we were lying a little bit. What we really meant is that these two expressions are equal *where both are defined*. The functions $(x^3 - 8)/(x - 2)$ and $x^2 + 2x + 4$ are different functions. The difference is that $x = 2$ is not in the domain of $(x^3 - 8)/(x - 2)$ and is in the domain of $x^2 + 2x + 4$. Since $\lim_{x \to 2} f(x)$ explicitly *ignores* any value that f may have at 2, this does not affect the limit. From the point of view of the limit at 2, these functions *are* equal. In general we have the following rule.

Functions with Equal Limits

If $f(x) = g(x)$ for all x except possibly $x = a$, then

$$\lim_{x\to a} f(x) = \lim_{x\to a} g(x).$$

Quick Example

$\dfrac{x^2-1}{x-1} = x+1$ for all x except $x = 1$. Write $\dfrac{x^2-1}{x-1}$ as $\dfrac{(x+1)(x-1)}{x-1}$ and cancel the $(x-1)$

Therefore,

$$\lim_{x\to 1}\frac{x^2-1}{x-1} = \lim_{x\to 1}(x+1) = 1+1 = 2.$$

Q: *How do we find $\lim_{x\to a} f(x)$ when $x = a$ is not in the domain of the function f, and we cannot simplify the given function to make a a point of the domain?*

A: In such a case, it might be necessary to analyze the function by some other method, such as numerically or graphically. However, if we do not obtain the indeterminate form 0/0 upon substitution, we can often say what the limit is, as the following example shows.

EXAMPLE 3 Limit of a Closed-Form Function at a Point Not in Its Domain: The Determinate Form *k*/0

Evaluate the following limits, if they exist:

a. $\displaystyle\lim_{x\to 1^+}\frac{x^2-4x+1}{x-1}$ **b.** $\displaystyle\lim_{x\to 1}\frac{x^2-4x+1}{x-1}$ **c.** $\displaystyle\lim_{x\to 1}\frac{x^2-4x+1}{x^2-2x+1}$

Solution

a. Although the function $f(x) = \dfrac{x^2-4x+1}{x-1}$ is a closed-form function, $x = 1$ is not in its domain. Notice that substituting $x = 1$ gives

$$\frac{x^2-4x+1}{x-1} = \frac{1^2-4+1}{1-1} = \frac{-2}{0}$$ The determinate form $\frac{k}{0}$

which, although not defined, conveys important information to us: As x gets closer and closer to 1, the numerator approaches −2 and the denominator gets closer and closer to 0. Now, if we divide a number close to −2 by a number close to zero, we get a number of large absolute value; for instance

$$\frac{-2.1}{0.0001} = -21{,}000 \quad\text{and}\quad \frac{-2.1}{-0.0001} = 21{,}000$$

$$\frac{-2.01}{0.00001} = -201{,}000 \quad\text{and}\quad \frac{-2.01}{-0.00001} = 201{,}000.$$

(Compare Example 5 in Section 3.1.) In our limit for part (a), x is approaching 1 from the right, so the denominator $x - 1$ is positive (as x is to the right of 1). Thus we have the scenario illustrated previously on the left, and we can conclude that

$$\lim_{x\to1^+} \frac{x^2 - 4x + 1}{x - 1} = -\infty. \qquad \text{Think of this as } \frac{-2}{0^+} = -\infty.$$

b. This time, x could be approaching 1 from either side. We already have, from part (a)

$$\lim_{x\to1^+} \frac{x^2 - 4x + 1}{x - 1} = -\infty.$$

The same reasoning we used in part (a) gives

$$\lim_{x\to1^-} \frac{x^2 - 4x + 1}{x - 1} = +\infty \qquad \text{Think of this as } \frac{-2}{0^-} = +\infty.$$

because now the denominator is negative and still approaching zero while the numerator still approaches -2 and therefore is also negative. (See the numerical calculations above on the right.) Because the left and right limits do not agree, we conclude that

$$\lim_{x\to1} \frac{x^2 - 4x + 1}{x - 1} \text{ does not exist.}$$

c. First notice that the denominator factors:

$$\lim_{x\to1} \frac{x^2 - 4x + 1}{x^2 - 2x + 1} = \lim_{x\to1} \frac{x^2 - 4x + 1}{(x - 1)^2}$$

As x approaches 1, the numerator approaches -2 as before, and the denominator approaches 0. However, this time, the denominator $(x - 1)^2$, being a square, is ≥ 0, regardless of from which side x is approaching 1. Thus, the entire function is negative as x approaches 1, and

$$\lim_{x\to1} \frac{x^2 - 4x + 1}{(x - 1)^2} = -\infty. \qquad \frac{-2}{0^+} = -\infty$$

➡ **Before we go on...** In general, the determinate forms $\frac{k}{0^+}$ and $\frac{k}{0^-}$ will always yield $\pm\infty$, with the sign depending on the sign of the overall expression as $x \to a$. (When we write the form $\frac{k}{0}$ we always mean $k \neq 0$.) This and other determinate forms are discussed further after Example 5.

Figure 16 shows the graphs of $\frac{x^2 - 4x + 1}{x - 1}$ and $\frac{x^2 - 4x + 1}{(x - 1)^2}$ from Example 3. You should check that results we obtained above agree with a geometric analysis of these graphs near $x = 1$.

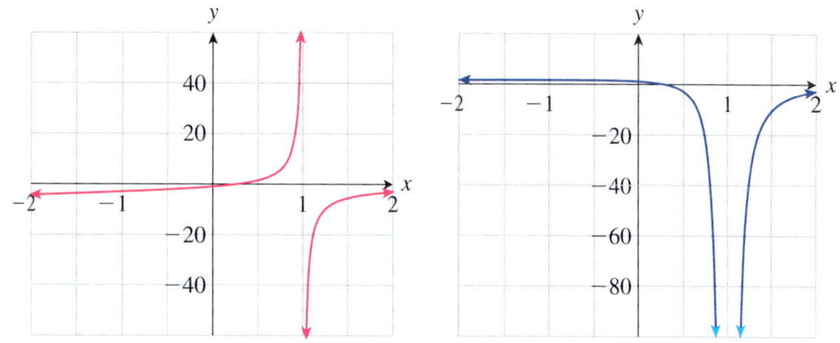

$y = \dfrac{x^2 - 4x + 1}{x - 1}$ $\qquad$ $y = \dfrac{x^2 - 4x + 1}{(x - 1)^2}$

Figure 16

We can also use algebraic techniques to analyze functions that are not given in closed form.

EXAMPLE 4 Functions Not Written in Closed Form

For which values of x are the following piecewise defined functions continuous?

a. $f(x) = \begin{cases} x^2 + 2 & \text{if } x < 1 \\ 2x - 1 & \text{if } x \geq 1 \end{cases}$ $\qquad$ **b.** $g(x) = \begin{cases} x^2 - x + 1 & \text{if } x \leq 0 \\ 1 - x & \text{if } 0 < x \leq 1 \\ x - 3 & \text{if } x > 1 \end{cases}$

Solution

a. The function $f(x)$ is given in closed form over the intervals $(-\infty, 1)$ and $[1, +\infty)$. At $x = 1$, $f(x)$ suddenly switches from one closed-form formula to another, so $x = 1$ is the only place where there is a potential problem with continuity. To investigate the continuity of $f(x)$ at $x = 1$, let's calculate the limit there:

$$\lim_{x \to 1^-} f(x) = \lim_{x \to 1^-} (x^2 + 2) \qquad f(x) = x^2 + 2 \text{ for } x < 1.$$

$$= (1)^2 + 2 = 3 \qquad x^2 + 2 \text{ is closed-form.}$$

$$\lim_{x \to 1^+} f(x) = \lim_{x \to 1^+} (2x - 1) \qquad f(x) = 2x - 1 \text{ for } x > 1.$$

$$= 2(1) - 1 = 1. \qquad 2x - 1 \text{ is closed-form.}$$

Because the left and right limits are different, $\lim_{x \to 1} f(x)$ does not exist, and so $f(x)$ is discontinuous at $x = 1$.

b. The only potential points of discontinuity for $g(x)$ occur at $x = 0$ and $x = 1$:

$$\lim_{x \to 0^-} g(x) = \lim_{x \to 0} (x^2 - x + 1) = 1$$

$$\lim_{x \to 0^+} g(x) = \lim_{x \to 0^+} (1 - x) = 1.$$

Thus, $\lim_{x\to 0} g(x) = 1$. Further, $g(0) = 0^2 - 0 + 1 = 1$ from the formula, and so

$$\lim_{x\to 0} g(x) = g(0),$$

which shows that $g(x)$ is continuous at $x = 0$. At $x = 1$ we have

$$\lim_{x\to 1^-} g(x) = \lim_{x\to 1^-} (1 - x) = 0$$

$$\lim_{x\to 1^+} g(x) = \lim_{x\to 1^+} (x - 3) = -2$$

so that $\lim_{x\to 1} g(x)$ does not exist. Thus, $g(x)$ is discontinuous at $x = 1$. We conclude that $g(x)$ is continuous at every real number x except $x = 1$.

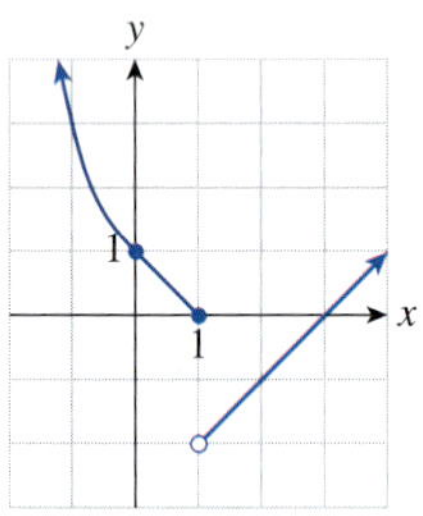

$y = g(x)$

Figure 17

Before we go on... Figure 17 shows the graph of g from Example 4(b). Notice how the discontinuity at $x = 1$ shows up as a break in the graph, whereas at $x = 0$ the two pieces "fit together" at the point (0, 1). ■

Limits at Infinity

Let's look once again at some limits similar to those in Examples 3 and 6 in Section 3.1.

EXAMPLE 5 Limits at Infinity

Compute the following limits, if they exist:

a. $\displaystyle\lim_{x\to+\infty} \frac{2x^2 - 4x}{x^2 - 1}$ **b.** $\displaystyle\lim_{x\to-\infty} \frac{2x^2 - 4x}{x^2 - 1}$

c. $\displaystyle\lim_{x\to+\infty} \frac{-x^3 - 4x}{2x^2 - 1}$ **d.** $\displaystyle\lim_{x\to+\infty} \frac{2x^2 - 4x}{5x^3 - 3x + 5}$

e. $\displaystyle\lim_{t\to+\infty} e^{0.1t} - 20$ **f.** $\displaystyle\lim_{t\to+\infty} \frac{80}{1 + 2.2(3.68)^{-t}}$

Solution **a** and **b.** While calculating the values for the tables used in Example 3 in Section 3.1, you might have noticed that the highest power of x in both the numerator and denominator dominated the calculations. For instance, when $x = 100{,}000$, the term $2x^2$ in the numerator has the value of 20,000,000,000, whereas the term $4x$ has the comparatively insignificant value of 400,000. Similarly, the term x^2 in the denominator overwhelms the term -1. In other words, for large values of x (or negative values with large magnitude),

$$\frac{2x^2 - 4x}{x^2 - 1} \approx \frac{2x^2}{x^2} \quad \text{Use only the highest powers top and bottom.}$$

$$= 2.$$

Therefore,

$$\lim_{x\to\pm\infty} \frac{2x^2 - 4x}{x^2 - 1} = \lim_{x\to\pm\infty} \frac{2x^2}{x^2}$$

$$= \lim_{x\to\pm\infty} 2 = 2.$$

The procedure of using only the highest powers of x to compute the limit is stated formally and justified after this example.

c. Applying the previous technique of looking only at highest powers gives

$$\lim_{x\to+\infty}\frac{-x^3-4x}{2x^2-1}=\lim_{x\to+\infty}\frac{-x^3}{2x^2}$$ Use only the highest powers top and bottom.

$$=\lim_{x\to+\infty}\frac{-x}{2}.$$ Simplify.

As x gets large, $-x/2$ gets large in magnitude but negative, so the limit is

$$\lim_{x\to+\infty}\frac{-x}{2}=-\infty.$$ $\frac{-\infty}{2}=-\infty$ (See below.)

d. $$\lim_{x\to+\infty}\frac{2x^2-4x}{5x^3-3x+5}=\lim_{x\to+\infty}\frac{2x^2}{5x^3}$$ Use only the highest powers top and bottom.

$$=\lim_{x\to+\infty}\frac{2}{5x}.$$

As x gets large, $2/(5x)$ gets close to zero, so the limit is

$$\lim_{x\to+\infty}\frac{2}{5x}=0.$$ $\frac{2}{\infty}=0$ (See below.)

e. Here we do not have a ratio of polynomials. However, we know that, as t becomes large and positive, so does $e^{0.1t}$, and hence also $e^{0.1t}-20$. Thus,

$$\lim_{t\to+\infty}e^{0.1t}-20=+\infty$$ $e^{+\infty}=+\infty$ (See below.)

f. As $t\to+\infty$, the term $(3.68)^{-t}=\dfrac{1}{3.68^t}$ in the denominator, being 1 divided by a very large number, approaches zero. Hence the denominator $1+2.2(3.68)^{-t}$ approaches $1+2.2(0)=1$ as $t\to+\infty$. Thus,

$$\lim_{t\to+\infty}\frac{80}{1+2.2(3.68)^{-t}}=\frac{80}{1+2.2(0)}=80$$ $(3.68)^{-\infty}=0$ (See below.)

Before we go on... Let's now look at the graph of the function $\dfrac{2x^2-4x}{x^2-1}$ in parts (a) and (b) of Example 5. We say that the graph of f has a **horizontal asymptote** at $y=2$ because of the limits we have just calculated. This means that the graph approaches the horizontal line $y=2$ far to the right or left (in this case, to both the right and left). Figure 18 shows the graph of f together with the line $y=2$.

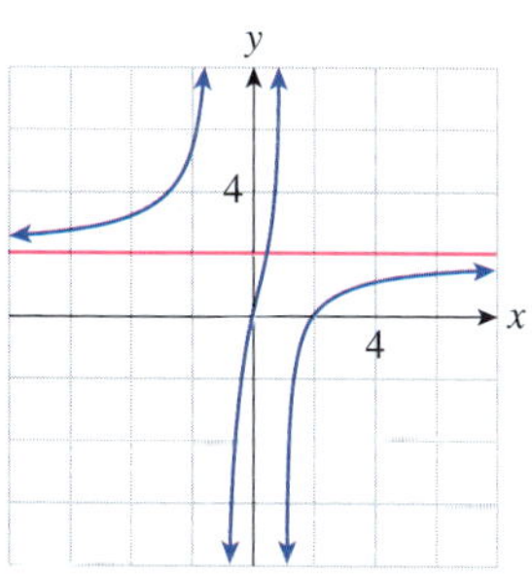

Figure 18

The graph reveals some additional interesting information: as $x\to1^+$, $f(x)\to-\infty$, and as $x\to1^-$, $f(x)\to+\infty$. Thus,

$$\lim_{x\to1}f(x)\text{ does not exist.}$$

See if you can determine what happens as $x\to-1$.

If you graph the functions in parts (d) and (f) of Example 5, you will again see a horizontal asymptote. Do the limits in parts (c) and (e) show horizontal asymptotes? ■

It is worthwhile looking again at what we did in each of the limits in Example 5:

a and **b.** We saw that $\dfrac{2x^2-4x}{x^2-1}\approx\dfrac{2x^2}{x^2}$, and then we canceled the x^2. Notice that, before we cancel, letting x approach $\pm\infty$ in the numerator and denominator yields the ratio $\frac{\infty}{\infty}$ which, like $\frac{0}{0}$, is another *indeterminate form*, and indicates to us that further work is needed—in this case cancellation—before we can write down the limit.

c. We obtained $\frac{-x^3-4x}{2x^2-1} \approx \frac{-x^3}{2x^2}$ which results in another indeterminate form, $\frac{-\infty}{\infty}$, as $x \to +\infty$. Cancellation of the x^2 gave us $\frac{-x}{2}$, resulting in the *determinate* form $\frac{-\infty}{2} = -\infty$ (a very large number divided by 2 is again a very large number).

d. Here, $\frac{2x^2-4x}{5x^3-3x+5} \approx \frac{2x^2}{5x^3} = \frac{2}{5x}$, and the cancellation step turns the indeterminate form $\frac{\infty}{\infty}$ into the determinate form $\frac{2}{\infty} = 0$ (dividing 2 by a very large number yields a very small number).

e. We reasoned that e raised to a large positive number is large and positive. Putting $t = +\infty$ gives us the determinate form $e^{+\infty} = +\infty$.

f. Here we reasoned that 3.68 raised to a large *negative* number is close to zero. Putting $t = +\infty$ gives us the determinate form $3.68^{-\infty} = 1/3.68^{+\infty} = 1/\infty = 0$ (see (d)).

In parts (a)–(d) of Example 5, $f(x)$ was a **rational function**: a quotient of polynomial functions. We calculated the limit of $f(x)$ at $\pm\infty$ by ignoring all powers of x in both the numerator and denominator except for the largest. Following is a theorem that justifies this procedure.

Theorem 3.2 Evaluating the Limit of a Rational Function at $\pm\infty$

If $f(x)$ has the form

$$f(x) = \frac{c_n x^n + c_{n-1}x^{n-1} + \cdots + c_1 x + c_0}{d_m x^m + d_{m-1}x^{m-1} + \cdots + d_1 x + d_0}$$

with the c_i and d_i constants ($c_n \neq 0$ and $d_m \neq 0$), then we can calculate the limit of $f(x)$ as $x \to \pm\infty$ by ignoring all powers of x except the highest in both the numerator and denominator. Thus,

$$\lim_{x\to\pm\infty} f(x) = \lim_{x\to\pm\infty} \frac{c_n x^n}{d_m x^m}.$$

Quick Examples

(See Example 5.)

1. $\lim_{x\to+\infty} \frac{2x^2-4x}{x^2-1} = \lim_{x\to+\infty} \frac{2x^2}{x^2} = \lim_{x\to+\infty} 2 = 2$
2. $\lim_{x\to+\infty} \frac{-x^3-4x}{2x^2-1} = \lim_{x\to+\infty} \frac{-x^3}{2x^2} = \lim_{x\to+\infty} \frac{-x}{2} = -\infty$
3. $\lim_{x\to+\infty} \frac{2x^2-4x}{5x^3-3x+5} = \lim_{x\to+\infty} \frac{2x^2}{5x^3} = \lim_{x\to+\infty} \frac{2}{5x} = 0$

Proof Our function $f(x)$ is a polynomial of degree n divided by a polynomial of degree m. If n happens to be larger than m, then dividing the top and bottom by the largest power x^n of x gives

$$f(x) = \frac{c_n x^n + c_{n-1}x^{n-1} + \cdots + c_1 x + c_0}{d_m x^m + d_{m-1}x^{m-1} + \cdots + d_1 x + d_0}$$

$$= \frac{c_n x^n/x^n + c_{n-1}x^{n-1}/x^n + \cdots + c_1 x/x^n + c_0/x^n}{d_m x^m/x^n + d_{m-1}x^{m-1}/x^n + \cdots + d_1 x/x^n + d_0/x^n}.$$

Canceling powers of x in each term and remembering that $n > m$ leaves us with

$$f(x) = \frac{c_n + c_{n-1}/x + \cdots + c_1/x^{n-1} + c_0/x^n}{d_m/x^{n-m} + d_{m-1}/x^{n-m+1} + \cdots + d_1/x^{n-1} + d_0/x^n}.$$

As $x \to \pm\infty$, all the terms shown in red approach 0, so we can ignore them in taking the limit. (The first term in the denominator happens to approach 0 as well, but we retain it for convenience.) Thus,

$$\lim_{x\to\pm\infty} f(x) = \lim_{x\to\pm\infty} \frac{c_n}{d_m/x^{n-m}} = \lim_{x\to\pm\infty} \frac{c_n x^n}{d_m x^m},$$

as required. The cases when n is smaller than m and $m = n$ are proved similarly by dividing top and bottom by the largest power of x in each case.

Some Determinate and Indeterminate Forms

The following table brings these ideas together with our observations in Example 3.

Some Determinate and Indeterminate Forms

$\frac{0}{0}$ and $\pm\frac{\infty}{\infty}$ are **indeterminate**; evaluating limits in which these arise requires simplification or further analysis.

The following are **determinate** forms for any nonzero number k:

$\frac{k}{0^\pm} = \pm\infty$ $\qquad$ $\frac{k}{\text{Small}} = \text{Big}^*$ (See Example 3.)

$k(\pm\infty) = \pm\infty$ $\qquad$ $k \times \text{Big} = \text{Big}^*$

$k \pm \infty = \pm\infty$ $\qquad$ $k \pm \text{Big} = \pm\text{Big}^*$

$\pm\frac{\infty}{k} = \pm\infty$ $\qquad$ $\frac{\text{Big}}{k} = \text{Big}^*$

$\pm\frac{k}{\infty} = 0$ $\qquad$ $\frac{k}{\text{Big}} = \text{Small}$

and, if k is positive, then

$k^{+\infty} = +\infty$ $\qquad$ $k^{\text{Big positive}} = \text{Big}$

$k^{-\infty} = 0$ $\qquad$ $k^{\text{Big negative}} = \text{Small}$

* **NOTE** The sign gets switched in these forms if k is negative.

Quick Examples

1. $\lim_{x\to 0} \frac{60}{2x^2} = +\infty$ $\quad \frac{k}{0^+} = +\infty$

2. $\lim_{x\to -1^-} \frac{2x-6}{x+1} = +\infty$ $\quad \frac{-4}{0^-} = +\infty$

3. $\lim_{x\to -\infty} 3x - 5 = -\infty$ $\quad 3(-\infty) - 5 = -\infty - 5 = -\infty$

4. $\lim_{x\to +\infty} \frac{2x}{60} = +\infty$ $\quad \frac{2(\infty)}{60} = \infty$

5. $\lim_{x\to -\infty} \frac{60}{2x} = 0$ $\quad \frac{60}{2(-\infty)} = 0$

6. $\lim_{x\to +\infty} \frac{60x}{2x} = 30$ $\quad \frac{\infty}{\infty}$ is indeterminate but we can cancel.

7. $\lim_{x\to -\infty} \frac{60}{e^x - 1} = \frac{60}{0-1} = -60$ $\quad e^{-\infty} = 0$

FAQs

Strategy for Evaluating Limits Algebraically

Q: *Is there a systematic way to evaluate a limit* $\lim_{x\to a} f(x)$ *algebraically?*

A: The following approach is often successful:

Case 1: a Is a Finite Number (Not $\pm\infty$)

1. Decide whether f is a closed-form function. If it is not, then find the left and right limits at the values of x where the function changes from one formula to another.

2. If f *is* a closed-form function, try substituting $x = a$ in the formula for $f(x)$. Then one of the following three things may happen:

   $f(a)$ is defined. Then $\lim_{x\to a} f(x) = f(a)$.

   $f(a)$ is not defined and has the indeterminate form 0/0. Try to simplify the expression for f to cancel one of the terms that gives 0.

   $f(a)$ is not defined and has one of the determinate forms listed above in the above table. Use the table to determine the limit as in the Quick Examples

Case 2: $a = \pm\infty$

Remember that we can use the determinate forms $k^{+\infty} = \infty$ and $k^{-\infty} = 0$ if k is positive. Further. if the given function is a polynomial or ratio of polynomials, use the technique of Example 5: Focus only on the highest powers of x and then simplify to obtain either a number L, in which case the limit exists and equals L, or one of the determinate forms $\pm\infty/k = \pm\infty$ or $\pm k/\infty = 0$.

There is another technique for evaluating certain difficult limits, called *l'Hospital's Rule*, but this uses derivatives, so we'll have to wait to discuss it until Section 4.1.

3.3 EXERCISES

▼ more advanced ◆ challenging

T indicates exercises that should be solved using technology

In Exercises 1–4 complete the given sentence.

1. The closed-form function $f(x) = \frac{1}{x-1}$ is continuous for all x except ______. HINT [See Quick Example on page 211.]

2. The closed-form function $f(x) = \frac{1}{x^2-1}$ is continuous for all x except ________. HINT [See Quick Example on page 211.]

3. The closed-form function $f(x) = \sqrt{x+1}$ has $x = 3$ in its domain. Therefore, $\lim_{x\to 3}\sqrt{x+1} =$ ___. HINT [See Example 1.]

4. The closed-form function $f(x) = \sqrt{x-1}$ has $x = 10$ in its domain. Therefore, $\lim_{x\to 10}\sqrt{x-1} =$ ___. HINT [See Example 1.]

In Exercises 5–20 determine if the given limit leads to a determinate or indeterminate form. Evaluate the limit if it exists, or say why if not. HINT [See Example 3 and Quick Examples on page 220.]

5. $\lim_{x\to 0}\frac{60}{x^4}$

6. $\lim_{x\to 0}\frac{2x^2}{x^2}$

7. $\lim_{x\to 0}\frac{x^3-1}{x^3}$

8. $\lim_{x\to 0}\frac{-2}{x^2}$

9. $\lim_{x\to -\infty}(-x^2+5)$

10. $\lim_{x\to 0}\frac{2x^2+4}{x}$

11. $\lim_{x\to +\infty}4^{-x}$

12. $\lim_{x\to +\infty}\frac{60+e^{-x}}{2-e^{-x}}$

13. $\lim_{x\to 0}\frac{-x^3}{3x^3}$

14. $\lim_{x\to -\infty}3x^2+6$

15. $\lim_{x\to -\infty}\frac{-x^3}{3x^6}$

16. $\lim_{x\to +\infty}\frac{-x^6}{3x^3}$

17. $\lim_{x\to -\infty}\frac{4}{-x+2}$

18. $\lim_{x\to -\infty}e^x$

19. $\lim_{x\to -\infty}\frac{60}{e^x-1}$

20. $\lim_{x\to -\infty}\frac{2}{2x^2+3}$

Calculate the limits in Exercises 21–72 algebraically. If a limit does not exist, say why.

21. $\lim_{x\to 0}(x+1)$ HINT [See Example 1.]

22. $\lim_{x\to 0}(2x-4)$ HINT [See Example 1.]

23. $\lim_{x\to 2}\frac{2+x}{x}$

24. $\lim_{x\to -1}\frac{4x^2+1}{x}$

25. $\lim_{x\to -1}\frac{x+1}{x}$

26. $\lim_{x\to 4}(x+\sqrt{x})$

27. $\lim_{x\to 8}(x-\sqrt[3]{x})$

28. $\lim_{x\to 1}\frac{x-2}{x+1}$

29. $\lim_{h\to 1}(h^2+2h+1)$

30. $\lim_{h\to 0}(h^3-4)$

31. $\lim_{h\to 3}2$

32. $\lim_{h\to 0}-5$

33. $\lim_{h\to 0}\frac{h^2}{h+h^2}$ HINT [See Example 2.]

34. $\lim_{h\to 0}\frac{h^2+h}{h^2+2h}$ HINT [See Example 2.]

35. $\lim_{x\to 1}\frac{x^2-2x+1}{x^2-x}$

36. $\lim_{x\to -1}\frac{x^2+3x+2}{x^2+x}$

37. $\lim_{x\to 2}\frac{x^3-8}{x-2}$

38. $\lim_{x\to -2}\frac{x^3+8}{x^2+3x+2}$

39. $\lim_{x\to 0^+}\frac{1}{x^2}$ HINT [See Example 3.]

40. $\lim_{x\to 0^+}\frac{1}{x^2-x}$ HINT [See Example 3.]

41. $\lim_{x\to -1}\frac{x^2+1}{x+1}$

42. $\lim_{x\to -1^-}\frac{x^2+1}{x+1}$

43. $\lim_{x\to -2^+}\frac{x^2+8}{x^2+3x+2}$

44. $\lim_{x\to -1}\frac{x^2+3x}{x^2+x}$

45. $\lim_{x\to -2}\frac{x^2+8}{x^2+3x+2}$

46. $\lim_{x\to -1}\frac{x^2+3x}{x^2+2x+1}$

47. $\lim_{x\to 2}\frac{x^2+8}{x^2-4x+4}$

48. $\lim_{x\to -1}\frac{x^2+3x}{x^2+3x+2}$

49. $\lim_{x\to +\infty}\frac{3x^2+10x-1}{2x^2-5x}$ HINT [See Example 5.]

50. $\lim_{x\to +\infty}\frac{6x^2+5x+100}{3x^2-9}$ HINT [See Example 5.]

51. $\lim_{x\to +\infty}\frac{x^5-1{,}000x^4}{2x^5+10{,}000}$

52. $\lim_{x\to +\infty}\frac{x^6+3{,}000x^3+1{,}000{,}000}{2x^6+1{,}000x^3}$

53. $\lim_{x\to +\infty}\frac{10x^2+300x+1}{5x+2}$

54. $\lim_{x\to +\infty}\frac{2x^4+20x^3}{1{,}000x^3+6}$

55. $\lim_{x\to +\infty}\frac{10x^2+300x+1}{5x^3+2}$

56. $\lim_{x\to +\infty}\frac{2x^4+20x^3}{1{,}000x^6+6}$

57. $\lim_{x\to -\infty}\frac{3x^2+10x-1}{2x^2-5x}$

58. $\lim_{x\to -\infty}\frac{6x^2+5x+100}{3x^2-9}$

59. $\lim_{x\to -\infty}\frac{x^5-1{,}000x^4}{2x^5+10{,}000}$

60. $\lim_{x\to -\infty}\frac{x^6+3000x^3+1{,}000{,}000}{2x^6+1{,}000x^3}$

61. $\lim_{x\to -\infty}\frac{10x^2+300x+1}{5x+2}$

62. $\lim_{x\to -\infty}\frac{2x^4+20x^3}{1{,}000x^3+6}$

63. $\lim_{x\to -\infty}\frac{10x^2+300x+1}{5x^3+2}$

64. $\lim_{x\to -\infty}\frac{2x^4+20x^3}{1{,}000x^6+6}$

65. $\lim_{x\to +\infty}(4e^{-3x}+12)$

66. $\lim_{x\to +\infty}\frac{2}{5-5.3e^{-3x}}$

67. $\lim_{t\to +\infty}\frac{2}{5-5.3(3^{3t})}$

68. $\lim_{t\to +\infty}(4.1-2e^{3t})$

69. $\lim_{t\to+\infty} \dfrac{2^{3t}}{1+5.3e^{-t}}$

70. $\lim_{x\to-\infty} \dfrac{4.2}{2-3^{2x}}$

71. $\lim_{x\to-\infty} \dfrac{-3^{2x}}{2+e^{x}}$

72. $\lim_{x\to+\infty} \dfrac{2^{-3x}}{1+5.3e^{-x}}$

In each of Exercises 73–80, find all points of discontinuity of the given function. HINT [See Example 4.]

73. $f(x) = \begin{cases} x+2 & \text{if } x < 0 \\ 2x-1 & \text{if } x \geq 0 \end{cases}$

74. $g(x) = \begin{cases} 1-x & \text{if } x \leq 1 \\ x-1 & \text{if } x > 1 \end{cases}$

75. $g(x) = \begin{cases} x+2 & \text{if } x < 0 \\ 2x+2 & \text{if } 0 \leq x < 2 \\ x^2+2 & \text{if } x \geq 2 \end{cases}$

76. $f(x) = \begin{cases} 1-x & \text{if } x \leq 1 \\ x+2 & \text{if } 1 < x < 3 \\ x^2-4 & \text{if } x \geq 3 \end{cases}$

77. ▼ $h(x) = \begin{cases} x+2 & \text{if } x < 0 \\ 0 & \text{if } x = 0 \\ 2x+2 & \text{if } x > 0 \end{cases}$

78. ▼ $h(x) = \begin{cases} 1-x & \text{if } x < 1 \\ 1 & \text{if } x = 1 \\ x+2 & \text{if } x > 1 \end{cases}$

79. ▼ $f(x) = \begin{cases} 1/x & \text{if } x < 0 \\ x & \text{if } 0 \leq x \leq 2 \\ 2^{x-1} & \text{if } x > 2 \end{cases}$

80. ▼ $f(x) = \begin{cases} x^3+2 & \text{if } x \leq -1 \\ x^2 & \text{if } -1 < x < 0 \\ x & \text{if } x \geq 0 \end{cases}$

APPLICATIONS

81. *Employment in Mexico* The number N of workers employed in manufacturing jobs in Mexico between 1995 and 2004 can be modeled by

$$N(t) = \begin{cases} 0.22t+3 & \text{if } 0 \leq t \leq 5 \\ -0.15t+4.85 & \text{if } 5 < t \leq 9 \end{cases} \text{ million jobs}$$

where t is time in years since 1995.[12]

a. Compute $\lim_{t\to5^-} N(t)$ and $\lim_{t\to5^+} N(t)$, and interpret each answer. HINT [See Example 4.]

b. Is the function N continuous at $t = 5$? Did the number of workers employed in manufacturing jobs in Mexico experience any abrupt changes during the period 1995–2004?

82. *Mortgage Delinquencies* The percentage P of borrowers in the highest risk category who were delinquent on their payments between 2001 and 2007 can be approximated by

$$P = \begin{cases} -1.8t+9.7 & \text{if } 0 \leq t \leq 3 \\ 2t-1.7 & \text{if } 3 < t \leq 6 \end{cases} \text{ percent}$$

[12] Source: *New York Times*, February 18, 2007, p. WK4.

where t is time in years since 2001.[13]

a. Compute $\lim_{t\to3^-} P(t)$ and $\lim_{t\to3^+} P(t)$, and interpret each answer.

b. Is the function P continuous at $t = 3$? Did the percentage of borrowers who were delinquent on their payments experience any abrupt changes during the period 2001–2007?

83. *Movie Advertising* Movie expenditures, in billions of dollars, on advertising in newspapers from 1995 to 2004 can be approximated by

$$f(t) = \begin{cases} 0.04t+0.33 & \text{if } t \leq 4 \\ -0.01t+1.2 & \text{if } t > 4 \end{cases}$$

where t is time in years since 1995.[14]

a. Compute $\lim_{t\to4^-} f(t)$ and $\lim_{t\to4^+} f(t)$, and interpret each answer. HINT [See Example 4.]

b. Is the function f continuous at $t = 4$? What does the answer tell you about movie advertising expenditures?

84. *Movie Advertising* The percentage of movie advertising as a share of newspapers' total advertising revenue from 1995 to 2004 can be approximated by

$$p(t) = \begin{cases} -0.07t+6.0 & \text{if } t \leq 4 \\ 0.3t+17.0 & \text{if } t > 4 \end{cases}$$

where t is time in years since 1995.[15]

a. Compute $\lim_{t\to4^-} p(t)$ and $\lim_{t\to4^+} p(t)$, and interpret each answer. HINT [See Example 4.]

b. Is the function p continuous at $t = 4$? What does the answer tell you about newspaper revenues?

85. *Law Enforcement* The cost of fighting crime in the United States increased steadily in the period 1982–1999. Total spending on police and courts can be approximated, respectively, by[16]

$$P(t) = 1.745t + 29.84 \text{ billion dollars} \quad (2 \leq t \leq 19)$$
$$C(t) = 1.097t + 10.65 \text{ billion dollars} \quad (2 \leq t \leq 19)$$

where t is time in years since 1980. Compute $\lim_{t\to+\infty} \dfrac{P(t)}{C(t)}$ to two decimal places and interpret the result. HINT [See Example 5.]

86. *Law Enforcement* Refer to Exercise 85. Total spending on police, courts, and prisons can be approximated, respectively, by[17]

$$P(t) = 1.745t + 29.84 \text{ billion dollars} \quad (2 \leq t \leq 19)$$
$$C(t) = 1.097t + 10.65 \text{ billion dollars} \quad (2 \leq t \leq 19)$$
$$J(t) = 1.919t + 12.36 \text{ billion dollars} \quad (2 \leq t \leq 19)$$

[13] Ibid.

[14] Model by the authors. Source for data: Newspaper Association of America Business Analysis and Research/*New York Times*, May 16, 2005.

[15] Ibid.

[16] Spending is adjusted for inflation and shown in 1999 dollars. Models are based on a linear regression. Source for data: Bureau of Justice Statistics/*New York Times*, February 11, 2002, p. A14.

[17] Ibid.

where t is time in years since 1980. Compute $\lim_{t\to+\infty} \frac{P(t)}{P(t)+C(t)+J(t)}$ to two decimal places and interpret the result.

87. ***Casino Revenues*** Annual revenues in 1996–2006 from slot machines and video poker machines at **Foxwoods** casino in Connecticut can be modeled by

$$R(t) = 825 - 240e^{-0.4t} \text{ million dollars} \quad (0 \le t \le 10)$$

where t is time in years since 1996.[18] If we extrapolate this model into the indefinite future, compute and interpret $\lim_{t\to+\infty} R(t)$. HINT [See Quick Example on page 220.]

88. ***Casino Revenues*** Annual revenues in 1996–2006 from slot machines and video poker machines at the **Mohegan Sun** casino in Connecticut can be modeled by

$$R(t) = 1{,}260 - 1{,}030e^{-0.1t} \text{ million dollars} \quad (0 \le t \le 10)$$

where t is time in years since 1996.[19] If we extrapolate this model into the indefinite future, compute and interpret $\lim_{t\to+\infty} R(t)$. HINT [See Quick Example on page 220.]

Foreign Trade *Annual U.S. imports from China in the years 1996 through 2003 can be approximated by*

$$I(t) = t^2 + 3.5t + 50 \quad (1 \le t \le 8)$$

billion dollars, where t represents time in years since 1995. Annual U.S. exports to China in the same years can be approximated by

$$E(t) = 0.4t^2 + 1.6t + 14 \quad (0 \le t \le 10)$$

billion dollars.[20] *Exercises 89 and 90 are based on these models.*

89. Assuming that the trends shown in the above models continue indefinitely, calculate the limits

$$\lim_{t\to+\infty} I(t) \text{ and } \lim_{t\to+\infty} \frac{I(t)}{E(t)}$$

algebraically, interpret your answers, and comment on the results. HINT [See Example 5.]

90. Repeat Exercise 89, this time calculating

$$\lim_{t\to+\infty} E(t) \text{ and } \lim_{t\to+\infty} \frac{E(t)}{I(t)} \quad \text{HINT [See Example 5.]}$$

91. ▼ ***Acquisition of Language*** The percentage $p(t)$ of children who can speak in at least single words by the age of t months can be approximated by the equation[21]

$$p(t) = 100\left(\frac{1 - 12{,}200}{t^{4.48}}\right). \quad (t \ge 8.5)$$

Calculate $\lim_{t\to+\infty} p(t)$ and interpret the results. HINT [See Example 5(e), (f).]

92. ▼ ***Acquisition of Language*** The percentage $q(t)$ of children who can speak in sentences of five or more words by the age of t months can be approximated by the equation[22]

$$q(t) = 100\left(1 - \frac{5.27 \times 10^{17}}{t^{12}}\right). \quad (t \ge 30)$$

If p is the function referred to in the preceding exercise, calculate $\lim_{t\to+\infty}[p(t) - q(t)]$ and interpret the result. HINT [See Example 5(e), (f).]

COMMUNICATION AND REASONING EXERCISES

93. Describe the algebraic method of evaluating limits as discussed in this section and give at least one disadvantage of this method.

94. What is a closed-form function? What can we say about such functions?

95. ▼ Your friend Karin tells you that $f(x) = 1/(x-2)^2$ cannot be a closed-form function because it is not continuous at $x = 2$. Comment on her assertion.

96. ▼ Give an example of a function f specified by means of algebraic formulas such that the domain of f consists of all real numbers and f is not continuous at $x = 2$. Is f a closed-form function?

97. Give examples of two limits that lead to two different indeterminate forms, but where both limits exist.

98. Give examples of two limits; one that leads to a determinate form and another that leads to an indeterminate form, but where neither limit exists.

99. ▼ What is wrong with the following statement? If $f(x)$ is specified algebraically and $f(a)$ is defined, then $\lim_{x\to a} f(x)$ exists and equals $f(a)$.

100. ▼ What is wrong with the following statement? $\lim_{x\to-2} \frac{x^2-4}{x+2}$ does not exist because substituting $x = -2$ yields $0/0$, which is undefined.

101. ▼ Give the formula for a function that is continuous everywhere except at two points.

102. ▼ Give the formula for a function that is continuous everywhere except at three points.

103. ◆ ***The Indeterminate Form*** $\infty - \infty$ An indeterminate form not mentioned in Section 3.3 is $\infty - \infty$. Give examples of three limits that lead to this indeterminate form, and where the first limit exists and equals 5, where the second limit diverges to $+\infty$, and where the third exists and equals -5.

104. ◆ ***The Indeterminate Form*** 1^∞ An indeterminate form not mentioned in Section 3.3 is 1^∞. Give examples of three limits that lead to this indeterminate form, and where the first limit exists and equals 1, where the second limit exists and equals e, and where the third diverges to $+\infty$. HINT [For the third, consider modifying the second.]

[18] Source for data: Connecticut Division of Special Revenue/*New York Times*, Sept. 23, 2007, p. C1.

[19] Ibid.

[20] Based on quadratic regression using data from the U.S. Census Bureau Foreign Trade Division Web site www.census.gov/foreign-trade/sitc1/ as of December 2004.

[21] The model is the authors' and is based on data presented in the article *The Emergence of Intelligence* by William H. Calvin, *Scientific American*, October, 1994, pp. 101–107.

[22] Ibid.

3.4 Average Rate of Change

Calculus is the mathematics of change, inspired largely by observation of continuously changing quantities around us in the real world. As an example, the New York metro area consumer confidence index C decreased from 100 points in January 2000 to 80 points in January 2002.[23] As we saw in Chapter 1, the **change** in this index can be measured as the difference:

$$\Delta C = \text{Second value} - \text{First value} = 80 - 100 = -20 \text{ points}$$

(The fact that the confidence index decreased is reflected in the negative sign of the change.) The kind of question we will concentrate on is *how fast* the confidence index was dropping. Because C decreased by 20 points in 2 years, we say it averaged a $20/2 = 10$ point drop each year. (It actually dropped less than 10 points the first year and more the second, giving an average drop of 10 points each year.)

Alternatively, we might want to measure this rate in points per month rather than points per year. Because C decreased by 20 points in 24 months, it went down at an average rate of $20/24 \approx 0.833$ points per month.

In both cases, we obtained the average rate of change by dividing the change by the corresponding length of time:

$$\text{Average rate of change} = \frac{\text{Change in } C}{\text{Change in time}} = \frac{-20}{2} = -10 \text{ points per year}$$

$$\text{Average rate of change} = \frac{\text{Change in } C}{\text{Change in time}} = \frac{-20}{24} \approx -0.833 \text{ points per month}$$

Average Rate of Change of a Function Numerically and Graphically

LWA-Dann Tardif/CORBIS

EXAMPLE 1 Standard and Poor's 500

The following table lists the approximate value of Standard and Poor's 500 stock market index (S&P) during the period 2000–2008* ($t = 0$ represents 2000):

t (year)	0	1	2	3	4	5	6	7	8
$S(t)$ S&P 500 Index (points)	1,450	1,200	950	1,000	1,100	1,200	1,250	1,500	1,300

a. What was the average rate of change in the S&P over the 2-year period 2005–2007 (the period $5 \le t \le 7$ or [5, 7] in interval notation); over the 4-year period 2000–2004 (the period $0 \le t \le 4$ or [0, 4]); and over the period [1, 5]?

b. Graph the values shown in the table. How are the rates of change reflected in the graph?

* The values are approximate values midway through the given year. Source: www.finance.google.com.

[23] Figures are approximate. Source: Siena College Research Institute/*New York Times*, February 10, 2002, p. LI.

Solution

a. During the 2-year period [5, 7], the S&P changed as follows:

Start of the period ($t = 5$):	$S(5) = 1{,}200$
End of the period ($t = 7$):	$S(7) = 1{,}500$
Change during the period [5, 7]:	$S(7) - S(5) = 300$

Thus, the S&P increased by 300 points in 2 years, giving an average rate of change of $300/2 = 150$ points per year. We can write the calculation this way:

$$\begin{aligned}\text{Average rate of change of } S &= \frac{\text{Change in } S}{\text{Change in } t}\\ &= \frac{\Delta S}{\Delta t}\\ &= \frac{S(7) - S(5)}{7 - 5}\\ &= \frac{1{,}500 - 1{,}200}{7 - 5} = \frac{300}{2} = 150 \text{ points per year}\end{aligned}$$

Interpreting the result: During the period [5, 7] (that is, 2005–2007), the S&P increased at an average rate of 150 points per year.

Similarly, the average rate of change during the period [0, 4] was

$$\begin{aligned}\text{Average rate of change of } S &= \frac{\Delta S}{\Delta t} = \frac{S(4) - S(0)}{4 - 0} = \frac{1{,}100 - 1{,}450}{4 - 0}\\ &= -\frac{350}{4} = -87.5 \text{ points per year.}\end{aligned}$$

Interpreting the result: During the period [0, 4] the S&P *decreased* at an average rate of 87.5 points per year.

Finally, during the period [1, 5], the average rate of change was

$$\begin{aligned}\text{Average rate of change of } S &= \frac{\Delta S}{\Delta t} = \frac{S(5) - S(1)}{5 - 1} = \frac{1{,}200 - 1{,}200}{5 - 1}\\ &= \frac{0}{4} = 0 \text{ points per year.}\end{aligned}$$

Interpreting the result: During the period [1, 5] the average rate of change of the S&P was zero points per year (even though its value did fluctuate during that period).

b. In Chapter 1, we saw that the rate of change of a quantity that changes linearly with time is measured by the slope of its graph. However, the S&P index does not change linearly with time. Figure 19 shows the data plotted two different ways: (a) as a bar chart and (b) as a piecewise linear graph. Bar charts are more commonly used in the media, but Figure 19(b) illustrates the changing index more clearly.

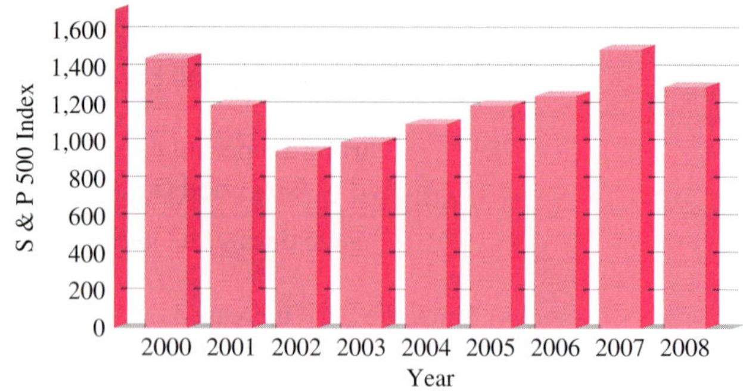

Figure 19(a)

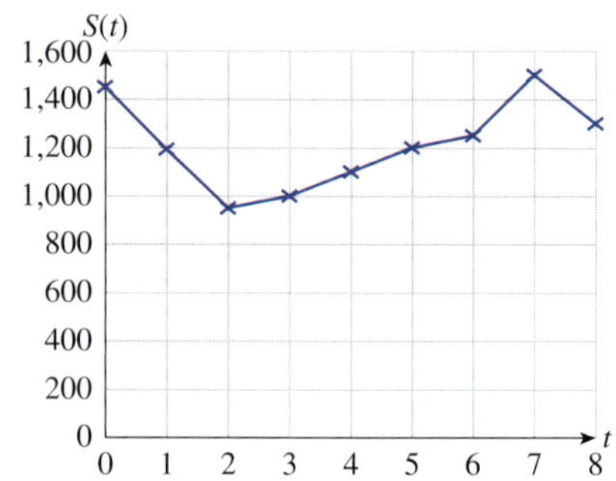

Figure 19(b)

We saw in part (a) that the average rate of change of S over the interval [5, 7] is the ratio

$$\text{Rate of change of } S = \frac{\Delta S}{\Delta t} = \frac{\text{Change in } S}{\text{Change in } t} = \frac{S(7) - S(5)}{7 - 5}.$$

Notice that this rate of change is also the slope of the line through P and Q shown in Figure 20, and we can estimate this slope directly from the graph as shown.

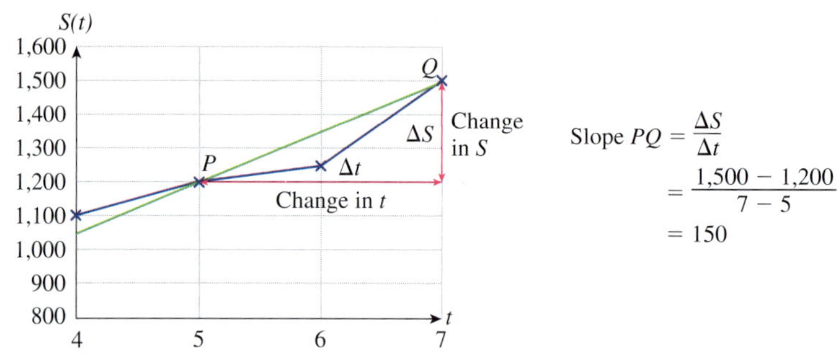

Figure 20

Average Rate of Change as Slope: The average rate of change of the S&P over the interval [5, 7] is the slope of the line passing through the points on the graph where $t = 5$ and $t = 7$.

Similarly, the average rates of change of the S&P over the intervals [0, 4] and [1, 5] are the slopes of the lines through pairs of corresponding points.

Here is the formal definition of the average rate of change of a function over an interval.

Change and Average Rate of Change of *f* over [*a*, *b*]: Difference Quotient

The **change** in $f(x)$ over the interval $[a, b]$ is

$$\begin{aligned}\text{Change in } f &= \Delta f \\ &= \text{Second value} - \text{First value} \\ &= f(b) - f(a).\end{aligned}$$

The **average rate of change** of $f(x)$ over the interval $[a, b]$ is

$$\begin{aligned}\text{Average rate of change of } f &= \frac{\text{Change in } f}{\text{Change in } x} \\ &= \frac{\Delta f}{\Delta x} = \frac{f(b) - f(a)}{b - a} \\ &= \text{Slope of line through points } P \text{ and } Q \\ &\quad \text{(see figure).}\end{aligned}$$

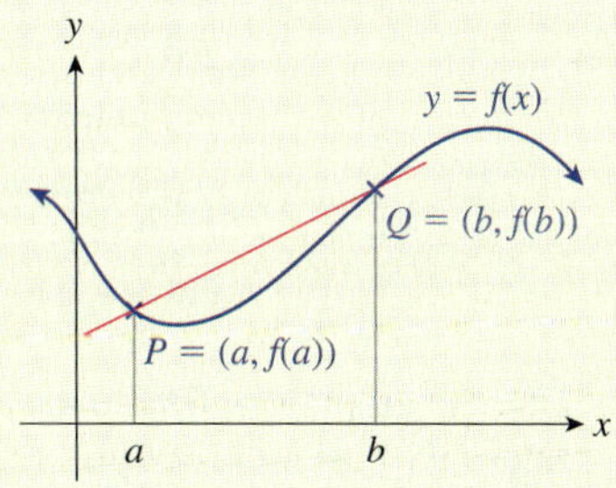

Average rate of change = Slope of PQ

We also call this average rate of change the **difference quotient** of f over the interval $[a, b]$. (It is the *quotient* of the *differences* $f(b) - f(a)$ and $b - a$.) A line through two points of a graph like P and Q is called a **secant line** of the graph.

Units

The units of the change in f are the units of $f(x)$.
The units of the average rate of change of f are units of $f(x)$ per unit of x.

Quick Example

If $f(3) = -1$ billion dollars, $f(5) = 0.5$ billion dollars, and x is measured in years, then the change and average rate of change of f over the interval [3, 5] are given by

$$\text{Change in } f = f(5) - f(3) = 0.5 - (-1) = 1.5 \text{ billion dollars}$$

$$\begin{aligned}\text{Average rate of change} &= \frac{f(5) - f(3)}{5 - 3} = \frac{0.5 - (-1)}{2} \\ &= 0.75 \text{ billion dollars/year.}\end{aligned}$$

Alternative Formula: Average Rate of Change of *f* over [*a*, *a* + *h*]

(Replace b above by $a + h$.) The average rate of change of f over the interval $[a, a + h]$ is

$$\text{Average rate of change of } f = \frac{f(a + h) - f(a)}{h}.$$

In Example 1 we saw that the average rate of change of a quantity can be estimated directly from a graph. Here is an example that further illustrates the graphical approach.

EXAMPLE 2 Freon Production

Figure 21 shows a function we looked at in Section 1.1: Let $f(t)$ be the number of tons of ozone-layer damaging Freon 22 (chlorodifluoromethane) produced in year t in developing countries, where $t = 0$ represents the year 2000.*

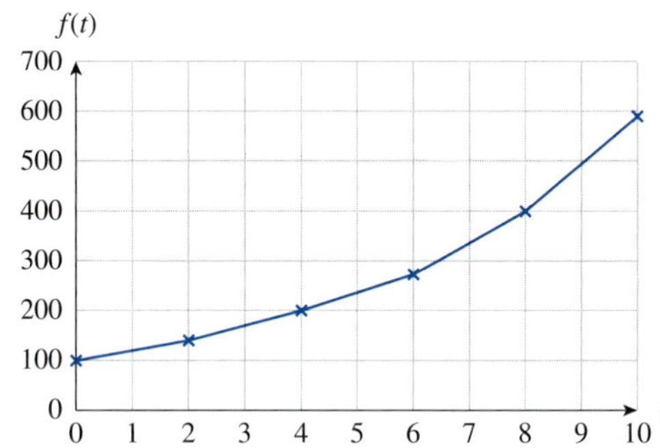

Figure 21

a. Use the graph to estimate the average rate of change of $f(t)$ with respect to t over the interval [4, 8] and interpret the result.

b. Over which 2-year period(s) was the average rate of change of Freon production the greatest?

c. Multiple choice: For the period of time under consideration, Freon production was

(A) increasing at a faster and faster rate.
(B) increasing at a slower and slower rate.
(C) decreasing at a faster and faster rate.
(D) decreasing at a slower and slower rate.

Solution

a. The average rate of change of f over the interval [4, 8] is given by the slope of the line through the points P and Q shown in Figure 22.

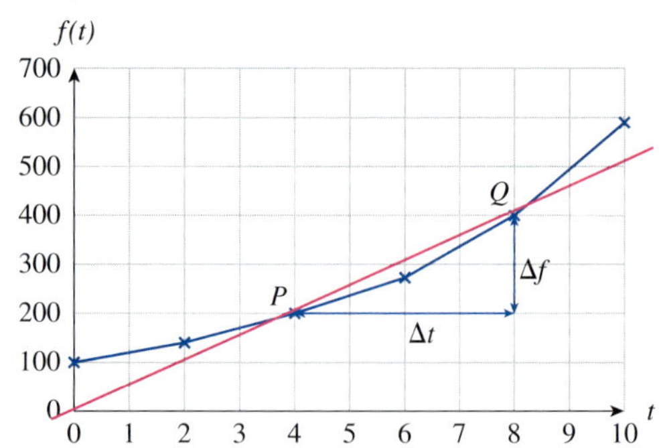

Figure 22

*Source: *New York Times*, February 23, 2007, p. C1.

From the figure,

$$\text{Average rate of change of } f = \frac{\Delta f}{\Delta t} = \text{slope } PQ \approx \frac{400 - 200}{8 - 4} = \frac{200}{4} = 50.$$

Thus, the rate of change of f over the interval [4, 8] is approximately 50.

Q: *How do we interpret the result?*

A: A clue is given by the units of the average rate of change: units of f per unit of t. The units of f are tons of Freon and the units of t are years. Thus, the average rate of change of f is measured in tons of Freon per year, and we can now interpret the result as follows:

Interpreting the average rate of change: Annual production of Freon was increasing at an average rate of 50 tons of Freon per year from 2004 to 2008.

b. The rates of change of Freon production over successive 2-year periods are given by the slopes of the individual line segments that make up the graph in Figure 21. Thus, the greatest average rate of change in a single two-year period corresponds to the segment(s) with the largest slope. If you look at the figure, you will notice that the segment corresponding to [8, 10] is the steepest. Thus, the average rate of change of Freon production was largest over the 2-year period from 2008 to 2010.

c. Looking again at the figure, notice that the graph rises as we go from left to right; that is, the value of the function (Freon production) is increasing with increasing t. At the same time, the fact that the curve bends up (is concave up) with increasing t tells us that the successive slopes get steeper, and so the average rates of change increase as well (Choice (A)).

➡ **Before we go on...** Notice in Example 2 that we do not get exact answers from a graph; the best we can do is *estimate* the rates of change: Was the exact answer to part (a) closer to 49 or 51? Two people can reasonably disagree about results read from a graph, and you should bear this in mind when you check the answers to the exercises. ■

Perhaps the most sophisticated way to compute the average rate of change of a quantity is through the use of a mathematical formula or model for the quantity in question.

Average Rate of Change of a Function Using Algebraic Data

EXAMPLE 3 Average Rate of Change from a Formula

You are a commodities trader and you monitor the price of gold on the New York Spot Market very closely during an active morning. Suppose you find that the price of an ounce of gold can be approximated by the function

$$G(t) = -8t^2 + 144t + 150 \text{ dollars} \qquad (7.5 \le t \le 10.5)$$

where t is time in hours. (See Figure 23. $t = 8$ represents 8:00 AM.)

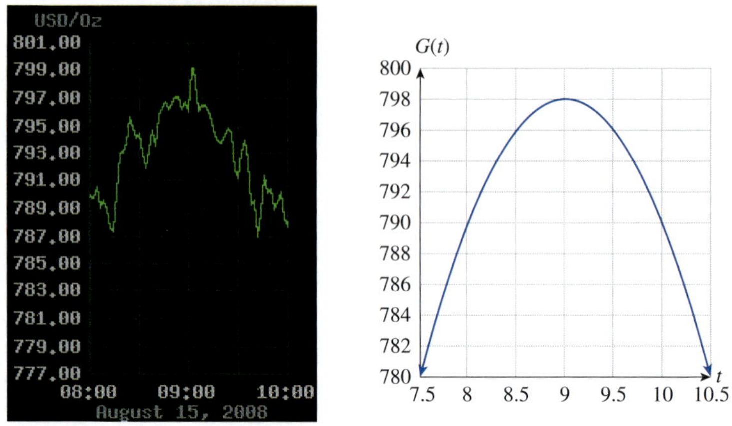

Source: www.kitco.com August 15, 2008 $G(t) = -8t^2 + 144t + 150$

Figure 23

Looking at the graph on the right, we can see that the price of gold rose rather rapidly at the beginning of the time period, but by $t = 8.5$ the rise had slowed, until the market faltered and the price began to fall more and more rapidly toward the end of the period. What was the average rate of change of the price of gold over the $1\frac{1}{2}$-hour period starting at 8:00 AM (the interval $[8, 9.5]$ on the t-axis)?

Solution We have

$$\text{Average rate of change of } G \text{ over } [8, 9.5] = \frac{\Delta G}{\Delta t} = \frac{G(9.5) - G(8)}{9.5 - 8}.$$

From the formula for $G(t)$, we find

$$G(9.5) = -8(9.5)^2 + 144(9.5) + 150 = 796$$
$$G(8) = -8(8)^2 + 144(8) + 150 = 790.$$

Thus, the average rate of change of G is given by

$$\frac{G(9.5) - G(8)}{9.5 - 8} = \frac{796 - 790}{1.5} = \frac{6}{1.5} = \$4 \text{ per hour.}$$

In other words, the price of gold was increasing at an average rate of \$4 per hour over the given $1\frac{1}{2}$-hour period.

using Technology

See the Technology Guide at the end of the chapter for detailed instructions on how to calculate the average rate of change of the function in Example 3 using a TI-83/84 Plus or Excel. Here is an outline:

TI-83/84 Plus

`Y1=-8X^2+144X+150`

Home screen: `(Y1(9.5)-Y1(8))/(9.5-8)`

[More details on page 280.]

Excel

Enter the headings t, $G(t)$, Rate of Change in A1–C1

t-values 8, 9.5 in A2–A3

`=-8*A2^2+144*A2+150` in B2 and copy down to B3

`=(B3-B2)/(A3-A2)` in C2.

[More details on page 282.]

EXAMPLE 4 Rates of Change over Shorter Intervals

Continuing with Example 3, use technology to compute the average rate of change of

$$G(t) = -8t^2 + 144t + 150 \qquad (7.5 \le t \le 10.5)$$

over the intervals $[8, 8 + h]$, where $h = 1, 0.1, 0.01, 0.001$, and 0.0001. What do the answers tell you about the price of gold?

Solution

We use the alternative formula

$$\text{Average rate of change of } G \text{ over } [a, a+h] = \frac{G(a+h) - G(a)}{h},$$

so

$$\text{Average rate of change of } G \text{ over } [8, 8+h] = \frac{G(8+h) - G(8)}{h}.$$

Let us calculate this average rate of change for some of the values of h listed:

$h = 1$: $G(8+h) = G(8+1) = G(9) = -8(9)^2 + 144(9) + 150 = 798$

$G(8) = -8(8)^2 + 144(8) + 150 = 790$

$$\text{Average rate of change of } G = \frac{G(9) - G(8)}{1} = \frac{798 - 790}{1} = 8$$

$h = 0.1$: $G(8+h) = G(8+0.1) = G(8.1) = -8(8.1)^2 + 144(8.1) + 150 = 791.52$

$G(8) = -8(8)^2 + 144(8) + 150 = 790$

$$\text{Average rate of change of } G = \frac{G(8.1) - G(8)}{0.1} = \frac{791.52 - 790}{0.1} = 15.2$$

$h = 0.01$: $G(8+h) = G(8+0.01) = G(8.01) = -8(8.01)^2 + 144(8.01) + 150$

$= 790.1592$

$G(8) = -8(8)^2 + 144(8) + 150 = 790$

$$\text{Average rate of change of } G = \frac{G(8.01) - G(8)}{0.01} = \frac{790.1592 - 790}{0.01}$$

$$= 15.92$$

Continuing in this way, we get the values in the following table:

h	1	0.1	0.01	0.001	0.0001
Ave. Rate of Change $\frac{G(8+h) - G(8)}{h}$	8	15.2	15.92	15.992	15.9992

Each value is an average rate of change of G. For example, the value corresponding to $h = 0.01$ is 15.92, which tells us:

Over the interval [8, 8.01] *the price of gold was increasing at an average rate of* \$15.92 *per hour.*

In other words, during the first one-hundredth of an hour (or 36 seconds) starting at $t = 8{:}00$ AM, the price of gold was increasing at an average rate of \$15.92 per hour. Put another way, in those 36 seconds, the price of gold increased at a rate that, if continued, would have produced an increase of \$15.92 in the price of gold during the next hour. We will return to this example at the beginning of Section 3.5.

using Technology

Example 4 is the kind of example where the use of technology can make a huge difference. See the Technology Guides at the end of the chapter to find out how to do these computations almost effortlessly using a TI-83/84 Plus or Excel. Here is an outline:

TI-83/84 Plus

`Y1=-8X^2+144X+150`

Home screen:

```
(Y1(8+1)-Y1(8))/1
(Y1(8+0.1)-Y1(8))/0.1
(Y1(8+0.01)-Y1(8))/0.01
```

etc.

[More details on page 280.]

Excel

Headings a, h, t, $G(t)$, Rate of Change in A1–E1

`8` in A2, `1` in B2,

`=A2` in C2, `=A2+B2` in C3

`=-8*C2^2+144*C2+150` in D2; copy down to D3

`=(D3-D2)/(C3-C2)` in E2

[More details on page 282.]

FAQs

Recognizing When and How to Compute the Average Rate of Change and How to Interpret the Answer

Q: *How do I know, by looking at the wording of a problem, that it is asking for an average rate of change?*

A: If a problem does not ask for an average rate of change directly, it might do so indirectly, as in "On average, how fast is quantity q increasing?"

Q: *If I know that a problem calls for computing an average rate of change, how should I compute it? By hand or using technology?*

A: All the computations can be done by hand, but when hand calculations are not called for, using technology might save time.

Q: *Lots of problems ask us to "interpret" the answer. How do I do that for questions involving average rates of change?*

A: The *units* of the average rate of change are often the key to interpreting the results:

The units of the average rate of change of f(x) are units of f(x) per unit of x.

Thus, for instance, if $f(x)$ is the cost, in dollars, of a trip of x miles in length, and the average rate of change of f is calculated to be 10, then the units of the average rate of change are dollars per mile, and so we can interpret the answer by saying that the cost of a trip rises an average of $10 for each additional mile.

3.4 EXERCISES

▼ more advanced ◆ challenging

T indicates exercises that should be solved using technology

In Exercises 1–18, calculate the average rate of change of the given function over the given interval. Where appropriate, specify the units of measurement. HINT [See Example 1.]

1.

x	0	1	2	3
$f(x)$	3	5	2	−1

Interval: [1, 3]

2.

x	0	1	2	3
$f(x)$	−1	3	2	1

Interval: [0, 2]

3.

x	−3	−2	−1	0
$f(x)$	−2.1	0	−1.5	0

Interval: [−3, −1]

4.

x	−2	−1	0	1
$f(x)$	−1.5	−0.5	4	6.5

Interval: [−1, 1]

5.

t (months)	2	4	6
$R(t)$ ($ millions)	20.2	24.3	20.1

Interval: [2, 6]

6.

x (kilos)	1	2	3
$C(x)$ (£)	2.20	3.30	4.00

Interval: [1, 3]

7.

p ($)	5.00	5.50	6.00
$q(p)$ (items)	400	300	150

Interval: [5, 5.5]

8.

t (hours)	0	0.1	0.2
$D(t)$ (miles)	0	3	6

Interval: [0.1, 0.2]

9.

Apple Computer Stock Price ($)

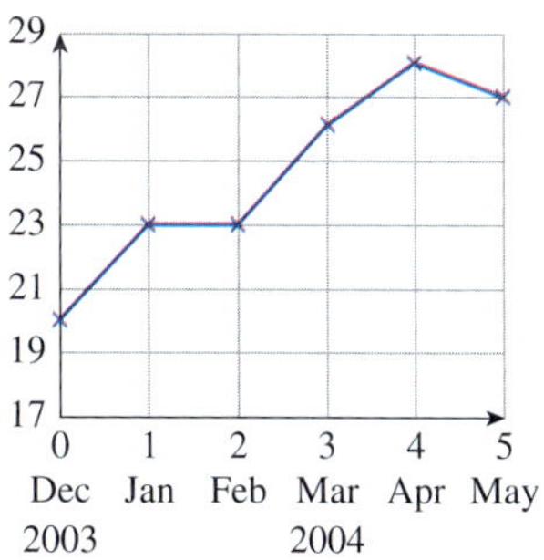

Interval: [2, 5]
HINT [See Example 2.]

10.

Cisco Systems Stock Price ($)

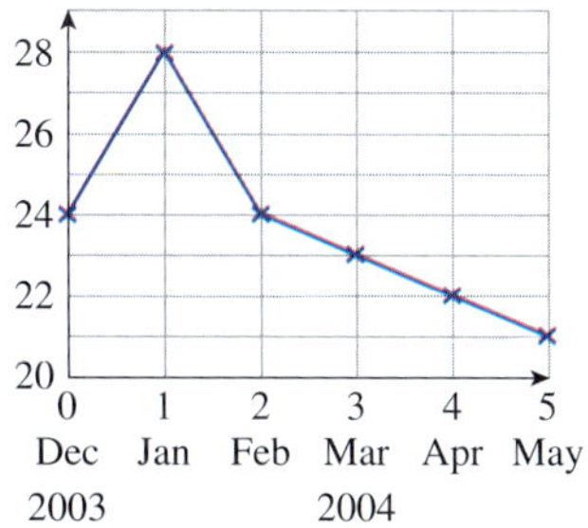

Interval: [1, 5]
HINT [See Example 2.]

11.

Unemployment (%)

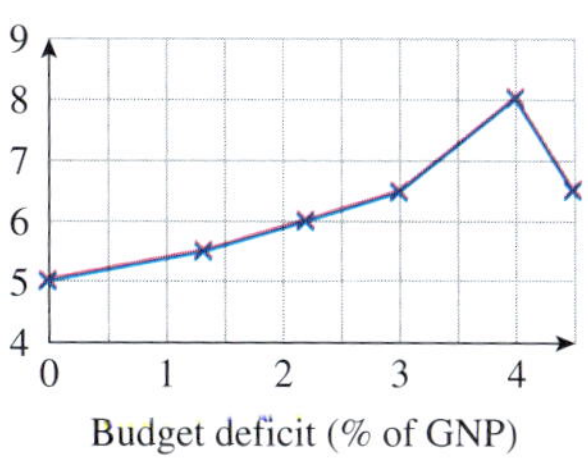

Interval: [0, 4]

12.

Inflation (%)

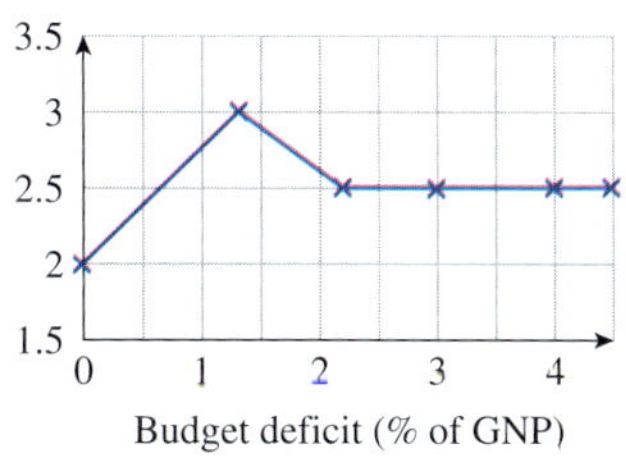

Interval: [0, 4]

13. $f(x) = x^2 - 3$; [1, 3] HINT [See Example 3.]

14. $f(x) = 2x^2 + 4$; [−1, 2] HINT [See Example 3.]

15. $f(x) = 2x + 4$; [−2, 0]

16. $f(x) = \frac{1}{x}$; [1, 4]

17. $f(x) = \frac{x^2}{2} + \frac{1}{x}$; [2, 3]

18. $f(x) = 3x^2 - \frac{x}{2}$; [3, 4]

In Exercises 19–24, calculate the average rate of change of the given function f over the intervals [a, a + h] where h = 1, 0.1, 0.01, 0.001, and 0.0001. (Technology is recommended for the cases h = 0.01, 0.001, and 0.0001) HINT [See Example 4.]

19. $f(x) = 2x^2$; $a = 0$

20. $f(x) = \frac{x^2}{2}$; $a = 1$

21. $f(x) = \frac{1}{x}$; $a = 2$

22. $f(x) = \frac{2}{x}$; $a = 1$

23. $f(x) = x^2 + 2x$; $a = 3$

24. $f(x) = 3x^2 - 2x$; $a = 0$

APPLICATIONS

25. ***World Military Expenditure*** The following table shows total military and arms trade expenditure in 1994, 1998, and 2006 ($t = 0$ represents 2000).[24]

Year t	−6	−2	6
Military Expenditure $C(t)$ ($ billion)	900	800	1,200

Compute and interpret the average rate of change of $C(t)$ **(a)** over the period 1994–2006 (that is, [−6, 6]), and **(b)** over the period [−2, 6]. Be sure to state the units of measurement. HINT [See Example 1.]

26. ***Education Expenditure*** The following table shows the percentage of the U.S. Discretionary Budget allocated to education in 2003, 2005, and 2009 ($t = 0$ represents 2000).[25]

Year t	3	5	9
Percentage $P(t)$	6.8	7	6.2

Compute and interpret the average rate of change of $P(t)$ **(a)** over the period 2003–2009 (that is, [3, 9]), and **(b)** over the period [5, 9]. Be sure to state the units of measurement. HINT [See Example 1.]

27. ***Oil Production in Mexico*** The following table shows approximate daily oil production by **Pemex**, Mexico's national oil company, for 2001–2009 ($t = 1$ represents the start of 2001):[26]

t (year since 2000)	1	2	3	4	5	6	7	8	9
$P(t)$ (million barrels)	3.1	3.3	3.4	3.4	3.4	3.3	3.2	3.1	3.0

a. Compute the average rate of change of $P(t)$ over the period 2002–2007. Interpret the result. HINT [See Example 1.]

b. Which of the following is true? From 2001 to 2008, the one-year average rates of change of oil production by **Pemex**

(A) increased in value
(B) decreased in value
(C) never increased in value
(D) never decreased in value

HINT [See Example 2.]

28. ***Oil Imports from Mexico*** The following table shows U.S. daily oil imports from Mexico, for 2000–2008 ($t = 0$ represents the start of 2000):[27]

t (year since 2000)	0	1	2	3	4	5	6	7	8
$I(t)$ (million barrels)	1.4	1.35	1.5	1.55	1.6	1.5	1.5	1.5	1.2

[24] Source: www.globalissues.org/Geopolitics/ArmsTrade/Spending.asp.

[25] Ibid.

[26] Figures are approximate, and 2008–2009 figures are projections by the Department of Energy. Source: Energy Information Administration/Pemex (http://www.eia.doe.gov).

[27] Ibid.

a. Use the data in the table to compute the average rate of change of $I(t)$ over the period 2001–2006. Interpret the result.

b. Which of the following is true? From 2000 to 2003, the one-year rate of change of oil imports from Mexico

(A) increased in value
(B) decreased in value
(C) never increased in value
(D) never decreased in value

29. ***Subprime Mortgages*** The following graph shows the approximate percentage $P(t)$ of mortgages issued in the U.S. that are subprime (normally classified as risky):[28]

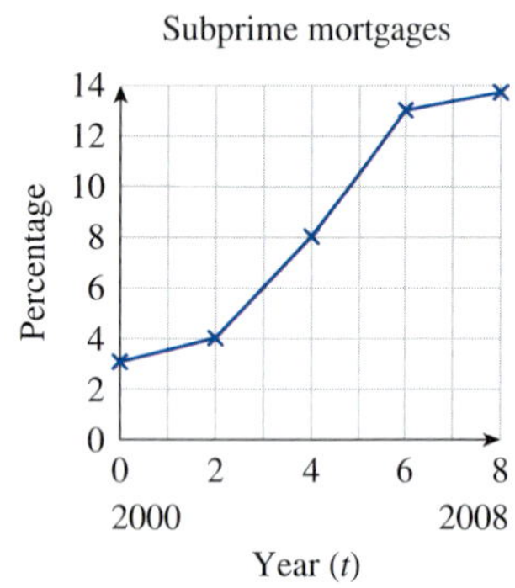

a. Use the graph to estimate, to one decimal place, the average rate of change of $P(t)$ with respect to t over the interval [0, 6] and interpret the result.

b. Over which 2-year period(s) was the average rate of change of $P(t)$ the greatest? HINT [See Example 2.]

30. ***Subprime Mortgage Debt*** The following graph shows the approximate value $V(t)$ of subprime (normally classified as risky) mortgage debt outstanding in the United States:[29]

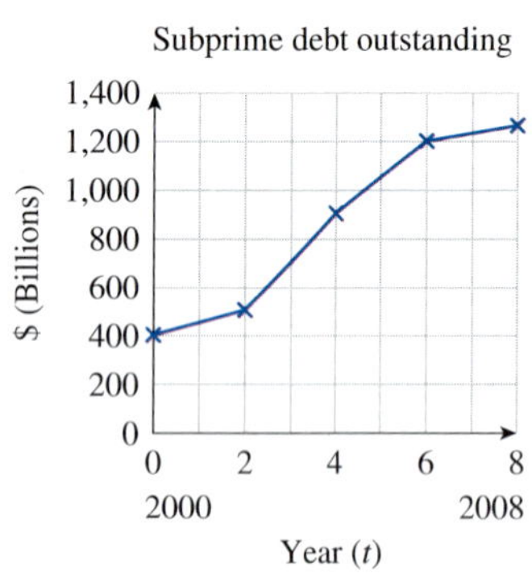

a. Use the graph to estimate, to one decimal place, the average rate of change of $V(t)$ with respect to t over the interval [2, 6] and interpret the result.

b. Over which 2-year period(s) was the average rate of change of $V(t)$ the least? HINT [See Example 2.]

[28] Sources: Mortgage Bankers Association, UBS.

[29] 2008 figure is an estimate. Source: www.data360.org/dataset.aspx?Data_Set_Id=9549.

31. ***iPhone Sales*** The following graph shows approximate Apple iPhone sales from the 2nd quarter in 2007 through the second quarter in 2008 ($t = 2$ represents the start of the second quarter of 2007):[30]

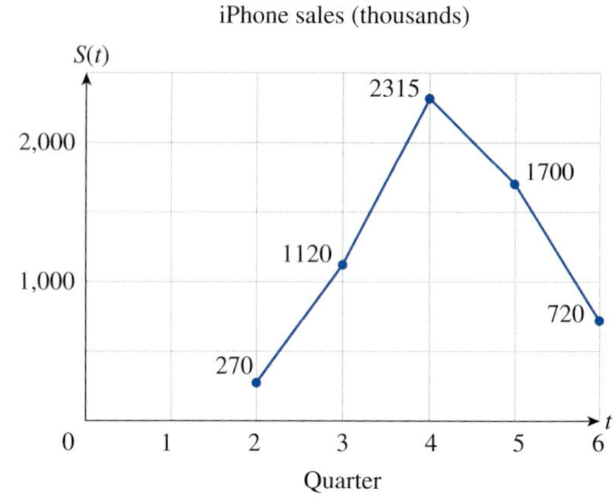

During which 2-quarter interval(s) was the average rate of change of $S(t)$ **(a)** greatest **(b)** least? Interpret your answers by referring to the rates of change. HINT [See Example 2.]

32. ***Facebook Membership*** The following graph gives the approximate number of Facebook users at various times since its establishment ($t = 0$ represents the start of 2004).[31]

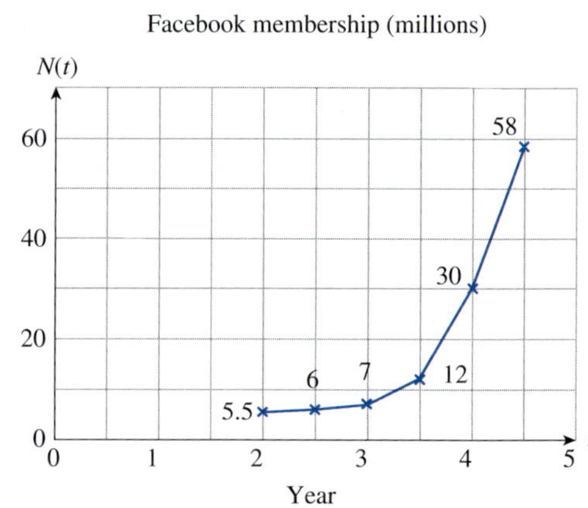

During which 1-year interval(s) was the average rate of change of $N(t)$ **(a)** greatest **(b)** least? Interpret your answers by referring to the rates of change. HINT [See Example 2.]

33. ▼ ***Physics Research in the U.S.*** The following table shows the number of research articles in the journal *Physical Review* authored by U.S researchers during the period 1993–2003 ($t = 3$ represents 1993):[32]

***t* (year since 1990)**	3	5	7	9	11	13
***N(t)* (articles, thousands)**	5.1	4.6	4.3	4.3	4.5	4.2

[30] Source: Apple financial statements (www.apple.com).

[31] Sources: Some data are interpolated. (www.facebook.com/, www.insidehighered.com).

[32] Source: The Americal Physical Society/*New York Times*, May 3, 2003, p. A1.

a. Find the interval(s) over which the average rate of change of N was the most negative. What was that rate of change? Interpret your answer.

b. The **percentage change of N over the interval $[a, b]$** is defined to be

$$\text{Percentage change of } N = \frac{\text{Change in } N}{\text{First value}} = \frac{N(b) - N(a)}{N(a)}.$$

Compute the percentage change of N over the interval $[3, 13]$ and also the average rate of change. Interpret the answers.

34. ▼ ***Physics Research in Europe*** The following table shows the number of research articles in the journal *Physical Review* authored by researchers in Europe during the period 1993–2003 ($t = 3$ represents 1993):[33]

***t* (year since 1990)**	3	5	7	9	11	13
***N(t)* (articles, thousands)**	3.8	4.6	5.0	5.0	6.0	5.7

a. Find the interval(s) over which the average rate of change of N was the most positive. What was that rate of change? Interpret your answer.

b. The **percentage change of N over the interval $[a, b]$** is defined to be

$$\text{Percentage change of } N = \frac{\text{Change in } N}{\text{First value}} = \frac{N(b) - N(a)}{N(a)}.$$

Compute the percentage change of N over the interval $[7, 13]$ and also the average rate of change. Interpret the answers.

35. ***College Basketball: Men*** The following chart shows the number of NCAA men's college basketball teams in the United States for various years since 2000.[34]

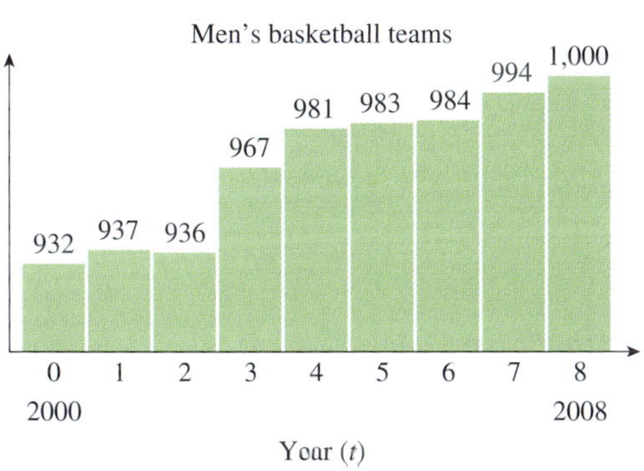

a. On average, how fast was the number of men's college basketball teams growing over the four-year period beginning in 2002?

b. By inspecting the graph, determine whether the four-year average rates of change increased or decreased beginning in 2002. HINT [See Example 2.]

36. ***College Basketball: Women*** The following chart shows the number of NCAA women's college basketball teams in the United States for various years since 2000.[35]

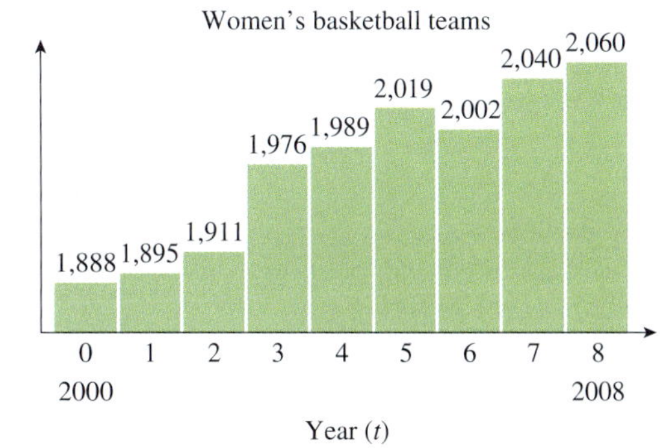

a. On average, how fast was the number of women's college basketball teams growing over the three-year period beginning in 2004?

b. By inspecting the graph, find the three-year period over which the average rate of change was largest. HINT [See Example 2.]

37. ▼ ***Funding for the Arts*** The following chart shows the total annual support for the arts in the U.S. by federal, state, and local government in 1995–2003 as a function of time in years ($t = 0$ represents 1995) together with the regression line:[36]

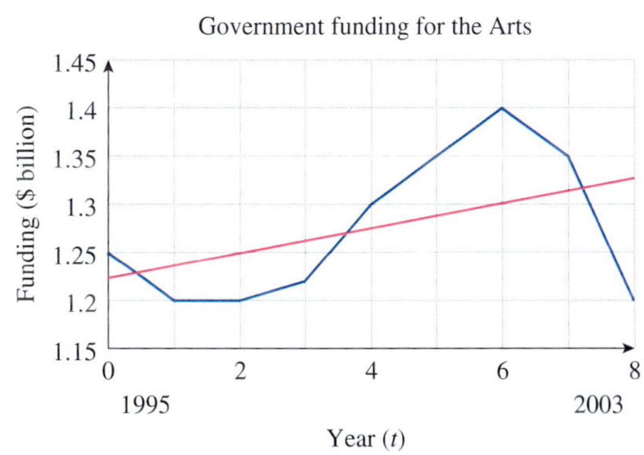

Multiple choice:

a. Over the period [0, 4] the average rate of change of government funding for the arts was
(A) less than **(B)** greater than **(C)** approximately equal
to the rate predicted by the regression line.

b. Over the period [4, 8] the average rate of change of government funding for the arts was
(A) less than **(B)** greater than **(C)** approximately equal
to the rate predicted by the regression line.

---

[33]Source: The Americal Physical Society/*New York Times*, May, 3, 2003, p. A1.

[34]2007 and 2008 figures are estimates. Source: The 2008 Statistical Abstract www.census.gov/.

[35]Ibid.

[36]Figures are adjusted for inflation. Sources: Giving USA, The Foundation Center, Americans for the Arts/*New York Times*, June 19, 2004, p. B7.

c. Over the period [3, 6] the average rate of change of government funding for the arts was
(A) less than **(B)** greater than **(C)** approximately equal to the rate predicted by the regression line.

d. Estimate, to two significant digits, the average rate of change of government funding for the arts over the period [0, 8]. (Be careful to state the units of measurement.) How does it compare to the slope of the regression line?

38. ▼ ***Funding for the Arts*** The following chart shows the total annual support for the arts in the U.S. by foundation endowments in 1995–2002 as a function of time in years ($t = 0$ represents 1995) together with the regression line:[37]

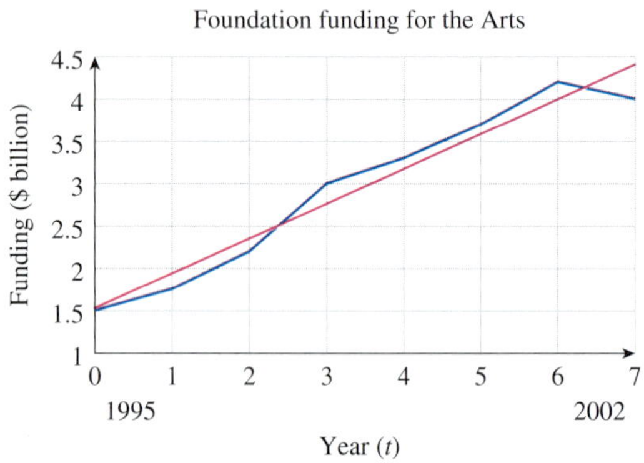

Multiple choice:

a. Over the period [0, 2.5] the average rate of change of government funding for the arts was
(A) less than **(B)** greater than **(C)** approximately equal to the rate predicted by the regression line.

b. Over the period [2, 6] the average rate of change of government funding for the arts was
(A) less than **(B)** greater than **(C)** approximately equal to the rate predicted by the regression line.

c. Over the period [3, 7] the average rate of change of government funding for the arts was
(A) less than **(B)** greater than **(C)** approximately equal to the rate predicted by the regression line.

d. Estimate, to two significant digits, the average rate of change of foundation funding for the arts over the period [0, 7]. (Be careful to state the units of measurement.) How does it compare to the slope of the regression line?

39. ▼ ***Market Volatility During the Dot-com Boom*** A volatility index generally measures the extent to which a market undergoes sudden changes in value. The volatility of the S&P 500 (as measured by one such index) was decreasing at an average rate of 0.2 points per year during 1991–1995, and was increasing at an average rate of about 0.3 points per year during 1995–1999. In 1995, the volatility of the S&P was 1.1.[38] Use this information to give a rough sketch of the volatility of the S&P 500 as a function of time, showing its values in 1991 and 1999.

40. ▼ ***Market Volatility During the Dot-com Boom*** The volatility (see the preceding exercise) of the NASDAQ had an average rate of change of 0 points per year during 1992–1995, and increased at an average rate of 0.2 points per year during 1995–1998. In 1995, the volatility of the NASDAQ was 1.1.[39] Use this information to give a rough sketch of the volatility of the NASDAQ as a function of time.

41. ***Market Index*** Joe Downs runs a small investment company from his basement. Every week he publishes a report on the success of his investments, including the progress of the "Joe Downs Index." At the end of one particularly memorable week, he reported that the index for that week had the value $I(t) = 1{,}000 + 1{,}500t - 800t^2 + 100t^3$ points, where t represents the number of business days into the week; t ranges from 0 at the beginning of the week to 5 at the end of the week. The graph of I is shown below.

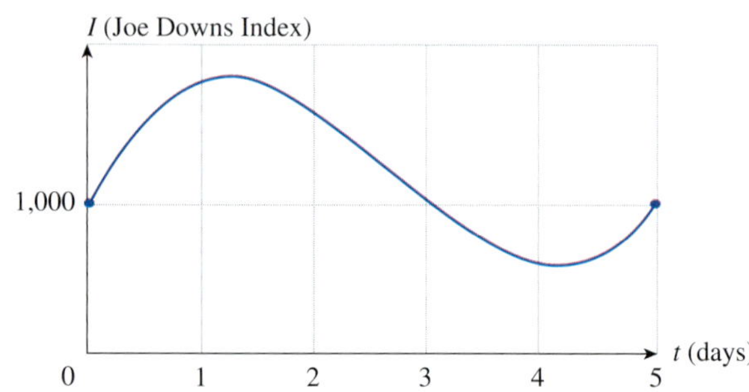

On average, how fast, and in what direction, was the index changing over the first two business days (the interval [0, 2])? HINT [See Example 3.]

42. ***Market Index*** Refer to the Joe Downs Index in the preceding exercise. On average, how fast, and in which direction, was the index changing over the last three business days (the interval [2, 5])? HINT [See Example 3.]

43. ***Crude Oil Prices*** The price per barrel of crude oil in constant 2008 dollars can be approximated by

$$P(t) = 0.45t^2 - 12t + 105 \text{ dollars} \quad (0 \le t \le 28)$$

where t is time in years since the start of 1980.[40]

a. What, in constant 2008 dollars, was the average rate of change of the price of oil from the start of 1981 ($t = 1$) to the start of 2006 ($t = 26$)? HINT [See Example 3.]

b. Your answer to part (a) is quite small. Can you conclude that the price of oil hardly changed at all over the 25-year period 1981 to 2006? Explain.

[37] Figures are adjusted for inflation. Sources: Giving USA, The Foundation Center, Americans for the Arts/*New York Times*, June 19, 2004, p. B7.

[38] Source for data: Sanford C. Bernstein Company/*New York Times*, March 24, 2000, p. C1.

[39] Ibid.

[40] Source for data: www.inflationdata.com

44. ***Median Home Price*** The median home price in the U.S. over the period 2004–2009 can be approximated by

$P(t) = -5t^2 + 75t - 30$ thousand dollars $\quad (4 \le t \le 9)$

where t is time in years since the start of 2000.[41]

a. What was the average rate of change of the median home price from the start of 2007 to the start of 2009? HINT [See Example 3.]

b. What, if anything, does your answer to part (a) say about the median home price in 2008? Explain.

45. ***SARS*** In the early stages of the deadly SARS (Severe Acute Respiratory Syndrome) epidemic in 2003, the number of reported cases can be approximated by

$$A(t) = 167(1.18)^t \quad (0 \le t \le 20)$$

t days after March 17, 2003 (the first day for which statistics were reported by the World Health Organization).

a. What was the average rate of change of $A(t)$ from March 17 to March 23? Interpret the result.

b. Which of the following is true? For the first 20 days of the epidemic, the number of reported cases

(A) increased at a faster and faster rate

(B) increased at a slower and slower rate

(C) decreased at a faster and faster rate

(D) decreased at a slower and slower rate HINT [See Example 2.]

46. ***SARS*** A few weeks into the deadly SARS (Severe Acute Respiratory Syndrome) epidemic in 2003, the number of reported cases can be approximated by

$$A(t) = 1{,}804(1.04)^t \quad (0 \le t \le 30)$$

t days after April 1, 2003.

a. What was the average rate of change of $A(t)$ from April 19 ($t = 18$) to April 29? Interpret the result.

b. Which of the following is true? During the 30-day period beginning April 1, the number of reported cases

(A) increased at a faster and faster rate

(B) increased at a slower and slower rate

(C) decreased at a faster and faster rate

(D) decreased at a slower and slower rate HINT [See Example 2.]

47. ▼ ***Ecology*** Increasing numbers of manatees ("sea sirens") have been killed by boats off the Florida coast. The following graph shows the relationship between the number of boats registered in Florida and the number of manatees killed each year.

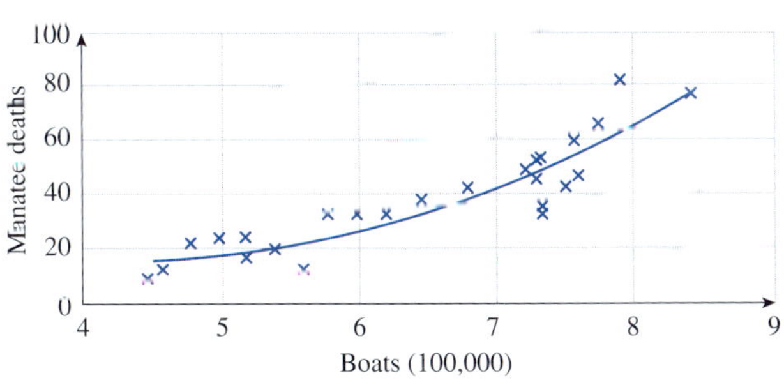

[41]Source for data: www.investmenttools.com.

The regression curve shown is given by

$f(x) = 3.55x^2 - 30.2x + 81$ manatees $\quad (4.5 \le x \le 8.5)$

where x is the number of boats (in hundreds of thousands) registered in Florida in a particular year, and $f(x)$ is the number of manatees killed by boats in Florida that year.[42]

a. Compute the average rate of change of f over the intervals [5, 6] and [7, 8].

b. What does the answer to part (a) tell you about the manatee deaths per boat?

48. ▼ ***Ecology*** Refer to Exercise 47.

a. Compute the average rate of change of f over the intervals [5, 7] and [6, 8].

b. Had we used a linear model instead of a quadratic one, how would the two answers in part (a) be related to each other?

49. T ▼ ***Poverty vs. Income*** Based on data from 1988 through 2003, the poverty rate (percentage of households with incomes below the poverty threshold) in the U.S. can be approximated by

$p(x) = 0.092x^2 - 8.1x + 190$ percentage points $\quad (38 \le x \le 44)$

where x is the U.S. median household income in thousands of dollars.[43]

a. Use technology to complete the following table which shows the average rate of change of p over successive intervals of length $\frac{1}{2}$. (Round all answers to two decimal places.) HINT [See Example 4.]

Interval	[39, 39.5]	[39.5, 40]	[40, 40.5]	[40.5, 41]	[41, 41.5]	[41.5, 42]
Average Rate of Change of p						

b. Interpret your answer for the interval [40, 40.5], being sure to indicate the direction of change and the units of measurement.

c. Multiple choice: As the median household income rises, the poverty rate

(A) increases

(B) decreases

(C) increases, then decreases

(D) decreases, then increases

d. Multiple choice: As the median income increases, the effect on the poverty rate is

(A) more pronounced

(B) less pronounced

[42]Regression model is based on data from 1976 to 2000. Sources for data: Florida Department of Highway Safety & Motor Vehicles, Florida Marine Institute/*New York Times*, February 12, 2002, p. F4.

[43]The model is based on a quadratic regression. Household incomes are in constant 2002 dollars. The poverty threshold is approximately \$18,000 for a family of four and \$9,200 for an individual. Sources: Census Bureau Current Population Survey/*New York Times*, Sept 27, 2003, p. A10/ U.S. Department of Labor Bureau of Labor Statistics www.stats.bls.gov June 17, 2004.

50. **T** ▼ ***Poverty vs. Unemployment*** Based on data from 1988 through 2003, the poverty rate (percentage of households with incomes below the poverty threshold) in the U.S. can be approximated by

$$p(x) = -0.12x^2 + 2.4x + 3.2 \text{ percentage points} \quad (4 \le x \le 8)$$

where x is the unemployment rate in percentage points.[44]

a. Use technology to complete the following table which shows the average rate of change of p over successive intervals of length $\frac{1}{2}$. (Round all answers to two decimal places.)

Interval	[5.0, 5.5]	[5.5, 6.0]	[6.0, 6.5]	[6.5, 7.0]	[7.0, 7.5]	[7.5, 8.0]
Average Rate of Change of *p*						

b. Interpret your answer for the interval [5.0, 5.5], being sure to indicate the direction of change and the units of measurement. HINT [See Example 4.]

c. Multiple choice: As the median household income rises, the poverty rate
(A) increases
(B) decreases
(C) increases, then decreases
(D) decreases, then increases

d. Multiple choice: As the unemployment rate increases, the effect on the poverty rate is
(A) more pronounced
(B) less pronounced

COMMUNICATION AND REASONING EXERCISES

51. Describe three ways we have used to determine the average rate of change of f over an interval $[a, b]$. Which of the three ways is *least* precise? Explain.

52. If f is a linear function of x with slope m, what is its average rate of change over any interval $[a, b]$?

53. Is the average rate of change of a function over $[a, b]$ affected by the values of the function between a and b? Explain.

54. If the average rate of change of a function over $[a, b]$ is zero, this means that the function has not changed over $[a, b]$, right?

55. Sketch the graph of a function whose average rate of change over [0, 3] is negative but whose average rate of change over [1, 3] is positive.

56. Sketch the graph of a function whose average rate of change over [0, 2] is positive but whose average rate of change over [0, 1] is negative.

[44]The model is based on a quadratic regression. Household incomes are in constant 2002 dollars. The poverty threshold is approximately $18,000 for a family of four and $9,200 for an individual. Sources: Census Bureau Current Population Survey/*New York Times*, Sept 27, 2003, p. A10; U.S. Department of Labor Bureau of Labor Statistics www.stats.bls.gov June 17, 2004.

57. ▼ If the rate of change of quantity A is 2 units of quantity A per unit of quantity B, and the rate of change of quantity B is 3 units of quantity B per unit of quantity C, what is the rate of change of quantity A with respect to quantity C?

58. ▼ If the rate of change of quantity A is 2 units of quantity A per unit of quantity B, what is the rate of change of quantity B with respect to quantity A?

59. ▼ A certain function has the property that its average rate of change over the interval $[1, 1+h]$ (for positive h) increases as h decreases. Which of the following graphs could be the graph of f?

(A)

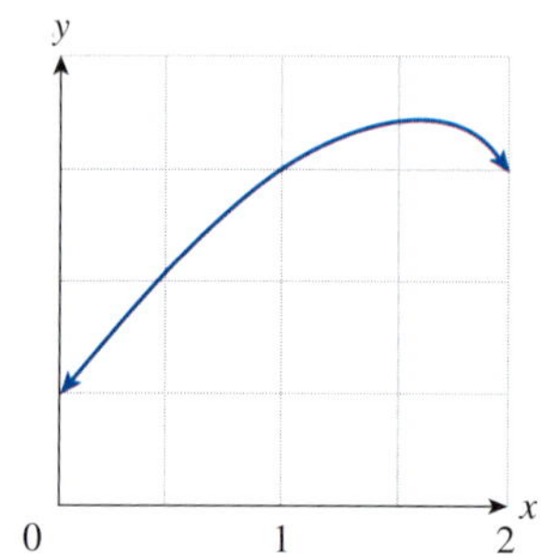

(B)

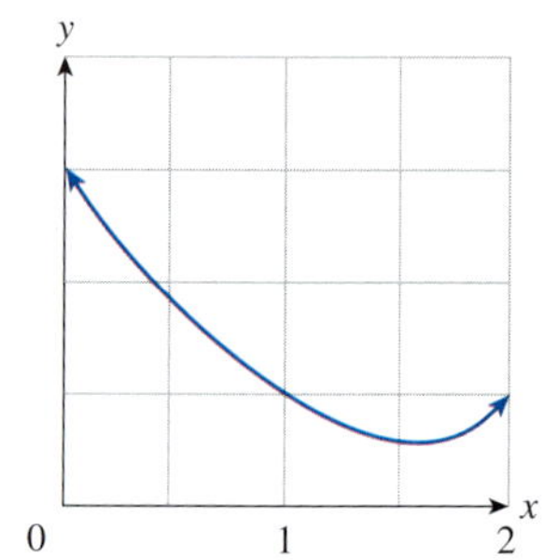

(C)

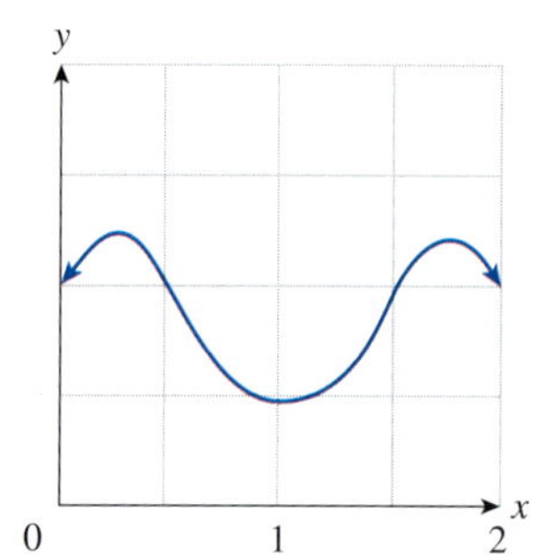

60. ▼ A certain function has the property that its average rate of change over the interval $[1, 1 + h]$ (for positive h) decreases as h decreases. Which of the following graphs could be the graph of f?

(A)

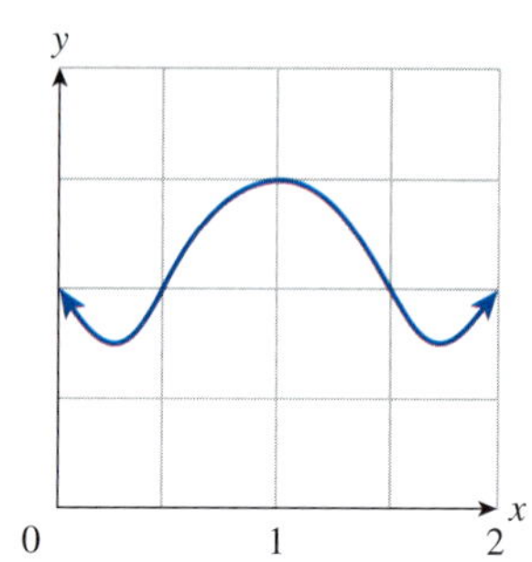

(B)

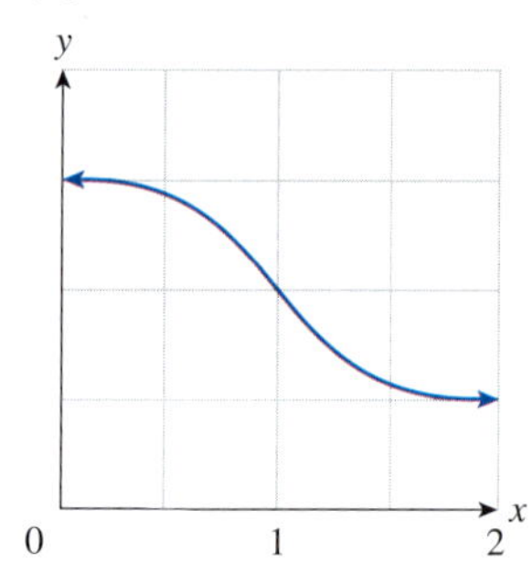

(C)

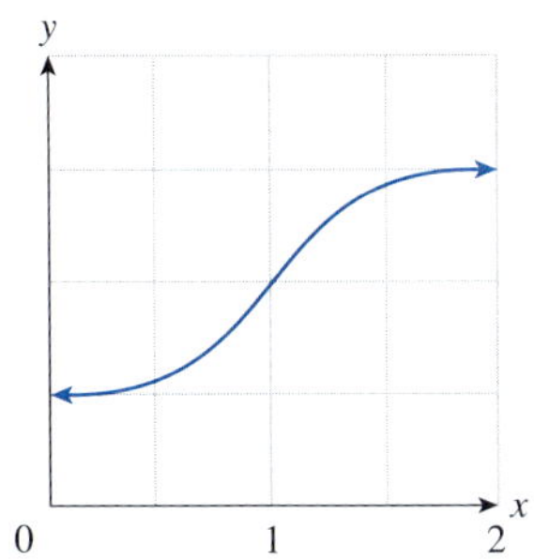

61. ▼ Is it possible for a company's revenue to have a negative 3-year average rate of growth, but a positive average rate of growth in 2 of the 3 years? (If not, explain; if so, illustrate with an example.)

62. ▼ Is it possible for a company's revenue to have a larger 2-year average rate of change than either of the 1-year average rates of change? (If not, explain why with the aid of a graph; if so, illustrate with an example.)

63. ◆ The average rate of change of f over $[1, 3]$ is

(A) always equal to
(B) never equal to
(C) sometimes equal to

the average of its average rates of change over $[1, 2]$ and $[2, 3]$.

64. ◆ The average rate of change of f over $[1, 4]$ is

(A) always equal to
(B) never equal to
(C) sometimes equal to

the average of its average rates of change over $[1, 2]$, $[2, 3]$, and $[3, 4]$.

3.5 Derivatives: Numerical and Graphical Viewpoints

In Example 4 of Section 3.4, we looked at the average rate of change of the function $G(t) = -8t^2 + 144t + 150$, approximating the price of gold on the New York Spot Market, over the intervals $[8, 8 + h]$ for successively smaller values of h. Here are some values we got:

h getting smaller; interval $[8, 8 + h]$ getting smaller →

h	1	0.1	0.01	0.001
Average Rate of Change Over [8, 8 + *h*]	8	15.2	15.92	15.992

Rate of change approaching \$16 per hour →

The average rates of change of the price of gold over smaller and smaller periods of time, starting at the instant $t = 8$ (8:00 AM), appear to be getting closer and closer to \$16 per hour. As we look at these shrinking periods of time, we are getting closer to looking at what happens at the *instant* $t = 8$. So it seems reasonable to say that the average rates of change are approaching the **instantaneous rate of change** at $t = 8$, which the table suggests is \$16 per hour. This is how fast the price of gold was changing *exactly* at 8:00 AM.

At $t = 8$, the instantaneous rate of change of $G(t)$ is 16.

We express this fact mathematically by writing $G'(8) = 16$ (which we read as "G prime of 8 equals 16"). Thus,

$G'(8) = 16$ *means that, at $t = 8$, the instantaneous rate of change of $G(t)$ is* 16.

The process of letting h get smaller and smaller is called taking the **limit** as h approaches 0 (as you recognize if you've done the sections on limits). We write $h \to 0$ as shorthand for "h approaches 0." Thus, taking the limit of the average rates of change as $h \to 0$ gives us the instantaneous rate of change.

Q: *All these intervals [8, 8 + h] are intervals to the right of 8. What about small intervals to the left of 8, such as [7.9, 8]?*

A: We can compute the average rate of change of our function for such intervals by choosing h to be negative ($h = -0.1, -0.01$, etc.) and using the same difference quotient formula we used for positive h:

$$\text{Average rate of change of } G \text{ over } [8+h, 8] = \frac{G(8) - G(8+h)}{8-(8+h)} = \frac{G(8+h) - G(8)}{h}.$$

Here are the results we get using negative h:

h getting closer to 0; interval [8 + h, 8] getting smaller →

h	−0.1	−0.01	−0.001	−0.0001
Average Rate of Change Over [8 + *h*, 8]	16.8	16.08	16.008	16.0008

Rate of change approaching \$16 per hour →

Notice that the average rates of change are again getting closer and closer to 16 as h approaches 0, suggesting once again that the instantaneous rate of change is \$16 per hour.

Instantaneous Rate of Change of *f*(*x*) at *x* = *a*: Derivative

The **instantaneous rate of change** of $f(x)$ at $x = a$ is defined as

$$f'(a) = \lim_{h\to 0} \frac{f(a+h) - f(a)}{h}.$$

f prime of a equals the limit as h approaches 0, of the ratio $\frac{f(a+h) - f(a)}{h}$

The quantity $f'(a)$ is also called the **derivative of *f*(*x*) at *x* = *a*.** Finding the derivative of f is called **differentiating *f*.**

Units

The units of $f'(a)$ are the same as the units of the average rate of change: units of f per unit of x.

Quick Examples

1. If $f(x) = -8x^2 + 144x + 150$, then the two previous tables suggest that
$$f'(8) = \lim_{h\to 0} \frac{f(8+h) - f(8)}{h} = 16.$$
2. If $f(t)$ is the number of insects in your room at time t hours, and we know that $f(3) = 5$ and $f'(3) = 8$, this means that, at time $t = 3$ hours, there are 5 insects in your room, and this number is growing at an instantaneous rate of 8 insects per hour.

IMPORTANT NOTES

1. Sections 3.1–3.3 discuss limits in some detail. If you have not (yet) covered those sections, you can trust to your intuition.
2. The formula for the derivative tells us that the instantaneous rate of change is the limit of the average rates of change $[f(a+h)-f(a)]/h$ over smaller and smaller intervals. Thus, the value of $f'(a)$ can be approximated by computing the average rate of change for smaller and smaller values of h, both positive and negative.*
3. In this section we will only *approximate* derivatives. In Section 3.6 we will begin to see how we find the *exact* values of derivatives.
4. $f'(a)$ is a number we can calculate, or at least approximate, for various values of a, as we have done in the earlier example. Since $f'(a)$ depends on the value of a, we can think of f' as *a function of* a. (We return to this idea at the end of this section.) An old name for f' is "the function *derived from* f," which has been shortened to the *derivative* of f.
5. It is because f' is a function that we sometimes refer to $f'(a)$ as "the derivative of f *evaluated* at a," or the "derivative of $f(x)$ evaluated at $x = a$."

* **NOTE** If a happens to be an endpoint of the domain of f, then $f'(a)$ can still exist, as the defining limit becomes a one-sided limit; for instance, if f has domain $[2, +\infty)$ then $f'(2)$ would be defined as

$$f'(2) = \lim_{h\to 0^+} \frac{f(2+h)-f(2)}{h}.$$

It may happen that the average rates of change $[f(a+h)-f(a)]/h$ do not approach any fixed number at all as h approaches zero, or that they approach one number on the intervals using positive h, and another on those using negative h. If this happens, $\lim_{h\to 0}[f(a+h)-f(a)]/h$ does not exist, and we say that f is **not differentiable** at $x = a$, or $f'(a)$ **does not exist**. When the limit *does* exist, we say that f is **differentiable** at the point $x = a$, or $f'(a)$ **exists**. It is comforting to know that all polynomials and exponential functions are differentiable at *every* point. On the other hand, certain functions are not differentiable. Examples are $f(x) = |x|$ and $f(x) = x^{1/3}$, neither of which is differentiable at $x = 0$. (See Section 4.1.)

EXAMPLE 1 Instantaneous Rate of Change: Numerically and Graphically

The air temperature one spring morning, t hours after 7:00 AM, was given by the function $f(t) = 50 + 0.1t^4$ degrees Fahrenheit $(0 \le t \le 4)$.

a. How fast was the temperature rising at 9:00 AM?

b. How is the instantaneous rate of change of temperature at 9:00 AM reflected in the graph of temperature vs. time?

Solution

a. We are being asked to find the instantaneous rate of change of the temperature at $t = 2$, so we need to find $f'(2)$. To do this we examine the average rates of change

$$\frac{f(2+h)-f(2)}{h}$$ Average rate of change = difference quotient

for values of h approaching 0. Calculating the average rate of change over $[2, 2+h]$ for $h = 1, 0.1, 0.01, 0.001$, and 0.0001 we get the following values (rounded to four decimal places):*

h	1	0.1	0.01	0.001	0.0001
Average Rate of Change Over [2, 2 + h]	6.5	3.4481	3.2241	3.2024	3.2002

* **NOTE** We can quickly compute these values using technology as in Example 4 in Section 3.4. (See the Technology Guides at the end of the chapter.)

Here are the values we get using negative values of h:

h	−1	−0.1	−0.01	−0.001	−0.0001
Average Rate of Change Over [2 + h, 2]	1.5	2.9679	3.1761	3.1976	3.1998

The average rates of change are clearly approaching the number 3.2, so we can say that $f'(2) = 3.2$. Thus, at 9:00 in the morning, the temperature was rising at the rate of 3.2 degrees per hour.

b. We saw in Section 3.4 that the average rate of change of f over an interval is the slope of the secant line through the corresponding points on the graph of f. Figure 24 illustrates this for the intervals $[2, 2 + h]$ with $h = 1, 0.5$, and 0.1.

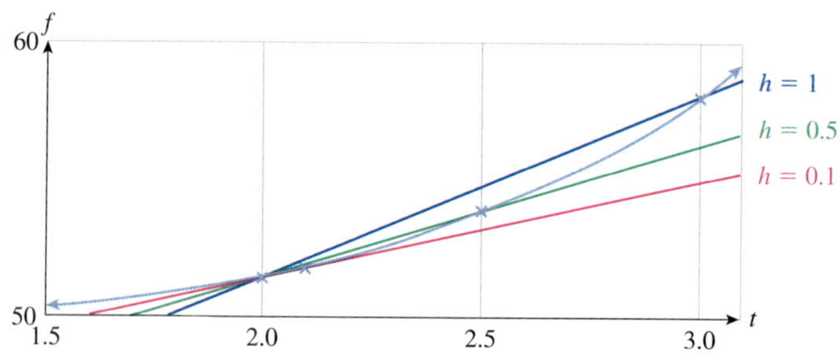

Figure 24

All three secant lines pass though the point $(2, f(2)) = (2, 51.6)$ on the graph of f. Each of them passes through a second point on the curve (the second point is different for each secant line) and this second point gets closer and closer to $(2, 51.6)$ as h gets closer to 0. What seems to be happening is that the secant lines are getting closer and closer to a line that just touches the curve at $(2, 51.6)$: the **tangent line** at $(2, 51.6)$, shown in Figure 25.

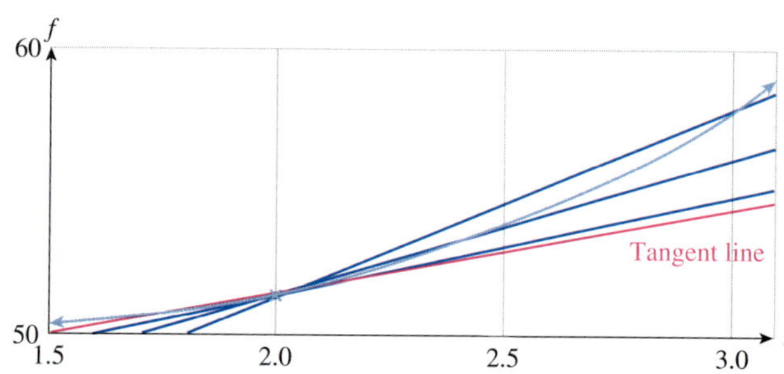

Figure 25

Q: *What is the slope of this tangent line?*

A: Because the slopes of the secant lines are getting closer and closer to 3.2, and because the secant lines are approaching the tangent line, the tangent line must have slope 3.2. In other words,

At the point on the graph where $x = 2$, the slope of the tangent line is $f'(2)$.

Be sure you understand the difference between $f(2)$ and $f'(2)$: Briefly, $f(2)$ is the *value of f* when $t = 2$, while $f'(2)$ is the *rate at which f is changing* when $t = 2$. Here,

$$f(2) = 50 + 0.1(2)^4 = 51.6 \text{ degrees.}$$

Thus, at 9:00 AM ($t = 2$), the temperature was 51.6 degrees. On the other hand,

$$f'(2) = 3.2 \text{ degrees per hour.}$$

Units of slope are units of f per unit of t.

This means that, at 9:00 AM ($t = 2$), the temperature was increasing at a rate of 3.2 degrees per hour.

Because we have been talking about tangent lines, we should say more about what they *are*. A tangent line to a *circle* is a line that touches the circle in just one point. A tangent line gives the circle "a glancing blow," as shown in Figure 26.

For a smooth curve other than a circle, a tangent line may touch the curve at more than one point, or pass through it (Figure 27).

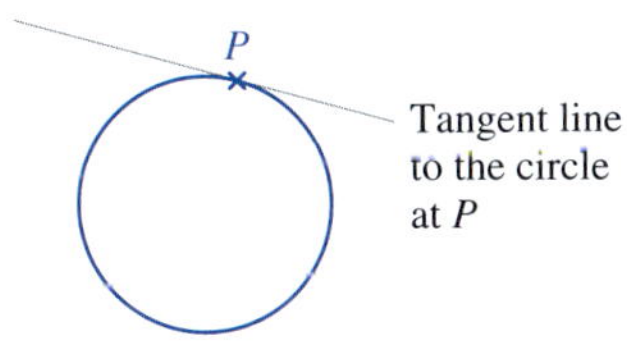

Figure 26

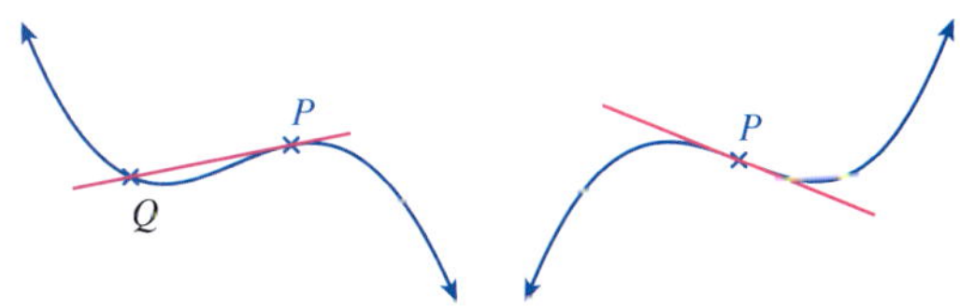

Tangent line at P intersects graph at Q

Tangent line at P passes through curve at P

Figure 27

However, all tangent lines have the following interesting property in common: If we focus on a small portion of the curve very close to the point P—in other words, if we "zoom in" to the graph near the point P—the curve will appear almost straight, and almost indistinguishable from the tangent line (Figure 28).

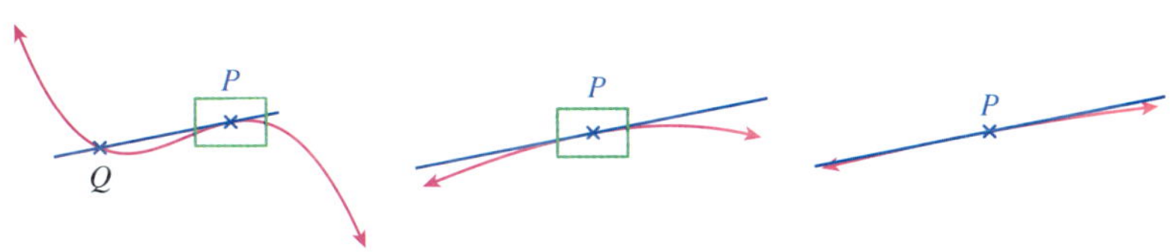

Original curve

Zoomed-in once

Zoomed-in twice

Figure 28

You can check this property by zooming in on the curve shown in Figures 24 and 25 in the previous example near the point where $x = 2$.

Secant and Tangent Lines

The *slope of the secant line* through the points on the graph of f where $x = a$ and $x = a + h$ is given by the average rate of change, or difference quotient,

$$m_{\text{sec}} = \text{slope of secant} = \text{average rate of change} = \frac{f(a+h) - f(a)}{h}.$$

The *slope of the tangent line* through the point on the graph of f where $x = a$ is given by the instantaneous rate of change, or derivative

$$m_{\tan} = \text{slope of tangent} = \text{derivative} = f'(a) = \lim_{h\to 0} \frac{f(a+h) - f(a)}{h}.$$

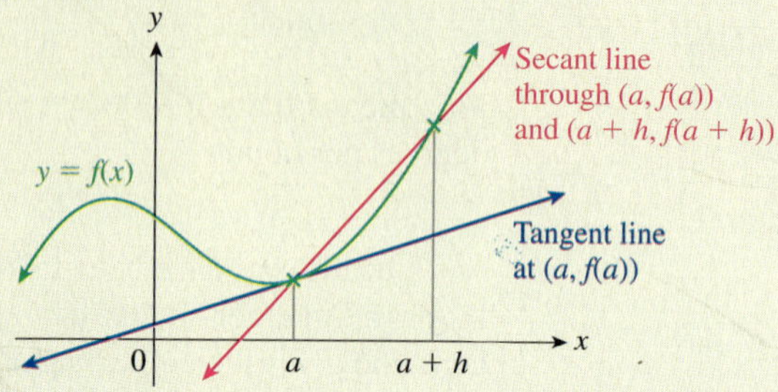

Quick Example

In the following graph, the tangent line at the point where $x = 2$ has slope 3. Therefore, the derivative at $x = 2$ is 3. That is, $f'(2) = 3$.

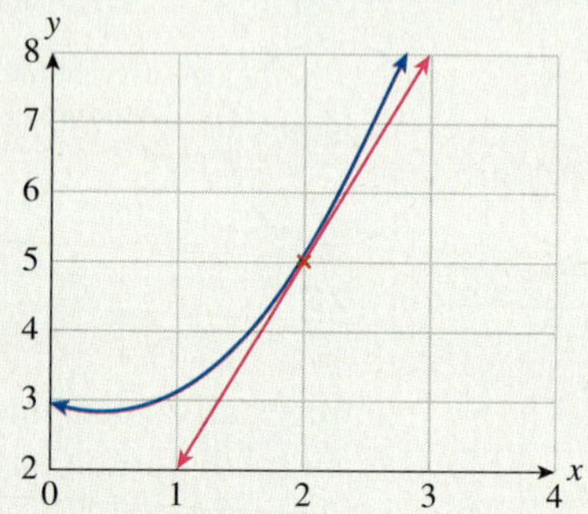

Note It might happen that the tangent line is vertical at some point or does not exist at all. These are the cases in which f is not differentiable at the given point. (See Section 3.6.) ■

We can now give a more precise definition of what we mean by the tangent line to a point P on the graph of f at a given point: The **tangent line** to the graph of f at the point $P(a, f(a))$ is the straight line passing through P with slope $f'(a)$.

Quick Approximation of the Derivative

Q: *Do we always need to make tables of difference quotients as above in order to calculate an approximate value for the derivative?*

A: We can usually *approximate* the value of the derivative by using a single, small value of h. In the example above, the value $h = 0.0001$ would have given a pretty good approximation. The problems with using a fixed value of h are that (1) we do not get an *exact* answer, only an *approximation* of the derivative, and (2) how good an approximation it is depends on the function we're differentiating.* However, with most of the functions we'll be considering, setting $h = 0.0001$ does give us a good approximation.

* **NOTE** In fact, no matter how small the value we decide to use for h, it is possible to craft a function f for which the difference quotient at a is not even close to $f'(a)$.

Calculating a Quick Approximation of the Derivative

We can calculate an approximate value of $f'(a)$ by using the formula

$$f'(a) \approx \frac{f(a+h) - f(a)}{h} \qquad \text{Rate of change over } [a,\ a+h]$$

with a small value of h. The value $h = 0.0001$ often works (but see the next example for a graphical way of determining a good value to use).

Alternative Formula: The Balanced Difference Quotient

The following alternative formula, which measures the rate of change of f over the interval $[a-h, a+h]$, often gives a more accurate result, and is the one used in many calculators:

$$f'(a) \approx \frac{f(a+h) - f(a-h)}{2h}. \qquad \text{Rate of change over } [a-h,\ a+h]$$

Note For the quick approximations to be valid, the function f must be differentiable; that is, $f'(a)$ must exist. ■

EXAMPLE 2 Quick Approximation of the Derivative

a. Calculate an approximate value of $f'(1.5)$ if $f(x) = x^2 - 4x$.

b. Find the equation of the tangent line at the point on the graph where $x = 1.5$.

Solution

a. We shall compute both the ordinary difference quotient and the balanced difference quotient.

Ordinary Difference Quotient: Using $h = 0.0001$, the ordinary difference quotient is:

$$f'(1.5) \approx \frac{f(1.5 + 0.0001) - f(1.5)}{0.0001} \qquad \text{Usual difference quotient}$$

$$= \frac{f(1.5001) - f(1.5)}{0.0001}$$

$$= \frac{(1.5001^2 - 4 \times 1.5001) - (1.5^2 - 4 \times 1.5)}{0.0001} = -0.9999$$

This answer is accurate to 0.0001; in fact, $f'(1.5) = -1$.

Graphically, we can picture this approximation as follows: Zoom in on the curve using the window $1.5 \le x \le 1.5001$ and measure the slope of the secant line joining both ends of the curve segment. Figure 29 shows close-up views of the curve and tangent line near the point P in which we are interested, the third view being the zoomed-in view used for this approximation.

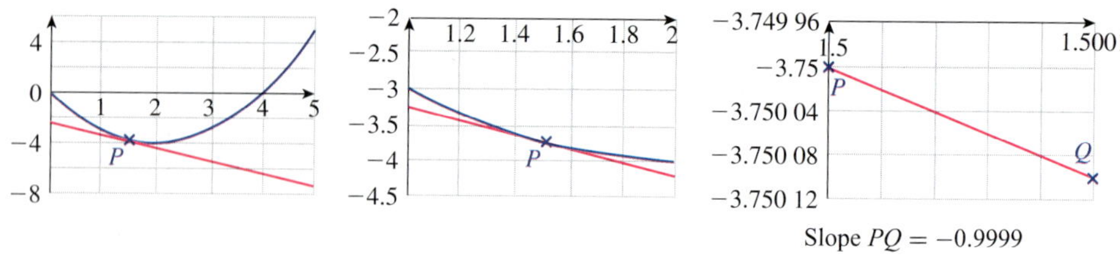

Figure 29

Notice that in the third window the tangent line and curve are indistinguishable. Also, the point P in which we are interested is on the left edge of the window.

Balanced Difference Quotient: For the balanced difference quotient, we get

$$f'(1.5) \approx \frac{f(1.5 + 0.0001) - f(1.5 - 0.0001)}{2(0.0001)} \qquad \text{Balanced difference quotient}$$

$$= \frac{f(1.5001) - f(1.4999)}{0.0002}$$

$$= \frac{(1.5001^2 - 4 \times 1.5001) - (1.4999^2 - 4 \times 1.4999)}{0.0002} = -1.$$

*** NOTE** The balanced difference quotient always gives the exact derivative for a quadratic function.

This balanced difference quotient gives the exact answer in this case!* Graphically, it is as though we have zoomed in using a window that puts the point P in the *center* of the screen (Figure 30) rather than at the left edge.

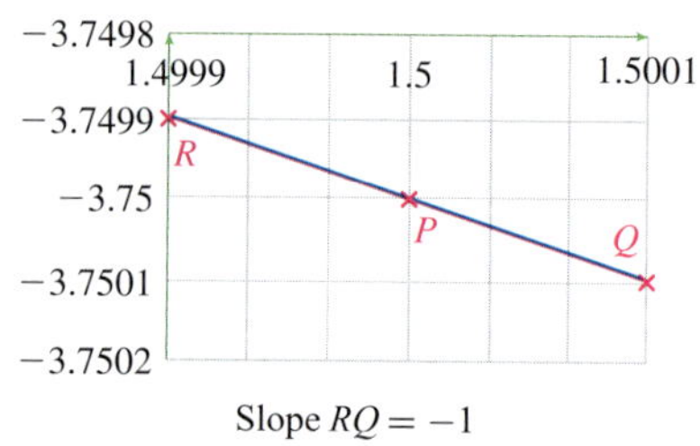

Figure 30

b. We find the equation of the tangent line from a point on the line and its slope, as we did in Chapter 1:

- ***Point*** $(1.5, f(1.5)) = (1.5, -3.75)$.
- ***Slope*** $m = f'(1.5) = -1$. Slope of the tangent line = derivative.

The equation is

$$y = mx + b$$

where $m = -1$ and $b = y_1 - mx_1 = -3.75 - (-1)(1.5) = -2.25$. Thus, the equation of the tangent line is

$$y = -x - 2.25.$$

using Technology

See the Technology Guides at the end of the chapter to find out how to calculate the quick approximations to the derivative in Example 2 using a TI-83/84 Plus or Excel. Here is an outline:

TI-83/84 Plus

`Y1=X^2-4*X`

Home screen:

`(Y1(1.5001)-Y1(1.5))/0.0001`

`(Y1(1.5001)-Y1(1.4999))/0.0002`

[More details on page 281.]

Excel

Headings *a*, *h*, *x*, *f*(*x*), Diff Quotient, Balanced Diff Quotient in A1–F1

`1.5` in A2, `0.0001` in B2, `=A2-B2` in C2, `=A2` in C3, `=A2+B2` in C4

`=C2^2-4*C2` in D2; copy down to D4

`=(D3-D2)/(C3-C2)` in E2

`=(D4-D2)/(C4-C2)` in E3

[More details on page 283.]

Q: *Why can't we simply put* $h = 0.000\,000\,000\,000\,000\,000\,01$ *for an incredibly accurate approximation to the instantaneous rate of change and be done with it?*

A: This approach would certainly work if you were patient enough to do the (thankless) calculation by hand! However, doing it with the help of technology—even an ordinary calculator—will cause problems: The issue is that calculators and spreadsheets represent numbers with a maximum number of significant digits (15 in the case of Excel). As the value of h gets smaller, the value of $f(a+h)$ gets closer and closer to the value of $f(a)$. For example, if $f(x) = 50 + 0.1x^4$, Excel might compute

$$\begin{aligned} &f(2 + 0.000\,000\,000\,000\,1) - f(2) \\ &\quad = 51.600\,000\,000\,000\,3 - 51.6 \qquad \text{Rounded to 15 digits} \\ &\quad = 0.000\,000\,000\,000\,3 \end{aligned}$$

and the corresponding difference quotient would be 3, not 3.2 as it should be. If h gets even smaller, Excel will not be able to distinguish between $f(a+h)$ and $f(a)$ at all, in which case it will compute 0 for the rate of change. This loss in accuracy when subtracting two very close numbers is called **subtractive error**.

Thus, there is a trade-off in lowering the value of h: smaller values of h yield *mathematically* more accurate approximations of the derivative, but if h gets too small, subtractive error becomes a problem and decreases the accuracy of computations that use technology.

Leibniz *d* Notation

We introduced the notation $f'(x)$ for the derivative of f at x, but there is another interesting notation. We have written the average rate of change as

$$\text{Average rate of change} = \frac{\Delta f}{\Delta x}. \qquad \frac{\text{Change in } f}{\text{Change in } x}$$

As we use smaller and smaller values for Δx, we approach the instantaneous rate of change, or derivative, for which we also have the notation df/dx, due to Leibniz:

$$\text{Instantaneous rate of change} = \lim_{\Delta x \to 0} \frac{\Delta f}{\Delta x} = \frac{df}{dx}.$$

That is, df/dx is just another notation for $f'(x)$. Do not think of df/dx as an actual quotient of two numbers: remember that we only use an actual quotient $\Delta f/\Delta x$ to *approximate* the value of df/dx.

In Example 3, we apply the quick approximation method of estimating the derivative.

EXAMPLE 3 Velocity

My friend Eric, an enthusiastic baseball player, claims he can "probably" throw a ball upward at a speed of 100 feet per second (ft/s).* Our physicist friends tell us that its height s (in feet) t seconds later would be $s = 100t - 16t^2$. Find its average velocity over the interval [2, 3] and its instantaneous velocity exactly 2 seconds after Eric throws it.

* **NOTE** Eric's claim is difficult to believe; 100 ft/s corresponds to around 68 mph, and professional pitchers can throw *forward* at about 100 mph.

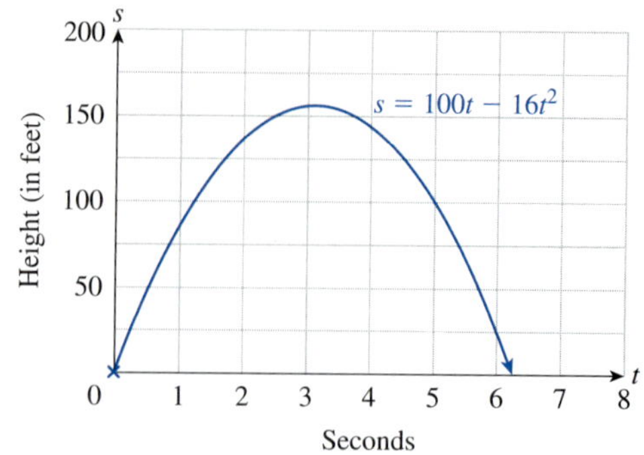

Figure 31

Solution The graph of the ball's height as a function of time is shown in Figure 31. Asking for the velocity is really asking for the rate of change of height with respect to time. (Why?) Consider average velocity first. To compute the **average velocity** of the ball from time 2 to time 3, we first compute the change in height:

$$\Delta s = s(3) - s(2) = 156 - 136 = 20 \text{ ft}$$

Since it rises 20 feet in $\Delta t = 1$ second, we use the defining formula *speed* = *distance*/*time* to get the average velocity:

$$\text{Average velocity} = \frac{\Delta s}{\Delta t} = \frac{20}{1} = 20 \text{ ft/s}$$

from time $t = 2$ to $t = 3$. This is just the difference quotient, so

The average velocity is the average rate of change of height.

To get the **instantaneous velocity** at $t = 2$, we find the instantaneous rate of change of height. In other words, we need to calculate the derivative ds/dt at $t = 2$. Using the balanced quick approximation described earlier, we get

$$\begin{aligned}\frac{ds}{dt} &\approx \frac{s(2+0.0001) - s(2-0.0001)}{2(0.0001)} \\ &= \frac{s(2.0001) - s(1.9999)}{0.0002} \\ &= \frac{100(2.0001) - 16(2.0001)^2 - (100(1.9999) - 16(1.9999)^2)}{0.0002} \\ &= 36 \text{ ft/s}.\end{aligned}$$

In fact, this happens to be the exact answer; the instantaneous velocity at $t = 2$ is exactly 36 ft/s. (Try an even smaller value of h to persuade yourself.)

➡ Before we go on... If we repeat the calculation in Example 3 at time $t = 5$, we get

$$\frac{ds}{dt} = -60 \text{ ft/s}$$

The negative sign tells us that the ball is *falling* at a rate of 60 feet per second at time $t = 5$. (How does the fact that it is falling at $t = 5$ show up on the graph?) ■

The preceding example gives another interpretation of the derivative.

Average and Instantaneous Velocity

For an object moving in a straight line with position $s(t)$ at time t, the **average velocity** from time t to time $t+h$ is the average rate of change of position with respect to time:

$$v_{ave} = \frac{s(t+h)-s(t)}{h} = \frac{\Delta s}{\Delta t}$$

Average velocity = Average rate of change of position

The **instantaneous velocity** at time t is

$$v = \lim_{h \to 0} \frac{s(t+h)-s(t)}{h} = \frac{ds}{dt}$$

Instantaneous velocity = Instantaneous rate of change of position

In other words, *instantaneous velocity is the derivative of position with respect to time.*

Here is one last comment on Leibniz notation. In Example 3, we could have written the velocity either as s' or as ds/dt, as we chose to do. To write the answer to the question, that the velocity at $t = 2$ sec was 36 ft/s, we can write either

$$s'(2) = 36$$

or

$$\left.\frac{ds}{dt}\right|_{t=2} = 36.$$

The notation "$|_{t=2}$" is read "evaluated at $t = 2$." Similarly, if $y = f(x)$, we can write the instantaneous rate of change of f at $x = 5$ in either functional notation as

$$f'(5)$$ The derivative of f, evaluated at $x = 5$

or in Leibniz notation as

$$\left.\frac{dy}{dx}\right|_{x=5}.$$ The derivative of y, evaluated at $x = 5$

The latter notation is obviously more cumbersome than the functional notation $f'(5)$, but the notation dy/dx has compensating advantages. You should practice using both notations.

The Derivative Function

The derivative $f'(x)$ is a number we can calculate, or at least approximate, for various values of x. Because $f'(x)$ depends on the value of x, we may think of f' as a function of x. This function is the **derivative function**.

Derivative Function

If f is a function, its **derivative function** f' is the function whose value $f'(x)$ is the derivative of f at x. Its domain is the set of all x at which f is differentiable. Equivalently, f' associates to each x the slope of the tangent to the graph of the function f at x, or the rate of change of f at x. The formula for the derivative function is

$$f'(x) = \lim_{h \to 0} \frac{f(x+h) - f(x)}{h}. \qquad \text{Derivative function}$$

Quick Examples

1. Let $f(x) = 3x - 1$. The graph of f is a straight line that has slope 3 everywhere. In other words, $f'(x) = 3$ for every choice of x; that is, f' is a constant function.

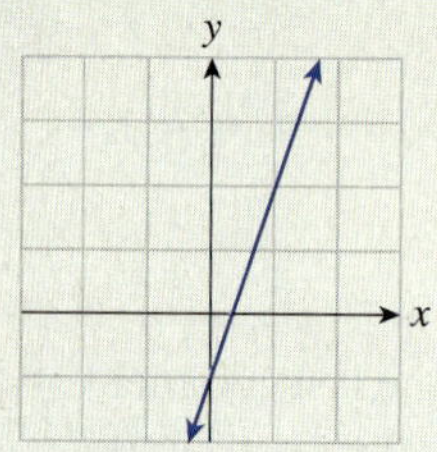

Original Function f

$f(x) = 3x - 1$

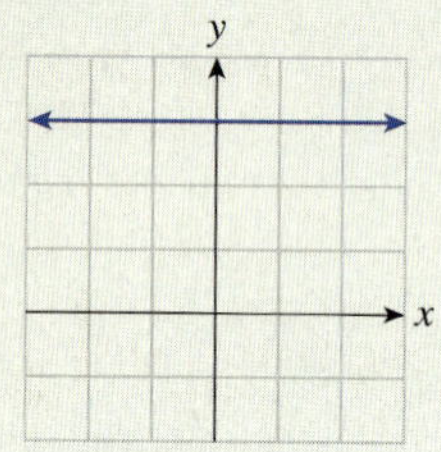

Derivative Function f'

$f'(x) = 3$

2. Given the graph of a function f, we can get a rough sketch of the graph of f' by estimating the slope of the tangent to the graph of f at several points, as illustrated below.*

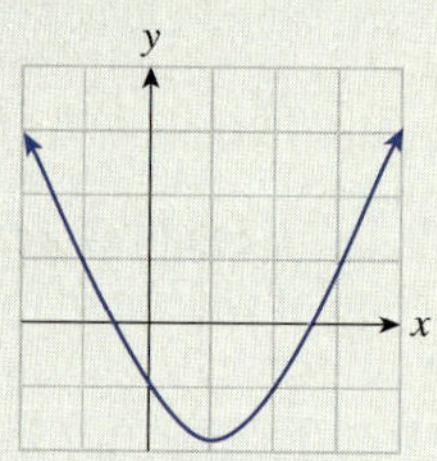

Original Function f

$y = f(x)$

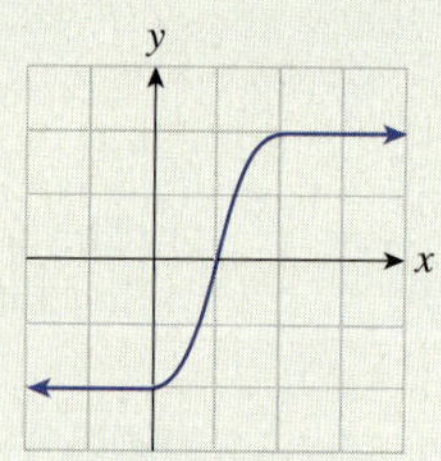

Derivative Function f'

$y = f'(x)$

For x between -2 and 0, the graph of f is linear with slope -2. As x increases from 0 to 2, the slope increases from -2 to 2. For x larger than 2, the graph of f is linear with slope 2. (Notice that, when $x = 1$, the graph of f has a horizontal tangent, so $f'(1) = 0$.)

3. Look again at the graph on the left in Quick Example 2. When $x < 1$ the derivative $f'(x)$ is negative, so the graph has negative slope and f is **decreasing**; its values are going down as x increases. When $x > 1$ the derivative $f'(x)$ is positive, so the graph has positive slope and f is **increasing**; its values are going up as x increases.

* **NOTE** This method is discussed in detail on the Web site at Web site → Online Text → Sketching the Graph of the Derivative.

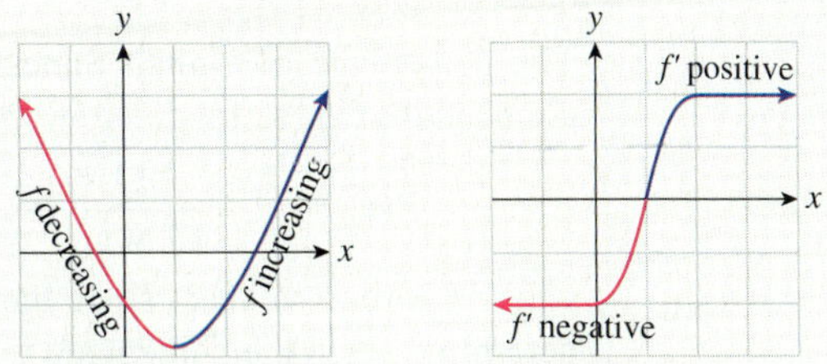

f decreasing for $x < 1$
f increasing for $x > 1$

f' negative for $x < 1$
f' positive for $x > 1$

The following example shows how we can use technology to graph the (approximate) derivative of a function, where it exists.

using Technology

See the Technology Guides at the end of the chapter to find out how to obtain a table of values of and graph the derivative in Example 4 using a TI-83/84 Plus or Excel. Here is an outline:

TI-83/84 Plus

```
Y1=-2X^2+6X+5
Y2=nDeriv(Y1,X,X)
```

[More details on page 281.]

Excel

Value of h in E2
Values of x from A2 down increasing by h
`-2*A2^2+6*A2+5` from B2 down `=(B3-B2)/$E$2` from C2 down
Insert scatter chart using columns A and C [More details on page 283.]

Web Site

www.AppliedCalc.org
Web grapher:
Student Home→Online Utilities→ Function Evaluator and Grapher
Enter `deriv(-2*x^2+6*x+5)` for y_1. Alternatively, enter `-2*x^2+6*x+5` for y_1 and `deriv(y1)` for y_2.
Excel grapher:
Student Home→Online Utilities→Excel First and Second Derivative Graphing Utility
Function: `-2*x^2+6*x+5`

EXAMPLE 4 Graphing the Derivative with Technology

Use technology to graph of the derivative of $f(x) = -2x^2 + 6x + 5$ for values of x starting at -5.

Solution The TI-83/84 Plus has a built-in function that approximates the derivative, and we can use it to graph the derivative of a function. In Excel, we need to create the approximation using one of the quick approximation formulas and we can then graph a table of its values. See the Technology Guides at the end of the chapter to find out how to graph the derivative (Figure 32) using the TI-83/84 Plus and Excel.

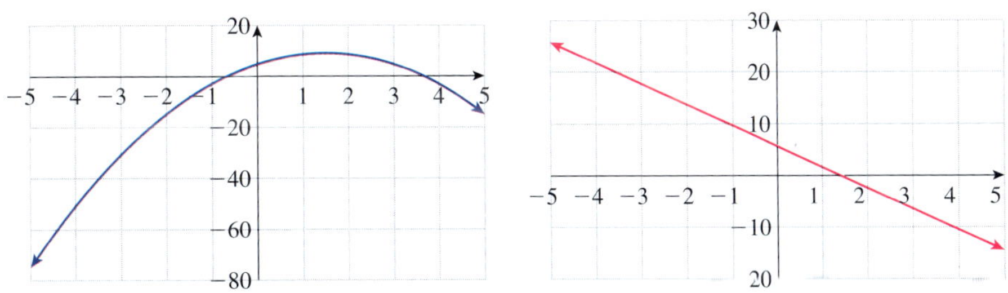

Graph of f Graph of f'

Figure 32

We said that f' records the slope of (the tangent line to) the function f at each point. Notice that the graph of f' confirms that the slope of the graph of f is decreasing as x increases from -5 to 5. Note also that the graph of f reaches a high point at $x = 1.5$ (the vertex of the parabola). At that point, the slope of the tangent is zero; that is, $f'(1.5) = 0$ as we see in the graph of f'.

EXAMPLE 5 T An Application: Broadband

The percentage of United States Internet-connected households that have broadband connections can be modeled by the logistic function

$$P(t) = \frac{100}{1 + 10.8(1.7)^{-t}} \qquad (0 \le t \le 9)$$

where t is time in years since the start of 2000.* Graph both P and its derivative, and determine when the percentage was growing most rapidly.

Solution Using one of the methods in Example 4, we obtain the graphs shown in Figure 33.

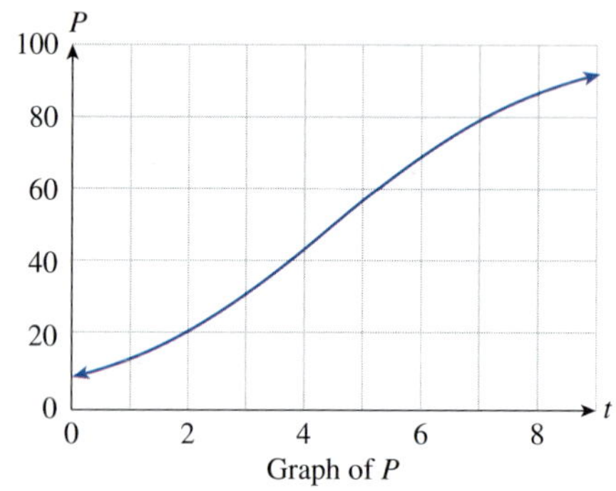

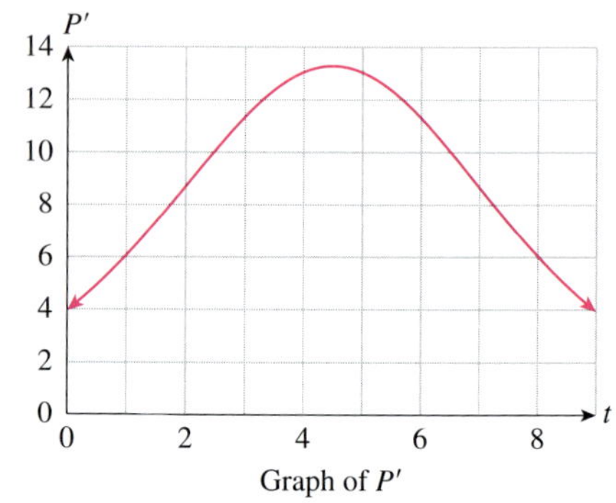

Figure 33
Technology formula: `100/(1+10.8*1.7^(-x))`

From the graph on the right, we see that P' reaches a peak somewhere between $t = 4$ and $t = 5$ (sometime during 2004). Recalling that P' measures the *slope* of the graph of P, we can conclude that the graph of P is steepest between $t = 4$ and $t = 5$, indicating that, according to the model, the percentage of Internet-connected households with broadband was growing most rapidly during 2004. Notice that this is not so easy to see directly on the graph of P.

To determine the point of maximum growth more accurately, we can zoom in on the graph of P' using the range $4 \le t \le 5$ (Figure 34).

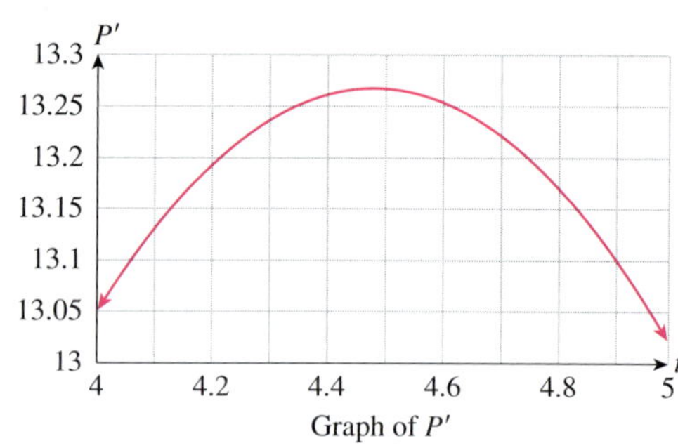

Figure 34

We can now see that P' reaches its highest point around $t = 4.5$, so we conclude that the percentage of Internet-connected households with broadband was growing most rapidly midway through 2004.

*See Example 2 in Section 2.4. Source for data: www.Nielsen.com.

➡ **Before we go on...** Besides helping us to determine the point of maximum growth, the graph of P' in Example 5 gives us a great deal of additional information. As just one example, in Figure 34 we can see that the maximum value of P' is just under 13.3, indicating that the percentage of Internet-connected households that have broadband never grew at a fastest rate of about 13.3 percentage points per quarter. ■

FAQs

Recognizing When and How to Compute the Instantaneous Rate of Change

Q: *How do I know, by looking at the wording of a problem, that it is asking for an instantaneous rate of change?*

A: If a problem does not ask for an instantaneous rate of change directly, it might do so indirectly, as in "How fast is quantity q increasing?" or "Find the rate of increase of q."

Q: *If I know that a problem calls for estimating an instantaneous rate of change, how should I estimate it: with a table showing smaller and smaller values of h, or by using a quick approximation?*

A: For most practical purposes, a quick approximation is accurate enough. Use a table showing smaller and smaller values of h when you would like to check the accuracy.

Q: *Which should I use in computing a quick approximation: the balanced difference quotient or the ordinary difference quotient?*

A: In general, the balanced difference quotient gives a more accurate answer.

Web Site

At the student Web site at www.AppliedCalc.org you can find the following optional interactive online sections:

- **Continuity and Differentiability**
- **Sketching the Graph of the Derivative**

You can find these sections by following

Chapter 2 → Online Text

3.5 EXERCISES

▼ more advanced ◆ challenging

T indicates exercises that should be solved using technology

In Exercises 1–4, estimate the derivative from the table of average rates of change. HINT [See discussion at the beginning of the section.]

1.

h	1	0.1	0.01	0.001	0.0001
Average Rate of Change of f Over $[5, 5+h]$	12	6.4	6.04	6.004	6.0004
h	−1	−0.1	−0.01	−0.001	−0.0001
Average Rate of Change of f Over $[5+h, 5]$	3	5.6	5.96	5.996	5.9996

Estimate $f'(5)$

2.

h	1	0.1	0.01	0.001	0.0001
Average Rate of Change of g Over $[7, 7+h]$	4	4.8	4.98	4.998	4.9998
h	−1	−0.1	−0.01	−0.001	−0.0001
Average Rate of Change of g Over $[7+h, 7]$	5	5.3	5.03	5.003	5.0003

Estimate $g'(7)$.

3.

h	1	0.1	0.01	0.001	0.0001
Average Rate of Change of r Over $[-6, -6+h]$	−5.4	−5.498	−5.4998	−5.499982	−5.49999822
h	−1	−0.1	−0.01	−0.001	−0.0001
Average Rate of Change of r Over $[-6+h, -6]$	−7.52	−6.13	−5.5014	−5.5000144	−5.500001444

Estimate $r'(-6)$.

4.

h	1	0.1	0.01	0.001	0.0001
Average Rate of Change of s Over $[0, h]$	−2.52	−1.13	−0.6014	−0.6000144	−0.600001444
h	−1	−0.1	−0.01	−0.001	−0.0001
Average Rate of Change of s Over $[h, 0]$	−0.4	−0.598	−0.5998	−0.599982	−0.59999822

Estimate $s'(0)$.

Consider the functions in Exercises 5–8 as representing the value of an ounce of palladium in U.S. dollars as a function of the time t in days.[45] *Find the average rates of change of R(t) over the time intervals [t, t + h], where t is as indicated and h = 1, 0.1, and 0.01 days. Hence, estimate the instantaneous rate of change of R at time t, specifying the units of measurement. (Use smaller values of h to check your estimates.)* HINT [See Example 1.]

5. $R(t) = 60t + 50t - t^2; t = 5$

6. $R(t) = 60t - 2t^2; t = 3$

7. $R(t) = 270 + 20t^3; t = 1$

8. $R(t) = 200 + 50t - t^3; t = 2$

Each of the functions in Exercises 9–12 gives the cost to manufacture x items. Find the average cost per unit of manufacturing h more items (i.e., the average rate of change of the total cost) at a production level of x, where x is as indicated and h = 10 and 1. Hence, estimate the instantaneous rate of change of the total cost at the given production level x, specifying the units of measurement. (Use smaller values of h to check your estimates.) HINT [See Example 1.]

9. $C(x) = 10{,}000 + 5x - \frac{x^2}{10{,}000}; x = 1{,}000$

10. $C(x) = 20{,}000 + 7x - \frac{x^2}{20{,}000}; x = 10{,}000$

11. $C(x) = 15{,}000 + 100x + \frac{1{,}000}{x};\ x = 100$

12. $C(x) = 20{,}000 + 50x + \frac{10{,}000}{x}; x = 100$

In Exercises 13–16, the graph of a function is shown together with the tangent line at a point P. Estimate the derivative of f at the corresponding x value. HINT [See Quick Example page 244.]

13.

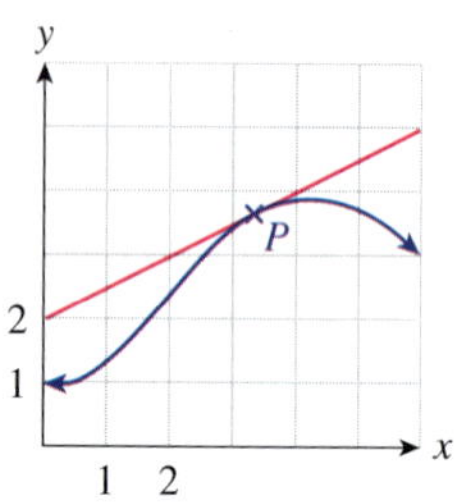

14.

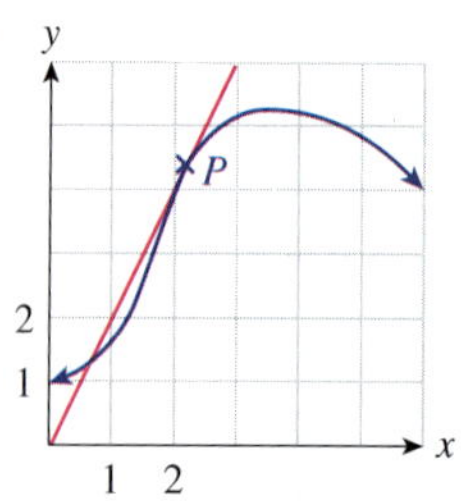

15.

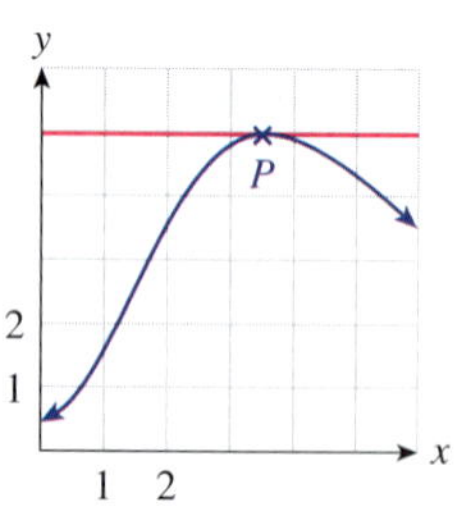

16.

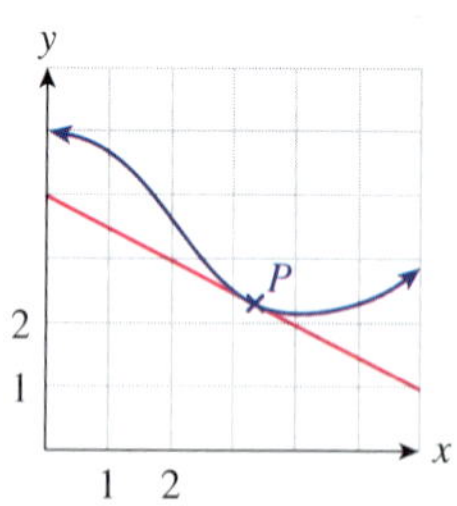

In each of the graphs given in Exercises 17–22, say at which labeled point the slope of the tangent is ***(a)*** *greatest and* ***(b)*** *least (in the sense that −7 is less than 1).* HINT [See Quick Example page 244.]

17.

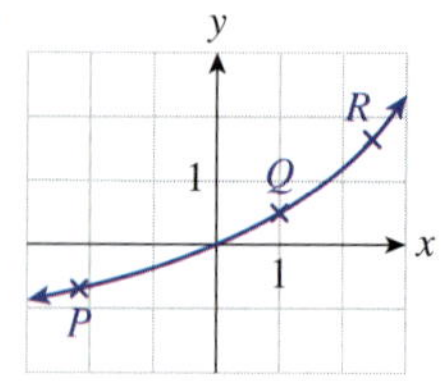

18.

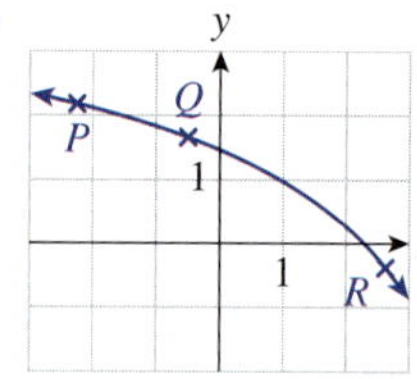

19.

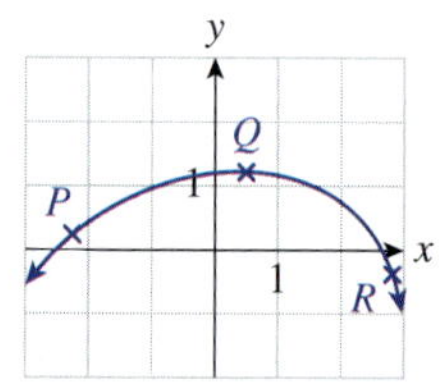

20.

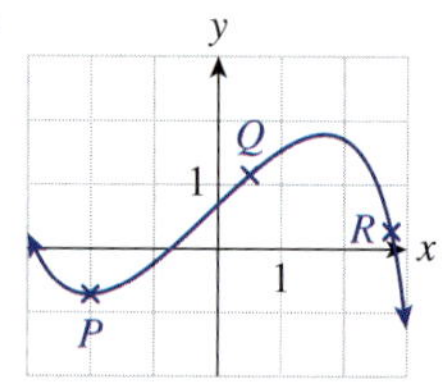

21.

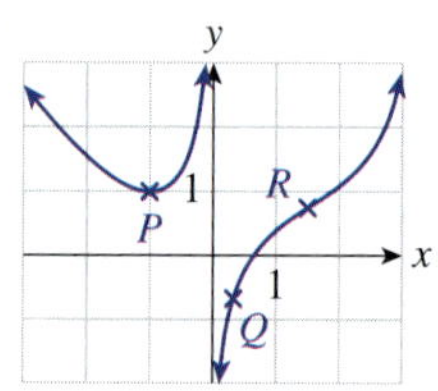

22.

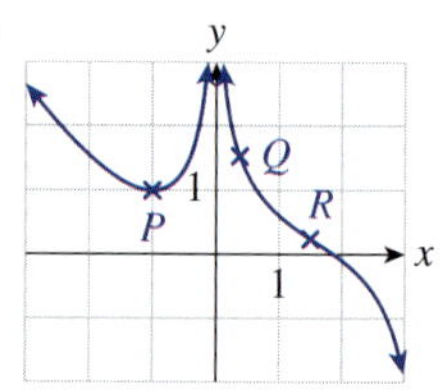

In each of Exercises 23–26, three slopes are given. For each slope, determine at which of the labeled points on the graph the tangent line has that slope.

23. a. 0 **b.** 4 **c.** −1

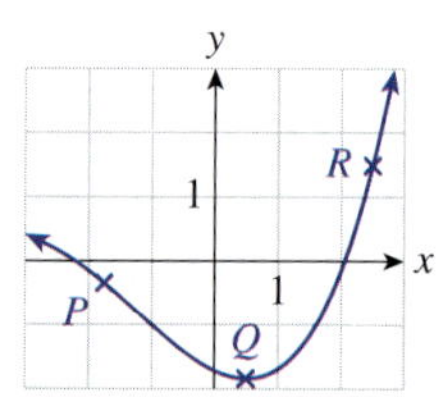

24. a. 0 **b.** 1 **c.** −1

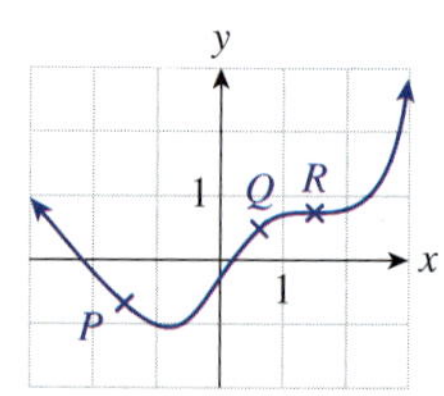

25. a. 0 **b.** 3 **c.** −3

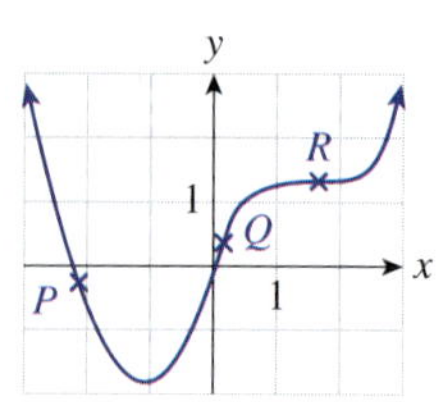

26. a. 0 **b.** 3 **c.** 1

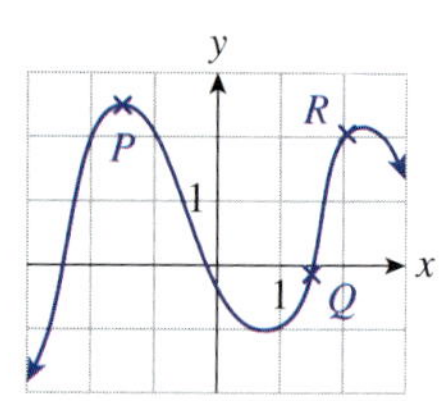

[45] Palladium was trading at around $290 in August 2008.

In each of Exercises 27–30, find the approximate coordinates of all points (if any) where the slope of the tangent is: ***(a)*** *0,* ***(b)*** *1,* ***(c)*** *−1.* HINT [See Quick Example page 244.]

27.

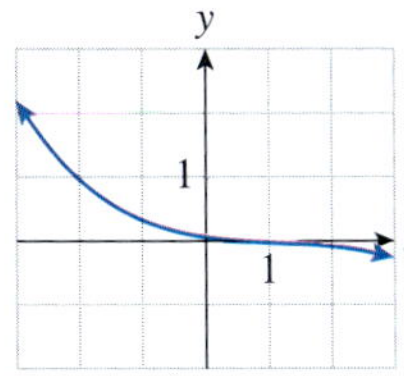

28.

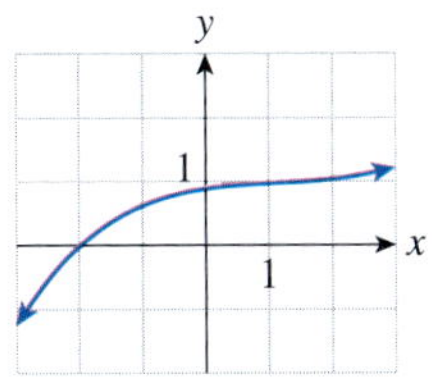

29.

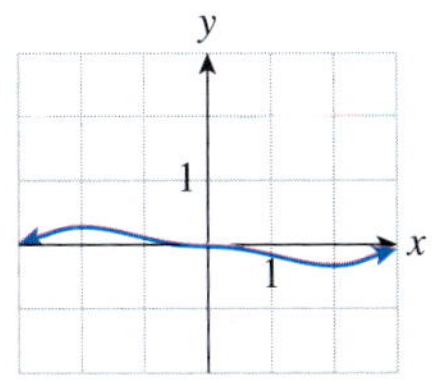

30.

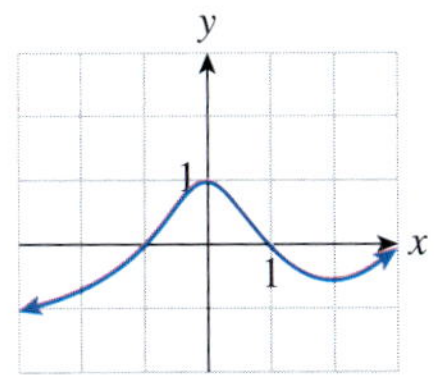

31. Complete the following: The tangent to the graph of the function f at the point where $x = a$ is the line passing through the point ____ with slope ____ .

32. Complete the following: The difference quotient for f at the point where $x = a$ gives the slope of the ____ line that passes through _____ .

33. Which is correct? The derivative function assigns to each value x

(A) the average rate of change of f at x
(B) the slope of the tangent to the graph of f at $(x, f(x))$
(C) the rate at which f is changing over the interval $[x, x + h]$ for $h = 0.0001$
(D) the balanced difference quotient $[f(x + h) - f(x - h)]/(2h)$ for $h \approx 0.0001$

34. Which is correct? The derivative function $f'(x)$ tells us
(A) the slope of the tangent line at each of the points $(x, f(x))$
(B) the approximate slope of the tangent line at each of the points $(x, f(x))$
(C) the slope of the secant line through $(x, f(x))$ and $(x + h, f(x + h))$ for $h = 0.0001$
(D) the slope of a certain secant line through each of the points $(x, f(x))$

35. ▼ Let f have the graph shown.

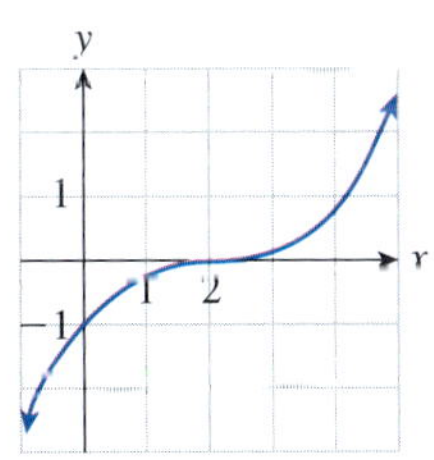

a. The average rate of change of f over the interval [2, 4] is
(A) greater than $f'(2)$ **(B)** less than $f'(2)$
(C) approximately equal to $f'(2)$

b. The average rate of change of f over the interval [−1, 1] is
(A) greater than $f'(0)$ **(B)** less than $f'(0)$
(C) approximately equal to $f'(0)$

c. Over the interval [0, 2], the instantaneous rate of change of f is
(A) increasing **(B)** decreasing **(C)** neither

d. Over the interval [0, 4], the instantaneous rate of change of f is
(A) increasing, then decreasing
(B) decreasing, then increasing
(C) always increasing
(D) always decreasing

e. When $x = 4$, $f(x)$ is
(A) approximately 0, and increasing at a rate of about 0.7 units per unit of x
(B) approximately 0, and decreasing at a rate of about 0.7 units per unit of x
(C) approximately 0.7, and increasing at a rate of about 1 unit per unit of x
(D) approximately 0.7, and increasing at a rate of about 3 units per unit of x

36. ▼ A function f has the following graph.

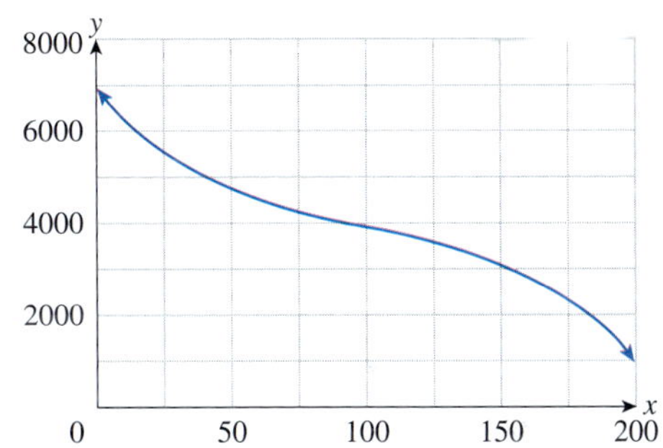

a. The average rate of change of f over [0, 200] is
(A) greater than
(B) less than
(C) approximately equal to
the instantaneous rate of change at $x = 100$.

b. The average rate of change of f over [0, 200] is
(A) greater than
(B) less than
(C) approximately equal to
the instantaneous rate of change at $x = 150$.

c. Over the interval [0, 50] the instantaneous rate of change of f is
(A) increasing, then decreasing
(B) decreasing, then increasing
(C) always increasing
(D) always decreasing

d. On the interval [0, 200], the instantaneous rate of change of f is
(A) always positive **(B)** always negative
(C) negative, positive, and then negative

e. $f'(100)$ is
(A) greater than $f'(25)$ **(B)** less than $f'(25)$
(C) approximately equal to $f'(25)$

In Exercises 37–40, use a quick approximation to estimate the derivative of the given function at the indicated point. HINT [See Example 2(a).]

37. $f(x) = 1 - 2x; x = 2$ **38.** $f(x) = \frac{x}{3} - 1; x = -3$

39. $f(x) = \frac{x^2}{4} - \frac{x^3}{3}; x = -1$ **40.** $f(x) = \frac{x^2}{2} + \frac{x}{4}; x = 2$

In Exercises 41–48, estimate the indicated derivative by any method. HINT [See Example 2.]

41. $g(t) = \frac{1}{t^5}$; estimate $g'(1)$

42. $s(t) = \frac{1}{t^3}$; estimate $s'(-2)$

43. $y = 4x^2$; estimate $\left.\frac{dy}{dx}\right|_{x=2}$

44. $y = 1 - x^2$; estimate $\left.\frac{dy}{dx}\right|_{x=-1}$

45. $s = 4t + t^2$; estimate $\left.\frac{ds}{dt}\right|_{t=-2}$

46. $s = t - t^2$; estimate $\left.\frac{ds}{dt}\right|_{t=2}$

47. $R = \frac{1}{p}$; estimate $\left.\frac{dR}{dp}\right|_{p=20}$

48. $R = \sqrt{p}$; estimate $\left.\frac{dR}{dp}\right|_{p=400}$

In Exercises 49–54, ***(a)*** *use any method to estimate the slope of the tangent to the graph of the given function at the point with the given x-coordinate and* ***(b)*** *find an equation of the tangent line in part (a). In each case, sketch the curve together with the appropriate tangent line.* HINT [See Example 2(b).]

49. $f(x) = x^3; x = -1$ **50.** $f(x) = x^2; x = 0$

51. $f(x) = x + \frac{1}{x}; x = 2$ **52.** $f(x) = \frac{1}{x^2}; x = 1$

53. $f(x) = \sqrt{x}; x = 4$ **54.** $f(x) = 2x + 4; x = -1$

In each of Exercises 55–58, estimate the given quantity.

55. $f(x) = e^x$; estimate $f'(0)$

56. $f(x) = 2e^x$; estimate $f'(1)$

57. $f(x) = \ln x$; estimate $f'(1)$

58. $f(x) = \ln x$; estimate $f'(2)$

In Exercises 59–64, match the graph of f to the graph of f′ (the graphs of f′ are shown after Exercise 64).

59. ▼

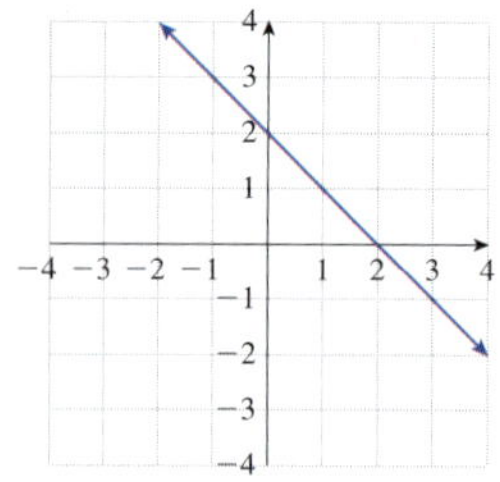

60. ▼

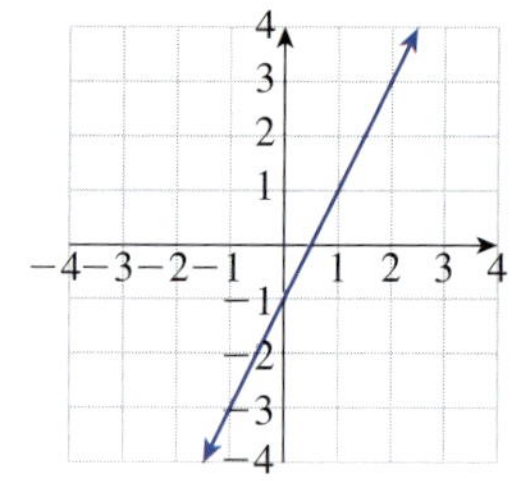

61. ▼

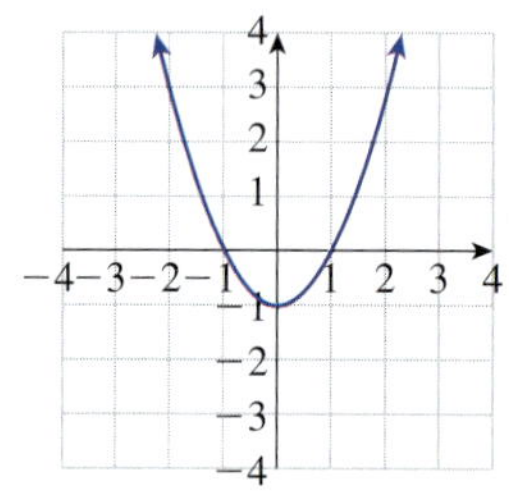

62. ▼

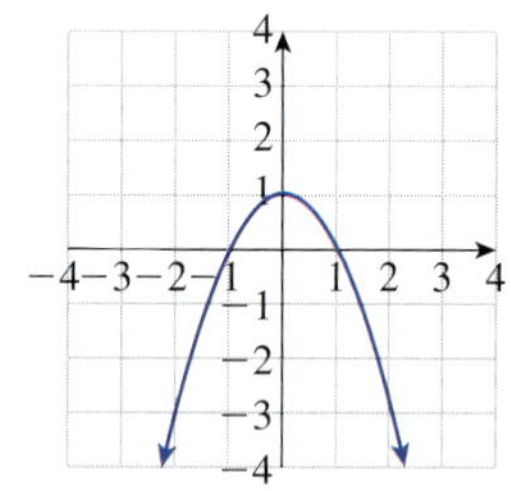

63. ▼

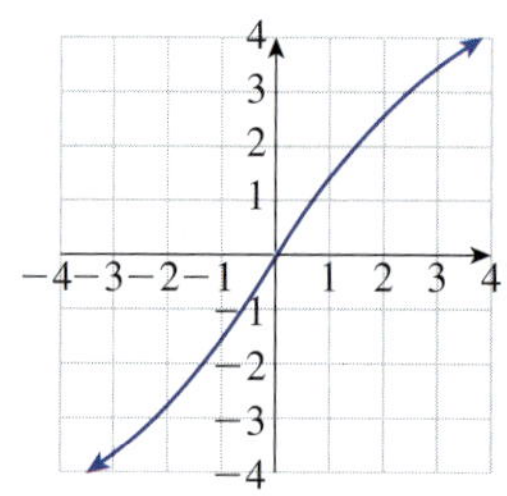

64. ▼

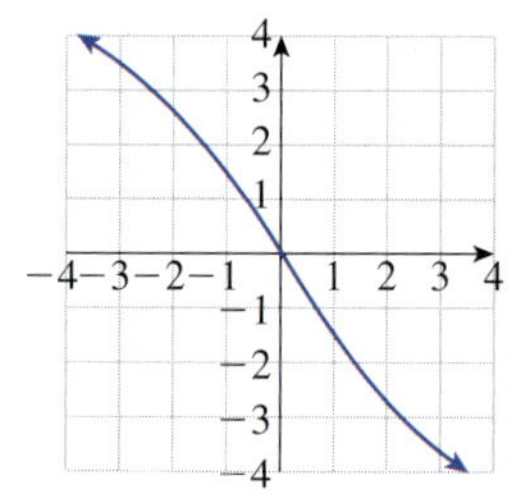

Graphs of derivatives for Exercises 59–64:

(A)

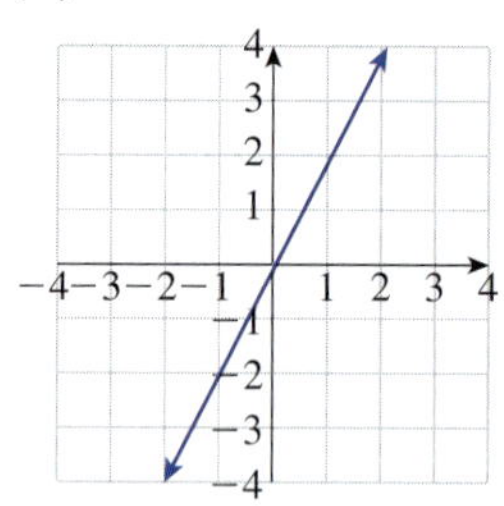

(B)

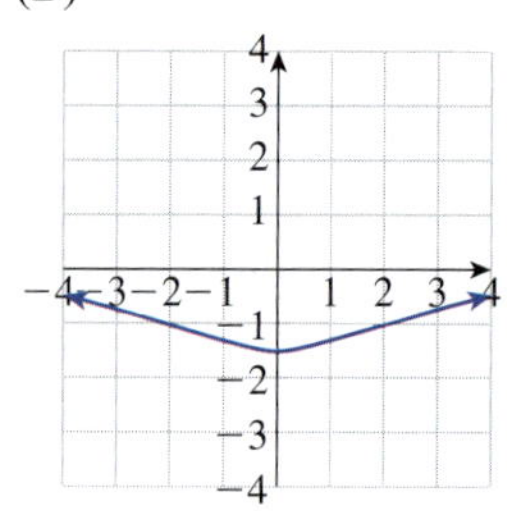

(C)

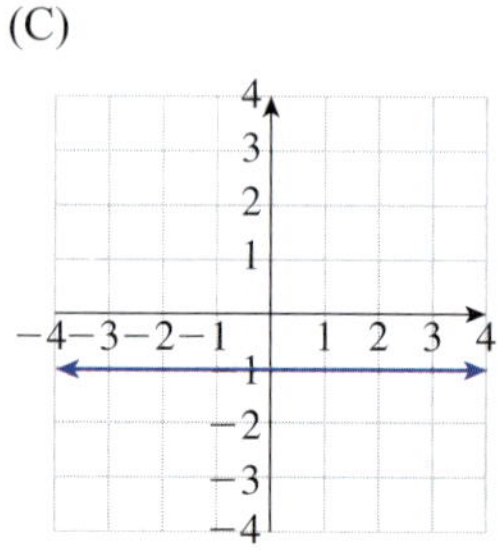

(D)

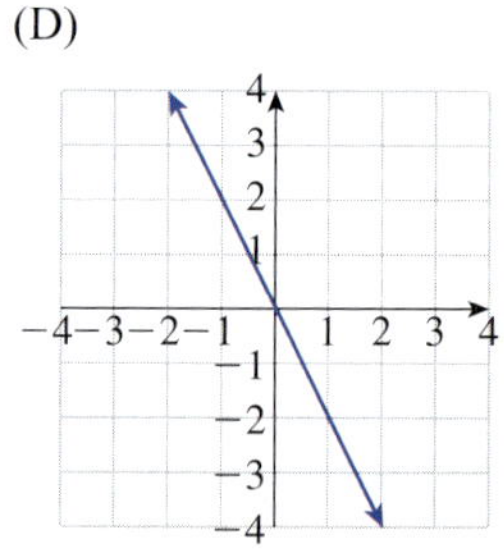

(E)

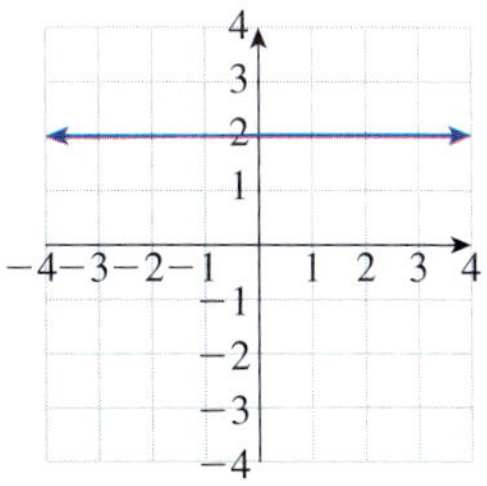

(F)

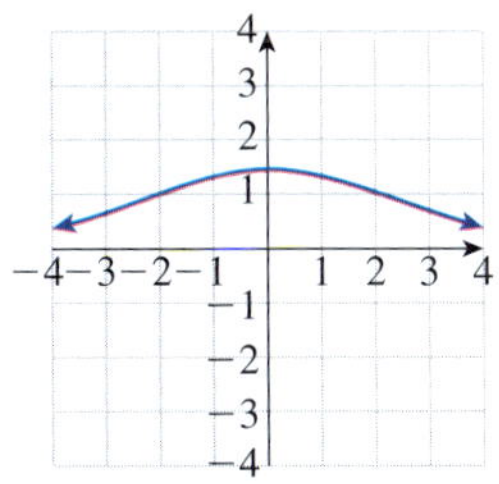

In Exercises 65–68 the graph of a function is given. For which x in the range shown is the function increasing? For which x is the function decreasing? HINT [See Quick Example 3 page 250.]

65.

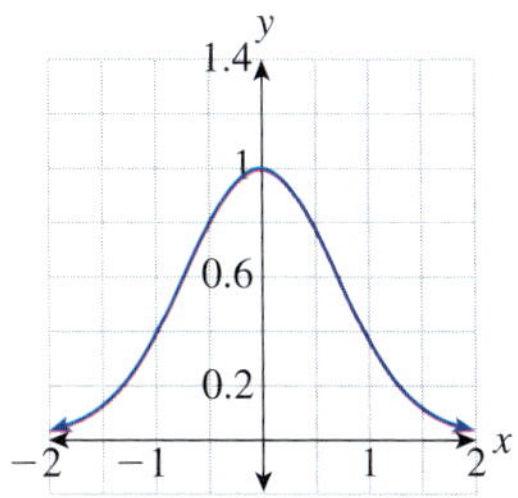

66.

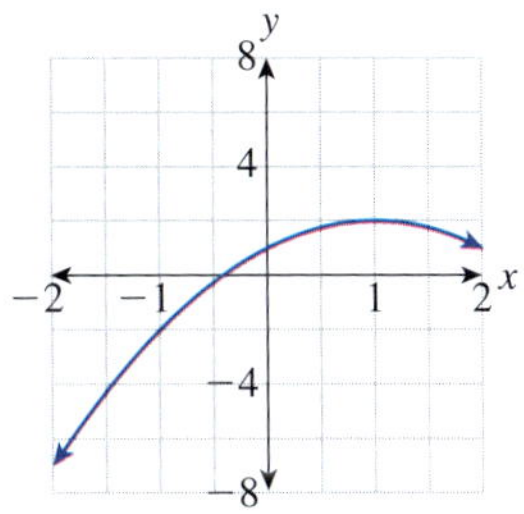

67.

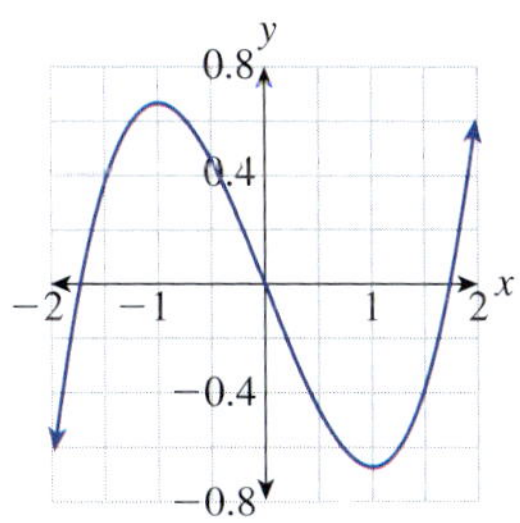

68.

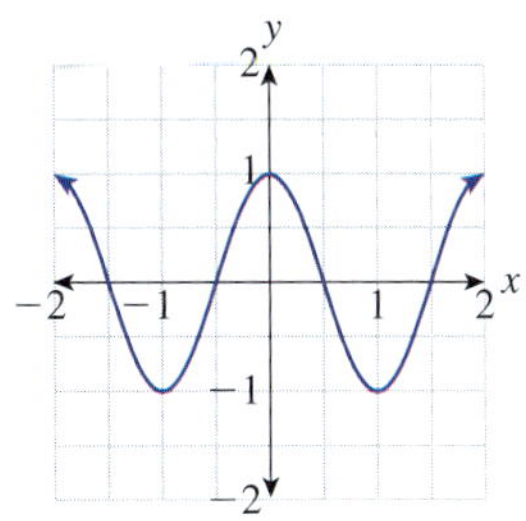

In Exercises 69–72 the graph of the derivative of a function is given. For which x is the (original) function increasing? For which x is the (original) function decreasing? HINT [See Quick Example 3 page 250.]

69.

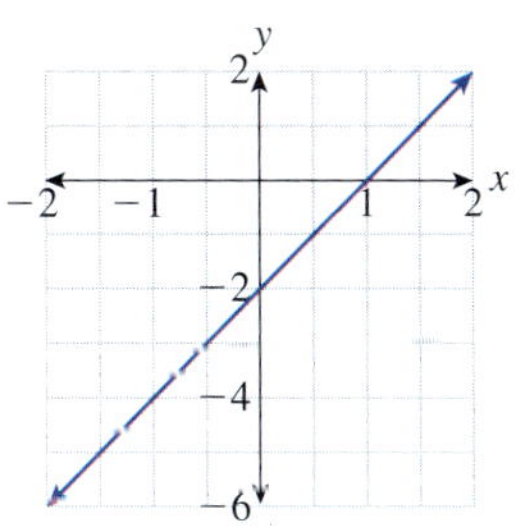

70.

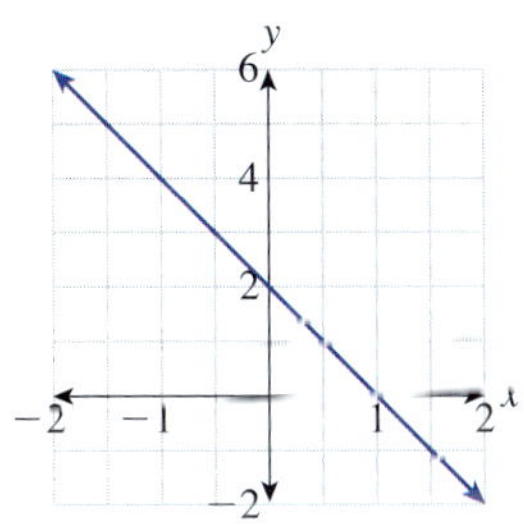

71.

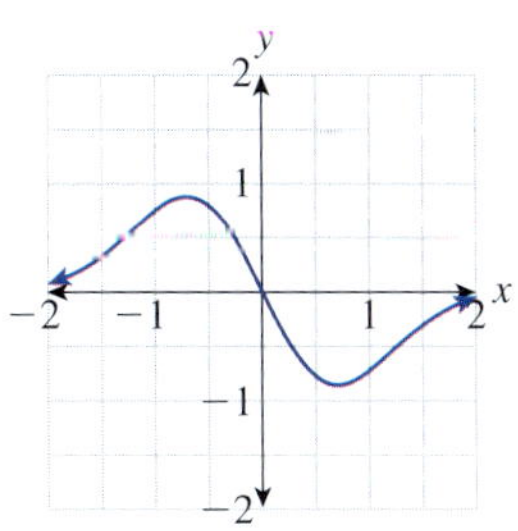

72.

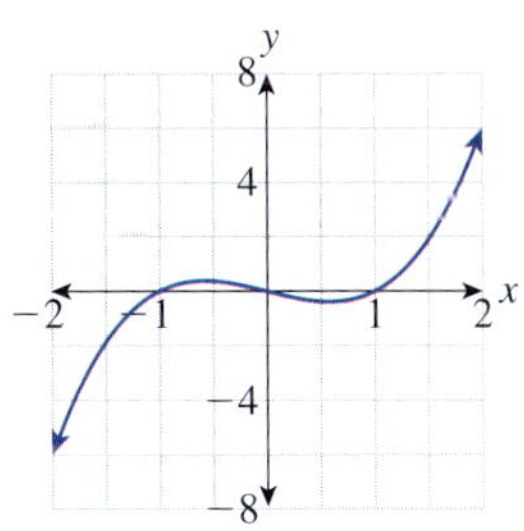

T *In Exercises 73 and 74, use technology to graph the derivative of the given function for the given range of values of x. Then use your graph to estimate all values of x (if any) where the tangent line to the graph of the given function is horizontal. Round answers to one decimal place.* HINT [See Example 4.]

73. $f(x) = x^4 + 2x^3 - 1; \quad -2 \le x \le 1$

74. $f(x) = -x^3 - 3x^2 - 1; \quad -3 \le x \le 1$

T *In Exercises 75 and 76, use the method of Example 4 to list approximate values of f′(x) for x in the given range. Graph f(x) together with f′(x) for x in the given range.*

75. $f(x) = \dfrac{x+2}{x-3}; \quad 4 \le x \le 5$

76. $f(x) = \dfrac{10x}{x-2}; \quad 2.5 \le x \le 3$

APPLICATIONS

77. *Demand* Suppose the demand for a new brand of sneakers is given by

$$q = \frac{5,000,000}{p}$$

where p is the price per pair of sneakers, in dollars, and q is the number of pairs of sneakers that can be sold at price p. Find $q(100)$ and estimate $q'(100)$. Interpret your answers. HINT [See Example 1.]

78. *Demand* Suppose the demand for an old brand of TV is given by

$$q = \frac{100,000}{p+10}$$

where p is the price per TV set, in dollars, and q is the number of TV sets that can be sold at price p. Find $q(190)$ and estimate $q'(190)$. Interpret your answers. HINT [See Example 1.]

79. *Oil Imports from Mexico* The following graph shows approximate daily oil imports to the U.S. from Mexico.[46] Also shown is the tangent line (and its slope) at the point corresponding to year 2005.

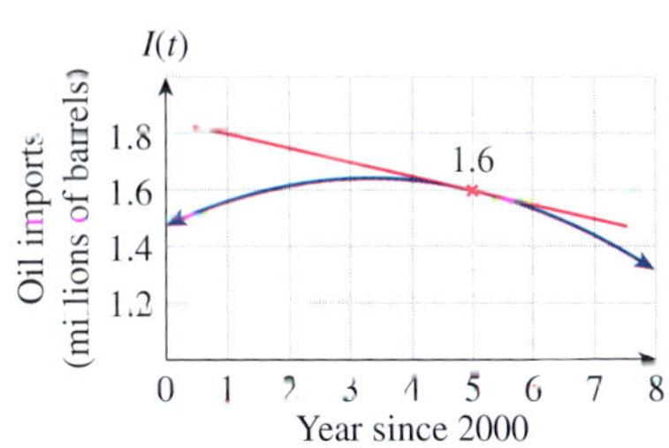

[46] Figures are approximate, and 2008–2009 figures are projections by the Department of Energy. Source: Energy Information Administration/ Pemex (http://www.eia.doe.gov).

a. Estimate the slope of the tangent line shown on the graph. What does the graph tell you about oil imports from Mexico in 2005? HINT [Identify two points on the tangent line. Then see Quick Example page 244.]

b. According to the graph, is the rate of change of oil imports from Mexico increasing, decreasing, or increasing then decreasing? Why?

80. ***Oil Production in Mexico*** The following graph shows approximate daily oil production by Pemex, Mexico's national oil company.[47] Also shown is the tangent line (and its slope) at the point corresponding to year 2003.

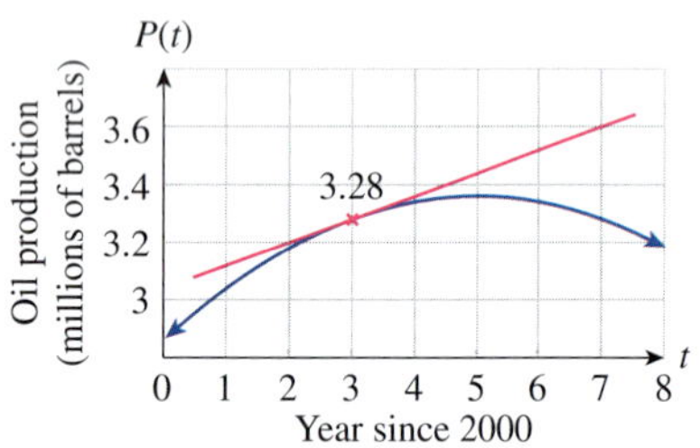

a. Estimate the slope of the tangent line shown on the graph. What does the graph tell you about oil production by Pemex in 2003? HINT [Identify two points on the tangent line. Then see Quick Example page 244.]

b. According to the graph, is the rate of change of oil production by Pemex increasing or decreasing over the range [0, 4]? Why?

81. ▼ ***Prison Population*** The following curve is a model of the total population in state prisons as a function of time in years ($t = 0$ represents 1980).[48]

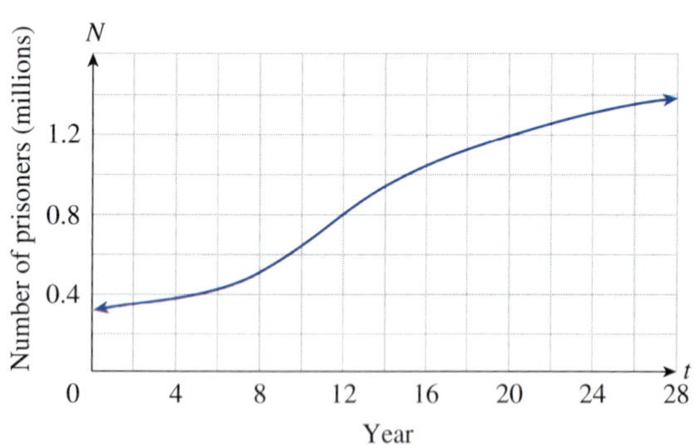

a. Which is correct? Over the period [16, 20] the instantaneous rate of change of N is

(A) increasing **(B)** decreasing

b. Which is correct? The instantaneous rate of change of prison population at $t = 12$ was

(A) less than **(B)** greater than

(C) approximately equal to

the average rate of change over the interval [0, 24].

c. Which is correct? Over the period [0, 28] the instantaneous rate of change of N is

(A) increasing, then decreasing

(B) decreasing, then increasing

(C) always increasing

(D) always decreasing

d. According to the model, the total state prison population was increasing fastest around what year?

e. Roughly estimate the instantaneous rate of change of N at $t = 16$ by using a balanced difference quotient with $h = 4$. Interpret the result.

82. ▼ ***Demand for Freon*** The demand for chlorofluorocarbon-12 (CFC-12)—the ozone-depleting refrigerant commonly known as Freon[49]—has been declining significantly in response to regulation and concern about the ozone layer. The graph below represents a model for the projected demand for CFC-12 as a function of time in years ($t = 0$ represents 1990).[50]

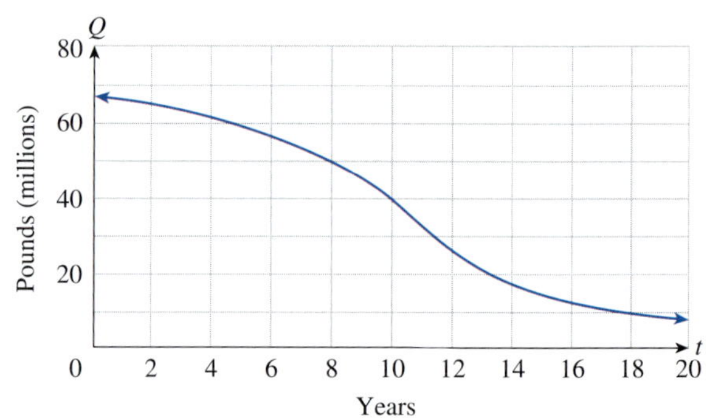

a. Which is correct? Over the period [12, 20] the instantaneous rate of change of Q is

(A) increasing **(B)** decreasing

b. Which is correct? The instantaneous rate of change of demand for Freon at $t = 10$ was

(A) less than **(B)** greater than

(C) approximately equal to

the average rate of change over the interval [0, 20].

c. Which is correct? Over the period [0, 20] the instantaneous rate of change of Q is

(A) increasing, then decreasing

(B) decreasing, then increasing

(C) always increasing

(D) always decreasing

d. According to the model, the demand for Freon was decreasing most rapidly around what year?

[47] Figures are approximate, and 2008–2009 figures are projections by the Department of Energy. Source: Energy Information Administration/Pemex (http://www.eia.doe.gov).

[48] The prison population represented excludes federal prisons. Source for 1980–2000 data: Bureau of Justice Statistics/*New York Times*, June 9, 2001, p. A10.

[49] The name given to it by DuPont.

[50] Source for data: The Automobile Consulting Group (*New York Times*, December 26, 1993, p. F23). The exact figures were not given, and the chart is a reasonable facsimile of the chart that appeared in *New York Times*.

e. Roughly estimate the instantaneous rate of change of Q at $t = 13$ by using a balanced difference quotient with $h = 5$. Interpret the result.

83. ***Velocity*** If a stone is dropped from a height of 400 feet, its height after t seconds is given by $s = 400 - 16t^2$.

a. Find its average velocity over the period [2, 4].

b. Estimate its instantaneous velocity at time $t = 4$. HINT [See Example 3.]

84. ***Velocity*** If a stone is thrown down at 120 ft/s from a height of 1,000 feet, its height after t seconds is given by $s = 1{,}000 - 120t - 16t^2$.

a. Find its average velocity over the period [1, 3].

b. Estimate its instantaneous velocity at time $t = 3$. HINT [See Example 3.]

85. ***Crude Oil Prices*** The price per barrel of crude oil in constant 2008 dollars can be approximated by

$$P(t) = 0.45t^2 - 12t + 105 \text{ dollars} \quad (0 \le t \le 28)$$

where t is time in years since the start of 1980.[51]

a. Compute the average rate of change of $P(t)$ over the interval [0, 28], and interpret your answer. HINT [See Section 3.4 Example 3.]

b. Estimate the instantaneous rate of change of $P(t)$ at $t = 0$, and interpret your answer. HINT [See Example 2(a).]

c. The answers to part (a) and part (b) have opposite signs. What does this indicate about the price of oil?

86. ***Median Home Price*** The median home price in the U.S. over the period 2004–2009 can be approximated by

$$P(t) = -5t^2 + 75t - 30 \text{ thousand dollars} \quad (4 \le t \le 9)$$

where t is time in years since the start of 2000.[52]

a. Compute the average rate of change of $P(t)$ over the interval [5, 9], and interpret your answer. HINT [See Section 3.4 Example 3.]

b. Estimate the instantaneous rate of change of $P(t)$ at $t = 5$, and interpret your answer. HINT [See Example 2(a).]

c. The answer to part (b) has larger absolute value than the answer to part (a). What does this indicate about the median home price?

87. ***SARS*** In the early stages of the deadly SARS (Severe Acute Respiratory Syndrome) epidemic in 2003, the number of reported cases can be approximated by

$$A(t) = 167(1.18)^t \quad (0 \le t \le 20)$$

t days after March 17, 2003 (the first day in which statistics were reported by the World Health Organization).

a. What, approximately, was the instantaneous rate of change of $A(t)$ on March 27 ($t = 10$)? Interpret the result.

b. Which of the following is true? For the first 20 days of the epidemic, the instantaneous rate of change of the number of cases

(A) increased **(B)** decreased

(C) increased and then decreased

(D) decreased and then increased

[51] Source for data: www.inflationdata.com.

[52] Source for data: www.investmenttools.com.

88. ***SARS*** A few weeks into the deadly SARS (Severe Acute Respiratory Syndrome) epidemic in 2003, the number of reported cases can be approximated by

$$A(t) = 1{,}804(1.04)^t \quad (0 \le t \le 30)$$

t days after April 1, 2003.

a. What, approximately, was the instantaneous rate of change of $A(t)$ on April 21 ($t = 20$)? Interpret the result.

b. Which of the following is true? During April, the instantaneous rate of change of the number of cases

(A) increased **(B)** decreased

(C) increased and then decreased

(D) decreased and then increased

89. ***Sales*** Weekly sales of a new brand of sneakers are given by

$$S(t) = 200 - 150e^{-t/10}$$

pairs sold per week, where t is the number of weeks since the introduction of the brand. Estimate $S(5)$ and $\left.\frac{dS}{dt}\right|_{t=5}$ and interpret your answers.

90. ***Sales*** Weekly sales of an old brand of TV are given by

$$S(t) = 100e^{-t/5}$$

sets per week, where t is the number of weeks after the introduction of a competing brand. Estimate $S(5)$ and $\left.\frac{dS}{dt}\right|_{t=5}$ and interpret your answers.

91. ***Early Internet Online Services*** On January 1, 1996, America Online was the biggest online service provider, with 4.5 million subscribers, and was adding new subscribers at a rate of 60,000 per week.[53] If $A(t)$ is the number of America Online subscribers t weeks after January 1, 1996, what do the given data tell you about values of the function A and its derivative? HINT [See Quick Example 2 on page 240.]

92. ***Early Internet Online Services*** On January 1, 1996, Prodigy was the third-biggest online service provider, with 1.6 million subscribers, but was losing subscribers.[54] If $P(t)$ is the number of Prodigy subscribers t weeks after January 1, 1996, what do the given data tell you about values of the function P and its derivative? HINT [See Quick Example 2 on page 240.]

93. ▼ ***Learning to Speak*** Let $p(t)$ represent the percentage of children who are able to speak at the age of t months.

a. It is found that $p(10) = 60$ and $\left.\frac{dp}{dt}\right|_{t=10} = 18.2$. What does this mean?[55] HINT [See Quick Example 2 on page 240.]

b. As t increases, what happens to p and $\frac{dp}{dt}$?

[53] Source: Information and Interactive Services Report/*New York Times*, January 2, 1996, p. C14.

[54] Ibid.

[55] Based on data presented in the article *The Emergence of Intelligence* by William H. Calvin, *Scientific American*, October, 1994, pp. 101–107.

94. ▼ ***Learning to Read*** Let $p(t)$ represent the number of children in your class who learned to read at the age of t years.

a. Assuming that everyone in your class could read by the age of 7, what does this tell you about $p(7)$ and $\left.\frac{dp}{dt}\right|_{t=7}$? HINT [See Quick Example 2 on page 240.]

b. Assuming that 25.0% of the people in your class could read by the age of 5, and that 25.3% of them could read by the age of 5 years and one month, estimate $\left.\frac{dp}{dt}\right|_{t=5}$. Remember to give its units.

95. ***Subprime Mortgages*** (Compare Exercise 29 of Section 3.4.) The percentage of mortgages issued in the United States that are subprime (normally classified as risky) can be approximated by

$$A(t) = \frac{15}{1 + 8.6(1.8)^{-t}} \quad (0 \le t \le 9)$$

where t is the number of years since the start of 2000.

a. Estimate $A(6)$ and $A'(6)$. (Round answers to two significant digits.) What do the answers tell you about subprime mortgages?

b. T Graph the extrapolated function and its derivative for $0 \le t \le 16$ and use your graphs to describe how the derivative behaves as t becomes large. (Express this behavior in terms of limits if you have studied the sections on limits.) What does this tell you about subprime mortgages? HINT [See Example 5.]

96. ***Subprime Mortgage Debt*** (Compare Exercise 30 of Section 3.4.) The value of subprime (normally classified as risky) mortgage debt outstanding in the U.S. can be approximated by

$$A(t) = \frac{1{,}350}{1 + 4.2(1.7)^{-t}} \text{ billion dollars} \quad (0 \le t \le 9)$$

where t is the number of years since the start of 2000.

a. Estimate $A(7)$ and $A'(7)$. (Round answers to three significant digits.) What do the answers tell you about subprime mortgages?

b. T Graph the function and its derivative and use your graphs to estimate when, to the nearest year, $A'(t)$ is greatest. What does this tell you about subprime mortgages? HINT [See Example 5.]

97. T ▼ ***Embryo Development*** The oxygen consumption of a turkey embryo increases from the time the egg is laid through the time the turkey chick hatches. In a brush turkey, the oxygen consumption (in milliliters per hour) can be approximated by

$$c(t) = -0.0012t^3 + 0.12t^2 - 1.83t + 3.97 \quad (20 \le t \le 50)$$

where t is the time (in days) since the egg was laid.[56] (An egg will typically hatch at around $t = 50$.) Use technology to graph $c'(t)$ and use your graph to answer the following questions. HINT [See Example 5.]

a. Over the interval [20, 32] the derivative c' is
(A) increasing, then decreasing
(B) decreasing, then increasing
(C) decreasing **(D)** increasing

b. When, to the nearest day, is the oxygen consumption increasing at the fastest rate?

c. When, to the nearest day, is the oxygen consumption increasing at the slowest rate?

98. T ▼ ***Embryo Development*** The oxygen consumption of a bird embryo increases from the time the egg is laid through the time the chick hatches. In a typical galliform bird, the oxygen consumption (in milliliters per hour) can be approximated by

$$c(t) = -0.0027t^3 + 0.14t^2 - 0.89t + 0.15 \quad (8 \le t \le 30)$$

where t is the time (in days) since the egg was laid.[57] (An egg will typically hatch at around $t = 28$.) Use technology to graph $c'(t)$ and use your graph to answer the following questions. HINT [See Example 5.]

a. Over the interval [8, 30] the derivative c' is
(A) increasing, then decreasing
(B) decreasing, then increasing
(C) decreasing **(D)** increasing

b. When, to the nearest day, is the oxygen consumption increasing the fastest?

c. When, to the nearest day, is the oxygen consumption increasing at the slowest rate?

The next two exercises are applications of Einstein's Special Theory of Relativity and relate to objects that are moving extremely fast. In science fiction terminology, a speed of warp 1 is the speed of light—about 3×10^8 meters per second. (Thus, for instance, a speed of warp 0.8 corresponds to 80% of the speed of light—about 2.4×10^8 meters per second.)

99. ◆ ***Lorentz Contraction*** According to Einstein's Special Theory of Relativity, a moving object appears to get shorter to a stationary observer as its speed approaches the speed of light. If a spaceship that has a length of 100 meters at rest travels at a speed of warp p, its length in meters, as measured by a stationary observer, is given by

$$L(p) = 100\sqrt{1 - p^2}$$

with domain [0, 1). Estimate $L(0.95)$ and $L'(0.95)$. What do these figures tell you?

100. ◆ ***Time Dilation*** Another prediction of Einstein's Special Theory of Relativity is that, to a stationary observer, clocks (as well as all biological processes) in a moving object appear to go more and more slowly as the speed of the object approaches that of light. If a spaceship travels at a speed of warp p, the time it takes for an onboard clock to register

[56] The model approximates graphical data published in the article *The Brush Turkey* by Roger S. Seymour, *Scientific American*, December, 1991, pp. 108–114.

[57] Ibid.

one second, as measured by a stationary observer, will be given by

$$T(p) = \frac{1}{\sqrt{1-p^2}} \text{ seconds}$$

with domain [0, 1). Estimate $T(0.95)$ and $T'(0.95)$. What do these figures tell you?

COMMUNICATION AND REASONING EXERCISES

101. Explain why we cannot put $h = 0$ in the approximation

$$f'(x) \approx \frac{f(x+h) - f(x)}{h}$$

for the derivative of f.

102. The balanced difference quotient

$$f'(a) \approx \frac{f(a+0.0001) - f(a-0.0001)}{0.0002}$$

is the average rate of change of f on what interval?

103. Let $H(t)$ represent the number of Handbook members in millions t years after its inception in 2020. It is found that $H(10) = 50$ and $H'(10) = -6$. This means that, in 2030 (Multiple Choice):

(A) There were 6 million members and this number was decreasing at a rate of 50 million per year

(B) There were –6 million members and this number was increasing at a rate of 50 million per year.

(C) Membership had dropped by 6 million since the previous year, but was now increasing at a rate of 50 million per year

(D) There were 50 million members and this number was decreasing at a rate of 6 million per year.

(E) There were 50 million members and membership had dropped by 6 million since the previous year.

104. Let $F(t)$ represent the net earnings of Footbook Inc. in millions of dollars t years after its inception in 3020. It is found that $F(100) = -10$ and $F'(100) = 60$. This means that, in 3120 (Multiple Choice):

(A) Footbook lost \$10 million but its net earnings were increasing at a rate of \$60 million per year.

(B) Footbook earned \$60 million but its earnings were decreasing at a rate of \$10 million per year.

(C) Footbook's net earnings had increased by \$60 million since the year before, but it still lost \$10 million.

(D) Footbook earned \$10 million but its net earnings were decreasing at a rate of \$60 million per year.

(E) Footbook's net earnings had decreased by \$10 million since the year before, but it still earned \$60 million.

105. It is now eight months since the Garden City lacrosse team won the national championship, and sales of team paraphernalia, while still increasing, have been leveling off. What does this tell you about the derivative of the sales curve?

106. Having been soundly defeated in the national lacrosse championships, Brakpan High has been faced with decreasing sales of its team paraphernalia. However, sales, while still decreasing, appear to be bottoming out. What does this tell you about the derivative of the sales curve?

107. ▼ Company A's profits are given by $P(0) = \$1$ million and $P'(0) = -\$1$ million/month. Company B's profits are given by $P(0) = -\$1$ million and $P'(0) = \$1$ million/month. In which company would you rather invest? Why?

108. ▼ Company C's profits are given by $P(0) = \$1$ million and $P'(0) = \$0.5$ million/month. Company D's profits are given by $P(0) = \$0.5$ million and $P'(0) = \$1$ million/month. In which company would you rather invest? Why?

109. ▼ During the one-month period starting last January 1, your company's profits increased at an average rate of change of \$4 million per month. On January 1, profits were increasing at an instantaneous rate of \$5 million per month. Which of the following graphs could represent your company's profits? Why?

(A)

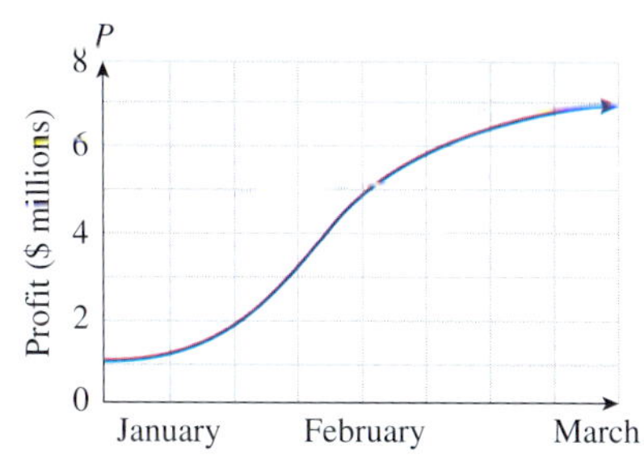

(B)

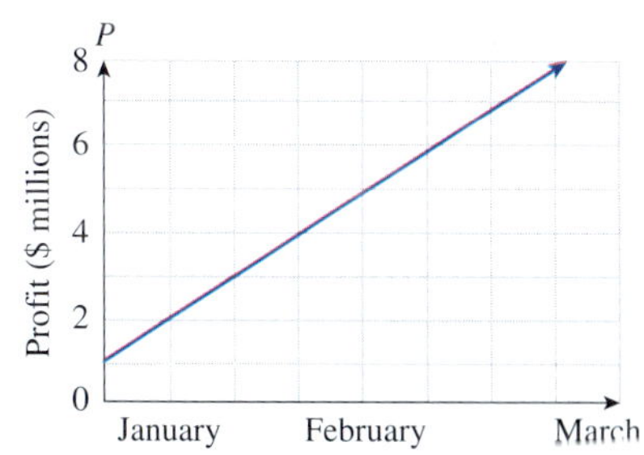

(C)

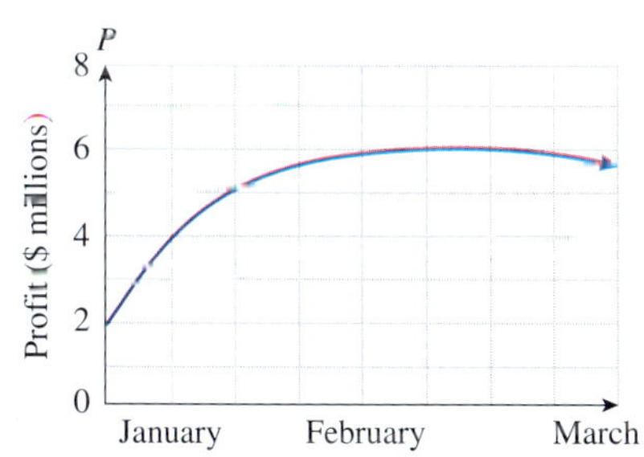

110. ▼ During the one-month period starting last January 1, your company's sales increased at an average rate of change of \$3,000 per month. On January 1, sales were changing at an instantaneous rate of –\$1,000 per month. Which of the following graphs could represent your company's sales? Why?

(A)

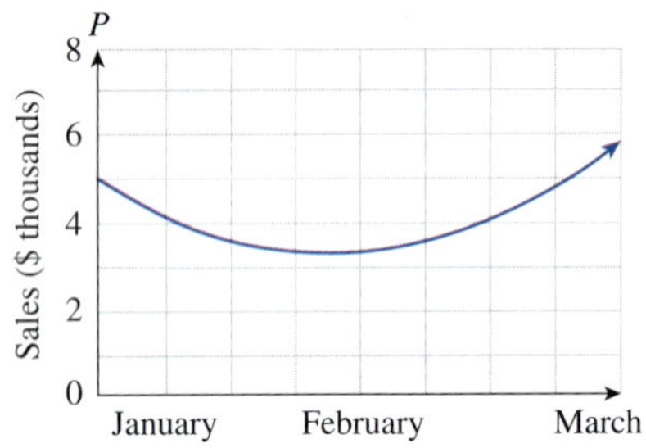

(B)

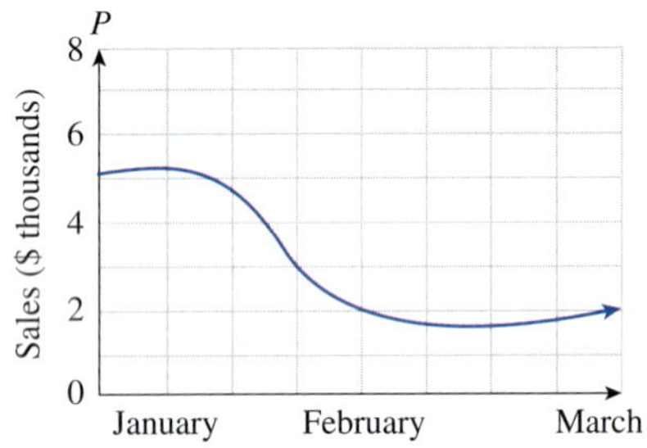

(C)

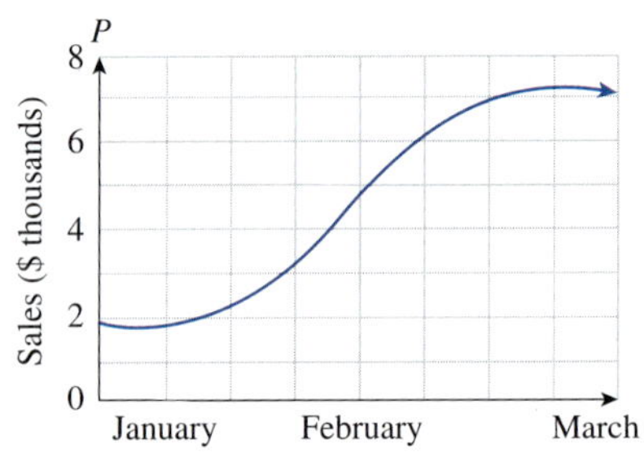

111. ▼ If the derivative of f is zero at a point, what do you know about the graph of f near that point?

112. ▼ Sketch the graph of a function whose derivative never exceeds 1.

113. ▼ Sketch the graph of a function whose derivative exceeds 1 at every point.

114. ▼ Sketch the graph of a function whose derivative is exactly 1 at every point.

115. ▼ Use the difference quotient to explain the fact that if f is a linear function, then the average rate of change over any interval equals the instantaneous rate of change at any point.

116. ▼ Give a numerical explanation of the fact that if f is a linear function, then the average rate of change over any interval equals the instantaneous rate of change at any point.

117. ◆ Consider the following values of the function f from Exercise 1.

h	0.1	0.01	0.001	0.0001
Average Rate of Change of f over $[5, 5+h]$	6.4	6.04	6.004	6.0004
h	−0.1	−0.01	−0.001	−0.0001
Average Rate of Change of f over $[5+h, 5]$	5.6	5.96	5.996	5.9996

Does the table suggests that the instantaneous rate of change of f is

(A) increasing **(B)** decreasing

as x increases toward 5?

118. ◆ Consider the following values of the function g from Exercise 2.

h	0.1	0.01	0.001	0.0001
Average Rate of Change of g over $[7, 7+h]$	4.8	4.98	4.998	4.9998
h	−0.1	−0.01	−0.001	−0.0001
Average Rate of Change of g over $[7+h, 7]$	5.3	5.03	5.003	5.0003

Does the table suggest that the instantaneous rate of change of g is

(A) increasing **(B)** decreasing

as x increases toward 7?

119. ▼ Sketch the graph of a function whose derivative is never zero but decreases as x increases.

120. ▼ Sketch the graph of a function whose derivative is never negative but is zero at exactly two points.

121. ◆ Here is the graph of the derivative f' of a function f. Give a rough sketch of the graph of f, given that $f(0) = 0$.

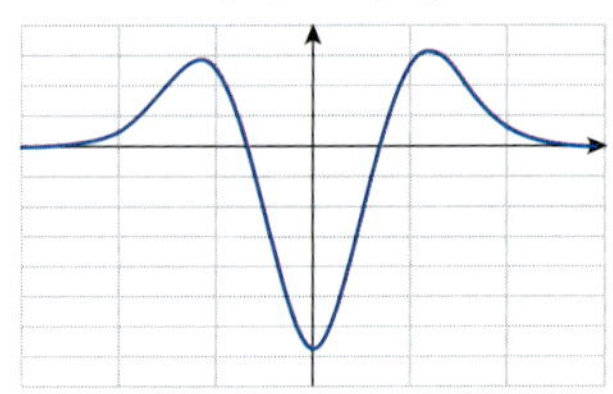

122. ◆ Here is the graph of the derivative f' of a function f. Give a rough sketch of the graph of f, given that $f(0) = 0$.

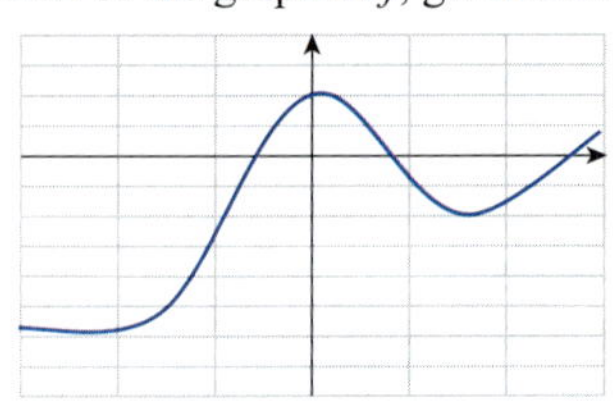

123. ◆ Professor Talker of the physics department drove a 60-mile stretch of road in exactly one hour. The speed limit along that stretch was 55 miles per hour. Which of the following must be correct:

(A) He exceeded the speed limit at no point of the journey.

(B) He exceeded the speed limit at some point of the journey.

(C) He exceeded the speed limit throughout the journey.
(D) He traveled slower than the speed limit at some point of the journey.

124. ◆ Professor Silent, another physics professor, drove a 50-mile stretch of road in exactly one hour. The speed limit along that stretch was 55 miles per hour. Which of the following must be correct:

(A) She exceeded the speed limit at no point of the journey.
(B) She exceeded the speed limit at some point of the journey.
(C) She traveled slower than the speed limit throughout the journey.
(D) She traveled slower than the speed limit at some point of the journey.

125. ◆ Draw the graph of a function f with the property that the balanced difference quotient gives a more accurate approximation of $f'(1)$ than the ordinary difference quotient.

126. ◆ Draw the graph of a function f with the property that the balanced difference quotient gives a less accurate approximation of $f'(1)$ than the ordinary difference quotient.

3.6 The Derivative: Algebraic Viewpoint

In Section 3.5 we saw how to estimate the derivative of a function using numerical and graphical approaches. In this section we use an algebraic approach that will give us the *exact value* of the derivative, rather than just an approximation, when the function is specified algebraically.

This algebraic approach is quite straightforward: Instead of subtracting numbers to estimate the average rate of change over smaller and smaller intervals, we subtract algebraic expressions. Our starting point is the definition of the derivative in terms of the difference quotient:

$$f'(a) = \lim_{h\to 0} \frac{f(a+h) - f(a)}{h}.$$

EXAMPLE 1 Calculating the Derivative at a Point Algebraically

Let $f(x) = x^2$. Use the definition of the derivative to compute $f'(3)$ algebraically.

Solution Substituting $a = 3$ into the definition of the derivative, we get:

$$f'(3) = \lim_{h\to 0} \frac{f(3+h) - f(3)}{h} \qquad \text{Formula for the derivative}$$

$$= \lim_{h\to 0} \frac{\overbrace{(3+h)^2}^{f(3+h)} - \overbrace{3^2}^{f(3)}}{h} \qquad \text{Substitute for } f(3) \text{ and } f(3+h).$$

$$= \lim_{h\to 0} \frac{(9 + 6h + h^2) - 9}{h} \qquad \text{Expand } (3+h)^2.$$

$$= \lim_{h\to 0} \frac{6h + h^2}{h} \qquad \text{Cancel the 9.}$$

$$= \lim_{h\to 0} \frac{h(6+h)}{h} \qquad \text{Factor out } h.$$

$$= \lim_{h\to 0} (6+h) \qquad \text{Cancel the } h.$$

Now we let h approach 0. As h gets closer and closer to 0, the sum $6 + h$ clearly gets closer and closer to $6 + 0 = 6$. Thus,

$$f'(3) = \lim_{h\to 0} (6+h) = 6. \qquad \text{As } h \to 0, (6+h) \to 6$$

(Calculations of limits like this are discussed and justified more fully in Sections 3.2 and 3.3.)

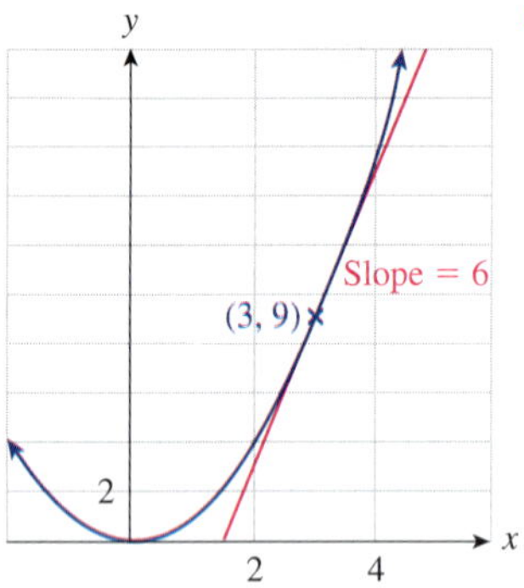

Figure 35

Before we go on... We did the following calculation in Example 1: If $f(x) = x^2$, then $f'(3) = 6$. In other words, the tangent to the graph of $y = x^2$ at the point (3, 9) has slope 6 (Figure 35). ■

There is nothing very special about $a = 3$ in Example 1. Let's try to compute $f'(x)$ for general x.

EXAMPLE 2 Calculating the Derivative Function Algebraically

Let $f(x) = x^2$.

a. Use the definition of the derivative to compute $f'(x)$ algebraically.
b. Use the answer to evaluate $f'(3)$.

Solution

a. Once again, our starting point is the definition of the derivative in terms of the difference quotient:

$$f'(x) = \lim_{h\to 0} \frac{f(x+h) - f(x)}{h} \qquad \text{Formula for the derivative}$$

$$= \lim_{h\to 0} \frac{\overbrace{(x+h)^2}^{f(x+h)} - \overbrace{x^2}^{f(x)}}{h} \qquad \text{Substitute for } f(x) \text{ and } f(x+h).$$

$$= \lim_{h\to 0} \frac{(x^2 + 2xh + h^2) - x^2}{h} \qquad \text{Expand } (x+h)^2.$$

$$= \lim_{h\to 0} \frac{2xh + h^2}{h} \qquad \text{Cancel the } x^2.$$

$$= \lim_{h\to 0} \frac{h(2x + h)}{h} \qquad \text{Factor out } h.$$

$$= \lim_{h\to 0} (2x + h) \qquad \text{Cancel the } h.$$

Now we let h approach 0. As h gets closer and closer to 0, the sum $2x + h$ clearly gets closer and closer to $2x + 0 = 2x$. Thus,

$$f'(x) = \lim_{h\to 0} (2x + h) = 2x.$$

This is the derivative function.

b. Now that we have a *formula* for the derivative of *f*, we can obtain $f'(a)$ for any value of a we choose by simply evaluating f' there. For instance,

$$f'(3) = 2(3) = 6$$

as we saw in Example 1.

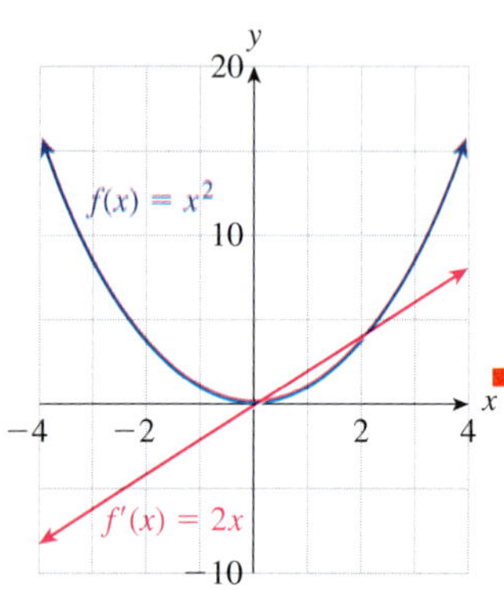

Figure 36

Before we go on... The graphs of $f(x) = x^2$ and $f'(x) = 2x$ from Example 2 are familiar. Their graphs are shown in Figure 36.

When $x < 0$, the parabola slopes downward, which is reflected in the fact that the derivative $2x$ is negative there. When $x > 0$, the parabola slopes upward, which is reflected in the fact that the derivative is positive there. The parabola has a horizontal tangent line at $x = 0$, reflected in the fact that $2x = 0$ there. ■

EXAMPLE 3 More Computations of Derivative Functions

Compute the derivative $f'(x)$ for each of the following functions:

a. $f(x) = x^3$ **b.** $f(x) = 2x^2 - x$ **c.** $f(x) = \dfrac{1}{x}$

Solution

a.

$$f'(x) = \lim_{h\to 0} \frac{f(x+h) - f(x)}{h} \qquad \text{Derivative formula}$$

$$= \lim_{h\to 0} \frac{\overbrace{(x+h)^3}^{f(x+h)} - \overbrace{x^3}^{f(x)}}{h} \qquad \text{Substitute for } f(x) \text{ and } f(x+h).$$

$$= \lim_{h\to 0} \frac{(x^3 + 3x^2h + 3xh^2 + h^3) - x^3}{h} \qquad \text{Expand } (x+h)^3.$$

$$= \lim_{h\to 0} \frac{3x^2h + 3xh^2 + h^3}{h} \qquad \text{Cancel the } x^3.$$

$$= \lim_{h\to 0} \frac{h(3x^2 + 3xh + h^2)}{h} \qquad \text{Factor out } h.$$

$$= \lim_{h\to 0} (3x^2 + 3xh + h^2) \qquad \text{Cancel the } h.$$

$$= 3x^2. \qquad \text{Let } h \text{ approach } 0.$$

b.

$$f'(x) = \lim_{h\to 0} \frac{f(x+h) - f(x)}{h} \qquad \text{Derivative formula}$$

$$= \lim_{h\to 0} \frac{\overbrace{(2(x+h)^2 - (x+h))}^{f(x+h)} - \overbrace{(2x^2 - x)}^{f(x)}}{h} \qquad \text{Substitute for } f(x) \text{ and } f(x+h).$$

$$= \lim_{h\to 0} \frac{(2x^2 + 4xh + 2h^2 - x - h) - (2x^2 - x)}{h} \qquad \text{Expand.}$$

$$= \lim_{h\to 0} \frac{4xh + 2h^2 - h}{h} \qquad \text{Cancel the } 2x^2 \text{ and } x.$$

$$= \lim_{h\to 0} \frac{h(4x + 2h - 1)}{h} \qquad \text{Factor out } h.$$

$$= \lim_{h\to 0} (4x + 2h - 1) \qquad \text{Cancel the } h.$$

$$= 4x - 1. \qquad \text{Let } h \text{ approach } 0.$$

c.

$$f'(x) = \lim_{h\to 0} \frac{f(x+h) - f(x)}{h} \qquad \text{Derivative formula}$$

$$= \lim_{h\to 0} \frac{\left[\overbrace{\frac{1}{x+h}}^{f(x+h)} - \overbrace{\frac{1}{x}}^{f(x)}\right]}{h} \qquad \text{Substitute for } f(x) \text{ and } f(x+h).$$

$$= \lim_{h\to 0} \frac{\left[\frac{x - (x+h)}{(x+h)x}\right]}{h} \qquad \text{Subtract the fractions.}$$

$$= \lim_{h\to 0} \frac{1}{h}\left[\frac{x-(x+h)}{(x+h)x}\right]$$ Dividing by h = Multiplying by $1/h$.

$$= \lim_{h\to 0} \left[\frac{-h}{h(x+h)x}\right]$$ Simplify.

$$= \lim_{h\to 0} \left[\frac{-1}{(x+h)x}\right]$$ Cancel the h.

$$= \frac{-1}{x^2}$$ Let h approach 0.

In Example 4, we redo Example 3 of Section 3.5, this time getting an exact, rather than approximate, answer.

Joseph Gareri/istockphoto.com

EXAMPLE 4 Velocity

My friend Eric, an enthusiastic baseball player, claims he can "probably" throw a ball upward at a speed of 100 feet per second (ft/s). Our physicist friends tell us that its height s (in feet) t seconds later would be $s(t) = 100t - 16t^2$. Find the ball's instantaneous velocity function and its velocity exactly 2 seconds after Eric throws it.

Solution The instantaneous velocity function is the derivative ds/dt, which we calculate as follow:

$$\frac{ds}{dt} = \lim_{h\to 0} \frac{s(t+h)-s(t)}{h}$$

Let us compute $s(t+h)$ and $s(t)$ separately:

$$\begin{aligned} s(t) &= 100t - 16t^2 \\ s(t+h) &= 100(t+h) - 16(t+h)^2 \\ &= 100t + 100h - 16(t^2 + 2th + h^2) \\ &= 100t + 100h - 16t^2 - 32th - 16h^2. \end{aligned}$$

Therefore,

$$\begin{aligned} \frac{ds}{dt} &= \lim_{h\to 0} \frac{s(t+h)-s(t)}{h} \\ &= \lim_{h\to 0} \frac{100t + 100h - 16t^2 - 32th - 16h^2 - (100t - 16t^2)}{h} \\ &= \lim_{h\to 0} \frac{100h - 32th - 16h^2}{h} \\ &= \lim_{h\to 0} \frac{h(100 - 32t - 16h)}{h} \\ &= \lim_{h\to 0} (100 - 32t - 16h) \\ &= 100 - 32t \text{ ft/s.} \end{aligned}$$

Thus, the velocity exactly 2 seconds after Eric throws it is

$$\left.\frac{ds}{dt}\right|_{t=2} = 100 - 32(2) = 36 \text{ ft/s.}$$

This verifies the accuracy of the approximation we made in Section 3.5.

➡ **Before we go on...** From the derivative function in Example 4, we can now describe the behavior of the velocity of the ball: Immediately on release ($t = 0$) the ball is traveling at 100 feet per second upward. The ball then slows down; precisely, it loses 32 feet per second of speed every second. When, exactly, does the velocity become zero and what happens after that? ■

Q: *Do we always have to calculate the limit of the difference quotient to find a formula for the derivative function?*

A: As it turns out, no. In Section 4.1 we will start to look at shortcuts for finding derivatives that allow us to bypass the definition of the derivative in many cases.

A Function Not Differentiable at a Point

Recall from Section 3.5 that a function is **differentiable** at a point a if $f'(a)$ exists; that is, if the difference quotient $[f(a+h) - f(a)]/h$ approaches a fixed value as h approaches 0. In Section 3.5, we mentioned that the function $f(x) = |x|$ is not differentiable at $x = 0$. In Example 5, we find out why.

EXAMPLE 5 A Function Not Differentiable at 0

Numerically, graphically, and algebraically investigate the differentiability of the function $f(x) = |x|$ at the points **(a)** $x = 1$ and **(b)** $x = 0$.

Solution

a. We compute

$$f'(1) = \lim_{h\to 0} \frac{f(1+h) - f(1)}{h} = \lim_{h\to 0} \frac{|1+h| - 1}{h}.$$

Numerically, we can make tables of the values of the average rate of change $(|1+h| - 1)/h$ for h positive or negative and approaching 0:

h	1	0.1	0.01	0.001	0.0001
Average Rate of Change Over [1, 1 + *h*]	1	1	1	1	1

h	−1	−0.1	−0.01	−0.001	−0.0001
Average Rate of Change Over [1 + *h*, 1]	1	1	1	1	1

From these tables it appears that $f'(1)$ is equal to 1. We can verify that algebraically: For h that is sufficiently small, $1 + h$ is positive (even if h is negative) and so

$$f'(1) = \lim_{h\to 0} \frac{1+h-1}{h}$$

$$= \lim_{h\to 0} \frac{h}{h} \qquad \text{Cancel the 1s.}$$

$$= \lim_{h\to 0} 1 \qquad \text{Cancel the } h.$$

$$= 1$$

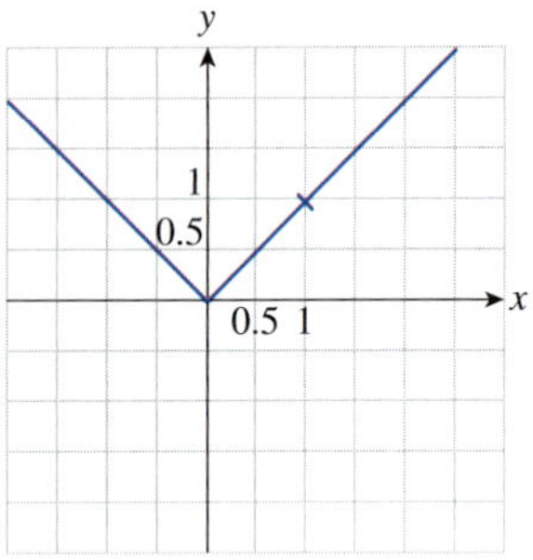

Figure 37

Graphically, we are seeing the fact that the tangent line at the point (1, 1) has slope 1 because the graph is a straight line with slope 1 near that point (Figure 37).

b. $$f'(0) = \lim_{h\to 0} \frac{f(0+h) - f(0)}{h} = \lim_{h\to 0} \frac{|0+h| - 0}{h} = \lim_{h\to 0} \frac{|h|}{h}$$

If we make tables of values in this case we get the following:

h	1	0.1	0.01	0.001	0.0001
Average Rate of Change over [0, 0 + h]	1	1	1	1	1

h	−1	−0.1	−0.01	−0.001	−0.0001
Average Rate of Change over [0 + h, 0]	−1	−1	−1	−1	−1

For the limit and hence the derivative $f'(0)$ to exist, the average rates of change should approach the same number for both positive and negative h. Because they do not, f is not differentiable at $x = 0$. We can verify this conclusion algebraically: If h is positive, then $|h| = h$, and so the ratio $|h|/h$ is 1, regardless of how small h is. Thus, according to the values of the difference quotients with $h > 0$, the limit should be 1. On the other hand if h is negative, then $|h| = -h$ (positive) and so $|h|/h = -1$, meaning that the limit should be -1. Because the limit cannot be both -1 and 1 (it must be a single number for the derivative to exist), we conclude that $f'(0)$ does not exist.

To see what is happening graphically, take a look at Figure 38, which shows zoomed-in views of the graph of f near $x = 0$.

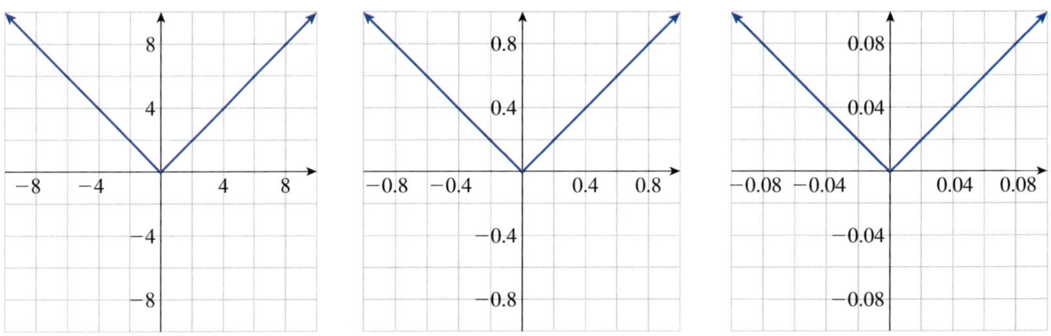

Figure 38

No matter what scale we use to view the graph, it has a sharp corner at $x = 0$ and hence has no tangent line there. Since there is no tangent line at $x = 0$, the function is not differentiable there.

Before we go on... Notice that $|x| = \begin{cases} -x & \text{if } x < 0 \\ x & \text{if } x \geq 0 \end{cases}$ is an example of a piecewise-linear function whose graph comes to a point at $x = 0$. In general, if $f(x)$ is any piecewise linear function whose graph comes to a point at $x = a$, it will be non-differentiable at $x = a$ for the same reason that $|x|$ fails to be differentiable at $x = 0$.

If we repeat the computation in Example 5(a) using any nonzero value for a in place of 1, we see that f is differentiable there as well. If a is positive, we find that $f'(a) = 1$ and, if a is negative, $f'(a) = -1$. In other words, the derivative function is

$$f'(x) = \begin{cases} -1 & \text{if } x < 0 \\ 1 & \text{if } x > 0 \end{cases}.$$

Immediately to the left of $x = 0$, we see that $f'(x) = -1$, immediately to the right, $f'(x) = 1$, and when $x = 0$, $f'(x)$ is not defined. ■

FAQ

Computing Derivatives Algebraically

Q: *The algebraic computation of $f'(x)$ seems to require a number of steps. How do I remember what to do, and when?*

A: If you examine the computations in the previous examples, you will find the following pattern:

1. Write out the formula for $f'(x)$, as the limit of the difference quotient, then substitute $f(x+h)$ and $f(x)$.
2. Expand and simplify the *numerator* of the expression, but not the denominator.
3. After simplifying the numerator, factor out an h to cancel with the h in the denominator. If h does not factor out of the numerator, you might have made an error. (A frequent error is a wrong sign.)
4. After canceling the h, you should be able to see what the limit is by letting $h \to 0$.

3.6 EXERCISES

▼ more advanced ◆ challenging

T indicates exercises that should be solved using technology

In Exercises 1–14, compute $f'(a)$ algebraically for the given value of a. HINT [See Example 1.]

1. $f(x) = x^2 + 1; a = 2$

2. $f(x) = x^2 - 3; a = 1$

3. $f(x) = 3x - 4; a = -1$

4. $f(x) = -2x + 4; a = -1$

5. $f(x) = 3x^2 + x; a = 1$

6. $f(x) = 2x^2 + x; a = -2$

7. $f(x) = 2x - x^2; a = -1$

8. $f(x) = -x - x^2; a = 0$

9. $f(x) = x^3 + 2x; a = 2$

10. $f(x) = x - 2x^3; a = 1$

11. $f(x) = \dfrac{-1}{x}; a = 1$ HINT [See Example 3.]

12. $f(x) = \dfrac{2}{x}; a = 5$ HINT [See Example 3.]

13. ▼ $f(x) = mx + b; a = 43$

14. ▼ $f(x) = \dfrac{x}{k} - b \ (k \neq 0); a = 12$

In Exercises 15–28, compute the derivative function $f'(x)$ algebraically. (Notice that the functions are the same as those in Exercises 1–14.) HINT [See Examples 2 and 3.]

15. $f(x) = x^2 + 1$

16. $f(x) = x^2 - 3$

17. $f(x) = 3x - 4$

18. $f(x) = -2x + 4$

19. $f(x) = 3x^2 + x$

20. $f(x) = 2x^2 + x$

21. $f(x) = 2x - x^2$

22. $f(x) = -x - x^2$

23. $f(x) = x^3 + 2x$

24. $f(x) = x - 2x^3$

25. ▼ $f(x) = \dfrac{-1}{x}$

26. ▼ $f(x) = \dfrac{2}{x}$

27. ▼ $f(x) = mx + b$

28. ▼ $f(x) = \dfrac{x}{k} - b \ (k \neq 0)$

In Exercises 29–38, compute the indicated derivative.

29. $R(t) = -0.3t^2; R'(2)$

30. $S(t) = 1.4t^2$; $S'(-1)$

31. $U(t) = 5.1t^2 + 5.1$; $U'(3)$

32. $U(t) = -1.3t^2 + 1.1$; $U'(4)$

33. $U(t) = -1.3t^2 - 4.5t$; $U'(1)$

34. $U(t) = 5.1t^2 - 1.1t$; $U'(1)$

35. $L(r) = 4.25r - 5.01$; $L'(1.2)$

36. $L(r) = -1.02r + 5.7$; $L'(3.1)$

37. ▼ $q(p) = \dfrac{2.4}{p} + 3.1$; $q'(2)$

38. ▼ $q(p) = \dfrac{1}{0.5p} - 3.1$; $q'(2)$

In Exercises 39–44, find the equation of the tangent to the graph at the indicated point. HINT [Compute the derivative algebraically; then see Example 2(b) in Section 3.5.]

39. ▼ $f(x) = x^2 - 3$; $a = 2$

40. ▼ $f(x) = x^2 + 1$; $a = 2$

41. ▼ $f(x) = -2x - 4$; $a = 3$

42. ▼ $f(x) = 3x + 1$; $a = 1$

43. ▼ $f(x) = x^2 - x$; $a = -1$

44. ▼ $f(x) = x^2 + x$; $a = -1$

APPLICATIONS

45. ***Velocity*** If a stone is dropped from a height of 400 feet, its height after t seconds is given by $s = 400 - 16t^2$. Find its instantaneous velocity function and its velocity at time $t = 4$. HINT [See Example 4.]

46. ***Velocity*** If a stone is thrown down at 120 feet per second from a height of 1,000 feet, its height after t seconds is given by $s = 1{,}000 - 120t - 16t^2$. Find its instantaneous velocity function and its velocity at time $t = 3$. HINT [See Example 4.]

47. ***Oil Imports from Mexico*** Daily oil imports to the United States from Mexico can be approximated by

$$I(t) = -0.015t^2 + 0.1t + 1.4 \text{ million barrels} \quad (0 \le t \le 8)$$

where t is time in years since the start of 2000.[58] Find the derivative function $\dfrac{dI}{dt}$. At what rate were oil imports changing at the start of 2007 ($t = 7$)? HINT [See Example 4.]

48. ***Oil Production in Mexico*** Daily oil production by **Pemex**, Mexico's national oil company, can be approximated by

$$P(t) = -0.022t^2 + 0.2t + 2.9 \text{ million barrels} \quad (1 \le t \le 9)$$

where t is time in years since the start of 2000.[59] Find the derivative function $\dfrac{dP}{dt}$. At what rate was oil production changing at the start of 2004 ($t = 4$)? HINT [See Example 4.]

49. ***Bottled Water Sales*** Annual U.S. sales of bottled water rose through the period 2000–2008 as shown in the following chart.[60]

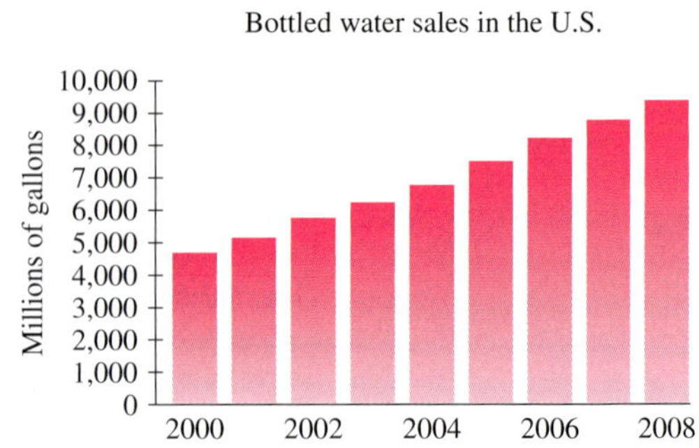

The function

$$R(t) = 12t^2 + 500t + 4{,}700 \text{ million gallons} \quad (0 \le t \le 8)$$

gives a good approximation, where t is time in years since 2000. Find the derivative function $R'(t)$. According to the model, how fast were annual sales of bottled water increasing in 2005?

50. ***Bottled Water Sales*** Annual U.S. per capita sales of bottled water rose through the period 2000–2008 as shown in the following chart.[61]

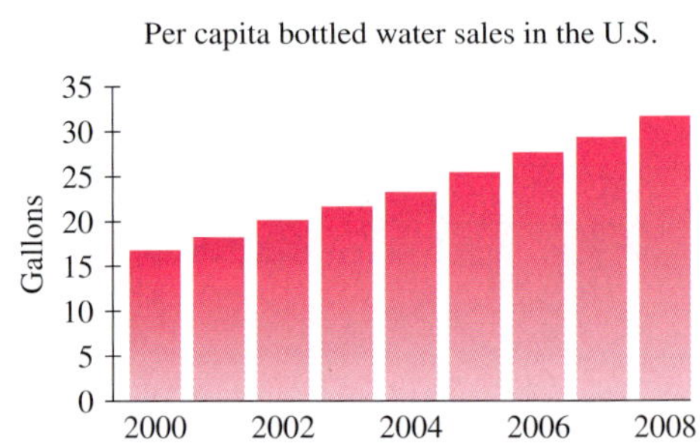

The function

$$Q(t) = 0.04t^2 + 1.5t + 17 \text{ gallons} \quad (0 \le t \le 8)$$

gives a good approximation, where t is the time in years since 2000. Find the derivative function $Q'(t)$. According to the model, how fast were annual per capita sales of bottled water increasing in 2008?

51. ▼ ***Ecology*** Increasing numbers of manatees ("sea sirens") have been killed by boats off the Florida coast. The following graph shows the relationship between the number of boats registered in Florida and the number of manatees killed each year.

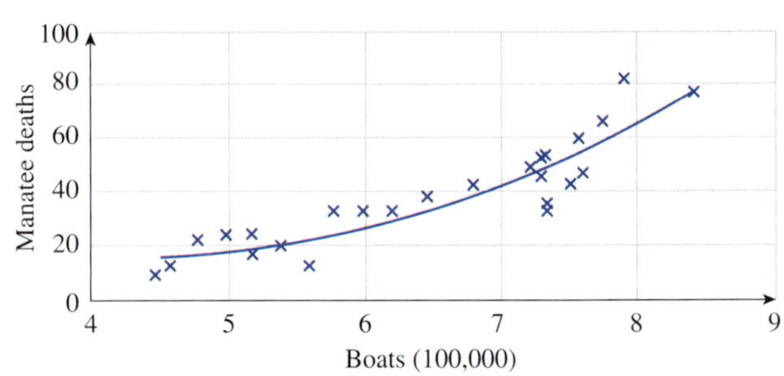

---

[58] Source for data: Energy Information Administration/Pemex (http://www.eia.doe.gov).

[59] Ibid.

[60] The 2008 figure is an estimate. Source: Beverage Marketing Corporation, www.bottledwater.org.

[61] Ibid.

The regression curve shown is given by

$$f(x) = 3.55x^2 - 30.2x + 81 \text{ manatee deaths}$$
$$(4.5 \le x \le 8.5)$$

where x is the number of boats (hundreds of thousands) registered in Florida in a particular year and $f(x)$ is the number of manatees killed by boats in Florida that year.[62] Compute and interpret $f'(8)$.

52. ▼ ***SAT Scores by Income*** The following graph shows U.S. verbal SAT scores as a function of parents' income level.[63]

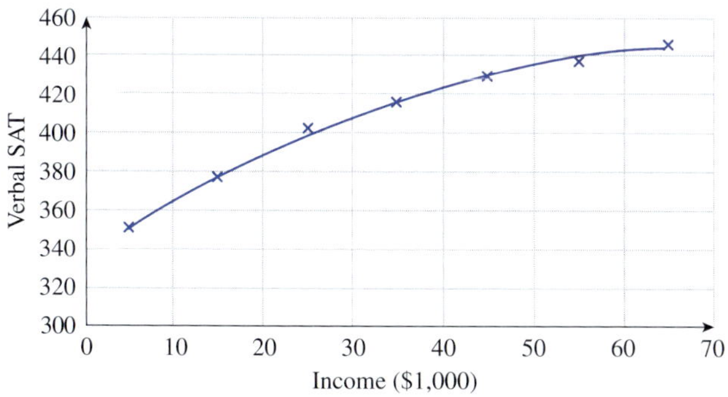

The regression curve shown is given by

$$f(x) = -0.021x^2 + 3.0x + 336 \quad (5 \le x \le 65)$$

where $f(x)$ is the average SAT verbal score of a student whose parents earn x thousand dollars per year. Compute and interpret $f'(30)$.

53. ▼ ***Television Advertising*** The cost, in millions of dollars, of a 30-second television ad during the Super Bowl in the years 1990 to 2007 can be approximated by the following piecewise linear function ($t = 0$ represents 1990):[64]

$$C(t) = \begin{cases} 0.08t + 0.6 & \text{if } 0 \le t < 8 \\ 0.13t + 0.20 & \text{if } 8 \le t \le 17 \end{cases}$$

a. Is C a continuous function of t? Why? HINT [See Example 4 of Section 3.3.]

b. Is C a differentiable function of t? Compute $\lim_{t\to 8^-} C'(t)$ and $\lim_{t\to 8^+} C'(t)$ and interpret the results. HINT [See *Before we go on...* after Example 5.]

54. ▼ ***Processor Speeds*** The processor speed, in megahertz (MHz), of **Intel** processors can be approximated by the following function of time t in years since the start of 1995:[65]

$$P(t) = \begin{cases} 180t + 200 & \text{if } 0 \le t \le 5 \\ 3{,}000t - 13{,}900 & \text{if } 5 < t \le 12 \end{cases}$$

a. Is P a continuous function of t? Why? HINT [See Example 4 of Section 3.3.]

b. Is P a differentiable function of t? Compute $\lim_{t\to 5^-} P'(t)$ and $\lim_{t\to 5^+} P'(t)$ and interpret the results. HINT [See *Before we go on...* after Example 5.]

COMMUNICATION AND REASONING EXERCISES

55. Of the three methods (numerical, graphical, algebraic) we can use to estimate the derivative of a function at a given value of x, which is always the most accurate? Explain.

56. Explain why we cannot put $h = 0$ in the formula

$$f'(a) = \lim_{h\to 0} \frac{f(a+h) - f(a)}{h}$$

for the derivative of f.

57. You just got your derivatives test back and you can't understand why that teacher of yours deducted so many points for what you thought was your best work:

$$\begin{aligned} &\lim_{h\to 0} \frac{f(x+h) - f(x)}{h} \\ &= \lim_{h\to 0} \frac{f(x) + h - f(x)}{h} \\ &= \lim_{h\to 0} \frac{h}{h} && \text{Canceled the } f(x) \\ &= 1 && \text{✗ WRONG } -10 \end{aligned}$$

What was wrong with your answer?

58. Your friend just got his derivatives test back and can't understand why that teacher of his deducted so many points for the following:

$$\begin{aligned} &\lim_{h\to 0} \frac{f(x+h) - f(x)}{h} \\ &= \lim_{h\to 0} \frac{f(x) + f(h) - f(x)}{h} \\ &= \lim_{h\to 0} \frac{f(h)}{h} && \text{Canceled the } f(x) \\ &= \lim_{h\to 0} \frac{f(\not h)}{\not h} && \text{Now cancel the } h. \\ &= f() && \text{✗ WRONG } -50 \end{aligned}$$

What was wrong with his answer?

59. Your other friend just got her derivatives test back and can't understand why that teacher of hers took off so many points for the following:

$$\begin{aligned} &\lim_{h\to 0} \frac{f(x+h) - f(x)}{h} \\ &= \lim_{h\to 0} \frac{f(x+\not h) - f(x)}{\not h} && \text{Now cancel the } h. \\ &= \lim_{h\to 0} f(x) - f(x) && \text{Canceled the } f(x) \\ &= 0 && \text{✗ WRONG } -15 \end{aligned}$$

What was wrong with his answer?

[62] Regression model is based on data from 1976 to 2000. Sources for data: Florida Department of Highway Safety & Motor Vehicles, Florida Marine Institute/*New York Times*, February 12, 2002, p. F4.

[63] Based on 1994 data. Source: The College Board/*New York Times*, March 5, 1995, p. E16.

[64] Sources for data: *New York Times*, January 26, 2001, p. C1., www.money.cnn.com.

[65] Sources for data: Sandpile.org/*New York Times*, May 17, 2004, p. C1 and www.Intel.com.

60. Your third friend just got her derivatives test back and can't understand why that teacher of hers took off so many points for the following:

$$\lim_{h\to 0} \frac{f(x+h)-f(x)}{h}$$
$$= \lim_{h\to 0} \frac{f(x)+h-f(x)}{h}$$
$$= \lim_{h\to 0} \frac{f(x)+\cancel{h}-f(x)}{\cancel{h}} \qquad \text{Now cancel the } h.$$
$$= \lim_{h\to 0} f(x)-f(x) \qquad \text{Canceled the } f(x)$$
$$= 0 \qquad \text{✗ WRONG} \quad -25$$

What was wrong with her answer?

61. Your friend Muffy claims that, because the balanced difference quotient is more accurate, it would be better to use that instead of the usual difference quotient when computing the derivative algebraically. Comment on this advice.

62. Use the balanced difference quotient formula,

$$f'(a) = \lim_{h\to 0} \frac{f(a+h)-f(a-h)}{2h}$$

to compute $f'(3)$ when $f(x) = x^2$. What do you find?

63. ▼ A certain function f has the property that $f'(a)$ does not exist. How is that reflected in the attempt to compute $f'(a)$ algebraically?

64. ▼ One cannot put $h = 0$ in the formula

$$f'(a) = \lim_{h\to 0} \frac{f(a+h)-f(a)}{h}$$

for the derivative of f. (See Exercise 56.) However, in the last step of each of the computations in the text, we are effectively setting $h = 0$ when taking the limit. What is going on here?

CHAPTER 3 REVIEW

KEY CONCEPTS

Web Site www.AppliedCalc.org

Go to the student Web site at www.AppliedCalc.org to find a comprehensive and interactive Web-based summary of Chapter 3.

3.1 Limits: Numerical and Graphical Approaches

$\lim_{x\to a} f(x) = L$ means that $f(x)$ approaches L as x approaches a *p. 194*

What it means for a limit to exist *p. 194*

Limits at infinity *p. 196*

Estimating limits graphically *p. 199*

Interpreting limits in real-world situations *p. 200*

3.2 Limits and Continuity

f is continuous at a if $\lim_{x\to a} f(x)$ exists and $\lim_{x\to a} f(x) = f(a)$ *p. 205*

Discontinuous, continuous on domain *p. 205*

Determining whether a given function is continuous *p. 205*

3.3 Limits and Continuity: Algebraic Approach

Closed-form function *p. 210*

Limits of closed form functions *p. 211*

Simplifying to obtain limits *p. 211*

The indeterminate form 0/0 *p. 212*

The determinate form k/0 *p. 213*

Limits of piecewise defined functions *p. 215*

Limits at infinity *p. 216*

Determinate and indeterminate forms *p. 219*

3.4 Average Rate of Change

Average rate of change of $f(x)$ over $[a, b]$: $\dfrac{\Delta f}{\Delta x} = \dfrac{f(b) - f(a)}{b - a}$ *p. 227*

Average rate of change as slope of the secant line *p. 227*

Computing the average rate of change from a graph *p. 228*

Computing the average rate of change from a formula *p. 229*

Computing the average rate of change over short intervals $[a, a + h]$ *p. 230*

3.5 The Derivative: Numerical and Graphical Viewpoints

Instantaneous rate of change of $f(x)$ (derivative of f at a);

$$f'(a) = \lim_{h\to 0} \frac{f(a + h) - f(a)}{h} \quad p.\ 240$$

The derivative as slope of the tangent line *p. 243*

Quick approximation of the derivative *p. 245*

$\dfrac{d}{dx}$ Notation *p. 247*

The derivative as velocity *p. 247*

Average and instantaneous velocity *p. 249*

The derivative function *p. 250*

Graphing the derivative function with technology *p. 251*

3.6 The Derivative: Algebraic Viewpoint

Derivative at the point $x = a$:

$$f'(a) = \lim_{h\to 0} \frac{f(a + h) - f(a)}{h} \quad p.\ 263$$

Derivative function:

$$f'(x) = \lim_{h\to 0} \frac{f(x + h) - f(x)}{h} \quad p.\ 264$$

Examples of the computation of $f'(x)$ *p. 265*

$f(x) = |x|$ is not differentiable at $x = 0$ *p. 267*

REVIEW EXERCISES

T indicates exercises that must be solved using technology

Numerically *estimate whether the limits in Exercises 1–4 exist. If a limit does exist, give its approximate value.*

1. $\lim_{x\to 3} \dfrac{x^2 - x - 6}{x - 3}$ **2.** $\lim_{x\to 3} \dfrac{x^2 - 2x - 6}{x - 3}$

3. $\lim_{x\to -1} \dfrac{|x + 1|}{x^2 - x - 2}$ **4.** $\lim_{x\to -1} \dfrac{|x + 1|}{x^2 + x - 2}$

In Exercises 5 and 6, the graph of a function f is shown. Graphically determine whether the given limits exist. If a limit does exist, give its approximate value.

5.

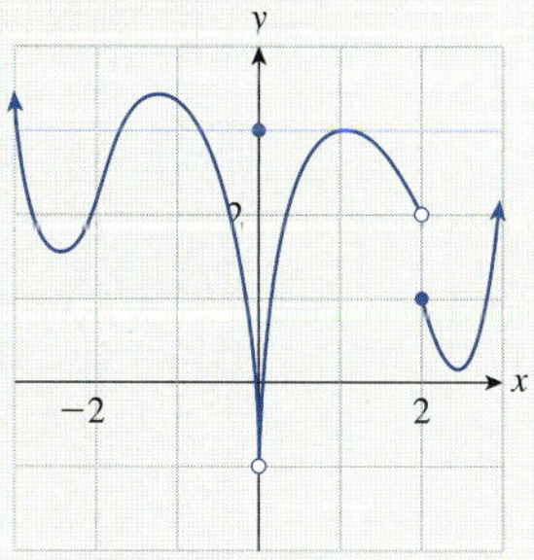

a. $\lim_{x\to 0} f(x)$ **b.** $\lim_{x\to 1} f(x)$

c. $\lim_{x\to 2} f(x)$

6.

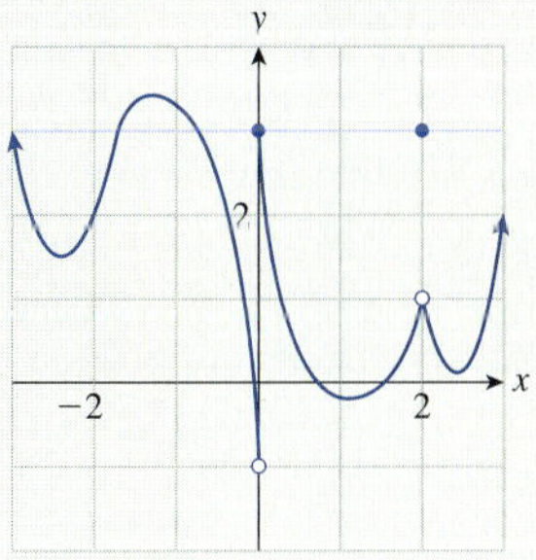

a. $\lim_{x\to 0} f(x)$ **b.** $\lim_{x\to -2} f(x)$

c. $\lim_{x\to 2} f(x)$

Calculate the limits in Exercises 7–24 algebraically. If a limit does not exist, say why.

7. $\lim_{x\to -2} \dfrac{x^2}{x - 3}$ **8.** $\lim_{x\to 3} \dfrac{x^2 - 9}{2x - 6}$

9. $\lim_{x\to 0} \dfrac{x}{2x^2 - x}$ **10.** $\lim_{x\to 1} \dfrac{x^2 - 9}{x - 1}$

11. $\lim_{x\to -1} \dfrac{x^2 + 3x}{x^2 - x - 2}$ **12.** $\lim_{x\to -1^+} \dfrac{x^2 + 1}{x^2 + 3x + 2}$

13. $\lim_{x\to 4} \dfrac{x^2 + 8}{x^2 - 2x - 8}$ **14.** $\lim_{x\to 4} \dfrac{x^2 + 3x}{x^2 - 8x + 16}$

15. $\lim_{x\to 1/2} \dfrac{x^2 + 8}{4x^2 - 4x + 1}$ **16.** $\lim_{x\to 1/2} \dfrac{x^2 + 3x}{2x^2 + 3x - 1}$

17. $\lim_{x\to -\infty} \dfrac{x^2 - x - 6}{x - 3}$ **18.** $\lim_{x\to \infty} \dfrac{x^2 - x - 6}{4x^2 - 3}$

19. $\lim_{t\to +\infty} \dfrac{-5}{5 + 5.3(3^{2t})}$ **20.** $\lim_{t\to +\infty} \left(3 + \dfrac{2}{e^{4t}}\right)$

21. $\lim_{x\to +\infty} \dfrac{2}{5 + 4e^{-3x}}$ **22.** $\lim_{x\to +\infty} (4e^{3x} + 12)$

23. $\lim_{t\to +\infty} \dfrac{1 + 2^{-3t}}{1 + 5.3e^{-t}}$ **24.** $\lim_{x\to -\infty} \dfrac{8 + 0.5^x}{2 - 3^{2x}}$

In Exercises 25–28, find the average rate of change of the given function over the interval $[a, a+h]$ *for* $h = 1, 0.01,$ *and* 0.001. *(Round answers to four decimal places.) Then estimate the slope of the tangent line to the graph of the function at a.*

25. $f(x) = \frac{1}{x+1}$; $a = 0$ **26.** $f(x) = x^x$; $a = 2$

27. $f(x) = e^{2x}$; $a = 0$ **28.** $f(x) = \ln(2x)$; $a = 1$

In Exercises 29–32 you are given the graph of a function with four points marked. Determine at which (if any) of these points the derivative of the function is: ***(i)*** *−1* ***(ii)*** *0* ***(iii)*** *1, and* ***(iv)*** *2.*

29.

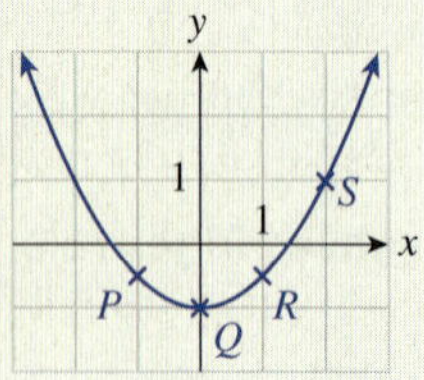

30.

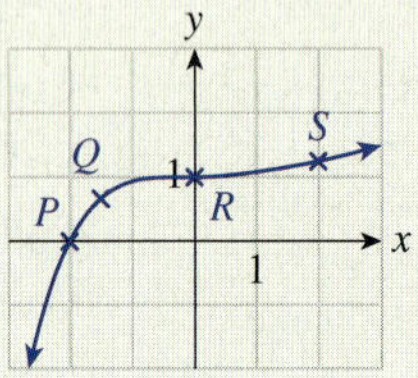

31.

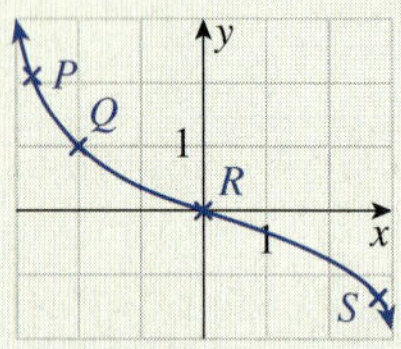

32.

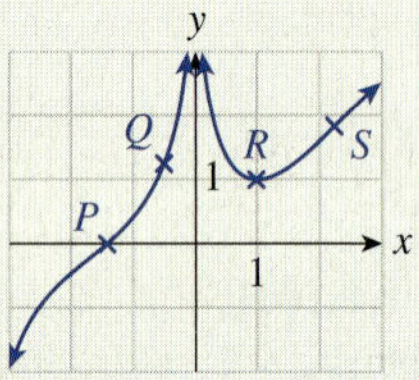

33. Let f have the graph shown.

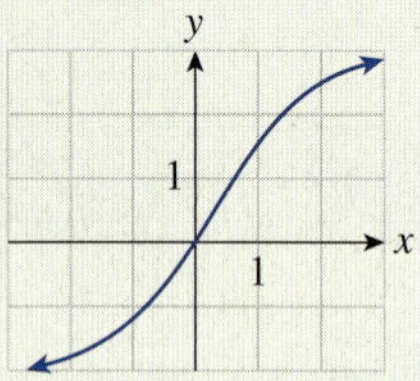

Select the correct answer.

a. The average rate of change of f over the interval [0, 2] is
(A) greater than $f'(0)$ **(B)** less than $f'(0)$
(C) approximately equal to $f'(0)$

b. The average rate of change of f over the interval $[-1, 1]$ is
(A) greater than $f'(0)$ **(B)** less than $f'(0)$
(C) approximately equal to $f'(0)$

c. Over the interval [0, 2], the instantaneous rate of change of f is
(A) increasing **(B)** decreasing
(C) neither increasing nor decreasing

d. Over the interval $[-2, 2]$, the instantaneous rate of change of f is
(A) increasing, then decreasing
(B) decreasing, then increasing
(C) approximately constant

e. When $x = 2$, $f(x)$ is
(A) approximately 1 and increasing at a rate of about 2.5 units per unit of x
(B) approximately 1.2 and increasing at a rate of about 1 unit per unit of x
(C) approximately 2.5 and increasing at a rate of about 0.5 units per unit of x
(D) approximately 2.5 and increasing at a rate of about 2.5 units per unit of x

34. Let f have the graph shown.

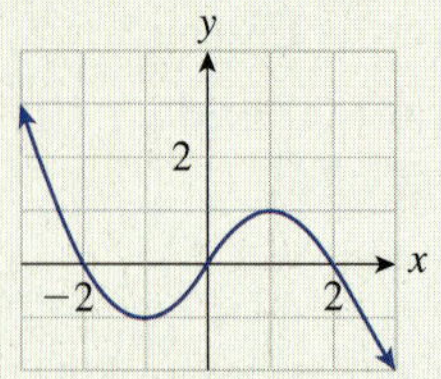

Select the correct answer.

a. The average rate of change of f over the interval [0, 1] is
(A) greater than $f'(0)$ **(B)** less than $f'(0)$
(C) approximately equal to $f'(0)$

b. The average rate of change of f over the interval [0, 2] is
(A) greater than $f'(1)$ **(B)** less than $f'(1)$
(C) approximately equal to $f'(1)$

c. Over the interval $[-2, 0]$, the instantaneous rate of change of f is
(A) increasing **(B)** decreasing
(C) neither increasing nor decreasing

d. Over the interval $[-2, 2]$, the instantaneous rate of change of f is
(A) increasing, then decreasing
(B) decreasing, then increasing
(C) approximately constant

e. When $x = 0$, $f(x)$ is
(A) approximately 0 and increasing at a rate of about 1.5 units per unit of x
(B) approximately 0 and decreasing at a rate of about 1.5 units per unit of x
(C) approximately 1.5 and neither increasing nor decreasing
(D) approximately 0 and neither increasing nor decreasing

In Exercises 35–38, use the definition of the derivative to calculate the derivative of each of the given functions algebraically.

35. $f(x) = x^2 + x$ **36.** $f(x) = 3x^2 - x + 1$

37. $f(x) = 1 - \frac{2}{x}$ **38.** $f(x) = \frac{1}{x} + 1$

T *In Exercises 39–42, use technology to graph the derivative of the given function. In each case, choose a range of x-values and y-values that shows the interesting features of the graph.*

39. $f(x) = 10x^5 + \frac{1}{2}x^4 - x + 2$

40. $f(x) = \frac{10}{x^5} + \frac{1}{2x^4} - \frac{1}{x} + 2$

41. $f(x) = 3x^3 + 3\sqrt[3]{x}$

42. $f(x) = \frac{2}{x^{2.1}} - \frac{x^{0.1}}{2}$

APPLICATIONS

43. ***Stock Investments*** OHaganBooks.com CEO John O'Hagan has terrible luck with stocks. The following graph shows the value of Fly-By-Night Airlines stock that he bought acting on a "hot tip" from Marjory Duffin (CEO of Duffin Press and a close business associate):

Fly-by-night stock

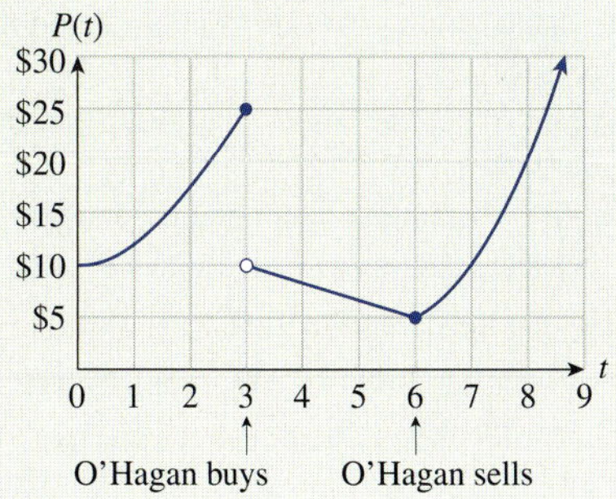

a. Compute $P(3)$, $\lim_{t \to 3^-} P(t)$ and $\lim_{t \to 3^+} P(t)$. Does $\lim_{t \to 3} P(t)$ exist? Interpret your answers in terms of Fly-By-Night stock.

b. Is P continuous at $t = 6$? Is P differentiable at $t = 6$? Interpret your answers in terms of Fly-By-Night stock.

44. ***Stock Investments*** John O'Hagan's golf partner Juan Robles seems to have had better luck with his investment in Gapple Gomputer Inc. stocks as shown in the following graph:

Gapple Inc. Stock

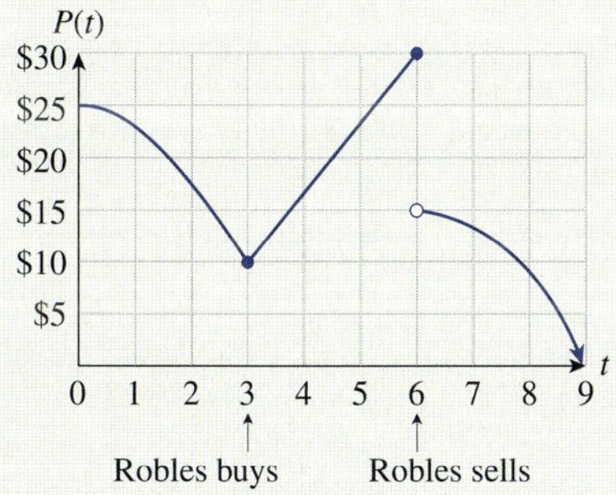

a. Compute $P(6)$, $\lim_{t \to 6^-} P(t)$ and $\lim_{t \to 6^+} P(t)$. Does $\lim_{t \to 6} P(t)$ exist? Interpret your answers in terms of Gapple stock.

b. Is P continuous at $t = 3$? Is P differentiable at $t = 3$? Interpret your answers in terms of Gapple stock.

45. ***Real Estate*** Marjory Duffin has persuaded John O'Hagan to consider investing a portion of OHaganBooks.com profits in real estate, now that the real estate market seems to have bottomed out. A real estate broker friend of hers e-mailed her the following (somewhat optimistic) graph from brokersadvocacy.com:[66]

Home price index

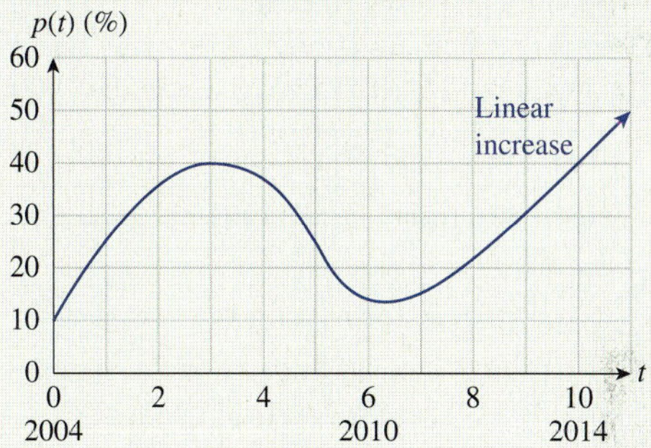

Here, $p(t)$ is the home price percentage over the 2003 level.

a. Assuming the trend shown in the graph were to continue indefinitely, estimate $\lim_{t \to 3} p(t)$ and $\lim_{t \to +\infty} p(t)$ and interpret the results.

b. Estimate $\lim_{t \to +\infty} p'(t)$ and interpret the result.

46. ***Advertising Costs*** OHaganBooks.com has (on further advice from Marjory Duffin) mounted an aggressive online marketing strategy. The following graph shows the weekly cost of this campaign for the six-week period since the start of July (t is time in weeks):

a. Assuming the trend shown in the graph were to continue indefinitely, estimate $\lim_{t \to 2} C(t)$ and $\lim_{t \to +\infty} C(t)$ and interpret the results.

b. Estimate $\lim_{t \to +\infty} C'(t)$ and interpret the result.

47. ***Sales*** Since the start of July, OHaganBooks.com has seen its weekly sales increase, as shown in the following table:

Week	1	2	3	4	5	6
Sales (books)	6,500	7,000	7,200	7,800	8,500	9,000

a. What was the average rate of increase of weekly sales over this entire period?

b. During which 1-week interval(s) did the rate of increase of sales exceed the average rate?

c. During which 2-week interval(s) did the weekly sales rise at the highest average rate, and what was that average rate?

[66] Authors' note: As of August 2008, brokersadvocacy.com is unregistered.

48. ***Rising Sea Level*** Marjory Duffin recently purchased a beachfront condominium in New York and is now in a panic, having just seen some disturbing figures about rising sea levels (sea levels as measured in New York relative to the 1900 level).[67]

Year since 1900	0	25	50	75	100	125
Sea Level (mm)	0	60	140	240	310	390

a. What was the average rate of increase of the sea level over this entire period?
b. During which 25-year interval(s) did the rate of increase of the sea level exceed the average rate?
c. Marjory Duffin's condominium is about 2 meters above sea level. Using the average rate of change from part (a), estimate how long she has before the sea rises to her condominium.

49. ***Real Estate*** The following graph (see Exercise 45) shows the home price index chart e-mailed to Marjory Duffin by a real estate broker:

Home price index

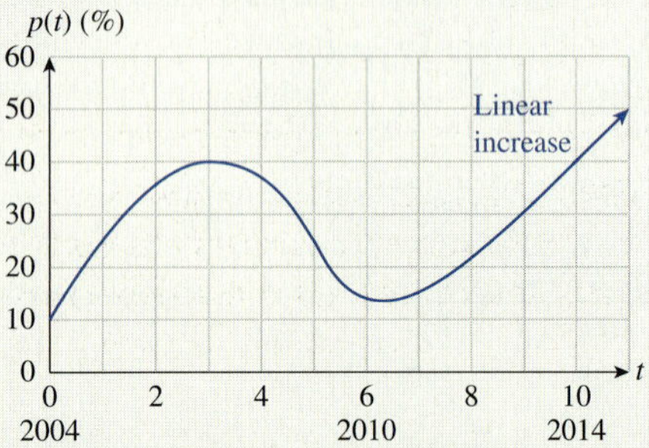

Use the graph to answer the following questions:
a. What was the average rate of change of the index over the 10-year period beginning 2004?
b. What was the average rate of change of the index over the period [3, 10]?
c. Which of the following is correct? Over the period [4, 6],
(A) The rate of change of the index increased.
(B) The rate of change of the index increased and then decreased.
(C) The rate of change of the index decreased.
(D) The rate of change of the index decreased and then increased.

50. ***Advertising Costs*** The following graph (see Exercise 46) shows the weekly cost of OHaganBooks.com's online ad campaign for the six-week period since the start of July (t is time in weeks).

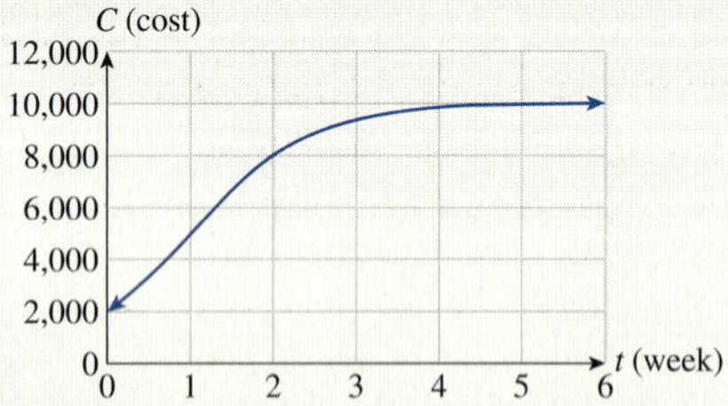

Use the graph to answer the following questions:
a. What was the average rate of change of cost over the entire six-week period?
b. What was the average rate of change of cost over the period [2, 6]?
c. Which of the following is correct? Over the period [2, 6],
(A) The rate of change of cost increased and the cost increased.
(B) The rate of change of cost decreased and the cost increased.
(C) The rate of change of cost increased and the cost decreased.
(D) The rate of change of cost decreased and the cost decreased.

51. ***Sales*** OHaganBooks.com fits the curve

$$w(t) = 36t^2 + 250t + 6240 \quad (0 \le t \le 6)$$

to its weekly sales figures from Exercise 47, as shown in the following graph:

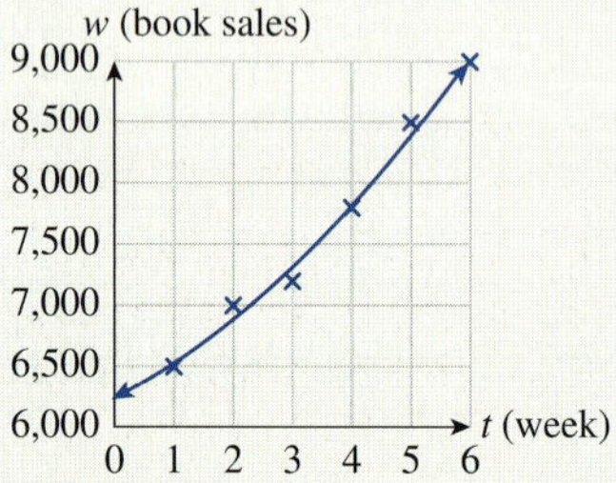

a. Compute the derivative function $w'(t)$.
b. According to the model, what was the rate of increase of sales at the beginning of the second week ($t = 1$)?
c. If we extrapolate the model, what would be the rate of increase of weekly sales at the beginning of the 8th week ($t = 7$)?

52. ***Sea Levels*** Marjory Duffin fit the curve

$$s(t) = 0.002t^2 + 3t - 6.4 \quad (0 \le t \le 125)$$

to her sea level figures from Exercise 48, as shown in the following graph:

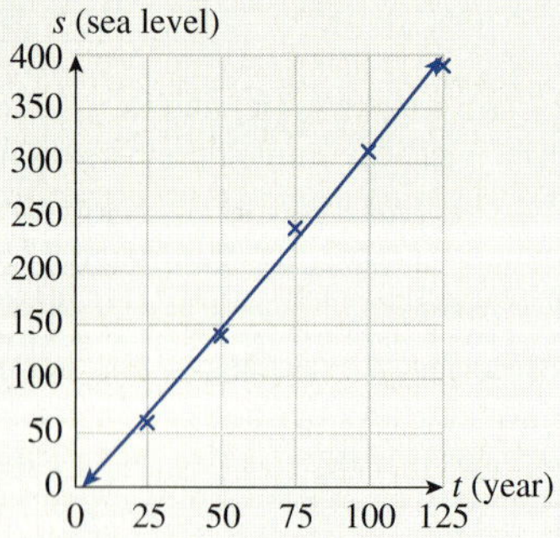

a. Compute the derivative function $s'(t)$.
b. According to the model, what was the rate of increase of the sea level in 2000 ($t = 100$)?
c. If we extrapolate the model, what would be the rate of increase of the sea level in 2100 ($t = 200$)?

[67] The 2025 level is a projection. Source: New England Integrated Science & Assessment (www.neisa.unh.edu/Climate/index.html).

Case Study Reducing Sulfur Emissions

Norbert Schaefer/CORBIS

The Environmental Protection Agency (EPA) wishes to formulate a policy that will encourage utilities to reduce sulfur emissions. Its goal is to reduce annual emissions of sulfur dioxide by a total of 10 million tons from the current level of 25 million tons by imposing a fixed charge for every ton of sulfur released into the environment per year. As a consultant to the EPA, you must determine the amount to be charged per ton of sulfur emissions.

You would like first to know the cost to the utility industry of reducing sulfur emissions. In other words, you would like to have a cost function of the form

$$C(q) = \text{Cost of removing } q \text{ tons of sulfur dioxide.}$$

Unfortunately, you do not have such a function handy. You do, however, have the following data, which show the *marginal* cost (that is, the *rate of change* of cost) to the utility industry of reducing sulfur emissions at several levels of reduction.[68]

Reduction (tons) q	8,000,000	10,000,000	12,000,000
Marginal Cost ($ per ton) $C'(q)$	270	360	779

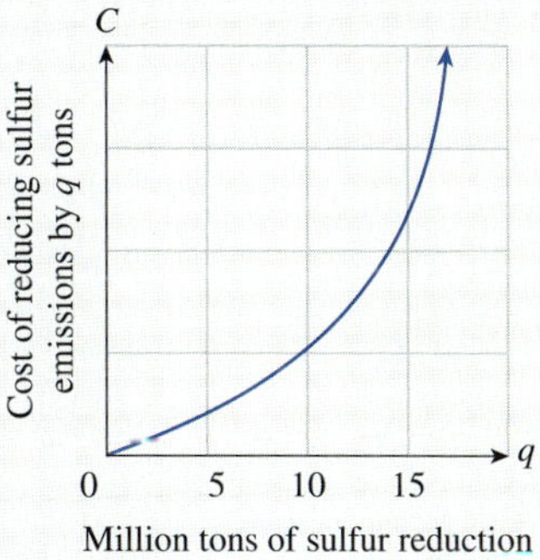

Figure 39

The table tells you that $C'(8{,}000{,}000) = \$270$ per ton, $C'(10{,}000{,}000) = \$360$ per ton, and $C'(12{,}000{,}000) = \$779$ per ton. Recalling that $C'(q)$ is the slope of the tangent to the graph of the cost function, you can see from the table that this slope is positive and increasing as q increases, so the graph of the cost function has the general shape shown in Figure 39.

Notice that the slope (additional cost) is increasing as you move to the right, so the utility industry has no cost incentive to reduce emissions further, as it costs the industry significantly more per ton for each additional ton of sulfur it removes. What you would like—if the goal of reducing total emissions by 10 million tons is to be reached—is that, somehow, the imposition of a fixed charge for every ton of sulfur dioxide released will *alter* the form of the cost curve so that it has the general shape shown in Figure 40. In this ideal curve, the cost D to utilities is lowest at a reduction level of 10 million tons, so if the utilities act to minimize cost, they can be expected to reduce emissions by 10 million tons, which is exactly the EPA goal! From the graph, you can see that the tangent line to the curve at the point where $q = 10$ million tons is horizontal, and thus has zero slope: $D'(10{,}000{,}000) = \$0$ per ton. Further, the slope $D'(q)$ is negative for values of q to the left of 10 million tons and positive for values to the right.

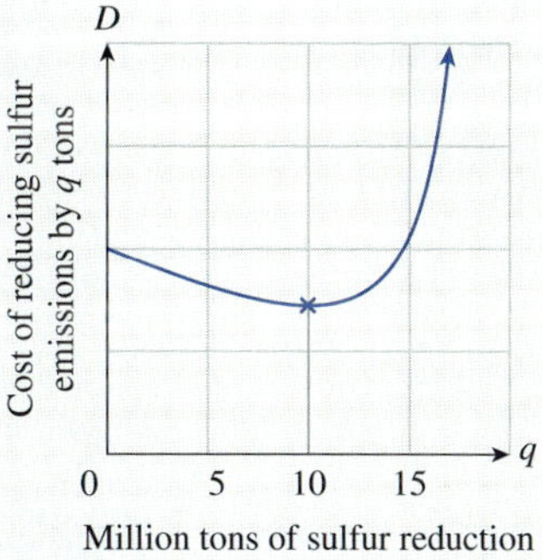

Figure 40

So, how much should the EPA charge per ton of sulfur released into the environment? Suppose the EPA charges $\$k$ per ton, so that

$$\text{Emission charge to utilities} = k \times \text{Sulfur emission}$$

It is your job to calculate k. Because you are working with q as the independent variable, you decide that it would be best to formulate the emission charge as a function of q. However, q represents the amount by which sulfur emissions have been reduced from the original 25 million tons; that is, the amount by which sulfur emissions are *lower than* the original 25 million tons:

$$q = 25{,}000{,}000 - \text{Sulfur emissions}$$

[68] These figures were produced in a computerized study of reducing sulfur emissions from the 1980 level by the given amounts. Source: Congress of the United States, Congressional Budget Office, *Curbing Acid Rain: Cost, Budget and Coal Market Effects* (Washington, DC: Government Printing Office, 1986): xx, xxii, 23, 80.

Thus, the total annual emission charge to the utilities is

$$k \times \text{Sulfur emission} = k(25{,}000{,}000 - q) = 25{,}000{,}000k - kq$$

This results in a total cost to the utilities of

$$\text{Total cost} = \text{Cost of reducing emissions} + \text{Emission charge}$$
$$D(q) = C(q) + 25{,}000{,}000k - kq$$

You now recall from calculus that the derivative of a sum of two functions is the sum of their derivatives (you will see why in Section 4.1*), so the derivative of D is given by

* **NOTE** This statement makes intuitive sense: For instance, if C is changing at a rate of 3 units per second and D is changing at a rate of 2 units per second, then their sum is changing at a rate of $3 + 2 = 5$ units per second.

$$D'(q) = \text{Derivative of } C + \text{Derivative of } (25{,}000{,}000k - kq)$$

The function $y = 25{,}000{,}000k - kq$ is a linear function of q with slope $-k$ and intercept $25{,}000{,}000k$. Thus its derivative is just its slope: $-k$. Therefore:

$$D'(q) = C'(q) - k$$

Remember that you want

$$D'(10{,}000{,}000) = 0.$$

Thus,

$$C'(10{,}000{,}000) - k = 0$$

Referring to the table, you see that

$$360 - k = 0$$

or

$$k = \$360 \text{ per ton.}$$

In other words, all you need to do is set the emission charge at $k = \$360$ per ton of sulfur emitted. Further, to ensure that the resulting curve will have the general shape shown in Figure 40, you would like to have $D'(q)$ negative for $q < 10{,}000{,}000$ and positive for $q > 10{,}000{,}000$. To check this, write

$$D'(q) = C'(q) - k$$
$$= C'(q) - 360$$

and refer to the table to obtain

$$D'(8{,}000{,}000) = 270 - 360 = -90 < 0 \checkmark$$

and

$$D'(12{,}000{,}000) = 779 - 360 = 419 > 0 \checkmark$$

Thus, based on the given data, the resulting curve will have the shape you require. You therefore inform the EPA that an annual emissions charge of \$360 per ton of sulfur released into the environment will create the desired incentive: to reduce sulfur emissions by 10 million tons per year.

One week later, you are informed that this charge would be unrealistic because the utilities cannot possibly afford such a cost. You are asked whether there is an alternative plan that accomplishes the 10-million-ton reduction goal and yet is cheaper to the utilities by \$5 billion per year. You then look at your expression for the emission charge

$$25{,}000{,}000k - kq$$

and notice that, if you decrease this amount by $5 billion, the derivative will not change at all because it will still have the same slope (only the intercept is affected). Thus, you propose the following revised formula for the emission charge:

$$\begin{aligned}&25{,}000{,}000k - kq - 5{,}000{,}000{,}000\\&= 25{,}000{,}000(360) - 360q - 5{,}000{,}000{,}000\\&= 4{,}000{,}000{,}000 - 360q\end{aligned}$$

At the expected reduction level of 10 million tons, the total amount paid by the utilities will then be

$$4{,}000{,}000{,}000 - 360(10{,}000{,}000) = \$400{,}000{,}000$$

Thus, your revised proposal is the following: Impose an annual emissions charge of $360 per ton of sulfur released into the environment and hand back $5 billion in the form of subsidies. The effect of this policy will be to cause the utilities industry to reduce sulfur emissions by 10 million tons per year and will result in $400 million in annual revenues to the government.

Notice that this policy also provides an incentive for the utilities to search for cheaper ways to reduce emissions. For instance, if they lowered costs to the point where they could achieve a reduction level of 12 million tons, they would have a total emission charge of

$$4{,}000{,}000{,}000 - 360(12{,}000{,}000) = -\$320{,}000{,}000.$$

The fact that this is negative means that the government would be paying the utilities industry $320 million more in annual subsidies than the industry is paying in per ton emission charges.

EXERCISES

1. Excluding subsidies, what should the annual emission charge be if the goal is to reduce sulfur emissions by 8 million tons?
2. Excluding subsidies, what should the annual emission charge be if the goal is to reduce sulfur emissions by 12 million tons?
3. What is the *marginal emission charge* (derivative of emission charge) in your revised proposal (as stated before the exercise set)? What is the relationship between the marginal cost of reducing sulfur emissions before emissions charges are implemented and the marginal emission charge, at the optimal reduction under your revised proposal?
4. We said that the revised policy provided an incentive for utilities to find cheaper ways to reduce emissions. How would $C(q)$ have to change to make 12 million tons the optimum reduction?
5. What change in $C(q)$ would make 8 million tons the optimum reduction?
6. If the scenario in Exercise 5 took place, what would the EPA have to do in order to make 10 million tons the optimal reduction once again?
7. Due to intense lobbying by the utility industry, you are asked to revise the proposed policy so that the utility industry will pay no charge if sulfur emissions are reduced by the desired 10 million tons. How can you accomplish this?
8. Suppose that instead of imposing a fixed charge per ton of emission, you decide to use a sliding scale, so that the total charge to the industry for annual emissions of x tons will be $\$kx^2$ for some k. What must k be to again make 10 million tons the optimum reduction? **HINT** [The derivative of kx^2 is $2kx$.]

TI-83/84 Plus Technology Guide

Section 3.1

Example 1 (page 193) Use a table to estimate the following limits.

a. $\lim_{x\to 2} \frac{x^3-8}{x-2}$ **b.** $\lim_{x\to 0} \frac{e^{2x}-1}{x}$

Solution with Technology

On the TI-83/84 Plus, use the table feature to automate these computations as follows:

1. Define `Y₁=(X^3-8)/(X-2)` for part (a) or `Y₁=(e^(2X)-1)/X` for part (b).
2. Press [2ND] [TABLE] to list its values for the given values of x. (If the calculator does not allow you to enter values of x, press [2ND] [TBLSET] and set `Indpnt` to `Ask`).

   Here is the table showing some of the values for part (a):

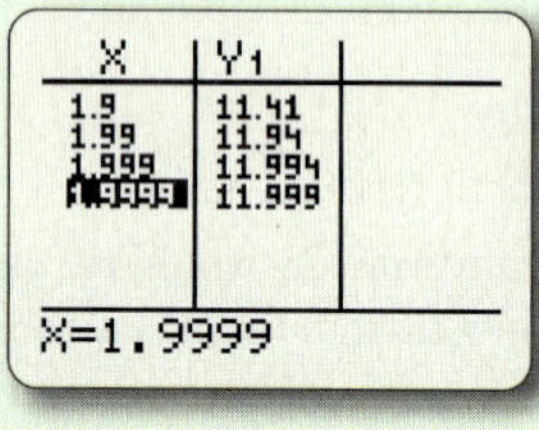

3. For part (b) use `Y₁=(e^(2X)-1)/X` and use values of x approaching 0 from either side.

Section 3.4

Example 3 (page 229) The price of an ounce of gold can be approximated by the function

$$G(t) = -8t^2 + 144t + 150 \text{ dollars} \quad (7.5 \le t \le 10.5)$$

where t is time in hours. What was the average rate of change of the price of gold over the $1\frac{1}{2}$-hour period starting at 8:00 AM (the interval [8, 9.5] on the t-axis)?

Solution with Technology

On the TI-83/84 Plus:

1. Enter the function G as `Y₁` (using X for t):

   `Y₁=-8X²+144X+150`

2. Now find the average rate of change over [8, 9.5] by evaluating the following on the home screen:

   `(Y₁(9.5)-Y₁(8))/(9.5-8)`

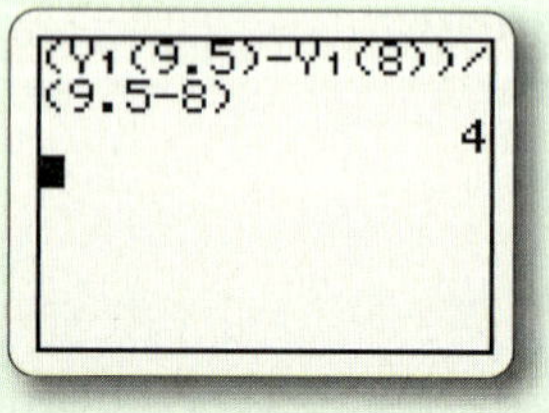

As shown on the screen, the average rate is of change is 4.

Example 4 (page 230) Continuing with Example 3, use technology to compute the average rate of change of

$$G(t) = -8t^2 + 144t + 150 \qquad (7.5 \le t \le 10.5)$$

over the intervals $[8, 8+h]$, where $h = 1, 0.1, 0.01, 0.001$, and 0.0001.

Solution with Technology

1. As in Example 3, enter the function G as `Y₁` (using X for t):

   `Y₁=-8X²+144X+150`

2. Now find the average rate of change for $h = 1$ by evaluating, on the home screen,

   `(Y₁(8+1)-Y₁(8))/1`

   which gives 8.

3. To evaluate for $h = 0.1$, recall the expression using [2ND] [ENTER] and then change the 1, both places it occurs, to 0.1, getting

   `(Y₁(8+0.1)-Y₁(8))/0.1`

   which gives 15.2.

4. Continuing, we can evaluate the average rate of change for all the desired values of h:

We get the values in the following table:

h	1	0.1	0.01	0.001	0.0001
Ave. Rate of Change $\dfrac{G(8+h)-G(8)}{h}$	8	15.2	15.92	15.992	15.9992

Section 3.5

Example 2 (page 245) Calculate an approximate value of $f'(1.5)$ if $f(x) = x^2 - 4x$, and then find the equation of the tangent line at the point on the graph where $x = 1.5$.

Solution with Technology

1. In the TI-83/84 Plus, enter the function f as Y_1

$$\mathrm{Y_1{=}X\hat{\ }2{-}4{*}X}$$

2. Go to the home screen to compute the approximations:

`(Y1(1.5001)-Y1(1.5))/0.0001`

Usual difference quotient

`(Y1(1.5001)-Y1(1.4999))/0.0002`

Balanced difference quotient

From the display on the right, we find that the difference quotient quick approximation is -0.9999 and the balanced difference quotient quick approximation is -1, which is in fact the exact value of $f'(1.5)$. See the discussion in the text for the calculation of the equation of the tangent line.

Example 4 (page 251) Use technology to graph of the derivative of $f(x) = -2x^2 + 6x + 5$ for values of x in starting at -5.

Solution with Technology

On the TI-83/84 Plus, the easiest way to obtain quick approximations of the derivative of a given function is to use the built-in `nDeriv` function, which calculates balanced difference quotients.

1. On the `Y=` screen, first enter the function:

$$\mathrm{Y_1{=}{-}2X\hat{\ }2{+}6X{+}5}$$

2. Then set

`Y2=nDeriv(Y1,X,X)` For `nDeriv` press [MATH] [8]

which is the TI-83's approximation of $f'(x)$.

Alternatively, we can enter the balanced difference quotient directly:

`Y2=(Y1(X+0.001)-Y1(X-0.001))/0.002`

(The TI-83 uses $h = 0.001$ by default in the balanced difference quotient when calculating `nDeriv`, but this can be changed by giving a value of h as a fourth argument, like `nDeriv(Y1,X,X,0.0001)`.) To see a table of approximate values of the derivative, we press 2ND TABLE and choose a collection of values for x:

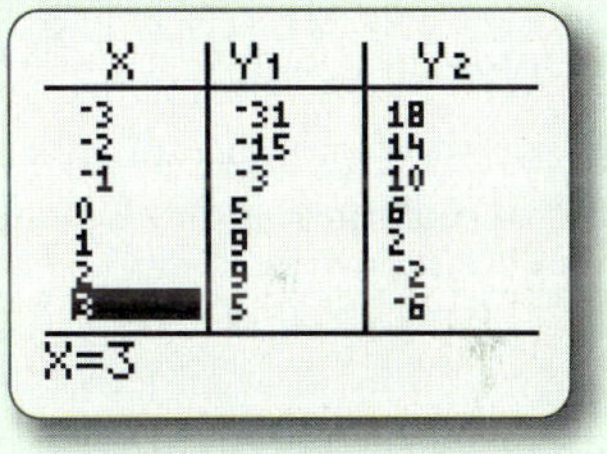

Here, Y_1 shows the value of f and Y_2 shows the values of f'.

To graph the function or its derivative, we can graph Y_1 or Y_2 in a window showing the given domain $[-5, 5]$:

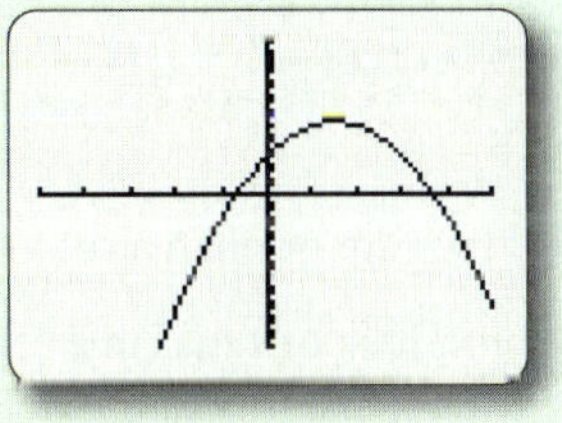

Graph of f

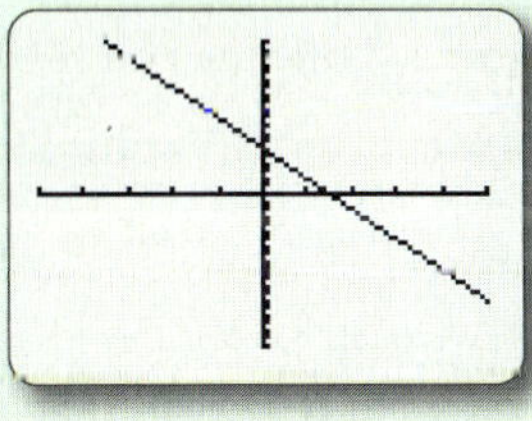

Graph of f'

EXCEL Technology Guide

Section 3.1

Example 1 (page 193) Use a table to estimate the following limits.

a. $\lim_{x \to 2} \frac{x^3 - 8}{x - 2}$ **b.** $\lim_{x \to 0} \frac{e^{2x} - 1}{x}$

Solution with Technology

1. Set up your spreadsheet to duplicate the table in part (a) as follows:

	A	B	C	D
1	x	f(x)	x	f(x)
2	1.9	=(A2^3-8)/(A2-2)	2.1	
3	1.99		2.01	
4	1.999		2.001	
5	1.9999		2.0001	

↓

	A	B	C	D
1	x	f(x)	x	f(x)
2	1.9	11.41	2.1	12.61
3	1.99	11.9401	2.01	12.0601
4	1.999	11.994001	2.001	12.006001
5	1.9999	11.99940001	2.0001	12.00060001

(The formula in cell B2 is copied to columns B and D as indicated by the shading.) The values of $f(x)$ will be calculated in columns B and D.

2. For part (b), use the formula `=(EXP(2*A2)-1)/A2` in cell B2 and, in columns A and C, use values of x approaching 0 from either side.

Section 3.4

Example 3 (page 229) The price of an ounce of gold can be approximated by the function

$$G(t) = -8t^2 + 144t + 150 \text{ dollars} \quad (7.5 \le t \le 10.5)$$

where t is time in hours. What was the average rate of change of the price of gold over the $1\frac{1}{2}$-hour period starting at 8:00 AM (the interval [8, 9.5] on the t-axis)?

Solution with Technology

To use Excel to compute the average rate of change of G:

1. Start with two columns, one for values of t and one for values of $G(t)$, which you enter using the formula for G:

	A	B
1	t	G(t)
2	8	=-8*A2^2+144*A2+150
3	9.5	

2. Next, calculate the average rate of change as shown here:

	A	B	C
1	t	G(t)	
2	8	790	Rate of change over [8, 9.5]:
3	9.5	796	=(B3-B2)/(A3-A2)

↓

	A	B	C	D
1	t	G(t)		
2	8	790	Rate of change over [8, 9.5]:	
3	9.5	796	4	

In Example 4, we describe another, more versatile Excel template for computing rates of change. ■

Example 4 (page 230) Continuing with Example 3, use technology to compute the average rate of change of

$$G(t) = -8t^2 + 144t + 150 \qquad (7.5 \le t \le 10.5)$$

over the intervals $[8, 8 + h]$, where $h = 1, 0.1, 0.01, 0.001$, and 0.0001.

Solution with Technology

The template we can use to compute the rates of change is an extension of what we used in Example 3:

	A	B	C	D	E
1	a	h	t	G(t)	Rate of Change
2	8	0.001	=A2	=-8*C2^2+144*C2+150	=(D3-D2)/(C3-C2)
3			=A2+B2		

↓

	A	B	C	D	E
1	a	h	t	G(t)	Rate of Change
2	8	0.001	8	790	15.992
3			8.001	790.015992	

1. Column C contains the values $t = a$ and $t = a + h$ we are using for the independent variable.
2. The formula in cell E2 is the average rate of change formula $\Delta G/\Delta t$. Entering the different values $h = 1, 0.1, 0.01, 0.001$, and 0.0001 in cell B2 gives the results shown in the following table:

h	1	0.1	0.01	0.001	0.0001
Ave. Rate of Change $\frac{G(8+h) - G(8)}{h}$	8	15.2	15.92	15.992	15.9992

Section 3.5

Example 2 (page 245) Calculate an approximate value of $f'(1.5)$ if $f(x) = x^2 - 4x$, and then find the equation of the tangent line at the point on the graph where $x = 1.5$.

Solution with Technology

You can compute both the difference quotient and the balanced difference quotient approximations in Excel using the following extension of the worksheet in Example 4 in Section 3.4.

	A	B	C	D	E	F
1	a	h	x	f(x)	Diff Quotients	Balanced Diff Quotient
2	1.5	0.0001	=A2-B2	=C2^2-4*C2	=(D3-D2)/(C3-C2)	=(D4-D2)/(C4-C2)
3			=A2			
4			=A2+B2			

Notice that we get two difference quotients in column E. The first uses $h = -0.0001$ while the second uses $h = 0.0001$ and is the one we use for our quick approximation. The balanced quotient is their average (column F). The results are as follows.

	A	B	C	D	E	F
1	a	h	x	f(x)	Diff Quotients	Balanced Diff Quotient
2	1.5	0.0001	1.4999	-3.7499	-1.0001	-1
3			1.5	-3.75	-0.9999	
4			1.5001	-3.7501		

From the results shown above, we find that the difference quotient quick approximation is -0.9999 and that the balanced difference quotient quick approximation is -1, which is in fact the exact value of $f'(1.5)$. See the discussion in the text for the calculation of the equation of the tangent line.

Example 4 (page 251) Use technology to graph of the derivative of $f(x) = -2x^2 + 6x + 5$ for values of x in starting at -5.

Solution with Technology

1. Start with a table of values for the function f:

	A	B	C	D	E
1	x	f(x)		Xmin	-5
2	=$E$1	=-2*A2^2+6*A2+5		h	0.1
3	=A2+$E$2				
4					
5					
101					
102					

2. Next, compute approximate derivatives in Column C:

	A	B	C	D	E
1	x	f(x)	f'(x)	Xmin	-5
2	-5	-75	=(B3-B2)/$E$2	h	0.1
3	-4.9	-72.42			
4	-4.8	-69.88			
5	-4.7	-67.38			
101	4.9	-13.62			
102	5	-15			

	A	B	C	D	E
1	x	f(x)	f'(x)	Xmin	-5
2	-5	-75	25.8	h	0.1
3	-4.9	-72.42	25.4		
4	-4.8	-69.88	25		
5	-4.7	-67.38	24.6		
101	4.9	-13.62	-13.8		
102	5	-15			

You cannot paste the difference quotient formula into cell C102. (Why?) Notice that this worksheet uses the ordinary difference quotients, $[f(x+h) - f(x)]/h$. If you prefer, you can use balanced difference quotients $[f(x+h) - f(x-h)]/(2h)$, in which case cells C2 and C102 would both have to be left blank.

We then graph the function and the derivative on different graphs as follows:

1. First, graph the function f in the usual way, using Columns A and B.
2. Make a copy of this graph and click on it once. Columns A and B should be outlined, indicating that these are the columns used in the graph.

3. By dragging from the center of the bottom edge of the box, move the Column B box over to Column C as shown:

	A	B	C
96	4.4	-7.32	-11.8
97	4.5	-8.5	-12.2
98	4.6	-9.72	-12.6
99	4.7	-10.98	-13
100	4.8	-12.28	-13.4
101	4.9	-13.62	-13.8
102	5	-15	

→

	A	B	C
96	4.4	-7.32	-11.8
97	4.5	-8.5	-12.2
98	4.6	-9.72	-12.6
99	4.7	-10.98	-13
100	4.8	-12.28	-13.4
101	4.9	-13.62	-13.8
102	5	-15	

The graph will then show the derivative (Columns A and C):

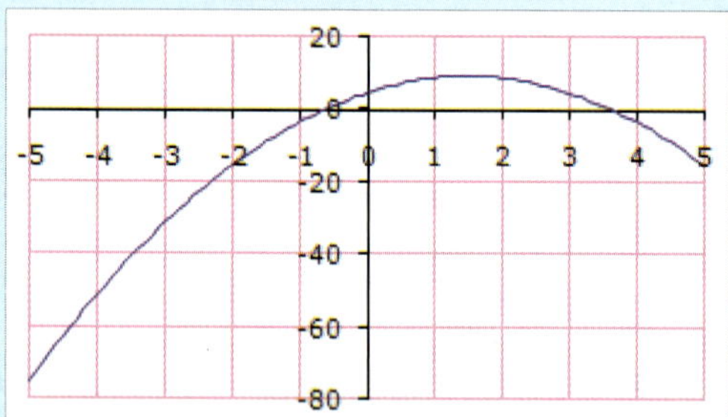

Graph of f

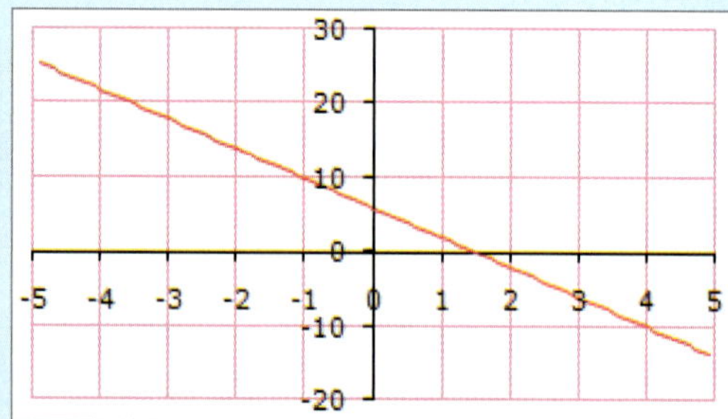

Graph of f'

4 Techniques of Differentiation with Applications

Case Study Projecting Market Growth

You are on the board of directors at Fullcourt Academic Press. The sales director of the high school division has just burst into your office with a proposal for an expansion strategy based on the assumption that the number of high school seniors in the United States will be growing at a rate of at least 5,600 per year through the year 2015. Because the figures actually appear to be leveling off, you are suspicious about this estimate. You would like to devise a model that predicts this trend before tomorrow's scheduled board meeting. **How do you go about doing this?**

Web Site

At the Web site you will find:

- Section by section tutorials, including game tutorials with randomized quizzes
- A detailed chapter summary
- A true/false quiz
- Additional review exercises
- Graphers, Excel tutorials, and other resources
- The following extra topic:

  Linear Approximation and Error Estimation

Introduction

In Chapter 3 we studied the concept of the derivative of a function, and we saw some of the applications for which derivatives are useful. However, computing the derivative of a function algebraically seemed to be a time-consuming process, forcing us to restrict attention to fairly simply functions.

In this chapter we develop shortcut techniques that will allow us to write down the derivative of a function directly without having to calculate any limit. These techniques will also enable us to differentiate any closed-form function—that is, any function, no matter how complicated, that can be specified by a formula involving powers, radicals, absolute values, exponents, and logarithms. (In a later chapter, we will discuss how to add trigonometric functions to this list.) We also show how to find the derivatives of functions that are only specified *implicitly*—that is, functions for which we are not given an explicit formula for y in terms of x but only an equation relating x and y.

algebra Review
For this chapter, you should be familiar with the algebra reviewed in **Chapter 0, Sections 3 and 4.**

4.1 Derivatives of Powers, Sums, and Constant Multiples

Up to this point we have approximated derivatives using difference quotients, and we have done exact calculations using the definition of the derivative as the limit of a difference quotient. In general, we would prefer to have an exact calculation, and it is also very useful to have a formula for the derivative function when we can find one. However, the calculation of a derivative as a limit is often tedious, so it would be nice to have a quicker method. We discuss the first of the shortcut rules in this section. By the end of this chapter, we will be able to find fairly quickly the derivative of almost any function we can write.

Shortcut Formula: The Power Rule

If you look at Examples 2 and 3 in Section 3.6, you may notice a pattern:

$$f(x) = x^2 \quad \Rightarrow \quad f'(x) = 2x$$
$$f(x) = x^3 \quad \Rightarrow \quad f'(x) = 3x^2$$

This pattern generalizes to any power of x:

Theorem 4.1 The Power Rule

If n is any constant and $f(x) = x^n$, then

$$f'(x) = nx^{n-1}.$$

Quick Examples

1. If $f(x) = x^2$, then $f'(x) = 2x^1 = 2x$.
2. If $f(x) = x^3$, then $f'(x) = 3x^2$.
3. If $f(x) = x$, rewrite as $f(x) = x^1$, so $f'(x) = 1x^0 = 1$.
4. If $f(x) = 1$, rewrite as $f(x) = x^0$, so $f'(x) = 0x^{-1} = 0$.

Web Site
www.AppliedCalc.org
At the Web site you can find a proof of the power rule by following:

Everything for Calculus
→ Chapter 4
→ Proof of the Power Rule

The proof of the power rule involves first studying the case when n is a positive integer, and then studying the cases of other types of exponents (negative integer, rational number, irrational number). You can find a proof at the Web site.

EXAMPLE 1 Using the Power Rule for Negative and Fractional Exponents

Calculate the derivatives of the following:

a. $f(x) = \frac{1}{x}$ **b.** $f(x) = \frac{1}{x^2}$ **c.** $f(x) = \sqrt{x}$

Solution

a. Rewrite* as $f(x) = x^{-1}$. Then $f'(x) = (-1)x^{-2} = -\frac{1}{x^2}$.

b. Rewrite as $f(x) = x^{-2}$. Then $f'(x) = (-2)x^{-3} = -\frac{2}{x^3}$.

c. Rewrite as $f(x) = x^{0.5}$. Then $f'(x) = 0.5x^{-0.5} = \frac{0.5}{x^{0.5}}$. Alternatively, rewrite $f(x)$ as $x^{1/2}$, so that $f'(x) = \frac{1}{2}x^{-1/2} = \frac{1}{2x^{1/2}} = \frac{1}{2\sqrt{x}}$.

* **NOTE** See the section on exponents in the algebra review to brush up on negative and fractional exponents.

By rewriting the given functions in Example 1 before taking derivatives, we converted them from **rational** or **radical form** (as in, say, $\frac{1}{x^2}$ and $\sqrt{x}$) to **exponent form** (as in x^{-2} and $x^{0.5}$; see the Algebra Review, Section 0.2) to enable us to use the power rule. (See the Caution below.)

Caution

We cannot apply the power rule to terms in the denominators or under square roots. For example:

1. The derivative of $\frac{1}{x^2}$ is **NOT** $\frac{1}{2x}$; it is $-\frac{2}{x^3}$. See Example 1(b).

2. The derivative of $\sqrt{x^3}$ is **NOT** $\sqrt{3x^2}$; it is $1.5x^{0.5}$. Rewrite $\sqrt{x^3}$ as $x^{3/2}$ or $x^{1.5}$ and apply the power rule.

Table 1 Table of Derivative Formulas

$f(x)$	$f'(x)$
1	0
x	1
x^2	$2x$
x^3	$3x^2$
x^n	nx^{n-1}
$\frac{1}{x}$	$-\frac{1}{x^2}$
$\frac{1}{x^2}$	$-\frac{2}{x^3}$
$\sqrt{x}$	$\frac{1}{2\sqrt{x}}$

Some of the derivatives in Example 1 are very useful to remember, so we summarize them in Table 1. We suggest that you add to this table as you learn more derivatives. It is *extremely* helpful to remember the derivatives of common functions such as $1/x$ and $\sqrt{x}$, even though they can be obtained by using the power rule as in the above example.

Another Notation: Differential Notation

Here is a useful notation based on the "d notation" we discussed in Section 3.5. **Differential notation** is based on an abbreviation for the phrase "the derivative with respect to x." For example, we learned that if $f(x) = x^3$, then $f'(x) = 3x^2$. When we say "$f'(x) = 3x^2$," we mean the following:

The derivative of x^3 with respect to x equals $3x^2$.

You may wonder why we sneaked in the words "with respect to x." All this means is that the variable of the function is x, and not any other variable.* Because we use the phrase "the derivative with respect to x" often, we use the following abbreviation.

* **NOTE** This may seem odd in the case of $f(x) = x^3$ because there are no other variables to worry about. But in expressions like st^3 that involve variables other than x, it is necessary to specify just what the variable of the function is. This is the same reason that we write "$f(x) = x^3$" rather than just "$f = x^3$."

Differential Notation; Differentiation

$\frac{d}{dx}$ means "the derivative with respect to x."

Thus, $\frac{d}{dx}[f(x)]$ is the same thing as $f'(x)$, the derivative of $f(x)$ with respect to x. If y is a function of x, then the derivative of y with respect to x is

$$\frac{d}{dx}(y) \quad \text{or, more compactly,} \quad \frac{dy}{dx}$$

To **differentiate** a function $f(x)$ with respect to x means to take its derivative with respect to x.

Quick Examples

In Words	Formula
1. The derivative with respect to x of x^3 is $3x^2$.	$\frac{d}{dx}(x^3) = 3x^2$
2. The derivative with respect to t of $\frac{1}{t}$ is $-\frac{1}{t^2}$.	$\frac{d}{dt}\left(\frac{1}{t}\right) = -\frac{1}{t^2}$

3. If $y = x^4$, then $\frac{dy}{dx} = 4x^3$.

4. If $u = \frac{1}{t^2}$, then $\frac{du}{dt} = -\frac{2}{t^3}$.

Notes

1. $\frac{dy}{dx}$ is Leibniz' notation for the derivative we discussed in Section 3.5. (See the discussion before Example 3 there.)

2. Leibniz notation illustrates units nicely: units of $\frac{dy}{dx}$ are units of y per unit of x. ■

The Rules for Sums and Constant Multiples

We can now find the derivatives of more complicated functions, such as polynomials, using the following rules:

Theorem 4.2 Derivatives of Sums, Differences, and Constant Multiples

If $f(x)$ and $g(x)$ are any two differentiable functions, and if c is any constant, then the functions $f(x) + g(x)$ and $cf(x)$ are differentiable, and

$$[f(x) \pm g(x)]' = f'(x) \pm g'(x) \qquad \text{Sum Rule}$$

$$[cf(x)]' = cf'(x). \qquad \text{Constant Multiple Rule}$$

In Words:

- The derivative of a sum is the sum of the derivatives, and the derivative of a difference is the difference of the derivatives.
- The derivative of c times a function is c times the derivative of the function.

Differential Notation:

$$\frac{d}{dx}[f(x) \pm g(x)] = \frac{d}{dx}f(x) \pm \frac{d}{dx}g(x)$$

$$\frac{d}{dx}[cf(x)] = c\frac{d}{dx}f(x)$$

Quick Examples

1. $\frac{d}{dx}[x^2 - x^4] = \frac{d}{dx}[x^2] - \frac{d}{dx}[x^4] = 2x - 4x^3$
2. $\frac{d}{dx}[7x^3] = 7\frac{d}{dx}[x^3] = 7(3x^2) = 21x^3$

   In other words, we multiply the coefficient (7) by the exponent (3), and then decrease the exponent by 1.
3. $\frac{d}{dx}[12x] = 12\frac{d}{dx}[x] = 12(1) = 12$

   In other words, the derivative of a constant times x is that constant.
4. $\frac{d}{dx}[-x^{0.5}] = \frac{d}{dx}[(-1)x^{0.5}] = (-1)\frac{d}{dx}[x^{0.5}] = (-1)(0.5)x^{-0.5}$

   $= -0.5x^{0.5}$
5. $\frac{d}{dx}[12] = \frac{d}{dx}[12(1)] = 12\frac{d}{dx}[1] = 12(0) = 0.$

   In other words, the derivative of a constant is zero.
6. If my company earns twice as much (annual) revenue as yours and the derivative of your revenue function is the curve on the left, then the derivative of my revenue function is the curve on the right.

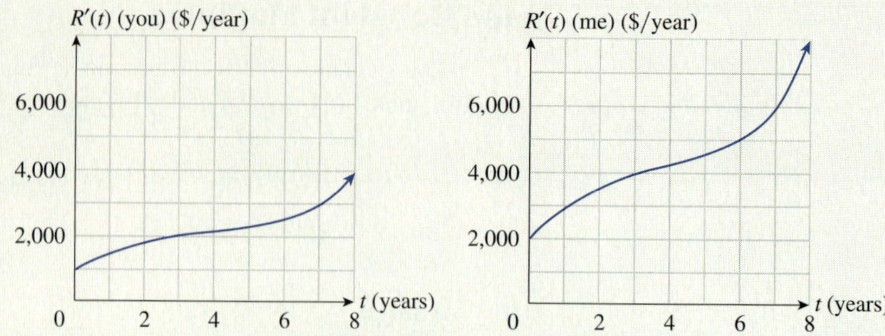

7. Suppose that a company's revenue R and cost C are changing with time. Then so is the profit, $P(t) = R(t) - C(t)$, and the rate of change of the profit is

$$P'(t) = R'(t) - C'(t).$$

In words: *The derivative of the profit is the derivative of revenue minus the derivative of cost.*

Proof of the Sum Rule

By the definition of the derivative of a function,

$$\begin{aligned}\frac{d}{dx}[f(x)+g(x)] &= \lim_{h\to 0}\frac{[f(x+h)+g(x+h)]-[f(x)+g(x)]}{h} \\ &= \lim_{h\to 0}\frac{[f(x+h)-f(x)]+[g(x+h)-g(x)]}{h} \\ &= \lim_{h\to 0}\left[\frac{f(x+h)-f(x)}{h}+\frac{g(x+h)-g(x)}{h}\right] \\ &= \lim_{h\to 0}\frac{f(x+h)-f(x)}{h}+\lim_{h\to 0}\frac{g(x+h)-g(x)}{h} \\ &= \frac{d}{dx}[f(x)]+\frac{d}{dx}[g(x)].\end{aligned}$$

The next-to-last step uses a property of limits: the limit of a sum is the sum of the limits. Think about why this should be true. The last step uses the definition of the derivative again (and the fact that the functions are differentiable).

The proof of the rule for constant multiples is similar.

EXAMPLE 2 Combining the Sum and Constant Multiple Rules, and Dealing with *x* in the Denominator

Find the derivatives of the following:

a. $f(x) = 3x^2 + 2x - 4$ **b.** $f(x) = \dfrac{2x}{3} - \dfrac{6}{x} + \dfrac{2}{3x^{0.2}} - \dfrac{x^4}{2}$

Solution

a. $$\frac{d}{dx}(3x^2 + 2x - 4) = \frac{d}{dx}(3x^2) + \frac{d}{dx}(2x - 4) \qquad \text{Rule for sums}$$

$$= \frac{d}{dx}(3x^2) + \frac{d}{dx}(2x) - \frac{d}{dx}(4) \qquad \text{Rule for differences}$$

$$= 3(2x) + 2(1) - 0 \qquad \text{See Quick Example 2.}$$

$$= 6x + 2$$

b. Notice that f has x and powers of x in the denominator. We deal with these terms the same way we did in Example 1, by rewriting them in exponent form (that is, in the form constant $\times$ power of x):

$$f(x) = \frac{2x}{3} - \frac{6}{x} + \frac{2}{3x^{0.2}} - \frac{x^4}{2} \qquad \text{Rational form}$$

$$= \frac{2}{3}x - 6x^{-1} + \frac{2}{3}x^{-0.2} - \frac{1}{2}x^4 \qquad \text{Exponent form}$$

We are now ready to take the derivative:

$$f'(x) = \frac{2}{3}(1) - 6(-1)x^{-2} + \frac{2}{3}(-0.2)x^{-1.2} - \frac{1}{2}(4x^3)$$

$$= \frac{2}{3} + 6x^{-2} - \frac{0.4}{3}x^{-1.2} - 2x^3 \qquad \text{Exponent form}$$

$$= \frac{2}{3} + \frac{6}{x^2} - \frac{0.4}{3x^{1.2}} - 2x^3 \qquad \text{Rational form}$$

Notice that in Example 2(a) we had three terms in the expression for $f(x)$, not just two. By applying the rule for sums and differences twice, we saw that the derivative of a sum or difference of three terms is the sum or difference of the derivatives of the terms. (One of those terms had zero derivative, so the final answer had only two terms.) In fact, the derivative of a sum or difference of any number of terms is the sum or difference of the derivatives of the terms. Put another way, to take the derivative of a sum or difference of any number of terms, we take derivatives term by term.

Note Nothing forces us to use only x as the independent variable when taking derivatives (although it is traditional to give x preference). For instance, part (a) in Example 2 can be rewritten as

$$\frac{d}{dt}(3t^2 + 2t - 4) = 6t + 2 \qquad \frac{d}{dt} \text{ means "derivative with respect to } t\text{."}$$

or

$$\frac{d}{du}(3u^2 + 2u - 4) = 6u + 2. \qquad \frac{d}{du} \text{ means "derivative with respect to } u\text{."}$$

■

In the previous examples, we saw instances of the following important facts. (Think about these graphically to see why they must be true.)

The Derivative of a Constant Times *x* and the Derivative of a Constant

If c is any constant, then:

Rule	Quick Examples	
$\frac{d}{dx}(cx) = c$	$\frac{d}{dx}(6x) = 6$	$\frac{d}{dx}(-x) = -1$
$\frac{d}{dx}(c) = 0$	$\frac{d}{dx}(5) = 0$	$\frac{d}{dx}(\pi) = 0$

In Example 5 of Section 3.6 we saw that $f(x) = |x|$ fails to be differentiable at $x = 0$. In the next example we use the power rule and find more functions not differentiable at a point.

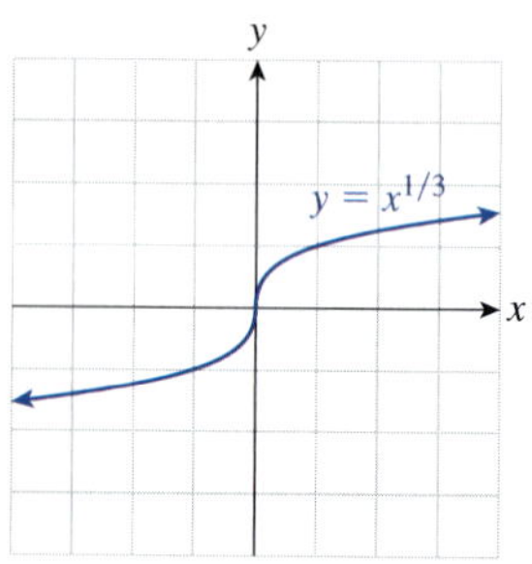

$f(x) = x^{1/3}$

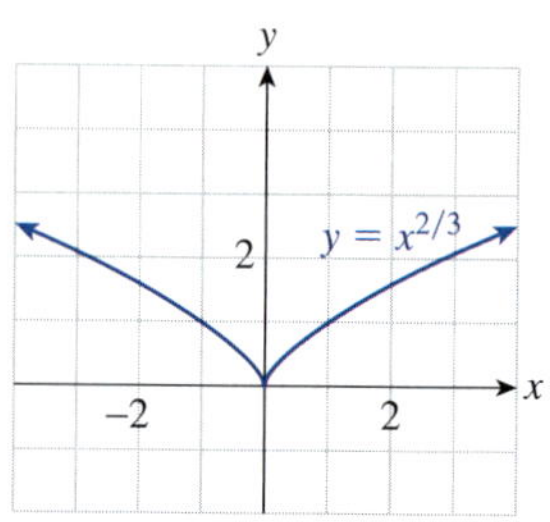

$g(x) = x^{2/3}$

Figure 1

EXAMPLE 3 Functions Not Differentiable at a Point

Find the natural domains of the derivatives of $f(x) = x^{1/3}$ and $g(x) = x^{2/3}$, and $h(x) = |x|$.

Solution Let's first look at the functions f and g. By the power rule,

$$f'(x) = \frac{1}{3}x^{-2/3} = \frac{1}{3x^{2/3}}$$

and

$$g'(x) = \frac{2}{3}x^{-1/3} = \frac{2}{3x^{1/3}}.$$

$f'(x)$ and $g'(x)$ are defined only for nonzero values of x, and their natural domains consist of all real numbers except 0. Thus, the derivatives f' and g' do not exist at $x = 0$. In other words, f and g are not differentiable at $x = 0$. If we look at Figure 1, we notice why these functions fail to be differentiable at $x = 0$: The graph of f has a vertical tangent line at 0. Because a vertical line has undefined slope, the derivative is undefined at that point. The graph of g comes to a sharp point (called a **cusp**) at 0, so it is not meaningful to speak about a tangent line at that point; therefore, the derivative of g is not defined there. (Actually, there is a reasonable candidate for the tangent line at $x = 0$, but it is the vertical line again.)

We can also detect this nondifferentiability by computing some difference quotients numerically. In the case of $f(x) = x^{1/3}$, we get the following table:

h	±1	±0.1	±0.01	±0.001	±0.0001
$\frac{f(0+h) - f(0)}{h}$	1	4.6416	21.544	100	464.16

suggesting that the difference quotients $[f(0 + h) - f(0)]/h$ grow large without bound rather than approach any fixed number as h approaches 0. (Can you see how the behavior of the difference quotients in the table is reflected in the graph?)

using Technology

If you try to graph the function $f(x) = x^{2/3}$ using the format

```
X^(2/3)
```

you may get only the right-hand portion of Figure 1 because graphing utilities are (often) not programmed to raise negative numbers to fractional exponents. (However, many will handle `X^(1/3)` correctly, as a special case they recognize.) To avoid this difficulty, you can take advantage of the identity

$$x^{2/3} = (x^2)^{1/3}$$

so that it is always a nonnegative number that is being raised to a fractional exponent. Thus, use the format

```
(X^2)^(1/3)
```

to obtain both portions of the graph.

Now we return to the function $h(x) = |x|$ discussed in Example 5 of Section 3.6. We saw there that $|x|$ is not differentiable at $x = 0$. What about values of x *other than* 0? For such values, we can write:

$$|x| = \begin{cases} -x & \text{if } x < 0 \\ x & \text{if } x > 0 \end{cases}$$

Hence, by the power rule (think of x as x^1):

$$f'(x) = \begin{cases} -1 & \text{if } x < 0 \\ 1 & \text{if } x > 0 \end{cases}$$

Q: *So does that mean there is no single formula for the derivative of $|x|$?*

A: Actually, there *is* a convenient formula. Consider the ratio

$$\frac{|x|}{x}.$$

If x is positive, then $|x| = x$, so $|x|/x = x/x = 1$. On the other hand, if x is negative then $|x| = -x$, so $|x|/x = -x/x = -1$. In other words,

$$\frac{|x|}{x} = \begin{cases} -1 & \text{if } x < 0 \\ 1 & \text{if } x > 0 \end{cases},$$

which is exactly the formula we obtained for $f'(x)$. In other words:

Derivative of |x|

$$\frac{d}{dx}|x| = \frac{|x|}{x}$$

Note that the derivative does not exist when $x = 0$.

Quick Example

$$\frac{d}{dx}[3|x| + x] = 3\frac{|x|}{x} + 1$$

Spencer Grant / PhotoEdit

APPLICATION

The next Example is similar to Example 3 in Section 3.4, but this time we analyze the curve using the derivative.

EXAMPLE 4 Gold Price

You are a commodities trader and you monitor the price of gold on the New York Spot Market very closely during an active morning. Suppose you find that the price of an ounce of gold can be approximated by the function

$$G(t) = -8t^2 + 144t + 150 \text{ dollars} \quad (7.5 \le t \le 10.5)$$

where t is time in hours. (See Figure 2. $t = 8$ represents 8:00 AM.)

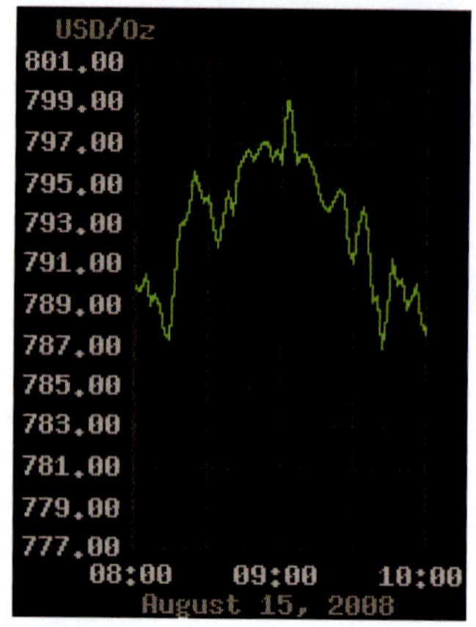

Source: www.kitco.com (August 15, 2008)

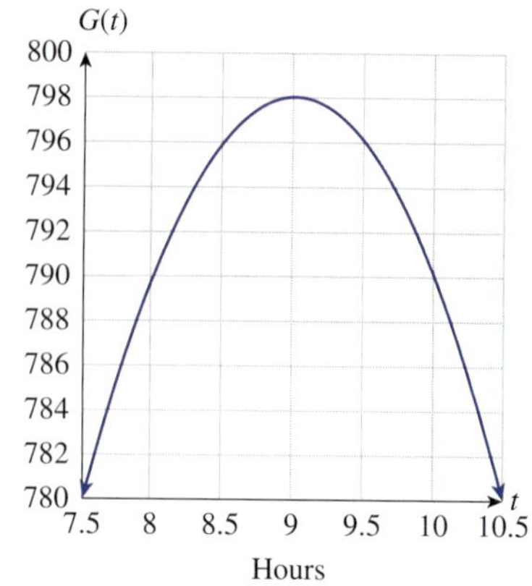

$G(t) = -8t^2 + 144t + 150$

Figure 2

a. According to the model, how fast was the price of gold changing at 10:00 AM?

b. According to the model, the price of gold

(A) increased at a faster and faster rate
(B) increased at a slower and slower rate
(C) decreased at a faster and faster rate
(D) decreased at a slower and slower rate

between 9:30 and 10:30 AM.

Solution

a. Differentiating the given function with respect to t gives

$$G'(t) = -16t + 144.$$

Because 10:00 AM corresponds to $t = 10$, we obtain

$$G'(10) = -16(10) + 144 = -16.$$

The units of the derivative are dollars per hour, so we conclude that, at 10:00 AM, the price of gold was dropping at a rate of \$16 per hour.

b. From the graph, we can see that, between 9:30 and 10:30 AM (the interval [9.5, 10.5]), the price of gold was decreasing. Also from the graph, we see that the slope of the tangent becomes more and more negative as t increases, so the price of gold is decreasing at a faster and faster rate (choice (C)).

We can also see this algebraically from the derivative, $G'(t) = -16t + 144$: For values of t larger than 9, $G'(t)$ is negative; that is, the rate of change of G is negative, so the price of gold is decreasing. Further, as t increases, $G'(t)$ becomes more and more negative, so the price of gold is decreasing at a faster and faster rate, confirming that choice (C) is the correct one.

An Application to Limits: L'Hospital's Rule

The limits that caused us some trouble in Sections 3.1–3.3 are those of the form $\lim_{x\to a} f(x)$ in which substituting $x = a$ gave us an indeterminate form, such as

$$\lim_{x\to 2} \frac{x^3 - 8}{x - 2}$$ Substituting $x = 2$ yields $\frac{0}{0}$.

$$\lim_{x\to +\infty} \frac{2x - 4}{x - 1}$$ Substituting $x = +\infty$ yields $\frac{\infty}{\infty}$.

L'Hospital's rule gives us an alternate way of computing limits such as these without the need to do any preliminary simplification. It also allows us to compute some limits for which algebraic simplification does not work.*

* **NOTE** Guillaume François Antoine, Marquis de l'Hospital (1661–1704) wrote the first textbook on calculus, *Analyse des infiniment petits pour l'intelligence des lignes courbes,* in 1692. The rule now known as l'Hospital's Rule appeared first in this book.

Theorem 4.3 L'Hospital's Rule

If f and g are two differentiable functions such that substituting $x = a$ in the expression $\frac{f(x)}{g(x)}$ gives the indeterminate form $\frac{0}{0}$ or $\frac{\infty}{\infty}$, then

$$\lim_{x\to a} \frac{f(x)}{g(x)} = \lim_{x\to a} \frac{f'(x)}{g'(x)}$$

That is, we can replace $f(x)$ and $g(x)$ with their *derivatives* and try again to take the limit.

Quick Examples

1. Substituting $x = 2$ in $\frac{x^3 - 8}{x - 2}$ yields $\frac{0}{0}$. Therefore, l'Hospital's rule applies and

$$\lim_{x\to 2} \frac{x^3 - 8}{x - 2} = \lim_{x\to 2} \frac{3x^2}{1} = \frac{3(2)^2}{1} = 12.$$

2. Substituting $x = +\infty$ in $\frac{2x - 4}{x - 1}$ yields $\frac{\infty}{\infty}$. Therefore, l'Hospital's rule applies and

$$\lim_{x\to +\infty} \frac{2x - 4}{x - 1} = \lim_{x\to +\infty} \frac{2}{1} = 2.$$

The proof of l'Hospital's rule is beyond the scope of this text.*

* **NOTE** A proof of l'Hospital's rule can be found in most advanced calculus textbooks.

EXAMPLE 5 Applying L'Hospital's Rule

Check whether l'Hospital's rule applies to each of the following limits. If it does, use it to evaluate the limit. Otherwise, use some other method to evaluate the limit.

a. $\lim_{x\to 1} \frac{x^2 - 2x + 1}{4x^3 - 3x^2 - 6x + 5}$ **b.** $\lim_{x\to +\infty} \frac{2x^2 - 4x}{5x^3 - 3x + 5}$

c. $\lim_{x \to 1} \frac{x - 1}{x^3 - 3x^2 + 3x - 1}$ **d.** $\lim_{x \to 1} \frac{x}{x^3 - 3x^2 + 3x - 1}$

Solution

a. Setting $x = 1$ yields

$$\frac{1 - 2 + 1}{4 - 3 - 6 + 5} = \frac{0}{0}.$$

Therefore, l'Hospital's rule applies and

$$\lim_{x \to 1} \frac{x^2 - 2x + 1}{4x^3 - 3x^2 - 6x + 5} = \lim_{x \to 1} \frac{2x - 2}{12x^2 - 6x - 6}.$$

We are left with a closed-form function. However, we cannot substitute $x = 1$ to find the limit because the function $(2x - 2)/(12x^2 - 6x - 6)$ is still not defined at $x = 1$. In fact, if we set $x = 1$, we again get $0/0$. Thus, l'Hospital's rule applies again, and

$$\lim_{x \to 1} \frac{2x - 2}{12x^2 - 6x - 6} = \lim_{x \to 1} \frac{2}{24x - 6}.$$

Once again we have a closed-form function, but this time it is defined when $x = 1$, giving

$$\frac{2}{24 - 6} = \frac{1}{9}.$$

Thus,

$$\lim_{x \to 1} \frac{x^2 - 2x + 1}{4x^3 - 3x^2 - 6x + 5} = \frac{1}{9}.$$

b. Setting $x = +\infty$ yields $\frac{\infty}{\infty}$, so

$$\lim_{x \to +\infty} \frac{2x^2 - 4x}{5x^3 - 3x + 5} = \lim_{x \to +\infty} \frac{4x - 4}{15x^2 - 3}.$$

Setting $x = +\infty$ again yields $\frac{\infty}{\infty}$, so we can apply the rule again to obtain

$$\lim_{x \to +\infty} \frac{4x - 4}{15x^2 - 3} = \lim_{x \to +\infty} \frac{4}{30x}.$$

Note that we cannot apply l'Hospital's rule a third time because setting $x = +\infty$ yields the *determinate* form $4/\infty = 0$ (see the discussion at the end of Section 3.3). Thus, the limit is 0.

c. Setting $x = 1$ yields $0/0$ so, by l'Hospital's rule,

$$\lim_{x \to 1} \frac{x - 1}{x^3 - 3x^2 + 3x - 1} = \lim_{x \to 1} \frac{1}{3x^2 - 6x + 3}.$$

We are left with a closed-form function that is still not defined at $x = 1$. Further, l'Hospital's rule no longer applies because putting $x = 1$ yields the determinate form $1/0$. To investigate this limit, we refer to the discussion at the end of Section 3.3 and find

$$\lim_{x \to 1} \frac{1}{3x^2 - 6x + 3} = \lim_{x \to 1} \frac{1}{3(x - 1)^2} = +\infty. \qquad \frac{1}{0^+} = +\infty$$

d. Setting $x = 1$ in the expression yields the determinate form $1/0$, so l'Hospital's rule does not apply here. Using the methods of Section 3.3 again, we find that the limit does not exist.

FAQs

Using the Rules and Recognizing When a Function is Not Differentiable

Q: I would *like* to say that the derivative of $5x^2 - 8x + 4$ is just $10x - 8$ without having to go through all that stuff about derivatives of sums and constant multiples. Can I simply forget about all the rules and write down the answer?

A: We developed the rules for sums and constant multiples precisely for that reason: so that we could simply write down a derivative without having to think about it too hard. So, you are perfectly justified in simply writing down the derivative without going through the rules, but bear in mind that what you are really doing is applying the power rule, the rule for sums, and the rule for multiples over and over.

Q: Is there a way of telling from its formula whether a function f is not differentiable at a point?

A: Here are some indicators to look for in the formula for f:

- The absolute value of some expression; f may not be differentiable at points where that expression is zero.

  Example: $f(x) = 3x^2 - |x - 4|$ is not differentiable at $x = 4$.

- A fractional power smaller than 1 of some expression; f may not be differentiable at points where that expression is zero.

  Example: $f(x) = (x^2 - 16)^{2/3}$ is not differentiable at $x = \pm 4$.

4.1 EXERCISES

▼ more advanced ◆ challenging

T indicates exercises that should be solved using technology

In Exercises 1–10, use the shortcut rules to ***mentally*** *calculate the derivative of the given function.* HINT [See Examples 1 and 2.]

1. $f(x) = x^5$
2. $f(x) = x^4$
3. $f(x) = 2x^{-2}$
4. $f(x) = 3x^{-1}$
5. $f(x) = -x^{0.25}$
6. $f(x) = -x^{-0.5}$
7. $f(x) = 2x^4 + 3x^3 - 1$
8. $f(x) = -x^3 - 3x^2 - 1$
9. $f(x) = -x + \frac{1}{x} + 1$
10. $f(x) = \frac{1}{x} + \frac{1}{x^2}$

In Exercises 11–16, obtain the derivative dy/dx and state the rules that you use. HINT [See Example 2.]

11. $y = 10$
12. $y = x^3$
13. $y = x^2 + x$
14. $y = x - 5$
15. $y = 4x^3 + 2x - 1$
16. $y = 4x^{-1} - 2x - 10$

In Exercises 17–40, find the derivative of each function. HINT [See Examples 1 and 2.]

17. $f(x) = x^2 - 3x + 5$
18. $f(x) = 3x^3 - 2x^2 + x$
19. $f(x) = x + x^{0.5}$
20. $f(x) = x^{0.5} + 2x^{-0.5}$
21. $g(x) = x^{-2} - 3x^{-1} - 2$
22. $g(x) = 2x^{-1} + 4x^{-2}$
23. $g(x) = \frac{1}{x} - \frac{1}{x^2}$
24. $g(x) = \frac{1}{x^2} + \frac{1}{x^3}$
25. $h(x) = \frac{2}{x^{0.4}}$
26. $h(x) = -\frac{1}{2x^{0.2}}$
27. $h(x) = \frac{1}{x^2} + \frac{2}{x^3}$
28. $h(x) = \frac{2}{x} - \frac{2}{x^3} + \frac{1}{x^4}$
29. $r(x) = \frac{2}{3x} - \frac{1}{2x^{0.1}}$
30. $r(x) = \frac{4}{3x^2} + \frac{1}{x^{3.2}}$
31. $r(x) = \frac{2x}{3} - \frac{x^{0.1}}{2} + \frac{4}{3x^{1.1}} - 2$
32. $r(x) = \frac{4x^2}{3} + \frac{x^{3.2}}{6} - \frac{2}{3x^2} + 4$
33. $t(x) = |x| + \frac{1}{x}$
34. $t(x) = 3|x| - \sqrt{x}$
35. $s(x) = \sqrt{x} + \frac{1}{\sqrt{x}}$
36. $s(x) = x + \frac{7}{\sqrt{x}}$

HINT [For Exercises 37–40, first expand the given function.]

37. ▼ $s(x) = x\left(x^2 - \frac{1}{x}\right)$ **38.** ▼ $s(x) = x^{-1}\left(x - \frac{2}{x}\right)$

39. ▼ $t(x) = \frac{x^2 - 2x^3}{x}$ **40.** ▼ $t(x) = \frac{2x + x^2}{x}$

In Exercises 41–46, evaluate the given expression.

41. $\frac{d}{dx}(2x^{1.3} - x^{-1.2})$ **42.** $\frac{d}{dx}(2x^{4.3} + x^{0.6})$

43. ▼ $\frac{d}{dx}[1.2(x - |x|)]$ **44.** ▼ $\frac{d}{dx}[4(x^2 + 3|x|)]$

45. ▼ $\frac{d}{dt}(at^3 - 4at)$; (a constant)

46. ▼ $\frac{d}{dt}(at^2 + bt + c)$; ($a$, b, c constant)

In Exercises 47–52, find the indicated derivative.

47. $y = \frac{x^{10.3}}{2} + 99x^{-1}$; $\frac{dy}{dx}$ **48.** $y = \frac{x^{1.2}}{3} - \frac{x^{0.9}}{2}$; $\frac{dy}{dx}$

49. $s = 2.3 + \frac{2.1}{t^{1.1}} - \frac{t^{0.6}}{2}$; $\frac{ds}{dt}$ **50.** $s = \frac{2}{t^{1.1}} + t^{-1.2}$; $\frac{ds}{dt}$

51. ▼ $V = \frac{4}{3}\pi r^3$; $\frac{dV}{dr}$ **52.** ▼ $A = 4\pi r^2$; $\frac{dA}{dr}$

In Exercises 53–58, find the slope of the tangent to the graph of the given function at the indicated point. HINT [Recall that the slope of the tangent to the graph of f at $x = a$ is $f'(a)$.]

53. $f(x) = x^3$; $(-1, -1)$ **54.** $g(x) = x^4$; $(-2, 16)$

55. $f(x) = 1 - 2x$; $(2, -3)$ **56.** $f(x) = \frac{x}{3} - 1$; $(-3, -2)$

57. $g(t) = \frac{1}{t^5}$; $(1, 1)$ **58.** $s(t) = \frac{1}{t^3}$; $\left(-2, -\frac{1}{8}\right)$

In Exercises 59–64, find the equation of the tangent line to the graph of the given function at the point with the indicated x-coordinate. In each case, sketch the curve together with the appropriate tangent line.

59. ▼ $f(x) = x^3$; $x = -1$ **60.** ▼ $f(x) = x^2$; $x = 0$

61. ▼ $f(x) = x + \frac{1}{x}$; $x = 2$ **62.** ▼ $f(x) = \frac{1}{x^2}$; $x = 1$

63. ▼ $f(x) = \sqrt{x}$; $x = 4$ **64.** ▼ $f(x) = 2x + 4$; $x = -1$

In Exercises 65–70, find all values of x (if any) where the tangent line to the graph of the given equation is horizontal. HINT [The tangent line is horizontal when its slope is zero.]

65. ▼ $y = 2x^2 + 3x - 1$ **66.** ▼ $y = -3x^2 - x$

67. ▼ $y = 2x + 8$ **68.** ▼ $y = -x + 1$

69. ▼ $y = x + \frac{1}{x}$ **70.** ▼ $y = x - \sqrt{x}$

71. ◆ Write out the proof that $\frac{d}{dx}(x^4) = 4x^3$.

72. ◆ Write out the proof that $\frac{d}{dx}(x^5) = 5x^4$.

T *In Exercises 73–76, use technology to graph the derivative of the given function for the given range of values of x. Then use your graph to estimate all values of x (if any) where **(a)** the given function is not differentiable, and **(b)** the tangent line to the graph of the given function is horizontal. Round answers to one decimal place.*

73. ▼ $h(x) = |x - 3|$; $-5 \le x \le 5$

74. ▼ $h(x) = 2x + (x - 3)^{1/3}$; $-5 \le x \le 5$

75. ▼ $f(x) = x - 5(x - 1)^{2/5}$; $-4 \le x \le 6$

76. ▼ $f(x) = |2x + 5| - x^2$; $-4 \le x \le 4$

T *In Exercises 77–80, investigate the differentiability of the given function at the given points numerically (that is, use a table of values). If $f'(a)$ exists, give its approximate value.* HINT [See Example 3.]

77. $f(x) = x^{1/3}$ **a.** $a = 1$ **b.** $a = 0$

78. $f(x) = x + |1 - x|$ **a.** $a = 1$ **b.** $a = 0$

79. ▼ $f(x) = [x(1 - x)]^{1/3}$ **a.** $a = 1$ **b.** $a = 0$

80. ▼ $f(x) = (1 - x)^{2/3}$ **a.** $a = -1$ **b.** $a = 1$

In Exercises 81–92 say whether l'Hospital's rule applies. If is does, use it to evaluate the given limit. If not, use some other method.

81. $\lim_{x\to 1} \frac{x^2 - 2x + 1}{x^2 - x}$ **82.** $\lim_{x\to -1} \frac{x^2 + 3x + 2}{x^2 + x}$

83. $\lim_{x\to 2} \frac{x^3 - 8}{x - 2}$ **84.** $\lim_{x\to 0} \frac{x^3 + 8}{x^2 + 3x + 2}$

85. $\lim_{x\to 1} \frac{x^2 + 3x + 2}{x^2 + x}$ **86.** $\lim_{x\to -2} \frac{x^3 + 8}{x^2 + 3x + 2}$

87. $\lim_{x\to -\infty} \frac{3x^2 + 10x - 1}{2x^2 - 5x}$ **88.** $\lim_{x\to -\infty} \frac{6x^2 + 5x + 100}{3x^2 - 9}$

89. $\lim_{x\to -\infty} \frac{10x^2 + 300x + 1}{5x + 2}$ **90.** $\lim_{x\to -\infty} \frac{2x^4 + 20x^3}{1000x^3 + 6}$

91. $\lim_{x\to -\infty} \frac{x^3 - 100}{2x^2 + 500}$ **92.** $\lim_{x\to -\infty} \frac{x^2 + 30x}{2x^6 + 10x}$

APPLICATIONS

93. ***Crude Oil Prices*** The price per barrel of crude oil in constant 2008 dollars can be approximated by

$$P(t) = 0.45t^2 - 12t + 105 \text{ dollars} \quad (0 \le t \le 28)$$

where t is time in years since the start of 1980.[1] Find $P'(t)$ and $P'(20)$. What does the answer tell you about the price of crude oil? HINT [See Example 2.]

94. ***Median Home Price*** The median home price in the United States over the period 2004–2009 can be approximated by

$$P(t) = -5t^2 + 75t - 30 \text{ thousand dollars} \quad (4 \le t \le 9)$$

where t is time in years since the start of 2000.[2] Find $P'(t)$ and $P'(6)$. What does the answer tell you about home prices? HINT [See Example 2.]

95. ***College Basketball: Men*** The number of NCAA men's college basketball teams in the United States can be modeled by:

$$n(t) = -0.56t^2 + 14t + 930 \quad (0 \le t \le 8)$$

where t is time in years since 2000.[3]

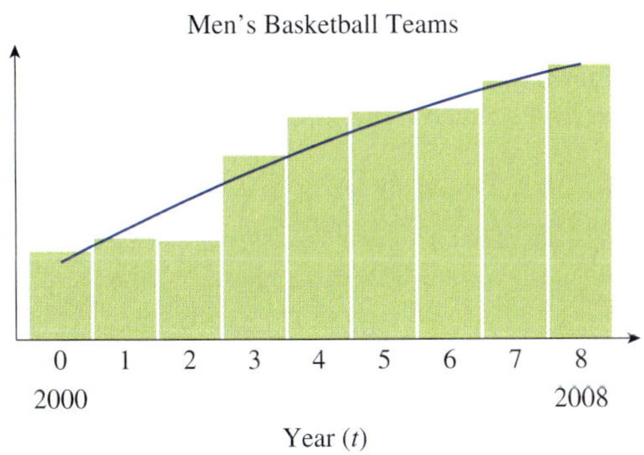

a. Find $n'(t)$. HINT [See Example 2.]

b. How fast (to the nearest whole number) was the number of men's college basketball teams increasing in 2006?

c. According to the model, did the rate of increase of the number of teams increase or decrease with time? Explain. HINT [See Example 4.]

96. ***College Basketball: Women*** The number of NCAA women's college basketball teams in the United States can be modeled by:

$$n(t) = -0.98t^2 + 32t + 1{,}850 \quad (0 \le t \le 8)$$

where t is time in years since 2000.[4]

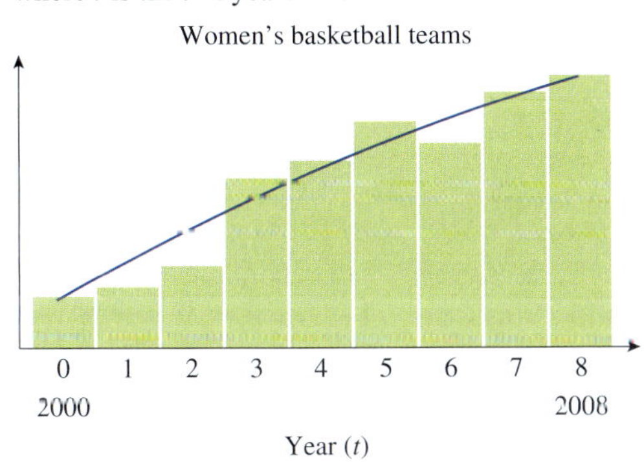

a. Find $n'(t)$. HINT [See Example 2.]

b. How fast (to the nearest whole number) was the number of women's college basketball teams increasing in 2005?

c. According to the model, did the rate of increase of the number of teams increase or decrease with time? Explain. HINT [See Example 4.]

97. ***Food Versus Education*** The following equation shows the approximate relationship between the percentage y of total personal consumption spent on food and the corresponding percentage x spent on education.[5]

$$y = \frac{35}{x^{0.35}} \text{ percentage points} \quad (6.5 \le x \le 17.5)$$

According to the model, spending on food is decreasing at a rate of _____ percentage points per one percentage point increase in spending on education when 10% of total consumption is spent on education. (Answer should be rounded to two significant digits.) HINT [See Example 2(b).]

98. ***Food Versus Recreation*** The following equation shows the approximate relationship between the percentage y of total personal consumption spent on food and the corresponding percentage x spent on recreation.[6]

$$y = \frac{33}{x^{0.63}} \text{ percentage points} \quad (2.5 \le x \le 4.5)$$

According to the model, spending on food is decreasing at a rate of _____ percentage points per one percentage point increase in spending on recreation when 3% of total consumption is spent on recreation. (Answer should be rounded to two significant digits.) HINT [See Example 2(b).]

99. ***Velocity*** If a stone is dropped from a height of 400 feet, its height s after t seconds is given by $s(t) = 400 - 16t^2$, with s in feet.

a. Compute $s'(t)$ and hence find its velocity at times $t = 0$, 1, 2, 3, and 4 seconds.

b. When does it reach the ground, and how fast is it traveling when it hits the ground? HINT [It reaches the ground when $s(t) = 0$.]

100. ***Velocity*** If a stone is thrown down at 120 ft/s from a height of 1,000 feet, its height s after t seconds is given by $s(t) = 1{,}000 - 120t - 16t^2$, with s in feet.

a. Compute $s'(t)$ and hence find its velocity at times $t = 0$, 1, 2, 3, and 4 seconds.

b. When does it reach the ground, and how fast is it traveling when it hits the ground? HINT [It reaches the ground when $s(t) = 0$.]

[1] Source for data: www.inflationdata.com.

[2] Source for data: www.investmenttools.com.

[3] 2007 and 2008 figures are estimates. Source: The 2008 Statistical Abstract, www.census.gov/.

[4] Ibid.

[5] Model based on historical and projected data from 1908–2010. Sources: Historical data, Bureau of Economic Analysis; projected data, Bureau of Labor Statistics/*New York Times*, December 1, 2003, p. C2.

[6] Ibid.

101. ***iPhone Sales*** iPhone sales from the 2nd quarter in 2007 through the 2nd quarter in 2008 can be approximated by

$$S(t) = -390t^2 + 3{,}300t - 4{,}800 \text{ thousand phones} \quad (2 \le t \le 6)$$

in quarter t. ($t = 1$ represents the start of the first quarter of 2007.)[7]

iPhone sales (thousands)

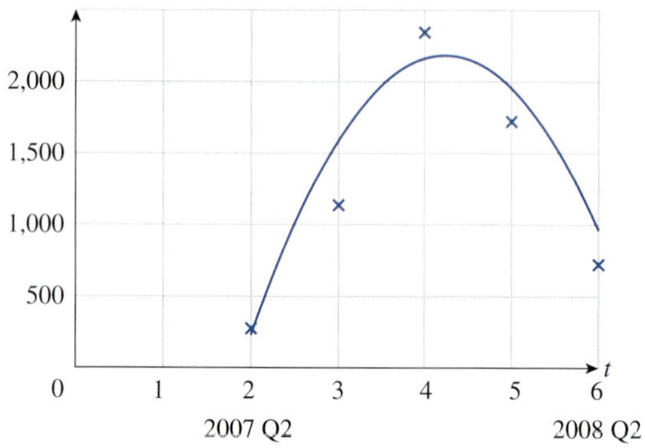

a. Compute $S'(t)$. How fast were iPhone sales changing in the first quarter of 2008 ($t = 5$)? (Be careful to give correct units of measurement.)

b. According to the model, iPhone sales

(A) increased at a faster and faster rate
(B) increased at a slower and slower rate
(C) decreased at a faster and faster rate
(D) decreased at a slower and slower rate

during the first two quarters shown (the interval [2, 4]). Justify your answer in two ways: geometrically, reasoning entirely from the graph; and algebraically, reasoning from the derivative of S. HINT [See Example 4.]

102. ***Facebook Membership*** The number of **Facebook** members from the start of 2006 to mid-2008 can be approximated by

$$S(t) = 15t^2 - 76t + 101 \text{ million members} \quad (2 \le t \le 4.5)$$

in year t ($t = 0$ represents the start of 2004).[8]

Facebook membership (millions)

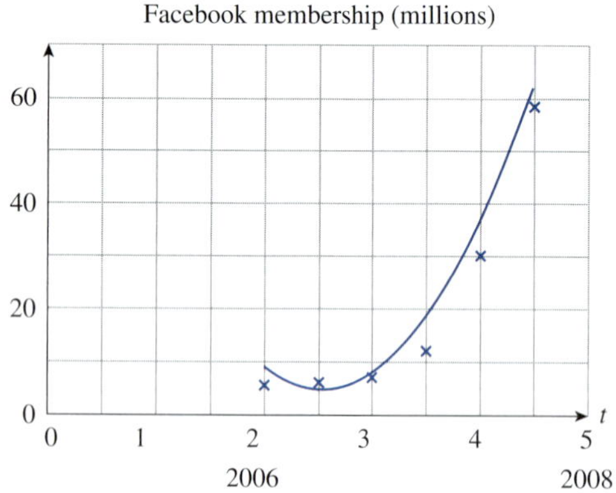

a. Compute $S'(t)$. According to the model, how fast was **Facebook** membership changing at the start of 2008 ($t = 4$)? (Be careful to give correct units of measurement.)

b. According to the model, **Facebook** membership

(A) increased at a faster and faster rate
(B) increased at a slower and slower rate
(C) decreased at a faster and faster rate
(D) decreased at a slower and slower rate

during 2007 (the interval [3, 4]). Justify your answer in two ways: geometrically, reasoning entirely from the graph; and algebraically, reasoning from the derivative of S. HINT [See Example 4.]

103. ***Ecology*** Increasing numbers of manatees ("sea sirens") have been killed by boats off the Florida coast. The following graph shows the relationship between the number of boats registered in Florida and the number of manatees killed each year.

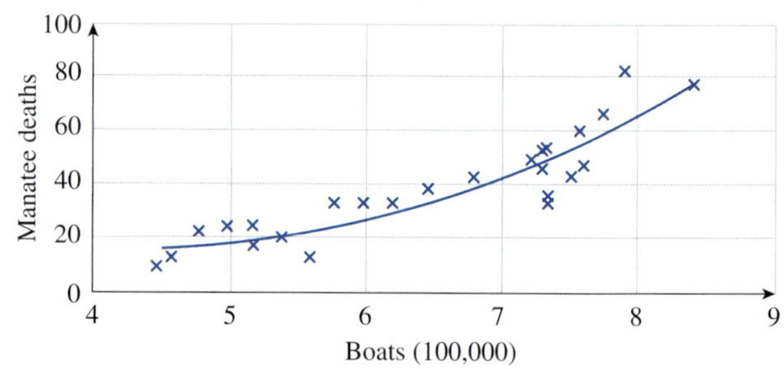

The regression curve shown is given by

$$f(x) = 3.55x^2 - 30.2x + 81 \quad (4.5 \le x \le 8.5)$$

where x is the number of boats (hundreds of thousands) registered in Florida in a particular year and $f(x)$ is the number of manatees killed by boats in Florida that year.[9]

a. Compute $f'(x)$. What are the units of measurement of $f'(x)$?

b. Is $f'(x)$ increasing or decreasing with increasing x? Interpret the answer. HINT [See Example 4.]

c. Compute and interpret $f'(8)$.

104. ***SAT Scores by Income*** The graph on the next page shows U.S. verbal SAT scores as a function of parents' income level.[10]

[7] The model is the authors'. Source for data: Apple financial statements, www.apple.com.

[8] The model is the authors'. Sources for data: www.facebook.com/, http://insidehighered.com (Some data are interpolated.)

[9] Regression model is based on data from 1976 to 2000. Sources for data: Florida Department of Highway Safety & Motor Vehicles, Florida Marine Institute/*New York Times*, February 12, 2002, p. F4.

[10] Based on 1994 data. Source: The College Board/*New York Times*, March 5, 1995, p. E16.

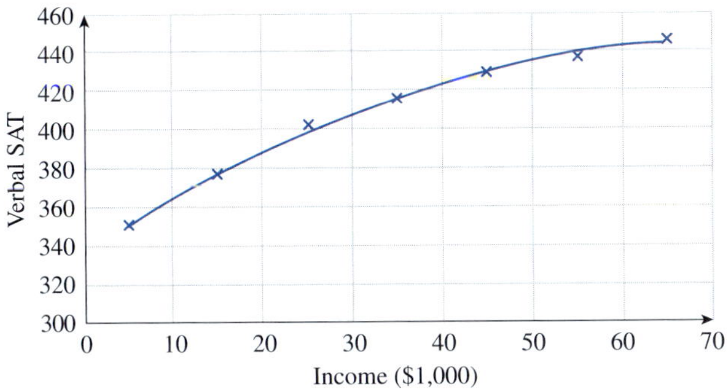

The regression curve shown is given by

$$f(x) = -0.021x^2 + 3.0x + 336 \quad (5 \le x \le 65)$$

where $f(x)$ is the average SAT verbal score of a student whose parents earn x thousand dollars per year.[11]

a. Compute $f'(x)$. What are the units of measurement of $f'(x)$?

b. Is $f'(x)$ increasing or decreasing with increasing x? Interpret the answer. HINT [See Example 4.]

c. Compute and interpret $f'(30)$.

105. ***ISP Market Share*** The following graph shows approximate market shares, in percentage points, of **Microsoft's MSN** Internet service provider, and the combined shares of **MSN**, **Comcast**, **Earthlink**, and **AOL** for the period 1999–2004.[12]

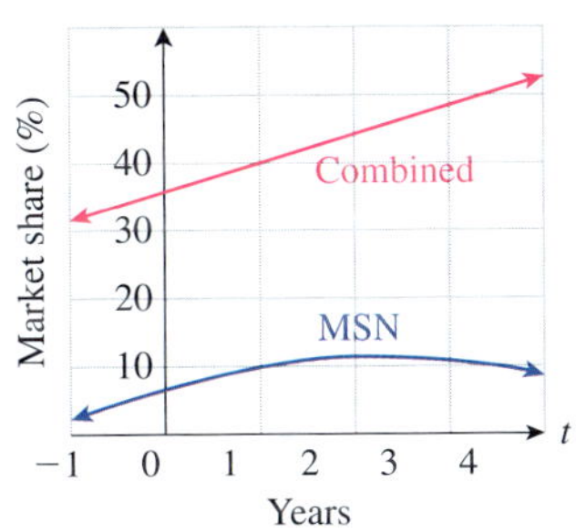

Here, t is time in years since June 2000. Let $c(t)$ be the combined market share at time t, and let $m(t)$ be **MSN**'s share at time t.

a. What does the function $c(t) - m(t)$ measure? What does $c'(t) - m'(t)$ measure?

b. Based on the graphs shown, $c(t) - m(t)$ is

(A) increasing
(B) decreasing
(C) increasing, then decreasing
(D) decreasing, then increasing

on the interval [3, 4].

c. Based on the graphs shown, $c'(t) - m'(t)$ is

(A) positive
(B) negative
(C) positive, then negative
(D) negative, then positive

on the interval [3, 4].

d. The two market shares are approximated by

MSN: $m(t) = -0.83t^2 + 3.8t + 6.8 \quad (-1 \le t \le 4)$

Combined: $c(t) = 4.2t + 36 \quad (-1 \le t \le 4)$

Compute $c'(2) - m'(2)$. Interpret your answer.

106. ▼ ***ISP Revenue*** The following graph shows the approximate total revenue, in millions of dollars, of **Microsoft's MSN** Internet service provider, as well as the portion of the revenue due to advertising for the period June 2001–January 2004.[13]

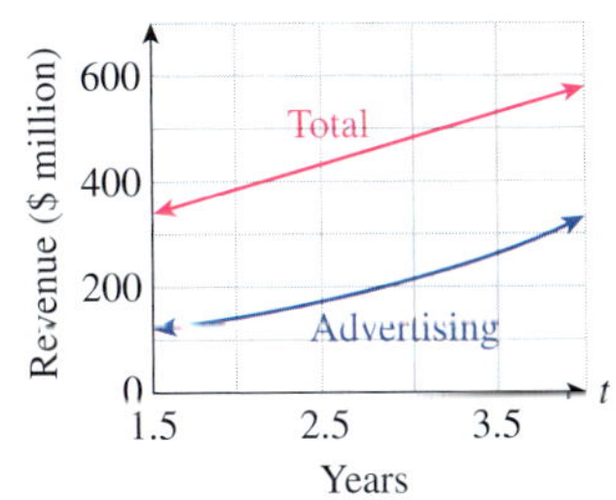

Here, t is time in years since January 2000. Let $s(t)$ be the total revenue at time t, and let $a(t)$ be revenue due to advertising at time t.

a. What does the function $s(t) - a(t)$ measure? What does $s'(t) - a'(t)$ measure?

b. Based on the graphs shown, $s(t) - a(t)$ is

(A) increasing
(B) decreasing
(C) increasing, then decreasing
(D) decreasing, then increasing

on the interval [2, 4].

c. Based on the graphs shown, $s'(t) - a'(t)$ is

(A) positive
(B) negative
(C) positive, then negative
(D) negative, then positive

on the interval [2, 4].

d. The two revenue curves are approximated by

Advertising: $a(t) = 20t^2 - 27t + 120 \quad (1.5 \le t \le 4)$

Total: $s(t) = 96t + 190 \quad (1.5 \le t \le 4)$

Compute $a'(2)$, $s'(2)$, and hence $s'(2) - a'(2)$. Interpret your answer.

[11] Regression model is based on 1994 data. Source: The College Board/*New York Times*, March 5, 1995, p. E16.

[12] The curves are regression models. Source for data: Solomon Research, Morgan Stanley/*New York Times*, July 19, 2004.

[13] Ibid.

COMMUNICATION AND REASONING EXERCISES

107. What instructions would you give to a fellow student who wanted to accurately graph the tangent line to the curve $y = 3x^2$ at the point $(-1, 3)$?

108. What instructions would you give to a fellow student who wanted to accurately graph a line at right angles to the curve $y = 4/x$ at the point where $x = 0.5$?

109. Consider $f(x) = x^2$ and $g(x) = 2x^2$. How do the slopes of the tangent lines of f and g at the same x compare?

110. Consider $f(x) = x^3$ and $g(x) = x^3 + 3$. How do the slopes of the tangent lines of f and g compare?

111. Suppose $g(x) = -f(x)$. How do the derivatives of f and g compare?

112. Suppose $g(x) = f(x) - 50$. How do the derivatives of f and g compare?

113. Following is an excerpt from your best friend's graded homework:

$$3x^4 + 11x^5 = 12x^3 + 55x^4 \quad \text{✗ WRONG} \quad -8$$

Why was it marked wrong? How would you correct it?

114. Following is an excerpt from your second best friend's graded homework:

$$f(x) = \frac{3}{4x^2};\ f'(x) = \frac{3}{8x} \quad \text{✗ WRONG} \quad -10$$

Why was it marked wrong? How would you correct it?

115. Following is an excerpt from your worst enemy's graded homework:

$$f(x) = 4x^2;\ f'(x) = (0)(2x) = 0 \quad \text{✗ WRONG} \quad -6$$

Why was it marked wrong? How would you correct it?

116. Following is an excerpt from your second worst enemy's graded homework:

$$f(x) = \frac{3}{4x};\ f'(x) = \frac{0}{4} = 0 \quad \text{✗ WRONG} \quad -10$$

Why was it marked wrong? How would you correct it?

117. One of the questions in your last calculus test was "**Question 1(a)** Give the definition of the derivative of a function f." Following is your answer and the grade you received:

$$nx^{n-1} \quad \text{✗ WRONG} \quad -10$$

Why was it marked wrong? What is the correct answer?

118. ▼ How would you respond to an acquaintance who says, "I finally understand what the derivative is: It is nx^{n-1}! Why weren't we taught that in the first place instead of the difficult way using limits?"

119. ▼ Sketch the graph of a function whose derivative is undefined at exactly two points but which has a tangent line at all but one point.

120. ▼ Sketch the graph of a function that has a tangent line at each of its points, but whose derivative is undefined at exactly two points.

4.2 A First Application: Marginal Analysis

In Chapter 1, we considered linear *cost functions* of the form $C(x) = mx + b$, where C is the total cost, x is the number of items, and m and b are constants. The slope m is the *marginal cost.* It measures the *cost of one more item.* Notice that the derivative of $C(x) = mx + b$ is $C'(x) = m$. In other words, for a linear cost function, *the marginal cost is the derivative of the cost function.*

In general, we make the following definition.

Marginal Cost

A **cost function** specifies the total cost C as a function of the number of items x. In other words, $C(x)$ is the total cost of x items. The **marginal cost function** is the derivative $C'(x)$ of the cost function $C(x)$. It measures the rate of change of cost with respect to x.

Units

The units of marginal cost are units of cost (dollars, say) per item.

Interpretation

*We interpret $C'(x)$ as the approximate cost of one more item.**

Quick Example

If $C(x) = 400x + 1{,}000$ dollars, then the marginal cost function is $C'(x) = \$400$ per item (a constant).

* **NOTE** See Example 1.

EXAMPLE 1 Marginal Cost

Suppose that the cost in dollars to manufacture portable CD players is given by

$$C(x) = 150{,}000 + 20x - 0.0001x^2$$

where x is the number of CD players manufactured.* Find the marginal cost function $C'(x)$ and use it to estimate the cost of manufacturing the 50,001st CD player.

* **NOTE** You might well ask where on Earth this formula came from. There are two approaches to obtaining cost functions in real life: analytical and empirical. The analytical approach is to calculate the cost function from scratch. For example, in the above situation, we might have fixed costs of \$150,000, plus a production cost of \$20 per CD player. The term $0.0001x^2$ may reflect a cost saving for high levels of production, such as a bulk discount in the cost of electronic components. In the empirical approach, we first obtain the cost at several different production levels by direct observation. This gives several points on the (as yet unknown) cost versus production level graph. Then find the equation of the curve that best fits these points, usually using regression.

Solution Since

$$C(x) = 150{,}000 + 20x - 0.0001x^2$$

the marginal cost function is

$$C'(x) = 20 - 0.0002x.$$

The units of $C'(x)$ are units of C (dollars) per unit of x (CD players). Thus, $C'(x)$ is measured in dollars per CD player.

The cost of the 50,001st CD player is the amount by which the total cost would rise if we increased production from 50,000 CD players to 50,001. Thus, we need to know the rate at which the total cost rises as we increase production. This rate of change is measured by the derivative, or marginal cost, which we just computed. At $x = 50{,}000$, we get

$$C'(50{,}000) = 20 - 0.0002(50{,}000) = \$10 \text{ per CD player.}$$

In other words, we estimate that the 50,001st CD player will cost approximately \$10.

➡ **Before we go on...** In Example 1, the marginal cost is really only an *approximation* to the cost of the 50,001st CD player:

$$\begin{aligned} C'(50{,}000) &\approx \frac{C(50{,}001) - C(50{,}000)}{1} && \text{Set } h = 1 \text{ in the definition of the derivative.} \\ &= C(50{,}001) - C(50{,}000) \\ &= \text{cost of the 50,001st CD player} \end{aligned}$$

The exact cost of the 50,001st CD player is

$$C(50{,}001) - C(50{,}000) = [150{,}000 + 20(50{,}001) - 0.0001(50{,}001)^2] - [150{,}000 + 20(50{,}000) - 0.0001(50{,}000)^2] = \$9.9999$$

So, the marginal cost is a good approximation to the actual cost.

Graphically, we are using the tangent line to approximate the cost function near a production level of 50,000. Figure 3 shows the graph of the cost function together with the tangent line at $x = 50{,}000$. Notice that the tangent line is essentially indistinguishable from the graph of the function for some distance on either side of 50,000.

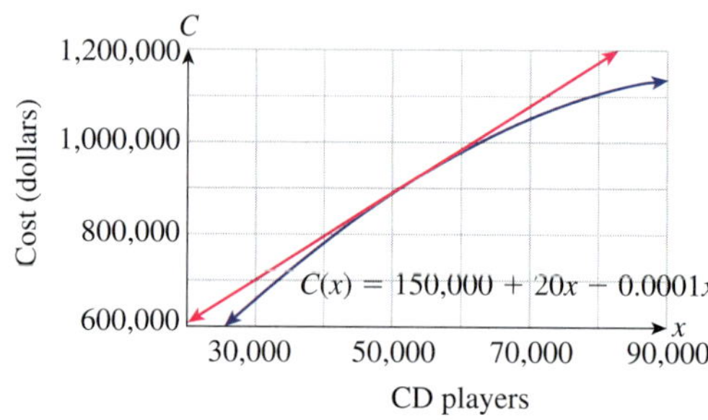

Figure 3

Notes

1. In general, the difference quotient $[C(x+h) - C(x)]/h$ gives the **average cost per item** to produce h more items at a current production level of x items. (Why?)
2. Notice that $C'(x)$ is much easier to calculate than $[C(x+h) - C(x)]/h$. (Try it.)

We can extend the idea of marginal cost to include other functions, like revenue and profit:

Marginal Revenue and Profit

A **revenue** or **profit function** specifies the total revenue R or profit P as a function of the number of items x. The derivatives, $R'(x)$ and $P'(x)$ of these functions are called the **marginal revenue** and **marginal profit** functions. They measure the rate of change of revenue and profit with respect to x.

Units

The units of marginal revenue and profit are the same as those of marginal cost: dollars (or euros, pesos, etc.) per item.

Interpretation

We interpret $R'(x)$ and $P'(x)$ as the approximate revenue and profit from the sale of one more item.

EXAMPLE 2 Marginal Revenue and Profit

You operate an *iPod* customizing service (a typical customized iPod might have a custom color case with blinking lights and a personalized logo). The cost to refurbish x iPods in a month is calculated to be

$$C(x) = 0.25x^2 + 40x + 1{,}000 \text{ dollars.}$$

You charge customers \$80 per iPod for the work.

a. Calculate the marginal revenue and profit functions. Interpret the results.

b. Compute the revenue and profit, and also the marginal revenue and profit, if you have refurbished 20 units this month. Interpret the results.

c. For which value of x is the marginal profit zero? Interpret your answer.

Solution

a. We first calculate the revenue and profit functions:

$$R(x) = 80x \qquad \text{Revenue} = \text{Price} \times \text{Quantity}$$

$$P(x) = R(x) - C(x) \qquad \text{Profit} = \text{Revenue} - \text{Cost}$$

$$= 80x - (0.25x^2 + 40x + 1{,}000)$$

$$P(x) = -0.25x^2 + 40x - 1{,}000.$$

The marginal revenue and profit functions are then the derivatives:

$$\text{Marginal revenue} = R'(x) = 80$$

$$\text{Marginal profit} = P'(x) = -0.5x + 40.$$

Interpretation: $R'(x)$ gives the approximate revenue from the refurbishing of one more item, and $P'(x)$ gives the approximate profit from the refurbishing of one more item. Thus, if x iPods have been refurbished in a month, you will earn a revenue of \$80 and make a profit of approximately $\$(-0.5x + 40)$ if you refurbish one more that month.

Notice that the marginal revenue is a constant, so you earn the same revenue (\$80) for each iPod you refurbish. However, the marginal profit, $\$(-0.5x + 40)$, decreases as x increases, so your additional profit is about 50¢ less for each additional iPod you refurbish.

b. From part (a), the revenue, profit, marginal revenue, and marginal profit functions are

$$R(x) = 80x$$

$$P(x) = -0.25x^2 + 40x - 1{,}000$$

$$R'(x) = 80$$

$$P'(x) = -0.5x + 40$$

Because you have refurbished $x = 20$ iPods this month, $x = 20$, so

$$R(20) = 80(20) = \$1{,}600 \qquad \text{Total revenue from 20 iPods}$$

$$P(20) = -0.25(20)^2 + 40(20) - 1{,}000 = -\$300 \qquad \text{Total profit from 20 iPods}$$

$$R'(20) = \$80 \text{ per unit} \qquad \text{Approximate revenue from the 21st iPod}$$

$$P'(20) = -0.5(20) + 40 = \$30 \text{ per unit} \qquad \text{Approximate profit from the 21st iPod}$$

Interpretation: If you refurbish 20 iPods in a month, you will earn a total revenue of \$160 and a profit of –\$300 (indicating a loss of \$300). Refurbishing one more iPod that month will earn you an additional revenue of \$80 and an additional profit of about \$30.

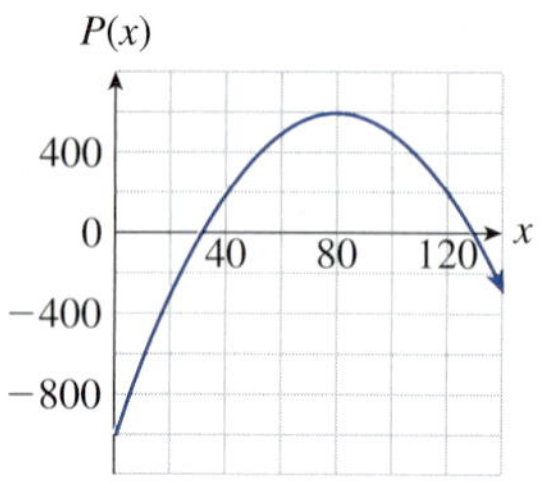

$P(x) = -0.25x^2 + 40x - 1{,}000$

Figure 4

c. The marginal profit is zero when $P'(x) = 0$:

$$-0.5x + 40 = 0$$

$$x = \frac{40}{0.5} = 80 \text{ iPods}$$

Thus, if you refurbish 80 iPods in a month, refurbishing one more will get you (approximately) zero additional profit. To understand this further, let us take a look at the graph of the profit function, shown in Figure 4. Notice that the graph is a parabola (the profit function is quadratic) with vertex at the point $x = 80$, where $P'(x) = 0$, so the profit is a maximum at this value of x.

➡ **Before we go on...** In general, setting $P'(x) = 0$ and solving for x will always give the exact values of x for which the profit peaks as in Figure 4, assuming there is such a value. We recommend that you graph the profit function to check whether the profit is indeed a maximum at such a point. ■

EXAMPLE 3 Marginal Product

A consultant determines that Precision Manufacturers' annual profit (in dollars) is given by

$$P(n) = -200{,}000 + 400{,}000n - 4{,}600n^2 - 10n^3 \qquad (10 \le n \le 50)$$

where n is the number of assembly-line workers it employs.

a. Compute $P'(n)$. $P'(n)$ is called the **marginal product** at the employment level of n assembly-line workers. What are its units?

b. Calculate $P(20)$ and $P'(20)$, and interpret the results.

c. Precision Manufacturers currently employs 20 assembly-line workers and is considering laying off some of them. What advice would you give the company's management?

Solution

a. Taking the derivative gives

$$P'(n) = 400{,}000 - 9{,}200n - 30n^2.$$

The units of $P'(n)$ are profit (in dollars) per worker.

b. Substituting into the formula for $P(n)$, we get

$$P(20) = -200{,}000 + 400{,}000(20) - 4{,}600(20)^2 - 10(20)^3 = \$5{,}880{,}000.$$

Thus, Precision Manufacturer will make an annual profit of \$5,880,000 if it employs 20 assembly-line workers. On the other hand,

$$P'(20) = 400{,}000 - 9{,}200(20) - 30(20)^2 = \$204{,}000/\text{worker}.$$

Thus, at an employment level of 20 assembly-line workers, annual profit is increasing at a rate of \$204,000 per additional worker. In other words, if the company were to employ one more assembly-line worker, its annual profit would increase by approximately \$204,000.

c. Because the marginal product is positive, profits will increase if the company increases the number of workers and will decrease if it decreases the number of workers, so your advice would be to hire additional assembly-line workers. Downsizing their assembly-line workforce would reduce their annual profits.

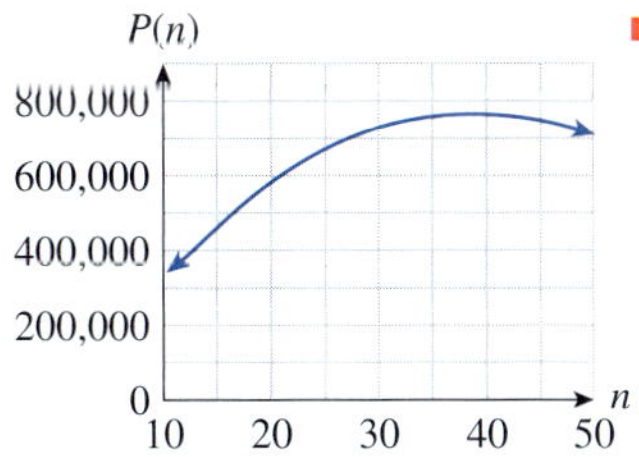

Figure 5

Before we go on... In Example 3, it would be interesting for Precision Manufacturers to ascertain how many additional assembly-line workers they should hire to obtain the *maximum* annual profit. Taking our cue from Example 2, we suspect that such a value of n would correspond to a point where $P'(n) = 0$. Figure 5 shows the graph of P, and on it we see that the highest point of the graph is indeed a point where the tangent line is horizontal; that is, $P'(n) = 0$, and occurs somewhere between $n = 35$ and 40. To compute this value of n more accurately, set $P'(n) = 0$ and solve for n:

$$P'(n) = 400{,}000 - 9{,}200n - 30n^2 = 0 \quad \text{or} \quad 40{,}000 - 920n - 3n^2 = 0.$$

We can now obtain n using the quadratic formula:

$$n = \frac{-b \pm \sqrt{b^2 - 4ac}}{2a} = \frac{920 \pm \sqrt{920^2 - 4(-3)(40{,}000)}}{2(-3)}$$

$$= \frac{920 \pm \sqrt{1{,}326{,}400}}{-6} \approx -345.3 \text{ or } 38.6.$$

The only meaningful solution is the positive one, $n \approx 38.6$ workers, and we conclude that the company should employ between 38 and 39 assembly-line workers for a maximum profit. To see which gives the larger profit, 38 or 39, we check:

$$P(38) = \$7{,}808{,}880$$

while

$$P(39) = \$7{,}810{,}210.$$

This tells us that the company should employ 39 assembly-line workers for a maximum profit. Thus, instead of laying off any of its 20 assembly-line workers, the company should hire 19 additional assembly line workers for a total of 39. ■

Average Cost

EXAMPLE 4 Average Cost

Suppose the cost in dollars to manufacture portable CD players is given by

$$C(x) = 150{,}000 + 20x - 0.0001x^2$$

where x is the number of CD players manufactured. (This is the cost equation we saw in Example 1.)

a. Find the average cost per CD player if 50,000 CD players are manufactured.

b. Find a formula for the average cost per CD player if x CD players are manufactured. This function of x is called the **average cost function, $\bar{C}(x)$**.

Solution

a. The total cost of manufacturing 50,000 CD players is given by

$$C(50{,}000) = 150{,}000 + 20(50{,}000) - 0.0001(50{,}000)^2$$
$$= \$900{,}000.$$

Because 50,000 CD players cost a total of $900,000 to manufacture, the average cost of manufacturing one CD player is this total cost divided by 50,000:

$$\bar{C}(50{,}000) = \frac{900{,}000}{50{,}000} = \$18.00 \text{ per CD player.}$$

Thus, if 50,000 CD players are manufactured, each CD player costs the manufacturer an average of $18.00 to manufacture.

b. If we replace 50,000 by x, we get the general formula for the average cost of manufacturing x CD players:

$$\begin{aligned}\bar{C}(x) &= \frac{C(x)}{x} \\ &= \frac{1}{x}(150{,}000 + 20x - 0.0001x^2) \\ &= \frac{150{,}000}{x} + 20 - 0.0001x. \qquad \text{Average cost function}\end{aligned}$$

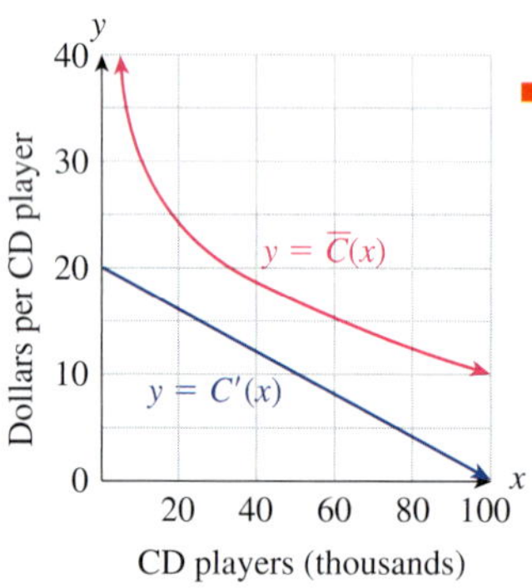

Figure 6

➡ **Before we go on...** Average cost and marginal cost convey different but related information. The average cost $\bar{C}(50{,}000) = \$18$ that we calculated in Example 4 is the cost per item of manufacturing the first 50,000 CD players, whereas the marginal cost $C'(50{,}000) = \$10$ that we calculated in Example 1 gives the (approximate) cost of manufacturing the *next* CD player. Thus, according to our calculations, the first 50,000 CD players cost an average of $18 to manufacture, but it costs only about $10 to manufacture the next one. Note that the marginal cost at a production level of 50,000 CD players is lower than the average cost. This means that the average cost to manufacture CDs is going down with increasing volume. (Think about why.)

Figure 6 shows the graphs of average and marginal cost. Notice how the decreasing marginal cost seems to pull the average cost down with it. ■

To summarize:

Average Cost

Given a cost function C, the **average cost** of the first x items is given by

$$\bar{C}(x) = \frac{C(x)}{x}.$$

The average cost is distinct from the **marginal cost** $C'(x)$, which tells us the approximate cost of the *next* item.

Quick Example

For the cost function $C(x) = 20x + 100$ dollars

Marginal Cost $= C'(x) = \$20$ per additional item.

Average Cost $= \bar{C}(x) = \dfrac{C(x)}{x} = \dfrac{20x + 100}{x} = \$(20 + 100/x)$ per item.

4.2 EXERCISES

▼ more advanced ◆ challenging

T indicates exercises that should be solved using technology

In Exercises 1–4, for each cost function, find the marginal cost at the given production level x, and state the units of measurement. (All costs are in dollars.) HINT [See Example 1.]

1. $C(x) = 10{,}000 + 5x - 0.0001x^2;\ x = 1{,}000$
2. $C(x) = 20{,}000 + 7x - 0.00005x^2;\ x = 10{,}000$
3. $C(x) = 15{,}000 + 100x + \dfrac{1{,}000}{x};\ x = 100$
4. $C(x) = 20{,}000 + 50x + \dfrac{10{,}000}{x};\ x = 100$

In Exercises 5 and 6, find the marginal cost, marginal revenue, and marginal profit functions, and find all values of x for which the marginal profit is zero. Interpret your answer. HINT [See Example 2.]

5. $C(x) = 4x;\quad R(x) = 8x - 0.001x^2$
6. $C(x) = 5x^2;\quad R(x) = x^3 + 7x + 10$
7. ▼ A certain cost function has the following graph:

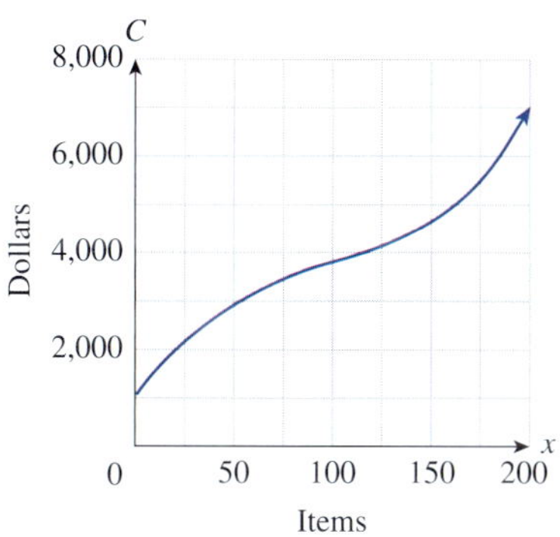

a. The associated marginal cost is
 (A) increasing, then decreasing
 (B) decreasing, then increasing
 (C) always increasing
 (D) always decreasing

b. The marginal cost is least at approximately
 (A) $x = 0$ (B) $x = 50$ (C) $x = 100$ (D) $x = 150$

c. The cost of 50 items is
 (A) approximately \$20, and increasing at a rate of about \$3,000 per item
 (B) approximately \$0.50, and increasing at a rate of about \$3,000 per item
 (C) approximately \$3,000, and increasing at a rate of about \$20 per item
 (D) approximately \$3,000, and increasing at a rate of about \$0.50 per item

8. ▼ A certain cost function has the following graph:

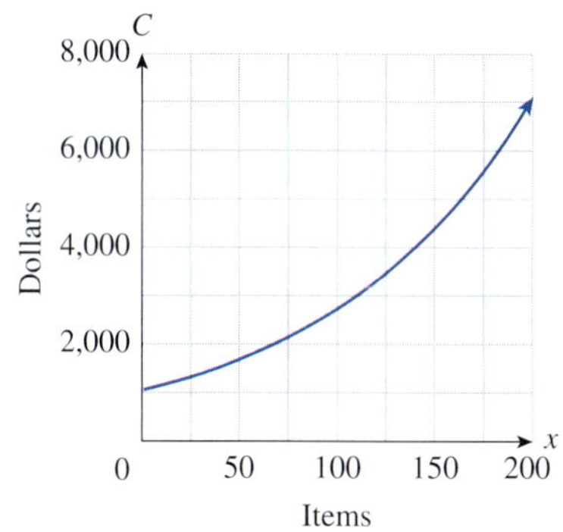

a. The associated marginal cost is
 (A) increasing, then decreasing
 (B) decreasing, then increasing
 (C) always increasing
 (D) always decreasing

b. When $x = 100$, the marginal cost is
 (A) greater than the average cost
 (B) less than the average cost
 (C) approximately equal to the average cost

c. The cost of 150 items is
 (A) approximately \$4,400, and increasing at a rate of about \$40 per item
 (B) approximately \$40, and increasing at a rate of about \$4,400 per item
 (C) approximately \$4,400, and increasing at a rate of about \$1 per item
 (D) approximately \$1, and increasing at a rate of about \$4,400 per item

APPLICATIONS

9. ***Advertising Costs*** The cost, in thousands of dollars, of airing x television commercials during a Super Bowl game is given by[14]

$$C(x) = 150 + 2{,}250x - 0.02x^2.$$

a. Find the marginal cost function and use it to estimate how fast the cost is increasing when $x = 4$. Compare this with the exact cost of airing the fifth commercial. HINT [See Example 1.]

b. Find the average cost function $\bar{C}$, and evaluate $\bar{C}(4)$. What does the answer tell you? HINT [See Example 4.]

[14] CBS charged an average of \$2.25 million per 30-second television spot during the 2004 Super Bowl game. This explains the coefficient of x in the cost function. Source: Advertising Age Research, www.AdAge.com/.

10. ***Marginal Cost and Average Cost*** The cost of producing x teddy bears per day at the Cuddly Companion Co. is calculated by their marketing staff to be given by the formula

$$C(x) = 100 + 40x - 0.001x^2.$$

a. Find the marginal cost function and use it to estimate how fast the cost is going up at a production level of 100 teddy bears. Compare this with the exact cost of producing the 101st teddy bear. HINT [See Example 1.]
b. Find the average cost function $\bar{C}$, and evaluate $\bar{C}(100)$. What does the answer tell you? HINT [See Example 4.]

11. ***Marginal Revenue and Profit*** Your college newspaper, *The Collegiate Investigator*, sells for 90¢ per copy. The cost of producing x copies of an edition is given by

$$C(x) = 70 + 0.10x + 0.001x^2 \text{ dollars.}$$

a. Calculate the marginal revenue and profit functions. HINT [See Example 2.]
b. Compute the revenue and profit, and also the marginal revenue and profit, if you have produced and sold 500 copies of the latest edition. Interpret the results.
c. For which value of x is the marginal profit zero? Interpret your answer.

12. ***Marginal Revenue and Profit*** The Audubon Society at Enormous State University (ESU) is planning its annual fund-raising "Eatathon." The society will charge students $1.10 per serving of pasta. The society estimates that the total cost of producing x servings of pasta at the event will be

$$C(x) = 350 + 0.10x + 0.002x^2 \text{ dollars.}$$

a. Calculate the marginal revenue and profit functions. HINT [See Example 2.]
b. Compute the revenue and profit, and also the marginal revenue and profit, if you have produced and sold 200 servings of pasta. Interpret the results.
c. For which value of x is the marginal profit zero? Interpret your answer.

13. ***Marginal Profit*** Suppose $P(x)$ represents the profit on the sale of x DVDs. If $P(1{,}000) = 3{,}000$ and $P'(1{,}000) = -3$, what do these values tell you about the profit?

14. ***Marginal Loss*** An automobile retailer calculates that its loss on the sale of type M cars is given by $L(50) = 5{,}000$ and $L'(50) = -200$, where $L(x)$ represents the loss on the sale of x type M cars. What do these values tell you about losses?

15. ***Marginal Profit*** Your monthly profit (in dollars) from selling magazines is given by

$$P = 5x + \sqrt{x}$$

where x is the number of magazines you sell in a month. If you are currently selling $x = 50$ magazines per month, find your profit and your marginal profit. Interpret your answers.

16. ***Marginal Profit*** Your monthly profit (in dollars) from your newspaper route is given by

$$P = 2n - \sqrt{n}$$

where n is the number of subscribers on your route. If you currently have 100 subscribers, find your profit and your marginal profit. Interpret your answers.

17. ▼ ***Marginal Revenue: Pricing Tuna*** Assume that the demand function for tuna in a small coastal town is given by

$$p = \frac{20{,}000}{q^{1.5}} \qquad (200 \le q \le 800)$$

where p is the price (in dollars) per pound of tuna, and q is the number of pounds of tuna that can be sold at the price p in one month.

a. Calculate the price that the town's fishery should charge for tuna in order to produce a demand of 400 pounds of tuna per month.
b. Calculate the monthly revenue R as a function of the number of pounds of tuna q.
c. Calculate the revenue and marginal revenue (derivative of the revenue with respect to q) at a demand level of 400 pounds per month, and interpret the results.
d. If the town fishery's monthly tuna catch amounted to 400 pounds of tuna, and the price is at the level in part (a), would you recommend that the fishery raise or lower the price of tuna in order to increase its revenue?

18. ▼ ***Marginal Revenue: Pricing Tuna*** Repeat Exercise 17, assuming a demand equation of

$$p = \frac{60}{q^{0.5}} \qquad (200 \le q \le 800).$$

19. ***Marginal Product*** A car wash firm calculates that its daily profit (in dollars) depends on the number n of workers it employs according to the formula

$$P = 400n - 0.5n^2.$$

Calculate the marginal product at an employment level of 50 workers, and interpret the result. HINT [See Example 3.]

20. ***Marginal Product*** Repeat the preceding exercise using the formula

$$P = -100n + 25n^2 - 0.005n^4.$$

HINT [See Example 3.]

21. ***Average and Marginal Cost*** The daily cost to manufacture generic trinkets for gullible tourists is given by the cost function

$$C(x) = -0.001x^2 + 0.3x + 500 \text{ dollars}$$

where x is the number of trinkets.

a. As x increases, the marginal cost
(A) increases **(B)** decreases **(C)** increases, then decreases **(D)** decreases, then increases

b. As x increases, the average cost
(A) increases (B) decreases (C) increases, then decreases (D) decreases, then increases

c. The marginal cost is
(A) greater than (B) equal to (C) less than
the average cost when $x = 100$. HINT [See Example 4.]

22. ***Average and Marginal Cost*** Repeat Exercise 21, using the following cost function for imitation oil paintings (x is the number of "oil paintings" manufactured):

$$C(x) = 0.1x^2 - 3.5x + 500 \text{ dollars}.$$

HINT [See Example 4.]

23. ***Advertising Cost*** Your company is planning to air a number of television commercials during the ABC Television Network's presentation of the Academy Awards. ABC is charging your company \$1.6 million per 30-second spot.[15] Additional fixed costs (development and personnel costs) amount to \$500,000, and the network has agreed to provide a discount of $\$10{,}000\sqrt{x}$ for x television spots.

a. Write down the cost function C, marginal cost function C', and average cost function $\bar{C}$.

b. Compute $C'(3)$ and $\bar{C}(3)$. (Round all answers to three significant digits.) Use these two answers to say whether the average cost is increasing or decreasing as x increases.

24. ***Housing Costs*** The cost C of building a house is related to the number k of carpenters used and the number x of electricians used by the formula[16]

$$C = 15{,}000 + 50k^2 + 60x^2.$$

a. Assuming that 10 carpenters are currently being used, find the cost function C, marginal cost function C', and average cost function $\bar{C}$, all as functions of x.

b. Use the functions you obtained in part (a) to compute $C'(15)$ and $\bar{C}(15)$. Use these two answers to say whether the average cost is increasing or decreasing as the number of electricians increases.

25. ▼ ***Emission Control*** The cost of controlling emissions at a firm rises rapidly as the amount of emissions reduced increases. Here is a possible model:

$$C(q) = 4{,}000 + 100q^2$$

where q is the reduction in emissions (in pounds of pollutant per day) and C is the daily cost (in dollars) of this reduction.

a. If a firm is currently reducing its emissions by 10 pounds each day, what is the marginal cost of reducing emissions further?

b. Government clean-air subsidies to the firm are based on the formula

$$S(q) = 500q$$

where q is again the reduction in emissions (in pounds per day) and S is the subsidy (in dollars). At what reduction level does the marginal cost surpass the marginal subsidy?

c. Calculate the net cost function, $N(q) = C(q) - S(q)$, given the cost function and subsidy above, and find the value of q that gives the lowest net cost. What is this lowest net cost? Compare your answer to that for part (b) and comment on what you find.

26. ▼ ***Taxation Schemes*** Here is a curious proposal for taxation rates based on income:

$$T(i) = 0.001i^{0.5}$$

where i represents total annual income in dollars and $T(i)$ is the income tax rate as a percentage of total annual income. (Thus, for example, an income of \$50,000 per year would be taxed at about 22%, while an income of double that amount would be taxed at about 32%.)[17]

a. Calculate the after-tax (net) income $N(i)$ an individual can expect to earn as a function of income i.

b. Calculate an individual's marginal after-tax income at income levels of \$100,000 and \$500,000.

c. At what income does an individual's marginal after-tax income become negative? What is the after-tax income at that level, and what happens at higher income levels?

d. What do you suspect is the most anyone can earn after taxes? (See NOTE at the bottom of this page.)

27. ▼ ***Fuel Economy*** Your Porsche's gas mileage (in miles per gallon) is given as a function $M(x)$ of speed x in miles per hour. It is found that

$$M'(x) = \frac{3{,}600x^{-2} - 1}{(3{,}600x^{-1} + x)^2}.$$

Estimate $M'(10)$, $M'(60)$, and $M'(70)$. What do the answers tell you about your car?

28. ▼ ***Marginal Revenue*** The estimated marginal revenue for sales of ESU soccer team T-shirts is given by

$$R'(p) = \frac{(8 - 2p)e^{-p^2+8p}}{10{,}000{,}000}$$

[15] ABC charged an average of \$1.6 million for a 30-second spot during the 2005 Academy Awards presentation. Source: CNN/Reuters, www.cnn.com, February 9, 2005.

[16] Based on an exercise in *Introduction to Mathematical Economics* by A. L. Ostrosky, Jr., and J. V. Koch (Waveland Press, Prospect Heights, Illinois, 1979).

[17] This model has the following interesting feature: An income of \$1 million per year would be taxed at 100%, leaving the individual penniless!

where p is the price (in dollars) that the soccer players charge for each shirt. Estimate $R'(3)$, $R'(4)$, and $R'(5)$. What do the answers tell you?

29. ◆ ***Marginal Cost*** *(from the GRE Economics Test)* In a multi-plant firm in which the different plants have different and continuous cost schedules, if costs of production for a given output level are to be minimized, which of the following is essential?

(A) Marginal costs must equal marginal revenue.
(B) Average variable costs must be the same in all plants.
(C) Marginal costs must be the same in all plants.
(D) Total costs must be the same in all plants.
(E) Output per worker per hour must be the same in all plants.

30. ◆ ***Study Time*** *(from the GRE economics test)* A student has a fixed number of hours to devote to study and is certain of the relationship between hours of study and the final grade for each course. Grades are given on a numerical scale (0 to 100), and each course is counted equally in computing the grade average. In order to maximize his or her grade average, the student should allocate these hours to different courses so that

(A) the grade in each course is the same.
(B) the marginal product of an hour's study (in terms of final grade) in each course is zero.
(C) the marginal product of an hour's study (in terms of final grade) in each course is equal, although not necessarily equal to zero.
(D) the average product of an hour's study (in terms of final grade) in each course is equal.
(E) the number of hours spent in study for each course is equal.

31. ◆ ***Marginal Product*** *(from the GRE Economics Test)* Assume that the marginal product of an additional senior professor is 50% higher than the marginal product of an additional junior professor and that junior professors are paid one-half the amount that senior professors receive. With a fixed overall budget, a university that wishes to maximize its quantity of output from professors should do which of the following?

(A) Hire equal numbers of senior professors and junior professors.
(B) Hire more senior professors and junior professors.
(C) Hire more senior professors and discharge junior professors.
(D) Discharge senior professors and hire more junior professors.
(E) Discharge all senior professors and half of the junior professors.

32. ◆ ***Marginal Product (based on a question from the GRE Economics Test)*** Assume that the marginal product of an additional senior professor is twice the marginal product of an additional junior professor and that junior professors are paid two-thirds the amount that senior professors receive. With a fixed overall budget, a university that wishes to maximize its quantity of output from professors should do which of the following?

(A) Hire equal numbers of senior professors and junior professors.
(B) Hire more senior professors and junior professors.
(C) Hire more senior professors and discharge junior professors.
(D) Discharge senior professors and hire more junior professors.
(E) Discharge all senior professors and half of the junior professors.

COMMUNICATION AND REASONING EXERCISES

33. The marginal cost of producing the 1,001st item is

(A) equal to
(B) approximately equal to
(C) always slightly greater than

the actual cost of producing the 1,001st item.

34. For the cost function $C(x) = mx + b$, the marginal cost of producing the 1,001st item is,

(A) equal to
(B) approximately equal to
(C) always slightly greater than

the actual cost of producing the 1,001st item.

35. What is a cost function? Carefully explain the difference between *average cost* and *marginal cost* in terms of **(a)** their mathematical definition, **(b)** graphs, and **(c)** interpretation.

36. The cost function for your grand piano manufacturing plant has the property that $\bar{C}(1{,}000) = \$3{,}000$ per unit and $C'(1{,}000) = \$2{,}500$ per unit. Will the average cost increase or decrease if your company manufactures a slightly larger number of pianos? Explain your reasoning.

37. If the average cost to manufacture one grand piano increases as the production level increases, which is greater, the marginal cost or the average cost?

38. If your analysis of a manufacturing company yielded positive marginal profit but negative profit at the company's current production levels, what would you advise the company to do?

39. ▼ If the marginal cost is decreasing, is the average cost necessarily decreasing? Explain.

40. ▼ If the average cost is decreasing, is the marginal cost necessarily decreasing? Explain.

41. ◆ If a company's marginal average cost is zero at the current production level, positive for a slightly higher production level, and negative for a slightly lower production level, what should you advise the company to do?

42. ◆ The **acceleration** of cost is defined as the derivative of the marginal cost function: that is, the derivative of the derivative—or *second derivative*—of the cost function. What are the units of acceleration of cost, and how does one interpret this measure?

4.3 The Product and Quotient Rules

We know how to find the derivatives of functions that are sums of powers, such as polynomials. In general, if a function is a sum or difference of functions whose derivatives we know, then we know how to find its derivative. But what about *products and quotients* of functions whose derivatives we know? For instance, how do we calculate the derivative of something like $x^2/(x+1)$? The derivative of $x^2/(x+1)$ is not, as one might suspect, $2x/1 = 2x$. That calculation is based on an assumption that the derivative of a quotient is the quotient of the derivatives. But it is easy to see that this assumption is false: For instance, the derivative of $1/x$ is not $0/1 = 0$, but $-1/x^2$. Similarly, the derivative of a product is not the product of the derivatives: For instance, the derivative of $x = 1 \cdot x$ is not $0 \cdot 1 = 0$, but 1.

To identify the correct method of computing the derivatives of products and quotients, let's look at a simple example. We know that the daily revenue resulting from the sale of q items per day at a price of p dollars per item is given by the product, $R = pq$ dollars. Suppose you are currently selling wall posters on campus. At this time your daily sales are 50 posters, and sales are increasing at a rate of 4 per day. Furthermore, you are currently charging \$10 per poster, and you are also raising the price at a rate of \$2 per day. Let's use this information to estimate how fast your daily revenue is increasing. In other words, let us estimate the rate of change, dR/dt, of the revenue R.

There are two contributions to the rate of change of daily revenue: the increase in daily sales and the increase in the unit price. We have

$$\frac{dR}{dt} \text{ due to increasing price:} \quad \$2 \text{ per day} \times 50 \text{ posters} = \$100 \text{ per day}$$

$$\frac{dR}{dt} \text{ due to increasing sales:} \quad \$10 \text{ per poster} \times 4 \text{ posters per day} = \$40 \text{ per day}$$

Thus, we estimate the daily revenue to be increasing at a rate of \$100 + \$40 = \$140 per day. Let us translate what we have said into symbols:

$$\frac{dR}{dt} \text{ due to increasing price:} \quad \frac{dp}{dt} \times q$$

$$\frac{dR}{dt} \text{ due to increasing sales:} \quad p \times \frac{dq}{dt}$$

Thus, the rate of change of revenue is given by

$$\frac{dR}{dt} = \frac{dp}{dt}q + p\frac{dq}{dt}.$$

Because $R = pq$, we have discovered the following rule for differentiating a product:

$$\frac{d}{dt}(pq) = \frac{dp}{dt}q + p\frac{dq}{dt}$$

The derivative of a product is the derivative of the first times the second, plus the first times the derivative of the second.

This rule and a similar rule for differentiating quotients are given next, and also a discussion of how these results are proved rigorously.

Product Rule

If $f(x)$ and $g(x)$ are differentiable functions of x, then so is their product $f(x)g(x)$, and

$$\frac{d}{dx}[f(x)g(x)] = f'(x)g(x) + f(x)g'(x).$$

Product Rule in Words

The derivative of a product is the derivative of the first times the second, plus the first times the derivative of the second.

Quick Example

$f(x) = x^2$ and $g(x) = 3x - 1$ are both differentiable functions of x, and so their product $x^2(3x - 1)$ is differentiable, and

$$\frac{d}{dx}[x^2(3x-1)] = \underset{\substack{\uparrow \\ \text{Derivative of first}}}{2x} \cdot \underset{\substack{\uparrow \\ \text{Second}}}{(3x-1)} + \underset{\substack{\uparrow \\ \text{First}}}{x^2} \cdot \underset{\substack{\uparrow \\ \text{Derivative of second}}}{(3)}.$$

Quotient Rule

If $f(x)$ and $g(x)$ are differentiable functions of x, then so is their quotient $f(x)/g(x)$ (provided $g(x) \neq 0$), and

$$\frac{d}{dx}\left(\frac{f(x)}{g(x)}\right) = \frac{f'(x)g(x) - f(x)g'(x)}{[g(x)]^2}.$$

Quotient Rule in Words

The derivative of a quotient is the derivative of the top times the bottom, minus the top times the derivative of the bottom, all over the bottom squared.

Quick Example

$f(x) = x^3$ and $g(x) = x^2 + 1$ are both differentiable functions of x, and so their quotient $x^3/(x^2 + 1)$ is differentiable, and

$$\frac{d}{dx}\left(\frac{x^3}{x^2+1}\right) = \frac{\overset{\substack{\text{Derivative of top} \\ \downarrow}}{3x^2}\overset{\substack{\text{Bottom} \\ \downarrow}}{(x^2+1)} - \overset{\substack{\text{Top} \\ \downarrow}}{x^3} \cdot \overset{\substack{\text{Derivative of bottom} \\ \downarrow}}{2x}}{\underset{\substack{\uparrow \\ \text{Bottom squared}}}{(x^2+1)^2}}$$

Notes

1. Don't try to remember the rules by the symbols we have used, but remember them in words. (The slogans are easy to remember, even if the terms are not precise.)
2. One more time: *The derivative of a product is* NOT *the product of the derivatives, and the derivative of a quotient is* NOT *the quotient of the derivatives.* To find the derivative of a product, you must use the product rule, and to find the derivative of a quotient, you must use the quotient rule.*

* **NOTE** Leibniz made this mistake at first, too, so you would be in good company if you forgot to use the product or quotient rule.

Q: *Wait a minute! The expression $2x^3$ is a product, and we already know that its derivative is $6x^2$. Where did we use the product rule?*

A: To differentiate functions such as $2x^3$, we have used the rule from Section 3.4:

The derivative of c times a function is c times the derivative of the function.

However, the product rule gives us the same result:

$$\frac{d}{dx}(2x^3) = \underset{\text{Derivative of first}}{(0)}\underset{\text{Second}}{(x^3)} + \underset{\text{First}}{(2)}\underset{\text{Derivative of second}}{(3x^2)} = 6x^2 \qquad \text{Product rule}$$

$$\frac{d}{dx}(2x^3) = (2)(3x^2) = 6x^2 \qquad \text{Derivative of a constant times a function}$$

We do not recommend that you use the product rule to differentiate functions such as $2x^3$; continue to use the simpler rule when one of the factors is a constant.

Derivation of the Product Rule

Before we look at more examples of using the product and quotient rules, let's see why the product rule is true. To calculate the derivative of the product $f(x)g(x)$ of two differentiable functions, we go back to the definition of the derivative:

$$\frac{d}{dx}[f(x)g(x)] = \lim_{h \to 0} \frac{f(x+h)g(x+h) - f(x)g(x)}{h}.$$

*** NOTE** Adding an appropriate form of zero is an age-old mathematical ploy.

We now rewrite this expression so that we can evaluate the limit: Notice that the numerator reflects a simultaneous change in f [from $f(x)$ to $f(x+h)$] and g [from $g(x)$ to $g(x+h)$]. To separate the two effects, we add and subtract a quantity in the numerator that reflects a change in only one of the functions:

$$\frac{d}{dx}[f(x)g(x)] = \lim_{h \to 0} \frac{f(x+h)g(x+h) - f(x)g(x)}{h}$$

$$= \lim_{h \to 0} \frac{f(x+h)g(x+h) - f(x)g(x+h) + f(x)g(x+h) - f(x)g(x)}{h} \qquad \text{We subtracted and added the quantity* } f(x)g(x+h).$$

$$= \lim_{h \to 0} \frac{[f(x+h) - f(x)]\,g(x+h) + f(x)[g(x+h) - g(x)]}{h} \qquad \text{Common factors}$$

$$= \lim_{h \to 0} \left(\frac{f(x+h) - f(x)}{h}\right) g(x+h) + \lim_{h \to 0} f(x)\left(\frac{g(x+h) - g(x)}{h}\right) \qquad \text{Limit of sum}$$

$$= \lim_{h \to 0} \left(\frac{f(x+h) - f(x)}{h}\right) \lim_{h \to 0} g(x+h) + \lim_{h \to 0} f(x) \lim_{h \to 0} \left(\frac{g(x+h) - g(x)}{h}\right) \qquad \text{Limit of product}$$

Web Site
www.AppliedCalc.org
For a proof of the fact that, if g is differentiable, it must be continuous, go to the Web site and follow the path
Everything for Calculus
→ Chapter 4
→ Continuity and Differentiability

Now we already know the following four limits:

$$\lim_{h \to 0} \frac{f(x+h) - f(x)}{h} = f'(x) \qquad \text{Definition of derivative of } f;\ f \text{ is differentiable.}$$

$$\lim_{h \to 0} \frac{g(x+h) - g(x)}{h} = g'(x) \qquad \text{Definition of derivative of } g;\ g \text{ is differentiable.}$$

$$\lim_{h \to 0} g(x+h) = g(x) \qquad \text{If } g \text{ is differentiable, it must be continuous.}$$

$$\lim_{h \to 0} f(x) = f(x) \qquad \text{Limit of a constant}$$

Web Site
www.AppliedCalc.org

The quotient rule can be proved in a very similar way. Go to the Web site and follow the path

Everything for Calculus
→ Chapter 4
→ Proof of Quotient Rule

Putting these limits into the one we're calculating, we get

$$\frac{d}{dx}[f(x)g(x)] = f'(x)g(x) + f(x)g'(x)$$

which is the product rule.

EXAMPLE 5 Using the Product Rule

Compute the following derivatives.

a. $\frac{d}{dx}[(x^{3.2}+1)(1-x)]$ Simplify the answer.

b. $\frac{d}{dx}[(x+1)(x^2+1)(x^3+1)]$ Do not expand the answer.

Solution

a. We can do the calculation in two ways.

Using the Product Rule: (Derivative of first ↓, Second ↓, First ↓, Derivative of second ↓)

$$\frac{d}{dx}[(x^{3.2}+1)(1-x)] = (3.2x^{2.2})(1-x) + (x^{3.2}+1)(-1)$$
$$= 3.2x^{2.2} - 3.2x^{3.2} - x^{3.2} - 1$$
$$= -4.2x^{3.2} + 3.2x^{2.2} - 1$$

Expand the answer.

Not Using the Product Rule: First, expand the given expression.

$$(x^{3.2}+1)(1-x) = -x^{4.2} + x^{3.2} - x + 1$$

Thus,

$$\frac{d}{dx}[(x^{3.2}+1)(1-x)] = \frac{d}{dx}(-x^{4.2} + x^{3.2} - x + 1)$$
$$= -4.2x^{3.2} + 3.2x^{2.2} - 1$$

In this example the product rule saves us little or no work, but in later sections we shall see examples that can be done in no other way. Learn how to use the product rule now!

b. Here we have a product of *three* functions, not just two. We can find the derivative by using the product rule twice:

$$\frac{d}{dx}[(x+1)(x^2+1)(x^3+1)]$$
$$= \frac{d}{dx}(x+1)\cdot[(x^2+1)(x^3+1)] + (x+1)\cdot\frac{d}{dx}[(x^2+1)(x^3+1)]$$
$$= (1)(x^2+1)(x^3+1) + (x+1)[(2x)(x^3+1) + (x^2+1)(3x^2)]$$
$$= (1)(x^2+1)(x^3+1) + (x+1)(2x)(x^3+1) + (x+1)(x^2+1)(3x^2)$$

We can see here a more general product rule:

$$(fgh)' = f'gh + fg'h + fgh'$$

Notice that every factor has a chance to contribute to the rate of change of the product. There are similar formulas for products of four or more functions.

EXAMPLE 6 Using the Quotient Rule

Compute the derivatives **a.** $\frac{d}{dx}\left[\frac{1-3.2x^{-0.1}}{x+1}\right]$ **b.** $\frac{d}{dx}\left[\frac{(x+1)(x+2)}{x-1}\right]$

Solution

a. $$\frac{d}{dx}\left[\frac{1-3.2x^{-0.1}}{x+1}\right] = \frac{\overset{\text{Derivative of top}}{(0.32x^{-1.1})}\overset{\text{Bottom}}{(x+1)} - \overset{\text{Top}}{(1-3.2x^{-0.1})}\overset{\text{Derivative of bottom}}{(1)}}{\underset{\text{Bottom squared}}{(x+1)^2}}$$

$$= \frac{0.32x^{-0.1} + 0.32x^{-1.1} - 1 + 3.2x^{-0.1}}{(x+1)^2} \quad \text{Expand the numerator.}$$

$$= \frac{3.52x^{-0.1} + 0.32x^{-1.1} - 1}{(x+1)^2}$$

b. Here we have both a product and a quotient. Which rule do we use, the product or the quotient rule? Here is a way to decide. Think about how we would calculate, step by step, the value of $(x+1)(x+2)/(x-1)$ for a specific value of x—say $x = 11$. Here is how we would probably do it:

1. Calculate $(x+1)(x+2) = (11+1)(11+2) = 156$.
2. Calculate $x - 1 = 11 - 1 = 10$.
3. Divide 156 by 10 to get 15.6.

Now ask: *What was the last operation we performed?* The last operation we performed was division, so we can regard the whole expression as a *quotient*—that is, as $(x+1)(x+2)$ *divided by* $(x-1)$. Therefore, we should use the quotient rule.

The first thing the quotient rule tells us to do is to take the derivative of the numerator. Now, the numerator is a product, so we must use the product rule to take its derivative. Here is the calculation:

$$\frac{d}{dx}\left[\frac{(x+1)(x+2)}{x-1}\right] = \frac{\overbrace{[(1)(x+2)+(x+1)(1)]}^{\text{Derivative of top}}\overset{\text{Bottom}}{(x-1)} - \overbrace{[(x+1)(x+2)]}^{\text{Top}}\overset{\text{Derivative of bottom}}{(1)}}{\underset{\text{Bottom squared}}{(x-1)^2}}$$

$$= \frac{(2x+3)(x-1) - (x+1)(x+2)}{(x-1)^2}$$

$$= \frac{x^2 - 2x - 5}{(x-1)^2}$$

What is important is to determine the *order of operations* and, in particular, to determine the last operation to be performed. Pretending to do an actual calculation reminds us of the order of operations; we call this technique the **calculation thought experiment**.

➡ **Before we go on...** We used the quotient rule in Example 6 because the function was a quotient; we used the product rule to calculate the derivative of the numerator because the numerator was a product. Get used to this: Differentiation rules usually must be used in combination.

Here is another way we could have done this problem: Our calculation thought experiment could have taken the following form.

1. Calculate $(x+1)/(x-1) = (11+1)/(11-1) = 1.2$.

2. Calculate $x+2 = 11+2 = 13$.

3. Multiply 1.2 by 13 to get 15.6.

We would have then regarded the expression as a *product*—the product of the factors $(x+1)/(x-1)$ and $(x+2)$ and used the product rule instead. We can't escape the quotient rule, however: We need to use it to take the derivative of the first factor, $(x+1)/(x-1)$. Try this approach for practice and check that you get the same answer. ■

Calculation Thought Experiment

The **calculation thought experiment** is a technique to determine whether to treat an algebraic expression as a product, quotient, sum, or difference. Given an expression, consider the steps you would use in computing its value. If the last operation is multiplication, treat the expression as a product; if the last operation is division, treat the expression as a quotient; and so on.

Quick Examples

1. $(3x^2-4)(2x+1)$ can be computed by first calculating the expressions in parentheses and then multiplying. Because the last step is multiplication, we can treat the expression as a product.

2. $\dfrac{2x-1}{x}$ can be computed by first calculating the numerator and denominator and then dividing one by the other. Because the last step is division, we can treat the expression as a quotient.

3. $x^2+(4x-1)(x+2)$ can be computed by first calculating x^2, then calculating the product $(4x-1)(x+2)$, and finally adding the two answers. Thus, we can treat the expression as a sum.

4. $(3x^2-1)^5$ can be computed by first calculating the expression in parentheses and then raising the answer to the fifth power. Thus, we can treat the expression as a power. (We shall see how to differentiate powers of expressions in Section 4.4.)

5. The expression $(x+1)(x+2)/(x-1)$ can be treated as either a quotient or a product: We can write it as a quotient: $\dfrac{(x+1)(x+2)}{x-1}$ or as a product: $(x+1)\left(\dfrac{x+2}{x-1}\right)$. (See Example 6(b).)

EXAMPLE 7 Using the Calculation Thought Experiment

Find $\frac{d}{dx}\left[6x^2 + 5\left(\frac{x}{x-1}\right)\right]$.

Solution The calculation thought experiment tells us that the expression we are asked to differentiate can be treated as a *sum*. Because the derivative of a sum is the sum of the derivatives, we get

$$\frac{d}{dx}\left[6x^2 + 5\left(\frac{x}{x-1}\right)\right] = \frac{d}{dx}(6x^2) + \frac{d}{dx}\left[5\left(\frac{x}{x-1}\right)\right].$$

In other words, we must take the derivatives of $6x^2$ and $5\left(\frac{x}{x-1}\right)$ separately and then add the answers. The derivative of $6x^2$ is $12x$. There are two ways of taking the derivative of $5\left(\frac{x}{x-1}\right)$: We could either first multiply the expression $\left(\frac{x}{x-1}\right)$ by 5 to get $\left(\frac{5x}{x-1}\right)$ and then take its derivative using the quotient rule, or we could pull the 5 out, as we do next.

$$\begin{aligned}
\frac{d}{dx}\left(6x^2 + 5\left(\frac{x}{x-1}\right)\right) &= \frac{d}{dx}(6x^2) + \frac{d}{dx}\left[5\left(\frac{x}{x-1}\right)\right] && \text{Derivative of sum} \\
&= 12x + 5\frac{d}{dx}\left(\frac{x}{x-1}\right) && \text{Constant} \times \text{Function} \\
&= 12x + 5\left(\frac{(1)(x-1) - (x)(1)}{(x-1)^2}\right) && \text{Quotient rule} \\
&= 12x + 5\left(\frac{-1}{(x-1)^2}\right) \\
&= 12x - \frac{5}{(x-1)^2}
\end{aligned}$$

APPLICATIONS

In the next example, we return to a scenario similar to the one discussed at the start of this section.

EXAMPLE 8 Applying the Product and Quotient Rules: Revenue and Average Cost

Sales of your newly launched miniature wall posters for college dorms, *iMiniPosters,* are really taking off. (Those old-fashioned large wall posters no longer fit in today's "downsized" college dorm rooms.) Monthly sales to students at the start of this year were 1,500 iMiniPosters, and since that time, sales have been increasing by 300 posters each month, even though the price you charge has also been going up.

a. The price you charge for iMiniPosters is given by:

$$p(t) = 10 + 0.05t^2 \text{ dollars per poster,}$$

where t is time in months since the start of January of this year. Find a formula for the monthly revenue, and then compute its rate of change at the beginning of March.

b. The number of students who purchase iMiniPosters in a month is given by

$$n(t) = 800 + 0.2t,$$

where t is as in part (a). Find a formula for the average number of posters each student buys, and hence estimate the rate at which this number was growing at the beginning of March.

Solution

a. To compute monthly revenue as a function of time t, we use

$$R(t) = p(t)q(t). \qquad \text{Revenue} = \text{Price} \times \text{Quantity}$$

We already have a formula for $p(t)$. The function $q(t)$ measures sales, which were 1,500 posters/month at time $t = 0$, and rising by 300 per month:

$$q(t) = 1{,}500 + 300t.$$

Therefore, the formula for revenue is

$$R(t) = p(t)q(t)$$
$$R(t) = (10 + 0.05t^2)(1{,}500 + 300t).$$

Rather than expand this expression, we shall leave it as a product so that we can use the product rule in computing its rate of change:

$$R'(t) = p'(t)q(t) + p(t)q'(t)$$
$$= [0.10t][1{,}500 + 300t] + [10 + 0.05t^2][300].$$

Because the beginning of March corresponds to $t = 2$, we have

$$R'(2) = [0.10(2)][1{,}500 + 300(2)] + [10 + 0.05(2)^2][300]$$
$$= (0.2)(2{,}100) + (10.2)(300) = \$3{,}480 \text{ per month.}$$

Therefore, your monthly revenue was increasing at a rate of $3,480 per month at the beginning of March.

b. The average number of posters sold to each student is

$$k(t) = \frac{\text{Number of posters}}{\text{Number of students}}$$
$$k(t) = \frac{q(t)}{n(t)} = \frac{1{,}500 + 300t}{800 + 0.2t}.$$

The rate of change of $k(t)$ is computed with the quotient rule:

$$k'(t) = \frac{q'(t)n(t) - q(t)n'(t)}{n(t)^2}$$
$$= \frac{(300)(800 + 0.2t) - (1{,}500 + 300t)(0.2)}{(800 + 0.2t)^2}$$

so that

$$k'(2) = \frac{(300)[800 + 0.2(2)] - [1{,}500 + 300(2)](0.2)}{[800 + 0.2(2)]^2}$$
$$= \frac{(300)(800.4) - (2{,}100)(0.2)}{800.4^2} \approx 0.37 \text{ posters/student per month.}$$

Therefore, the average number of posters sold to each student was increasing at a rate of about 0.37 posters/student per month.

4.3 EXERCISES

▼ more advanced ◆ challenging

T indicates exercises that should be solved using technology

In Exercises 1–12:

a. Calculate the derivative of the given function without using either the product or quotient rule.

b. Use the product or quotient rule to find the derivative. Check that you obtain the same answer. HINT [See Quick Examples on page 314.]

1. $f(x) = 3x$ **2.** $f(x) = 2x^2$

3. $g(x) = x \cdot x^2$ **4.** $g(x) = x \cdot x$

5. $h(x) = x(x+3)$ **6.** $h(x) = x(1+2x)$

7. $r(x) = 100x^{2.1}$ **8.** $r(x) = 0.2x^{-1}$ **9.** $s(x) = \frac{2}{x}$

10. $t(x) = \frac{x}{3}$ **11.** $u(x) = \frac{x^2}{3}$ **12.** $s(x) = \frac{3}{x^2}$

Calculate $\frac{dy}{dx}$ *in Exercises 13–20. Simplify your answer.* HINT [See Example 5.]

13. $y = 3x(4x^2 - 1)$ **14.** $y = 3x^2(2x+1)$

15. $y = x^3(1 - x^2)$ **16.** $y = x^5(1 - x)$

17. $y = (2x+3)^2$ **18.** $y = (4x-1)^2$

19. $x\sqrt{x}$ **20.** $x^2\sqrt{x}$

Calculate $\frac{dy}{dx}$ *in Exercises 21–50. You need not expand your answers.*

21. $y = (x+1)(x^2-1)$

22. $y = (4x^2 + x)(x - x^2)$

23. $y = (2x^{0.5} + 4x - 5)(x - x^{-1})$

24. $y = (x^{0.7} - 4x - 5)(x^{-1} + x^{-2})$

25. $y = (2x^2 - 4x + 1)^2$

26. $y = (2x^{0.5} - x^2)^2$

27. $y = \left(\frac{x}{3.2} + \frac{3.2}{x}\right)(x^2+1)$

28. $y = \left(\frac{x^{2.1}}{7} + \frac{2}{x^{2.1}}\right)(7x-1)$

29. $x^2(2x+3)(7x+2)$ HINT [See Example 5b.]

30. $x(x^2-3)(2x^2+1)$ HINT [See Example 5b.]

31. $(5.3x - 1)(1 - x^{2.1})(x^{-2.3} - 3.4)$

32. $(1.1x + 4)(x^{2.1} - x)(3.4 - x^{-2.1})$

33. ▼ $y = (\sqrt{x} + 1)\left(\sqrt{x} + \frac{1}{x^2}\right)$

34. ▼ $y = (4x^2 - \sqrt{x})\left(\sqrt{x} - \frac{2}{x^2}\right)$

35. $y = \frac{2x+4}{3x-1}$ HINT [See Example 6.]

36. $y = \frac{3x-9}{2x+4}$ HINT [See Example 6.]

37. $y = \frac{2x^2+4x+1}{3x-1}$ **38.** $y = \frac{3x^2-9x+11}{2x+4}$

39. $y = \frac{x^2-4x+1}{x^2+x+1}$ **40.** $y = \frac{x^2+9x-1}{x^2+2x-1}$

41. $y = \frac{x^{0.23} - 5.7x}{1 - x^{-2.9}}$ **42.** $y = \frac{8.43x^{-0.1} - 0.5x^{-1}}{3.2 + x^{2.9}}$

43. ▼ $y = \frac{\sqrt{x}+1}{\sqrt{x}-1}$ **44.** ▼ $y = \frac{\sqrt{x}-1}{\sqrt{x}+1}$

45. ▼ $y = \frac{\left(\frac{1}{x} + \frac{1}{x^2}\right)}{x + x^2}$ **46.** ▼ $y = \frac{\left(1 - \frac{1}{x^2}\right)}{x^2 - 1}$

47. $y = \frac{(x+3)(x+1)}{3x-1}$ HINT [See Example 6b.]

48. $y = \frac{x}{(x-5)(x-4)}$ HINT [See Example 6b.]

49. $y = \frac{(x+3)(x+1)(x+2)}{3x-1}$

50. $y = \frac{3x-1}{(x-5)(x-4)(x-1)}$

In Exercises 51–56, compute the derivatives.

51. $\frac{d}{dx}[(x^2+x)(x^2-x)]$

52. $\frac{d}{dx}[(x^2+x^3)(x+1)]$

53. $\frac{d}{dx}[(x^3+2x)(x^2-x)]\Big|_{x=2}$

54. $\frac{d}{dx}[(x^2+x)(x^2-x)]\Big|_{x=1}$

55. $\frac{d}{dt}[(t^2 - t^{0.5})(t^{0.5} + t^{-0.5})]\Big|_{t=1}$

56. $\frac{d}{dt}[(t^2 + t^{0.5})(t^{0.5} - t^{-0.5})]\Big|_{t=1}$

In Exercises 57–64 use the calculation thought experiment to say whether the expression is written as a sum, difference, scalar multiple, product, or quotient. Then use the appropriate rules to find its derivative. HINT [See Quick Examples on page 318 and Example 7.]

57. $y = x^4 - (x^2 + 120)(4x - 1)$

58. $y = x^4 - \dfrac{x^2 + 120}{4x - 1}$

59. $y = x + 1 + 2\left(\dfrac{x}{x+1}\right)$

60. $y = (x + 2) - 4(x^2 - x)\left(x + \dfrac{1}{x}\right)$

(Do not simplify the answer.)

61. $y = (x + 2)\left(\dfrac{x}{x+1}\right)$

(Do not simplify the answer.)

62. $y = \dfrac{(x+2)x}{x+1}$

(Do not simplify the answer.)

63. $y = (x + 1)(x - 2) - 2\left(\dfrac{x}{x+1}\right)$

64. $y = \dfrac{x+2}{x+1} + (x + 1)(x - 2)$

In Exercises 65–70, find the equation of the line tangent to the graph of the given function at the point with the indicated x-coordinate.

65. $f(x) = (x^2 + 1)(x^3 + x); x = 1$

66. $f(x) = (x^{0.5} + 1)(x^2 + x); x = 1$

67. $f(x) = \dfrac{x+1}{x+2}; x = 0$

68. $f(x) = \dfrac{\sqrt{x}+1}{\sqrt{x}+2}; x = 4$

69. $f(x) = \dfrac{x^2+1}{x}; x = -1$

70. $f(x) = \dfrac{x}{x^2+1}; x = 1$

APPLICATIONS

71. ***Revenue*** The monthly sales of **Sunny Electronics**' new sound system are given by $q(t) = 2{,}000t - 100t^2$ units per month, t months after its introduction. The price Sunny charges is $p(t) = 1{,}000 - t^2$ dollars per sound system, t months after introduction. Find the rate of change of monthly sales, the rate of change of the price, and the rate of change of monthly revenue five months after the introduction of the sound system. Interpret your answers. HINT [See Example 8(a).]

72. ***Revenue*** The monthly sales of **Sunny Electronics**' new *iSun* walkman is given by $q(t) = 2{,}000t - 100t^2$ units per month, t months after its introduction. The price Sunny charges is $p(t) = 100 - t^2$ dollars per *iSun*, t months after introduction. Find the rate of change of monthly sales, the rate of change of the price, and the rate of change of monthly revenue six months after the introduction of the *iSun*. Interpret your answers. HINT [See Example 8(a).]

73. ***Saudi Oil Revenues*** The spot price of crude oil during the period 2000–2005 can be approximated by

$$P(t) = 5t + 25 \text{ dollars per barrel} \quad (0 \le t \le 5)$$

in year t, where $t = 0$ represents 2000. Saudi Arabia's crude oil production over the same period can be approximated by

$$Q(t) = 0.082t^2 - 0.22t + 8.2 \text{ million barrels per day.}^{18} \quad (0 \le t \le 5)$$

Use these models to estimate Saudi Arabia's daily oil revenue and also its rate of change in 2001. (Round your answers to the nearest \$1 million.)

74. ***Russian Oil Revenues*** Russia's crude oil production during the period 2000–2005 can be approximated by

$$Q(t) = -0.066t^2 + 0.96t + 6.1 \text{ million barrels per day}^{19} \quad (0 \le t \le 5)$$

in year t, where $t = 0$ represents 2000. Use the model for the spot price in Exercise 73 to estimate Russia's daily oil revenue and also its rate of change in 2001.

75. ***Revenue*** Dorothy Wagner is currently selling 20 "I ♥ Calculus" T-shirts per day, but sales are dropping at a rate of 3 per day. She is currently charging \$7 per T-shirt, but to compensate for dwindling sales, she is increasing the unit price by \$1 per day. How fast, and in what direction is her daily revenue currently changing?

76. ***Pricing Policy*** Let us turn Exercise 75 around a little: Dorothy Wagner is currently selling 20 "I ♥ Calculus" T-shirts per day, but sales are dropping at a rate of 3 per day. She is currently charging \$7 per T-shirt, and she wishes to increase her daily revenue by \$10 per day. At what rate should she increase the unit price to accomplish this (assuming that the price increase does not affect sales)?

77. ***Bus Travel*** Thoroughbred Bus Company finds that its monthly costs for one particular year were given by $C(t) = 10{,}000 + t^2$ dollars after t months. After t months the company had $P(t) = 1{,}000 + t^2$ passengers per month. How fast is its cost per passenger changing after 6 months? HINT [See Example 8(b).]

78. ***Bus Travel*** Thoroughbred Bus Company finds that its monthly costs for one particular year were given by $C(t) = 100 + t^2$ dollars after t months. After t months, the company had $P(t) = 1{,}000 + t^2$ passengers per month. How fast is its cost per passenger changing after 6 months? HINT [See Example 8(b).]

79. ***Fuel Economy*** Your muscle car's gas mileage (in miles per gallon) is given as a function $M(x)$ of speed x in mph, where

$$M(x) = \frac{3{,}000}{x + 3{,}600x^{-1}}.$$

Calculate $M'(x)$, and then $M'(10)$, $M'(60)$, and $M'(70)$. What do the answers tell you about your car?

---

[18] Source for data: EIA/Saudi British Bank (www.sabb.com). 2004 figures are based on mid-year data, and 2005 data are estimates.

[19] Source for data: Energy Information Administration (www.eia.doe.gov), Pravda (http://english.pravda.ru). 2004 figures are based on mid-year data, and 2005 data are estimates.

80. ***Fuel Economy*** Your used Chevy's gas mileage (in miles per gallon) is given as a function $M(x)$ of speed x in mph, where

$$M(x) = \frac{4{,}000}{x + 3{,}025x^{-1}}.$$

Calculate $M'(x)$ and hence determine *the sign* of each of the following: $M'(40)$, $M'(55)$, and $M'(60)$. Interpret your results.

81. ▼ ***Oil Imports from Mexico*** Daily oil production in Mexico and daily U.S. oil imports from Mexico during 2005–2009 can be approximated by

$$P(t) = 3.9 - 0.10t \text{ million barrels} \quad (5 \le t \le 9)$$
$$I(t) = 2.1 - 0.11t \text{ million barrels} \quad (5 \le t \le 9)$$

where t is time in years since the start of 2000.[20]

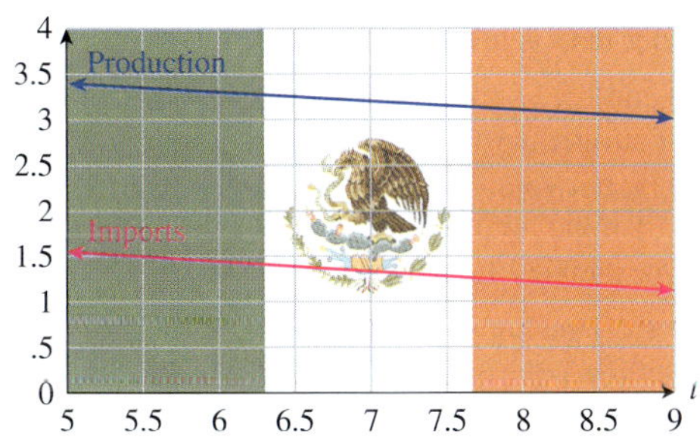

a. What are represented by the functions $P(t) - I(t)$ and $I(t)/P(t)$?

b. Compute $\frac{d}{dt}\left[\frac{I(t)}{P(t)}\right]\Big|_{t=8}$ to two significant digits. What does the answer tell you about oil imports from Mexico?

82. ▼ ***Oil Imports from Mexico*** Daily oil production in Mexico and daily U.S. oil imports from Mexico during 2000–2004 can be approximated by

$$P(t) = 3.0 + 0.13t \text{ million barrels} \quad (0 \le t \le 4)$$
$$I(t) = 1.4 + 0.06t \text{ million barrels} \quad (0 \le t \le 4)$$

where t is time in years since the start of 2000.[21]

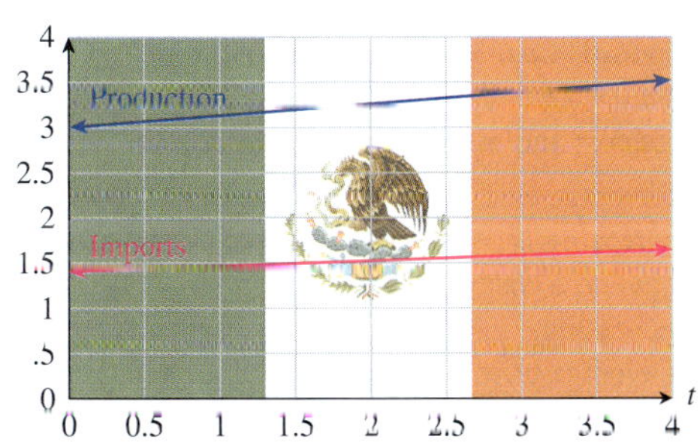

a. What are represented by the functions $P(t) - I(t)$ and $I(t)/P(t)$?

b. Compute $\frac{d}{dt}\left[\frac{I(t)}{P(t)}\right]\Big|_{t=3}$ to two significant digits. What does the answer tell you about oil imports from Mexico?

83. ▼ ***Military Spending*** The annual cost per active-duty armed service member in the United States increased from \$80,000 in 1995 to a projected \$120,000 in 2007. In 1995, there were 1.5 million armed service personnel, and this number was projected to decrease to 1.4 million in 2003.[22] Use linear models for annual cost and personnel to estimate, to the nearest \$10 million, the rate of change of total military personnel costs in 2002.

84. ▼ ***Military Spending in the 1990s*** The annual cost per active-duty armed service member in the United States increased from \$80,000 in 1995 to \$90,000 in 2000. In 1990, there were 2 million armed service personnel and this number decreased to 1.5 million in 2000.[23] Use linear models for annual cost and personnel to estimate, to the nearest \$10 million, the rate of change of total military personnel costs in 1995.

85. ▼ ***Biology—Reproduction*** The Verhulst model for population growth specifies the reproductive rate of an organism as a function of the total population according to the following formula:

$$R(p) = \frac{r}{1 + kp}$$

where p is the total population in thousands of organisms, r and k are constants that depend on the particular circumstances and the organism being studied, and $R(p)$ is the reproduction rate in thousands of organisms per hour.[24] If $k = 0.125$ and $r = 45$, find $R'(p)$ and then $R'(4)$. Interpret the result.

86. ▼ ***Biology—Reproduction*** Another model, the predator satiation model for population growth, specifies that the reproductive rate of an organism as a function of the total population varies according to the following formula:

$$R(p) = \frac{rp}{1 + kp}$$

where p is the total population in thousands of organisms, r and k are constants that depend on the particular circumstances and the organism being studied, and $R(p)$ is the reproduction rate in new organisms per hour.[25] Given that $k = 0.2$ and $r = 0.08$, find $R'(p)$ and $R'(2)$. Interpret the result.

[20] Source for data: Energy Information Administration (www.eia.doe.gov)/Pemex.

[21] Ibid.

[22] Annual costs are adjusted for inflation. Sources: Department of Defense, Stephen Daggett, military analyst, Congressional Research Service/*New York Times*, April 19, 2002, p. A21.

[23] Ibid.

[24] Source: *Mathematics in Medicine and the Life Sciences* by F. C. Hoppensteadt and C. S. Peskin (Springer-Verlag, New York, 1992) pp. 20–22.

[25] Ibid.

87. ▼ ***Embryo Development*** Bird embryos consume oxygen from the time the egg is laid through the time the chick hatches. For a typical galliform bird egg, the oxygen consumption (in milliliters) t days after the egg was laid can be approximated by[26]

$$C(t) = -0.016t^4 + 1.1t^3 - 11t^2 + 3.6t. \quad (15 \le t \le 30)$$

(An egg will usually hatch at around $t = 28$.) Suppose that at time $t = 0$ you have a collection of 30 newly laid eggs and that the number of eggs decreases linearly to zero at time $t = 30$ days. How fast is the total oxygen consumption of your collection of embryos changing after 25 days? (Round your answers to two significant digits.) Comment on the result. HINT [Total oxygen consumption = Oxygen consumption per egg × Number of eggs.]

88. ▼ ***Embryo Development*** Turkey embryos consume oxygen from the time the egg is laid through the time the chick hatches. For a brush turkey, the oxygen consumption (in milliliters) t days after the egg was laid can be approximated by[27]

$$C(t) = -0.0071t^4 + 0.95t^3 - 22t^2 + 95t. \quad (25 \le t \le 50)$$

(An egg will typically hatch at around $t = 50$.) Suppose that at time $t = 0$ you have a collection of 100 newly laid eggs and that the number of eggs decreases linearly to zero at time $t = 50$ days. How fast is the total oxygen consumption of your collection of embryos changing after 40 days? (Round your answer to two significant digits.) Interpret the result. HINT [Total oxygen consumption = Oxygen consumption per egg × Number of eggs.]

COMMUNICATION AND REASONING EXERCISES

89. If f and g are functions of time, and at time $t = 3$, f equals 5 and is rising at a rate of 2 units per second, and g equals 4 and is rising at a rate of 5 units per second, then the product fg equals ____ and is rising at a rate of ____ units per second.

90. If f and g are functions of time, and at time $t = 2$, f equals 3 and is rising at a rate of 4 units per second, and g equals 5 and is rising at a rate of 6 units per second, then fg equals ____ and is rising at a rate of ____ units per second.

91. If f and g are functions of time, and at time $t = 3$, f equals 5 and is rising at a rate of 2 units per second, and g equals 4 and is rising at a rate of 5 units per second, then f/g equals ____ and is changing at a rate of ____ units per second.

92. If f and g are functions of time, and at time $t = 2$, f equals 3 and is rising at a rate of 4 units per second, and g equals 5 and is rising at a rate of 6 units per second, then f/g equals ____ and is changing at a rate of ____ units per second.

93. You have come across the following in a newspaper article: "Revenues of HAL Home Heating Oil Inc. are rising by \$4.2 million per year. This is due to an annual increase of 70¢ per gallon in the price HAL charges for heating oil and an increase in sales of 6 million gallons of oil per year." Comment on this analysis.

94. Your friend says that because average cost is obtained by dividing the cost function by the number of units x, it follows that the derivative of average cost is the same as marginal cost because the derivative of x is 1. Comment on this analysis.

95. ▼ Find a demand function $q(p)$ such that, at a price per item of $p = \$100$, revenue will rise if the price per item is increased.

96. ▼ What must be true about a demand function $q(p)$ so that, at a price per item of $p = \$100$, revenue will decrease if the price per item is increased?

97. ▼ You and I are both selling a steady 20 T-shirts per day. The price I am getting for my T-shirts is increasing twice as fast as yours, but your T-shirts are currently selling for twice the price of mine. Whose revenue is increasing faster: yours, mine, or neither? Explain.

98. ▼ You and I are both selling T-shirts for a steady \$20 per shirt. Sales of my T-shirts are increasing at twice the rate of yours, but you are currently selling twice as many as I am. Whose revenue is increasing faster: yours, mine, or neither? Explain.

99. ◆ ***Marginal Product*** *(from the GRE Economics Test)* Which of the following statements about average product and marginal product is correct?

(A) If average product is decreasing, marginal product must be less than average product.
(B) If average product is increasing, marginal product must be increasing.
(C) If marginal product is decreasing, average product must be less than marginal product.
(D) If marginal product is increasing, average product must be decreasing.
(E) If marginal product is constant over some range, average product must be constant over that range.

100. ◆ ***Marginal Cost*** *(based on a question from the GRE Economics Test)* Which of the following statements about average cost and marginal cost is correct?

(A) If average cost is increasing, marginal cost must be increasing.
(B) If average cost is increasing, marginal cost must be decreasing.
(C) If average cost is increasing, marginal cost must be more than average cost.
(D) If marginal cost is increasing, average cost must be increasing.
(E) If marginal cost is increasing, average cost must be larger than marginal cost.

[26] The model is derived from graphical data published in the article "The Brush Turkey" by Roger S. Seymour, *Scientific American*, December, 1991, pp. 108–114.

[27] Ibid.

4.4 The Chain Rule

We can now find the derivatives of expressions involving powers of x combined using addition, subtraction, multiplication, and division, but we still cannot take the derivative of an expression like $(3x + 1)^{0.5}$. For this we need one more rule. The function $h(x) = (3x + 1)^{0.5}$ is not a sum, difference, product, or quotient. To find out what it is, we can use the calculation thought experiment and think about the last operation we would perform in calculating $h(x)$.

1. Calculate $3x + 1$.

2. Take the 0.5 power (square root) of the answer.

The last operation is "take the 0.5 power." We do not yet have a rule for finding the derivative of the 0.5 power of a quantity other than x.

There is a way to build $h(x) = (3x + 1)^{0.5}$ out of two simpler functions: $u(x) = 3x + 1$ (the function that corresponds to the first step in the calculation above) and $f(x) = x^{0.5}$ (the function that corresponds to the second step):

$$\begin{aligned} h(x) &= (3x + 1)^{0.5} \\ &= [u(x)]^{0.5} && u(x) = 3x + 1 \\ &= f(u(x)) && f(x) = x^{0.5} \end{aligned}$$

We say that h is the **composite** of f and u. We read $f(u(x))$ as "f of u of x."

To compute $h(1)$, say, we first compute $3 \cdot 1 + 1 = 4$ and then take the square root of 4, giving $h(1) = 2$. To compute $f(u(1))$ we follow exactly the same steps: First compute $u(1) = 4$ and then $f(u(1)) = f(4) = 2$. We always compute $f(u(x))$ numerically from the inside out: Given x, first compute $u(x)$ and then $f(u(x))$.

Now, f and u are functions *whose derivatives we know.* The *chain rule* allows us to use our knowledge of the derivatives of f and u to find the derivative of $f(u(x))$. For the purposes of stating the rule, let us avoid some of the nested parentheses by abbreviating $u(x)$ as u. Thus, we write $f(u)$ instead of $f(u(x))$ and remember that u is a function of x.

Chain Rule

If f is a differentiable function of u and u is a differentiable function of x, then the composite $f(u)$ is a differentiable function of x, and

$$\frac{d}{dx}[f(u)] = f'(u)\frac{du}{dx} \qquad \text{Chain Rule}$$

In words *The derivative of f(quantity) is the derivative of f, evaluated at that quantity, times the derivative of the quantity.*

Quick Examples

1. Take $f(u) = u^2$. Then

$$\frac{d}{dx}[u^2] = 2u\frac{du}{dx} \qquad \text{Because } f'(u) = 2u$$

The derivative of a quantity squared is two times the quantity, times the derivative of the quantity.

2. Take $f(u) = u^{0.5}$. Then

$$\frac{d}{dx}[u^{0.5}] = 0.5u^{-0.5}\frac{du}{dx} \qquad \text{Because } f'(u) = 0.5u^{-0.5}$$

The derivative of a quantity raised to the 0.5 is 0.5 times the quantity raised to the −0.5, times the derivative of the quantity.

As the quick examples illustrate, for every power of a function u whose derivative we know, we now get a "generalized" differentiation rule. The following table gives more examples.

Original Rule	***Generalized Rule***	***In Words***
$\frac{d}{dx}[x^2] = 2x$	$\frac{d}{dx}[u^2] = 2u\frac{du}{dx}$	*The derivative of a quantity squared is twice the quantity, times the derivative of the quantity.*
$\frac{d}{dx}[x^3] = 3x^2$	$\frac{d}{dx}[u^3] = 3u^2\frac{du}{dx}$	*The derivative of a quantity cubed is 3 times the quantity squared, times the derivative of the quantity.*
$\frac{d}{dx}\left(\frac{1}{x}\right) = -\frac{1}{x^2}$	$\frac{d}{dx}\left(\frac{1}{u}\right) = -\frac{1}{u^2}\frac{du}{dx}$	*The derivative of 1 over a quantity is negative 1 over the quantity squared, times the derivative of the quantity.*
Power Rule	***Generalized Power Rule***	***In Words***
$\frac{d}{dx}[x^n] = nx^{n-1}$	$\frac{d}{dx}[u^n] = nu^{n-1}\frac{du}{dx}$	*The derivative of a quantity raised to the n is n times the quantity raised to the n − 1, times the derivative of the quantity.*

To motivate the chain rule, let us see why it is true in the special case when $f(u) = u^3$, where the chain rule tells us that

$$\frac{d}{dx}[u^3] = 3u^2\frac{du}{dx} \qquad \text{Generalized Power Rule with } n = 3$$

But we could have done this using the product rule instead:

$$\frac{d}{dx}[u^3] = \frac{d}{dx}[u \cdot u \cdot u] = \frac{du}{dx}u \cdot u + u\frac{du}{dx}u + u \cdot u\frac{du}{dx} = 3u^2\frac{du}{dx}$$

which gives us the same result. A similar argument works for $f(u) = u^n$ where $n = 2, 3, 4, \ldots$ We can then use the quotient rule and the chain rule for positive powers to verify the generalized power rule for *negative* powers as well. For the case of a general differentiable function f, the proof of the chain rule is beyond the scope of this book, but you can find one on the Web site by the path shown in the margin.

Web Site
www.AppliedCalc.org
Everything for Calculus
→ Chapter 4
→ Proof of Chain Rule

EXAMPLE 1 Using the Chain Rule

Compute the following derivatives.

a. $\frac{d}{dx}[(2x^2 + x)^3]$ **b.** $\frac{d}{dx}[(x^3 + x)^{100}]$ **c.** $\frac{d}{dx}\sqrt{3x + 1}$

Solution

a. Using the calculation thought experiment, we see that the last operation we would perform in calculating $(2x^2 + x)^3$ is that of *cubing*. Thus we think of $(2x^2 + x)^3$ as *a quantity cubed*. There are two similar methods we can use to calculate its derivative.

Method 1: Using the formula We think of $(2x^2 + x)^3$ as u^3, where $u = 2x^2 + x$. By the formula,

$$\frac{d}{dx}[u^3] = 3u^2 \frac{du}{dx} \quad \text{Generalized Power Rule}$$

Now substitute for u:

$$\frac{d}{dx}[(2x^2 + x)^3] = 3(2x^2 + x)^2 \frac{d}{dx}(2x^2 + x)$$
$$= 3(2x^2 + x)^2(4x + 1)$$

Method 2: Using the verbal form If we prefer to use the verbal form, we get:

The derivative of $(2x^2 + x)$ *cubed is three times* $(2x^2 + x)$ *squared, times the derivative of* $(2x^2 + x)$.

In symbols,

$$\frac{d}{dx}[(2x^2 + x)^3] = 3(2x^2 + x)^2(4x + 1),$$

as we obtained above.

b. First, the calculation thought experiment: If we were computing $(x^3 + x)^{100}$, the last operation we would perform is *raising a quantity to the power* 100. Thus we are dealing with *a quantity raised to the power* 100, and so we must again use the generalized power rule. According to the verbal form of the generalized power rule, the derivative of a quantity raised to the power 100 is 100 times that quantity to the power 99, times the derivative of that quantity. In symbols,

$$\frac{d}{dx}[(x^3 + x)^{100}] = 100(x^3 + x)^{99}(3x^2 + 1).$$

c. We first rewrite the expression $\sqrt{3x + 1}$ as $(3x + 1)^{0.5}$ and then use the generalized power rule as in parts (a) and (b):

The derivative of a quantity raised to the 0.5 *is* 0.5 *times the quantity raised to the* −0.5, *times the derivative of the quantity.*

Thus,

$$\frac{d}{dx}[(3x + 1)^{0.5}] = 0.5(3x + 1)^{-0.5} \cdot 3 = 1.5(3x + 1)^{-0.5}.$$

➡ Before we go on... The following are examples of common errors in solving Example 1(b):

$$\text{“}\frac{d}{dx}[(x^3 + x)^{100}] = 100(3x^2 + 1)^{99}\text{”} \quad ✗ \text{ WRONG!}$$

$$\text{“}\frac{d}{dx}[(x^3 + x)^{100}] = 100(x^3 + x)^{99}\text{”} \quad ✗ \text{ WRONG!}$$

Remember that the generalized power rule says that the derivative of a quantity to the power 100 is 100 times *that same quantity* raised to the power 99, *times the derivative of that quantity.* ■

Q: *It seems that there are now two formulas for the derivative of an nth power:*

1. $\frac{d}{dx}[x^n] = nx^{n-1}$

2. $\frac{d}{dx}[u^n] = nu^{n-1}\frac{du}{dx}$

Which one do I use?

A: Formula 1 is the original power rule, which applies only to a power of x. For instance, it applies to x^{10}, but it does not apply to $(2x+1)^{10}$ because the quantity that is being raised to a power is not x. Formula 2 applies to a power of any *function of x*, such as $(2x+1)^{10}$. It can even be used in place of the original power rule. For example, if we take $u = x$ in Formula 2, we obtain

$$\frac{d}{dx}[x^n] = nx^{n-1}\frac{dx}{dx}$$

$$= nx^{n-1} \qquad \text{The derivative of } x \text{ with respect to } x \text{ is } 1.$$

Thus, the generalized power rule really *is* a generalization of the original power rule, as its name suggests.

EXAMPLE 2 More Examples Using the Chain Rule

Find: **a.** $\frac{d}{dx}[(2x^5+x^2-20)^{-2/3}]$ **b.** $\frac{d}{dx}\left[\frac{1}{\sqrt{x+2}}\right]$ **c.** $\frac{d}{dx}\left[\frac{1}{x^2+x}\right]$

Solution Each of the given functions is, or can be rewritten as, a power of a function whose derivative we know. Thus, we can use the method of Example 1.

a. $\frac{d}{dx}[(2x^5+x^2-20)^{-2/3}] = -\frac{2}{3}(2x^5+x^2-20)^{-5/3}(10x^4+2x)$

b. $\frac{d}{dx}\left[\frac{1}{\sqrt{x+2}}\right] = \frac{d}{dx}(x+2)^{-1/2} = -\frac{1}{2}(x+2)^{-3/2}\cdot 1 = -\frac{1}{2(x+2)^{3/2}}$

c. $\frac{d}{dx}\left[\frac{1}{x^2+x}\right] = \frac{d}{dx}(x^2+x)^{-1} = -(x^2+x)^{-2}(2x+1) = -\frac{2x+1}{(x^2+x)^2}$

➡ **Before we go on...** In Example 2(c), we could have used the quotient rule instead of the generalized power rule. We can think of the quantity $1/(x^2+x)$ in two different ways using the calculation thought experiment:

1. As 1 divided by something—in other words, as a quotient
2. As something raised to the -1 power

Of course, we get the same derivative using either approach. ■

We now look at some more complicated examples.

EXAMPLE 3 Harder Examples Using the Chain Rule

Find $\frac{dy}{dx}$ in each case. **a.** $y = [(x+1)^{-2.5} + 3x]^{-3}$ **b.** $y = (x+10)^3\sqrt{1-x^2}$

Solution

a. The calculation thought experiment tells us that the last operation we would perform in calculating y is raising the quantity $[(x+1)^{-2.5} + 3x]$ to the power -3. Thus, we use the generalized power rule.

$$\frac{dy}{dx} = -3[(x+1)^{-2.5} + 3x]^{-4}\frac{d}{dx}[(x+1)^{-2.5} + 3x]$$

We are not yet done; we must still find the derivative of $(x+1)^{-2.5} + 3x$. Finding the derivative of a complicated function in several steps helps to keep the problem manageable. Continuing, we have

$$\frac{dy}{dx} = -3[(x+1)^{-2.5} + 3x]^{-4}\frac{d}{dx}[(x+1)^{-2.5} + 3x]$$

$$= -3[(x+1)^{-2.5} + 3x]^{-4}\left[\frac{d}{dx}[(x+1)^{-2.5}] + \frac{d}{dx}(3x)\right] \quad \text{Derivative of a sum}$$

Now we have two derivatives left to calculate. The second of these we know to be 3, and the first is the derivative of a quantity raised to the -2.5 power. Thus

$$\frac{dy}{dx} = -3[(x+1)^{-2.5} + 3x]^{-4}[-2.5(x+1)^{-3.5} \cdot 1 + 3].$$

b. The expression $(x+10)^3\sqrt{1-x^2}$ is a product, so we use the product rule:

$$\frac{d}{dx}[(x+10)^3\sqrt{1-x^2}] = \left(\frac{d}{dx}[(x+10)^3]\right)\sqrt{1-x^2} + (x+10)^3\left(\frac{d}{dx}\sqrt{1-x^2}\right)$$

$$= 3(x+10)^2\sqrt{1-x^2} + (x+10)^3\frac{1}{2\sqrt{1-x^2}}(-2x)$$

$$= 3(x+10)^2\sqrt{1-x^2} - \frac{x(x+10)^3}{\sqrt{1-x^2}}$$

APPLICATIONS

The next example is a new treatment of Example 3 from Section 4.2.

EXAMPLE 4 Marginal Product

Precision Manufacturers is informed by a consultant that its annual profit is given by

$$P = -200{,}000 + 4{,}000q - 0.46q^2 - 0.00001q^3$$

where q is the number of surgical lasers it sells each year. The consultant also informs Precision that the number of surgical lasers it can manufacture each year depends on the number n of assembly line workers it employs according to the equation

$$q = 100n \quad \text{Each worker contributes 100 lasers per year.}$$

Use the chain rule to find the marginal product $\frac{dP}{dn}$.

Solution We could calculate the marginal product by substituting the expression for q in the expression for P to obtain P as a function of n (as given in Example 3 from Section 4.2) and then finding dP/dn. Alternatively—and this will simplify the calculation—we can use the chain rule. To see how the chain rule applies, notice that P is a function of q, where q in turn is given as a function of n. By the chain rule,

$$\frac{dP}{dn} = P'(q)\frac{dq}{dn} \qquad \text{Chain Rule}$$

$$= \frac{dP}{dq}\frac{dq}{dn} \qquad \text{Notice how the "quantities" } dq \text{ appear to cancel.}$$

Now we compute

$$\frac{dP}{dq} = 4{,}000 - 0.92q - 0.00003q^2$$

and $$\frac{dq}{dn} = 100.$$

Substituting into the equation for $\frac{dP}{dn}$ gives

$$\frac{dP}{dn} = (4{,}000 - 0.92q - 0.00003q^2)(100)$$

$$= 400{,}000 - 92q - 0.003q^2$$

Notice that the answer has q as a variable. We can express dP/dn as a function of n by substituting $100n$ for q:

$$\frac{dP}{dn} = 400{,}000 - 92(100n) - 0.003(100n)^2$$

$$= 400{,}000 - 9{,}200n - 30n^2$$

The equation

$$\frac{dP}{dn} = \frac{dP}{dq}\frac{dq}{dn}$$

in the example above is an appealing way of writing the chain rule because it suggests that the "quantities" dq cancel. In general, we can write the chain rule as follows.

Chain Rule in Differential Notation

If y is a differentiable function of u, and u is a differentiable function of x, then

$$\frac{dy}{dx} = \frac{dy}{du}\frac{du}{dx}$$

Notice how the units cancel:

$$\frac{\text{Units of } y}{\text{Units of } x} = \frac{\text{Units of } y}{\cancel{\text{Units of } u}}\,\frac{\cancel{\text{Units of } u}}{\text{Units of } x}$$

Quick Examples

1. If $y = u^3$, where $u = 4x + 1$, then

$$\frac{dy}{dx} = \frac{dy}{du}\frac{du}{dx} = 3u^2 \cdot 4 = 12u^2 = 12(4x + 1)^2.$$

2. If $q = 43p^2$ where p (and hence q also) is a differentiable function of t, then

$$\frac{dq}{dt} = \frac{dq}{dp}\frac{dp}{dt}$$

$$= 86p\frac{dp}{dt}.$$ p is not specified, so we leave dp/dt as is.

You can see one of the reasons we still use Leibniz differential notation: The chain rule looks like a simple "cancellation" of du terms.

EXAMPLE 5 Marginal Revenue

Suppose a company's weekly revenue R is given as a function of the unit price p, and p in turn is given as a function of weekly sales q (by means of a demand equation). If

$$\left.\frac{dR}{dp}\right|_{q=1,000} = \$40 \text{ per } \$1 \text{ increase in price}$$

and

$$\left.\frac{dp}{dq}\right|_{q=1,000} = -\$20 \text{ per additional item sold per week}$$

find the marginal revenue when sales are 1,000 items per week.

Solution The marginal revenue is $\frac{dR}{dq}$. By the chain rule, we have

$$\frac{dR}{dq} = \frac{dR}{dp}\frac{dp}{dq}$$ Units: Revenue per item = Revenue per \$1 price increase × price increase per additional item

Because we are interested in the marginal revenue at a demand level of 1,000 items per week, we have

$$\left.\frac{dR}{dq}\right|_{q=1,000} = (40)(-20) = -\$800 \text{ per additional item sold.}$$

Thus, if the price is lowered to increase the demand from 1,000 to 1,001 items per week, the weekly revenue will drop by approximately \$800.

Look again at the way the terms "du" appeared to cancel in the differential formula $\frac{dy}{dx} = \frac{dy}{du}\frac{du}{dx}$. In fact, the chain rule tells us more:

Manipulating Derivatives in Differential Notation

1. Suppose y is a function of x. Then, thinking of x as a function of y (as, for instance, when we can solve for x)* one has

$$\frac{dx}{dy} = \frac{1}{\left(\frac{dy}{dx}\right)}, \text{ provided } \frac{dy}{dx} \neq 0.$$

Notice again how $\frac{dy}{dx}$ behaves like a fraction.

* **NOTE** The notion of "thinking of x as a function of y" will be made more precise in Section 4.4.

Quick Example

In the demand equation $q = -0.2p - 8$, we have $\frac{dq}{dp} = -0.2$. Therefore,

$$\frac{dp}{dq} = \frac{1}{\left(\frac{dq}{dp}\right)} = \frac{1}{-0.2} = -5.$$

2. Suppose x and y are functions of t. Then, thinking of y as a function of x (as, for instance, when we can solve for t as a function of x, and hence obtain y as a function of x) one has

$$\frac{dy}{dx} = \frac{dy/dt}{dx/dt}.$$

The terms dt appear to cancel.

Quick Example

If $x = 3 - 0.2t$ and $y = 6 + 6t$, then

$$\frac{dy}{dx} = \frac{dy/dt}{dx/dt} = \frac{6}{-0.2} = -30.$$

To see why the above formulas work, notice that the second formula,

$$\frac{dy}{dx} = \frac{\left(\frac{dy}{dt}\right)}{\left(\frac{dx}{dt}\right)}$$

can be written as

$$\frac{dy}{dx}\frac{dx}{dt} = \frac{dy}{dt},$$

Multiply both sides by $\frac{dx}{dt}$.

which is just the differential form of the chain rule. For the first formula, use the second formula with y playing the role of t:

$$\frac{dy}{dx} = \frac{dy/dy}{dx/dy}$$

$$= \frac{1}{dx/dy}.$$

$\frac{dy}{dy} = \frac{d}{dy}[y] = 1$

FAQs

Using the Chain Rule

Q: *How do I decide whether or not to use the chain rule when taking a derivative?*

A: Use the calculation thought experiment (Section 4.3): Given an expression, consider the steps you would use in computing its value.

- If the last step is *raising a quantity to a power,* as in $\left(\dfrac{x^2-1}{x+4}\right)^4$, then the first step to use is the chain rule (in the form of the generalized power rule):

$$\frac{d}{dx}\left(\frac{x^2-1}{x+4}\right)^4 = 4\left(\frac{x^2-1}{x+4}\right)^3 \frac{d}{dx}\left(\frac{x^2-1}{x+4}\right).$$

Then use the appropriate rules to finish the computation. You may need to again use the calculation thought experiment to decide on the next step (here the quotient rule):

$$= 4\left(\frac{x^2-1}{x+4}\right)^3 \frac{(2x)(x+4)-(x^2-1)(1)}{(x+4)^2}.$$

- If the last step is *division,* as in $\dfrac{(x^2-1)}{(3x+4)^4}$, then the first step to use is the quotient rule:

$$\frac{d}{dx}\frac{(x^2-1)}{(3x+4)^4} = \frac{(2x)(3x+4)^4-(x^2-1)\dfrac{d}{dx}(3x+4)^4}{(3x+4)^8}.$$

Then use the appropriate rules to finish the computation (here the chain rule):

$$= \frac{(2x)(3x+4)^4-(x^2-1)4(3x+4)^3(3)}{(3x+4)^8}.$$

- If the last step is *multiplication, addition, subtraction, or multiplication by a constant,* then the first rule to use is the product rule, or the rule for sums, differences, or constant multiples as appropriate.

Q: *Every time I compute a derivative, I leave something out. How do I make sure I am really done when taking the derivative of a complicated-looking expression?*

A: Until you are an expert at taking derivatives, the key is to use one rule at a time and write out each step, rather than trying to compute the derivative in a single step. To illustrate this, try computing the derivative of $(x+10)^3\sqrt{1-x^2}$ in Example 3(b) in two ways: First try to compute it in a single step, and then compute it by writing out each step as shown in the example. How do your results compare? For more practice, try Exercises 83 and 84 following.

4.4 EXERCISES

▼ more advanced ◆ challenging

T indicates exercises that should be solved using technology

Calculate the derivatives of the functions in Exercises 1–46. HINT [See Example 1.]

1. $f(x) = (2x+1)^2$

2. $f(x) = (3x-1)^2$

3. $f(x) = (x-1)^{-1}$

4. $f(x) = (2x-1)^{-2}$

5. $f(x) = (2-x)^{-2}$

6. $f(x) = (1-x)^{-1}$

7. $f(x) = (2x+1)^{0.5}$

8. $f(x) = (-x+2)^{1.5}$

9. $f(x) = (4x-1)^{-1}$

10. $f(x) = (x+7)^{-2}$

11. $f(x) = \dfrac{1}{3x-1}$

12. $f(x) = \dfrac{1}{(x+1)^2}$

13. $f(x) = (x^2+2x)^4$

14. $f(x) = (x^3-x)^3$

15. $f(x) = (2x^2-2)^{-1}$

16. $f(x) = (2x^3+x)^{-2}$

17. $g(x) = (x^2-3x-1)^{-5}$

18. $g(x) = (2x^2+x+1)^{-3}$

19. $h(x) = \dfrac{1}{(x^2+1)^3}$ HINT [See Example 2.]

20. $h(x) = \dfrac{1}{(x^2+x+1)^2}$ HINT [See Example 2.]

21. $r(x) = (0.1x^2-4.2x+9.5)^{1.5}$

22. $r(x) = (0.1x-4.2x^{-1})^{0.5}$

23. $r(s) = (s^2-s^{0.5})^4$

24. $r(s) = (2s+s^{0.5})^{-1}$

25. $f(x) = \sqrt{1-x^2}$

26. $f(x) = \sqrt{x+x^2}$

27. $h(x) = 2[(x+1)(x^2-1)]^{-1/2}$ HINT [See Example 3.]

28. $h(x) = 3[(2x-1)(x-1)]^{-1/3}$ HINT [See Example 3.]

29. $h(x) = (3.1x-2)^2 - \dfrac{1}{(3.1x-2)^2}$

30. $h(x) = \left[3.1x^2 - 2 - \dfrac{1}{3.1x-2}\right]^2$

31. $f(x) = [(6.4x-1)^2 + (5.4x-2)^3]^2$

32. $f(x) = (6.4x-3)^{-2} + (4.3x-1)^{-2}$

33. $f(x) = (x^2-3x)^{-2}(1-x^2)^{0.5}$

34. $f(x) = (3x^2+x)(1-x^2)^{0.5}$

35. $s(x) = \left(\dfrac{2x+4}{3x-1}\right)^2$

36. $s(x) = \left(\dfrac{3x-9}{2x+4}\right)^3$

37. $g(z) = \left(\dfrac{z}{1+z^2}\right)^3$

38. $g(z) = \left(\dfrac{z^2}{1+z}\right)^2$

39. $f(x) = [(1+2x)^4 - (1-x)^2]^3$

40. $f(x) = [(3x-1)^2 + (1-x)^5]^2$

41. $t(x) = [2 + (x+1)^{-0.1}]^{4.3}$

42. $t(x) = [(x+1)^{0.1} - 4x]^{-5.1}$

43. ▼ $r(x) = \left(\sqrt{2x+1} - x^2\right)^{-1}$

44. ▼ $r(x) = \left(\sqrt{x+1} + \sqrt{x}\right)^3$

45. ▼ $f(x) = (1 + (1 + (1+2x)^3)^3)^3$

46. ▼ $f(x) = 2x + (2x + (2x+1)^3)^3$

Find the indicated derivatives in Exercises 47–54. In each case, the independent variable is a (unspecified) function of t. HINT [See Quick Example 2 on page 331.]

47. $y = x^{100} + 99x^{-1}$. Find $\dfrac{dy}{dt}$.

48. $y = x^{0.5}(1+x)$. Find $\dfrac{dy}{dt}$.

49. $s = \dfrac{1}{r^3} + r^{0.5}$. Find $\dfrac{ds}{dt}$.

50. $s = r + r^{-1}$. Find $\dfrac{ds}{dt}$.

51. $V = \dfrac{4}{3}\pi r^3$. Find $\dfrac{dV}{dt}$.

52. $A = 4\pi r^2$. Find $\dfrac{dA}{dt}$.

53. ▼ $y = x^3 + \dfrac{1}{x}$, $x = 2$ when $t = 1$, $\left.\dfrac{dx}{dt}\right|_{t=1} = -1$

Find $\left.\dfrac{dy}{dt}\right|_{t=1}$.

54. ▼ $y = \sqrt{x} + \dfrac{1}{\sqrt{x}}$, $x = 9$ when $t = 1$, $\left.\dfrac{dx}{dt}\right|_{t=1} = -1$

Find $\left.\dfrac{dy}{dt}\right|_{t=1}$.

In Exercises 55–60, compute the indicated derivative using the chain rule. HINT [See Quick Examples on page 332.]

55. $y = 3x - 2$; $\dfrac{dx}{dy}$

56. $y = 8x + 4$; $\dfrac{dx}{dy}$

57. $x = 2 + 3t$, $y = -5t$; $\dfrac{dy}{dx}$

58. $x = 1 - t/2$, $y = 4t - 1$; $\dfrac{dy}{dx}$

59. $y = 3x^2 - 2x$; $\left.\dfrac{dx}{dy}\right|_{x=1}$

60. $y = 3x - \dfrac{2}{x}$; $\left.\dfrac{dx}{dy}\right|_{x=2}$

APPLICATIONS

61. ***Marginal Product*** Paramount Electronics has an annual profit given by

$$P = -100{,}000 + 5{,}000q - 0.25q^2$$

where q is the number of laptop computers it sells each year. The number of laptop computers it can make and sell each year depends on the number n of electrical engineers Paramount employs, according to the equation

$$q = 30n + 0.01n^2.$$

Use the chain rule to find $\left.\frac{dP}{dn}\right|_{n=10}$ and interpret the result. HINT [See Example 3.]

62. *Marginal Product* Refer back to Exercise 61. The average profit $\bar{P}$ per computer is given by dividing the total profit P by q:

$$\bar{P} = -\frac{100{,}000}{q} + 5{,}000 - 0.25q.$$

Determine the **marginal average product**, $d\bar{P}/dn$ at an employee level of 10 engineers. Interpret the result. HINT [See Example 3.]

63. *Food Versus Education* The percentage y (of total personal consumption) an individual spends on food is approximately

$$y = 35x^{-0.25} \text{ percentage points} \quad (6.5 \le x \le 17.5)$$

where x is the percentage the individual spends on education.[28] An individual finds that she is spending

$$x = 7 + 0.2t$$

percent of her personal consumption on education, where t is time in months since January 1. Use direct substitution to express the percentage y as a function of time t (do not simplify the expression) and then use the chain rule to estimate how fast the percentage she spends on food is changing on November 1. Be sure to specify the units.

64. *Food Versus Recreation* The percentage y (of total personal consumption) an individual spends on food is approximately

$$y = 33x^{-0.63} \text{ percentage points} \quad (2.5 \le x \le 4.5)$$

where x is the percentage the individual spends on recreation.[29] A college student finds that he is spending

$$x = 3.5 + 0.1t$$

percent of his personal consumption on recreation, where t is time in months since January 1. Use direct substitution to express the percentage y as a function of time t (do not simplify the expression) and then use the chain rule to estimate how fast the percentage he spends on food is changing on November 1. Be sure to specify the units.

65. *Marginal Revenue* The weekly revenue from the sale of rubies at **Royal Ruby Retailers (RRR)** is increasing at a rate of \$40 per \$1 increase in price, and the price is decreasing at a rate of \$0.75 per additional ruby sold. What is the marginal revenue? (Be sure to state the units of measurement.) Interpret the result. HINT [See Example 5.]

66. *Marginal Revenue* The weekly revenue from the sale of emeralds at **Eduardo's Emerald Emporium (EEE)** is decreasing at a rate of €500 per €1 increase in price, and the price is decreasing at a rate of €0.45 per additional emerald sold. What is the marginal revenue? (Be sure to state the units of measurement.) Interpret the result. HINT [See Example 5.]

67. *Crime Statistics* The murder rate in large cities (over 1 million residents) can be related to that in smaller cities (500,000–1,000,000 residents) by the following linear model:[30]

$$y = 1.5x - 1.9 \quad (15 \le x \le 25)$$

where y is the murder rate (in murders per 100,000 residents each year) in large cities and x is the murder rate in smaller cities. During the period 1991–1998, the murder rate in small cities was decreasing at an average rate of 2 murders per 100,000 residents each year. Use the chain rule to estimate how fast the murder rate was changing in larger cities during that period. (Show how you used the chain rule in your answer.)

68. *Crime Statistics* Following is a quadratic model relating the murder rates described in the preceding exercise:

$$y = 0.1x^2 - 3x + 39 \quad (15 \le x \le 25)$$

In 1996, the murder rate in smaller cities was approximately 22 murders per 100,000 residents each year and was decreasing at a rate of approximately 2.5 murders per 100,000 residents each year. Use the chain rule to estimate how fast the murder rate was changing for large cities. (Show how you used the chain rule in your answer.)

69. *Existing Home Sales* The following graph shows the approximate value of home prices and existing home sales in 2006–2010 as a percentage change from 2003, together with quadratic approximations.[31]

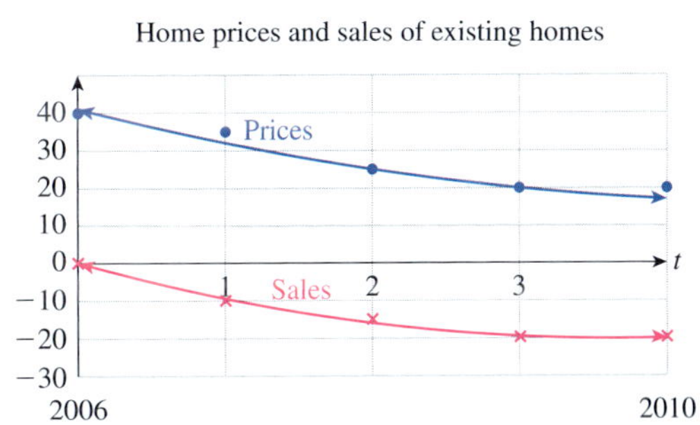

The quadratic approximations are given by

$$\text{Home Prices:} \quad P(t) = t^2 - 10t + 41 \quad (0 \le t \le 4)$$

$$\text{Existing Home Sales:} \quad S(t) = 1.5t^2 - 11t \quad (0 \le t \le 4)$$

where t is time in years since the start of 2006. Use the chain rule to estimate $\left.\frac{dS}{dP}\right|_{t=2}$. What does the answer tell you about home sales and prices? HINT [See Quick Example 2 on page 332.]

---

[28] Model based on historical and projected data from 1908–2010. Sources: Historical data, Bureau of Economic Analysis; projected data, Bureau of Labor Statistics/*New York Times*, December 1, 2003, p. C2.

[29] Ibid.

[30] The model is a linear regression model. Source for data: Federal Bureau of Investigation, Supplementary Homicide Reports/*New York Times*, May 29, 2000, p. A12.

[31] Sources: Standard & Poors/Bloomberg Financial Markets/*New York Times*, September 29, 2007, p. C3. Projection is the authors'.

70. ***Existing Home Sales*** The following graph shows the approximate value of home prices and existing home sales in 2004–2007 as a percentage change from 2003, together with quadratic approximations.[32]

Home prices and sales of existing homes

The quadratic approximations are given by

Home Prices: $P(t) = -6t^2 + 27t + 10 \quad (0 \le t \le 3)$

Existing Home Sales: $S(t) = -4t^2 + 4t + 11 \quad (0 \le t \le 3)$

where t is time in years since the start of 2004. Use the chain rule to estimate $\left.\frac{dS}{dP}\right|_{t=2}$. What does the answer tell you about home sales and prices? HINT [See Quick Example 2 on page 332.]

71. ▼ ***Pollution*** An offshore oil well is leaking oil and creating a circular oil slick. If the radius of the slick is growing at a rate of 2 miles/hour, find the rate at which the area is increasing when the radius is 3 miles. (The area of a disc of radius r is $A = \pi r^2$.) HINT [See Quick Example 2 on page 332.]

72. ▼ ***Mold*** A mold culture in a dorm refrigerator is circular and growing. The radius is growing at a rate of 0.3 cm/day. How fast is the area growing when the culture is 4 centimeters in radius? (The area of a disc of radius r is $A = \pi r^2$.) HINT [See Quick Example 2 on page 332.]

73. ▼ ***Budget Overruns*** The Pentagon is planning to build a new, spherical satellite. As is typical in these cases, the specifications keep changing, so that the size of the satellite keeps growing. In fact, the radius of the planned satellite is growing 0.5 feet per week. Its cost will be $1,000 per cubic foot. At the point when the plans call for a satellite 10 feet in radius, how fast is the cost growing? (The volume of a solid sphere of radius r is $V = \frac{4}{3}\pi r^3$.)

74. ▼ ***Soap Bubbles*** The soap bubble I am blowing has a radius that is growing at a rate of 4 cm/s. How fast is the surface area growing when the radius is 10 cm? (The surface area of a sphere of radius r is $S = 4\pi r^2$.)

75. T ▼ ***Revenue Growth*** The demand for the Cyberpunk II arcade video game is modeled by the logistic curve

$$q(t) = \frac{10{,}000}{1 + 0.5e^{-0.4t}}$$

where $q(t)$ is the total number of units sold t months after its introduction.

a. Use technology to estimate $q'(4)$.

b. Assume that the manufacturers of Cyberpunk II sell each unit for $800. What is the company's marginal revenue dR/dq?

c. Use the chain rule to estimate the rate at which revenue is growing 4 months after the introduction of the video game.

76. T ▼ ***Information Highway*** The amount of information transmitted each month in the early years of the Internet (1988 to 1994) can be modeled by the equation

$$q(t) = \frac{2e^{0.69t}}{3 + 1.5e^{-0.4t}} \qquad (0 \le t \le 6)$$

where q is the amount of information transmitted each month in billions of data packets and t is the number of years since the start of 1988.[33]

a. Use technology to estimate $q'(2)$.

b. Assume that it costs $5 to transmit a million packets of data. What is the marginal cost $C'(q)$?

c. How fast was the cost increasing at the start of 1990?

Money Stock *Exercises 77–80 are based on the following demand function for money (taken from a question on the GRE Economics Test):*

$$M_d = 2 \times y^{0.6} \times r^{-0.3} \times p$$

where

M_d = *demand for nominal money balances (money stock)*
y = *real income*
r = *an index of interest rates*
p = *an index of prices*

These exercises also use the idea of ***percentage rate of growth****:*

$$\text{Percentage Rate of Growth of } M = \frac{\text{Rate of Growth of } M}{M} = \frac{dM/dt}{M}$$

77. ◆ (from the GRE Economics Test) If the interest rate and price level are to remain constant while real income grows at 5 percent per year, the money stock must grow at what percent per year?

78. ◆ (from the GRE Economics Test) If real income and price level are to remain constant while the interest rate grows at 5 percent per year, the money stock must change by what percent per year?

79. ◆ (from the GRE Economics Test) If the interest rate is to remain constant while real income grows at 5 percent per year

[32] Sources: Standard & Poors /Bloomberg Financial Markets/*New York Times*, September 29, 2007, p. C3. Projection is the authors'.

[33] This is the authors' model, based on figures published in *New York Times*, Nov. 3, 1993.

and the price level rises at 5 percent per year, the money stock must grow at what percent per year?

80. ◆ (from the GRE Economics Test) If real income grows by 5 percent per year, the interest rate grows by 2 percent per year, and the price level drops by 3 percent per year, the money stock must change by what percent per year?

COMMUNICATION AND REASONING EXERCISES

81. Complete the following: The derivative of 1 over a glob is −1 over . . . .

82. Complete the following: The derivative of the square root of a glob is 1 over . . . .

83. Say why the following was marked wrong and give the correct answer.

$$\frac{d}{dx}[(3x^3 - x)^3] = 3(9x^2 - 1)^2 \quad \text{✗ WRONG!}$$

84. Say why the following was marked wrong and give the correct answer.

$$\frac{d}{dx}\left[\left(\frac{3x^2-1}{2x-2}\right)^3\right] = 3\left(\frac{3x^2-1}{2x-2}\right)^2\left(\frac{6x}{2}\right) \quad \text{✗ WRONG!}$$

85. Name two major errors in the following graded test question and give the correct answer.

$$\frac{d}{dx}\left[\left(\frac{3x^2-1}{2x-2}\right)^3\right] = 3\left(\frac{6x}{2}\right)^2 \quad \text{✗ WRONG! SEE ME!}$$

86. Name two major errors in the following graded test question and give the correct answer.

$$\frac{d}{dx}[(3x^3 - x)(2x+1)]^4 = 4[(9x^2 - 1)(2)]^3 \quad \text{✗ WRONG! SEE ME!}$$

87. ▼ Formulate a simple procedure for deciding whether to apply first the chain rule, the product rule, or the quotient rule when finding the derivative of a function.

88. ▼ Give an example of a function *f* with the property that calculating $f'(x)$ requires use of the following rules in the given order: (1) the chain rule, (2) the quotient rule, and (3) the chain rule.

89. ◆ Give an example of a function *f* with the property that calculating $f'(x)$ requires use of the chain rule five times in succession.

90. ◆ What can you say about composites of linear functions?

4.5 Derivatives of Logarithmic and Exponential Functions

At this point, we know how to take the derivative of any algebraic expression in x (involving powers, radicals, and so on). We now turn to the derivatives of logarithmic and exponential functions.

Derivative of the Natural Logarithm

$$\frac{d}{dx}[\ln x] = \frac{1}{x} \qquad \text{Recall that } \ln x = \log_e x$$

Quick Examples

1. $\frac{d}{dx}[3 \ln x] = 3 \cdot \frac{1}{x} = \frac{3}{x}$ — Derivative of a constant times a function

2. $\frac{d}{dx}[x \ln x] = 1 \cdot \ln x + x \cdot \frac{1}{x} = \ln x + 1.$ — Product rule, because $x \ln x$ is a product

The above simple formula works only for the natural logarithm (the logarithm with base e). For logarithms with bases other than e, we have the following:

Derivative of the Logarithm with Base *b*

$$\frac{d}{dx}[\log_b x] = \frac{1}{x \ln b}$$

Notice that, if $b = e$, we get the same formula as previously.

Quick Examples

1. $\frac{d}{dx}[\log_3 x] = \frac{1}{x \ln 3} \approx \frac{1}{1.0986x}$

2. $\frac{d}{dx}[\log_2(x^4)] = \frac{d}{dx}(4 \log_2 x)$ We used the logarithm identity $\log_b(x^r) = r \log_b x$.

$= 4 \cdot \frac{1}{x \ln 2} \approx \frac{4}{0.6931x}$

Derivation of the formulas $\frac{d}{dx}[\ln x] = \frac{1}{x}$ and $\frac{d}{dx}[\log_b x] = \frac{1}{x \ln b}$

To compute $\frac{d}{dx}[\ln x]$, we need to use the definition of the derivative. We also use properties of the logarithm to help evaluate the limit.

$$\frac{d}{dx}[\ln x] = \lim_{h\to 0} \frac{\ln(x+h) - \ln x}{h}$$ Definition of the derivative

$$= \lim_{h\to 0} \frac{1}{h}[\ln(x+h) - \ln x]$$ Algebra

$$= \lim_{h\to 0} \frac{1}{h} \ln\left(\frac{x+h}{x}\right)$$ Properties of the logarithm

$$= \lim_{h\to 0} \frac{1}{h} \ln\left(1 + \frac{h}{x}\right)$$ Algebra

$$= \lim_{h\to 0} \ln\left(1 + \frac{h}{x}\right)^{1/h}$$ Properties of the logarithm

which we rewrite as

$$\lim_{h\to 0} \ln\left[\left(1 + \frac{1}{(x/h)}\right)^{x/h}\right]^{1/x}.$$

As $h \to 0^+$, the quantity x/h is getting large and positive, and so the quantity in brackets is approaching e (see the definition of e in Section 2.2), which leaves us with

$$\ln[e]^{1/x} = \frac{1}{x}\ln e = \frac{1}{x}$$

which is the derivative we are after.* What about the limit as $h \to 0^-$? We will glide over that case and leave it for the interested reader to pursue.†

The rule for the derivative of $\log_b x$ follows from the fact that $\log_b x = \ln x/\ln b$.

If we were to take the derivative of the natural logarithm of a *quantity* (a function of x), rather than just x, we would need to use the chain rule:

*** NOTE** We actually used the fact that the logarithm function is continuous when we took the limit.

† NOTE Here is an outline of the argument for negative h. Because x must be positive for $\ln x$ to be defined, we find that $x/h \to -\infty$ as $h \to 0^-$, and so we must consider the quantity $(1 + 1/m)^m$ for large *negative* m. It turns out the limit is still e (check it numerically!) and so the computation above still works.

Derivatives of Logarithms of Functions

Original Rule	*Generalized Rule*	*In Words*
$\frac{d}{dx}[\ln x] = \frac{1}{x}$	$\frac{d}{dx}[\ln u] = \frac{1}{u}\frac{du}{dx}$	*The derivative of the natural logarithm of a quantity is 1 over that quantity, times the derivative of that quantity.*
$\frac{d}{dx}[\log_b x] = \frac{1}{x \ln b}$	$\frac{d}{dx}[\log_b u] = \frac{1}{u \ln b}\frac{du}{dx}$	*The derivative of the log to base b of a quantity is 1 over the product of* ln *b and that quantity, times the derivative of that quantity.*

Quick Examples

1. $$\frac{d}{dx}\ln[x^2+1] = \frac{1}{x^2+1}\frac{d}{dx}(x^2+1) \qquad u = x^2+1 \text{ (See the margin note.§)}$$
$$= \frac{1}{x^2+1}(2x) = \frac{2x}{x^2+1}$$

2. $$\frac{d}{dx}\log_2[x^3+x] = \frac{1}{(x^3+x)\ln 2}\frac{d}{dx}(x^3+x) \qquad u = x^3+x$$
$$= \frac{1}{(x^3+x)\ln 2}(3x^2+1) = \frac{3x^2+1}{(x^3+x)\ln 2}$$

§ NOTE If we were to evaluate $\ln(x^2+1)$, the last operation we would perform would be to take the natural logarithm of a quantity. Thus, the calculation thought experiment tells us that we are dealing with ln *of a quantity*, and so we need the generalized logarithm rule as stated above.

EXAMPLE 1 Derivative of Logarithmic Function

Find $\frac{d}{dx}[\ln\sqrt{x+1}]$.

Solution The calculation thought experiment tells us that we have the natural logarithm of a quantity, so

$$\frac{d}{dx}[\ln\sqrt{x+1}] = \frac{1}{\sqrt{x+1}}\frac{d}{dx}\sqrt{x+1} \qquad \frac{d}{dx}\ln u = \frac{1}{u}\frac{du}{dx}$$

$$= \frac{1}{\sqrt{x+1}} \cdot \frac{1}{2\sqrt{x+1}} \qquad \frac{d}{dx}\sqrt{u} = \frac{1}{2\sqrt{u}}\frac{du}{dx}$$

$$= \frac{1}{2(x+1)}.$$

Before we go on... What happened to the square root in Example 1? As with many problems involving logarithms, we could have done this one differently and with less bother if we had simplified the expression $\ln\sqrt{x+1}$ using the properties of logarithms *before* differentiating. Doing this, we get

$$\ln\sqrt{x+1} = \ln(x+1)^{1/2} = \frac{1}{2}\ln(x+1). \qquad \text{Simplify the logarithm first.}$$

Thus,

$$\frac{d}{dx}[\ln\sqrt{x+1}] = \frac{d}{dx}\left[\frac{1}{2}\ln(x+1)\right]$$

$$= \frac{1}{2}\left(\frac{1}{x+1}\right) \cdot 1 = \frac{1}{2(x+1)},$$

the same answer as above. ■

EXAMPLE 2 Derivative of a Logarithmic Function

Find $\frac{d}{dx}[\ln[(1+x)(2-x)]]$.

Solution This time, we simplify the expression $\ln[(1+x)(2-x)]$ before taking the derivative.

$$\ln[(1+x)(2-x)] = \ln(1+x) + \ln(2-x) \qquad \text{Simplify the logarithm first.}$$

Thus,

$$\frac{d}{dx}[\ln[(1+x)(2-x)]] = \frac{d}{dx}[\ln(1+x)] + \frac{d}{dx}\ln(2-x)]$$

$$= \frac{1}{1+x} - \frac{1}{2-x} \qquad \text{Because } \frac{d}{dx}\ln(2-x) = -\frac{1}{2-x}$$

$$= \frac{1-2x}{(1+x)(2-x)}.$$

➡ **Before we go on...** For practice, try doing Example 2 without simplifying first. What other differentiation rule do you need to use? ■

EXAMPLE 3 Logarithm of an Absolute Value

Find $\frac{d}{dx}[\ln|x|]$.

Solution Before we start, we note that ln x is defined only for positive values of x, so its domain is the set of positive real numbers. The domain of $\ln|x|$, on the other hand, is the set of *all* nonzero real numbers. For example, $\ln|-2| = \ln 2 \approx 0.6931$. For this reason, $\ln|x|$ often turns out to be more useful than the ordinary logarithm function.

Now we'll get to work. The calculation thought experiment tells us that $\ln|x|$ is the natural logarithm of a quantity, so we use the chain rule:

$$\frac{d}{dx}[\ln|x|] = \frac{1}{|x|}\frac{d}{dx}|x| \qquad u = |x|$$

$$= \frac{1}{|x|}\frac{|x|}{x} \qquad \text{Recall that } \frac{d}{dx}|x| = \frac{|x|}{x}.$$

$$= \frac{1}{x}.$$

➡ **Before we go on...** Figure 7(a) shows the graphs of $y = \ln|x|$ and $y = 1/x$. Figure 7(b) shows the graphs of $y = \ln|x|$ and $y = 1/|x|$. You should be able to see from these graphs why the derivative of $\ln|x|$ is $1/x$ and not $1/|x|$.

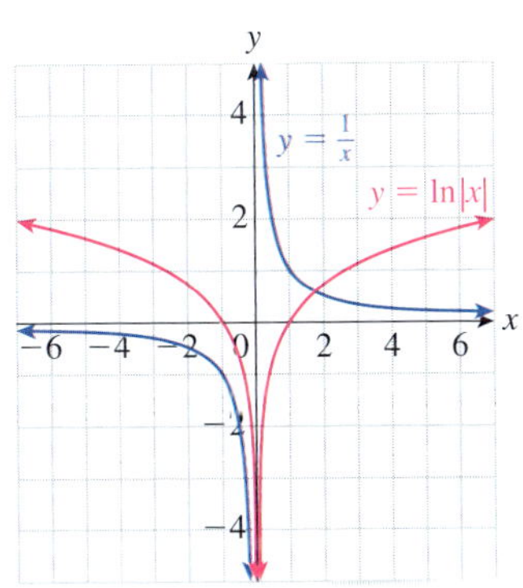

Figure 7(a)

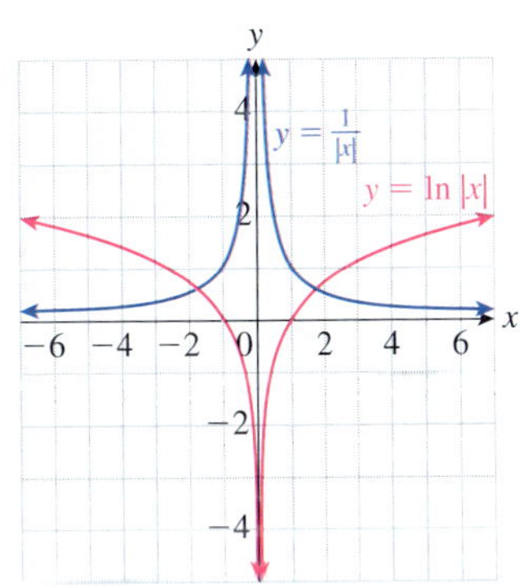

Figure 7(b) ■

This last example, in conjunction with the chain rule, gives us the following formulas.

Derivative of Logarithms of Absolute Values

Original Rule	*Generalized Rule*	*In Words*
$\frac{d}{dx}[\ln\|x\|] = \frac{1}{x}$	$\frac{d}{dx}[\ln\|u\|] = \frac{1}{u}\frac{du}{dx}$	*The derivative of the natural logarithm of the absolute value of a quantity is 1 over that quantity, times the derivative of that quantity.*
$\frac{d}{dx}[\log_b\|x\|] = \frac{1}{x\ln b}$	$\frac{d}{dx}[\log_b\|u\|] = \frac{1}{u\ln b}\frac{du}{dx}$	*The derivative of the log to base b of the absolute value of a quantity is 1 over the product of* ln *b and that quantity, times the derivative of that quantity.*

Note: Compare the above formulas with those on page 339. They tell us that we can simply ignore the absolute values in $\ln|u|$ or $\log_b|u|$ when taking the derivative.

Quick Examples

1. $\frac{d}{dx}[\ln|x^2+1|] = \frac{1}{x^2+1}\frac{d}{dx}(x^2+1)$ $\quad u = x^2+1$

$= \frac{1}{x^2+1}(2x) = \frac{2x}{x^2+1}$

2. $\frac{d}{dx}[\log_2|x^3+x|] = \frac{1}{(x^3+x)\ln 2}\frac{d}{dx}(x^3+x)$ $\quad u = x^3+x$

$= \frac{1}{(x^3+x)\ln 2}(3x^2+1) = \frac{3x^2+1}{(x^3+x)\ln 2}$

We now turn to the derivatives of *exponential* functions—that is, functions of the form $f(x) = b^x$. We begin by showing how *not* to differentiate them.

Caution The derivative of b^x is *not* xb^{x-1}. The power rule applies only to *constant* exponents. In this case the exponent is decidedly *not* constant, and so the power rule does not apply.

The following shows the correct way of differentiating b^x, beginning with a special case.

Derivative of e^x

$$\frac{d}{dx}[e^x] = e^x$$

Quick Examples

1. $\frac{d}{dx}[3e^x] = 3\frac{d}{dx}[e^x] = 3e^x$

2. $\frac{d}{dx}\left[\frac{e^x}{x}\right] = \frac{e^x x - e^x(1)}{x^2}$ Quotient rule

$= \frac{e^x(x-1)}{x^2}$

Thus, e^x has the amazing property that its derivative is itself!* For bases other than e, we have the following generalization:

* **NOTE** There is another—very simple—function that is its own derivative. What is it?

Derivative of b^x

If b is any positive number, then

$$\frac{d}{dx}[b^x] = b^x \ln b.$$

Note that if $b = e$, we obtain the previous formula.

Quick Example

$$\frac{d}{dx}[3^x] = 3^x \ln 3$$

Derivation of the formula $\frac{d}{dx}[e^x] = e^x$

To find the derivative of e^x we use a shortcut.* Write $g(x) = e^x$. Then

$$\ln g(x) = x.$$

Take the derivative of both sides of this equation to get

$$\frac{g'(x)}{g(x)} = 1$$

or

$$g'(x) = g(x) = e^x.$$

* **NOTE** This shortcut is an example of a technique called *logarithmic differentiation*, which is occasionally useful. We will see it again in the next section.

In other words, the exponential function with base e is its own derivative. The rule for exponential functions with other bases follows from the equality $b^x = e^{x \ln b}$ (why?) and the chain rule. (Try it.)

If we were to take the derivative of e raised to a *quantity*, not just x, we would need to use the chain rule, as follows.

Derivatives of Exponentials of Functions

Original Rule	*Generalized Rule*	*In Words*
$\frac{d}{d^x}[e^x] = e^x$	$\frac{d}{dx}[e^u] = e^u \frac{du}{dx}$	*The derivative of e raised to a quantity is e raised to that quantity, times the derivative of that quantity.*
$\frac{d}{dx}[b^x] = b^x \ln b$	$\frac{d}{dx}[b^u] = b^u \ln b \frac{du}{dx}$	*The derivative of b raised to a quantity is b raised to that quantity, times* ln *b, times the derivative of that quantity.*

Quick Examples

1. $\frac{d}{dx}\left[e^{x^2+1}\right] = e^{x^2+1}\frac{d}{dx}[x^2+1]$ $\qquad u = x^2 + 1$ (See margin note.*)

$$= e^{x^2+1}(2x) = 2x\, e^{x^2+1}$$

2. $\frac{d}{dx}[2^{3x}] = 2^{3x} \ln 2 \frac{d}{dx}[3x]$ $\qquad u = 3x$

$$= 2^{3x}(\ln 2)(3) = (3 \ln 2)2^{3x}$$

3. $\frac{d}{dt}[30e^{1.02t}] = 30e^{1.02t}(1.02) = 30.6e^{1.02t}$ $\qquad u = 1.02t$

4. If \$1,000 is invested in an account earning 5% per year compounded continuously, then the rate of change of the account balance after t years is

$$\frac{d}{dt}[1{,}000e^{0.05t}] = 1{,}000(0.05)e^{0.05t} = 50e^{0.05t} \text{ dollars/year.}$$

* **NOTE** The calculation thought experiment tells us that we have e raised to a quantity.

APPLICATIONS

EXAMPLE 4 Epidemics

In the early stages of the AIDS epidemic during the 1980s, the number of cases in the United States was increasing by about 50% every 6 months. By the start of 1983, there were approximately 1,600 AIDS cases in the United States.* Had this trend continued, how many new cases per year would have been occurring by the start of 1993?

*Data based on regression of 1982–1986 figures. Source for data: Centers for Disease Control and Prevention. HIV/AIDS Surveillance Report, 2000;12 (No. 2).

Solution To find the answer, we must first model this exponential growth using the methods of Chapter 2. Referring to Example 4 in Section 2.2, we find that t years after the start of 1983 the number of cases is

$$A = 1{,}600(2.25^t).$$

We are asking for the number of new cases each year. In other words, we want the rate of change, dA/dt:

$$\frac{dA}{dt} = 1{,}600(2.25)^t \ln 2.25 \text{ cases per year.}$$

At the start of 1993, $t = 10$, so the number of new cases per year is

$$\left.\frac{dA}{dt}\right|_{t=10} = 1{,}600(2.25)^{10} \ln 2.25 \approx 4{,}300{,}000 \text{ cases per year.}$$

➡ **Before we go on...** In Example 4, the figure for the number of new cases per year is so large because we assumed that exponential growth—the 50% increase every six months—would continue. A more realistic model for the spread of a disease is the logistic model. (See Section 2.4, as well as the next example.) ■

Photononstop/Superstock

EXAMPLE 5 Sales Growth

The sales of the Cyberpunk II video game can be modeled by the logistic curve

$$q(t) = \frac{10{,}000}{1 + 0.5e^{-0.4t}}$$

where $q(t)$ is the total number of units sold t months after its introduction. How fast is the game selling 2 years after its introduction?

Solution We are asked for $q'(24)$. We can find the derivative of $q(t)$ using the quotient rule, or we can first write

$$q(t) = 10{,}000(1 + 0.5e^{-0.4t})^{-1}$$

and then use the generalized power rule:

$$q'(t) = -10{,}000(1 + 0.5e^{-0.4t})^{-2}(0.5e^{-0.4t})(-0.4)$$
$$= \frac{2{,}000e^{-0.4t}}{(1 + 0.5e^{-0.4t})^2}.$$

Thus,

$$q'(24) = \frac{2{,}000e^{-0.4(24)}}{(1 + 0.5e^{-0.4(24)})^2} \approx 0.135 \text{ units per month.}$$

So, after 2 years, sales are quite slow.

➡ **Before we go on...** We can check the answer in Example 5 graphically. If we plot the total sales curve for $0 \le t \le 30$ and $6{,}000 \le q \le 10{,}000$, on a TI-83/84 Plus, for example, we get the graph shown in Figure 8. Notice that total sales level off at about

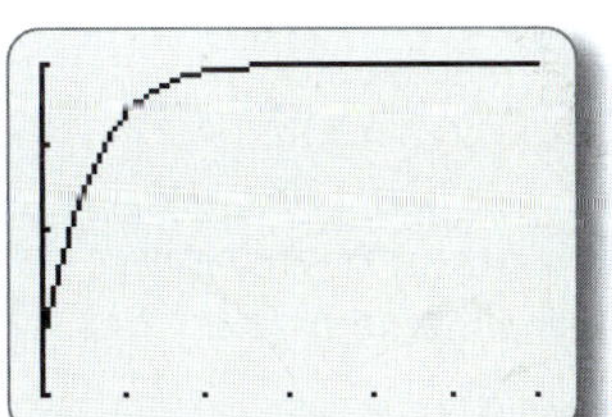

Figure 8

* **NOTE** We can also say this using limits: $\lim_{t \to +\infty} q(t) = 10{,}000$.

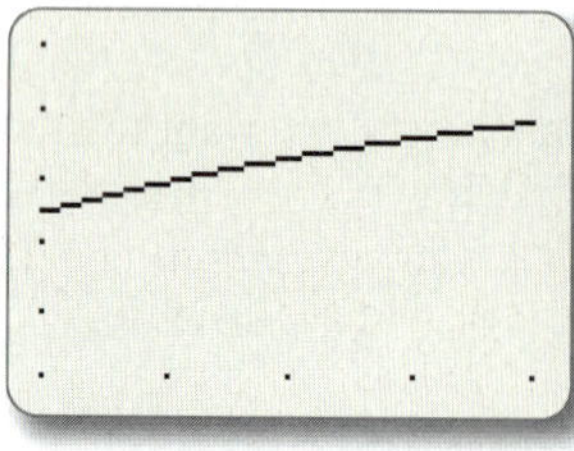

Figure 9

10,000 units.* We computed $q'(24)$, which is the slope of the curve at the point with t-coordinate 24. If we zoom in to the portion of the curve near $t = 24$, we obtain the graph shown in Figure 9, with $23 \le t \le 25$ and $9{,}999 \le q \le 10{,}000$. The curve is almost linear in this range. If we use the two endpoints of this segment of the curve, (23, 9,999.4948) and (25, 9,999.7730), we can approximate the derivative as

$$\frac{9{,}999.7730 - 9{,}999.4948}{25 - 23} = 0.1391$$

which is accurate to two decimal places. ■

4.5 EXERCISES

▼ more advanced ◆ challenging

T indicates exercises that should be solved using technology

Find the derivatives of the functions in Exercises 1–76. **HINT** [See Quick Examples on page 339.]

1. $f(x) = \ln(x - 1)$
2. $f(x) = \ln(x + 3)$
3. $f(x) = \log_2 x$
4. $f(x) = \log_3 x$
5. $g(x) = \ln|x^2 + 3|$
6. $g(x) = \ln|2x - 4|$
7. $h(x) = e^{x+3}$ **HINT** [See Quick Examples on page 344.]
8. $h(x) = e^{x^2}$ **HINT** [See Quick Examples on page 344.]
9. $f(x) = e^{-x}$
10. $f(x) = e^{1-x}$
11. $g(x) = 4^x$
12. $g(x) = 5^x$
13. $h(x) = 2^{x^2-1}$
14. $h(x) = 3^{x^2-x}$
15. $f(x) = x \ln x$
16. $f(x) = 3 \ln x$
17. $f(x) = (x^2 + 1) \ln x$
18. $f(x) = (4x^2 - x) \ln x$
19. $f(x) = (x^2 + 1)^5 \ln x$
20. $f(x) = (x + 1)^{0.5} \ln x$
21. $g(x) = \ln|3x - 1|$
22. $g(x) = \ln|5 - 9x|$
23. $g(x) = \ln|2x^2 + 1|$
24. $g(x) = \ln|x^2 - x|$
25. $g(x) = \ln(x^2 - 2.1x^{0.3})$
26. $g(x) = \ln(x - 3.1x^{-1})$
27. $h(x) = \ln[(-2x + 1)(x + 1)]$
28. $h(x) = \ln[(3x + 1)(-x + 1)]$
29. $h(x) = \ln\left(\frac{3x + 1}{4x - 2}\right)$
30. $h(x) = \ln\left(\frac{9x}{4x - 2}\right)$
31. $r(x) = \ln\left|\frac{(x + 1)(x - 3)}{-2x - 9}\right|$
32. $r(x) = \ln\left|\frac{-x + 1}{(3x - 4)(x - 9)}\right|$
33. $s(x) = \ln(4x - 2)^{1.3}$
34. $s(x) = \ln(x - 8)^{-2}$
35. $s(x) = \ln\left|\frac{(x + 1)^2}{(3x - 4)^3(x - 9)}\right|$
36. $s(x) = \ln\left|\frac{(x + 1)^2(x - 3)^4}{2x + 9}\right|$
37. $h(x) = \log_2(x + 1)$
38. $h(x) = \log_3(x^2 + x)$
39. $r(t) = \log_3(t + 1/t)$
40. $r(t) = \log_3\left(t + \sqrt{t}\right)$
41. $f(x) = (\ln|x|)^2$
42. $f(x) = \frac{1}{\ln|x|}$
43. $r(x) = \ln(x^2) - [\ln(x - 1)]^2$
44. $r(x) = (\ln(x^2))^2$
45. $f(x) = xe^x$
46. $f(x) = 2e^x - x^2e^x$
47. $r(x) = \ln(x + 1) + 3x^3e^x$
48. $r(x) = \ln|x + e^x|$
49. $f(x) = e^x \ln|x|$
50. $f(x) = e^x \log_2|x|$
51. $f(x) = e^{2x+1}$
52. $f(x) = e^{4x-5}$
53. $h(x) = e^{x^2-x+1}$
54. $h(x) = e^{2x^2-x+1/x}$
55. $s(x) = x^2e^{2x-1}$
56. $s(x) = \frac{e^{4x-1}}{x^3 - 1}$
57. $r(x) = (e^{2x-1})^2$
58. $r(x) = (e^{2x^2})^3$
59. $t(x) = 3^{2x-4}$
60. $t(x) = 4^{-x+5}$
61. $v(x) = 3^{2x+1} + e^{3x+1}$
62. $v(x) = e^{2x}4^{2x}$
63. $u(x) = \frac{3^{x^2}}{x^2 + 1}$
64. $u(x) = (x^2 + 1)4^{x^2-1}$
65. $g(x) = \frac{e^x + e^{-x}}{e^x - e^{-x}}$
66. $g(x) = \frac{1}{e^x + e^{-x}}$
67. ▼ $g(x) = e^{3x-1}e^{x-2}e^x$
68. ▼ $g(x) = e^{-x+3}e^{2x-1}e^{-x+11}$
69. ▼ $f(x) = \frac{1}{x \ln x}$
70. ▼ $f(x) = \frac{e^{-x}}{xe^x}$
71. ▼ $f(x) = [\ln(e^x)]^2 - \ln[(e^x)^2]$
72. ▼ $f(x) = e^{\ln x} - e^{2\ln(x^2)}$
73. ▼ $f(x) = \ln|\ln x|$
74. ▼ $f(x) = \ln|\ln|\ln x||$
75. ▼ $s(x) = \ln\sqrt{\ln x}$
76. ▼ $s(x) = \sqrt{\ln(\ln x)}$

Find the equations of the straight lines described in Exercises 77–82. Use graphing technology to check your answers by plotting the given curve together with the tangent line.

77. Tangent to $y = e^x \log_2 x$ at the point (1, 0)
78. Tangent to $y = e^x + e^{-x}$ at the point (0, 2)
79. Tangent to $y = \ln\sqrt{2x + 1}$ at the point where $x = 0$
80. Tangent to $y = \ln\sqrt{2x^2 + 1}$ at the point where $x = 1$

81. At right angles to $y = e^{x^2}$ at the point where $x = 1$

82. At right angles to $y = \log_2(3x + 1)$ at the point where $x = 1$

APPLICATIONS

83. ***Research and Development: Industry*** The total spent on research and development by industry in the United States during 1995–2007 can be approximated by

$$S(t) = 57.5 \ln t + 31 \text{ billion dollars} \quad (5 \le t \le 19)$$

where t is the year since 1990.[34] What was the total spent in 2000 ($t = 10$) and how fast was it increasing? HINT [See Quick Examples on page 337.]

84. ***Research and Development: Federal*** The total spent on research and development by the federal government in the United States during 1995–2007 can be approximated by

$$S(t) = 7.4 \ln t + 3 \text{ billion dollars} \quad (5 \le t \le 19)$$

where t is the year since 1990.[35] What was the total spent in 2005 ($t = 15$) and how fast was it increasing? HINT [See Quick Examples on page 337.]

85. ***Research and Development: Industry*** The function $S(t)$ in Exercise 83 can also be written (approximately) as

$$S(t) = 57.5 \ln(1.71t + 17.1) \text{ billion dollars} \quad (-5 \le t \le 9)$$

where this time t is the year since 2000. Use this alternative formula to estimate the amount spent in 2000 and its rate of change, and check your answers by comparing it with those in Exercise 83.

86. ***Research and Development: Federal*** The function $S(t)$ in Exercise 84 can also be written (approximately) as

$$S(t) = 7.4 \ln(1.5t + 15) \text{ billion dollars} \quad (-5 \le t \le 9)$$

where this time t is the year since 2000. Use this alternative formula to estimate the amount spent in 2005 and its rate of change, and check your answers by comparing it with those in Exercise 84.

87. ▼ ***Carbon Dating*** The age in years of a specimen that originally contained 10g of carbon 14 is given by

$$y = \log_{0.999879}(0.1x)$$

where x is the amount of carbon 14 it currently contains. Compute $\left.\frac{dy}{dx}\right|_{x=5}$ and interpret your answer. HINT [For the calculation, see Quick Examples on page 339.]

88. ▼ ***Iodine Dating*** The age in years of a specimen that originally contained 10g of iodine 131 is given by

$$y = \log_{0.999567}(0.1x)$$

where x is the amount of iodine 131 it currently contains. Compute $\left.\frac{dy}{dx}\right|_{x=8}$ and interpret your answer. HINT [For the calculation, see Quick Examples on page 339.]

89. ***New York City Housing Costs: Downtown*** The average price of a two-bedroom apartment in downtown New York City during the real estate boom from 1994 to 2004 can be approximated by

$$p(t) = 0.33e^{0.16t} \text{ million dollars} \quad (0 \le t \le 10)$$

where t is time in years ($t = 0$ represents 1994).[36] What was the average price of a two-bedroom apartment in downtown New York City in 2003, and how fast was it increasing? (Round your answers to two significant digits.) HINT [See Quick Example 3 on page 344.]

90. ***New York City Housing Costs: Uptown*** The average price of a two-bedroom apartment in uptown New York City during the real estate boom from 1994 to 2004 can be approximated by

$$p(t) = 0.14e^{0.10t} \text{ million dollars} \quad (0 \le t \le 10)$$

where t is time in years ($t = 0$ represents 1994).[37] What was the average price of a two-bedroom apartment in uptown New York City in 2002, and how fast was it increasing? (Round your answers to two significant digits.) HINT [See Quick Example 3 on page 344.]

91. ***Big Brother*** The following chart shows the total number of wiretaps authorized each year by U.S. state and federal courts from 1990 to 2007 ($t = 0$ represents 1990):[38]

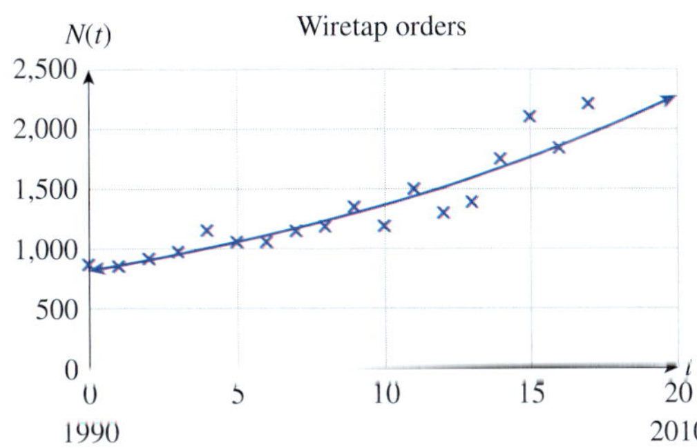

These data can be approximated with the model (shown on the graph)

$$N(t) = 820e^{0.051t}. \quad (0 \le t \le 17)$$

[34] Spending is in constant 2000 dollars. Source for data through 2006: National Science Foundation, Division of Science Resources Statistics, National Patterns of R&D Resources (www.nsf.gov/statistics) August 2008.

[35] Federal funding excluding grants to industry and nonprofit organizations. Spending is in constant 2000 dollars. Source for data through 2006: National Science Foundation, Division of Science Resources Statistics, National Patterns of R&D Resources (www.nsf.gov/statistics) August 2008.

[36] Model is based on a exponential regression. Source for data: Miller Samuel/*New York Times*, March 28, 2004, p. RE 11.

[37] Ibid.

[38] Source for data: 2007 Wiretap Report, Administrative Office of the United States Courts (www.uscourts.gov/wiretap07/2007WTText.pdf).

a. Find $N(15)$ and $N'(15)$. Be sure to state the units of measurement. To how many significant digits should we round the answers? Why?

b. The number of people whose communications are intercepted averages around 100 per wiretap order.[39] What does the answer to part (a) tell you about the number of people whose communications were intercepted?[40]

c. According to the model, the number of wiretaps orders each year (choose one):

(A) increased at a linear rate
(B) decreased at a quadratic rate
(C) increased at an exponential rate
(D) increased at a logarithmic rate

over the period shown.

92. ***Big Brother*** The following chart shows the number of wiretaps authorized each year by U.S. state courts from 1990 to 2007 ($t = 0$ represents 1990):[41]

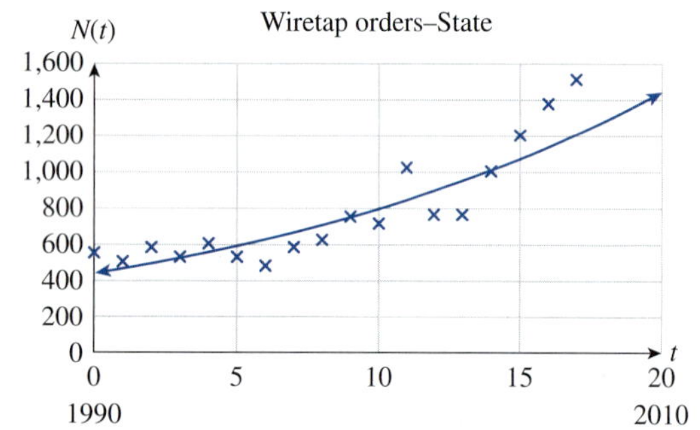

These data can be approximated with the model (shown on the graph)

$$N(t) = 440e^{0.06t}. \quad (0 \le t \le 17)$$

a. Find $N(10)$ and $N'(10)$. Be sure to state the units of measurement. To how many significant digits should we round the answers? Why?

b. The number of people whose communications are intercepted averages around 100 per wiretap order.[42] What does the answer to part (a) tell you about the number of people whose communications were intercepted?[43]

c. According to the model, the number of wiretaps orders each year (choose one)

(A) increased at a linear rate
(B) decreased at a quadratic rate
(C) increased at an exponential rate
(D) increased at a logarithmic rate

over the period shown.

93. ***Investments*** If \$10,000 is invested in a savings account offering 4% per year, compounded continuously, how fast is the balance growing after 3 years?

94. ***Investments*** If \$20,000 is invested in a savings account offering 3.5% per year, compounded continuously, how fast is the balance growing after 3 years?

95. ***Investments*** If \$10,000 is invested in a savings account offering 4% per year, compounded semiannually, how fast is the balance growing after 3 years?

96. ***Investments*** If \$20,000 is invested in a savings account offering 3.5% per year, compounded semiannually, how fast is the balance growing after 3 years?

97. ***SARS*** In the early stages of the deadly SARS (Severe Acute Respiratory Syndrome) epidemic in 2003, the number of cases was increasing by about 18% each day.[44] On March 17, 2003 (the first day for which statistics were reported by the World Health Organization) there were 167 cases. Find an exponential model that predicts the number of people infected t days after March 17, 2003, and use it to estimate how fast the epidemic was spreading on March 31, 2003. (Round your answer to the nearest whole number of new cases per day.) HINT [See Example 4.]

98. ***SARS*** A few weeks into the deadly SARS (Severe Acute Respiratory Syndrome) epidemic in 2003, the number of cases was increasing by about 4% each day.[45] On April 1, 2003 there were 1,804 cases. Find an exponential model that predicts the number $A(t)$ of people infected t days after April 1, 2003, and use it to estimate how fast the epidemic was spreading on April 30, 2003. (Round your answer to the nearest whole number of new cases per day.) HINT [See Example 4.]

99. ▼ ***SAT Scores by Income*** The following chart shows United State verbal SAT scores as a function of parents' income level:[46]

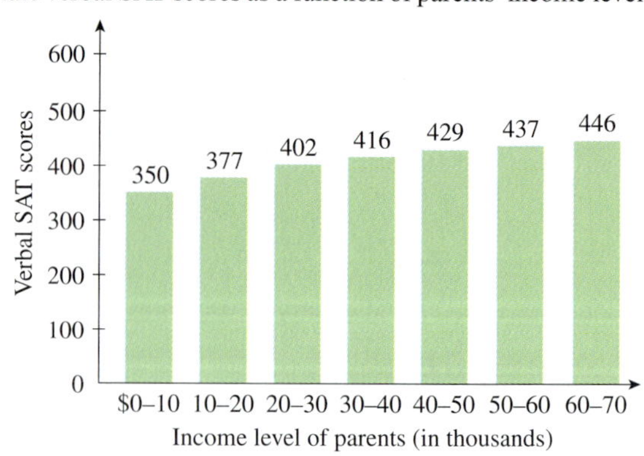

a. The data can best be modeled by which of the following?

(A) $S(x) = 470 - 136e^{-0.0000264x}$
(B) $S(x) = 136e^{-0.0000264x}$

[39] Source for data: 2007 Wiretap Report, Administrative Office of the United States Courts (www.uscourts.gov/wiretap07/2007WTText.pdf).

[40] Assume there is no significant overlap between the people whose communications are intercepted in different wiretap orders.

[41] Source for data: 2007 Wiretap Report, Administrative Office of the United States Courts (www.uscourts.gov/wiretap07/2007WTText.pdf).

[42] Ibid.

[43] Assume there is no significant overlap between the people whose communications are intercepted in different wiretap orders.

[44] World Health Organization (www.who.int).

[45] Ibid.

[46] Source: The College Board/*New York Times*, March 5, 1995, p. E16.

(C) $S(x) = 355(1.000004^x)$
(D) $S(x) = 470 - 355(1.000004^x)$

($S(x)$ is the average verbal SAT score of students whose parents earn \$$x$ per year.)

b. Use $S'(x)$ to predict how a student's verbal SAT score is affected by a \$1,000 increase in parents' income for a student whose parents earn \$45,000.

c. Does $S'(x)$ increase or decrease as x increases? Interpret your answer.

100. ▼ ***SAT Scores by Income*** The following chart shows U.S. average math SAT scores as a function of parents' income level:[47]

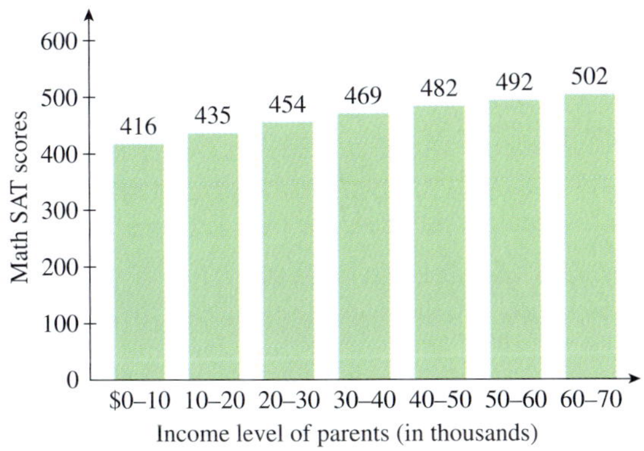

a. The data can best be modeled by which of the following?

(A) $S(x) = 535 - 415(1.000003^x)$
(B) $S(x) = 535 - 136e^{0.0000213x}$
(C) $S(x) = 535 - 136e^{-0.0000213x}$
(D) $S(x) = 415(1.000003^x)$

($S(x)$ is the average math SAT score of students whose parents earn \$$x$ per year.)

b. Use $S'(x)$ to predict how a student's math SAT score is affected by a \$1,000 increase in parents' income for a student whose parents earn \$45,000.

c. Does $S'(x)$ increase or decrease as x increases? Interpret your answer.

101. ▼ ***Demographics: Average Age and Fertility*** The following graph shows a plot of average age of a population versus fertility rate (the average number of children each woman has in her lifetime) in the United States and Europe over the period 1950–2005.[48]

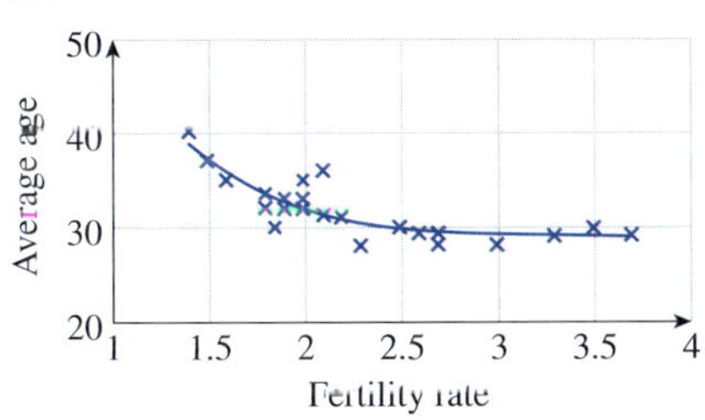

The equation of the accompanying curve is

$$a = 28.5 + 120(0.172)^x \quad (1.4 \le x \le 3.7)$$

where a is the average age (in years) of the population and x is the fertility rate.

a. Compute $a'(2)$. What does the answer tell you about average age and fertility rates?

b. Use the answer to part (a) to estimate how much the fertility rate would need to increase from a level of 2 children per woman to lower the average age of a population by about 1 year.

102. ▼ ***Demographics: Average Age and Fertility*** The following graph shows a plot of average age of a population versus fertility rate (the average number of children each woman has in her lifetime) in Europe over the period 1950–2005.[49]

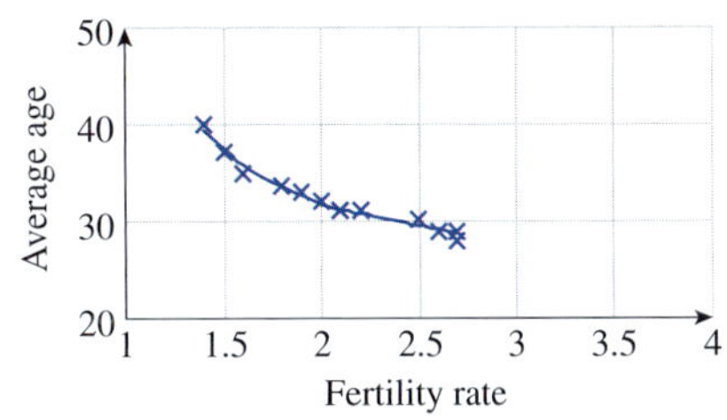

The equation of the accompanying curve is

$$g = 27.6 + 128(0.181)^x \quad (1.4 \le x \le 3.7)$$

where g is the average age (in years) of the population and x is the fertility rate.

a. Compute $g'(2.5)$. What does the answer tell you about average age and fertility rates?

b. Referring to the model that combines the data for Europe and the United States in Exercise 101, which population's average age is affected more by a changing fertility rate at the level of 2.5 children per woman?

103. ***Epidemics*** A flu epidemic described in Example 1 in Section 2.4 approximately followed the curve

$$P = \frac{150}{1 + 15{,}000e^{-0.35t}} \text{ million people}$$

where P is the number of people infected and t is the number of weeks after the start of the epidemic. How fast is the epidemic growing (that is, how many new cases are there each week) after 20 weeks? After 30 weeks? After 40 weeks? (Round your answers to two significant digits.) HINT [See Example 5.]

104. ***Epidemics*** Another epidemic follows the curve

$$P = \frac{200}{1 + 20{,}000e^{-0.549t}} \text{ million people}$$

[47] Source: The College Board/*New York Times*, March 5, 1995, p. E16.

[48] The separate data for Europe and the United States are collected in the same graph. 2005 figures are estimates. Source: United Nations World Population Division/*New York Times*, June 29, 2003, p. 3.

[49] All European countries including the Russian Federation. 2005 figures are estimates. Source: United Nations World Population Division/*New York Times*, June 29, 2003, p. 3.

where t is in years. How fast is the epidemic growing after 10 years? After 20 years? After 30 years? (Round your answers to two significant digits.) HINT [See Example 5.]

105. ***Subprime Mortgages*** The percentage of mortgages issued in the United States that are subprime (normally classified as risky) can be approximated by

$$A(t) = \frac{15.0}{1 + 8.6e^{-0.59t}} \text{ percent} \quad (0 \le t \le 8)$$

t years after the start of 2000.[50]

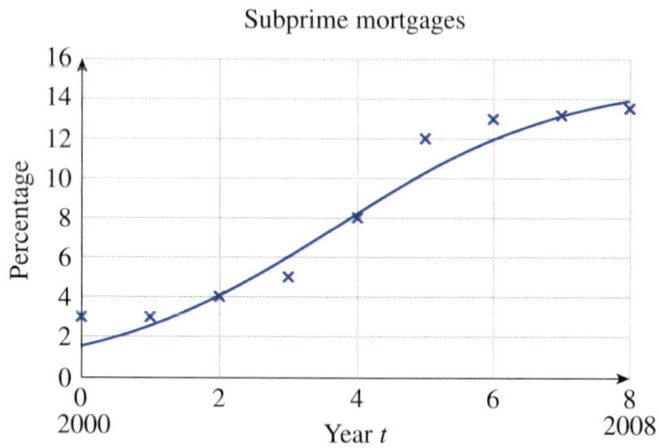

How fast, to the nearest 0.1 percent, was the percentage increasing at the start of 2003? How would you check that the answer is approximately correct by looking at the graph?

106. ***Subprime Mortgage Debt*** The approximate value of subprime (normally classified as risky) mortgage debt outstanding in the United States can be approximated by

$$A(t) = \frac{1,350}{1 + 4.2e^{-0.53t}} \text{ \$ billion} \quad (0 \le t \le 8)$$

t years after the start of 2000.[51]

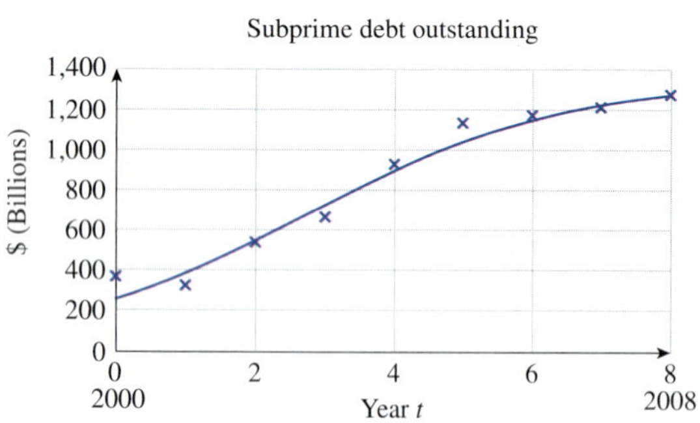

How fast, to the nearest \$1 billion, was the subprime mortgage debt increasing at the start of 2005? How would you check that the answer is approximately correct by looking at the graph?

107. ***Subprime Mortgages*** (Compare Exercise 105.) The percentage of mortgages issued in the United States that are subprime (normally classified as risky) can be approximated by

$$A(t) = \frac{15.0}{1 + 8.6(1.8)^{-t}} \text{ percent} \quad (0 \le t \le 8)$$

t years after the start of 2000.[52] How fast, to the nearest 0.1 percent, was the percentage increasing at the start of 2003?

108. ***Subprime Mortgage Debt*** (Compare Exercise 106.) The approximate value of subprime (normally classified as risky) mortgage debt outstanding in the United States can be approximated by

$$A(t) = \frac{1,350}{1 + 4.2(1.7)^{-t}} \text{ \$ billion} \quad (0 \le t \le 8)$$

t years after the start of 2000.[53] How fast, to the nearest \$1 billion, was the subprime mortgage debt increasing at the start of 2005?

109. ▼ ***Population Growth*** The population of Lower Anchovia was 4,000,000 at the start of 2010 and was doubling every 10 years. How fast was it growing per year at the start of 2010? (Round your answer to three significant digits.) HINT [Use the method of Example 2 in Section 2.2 to obtain an exponential model for the population.]

110. ▼ ***Population Growth*** The population of Upper Anchovia was 3,000,000 at the start of 2011 and doubling every 7 years. How fast was it growing per year at the start of 2011? (Round your answer to three significant digits.) HINT [Use the method of Example 2 in Section 2.2 to obtain an exponential model for the population.]

111. ▼ ***Radioactive Decay*** Plutonium 239 has a half-life of 24,400 years. How fast is a lump of 10 grams decaying after 100 years?

112. ▼ ***Radioactive Decay*** Carbon 14 has a half-life of 5,730 years. How fast is a lump of 20 grams decaying after 100 years?

113. T ▼ ***Diffusion of New Technology*** Numeric control is a technology whereby the operation of machines is controlled by numerical instructions on disks, tapes, or cards. In a study, E. Mansfield and associates[54] modeled the growth of this technology using the equation

$$p(t) = \frac{0.80}{1 + e^{4.46-0.477t}}$$

[50] 2009 figure is an estimate. Sources: Mortgage Bankers Association, UBS.

[51] 2008–2009 figures are estimates. Source: www.data360.org/dataset.aspx? Data_Set_Id=9549.

[52] 2009 figure is an estimate. Sources: Mortgage Bankers Association, UBS.

[53] 2008–2009 figures are estimates. Source: www.data360.org/dataset.aspx? Data_Set_Id=9549.

[54] Source: "The Diffusion of a Major Manufacturing Innovation," in *Research and Innovation in the Modern Corporation* (W.W. Norton and Company, Inc., New York, 1971, pp. 186-205).

where $p(t)$ is the fraction of firms using numeric control in year t.

a. Graph this function for $0 \le t \le 20$ and estimate $p'(10)$ graphically. Interpret the result.

b. Use your graph to estimate $\lim_{t\to+\infty} p(t)$ and interpret the result.

c. Compute $p'(t)$, graph it, and again find $p'(10)$.

d. Use your graph to estimate $\lim_{t\to+\infty} p'(t)$ and interpret the result.

114. T ▼ ***Diffusion of New Technology*** Repeat Exercise 113 using the revised formula

$$p(t) = \frac{0.90e^{-0.1t}}{1 + e^{4.50-0.477t}}$$

which takes into account that in the long run this new technology will eventually become outmoded and will be replaced by a newer technology. Draw your graphs using the range $0 \le t \le 40$.

115. ◆***Cell Phone Revenues*** The number of cell phone subscribers in China for the period 2000–2005 was projected to follow the equation[55]

$$N(t) = 39t + 68 \text{ million subscribers}$$

in year t ($t = 0$ represents 2000). The average annual revenue per cell phone user was \$350 in 2000. Assuming that, due to competition, the revenue per cell phone user decreases continuously at an annual rate of 10%, give a formula for the annual revenue in year t. Hence, project the annual revenue and its rate of change in 2002. Round all answers to the nearest billion dollars or billion dollars per year.

116. ◆***Cell Phone Revenues*** The annual revenue for cell phone use in China for the period 2000–2005 was projected to follow the equation[56]

$$R(t) = 14t + 24 \text{ billion dollars}$$

in year t ($t = 0$ represents 2000). At the same time, there were approximately 68 million subscribers in 2000. Assuming that the number of subscribers increases continuously at an annual rate of 10%, give a formula for the annual revenue per subscriber in year t. Hence, project to the nearest dollar the annual revenue per subscriber and its rate of change in 2002. (Be careful with units!)

[55] Based on a regression of projected figures (coefficients are rounded). Source: Intrinsic Technology/*New York Times*, Nov. 24, 2000, p. C1.

[56] Not allowing for discounting due to increased competition. Source: Ibid.

COMMUNICATION AND REASONING EXERCISES

117. Complete the following: The derivative of e raised to a glob is . . . .

118. Complete the following: The derivative of the natural logarithm of a glob is . . . .

119. Complete the following: The derivative of 2 raised to a glob is . . . .

120. Complete the following: The derivative of the base 2 logarithm of a glob is . . . .

121. What is wrong with the following?

$$\frac{d}{dx}\ln|3x+1| = \frac{3}{|3x+1|} \quad \text{✗ WRONG!}$$

122. What is wrong with the following?

$$\frac{d}{dx}2^{2x} = (2)2^{2x} \quad \text{✗ WRONG!}$$

123. What is wrong with the following?

$$\frac{d}{dx}3^{2x} = (2x)3^{2x-1} \quad \text{✗ WRONG!}$$

124. What is wrong with the following?

$$\frac{d}{dx}\ln(3x^2-1) = \frac{1}{6x} \quad \text{✗ WRONG!}$$

125. ▼ The number N of music downloads on campus is growing exponentially with time. Can $N'(t)$ grow linearly with time? Explain.

126. ▼ The number N of graphing calculators sold on campus is decaying exponentially with time. Can $N'(t)$ grow with time? Explain.

*The **percentage rate of change** or **fractional rate of change** of a function is defined to be the ratio $f'(x)/f(x)$. (It is customary to express this as a percentage when speaking about percentage rate of change.)*

127. ◆Show that the fractional rate of change of the exponential function e^{kx} is equal to k, which is often called its **fractional growth rate**.

128. ◆Show that the fractional rate of change of $f(x)$ is the rate of change of $\ln(f(x))$.

129. ◆Let $A(t)$ represent a quantity growing exponentially. Show that the percentage rate of change, $A'(t)/A(t)$, is constant.

130. ◆Let $A(t)$ be the amount of money in an account that pays interest which is compounded some number of times per year. Show that the percentage rate of growth, $A'(t)/A(t)$, is constant. What might this constant represent?

4.6 Implicit Differentiation

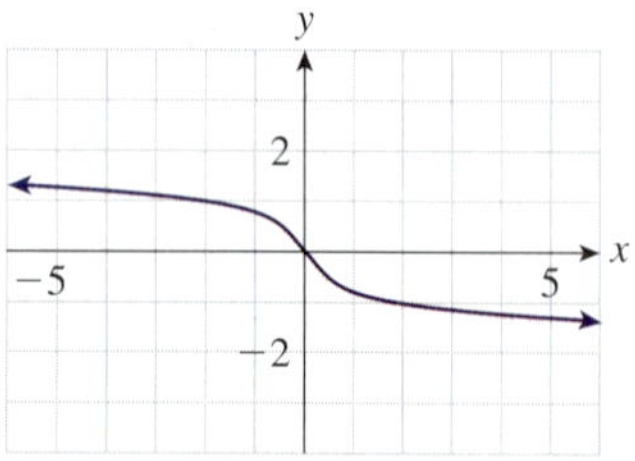

Figure 10

Consider the equation $y^5 + y + x = 0$, whose graph is shown in Figure 10.

How did we obtain this graph? We did not solve for y as a function of x; that is impossible. In fact, we solved for x in terms of y to find points to plot. Nonetheless, the graph in Figure 10 is the graph of a function because it passes the vertical line test: Every vertical line crosses the graph no more than once, so for each value of x there is no more than one corresponding value of y. Because we cannot solve for y explicitly in terms of x, we say that the equation $y^5 + y + x = 0$ determines y as an **implicit function** of x.

Now, suppose we want to find the slope of the tangent line to this curve at, say, the point $(2, -1)$ (which, you should check, is a point on the curve). In the following example we find, surprisingly, that it is possible to obtain a formula for dy/dx without having to first solve the equation for y.

EXAMPLE 1 Implicit Differentiation

Find $\dfrac{dy}{dx}$, given that $y^5 + y + x = 0$.

Solution We use the chain rule and a little cleverness. Think of y as a function of x and take the derivative with respect to x of both sides of the equation:

$$y^5 + y + x = 0 \qquad \text{Original equation}$$

$$\frac{d}{dx}[y^5 + y + x] = \frac{d}{dx}[0] \qquad \text{Derivative with respect to } x \text{ of both sides}$$

$$\frac{d}{dx}[y^5] + \frac{d}{dx}[y] + \frac{d}{dx}[x] = 0 \qquad \text{Derivative rules}$$

Now we must be careful. The derivative *with respect to x* of y^5 is *not* $5y^4$. Rather, because y is a function of x, we must use the chain rule, which tells us that

$$\frac{d}{dx}[y^5] = 5y^4\frac{dy}{dx}.$$

Thus, we get

$$5y^4\frac{dy}{dx} + \frac{dy}{dx} + 1 = 0.$$

We want to find dy/dx, so we *solve for it*:

$$(5y^4 + 1)\frac{dy}{dx} = -1 \qquad \text{Isolate } dy/dx \text{ on one side.}$$

$$\frac{dy}{dx} = -\frac{1}{5y^4 + 1} \qquad \text{Divide both sides by } 5y^4 + 1.$$

➡ **Before we go on...** Note that we should not expect to obtain dy/dx as an explicit function of x if y was not an explicit function of x to begin with. For example, the formula we found in Example 1 for dy/dx is not a function of x because there is a y in it. However, the result is still useful because we can evaluate the derivative at any point on the graph. For instance, at the point $(2, -1)$ on the graph, we get

$$\frac{dy}{dx} = -\frac{1}{5y^4 + 1} = -\frac{1}{5(-1)^4 + 1} = -\frac{1}{6}.$$

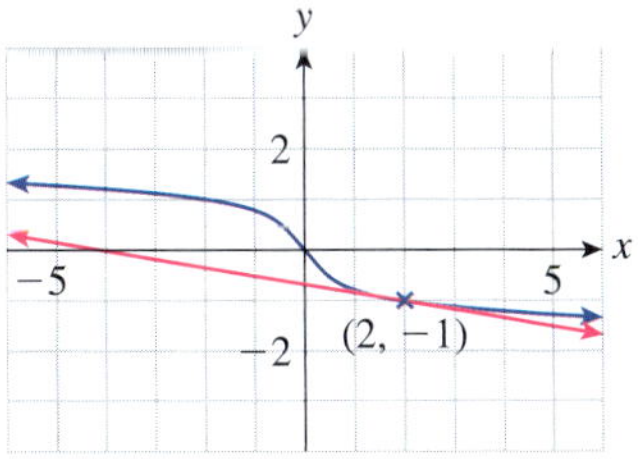

Figure 11

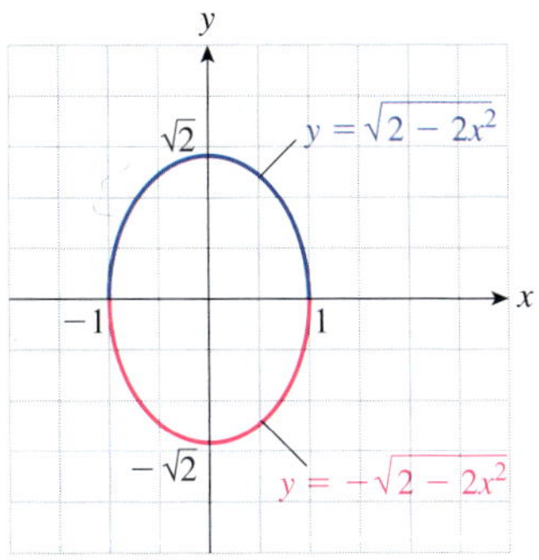

Figure 12

Thus, the slope of the tangent line to the curve $y^5 + y + x = 0$ at the point $(2, -1)$ is $-1/6$. Figure 11 shows the graph and this tangent line. ■

This procedure we just used—differentiating an equation to find dy/dx without first solving the equation for y—is called **implicit differentiation.**

In Example 1 we were given an equation in x and y that determined y as an (implicit) function of x, even though we could not solve for y. But an equation in x and y need not always determine y as a function of x. Consider, for example, the equation

$$2x^2 + y^2 = 2.$$

Solving for y yields $y = \pm\sqrt{2 - 2x^2}$. The $\pm$ sign reminds us that for some values of x there are two corresponding values for y. We can graph this equation by superimposing the graphs of

$$y = \sqrt{2 - 2x^2} \qquad \text{and} \qquad y = -\sqrt{2 - 2x^2}.$$

The graph, an *ellipse*, is shown in Figure 12.

The graph of $y = \sqrt{2 - 2x^2}$ constitutes the top half of the ellipse, and the graph of $y = -\sqrt{2 - 2x^2}$ constitutes the bottom half.

EXAMPLE 2 Slope of Tangent Line

Refer to Figure 12. Find the slope of the tangent line to the ellipse $2x^2 + y^2 = 2$ at the point $(1/\sqrt{2}, 1)$.

Solution Because $(1/\sqrt{2}, 1)$ is on the top half of the ellipse in Figure 12, we *could* differentiate the function $y = \sqrt{2 - 2x^2}$, to obtain the result, but it is actually easier to apply implicit differentiation to the original equation.

$$2x^2 + y^2 = 2 \qquad \text{Original equation}$$

$$\frac{d}{dx}[2x^2 + y^2] = \frac{d}{dx}[2] \qquad \text{Derivative with respect to } x \text{ of both sides}$$

$$4x + 2y\frac{dy}{dx} = 0$$

$$2y\frac{dy}{dx} = -4x \qquad \text{Solve for } dy/dx.$$

$$\frac{dy}{dx} = -\frac{4x}{2y} = -\frac{2x}{y}$$

To find the slope at $(1/\sqrt{2}, 1)$ we now substitute for x and y:

$$\left.\frac{dy}{dx}\right|_{(1/\sqrt{2}, 1)} = -\frac{2/\sqrt{2}}{1} = -\sqrt{2}.$$

Thus, the slope of the tangent to the ellipse at the point $(1/\sqrt{2}, 1)$ is $-\sqrt{2} \approx -1.414$.

EXAMPLE 3 Tangent Line for an Implicit Function

Find the equation of the tangent line to the curve $\ln y = xy$ at the point where $y = 1$.

Solution First, we use implicit differentiation to find dy/dx:

$$\frac{d}{dx}[\ln y] = \frac{d}{dx}[xy] \qquad \text{Take } d/dx \text{ of both sides.}$$

$$\frac{1}{y}\frac{dy}{dx} = (1)y + x\frac{dy}{dx}. \qquad \text{Chain rule on left, product rule on right}$$

To solve for dy/dx, we bring all the terms containing dy/dx to the left-hand side and all terms not containing it to the right-hand side:

$$\frac{1}{y}\frac{dy}{dx} - x\frac{dy}{dx} = y \qquad \text{Bring the terms with } dy/dx \text{ to the left.}$$

$$\frac{dy}{dx}\left(\frac{1}{y} - x\right) = y \qquad \text{Factor out } dy/dx.$$

$$\frac{dy}{dx}\left(\frac{1-xy}{y}\right) = y$$

$$\frac{dy}{dx} = y\left(\frac{y}{1-xy}\right) = \frac{y^2}{1-xy}. \qquad \text{Solve for } dy/dx.$$

The derivative gives the slope of the tangent line, so we want to evaluate the derivative at the point where $y = 1$. However, the formula for dy/dx requires values for both x and y. We get the value of x by substituting $y = 1$ in the original equation:

$$\ln y = xy$$

$$\ln(1) = x \cdot 1$$

But $\ln(1) = 0$, and so $x = 0$ for this point. Thus,

$$\left.\frac{dy}{dx}\right|_{(0,1)} = \frac{1^2}{1-(0)(1)} = 1.$$

Therefore, the tangent line is the line through $(x, y) = (0, 1)$ with slope 1, which is

$$y = x + 1.$$

➡ **Before we go on...** Example 3 presents an instance of an implicit function in which it is simply not possible to solve for y. Try it. ■

Sometimes, it is easiest to differentiate a complicated function of x by first taking the logarithm and then using implicit differentiation—a technique called **logarithmic differentiation**.

EXAMPLE 4 Logarithmic Differentiation

Find $\frac{d}{dx}\left[\frac{(x+1)^{10}(x^2+1)^{11}}{(x^3+1)^{12}}\right]$ without using the product or quotient rules.

Solution Write

$$y = \frac{(x+1)^{10}(x^2+1)^{11}}{(x^3+1)^{12}}$$

and then take the natural logarithm of both sides:

$$\ln y = \ln\left[\frac{(x+1)^{10}(x^2+1)^{11}}{(x^3+1)^{12}}\right].$$

We can use properties of the logarithm to simplify the right-hand side:

$$\begin{aligned}\ln y &= \ln(x+1)^{10} + \ln(x^2+1)^{11} - \ln(x^3+1)^{12} \\ &= 10\ln(x+1) + 11\ln(x^2+1) - 12\ln(x^3+1).\end{aligned}$$

Now we can find $\frac{dy}{dx}$ using implicit differentiation:

$$\frac{1}{y}\frac{dy}{dx} = \frac{10}{x+1} + \frac{22x}{x^2+1} - \frac{36x^2}{x^3+1} \qquad \text{Take } d/dx \text{ of both sides.}$$

$$\frac{dy}{dx} = y\left(\frac{10}{x+1} + \frac{22x}{x^2+1} - \frac{36x^2}{x^3+1}\right) \qquad \text{Solve for } dy/dx.$$

$$= \frac{(x+1)^{10}(x^2+1)^{11}}{(x^3+1)^{12}}\left(\frac{10}{x+1} + \frac{22x}{x^2+1} - \frac{36x^2}{x^3+1}\right). \qquad \text{Substitute for } y.$$

➡ **Before we go on...** Redo Example 4 using the product and quotient rules (and the chain rule) instead of logarithmic differentiation and compare the answers. Compare also the amount of work involved in both methods. ■

APPLICATION

Productivity usually depends on both labor and capital. Suppose, for example, you are managing a surfboard manufacturing company. You can measure its productivity by counting the number of surfboards the company makes each year. As a measure of labor, you can use the number of employees, and as a measure of capital you can use its operating budget. The so-called *Cobb-Douglas* model uses a function of the form:

$$P = Kx^a y^{1-a} \qquad \text{Cobb-Douglas model for productivity}$$

where P stands for the number of surfboards made each year, x is the number of employees, and y is the operating budget. The numbers K and a are constants that depend on the particular situation studied, with a between 0 and 1.

David Samuel Robbins/CORBIS

EXAMPLE 5 Cobb-Douglas Production Function

The surfboard company you own has the Cobb-Douglas production function

$$P = x^{0.3}y^{0.7}$$

where P is the number of surfboards it produces per year, x is the number of employees, and y is the daily operating budget (in dollars). Assume that the production level P is constant.

a. Find $\frac{dy}{dx}$.

b. Evaluate this derivative at $x = 30$ and $y = 10{,}000$, and interpret the answer.

Solution

a. We are given the equation $P = x^{0.3}y^{0.7}$, in which P is constant. We find $\frac{dy}{dx}$ by implicit differentiation

$$0 = \frac{d}{dx}[x^{0.3}y^{0.7}] \qquad d/dx \text{ of both sides}$$

$$0 = 0.3x^{-0.7}y^{0.7} + x^{0.3}(0.7)y^{-0.3}\frac{dy}{dx} \qquad \text{Product and chain rules}$$

$$-0.7x^{0.3}y^{-0.3}\frac{dy}{dx} = 0.3x^{-0.7}y^{0.7}$$ Bring term with dy/dx to left.

$$\frac{dy}{dx} = -\frac{0.3x^{-0.7}y^{0.7}}{0.7x^{0.3}y^{-0.3}}$$ Solve for dy/dx.

$$= -\frac{3y}{7x}.$$ Simplify.

b. Evaluating this derivative at $x = 30$ and $y = 10{,}000$ gives

$$\left.\frac{dy}{dx}\right|_{x=30,\ y=10{,}000} = -\frac{3(10{,}000)}{7(30)} \approx -143.$$

To interpret this result, first look at the units of the derivative: We recall that the units of dy/dx are units of y per unit of x. Because y is the daily budget, its units are dollars; because x is the number of employees, its units are employees. Thus,

$$\left.\frac{dy}{dx}\right|_{x=30,\ y=10{,}000} \approx -\$143 \text{ per employee.}$$

Next, recall that dy/dx measures the rate of change of y as x changes. Because the answer is negative, the daily budget to maintain production at the fixed level is decreasing by approximately \$143 per additional employee at an employment level of 30 employees and a daily operating budget of \$10,000. In other words, increasing the workforce by one worker will result in a savings of approximately \$143 per day. Roughly speaking, *a new employee is worth \$143 per day* at the current levels of employment and production.

4.6 EXERCISES

▼ more advanced ◆ challenging

T indicates exercises that should be solved using technology

In Exercises 1–10, find dy/dx, using implicit differentiation. In each case, compare your answer with the result obtained by first solving for y as a function of x and then taking the derivative. HINT [See Example 1.]

1. $2x + 3y = 7$

2. $4x - 5y = 9$

3. $x^2 - 2y = 6$

4. $3y + x^2 = 5$

5. $2x + 3y = xy$

6. $x - y = xy$

7. $e^x y = 1$

8. $e^x y - y = 2$

9. $y \ln x + y = 2$

10. $\dfrac{\ln x}{y} = 2 - x$

In Exercises 11–30, find the indicated derivative using implicit differentiation. HINT [See Example 1.]

11. $x^2 + y^2 = 5;\ \dfrac{dy}{dx}$

12. $2x^2 - y^2 = 4;\ \dfrac{dy}{dx}$

13. $x^2y - y^2 = 4;\ \dfrac{dy}{dx}$

14. $xy^2 - y = x;\ \dfrac{dy}{dx}$

15. $3xy - \dfrac{y}{3} = \dfrac{2}{x};\ \dfrac{dy}{dx}$

16. $\dfrac{xy}{2} - y^2 = 3;\ \dfrac{dy}{dx}$

17. $x^2 - 3y^2 = 8;\ \dfrac{dx}{dy}$

18. $(xy)^2 + y^2 = 8;\ \dfrac{dx}{dy}$

19. $p^2 - pq = 5p^2q^2;\ \dfrac{dp}{dq}$

20. $q^2 - pq = 5p^2q^2;\ \dfrac{dp}{dq}$

21. $xe^y - ye^x = 1;\ \dfrac{dy}{dx}$

22. $x^2e^y - y^2 = e^x;\ \dfrac{dy}{dx}$

23. ▼ $e^{st} = s^2;\ \dfrac{ds}{dt}$

24. ▼ $e^{s^2t} - st = 1;\ \dfrac{ds}{dt}$

25. ▼ $\dfrac{e^x}{y^2} = 1 + e^y;\ \dfrac{dy}{dx}$

26. ▼ $\dfrac{x}{e^y} + xy = 9y;\ \dfrac{dy}{dx}$

27. ▼ $\ln(y^2 - y) + x = y;\ \dfrac{dy}{dx}$

28. ▼ $\ln(xy) - x\ln y = y;\ \dfrac{dy}{dx}$

29. ▼ $\ln(xy + y^2) = e^y;\ \dfrac{dy}{dx}$

30. ▼ $\ln(1 + e^{xy}) = y;\ \dfrac{dy}{dx}$

In Exercises 31–42, use implicit differentiation to find ***(a)*** *the slope of the tangent line, and* ***(b)*** *the equation of the tangent line at the indicated point on the graph. (Round answers to four decimal places as needed.) If only the x-coordinate is given, you must also find the y-coordinate.)* HINT [See Examples 2, 3.]

31. $4x^2 + 2y^2 = 12, (1, -2)$

32. $3x^2 - y^2 = 11, (-2, 1)$

33. $2x^2 - y^2 = xy, (-1, 2)$

34. $2x^2 + xy = 3y^2, (-1, -1)$

35. $x^2y - y^2 + x = 1, (1, 0)$

36. $(xy)^2 + xy - x = 8, (-8, 0)$

37. $xy - 2000 = y, x = 2$

38. $x^2 - 10xy = 200, x = 10$

39. ▼ $\ln(x + y) - x = 3x^2, x = 0$

40. ▼ $\ln(x - y) + 1 = 3x^2, x = 0$

41. ▼ $e^{xy} - x = 4x, x = 3$

42. ▼ $e^{-xy} + 2x = 1, x = -1$

In Exercises 43–52, use logarithmic differentiation to find dy/dx. Do not simplify the result. HINT [See Example 4.]

43. $y = \dfrac{2x + 1}{4x - 2}$

44. $y = (3x + 2)(8x - 5)$

45. $y = \dfrac{(3x + 1)^2}{4x(2x - 1)^3}$

46. $y = \dfrac{x^2(3x + 1)^2}{(2x - 1)^3}$

47. $y = (8x - 1)^{1/3}(x - 1)$

48. $y = \dfrac{(3x + 2)^{2/3}}{3x - 1}$

49. $y = (x^3 + x)\sqrt{x^3 + 2}$

50. $y = \sqrt{\dfrac{x - 1}{x^2 + 2}}$

51. ▼ $y = x^x$

52. ▼ $y = x^{-x}$

APPLICATIONS

53. ***Productivity*** The number of CDs per hour that Snappy Hardware can manufacture at its plant is given by

$$P = x^{0.6}y^{0.4}$$

where x is the number of workers at the plant and y is the monthly budget (in dollars). Assume P is constant, and compute $\dfrac{dy}{dx}$ when $x = 100$ and $y = 200{,}000$. Interpret the result. HINT [See Example 5.]

54. ***Productivity*** The number of cell-phone accessory kits (neon lights, matching covers, and earpods) per day that USA Cellular Makeover Inc. can manufacture at its plant in Cambodia is given by

$$P = x^{0.5}y^{0.5}$$

where x is the number of workers at the plant and y is the monthly budget (in dollars). Assume P is constant, and compute $\dfrac{dy}{dx}$ when $x = 200$ and $y = 100{,}000$. Interpret the result. HINT [See Example 5.]

55. ***Demand*** The demand equation for soccer tournament T-shirts is

$$xy - 2{,}000 = y$$

where y is the number of T-shirts the Enormous State University soccer team can sell at a price of \$$x$ per shirt. Find $\left.\dfrac{dy}{dx}\right|_{x=5}$, and interpret the result.

56. ***Cost Equations*** The cost y (in cents) of producing x gallons of Ectoplasm hair gel is given by the cost equation

$$y^2 - 10xy = 200.$$

Evaluate $\dfrac{dy}{dx}$ at $x = 1$ and interpret the result.

57. ***Housing Costs***[57] The cost C (in dollars) of building a house is related to the number k of carpenters used and the number e of electricians used by the formula

$$C = 15{,}000 + 50k^2 + 60e^2.$$

If the cost of the house is fixed at \$200,000, find $\left.\dfrac{dk}{de}\right|_{e=15}$ and interpret your result.

58. ***Employment*** An employment research company estimates that the value of a recent MBA graduate to an accounting company is

$$V = 3e^2 + 5g^3$$

where V is the value of the graduate, e is the number of years of prior business experience, and g is the graduate school grade-point average. If V is fixed at 200, find $\dfrac{de}{dg}$ when $g = 3.0$ and interpret the result.

59. ▼ ***Grades***[58] A productivity formula for a student's performance on a difficult English examination is

$$g = 4tx - 0.2t^2 - 10x^2 \quad (t < 30)$$

where g is the score the student can expect to obtain, t is the number of hours of study for the examination, and x is the student's grade-point average.

a. For how long should a student with a 3.0 grade-point average study in order to score 80 on the examination?

b. Find $\dfrac{dt}{dx}$ for a student who earns a score of 80, evaluate it when $x = 3.0$, and interpret the result.

60. ▼ ***Grades*** Repeat the preceding exercise using the following productivity formula for a basket-weaving examination:

$$g = 10tx - 0.2t^2 - 10x^2 \quad (t < 10).$$

Comment on the result.

[57] Based on an Exercise in *Introduction to Mathematical Economics* by A. L. Ostrosky Jr., and J. V. Koch (Waveland Press, Springfield, Illinois, 1979).

[58] Ibid.

Exercises 61 and 62 are based on the following demand function for money (taken from a question on the GRE Economics Test):

$$M_d = (2) \times (y)^{0.6} \times (r)^{-0.3} \times (p)$$

where

M_d = *demand for nominal money balances (money stock)*

y = *real income*

r = *an index of interest rates*

p = *an index of prices.*

61. ◆ ***Money Stock*** If real income grows while the money stock and the price level remain constant, the interest rate must change at what rate? (First find dr/dy, then dr/dt; your answers will be expressed in terms of r, y, and $\frac{dy}{dt}$.)

62. ◆ ***Money Stock*** If real income grows while the money stock and the interest rate remain constant, the price level must change at what rate?

COMMUNICATION AND REASONING EXERCISES

63. Fill in the missing terms: The equation $x = y^3 + y - 3$ specifies ___ as a function of ___, and ___ as an implicit function of ___.

64. Fill in the missing terms: When $x \neq 0$ in the equation $xy = x^3 + 4$, it is possible to specify ___ as a function of ___. However, ___ is only an implicit function of ___.

65. ▼ Use logarithmic differentiation to give another proof of the product rule.

66. ▼ Use logarithmic differentiation to give a proof of the quotient rule.

67. ▼ If y is given explicitly as a function of x by an equation $y = f(x)$, compare finding dy/dx by implicit differentiation to finding it explicitly in the usual way.

68. ▼ Explain why one should not expect dy/dx to be a function of x if y is not a function of x.

69. ◆ If y is a function of x and $dy/dx \neq 0$ at some point, regard x as an implicit function of y and use implicit differentiation to obtain the equation

$$\frac{dx}{dy} = \frac{1}{dy/dx}.$$

70. ◆ If you are given an equation in x and y such that dy/dx is a function of x only, what can you say about the graph of the equation?

CHAPTER 4 REVIEW

KEY CONCEPTS

Web Site www.AppliedCalc.org

Go to the student Web site at www.AppliedCalc.org to find a comprehensive and interactive Web-based summary of Chapter 4.

4.1 Derivatives of Powers, Sums, and Constant Multiples

Power Rule: If n is any constant and $f(x) = x^n$, then $f'(x) = nx^{n-1}$. *p. 286*

Using the power rule for negative and fractional exponents *p. 287*

Sums, differences, and constant multiples *p. 289*

Combining the rules *p. 290*

$\frac{d}{dx}(cx) = c,\ \frac{d}{dx}(c) = 0$ *p. 292*

$f(x) = x^{1/3}$ and $g(x) = x^{2/3}$ are not differentiable at $x = 0$. *p. 292*

Derivative of $f(x) = |x|$: $\frac{d}{dx}|x| = \frac{|x|}{x}$ *p. 293*

L'Hospital's Rule *p. 295*

4.2 A First Application: Marginal Analysis

Marginal cost function $C'(x)$ *p. 302*

Marginal revenue and profit functions $R'(x)$ and $P'(x)$ *p. 304*

What it means when the marginal profit is zero *p. 305*

Marginal product *p. 306*

Average cost of the first x items:

$\bar{C}(x) = \frac{C(x)}{x}$ *p. 307*

4.3 The Product and Quotient Rules

Product rule: $\frac{d}{dx}[f(x)g(x)] = f'(x)g(x) + f(x)g'(x)$ *p. 313*

Quotient rule: $\frac{d}{dx}\left[\frac{f(x)}{g(x)}\right] = \frac{f'(x)g(x) - f(x)g'(x)}{[g(x)]^2}$ *p. 314*

Using the product rule *p. 316*

Using the quotient rule *p. 317*

Calculation thought experiment *p. 318*

Application to revenue and average cost *p. 319*

4.4 The Chain Rule

Chain rule: $\frac{d}{dx}[f(u)] = f'(u)\frac{du}{dx}$ *p. 325*

Generalized power rule:

$\frac{d}{dx}[u^n] = nu^{n-1}\frac{du}{dx}$ *p. 326*

Using the chain rule *p. 326*

Application to marginal product *p. 329*

Chain rule in differential notation:

$\frac{dy}{dx} = \frac{dy}{du}\frac{du}{dx}$ *p. 330*

Manipulating derivatives in differential notation *p. 332*

4.5 Derivatives of Logarithmic and Exponential Functions

Derivative of the natural logarithm:

$\frac{d}{dx}[\ln x] = \frac{1}{x}$ *p. 337*

Derivative of logarithm with base b:

$\frac{d}{dx}[\log_b x] = \frac{1}{x \ln b}$ *p. 338*

Derivatives of logarithms of functions:

$\frac{d}{dx}[\ln u] = \frac{1}{u}\frac{du}{dx}$

$\frac{d}{dx}[\log_b u] = \frac{1}{u \ln b}\frac{du}{dx}$ *p. 339*

Derivatives of logarithms of absolute values:

$\frac{d}{dx}[\ln|x|] = \frac{1}{x}$ $\quad \frac{d}{dx}[\ln|u|] = \frac{1}{u}\frac{du}{dx}$

$\frac{d}{dx}[\log_b |x|] = \frac{1}{x \ln b}$

$\frac{d}{dx}[\log_b |u|] = \frac{1}{u \ln b}\frac{du}{dx}$ *p. 342*

Derivative of e^x: $\frac{d}{dx}[e^x] = e^x$ *p. 343*

Derivative of b^x: $\frac{d}{dx}[b^x] = b^x \ln b$ *p. 343*

Derivatives of exponential functions *p. 344*

Application to epidemics *p. 344*

Application to sales growth (logistic function) *p. 345*

4.6 Implicit Differentiation

Implicit function of x *p. 352*

Implicit differentiation *p. 352*

Using implicit differentiation *p. 352*

Finding a tangent line *p. 353*

Logarithmic differentiation *p. 354*

REVIEW EXERCISES

In Exercises 1–20 find the derivative of the given function.

1. $f(x) = 10x^5 + \frac{1}{2}x^4 - x + 2$
2. $f(x) = \frac{10}{x^5} + \frac{1}{2x^4} - \frac{1}{x} + 2$
3. $f(x) = 3x^3 + 3\sqrt[3]{x}$
4. $f(x) = \frac{2}{x^{2.1}} - \frac{x^{0.1}}{2}$
5. $f(x) = x + \frac{1}{x^2}$
6. $f(x) = 2x - \frac{1}{x}$
7. $f(x) = \frac{4}{3x} - \frac{2}{x^{0.1}} + \frac{x^{1.1}}{3.2} - 4$
8. $f(x) = \frac{4}{x} + \frac{x}{4} - |x|$
9. $f(x) = e^x(x^2 - 1)$
10. $f(x) = \frac{x^2 + 1}{x^2 - 1}$
11. $f(x) = (x^2 - 1)^{10}$
12. $f(x) = \frac{1}{(x^2 - 1)^{10}}$
13. $f(x) = e^x(x^2 + 1)^{10}$
14. $f(x) = \left[\frac{x - 1}{3x + 1}\right]^3$
15. $f(x) = \frac{3^x}{x - 1}$
16. $f(x) = 4^{-x}(x + 1)$
17. $f(x) = e^{x^2 - 1}$
18. $f(x) = (x^2 + 1)e^{x^2 - 1}$
19. $f(x) = \ln(x^2 - 1)$
20. $f(x) = \frac{\ln(x^2 - 1)}{x^2 - 1}$

In Exercises 21–28 find all values of x (if any) where the tangent line to the graph of the given equation is horizontal.

21. $y = -3x^2 + 7x - 1$ **22.** $y = 5x^2 - 2x + 1$

23. $y = \frac{x}{2} + \frac{2}{x}$ **24.** $y = \frac{x^2}{2} - \frac{8}{x^2}$

25. $y = x - e^{2x-1}$ **26.** $y = e^{x^2}$

27. $y = \frac{x}{x+1}$ **28.** $y = \sqrt{x}(x - 1)$

In Exercises 29–34, find dy/dx for the given equation.

29. $x^2 - y^2 = x$ **30.** $2xy + y^2 = y$

31. $e^{xy} + xy = 1$ **32.** $\ln\left(\frac{y}{x}\right) = y$

33. $y = \frac{(2x-1)^4(3x+4)}{(x+1)(3x-1)^3}$ **34.** $y = x^{x-1}3^x$

In Exercises 35–40 find the equation of the tangent line to the graph of the given equation at the specified point.

35. $y = (x^2 - 3x)^{-2}$; $x = 1$ **36.** $y = (2x^2 - 3)^{-3}$; $x = -1$

37. $y = x^2 e^{-x}$; $x = -1$ **38.** $y = \frac{x}{1 + e^x}$; $x = 0$

39. $xy - y^2 = x^2 - 3$; $(-1, 1)$ **40.** $\ln(xy) + y^2 = 1$; $(-1, -1)$

APPLICATIONS

41. *Sales* OHaganBooks.com fits the cubic curve

$$w(t) = -3.7t^3 + 74.6t^2 + 135.5t + 6{,}300$$

to its weekly sales figures (see Chapter 3 Review Exercise 47), as shown in the following graph:

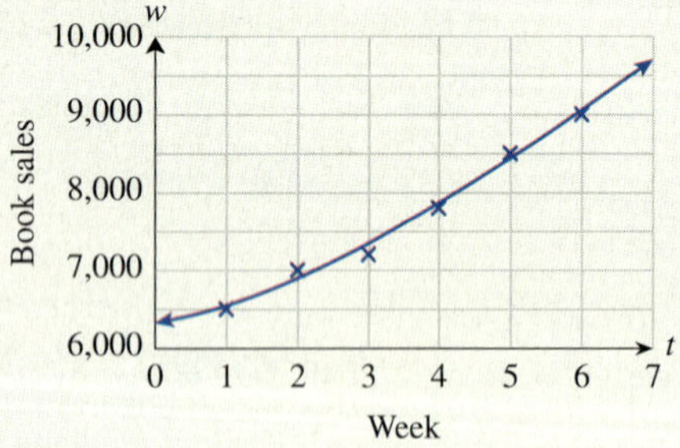

a. According to the cubic model, what was the rate of increase of sales at the beginning of the second week ($t = 1$)? (Round your answer to the nearest unit.)

b. If we extrapolate the model, what would be the rate of increase of weekly sales at the beginning of the 8th week ($t = 7$)?

c. Graph the function w for $0 \le t \le 20$. Would it be realistic to use the function to predict sales through week 20? Why?

d. By examining the graph, say why the choice of a quadratic model would result in radically different long-term predictions of sales.

42. *Rising Sea Level* Marjory Duffin fit the cubic curve

$$L(t) = -0.0001t^3 + 0.02t^2 + 2.2t \text{ mm}$$

to the New York sea level figures she had seen after purchasing a beachfront condominium in New York (see Chapter 3 Review Exercise 48; t is time in years since 1900). The curve and data are shown in the following graph:

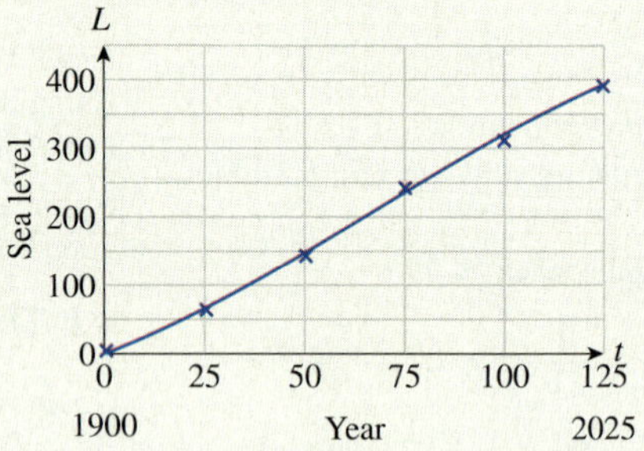

Sea Level Change since 1900

a. According to the cubic model, what was the rate at which the sea level was rising in 2000 ($t = 100$)? (Round your answer to two significant digits.)

b. If we extrapolate the model, what would be the rate at which the sea level is rising in 2025 ($t = 125$)?

c. Graph the function L for $0 \le t \le 200$. Why is it not realistic to use the function to predict the sea level through 2100?

d. James Stewart, a summer intern at Duffin House Publishers, differs. As he puts it, "The cubic curve came from doing regression on the actual data, and thus reflects the actual trend of the data. We can't argue against reality!" Comment on this assertion.

43. *Cost* As OHaganBooks.com's sales increase, so do its costs. If we take into account volume discounts from suppliers and shippers, the weekly cost of selling x books is

$$C(x) = -0.00002x^2 + 3.2x + 5{,}400 \text{ dollars}$$

a. What is the marginal cost at a sales level of 8,000 books per week?

b. What is the average cost per book at a sales level of 8,000 books per week?

c. What is the marginal average cost ($d\bar{C}/dx$) at a sales level of 8,000 books per week?

d. Interpret the results of parts (a)–(c).

44. *Cost* OHaganBooks.com has been experiencing a run of bad luck with its summer college intern program in association with Party Central University (begun as a result of a suggestion by Marjory Duffin over dinner one evening). The frequent errors in filling orders, charges from movie download sites and dating sites, and beverages spilled on computer equipment have resulted in an estimated weekly cost to the company of

$$C(x) = 25x^2 - 5.2x + 4{,}000 \text{ dollars}$$

where x is the number of college interns employed.

a. What is the marginal cost at a level of 10 interns?
b. What is the average cost per intern at a level of 10 interns?
c. What is the marginal average cost at a level of 10 interns?
d. Interpret the results of parts (a)–(c).

45. ***Revenue*** At the moment, OHaganBooks.com is selling 1,000 books per week and its sales are rising at a rate of 200 books per week. Also, it is now selling all its books for \$20 each, but its price is dropping at a rate of \$1 per week.

a. At what rate is OHaganBooks.com's weekly revenue rising or falling?
b. John O'Hagan would like to see the company's weekly revenue increase at a rate of \$5,000 per week. At what rate would sales have to have been increasing to accomplish that goal, assuming all the other information is as given above?

46. ***Revenue*** Due to ongoing problems with its large college intern program, OHaganBooks.com has decided to offer to transfer its interns to its competitor JungleBooks.com (whose headquarters happens to be across the road) for a small fee. At the moment, it is transferring 5 students per week, and this number is rising at a rate of 4 students per week. Also, it is now charging JungleBooks \$400 per intern, but this amount is decreasing at a rate of \$20 per week.

a. At what rate is OHaganBooks.com's weekly revenue from this transaction rising or falling?
b. Flush with success of the transfer program, John O'Hagan would like to see the company's resulting revenue increase at a rate of \$3,900 per week. At what rate would the transfer of interns have to increase to accomplish that goal, assuming all the other information is as given above?

47. ***Percentage Rate of Change of Revenue*** The percentage rate of change of a quantity Q is Q'/Q. Why is the percentage rate of change of revenue always equal to the sum of the percentage rates of change of unit price and weekly sales?

48. ***P/E Ratios*** At the beginning of last week, OHaganBooks.com stock was selling for \$100 per share, rising at a rate of \$50 per year. Its earnings amounted to \$1 per share, rising at a rate of \$0.10 per year. At what rate was its price-to-earnings (P/E) ratio, the ratio of its stock price to its earnings per share, rising or falling?

49. ***P/E Ratios*** Refer to Exercise 48. Curt Hinrichs, who recently invested in OHaganBooks.com stock, would have liked to see the P/E ratio increase at a rate of 100 points per year. How fast would the stock have to have been rising, assuming all the other information is as given in Exercise 48?

50. ***Percentage Rate of Change of P/E Ratios*** The percentage rate of change of a quantity Q is Q'/Q. Why is the percentage rate of change of P/E always equal to the percentage rate of change of unit price minus the percentage rate of change of earnings?

51. ***Sales*** OHaganBooks.com decided that the cubic curve in Exercise 41 was not suitable for extrapolation, so instead it tried

$$s(t) = 6{,}000 + \frac{4{,}500}{1 + e^{-0.55(t-4.8)}}$$

as shown in the following graph:

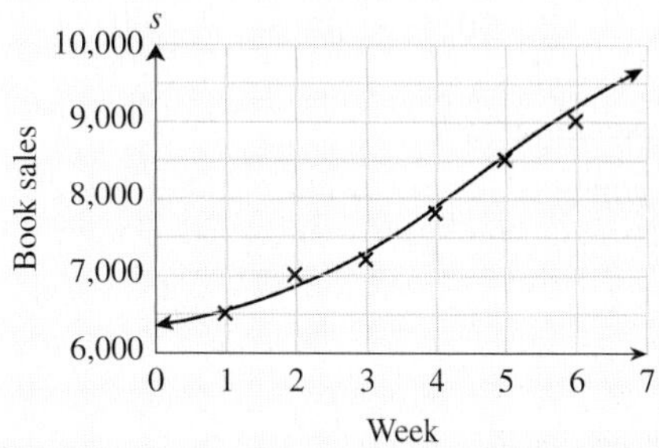

a. Compute $s'(t)$ and use the answer to estimate the rate of increase of weekly sales at the beginning of the 7th week ($t = 6$). (Round your answer to the nearest unit.)
b. Compute $\lim_{t \to +\infty} s'(t)$ and interpret the answer.

52. ***Rising Sea Level*** Upon some reflection, Marjory Duffin decided that the curve in Exercise 42 was not suitable for extrapolation, so instead she tried

$$L(t) = \frac{418}{1 + 17.2e^{-0.041t}} \qquad (0 \le t \le 125)$$

(t is time in years since 1900) as shown in the following graph:

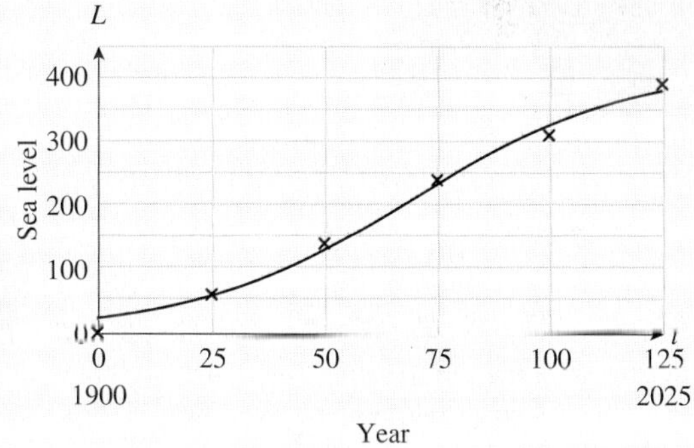

a. Compute $L'(t)$ and use the answer to estimate the rate at which the sea level was rising in 2000 ($t = 100$). (Round your answer to two decimal places.)
b. Compute $\lim_{t \to +\infty} L'(t)$ and interpret the answer.

53. ***Web Site Activity*** The number of "hits" on OHaganBooks.com's Web site was 1,000 per day at the beginning of the year, and was growing at a rate of 5% per week. If this growth rate continued for the whole year (52 weeks), find the rate of increase (in hits per day per week) at the end of the year.

54. ***Web Site Activity*** The number of "hits" on ShadyDownload.net during the summer intern program at OHaganBooks.com was 100 per day at the beginning of the intern program, and was growing at a rate of 15% per day. If this growth rate continued for the duration of the whole summer intern program (85 days), find the rate of increase (in hits per day per day) at the end of the program.

55. ***Demand and Revenue*** The price p that OHaganBooks.com charges for its latest leather-bound gift edition of *The Complete Harry Potter* is related to the demand q in weekly sales by the equation

$$250pq + q^2 = 13,500,000$$

Suppose the price is set at \$50, which would make the demand 1,000 copies per week.

a. Using implicit differentiation, compute the rate of change of demand with respect to price, and interpret the result. (Round the answer to two decimal places.)

b. Use the result of part (a) to compute the rate of change of revenue with respect to price. Should the price be raised or lowered to increase revenue?

56. ***Demand and Revenue*** The price p that OHaganBooks.com charges for its latest leather-bound gift edition of *The Lord of the Rings* is related to the demand q in weekly sales by the equation

$$100pq + q^2 = 5,000,000.$$

Suppose the price is set at \$40, which would make the demand 1,000 copies per week.

a. Using implicit differentiation, compute the rate of change of demand with respect to price, and interpret the result. (Round the answer to two decimal places.)

b. Use the result of part (a) to compute the rate of change of revenue with respect to price. Should the price be raised or lowered to increase revenue?

Case Study — Projecting Market Growth

Yuri Arcurs, 2009/Used under license from Shutterstock.com

You are on the board of directors at Fullcourt Academic Press, a major textbook supplier to private schools, and TJM, the sales director of the high school division, has just burst into your office with data showing the number of private high school graduates each year over the past 14 years (Figure 13).[59]

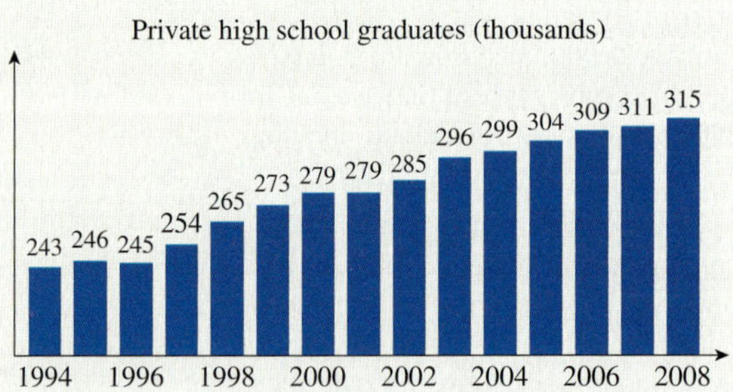

Figure 13

TJM is pleased that the figures appear to support a basic premise of his recent proposal for an expansion strategy: The number of high school seniors in private schools in the United States will be growing at a rate of at least 5,600 per year through the year 2015. TJM points out that the rate of increase predicted by the regression line (see Figure 14) based on the data since 1994 is around 5.6 thousand students per year, so it would not be overly optimistic to assume that the trend will continue—at least for the next 7 years.

[59] Source: National Center for Educational Statistics (http://nces.ed.gov/).

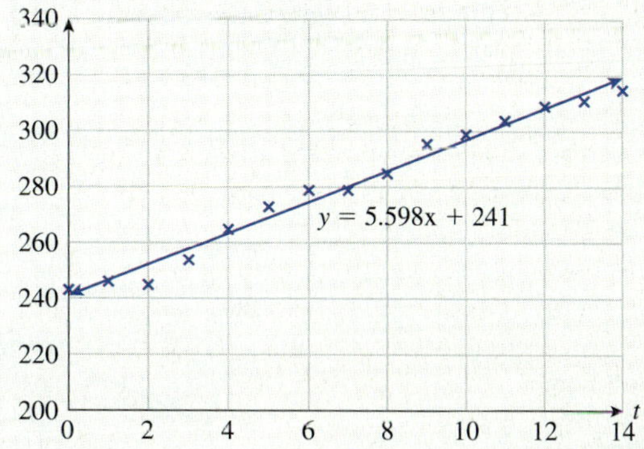

Figure 14

Although you are tempted to support TJM's proposal at the next board meeting, you would like to estimate first whether the 5,600 figure is a realistic expectation, especially because the graph suggests that the number of graduates began to "level off" (in the language of calculus, the *derivative appears to be decreasing*) toward the end of the period. Moreover, you recall reading somewhere that the numbers of students in the lower grades have also begun to level off, so it is safe to predict that the slowing of growth in the senior class will continue over the next few years. You really need precise data about numbers in the lower grades in order to make a meaningful prediction, but TJM's report is scheduled to be presented tomorrow and you would like a quick and easy way of "extending the curve to the right" by then.

It would certainly be helpful if you had a mathematical model of the data in Figure 13 that you could use to project the current trend. But what kind of model should you use? A linear model would be no good because it would not show any change in the derivative (the derivative of a linear function is constant). In addition, best-fit polynomial and exponential functions do not accurately reflect the leveling off, as you realize after trying to fit a few of them (Figure 15).

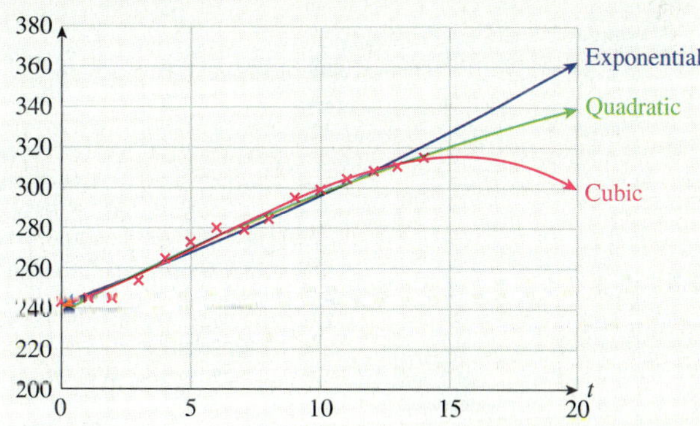

Figure 15

You then recall that a logistic curve can model the leveling-off property you desire, and so you try fitting a curve of the form

$$y = \frac{N}{1 + Ab^{-t}}.$$

Figure 16 shows the best-fit logistic curve, which eventually levels off at around $N = 390$.

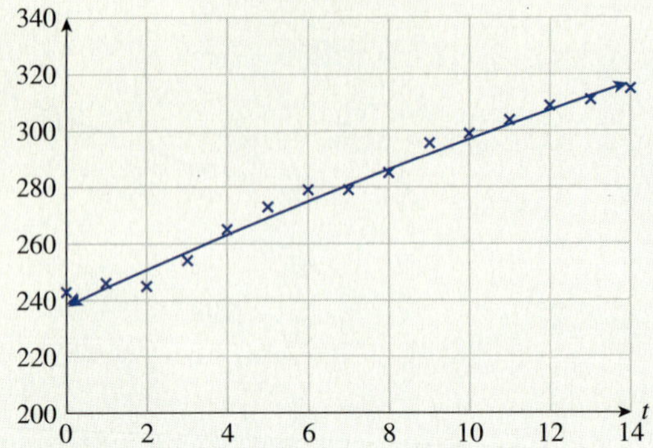

Figure 16

*** NOTE** There is another, more mathematical, reason for not using logistic regression to predict long-term leveling off: The regression value of the long-term level *N* is extremely sensitive to the values of the other coefficients. As a result, there can be good fits to the same set of data with a wide variety of values of *N*.

Web Site
www.AppliedCalc.org
Follow the path
Chapter 1
→ New Functions from Old: Scaled and Shifted Functions
to find a detailed treatment of scaled and shifted functions.

You are slightly troubled by the shape of the regression curve: It doesn't seem to "follow the s-shape" of the data very convincingly. Moreover, the curve doesn't appear to fit the data significantly more snugly than the quadratic or cubic models.* To reassure yourself, you decide to look for another kind of s-shaped model as a backup.

After flipping through a calculus book, you stumble across a function that is slightly more general than the one you have:

$$y = c + \frac{N}{1 + Ab^{-t}} \qquad \text{Shifted Logistic Curve}$$

The added term c has the effect of shifting the graph up c units. Turning once again to your calculus book (see the discussion of logistic regression in Section 2.4), you see that a best-fit curve is one that minimizes the sum-of-squares error. Here is an Excel spreadsheet showing the errors for $c = 200, N = 100, A = 3$, and $b = 1.2$.

	A	B	C	D	E	F	G	H
1	t (Year)	y (Observed)	y (Predicted)	residual^2	N	A	b	c
2	0	243	225	324	100	3	1.2	200
3	1	246	228.5714286	303.755102				
4	2	245	232.4324324	157.9437546	SSE:	11120.4856		
5	3	254	236.5482234	304.5645082				
6	4	265	240.8703879	582.2381806				
7	5	273	245.3384642	765.1605612				
8	6	279	249.8829265	847.8039683				
9	7	279	254.4293239	603.7181244				
10	8	285	258.9030791	681.0492791				
11	9	296	263.2343045	1073.590803				
12	10	299	267.3619884	1000.963778				
13	11	304	271.2370477	1073.411042				
14	12	309	274.8239786	1168.000439				
15	13	311	278.1011037	1082.337379				
16	14	315	281.0596311	1151.94864				

The formula for the function in cell C2 is

```
=$H$2+$E$2/(1+$F$2*$G$2^(-A2))
```

$$c + \frac{N}{1 + Ab^{-t}}$$

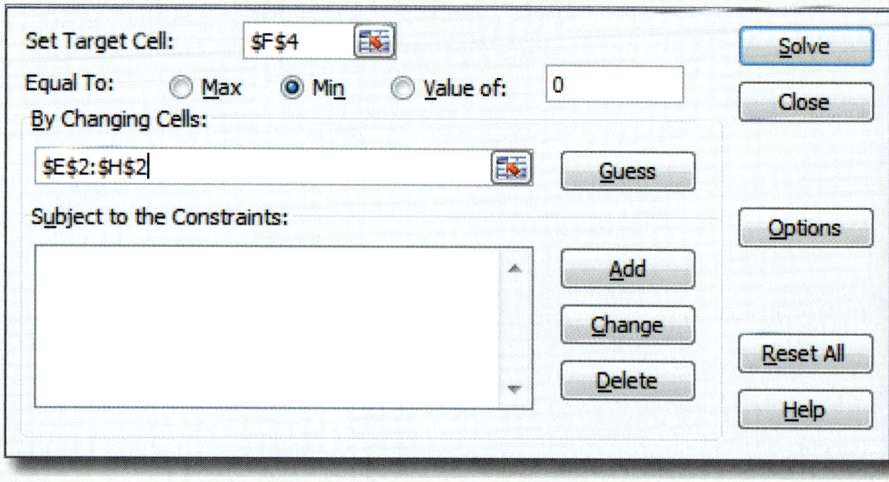

Figure 17

Figure 17 shows how to set up Solver to find the best values for N, A, b, and c for the setup used in this spreadsheet. The Target Cell, $F$4, contains the value of SSE, which is to be minimized. The Changing Cells are the cells containing the values of the constants N, A, b, and c that we want to change. That's it.

Now click "Solve." After thinking about it for a few seconds, Excel gives the optimal values of N, A, b, and c in cells E2–H2, and the minimum value of SSE in cell F4.* You find

*** NOTE** Depending on the settings in Solver, you may need to run the utility twice in succession to reach the minimum value of SSE.

$$N = 108.248027, \quad A = 3.85006915, \quad b = 1.27688839, \quad c = 218.351709$$

with SSE $\approx$ 101.2, which is a better fit than the unshifted logistic regression curve (SSE $\approx$ 129.8). Figure 18 shows that not only does this choice of model and constants give a good fit, but that the curve seems to follow the "s-shape" more convincingly than the unshifted logistic curve.

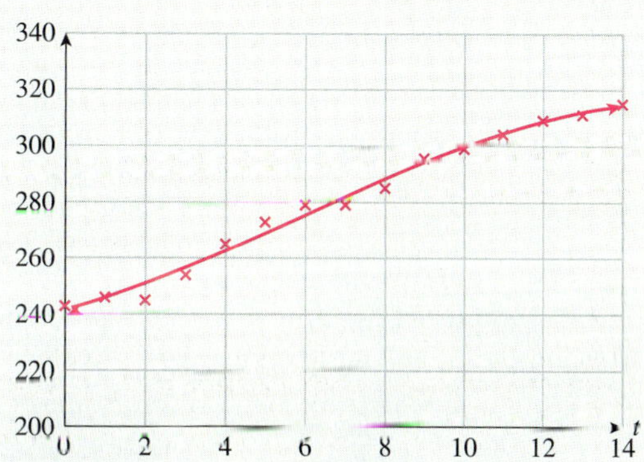

Figure 18

The derivative, dy/dt, will represent the rate of increase of high school graduates, which is exactly what you wish to estimate:

$$y = c + \frac{N}{1 + Ab^{-t}}$$

$$\frac{dy}{dt} = -\frac{N}{(1 + Ab^{-t})^2}\frac{d}{dt}\left[1 + Ab^{-t}\right]$$

$$= \frac{NAb^{-t}\ln b}{(1 + Ab^{-t})^2}$$

The rate of increase in the number of high school students in 2015 ($t = 21$) is given by

$$\left.\frac{dy}{dt}\right|_{t=21} = \frac{(108.248027)(3.85006915)1.27688839^{-21}\ln 1.27688839}{(1 + (3.85006915)1.27688839^{-21})^2}$$

$$\approx 0.5745 \text{ thousand students per year}$$

or 570 students per year—far less than the optimistic estimate of 5,600 in the proposal!

You now conclude that TJM's prediction is suspect and that further research will have to be done before the board can support the proposal.

Q: *How accurately does the model predict the number of high school graduates?*

A: Using a regression curve-fitting model to make long-term predictions is always risky. A more accurate model would have to take into account such factors as the birth rate and current school populations at all levels.

Q: *Which values of the constants should I use as starting values when using Excel to find the best-fit curve?*

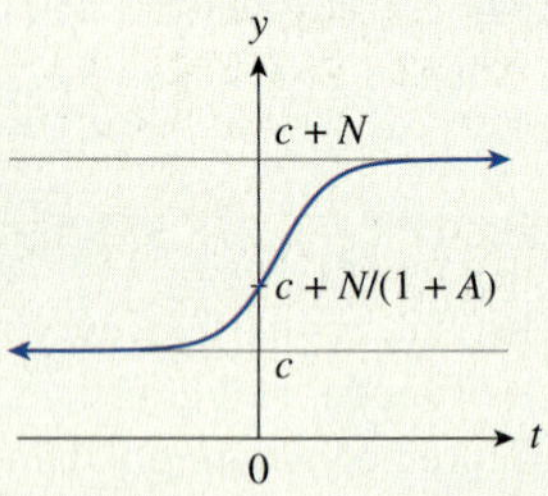

Figure 19

A: If the starting values of the constants are far from the optimal values, Solver may find a nonoptimal solution. Thus, you need to obtain some rough initial estimate of the constants by examining the graph. Figure 19 shows some important features of the curve that you can use to obtain estimates of N, A, b, and c by inspecting the graph.

From the graph, c is the lower asymptote (in a graph showing only the right-hand portion of the curve in Figure 19 we need to mentally extend the curve to the left), and N is the vertical distance between the upper and lower asymptotes. Once we have estimates for c and N, we can estimate A from the y-intercept m using

$$A = 1 - \frac{N}{m - c}. \qquad m = y\text{-intercept}$$

As in the discussion of logistic curves in Chapter 2, b is the base of exponential growth for values of t close to zero.

EXERCISES

1. In 1993 there were 247,000 private high school graduates. What does the regression model "predict" for 1993? (Round answer to the nearest 1,000.) What is the residual ($y_{\text{observed}} - y_{\text{predicted}}$)?
2. What is the long-term prediction of the model? (Round answer to the nearest 1,000.)
3. Find $\lim_{t\to\infty} \frac{dy}{dt}$ and interpret the result.
4. T You receive a memo to the effect that the 2007 and 2008 figures are not accurate. Use Excel Solver to re-estimate the best-fit constants N, a, b, and c in the absence of this data and obtain new estimates for the 2007 and 2008 data. What does the new model predict the rate of change in the number of high school seniors will be in 2015?

5. **Another Model** Using the original data, find the best-fit shifted *predator-prey satiation curve* of the form

$$f(t) = y = c + b\frac{a(t-m)}{1+a|t-m|}. \quad (a, b, c, m \text{ constant})$$

Its graph is shown below:

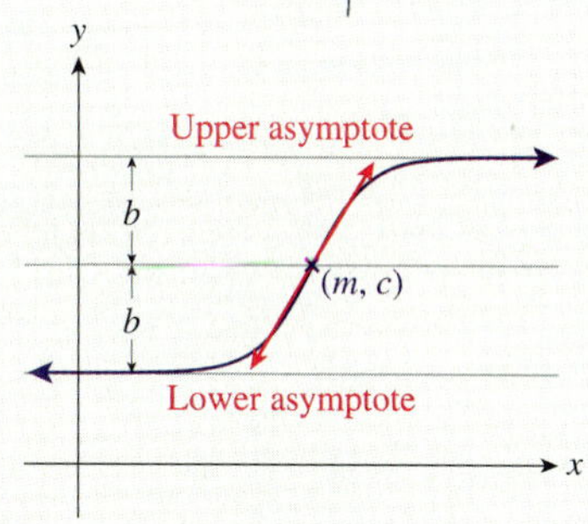

$$a = \frac{\text{Slope of tangent}}{b}$$

(Start with the following values: $a = 0.05$, $b = 160$, $c = 250$, $m = 5$. You might have to run Solver twice in succession to minimize SSE.) Graph the data together with the model. What is SSE? Is the model as accurate a fit as the model used in the text? What does this model predict will be the growth rate of the number of high school graduates in 2015? Comment on the answer. (Round the coefficients in the model and all answers to four decimal places.)

6. **Demand for Freon** The demand for chlorofluorocarbon-12 (CFC-12)—the ozone-depleting refrigerant commonly known as Freon[60]—has been declining significantly in response to regulation and concern about the ozone layer. The chart below shows the projected demand for CFC-12 for the period 1994–2005.[61]

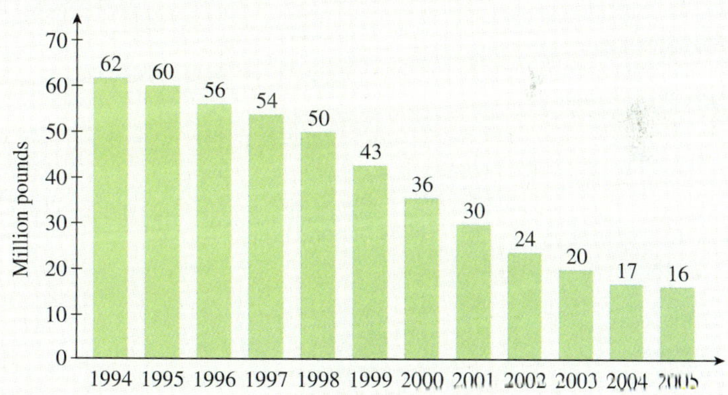

a. Use Excel Solver to obtain the best-fit equation of the form

$$f(t) = c + b\frac{a(t-m)}{1+a|t-m|}$$

where t = years since 1990. Use your function to estimate the total demand for CFC-12 from the start of the year 2000 to the start of 2010. (Start with the following values: $a = 1$, $b = -25$, $c = 35$, and $m = 10$, and round your answers to four decimal places.)

b. According to your model, how fast is the demand for Freon declining in 2000?

---

[60] The name given to it by Du Pont.

[61] Source: The Automobile Consulting Group (*New York Times*, December 26, 1993, p. F23). The exact figures were not given, and the chart is a reasonable facsimile of the chart that appeared in *New York Times*.

5 Further Applications of the Derivative

Case Study Production Lot Size Management

Your publishing company is planning the production of its latest best seller, which it predicts will sell 100,000 copies each month over the coming year. The book will be printed in several batches of the same number, evenly spaced throughout the year. Each print run has a setup cost of \$5,000, a single book costs \$1 to produce, and monthly storage costs for books awaiting shipment average 1¢ per book. **To meet the anticipated demand at minimum total cost to your company, how many printing runs should you plan?**

SUNNYphotography.com/Alamy

Web Site

At the Web site you will find:

- Section by section tutorials, including game tutorials with randomized quizzes
- A detailed chapter summary
- A true/false quiz
- Additional review exercises
- Graphers, Excel tutorials, and other resources
- The following extra topic:

  Linear Approximation and Error Estimation

Introduction

In this chapter we begin to see the power of calculus as an optimization tool. In Chapter 2 we saw how to price an item in order to get the largest revenue when the demand function is linear. Using calculus, we can handle nonlinear functions, which are much more general. In Section 5.1 we show how calculus can be used to solve the problem of finding the values of a variable that lead to a maximum or minimum value of a given function. In Section 5.2 we show how this helps us in various real-world applications.

Another theme in this chapter is that calculus can help us to draw and understand the graph of a function. By the time you have completed the material in Section 5.1, you will be able to locate and sketch some of the important features of a graph, such as where it rises and where it falls. In Section 5.3 we look at the *second derivative,* the derivative of the derivative function, and what it tells us about how the graph *curves*. In Section 5.4 we put a number of ideas together that help to explain what you see in a graph (drawn, for example, using graphing technology) and to locate its most important points.

We also include sections on related rates and elasticity of demand. The first of these (Section 5.5) examines further the concept of the derivative as a rate of change. The second (Section 5.6) returns to the problem of optimizing revenue based on the demand equation, looking at it in a new way that leads to an important idea in economics—elasticity.

algebra Review

For this chapter, you should be familiar with the algebra reviewed in **Chapter 0, sections 5 and 6**.

5.1 Maxima and Minima

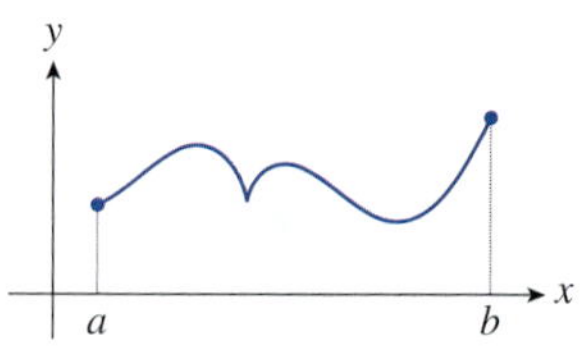

Figure 1

Figure 1 shows the graph of a function f whose domain is the closed interval $[a, b]$. A mathematician sees lots of interesting things going on here. There are hills and valleys, and even a small chasm (called a *cusp*) near the center. For many purposes, the important features of this curve are the highs and lows. Suppose, for example, you know that the price of the stock of a certain company will follow this graph during the course of a week. Although you would certainly make a handsome profit if you bought at time a and sold at time b, your best strategy would be to follow the old adage to "buy low and sell high," buying at all the lows and selling at all the highs.

Figure 2 shows the graph once again with the highs and lows marked. Mathematicians have names for these points: the highs (at the x-values p, r, and b) are referred to as **relative maxima**, and the lows (at the x-values a, q, and s) are referred to as **relative minima**. Collectively, these highs and lows are referred to as **relative extrema**. (A point of language: The singular forms of the plurals *minima, maxima,* and *extrema* are *minimum, maximum,* and *extremum.*)

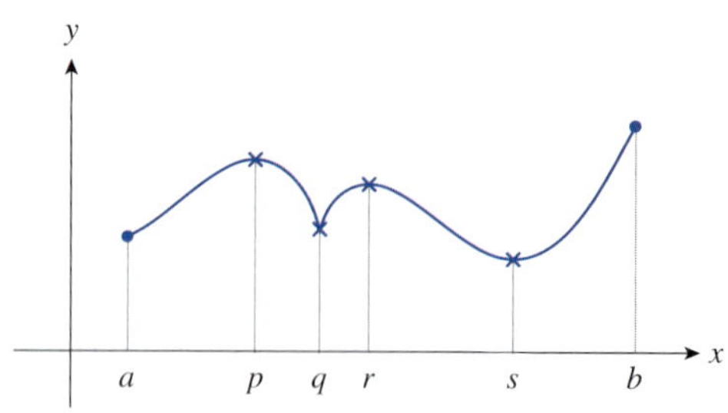

Figure 2

Why do we refer to these points as relative extrema? Take a look at the point corresponding to $x = r$. It is the highest point of the graph *compared to other points nearby.*

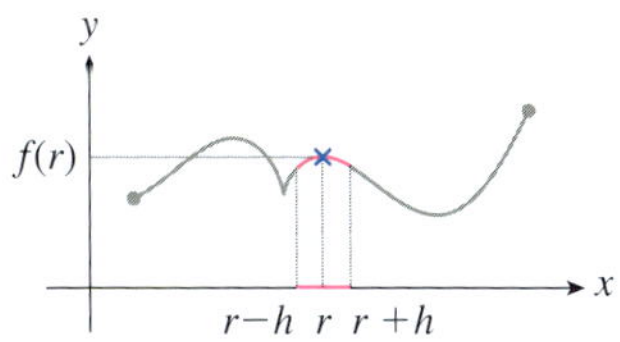

Figure 3

If you were an extremely nearsighted mountaineer standing at the point where $x = r$, you would *think* that you were at the highest point of the graph, not being able to see the distant peaks at $x = p$ and $x = b$.

Let's translate into mathematical terms. We are talking about the heights of various points on the curve. The height of the curve at $x = r$ is $f(r)$, so we are saying that $f(r)$ is greater than or equal to $f(x)$ for every x near r. In other words, *$f(r)$ is the greatest value that $f(x)$ has for all choices of x between $r - h$ and $r + h$* for some (possibly small) h. (See Figure 3.)

We can phrase the formal definition as follows.

Relative Extrema

f has a **relative maximum** at $x = r$ if there is some interval $(r - h, r + h)$ (even a very small one) for which $f(r) \geq f(x)$ for all x in $(r - h, r + h)$ for which $f(x)$ is defined.

f has a **relative minimum** at $x = r$ if there is some interval $(r - h, r + h)$ (even a very small one) for which $f(r) \leq f(x)$ for all x in $(r - h, r + h)$ for which $f(x)$ is defined.

Quick Examples

In Figure 2, f has the following relative extrema:

1. Relative maxima at p and r.
2. A relative maximum at b. (See Figure 4.) Note that $f(x)$ is not defined for $x > b$. However, $f(b) \geq f(x)$ for every x in the interval $(b - h, b + h)$ *for which $f(x)$ is defined*—that is, for every x in $(b - h, b]$.
3. Relative minima at a, q, and s.

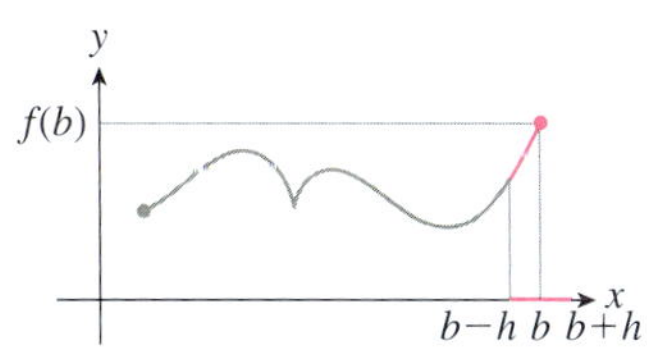

Figure 4

Note Our definition of relative extremum allows f to have a relative extremum at an endpoint of its domain; the definitions used in some books do not. In view of examples like the stock-market investing strategy mentioned above, we find it more useful to allow endpoints as relative extrema. ■

Looking carefully at Figure 2, we can see that the lowest point on the whole graph is where $x = s$ and the highest point is where $x = b$. This means that $f(s)$ is the least value of f on the whole domain of f (the interval $[a, b]$) and $f(b)$ is the greatest value. We call these the *absolute* minimum and maximum.

Absolute Extrema

f has an **absolute maximum** at r if $f(r) \geq f(x)$ for every x in the domain of f.

f has an **absolute minimum** at r if $f(r) \leq f(x)$ for every x in the domain of f.

Quick Examples

1. In Figure 2, f has an absolute maximum at b and an absolute minimum at s.
2. If $f(x) = x^2$ then $f(x) \geq f(0)$ for every real number x. Therefore, $f(x) = x^2$ has an absolute minimum at $x = 0$. (See the figure.)

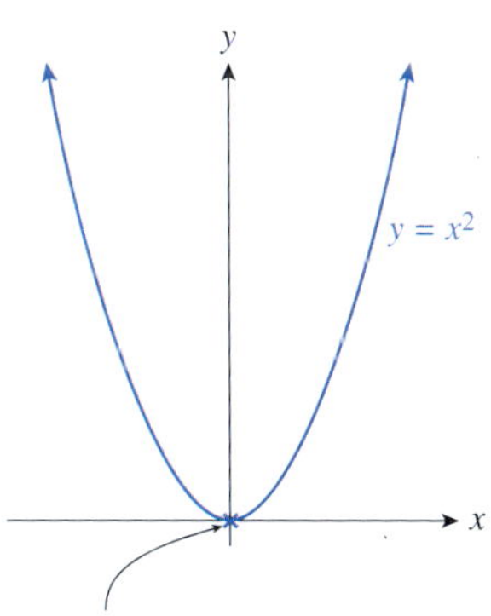

Absolute minimum at $x = 0$

> **3.** Generalizing (2), every quadratic function $f(x) = ax^2 + bx + c$ has an absolute extremum at its vertex $x = -b/(2a)$, an absolute minimum if $a > 0$, and an absolute maximum if $a < 0$.

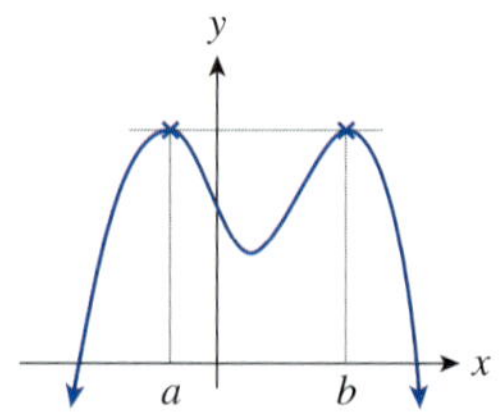

Absolute maxima at $x = a$ and $x = b$

Figure 5

Some graphs have no absolute extrema at all (think of the graph of $y = x$), while others might have an absolute maximum but no absolute minimum (like $y = x^2$), or vice versa. When f does have an absolute maximum, there is only one absolute maximum *value* of f, but this value may occur at different values of x. (See Figure 5.)

Q: *At how many different values of x can f take on its absolute maximum value?*

A: An extreme case is that of a constant function; because we use $\geq$ in the definition of absolute maximum, a constant function has an absolute maximum (and minimum) at every point in its domain.

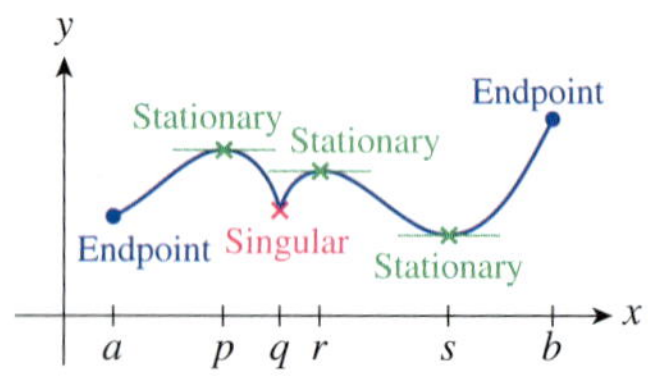

Figure 6

Now, how do we go about locating extrema? In many cases we can get a good idea by using graphing technology to zoom in on a maximum or minimum and approximate its coordinates. However, calculus gives us a way to find the exact locations of the extrema and at the same time to understand why the graph of a function behaves the way it does. In fact, it is often best to combine the powers of graphing technology with those of calculus, as we shall see.

In Figure 6 we see the graph from Figure 1 once more, but we have labeled each extreme point as one of three types. At the points labeled "Stationary," the tangent lines to the graph are horizontal, and so have slope 0, so f' (which gives the slope) is 0. Any time $f'(x) = 0$, we say that f has a **stationary point** at x because the rate of change of f is zero there. We call an extremum that occurs at a stationary point a **stationary extremum**. In general, to find the exact location of each stationary point, we need to solve the equation $f'(x) = 0$.

There is a relative minimum in Figure 6 at $x = q$, but there is no horizontal tangent there. In fact, there is no tangent line at all; $f'(q)$ is not defined. (Recall a similar situation with the graph of $f(x) = |x|$ at $x = 0$.) When $f'(x)$ does not exist for some x in the domain of f, we say that f has a **singular point** at x. We shall call an extremum that occurs at a singular point a **singular extremum**. The points that are either stationary or singular we call collectively the **critical points** of f.

The remaining two extrema are at the **endpoints** of the domain.* As we see in the figure, they are (almost) always either relative maxima or relative minima.

*** NOTE** Remember that we do allow relative extrema at endpoints.

We bring all the above information together in Figure 7:

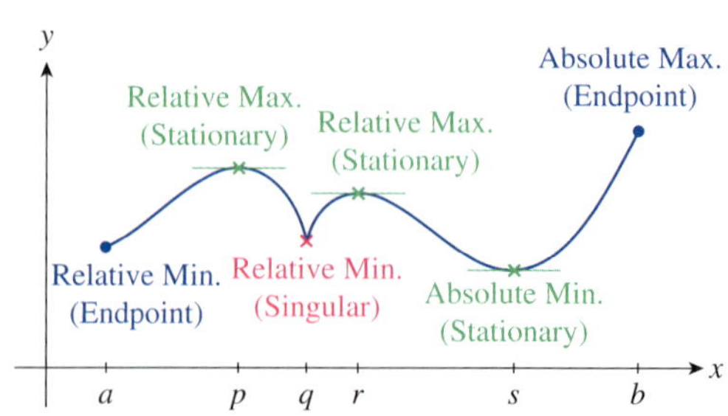

Figure 7

Q: *Are there any other types of relative extrema?*

A: No; relative extrema of a function always occur at critical points or endpoints. (A rigorous proof is beyond the scope of this book.)*

* **NOTE** Here is an outline of the argument. Suppose f has a relative maximum, say, at $x = a$, at a point other than an endpoint of the domain. Then either f is differentiable there, or it is not. If it is not, then we have a singular point. If f is differentiable at $x = a$, then consider the slope of the secant line through the points where $x = a$ and $x = a + h$ for small positive h. Because f has a relative maximum at $x = a$, it is falling (or level) to the right of $x = a$, and so the slope of this secant line must be ≤ 0. Thus, we must have $f'(a) \leq 0$ in the limit as $h \to 0$. On the other hand, if h is small and *negative,* then the corresponding secant line must have slope ≥ 0 because f is also falling (or level) as we move left from $x = a$, and so $f'(a) \geq 0$. Because $f'(a)$ is both ≥ 0 and ≤ 0, it must be zero, and so we have a stationary point at $x = a$.

Locating Candidates for Relative Extrema

If f is a real valued function, then its relative extrema occur among the following types of points:

1. **Stationary Points:** f has a stationary point at x if x is in the domain and $f'(x) = 0$. To locate stationary points, set $f'(x) = 0$ and solve for x.
2. **Singular Points:** f has a singular point at x if x is in the domain and $f'(x)$ is not defined. To locate singular points, find values of x where $f'(x)$ is *not* defined, but $f(x)$ *is* defined.
3. **Endpoints:** The x-coordinates of endpoints are endpoints of the domain, if any. Recall that closed intervals contain endpoints, but open intervals do not.

Once we have the x-coordinates of a candidate for a relative extremum, we find the corresponding y-coordinate using $y = f(x)$.

Quick Examples

1. **Stationary Points:** Let $f(x) = x^3 - 12x$. Then to locate the stationary points, set $f'(x) = 0$ and solve for x. This gives $3x^2 - 12 = 0$, so f has stationary points at $x = \pm 2$. Thus, the stationary points are $(-2, f(-2)) = (-2, 16)$ and $(2, f(2)) = (2, -16)$.
2. **Singular points:** Let $f(x) = 3(x-1)^{1/3}$. Then $f'(x) = (x-1)^{-2/3} = 1/(x-1)^{2/3}$. $f'(1)$ is not defined, although $f(1)$ *is* defined. Thus, the (only) singular point occurs at $x = 1$. Its coordinates are $(1, f(1)) = (1, 0)$.
3. **Endpoints:** Let $f(x) = 1/x$, with domain $(-\infty, 0) \cup [1, +\infty)$. Then the only endpoint in the domain of f occurs when $x = 1$ and has coordinates $(1, 1)$. The natural domain of $1/x$ has no endpoints.

Remember, though, that these are only *candidates* for relative extrema. It is quite possible, as we shall see, to have a stationary point or a singular point that is neither a relative maximum nor a relative minimum. (It is also possible for an endpoint to be neither a maximum nor a minimum, but only in functions whose graphs are rather bizarre—see Exercise 65.)

Now let's look at some examples of finding maxima and minima. In all of these examples, we will use the following procedure: First, we find the derivative, which we examine to find the stationary points and singular points. Next, we make a table listing the x-coordinates of the critical points and endpoints, together with their y-coordinates. We use this table to make a rough sketch of the graph. From the table and rough sketch, we usually have enough data to be able to say where the extreme points are and what kind they are.

EXAMPLE 1 Maxima and Minima

Find the relative and absolute maxima and minima of

$$f(x) = x^2 - 2x$$

on the interval [0, 4].

Solution We first calculate $f'(x) = 2x - 2$. We use this derivative to locate the critical points (stationary and singular points).

Stationary Points To locate the stationary points, we solve the equation $f'(x) = 0$, or

$$2x - 2 = 0,$$

getting $x = 1$. The domain of the function is [0, 4], so $x = 1$ is in the domain. Thus, the only candidate for a stationary relative extremum occurs when $x = 1$.

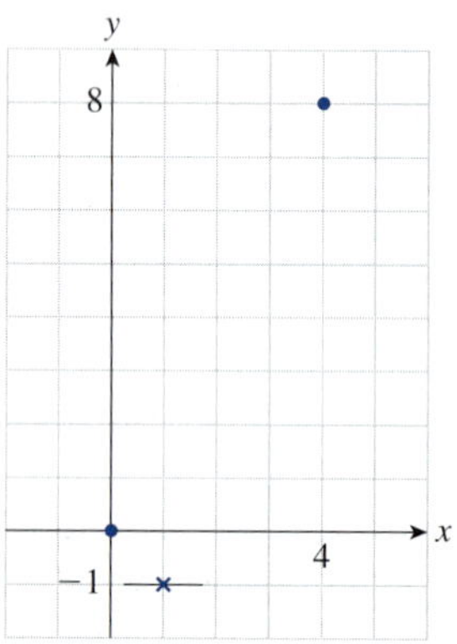

Figure 8

Singular Points We look for points where the derivative is not defined. However, the derivative is $2x - 2$, which is defined for every x. Thus, there are no singular points and hence no candidates for singular relative extrema.

Endpoints The domain is [0, 4], so the endpoints occur when $x = 0$ and $x = 4$.

We record these values of x in a table, together with the corresponding y-coordinates (values of f):

x	0	1	4
$f(x) = x^2 - 2x$	0	−1	8

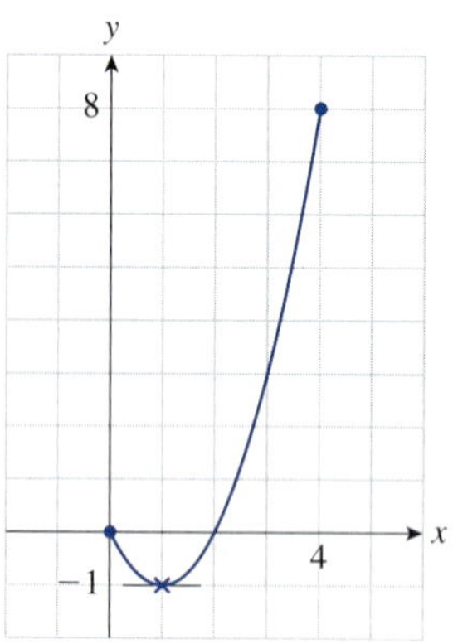

Figure 9

This gives us three points on the graph, (0, 0), (1, −1), and (4, 8), which we plot in Figure 8. We remind ourselves that the point (1, −1) is a stationary point of the graph by drawing in a part of the horizontal tangent line. Connecting these points must give us a graph something like that in Figure 9. Notice that the graph has a horizontal tangent line at $x = 1$ but not at either of the endpoints because the endpoints are not stationary points.

From Figure 9 we can see that f has the following extrema:

x	$y = x^2 - 2x$	*Classification*
0	0	Relative maximum (endpoint)
1	−1	Absolute minimum (stationary point)
4	8	Absolute maximum (endpoint)

➡ **Before we go on...** A little terminology: If the point (a, b) on the graph of f represents a maximum (or minimum) of f, we sometimes say that **f has a maximum (or minimum) value of b at $x = a$**. Thus, in the above example, we could have said the following:

- f has a relative maximum value of 0 at $x = 0$.
- f has an absolute minimum value of −1 at $x = 1$.
- f has an absolute maximum value of 8 at $x = 4$.

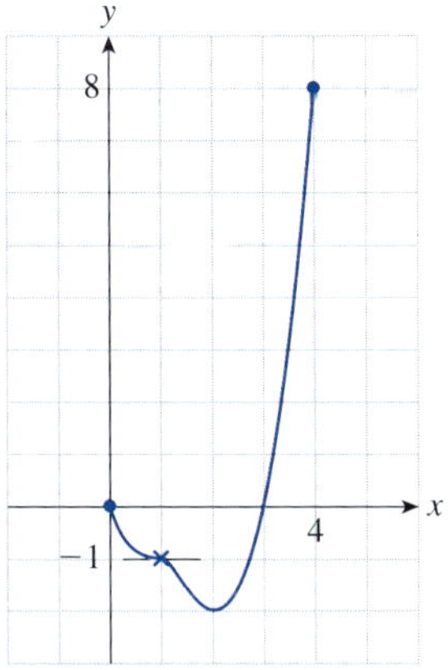

Figure 10

Q: *How can we be sure that the graph in Example 1 doesn't look like Figure 10?*

A: If it did, there would be another critical point somewhere between $x = 1$ and $x = 4$. But we already know that there aren't any other critical points. The table we made listed all of the possible extrema; there can be no more.

First Derivative Test

The **first derivative test*** gives another, very systematic, way of checking whether a critical point is a relative maximum or minimum. To motivate the first derivative test, consider again the critical point $x = 1$ in Example 1. If we look at some values of $f'(x)$ to the left and right of the critical point, we obtain the information shown in the following table:

* **NOTE** Why "first" derivative test? To distinguish it from a test based on the **second derivative** of a function, which we shall discuss in Section 5.3.

	Point to the Left	Critical Point	Point to the Right
x	0	1	2
$f'(x) = 2x - 2$	−2	0	2
Direction of Graph	↘	→	↗

At $x = 0$ (to the left of the critical point) we see that $f'(0) = -2 < 0$, so the graph has negative slope and f is decreasing. We note this with the downward pointing arrow. At $x = 2$ (to the right of the critical point), we find $f'(2) = 2 > 0$, so the graph has positive slope and f is increasing. In fact, because $f'(x) = 0$ only at $x = 1$, we know that $f'(x) < 0$ for all x in $[0, 1)$, and we can say that f is decreasing on the interval $[0, 1]$. Similarly, f is increasing on $[1, 4]$.

So, starting at $x = 0$, the graph of f goes down until we reach $x = 1$ and then it goes back up, telling us that $x = 1$ must be a relative minimum. Notice how the relative minimum is suggested by the arrows to the left and right.

First Derivative Test for Relative Extrema

Suppose that c is a critical point of the continuous function f, and that its derivative is defined for x close to, and on both sides of, $x = c$. Then, determine the sign of the derivative to the left and right of $x = c$.

1. If $f'(x)$ is positive to the left of $x = c$ and negative to the right, then f has a relative maximum at $x = c$.
2. If $f'(x)$ is negative to the left of $x = c$ and positive to the right, then f has a relative minimum at $x = c$.
3. If $f'(x)$ has the same sign on both sides of $x = c$, then f has neither a relative maximum nor a relative minimum at $x = c$.

Quick Examples

1. In Example 1 above, we saw that $f(x) = x^2 - 2x$ has a critical point at $x = 1$ with $f'(x)$ negative to the left of $x = 1$ and positive to the right (see the table). Therefore, f has a relative minimum at $x = 1$.

2. Here is a graph showing a function f with a singular point at $x = 1$:

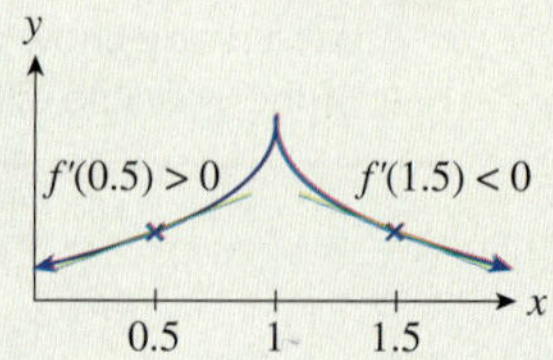

The graph gives us the information shown in the table:

	Point to the Left	Critical Point	Point to the Right
x	0.5	1	1.5
$f'(x)$	$+$	Undefined	$-$
Direction of Graph	↗		↘

Since $f'(x)$ is positive to the left of $x = 1$ and negative to the right, we see that f has a relative maximum at $x = 1$. (Notice again how this is suggested by the direction of the arrows.)

EXAMPLE 2 Unbounded Interval

Find all extrema of $f(x) = 3x^4 - 4x^3$ on $[-1, \infty)$.

Solution We first calculate $f'(x) = 12x^3 - 12x^2$.

Stationary points We solve the equation $f'(x) = 0$, which is

$$12x^3 - 12x^2 = 0 \text{ or}$$
$$12x^2(x - 1) = 0.$$

There are two solutions, $x = 0$ and $x = 1$, and both are in the domain. These are our candidates for the x-coordinates of stationary relative extrema.

Singular points There are no points where $f'(x)$ is not defined, so there are no singular points.

Endpoints The domain is $[-1, \infty)$, so there is one endpoint, at $x = -1$.

We record these points in a table with the corresponding y-coordinates:

x	-1	0	1
$f(x) = 3x^4 - 4x^3$	7	0	-1

We will illustrate three methods we can use to determine which are minima, which are maxima, and which are neither:

1. Plot these points and sketch the graph by hand.

2. Use the First Derivative Test.

3. Use technology to help us.

Use the method you find most convenient.

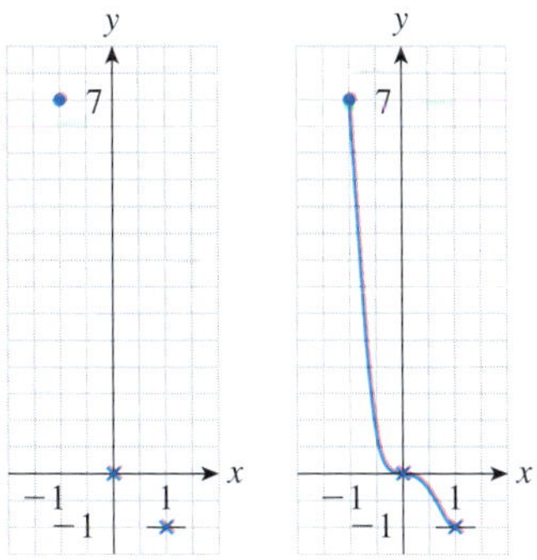

(a) (b)

Figure 11

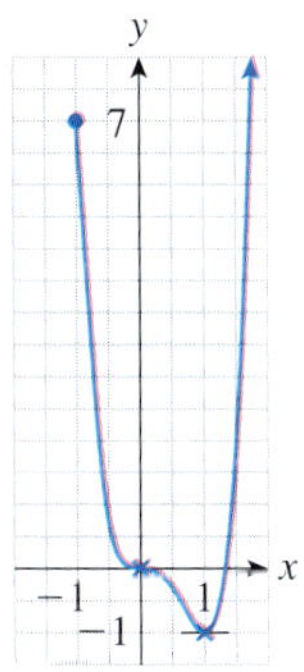

Figure 12

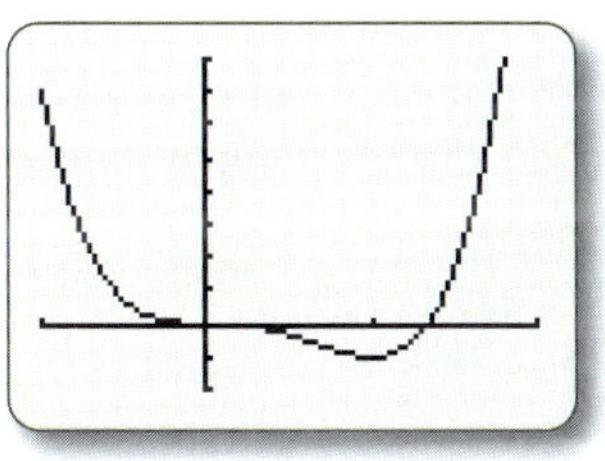
Figure 13

Using a Hand Plot: If we plot these points by hand, we obtain Figure 11(a), which suggests Figure 11(b).

We can't be sure what happens to the right of $x = 1$. Does the curve go up, or does it go down? To find out, let's plot a "test point" to the right of $x = 1$. Choosing $x = 2$, we obtain $y = 3(2)^4 - 4(2)^3 = 16$, so (2, 16) is another point on the graph. Thus, it must turn upward to the right of $x = 1$, as shown in Figure 12.

From the graph, we find that f has the following extrema:

A relative (endpoint) maximum at $(-1, 7)$

An absolute (stationary) minimum at $(1, -1)$

Using the First Derivative Test: List the critical and endpoints in a table, and add additional points as necessary so that each critical point has a noncritical point on either side. Then compute the derivative at each of these points, and draw an arrow to indicate the direction of the graph.

	End Point	Critical Point		Critical Point	
x	−1	0	0.5	1	2
$f'(x) = 12x^3 - 12x^2$	−24	0	−1.5	0	48
Direction of Graph	↘	→	↘	→	↗

Notice that the arrows now suggest the shape of the curve in Figure 12. The first derivative test tells us that the function has a relative maximum at $x = -1$, neither a maximum nor a minimum at $x = 0$, and a relative minimum at $x = 1$. Deciding which of these extrema are absolute and which are relative requires us to compute y-coordinates and plot the corresponding points on the graph by hand, as we did in the first method.

using Technology

If we use technology to show the graph, we should choose the viewing window so that it contains the three interesting points we found: $x = -1$, $x = 0$, and $x = 1$. Again, we can't be sure yet what happens to the right of $x = 1$; does the graph go up or down from that point? If we set the viewing window to an interval of $[-1, 2]$ for x and $[-2, 8]$ for y, we will leave enough room to the right of $x = 1$ and below $y = -1$ to see what the graph will do. The result will be something like Figure 13.

Now we can tell what happens to the right of $x = 1$: the function increases. We know that it cannot later decrease again because if it did, there would have to be another critical point where it turns around, and we found that there are no other critical points. ■

➡ **Before we go on...** Notice that the stationary point at $x = 0$ in Example 2 is neither a relative maximum nor a relative minimum. It is simply a place where the graph of f flattens out for a moment before it continues to fall. Notice also that f has no absolute maximum because $f(x)$ increases without bound as x gets large. ■

EXAMPLE 3 Singular Point

Find all extrema of $f(t) = t^{2/3}$ on $[-1, 1]$.

Solution First, $f'(t) = \frac{2}{3}t^{-1/3}$.

Stationary points We need to solve

$$\frac{2}{3}t^{-1/3} = 0.$$

We can rewrite this equation without the negative exponent:

$$\frac{2}{3t^{1/3}} = 0.$$

Now, the only way that a fraction can equal 0 is if the numerator is 0, so this fraction can never equal 0. Thus, there are no stationary points.

Singular points The derivative

$$f'(t) = \frac{2}{3t^{1/3}}$$

is not defined for $t = 0$. However, f itself *is* defined at $t = 0$, so 0 is in the domain. Thus, f has a singular point at $t = 0$.

Endpoints There are two endpoints, -1 and 1.

We now put these three points in a table with the corresponding y-coordinates:

t	-1	0	1
$f(t)$	1	0	1

Using a Hand Plot: The derivative, $f'(t) = 2/(3t^{1/3})$, is not defined at the singular point $t = 0$. To help us sketch the graph, let's use limits to investigate what happens to the derivative as we approach 0 from either side:

$$\lim_{t\to 0^-} f'(t) = \lim_{t\to 0^-} \frac{2}{3t^{1/3}} = -\infty$$

$$\lim_{t\to 0^+} f'(t) = \lim_{t\to 0^+} \frac{2}{3t^{1/3}} = +\infty.$$

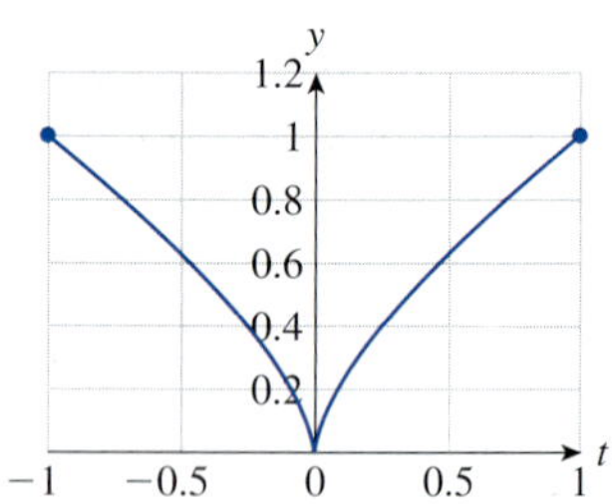

Figure 14

Thus, the graph decreases very steeply, approaching $t = 0$ from the left, and then rises very steeply as it leaves to the right. It would make sense to say that the tangent line at $x = 0$ is vertical, as seen in Figure 14.

From this graph, we find the following extrema for f:

An absolute (endpoint) maximum at $(-1, 1)$

An absolute (singular) minimum at $(0, 0)$

An absolute (endpoint) maximum at $(1, 1)$.

Notice that the absolute maximum value of f is achieved at two values of t: $t = -1$ and $t = 1$.

First Derivative Test: Here is the corresponding table for the first derivative test.

t	-1	0	1
$f'(t) = \dfrac{2}{3t^{1/3}}$	$-\dfrac{2}{3}$	Undefined	$\dfrac{2}{3}$
Direction of Graph	↘	↕	↗

(We drew a vertical arrow at $t = 0$ to indicate a vertical tangent.) Again, notice how the arrows suggest the shape of the curve in Figure 14, and the first derivative test confirms that we have a relative minimum at $x = 0$.

using Technology

Because there is only one critical point, at $t = 0$, it is clear from this table that f must decrease from $t = -1$ to $t = 0$ and then increase from $t = 0$ to $t = 1$. To graph f using technology, choose a viewing window with an interval of $[-1, 1]$ for t and $[0, 1]$ for y. The result will be something like Figure 14.* ■

* **NOTE** Many graphing calculators will give you only the right-hand half of the graph shown in Figure 14 because fractional powers of negative numbers are not, in general, real numbers. To obtain the whole curve, enter the formula as `Y=(x^2)^(1/3)`, a fractional power of the non-negative function x^2.

In Examples 1 and 3, we could have found the absolute maxima and minima without doing any graphing. In Example 1, after finding the critical points and endpoints, we created the following table:

x	0	1	4
$f(x)$	0	-1	8

From this table we can see that f must decrease from its value of 0 at $x = 0$ to -1 at $x = 1$, and then increase to 8 at $x = 4$. The value of 8 must be the largest value it takes on, and the value of -1 must be the smallest, on the interval $[0, 4]$. Similarly, in Example 3 we created the following table:

t	-1	0	1
$f(t)$	1	0	1

From this table we can see that the largest value of f on the interval $[-1, 1]$ is 1 and the smallest value is 0. We are taking advantage of the following fact, the proof of which uses some deep and beautiful mathematics (alas, beyond the scope of this book):

Extreme Value Theorem

If f is *continuous* on a *closed interval* $[a, b]$, then it will have an absolute maximum and an absolute minimum value on that interval. Each absolute extremum must occur at either an endpoint or a critical point. Therefore, the absolute maximum is the largest value in a table of the values of f at the endpoints and critical points, and the absolute minimum is the smallest value.

Quick Example

The function $f(x) = 3x - x^3$ on the interval $[0, 2]$ has one critical point at $x = 1$. The values of f at the critical point and the endpoints of the interval are given in the following table:

	Endpoint	Critical point	Endpoint
x	0	1	2
$f(x)$	0	2	-2

From this table we can say that the absolute maximum value of f on $[0, 2]$ is 2, which occurs at $x = 1$, and the absolute minimum value of f is -2, which occurs at $x = 2$.

As we can see in Example 2 and the following examples, if the domain is not a closed interval then f may not have an absolute maximum and minimum, and a table of values as above is of little help in determining whether it does.

EXAMPLE 4 Domain Not a Closed Interval

Find all extrema of $f(x) = x + \frac{1}{x}$.

Solution Because no domain is specified, we take the domain to be as large as possible. The function is not defined at $x = 0$ but is at all other points, so we take its domain to be $(-\infty, 0) \cup (0, +\infty)$. We calculate

$$f'(x) = 1 - \frac{1}{x^2}.$$

Stationary Points Setting $f'(x) = 0$, we solve

$$1 - \frac{1}{x^2} = 0$$

to find $x = \pm 1$. Calculating the corresponding values of f, we get the two stationary points $(1, 2)$ and $(-1, -2)$.

Singular Points The only value of x for which $f'(x)$ is not defined is $x = 0$, but then f is not defined there either, so there are no singular points in the domain.

Endpoints The domain, $(-\infty, 0) \cup (0, +\infty)$, has no endpoints.

From this scant information, it is hard to tell what f does. If we are sketching the graph by hand, or using the first derivative test, we will need to plot additional "test points" to the left and right of the stationary points $x = \pm 1$.

using Technology

For the technology approach, let's choose a viewing window with an interval of $[-3, 3]$ for x and $[-4, 4]$ for y, which should leave plenty of room to see how f behaves near the stationary points. The result is something like Figure 15.

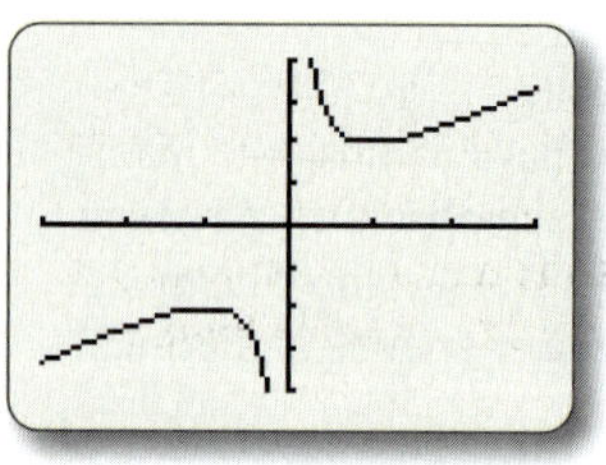

Figure 15

From this graph we can see that f has:

A relative (stationary) maximum at $(-1, -2)$

A relative (stationary) minimum at $(1, 2)$

Curiously, the relative maximum is lower than the relative minimum! Notice also that, because of the break in the graph at $x = 0$, the graph did not need to rise to get from $(-1, -2)$ to $(1, 2)$. ■

So far we have been solving the equation $f'(x) = 0$ to obtain our candidates for stationary extrema. However, it is often not easy—or even possible—to solve equations analytically. In the next example, we show a way around this problem by using graphing technology.

EXAMPLE 5 T Finding Approximate Extrema Using Technology

Graph the function $f(x) = (x - 1)^{2/3} - \frac{x^2}{2}$ with domain $[-2, +\infty)$. Also graph its derivative and hence locate and classify all extrema of f, with coordinates accurate to two decimal places.

Solution In Example 4 of Section 3.5, we saw how to draw the graphs of f and f' using technology. Note that the technology formula to use for the graph of f is

```
((x-1)^2)^(1/3)-0.5*x^2
```

instead of

```
(x-1)^(2/3)-0.5*x^2.
```

(Why?)

Figure 16 shows the resulting graphs of f and f'.

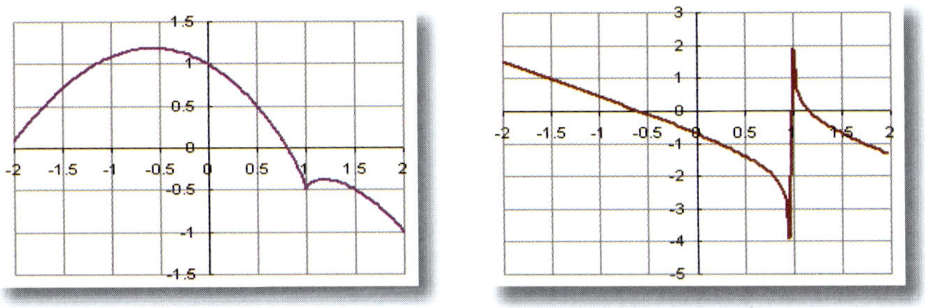

Graph of f — Graph of f'

Figure 16 **(a)** **(b)**

If we extend Xmax beyond $x = 2$, we find that the graph continues downward, apparently without any further interesting behavior.

Stationary Points The graph of f shows two stationary points, both maxima, at around $x = -0.6$ and $x = 1.2$. Notice that the graph of f' is zero at precisely these points. Moreover, it is easier to locate these values accurately on the graph of f' because it is easier to pinpoint where a graph crosses the x-axis than to locate a stationary point. Zooming in to the stationary point at $x \approx -0.6$ results in Figure 17.

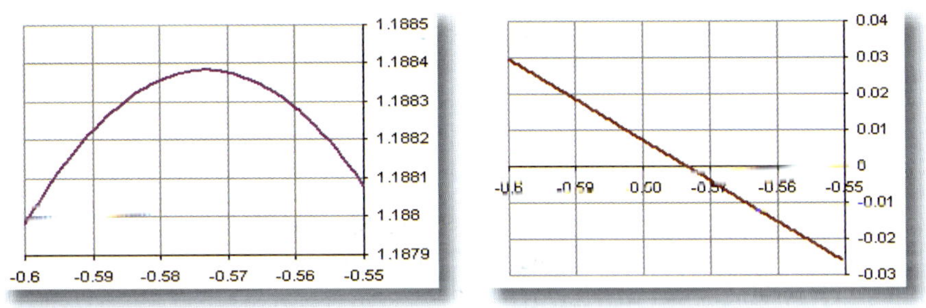

Graph of f — Graph of f'

Figure 17 **(a)** **(b)**

From the graph of f, we can see that the stationary point is somewhere between -0.58 and -0.57. The graph of f' shows more clearly that the zero of f', hence the stationary point of f lies somewhat closer to -0.57 than to -0.58. Thus, the stationary point occurs at $x \approx -0.57$, rounded to two decimal places.

In a similar way, we find the second stationary point at $x \approx 1.18$.

Singular Points Going back to Figure 16, we notice what appears to be a cusp (singular point) at the relative minimum around $x = 1$, and this is confirmed by a glance at the graph of f', which seems to take a sudden jump at that value. Zooming in closer suggests that the singular point occurs at exactly $x = 1$. In fact, we can calculate

$$f'(x) = \frac{2}{3(x-1)^{1/3}} - x.$$

From this formula we see clearly that $f'(x)$ is defined everywhere except at $x = 1$.

Endpoints The only endpoint in the domain is $x = -2$, which gives a relative minimum.

Thus, we have found the following approximate extrema for f:

A relative (endpoint) minimum at $(-2, 0.08)$

An absolute (stationary) maximum at $(-0.57, 1.19)$

A relative (singular) minimum at $(1, -0.5)$

A relative (stationary) maximum at $(1.18, -0.38)$.

5.1 EXERCISES

▼ more advanced ◆ challenging

T indicates exercises that should be solved using technology

In Exercises 1–12, locate and classify all extrema in each graph. (By classifying the extrema, we mean listing whether each extremum is a relative or absolute maximum or minimum.) Also, locate any stationary points or singular points that are not relative extrema. HINT [See Figure 7.]

1.

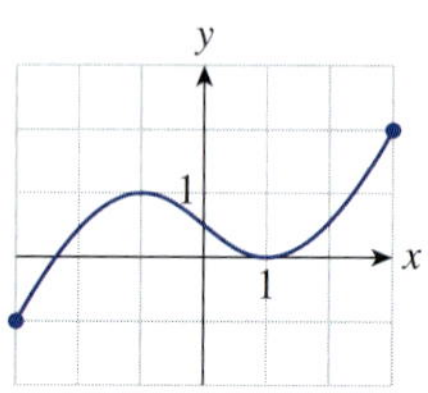

2.

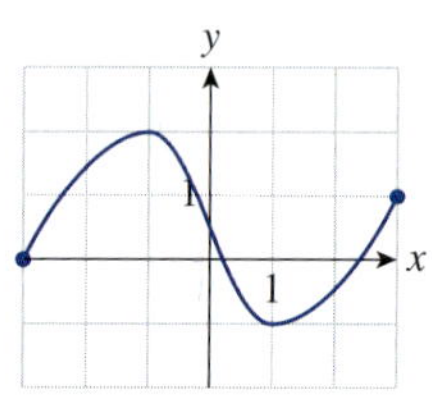

3.

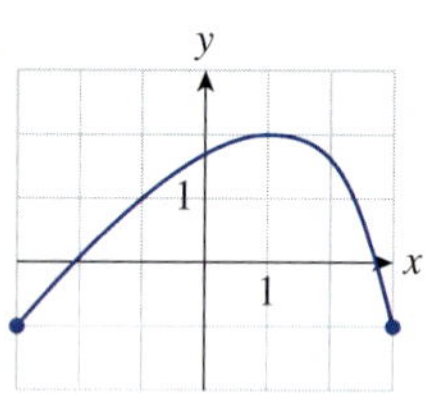

4.

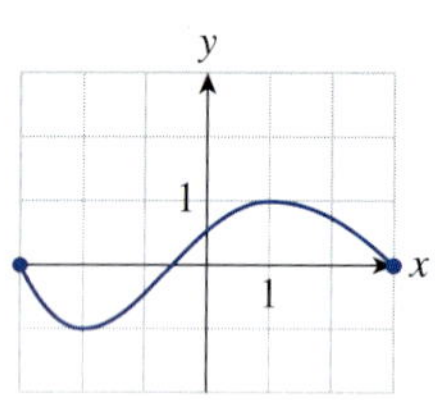

5.

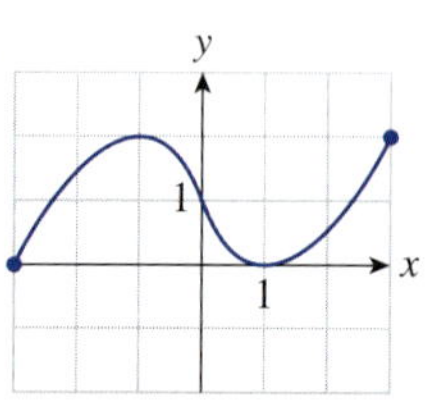

6.

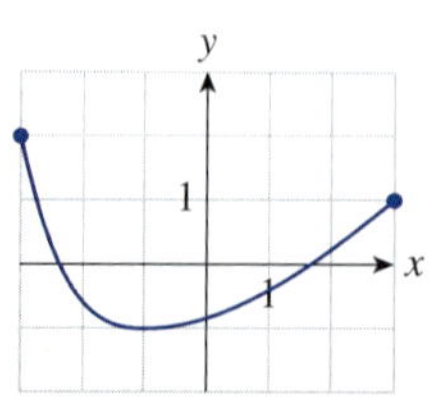

7.

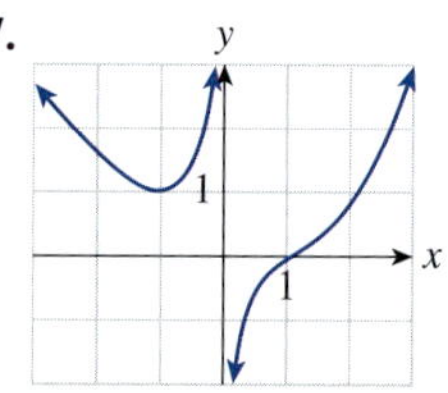

8.

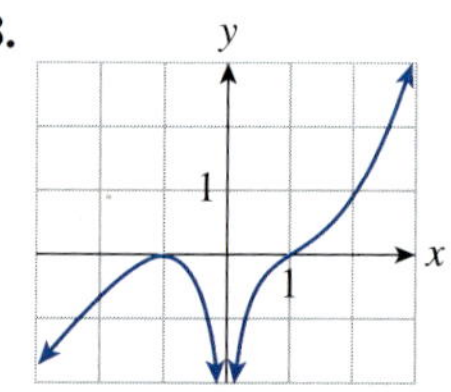

9.

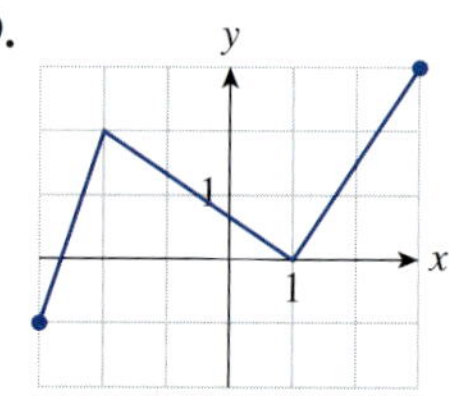

10.

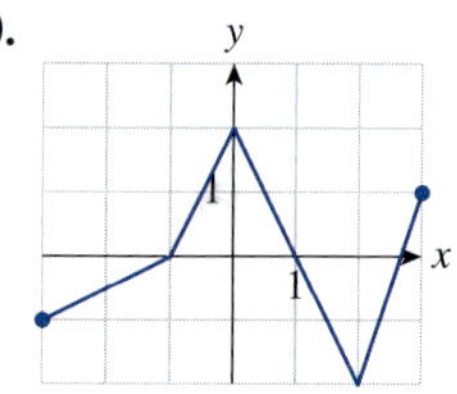

11.

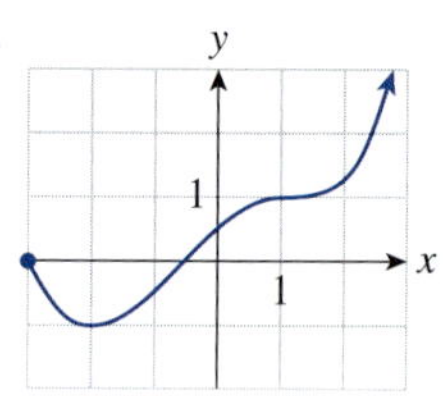

12.

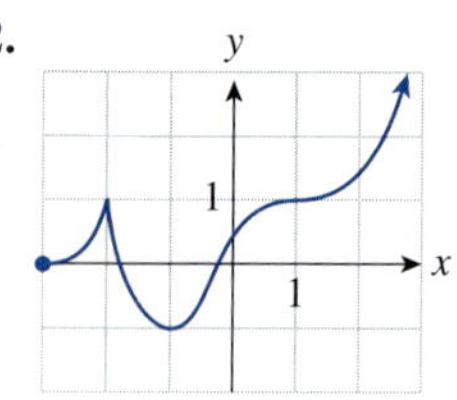

Find the exact location of all the relative and absolute extrema of each function in Exercises 13–44. HINT [See Example 1.]

13. $f(x) = x^2 - 4x + 1$ with domain $[0, 3]$

14. $f(x) = 2x^2 - 2x + 3$ with domain $[0, 3]$

15. $g(x) = x^3 - 12x$ with domain $[-4, 4]$

16. $g(x) = 2x^3 - 6x + 3$ with domain $[-2, 2]$

17. $f(t) = t^3 + t$ with domain $[-2, 2]$

18. $f(t) = -2t^3 - 3t$ with domain $[-1, 1]$

19. $h(t) = 2t^3 + 3t^2$ with domain $[-2, +\infty)$ HINT [See Example 2.]

20. $h(t) = t^3 - 3t^2$ with domain $[-1, +\infty)$ HINT [See Example 2.]

21. $f(x) = x^4 - 4x^3$ with domain $[-1, +\infty)$

22. $f(x) = 3x^4 - 2x^3$ with domain $[-1, +\infty)$

23. $g(t) = \frac{1}{4}t^4 - \frac{2}{3}t^3 + \frac{1}{2}t^2$ with domain $(-\infty, +\infty)$

24. $g(t) = 3t^4 - 16t^3 + 24t^2 + 1$ with domain $(-\infty, +\infty)$

25. $h(x) = (x-1)^{2/3}$ with domain $[0, 2]$ HINT [See Example 3.]

26. $h(x) = (x+1)^{2/5}$ with domain $[-2, 0]$ HINT [See Example 3.]

27. $k(x) = \frac{2x}{3} + (x+1)^{2/3}$ with domain $(-\infty, 0]$

28. $k(x) = \frac{2x}{5} - (x-1)^{2/5}$ with domain $[0, +\infty)$

29. ▼ $f(t) = \frac{t^2+1}{t^2-1}; -2 \le t \le 2, t \ne \pm 1$

30. ▼ $f(t) = \frac{t^2-1}{t^2+1}$ with domain $[-2, 2]$

31. ▼ $f(x) = \sqrt{x}(x-1); x \ge 0$

32. ▼ $f(x) = \sqrt{x}(x+1); x \ge 0$

33. ▼ $g(x) = x^2 - 4\sqrt{x}$

34. ▼ $g(x) = \frac{1}{x} - \frac{1}{x^2}$

35. ▼ $g(x) = \frac{x^3}{x^2+3}$

36. ▼ $g(x) = \frac{x^3}{x^2-3}$

37. ▼ $f(x) = x - \ln x$ with domain $(0, +\infty)$

38. ▼ $f(x) = x - \ln x^2$ with domain $(0, +\infty)$

39. ▼ $g(t) = e^t - t$ with domain $[-1, 1]$

40. ▼ $g(t) = e^{-t^2}$ with domain $(-\infty, +\infty)$

41. ▼ $f(x) = \frac{2x^2 - 24}{x+4}$

42. ▼ $f(x) = \frac{x-4}{x^2+20}$

43. ▼ $f(x) = xe^{1-x^2}$

44. ▼ $f(x) = x \ln x$ with domain $(0, +\infty)$

In Exercises 45–48, use graphing technology and the method in Example 5 to find the x-coordinates of the critical points, accurate to two decimal places. Find all relative and absolute maxima and minima. HINT [See Example 5.]

45. T $y = x^2 + \frac{1}{x-2}$ with domain $(-3, 2) \cup (2, 6)$

46. T $y = x^2 - 10(x-1)^{2/3}$ with domain $(-4, 4)$

47. T $f(x) = (x-5)^2(x+4)(x-2)$ with domain $[-5, 6]$

48. T $f(x) = (x+3)^2(x-2)^2$ with domain $[-5, 5]$

In Exercises 49–56, the graph of the derivative *of a function f is shown. Determine the x-coordinates of all stationary and singular points of f, and classify each as a relative maximum, relative minimum, or neither. (Assume that f(x) is defined and continuous everywhere in* $[-3, 3]$*.)* HINT [See Example 5.]

49. ▼

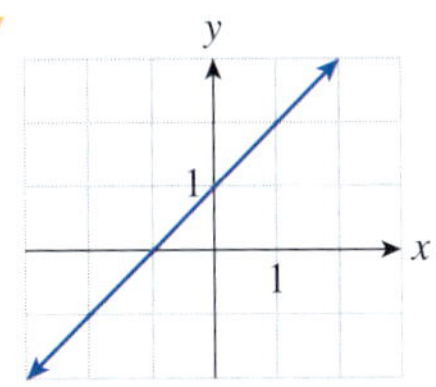

50. ▼

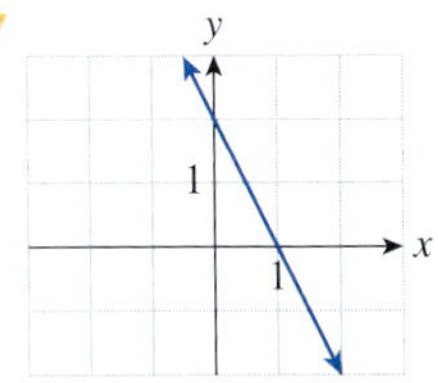

51. ▼

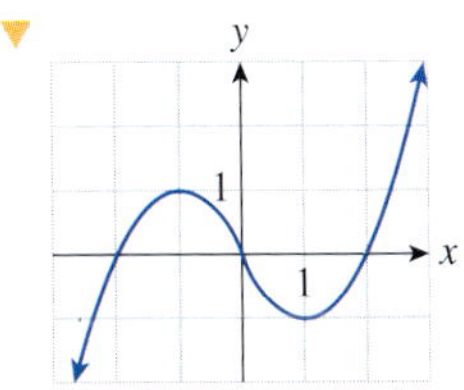

52. ▼

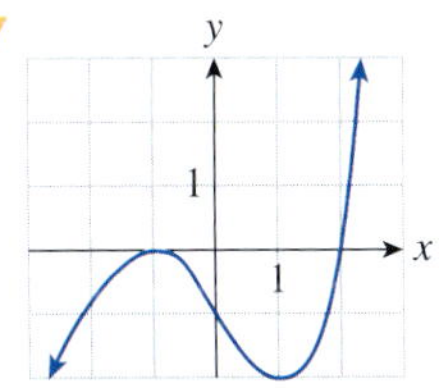

53. ▼

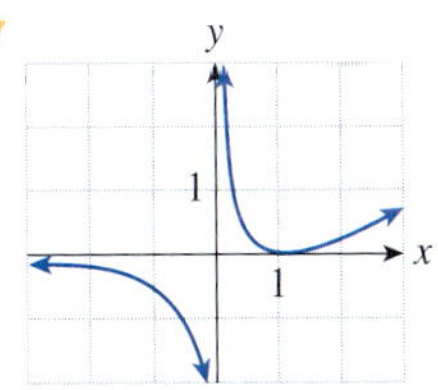

54. ▼

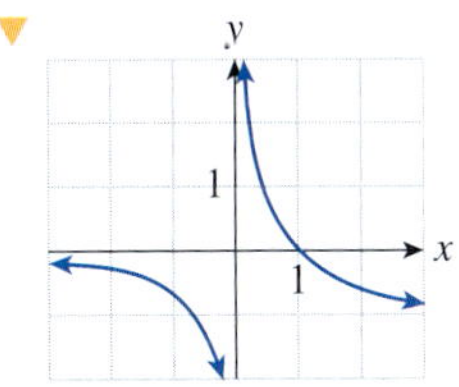

55. ▼

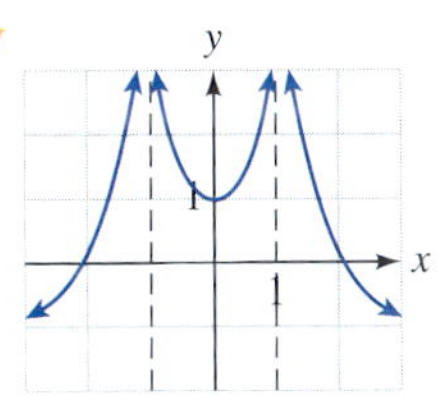

56. ▼

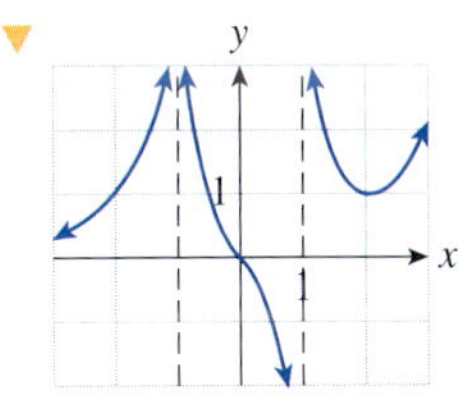

COMMUNICATION AND REASONING EXERCISES

57. Draw the graph of a function f with domain the set of all real numbers, such that f is not linear and has no relative extrema.

58. Draw the graph of a function g with domain the set of all real numbers, such that g has a relative maximum and minimum but no absolute extrema.

59. Draw the graph of a function that has stationary and singular points but no relative extrema.

60. Draw the graph of a function that has relative, not absolute, maxima and minima, but has no stationary or singular points.

61. If a stationary point is not a relative maximum, then must it be a relative minimum? Explain your answer.

62. If one endpoint is a relative maximum, must the other be a relative minimum? Explain your answer.

63. ▼ We said that if f is continuous on a closed interval $[a, b]$, then it will have an absolute maximum and an absolute minimum. Draw the graph of a function with domain $[0, 1]$ having an absolute maximum but no absolute minimum.

64. ▼ Refer to Exercise 63. Draw the graph of a function with domain $[0, 1]$ having no absolute extrema.

65. T ▼ Must endpoints always be extrema? Consider the following function (based on the trigonometric sine function—see Chapter 9 for a discussion of its properties):

$$f(x) = \begin{cases} x \sin\left(\frac{1}{x}\right) & \text{if } x > 0 \\ 0 & \text{if } x = 0 \end{cases}.$$

Technology formula: `x*sin(1/x)`

Graph this function using the technology formula above for $0 \le x \le h$, choosing smaller and smaller values of h, and decide whether f has a either a relative maximum or relative minimum at the endpoint $x = 0$. Explain your answer. (Note: Very few graphers can draw this curve accurately; the grapher that comes with Mac computers is probably among the best, while the TI-83/84 Plus is probably among the worst.)

66. T ▼ Refer to the preceding exercise, and consider the function

$$f(x) = \begin{cases} x^2 \sin\left(\frac{1}{x}\right) & \text{if } x \ne 0 \\ 0 & \text{if } x = 0 \end{cases}.$$

Technology formula: `x^2*sin(1/x)`

Graph this function using the technology formula above for $0 \le x \le h$, choosing smaller and smaller values of h, and decide **(a)** whether $x = 0$ is a stationary point, and **(b)** whether f has either a relative maximum or a relative minimum at $x = 0$. [For part (a), use technology to estimate the derivative at $x = 0$.] Explain your answers.

5.2 Applications of Maxima and Minima

In many applications we would like to find the largest or smallest possible value of some quantity—for instance, the greatest possible profit or the lowest cost. We call this the *optimal* (best) value. In this section we consider several such examples and use calculus to find the optimal value in each.

In all applications the first step is to translate a written description into a mathematical problem. In the problems we look at in this section, there are *unknowns* that we are asked to find, there is an expression involving those unknowns that must be made as large or as small as possible—the **objective function**—and there may be **constraints**—equations or inequalities relating the variables.*

* **NOTE** If you have studied linear programming, you will notice a similarity here, but unlike the situation in linear programming, neither the objective function nor the constraints need be linear.

EXAMPLE 1 Minimizing Average Cost

Gymnast Clothing manufactures expensive hockey jerseys for sale to college bookstores in runs of up to 500. Its cost (in dollars) for a run of x hockey jerseys is

$$C(x) = 2{,}000 + 10x + 0.2x^2.$$

How many jerseys should Gymnast produce per run in order to minimize average cost?*

* **NOTE** Why don't we seek to minimize total cost? The answer would be uninteresting; to minimize total cost, we would make *no* jerseys at all. Minimizing the average cost is a more practical objective.

Solution Here is the procedure we will follow to solve problems like this.

1. ***Identify the unknown(s).*** There is one unknown: x, the number of hockey jerseys Gymnast should produce per run. (We know this because the question is, How many jerseys . . . ?)
2. ***Identify the objective function.*** The objective function is the quantity that must be made as small (in this case) as possible. In this example it is the average cost, which is given by

$$\bar{C}(x) = \frac{C(x)}{x} = \frac{2{,}000 + 10x + 0.2x^2}{x}$$
$$= \frac{2{,}000}{x} + 10 + 0.2x \text{ dollars/jersey.}$$

3. ***Identify the constraints (if any).*** At most 500 jerseys can be manufactured in a run. Also, $\bar{C}(0)$ is not defined. Thus, x is constrained by

$$0 < x \le 500.$$

Put another way, the domain of the objective function $\bar{C}(x)$ is (0, 500].

4. ***State and solve the resulting optimization problem.*** Our optimization problem is:

$$\text{Minimize } \bar{C}(x) = \frac{2{,}000}{x} + 10 + 0.2x \qquad \text{Objective function}$$

$$\text{subject to } 0 < x \le 500. \qquad \text{Constraint}$$

We now solve this problem as in Section 5.1. We first calculate

$$\bar{C}'(x) = -\frac{2{,}000}{x^2} + 0.2.$$

We solve $\bar{C}'(x) = 0$ to find $x = \pm 100$. We reject $x = -100$ because -100 is not in the domain of $\bar{C}$ (and makes no sense), so we have one stationary point, at $x = 100$. There, the average cost is $\bar{C}(100) = \$50$ per jersey.

The only point at which the formula for $\bar{C}'$ is not defined is $x = 0$, but that is not in the domain of $\bar{C}$, so we have no singular points. We have one endpoint in the domain, at $x = 500$. There, the average cost is $\bar{C}(500) = \$114$.

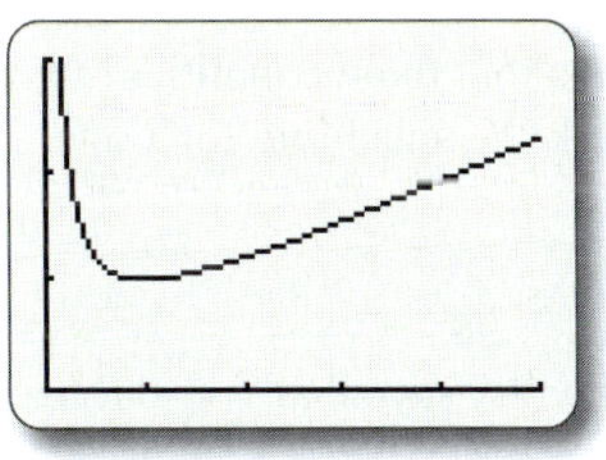

Figure 18

using Technology

Let's plot $\bar{C}$ in a viewing window with the intervals [0, 500] for x and [0, 150] for y, which will show the whole domain and the two interesting points we've found so far. The result is Figure 18.

From the graph of $\bar{C}$, we can see that the stationary point at $x = 100$ gives the absolute minimum. We can therefore say that Gymnast Clothing should produce 100 jerseys per run, for a lowest possible average cost of \$50 per jersey. ■

EXAMPLE 2 Maximizing Area

Slim wants to build a rectangular enclosure for his pet rabbit, Killer, against the side of his house, as shown in Figure 19. He has bought 100 feet of fencing. What are the dimensions of the largest area that he can enclose?

Figure 19

Solution

1. ***Identify the unknown(s).*** To identify the unknown(s), we look at the question: What are the *dimensions* of the largest area he can enclose? Thus, the unknowns are the dimensions of the fence. We call these x and y, as shown in Figure 20.

Figure 20

2. ***Identify the objective function.*** We look for what it is that we are trying to maximize (or minimize). The phrase "largest area" tells us that our object is to *maximize the area*, which is the product of length and width, so our objective function is

$$A = xy, \text{ where } A \text{ is the area of the enclosure.}$$

3. ***Identify the constraints (if any).*** What stops Slim from making the area as large as he wants? He has only 100 feet of fencing to work with. Looking again at Figure 20, we see that the sum of the lengths of the three sides must equal 100, so

$$x + 2y = 100.$$

One more point: Because x and y represent the lengths of the sides of the enclosure, neither can be a negative number.

4. ***State and solve the resulting optimization problem.*** Our mathematical problem is:

Maximize $A = xy$ — Objective function

subject to $x + 2y = 100$, $x \geq 0$, and $y \geq 0$. — Constraints

We know how to find maxima and minima of a function of one variable, but A appears to depend on two variables. We can remedy this by using a constraint to express one variable in terms of the other. Let's take the constraint $x + 2y = 100$ and solve for x in terms of y:

$$x = 100 - 2y.$$

Substituting into the objective function gives

$$A = xy = (100 - 2y)y = 100y - 2y^2$$

and we have eliminated x from the objective function. What about the inequalities? One says that $x \geq 0$, but we want to eliminate x from this as well. We substitute for x again, getting

$$100 - 2y \geq 0.$$

Solving this inequality for y gives $y \leq 50$. The second inequality says that $y \geq 0$. Now, we can restate our problem with x eliminated:

Maximize $A(y) = 100y - 2y^2$ subject to $0 \leq y \leq 50$.

We now proceed with our usual method of solving such problems. We calculate $A'(y) = 100 - 4y$. Solving $100 - 4y = 0$, we get one stationary point at $y = 25$. There, $A(25) = 1{,}250$. There are no points at which $A'(y)$ is not defined, so there are no singular points. We have two endpoints, at $y = 0$ and $y = 50$. The corresponding areas are $A(0) = 0$ and $A(50) = 0$. We record the three points we found in a table:

y	0	25	50
$A(y)$	0	1,250	0

It's clear now how A must behave: It increases from 0 at $y = 0$ to 1,250 at $y = 25$ and then decreases back to 0 at $y = 50$. Thus, the largest possible value of A is 1,250 square feet, which occurs when $y = 25$. To completely answer the question that was asked, we need to know the corresponding value of x. We have $x = 100 - 2y$, so $x = 50$ when $y = 25$. Thus, Slim should build his enclosure 50 feet across and 25 feet deep (with the "missing" 50-foot side being formed by part of the house).

➡ **Before we go on...** Notice that the problem in Example 2 came down to finding the absolute maximum value of A on the closed and bounded interval [0, 50]. As we noted in the preceding section, the table of values of A at its critical points and the endpoints of the interval gives us enough information to find the absolute maximum. ■

Let's stop for a moment and summarize the steps we've taken in these two examples.

Solving an Optimization Problem

1. **Identify the unknown(s), possibly with the aid of a diagram.** These are usually the quantities asked for in the problem.
2. **Identify the objective function.** This is the quantity you are asked to maximize or minimize. You should name it explicitly, as in "Let S = surface area."
3. **Identify the constraint(s).** These can be equations relating variables or inequalities expressing limitations on the values of variables.
4. **State the optimization problem.** This will have the form "Maximize [minimize] the objective function subject to the constraint(s)."
5. **Eliminate extra variables.** If the objective function depends on several variables, solve the constraint equations to express all variables in terms of one particular variable. Substitute these expressions into the objective function to rewrite it as a function of a single variable. Substitute the expressions into any inequality constraints to help determine the domain of the objective function.
6. **Find the absolute maximum (or minimum) of the objective function.** Use the techniques of the preceding section.

Now for some further examples.

EXAMPLE 3 Maximizing Revenue

Cozy Carriage Company builds baby strollers. Using market research, the company estimates that if it sets the price of a stroller at p dollars, then it can sell $q = 300{,}000 - 10p^2$ strollers per year. What price will bring in the greatest annual revenue?

Solution The question we are asked identifies our main unknown, the price p. However, there is another quantity that we do not know, q, the number of strollers the company will sell per year. The question also identifies the objective function, revenue, which is

$$R = pq.$$

Including the equality constraint given to us, that $q = 300{,}000 - 10p^2$, and the "reality" inequality constraints $p \geq 0$ and $q > 0$, we can write our problem as

$$\text{Maximize } R = pq \text{ subject to } q = 300{,}000 - 10p^2,\ p \geq 0, \text{ and } q \geq 0.$$

We are given q in terms of p, so let's substitute to eliminate q:

$$R = pq = p(300{,}000 - 10p^2) = 300{,}000p - 10p^3$$

Substituting in the inequality $q \geq 0$, we get

$$300{,}000 - 10p^2 \geq 0.$$

Thus, $p^2 \leq 30{,}000$, which gives $-100\sqrt{3} \leq p \leq 100\sqrt{3}$. When we combine this with $p \geq 0$, we get the following restatement of our problem:

$$\text{Maximize } R(p) = 300{,}000p - 10p^3 \text{ such that } 0 \leq p \leq 100\sqrt{3}.$$

We solve this problem in much the same way we did the preceding one. We calculate $R'(p) = 300{,}000 - 30p^2$. Setting $300{,}000 - 30p^2 = 0$, we find one stationary point at $p = 100$. There are no singular points and we have the endpoints $p = 0$ and $p = 100\sqrt{3}$. Putting these points in a table and computing the corresponding values of R, we get the following:

p	0	100	$100\sqrt{3}$
$R(p)$	0	20,000,000	0

Thus, Cozy Carriage should price its strollers at \$100 each, which will bring in the largest possible revenue of \$20,000,000.

Figure 21

EXAMPLE 4 Optimizing Resources

The Metal Can Company has an order to make cylindrical cans with a volume of 250 cubic centimeters. What should be the dimensions of the cans in order to use the least amount of metal in their production?

Solution We are asked to find the dimensions of the cans. It is traditional to take as the dimensions of a cylinder the height h and the radius of the base r, as in Figure 21.

We are also asked to minimize the amount of metal used in the can, which is the area of the surface of the cylinder. We can look up the formula or figure it out ourselves: Imagine removing the circular top and bottom and then cutting vertically and flattening out the hollow cylinder to get a rectangle, as shown in Figure 22.

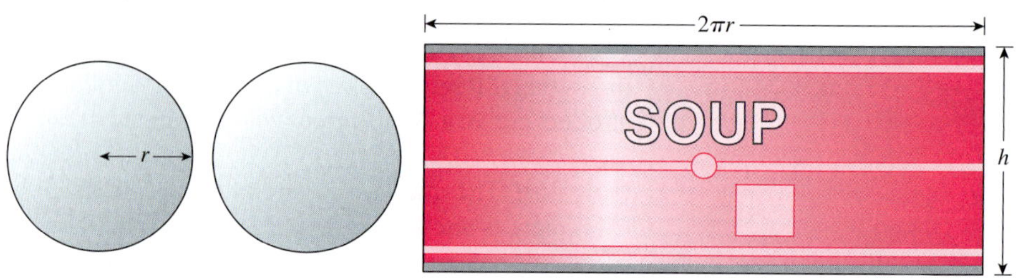

Figure 22

Our objective function is the (total) surface area S of the can. The area of each disc is πr^2, while the area of the rectangular piece is $2\pi rh$. Thus, our objective function is

$$S = 2\pi r^2 + 2\pi rh.$$

As usual, there is a constraint: The volume must be exactly 250 cubic centimeters. The formula for the volume of a cylinder is $V = \pi r^2 h$, so

$$\pi r^2 h = 250.$$

It is easiest to solve this constraint for h in terms of r:

$$h = \frac{250}{\pi r^2}.$$

Substituting in the objective function, we get

$$S = 2\pi r^2 + 2\pi r\frac{250}{\pi r^2} = 2\pi r^2 + \frac{500}{r}.$$

Now r cannot be negative or 0, but it can become very large (a very wide but very short can could have the right volume). We therefore take the domain of $S(r)$ to be $(0, +\infty)$, so our mathematical problem is as follows:

$$\text{Minimize } S(r) = 2\pi r^2 + \frac{500}{r} \text{ subject to } r > 0.$$

Now we calculate

$$S'(r) = 4\pi r - \frac{500}{r^2}.$$

To find stationary points, we set this equal to 0 and solve:

$$\begin{aligned} 4\pi r - \frac{500}{r^2} &= 0 \\ 4\pi r &= \frac{500}{r^2} \\ 4\pi r^3 &= 500 \\ r^3 &= \frac{125}{\pi}. \end{aligned}$$

So

$$r = \sqrt[3]{\frac{125}{\pi}} = \frac{5}{\sqrt[3]{\pi}} \approx 3.41.$$

The corresponding surface area is approximately $S(3.41) \approx 220$. There are no singular points or endpoints in the domain.

using Technology

To see how S behaves near the one stationary point, let's graph it in a viewing window with interval $[0, 5]$ for r and $[0, 300]$ for S. The result is Figure 23.

From the graph we can clearly see that the smallest surface area occurs at the stationary point at $r \approx 3.41$. The height of the can will be

$$h = \frac{250}{\pi r^2} \approx 6.83.$$

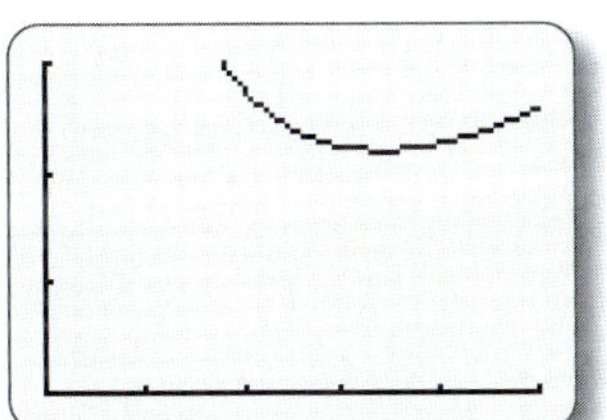

Figure 23

Thus, the can that uses the least amount of metal has a height of approximately 6.83 centimeters and a radius of approximately 3.41 centimeters. Such a can will use approximately 220 square centimeters of metal.

Before we go on... We obtained the value of r in Example 4 by solving the equation

$$4\pi r - \frac{500}{r^2}$$

This time, let us do things differently: Divide both sides by 4π to obtain

$$r = \frac{500}{4\pi r^2} = \frac{125}{\pi r^2}$$

and compare what we got with the expression for h:

$$h = \frac{250}{\pi r^2}$$

which we see is exactly twice the expression for r. Put another way, the height is exactly equal to the diameter so that the can looks square when viewed from the side. Have you ever seen cans with that shape? Why do you think most cans do not have this shape? ■

EXAMPLE 5 Allocation of Labor

The Gym Sock Company manufactures cotton athletic socks. Production is partially automated through the use of robots. Daily operating costs amount to \$50 per laborer and \$30 per robot. The number of pairs of socks the company can manufacture in a day is given by a Cobb-Douglas* production formula

$$q = 50n^{0.6}r^{0.4}$$

where q is the number of pairs of socks that can be manufactured by n laborers and r robots. Assuming that the company wishes to produce 1,000 pairs of socks per day at a minimum cost, how many laborers and how many robots should it use?

* **NOTE** Cobb-Douglas production formulas were discussed in Section 4.6.

Solution The unknowns are the number of laborers n and the number of robots r. The objective is to minimize the daily cost:

$$C = 50n + 30r.$$

The constraints are given by the daily quota

$$1{,}000 = 50n^{0.6}r^{0.4}$$

and the fact that n and r are nonnegative. We solve the constraint equation for one of the variables; let's solve for n:

$$n^{0.6} = \frac{1{,}000}{50r^{0.4}} = \frac{20}{r^{0.4}}.$$

Taking the $1/0.6$ power of both sides gives

$$n = \left(\frac{20}{r^{0.4}}\right)^{1/0.6} = \frac{20^{1/0.6}}{r^{0.4/0.6}} = \frac{20^{5/3}}{r^{2/3}} \approx \frac{147.36}{r^{2/3}}.$$

Substituting in the objective equation gives us the cost as a function of r:

$$C(r) \approx 50\left(\frac{147.36}{r^{2/3}}\right) + 30r$$
$$= 7{,}368r^{-2/3} + 30r.$$

The only remaining constraint on r is that $r > 0$. To find the minimum value of $C(r)$, we first take the derivative:

$$C'(r) \approx -4{,}912r^{-5/3} + 30.$$

Setting this equal to zero, we solve for r:

$$r^{-5/3} \approx 0.006107$$
$$r \approx (0.006107)^{-3/5} \approx 21.3.$$

The corresponding cost is $C(21.3) \approx \$1{,}600$. There are no singular points or endpoints in the domain of C.

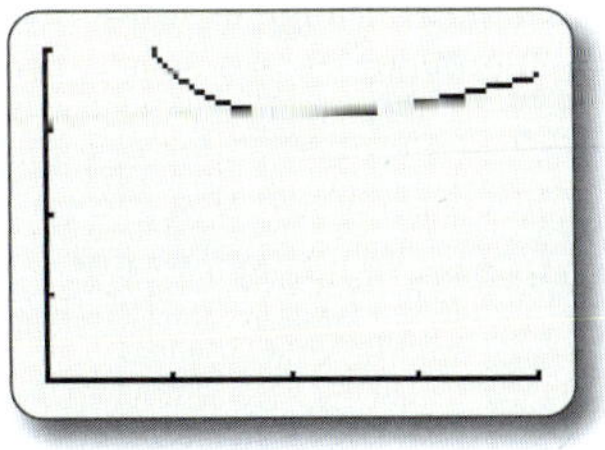

Figure 24

using Technology

To see how C behaves near its stationary point, let's draw its graph in a viewing window with an interval of [0, 40] for r and [0, 2,000] for C. The result is Figure 24.

From the graph we can see that C does have its minimum at the stationary point. The corresponding value of n is

$$n \approx \frac{147.36}{r^{2/3}} \approx 19.2.$$

At this point, our solution appears to be this: Use (approximately) 19.2 laborers and (approximately) 21.3 robots to meet the manufacturing quota at a minimum cost. However, we are not interested in fractions of robots or people, so we need to find integer solutions for n and r. If we round these numbers, we get the solution $(n, r) = (19, 21)$. However, a quick calculation shows that

$$q = 50(19)^{0.6}(21)^{0.4} \approx 989 \text{ pairs of socks,}$$

which fails to meet the quota of 1,000. Thus, we need to round at least one of the quantities n and r *upward* in order to meet the quota. The three possibilities, with corresponding values of q and C, are as follows:

$$(n, r) = (20, 21), \text{ with } q \approx 1{,}020 \text{ and } C = \$1{,}630$$
$$(n, r) = (19, 22), \text{ with } q \approx 1{,}007 \text{ and } C = \$1{,}610$$
$$(n, r) = (20, 22), \text{ with } q \approx 1{,}039 \text{ and } C = \$1{,}660.$$

Of these, the solution that meets the quota at a minimum cost is $(n, r) = (19, 22)$. Thus, the Gym Sock Co. should use 19 laborers and 22 robots, at a cost of $50 \times 19 + 30 \times 22 = \$1{,}610$, to manufacture $50 \times 19^{0.6} \times 22^{0.4} \approx 1{,}007$ pairs of socks.

5.2 EXERCISES

▼ more advanced ◆ challenging

T indicates exercises that should be solved using technology

Solve the optimization problems in Exercises 1–8. HINT [See Example 2.]

1. Maximize $P = xy$ with $x + y = 10$.

2. Maximize $P = xy$ with $x + 2y = 40$.

3. Minimize $S = x + y$ with $xy = 9$ and both x and $y > 0$.

4. Minimize $S = x + 2y$ with $xy = 2$ and both x and $y > 0$.

5. Minimize $F = x^2 + y^2$ with $x + 2y = 10$.

6. Minimize $F = x^2 + y^2$ with $xy^2 = 16$.

7. Maximize $P = xyz$ with $x + y = 30$ and $y + z = 30$, and x, y, and $z \geq 0$.

8. Maximize $P = xyz$ with $x + z = 12$ and $y + z = 12$, and x, y, and $z \geq 0$.

9. For a rectangle with perimeter 20 to have the largest area, what dimensions should it have?

10. For a rectangle with area 100 to have the smallest perimeter, what dimensions should it have?

APPLICATIONS

11. ***Average Cost: iPods*** Assume that it costs **Apple** approximately

$$C(x) = 22{,}500 + 100x + 0.01x^2$$

dollars to manufacture x 30-gigabyte video iPods in a day.[1] How many iPods should be manufactured in order to minimize average cost? What is the resulting average cost of an iPod? (Give your answer to the nearest dollar.) HINT [See Example 1.]

12. ***Average Cost: Xboxes*** Assume that it costs **Microsoft** approximately

$$C(x) = 14{,}400 + 550x + 0.01x^2$$

[1] Not the actual cost equation; the authors do not know Apple's actual cost equation. The marginal cost in the model given is in rough agreement with the actual marginal cost for reasonable values of x. Source for cost data: *Manufacturing & Technology News*, July 31, 2007 Volume 14, No. 14 (www.manufacturingnews.com).

dollars to manufacture x Xbox 360s in a day.[2] How many Xboxes should be manufactured in order to minimize average cost? What is the resulting average cost of an Xbox? (Give your answer to the nearest dollar.) HINT [See Example 1.]

13. *Pollution Control* The cost of controlling emissions at a firm rises rapidly as the amount of emissions reduced increases. Here is a possible model:

$$C(q) = 4{,}000 + 100q^2$$

where q is the reduction in emissions (in pounds of pollutant per day) and C is the daily cost to the firm (in dollars) of this reduction. What level of reduction corresponds to the lowest average cost per pound of pollutant, and what would be the resulting average cost to the nearest dollar?

14. *Pollution Control* Repeat the preceding exercise using the following cost function:

$$C(q) = 2{,}000 + 200q^2.$$

15. *Pollution Control* (Compare Exercise 13.) The cost of controlling emissions at a firm is given by

$$C(q) = 4{,}000 + 100q^2$$

where q is the reduction in emissions (in pounds of pollutant per day) and C is the daily cost to the firm (in dollars) of this reduction. Government clean-air subsidies amount to \$500 per pound of pollutant removed. How many pounds of pollutant should the firm remove each day in order to minimize *net* cost (cost minus subsidy)?

16. *Pollution Control* (Compare Exercise 14.) Repeat the preceding exercise, using the following cost function:

$$C(q) = 2{,}000 + 200q^2$$

with government subsidies amounting to \$100 per pound of pollutant removed per day.

17. *Fences* I would like to create a rectangular vegetable patch. The fencing for the east and west sides costs \$4 per foot, and the fencing for the north and south sides costs only \$2 per foot. I have a budget of \$80 for the project. What are the dimensions of the vegetable patch with the largest area I can enclose? HINT [See Example 2.]

18. *Fences* I would like to create a rectangular orchid garden that abuts my house so that the house itself forms the northern boundary. The fencing for the southern boundary costs \$4 per foot, and the fencing for the east and west sides costs \$2 per foot. If I have a budget of \$80 for the project, what are the dimensions of the garden with the largest area I can enclose? HINT [See Example 2.]

[2]Not the actual cost equation; the authors do not know Microsoft's actual cost equation. The marginal cost in the model given is in rough agreement with the actual marginal cost for reasonable values of x. Source for estimate of marginal cost: iSuppli (www.isuppli.com).

19. *Fences* You are building a right-angled triangular flower garden along a stream as shown in the figure.

The fencing of the left border costs \$5 per foot, while the fencing of the lower border costs \$1 per foot. (No fencing is required along the river.) You want to spend \$100 and enclose as much area as possible. What are the dimensions of your garden, and what area does it enclose? [The area of a right-triangle is given by $A = xy/2$.]

20. *Fences* Repeat Exercise 19, this time assuming that the fencing of the left border costs \$8 per foot, while the fencing of the lower border costs \$2 per foot, and that you can spend \$400.

21. ▼ ***Fences*** (Compare Exercise 17.) For tax reasons, I need to create a rectangular vegetable patch with an area of exactly 242 sq. ft. The fencing for the east and west sides costs \$4 per foot, and the fencing for the north and south sides costs only \$2 per foot. What are the dimensions of the vegetable patch with the least expensive fence? HINT [Compare Exercise 3.]

22. ▼ ***Fences*** (Compare Exercise 18.) For reasons too complicated to explain, I need to create a rectangular orchid garden with an area of exactly 324 sq. ft. abutting my house so that the house itself forms the northern boundary. The fencing for the southern boundary costs \$4 per foot, and the fencing for the east and west sides costs \$2 per foot. What are the dimensions of the orchid garden with the least expensive fence? HINT [Compare Exercise 4.]

23. *Revenue* Hercules Films is deciding on the price of the video release of its film *Son of Frankenstein*. Its marketing people estimate that at a price of p dollars, it can sell a total of $q = 200{,}000 - 10{,}000p$ copies. What price will bring in the greatest revenue? HINT [See Example 3.]

24. *Profit* Hercules Films is also deciding on the price of the video release of its film *Bride of the Son of Frankenstein*. Again, marketing estimates that at a price of p dollars, it can sell $q = 200{,}000 - 10{,}000p$ copies, but each copy costs \$4 to make. What price will give the greatest *profit*?

25. *Revenue: Cell Phones* Worldwide quarterly sales of Nokia cell phones were approximately $q = -p + 156$ million phones when the wholesale price was $\$p$. At what wholesale price should Nokia have sold its phones to maximize its quarterly revenue? What would have been the resulting revenue?[3]

26. ***Revenue: Cell Phones*** Worldwide annual sales of all cell phones were approximately $-10p + 1{,}600$ million phones when the wholesale price was $\$p$. At what wholesale price should cell phones have been sold to maximize annual revenue? What would have been the resulting revenue?[4]

27. ***Revenue: Monorail Service*** The demand, in rides per day, for monorail service in Las Vegas in 2005 can be approximated by $q = -4{,}500p + 41{,}500$ when the fare was $\$p$. What price should have been charged to maximize total revenue?[5]

28. ***Demand for Monorail Service, Mars*** The demand, in rides per day, for monorail service in the three urbynes (or districts) of Utarek, Mars, can be approximated by $q = -2p + 24$ million riders when the fare is $\overline{\overline{Z}}p$. What price should be charged to maximize total revenue?[6]

29. ▼ ***Revenue*** Assume that the demand for tuna in a small coastal town is given by

$$p = \frac{500{,}000}{q^{1.5}}$$

where q is the number of pounds of tuna that can be sold in a month at p dollars per pound. Assume that the town's fishery wishes to sell at least 5,000 pounds of tuna per month.

a. How much should the town's fishery charge for tuna in order to maximize monthly revenue? HINT [See Example 3, and don't neglect endpoints.]

b. How much tuna will it sell per month at that price?

c. What will be its resulting revenue?

30. ▼ ***Revenue*** Economist Henry Schultz devised the following demand function for corn:

$$p = \frac{6{,}570{,}000}{q^{1.3}}$$

where q is the number of bushels of corn that could be sold at p dollars per bushel in one year.[7] Assume that at least 10,000 bushels of corn per year must be sold.

a. How much should farmers charge per bushel of corn to maximize annual revenue? HINT [See Example 3, and don't neglect endpoints.]

b. How much corn can farmers sell per year at that price?

c. What will be the farmers' resulting revenue?

31. ▼ ***Revenue*** The wholesale price for chicken in the United States fell from 25¢ per pound to 14¢ per pound, while per capita chicken consumption rose from 22 pounds per year to 27.5 pounds per year.[8] Assuming that the demand for chicken depends linearly on the price, what wholesale price for chicken maximizes revenues for poultry farmers, and what does that revenue amount to?

32. ▼ ***Revenue*** Your underground used-book business is booming. Your policy is to sell all used versions of *Calculus and You* at the same price (regardless of condition). When you set the price at \$10, sales amounted to 120 volumes during the first week of classes. The following semester, you set the price at \$30 and sold not a single book. Assuming that the demand for books depends linearly on the price, what price gives you the maximum revenue, and what does that revenue amount to?

33. ***Profit: Cell Phones*** (Compare Exercise 25.) Worldwide quarterly sales of Nokia cell phones were approximately $q = -p + 156$ million phones when the wholesale price was $\$p$. Assuming that it cost Nokia \$40 to manufacture each cell phone, at what wholesale price should Nokia have sold its phones to maximize its quarterly profit? What would have been the resulting profit?[9] (The actual wholesale price was \$105 in the fourth quarter of 2004.) HINT [See Example 3, and recall that Profit = Revenue − Cost.]

34. ***Profit: Cell Phones*** (Compare Exercise 26.) Worldwide annual sales of all cell phones were approximately $-10p + 1{,}600$ million phones when the wholesale price was $\$p$. Assuming that it costs \$30 to manufacture each cell phone, at what wholesale price should cell phones have been sold to maximize annual profit? What would have been the resulting profit?[10] HINT [See Example 3, and recall that Profit = Revenue − Cost.]

35. ▼ ***Profit*** The demand equation for your company's virtual reality video headsets is

$$p = \frac{1{,}000}{q^{0.3}}$$

where q is the total number of headsets that your company can sell in a week at a price of p dollars. The total manufacturing and shipping cost amounts to \$100 per headset.

a. What is the greatest profit your company can make in a week, and how many headsets will your company sell at this level of profit? (Give answers to the nearest whole number.)

b. How much, to the nearest \$1, should your company charge per headset for the maximum profit?

[3] Demand equation based on second- and fourth-quarter sales. Source: Embedded.com/Company reports December, 2004.

[4] Demand equation based on estimated 2004 sales and projected 2008 sales. Source: I-Stat/NDR, December 2004.

[5] Source for ridership data: *New York Times*, February 10, 2007, p. A9.

[6] $\overline{\overline{Z}}$ designates Zonars, the official currency in Mars. See www.marsnext.com for details of the Mars colony, its commerce, and its culture.

[7] Based on data for the period 1915–1929. Source: Henry Schultz, *The Theory and Measurement of Demand*, (as cited in *Introduction to Mathematical Economics* by A. L. Ostrosky, Jr., and J. V. Koch, Waveland Press, Prospect Heights, Illinois, 1979).

[8] Data are provided for the years 1951–1958. Source: U.S. Department of Agriculture, *Agricultural Statistics*.

[9] Source: Embedded.com/Company reports, December 2004.

[10] Wholesale price projections are the authors'. Source for sales prediction: I-Stat/NDR, December 2004.

36. ▼ ***Profit*** Due to sales by a competing company, your company's sales of virtual reality video headsets have dropped, and your financial consultant revises the demand equation to

$$p = \frac{800}{q^{0.35}}$$

where q is the total number of headsets that your company can sell in a week at a price of p dollars. The total manufacturing and shipping cost still amounts to \$100 per headset.

a. What is the greatest profit your company can make in a week, and how many headsets will your company sell at this level of profit? (Give answers to the nearest whole number.)

b. How much, to the nearest \$1, should your company charge per headset for the maximum profit?

37. ***Paint Cans*** A company manufactures cylindrical paint cans with open tops with a volume of 27,000 cubic centimeters. What should be the dimensions of the cans in order to use the least amount of metal in their production? HINT [See Example 4.]

38. ***Metal Drums*** A company manufactures cylindrical metal drums with open tops with a volume of 1 cubic meter. What should be the dimensions of the drum in order to use the least amount of metal in their production? HINT [See Example 4.]

39. ***Tin Cans*** A company manufactures cylindrical tin cans with closed tops with a volume of 250 cubic centimeters. The metal used to manufacture the cans costs \$0.01 per square cm for the sides and \$0.02 per square cm for the (thicker) top and bottom. What should be the dimensions of the cans in order to minimize the cost of metal in their production? What is the ratio height/radius? HINT [See Example 4.]

40. ***Metal Drums*** A company manufactures cylindrical metal drums with open tops with a volume of 2 cubic meters. The metal used to manufacture the cans costs \$2 per square meter for the sides and \$3 per square meter for the (thicker) bottom. What should be the dimensions of the drums in order to minimize the cost of metal in their production? What is the ratio height/radius? HINT [See Example 4.]

41. ▼ ***Box Design*** Chocolate Box Company is going to make open-topped boxes out of 6×16-inch rectangles of cardboard by cutting squares out of the corners and folding up the sides. What is the largest volume box it can make this way?

42. ▼ ***Box Design*** Vanilla Box Company is going to make open-topped boxes out of 12×12-inch rectangles of cardboard by cutting squares out of the corners and folding up the sides. What is the largest volume box it can make this way?

43. ▼ ***Box Design*** A packaging company is going to make closed boxes, with square bases, that hold 125 cubic centimeters. What are the dimensions of the box that can be built with the least material?

44. ▼ ***Box Design*** A packaging company is going to make open-topped boxes, with square bases, that hold 108 cubic centimeters. What are the dimensions of the box that can be built with the least material?

45. ▼ ***Luggage Dimensions*** American Airlines requires that the total outside dimensions (length + width + height) of a checked bag not exceed 62 inches.[11] Suppose you want to check a bag whose height equals its width. What is the largest volume bag of this shape that you can check on an American flight?

46. ▼ ***Luggage Dimensions*** American Airlines requires that the total outside dimensions (length + width + height) of a carry-on bag not exceed 45 inches.[12] Suppose you want to carry on a bag whose length is twice its height. What is the largest volume bag of this shape that you can carry on an American flight?

47. ▼ ***Luggage Dimensions*** Fly-by-Night Airlines has a peculiar rule about luggage: The length and width of a bag must add up to at most 45 inches, and the width and height must also add up to 45 inches. What are the dimensions of the bag with the largest volume that Fly-by-Night will accept?

48. ▼ ***Luggage Dimensions*** Fair Weather Airlines has a similar rule. It will accept only bags for which the sum of the length and width is at most 36 inches, while the sum of length, height, and twice the width is at most 72 inches. What are the dimensions of the bag with the largest volume that Fair Weather will accept?

49. ▼ ***Package Dimensions*** The U.S. Postal Service (USPS) will accept packages only if the length plus girth is no more than 108 inches.[13] (See the figure.)

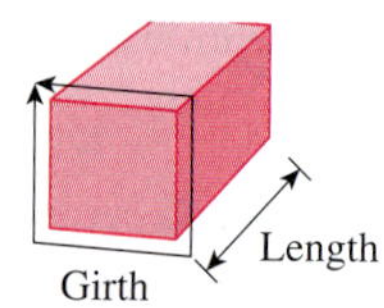

Assuming that the front face of the package (as shown in the figure) is square, what is the largest volume package that the USPS will accept?

50. ▼ ***Package Dimensions*** United Parcel Service (UPS) will accept only packages with a length of no more than 108 inches and length plus girth of no more than 165 inches.[14] (See figure for the preceding exercise.) Assuming that the front face of the package (as shown in the figure) is square, what is the largest volume package that UPS will accept?

51. ▼ ***Cell Phone Revenues*** The number of cell phone subscribers in China in the years 2000–2005 was projected to follow the equation $N(t) = 39t + 68$ million subscribers in year t ($t = 0$ represents January 2000). The average annual revenue per cell phone user was \$350 in 2000.[15] If we assume

[11] According to information on its Web site (www.aa.com/).

[12] Ibid.

[13] The requirement for packages sent other than Parcel Post, as of September 2008 (www.usps.com/).

[14] The requirement as of September 2008 (www.ups.com/).

[15] Based on a regression of projected figures (coefficients are rounded). Source: Intrinsic Technology/*New York Times*, Nov. 24, 2000, p. C1.

that due to competition the revenue per cell phone user decreases continuously at an annual rate of 30%, we can model the annual revenue as

$$R(t) = 350(39t + 68)e^{-0.3t} \text{ million dollars.}$$

Determine **a.** when to the nearest 0.1 year the revenue was projected to peak and **b.** the revenue, to the nearest \$1 million, at that time.

52. ▼ ***Cell Phone Revenues*** (Refer to Exercise 51.) If we assume instead that the revenue per cell phone user decreases continuously at an annual rate of 20%, we obtain the revenue model

$$R(t) = 350(39t + 68)e^{-0.2t} \text{ million dollars.}$$

Determine **a.** when to the nearest 0.1 year the revenue was projected to peak and **b.** the revenue, to the nearest \$1 million, at that time.

53. ▼ ***Research and Development*** Spending on research and development by drug companies in the United States t years after 1970 can be modeled by

$$S(t) = 2.5e^{0.08t} \text{ billion dollars.} \quad (0 \le t \le 31)$$

The number of new drugs approved by the Federal Drug Administration (FDA) over the same period can be modeled by

$$D(t) = 10 + t \text{ drugs per year.}^{16} \quad (0 \le t \le 31)$$

When was the function $D(t)/S(t)$ at a maximum? What is the maximum value of $D(t)/S(t)$? What does the answer tell you about the cost of developing new drugs?

54. ▼ ***Research and Development*** (Refer to Exercise 53.) If the number of new drugs approved by the FDA had been $10 + 2t$ new drugs each year, when would the function $D(t)/S(t)$ have reached a maximum? What does the answer tell you about the cost of developing new drugs?

55. ▼ ***Asset Appreciation*** As the financial consultant to a classic auto dealership, you estimate that the total value (in dollars) of its collection of 1959 Chevrolets and Fords is given by the formula

$$v = 300{,}000 + 1{,}000t^2 \quad (t \ge 5)$$

where t is the number of years from now. You anticipate a continuous inflation rate of 5% per year, so that the discounted (present) value of an item that will be worth \$$v$ in t years' time is

$$p = ve^{-0.05t}.$$

When would you advise the dealership to sell the vehicles to maximize their discounted value?

56. ▼ ***Plantation Management*** The value of a fir tree in your plantation increases with the age of the tree according to the formula

$$v = \frac{20t}{1 + 0.05t}$$

where t is the age of the tree in years. Given a continuous inflation rate of 5% per year, the discounted (present) value of a newly planted seedling is

$$p = ve^{-0.05t}.$$

At what age (to the nearest year) should you harvest your trees in order to ensure the greatest possible discounted value?

57. ▼ ***Marketing Strategy*** FeatureRich Software Company has a dilemma. Its new program, Doors-X 10.27, is almost ready to go on the market. However, the longer the company works on it, the better it can make the program and the more it can charge for it. The company's marketing analysts estimate that if it delays t days, it can set the price at $100 + 2t$ dollars. On the other hand, the longer it delays, the more market share they will lose to their main competitor (see the next exercise) so that if it delays t days it will be able to sell $400{,}000 - 2{,}500t$ copies of the program. How many days should FeatureRich delay the release in order to get the greatest revenue?

58. ▼ ***Marketing Strategy*** FeatureRich Software's main competitor (see previous exercise) is Moon Systems, and Moon is in a similar predicament. Its product, Walls-Y 11.4, could be sold now for \$200, but for each day Moon delays, it could increase the price by \$4. On the other hand, it could sell 300,000 copies now, but each day it waits will cut sales by 1,500. How many days should Moon delay the release in order to get the greatest revenue?

59. ▼ ***Average Profit*** The FeatureRich Software Company sells its graphing program, Dogwood, with a volume discount. If a customer buys x copies, then he or she pays[17] \$$500\sqrt{x}$. It cost the company \$10,000 to develop the program and \$2 to manufacture each copy. If a single customer were to buy all the copies of Dogwood, how many copies would the customer have to buy for FeatureRich Software's average profit per copy to be maximized? How are average profit and marginal profit related at this number of copies?

60. ▼ ***Average Profit*** Repeat the preceding exercise with the charge to the customer \$$600\sqrt{x}$ and the cost to develop the program \$9,000.

61. ***Resource Allocation*** Your company manufactures automobile alternators, and production is partially automated through the use of robots. Daily operating costs amount to \$100 per laborer and \$16 per robot. In order to meet production deadlines, the company calculates that the numbers of laborers and robots must satisfy the constraint

$$xy = 10{,}000$$

[16] The exponential model for R&D is based on the 1970 and 2001 spending in constant 2001 dollars, while the linear model for new drugs approved is based on the 6-year moving average from data from 1970–2000. Source for data: Pharmaceutical Research and Manufacturers of America, FDA/*New York Times*, April 19, 2002, p. C1.

[17] This is similar to the way site licenses have been structured for the program Maple®.

where x is the number of laborers and y is the number of robots. Assuming that the company wishes to meet production deadlines at a minimum cost, how many laborers and how many robots should it use? HINT [See Example 5.]

62. ***Resource Allocation*** Your company is the largest sock manufacturer in the solar system, and production is automated through the use of androids and robots. Daily operating costs amount to Ʉ200 per android and Ʉ8 per robot.[18] In order to meet production deadlines, the company calculates that the numbers of androids and robots must satisfy the constraint

$$xy = 1{,}000{,}000$$

where x is the number of androids and y is the number of robots. Assuming that the company wishes to meet production deadlines at a minimum cost, how many androids and how many robots should it use? HINT [See Example 5.]

63. ▼ ***Resource Allocation*** Your automobile assembly plant has a Cobb-Douglas production function given by

$$q = x^{0.4}y^{0.6}$$

where q is the number of automobiles it produces per year, x is the number of employees, and y is the daily operating budget (in dollars). Annual operating costs amount to an average of \$20,000 per employee plus the operating budget of \$365y. Assume that you wish to produce 1,000 automobiles per year at a minimum cost. How many employees should you hire? HINT [See Example 5.]

64. ▼ ***Resource Allocation*** Repeat the preceding exercise using the production formula

$$q = x^{0.5}y^{0.5}.$$

HINT [See Example 5.]

65. ▼ ***Incarceration Rate*** The incarceration rate (the number of persons in prison per 100,000 residents) in the United States can be approximated by

$$N(t) = 0.04t^3 - 2t^2 + 40t + 460 \quad (0 \le t \le 18)$$

(t is the year since 1990).[19] When, to the nearest year, was the incarceration rate increasing most rapidly? When was it increasing least rapidly? HINT [You are being asked to find the extreme values of the rate of change of the incarceration rate.]

66. ▼ ***Prison Population*** The prison population in the United States can be approximated by

$$N(t) = 0.02t^3 - 2t^2 + 100t + 1{,}100 \text{ thousand people} \quad (0 \le t \le 18)$$

(t is the year since 1990).[20] When, to the nearest year, was the prison population increasing most rapidly? When was it increasing least rapidly? HINT [You are being asked to find the extreme values of the rate of change of the prison population.]

[18] Ʉ are Standard Solar Units of currency.

[19] Source for data: Sourcebook of Criminal Justice Statistics Online (www.albany.edu/sourcebook).

[20] Ibid.

67. ▼ ***Embryo Development*** The oxygen consumption of a bird embryo increases from the time the egg is laid through the time the chick hatches. In a typical galliform bird, the oxygen consumption can be approximated by

$$c(t) = -0.065t^3 + 3.4t^2 - 22t + 3.6 \text{ milliliters per day} \quad (8 \le t \le 30)$$

where t is the time (in days) since the egg was laid.[21] (An egg will typically hatch at around $t = 28$.) When, to the nearest day, is $c'(t)$ a maximum? What does the answer tell you?

68. ▼ ***Embryo Development*** The oxygen consumption of a turkey embryo increases from the time the egg is laid through the time the chick hatches. In a brush turkey, the oxygen consumption can be approximated by

$$c(t) = -0.028t^3 + 2.9t^2 - 44t + 95 \text{ milliliters per day} \quad (20 \le t \le 50)$$

where t is the time (in days) since the egg was laid.[22] (An egg will typically hatch at around $t = 50$.) When, to the nearest day, is $c'(t)$ a maximum? What does the answer tell you?

69. T ▼ ***Subprime Mortgages*** The percentage of U.S.-issued mortgages that are subprime can be approximated by

$$A(t) = \frac{15.0}{1 + 8.6(1.8)^{-t}} \text{ percent} \quad (0 \le t \le 8)$$

t years after the start of 2000.[23] Graph the *derivative* of $A(t)$ and determine the year during which this derivative had an absolute maximum and also its value at that point. What does the answer tell you?

70. T ▼ ***Subprime Mortgage Debt*** The approximate value of subprime (normally classified as risky) mortgage debt outstanding in the United States can be approximated by

$$A(t) = \frac{1{,}350}{1 + 4.2(1.7)^{-t}} \text{ billion dollars} \quad (0 \le t \le 8)$$

t years after the start of 2000.[24] Graph the *derivative* of $A(t)$ and determine the year during which this derivative had an absolute maximum and also its value at that point. What does the answer tell you?

71. T ▼ ***Asset Appreciation*** You manage a small antique company that owns a collection of Louis XVI jewelry boxes. Their value v is increasing according to the formula

$$v = \frac{10{,}000}{1 + 500e^{-0.5t}}$$

[21] The model approximates graphical data published in the article "The Brush Turkey" by Roger S. Seymour, *Scientific American*, December, 1991, pp. 108–114.

[22] Ibid.

[23] Sources: Mortgage Bankers Association, UBS.

[24] Source: www.data360.org/dataset.aspx?Data_Set_Id=9549.

where t is the number of years from now. You anticipate an inflation rate of 5% per year, so that the present value of an item that will be worth \$$v$ in t years' time is given by

$$p = v(1.05)^{-t}.$$

When (to the nearest year) should you sell the jewelry boxes to maximize their present value? How much (to the nearest constant dollar) will they be worth at that time?

72. **T** ▼ ***Harvesting Forests*** The following equation models the approximate volume in cubic feet of a typical Douglas fir tree of age t years.[25]

$$V = \frac{22{,}514}{1 + 22{,}514t^{-2.55}}$$

The lumber will be sold at \$10 per cubic foot, and you do not expect the price of lumber to appreciate in the foreseeable future. On the other hand, you anticipate a general inflation rate of 5% per year, so that the present value of an item that will be worth \$$v$ in t years' time is given by

$$p = v(1.05)^{-t}.$$

At what age (to the nearest year) should you harvest a Douglas fir tree in order to maximize its present value? How much (to the nearest constant dollar) will a Douglas fir tree be worth at that time?

73. ◆ ***Agriculture*** The fruit yield per tree in an orchard containing 50 trees is 100 pounds per tree each year. Due to crowding, the yield decreases by 1 pound per season for every additional tree planted. How may additional trees should be planted for a maximum total annual yield?

74. ◆ ***Agriculture*** Two years ago your orange orchard contained 50 trees and the total yield was 75 bags of oranges. Last year you removed ten of the trees and noticed that the total yield increased to 80 bags. Assuming that the yield per tree depends linearly on the number of trees in the orchard, what should you do this year to maximize your total yield?

75. ◆ ***Revenue*** (based on a question on the GRE Economics Test[26]) If total revenue (*TR*) is specified by $TR = a + bQ - cQ^2$, where Q is quantity of output and a, b, and c are positive parameters, then *TR* is maximized for this firm when it produces Q equal to:

(A) $b/2ac$ **(B)** $b/4c$ **(C)** $(a+b)/c$ **(D)** $b/2c$ **(E)** $c/2b$

76. ◆ ***Revenue*** (based on a question on the GRE Economics Test) If total demand (Q) is specified by $Q = -aP + b$, where P is unit price and a and b are positive parameters, then total revenue is maximized for this firm when it charges P equal to:

(A) $b/2a$ **(B)** $b/4a$ **(C)** a/b **(D)** $a/2b$ **(E)** $-b/2a$

COMMUNICATION AND REASONING EXERCISES

77. You are interested in knowing the height of the tallest condominium complex that meets the city zoning requirements that the height H should not exceed eight times the distance D from the road and that it must provide parking for at least 50 cars. The objective function of the associated optimization problem is then:

(A) H **(B)** $H - 8D$ **(C)** D **(D)** $D - 8H$

One of the constraints is:

(A) $8H = D$ **(B)** $8D = H$
(C) $H'(D) = 0$ **(D)** $D'(H) = 0$

78. You are interested in building a condominium complex with a height H of at least 8 times the distance D from the road and parking area of at least 1,000 sq ft. at the cheapest cost C. The objective function of the associated optimization problem is then:

(A) H **(B)** D **(C)** C **(D)** $H + D - C$

One of the constraints is:

(A) $H - 8D = 0$ **(B)** $H + D - C = 0$
(C) $C'(D) = 0$ **(D)** $8H = D$

79. Explain why the following problem is uninteresting: A packaging company wishes to make cardboard boxes with open tops by cutting square pieces from the corners of a square sheet of cardboard and folding up the sides. What is the box with the least surface area it can make this way?

80. Explain why finding the production level that minimizes a cost function is frequently uninteresting. What would a more interesting objective be?

81. Your friend Margo claims that all you have to do to find the absolute maxima and minima in applications is set the derivative equal to zero and solve. "All that other stuff about endpoints and so-on is a waste of time just to make life hard for us," according to Margo. Explain why she is wrong, and find at least one exercise in this exercise set to illustrate your point.

82. You are having a hard time persuading your friend Marco that maximizing revenue is not the same as maximizing profit. "How on earth can you expect to obtain the largest profit if you are not taking in the largest revenue?" Explain why he is wrong, and find at least one exercise in this exercise set to illustrate your point.

83. ▼ If demand q decreases as price p increases, what does the minimum value of dq/dp measure?

84. ▼ Explain how you would solve an optimization problem of the following form. Maximize $P = f(x, y, z)$ subject to $z = g(x, y)$ and $y = h(x)$.

[25] The model is the authors' and is based on data in *Environmental and Natural Resource Economics* by Tom Tietenberg, Third Edition, (New York: HarperCollins, 1992), p. 282.

[26] Source: GRE Economics Test, by G. Gallagher, G. E. Pollock, W. J. Simeone, G. Yohe (Piscataway, NJ: Research and Education Association, 1989).

5.3 Higher Order Derivatives: Acceleration and Concavity

The **second derivative** is simply the derivative of the derivative function. To explain why we would be interested in such a thing, we start by discussing one of its interpretations.

Acceleration

Recall that if $s(t)$ represents the position of a car at time t, then its velocity is given by the derivative: $v(t) = s'(t)$. But one rarely drives a car at a constant speed; the velocity itself is changing. The rate at which the velocity is changing is the **acceleration**. Because the derivative measures the rate of change, acceleration is the derivative of velocity: $a(t) = v'(t)$. Because v is the derivative of s, we can express the acceleration in terms of s:

$$a(t) = v'(t) = (s')'(t) = s''(t)$$

That is, a is the derivative of the derivative of s, in other words, the second derivative of s, which we write as s''. (In this context you will often hear the derivative s' referred to as the **first derivative**.)

Second Derivative, Acceleration

If a function f has a derivative that is in turn differentiable, then its **second derivative** is the derivative of the derivative of f, written as f''. If $f''(a)$ exists, we say that f is **twice differentiable at** $x = a$.

Quick Examples

1. If $f(x) = x^3 - x$, then $f'(x) = 3x^2 - 1$, so $f''(x) = 6x$ and $f''(-2) = -12$.
2. If $f(x) = 3x + 1$, then $f'(x) = 3$, so $f''(x) = 0$.
3. If $f(x) = e^x$, then $f'(x) = e^x$, so $f''(x) = e^x$ as well.

The **acceleration** of a moving object is the derivative of its velocity—that is, the second derivative of the position function.

Quick Example

If t is time in hours and the position of a car at time t is $s(t) = t^3 + 2t^2$ miles, then the car's velocity is $v(t) = s'(t) = 3t^2 + 4t$ miles per hour and its acceleration is $a(t) = s''(t) = v'(t) = 6t + 4$ miles per hour per hour.

Differential Notation for the Second Derivative

We have written the second derivative of $f(x)$ as $f''(x)$. We could also use differential notation:

$$f''(x) = \frac{d^2f}{dx^2}$$

This notation comes from writing the second derivative as the derivative of the derivative in differential notation:

$$f''(x) = \frac{d}{dx}\left[\frac{df}{dx}\right] = \frac{d^2f}{dx^2}$$

Similarly, if $y = f(x)$, we write $f''(x)$ as $\frac{d}{dx}\left[\frac{dy}{dx}\right] = \frac{d^2y}{dx^2}$. For example, if $y = x^3$, then $\frac{d^2y}{dx^2} = 6x$.

An important example of acceleration is the acceleration due to gravity.

EXAMPLE 1 Acceleration Due to Gravity

According to the laws of physics, the height of an object near the surface of the earth falling in a vacuum from an initial rest position 100 feet above the ground under the influence of gravity is approximately

$$s(t) = 100 - 16t^2 \text{ feet}$$

in t seconds. Find its acceleration.

Solution The velocity of the object is

$$v(t) = s'(t) = -32t \text{ ft/s}.$$

Differential notation: $v = \frac{ds}{dt} = -32t$ ft/s.

The reason for the negative sign is that the height of the object is decreasing with time, so its velocity is negative. Hence, the acceleration is

$$a(t) = s''(t) = -32 \text{ ft/s}^2.$$

Differential notation: $a = \frac{d^2s}{dt^2} = -32$ ft/s^2.

(We write ft/s^2 as an abbreviation for feet/second/second—that is, feet per second per second. It is often read "feet per second squared.") Thus, the *downward* velocity is increasing by 32 ft/s every second. We say that 32 ft/s^2 is the **acceleration due to gravity**. If we ignore air resistance, all falling bodies near the surface of the earth, no matter what their weight, will fall with this acceleration.*

Before we go on... In very careful experiments using balls rolling down inclined planes, Galileo made one of his most important discoveries—that the acceleration due to gravity is constant and does not depend on the weight or composition of the object falling.† A famous, though probably apocryphal, story has him dropping cannonballs of different weights off the Leaning Tower of Pisa to prove his point.§ ■

* **NOTE** On other planets the acceleration due to gravity is different. For example, on Jupiter, it is about three times as large as on Earth.

† **NOTE** An interesting aside: Galileo's experiments depended on getting extremely accurate timings. Because the timepieces of his day were very inaccurate, he used the most accurate time measurement he could: He sang and used the beat as his stopwatch.

§ **NOTE** A true story: The point was made again during the Apollo 15 mission to the moon (July 1971) when astronaut David R. Scott dropped a feather and a hammer from the same height. The moon has no atmosphere, so the two hit the surface of the moon simultaneously.

EXAMPLE 2 Acceleration of Sales

For the first 15 months after the introduction of a new video game, the total sales can be modeled by the curve

$$S(t) = 20e^{0.4t} \text{ units sold}$$

where t is the time in months since the game was introduced. After about 25 months total sales follow more closely the curve

$$S(t) = 100{,}000 - 20e^{17-0.4t}$$

How fast are total sales accelerating after 10 months? How fast are they accelerating after 30 months? What do these numbers mean?

Solution By acceleration we mean the rate of change of the rate of change, which is the second derivative. During the first 15 months, the first derivative of sales is

$$\frac{dS}{dt} = 8e^{0.4t}$$

and so the second derivative is

$$\frac{d^2S}{dt^2} = 3.2e^{0.4t}$$

Thus, after 10 months the acceleration of sales is

$$\left.\frac{d^2S}{dt^2}\right|_{t=10} = 3.2e^4 \approx 175 \text{ units/month/month, or units/month}^2$$

We can also compute total sales

$$S(10) = 20e^4 \approx 1{,}092 \text{ units}$$

and the rate of change of sales

$$\left.\frac{dS}{dt}\right|_{t=10} = 8e^4 \approx 437 \text{ units/month.}$$

What do these numbers mean? By the end of the tenth month, a total of 1,092 video games have been sold. At that time the game is selling at the rate of 437 units per month. This rate of sales is increasing by 175 units per month per month. More games will be sold each month than the month before.

To analyze the sales after 30 months is similar, using the formula

$$S(t) = 100{,}000 - 20e^{17-0.4t}.$$

The derivative is

$$\frac{dS}{dt} = 8e^{17-0.4t}$$

and the second derivative is

$$\frac{d^2S}{dt^2} = -3.2e^{17-0.4t}.$$

After 30 months,

$$S(30) = 100{,}000 - 20e^{17-12} \approx 97{,}032 \text{ units}$$

$$\left.\frac{dS}{dt}\right|_{t=30} = 8e^{17-12} \approx 1{,}187 \text{ units/month}$$

$$\left.\frac{d^2S}{dt^2}\right|_{t=30} = -3.2e^{17-12} \approx -475 \text{ units/month}^2.$$

By the end of the 30th month, 97,032 video games have been sold, the game is selling at a rate of 1,187 units per month, and the rate of sales is *decreasing* by 475 units per month. Fewer games are sold each month than the month before.

Geometric Interpretation of Second Derivative: Concavity

The first derivative of f tells us where the graph of f is rising [where $f'(x) > 0$] and where it is falling [where $f'(x) < 0$]. The second derivative tells in what direction the graph of f *curves* or *bends*. Consider the graphs in Figures 25 and 26.

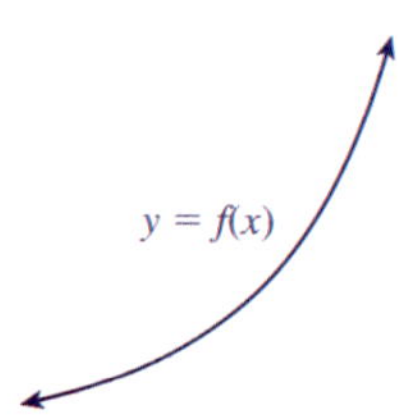

Figure 25

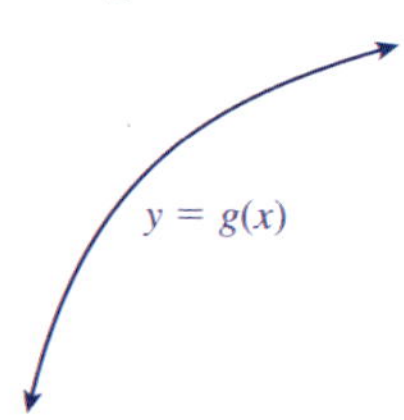

Figure 26

Think of a car driving from left to right along each of the roads shown in the two figures. A car driving along the graph of f in Figure 25 will turn to the left (upward); a car driving along the graph of g in Figure 26 will turn to the right (downward). We say that the graph of f is **concave up** and the graph of g is **concave down**. Now think about the derivatives of f and g. The derivative $f'(x)$ starts small but *increases* as the graph gets steeper. Because $f'(x)$ is increasing, its derivative $f''(x)$ must be positive. On the other hand, $g'(x)$ *decreases* as we go to the right. Because $g'(x)$ is decreasing, its derivative $g''(x)$ must be negative. Summarizing, we have the following.

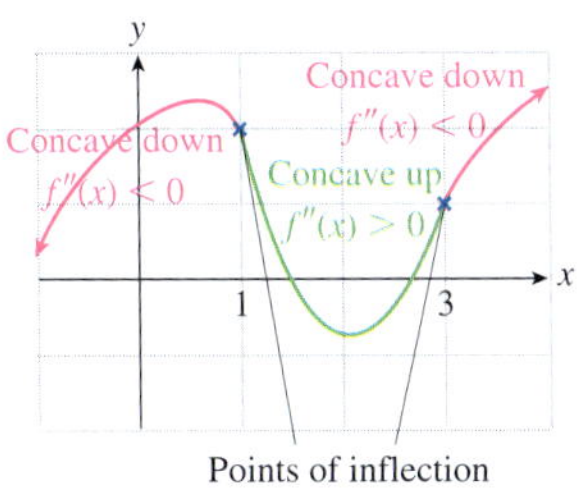

Figure 27

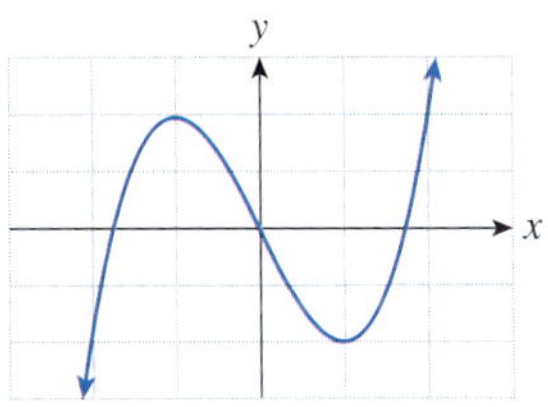

Figure 28

Concavity and the Second Derivative

A curve is **concave up** if its slope is increasing, in which case the second derivative is positive. A curve is **concave down** if its slope is decreasing, in which case the second derivative is negative. A point in the domain of f where the graph of f changes concavity, from concave up to concave down or vice versa, is called a **point of inflection**. At a point of inflection, the second derivative is either zero or undefined.

Locating Points of Inflection

To locate possible points of inflection, list points where $f''(x) = 0$ and also points where $f''(x)$ is not defined.

Quick Examples

1. The graph of the function f shown in Figure 27 is concave up when $1 < x < 3$, so $f''(x) > 0$ for $1 < x < 3$. It is concave down when $x < 1$ and $x > 3$, so $f''(x) < 0$ when $x < 1$ and $x > 3$. It has points of inflection at $x = 1$ and $x = 3$.
2. Consider $f(x) = x^3 - 3x$, whose graph is shown in Figure 28. $f''(x) = 6x$ is negative when $x < 0$ and positive when $x > 0$. The graph of f is concave down when $x < 0$ and concave up when $x > 0$. f has a point of inflection at $x = 0$, where the second derivative is 0.

The following example shows one of the reasons it's useful to look at concavity.

EXAMPLE 3 Inflation

Figure 29 shows the value of the U.S. Consumer Price Index (CPI) from January 2007 through June 2008.*

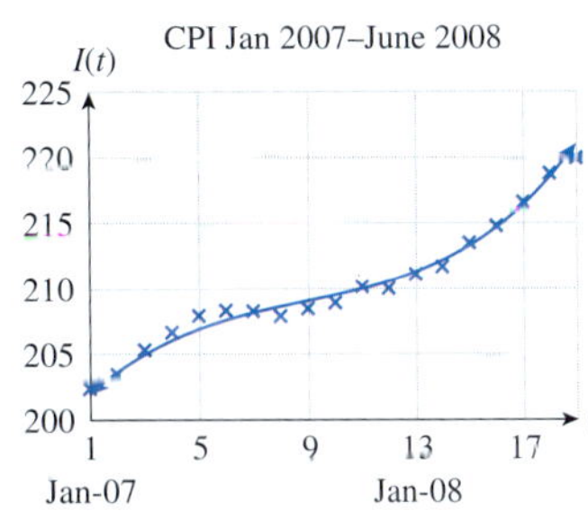

Figure 29

The approximating curve shown on the figure is given by

$$I(t) = 0.0075t^3 - 0.2t^2 + 2.2t + 200 \qquad (1 \le t \le 19)$$

*The CPI is compiled by the Bureau of Labor Statistics and is based upon a 1982 value of 100. For instance, a CPI of 200 means the CPI has doubled since 1982. Source: InflationData.com (www.inflationdata.com).

where t is time in months ($t = 1$ represents January 2007). When the CPI is increasing, the U.S. economy is **experiencing inflation**. In terms of the model, this means that the derivative is positive: $I'(t) > 0$. Notice that $I'(t) > 0$ for the entire period shown (the graph is sloping upward), so the U.S. economy experienced inflation for $1 \le t \le 19$. We could measure **inflation** by the first derivative $I'(t)$ of the CPI, but we traditionally measure it as a ratio:

$$\text{Inflation rate} = \frac{I'(t)}{I(t)}, \qquad \text{Relative rate of change of the CPI}$$

expressed as a percentage per unit time (per month in this case).

a. Use the model to estimate the inflation rate in January 2008.

b. Was inflation slowing or speeding up in January 2008?

c. When was inflation slowing? When was inflation speeding up? When was inflation slowest?

Solution

a. We need to compute $I'(t)$:

$$I'(t) = 0.0225t^2 - 0.4t + 2.2$$

Thus, the inflation rate in January 2008 was given by

$$\text{Inflation rate} = \frac{I'(13)}{I(13)} = \frac{0.0225(13)^2 - 0.4(13) + 2.2}{0.0075(13)^3 - 0.2(13)^2 + 2.2(13) + 200}$$

$$= \frac{0.8025}{211.2775} \approx 0.00380,$$

or 0.38% per month.*

* **NOTE** The 0.38% monthly inflation rate corresponds to a $12 \times 0.38 = 4.56\%$ annual inflation rate. This result could be obtained directly by changing the units of the t-axis from months to years and then redoing the calculation.

b. We say that inflation is "slowing" when the CPI is decelerating ($I''(t) < 0$; the index rises at a slower rate). Similarly, inflation is "speeding up" when the CPI is accelerating ($I''(t) > 0$; the index rises at a faster rate). From the formula for $I'(t)$, the second derivative is

$$I''(t) = 0.045t - 0.4$$
$$I''(13) = 0.045(13) - 0.4 = 0.185.$$

Because this quantity is positive, we conclude that inflation was speeding up in January 2008.

c. When inflation is slowing, $I''(t)$ is negative, so the graph of the CPI is concave down. When inflation is speeding up, it is concave up. At the point at which it switches, there is point of inflection (Figure 30).

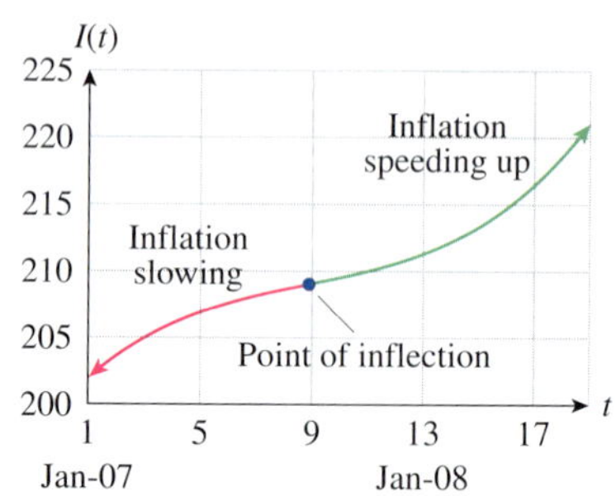

Figure 30

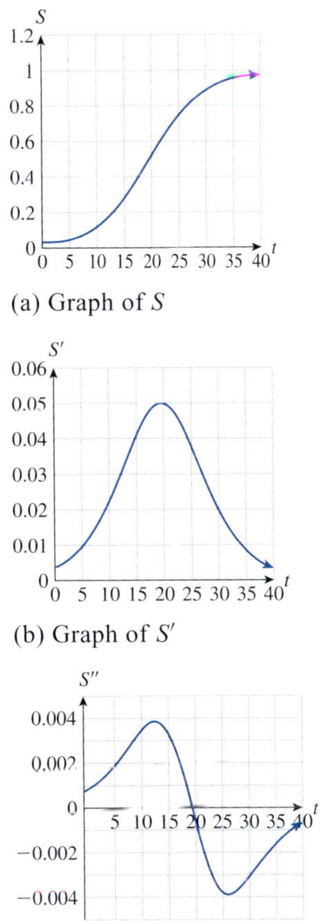

(a) Graph of S

(b) Graph of S'

(c) Graph of S''

Figure 31

The point of inflection occurs when $I''(t) = 0$; that is,

$$0.045t - 0.4 = 0$$

$$t = \frac{0.4}{0.45} \approx 8.9.$$

Thus, inflation was slowing when $t < 8.9$ (that is, until the end of August), and speeding up when $t > 8.9$ (after that time). Inflation was slowest at the point when it stopped slowing down and began to speed up, $t \approx 8.9$; notice that the graph has the least slope at that point.

EXAMPLE 4 The Point of Diminishing Returns

After the introduction of a new video game, the total worldwide sales are modeled by the curve

$$S(t) = \frac{1}{1 + 50e^{-0.2t}} \text{ million units sold}$$

where t is the time in months since the game was introduced (compare Example 2). The graphs of $S(t)$, $S'(t)$, and $S''(t)$ are shown in Figure 31. Where is the graph of S concave up, and where is it concave down? Where are any points of inflection? What does this all mean?

Solution Look at the graph of S. We see that the graph of S is concave up in the early months and then becomes concave down later. The point of inflection, where the concavity changes, is somewhere between 15 and 25 months.

Now look at the graph of S''. This graph crosses the t-axis very close to $t = 20$, is positive before that point, and negative after that point. Because positive values of S'' indicate S is concave up and negative values concave down, we conclude that the graph of S is concave up for about the first 20 months; that is, for $0 < t < 20$ and concave down for $20 < t < 40$. The concavity switches at the point of inflection, which occurs at about $t = 20$ (when $S''(t) = 0$; a more accurate answer is $t \approx 19.56$).

What does this all mean? Look at the graph of S', which shows sales per unit time, or monthly sales. From this graph we see that monthly sales are increasing for $t < 20$: more units are being sold each month than the month before. Monthly sales reach a peak of 0.05 million = 50,000 games per month at the point of inflection $t = 20$ and then begin to drop off. Thus, the point of inflection occurs at the time when monthly sales stop increasing and start to fall off; that is, the time when monthly sales peak. The point of inflection is sometimes called the **point of diminishing returns**. Although the total sales figure continues to rise (see the graph of S: game units continue to be sold), the *rate* at which units are sold starts to drop. (See Figure 32.)

Figure 32

using Technology

We can use a TI-83/84 Plus or a downloadable Excel sheet at the Web site to graph the second derivative of the function in Example 4:

TI-83/84 Plus

```
Y1=1/(1+50*e^(-0.2X))
Y2=nDeriv(Y1,X,X)
Y3=nDeriv(Y2,X,X)
```

Web Site

www.AppliedCalc.org

Student Web Site

→ Online Utilities

→ Excel First and Second Derivative Graphing Utility

Function:

```
1/(1+50*exp(-0.2*x))
```

The Second Derivative Test for Relative Extrema

The second derivative often gives us a way of knowing whether or not a stationary point is a relative extremum. Figure 33 shows a graph with two stationary points: a relative maximum at $x = a$ and a relative minimum at $x = b$.

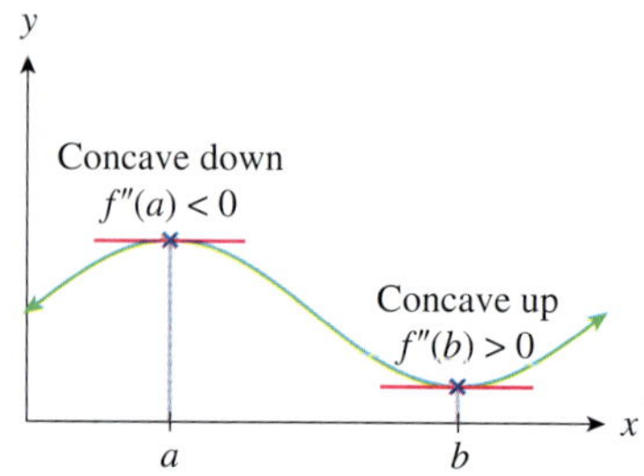

Figure 33

Notice that the curve is *concave down* at the relative maximum ($x = a$), so that $f''(a) < 0$, and *concave up* at the relative minimum ($x = b$), so that $f''(b) > 0$. This suggests the following (compare the First Derivative Test in Section 5.1).

Second Derivative Test for Relative Extrema

Suppose that the function f has a stationary point at $x = c$, and that $f''(c)$ exists. Determine the sign of $f''(c)$.

1. If $f''(c) > 0$ then f has a relative minimum at $x = c$.

2. If $f''(c) < 0$ then f has a relative maximum at $x = c$.

If $f''(c) = 0$ then the test is inconclusive and you need to use one of the methods of Section 5.1 (such as the first derivative test) to determine whether or not f has a relative extremum at $x = c$.

Quick Examples

1. $f(x) = x^2 - 2x$ has $f'(x) = 2x - 2$ and hence a stationary point at $x = 1$. $f''(x) = 2$, and so $f''(1) = 2$, which is positive, so f has a relative minimum at $x = 1$.

2. Let $f(x) = x^3 - 3x^2 - 9x$. Then

$f'(x) = 3x^2 - 6x - 9 = 3(x + 1)(x - 3)$

Stationary points at $x = -1, x = 3$

$f''(x) = 6x - 6$

$f''(-1) = -12$, so there is a relative maximum at $x = -1$

$f''(3) = 12$, so there is a relative minimum at $x = 3$.

3. $f(x) = x^4$ has $f'(x) = 4x^3$ and hence a stationary point at $x = 0$. $f''(x) = 12x^2$ and so $f''(0) = 0$, telling us that the second derivative test is inconclusive. However, we can see from the graph of f or the first derivative test that f has a minimum at $x = 0$.

Higher Order Derivatives

There is no reason to stop at the second derivative; we could once again take the *derivative* of the second derivative to obtain the **third derivative**, f''', and we could take the derivative once again to obtain the **fourth derivative**, written $f^{(4)}$, and then continue to obtain $f^{(5)}$, $f^{(6)}$, and so on (assuming we get a differentiable function at each stage).

Higher Order Derivatives

We define

$$f'''(x) = \frac{d}{dx}[f''(x)]$$

$$f^{(4)}(x) = \frac{d}{dx}[f'''(x)]$$

$$f^{(5)}(x) = \frac{d}{dx}[f^{(4)}(x)],$$

and so on, assuming all these derivatives exist.

Different Notations

$$f'(x),\ f''(x),\ f'''(x),\ f^{(4)}(x),\ \dots,\ f^{(n)}(x),\ \dots$$

$$\frac{df}{dx},\ \frac{d^2f}{dx^2},\ \frac{d^3f}{dx^3},\ \frac{d^4f}{dx^4},\ \dots,\ \frac{d^nf}{dx^n},\ \dots$$

$$\frac{dy}{dx},\ \frac{d^2y}{dx^2},\ \frac{d^3y}{dx^3},\ \frac{d^4y}{dx^4},\ \dots,\ \frac{d^ny}{dx^n},\ \dots \qquad \text{When } y = f(x)$$

$$y,\ y',\ y'',\ y''',\ y^{(4)},\ \dots,\ y^{(n)},\ \dots \qquad \text{When } y = f(x)$$

Quick Examples

1. If $f(x) = x^3 - x$, then $f'(x) = 3x^2 - 1$, $f''(x) = 6x$, $f'''(x) = 6$, $f^{(4)}(x) = f^{(5)}(x) = \cdots = 0$.
2. If $f(x) = e^x$, then $f'(x) = e^x$, $f''(x) = e^x$, $f'''(x) = e^x$, $f^{(4)}(x) = f^{(5)}(x) = \cdots = e^x$.

Q: *We know that the second derivative can be interpreted as acceleration. How do we interpret the third derivative; and the fourth, fifth, and so on?*

A: Think of a car traveling down the road (with position $s(t)$ at time t) in such a way that its acceleration $\frac{d^2s}{dt^2}$ is changing with time (for instance, the driver may be slowly increasing pressure on the accelerator, causing the car to accelerate at a greater and greater rate). Then $\frac{d^3s}{dt^3}$ is the rate of change of acceleration. $\frac{d^4s}{dt^4}$ would then be the *acceleration* of the acceleration, and so on.

Q: *How are these higher order derivatives reflected in the graph of a function f?*

A: Because the concavity is measured by f'', its derivative f''' tells us the rate of change of concavity. Similarly, $f^{(4)}$ would tell us the *acceleration* of concavity, and so on. These properties are very subtle and hard to discern by simply looking at the curve; the higher the order, the more subtle the property. There is a remarkable theorem by Taylor* that tells us that, for a large class of functions (including polynomial, exponential, logarithmic, and trigonometric functions) the values of all orders of derivative $f(a)$, $f'(a)$, $f''(a)$, $f'''(a)$, and so on at the single point $x = a$ are enough to describe the entire graph (even at points very far from $x = a$)! In other words, the smallest piece of a graph near any point a contains sufficient information to "clone" the entire graph!

* **NOTE** Brook Taylor (1685–1731) was an English mathematician.

FAQs

Interpreting Points of Inflection and Using the Second Derivative Test

Q: *It says in Example 4 that monthly sales reach a maximum at the point of inflection (second derivative is zero), but the Second Derivative test says that, for a maximum, the second derivative must be positive. What is going on here?*

A: What is a maximum in Example 4 is the *rate of change of* sales: which is measured in sales per unit time (monthly sales in the example). In other words, it is the *derivative* of the total sales function that is a maximum, so we located the maximum by setting its derivative (which is the *second* derivative of total sales) equal to zero. In general: To find relative (stationary) extrema of the *original* function, set $f'(x)$ equal to zero and solve for x as usual. The second derivative test can then be used to test the stationary point obtained. To find relative (stationary) extrema of the *rate of change of* f, set $f''(x) = 0$ and solve for x.

Q: *I used the second derivative test and it was inconclusive. That means that there is neither a relative maximum nor a relative minimum at x = a, right?*

A: Wrong. If (as is often the case) the second derivative is zero at a stationary point, all it means is that the second derivative test itself cannot determine whether the given point is a relative maximum, minimum, or neither. For instance, $f(x) = x^4$ has a stationary minimum at $x = 0$, but the second derivative test is inconclusive. In such cases, one should use another test (such as the first derivative test) to decide if the point is a relative maximum, minimum, or neither.

5.3 EXERCISES

▼ more advanced ◆ challenging

T indicates exercises that should be solved using technology

In Exercises 1–10, calculate $\frac{d^2y}{dx^2}$. HINT [See Quick Examples on page 398.]

1. $y = 3x^2 - 6$

2. $y = -x^2 + x$

3. $y = \frac{2}{x}$

4. $y = -\frac{2}{x^2}$

5. $y = 4x^{0.4} - x$

6. $y = 0.2x^{-0.1}$

7. $y = e^{-(x-1)} - x$

8. $y = e^{-x} + e^x$

9. $y = \frac{1}{x} - \ln x$

10. $y = x^{-2} + \ln x$

In Exercises 11–16, the position s of a point (in feet) is given as a function of time t (in seconds). Find (a) its acceleration as a function of t and (b) its acceleration at the specified time. HINT [See Example 1.]

11. $s = 12 + 3t - 16t^2;\ t = 2$

12. $s = -12 + t - 16t^2;\ t = 2$

13. $s = \dfrac{1}{t} + \dfrac{1}{t^2};\ t = 1$

14. $s = \dfrac{1}{t} - \dfrac{1}{t^2};\ t = 2$

15. $s = \sqrt{t} + t^2;\ t = 4$

16. $s = 2\sqrt{t} + t^3;\ t = 1$

In Exercises 17–24, the graph of a function is given. Find the approximate coordinates of all points of inflection of each function (if any). HINT [See Quick Examples on page 401.]

17.

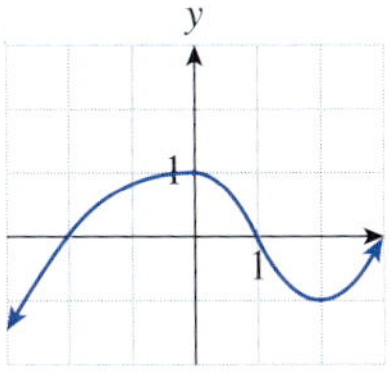

18.

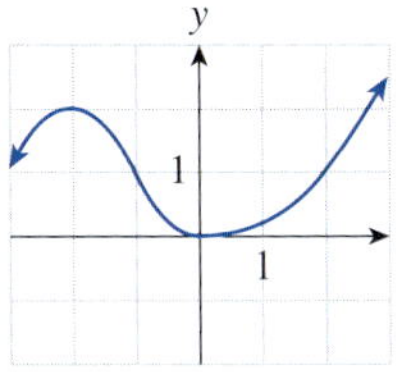

19.

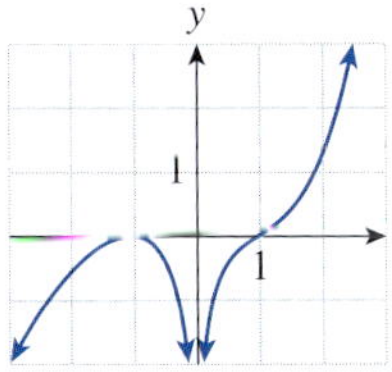

20.

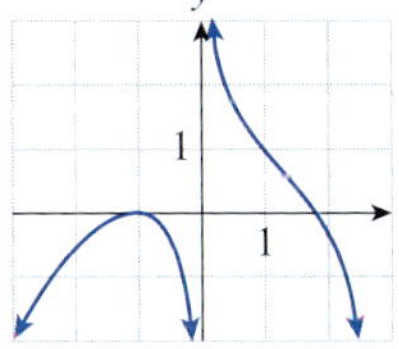

21.

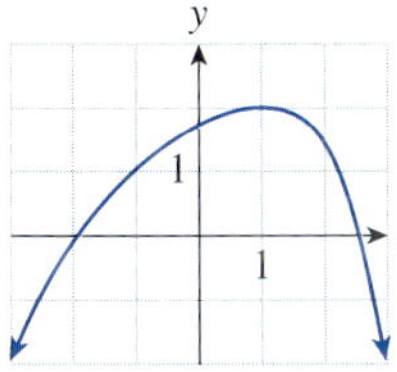

22.

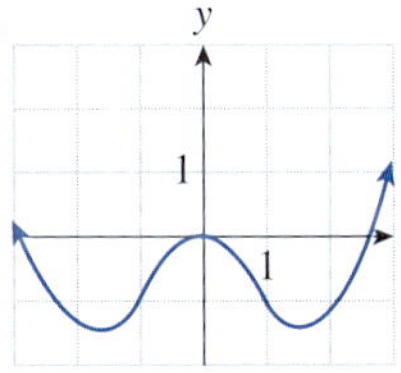

23.

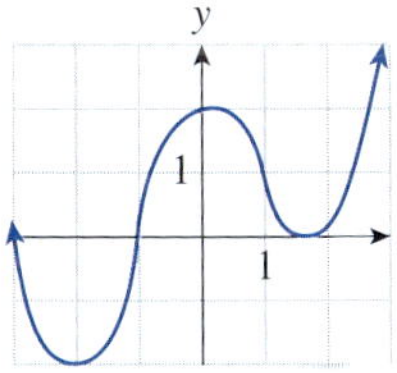

24.

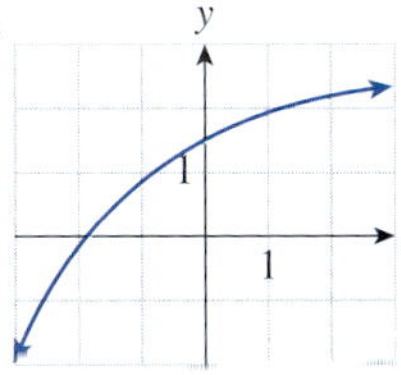

In Exercises 25–28, the graph of the derivative, $f'(x)$, is given. Determine the x-coordinates of all points of inflection of $f(x)$, if any. (Assume that $f(x)$ is defined and continuous everywhere in $[-3, 3]$.) HINT [See the **Before we go on** discussion in Example 4.]

25.

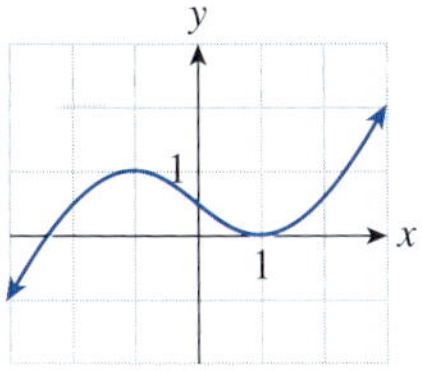

26.

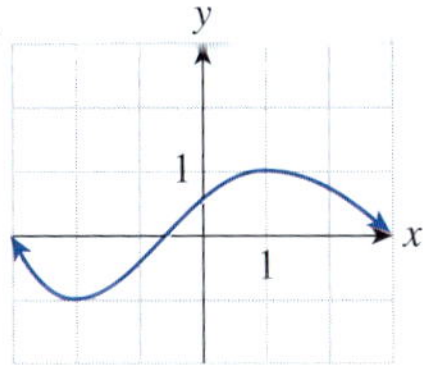

27.

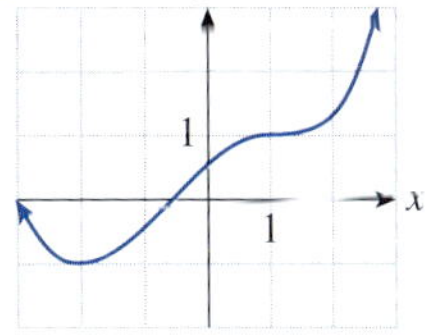

28.

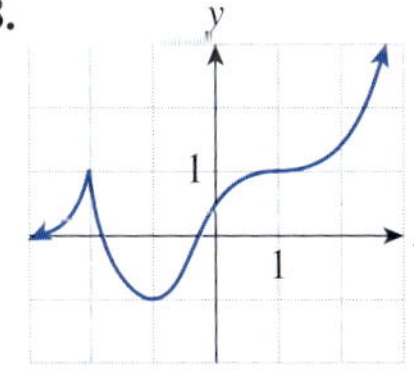

In Exercises 29–32, the graph of the second derivative, *$f''(x)$, is given. Determine the x-coordinates of all points of inflection of $f(x)$, if any. (Assume that $f(x)$ is defined and continuous everywhere in $[-3, 3]$.)*

29.

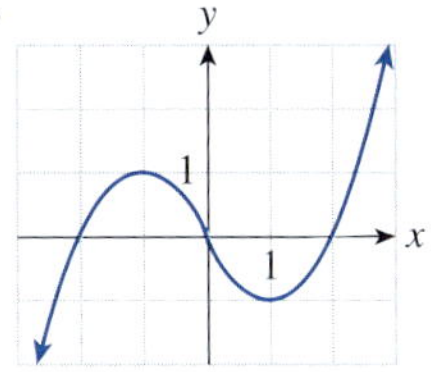

30.

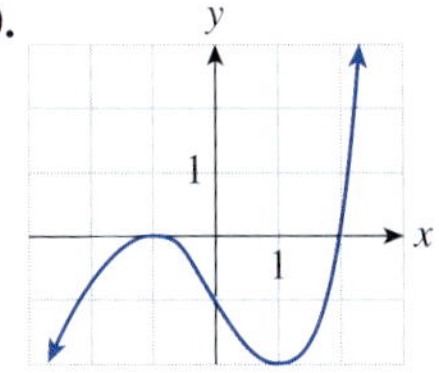

31.

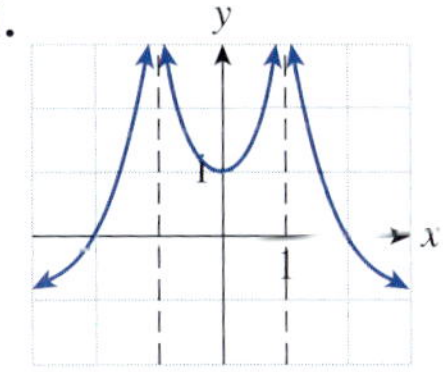

32.

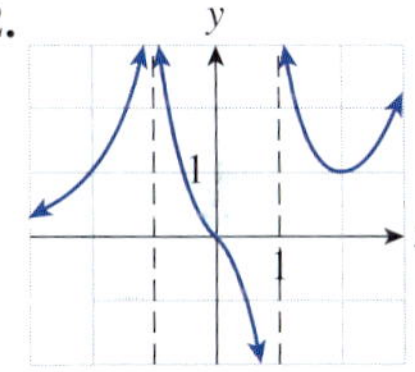

In Exercises 33–44, find the x-coordinates of all critical points of the given function. Determine whether each critical point is a relative maximum, minimum, or neither by first applying the second derivative test, and, if the test fails, by some other method. HINT [See Quick Examples on page 404.]

33. $f(x) = x^2 - 4x + 1$

34. $f(x) = 2x^2 - 2x + 3$

35. $g(x) = x^3 - 12x$

36. $g(x) = 2x^3 - 6x + 3$

37. $f(t) = t^3 - t$

38. $f(t) = -2t^3 + 3t$

39. $f(x) = x^4 - 4x^3$

40. $f(x) = 3x^4 - 2x^3$

41. $f(x) = e^{-x^2}$

42. $f(x) = e^{2-x^2}$

43. $f(x) = xe^{1-x^2}$

44. $f(x) = xe^{-x^2}$

In Exercises 45–54, calculate the derivatives of all orders: $f'(x), f''(x), f'''(x), f^{(4)}(x), \ldots, f^{(n)}(x), \ldots$ HINT [See Quick Examples on page 405.]

45. $f(x) = 4x^2 - x + 1$

46. $f(x) = -3x^3 + 4x$

47. $f(x) = -x^4 + 3x^2$

48. $f(x) = x^4 + x^3$

49. $f(x) = (2x + 1)^4$

50. $f(x) = (-2x + 1)^3$

51. $f(x) = e^{-x}$

52. $f(x) = e^{2x}$

53. $f(x) = e^{3x-1}$

54. $f(x) = 2e^{-x+3}$

APPLICATIONS

55. ***Acceleration on Mars*** If a stone is dropped from a height of 40 meters above the Martian surface, its height in meters after t seconds is given by $s = 40 - 1.9t^2$. What is its acceleration? HINT [See Example 1.]

56. ***Acceleration on the Moon*** If a stone is thrown up at 10 m per second from a height of 100 meters above the surface of the Moon, its height in meters after t seconds is given by $s = 100 + 10t - 0.8t^2$. What is its acceleration? HINT [See Example 1.]

57. ***Motion in a Straight Line*** The position of a particle moving in a straight line is given by $s = t^3 - t^2$ ft after t seconds. Find an expression for its acceleration after a time t. Is its velocity increasing or decreasing when $t = 1$?

58. ***Motion in a Straight Line*** The position of a particle moving in a straight line is given by $s = 3e^t - 8t^2$ ft after t seconds. Find an expression for its acceleration after a time t. Is its velocity increasing or decreasing when $t = 1$?

59. ***Bottled Water Sales*** Annual sales of bottled water in the United States in the period 2000–2008 can be approximated by

$$R(t) = 12t^2 + 500t + 4{,}700 \text{ million gallons} \quad (0 \le t \le 8)$$

where t is time in years since 2000.[27] Were sales of bottled water accelerating or decelerating in 2004? How quickly? HINT [See Example 2.]

60. ***Bottled Water Sales*** Annual U.S. per capita sales of bottled water through the period 2000–2008 can be approximated by

$$Q(t) = 0.04t^2 + 1.5t + 17 \text{ gallons} \quad (0 \le t \le 8)$$

where t is time in years since 2000.[28] Were U.S. per capita sales of bottled water accelerating or decelerating in 2006? How quickly?

61. ***Embryo Development*** The daily oxygen consumption of a bird embryo increases from the time the egg is laid through the time the chick hatches. In a typical galliform bird, the oxygen consumption can be approximated by

$$c(t) = -0.065t^3 + 3.4t^2 - 22t + 3.6 \text{ ml} \quad (8 \le t \le 30)$$

where t is the time (in days) since the egg was laid.[29] (An egg will typically hatch at around $t = 28$.) Use the model to estimate the following (give the units of measurement for each answer and round all answers to two significant digits):

a. The daily oxygen consumption 20 days after the egg was laid
b. The rate at which the oxygen consumption is changing 20 days after the egg was laid
c. The rate at which the oxygen consumption is accelerating 20 days after the egg was laid

62. ***Embryo Development*** The daily oxygen consumption of a turkey embryo increases from the time the egg is laid through the time the chick hatches. In a brush turkey, the oxygen consumption can be approximated by

$$c(t) = -0.028t^3 + 2.9t^2 - 44t + 95 \text{ ml} \quad (20 \le t \le 50)$$

where t is the time (in days) since the egg was laid.[30] (An egg will typically hatch at around $t = 50$.) Use the model to estimate the following (give the units of measurement for each answer and round all answers to two significant digits):

a. The daily oxygen consumption 40 days after the egg was laid
b. The rate at which the oxygen consumption is changing 40 days after the egg was laid
c. The rate at which the oxygen consumption is accelerating 40 days after the egg was laid

63. ***Inflation*** The following graph shows the approximate value of the United States Consumer Price Index (CPI) from December 2006 through July 2007.[31]

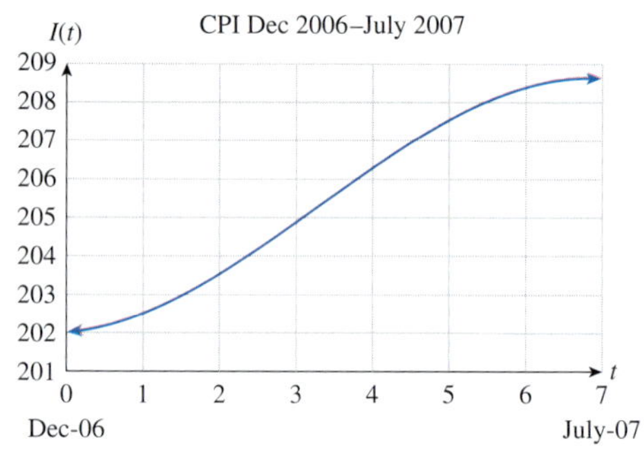

The approximating curve shown on the figure is given by

$$I(t) = -0.04t^3 + 0.4t^2 + 0.1t + 202 \quad (0 \le t \le 7)$$

where t is time in months ($t = 0$ represents December 2006).

a. Use the model to estimate the monthly inflation rate in February 2007 ($t = 2$). [Recall that the inflation *rate* is $I'(t)/I(t)$.]
b. Was inflation slowing or speeding up in February 2007?
c. When was inflation speeding up? When was inflation slowing? HINT [See Example 3.]

64. ***Inflation*** The following graph shows the approximate value of the U.S. Consumer Price Index (CPI) from September 2004 through November 2005.[32]

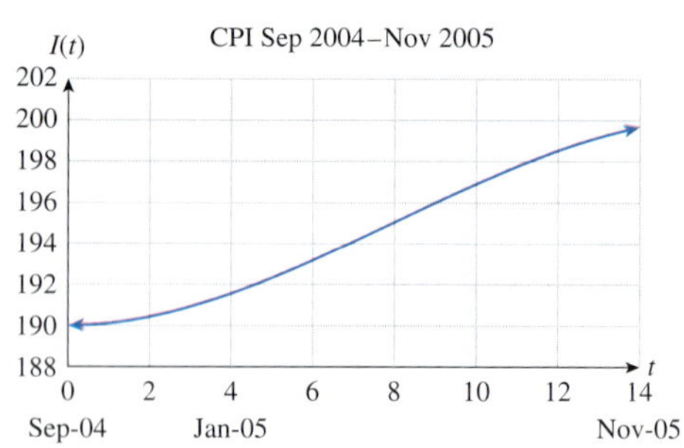

[27] Source for data: Beverage Marketing Corporation (www.bottledwater.org).

[28] Ibid.

[29] The model approximates graphical data published in the article "The Brush Turkey" by Roger S. Seymour, *Scientific American*, December, 1991, pp. 108–114.

[30] Ibid.

[31] The CPI is compiled by the Bureau of Labor Statistics and is based upon a 1982 value of 100. For instance, a CPI of 200 means the CPI has doubled since 1982. Source: InflationData.com (www.inflationdata.com).

[32] Ibid.

The approximating curve shown on the figure is given by

$$I(t) = -0.005t^3 + 0.12t^2 - 0.01t + 190 \quad (0 \le t \le 14)$$

where t is time in months ($t = 0$ represents September 2004).

a. Use the model to estimate the monthly inflation rate in July 2005 ($t = 10$). [Recall that the inflation *rate* is $I'(t)/I(t)$.]
b. Was inflation slowing or speeding up in July 2005?
c. When was inflation speeding up? When was inflation slowing? HINT [See Example 3.]

65. *Inflation* The following graph shows the approximate value of the U.S. Consumer Price Index (CPI) from July 2005 through March 2006.[33]

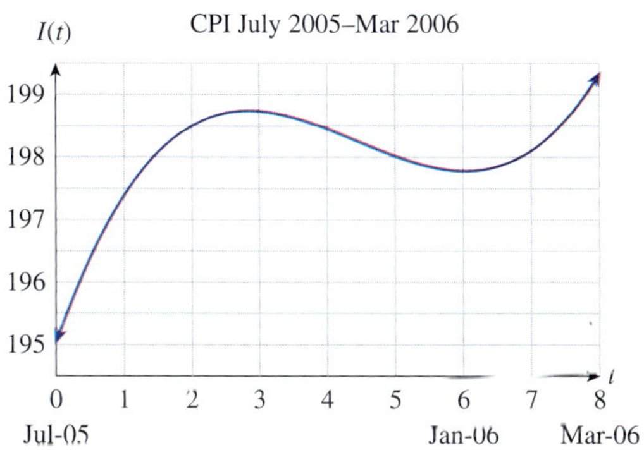

The approximating curve shown on the figure is given by

$$I(t) = 0.06t^3 - 0.8t^2 + 3.1t + 195 \quad (0 \le t \le 8)$$

where t is time in months ($t = 0$ represents July 2005).

a. Use the model to estimate the monthly inflation rates in December 2005 and February 2006 ($t = 5$ and $t = 7$).
b. Was inflation slowing or speeding up in February 2006?
c. When was inflation decreasing? When was inflation increasing?

66. *Inflation* The following graph shows the approximate value of the U.S. Consumer Price Index (CPI) from March 2006 through May 2007.[34]

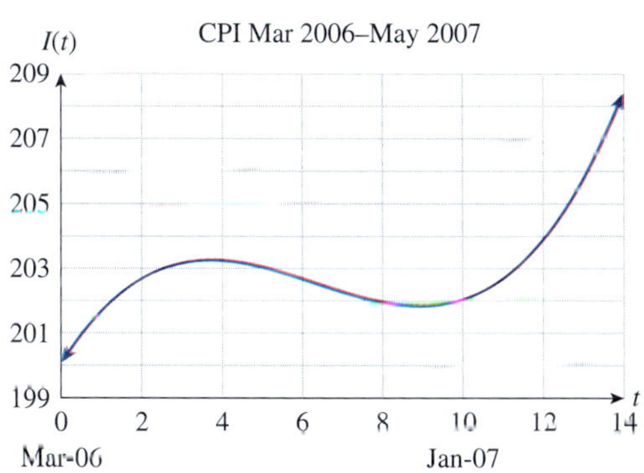

[33] The CPI is compiled by the Bureau of Labor Statistics and is based upon a 1982 value of 100. For instance, a CPI of 200 means the CPI has doubled since 1982. Source: InflationData.com (www.inflationdata.com).

[34] Ibid.

The approximating curve shown on the figure is given by

$$I(t) = 0.02t^3 - 0.38t^2 + 2t + 200 \quad (0 \le t \le 14)$$

where t is time in months ($t = 0$ represents March, 2006).

a. Use the model to estimate the monthly inflation rates in September 2006 and January 2007 ($t = 6$ and $t = 10$).
b. Was inflation slowing or speeding up in January 2007?
c. When was inflation decreasing? When was inflation increasing?

67. *Scientific Research* The percentage of research articles in the prominent journal *Physical Review* that were written by researchers in the United States during the years 1983–2003 can be modeled by

$$P(t) = 25 + \frac{36}{1 + 0.06(0.7)^{-t}}$$

where t is time in years since 1983.[35] The graphs of P, P', and P'' are shown here:

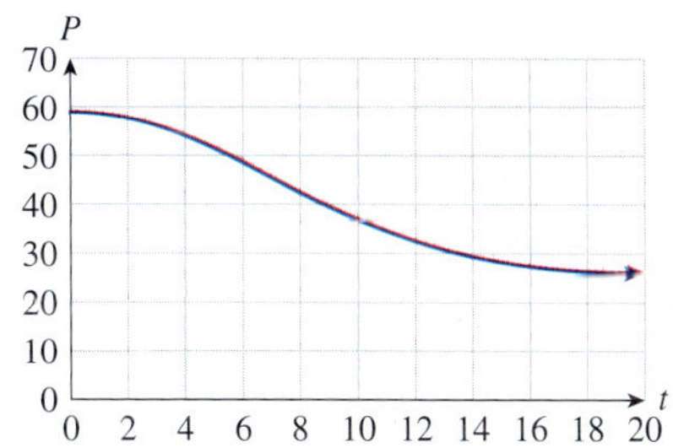

Graph of P

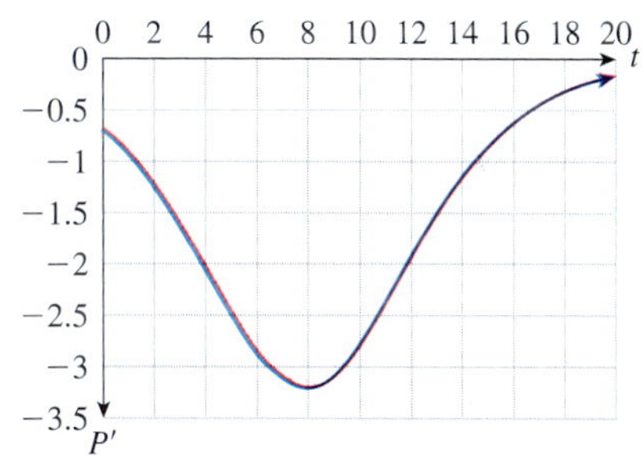

Graph of P'

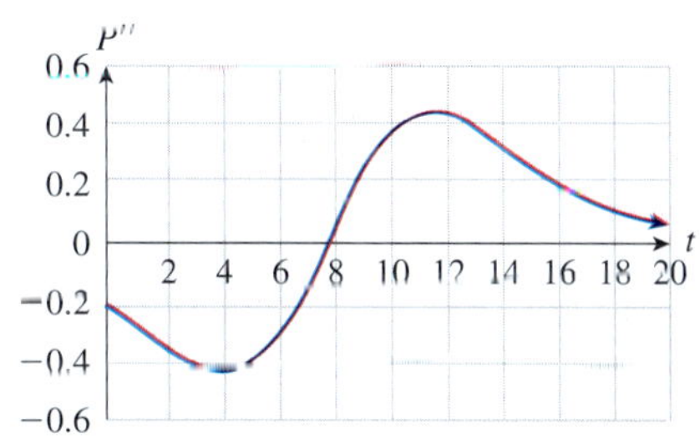

Graph of P''

[35] Source: The American Physical Society/*New York Times*, May 3, 2003, p. A1.

Determine, to the nearest whole number, the values of t for which the graph of P is concave up, and where it is concave down, and locate any points of inflection. What does the point of inflection tell you about science articles? HINT [See Example 4.]

68. ***Scientific Research*** The number of research articles in the prominent journal *Physical Review* that were written by researchers in Europe during the years 1983–2003 can be modeled by

$$P(t) = \frac{7.0}{1 + 5.4(1.2)^{-t}}$$

where t is time in years since 1983.[36] The graphs of P, P', and P'' are shown here:

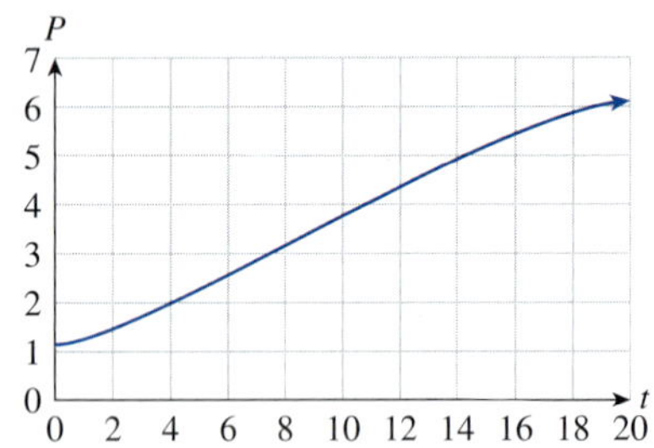

Graph of P

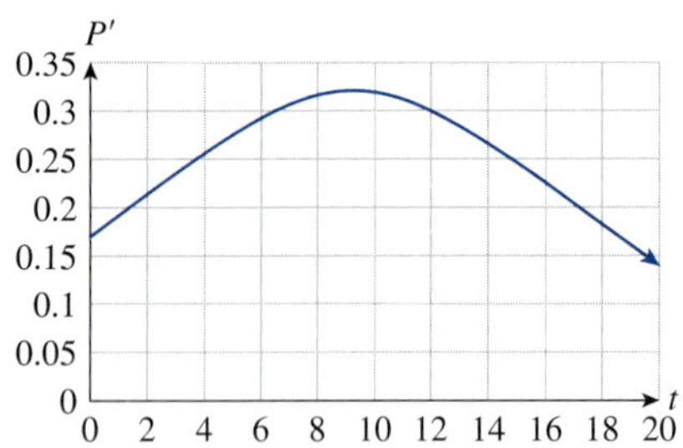

Graph of P'

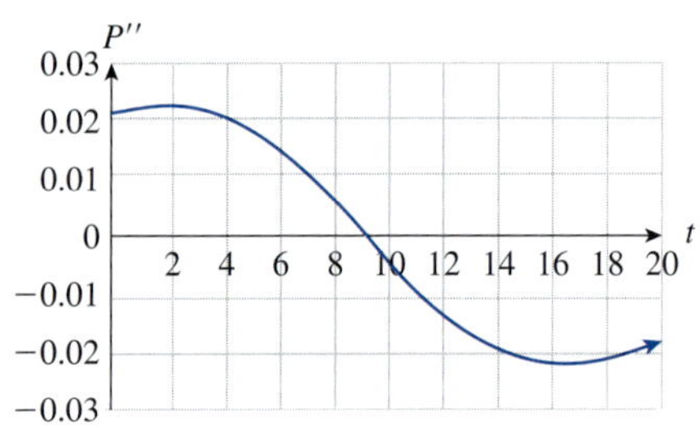

Graph of P''

Determine, to the nearest whole number, the values of t for which the graph of P is concave up, and where it is concave down, and locate any points of inflection. What does the point of inflection tell you about science articles?

[36] Source: The American Physical Society/*New York Times*, May 3, 2003, p. A1.

69. ***Embryo Development*** Here are sketches of the graphs of c, c', and c'' from Exercise 61:

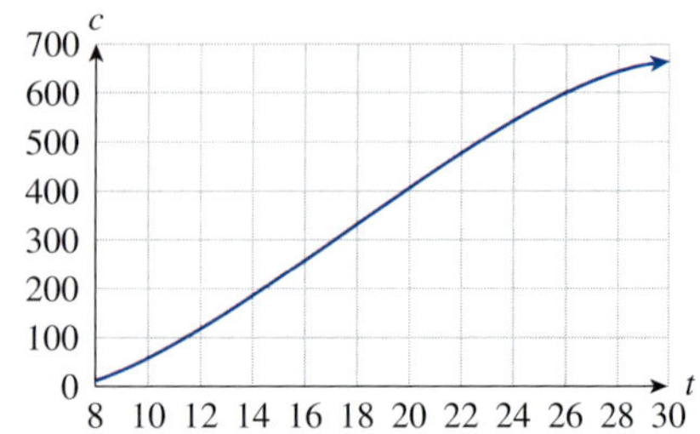

Graph of c

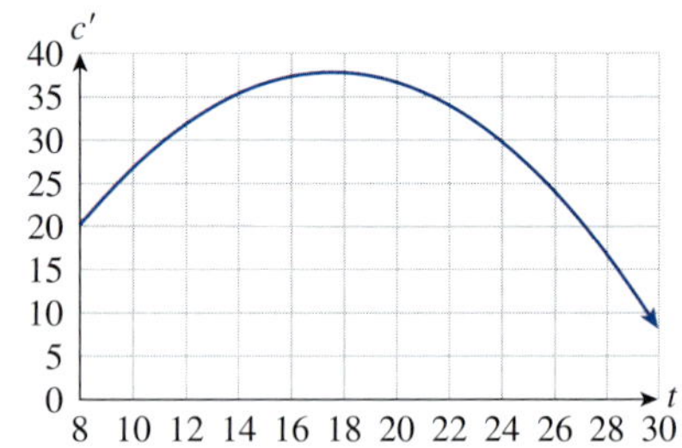

Graph of c'

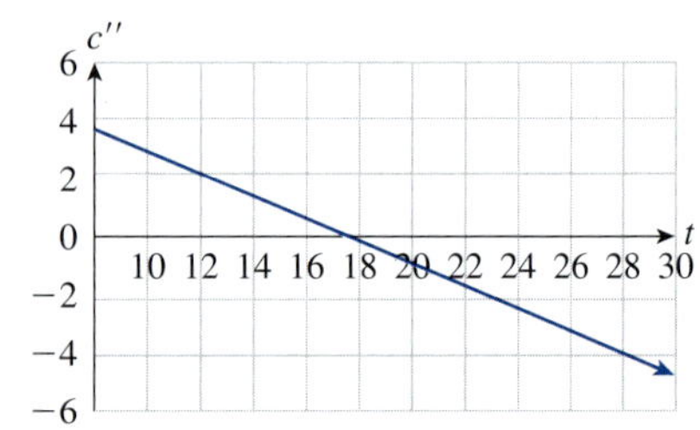

Graph of c''

Multiple choice:

a. The graph of c' **(A)** has a point of inflection **(B)** has no points of inflection in the range shown.

b. At around 18 days after the egg is laid, daily oxygen consumption is: **(A)** at a maximum, **(B)** increasing at a maximum rate, or **(C)** just beginning to decrease.

c. For $t > 18$ days, the oxygen consumption is **(A)** increasing at a decreasing rate, **(B)** decreasing at an increasing rate, or **(C)** increasing at an increasing rate.

70. ***Embryo Development*** Here are sketches of the graphs of c, c', and c'' from Exercise 62:

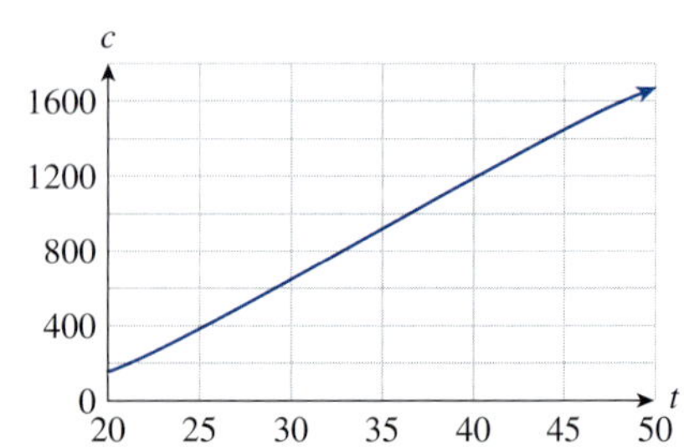

Graph of c

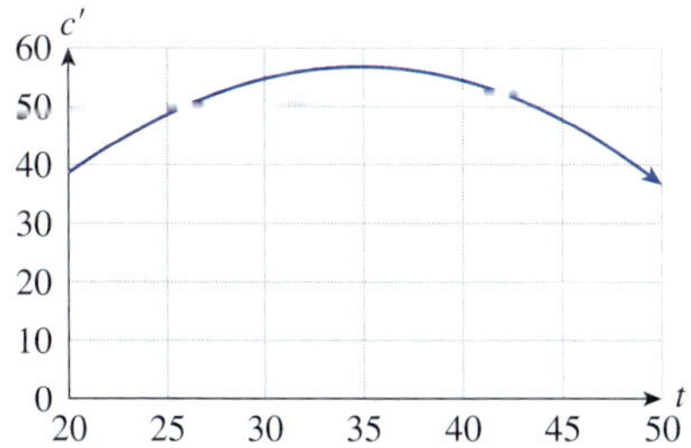

Graph of c'

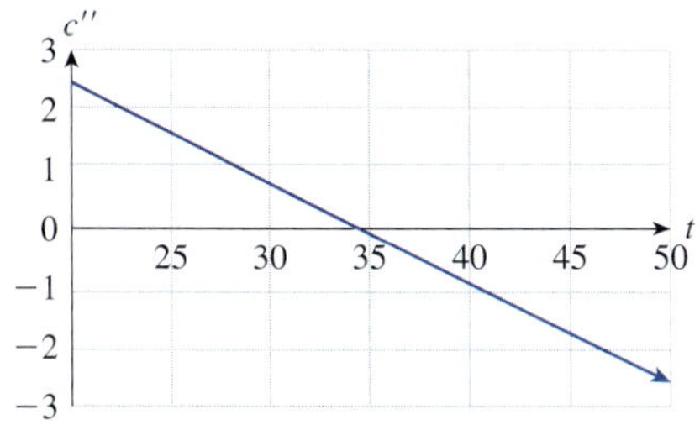

Graph of c''

Multiple choice:

a. The graph of c: **(A)** has points of inflection, **(B)** has no points of inflection, or **(C)** may or may not have a point of inflection, but the graphs do not provide enough information.

b. At around 35 days after the egg is laid, the rate of change of daily oxygen consumption is: **(A)** at a maximum, **(B)** increasing at a maximum rate, or **(C)** just becoming negative.

c. For $t < 35$ days, the oxygen consumption is: **(A)** increasing at an increasing rate, **(B)** increasing at a decreasing rate, or **(C)** decreasing at an increasing rate.

71. T ***Subprime Mortgages*** The percentage of U.S.-issued mortgages that are subprime can be approximated by

$$A(t) = \frac{15.0}{1 + 8.6(1.8)^{-t}} \text{ percent } \quad (0 \le t \le 8)$$

t years after the start of 2000.[37] Graph the function as well as its first and second derivatives. Determine, to the nearest whole number, the values of t for which the graph of A is concave up and concave down, and the t-coordinate of any points of inflection. What does the point of inflection tell you about subprime mortgages? HINT [To graph the second derivative, see the note in the margin on page 403.]

72. T ***Subprime Mortgage Debt*** The approximate value of subprime (normally classified as risky) mortgage debt outstanding in the United States can be approximated by

$$A(t) = \frac{1{,}350}{1 + 4.2(1.7)^{-t}} \text{ billion dollars } \quad (0 \le t \le 8)$$

t years after the start of 2000.[38] Graph the function as well as its first and second derivatives. Determine, to the nearest whole number, the values of t for which the graph of A is concave up and concave down, and the t-coordinate of any points of inflection. What does the point of inflection tell you about subprime mortgages? HINT [To graph the second derivative, see the note in the margin on page 403.]

73. ***Epidemics*** The following graph shows the total number n of people (in millions) infected in an epidemic as a function of time t (in years):

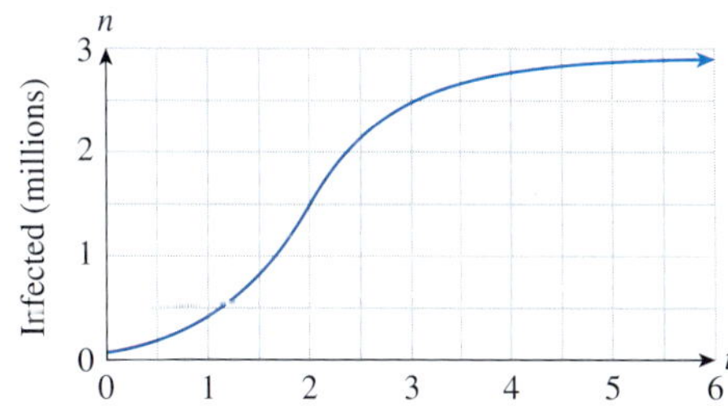

a. When to the nearest year was the rate of new infection largest?

b. When could the Centers for Disease Control and Prevention announce that the rate of new infection was beginning to drop? HINT [See Example 4.]

74. ***Sales*** The following graph shows the total number of Pomegranate Q4 computers sold since their release (t is in years):

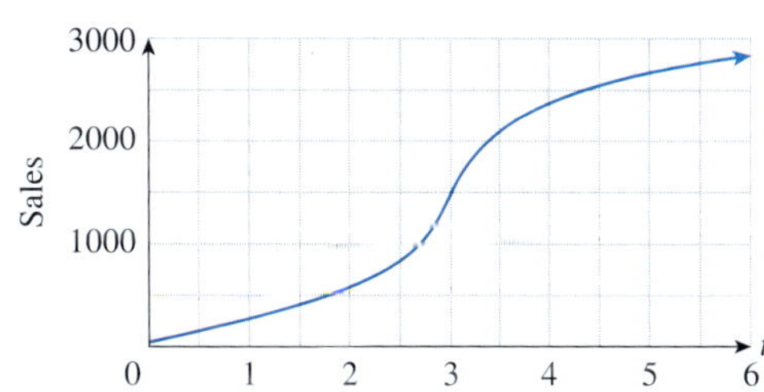

a. When were the computers selling fastest?

b. Explain why this graph might look as it does. HINT [See Example 4.]

[37] Sources: Mortgage Bankers Association, UBS.

[38] 2008 figure is an estimate.
Source: www.data360.org/dataset.aspx?Data_Set_Id=9549.

75. ***Industrial Output*** The following graph shows the yearly industrial output (measured in billions of zonars) of the Republic of Mars over a seven-year period:

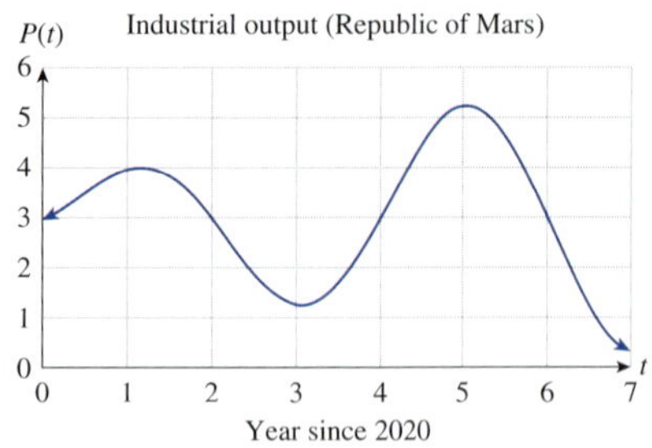

a. When to the nearest year did the rate of change of yearly industrial output reach a maximum?
b. When to the nearest year did the rate of change of yearly industrial output reach a minimum?
c. When to the nearest year does the graph first change from concave down to concave up? The result tells you that:

(A) In that year the rate of change of industrial output reached a minimum compared with nearby years.
(B) In that year the rate of change of industrial output reached a maximum compared with nearby years.

76. ***Profits*** The following graph shows the yearly profits of Gigantic Conglomerate, Inc. (GCI) from 2020 to 2035:

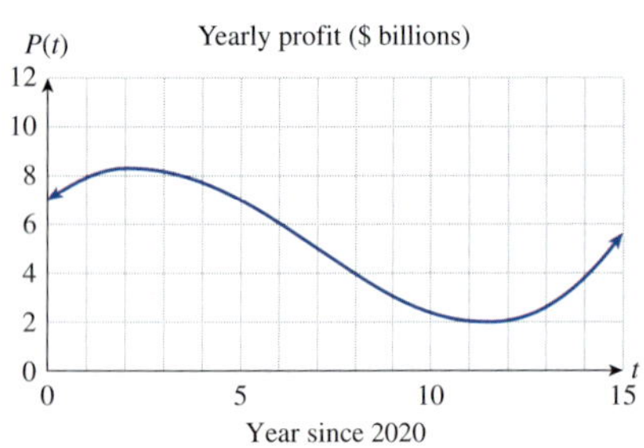

a. Approximately when were the profits rising most rapidly?
b. Approximately when were the profits falling most rapidly?
c. Approximately when could GCI's board of directors legitimately tell stockholders that they had "turned the company around"?

77. ▼ ***Education and Crime*** The following graph shows a striking relationship between the total prison population and the average combined SAT score in the United States:

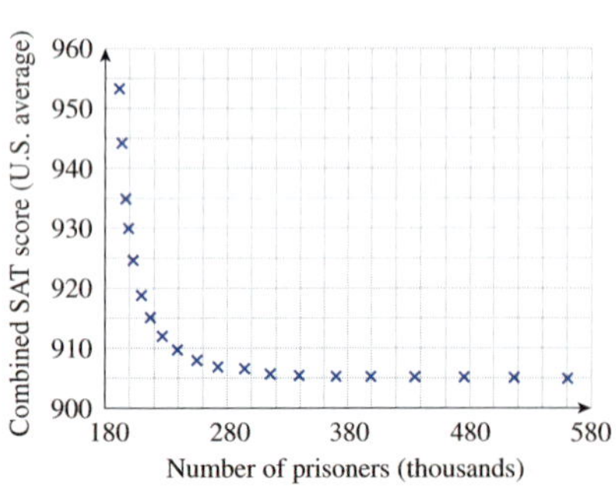

These data can be accurately modeled by

$$S(n) = 904 + \frac{1{,}326}{(n - 180)^{1.325}}. \quad (192 \le n \le 563)$$

Here, $S(n)$ is the combined U.S. average SAT score at a time when the total U.S. prison population was n thousand.[39]

a. Are there any points of inflection on the graph of S?
b. What does the concavity of the graph of S tell you about prison populations and SAT scores?

78. ▼ ***Education and Crime*** Refer back to the model in the preceding exercise.

a. Are there any points of inflection on the graph of S'?
b. When is S'' a maximum? Interpret your answer in terms of prisoners and SAT scores.

79. ▼ ***Patents*** In 1965, the economist F. M. Scherer modeled the number, n, of patents produced by a firm as a function of the size, s, of the firm (measured in annual sales in millions of dollars). He came up with the following equation based on a study of 448 large firms:[40]

$$n = -3.79 + 144.42s - 23.86s^2 + 1.457s^3.$$

a. Find $\left.\frac{d^2n}{ds^2}\right|_{s=3}$. Is the rate at which patents are produced as the size of a firm goes up increasing or decreasing with size when $s = 3$? Comment on Scherer's words, ". . . we find diminishing returns dominating."

b. Find $\left.\frac{d^2n}{ds^2}\right|_{s=7}$ and interpret the answer.

c. Find the s-coordinate of any points of inflection and interpret the result.

80. ▼ ***Returns on Investments*** A company finds that the number of new products it develops per year depends on the size of its annual R&D budget, x (in thousands of dollars), according to the formula

$$n(x) = -1 + 8x + 2x^2 - 0.4x^3.$$

a. Find $n''(1)$ and $n''(3)$, and interpret the results.
b. Find the size of the budget that gives the largest rate of return as measured in new products per dollar (again, called the point of diminishing returns).

81. T ▼ ***Oil Imports from Mexico*** Daily oil production in Mexico and daily U.S. oil imports from Mexico during 2005–2009 can be approximated by

$$P(t) = 3.9 - 0.10t \text{ million barrels} \quad (5 \le t \le 9)$$
$$I(t) = 2.1 - 0.11t \text{ million barrels} \quad (5 \le t \le 9)$$

[39] The model is the authors' based on data for the years 1967–1989. Sources: *Sourcebook of Criminal Justice Statistics*, 1990, p. 604/ Educational Testing Service.

[40] Source: F. M. Scherer, "Firm Size, Market Structure, Opportunity, and the Output of Patented Inventions," *American Economic Review* 55 (December 1965): pp. 1097–1125.

where t is time in years since the start of 2000.[41]

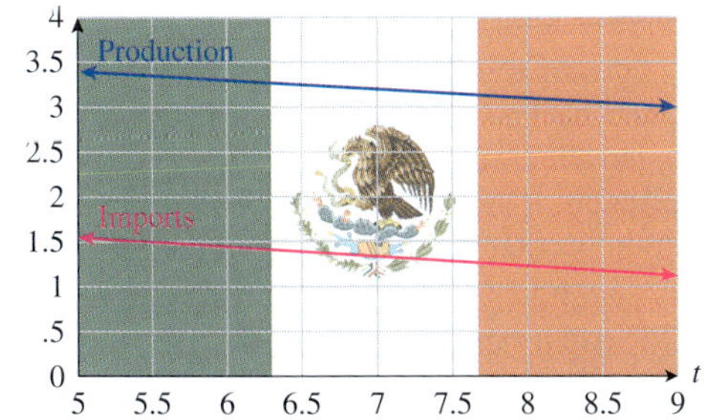

Graph the function $I(t)/P(t)$ and its derivative. Is the graph of $I(t)/P(t)$ concave up or concave down? The concavity of $I(t)/P(t)$ tells you that:

(A) The percentage of oil produced in Mexico that was exported to the United States was decreasing.
(B) The percentage of oil produced in Mexico that was not exported to the United States was increasing.
(C) The percentage of oil produced in Mexico that was exported to the United States was decreasing at a slower rate.
(D) The percentage of oil produced in Mexico that was exported to the United States was decreasing at a faster rate.

82. T ▼ ***Oil Imports from Mexico*** Repeat Exercise 81 using instead the models for 2000–2004 shown below:

$$P(t) = 3.0 + 0.13t \text{ million barrels} \quad (0 \le t \le 4)$$
$$I(t) = 1.4 + 0.06t \text{ million barrels} \quad (0 \le t \le 4)$$

(t is time in years since the start of 2000).[42]

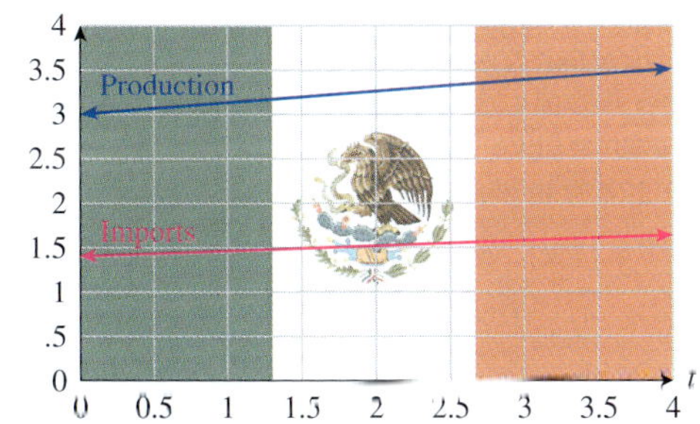

83. ◆ ***Logistic Models*** Let

$$f(x) = \frac{N}{1 + Ab^{-x}}$$

for constants N, A, and b (A and b positive and $b \ne 1$). Show that f has a single point of inflection at $x = \ln A/\ln b$.

84. ◆ ***Logistic Models*** Let

$$f(x) = \frac{N}{1 + Ae^{-kx}}$$

for constants N, A, and k (A and k positive). Show that f has a single point of inflection at $x = \ln A/k$.

85. T ***Population: Puerto Rico*** The population of Puerto Rico in 1950–2025 can be approximated by

$$P(t) = \frac{4{,}500}{1 + 1.1466\,(1.0357)^{-t}} \text{ thousand people} \quad (0 \le t \le 75)$$

(t is the year since 1950).[43] Use the result of Exercise 83 to find the location of the point of inflection in the graph of P. What does the result tell you about the population of Puerto Rico?

86. T ***Population: Virgin Islands*** The population of the Virgin Islands can be approximated by

$$P(t) = \frac{110}{1 + 2.3596\,(1.0767)^{-t}} \text{ thousand people} \quad (0 \le t \le 75)$$

(t is the year since 1950).[44] Use the result of Exercise 83 to find the location of the point of inflection in the graph of P. What does the result tell you about the population of the Virgin Islands?

87. T ▼ ***Asset Appreciation*** You manage a small antique store that owns a collection of Louis XVI jewelry boxes. Their value v is increasing according to the formula

$$v = \frac{10{,}000}{1 + 500e^{-0.5t}}$$

where t is the number of years from now. You anticipate an inflation rate of 5% per year, so that the present value of an item that will be worth \$$v$ in t years' time is given by

$$p = v(1.05)^{-t}.$$

What is the greatest rate of increase of the value of your antiques, and when is this rate attained?

88. T ▼ ***Harvesting Forests*** The following equation models the approximate volume in cubic feet of a typical Douglas fir tree of age t years[45]:

$$V = \frac{22{,}514}{1 + 22{,}514t^{-2.55}}.$$

The lumber will be sold at \$10 per cubic foot, and you do not expect the price of lumber to appreciate in the foreseeable future. On the other hand, you anticipate a general inflation

[41] Source for data: Energy Information Administration/Pemex (http://www.eia.doe.gov)

[42] Ibid.

[43] Figures from 2010 on are U.S. census projections. Source for data: The 2008 Statistical Abstract (www.census.gov/).

[44] Ibid.

[45] The model is the authors', and is based on data in *Environmental and Natural Resource Economics* by Tom Tietenberg, Third Edition (New York: HarperCollins, 1992), p. 282.

rate of 5% per year, so that the present value of an item that will be worth $\$v$ in t years time is given by

$$p = v(1.05)^{-t}.$$

What is the largest rate of increase of the value of a fir tree, and when is this rate attained?

89. T ▼ ***Asset Appreciation*** As the financial consultant to a classic auto dealership, you estimate that the total value of its collection of 1959 Chevrolets and Fords is given by the formula

$$v = 300{,}000 + 1{,}000t^2$$

where t is the number of years from now. You anticipate a continuous inflation rate of 5% per year, so that the discounted (present) value of an item that will be worth $\$v$ in t years' time is given by

$$p = ve^{-0.05t}.$$

When is the value of the collection of classic cars increasing most rapidly? When is it decreasing most rapidly?

90. T ▼ ***Plantation Management*** The value of a fir tree in your plantation increases with the age of the tree according to the formula

$$v = \frac{20t}{1 + 0.05t}$$

where t is the age of the tree in years. Given a continuous inflation rate of 5% per year, the discounted (present) value of a newly planted seedling is

$$p = ve^{-0.05t}.$$

When is the discounted value of a tree increasing most rapidly? Decreasing most rapidly?

COMMUNICATION AND REASONING EXERCISES

91. Complete the following: If the graph of a function is concave up on its entire domain, then its second derivative is ______ on the domain.

92. Complete the following: If the graph of a function is concave up on its entire domain, then its first derivative is ______ on the domain.

93. Daily sales of Kent's Tents reached a maximum in January 2002 and declined to a minimum in January 2003 before starting to climb again. The graph of daily sales shows a point of inflection at June 2002. What is the significance of the point of inflection?

94. The graph of daily sales of Luddington's Wellington boots is concave down, although sales continue to increase. What properties of the graph of daily sales versus time are reflected in the following behaviors?

a. a point of inflection next year
b. a horizontal asymptote

95. ▼ Company A's profits satisfy $P(0) = \$1$ million, $P'(0) = \$1$ million per year, and $P''(0) = -\$1$ million per year per year. Company B's profits satisfy $P(0) = \$1$ million, $P'(0) = -\$1$ million per year, and $P''(0) = \$1$ million per year per year. There are no points of inflection in either company's profit curve. Sketch two pairs of profit curves: one in which Company A ultimately outperforms Company B and another in which Company B ultimately outperforms Company A.

96. ▼ Company C's profits satisfy $P(0) = \$1$ million, $P'(0) = \$1$ million per year, and $P''(0) = -\$1$ million per year per year. Company D's profits satisfy $P(0) = \$0$ million, $P'(0) = \$0$ million per year, and $P''(0) = \$1$ million per year per year. There are no points of inflection in either company's profit curve. Sketch two pairs of profit curves: one in which Company C ultimately outperforms Company D and another in which Company D ultimately outperforms Company C.

97. ▼ Explain geometrically why the derivative of a function has a relative extremum at a point of inflection, if it is defined there. Which points of inflection give rise to relative maxima in the derivative?

98. ▼ If we regard position, s, as a function of time, t, what is the significance of the *third* derivative, $s'''(t)$? Describe an everyday scenario in which this arises.

5.4 Analyzing Graphs

Mathematical curves are beautiful—their subtle form can be imitated by only the best of artists—and calculus gives us the tools we need to probe their secrets. While it is easy to use graphing technology to draw a graph, we must use calculus to understand what we are seeing. Following is a list of some of the most interesting features of the graph of a function.

Features of a Graph

1. ***The x- and y-intercepts:*** If $y = f(x)$, find the x-intercept(s) by setting $y = 0$ and solving for x; find the y-intercept by setting $x = 0$ and solving for y:

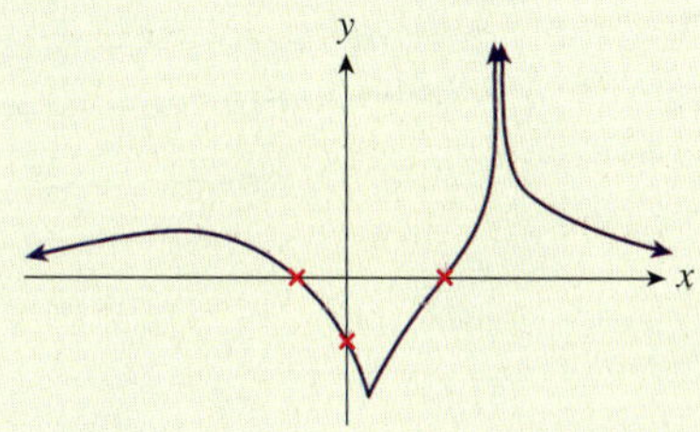

2. ***Extrema:*** Use the techniques of Section 5.1 to locate the maxima and minima:

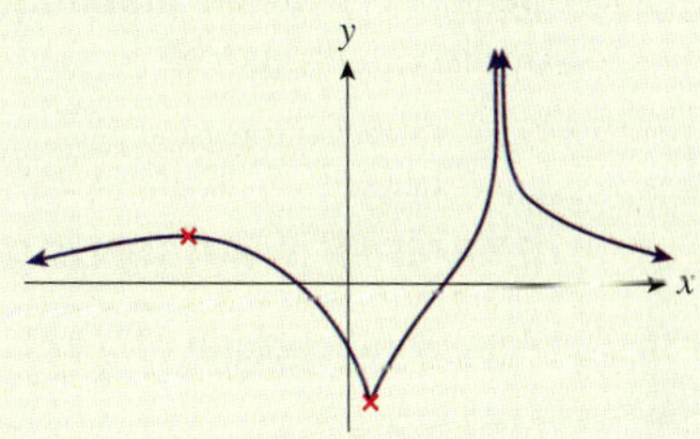

3. ***Points of inflection:*** Use the techniques of Section 5.2 to locate the points of inflection:

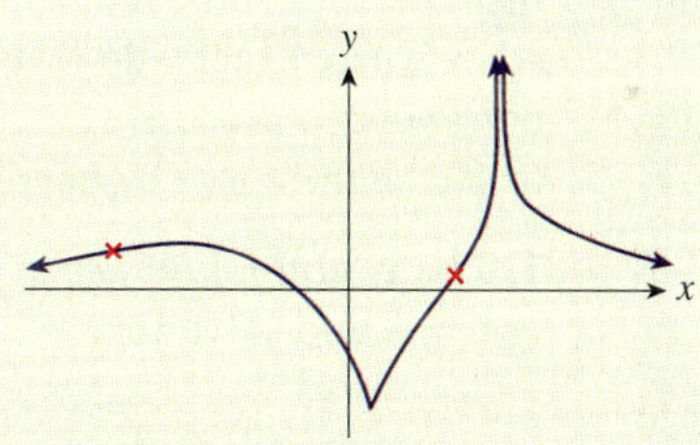

4. ***Behavior near points where the function is not defined:*** If $f(x)$ is not defined at $x = a$, consider $\lim_{x\to a^-} f(x)$ and $\lim_{x\to a^+} f(x)$ to see how the graph of f behaves as x approaches a:

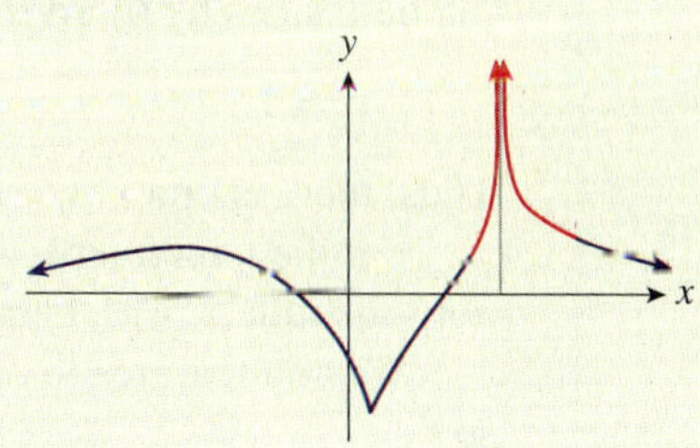

5. ***Behavior at infinity:*** Consider $\lim_{x\to-\infty} f(x)$ and $\lim_{x\to+\infty} f(x)$ if appropriate, to see how the graph of f behaves far to the left and right:

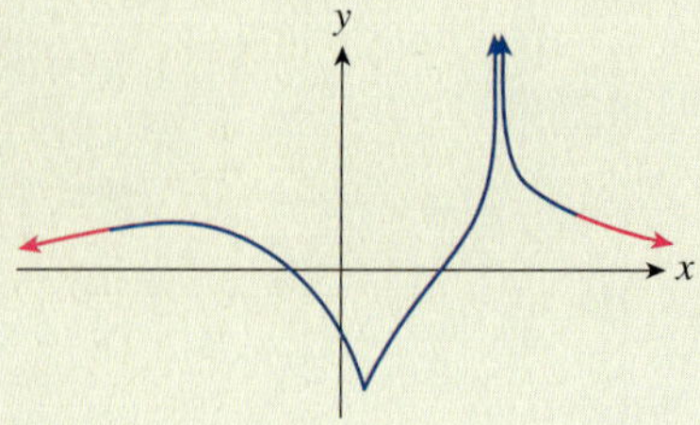

Note It is sometimes difficult or impossible to solve all of the equations that come up in Steps 1, 2, and 3 of the previous analysis. As a consequence, we might not be able to say exactly where the x-intercept, extrema, or points of inflection are. When this happens, we will use graphing technology to assist us in determining accurate numerical approximations. ■

EXAMPLE 1 Analyzing a Graph

Analyze the graph of $f(x) = \dfrac{1}{x} - \dfrac{1}{x^2}$.

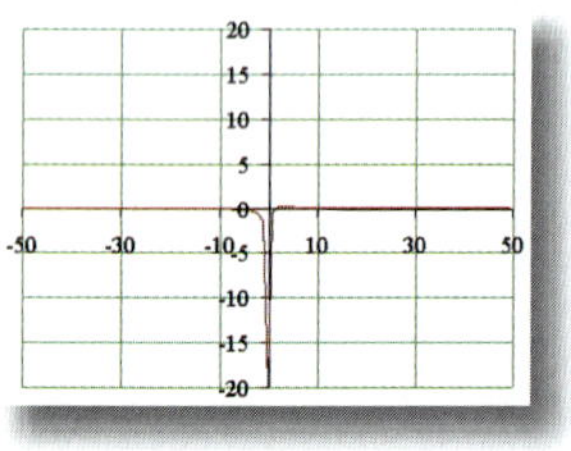

$-50 \le x \le 50, -20 \le y \le 20$

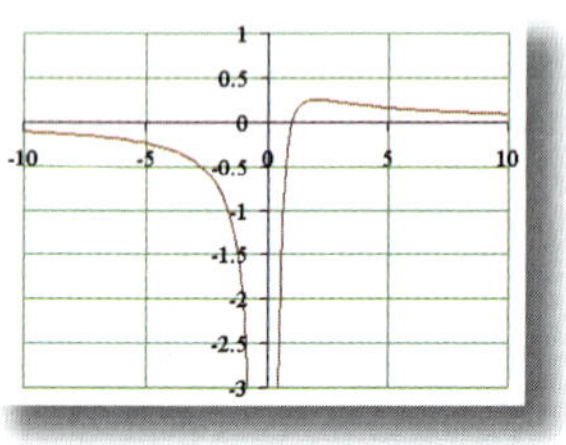

$-10 \le x \le 10, -3 \le y \le 1$

Figure 34

Solution The graph, as drawn using graphing technology, is shown in Figure 34, using two different viewing windows. (Note that $x = 0$ is not in the domain of f.) The second window in Figure 34 seems to show the features of the graph better than the first. Does the second viewing window include *all* the interesting features of the graph? Or are there perhaps some interesting features to the right of $x = 10$ or to the left of $x = -10$? Also, where exactly do features like maxima, minima, and points of inflection occur? In our five-step process of analyzing the interesting features of the graph, we will be able to sketch the curve by hand, and also answer these questions.

1. ***The x- and y-intercepts:*** We consider $y = \dfrac{1}{x} - \dfrac{1}{x^2}$. To find the x-intercept(s), we set $y = 0$ and solve for x:

$$0 = \frac{1}{x} - \frac{1}{x^2}$$
$$\frac{1}{x} = \frac{1}{x^2}.$$

Multiplying both sides by x^2 (we know that x cannot be zero, so we are not multiplying both sides by 0) gives

$$x = 1.$$

Thus, there is one x-intercept (which we can see in Figure 34) at $x = 1$.

For the y-intercept, we would substitute $x = 0$ and solve for y. However, we cannot substitute $x = 0$; because $f(0)$ is not defined, the graph does not meet the y-axis.

We add features to our freehand sketch as we go. Figure 35 shows what we have so far.

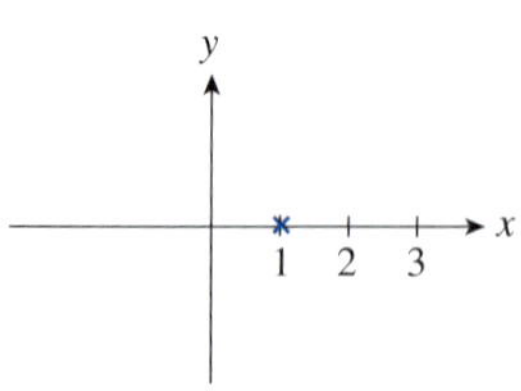

Figure 35

2. ***Relative extrema:*** We calculate $f'(x) = -\frac{1}{x^2} + \frac{2}{x^3}$. To find any stationary points, we set the derivative equal to 0 and solve for x:

$$-\frac{1}{x^2} + \frac{2}{x^3} = 0$$
$$\frac{1}{x^2} = \frac{2}{x^3}$$
$$x = 2.$$

Thus, there is one stationary point, at $x = 2$. We can use a test point to the right to determine that this stationary point is a relative maximum:

x	1 (Intercept)	2	3 (Test point)
$y = \frac{1}{x} - \frac{1}{x^2}$	0	$\frac{1}{4}$	$\frac{2}{9}$

The only possible singular point is at $x = 0$ because $f'(0)$ is not defined. However, $f(0)$ is not defined either, so there are no singular points. Figure 36 shows our graph so far.

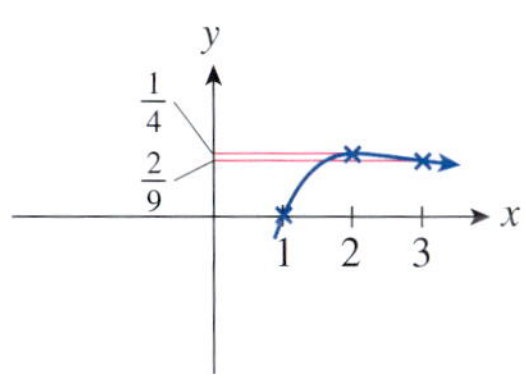

Figure 36

3. ***Points of inflection:*** We calculate $f''(x) = \frac{2}{x^3} - \frac{6}{x^4}$. To find points of inflection, we set the second derivative equal to 0 and solve for x:

$$\frac{2}{x^3} - \frac{6}{x^4} = 0$$
$$\frac{2}{x^3} = \frac{6}{x^4}$$
$$2x = 6$$
$$x = 3.$$

Figure 34 confirms that the graph of f changes from being concave down to being concave up at $x = 3$, so this is a point of inflection. $f''(x)$ is not defined at $x = 0$, but that is not in the domain, so there are no other points of inflection. In particular, the graph must be concave down in the whole region $(-\infty, 0)$, as we can see by calculating the second derivative at any one point in that interval: $f''(-1) = -8 < 0$.

Figure 37 shows our graph so far (we extended the curve near $x = 3$ to suggest a point of inflection at $x = 3$).

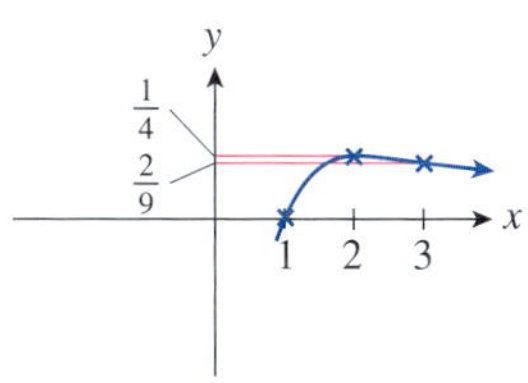

Figure 37

4. ***Behavior near points where f is not defined:*** The only point where $f(x)$ is not defined is $x = 0$. From the graph, $f(x)$ appears to go to $-\infty$ as x approaches 0 from either side. To calculate these limits, we rewrite $f(x)$:

$$f(x) = \frac{1}{x} - \frac{1}{x^2} = \frac{x-1}{x^2}$$

Now, if x is close to 0 (on either side), the numerator $x - 1$ is close to -1 and the denominator is a very small but positive number. The quotient is therefore a negative number of very large magnitude. Therefore,

$$\lim_{x \to 0^-} f(x) = -\infty$$

and

$$\lim_{x \to 0^+} f(x) = -\infty.$$

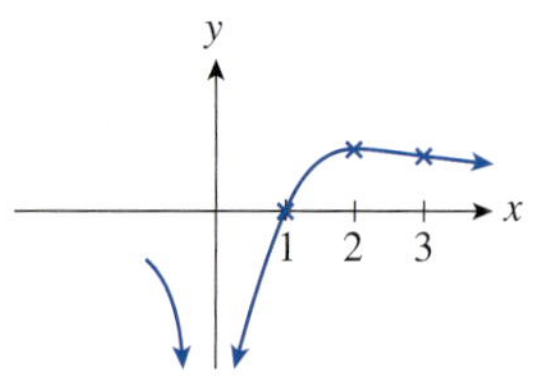

Figure 38

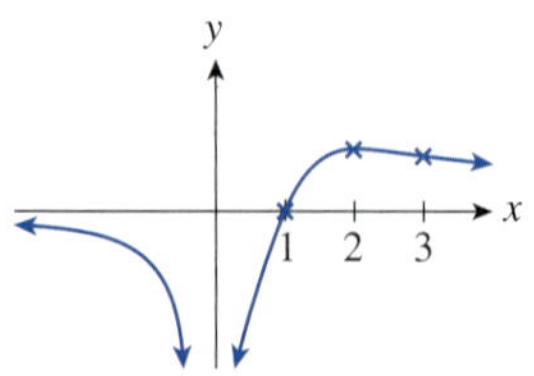

Figure 39

From these limits, we see the following:

(1) Immediately to the *left* of $x = 0$, the graph plunges down toward $-\infty$.

(2) Immediately to the *right* of $x = 0$, the graph also plunges down toward $-\infty$.

Figure 38 shows our graph with these features added. We say that f has a **vertical asymptote** at $x = 0$, meaning that the points on the graph of f get closer and closer to points on a vertical line (the y-axis in this case) further and further from the origin.

5. ***Behavior at infinity:*** Both $1/x$ and $1/x^2$ go to 0 as x goes to $-\infty$ or $+\infty$; that is,

$$\lim_{x \to -\infty} f(x) = 0$$

and

$$\lim_{x \to +\infty} f(x) = 0.$$

Thus, on the extreme left and right of our picture, the height of the curve levels off toward zero. Figure 39 shows the completed freehand sketch of the graph.

We say that f has a **horizontal asymptote** at $y = 0$. (Notice another thing: We haven't plotted a single point to the left of the y-axis, and yet we have a pretty good idea of what the curve looks like there! Compare the technology-drawn curve in Figure 34.)

In summary, there is one x-intercept at $x = 1$; there is one relative maximum (which, we can now see, is also an absolute maximum) at $x = 2$; there is one point of inflection at $x = 3$, where the graph changes from being concave down to concave up. There is a vertical asymptote at $x = 0$, on both sides of which the graph goes down toward $-\infty$, and a horizontal asymptote at $y = 0$.

EXAMPLE 2 Analyzing a Graph

Analyze the graph of $f(x) = \dfrac{2x}{3} - (x - 2)^{2/3}$.

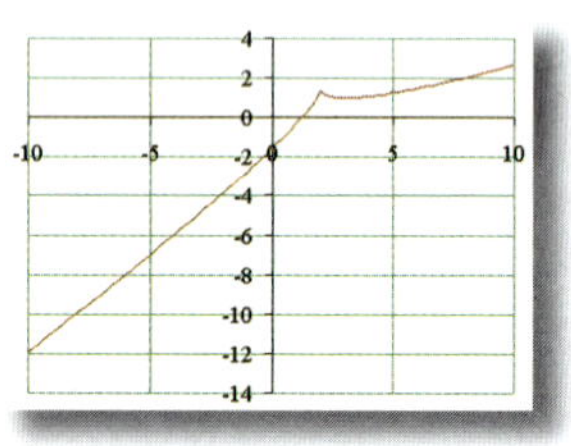

Technology:
`2*x/3-((x-2)^2)^(1/3)`

Figure 40

Solution Figure 40 shows a technology-generated version of the graph. Note that in the technology formulation $(x - 2)^{2/3}$ is written as $[(x - 2)^2]^{1/3}$ to avoid problems with some graphing calculators and Excel.

Let us now recreate this graph by hand, and in the process identify the features we see in Figure 40.

1. ***The x- and y-intercepts:*** We consider $y = \dfrac{2x}{3} - (x - 2)^{2/3}$. For the y-intercept, we set $x = 0$ and solve for y:

$$y = \frac{2(0)}{3} - (0 - 2)^{2/3} = -2^{2/3} \approx -1.59.$$

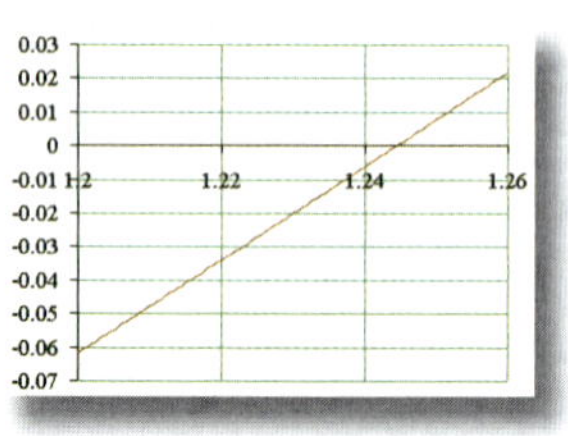

Figure 41

To find the x-intercept(s), we set $y = 0$ and solve for x. However, if we attempt this, we will find ourselves with a cubic equation that is hard to solve. (Try it!) Following the advice in the note on page 416, we use graphing technology to locate the x-intercept we see in Figure 40 by zooming in (Figure 41). From Figure 41, we find $x \approx 1.24$. We shall see in the discussion to follow that there can be no other x-intercepts.

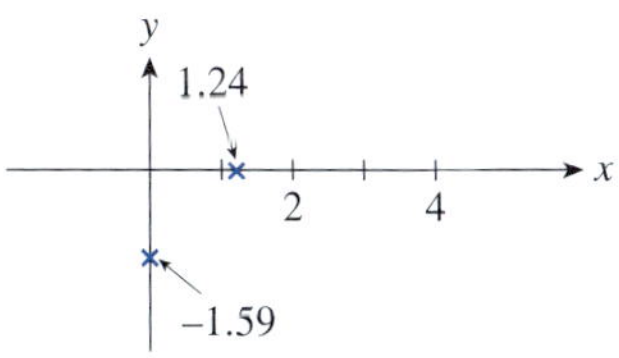

Figure 42

Figure 42 shows our freehand sketch so far.

2. ***Relative extrema:*** We calculate

$$f'(x) = \frac{2}{3} - \frac{2}{3}(x-2)^{-1/3}$$
$$= \frac{2}{3} - \frac{2}{3(x-2)^{1/3}}.$$

To find any stationary points, we set the derivative equal to 0 and solve for x:

$$\frac{2}{3} - \frac{2}{3(x-2)^{1/3}} = 0$$
$$(x-2)^{1/3} = 1$$
$$x - 2 = 1^3 = 1$$
$$x = 3.$$

To check for singular points, look for points where $f(x)$ is defined and $f'(x)$ is not defined. The only such point is $x = 2$: $f'(x)$ is not defined at $x = 2$, whereas $f(x)$ is defined there, so we have a singular point at $x = 2$.

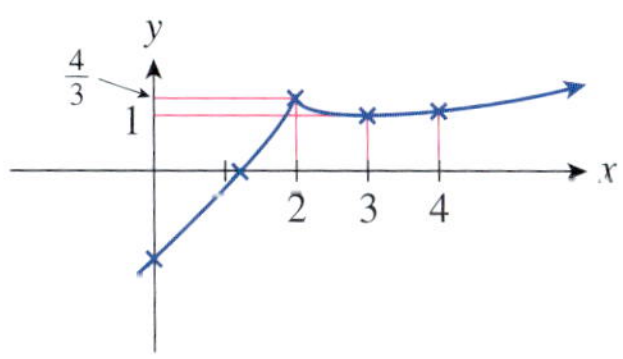

Figure 43

x	2 (Singular point)	3 (Stationary point)	4 (Test point)
$y = \frac{2x}{3} - (x-2)^{2/3}$	$\frac{4}{3}$	1	1.079

Figure 43 shows our graph so far.

We see that there is a singular relative maximum at $(2, 4/3)$ (we will confirm that the graph eventually gets higher on the right) and a stationary relative minimum at $x = 3$.

3. ***Points of inflection:*** We calculate

$$f''(x) = \frac{2}{9(x-2)^{4/3}}.$$

To find points of inflection, we set the second derivative equal to 0 and solve for x. But the equation

$$0 = \frac{2}{9(x-2)^{4/3}}$$

has no solution for x, so there are no points of inflection on the graph.

4. ***Behavior near points where f is not defined:*** Because $f(x)$ is defined everywhere, there are no such points to consider. In particular, there are no vertical asymptotes.

5. ***Behavior at infinity:*** We estimate the following limits numerically:

$$\lim_{x \to -\infty} \left[\frac{2x}{3} - (x-2)^{2/3}\right] = -\infty$$

and

$$\lim_{x \to +\infty} \left[\frac{2x}{3} - (x-2)^{2/3}\right] = +\infty.$$

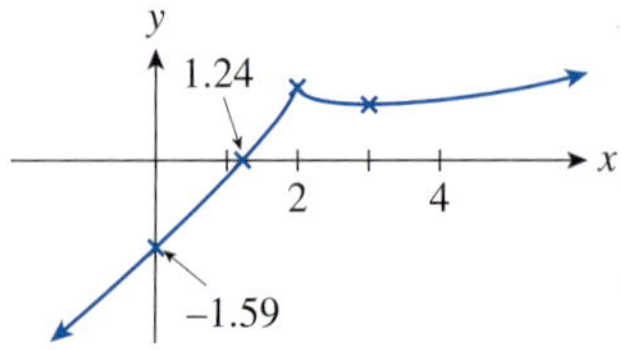

Figure 44

Thus, on the extreme left the curve goes down toward $-\infty$, and on the extreme right the curve rises toward $+\infty$. In particular, there are no horizontal asymptotes. (There can also be no other x-intercepts.)

Figure 44 shows the completed graph.

5.4 EXERCISES

▼ more advanced ◆ challenging

T indicates exercises that should be solved using technology

In Exercises 1–26, sketch the graph of the given function, indicating (a) x- and y-intercepts, (b) extrema, (c) points of inflection, (d) behavior near points where the function is not defined, and (e) behavior at infinity. Where indicated, technology should be used to approximate the intercepts, coordinates of extrema, and/or points of inflection to one decimal place. Check your sketch using technology. HINT [See Example 1.]

1. $f(x) = x^2 + 2x + 1$

2. $f(x) = -x^2 - 2x - 1$

3. $g(x) = x^3 - 12x$, domain $[-4, 4]$

4. $g(x) = 2x^3 - 6x$, domain $[-4, 4]$

5. $h(x) = 2x^3 - 3x^2 - 36x$ [Use technology for x-intercepts.]

6. $h(x) = -2x^3 - 3x^2 + 36x$ [Use technology for x-intercepts.]

7. $f(x) = 2x^3 + 3x^2 - 12x + 1$ [Use technology for x-intercepts.]

8. $f(x) = 4x^3 + 3x^2 + 2$ [Use technology for x-intercepts.]

9. $k(x) = -3x^4 + 4x^3 + 36x^2 + 10$ [Use technology for x-intercepts.]

10. $k(x) = 3x^4 + 4x^3 - 36x^2 - 10$ [Use technology for x-intercepts.]

11. $g(t) = \frac{1}{4}t^4 - \frac{2}{3}t^3 + \frac{1}{2}t^2$

12. $g(t) = 3t^4 - 16t^3 + 24t^2 + 1$

13. $f(x) = x + \frac{1}{x}$

14. $f(x) = x^2 + \frac{1}{x^2}$

15. $g(x) = x^3/(x^2 + 3)$

16. $g(x) = x^3/(x^2 - 3)$

17. $f(t) = \frac{t^2 + 1}{t^2 - 1}$, domain $[-2, 2]$, $t \neq \pm 1$

18. $f(t) = \frac{t^2 - 1}{t^2 + 1}$, domain $[-2, 2]$

19. $k(x) = \frac{2x}{3} + (x + 1)^{2/3}$ [Use technology for x-intercepts. HINT [See Example 2.]

20. $k(x) = \frac{2x}{5} - (x - 1)^{2/5}$ [Use technology for x-intercepts. HINT [See Example 2.]

21. $f(x) = x - \ln x$, domain $(0, +\infty)$

22. $f(x) = x - \ln x^2$, domain $(0, +\infty)$

23. $f(x) = x^2 + \ln x^2$ [Use technology for x-intercepts.]

24. $f(x) = 2x^2 + \ln x$ [Use technology for x-intercepts.]

25. $g(t) = e^t - t$, domain $[-1, 1]$

26. $g(t) = e^{-t^2}$

T *In Exercises 27–30, use technology to sketch the graph of the given function, labeling all relative and absolute extrema and points of inflection, and vertical and horizontal asymptotes. The coordinates of the extrema and points of inflection should be accurate to two decimal places.* HINT [To locate extrema accurately, plot the first derivative; to locate points of inflection accurately, plot the second derivative.]

27. ▼ $f(x) = x^4 - 2x^3 + x^2 - 2x + 1$

28. ▼ $f(x) = x^4 + x^3 + x^2 + x + 1$

29. ▼ $f(x) = e^x - x^3$

30. ▼ $f(x) = e^x - \frac{x^4}{4}$

APPLICATIONS

31. ***Home Prices*** The following graph shows the approximate value of the home price index as a percentage change from 2003. The locations of the maximum and the point of inflection are indicated on the graph (t is time in years since the start of 2004).[46]

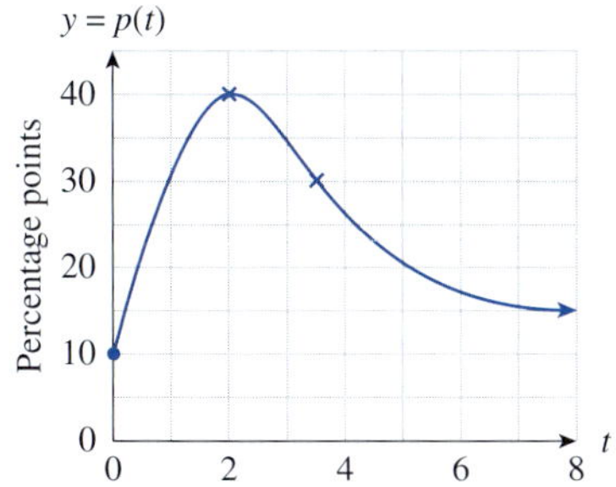

Analyze the graph's important features and interpret each feature in terms of the home price index.

32. ***Existing Home Sales*** The following graph shows the approximate value of existing home sales as a percentage change from 2003. The locations of the maximum and the point of inflection are indicated on the graph (t is time in years since the start of 2004).[47]

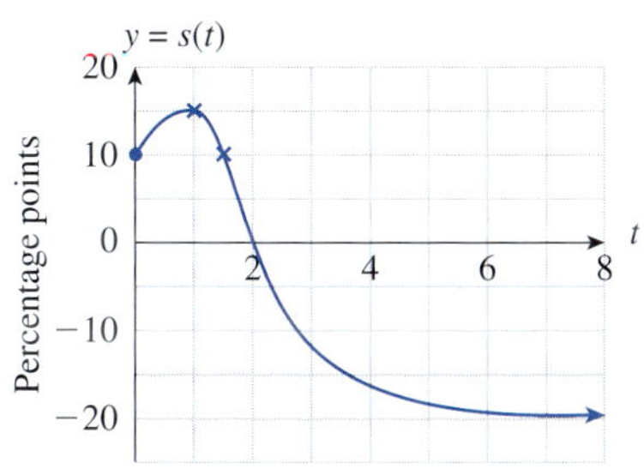

Analyze the graph's important features and interpret each feature in terms of the existing home sales.

33. ***Consumer Price Index*** The following graph shows the approximate value of the U.S. Consumer Price Index (CPI) from July 2005 through March 2006.[48]

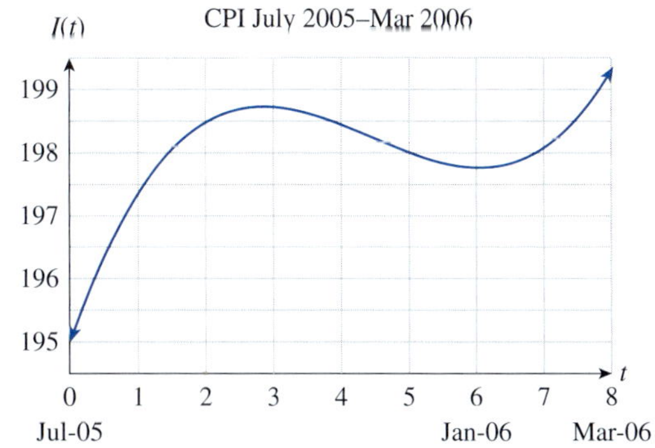

The approximating curve shown on the figure is given by

$$I(t) = 0.06t^3 - 0.8t^2 + 3.1t + 195 \quad (0 \le t \le 8)$$

where t is time in months ($t = 0$ represents July 2005).

a. Locate the intercepts, extrema, and points of inflection of the curve and interpret each feature in terms of the CPI. (Approximate all coordinates to one decimal place.) HINT [See Example 1.]

b. Recall from Section 5.2 that the inflation rate is defined to be $\frac{I'(t)}{I(t)}$. What do the stationary extrema of the curve shown above tell you about the inflation rate?

34. ***Inflation*** The following graph shows the approximate value of the U.S. Consumer Price Index (CPI) from March 2006 through May 2007.[49]

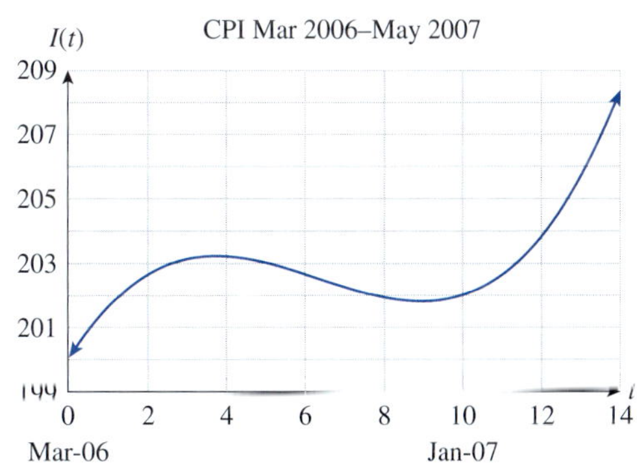

The approximating curve shown on the figure is given by

$$I(t) = 0.02t^3 - 0.38t^2 + 2t + 200 \quad (0 \le t \le 14)$$

where t is time in months ($t = 0$ represents March, 2006).

[46] 2008–2012 data are the authors' projections. Source for data: S&P/Case-Shiller Home Price Index. Source: Standard & Poors/*New York Times*, September 29, 2007, p. C3.

[47] 2008–2012 data are the authors' projections. Source for data: Bloomberg Finiancial Markets/*New York Times*, September 29, 2007, p. C3.

[48] The CPI is compiled by the Bureau of Labor Statistics and is based upon a 1982 value of 100. For instance, a CPI of 200 means the CPI has doubled since 1982. Source: InflationData.com (www.inflationdata.com).

[49] Ibid.

a. Locate the intercepts, extrema, and points of inflection of the curve and interpret each feature in terms of the CPI. (Approximate all coordinates to one decimal place.) HINT [See Example 1.]

b. Recall from Section 5.2 that the inflation rate is defined to be $\frac{I'(t)}{I(t)}$. What do the stationary extrema of the curve shown above tell you about the inflation rate?

35. ***Motion in a Straight Line*** The distance of a UFO from an observer is given by $s = 2t^3 - 3t^2 + 100$ feet after t seconds $(t \geq 0)$. Obtain the extrema, points of inflection, and behavior at infinity. Sketch the curve and interpret these features in terms of the movement of the UFO.

36. ***Motion in a Straight Line*** The distance of the Mars orbiter from your location in Utarek, Mars is given by $s = 2(t-1)^3 - 3(t-1)^2 + 100$ km after t seconds $(t \geq 0)$. Obtain the extrema, points of inflection, and behavior at infinity. Sketch the curve and interpret these features in terms of the movement of the Mars orbiter.

37. ***Average Cost: iPods*** Assume that it costs **Apple** approximately

$$C(x) = 22{,}500 + 100x + 0.01x^2$$

dollars to manufacture x 30-gigabyte video iPods in a day.[50] Obtain the average cost function, sketch its graph, and analyze the graph's important features. Interpret each feature in terms of iPods. HINT [Recall that the average cost function is $\bar{C}(x) = C(x)/x$.]

38. ***Average Cost: Xboxes*** Assume that it costs **Microsoft** approximately

$$C(x) = 14{,}400 + 550x + 0.01x^2$$

dollars to manufacture x Xbox 360s in a day.[51] Obtain the average cost function, sketch its graph, and analyze the graph's important features. Interpret each feature in terms of Xboxes. HINT [Recall that the average cost function is $\bar{C}(x) = C(x)/x$.]

39. T ▼ ***Subprime Mortgages*** The percentage of U.S.-issued mortgages that were subprime can be approximated by

$$A(t) = \frac{15.0}{1 + 8.6(1.8)^{-t}} \text{ percent} \qquad (0 \leq t \leq 8)$$

t years after the start of 2000.[52] Graph the *derivative* $A'(t)$ of $A(t)$ using an extended domain of $0 \leq t \leq 15$. Determine the approximate coordinates of the maximum and determine the behavior of $A'(t)$ at infinity. What do the answers tell you?

40. T ▼ ***Subprime Mortgage Debt*** The approximate value of subprime (normally classified as risky) mortgage debt outstanding in the United States can be approximated by

$$A(t) = \frac{1{,}350}{1 + 4.2(1.7)^{-t}} \text{ billion dollars} \qquad (0 \leq t \leq 8)$$

t years after the start of 2000.[53] Graph the *derivative* $A'(t)$ of $A(t)$ using an extended domain of $0 \leq t \leq 15$. Determine the approximate coordinates of the maximum and determine the behavior of $A'(t)$ at infinity. What do the answers tell you?

COMMUNICATION AND REASONING EXERCISES

41. A function is *bounded* if its entire graph lies between two horizontal lines. Can a bounded function have vertical asymptotes? Can a bounded function have horizontal asymptotes? Explain.

42. A function is *bounded above* if its entire graph lies below some horizontal line. Can a bounded above function have vertical asymptotes? Can a bounded above function have horizontal asymptotes? Explain.

43. If the graph of a function has a vertical asymptote at $x = a$ in such a way that y increases to $+\infty$ as $x \to a$, what can you say about the graph of its derivative? Explain.

44. If the graph of a function has a horizontal asymptote at $y = a$ in such a way that y decreases to a as $x \to +\infty$, what can you say about the graph of its derivative? Explain.

45. Your friend tells you that he has found a continuous function defined on $(-\infty, +\infty)$ with exactly two critical points, each of which is a relative maximum. Can he be right?

46. Your other friend tells you that she has found a continuous function with two critical points, one a relative minimum and one a relative maximum, and no point of inflection between them. Can she be right?

47. ▼ By thinking about extrema, show that, if $f(x)$ is a polynomial, then between every pair of zeros (x-intercepts) of $f(x)$ there is a zero of $f'(x)$.

48. ▼ If $f(x)$ is a polynomial of degree 2 or higher, show that between every pair of relative extrema of $f(x)$ there is a point of inflection of $f(x)$.

[50] Not the actual cost equation; the authors do not know Apple's actual cost equation. The marginal cost in the model given is in rough agreement with the actual marginal cost for reasonable values of x. Source for cost data: Manufacturing & Technology News, July 31, 2007 Volume 14, No. 14 (www.manufacturingnews.com).

[51] Not the actual cost equation; the authors do not know Microsoft's actual cost equation. The marginal cost in the model given is in rough agreement with the actual marginal cost for reasonable values of x. Source for estimate of marginal cost: iSuppli: (www.isuppli.com/news/xbox/).

[52] 2009 figure is an estimate. Sources: Mortgage Bankers Association, UBS.

[53] 2008–2009 figure are estimates. Source: www.data360.org/dataset.aspx?Data_Set_Id=9549.

5.5 Related Rates

We start by recalling some basic facts about the rate of change of a quantity:

Rate of Change of Q

If Q is a quantity changing over time t, then the derivative dQ/dt is the rate at which Q changes over time.

Quick Examples

1. If A is the area of an expanding circle, then dA/dt is the rate at which the area is increasing.
2. *Words:* The radius r of a sphere is currently 3 cm and increasing at a rate of 2 cm/s.

   Symbols: $r = 3$ cm and $dr/dt = 2$ cm/s.

In this section we are concerned with what are called **related rates** problems. In such a problem we have two (sometimes more) related quantities, we know the rate at which one is changing, and we wish to find the rate at which another is changing. A typical example is the following.

EXAMPLE 1 The Expanding Circle

The radius of a circle is increasing at a rate of 10 cm/s. How fast is the area increasing at the instant when the radius has reached 5 cm?

Solution We have two related quantities: the radius of the circle, r, and its area, A. The first sentence of the problem tells us that r is increasing at a certain rate. When we see a sentence referring to speed or change, it is very helpful to rephrase the sentence using the phrase "the rate of change of." Here, we can say

The rate of change of r is 10 cm/s.

Because the rate of change is the derivative, we can rewrite this sentence as the equation

$$\frac{dr}{dt} = 10.$$

Similarly, the second sentence of the problem asks how A is changing. We can rewrite that question:

What is the rate of change of A when the radius is 5 cm?

Using mathematical notation, the question is:

$$\text{What is } \frac{dA}{dt} \text{ when } r = 5?$$

Thus, knowing one rate of change, dr/dt, we wish to find a related rate of change, dA/dt. To find exactly how these derivatives are related, we need the equation relating the variables, which is

$$A = \pi r^2.$$

To find the relationship between the derivatives, we take the derivative of both sides of this equation *with respect to t*. On the left we get dA/dt. On the right we need to remember that r is a function of t and use the chain rule. We get

$$\frac{dA}{dt} = 2\pi r \frac{dr}{dt}.$$

Now we substitute the given values $r = 5$ and $dr/dt = 10$. This gives

$$\left.\frac{dA}{dt}\right|_{r=5} = 2\pi(5)(10) = 100\pi \approx 314 \text{ cm}^2/\text{s}.$$

Thus, the area is increasing at the rate of 314 cm^2/s when the radius is 5 cm.

We can organize our work as follows:

Solving a Related Rates Problem

A. The Problem

1. List the related, changing quantities.
2. Restate the problem in terms of rates of change. Rewrite the problem using mathematical notation for the changing quantities and their derivatives.

B. The Relationship

1. Draw a diagram, if appropriate, showing the changing quantities.
2. Find an equation or equations relating the changing quantities.
3. Take the derivative with respect to time of the equation(s) relating the quantities to get the **derived equation(s)**, which relate the rates of change of the quantities.

C. The Solution

1. Substitute into the derived equation(s) the given values of the quantities and their derivatives.
2. Solve for the derivative required.

We can illustrate the procedure with the "ladder problem" found in almost every calculus textbook.

EXAMPLE 2 The Falling Ladder

Jane is at the top of a 5-foot ladder when it starts to slide down the wall at a rate of 3 feet per minute. Jack is standing on the ground behind her. How fast is the base of the ladder moving when it hits him if Jane is 4 feet from the ground at that instant?

Solution The first sentence talks about (the top of) the ladder sliding down the wall. Thus, one of the changing quantities is the height of the top of the ladder. The question asked refers to the motion of the base of the ladder, so another changing quantity is the distance of the base of the ladder from the wall. Let's record these variables and follow the outline above to obtain the solution.

A. The Problem

1. The changing quantities are

$h =$ height of the top of the ladder
$b =$ distance of the base of the ladder from the wall

2. We rephrase the problem in words, using the phrase "rate of change":

The rate of change of the height of the top of the ladder is -3 feet per minute. What is the rate of change of the distance of the base from the wall when the top of the ladder is 4 feet from the ground?

We can now rewrite the problem mathematically:

$$\frac{dh}{dt} = -3. \text{ Find } \frac{db}{dt} \text{ when } h = 4.$$

Figure 45

B. The Relationship

1. Figure 45 shows the ladder and the variables h and b. Notice that we put in the figure the fixed length, 5, of the ladder, but any changing quantities, like h and b, we leave as variables. We shall not use any specific values for h or b until the very end.

2. From the figure, we can see that h and b are related by the Pythagorean theorem:

$$h^2 + b^2 = 25.$$

3. Taking the derivative with respect to time of the equation above gives us the derived equation:

$$2h\frac{dh}{dt} + 2b\frac{db}{dt} = 0.$$

C. The Solution

1. We substitute the known values $dh/dt = -3$ and $h = 4$ into the derived equation:

$$2(4)(-3) + 2b\frac{db}{dt} = 0.$$

We would like to solve for db/dt, but first we need the value of b, which we can determine from the equation $h^2 + b^2 = 25$, using the value $h = 4$:

$$\begin{aligned} 16 + b^2 &= 25 \\ b^2 &= 9 \\ b &= 3. \end{aligned}$$

Substituting into the derived equation, we get

$$-24 + 2(3)\frac{db}{dt} = 0.$$

2. Solving for db/dt gives

$$\frac{db}{dt} = \frac{24}{6} = 4.$$

Thus, the base of the ladder is sliding away from the wall at 4 ft/min when it hits Jack.

AFP/Getty Images

EXAMPLE 3 Average Cost

The cost to manufacture x cell phones in a day is

$$C(x) = 10{,}000 + 20x + \frac{x^2}{10{,}000} \text{ dollars.}$$

The daily production level is currently $x = 5{,}000$ cell phones and is increasing at a rate of 100 units per day. How fast is the average cost changing?

Solution

A. The Problem

1. The changing quantities are the production level x and the average cost, $\bar{C}$.

2. We rephrase the problem as follows:

The daily production level is $x = 5{,}000$ *units and the rate of change of* x *is* 100 *units/day. What is the rate of change of the average cost,* $\bar{C}$*?*

In mathematical notation,

$$x = 5{,}000 \text{ and } \frac{dx}{dt} = 100. \text{ Find } \frac{d\bar{C}}{dt}.$$

B. The Relationship

1. In this example the changing quantities cannot easily be depicted geometrically.

2. We are given a formula for the *total* cost. We get the *average* cost by dividing the total cost by x:

$$\bar{C} = \frac{C}{x}.$$

So,

$$\bar{C} = \frac{10{,}000}{x} + 20 + \frac{x}{10{,}000}.$$

3. Taking derivatives with respect to t of both sides, we get the derived equation:

$$\frac{d\bar{C}}{dt} = \left(-\frac{10{,}000}{x^2} + \frac{1}{10{,}000}\right)\frac{dx}{dt}.$$

C. The Solution

Substituting the values from part A into the derived equation, we get

$$\frac{d\bar{C}}{dt} = \left(-\frac{10{,}000}{5{,}000^2} + \frac{1}{10{,}000}\right)100$$

$$= -0.03 \text{ dollars/day.}$$

Thus, the average cost is decreasing by 3¢ per day.

The scenario in the following example is similar to Example 5 in Section 5.2.

EXAMPLE 4 Allocation of Labor

The Gym Sock Company manufactures cotton athletic socks. Production is partially automated through the use of robots. The number of pairs of socks the company can manufacture in a day is given by a Cobb-Douglas production formula:

$$q = 50n^{0.6}r^{0.4}$$

where q is the number of pairs of socks that can be manufactured by n laborers and r robots. The company currently produces 1,000 pairs of socks each day and employs 20 laborers. It is bringing one new robot on line every month. At what rate are laborers being laid off, assuming that the number of socks produced remains constant?

Solution

A. The Problem

1. The changing quantities are the number of laborers n and the number of robots r.

2. $\frac{dr}{dt} = 1$. Find $\frac{dn}{dt}$ when $n = 20$.

B. The Relationship

1. No diagram is appropriate here.

2. The equation relating the changing quantities:

$$1{,}000 = 50n^{0.6}r^{0.4}$$

or

$$20 = n^{0.6}r^{0.4}.$$

(Productivity is constant at 1,000 pairs of socks each day.)

3. The derived equation is

$$\begin{aligned} 0 &= 0.6n^{-0.4}\left(\frac{dn}{dt}\right)r^{0.4} + 0.4n^{0.6}r^{-0.6}\left(\frac{dr}{dt}\right) \\ &= 0.6\left(\frac{r}{n}\right)^{0.4}\left(\frac{dn}{dt}\right) + 0.4\left(\frac{n}{r}\right)^{0.6}\left(\frac{dr}{dt}\right). \end{aligned}$$

We solve this equation for dn/dt because we shall want to find dn/dt below and because the equation becomes simpler when we do this:

$$0.6\left(\frac{r}{n}\right)^{0.4}\left(\frac{dn}{dt}\right) = -0.4\left(\frac{n}{r}\right)^{0.6}\left(\frac{dr}{dt}\right)$$

$$\begin{aligned} \frac{dn}{dt} &= -\frac{0.4}{0.6}\left(\frac{n}{r}\right)^{0.6}\left(\frac{n}{r}\right)^{0.4}\left(\frac{dr}{dt}\right) \\ &= -\frac{2}{3}\left(\frac{n}{r}\right)\left(\frac{dr}{dt}\right). \end{aligned}$$

C. The Solution

Substituting the numbers in A into the last equation in B, we get

$$\frac{dn}{dt} = -\frac{2}{3}\left(\frac{20}{r}\right)(1).$$

We need to compute r by substituting the known value of n in the original formula:

$$20 = n^{0.6}r^{0.4}$$
$$20 = 20^{0.6}r^{0.4}$$
$$r^{0.4} = \frac{20}{20^{0.6}} = 20^{0.4}$$
$$r = 20.$$

Thus,

$$\frac{dn}{dt} = -\frac{2}{3}\left(\frac{20}{20}\right)(1) = -\frac{2}{3} \text{ laborers per month.}$$

The company is laying off laborers at a rate of 2/3 per month, or two every three months.

We can interpret this result as saying that, at the current level of production and number of laborers, one robot is as productive as 2/3 of a laborer, or 3 robots are as productive as 2 laborers.

5.5 EXERCISES

▼ more advanced ◆ challenging

T indicates exercises that should be solved using technology

Rewrite the statements and questions in Exercises 1–8 in mathematical notation. HINT [See Quick Examples on page 423.]

1. The population P is currently 10,000 and growing at a rate of 1,000 per year.
2. There are presently 400 cases of Bangkok flu, and the number is growing by 30 new cases every month.
3. The annual revenue of your tie-dye T-shirt operation is currently \$7,000 but is decreasing by \$700 each year. How fast are annual sales changing?
4. A ladder is sliding down a wall so that the distance between the top of the ladder and the floor is decreasing at a rate of 3 feet per second. How fast is the base of the ladder receding from the wall?
5. The price of shoes is rising \$5 per year. How fast is the demand changing?
6. Stock prices are rising \$1,000 per year. How fast is the value of your portfolio increasing?
7. The average global temperature is 60°F and rising by 0.1°F per decade. How fast are annual sales of Bermuda shorts increasing?
8. The country's population is now 260,000,000 and is increasing by 1,000,000 people per year. How fast is the annual demand for diapers increasing?

APPLICATIONS

9. ***Sun Spots*** The area of a circular sun spot is growing at a rate of 1,200 km^2/s.
   a. How fast is the radius growing at the instant when it equals 10,000 km? HINT [See Example 1.]
   b. How fast is the radius growing at the instant when the sun spot has an area of 640,000 km^2? HINT [Use the area formula to determine the radius at that instant.]
10. ***Puddles*** The radius of a circular puddle is growing at a rate of 5 cm/s.
    a. How fast is its area growing at the instant when the radius is 10 cm? HINT [See Example 1.]
    b. How fast is the area growing at the instant when it equals 36 cm^2? HINT [Use the area formula to determine the radius at that instant.]
11. ***Balloons*** A spherical party balloon is being inflated with helium pumped in at a rate of 3 cubic feet per minute. How fast is the radius growing at the instant when the radius has reached 1 foot? (The volume of a sphere of radius r is $V = \frac{4}{3}\pi r^3$.) HINT [See Example 1.]
12. ***More Balloons*** A rather flimsy spherical balloon is designed to pop at the instant its radius has reached 10 centimeters. Assuming the balloon is filled with helium at a rate of 10 cubic centimeters per second, calculate how fast the radius is growing at the instant it pops. (The volume of a sphere of radius r is $V = \frac{4}{3}\pi r^3$.) HINT [See Example 1.]

13. ***Sliding Ladders*** The base of a 50-foot ladder is being pulled away from a wall at a rate of 10 feet per second. How fast is the top of the ladder sliding down the wall at the instant when the base of the ladder is 30 feet from the wall? HINT [See Example 2.]

14. ***Sliding Ladders*** The top of a 5-foot ladder is sliding down a wall at a rate of 10 feet per second. How fast is the base of the ladder sliding away from the wall at the instant when the top of the ladder is 3 feet from the ground? HINT [See Example 2.]

15. ***Average Cost*** The average cost function for the weekly manufacture of portable CD players is given by

$$\bar{C}(x) = 150{,}000x^{-1} + 20 + 0.0001x \text{ dollars per player,}$$

where x is the number of CD players manufactured that week. Weekly production is currently 3,000 players and is increasing at a rate of 100 players per week. What is happening to the average cost? HINT [See Example 3.]

16. ***Average Cost*** Repeat the preceding exercise, using the revised average cost function

$$\bar{C}(x) = 150{,}000x^{-1} + 20 + 0.01x \text{ dollars per player.}$$

HINT [See Example 3.]

17. ***Demand*** Demand for your tie-dyed T-shirts is given by the formula

$$q = 500 - 100p^{0.5}$$

where q is the number of T-shirts you can sell each month at a price of p dollars. If you currently sell T-shirts for \$15 each and you raise your price by \$2 per month, how fast will the demand drop? (Round your answer to the nearest whole number.)

18. ***Supply*** The number of portable CD players you are prepared to supply to a retail outlet every week is given by the formula

$$q = 0.1p^2 + 3p$$

where p is the price it offers you. The retail outlet is currently offering you \$40 per CD player. If the price it offers decreases at a rate of \$2 per week, how will this affect the number you supply?

19. ***Revenue*** You can now sell 50 cups of lemonade per week at 30¢ per cup, but demand is dropping at a rate of 5 cups per week each week. Assuming that raising the price does not affect demand, how fast do you have to raise your price if you want to keep your weekly revenue constant? HINT [Revenue = Price × Quantity.]

20. ***Revenue*** You can now sell 40 cars per month at \$20,000 per car, and demand is increasing at a rate of 3 cars per month each month. What is the fastest you could drop your price before your monthly revenue starts to drop? HINT [Revenue = Price × Quantity.]

21. ▼ ***Oil Revenues*** Daily oil production by **Pemex**, Mexico's national oil company, can be approximated by

$$q(t) = -0.022t^2 + 0.2t + 2.9 \text{ million barrels} \quad (1 \le t \le 9)$$

where t is time in years since the start of 2000.[54]

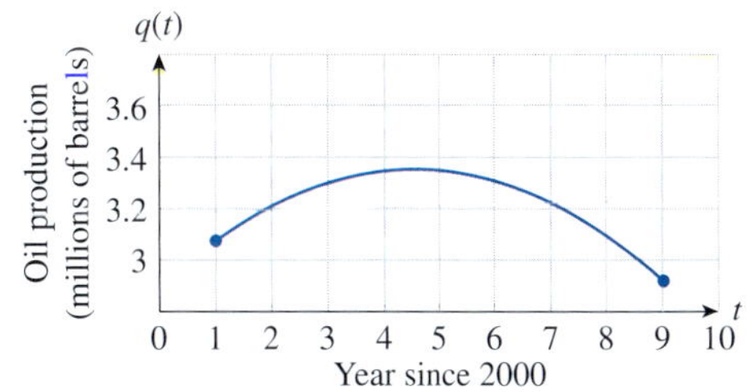

At the start of 2008 the price of oil was \$90 per barrel and increasing at a rate of \$80 per year.[55] How fast was **Pemex**'s oil (daily) revenue changing at that time?

22. ▼ ***Oil Expenditures*** Daily oil imports to the United States from Mexico can be approximated by

$$q(t) = -0.015t^2 + 0.1t + 1.4 \text{ million barrels} \quad (0 \le t \le 8)$$

where t is time in years since the start of 2000.[56]

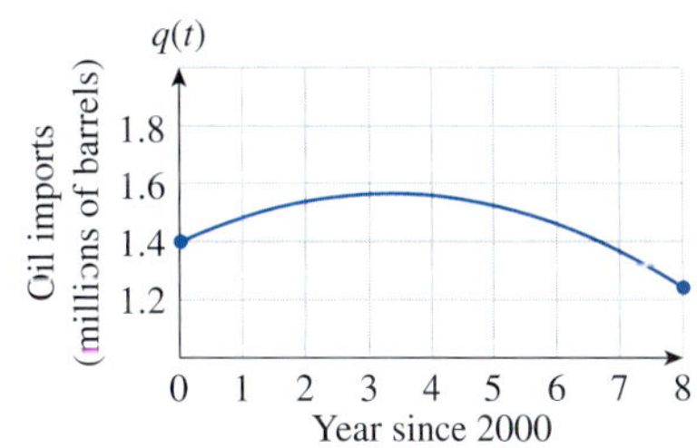

At the start of 2004 the price of oil was \$30 per barrel and increasing at a rate of \$40 per year.[57] How fast was (daily) oil expenditure for imports from Mexico changing at that time?

23. ***Resource Allocation*** Your company manufactures automobile alternators, and production is partially automated through the use of robots. In order to meet production deadlines, your company calculates that the numbers of laborers and robots must satisfy the constraint

$$xy = 10{,}000$$

where x is the number of laborers and y is the number of robots. Your company currently uses 400 robots and is increasing robot deployment at a rate of 16 per month. How fast is it laying off laborers? HINT [See Example 4.]

24. ***Resource Allocation*** Your company is the largest sock manufacturer in the solar system, and production is automated through the use of androids and robots. In order to meet production deadlines, your company calculates that the numbers of androids and robots must satisfy the constraint

$$xy = 1{,}000{,}000$$

[54] Source for data: Energy Information Administration/Pemex (http://www.eia.doe.gov).

[55] Based on NYMEX crude oil futures; average rate of change during January–June, 2008.

[56] Source for data: Energy Information Administration/Pemex (http://www.eia.doe.gov).

[57] Based on NYMEX crude oil futures; average rate of change during 2004–2005.

where x is the number of androids and y is the number of robots. Your company currently uses 5000 androids and is increasing android deployment at a rate of 200 per month. How fast is it scrapping robots? HINT [See Example 4.]

25. ***Production*** The automobile assembly plant you manage has a Cobb-Douglas production function given by

$$P = 10x^{0.3}y^{0.7}$$

where P is the number of automobiles it produces per year, x is the number of employees, and y is the daily operating budget (in dollars). You maintain a production level of 1,000 automobiles per year. If you currently employ 150 workers and are hiring new workers at a rate of 10 per year, how fast is your daily operating budget changing? HINT [See Example 4.]

26. ***Production*** Refer back to the Cobb-Douglas production formula in the preceding exercise. Assume that you maintain a constant work force of 200 workers and wish to increase production in order to meet a demand that is increasing by 100 automobiles per year. The current demand is 1000 automobiles per year. How fast should your daily operating budget be increasing? HINT [See Example 4.]

27. ***Demand*** Assume that the demand equation for tuna in a small coastal town is

$$pq^{1.5} = 50{,}000$$

where q is the number of pounds of tuna that can be sold in one month at the price of p dollars per pound. The town's fishery finds that the demand for tuna is currently 900 pounds per month and is increasing at a rate of 100 pounds per month each month. How fast is the price changing?

28. ***Demand*** The demand equation for rubies at Royal Ruby Retailers is

$$q + \frac{4}{3}p = 80$$

where q is the number of rubies RRR can sell per week at p dollars per ruby. RRR finds that the demand for its rubies is currently 20 rubies per week and is dropping at a rate of one ruby per week. How fast is the price changing?

29. ▼ ***Ships Sailing Apart*** The H.M.S. Dreadnaught is 40 miles north of Montauk and steaming due north at 20 miles/hour, while the U.S.S. Mona Lisa is 50 miles east of Montauk and steaming due east at an even 30 miles/hour. How fast is their distance apart increasing?

30. ▼ ***Near Miss*** My aunt and I were approaching the same intersection, she from the south and I from the west. She was traveling at a steady speed of 10 miles/hour, while I was approaching the intersection at 60 miles/hour. At a certain instant in time, I was one-tenth of a mile from the intersection, while she was one-twentieth of a mile from it. How fast were we approaching each other at that instant?

31. ▼ ***Baseball*** A baseball diamond is a square with side 90 ft.

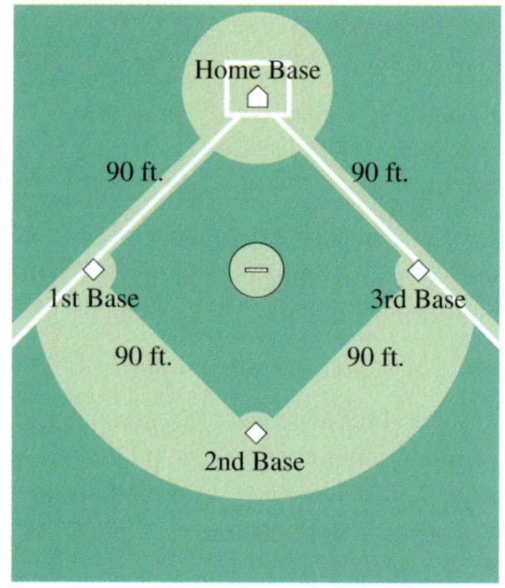

A batter at home base hits the ball and runs toward first base with a speed of 24 ft/s. At what rate is his distance from third base increasing when he is halfway to first base?

32. ▼ ***Baseball*** Refer to Exercise 31. Another player is running from third base to home at 30 ft/s. How fast is her distance from second base increasing when she is 60 feet from third base?

33. ▼ ***Movement along a Graph*** A point on the graph of $y = 1/x$ is moving along the curve in such a way that its x-coordinate is increasing at a rate of 4 units per second. What is happening to the y-coordinate at the instant the y-coordinate is equal to 2?

34. ▼ ***Motion around a Circle*** A point is moving along the circle $x^2 + (y-1)^2 = 8$ in such a way that its x-coordinate is decreasing at a rate of 1 unit per second. What is happening to the y-coordinate at the instant when the point has reached $(-2, 3)$?

35. ▼ ***Education*** In 1991, the expected income of an individual depended on his or her educational level according to the following formula:

$$I(n) = 2928.8n^3 - 115{,}860n^2 + 1{,}532{,}900n - 6{,}760{,}800. \quad (12 \le n \le 15)$$

Here, n is the number of school years completed and $I(n)$ is the individual's expected income.[58] You have completed 13 years of school and are currently a part-time student. Your schedule is such that you will complete the equivalent of one year of college every three years. Assuming that your salary is linked to the above model, how fast is your income going up? (Round your answer to the nearest $1.)

[58] The model is a best-fit cubic based on Table 358, U.S. Department of Education, *Digest of Education Statistics, 1991*, Washington, DC: Government Printing Office, 1991.

36. ▼ ***Education*** Refer back to the model in the preceding exercise. Assume that someone has completed 14 years of school and that her income is increasing by \$10,000 per year. How much schooling per year is this rate of increase equivalent to?

37. ▼ ***Employment*** An employment research company estimates that the value of a recent MBA graduate to an accounting company is

$$V = 3e^2 + 5g^3$$

where V is the value of the graduate, e is the number of years of prior business experience, and g is the graduate school grade point average. A company that currently employs graduates with a 3.0 average wishes to maintain a constant employee value of $V = 200$, but finds that the grade point average of its new employees is dropping at a rate of 0.2 per year. How fast must the experience of its new employees be growing in order to compensate for the decline in grade point average?

38. ▼ ***Grades***[59] A production formula for a student's performance on a difficult English examination is given by

$$g = 4hx - 0.2h^2 - 10x^2$$

where g is the grade the student can expect to obtain, h is the number of hours of study for the examination, and x is the student's grade point average. The instructor finds that students' grade point averages have remained constant at 3.0 over the years, and that students currently spend an average of 15 hours studying for the examination. However, scores on the examination are dropping at a rate of 10 points per year. At what rate is the average study time decreasing?

39. ▼ ***Cones*** A right circular conical vessel is being filled with green industrial waste at a rate of 100 cubic meters per second. How fast is the level rising after 200π cubic meters have been poured in? The cone has a height of 50 m and a radius of 30 m at its brim. (The volume of a cone of height h and cross-sectional radius r at its brim is given by $V = \frac{1}{3}\pi r^2 h$.)

40. ▼ ***More Cones*** A circular conical vessel is being filled with ink at a rate of 10 cm^3/s. How fast is the level rising after 20 cm^3 have been poured in? The cone has height 50 cm and radius 20 cm at its brim. (The volume of a cone of height h and cross-sectional radius r at its brim is given by $V = \frac{1}{3}\pi r^2 h$.)

41. ▼ ***Cylinders*** The volume of paint in a right cylindrical can is given by $V = 4t^2 - t$ where t is time in seconds and V is the volume in cm^3. How fast is the level rising when the height is 2 cm? The can has a height of 4 cm and a radius of 2 cm. HINT [To get h as a function of t, first solve the volume $V = \pi r^2 h$ for h.]

42. ▼ ***Cylinders*** A cylindrical bucket is being filled with paint at a rate of 6 cm^3 per minute. How fast is the level rising when the bucket starts to overflow? The bucket has a radius of 30 cm and a height of 60 cm.

[59] Based on an Exercise in *Introduction to Mathematical Economics* by A.L. Ostrosky Jr. and J.V. Koch (Waveland Press, Illinois, 1979.)

43. ▼ ***Computers vs. Income*** The demand for personal computers in the home goes up with household income. For a given community, we can approximate the average number of computers in a home as

$$q = 0.3454 \ln x - 3.047 \qquad 10{,}000 \le x \le 125{,}000$$

where x is mean household income.[60] Your community has a mean income of \$30,000, increasing at a rate of \$2,000 per year. How many computers per household are there, and how fast is the number of computers in a home increasing? (Round your answer to four decimal places.)

44. ▼ ***Computers vs. Income*** Refer back to the model in the preceding exercise. The average number of computers per household in your town is 0.5 and is increasing at a rate of 0.02 computers per household per year. What is the average household income in your town, and how fast is it increasing? (Round your answers to the nearest \$10).

Education and Crime *The following graph shows a striking relationship between the total prison population and the average combined SAT score in the U.S.*

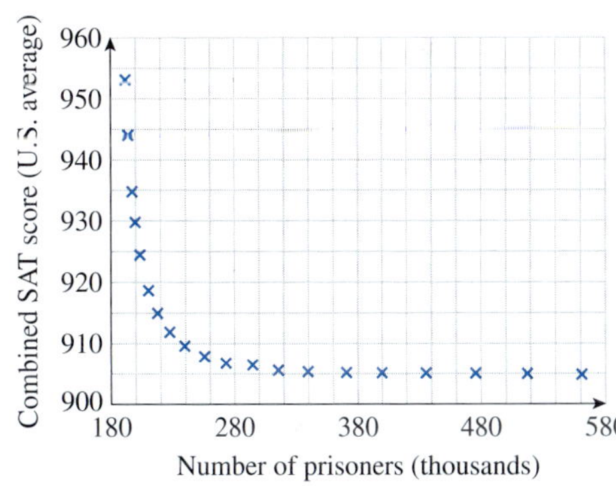

Exercises 45 and 46 are based on the following model for these data:

$$S(n) = 904 + \frac{1{,}326}{(n - 180)^{1.325}}. \qquad (192 \le n \le 563)$$

Here, S(n) is the combined average SAT score at a time when the total prison population is n thousand.[61]

45. ▼ In 1985, the U.S. prison population was 475,000 and increasing at a rate of 35,000 per year. What was the average SAT score, and how fast, and in what direction, was it changing? (Round your answers to two decimal places.)

[60] The model is a regression model. Source for data: Income distribution: Computer data: Forrester Research/*The New York Times*, August 8, 1999, p. BU4.

[61] The model is the authors' based on data for the years 1967–1989. Sources: Sourcebook of Criminal Justice Statistics, 1990, p. 604/ Educational Testing Service.

46. ▼ In 1970, the U.S. combined SAT average was 940 and dropping by 10 points per year. What was the U.S. prison population, and how fast, and in what direction, was it changing? (Round your answers to the nearest 100.)

Divorce Rates *A study found that the divorce rate d (given as a percentage) appears to depend on the ratio r of available men to available women.*[62] *This function can be approximated by*

$$d(r) = \begin{cases} -40r + 74 & \text{if } r \le 1.3 \\ \dfrac{130r}{3} - \dfrac{103}{3} & \text{if } r > 1.3 \end{cases}.$$

Exercises 47 and 48 are based on this model.

47. ◆ There are currently 1.1 available men per available woman in Littleville, and this ratio is increasing by 0.05 per year. What is happening to the divorce rate?

48. ◆ There are currently 1.5 available men per available woman in Largeville, and this ratio is decreasing by 0.03 per year. What is happening to the divorce rate?

COMMUNICATION AND REASONING EXERCISES

49. Why is this section titled "related rates"?

50. If you know how fast one quantity is changing and need to compute how fast a second quantity is changing, what kind of information do you need?

51. In a related rates problem, there is no limit to the number of changing quantities we can consider. Illustrate this by creating a related rates problem with four changing quantities.

[62] The cited study, by Scott J. South and associates, appeared in the *American Sociological Review* (February, 1995). Figures are rounded. Source: *The New York Times*, February 19, 1995, p. 40.

52. If three quantities are related by a single equation, how would you go about computing how fast one of them is changing based on a knowledge of the other two?

53. ▼ The demand and unit price for your store's checkered T-shirts are changing with time. Show that the percentage rate of change of revenue equals the sum of the percentage rates of change of price and demand. (The percentage rate of change of a quantity Q is $Q'(t)/Q(t)$.)

54. ▼ The number N of employees and the total floor space S of your company are both changing with time. Show that the percentage rate of change of square footage per employee equals the percentage rate of change of S minus the percentage rate of change of N. (The percentage rate of change of a quantity Q is $Q'(t)/Q(t)$.)

55. ▼ In solving a related rates problem, a key step is solving the derived equation for the unknown rate of change (once we have substituted the other values into the equation). Call the unknown rate of change X. The derived equation is what kind of equation in X?

56. ▼ On a recent exam, you were given a related rates problem based on an algebraic equation relating two variables x and y. Your friend told you that the correct relationship between dx/dt and dy/dt was given by

$$\left(\frac{dx}{dt}\right) = \left(\frac{dy}{dt}\right)^2.$$

Could he be correct?

57. ▼ Transform the following into a mathematical statement about derivatives: If my grades are improving at twice the speed of yours, then your grades are improving at half the speed of mine.

58. ▼ If two quantities x and y are related by a linear equation, how are their rates of change related?

5.6 Elasticity

You manufacture an extremely popular brand of sneakers and want to know what will happen if you increase the selling price. Common sense tells you that demand will drop as you raise the price. But will the drop in demand be enough to cause your revenue to fall? Or will it be small enough that your revenue will rise because of the higher selling price? For example, if you raise the price by 1%, you might suffer only a 0.5% loss in sales. In this case, the loss in sales will be more than offset by the increase in price and your revenue will rise. In such a case, we say that the demand is **inelastic**, because it is not very sensitive to the increase in price. On the other hand, if your 1% price increase results in a 2% drop in demand, then raising the price will cause a drop in revenues. We then say that the demand is **elastic** because it reacts strongly to a price change.

*** NOTE** Coming up with a good demand equation is not always easy. We saw in Chapter 1 that it is possible to find a linear demand equation if we know the sales figures at two different prices. However, such an equation is only a first approximation. To come up with a more accurate demand equation, we might need to gather data corresponding to sales at several different prices and use curve-fitting techniques like regression. Another approach would be an analytic one, based on mathematical modeling techniques that an economist might use.

We can use calculus to measure the response of demand to price changes if we have a demand equation for the item we are selling.* We need to know the *percentage drop in demand per percentage increase in price.* This ratio is called the **elasticity of demand**, or **price elasticity of demand**, and is usually denoted by E. Let's derive a formula for E in terms of the demand equation.

Assume that we have a demand equation

$$q = f(p)$$

where q stands for the number of items we would sell (per week, per month, or what have you) if we set the price per item at p. Now suppose we increase the price p by a very small amount, Δp. Then our percentage increase in price is $(\Delta p/p) \times 100\%$. This increase in p will presumably result in a decrease in the demand q. Let's denote this corresponding decrease in q by $-\Delta q$ (we use the minus sign because, by convention, Δq stands for the *increase* in demand). Thus, the percentage decrease in demand is $(-\Delta q/q) \times 100\%$.

Now E is the ratio

$$E = \frac{\text{Percentage decrease in demand}}{\text{Percentage increase in price}}$$

so

$$E = \frac{-\dfrac{\Delta q}{q} \times 100\%}{\dfrac{\Delta p}{p} \times 100\%}.$$

Canceling the 100%s and reorganizing, we get

$$E = -\frac{\Delta q}{\Delta p} \cdot \frac{p}{q}.$$

Q: *What small change in price will we use for Δp?*

A: It should probably be pretty small. If, say, we increased the price of sneakers to $1 million per pair, the sales would likely drop to zero. But knowing this tells us nothing about how the market would respond to a modest increase in price. In fact, we'll do the usual thing we do in calculus and let Δp approach 0.

In the expression for E, if we let Δp go to 0, then the ratio $\Delta q/\Delta p$ goes to the derivative dq/dp. This gives us our final and most useful definition of the elasticity.

Price Elasticity of Demand

The **price elasticity of demand *E*** is the percentage rate of decrease of demand per percentage increase in price. E is given by the formula

$$E = -\frac{dq}{dp} \cdot \frac{p}{q}.$$

We say that the demand is **elastic** if $E > 1$, is **inelastic** if $E < 1$, and has **unit elasticity** if $E = 1$.

Quick Example

Suppose that the demand equation is $q = 20{,}000 - 2p$ where p is the price in dollars. Then

$$E = -(-2)\frac{p}{20{,}000 - 2p} = \frac{p}{10{,}000 - p}.$$

If $p = \$2{,}000$, then $E = 1/4$, and demand is inelastic at this price.
If $p = \$8{,}000$, then $E = 4$, and demand is elastic at this price.
If $p = \$5{,}000$, then $E = 1$, and the demand has unit elasticity at this price.

* **NOTE** For another—more rigorous—argument, see Exercise 27.

* **NOTE** See, for example, Exercise 53 in Section 5.4.

We are generally interested in the price that maximizes revenue and, in ordinary cases, the price that maximizes revenue must give unit elasticity. One way of seeing this is as follows:* If the demand is inelastic (which ordinarily occurs at a low unit price) then raising the price by a small percentage—1% say—results in a smaller percentage drop in demand. For example, in the Quick Example above, if $p = \$2{,}000$,d then the demand would drop by only $\frac{1}{4}\%$ for every 1% increase in price. To see the effect on revenue, we use the fact* that, for small changes in price,

$$\begin{aligned}\text{Percentage change in revenue} &\approx \text{Percentage change in price} \\ &\quad + \text{Percentage change in demand} \\ &= 1 + \left(-\frac{1}{4}\right) = \frac{3}{4}\%.\end{aligned}$$

Thus, the revenue will increase by about 3/4%. Put another way:

If the demand is inelastic, raising the price increases revenue.

On the other hand, if the price is elastic (which ordinarily occurs at a high unit price), then increasing the price slightly will lower the revenue, so:

If the demand is elastic, lowering the price increases revenue.

The price that results in the largest revenue must therefore be at unit elasticity.

EXAMPLE 1 Price Elasticity of Demand: Dolls

Suppose that the demand equation for Bobby Dolls is given by $q = 216 - p^2$, where p is the price per doll in dollars and q is the number of dolls sold per week.

a. Compute the price elasticity of demand when $p = \$5$ and $p = \$10$, and interpret the results.

b. Find the ranges of prices for which the demand is elastic and the range for which the demand is inelastic

c. Find the price at which the weekly revenue is maximized. What is the maximum weekly revenue?

Solution

a. The price elasticity of demand is

$$E = -\frac{dq}{dp} \cdot \frac{p}{q}.$$

Taking the derivative and substituting for q gives

$$E = 2p \cdot \frac{p}{216 - p^2} = \frac{2p^2}{216 - p^2}.$$

When $p = \$5$,

$$E = \frac{2(5)^2}{216 - 5^2} = \frac{50}{191} \approx 0.26.$$

Thus, when the price is set at \$5, the demand is dropping at a rate of 0.26% per 1% increase in the price. Because $E < 1$, the demand is inelastic at this price, so raising the price will increase revenue.

When $p = \$10$,

$$E = \frac{2(10)^2}{216 - 10^2} = \frac{200}{116} \approx 1.72.$$

Thus, when the price is set at \$10, the demand is dropping at a rate of 1.72% per 1% increase in the price. Because $E > 1$, demand is elastic at this price, so raising the price will decrease revenue; lowering the price will increase revenue.

b. and **c.** We answer part (c) first. Setting $E = 1$, we get

$$\frac{2p^2}{216 - p^2} = 1$$

$$p^2 = 72.$$

Thus, we conclude that the maximum revenue occurs when $p = \sqrt{72} \approx \$8.49$. We can now answer part (b): The demand is elastic when $p > \$8.49$ (the price is too high), and the demand is inelastic when $p < \$8.49$ (the price is too low). Finally, we calculate the maximum weekly revenue, which equals the revenue corresponding to the price of \$8.49:

$$R = qp = (216 - p^2)p = (216 - 72)\sqrt{72} = 144\sqrt{72} \approx \$1,222.$$

using Technology

See the Technology Guides at the end of the chapter to find out how to automate computations like those in part (a) of Example 1 using a graphing calculator or Excel. Here is an outline for the TI-83/84 Plus:

TI-83/84 Plus

```
Y1=216-X^2
Y2=-nDeriv(Y1,X,X)*X/Y1
```

[2ND] [TABLE] Enter $x = 5$

[More details on page 448.]

Excel

Enter values of p: 4.9, 4.91, . . . , 5.0, 5.01, . . . , 5.1 in A5–A25.

In B5 enter `216-A5^2` and copy down to B25.

In C5 enter `=(A6-A5)/A5` and paste the formula in C5–D24.

In E5 enter = −D5/C5 and copy down to E24. This column contains the values of E for the values of p in column A.

[More details on page 448.]

The concept of elasticity can be applied in other situations. In the following example we consider *income* elasticity of demand—the percentage increase in demand for a particular item per percentage increase in personal income.

EXAMPLE 2 Income Elasticity of Demand: Porsches

You are the sales director at Suburban Porsche and have noticed that demand for Porsches depends on income according to

$$q = 0.005e^{-0.05x^2+x}. \qquad (1 \le x \le 10)$$

Here, x is the income of a potential customer in hundreds of thousands of dollars and q is the probability that the person will actually purchase a Porsche.* The **income elasticity of demand** is

$$E = \frac{dq}{dx}\frac{x}{q}.$$

Compute and interpret E for $x = 2$ and 9.

* **NOTE** In other words, q is the fraction of visitors to your showroom having income x who actually purchase a Porsche.

Solution

Q: *Why is there no negative sign in the formula?*

A: Because we anticipate that the demand will increase as income increases, the ratio

$$\frac{\text{Percentage increase in demand}}{\text{Percentage increase in income}}$$

will be positive, so there is no need to introduce a negative sign.

Turning to the calculation, since $q = 0.005e^{-0.05x^2+x}$,

$$\frac{dq}{dx} = 0.005e^{-0.05x^2+x}(-0.1x + 1)$$

and so

$$E = \frac{dq}{dx}\frac{x}{q}$$

$$= 0.005e^{-0.05x^2+x}(-0.1x + 1)\frac{x}{0.005e^{-0.05x^2+x}}$$

$$= x(-0.1x + 1).$$

When $x = 2$, $E = 2[-0.1(2) + 1)] = 1.6$. Thus, at an income level of \$200,000, the probability that a customer will purchase a Porsche increases at a rate of 1.6% per 1% increase in income.

When $x = 9$, $E = 9[-0.1(9) + 1)] = 0.9$. Thus, at an income level of \$900,000, the probability that a customer will purchase a Porsche increases at a rate of 0.9% per 1% increase in income.

5.6 EXERCISES

▼ more advanced ◆ challenging

T indicates exercises that should be solved using technology

APPLICATIONS

1. ***Demand for Oranges*** The weekly sales of Honolulu Red Oranges is given by $q = 1{,}000 - 20p$. Calculate the price elasticity of demand when the price is \$30 per orange (yes, \$30 per orange[63]). Interpret your answer. Also, calculate the price that gives a maximum weekly revenue, and find this maximum revenue. HINT [See Example 1.]

2. ***Demand for Oranges*** Repeat the preceding exercise for weekly sales of $1{,}000 - 10p$. HINT [See Example 1.]

3. ***Tissues*** The consumer demand equation for tissues is given by $q = (100 - p)^2$, where p is the price per case of tissues and q is the demand in weekly sales.

   a. Determine the price elasticity of demand E when the price is set at \$30, and interpret your answer.

   b. At what price should tissues be sold in order to maximize the revenue?

   c. Approximately how many cases of tissues would be demanded at that price?

4. ***Bodybuilding*** The consumer demand curve for Professor Stefan Schwarzenegger dumbbells is given by $q = (100 - 2p)^2$, where p is the price per dumbbell, and q is the demand in weekly sales. Find the price Professor Schwarzenegger should charge for his dumbbells in order to maximize revenue.

5. ***T-Shirts*** The Physics Club sells $E = mc^2$ T-shirts at the local flea market. Unfortunately, the club's previous administration has been losing money for years, so you decide to do an analysis of the sales. A quadratic regression based on old sales data reveals the following demand equation for the T-shirts:

$$q = -2p^2 + 33p. \quad (9 \le p \le 15)$$

[63] They are very hard to find, and their possession confers considerable social status.

Here, p is the price the club charges per T-shirt, and q is the number it can sell each day at the flea market.

a. Obtain a formula for the price elasticity of demand for $E = mc^2$ T-shirts.
b. Compute the elasticity of demand if the price is set at \$10 per shirt. *Interpret the result.*
c. How much should the Physics Club charge for the T-shirts in order to obtain the maximum daily revenue? What will this revenue be?

6. ***Comics*** The demand curve for original *Iguanawoman* comics is given by

$$q = \frac{(400 - p)^2}{100} \qquad (0 \le p \le 400)$$

where q is the number of copies the publisher can sell per week if it sets the price at \$$p$.

a. Find the price elasticity of demand when the price is set at \$40 per copy.
b. Find the price at which the publisher should sell the books in order to maximize weekly revenue.
c. What, to the nearest \$1, is the maximum weekly revenue the publisher can realize from sales of *Iguanawoman* comics?

7. ***College Tuition*** A study of about 1,800 U.S. colleges and universities resulted in the demand equation $q = 9{,}900 - 2.2p$, where q is the enrollment at a college or university, and p is the average annual tuition (plus fees) it charges.[64]

a. The study also found that the average tuition charged by universities and colleges was \$2,900. What is the corresponding price elasticity of demand? Is the price elastic or inelastic? Should colleges charge more or less on average to maximize revenue?
b. Based on the study, what would you advise a college to charge its students in order to maximize total revenue, and what would the revenue be?

8. ***Demand for Fried Chicken*** A fried chicken franchise finds that the demand equation for its new roast chicken product, "Roasted Rooster," is given by

$$p = \frac{40}{q^{1.5}}$$

where p is the price (in dollars) per quarter-chicken serving and q is the number of quarter-chicken servings that can be sold per hour at this price. Express q as a function of p and find the price elasticity of demand when the price is set at \$4 per serving. Interpret the result.

9. ***Paint-By-Number*** The estimated monthly sales of *Mona Lisa* paint-by-number sets is given by the formula $q = 100e^{-3p^2+p}$, where q is the demand in monthly sales and p is the retail price in yen.

a. Determine the price elasticity of demand E when the retail price is set at ¥3 and interpret your answer.
b. At what price will revenue be a maximum?
c. Approximately how many paint-by-number sets will be sold per month at the price in part (b)?

10. ***Paint-By-Number*** Repeat the previous exercise using the demand equation $q = 100e^{p-3p^2/2}$.

11. ▼ ***Linear Demand Functions*** A general linear demand function has the form $q = mp + b$ (m and b constants, $m \neq 0$).

a. Obtain a formula for the price elasticity of demand at a unit price of p.
b. Obtain a formula for the price that maximizes revenue.

12. ▼ ***Exponential Demand Functions*** A general exponential demand function has the form $q = Ae^{-bp}$ (A and b nonzero constants).

a. Obtain a formula for the price elasticity of demand at a unit price of p.
b. Obtain a formula for the price that maximizes revenue.

13. ▼ ***Hyperbolic Demand Functions*** A general hyperbolic demand function has the form $q = \dfrac{k}{p^r}$ (r and k nonzero constants).

a. Obtain a formula for the price elasticity of demand at unit price p.
b. How does E vary with p?
c. What does the answer to part (b) say about the model?

14. ▼ ***Quadratic Demand Functions*** A general quadratic demand function has the form $q = ap^2 + bp + c$ (a, b, and c constants with $a \neq 0$).

a. Obtain a formula for the price elasticity of demand at a unit price p.
b. Obtain a formula for the price or prices that could maximize revenue.

15. ▼ ***Modeling Linear Demand*** You have been hired as a marketing consultant to Johannesburg Burger Supply, Inc., and you wish to come up with a unit price for its hamburgers in order to maximize its weekly revenue. To make life as simple as possible, you assume that the demand equation for Johannesburg hamburgers has the linear form $q = mp + b$, where p is the price per hamburger, q is the demand in weekly sales, and m and b are certain constants you must determine.

a. Your market studies reveal the following sales figures: When the price is set at \$2.00 per hamburger, the sales amount to 3,000 per week, but when the price is set at \$4.00 per hamburger, the sales drop to zero. Use these data to calculate the demand equation.
b. Now estimate the unit price that maximizes weekly revenue and predict what the weekly revenue will be at that price.

[64] Based on a study by A.L. Ostrosky Jr. and J.V. Koch , as cited in their book, *Introduction to Mathematical Economics* (Waveland Press, Illinois, 1979) p. 133.

16. ▼ ***Modeling Linear Demand*** You have been hired as a marketing consultant to Big Book Publishing, Inc., and you have been approached to determine the best selling price for the hit calculus text by Whiner and Istanbul entitled *Fun with Derivatives*. You decide to make life easy and assume that the demand equation for *Fun with Derivatives* has the linear form $q = mp + b$, where p is the price per book, q is the demand in annual sales, and m and b are certain constants you'll have to figure out.

a. Your market studies reveal the following sales figures: when the price is set at \$50.00 per book, the sales amount to 10,000 per year; when the price is set at \$80.00 per book, the sales drop to 1000 per year. Use these data to calculate the demand equation.

b. Now estimate the unit price that maximizes annual revenue and predict what Big Book Publishing, Inc.'s annual revenue will be at that price.

17. ***Income Elasticity of Demand: Live Drama*** The likelihood that a child will attend a live theatrical performance can be modeled by

$$q = 0.01(-0.0078x^2 + 1.5x + 4.1) \qquad (15 \le x \le 100)$$

Here, q is the fraction of children with annual household income x thousand dollars who will attend a live dramatic performance at a theater during the year.[65] Compute the income elasticity of demand at an income level of \$20,000 and interpret the result. (Round your answer to two significant digits.) HINT [See Example 2.]

18. ***Income Elasticity of Demand: Live Concerts*** The likelihood that a child will attend a live musical performance can be modeled by

$$q = 0.01(0.0006x^2 + 0.38x + 35). \quad (15 \le x \le 100)$$

Here, q is the fraction of children with annual household income x who will attend a live musical performance during the year.[66] Compute the income elasticity of demand at an income level of \$30,000 and interpret the result. HINT [See Example 2.]

19. ***Income Elasticity of Demand: Computer Usage*** The demand for personal computers in the home goes up with household income. The following graph shows some data on computer usage together with the logarithmic model $q = 0.3454\ \ln(x) - 3.047$, where q is the probability that a household with annual income x will have a computer.[67]

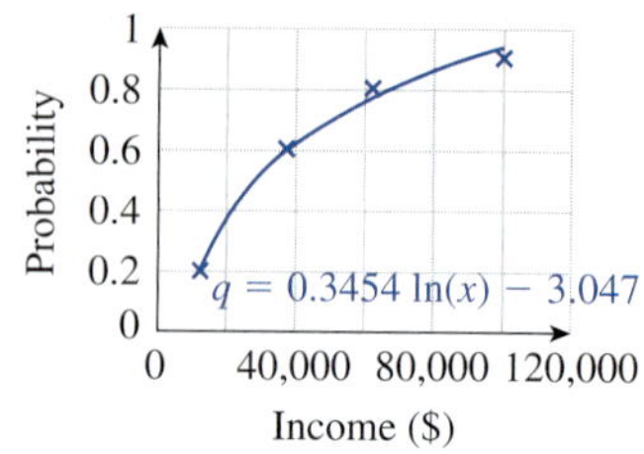

a. Compute the income elasticity of demand for computers, to two decimal places, for a household income of \$60,000 and interpret the result.

b. As household income increases, how is income elasticity of demand affected?

c. How reliable is the given model of demand for incomes well above \$120,000? Explain.

d. What can you say about E for incomes much larger than those shown?

20. ***Income Elasticity of Demand: Internet Usage*** The demand for Internet connectivity also goes up with household income. The following graph shows some data on Internet usage, together with the logarithmic model $q = 0.2802\ \ln(x) - 2.505$, where q is the probability that a home with annual household income x will have an Internet connection.[68]

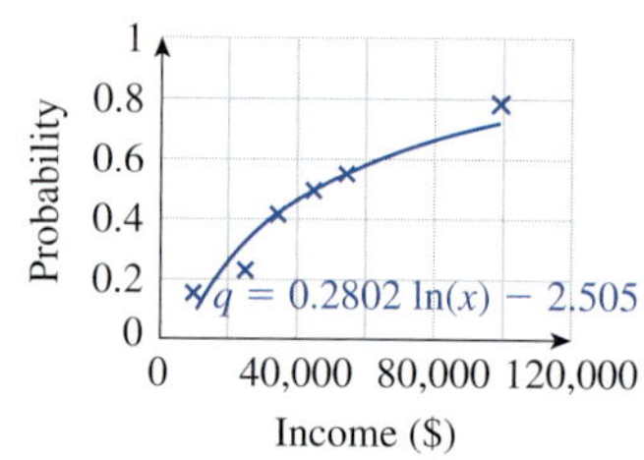

a. Compute the income elasticity of demand to two decimal places for a household income of \$60,000 and interpret the result.

b. As household income increases, how is income elasticity of demand affected?

c. The logarithmic model shown above is not appropriate for incomes well above \$100,000. Suggest a model that might be more appropriate.

d. In the model you propose, how does E behave for very large incomes?

21. ▼ ***Income Elasticity of Demand*** *(based on a question on the GRE Economics Test)* If $Q = aP^{\alpha}Y^{\beta}$ is the individual's demand function for a commodity, where P is the (fixed) price of the commodity, Y is the individual's income, and a, α, and β are parameters, explain why β can be interpreted as the income elasticity of demand.

---

[65] Based on a quadratic regression of data from a 2001 survey. Source for data: New York Foundation of the Arts (www.nyfa.org/culturalblueprint/).

[66] Ibid.

[67] All figures are approximate. The model is a regression model, and x measures the probability that a given household will have one or more computers. Source: Income distribution computer data: Forrester Research/*The New York Times*, August 8, 1999, p. BU4.

[68] All figures are approximate, and the model is a regression model. The Internet connection figures were actually quoted as "share of consumers who use the Internet, by household income." Sources: Luxembourg Income Study/*The New York Times*, August 14, 1995, p. A9, Commerce Department, Deloitte & Touche Survey/*The New York Times*, November 24, 1999, p. C1.

22. ▼ ***College Tuition*** *(from the GRE Economics Test)* A time-series study of the demand for higher education, using tuition charges as a price variable, yields the following result:

$$\frac{dq}{dp} \cdot \frac{p}{q} = -0.4$$

where p is tuition and q is the quantity of higher education. Which of the following is suggested by the result?

(A) As tuition rises, students want to buy a greater quantity of education.
(B) As a determinant of the demand for higher education, income is more important than price.
(C) If colleges lowered tuition slightly, their total tuition receipts would increase.
(D) If colleges raised tuition slightly, their total tuition receipts would increase.
(E) Colleges cannot increase enrollments by offering larger scholarships.

23. ▼ ***Modeling Exponential Demand*** As the new owner of a supermarket, you have inherited a large inventory of unsold imported Limburger cheese, and you would like to set the price so that your revenue from selling it is as large as possible. Previous sales figures of the cheese are shown in the following table:

Price per Pound, *p*	\$3.00	\$4.00	\$5.00
Monthly Sales in Pounds, *q*	407	287	223

a. Use the sales figures for the prices \$3 and \$5 per pound to construct a demand function of the form $q = Ae^{-bp}$, where A and b are constants you must determine. (Round A and b to two significant digits.)
b. Use your demand function to find the price elasticity of demand at each of the prices listed.
c. At what price should you sell the cheese in order to maximize monthly revenue?
d. If your total inventory of cheese amounts to only 200 pounds, and it will spoil one month from now, how should you price it in order to receive the greatest revenue? Is this the same answer you got in part (c)? If not, give a brief explanation.

24. ▼ ***Modeling Exponential Demand*** Repeat the preceding exercise, but this time use the sales figures for \$4 and \$5 per pound to construct the demand function.

COMMUNICATION AND REASONING EXERCISES

25. Complete the following: When demand is inelastic, revenue will decrease if ____ .

26. Complete the following: When demand has unit elasticity, revenue will decrease if ____ .

27. ▼ Given that the demand q is a differentiable function of the unit price p, show that the revenue $R = pq$ has a stationary point when

$$q + p\frac{dq}{dp} = 0.$$

Deduce that the stationary points of R are the same as the points of unit price elasticity of demand. (Ordinarily, there is only one such stationary point, corresponding to the absolute maximum of R.) HINT [Differentiate R with respect to p.]

28. ▼ Given that the demand q is a differentiable function of income x, show that the quantity $R = q/x$ has a stationary point when

$$q - x\frac{dq}{dx} = 0.$$

Deduce that stationary points of R are the same as the points of unit income elasticity of demand. HINT [Differentiate R with respect to x.]

29. ◆ Your calculus study group is discussing price elasticity of demand, and a member of the group asks the following question: "Since elasticity of demand measures the response of demand to change in unit price, what is the difference between elasticity of demand and the quantity $-dq/dp$?" How would you respond?

30. ◆ Another member of your study group claims that unit price elasticity of demand need not always correspond to maximum revenue. Is he correct? Explain your answer.

CHAPTER 5 REVIEW

KEY CONCEPTS

Web Site www.AppliedCalc.org
Go to the student Web site at www.AppliedCalc.org to find a comprehensive and interactive Web-based summary of Chapter 5.

5.1 Maxima and Minima

Relative maximum, relative minimum *p. 371*
Absolute maximum, absolute minimum *p. 371*
Stationary points, singular points, endpoints *p. 373*
Finding and classifying maxima and minima *p. 373*
First derivative test for relative extrema *p. 375*
Extreme value theorem *p. 379*
Using technology to locate approximate extrema *p. 380*

5.2 Applications of Maxima and Minima

Minimizing average cost *p. 384*
Maximizing area *p. 385*
Steps in solving optimization problems *p. 387*
Maximizing revenue *p. 387*
Optimizing resources *p. 388*
Allocation of labor *p. 390*

5.3 Higher Order Derivatives: Acceleration and Concavity

The second derivative of a function f is the derivative of the derivative of f, written as f'' *p. 398*
The acceleration of a moving object is the second derivative of the position function *p. 398*
Acceleration due to gravity *p. 399*
Acceleration of sales *p. 399*
Concave up, concave down, point of inflection *p. 400*
Locating points of inflection *p. 401*
Application to inflation *p. 401*
Second derivative test for relative extrema *p. 404*
Higher order derivatives *p. 405*

5.4 Analyzing Graphs

Features of a graph: x- and y-intercepts, relative extrema, points of inflection; behavior near points where the function is not defined, behavior at infinity *p. 415*
Analyzing a graph *p. 416*

5.5 Related Rates

If Q is a quantity changing over time t, then the derivative dQ/dt is the rate at which Q changes over time *p. 423*
The expanding circle *p. 423*
Steps in solving related rates problems *p. 424*
The falling ladder *p. 424*
Average cost *p. 426*
Allocation of labor *p. 427*

5.6 Elasticity

Price elasticity of demand $E = -\frac{dq}{dp} \cdot \frac{p}{q}$; demand is elastic if $E > 1$, inelastic if $E < 1$, has unit elasticity if $E = 1$ *p. 433*
Computing and interpreting elasticity, and maximizing revenue *p. 433*
Using technology to compute elasticity *p. 435*
Income elasticity of demand *p. 435*

REVIEW EXERCISES

In Exercises 1–8, find all the relative and absolute extrema of the given functions on the given domain (if supplied) or on the largest possible domain (if no domain is supplied).

1. $f(x) = 2x^3 - 6x + 1$ on $[-2, +\infty)$

2. $f(x) = x^3 - x^2 - x - 1$ on $(-\infty, \infty)$

3. $g(x) = x^4 - 4x$ on $[-1, 1]$

4. $f(x) = \dfrac{x+1}{(x-1)^2}$ on $[-2, 1) \cup (1, 2]$

5. $g(x) = (x-1)^{2/3}$ **6.** $g(x) = x^2 + \ln x$ on $(0, +\infty)$

7. $h(x) = \dfrac{1}{x} + \dfrac{1}{x^2}$ **8.** $h(x) = e^{x^2} + 1$

In Exercises 9–12, the graph of the function f or its derivative is given. Find the approximate x-coordinates of all relative extrema and points of inflection of the original function f (if any).

9. Graph of f:

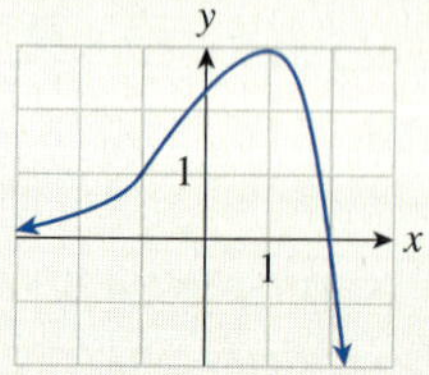

10. Graph of f:

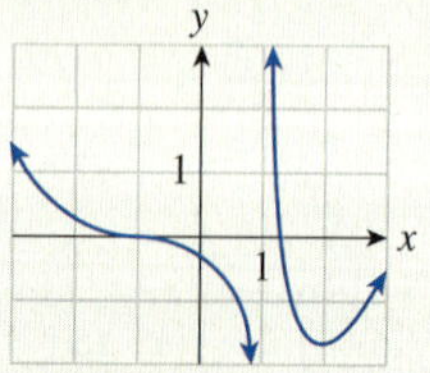

11. Graph of f':

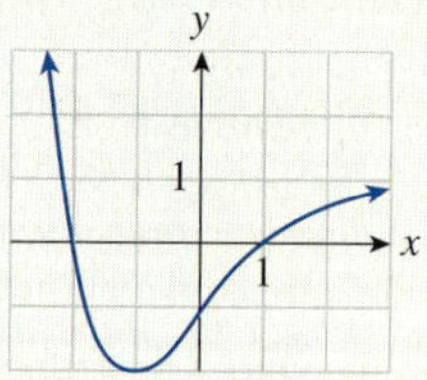

12. Graph of f':

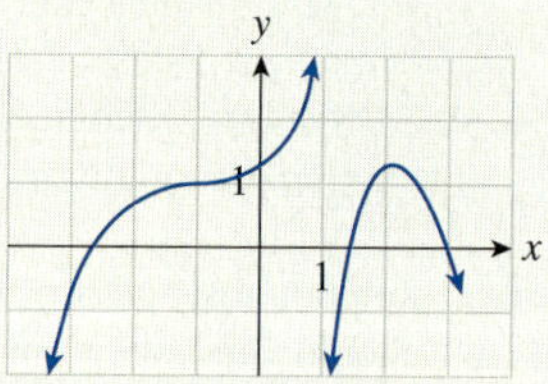

In Exercises 13 and 14, the graph of the second derivative of a function f is given. Find the approximate x-coordinates of all points of inflection of the original function f (if any).

13. Graph of f''

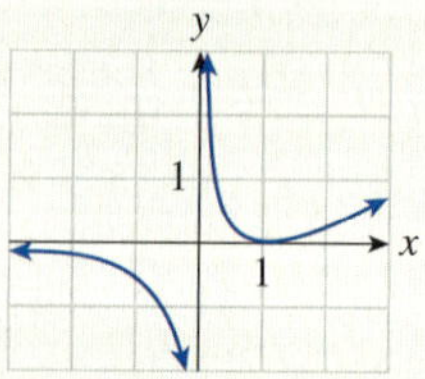

14. Graph of f''

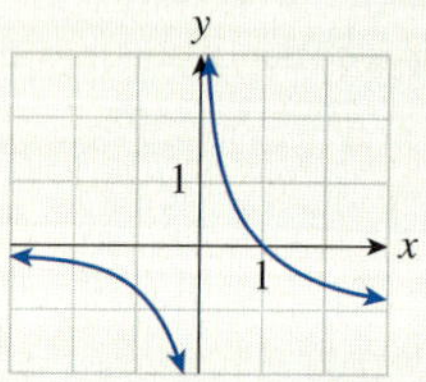

*In Exercises 15 and 16, the position s of a point (in meters) is given as a function of time t (in seconds). Find **(a)** its acceleration as a function of t and **(b)** its acceleration at the specified time.*

15. $s = \frac{2}{3t^2} - \frac{1}{t}; t = 1$

16. $s = \frac{4}{t^2} - \frac{3t}{4}; t = 2$

In Exercises 17–22, sketch the graph of the given function, indicating all relative and absolute extrema and points of inflection. Find the coordinates of these points exactly, where possible. Also indicate any horizontal and vertical asymptotes.

17. $f(x) = x^3 - 12x$ on $[-2, +\infty)$

18. $g(x) = x^4 - 4x$ on $[-1, 1]$

19. $f(x) = \frac{x^2 - 3}{x^3}$

20. $f(x) = (x-1)^{2/3} + \frac{2x}{3}$

21. $g(x) = (x-3)\sqrt{x}$

22. $g(x) = (x+3)\sqrt{x}$

APPLICATIONS

23. ***Revenue*** Demand for the latest best-seller at OHaganBooks.com, *A River Burns Through It*, is given by

$$q = -p^2 + 33p + 9 \qquad (18 \le p \le 28)$$

copies sold per week when the price is p dollars. What price should the company charge to obtain the largest revenue?

24. ***Revenue*** Demand for *The Secret Loves of John O*, a romance novel by Margó Dufón that flopped after two weeks on the market, is given by

$$q = -2p^2 + 5p + 6 \qquad (0 \le p \le 3.3)$$

copies sold per week when the price is p dollars. What price should OHaganBooks charge to obtain the largest revenue?

25. ***Profit*** Taking into account storage and shipping, it costs OHaganBooks.com

$$C = 9q + 100$$

dollars to sell q copies of *A River Burns Through It* in a week (see Exercise 23).

a. If demand is as in Exercise 23, express the weekly profit earned by OHaganBooks.com from the sale of *A River Burns Through It* as a function of unit price p.

b. What price should the company charge to get the largest weekly profit? What is the maximum possible weekly profit?

c. Compare your answer in part (b) with the price the company should charge to obtain the largest revenue (Exercise 23). Explain any difference.

26. ***Profit*** Taking into account storage and shipping, it costs OHaganBooks.com

$$C = 3q$$

dollars to sell q copies of Margó Dufón's *The Secret Loves of John O* in a week (see Exercise 24).

a. If demand is as in Exercise 24, express the weekly profit earned by OHaganBooks.com from the sale of *The Secret Loves of John O* as a function of unit price p.

b. What price should the company charge to get the largest weekly profit? What is the maximum possible weekly profit?

c. Compare your answer in part (b) with the price the company should charge to obtain the largest revenue (Exercise 24). Explain any difference.

27. ***Box Design*** The sales department at OHaganBooks.com, which has decided to send chocolate lobsters to each of its customers, is trying to design a shipping box with a square base. It has a roll of cardboard 36 inches wide from which to make the boxes. Each box will be obtained by cutting out corners from a rectangle of cardboard as shown in the following diagram:

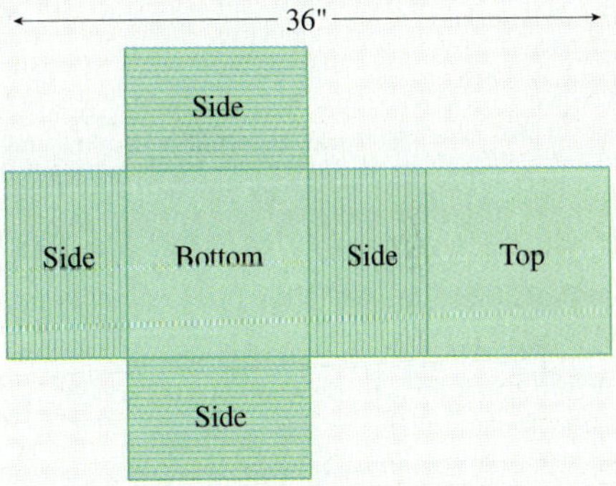

(Notice that the top and bottom of each box will be square, but the sides will not necessarily be square.) What are the dimensions of the boxes with the largest volume that can be made in this way? What is the maximum volume?

28. ***Box Redesign*** The sales department at OHaganBooks.com was not pleased with the result of the box design in the preceding exercise; the resulting box was too large for the chocolate lobsters, so, following a suggestion by a math major student intern, the department decided to redesign the boxes to meet the following specifications: As in Exercise 27, each box would be obtained by cutting out corners from a rectangle of cardboard as shown in the following diagram:

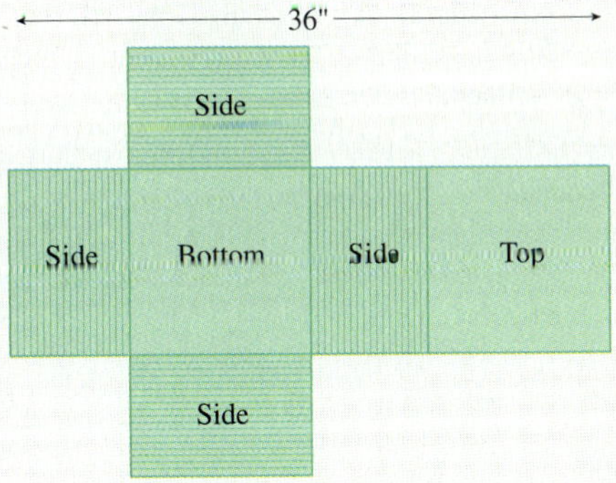

(Notice that the top and bottom of each box would be square, but not necessarily the sides.) The dimensions would be such that the total surface area of the sides plus the bottom of the box would be as large as possible. What are the dimensions of the boxes with the largest area that can be made in this way? How does this box compare with that obtained in Exercise 27?

29. ***Elasticity of Demand*** (Compare Exercise 23). Demand for the latest best-seller at OHaganBooks.com, *A River Burns Through It*, is given by

$$q = -p^2 + 33p + 9 \qquad (18 \le p \le 28)$$

copies sold per week when the price is p dollars.

a. Find the price elasticity of demand as a function of p.
b. Find the elasticity of demand for this book at a price of \$20 and at a price of \$25. (Round your answers to two decimal places.) Interpret the answers.
c. What price should the company charge to obtain the largest revenue?

30. ***Elasticity of Demand*** (Compare Exercise 24). Demand for *The Secret Loves of John O*, a romance novel by Margó Dufón that flopped after two weeks on the market, is given by

$$q = -2p^2 + 5p + 6 \qquad (0 \le p \le 3.3)$$

copies sold per week when the price is p dollars.

a. Find the price elasticity of demand as a function of p.
b. Find the elasticity of demand for this book at a price of \$1 and at a price of \$3. (Round your answers to two decimal places.) Interpret the answers.
c. What price should the company charge to obtain the largest revenue?

31. ***Elasticity of Demand*** Last year OHaganBooks.com experimented with an online subscriber service, Red On Line (ROL), for its electronic book service. The consumer demand for ROL was modeled by the equation

$$q = 1{,}000e^{-p^2+p}$$

where p was the monthly access charge and q is the number of subscribers.

a. Obtain a formula for the price elasticity of demand, E, for ROL services.
b. Compute the elasticity of demand if the monthly access charge is set at \$2 per month. Interpret the result.
c. How much should the company have charged in order to obtain the maximum monthly revenue? What would this revenue have been?

32. ***Elasticity of Demand*** JungleBooks.com (one of OHaganBooks' main competitors) responded with its own online subscriber service, Better On Line (BOL), for its electronic book service. The consumer demand for BOL was modeled by the equation

$$q = 2{,}000e^{-3p^2+2p}$$

where p was the monthly access charge and q is the number of subscribers.

a. Obtain a formula for the price elasticity of demand, E, for BOL services.
b. Compute the elasticity of demand if the monthly access charge is set at \$2 per month. Interpret the result.
c. How much should the company have charged in order to obtain the maximum monthly revenue? What would this revenue have been?

33. ***Sales*** OHaganBooks.com modeled its weekly sales over a period of time with the function

$$s(t) = 6{,}053 + \frac{4{,}474}{1 + e^{-0.55(t-4.8)}}$$

where t is the time in weeks. Following are the graphs of s, s', and s'':

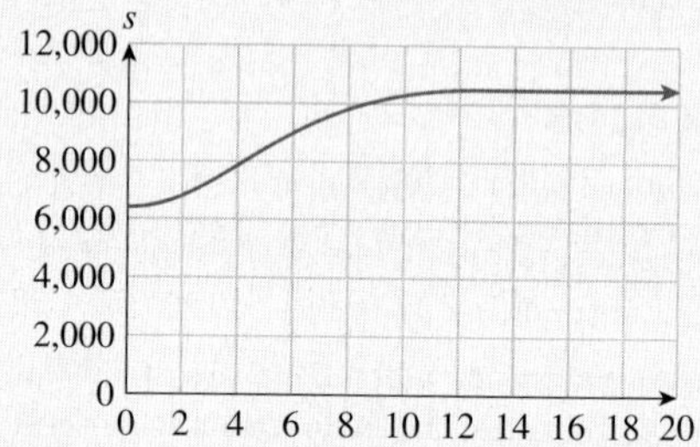

Graph of s

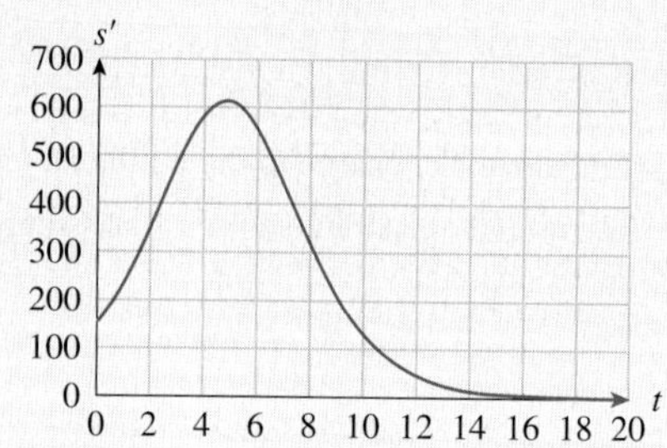

Graph of s'

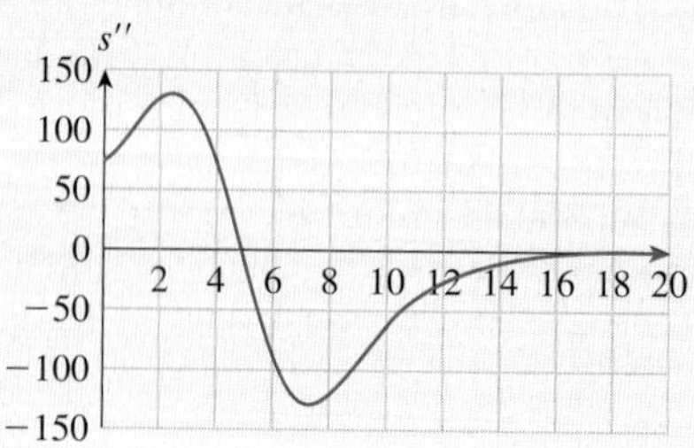

Graph of s''

a. Estimate when, to the nearest week, the weekly sales were growing fastest.

b. To what features on the graphs of s, s', and s'' does your answer to part (a) correspond?

c. The graph of s has a horizontal asymptote. What is the approximate value (s-coordinate) of this asymptote, and what is its significance in terms of weekly sales at OHaganBooks.com?

d. The graph of s' has a horizontal asymptote. What is the value (s'-coordinate) of this asymptote, and what is its significance in terms of weekly sales at OHaganBooks.com?

34. ***Sales*** The quarterly sales of OHagan χPods (OHaganBooks' answer to the *iPod*; a portable audio book unit with an incidental music feature) from the fourth quarter of 2009 can be roughly approximated by the function

$$N(t) = \frac{1{,}100}{1 + 9(1.8)^{-t}} \quad (t \geq 0)$$

where t is time in quarters since the fourth quarter of 2009. Following are the graphs of N, N', and N'':

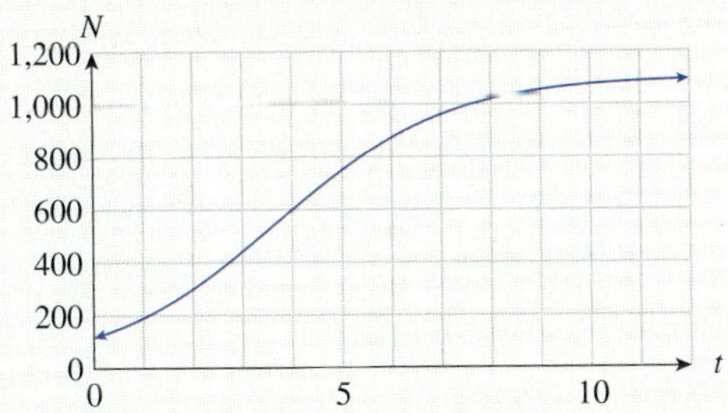

Graph of N

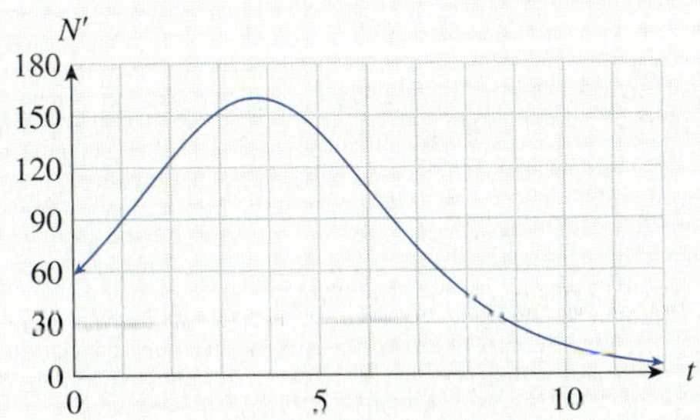

Graph of N'

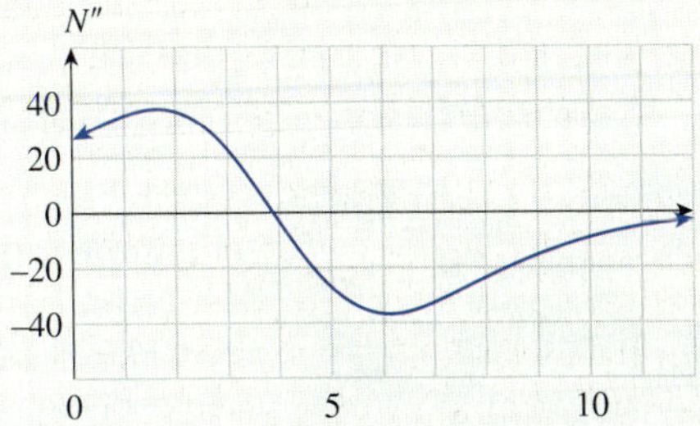

Graph of N''

a. Estimate when, to the nearest quarter, the quarterly sales were growing fastest.

b. To what features on the graphs of N, N', and N'' does your answer to part (a) correspond?

c. The graph of N has a horizontal asymptote. What is the approximate value (N-coordinate) of this asymptote, and what is its significance in terms of quarterly sales of χPods?

d. The graph of N' has a horizontal asymptote. What is the value (N'-coordinate) of this asymptote, and what is its significance in terms of quarterly sales of χPods?

35. ***Chance Encounter*** Marjory Duffin is walking north towards the corner entrance of OHaganBooks.com company headquarters at 5 ft/s, while John O'Hagan is walking west toward the same entrance, also at 5 ft/s. How fast is their distance apart decreasing when:

a. Each of them is 2 ft from the corner?

b. Each of them is 1 ft. from the corner?

c. Each of them is h ft. from the corner?

d. They collide on the corner?

36. ***Company Logos*** OHaganBooks.com's Web site has an animated graphic with its name in a rectangle whose height and width change; on either side of the rectangle are semicircles, as in the figure, whose diameters are the same as the height of the rectangle.

For reasons too complicated to explain, the designer wanted the combined area of the rectangle and semicircles to remain constant. At one point during the animation, the width of the rectangle is 1 inch, growing at a rate of 0.5 inches per second, while the height is 3 inches. How fast is the height changing?

Case Study Production Lot Size Management

SUNNYphotography.com/Alamy

Your publishing company, Knockem Dead Paperbacks, Inc., is about to release its next best-seller, *Henrietta's Heaving Heart* by Celestine A. Lafleur. The company expects to sell 100,000 books each month in the next year. You have been given the job of scheduling print runs to meet the anticipated demand and minimize total costs to the company. Each print run has a setup cost of $5,000, each book costs $1 to produce, and monthly storage costs for books awaiting shipment average 1¢ per book. What will you do?

If you decide to print all 1,200,000 books (the total demand for the year, 100,000 books per month for 12 months) in a single run at the start of the year and sales run as predicted, then the number of books in stock would begin at 1,200,000 and decrease to zero by the end of the year, as shown in Figure 46.

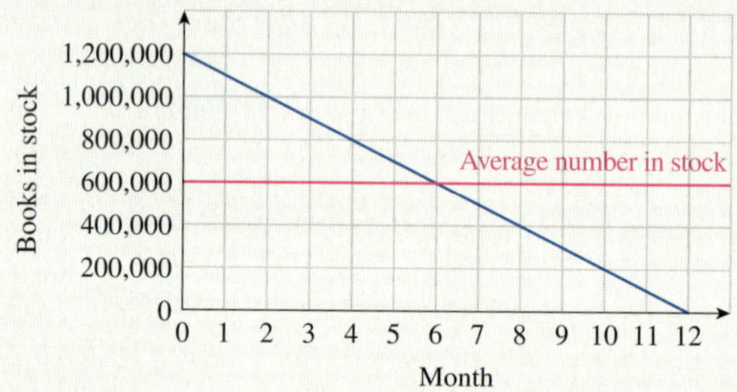

Figure 46

On average, you would be storing 600,000 books for 12 months at 1¢ per book, giving a total storage cost of $600{,}000 \times 12 \times .01 = \$72{,}000$. The setup cost for the single print run would be $5,000. When you add to these the total cost of producing 1,200,000 books at $1 per book, your total cost would be $1,277,000.

If, on the other hand, you decide to cut down on storage costs by printing the book in two runs of 600,000 each, you would get the picture shown in Figure 47.

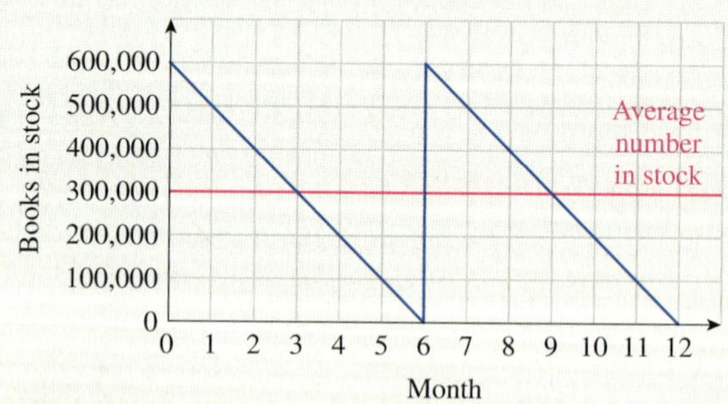

Figure 47

Now, the storage cost would be cut in half because on average there would be only 300,000 books in stock. Thus, the total storage cost would be $36,000, and the setup cost would double to $10,000 (because there would now be two runs). The production costs would be the same: 1,200,000 books @ $1 per book. The total cost would therefore be reduced to $1,246,000, a savings of $31,000 compared to your first scenario.

"Aha!" you say to yourself, after doing these calculations. "Why not drastically cut costs by setting up a run every month?" You calculate that the setup costs alone would be $12 \times \$5{,}000 = \$60{,}000$, which is already more than the setup plus storage costs for two runs, so a run every month will cost too much. Perhaps, then, you should investigate three runs, four runs, and so on, until you find the lowest cost. This strikes you as too laborious a process, especially considering that you will have to do it all over again when planning for Lafleur's sequel, *Lorenzo's Lost Love,* due to be released next year. Realizing that this is an optimization problem, you decide to use some calculus to help you come up with a *formula* that you can use for all future plans. So you get to work.

Instead of working with the number 1,200,000, you use the letter N so that you can be as flexible as possible. (What if *Lorenzo's Lost Love* sells more copies?) Thus, you have a total of N books to be produced for the year. You now calculate the total cost of using x print runs per year. Because you are to produce a total of N books in x print runs, you will have to produce N/x books in each print run. N/x is called the **lot size**. As you can see from the diagrams above, the average number of books in storage will be half that amount, $N/(2x)$.

Now you can calculate the total cost for a year. Write P for the setup cost of a single print run ($P = \$5{,}000$ in your case) and c for the *annual* cost of storing a book (to convert all of the time measurements to years; $c = \$0.12$ here). Finally, write b for the cost of producing a single book ($b = \$1$ here). The costs break down as follows.

Setup Costs: x print runs @ P dollars per run:	Px
Storage Costs: $N/(2x)$ books stored @ c dollars per year:	$cN/(2x)$
Production Costs: N books @ b dollars per book:	Nb
Total Cost:	$Px + \dfrac{cN}{2x} + Nb$

Remember that P, N, c, and b are all constants and x is the only variable. Thus, your cost function is

$$C(x) = Px + \frac{cN}{2x} + Nb$$

and you need to find the value of x that will minimize $C(x)$. But that's easy! All you need to do is find the relative extrema and select the absolute minimum (if any).

The domain of $C(x)$ is $(0, +\infty)$ because there is an x in the denominator and x can't be negative. To locate the extrema, you start by locating the critical points:

$$C'(x) = P - \frac{cN}{2x^2}.$$

The only singular point would be at $x = 0$, but 0 is not in the domain. To find stationary points, you set $C'(x) = 0$ and solve for x:

$$P - \frac{cN}{2x^2} = 0$$

$$2x^2 = \frac{cN}{P}$$

so

$$x = \sqrt{\frac{cN}{2P}}.$$

There is only one stationary point, and there are no singular points or endpoints. To graph the function you will need to put in numbers for the various constants. Substituting $N = 1{,}200{,}000$, $P = 5{,}000$, $c = 0.12$, and $b = 1$, you get

$$C(x) = 5{,}000x + \frac{72{,}000}{x} + 1{,}200{,}000$$

with the stationary point at

$$x = \sqrt{\frac{(0.12)(1{,}200{,}000)}{2(5000)}} \approx 3.79.$$

The total cost at the stationary point is

$$C(3.79) \approx 1{,}240{,}000.$$

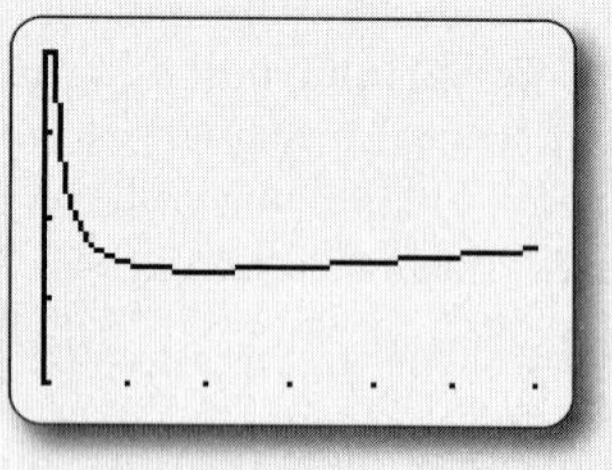

Figure 48

You now graph $C(x)$ in a window that includes the stationary point, say, $0 \le x \le 12$ and $1{,}100{,}000 \le C \le 1{,}500{,}000$, getting Figure 48.

From the graph, you can see that the stationary point is an absolute minimum. In the graph it appears that the graph is always concave up, which also tells you that your stationary point is a minimum. You can check the concavity by computing the second derivative:

$$C''(x) = \frac{cN}{x^3} > 0.$$

The second derivative is always positive because c, N, and x are all positive numbers, so indeed the graph is always concave up. Now you also know that it works regardless of the particular values of the constants.

So now you are practically done! You know that the absolute minimum cost occurs when you have $x \approx 3.79$ print runs per year. Don't be disappointed that the answer is not a whole number; whole number solutions are rarely found in real scenarios. What the answer (and the graph) do indicate is that either 3 or 4 print runs per year will cost the least money. If you take $x = 3$, you get a total cost of

$$C(3) = \$1{,}239{,}000$$

If you take $x = 4$, you get a total cost of

$$C(4) = \$1{,}238{,}000$$

So, four print runs per year will allow you to minimize your total costs.

EXERCISES

1. *Lorenzo's Lost Love* will sell 2,000,000 copies in a year. The remaining costs are the same. How many print runs should you use now?
2. In general, what happens to the number of runs that minimizes cost if both the setup cost and the total number of books are doubled?
3. In general, what happens to the number of runs that minimizes cost if the setup cost increases by a factor of 4?
4. Assuming that the total number of copies and storage costs are as originally stated, find the setup cost that would result in a single print run.
5. Assuming that the total number of copies and setup cost are as originally stated, find the storage cost that would result in a print run each month.

6. In Figure 47 we assumed that all the books in each run were manufactured in a very short time; otherwise the figure might have looked more like Figure 49, which shows the inventory, assuming a slower rate of production.

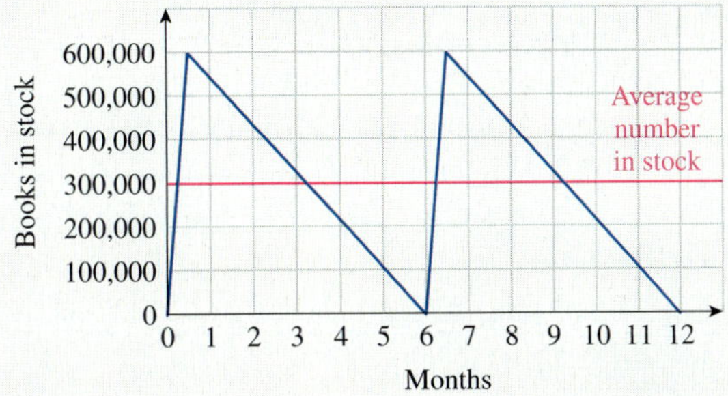

Figure 49

How would this affect the answer?

7. Referring to the general situation discussed in the text, find the cost as a function of the total number of books produced, assuming that the number of runs is chosen to minimize total cost. Also find the average cost per book.
8. Let $\bar{C}$ be the average cost function found in the preceding exercise. Calculate $\lim_{N \to +\infty} \bar{C}(N)$ and interpret the result.

TECHNOLOGY GUIDE

TI-83/84 Plus Technology Guide

Section 5.6

Example 1(a) (page 434) Suppose that the demand equation for Bobby Dolls is given by $q = 216 - p^2$, where p is the price per doll in dollars and q is the number of dolls sold per week. Compute the price elasticity of demand when $p = \$5$ and $p = \$10$, and interpret the results.

Solution with Technology

The TI-83/84 Plus function `nDeriv` can be used to compute approximations of the elasticity E at various prices.

1. Set

```
Y1= 216-X^2                     Demand equation
Y2= -nDeriv(Y1,X,X)* X/Y1       Formula for E
```

2. Use the table feature to list the values of elasticity for a range of prices. For part (a) we chose values of `X` close to 5:

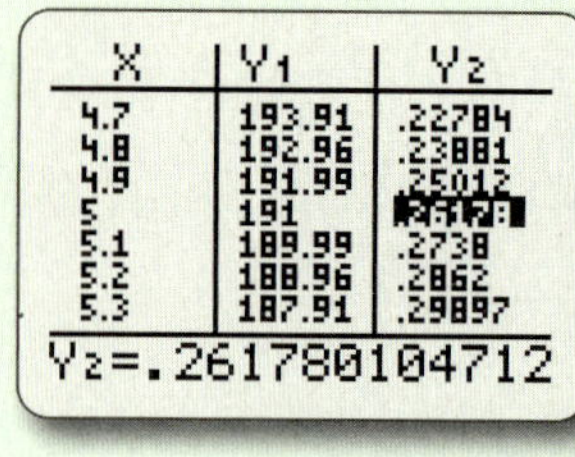

EXCEL Technology Guide

Section 5.6

Example 1(a) (page 434) Suppose that the demand equation for Bobby Dolls is given by $q = 216 - p^2$, where p is the price per doll in dollars and q is the number of dolls sold per week. Compute the price elasticity of demand when $p = \$5$ and $p = \$10$, and interpret the results.

Solution with Technology

To approximate E in Excel, we can use the following approximation of E.

$$E \approx \frac{\text{Percentage decrease in demand}}{\text{Percentage increase in price}} \approx -\frac{\left(\dfrac{\Delta q}{q}\right)}{\left(\dfrac{\Delta p}{p}\right)}$$

The smaller Δp is, the better the approximation. Let's use $\Delta p = 1¢$, or 0.01 (which is small compared with the typical prices we consider—around \$5 to \$10).

1. We start by setting up our worksheet to list a range of prices, in increments of Δp, on either side of a price in which we are interested, such as $p_0 = \$5$:

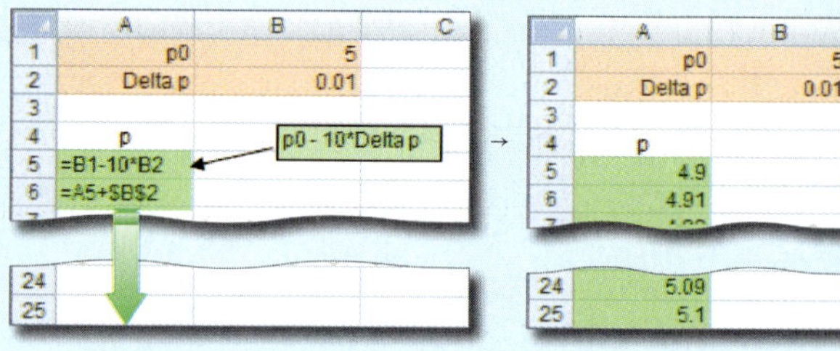

We start in cell A5 with the formula for $p_0 - 10\Delta p$ and then successively add Δp going down column A. You will find that the value $p_0 = 5$ appears midway down the list.

2. Next, we compute the corresponding values for the demand q in Column B.

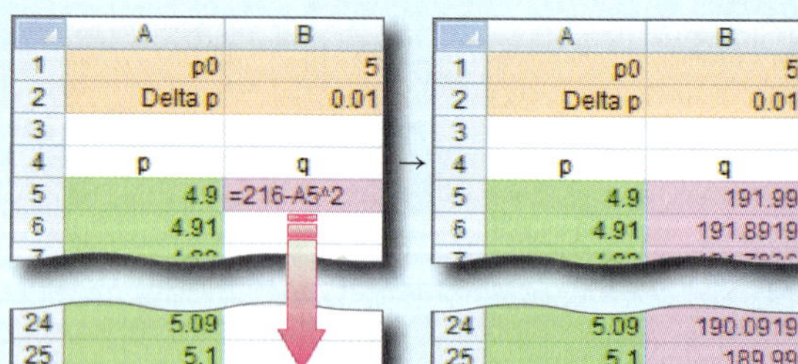

3. We add two new columns for the percentage changes in p and q. The formula shown in cell C5 is copied down columns C and D, to Row 24. (Why not Row 25?)

	A	B	C	D
1	p0	5		
2	Delta p	0.01		
3				
4	p	q	Δp/p	Δq/q
5	4.9	191.99	=(A6-A5)/A5	
6	4.91	191.8919		
24	5.09	190.0919		
25	5.1	189.99		

4. The elasticity can now be computed in column E as shown:

	A	B	C	D	E
1	p0	5			
2	Delta p	0.01			
3					
4	p	q	Δp/p	Δq/q	E
5	4.9	191.99	0.00204082	-0.00051096	= -D5/C5
6	4.91	191.8919	0.00203666	-0.00051227	
24	5.09	190.0919	0.00196464	-0.00053606	
25	5.1	189.99			

	A	B	C	D	E
1	p0	5			
2	Delta p	0.01			
3					
4	p	q	Δp/p	Δq/q	E
5	4.9	191.99	0.00204082	-0.00051096	0.25037242
14	4.99	191.0999	[illegible]	[illegible]	[illegible]
15	5	191	0.002	-0.00052408	0.26204188
16	5.01	190.8999	0.00199601	-0.00052541	0.26322853
17	5.02	190.7996	0.00199203	-0.00052673	0.26441879

6 The Integral

Case Study Spending on Housing Construction

As a summer intern at the U.S. Department of Housing and Urban Development, you have been asked to find formulas for monthly spending on housing construction in the United States and for the average spent per month over a given period. You have data about percentage spending changes. **How will you model the trend and estimate the total?**

Bill Varie/CORBIS

Web Site

www.AppliedCalc.org

At the Web site you will find:

- Section by section tutorials, including game tutorials with randomized quizzes
- A detailed chapter summary
- A true/false quiz
- Additional review exercises
- A numerical integration utility
- Graphing calculator programs for numerical integration
- Graphers, Excel tutorials, and other resources
- The following extra topic:

  Numerical Integration

Introduction

Roughly speaking, calculus is divided into two parts: **differential calculus** (the calculus of derivatives) and **integral calculus**, which is the subject of this chapter and the next. Integral calculus is concerned with problems that are in some sense the reverse of the problems seen in differential calculus. For example, where differential calculus shows how to compute the rate of change of a quantity, integral calculus shows how to find the quantity if we know its rate of change. This idea is made precise in the **Fundamental Theorem of Calculus**. Integral calculus and the Fundamental Theorem of Calculus allow us to solve many problems in economics, physics, and geometry, including one of the oldest problems in mathematics—computing areas of regions with curved boundaries.

6.1 The Indefinite Integral

Suppose that we knew the marginal cost to manufacture an item and we wanted to reconstruct the cost function. We would have to *reverse* the process of differentiation, to go from the derivative (the marginal cost function) back to the original function (the total cost). We'll first discuss how to do that and then look at some applications.

Here is an example: If the derivative of $F(x)$ is $4x^3$, what was $F(x)$? We recognize $4x^3$ as the derivative of x^4. So, we might have $F(x) = x^4$. However, $F(x) = x^4 + 7$ works just as well. In fact, $F(x) = x^4 + C$ works for any number C. Thus, there are *infinitely many* possible answers to this question.

In fact, we will see shortly that the formula $F(x) = x^4 + C$ covers *all* possible answers to the question. Let's give a name to what we are doing.

Antiderivative

An **antiderivative** of a function f is a function F such that $F' = f$.

Quick Examples

1. An antiderivative of $4x^3$ is x^4. — Because the derivative of x^4 is $4x^3$
2. Another antiderivative of $4x^3$ is $x^4 + 7$. — Because the derivative of $x^4 + 7$ is $4x^3$
3. An antiderivative of $2x$ is $x^2 + 12$. — Because the derivative of $x^2 + 12$ is $2x$

Thus,

If the derivative of $A(x)$ is $B(x)$, then an antiderivative of $B(x)$ is $A(x)$.

We call the set of *all* antiderivatives of a function the **indefinite integral** of the function.

Indefinite Integral

$$\int f(x)\,dx$$

is read "the **indefinite integral** of $f(x)$ with respect to x" and stands for the set of all antiderivatives of f. Thus, $\int f(x)\,dx$ is a *collection of functions*; it is not a single function or a number. The function f that is being **integrated** is called the **integrand**, and the variable x is called the **variable of integration**.

Quick Examples

1. $\int 4x^3\,dx = x^4 + C$ Every possible antiderivative of $4x^3$ has the form $x^4 + C$.
2. $\int 2x\,dx = x^2 + C$ Every possible antiderivative of $2x$ has the form $x^2 + C$.

The **constant of integration** C reminds us that we can add any constant and get a different antiderivative.

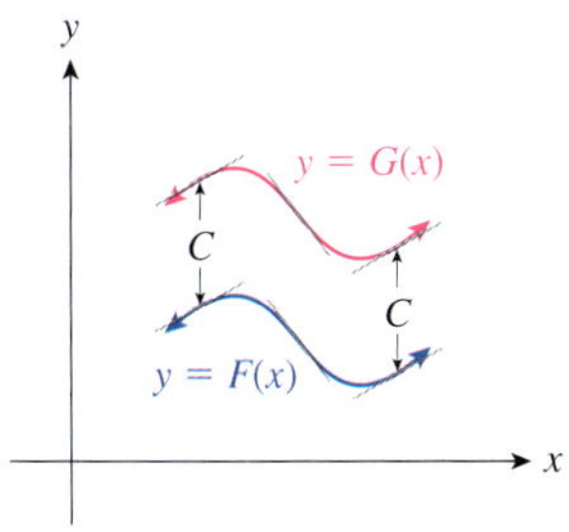

Figure 1

Q: *If $F(x)$ is one antiderivative of $f(x)$, why must all other antiderivatives have the form $F(x) + C$?*

A: Suppose $F(x)$ and $G(x)$ are both antiderivatives of $f(x)$, so that $F'(x) = G'(x)$. Consider what this means by looking at Figure 1. If $F'(x) = G'(x)$ for all x, then F and G have the *same slope* at each value of x. This means that their graphs must be *parallel* and hence remain exactly the same vertical distance apart. But that is the same as saying that the functions differ by a constant—that is, that $G(x) = F(x) + C$ for some constant C.*

* **NOTE** This argument can be turned into a more rigorous proof—that is, a proof that does not rely on geometric concepts such as "parallel graphs." We should also say that the result requires that F and G have the same derivative *on an interval* $[a, b]$.

EXAMPLE 1 Indefinite Integral

Check that $\int x\,dx = \frac{x^2}{2} + C$.

Solution We check our answer by taking the derivative of the right-hand side:

$$\frac{d}{dx}\left(\frac{x^2}{2} + C\right) = \frac{2x}{2} + 0 = x \quad ✔$$

Because the derivative of the right-hand side is the integrand x, we can conclude that $\int x\,dx = \frac{x^2}{2} + C$, as claimed.

Now, we would like to make the process of finding indefinite integrals (antiderivatives) more mechanical. For example, it would be nice to have a power rule for indefinite integrals similar to the one we already have for derivatives. Two cases suggested by the examples above are:

$$\int x\,dx = \frac{x^2}{2} + C \qquad \int x^3\,dx = \frac{x^4}{4} + C$$

You should check the last equation by taking the derivative of its right-hand side. These cases suggest the following general statement:

Power Rule for the Indefinite Integral, Part I

$$\int x^n\,dx = \frac{x^{n+1}}{n+1} + C \qquad (\text{if } n \neq -1)$$

In Words To find the integral of x^n, add 1 to the exponent, and then divide by the new exponent. This rule works provided that n is not -1.

Quick Examples

1. $\int x^{55}\,dx = \frac{x^{56}}{56} + C$

2. $\int \frac{1}{x^{55}}\,dx = \int x^{-55}\,dx$ — Exponent form

$= \frac{x^{-54}}{-54} + C$ — When we add 1 to -55, we get -54, *not* -56.

$= -\frac{1}{54x^{54}} + C$ — See marginal note.*

3. $\int 1\,dx = x + C$ — Because $1 = x^0$. This is an important special case.

4. $\int \sqrt{x}\,dx = \int x^{1/2}dx$ — Exponent form

$= \frac{x^{3/2}}{3/2} + C$

$= \frac{2x^{3/2}}{3} + C$

* **NOTE** We are glossing over a subtlety here: The constant of integration C can be different for $x < 0$ and $x > 0$ because the graph breaks at $x = 0$. In general, our understanding will be that the constant of integration may be different on disconnected intervals of the domain.

Notes

1. The integral $\int 1\,dx$ is commonly written as $\int dx$. Similarly, the integral $\int \frac{1}{x^{55}}\,dx$ may be written as $\int \frac{dx}{x^{55}}$.

2. We can easily check the power rule formula by taking the derivative of the right-hand side:

$$\frac{d}{dx}\left(\frac{x^{n+1}}{n+1} + C\right) = \frac{(n+1)x^n}{n+1} = x^n \quad ✔$$

■

Q: *What is the reason for the restriction $n \neq -1$?*

A: The right-hand side of the power rule formula has $n+1$ in the denominator, and thus makes no sense if $n = -1$. This leaves us not yet knowing how to compute

$$\int x^{-1}\,dx = \int \frac{1}{x}\,dx.$$

Computing this derivative amounts to finding a function whose derivative is $1/x$. Prodding our memories a little, we recall that $\ln x$ has derivative $1/x$. In fact, as we pointed out when we first discussed it, $\ln|x|$ also has derivative $1/x$, but it has the advantage that its domain is the same as that of $1/x$. Thus, we can fill in the missing case as follows:

Power Rule for the Indefinite Integral, Part II

$$\int x^{-1}\,dx = \ln|x| + C \qquad \text{Equivalently, } \int \frac{1}{x}\,dx = \ln|x| + C.$$

Two other indefinite integrals that come from formulas for differentiation are the following:

Indefinite Integral of e^x and b^x

$$\int e^x\,dx = e^x + C \qquad \text{Because } \frac{d}{dx}(e^x) = e^x$$

If b is any positive number other than 1, then

$$\int b^x\,dx = \frac{b^x}{\ln b} + C \qquad \text{Because } \frac{d}{dx}\left(\frac{b^x}{\ln b}\right) = \frac{b^x \ln b}{\ln b} = b^x$$

Quick Example

$$\int 2^x\,dx = \frac{2^x}{\ln 2} + C$$

For more complicated functions, like $2x^3 + 6x^5 - 1$, we need the following rules for integrating sums, differences, and constant multiples.

Sums, Differences, and Constant Multiples

Sum and Difference Rules

$$\int [f(x) \pm g(x)]\,dx = \int f(x)\,dx \pm \int g(x)\,dx$$

In Words: The integral of a sum is the sum of the integrals, and the integral of a difference is the difference of the integrals.

Constant Multiple Rule

$$\int kf(x)\,dx = k\int f(x)\,dx \quad (k\ \textit{constant})$$

In Words: The integral of a constant times a function is the constant times the integral of the function. (In other words, the constant "goes along for the ride.")

Quick Examples

Sum Rule: $\int (x^3+1)\,dx = \int x^3\,dx + \int 1\,dx = \frac{x^4}{4} + x + C$ $\quad f(x) = x^3;\ g(x) = 1$

Constant Multiple Rule: $\int 5x^3\,dx = 5\int x^3\,dx = 5\frac{x^4}{4} + C$ $\quad k = 5;\ f(x) = x^3$

Constant Multiple Rule: $\int 4\,dx = 4\int 1\,dx = 4x + C$ $\quad k = 4;\ f(x) = 1$

Constant Multiple Rule: $\int 4e^x\,dx = 4\int e^x\,dx = 4e^x + C$ $\quad k = 4;\ f(x) = e^x$

Proof of the Sum Rule

We saw above that if two functions have the same derivative, they differ by a (possibly zero) constant. Look at the rule for sums:

$$\int [f(x) + g(x)]\,dx = \int f(x)\,dx + \int g(x)\,dx$$

If we take the derivative of the left-hand side with respect to x, we get the integrand, $f(x) + g(x)$. If we take the derivative of the right-hand side, we get

$$\frac{d}{dx}\left[\int f(x)\,dx + \int g(x)\,dx\right] = \frac{d}{dx}\left[\int f(x)\,dx\right] + \frac{d}{dx}\left[\int g(x)\,dx\right]$$

Derivative of a sum = Sum of derivatives.

$$= f(x) + g(x)$$

Because the left- and right-hand sides have the same derivative, they differ by a constant. But, because both expressions are indefinite integrals, adding a constant does not affect their value, so they are the same as indefinite integrals.

Notice that a key step in the proof was the fact that the derivative of a sum is the sum of the derivatives.

A similar proof works for the difference and constant multiple rule.

EXAMPLE 2 Using the Sum and Difference Rules

Find the integrals.

a. $\int (x^3 + x^5 - 1)\,dx$ **b.** $\int \left(x^{2.1} + \frac{1}{x^{1.1}} + \frac{1}{x} + e^x\right) dx$ **c.** $\int (e^x + 3^x - 1)\,dx$

Solution

a.
$$\int (x^3 + x^5 - 1)\,dx = \int x^3\,dx + \int x^5\,dx - \int 1\,dx \quad \text{Sum/difference rule}$$
$$= \frac{x^4}{4} + \frac{x^6}{6} - x + C \quad \text{Power rule}$$

b.
$$\int \left(x^{2.1} + \frac{1}{x^{1.1}} + \frac{1}{x} + e^x\right) dx$$
$$= \int (x^{2.1} + x^{-1.1} + x^{-1} + e^x)\,dx \quad \text{Exponent form}$$
$$= \int x^{2.1}\,dx + \int x^{-1.1}\,dx + \int x^{-1}\,dx + \int e^x\,dx \quad \text{Sum rule}$$
$$= \frac{x^{3.1}}{3.1} + \frac{x^{-0.1}}{-0.1} + \ln|x| + e^x + C \quad \text{Power rule and exponential rule}$$
$$= \frac{x^{3.1}}{3.1} - \frac{10}{x^{0.1}} + \ln|x| + e^x + C$$

c.
$$\int (e^x + 3^x - 1)\,dx = \int e^x\,dx + \int 3^x\,dx - \int 1\,dx \quad \text{Sum/difference rule}$$
$$= e^x + \frac{3^x}{\ln 3} - x + C \quad \text{Power rule and exponential rule}$$

➡ **Before we go on...** You should check each of the answers in Example 2 by differentiating.

Q: *Why is there only a single arbitrary constant C in each of the answers?*

A: We could have written the answer to part (a) as

$$\frac{x^4}{4} + D + \frac{x^6}{6} + E - x + F$$

where D, E, and F are all arbitrary constants. Now suppose, for example, we set $D = 1$, $E = -2$, and $F = 6$. Then the particular antiderivative we get is $x^4/4 + x^6/6 - x + 5$, which has the form $x^4/4 + x^6/6 - x + C$. Thus, we could have chosen the single constant C to be 5 and obtained the same answer. In other words, the answer $x^4/4 + x^6/6 - x + C$ is just as general as the answer $x^4/4 + D + x^6/6 + E - x + F$, but simpler.

■

In practice we do not explicitly write the integral of a sum as a sum of integrals but just "integrate term by term," much as we learned to differentiate term by term.

EXAMPLE 3 Combining the Rules

Find the integrals.

a. $\int (10x^4 + 2x^2 - 3e^x)\,dx$ **b.** $\int \left(\frac{2}{x^{0.1}} + \frac{x^{0.1}}{2} - \frac{3}{4x}\right) dx$ **c.** $\int [3e^x - 2(1.2^x)]\,dx$

Solution

a. We need to integrate separately each of the terms $10x^4$, $2x^2$, and $3e^x$. To integrate $10x^4$ we use the rules for constant multiples and powers:

$$\int 10x^4\,dx = 10\int x^4\,dx = 10\frac{x^5}{5} + C = 2x^5 + C.$$

The other two terms are similar. We get

$$\int (10x^4 + 2x^2 - 3e^x)\,dx = 10\frac{x^5}{5} + 2\frac{x^3}{3} - 3e^x + C = 2x^5 + \frac{2}{3}x^3 - 3e^x + C.$$

b. We first convert to exponent form and then integrate term by term:

$$\int \left(\frac{2}{x^{0.1}} + \frac{x^{0.1}}{2} - \frac{3}{4x}\right) dx = \int \left(2x^{-0.1} + \frac{1}{2}x^{0.1} - \frac{3}{4}x^{-1}\right) dx$$ Exponent form

$$= 2\frac{x^{0.9}}{0.9} + \frac{1}{2}\frac{x^{1.1}}{1.1} - \frac{3}{4}\ln|x| + C$$ Integrate term by term.

$$= \frac{20x^{0.9}}{9} + \frac{x^{1.1}}{2.2} - \frac{3}{4}\ln|x| + C.$$ Back to rational form

c. $\int [3e^x - 2(1.2^x)]\,dx = 3e^x - 2\frac{1.2^x}{\ln(1.2)} + C$

EXAMPLE 4 Different Variable Name

Find $\int \left(\frac{1}{u} + \frac{1}{u^2}\right) du$.

Solution This integral may look a little strange because we are using the letter u instead of x, but there is really nothing special about x. Using u as the variable of integration, we get

$$\int \left(\frac{1}{u} + \frac{1}{u^2}\right) du = \int (u^{-1} + u^{-2})\,du$$ Exponent form.

$$= \ln|u| + \frac{u^{-1}}{-1} + C$$ Integrate term by term.

$$= \ln|u| - \frac{1}{u} + C.$$ Simplify the result.

➡ **Before we go on...** When we compute an indefinite integral, we want the independent variable in the answer to be the same as the variable of integration. Thus, if the integral in Example 4 had been written in terms of x rather than u, we would have written

$$\int \left(\frac{1}{x} + \frac{1}{x^2}\right) dx = \ln|x| - \frac{1}{x} + C.$$ ■

APPLICATIONS

Rudi Von Briel/PhotoEdit

EXAMPLE 5 Finding Cost from Marginal Cost

The marginal cost to produce baseball caps at a production level of x caps is $4 - 0.001x$ dollars per cap, and the cost of producing 100 caps is \$500. Find the cost function.

Solution We are asked to find the cost function $C(x)$, given that the *marginal* cost function is $4 - 0.001x$. Recalling that the marginal cost function is the derivative of the cost function, we can write

$$C'(x) = 4 - 0.001x$$

and must find $C(x)$. Now $C(x)$ must be an antiderivative of $C'(x)$, so

$$\begin{aligned} C(x) &= \int (4 - 0.001x)\,dx \\ &= 4x - 0.001\frac{x^2}{2} + K \qquad \text{K is the constant of integration.*} \\ &= 4x - 0.0005x^2 + K. \end{aligned}$$

* **NOTE** We used K and not C for the constant of integration because we are using C for cost.

Now, unless we have a value for K, we don't really know what the cost function is. However, there is another piece of information we have ignored: The cost of producing 100 baseball caps is \$500. In symbols

$$C(100) = 500.$$

Substituting in our formula for $C(x)$, we have

$$\begin{aligned} C(100) &= 4(100) - 0.0005(100)^2 + K \\ 500 &= 395 + K \\ K &= 105. \end{aligned}$$

Now that we know what K is, we can write down the cost function:

$$C(x) = 4x - 0.0005x^2 + 105.$$

➡ **Before we go on...** Let us consider the significance of the constant term 105 in Example 5. If we substitute $x = 0$ into the cost function, we get

$$C(0) = 4(0) - 0.0005(0)^2 + 105 = 105.$$

Thus, \$105 is the cost of producing zero items; in other words, it is the **fixed cost**. ■

EXAMPLE 6 Total Sales from Annual Sales

By the start of 2004, **Apple** had sold a total of about 2 million iPods. From the start of 2004 through mid-2008, sales of iPods were approximately

$$s(t) = -3t^2 + 28t - 5.4 \text{ million iPods per year} \qquad (0 \le t \le 4.5)$$

where t is time in years since the start of 2004.*

a. Find an expression for the total sales of iPods up to time t.

b. Use the answer to part (a) to estimate the total sales of iPods by the start of 2008. (The actual figure was 141 million.)

Solution

a. Let $S(t)$ be the total sales of iPods up to time t, where t is measured in years since the start of 2004, so we know that $S(0) = 2$. We are also given an expression for the number of iPods sold per year. This function is the *derivative* of $S(t)$:

$$S'(t) = -3t^2 + 28t - 5.4.$$

Thus, the desired total sales function must be an antiderivative of $S'(t)$:

$$\begin{aligned} S(t) &= \int (-3t^2 + 28t - 5.4)\, dt \\ &= -\frac{3t^2}{3} + \frac{28t^2}{2} - 5.4t + C \\ &= -t^3 + 14t^2 - 5.4t + C. \end{aligned}$$

To calculate the value of the constant C, we can, as in the preceding example, use the known value of S: $S(0) = 2$

$$S(0) = -(0)^3 + 14(0)^2 - 5.4(0) + C = 2$$

so

$$C = 2.$$

We can now write down the total sales function:

$$S(t) = -t^3 + 14t^2 - 5.4t + 2 \text{ million iPods.}$$

b. Because the start of 2008 corresponds to $t = 4$, we calculate total sales as

$$S(4) = -(4)^3 + 14(4)^2 - 5.4(4) + 2 \approx 140 \text{ million iPods.}$$

*Source for data: Apple quarterly earnings reports (www.apple.com/investor/).

Motion in a Straight Line

An important application of the indefinite integral is to the study of motion. The application of calculus to problems about motion is an example of the intertwining of mathematics and physics. We begin by bringing together some facts, scattered through the last several chapters, that have to do with an object moving in a straight line, and then restating them in terms of antiderivatives.

Position, Velocity, and Acceleration: Derivative Form

If $s = s(t)$ is the **position** of an object at time t, then its **velocity** is given by the derivative

$$v = \frac{ds}{dt}.$$

In Words: Velocity is the derivative of position.

The **acceleration** of an object is given by the derivative

$$a = \frac{dv}{dt}.$$

In Words: Acceleration is the derivative of velocity.

Position, Velocity, and Acceleration: Integral Form

$$s(t) = \int v(t)\,dt \qquad \text{Because } v = \frac{ds}{dt}$$

$$v(t) = \int a(t)\,dt \qquad \text{Because } a = \frac{dv}{dt}$$

Quick Examples

1. If the velocity of a particle moving in a straight line is given by $v(t) = 4t + 1$, then its position after t seconds is given by $s(t) = \int v(t)\,dt = \int (4t + 1)\,dt = 2t^2 + t + C$.
2. If sales are accelerating at 2 golf balls/day^2, then the rate of change of sales ("velocity of sales") is $v(t) = \int a(t)\,dt = \int 2\,dt = 2t + C$ golf balls/day.
3. If the rate of change of sales is $v(t) = 2t + 5$ golf balls per day, then the total sales are $s(t) = \int v(t)\,dt = \int (2t + 5)\,dt = t^2 + 5t + C$ golf balls sold through time t.

EXAMPLE 7 Motion in a Straight Line

a. The velocity of a particle moving along in a straight line is given by $v(t) = 4t + 1$ m/s. Given that the particle is at position $s = 2$ meters at time $t = 1$, find an expression for s in terms of t.

b. For a freely falling body experiencing no air resistance and zero initial velocity, find an expression for the velocity v in terms of t. [Note: On Earth, a freely falling body experiencing no air resistance accelerates downward at approximately 9.8 meters per second per second, or 9.8 m/s^2 (or 32 ft/s^2).]

Solution

a. As we saw in the Quick Example above, the position of the particle after t seconds is given by

$$\begin{aligned} s(t) &= \int v(t)\,dt \\ &= \int (4t + 1)\,dt = 2t^2 + t + C. \end{aligned}$$

But what is the value of C? Now, we are told that the particle is at position $s = 2$ at time $t = 1$. In other words, $s(1) = 2$. Substituting this into the expression for $s(t)$ gives

$$2 = 2(1)^2 + 1 + C$$

so

$$C = -1.$$

Hence the position after t seconds is given by

$$s(t) = 2t^2 + t - 1 \text{ meters.}$$

b. Let's measure heights above the ground as positive, so that a rising object has positive velocity and the acceleration due to gravity is negative. (It causes the upward velocity to decrease in value.) Thus, the acceleration of the stone is given by

$$a(t) = -9.8 \text{ m/s}^2.$$

We wish to know the velocity, which is an antiderivative of acceleration, so we compute

$$v(t) = \int a(t)\, dt = \int (-9.8)\, dt = -9.8t + C.$$

To find the value of C, we use the given information that at time $t = 0$ the velocity is 0: $v(0) = 0$. Substituting this into the expression for $v(t)$ gives

$$0 = -9.8(0) + C$$

so

$$C = 0.$$

Hence, the velocity after t seconds is given by

$$v(t) = -9.8t \text{ m/s.}$$

EXAMPLE 8 Vertical Motion Under Gravity

You are standing on the edge of a cliff and toss a stone upward at a speed of $v_0 = 30$ feet per second (v_0 is called the *initial velocity*).

a. Find the stone's velocity as a function of time. How fast and in what direction is it going after 5 seconds? (Neglect the effects of air resistance.)

b. Find the position of the stone as a function of time. Where will it be after 5 seconds?

c. When and where will the stone reach its zenith, its highest point?

Solution

a. This is similar to Example 7(b): Measuring height above the ground as positive, the acceleration of the stone is given by $a(t) = -32$ ft/s^2, and so

$$v(t) = \int (-32)\, dt = -32t + C.$$

To obtain C, we use the fact that you tossed the stone upward at 30 ft/s; that is, when $t = 0$, $v = 30$, or $v(0) = 30$. Thus,

$$30 = v(0) = -32(0) + C.$$

So, $C = 30$ and the formula for velocity is

$$v(t) = -32t + 30 \text{ ft.}$$ $v(t) = -32t + v_0$

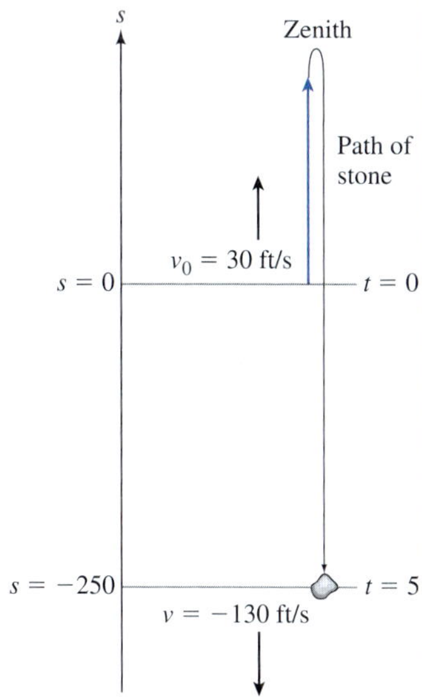

Figure 2

In particular, after 5 seconds the velocity will be

$$v(5) = -32(5) + 30 = -130 \text{ ft/s}.$$

After 5 seconds the stone is *falling* with a speed of 130 ft/s.

b. We wish to know the position, but position is an antiderivative of velocity. Thus,

$$s(t) = \int v(t)\, dt = \int (-32t + 30)\, dt = -16t^2 + 30t + C.$$

Now to find C, we need to know the initial position $s(0)$. We are not told this, so let's measure heights so that the initial position is zero. Then

$$0 = s(0) = C$$

and

$$s(t) = -16t^2 + 30t.$$

$s(t) = -16t^2 + v_0t + s_0$
s_0 = initial position

In particular, after 5 seconds the stone has a height of

$$s(5) = -16(5)^2 + 30(5) = -250 \text{ ft}.$$

In other words, the stone is now 250 ft *below* where it was when you first threw it, as shown in Figure 2.

c. The stone reaches its zenith when its height $s(t)$ is at its maximum value, which occurs when $v(t) = s'(t)$ is zero. So we solve

$$v(t) = -32t + 30 = 0$$

getting $t = 30/32 = 15/16 = 0.9375$ s. This is the time when the stone reaches its zenith. The height of the stone at that time is

$$s(15/16) = -16(15/16)^2 + 30(15/16) = 14.0625 \text{ ft}.$$

➡ Before we go on... Here again are the formulas we obtained in Example 8, together with their metric equivalents:

Vertical Motion Under Gravity: Velocity and Position

If one ignores air resistance, the vertical velocity and position of an object moving under gravity are given by

	British Units	Metric Units
Velocity:	$v(t) = -32t + v_0$ ft/s	$v(t) = -9.8t + v_0$ m/s
Position:	$s(t) = -16t^2 + v_0t + s_0$ ft	$s(t) = -4.9t^2 + v_0t + s_0$ m

v_0 = initial velocity = velocity at time 0
s_0 = initial position = position at time 0

Quick Example

If a ball is thrown down at 2 ft/s from a height of 200 ft, then its velocity and position after t seconds are $v(t) = -32t - 2$ ft/s and $s(t) = -16t^2 - 2t + 200$ ft. ■

6.1 EXERCISES

▼ more advanced ◆ challenging

T indicates exercises that should be solved using technology

Evaluate the integrals in Exercises 1–40.
HINT [for 1–6: See Quick Examples on page 453.]

1. $\int x^5\,dx$ **2.** $\int x^7\,dx$ **3.** $\int 6\,dx$

4. $\int (-5)\,dx$ **5.** $\int x\,dx$ **6.** $\int (-x)\,dx$

HINT [for 7–18: See Example 2.]

7. $\int (x^2 - x)\,dx$ **8.** $\int (x + x^3)\,dx$

9. $\int (1 + x)\,dx$ **10.** $\int (4 - x)\,dx$

11. $\int x^{-5}\,dx$ **12.** $\int x^{-7}\,dx$

13. $\int (x^{2.3} + x^{-1.3})\,dx$ **14.** $\int (x^{-0.2} - x^{0.2})\,dx$

15. $\int (u^2 - 1/u)\,du$ HINT [See Example 4.]

16. $\int (1/v^2 + 2/v)\,dv$ HINT [See Example 4.]

17. $\int \sqrt[4]{x}\,dx$ **18.** $\int \sqrt[3]{x}\,dx$

HINT [for 19–40: See Example 3.]

19. $\int (3x^4 - 2x^{-2} + x^{-5} + 4)\,dx$ **20.** $\int (4x^7 - x^{-3} + 1)\,dx$

21. $\int \left(\frac{2}{u} + \frac{u}{4}\right) du$ **22.** $\int \left(\frac{2}{u^2} + \frac{u^2}{4}\right) du$

23. $\int \left(\frac{1}{x} + \frac{2}{x^2} - \frac{1}{x^3}\right) dx$ **24.** $\int \left(\frac{3}{x} - \frac{1}{x^5} + \frac{1}{x^7}\right) dx$

25. $\int (3x^{0.1} - x^{4.3} - 4.1)\,dx$ **26.** $\int \left(\frac{x^{2.1}}{2} - 2.3\right) dx$

27. $\int \left(\frac{3}{x^{0.1}} - \frac{4}{x^{1.1}}\right) dx$ **28.** $\int \left(\frac{1}{x^{1.1}} - \frac{1}{x}\right) dx$

29. $\int \left(5.1t - \frac{1.2}{t} + \frac{3}{t^{1.2}}\right) dt$

30. $\int \left(3.2 + \frac{1}{t^{0.9}} + \frac{t^{1.2}}{3}\right) dt$

31. $\int (2e^x + 5/x + 1/4)\,dx$

32. $\int (-e^x + x^{-2} - 1/8)\,dx$

33. $\int \left(\frac{6.1}{x^{0.5}} + \frac{x^{0.5}}{6} - e^x\right) dx$

34. $\int \left(\frac{4.2}{x^{0.4}} + \frac{x^{0.4}}{3} - 2e^x\right) dx$

35. $\int (2^x - 3^x)\,dx$ **36.** $\int (1.1^x + 2^x)\,dx$

37. $\int 100(1.1^x)\,dx$ **38.** $\int 1{,}000(0.9^x)\,dx$

39. ▼ $\int \frac{x+2}{x^3} dx$ **40.** ▼ $\int \frac{x^2-2}{x} dx$

41. Find $f(x)$ if $f(0) = 1$ and the tangent line at $(x, f(x))$ has slope x. HINT [See Example 5.]

42. Find $f(x)$ if $f(1) = 1$ and the tangent line at $(x, f(x))$ has slope $\frac{1}{x}$. HINT [See Example 5.]

43. Find $f(x)$ if $f(0) = 0$ and the tangent line at $(x, f(x))$ has slope $e^x - 1$.

44. Find $f(x)$ if $f(1) = -1$ and the tangent line at $(x, f(x))$ has slope $2e^x + 1$.

45. ***Antiderivative of*** $|x|$ Show that an antiderivative of $|x|$ is $\frac{1}{2}x|x|$. Hence evaluate $\int (3|x| - x)\,dx$.

46. ***Antiderivative of*** $|x|$ Use the result of Exercise 45 to evaluate $\int \left(\frac{|x|}{3} - \frac{1}{x}\right) dx$.

APPLICATIONS

47. ***Marginal Cost*** The marginal cost of producing the xth box of light bulbs is $5 - \frac{x}{10{,}000}$ and the fixed cost is \$20,000. Find the cost function $C(x)$. HINT [See Example 5.]

48. ***Marginal Cost*** The marginal cost of producing the xth box of Zip disks is $10 + \frac{x^2}{100{,}000}$ and the fixed cost is \$100,000. Find the cost function $C(x)$. HINT [See Example 5.]

49. ***Marginal Cost*** The marginal cost of producing the xth roll of film is $5 + 2x + \frac{1}{x}$. The total cost to produce one roll is \$1,000. Find the cost function $C(x)$. HINT [See Example 5.]

50. ***Marginal Cost*** The marginal cost of producing the xth box of CDs is $10 + x + \frac{1}{x^2}$. The total cost to produce 100 boxes is \$10,000. Find the cost function $C(x)$. HINT [See Example 5.]

51. **Facebook *Membership*** At the start of 2005, **Facebook** had 1 million members. Since that time, new members joined at a rate of approximately

$$m(t) = 12t^2 - 20t + 10 \text{ million members per year} \quad (0 \le t \le 3.5)$$

where t is time in years since the start of 2005.[1]

[1] Sources for data: www.facebook.com, www.insidehighered.com (Some data are interpolated.)

a. Find an expression for total **Facebook** membership $M(t)$ at time t. HINT [See Example 6.]

b. Use the answer to part (a) to estimate **Facebook** membership midway through 2008. (The actual figure was 80 million.)

52. **MySpace** ***Membership*** At the start of 2004, **MySpace** had 0.6 million members. Since that time, new members joined at a rate of approximately

$$m(t) = 10.5t^2 + 14t - 6 \text{ million members per year} \quad (0 \le t \le 3.5)$$

where t is time in years since the start of 2004.[2]

a. Find an expression for total **MySpace** membership $M(t)$ at time t. HINT [See Example 6.]

b. Use the answer to part (a) to estimate **MySpace** membership in January 2007. (The actual figure was 141 million.)

53. ***Median Household Income*** From 2000 to 2007, median household income in the United States rose by an average of approximately \$1,200 per year.[3] Given that the median household income in 2000 was approximately \$42,000, use an indefinite integral to find a formula for median household income I as a function of the year t ($t = 0$ represents 2000), and use your formula to estimate the median household income in 2005. HINT [See Example 6.]

54. ***Mean Household Income*** From 2000 to 2007, the mean household income in the United States rose by an average of approximately \$1,500 per year.[4] Given that the mean household income in 2000 was approximately \$57,000, use an indefinite integral to find a formula for the mean household income I as a function of the year t ($t = 0$ represents 2000), and use your formula to estimate the mean household income in 2006. HINT [See Example 6.]

55. ***Bottled Water Sales*** The rate of U.S. sales of bottled water for the period 2000–2008 can be approximated by

$$s(t) = 12t^2 + 500t + 4{,}700 \text{ million gallons per year} \quad (0 \le t \le 8)$$

where t is time in years since the start of 2000.[5] Use an indefinite integral to approximate the total sales $S(t)$ of bottled water since the start of 2000. Approximately how much bottled water was sold from the start of 2000 to the end of 2005? HINT [See Example 6. At the start of 2000, sales since that time are zero.]

56. ***Bottled Water Sales*** The rate of U.S. per capita sales of bottled water for the period 2000–2008 can be approximated by

$$s(t) = 0.04t^2 + 1.5t + 17 \text{ gallons per year} \quad (0 \le t \le 8)$$

where t is the time in years since the start of 2000.[6] Use an indefinite integral to approximate the total per capita sales $S(t)$ of bottled water since the start of 2000. Approximately how much bottled water was sold, per capita, from the start of 2000 to the end of 2007? HINT [See Example 6. At the start of 2000, sales since that time are zero.]

57. ▼ ***Health Care Spending*** Write $H(t)$ for the amount spent in the United States on health care in year t, where t is measured in years since 1990. The rate of increase of $H(t)$ was approximately \$65 billion per year in 1990 and rose to \$100 billion per year in 2000.[7]

a. Find a linear model for the rate of change $H'(t)$.

b. Given that \$700 billion was spent on health care in the United States in 1990, find the function $H(t)$.

58. ▼ ***Health Care Spending*** Write $H(t)$ for the amount spent in the United States on health care in year t, where t is measured in years since 2000. The rate of increase of $H(t)$ was projected to rise from \$100 billion per year in 2000 to approximately \$190 billion per year in 2010.[8]

a. Find a linear model for the rate of change $H'(t)$.

b. Given that \$1,300 billion was spent on health care in the United States in 2000, find the function $H(t)$.

59. ▼ ***Subprime Mortgage Debt*** At the start of 2007, the percentage of U.S. mortgages that were subprime was about 13%, was increasing at a rate of 1 percentage point per year, but was decelerating at 0.4 percentage points per year per year.[9]

a. Find an expression for the rate of change (velocity) of this percentage at time t in years since the start of 2007.

b. Use the result of part (a) to find an expression for the percentage of mortgages that were subprime at time t, and use it to estimate the rate at the start of 2008. HINT [See Quick Examples 2 and 3 on page 461.]

60. ▼ ***Subprime Mortgage Debt*** At the start of 2008, the value of subprime mortgage debt outstanding in the U.S. was about \$1,300 billion, was increasing at a rate of 40 billion dollars per year, but was decelerating at 20 billion dollars points per year per year.[10]

a. Find an expression for the rate of change (velocity) of the value of subprime mortgage debt at time t in years since the start of 2008.

---

[2] Source for data: www.swivel.com/data_sets.

[3] In current dollars, unadjusted for inflation. Source: U.S. Census Bureau; "Table H-5. Race and Hispanic Origin of Householder—Households by Median and Mean Income: 1967 to 2007" (www.census.gov/hhes/www/income/histinc/h05.html).

[4] Ibid.

[5] The 2008 figure is an estimate. Source: Beverage Marketing Corporation (www.bottledwater.org).

[6] Ibid.

[7] Source: Centers for Medicare and Medicaid Services, "National Health Expenditures," 2002 version, released January 2004 (www.cms.hhs.gov/statistics/nhe/).

[8] Source: Centers for Medicare and Medicaid Services, "National Health Expenditures 1965–2013, History and Projections" (www.cms.hhs.gov/statistics/nhe/).

[9] Sources: Mortgage Bankers Association, UBS.

[10] Source: www.data360.org/dataset.aspx?Data_Set_Id=9549.

b. Use the result of part (a) to find an expression for the value of subprime mortgage debt at time t, and use it to estimate the value at the start of 2009. HINT [See Quick Examples 2 and 3 on page 461.]

61. ***Motion in a Straight Line*** The velocity of a particle moving in a straight line is given by $v(t) = t^2 + 1$.

a. Find an expression for the position s after a time t.

b. Given that $s = 1$ at time $t = 0$, find the constant of integration C, and hence find an expression for s in terms of t without any unknown constants. HINT [See Example 7.]

62. ***Motion in a Straight Line*** The velocity of a particle moving in a straight line is given by $v = 3e^t + t$.

a. Find an expression for the position s after a time t.

b. Given that $s = 3$ at time $t = 0$, find the constant of integration C, and hence find an expression for s in terms of t without any unknown constants. HINT [See Example 7.]

63. ***Vertical Motion under Gravity*** If a stone is dropped from a rest position above the ground, how fast (in feet per second) and in what direction will it be traveling after 10 seconds? (Neglect the effects of air resistance.) HINT [See Example 8.]

64. ***Vertical Motion under Gravity*** If a stone is thrown upward at 10 feet per second, how fast (in feet per second) and in what direction will it be traveling after 10 seconds? (Neglect the effects of air resistance.) HINT [See Example 8.]

65. ***Vertical Motion under Gravity*** Your name is Galileo Galilei and you toss a weight upward at 16 feet per second from the top of the Leaning Tower of Pisa (height 185 ft).

a. Neglecting air resistance, find the weight's velocity as a function of time t in seconds.

b. Find the height of the weight above the ground as a function of time. Where and when will it reach its zenith? HINT [See Example 8.]

66. ***Vertical Motion under Gravity*** Your name is Spaghettini Bologna (an assistant of Galileo Galilei) and, to impress your boss, you toss a weight upward at 24 feet per second from the top of the Leaning Tower of Pisa (height 185 ft).

a. Neglecting air resistance, find the weight's velocity as a function of time t in seconds.

b. Find the height of the weight above the ground as a function of time. Where and when will it reach its zenith? HINT [See Example 8.]

67. ▼ ***Tail Winds*** The ground speed of an airliner is obtained by adding its air speed and the tail-wind speed. On your recent trip from Mexico to the United States your plane was traveling at an air speed of 500 miles per hour and experienced tail winds of $25 + 50t$ miles per hour, where t is the time in hours since takeoff.

a. Obtain an expression for the distance traveled in terms of the time since takeoff. HINT [Ground speed = Air speed + Tail-wind speed.]

b. Use the result of part (a) to estimate the time of your 1,800-mile trip.

c. The equation solved in part (b) leads mathematically to two solutions. Explain the meaning of the solution you rejected.

68. ▼ ***Head Winds*** The ground speed of an airliner is obtained by subtracting its head-wind speed from its air speed. On your recent trip to Mexico from the United States your plane was traveling at an air speed of 500 miles per hour and experienced head winds of $25 + 50t$ miles per hour, where t is the time in hours since takeoff. HINT [Ground speed = Air speed − Head-wind speed.]

a. Obtain an expression for the distance traveled in terms of the time since takeoff.

b. Use the result of part (a) to estimate the time of your 1,500-mile trip.

c. The equation solved in part (b) leads mathematically to two solutions. Explain the meaning of the solution you rejected.

69. ▼ ***Vertical Motion*** Show that if a projectile is thrown upward with a velocity of v_0 ft/s, then (neglecting air resistance) it will reach its highest point after $v_0/32$ seconds. HINT [See the formulas after Example 8.]

70. ▼ ***Vertical Motion*** Use the result of the preceding exercise to show that if a projectile is thrown upward with a velocity of v_0 ft/s, its highest point will be $v_0^2/64$ feet above the starting point (if we neglect the effects of air resistance).

Exercises 71–76 use the results in the preceding two exercises.

71. ▼ I threw a ball up in the air to a height of 20 feet. How fast was the ball traveling when it left my hand?

72. ▼ I threw a ball up in the air to a height of 40 feet. How fast was the ball traveling when it left my hand?

73. ▼ A piece of chalk is tossed vertically upward by Prof. Schwarzenegger and hits the ceiling 100 feet above with a *BANG.*

a. What is the minimum speed the piece of chalk must have been traveling to enable it to hit the ceiling?

b. Assuming that Prof. Schwarzenegger in fact tossed the piece of chalk up at 100 ft/s, how fast was it moving when it struck the ceiling?

c. Assuming that Prof. Schwarzenegger tossed the chalk up at 100 ft/s, and that it recoils from the ceiling with the same speed it had at the instant it hit, how long will it take the chalk to make the return journey and hit the ground?

74. ▼ A projectile is fired vertically upward from ground level at 16,000 feet per second.

a. How high does the projectile go?

b. How long does it take to reach its zenith (highest point)?

c. How fast is it traveling when it hits the ground?

75. ▼ ***Strength*** Prof. Strong can throw a 10-pound dumbbell twice as high as Professor Weak can. How much faster can Prof. Strong throw it?

76. ▼ ***Weakness*** Prof. Weak can throw a computer disk three times as high as Professor Strong can. How much faster can Prof. Weak throw it?

COMMUNICATION AND REASONING EXERCISES

77. Why is this section called "The *Indefinite* Integral?"

78. If the derivative of Julius is Augustus, then Augustus is _________ of Julius.

79. Linear functions are antiderivatives of what kind of function? Explain.

80. Constant functions are antiderivatives of what kind of function? Explain.

81. If we know the *derivative* of a function, do we know the function? Explain. If not, what further informaion will suffice?

82. If we know an *antiderivative* of a function, do we know the function? Explain. If not, what further informaion will suffice?

83. If $F(x)$ and $G(x)$ are both antiderivatives of $f(x)$, how are $F(x)$ and $G(x)$ related?

84. Your friend Marco claims that once you have one antiderivative of $f(x)$, you have all of them. Explain what he means.

85. Complete the following: The total cost function is a(n) _____ of the _____ cost function.

86. Complete the following: The distance covered is an antiderivative of the _____ function, and the velocity is an antiderivative of the _____ function.

87. If x represents the number of items manufactured and $f(x)$ represents dollars per item, what does $\int f(x)\,dx$ represent? In general, how are the units of $f(x)$ and the units of $\int f(x)\,dx$ related?

88. Complete the following: $-\frac{1}{x}$ is a(n) _____ of $\frac{1}{x^2}$, whereas $\ln x^2$ is not. Also, $-\frac{1}{x}+C$ is the _____ of $\frac{1}{x^2}$, because the _____ of $-\frac{1}{x}+C$ is _____.

89. ▼ Give an argument for the rule that the integral of a sum is the sum of the integrals.

90. ▼ Is it true that $\int \frac{1}{x^3}\,dx = \ln(x^3)+C$? Give a reason for your answer.

91. ▼ Give an example to show that the integral of a product is not the product of the integrals.

92. ▼ Give an example to show that the integral of a quotient is not the quotient of the integrals.

93. ▼ Complete the following: If you take the _____ of the _____ of $f(x)$, you obtain $f(x)$ back. On the other hand, if you take the _____ of the _____ of $f(x)$, you obtain $f(x)+C$.

94. ▼ If a Martian told you that the Institute of Alien Mathematics, after a long and difficult search, has announced the discovery of a new antiderivative of x 1 called $M(x)$ [the formula for $M(x)$ is classified information and cannot be revealed here], how would you respond?

6.2 Substitution

The chain rule for derivatives gives us an extremely useful technique for finding antiderivatives. This technique is called **change of variables** or **substitution**.

Recall that to differentiate a function like $(x^2+1)^6$, we first think of the function as $g(u)$ where $u = x^2+1$ and $g(u) = u^6$. We then compute the derivative, using the chain rule, as

$$\frac{d}{dx}g(u) = g'(u)\frac{du}{dx}.$$

Any rule for derivatives can be turned into a technique for finding antiderivatives by writing it in integral form. The integral form of the above formula is

$$\int g'(u)\frac{du}{dx}\,dx = g(u)+C$$

But, if we write $g(u)+C = \int g'(u)\,du$, we get the following interesting equation:

$$\int g'(u)\frac{du}{dx}\,dx = \int g'(u)\,du.$$

This equation is the one usually called the *change of variables formula.* We can turn it into a more useful integration technique as follows. Let $f = g'(u)(du/dx)$. We can rewrite the above change of variables formula using f:

$$\int f\,dx = \int \left(\frac{f}{du/dx}\right) du.$$

In essence, we are making the formal substitution

$$dx = \frac{1}{du/dx}\,du.$$

Here's the technique:

Substitution Rule

If u is a function of x, then we can use the following formula to evaluate an integral:

$$\int f\,dx = \int \left(\frac{f}{du/dx}\right) du.$$

Rather than use the formula directly, we use the following step-by-step procedure:

1. Write u as a function of x.
2. Take the derivative du/dx and solve for the quantity dx in terms of du.
3. Use the expression you obtain in step 2 to substitute for dx in the given integral and substitute u for its defining expression.

Now let's see how this procedure works in practice.

EXAMPLE 1 Substitution

Find $\int 4x(x^2+1)^6\,dx$.

Solution To use substitution we need to choose an expression to be u. There is no hard and fast rule, but here is one hint that often works:

Take u to be an expression that is being raised to a power.

In this case, let's set $u = x^2 + 1$. Continuing the procedure above, we place the calculations for step 2 in a box.

$u = x^2 + 1$	Write u as a function of x.
$\frac{du}{dx} = 2x$	Take the derivative of u with respect to x.
$dx = \frac{1}{2x}\,du$	Solve for dx: $dx = \frac{1}{du/dx}\,du$.

Now we *substitute u for its defining expression and substitute for dx* in the original integral:

$$\int 4x(x^2+1)^6\,dx = \int 4xu^6 \frac{1}{2x}\,du \qquad \text{Substitute* for } u \text{ and } dx.$$

$$= \int 2u^6\,du. \qquad \text{Cancel the } x\text{s and simplify.}$$

*** NOTE** This step is equivalent to using the formula stated in the Substitution Rule box. If it should bother you that the integral contains both x and u, note that x is now a function of u.

We have boiled the given integral down to the much simpler integral $\int 2u^6\,du$, and we can now write down the solution:

$$2\frac{u^7}{7} + C = \frac{2(x^2+1)^7}{7} + C. \qquad \text{Substitute } (x^2+1) \text{ for } u \text{ in the answer.}$$

Before we go on... There are two points to notice in Example 1. First, before we can actually integrate with respect to u, *we must eliminate all* x*s from the integrand.* If we cannot, we may have chosen the wrong expression for u. Second, after integrating, we must substitute back to obtain an expression involving x.

It is easy to check our answer. We differentiate:

$$\frac{d}{dx}\left[\frac{2(x^2+1)^7}{7}\right] = \frac{2(7)(x^2+1)^6(2x)}{7} = 4x(x^2+1)^6. \quad ✓$$

Notice how we used the chain rule to check the result obtained by substitution. ■

When we use substitution, the first step is always to decide what to take as u. Again, there are no set rules, but we see some common cases in the examples.

EXAMPLE 2 More Substitution

Calculate $\int x^2(x^3+1)^2\,dx$.

Solution As we said in Example 1, it often works to take u to be an expression that is being raised to a power. We usually also want to see the derivative of u as a factor in the integrand so that we can cancel terms involving x. In this case, x^3+1 is being raised to a power, so let's set $u = x^3+1$. Its derivative is $3x^2$; in the integrand is x^2, which is missing the factor 3, but missing or incorrect constant factors are not a problem.

$u = x^3 + 1$	Write u as a function of x.
$\dfrac{du}{dx} = 3x^2$	Take the derivative of u with respect to x.
$dx = \dfrac{1}{3x^2}\,du$	Solve for dx: $dx = \dfrac{1}{du/dx}\,du$.

$$\int x^2(x^3+1)^2\,dx = \int x^2u^2\frac{1}{3x^2}du \qquad \text{Substitute for } u \text{ and } dx.$$

$$= \int \frac{1}{3}u^2\,du \qquad \text{Cancel the terms with } x.$$

$$= \frac{1}{9}u^3 + C \qquad \text{Take the antiderivative.}$$

$$= \frac{1}{9}(x^3+1)^3 + C \qquad \text{Substitute for } u \text{ in the answer.}$$

EXAMPLE 3 An Expression in the Exponent

Evaluate $\int 3xe^{x^2}\,dx$.

Solution When we have an exponential with an expression in the exponent, it often works to substitute u for that expression. In this case, let's set $u = x^2$.

$$u = x^2$$
$$\frac{du}{dx} = 2x$$
$$dx = \frac{1}{2x}du$$

Substituting into the integral, we have

$$\int 3xe^{x^2}dx = \int 3xe^u\frac{1}{2x}\,du = \int \frac{3}{2}e^u du$$
$$= \frac{3}{2}e^u + C = \frac{3}{2}e^{x^2} + C.$$

EXAMPLE 4 A Special Power

Evaluate: **a.** $\int \frac{1}{2x+5}\,dx$ **b.** $\int \left(\frac{1}{2x+5} + 4x^2 + 1\right)dx$

Solution

a. We begin by rewriting the integrand as a power.

$$\int \frac{1}{2x+5}\,dx = \int (2x+5)^{-1}\,dx$$

Now we take our earlier advice and set u equal to the expression that is being raised to a power:

$$u = 2x + 5$$
$$\frac{du}{dx} = 2$$
$$dx = \frac{1}{2}du$$

Substituting into the integral, we have

$$\int \frac{1}{2x+5}\,dx = \int \frac{1}{2}u^{-1}\,du = \frac{1}{2}\ln|u| + C$$
$$= \frac{1}{2}\ln|2x+5| + C.$$

b. Here, the substitution $u = 2x + 5$ works for the first part of the integrand, $1/(2x + 5)$, but not for the rest of it, so we break up the integral:

$$\int \left(\frac{1}{2x+5} + 4x^2 + 1 \right) dx = \int \frac{1}{2x+5}\, dx + \int (4x^2 + 1)\, dx.$$

For the first integral we can use the substitution $u = 2x + 5$ [which we did in part (a)], and for the second, no substitution is necessary:

$$\int \frac{1}{2x+5}\, dx + \int (4x^2 + 1)\, dx = \frac{1}{2} \ln|2x + 5| + \frac{4x^3}{3} + x + C.$$

EXAMPLE 5 Choosing *u*

Evaluate $\int (x + 3)\sqrt{x^2 + 6x}\, dx$.

Solution There are two parenthetical expressions. Notice, however, that the derivative of the expression $(x^2 + 6x)$ is $2x + 6$, which is twice the term $(x + 3)$ in front of the radical. Recall that we would like the derivative of u to appear as a factor. Thus, let's take $u = x^2 + 6x$.

$$u = x^2 + 6x$$
$$\frac{du}{dx} = 2x + 6 = 2(x + 3)$$
$$dx = \frac{1}{2(x+3)}\, du$$

Substituting into the integral, we have

$$\int (x + 3)\sqrt{x^2 + 6x}\, dx$$
$$= \int (x + 3)\sqrt{u} \left(\frac{1}{2(x+3)} \right) du$$
$$= \int \frac{1}{2}\sqrt{u}\, du = \frac{1}{2} \int u^{1/2} du$$
$$= \frac{1}{2}\frac{2}{3} u^{3/2} + C = \frac{1}{3}(x^2 + 6x)^{3/2} + C.$$

Some cases require a little more work.

EXAMPLE 6 When the *x* Terms Do Not Cancel

Evaluate $\int \frac{2x}{(x - 5)^2}\, dx$.

Solution We first rewrite

$$\int \frac{2x}{(x-5)^2}\, dx = \int 2x(x - 5)^{-2}\, dx.$$

This suggests that we should set $u = x - 5$.

$$u = x - 5$$
$$\frac{du}{dx} = 1$$
$$dx = du$$

Substituting, we have

$$\int \frac{2x}{(x-5)^2}\,dx = \int 2xu^{-2}\,du.$$

Now, there is nothing in the integrand to cancel the x that appears. If, as here, there is still an x in the integrand after substituting, we go back to the expression for u, solve for x, and substitute the expression we obtain for x in the integrand. So, we take $u = x - 5$ and solve for $x = u + 5$. Substituting, we get

$$\begin{aligned}\int 2xu^{-2}\,du &= \int 2(u+5)u^{-2}\,du \\ &= 2\int (u^{-1} + 5u^{-2})\,du \\ &= 2\ln|u| - \frac{10}{u} + C \\ &= 2\ln|x-5| - \frac{10}{x-5} + C.\end{aligned}$$

EXAMPLE 7 Application: Bottled Water for Pets

Annual sales of bottled spring water for pets can be modeled by the logistic function

$$s(t) = \frac{3{,}000e^{0.5t}}{3 + e^{0.5t}} \text{ million gallons per year} \qquad (0 \le t \le 8)$$

where t is time in years since the start of 2000.*

a. Find an expression for the total amount of bottled spring water for pets sold since the start of 2000.

b. How much bottled spring water for pets was sold from the start of 2005 to the start of 2008?

Solution

a. If we write the total amount of pet spring water sold since the start of 2000 as $S(t)$, then the information we are given says that

$$S'(t) = s(t) = \frac{3{,}000e^{0.5t}}{3 + e^{0.5t}}.$$

Thus,

$$S(t) = \int \frac{3{,}000e^{0.5t}}{3 + e^{0.5t}}\,dt$$

* Source for data: Beverage Marketing Corporation (www.beveragemarketing.com).

is the function we are after. To integrate the expression, take u to be the denominator of the integrand:

$$u = 3 + e^{0.5t}$$
$$\frac{du}{dt} = 0.5e^{0.5t}$$
$$dt = \frac{1}{0.5e^{0.5t}}\,du$$

$$\begin{aligned} S(t) &= \int \frac{3{,}000e^{0.5t}}{3 + e^{0.5t}}\,dt \\ &= \int \frac{3{,}000e^{0.5t}}{u} \cdot \frac{1}{0.5e^{0.5t}}\,du \\ &= \frac{3{,}000}{0.5} \int \frac{1}{u}\,du \\ &= 6{,}000 \ln|u| + C = 6{,}000 \ln(3 + e^{0.5t}) + C. \end{aligned}$$

(Why could we drop the absolute value in the last step?)

Now what is C? Because $S(t)$ represents the total amount of bottled spring water for pets sold *since time* $t = 0$, we have $S(0) = 0$ (because that is when we started counting). Thus,

$$\begin{aligned} 0 &= 6{,}000 \ln\left(3 + e^{0.5(0)}\right) + C \\ &= 6{,}000 \ln(4) + C \\ C &= -6{,}000 \ln(4) \approx -8{,}318. \end{aligned}$$

Therefore, the total sales from the start of 2000 is approximately

$$S(t) = 6{,}000 \ln(3 + e^{0.5t}) - 8{,}318 \text{ million gallons.}$$

b. The period from the start of 2005 to the start of 2008 is represented by the interval [5, 8]. From part (a):

$$\begin{aligned} \text{Sales through the start of 2005} &= S(5) \\ &= 6{,}000 \ln\left(3 + e^{0.5(5)}\right) - 8{,}318 \approx 8{,}003 \text{ million gallons.} \end{aligned}$$

$$\begin{aligned} \text{Sales through the start of 2008} &= S(8) \\ &= 6{,}000 \ln\left(3 + e^{0.5(8)}\right) - 8{,}318 \approx 16{,}003 \text{ million gallons.} \end{aligned}$$

Therefore, sales over the period were about $16{,}003 - 8{,}003 = 8{,}000$ million gallons.

➡ **Before we go on...** You might wonder why we are writing a logistic function in the form we used in Example 7 rather than in one of the "standard" forms $\frac{N}{1 + Ab^{-t}}$ or $\frac{N}{1 + Ae^{-kt}}$. Our only reason for doing this is to make the substitution work. To convert from the second "standard" form to the form we used in the example, multiply top and bottom by e^{kt}. (See Exercises 81 and 82 in Section 6.4 for further discussion.) ■

Shortcuts

The following shortcuts allow us to simply write down the antiderivative in cases where we would otherwise need the substitution $u = ax + b$, as in Example 4. (a and b are constants with $a \neq 0$.) All of the shortcuts can be obtained using the substitution $u = ax + b$. Their derivation will appear in the exercises.

Shortcuts: Integrals of Expressions Involving $(ax + b)$

Rule	Quick Example
$\int (ax+b)^n\,dx = \frac{(ax+b)^{n+1}}{a(n+1)} + C$ (if $n \neq -1$)	$\int (3x-1)^2\,dx = \frac{(3x-1)^3}{3(3)} + C = \frac{(3x-1)^3}{9} + C$
$\int (ax+b)^{-1}\,dx = \frac{1}{a}\ln\|ax+b\| + C$	$\int (3-2x)^{-1}\,dx = \frac{1}{(-2)}\ln\|3-2x\| + C = -\frac{1}{2}\ln\|3-2x\| + C$
$\int e^{ax+b}\,dx = \frac{1}{a}e^{ax+b} + C$	$\int e^{-x+4}\,dx = \frac{1}{(-1)}e^{-x+4} + C = -e^{-x+4} + C$
$\int c^{ax+b}\,dx = \frac{1}{a\ln c}c^{ax+b} + C$	$\int 2^{-3x+4}\,dx = \frac{1}{(-3\ln 2)}2^{-3x+4} + C = -\frac{1}{3\ln 2}2^{-3x+4} + C$
$\int \|ax+b\|\,dx = \frac{1}{2a}(ax+b)\|ax+b\| + C$*	$\int \|2x-1\|\,dx = \frac{1}{4}(2x-1)\|2x-1\| + C$

* **NOTE** Recall that $\int |x|\,dx = \frac{1}{2}x|x| + C.$ (See Exercise 45 of Section 6.1 on page 464.)

FAQs

When to Use Substitution and What to Use for *u*

Q: *If I am asked to calculate an antiderivative, how do I know when to use a substitution and when not to use one?*

A: Do *not* use substitution when integrating sums, differences, and/or constant multiples of powers of x and exponential functions, such as $2x^3 - \frac{4}{x^2} + \frac{1}{2x} + 3^x + \frac{2^x}{3}$.

To recognize when you should try a substitution, pretend that you are *differentiating* the given expression instead of integrating it. If differentiating the expression would require use of the chain rule, then integrating that expression may well require a substitution, as in, say, $x(3x^2-4)^3$ or $(x+1)e^{x^2+2x-1}$. (In the first we have a *quantity* cubed, and in the second we have e raised to a *quantity*.)

Q: *If an integral seems to call for a substitution, what should I use for u?*

A: There are no set rules for deciding what to use for u, but the preceding examples show some common patterns:

- If you see a linear expression raised to a power, try setting u equal to that linear expression. For example, in $(3x - 2)^{-3}$, set $u = 3x - 2$. (Alternatively, try using the shortcuts above.)
- If you see a constant raised to a linear expression, try setting u equal to that linear expression. For example, in $3^{(2x+1)}$, set $u = 2x + 1$. (Alternatively, try a shortcut.)
- If you see an expression raised to a power multiplied by the derivative of that expression (or a constant multiple of the derivative), try setting u equal to that expression. For example, in $x^2(3x^3 - 4)^{-1}$, set $u = 3x^3 - 4$.
- If you see a constant raised to an expression, multiplied by the derivative of that expression (or a constant multiple of its derivative), try setting u equal to that expression. For example, in $5(x + 1)e^{x^2+2x-1}$, set $u = x^2 + 2x - 1$.
- If you see an expression in the denominator and its derivative (or a constant multiple of its derivative) in the numerator, try setting u equal to that expression. For example, in $\dfrac{2^{3x}}{3 - 2^{3x}}$, set $u = 3 - 2^{3x}$.

Persistence often pays off: If a certain substitution does not work, try another approach or a different substitution.

6.2 EXERCISES

▼ more advanced ◆ challenging

T indicates exercises that should be solved using technology

In Exercises 1–10, evaluate the given integral using the substitution (or method) indicated.

1. $\int (3x - 5)^3\, dx;\ u = 3x - 5$

2. $\int (2x + 5)^{-2}\, dx;\ u = 2x + 5$

3. $\int (3x - 5)^3\, dx$; shortcut page 474

4. $\int (2x + 5)^{-2}\, dx$; shortcut page 474

5. $\int e^{-x}\, dx;\ u = -x$

6. $\int e^{x/2}\, dx;\ u = x/2$

7. $\int e^{-x}\, dx$; shortcut page 474

8. $\int e^{x/2}\, dx$; shortcut page 474

9. $\int (x + 1)e^{(x+1)^2}\, dx;\ u = (x + 1)^2$

10. $\int (x - 1)^2 e^{(x-1)^3}\, dx;\ u = (x - 1)^3$

In Exercises 11–48, decide on what substitution to use, and then evaluate the given integral using a substitution. HINT [See Example 1.]

11. $\int (3x + 1)^5\, dx$

12. $\int (-x - 1)^7\, dx$

13. $\int (-2x + 2)^{-2}\, dx$

14. $\int (2x)^{-1}\, dx$

15. $\int 7.2\sqrt{3x - 4}\, dx$

16. $\int 4.4e^{(-3x+4)}\, dx$

17. $\int 1.2e^{(0.6x+2)}\, dx$

18. $\int 8.1\sqrt{-3x + 4}\, dx$

19. $\int x(3x^2 + 3)^3\, dx$

20. $\int x(-x^2 - 1)^3\, dx$

21. $\int x(x^2 + 1)^{1.3}\, dx$

22. $\int \dfrac{x}{(3x^2 - 1)^{0.4}}\, dx$

23. $\int (1 + 9.3e^{3.1x-2})\, dx$

24. $\int (3.2 - 4e^{1.2x-3})\, dx$

25. $\int 2x\sqrt{3x^2-1}\,dx$

26. $\int 3x\sqrt{-x^2+1}\,dx$

27. $\int xe^{-x^2+1}\,dx$

28. $\int xe^{2x^2-1}\,dx$

29. $\int (x+1)e^{-(x^2+2x)}\,dx$ HINT [See Example 5.]

30. $\int (2x-1)e^{2x^2-2x}\,dx$ HINT [See Example 5.]

31. $\int \frac{-2x-1}{(x^2+x+1)^3}\,dx$

32. $\int \frac{x^3-x^2}{3x^4-4x^3}\,dx$

33. $\int \frac{x^2+x^5}{\sqrt{2x^3+x^6-5}}\,dx$ HINT [See Example 6.]

34. $\int \frac{2(x^3-x^4)}{(5x^4-4x^5)^5}\,dx$ HINT [See Example 6.]

35. $\int x(x-2)^5\,dx$

36. $\int x(x-2)^{1/3}\,dx$

37. $\int 2x\sqrt{x+1}\,dx$

38. $\int \frac{x}{\sqrt{x+1}}\,dx$

39. $\int \frac{e^{-0.05x}}{1-e^{-0.05x}}\,dx$ HINT [See Example 7.]

40. $\int \frac{3e^{1.2x}}{2+e^{1.2x}}\,dx$ HINT [See Example 7.]

41. ▼ $\int \frac{3e^{-1/x}}{x^2}\,dx$

42. ▼ $\int \frac{2e^{2/x}}{x^2}\,dx$

43. ▼ $\int \frac{e^x+e^{-x}}{2}\,dx$ HINT [See Example 4(b).]

44. ▼ $\int (e^{x/2}+e^{-x/2})\,dx$ HINT [See Example 4(b).]

45. ▼ $\int \frac{e^x-e^{-x}}{e^x+e^{-x}}\,dx$

46. ▼ $\int \frac{e^{x/2}+e^{-x/2}}{e^{x/2}-e^{-x/2}}\,dx$

47. ▼ $\int \left((2x-1)e^{2x^2-2x}+xe^{x^2}\right)dx$

48. ▼ $\int \left(xe^{-x^2+1}+e^{2x}\right)dx$

In Exercises 49–52, derive the given equation, where a and b are constants with $a \neq 0$.

49. $\int (ax+b)^n\,dx = \frac{(ax+b)^{n+1}}{a(n+1)}+C \quad (\text{if } n \neq -1)$

50. $\int (ax+b)^{-1}\,dx = \frac{1}{a}\ln|ax+b|+C$

51. $\int |ax+b|\,dx = \frac{1}{2a}(ax+b)|ax+b|+C$

52. $\int e^{ax+b}\,dx = \frac{1}{a}e^{ax+b}+C$

In Exercises 53–68, use the shortcut formulas (see page 474 and Exercises 49–52) to calculate the given integral.

53. $\int e^{-x}\,dx$

54. $\int e^{x-1}\,dx$

55. $\int e^{2x-1}\,dx$

56. $\int e^{-3x}\,dx$

57. $\int (2x+4)^2\,dx$

58. $\int (3x-2)^4\,dx$

59. $\int \frac{1}{5x-1}\,dx$

60. $\int (x-1)^{-1}\,dx$

61. $\int (1.5x)^3\,dx$

62. $\int e^{2.1x}\,dx$

63. $\int 1.5^{3x}\,dx$

64. $\int 4^{-2x}\,dx$

65. $\int |2x+4|\,dx$

66. $\int |3x-2|\,dx$

67. $\int (2^{3x+4}+2^{-3x+4})\,dx$ HINT [See Example 4(b).]

68. $\int (1.1^{-x+4}+1.1^{x+4})\,dx$ HINT [See Example 4(b).]

69. Find $f(x)$ if $f(0)=0$ and the tangent line at $(x, f(x))$ has slope $x(x^2+1)^3$.

70. Find $f(x)$ if $f(1)=0$ and the tangent line at $(x, f(x))$ has slope $\frac{x}{x^2+1}$.

71. Find $f(x)$ if $f(1)=1/2$ and the tangent line at $(x, f(x))$ has slope xe^{x^2-1}.

72. Find $f(x)$ if $f(2)=1$ and the tangent line at x has slope $(x-1)e^{x^2-2x}$.

APPLICATIONS

73. ***Economic Growth*** The value of sold goods in Mexico can be approximated by

$$v(t) = 210 - 62e^{-0.05t} \text{ trillion pesos per month} \quad (t \geq 0)$$

where t is time in months since January 2005.[11] Find an expression for the total value $V(t)$ of sold goods in Mexico from January 2005 to time t. HINT [Use the shortcut on page 474.]

74. ***Economic Contraction*** The number of housing starts in the United States can be approximated by

$$n(t) = \frac{1}{12}(1.1 + 1.2e^{-0.08t}) \text{ million homes per month} \quad (t \geq 0)$$

where t is time in months from the start of 2006.[12] Find an expression for the total number $N(t)$ of housing starts in the United States from January 2006 to time t. HINT [Use the shortcut on page 474.]

[11] Source: Instituto Nacional de Estadística y Geografía (INEGI) (www.inegi.gob.mx).

[12] Source for data: *New York Times*, February 17, 2007, p. C3.

75. ***Casino Revenues*** Annual revenues in 1996–2006 from slot machines and video poker machines at **Foxwoods** casino in Connecticut can be modeled by

$$R(t) = 825 - 240e^{-0.4t} \text{ million dollars} \quad (0 \le t \le 10)$$

where t is time in years since 1996.[13] Find an expression for the total revenue earned since 1996, and hence estimate the total revenue earned over the period 1996–2006. (Round your answer to the nearest $10 million.)

76. ***Casino Revenues*** Annual revenues in 1996–2006 from slot machines and video poker machines at the **Mohegan Sun** casino in Connecticut can be modeled by

$$R(t) = 1{,}260 - 1{,}030e^{-0.1t} \text{ million dollars} \quad (0 \le t \le 10)$$

where t is time in years since 1996.[14] Find an expression for the total revenue earned since 1996, and hence estimate the total revenue earned over the period 1996–2006. (Round your answer to the nearest $10 million.)

77. ***Cost*** The marginal cost of producing the xth roll of film is given by $5 + 1/(x + 1)^2$. The total cost to produce one roll is $1,000. Find the total cost function $C(x)$.

78. ***Cost*** The marginal cost of producing the xth box of CDs is given by $10 - x/(x^2 + 1)^2$. The total cost to produce 2 boxes is $1,000. Find the total cost function $C(x)$.

79. ***Scientific Research*** The number of research articles in the prominent journal *Physical Review* written by researchers in Europe can be approximated by

$$E(t) = \frac{7e^{0.2t}}{5 + e^{0.2t}} \text{ thousand articles per year} \quad (t \ge 0)$$

where t is time in years ($t = 0$ represents 1983).[15]

a. Find an (approximate) expression for the total number of articles written by researchers in Europe since 1983 ($t = 0$). HINT [See Example 7.]

b. Roughly how many articles were written by researchers in Europe from 1983 to 2003? (Round your answer to the nearest 1,000 articles.)

80. ***Scientific Research*** The number of research articles in the prominent journal *Physical Review* written by researchers in the United States can be approximated by

$$U(t) = \frac{4.6e^{0.6t}}{0.4 + e^{0.6t}} \text{ thousand articles per year} \quad (t \ge 0)$$

where t is time in years ($t = 0$ represents 1983).[16]

a. Find an (approximate) expression for the total number of articles written by researchers in the United States since 1983 ($t = 0$). HINT [See Example 7.]

b. Roughly how many articles were written by researchers in the United States from 1983 to 2003?

81. ***Sales*** The rate of sales of your company's Jackson Pollock Advanced Paint-by-Number sets can be modeled by

$$s(t) = \frac{900e^{0.25t}}{3 + e^{0.25t}} \text{ sets per month}$$

t months after their introduction. Find an expression for the total number of paint-by-number sets $S(t)$ sold t months after their introduction, and use it to estimate the total sold in the first 12 months. HINT [See Example 7.]

82. ***Sales*** The rate of sales of your company's Jackson Pollock Beginners Paint-by-Number sets can be modeled by

$$s(t) = \frac{1{,}800e^{0.75t}}{10 + e^{0.75t}} \text{ sets per month}$$

t months after their introduction. Find an expression for the total number of paint-by-number sets $S(t)$ sold t months after their introduction, and use it to estimate the total sold in the first 12 months. HINT [See Example 7.]

83. ***Motion in a Straight Line*** The velocity of a particle moving in a straight line is given by $v = t(t^2 + 1)^4 + t$.

a. Find an expression for the position s after a time t. HINT [See Example 4(b).]

b. Given that $s = 1$ at time $t = 0$, find the constant of integration C and hence an expression for s in terms of t without any unknown constants.

84. ***Motion in a Straight Line*** The velocity of a particle moving in a straight line is given by $v = 3te^{t^2} + t$.

a. Find an expression for the position s after a time t. HINT [See Example 4(b).]

b. Given that $s = 3$ at time $t = 0$, find the constant of integration C and hence an expression for s in terms of t without any unknown constants.

85. ***Bottled Water Sales*** (Compare Exercise 55 in Section 6.1.) The rate of U.S. sales of bottled water for the period 2000–2008 can be approximated by

$$s(t) = 12(t - 2000)^2 + 500(t - 2000) + 4{,}700 \text{ million gallons per year} \quad (2000 \le t \le 2008)$$

where t is the year.[17] Use an indefinite integral to approximate the total sales $S(t)$ of bottled water since 2003 ($t = 2003$). Approximately how much bottled water was sold from $t = 2003$ to $t = 2008$?

[13] Source for data: Connecticut Division of Special Revenue/*New York Times*, Sept. 23, 2007, p. C1.

[14] Ibid.

[15] Based on data from 1983 to 2003. Source: The American Physical Society/*New York Times*, May 3, 2003, p. A1.

[16] Ibid.

[17] The 2008 figure is an estimate. Source: Beverage Marketing Corporation (www.bottledwater.org).

86. ***Bottled Water Sales*** (Compare Exercise 56 in Section 6.1.) The rate of U.S. per capita sales of bottled water for the period 2000–2008 can be approximated by

$$s(t) = 0.04(t - 2000)^2 + 1.5(t - 2000) + 17 \text{ gallons per year} \quad (2000 \le t \le 2008)$$

where t is the year.[18] Use an indefinite integral to approximate the total per capita sales $S(t)$ of bottled water since the start of 2006. Approximately how much bottled water was sold, per capita, from $t = 2006$ to $t = 2008$?

COMMUNICATION AND REASONING EXERCISES

87. Are there any circumstances in which you should use the substitution $u = x$? Illustrate your answer by giving an example that shows the effect of this substitution.

88. At what stage of a calculation using a u substitution should you substitute back for u in terms of x: before or after taking the antiderivative?

89. Consider $\int \left(\frac{x}{x^2-1} + \frac{3x}{x^2+1}\right) dx$. To compute it, you should use which of the following?

(A) $u = x^2 - 1$ **(B)** $u = x^2 + 1$ **(C)** Neither **(D)** Both

Explain your answer.

[18]The 2008 figure is an estimate. Source: Beverage Marketing Corporation (www.bottledwater.org).

90. If the substitution $u = x^2 - 1$ works in $\int \frac{x}{x^2-1}\,dx$, why does it not work nearly so easily in $\int \frac{x^2-1}{x}\,dx$? How would you do the second integral most simply?

91. You are asked to calculate $\int \frac{u}{u^2+1}\,du$. What is wrong with the substitution $u = u^2 + 1$?

92. What is wrong with the following "calculation" of $\int \frac{1}{x^2-1}\,dx$?

$$\int \frac{1}{x^2-1} = \int \frac{1}{u} \qquad \text{Using the substitution } u = x^2 - 1$$
$$= \ln|u| + C$$
$$= \ln|x^2 - 1| + C$$

93. Give an example of an integral that can be calculated by using the substitution $u = x^2 + 1$, and then carry out the calculation.

94. ▼ Give an example of an integral that can be calculated either by using the power rule for antiderivatives or by using the substitution $u = x^2 + x$, and then carry out the calculations.

95. ▼ Show that *none* of the following substitutions work for $\int e^{-x^2}\,dx$: $u = -x$, $u = x^2$, $u = -x^2$. (The antiderivative of e^{-x^2} involves the *error function* erf(x).)

96. ▼ Show that *none* of the following substitutions work for $\int \sqrt{1-x^2}\,dx$: $u = 1 - x^2$, $u = x^2$, and $u = -x^2$. (The antiderivative of $\sqrt{1-x^2}$ involves inverse trigonometric functions, discussion of which is beyond the scope of this book.)

6.3 The Definite Integral: Numerical and Graphical Approaches

In Sections 6.1 and 6.2, we discussed the indefinite integral. There is an older, related concept called the **definite integral**. Let's introduce this new idea with an example. (We'll drop hints now and then about how the two types of integral are related. In Section 6.4 we discuss the exact relationship, which is one of the most important results in calculus.)

In Section 6.1, we used antiderivatives to answer questions of the form "Given the marginal cost, compute the total cost." (See Example 5 in Section 6.1.) In this section we approach such questions more directly, and we will forget about antiderivatives for now.

CORBIS

EXAMPLE 1 Total Cost

Your cell phone company offers you an innovative pricing scheme. When you make a call, the *marginal* cost is

$$c(t) = \frac{5}{10t + 1} \text{ dollars per hour.}$$

Use a numerical calculation to estimate the total cost of a 2-hour phone call.

Solution The graph of $c(t)$ is shown in Figure 3.

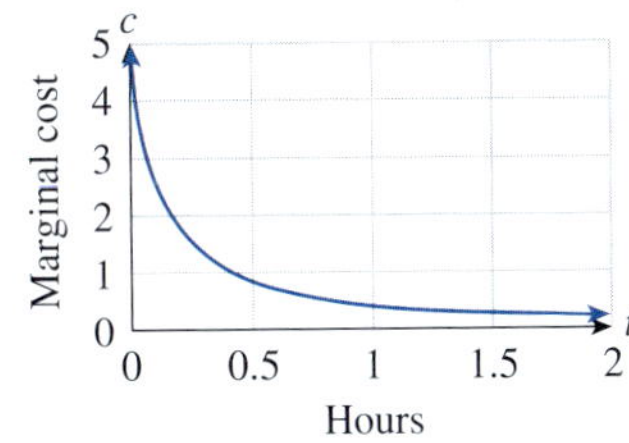

Figure 3

Let's start with a very crude estimate of the total cost, using the graph as a guide. The marginal cost at the beginning of your call is $c(0) = 5/(0+1) = \$5$ per hour. If this cost were to remain constant for the length of your call, the total cost of the call would be

$$\text{Cost of call} = \text{Cost per hour} \times \text{Number of hours} = 5 \times 2 = \$10.$$

Figure 4 shows how we can represent this calculation on the graph of $c(t)$.

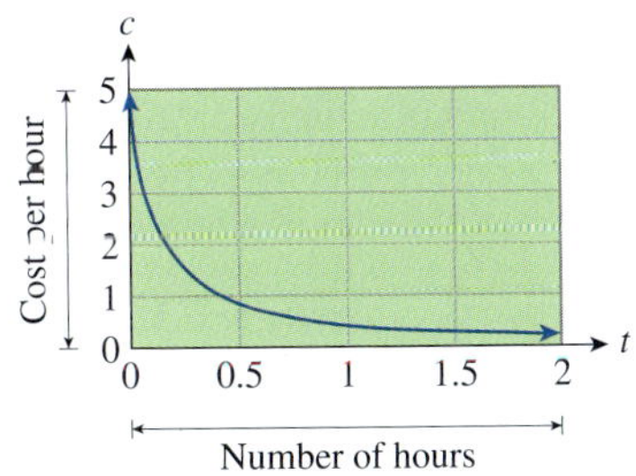

Figure 4

The cost per hour based on $c(0) = 5$ is represented by the y-coordinate of the graph at its left edge, while the number of hours is represented by the width of the interval $[0, 2]$ on the x-axis. Therefore, computing the area of the shaded rectangle in the figure gives the same calculation:

$$\begin{aligned}\text{Area of rectangle} &= \text{Cost per hour} \times \text{Number of hours}\\ &= 5 \times 2 = \$10 = \text{Cost of call.}\end{aligned}$$

But, as we see in the graph, the marginal cost does not remain constant, but goes down quite dramatically over the course of the call. We can obtain a somewhat more accurate estimate of the total cost by looking at the call hour by hour—that is, by dividing the length of the call into two equal intervals, or subdivisions. We estimate the cost of each one-hour subdivision, using the marginal cost at the beginning of that hour.

$$\begin{aligned}\text{Cost of first hour} &= \text{Cost per hour} \times \text{Number of hours}\\ &= c(0) \times 1 = 5 \times 1 = \$5\\ \text{Cost of second hour} &= \text{Cost per hour} \times \text{Number of hours}\\ &= c(1) \times 1 = 5/11 \times 1 \approx \$0.45\end{aligned}$$

Adding these costs gives us the more accurate estimate

$$c(0) \times 1 + c(1) \times 1 = \$5.45. \qquad \text{Calculation using 2 subdivisions}$$

In Figure 5 we see that we are computing the combined area of two rectangles, each of whose heights is determined by the height of the graph at its left edge:

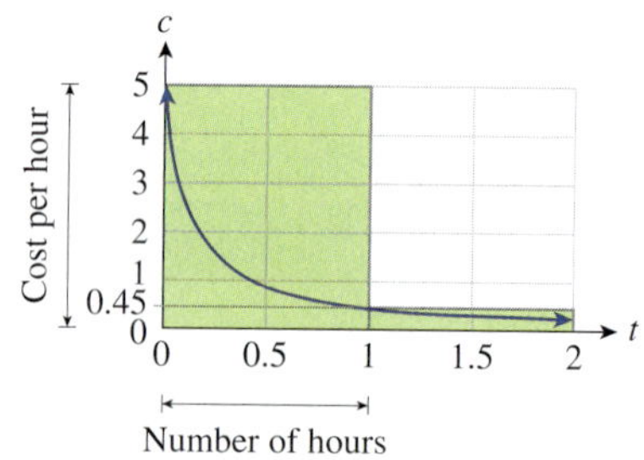

The areas of the rectangles are estimates of the costs for successive one-hour periods.

Figure 5

$$\begin{aligned}\text{Area of first rectangle} &= \text{Cost per hour} \times \text{Number of hours}\\ &= c(0) \times 1 = \$5 = \text{Cost of first hour}\\ \text{Area of second rectangle} &= \text{Cost per hour} \times \text{Number of hours}\\ &= c(1) \times 1 \approx \$0.45 = \text{Cost of second hour}\end{aligned}$$

If we assume that the phone company is honest about $c(t)$ being the marginal cost and is actually calculating your cost more than once an hour, we get an even better estimate of the cost by looking at the call by using four divisions of a half-hour each. The calculation for this estimate is

$$\begin{aligned}&c(0) \times 0.5 + c(0.5) \times 0.5 + c(1) \times 0.5 + c(1.5) \times 0.5\\ &\approx 2.500 + 0.417 + 0.227 + 0.156 = \$3.30.\end{aligned}$$

Calculation using 4 subdivisions

As we see in Figure 6, we have now computed the combined area of *four* rectangles, each of whose heights is again determined by the height of the graph at its left edge.

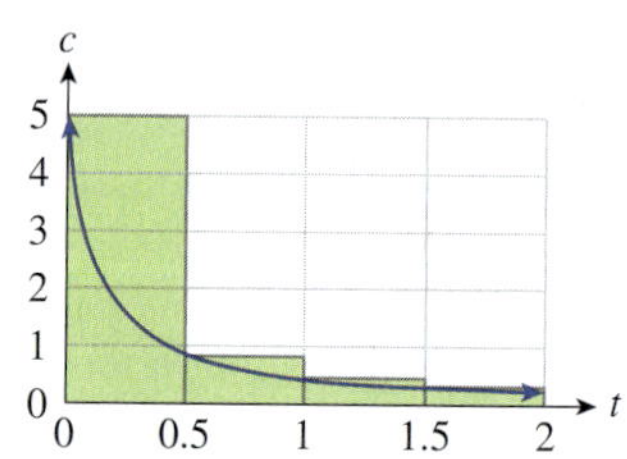

Estimated Cost Using 4 Subdivisions.
The areas of the rectangles are estimates of the costs for successive half-hour periods.

Figure 6

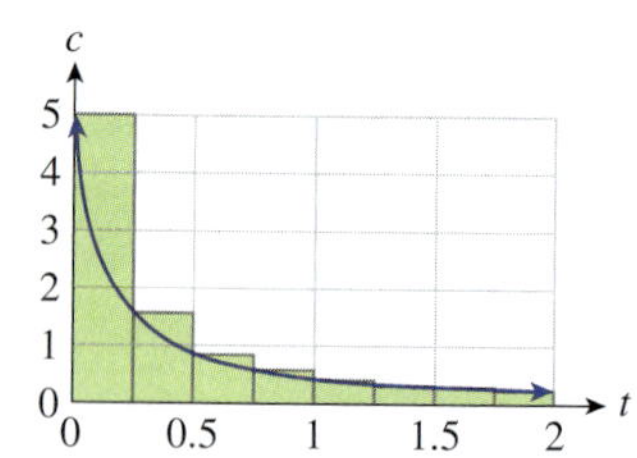

Estimated Cost Using 8 Subdivisions.
The areas of the rectangles are estimates of the costs for successive quarter-hour periods.

Figure 7

Notice how the cost seems to be decreasing as we use more subdivisions. More importantly, total cost seems to be getting closer to the area under the graph. Figure 7 illustrates the calculation for 8 equal subdivisions. The approximate total cost using 8 subdivisions is the total area of the shaded region in Figure 7:

$$c(0) \times 0.25 + c(0.25) \times 0.25 + c(0.5) \times 0.25 + \cdots + c(1.75) \times 0.25 \approx \$2.31.$$

Calculation using 8 subdivisions

Looking at Figure 7, one still gets the impression that we are being overcharged, especially for the first period. If the phone company wants to be *really* honest about $c(t)$ being the marginal cost, it should really be calculating your cost *continuously*, minute by minute or, better yet, second-by-second, as illustrated in Figure 8.

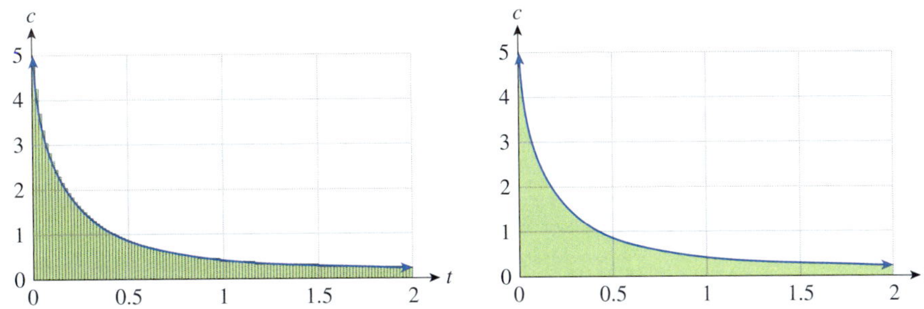

Minute-by-Minute (120 subdivisions)
Cost ≈ \$1.56

Second-by-Second (7,200 subdivisions)
Cost ≈ \$1.52

Figure 8

Figure 8 strongly suggests that the more accurately we calculate the total cost, the closer the answer gets to the exact area under the portion of the graph of $c(t)$ with $0 \le t \le 2$, and leads us to the conclusion that the *exact* total cost is the exact area under the marginal cost curve for $0 \le t \le 2$. In other words, we have made the following remarkable discovery:

Total cost is the area under the marginal cost curve!

➡ **Before we go on...** The minute-by-minute calculation in Example 1 is tedious to do by hand, and no one in his or her right mind would even *attempt* to do the second-by-second calculation by hand! Below we discuss ways of doing these calculations with the aid of technology. ■

The type of calculation done in Example 1 is useful in many applications. Let's look at the general case and give the result a name.

In general, we have a function f (such as the function c in the example), and we consider an interval $[a, b]$ of possible values of the independent variable x. We subdivide the interval $[a, b]$ into some number of segments of equal length. Write n for the number of segments, or **subdivisions**.

Next, we label the endpoints of these subdivisions x_0 for a, x_1 for the end of the first subdivision, x_2 for the end of the second subdivision, and so on until we get to x_n, the end of the nth subdivision, so that $x_n = b$. Thus,

$$a = x_0 < x_1 < \cdots < x_n = b.$$

The first subdivision is the interval $[x_0, x_1]$, the second subdivision is $[x_1, x_2]$, and so on until we get to the last subdivision, which is $[x_{n-1}, x_n]$. We are dividing the interval $[a, b]$ into n subdivisions of equal length, so each segment has length $(b - a)/n$. We write Δx for $(b - a)/n$ (Figure 9).

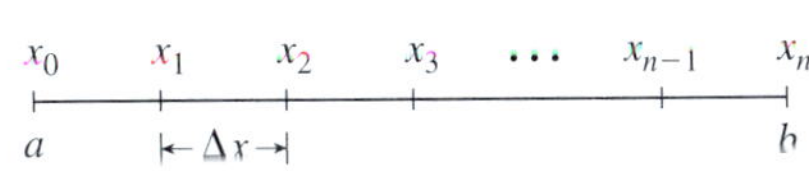

Figure 9

Having established this notation, we can write the calculation that we want to do as follows: For each subdivision $[x_{k-1}, x_k]$, compute $f(x_{k-1})$, the value of the function f at the left endpoint. Multiply this value by the length of the interval, which is Δx. Then add together all n of these products to get the number

$$f(x_0)\Delta x + f(x_1)\Delta x + \cdots + f(x_{n-1})\Delta x.$$

* **NOTE** After Georg Friedrich Bernhard Riemann (1826–1866).

This sum is called a **(left) Riemann*** **sum** for f. In Example 1 we computed several different Riemann sums. Here is the computation for $n = 4$ we used in the cell phone example (see Figure 10):

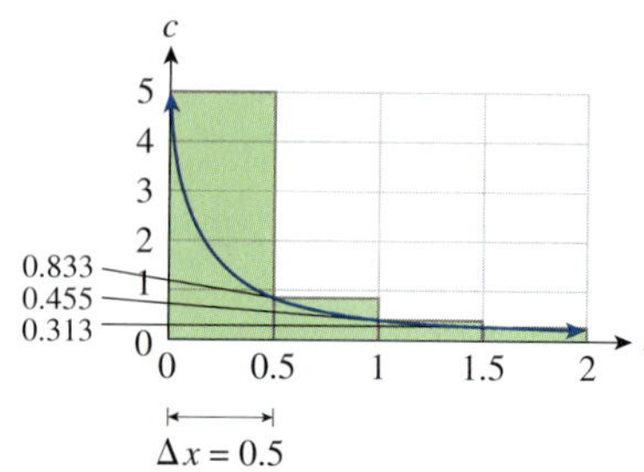

Figure 10

$$\begin{aligned}\text{Left Riemann sum} &= f(x_0)\Delta x + f(x_1)\Delta x + \cdots + f(x_{n-1})\Delta x \\ &= f(0)(0.5) + f(0.5)(0.5) + f(1)(0.5) + f(1.5)(0.5) \\ &\approx (5)(0.5) + (0.833)(0.5) + (0.455)(0.5) + (0.313)(0.5) \approx 3.30\end{aligned}$$

Because sums are often used in mathematics, mathematicians have developed a shorthand notation for them. We write

$$f(x_0)\Delta x + f(x_1)\Delta x + \cdots + f(x_{n-1})\Delta x \text{ as } \sum_{k=0}^{n-1} f(x_k)\Delta x.$$

The symbol $\sum$ is the Greek letter sigma and stands for **summation**. The letter k here is called the index of summation, and we can think of it as counting off the segments. We read the notation as "the sum from $k = 0$ to $n - 1$ of the quantities $f(x_k)\Delta x$." Think of it as a set of instructions:

Set $k = 0$, and calculate $f(x_0)\Delta x$.   *$f(0)(0.5)$ in the above calculation*

Set $k = 1$, and calculate $f(x_1)\Delta x$.   *$f(0.5)(0.5)$ in the above calculation*

...

Set $k = n - 1$, and calculate $f(x_{n-1})\Delta x$.   *$f(1.5)(0.5)$ in the above calculation*

Then sum all the quantities so calculated.

Riemann Sum

If f is a continuous function, the **left Riemann sum** with n equal subdivisions for f over the interval $[a, b]$ is defined to be

$$\begin{aligned}\text{Left Riemann sum} &= \sum_{k=0}^{n-1} f(x_k)\Delta x \\ &= f(x_0)\Delta x + f(x_1)\Delta x + \cdots + f(x_{n-1})\Delta x \\ &= [f(x_0) + f(x_1) + \cdots + f(x_{n-1})]\Delta x\end{aligned}$$

where $a = x_0 < x_1 < \cdots < x_n = b$ are the subdivisions, and $\Delta x = (b - a)/n$.

Interpretation of the Riemann Sum

If f is the rate of change of a quantity F (that is, $f = F'$), then the Riemann sum of f approximates the total change of F from $x = a$ to $x = b$. The approximation improves as the number of subdivisions increases toward infinity.

Quick Example

If $f(t)$ is the rate of change in the number of bats in a belfry and $[a, b] = [2, 3]$, then the Riemann sum approximates the total change in the number of bats in the belfry from time $t = 2$ to time $t = 3$.

Visualizing a Left Riemann Sum (Non-negative Function)

Graphically, we can represent a left Riemann sum of a non-negative function as an approximation of the area under a curve:

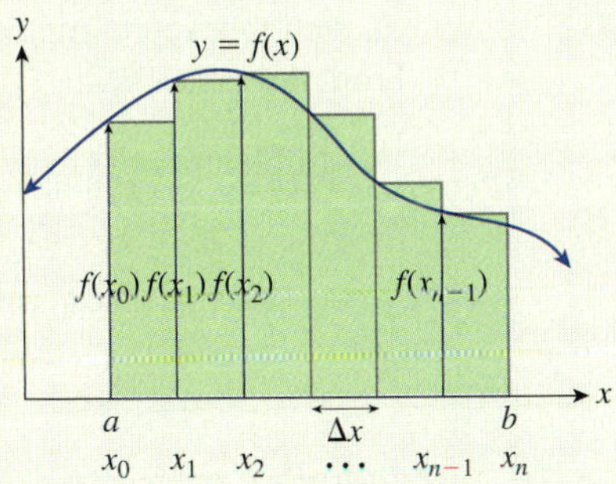

Riemann sum = Shaded area = Area of first rectangle + Area of second rectangle + ⋯ + Area of nth rectangle $= f(x_0)\Delta x + f(x_1)\Delta x + f(x_2)\Delta x + \cdots + f(x_{n-1})\Delta x$

Quick Example

In Example 1 we computed several Riemann sums, including these:

$n = 1$: Riemann sum $= c(0)\Delta t = 5 \times 2 = \$10$

$n = 2$: Riemann sum $= [c(t_0) + c(t_1)]\Delta t$
$= [c(0) + c(1)] \cdot (1) \approx \$5.45$

$n = 4$: Riemann sum $= [c(t_0) + c(t_1) + c(t_2) + c(t_3)]\Delta t$
$= [c(0) + c(0.5) + c(1) + c(1.5)] \cdot (0.5) \approx \$3.30$

$n = 8$: Riemann sum $= [c(t_0) + c(t_1) + \cdots + c(t_7)]\Delta t$
$= [c(0) + c(0.25) + \cdots + c(1.75)] \cdot (0.25) \approx \$2.31$

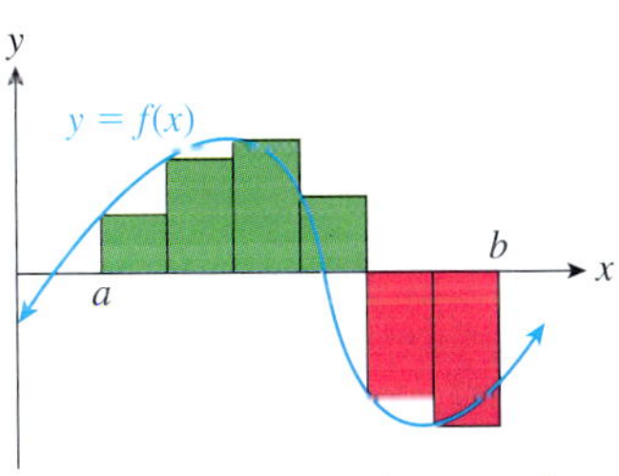

Riemann sum = Area above x-axis − Area below x-axis

Figure 11

Note To visualize the Riemann sum of a function that is negative, look again at the formula $f(x_0)\Delta x + f(x_1)\Delta x + f(x_2)\Delta x + \cdots + f(x_{n-1})\Delta x$ for the Riemann sum. Each term $f(x_k)\Delta x_k$ represents the area of one rectangle in the figure above. So, the areas of the rectangles with negative values of $f(x_k)$ are automatically counted as negative. They appear as red rectangles in Figure 11. ■

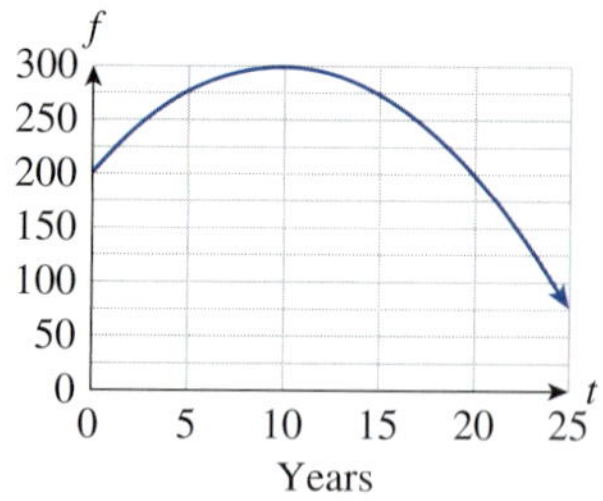

Figure 12

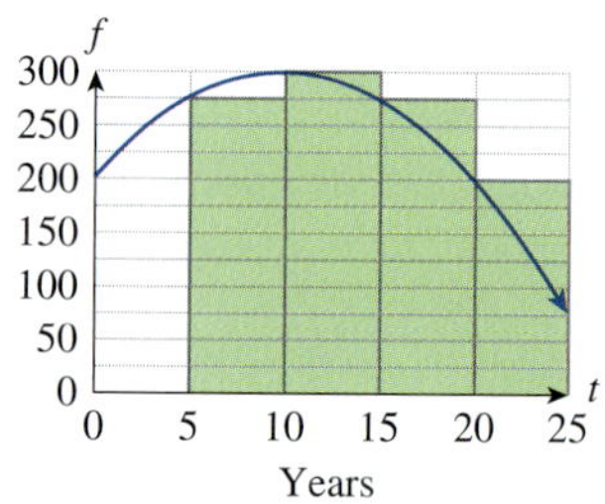

Figure 13

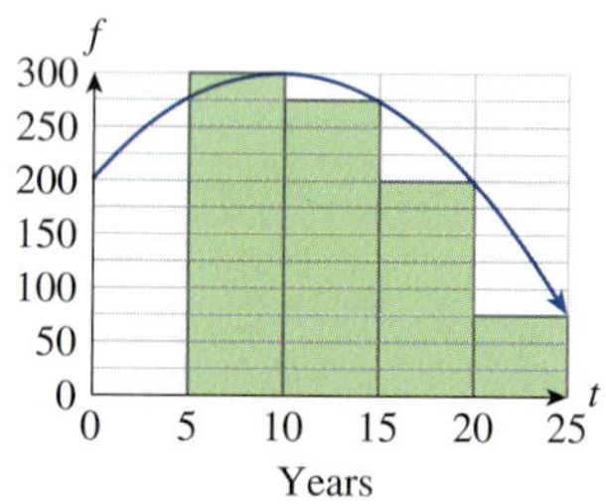

Right Riemann Sum = (300)(5) + (275)(5) + (200)(5) + (75)(5) = 4,250

Height of each rectangle is determined by height of graph at right edge.

Figure 14

EXAMPLE 2 Computing a Riemann Sum From a Graph

Figure 12 shows the approximate rate f at which convicts have been given the death sentence in the United States since 1980 (t is time in years since 1980.)* Use a left Riemann sum with 4 subdivisions to estimate the total number of death sentences handed down from 1985 to 2005.

Solution Let us represent the total number of death sentences handed down since the institution of the death sentence up to time t (measured in years since 1980) by $F(t)$. The total number of death sentences handed down from 1985 to 2005 is then the total change in F over the interval [5, 25]. In view of the above discussion, we can approximate the total change in F using a Riemann sum of its rate of change f. Because $n = 4$ subdivisions are specified, the width of each subdivision is

$$\Delta x = \frac{b-a}{n} = \frac{25-5}{4} = 5.$$

We can therefore represent the left Riemann sum by the shaded area shown in Figure 13: From the graph,

$$\begin{aligned}\text{Left sum} &= f(5)\Delta t + f(10)\Delta t + f(15)\Delta t + f(20)\Delta t \\ &= (275)(5) + (300)(5) + (275)(5) + (200)(5) = 5{,}250.\end{aligned}$$

So, we estimate that a total of 5,250 death sentences were handed down during the given period.

*The death penalty was reinstated by the U.S. Supreme Court in 1976. Source for data through 2003: Bureau of Justice Statistics, NAACP Defense Fund Inc./*New York Times*, September 15, 2004, p. A16.

Before we go on... Although in this section we focus primarily on left Riemann sums, we could also approximate the total in Example 2 using a **right Riemann sum**, as shown in Figure 14. For continuous functions, the distinction between these two types of Riemann sums approaches zero as the number of subdivisions approaches infinity. (See below.) ■

EXAMPLE 3 Computing a Riemann Sum From a Formula

Compute the left Riemann sum for $f(x) = x^2 + 1$ over the interval $[-1, 1]$, using $n = 5$ subdivisions.

Solution Because the interval is $[a, b] = [-1, 1]$ and $n = 5$, we have

$$\Delta x = \frac{b-a}{n} = \frac{1-(-1)}{5} = 0.4.$$

Width of subdivisions

Thus, the subdivisions of $[-1, 1]$ are given by

$$-1 < -0.6 < -0.2 < 0.2 < 0.6 < 1.$$

Start with -1 and keep adding $\Delta x = 0.4$.

Figure 15 shows the graph with a representation of the Riemann sum.

The Riemann sum we want is

$$\begin{aligned}&[f(x_0) + f(x_1) + \cdots + f(x_4)]\Delta x \\ &= [f(-1) + f(-0.6) + f(-0.2) + f(0.2) + f(0.6)]0.4.\end{aligned}$$

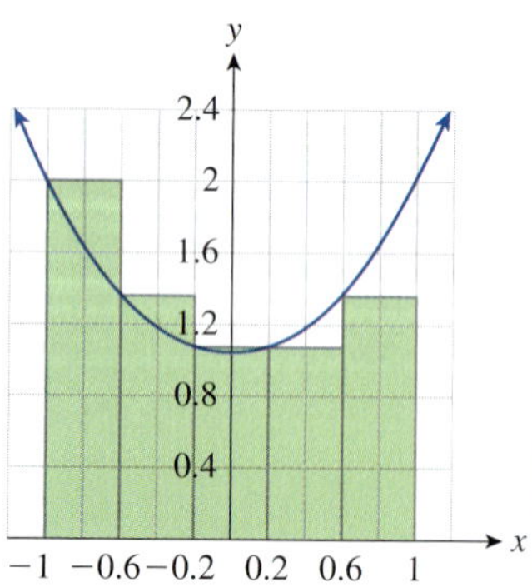

Figure 15

We can conveniently organize this calculation in a table as follows:

x	-1	-0.6	-0.2	0.2	0.6	**Total**
$f(x) = x^2 + 1$	2	1.36	1.04	1.04	1.36	6.8

The Riemann sum is therefore

$$6.8\Delta x = 6.8 \times 0.4 = 2.72.$$

As in Example 1, we're most interested in what happens to the Riemann sum when we let n get very large. When f is continuous,* its Riemann sums will always approach a limit as n goes to infinity. (This is not meant to be obvious. Proofs may be found in advanced calculus texts.) We give the limit a name.

* **NOTE** This applies to some other functions as well, including "piecewise continuous" functions discussed in the next section.

The Definite Integral

If f is a continuous function, the **definite integral of f from a to b** is defined to be the limit of the Riemann sums as the number of subdivisions approaches infinity:

$$\int_a^b f(x)\,dx = \lim_{n\to\infty} \sum_{k=0}^{n-1} f(x_k)\,\Delta x.$$

In Words: The integral, from a to b, of $f(x)\,dx$ equals the limit, as $n \to \infty$, of the Riemann Sum with a partition of n subdivisions.

The function f is called the **integrand**, the numbers a and b are the **limits of integration**, and the variable x is the **variable of integration**. A Riemann sum with a large number of subdivisions may be used to approximate the definite integral.

Interpretation of the Definite Integral

If f is the rate of change of a quantity F (that is, $f = F'$), then $\int_a^b f(x)\,dx$ is the (exact) total change of F from $x = a$ to $x = b$.

Quick Examples

1. If $f(t)$ is the rate of change in the number of bats in a belfry and $[a, b] = [2, 3]$, then $\int_2^3 f(t)\,dt$ is the total change in the number of bats in the belfry from time $t = 2$ to time $t = 3$.
2. If, at time t hours, you are selling wall posters at a rate of $s(t)$ posters per hour, then

$$\text{Total number of posters sold from hour 3 to hour 5} = \int_3^5 s(t)\,dt.$$

Visualizing the Definite Integral

Non-negative Functions: If $f(x) \geq 0$ for all x in $[a, b]$, then $\int_a^b f(x)\,dx$ is the area under the graph of f over the interval $[a, b]$, as shaded in the figure.

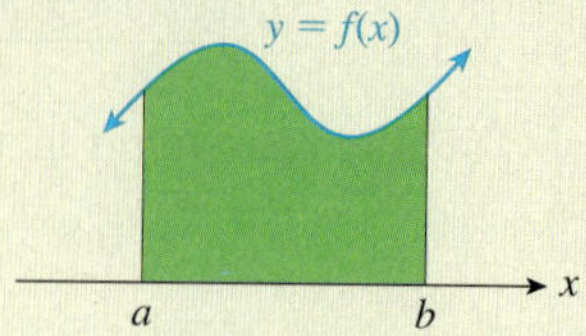

General Functions: $\int_a^b f(x)\,dx$ is the area between $x = a$ and $x = b$ that is above the x-axis and below the graph of f, minus the area that is below the x-axis and above the graph of f:

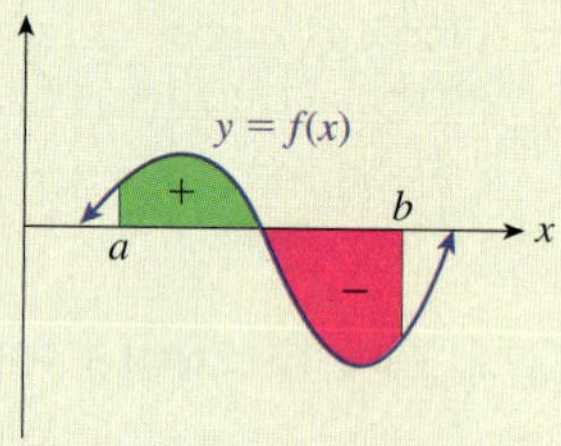

$$\int_a^b f(x)\,dx = \text{Area above } x\text{-axis} - \text{Area below } x\text{-axis}$$

Quick Examples

1.

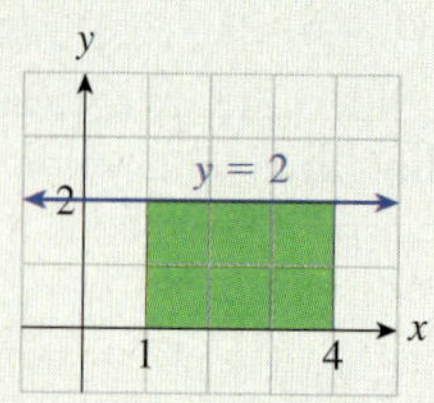

$$\int_1^4 2\,dx = \text{Area of rectangle} = 6$$

2.

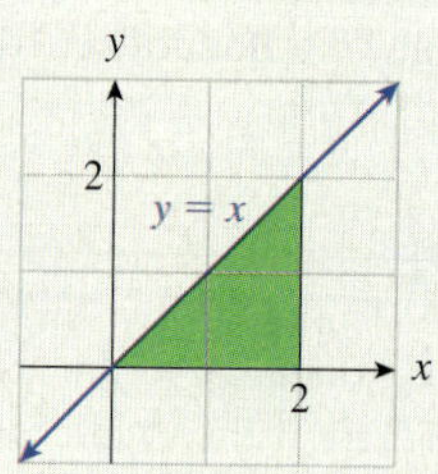

$$\int_0^2 x\,dx = \text{Area of triangle} = \frac{1}{2}\text{ base} \times \text{height} = 2$$

3.

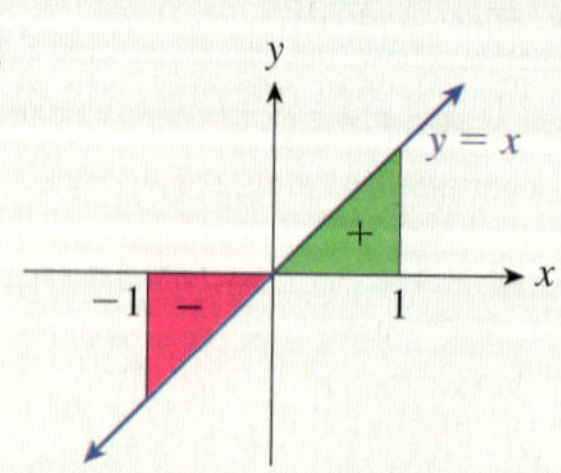

$$\int_{-1}^1 x\,dx = 0$$

The areas above and below the x-axis are equal.

Notes

1. Remember that $\int_a^b f(x)\,dx$ stands for a number that depends on f, a, and b. The variable x that appears is called a **dummy variable** because it has no effect on the answer. In other words,

$$\int_a^b f(x)\,dx = \int_a^b f(t)\,dt. \qquad x \text{ or } t \text{ is just a name we give the variable.}$$

2. The notation for the definite integral (due to Leibniz) comes from the notation for the Riemann sum. The integral sign $\int$ is an elongated S, the Roman equivalent of the Greek $\sum$. The d in dx is the lowercase Roman equivalent of the Greek Δ.
3. The definition above is adequate for continuous functions, but more complicated definitions are needed to handle other functions. For example, we broke the interval $[a, b]$ into n subdivisions of equal length, but other definitions allow a **partition** of the interval into subdivisions of possibly unequal lengths. We have evaluated f at the left endpoint of each subdivision, but we could equally well have used the right endpoint or any other point in the subdivision. All of these variations lead to the same answer when f is continuous.
4. The similarity between the notations for the definite integral and the indefinite integral is no mistake. We will discuss the exact connection in the next section. ■

Computing Definite Integrals

In some cases, we can compute the definite integral directly from the graph (see the quick examples above and the next example below). In general, the only method of computing definite integrals we have discussed so far is numerical estimation: compute the Riemann sums for larger and larger values of n and then estimate the number it seems to be approaching as we did in Example 1. (In the next section we will discuss an algebraic method for computing them.)

EXAMPLE 4 Estimating a Definite Integral From a Graph

Figure 16 shows the graph of the (approximate) rate $f'(t)$, at which the United States has been consuming gasoline from 2000 through 2008. (t is time in years since 2000.)*

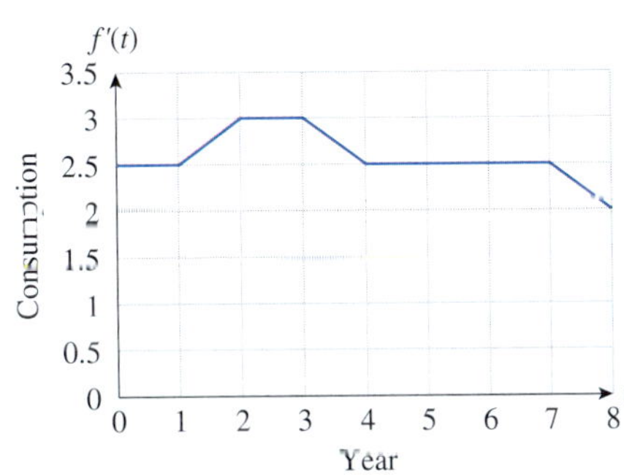

$f'(t)$ = Gasoline consumption (100 million gals per year)

Figure 16

Use the graph to estimate the total U.S. consumption of gasoline over the period shown.

* Source: Energy Information Administration (Department of Energy), (http://www.eia.doe.gov).

Solution The derivative $f'(t)$ represents the rate of change of the total U.S. consumption of gasoline, and so the total U.S. consumption of gasoline over the given period $[0, 8]$ is given by the definite integral:

$$\text{Total U.S. consumption of gasoline} = \text{Total change in } f(t) = \int_0^8 f'(t)\,dt$$

and is given by the area under the graph (Figure 17).

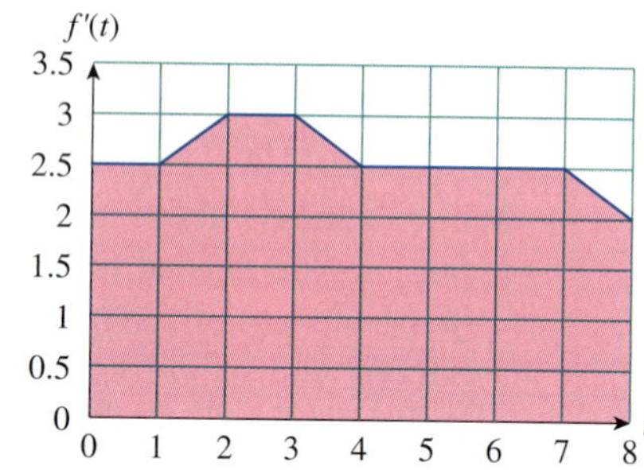

Figure 17

One way to determine the area is to count the number of filled rectangles as defined by the grid. Each rectangle has an area of $1 \times 0.5 = 0.5$ units (and the half-rectangles determined by diagonal portions of the graph have half that area). Counting rectangles, we find a total of 41.5 complete rectangles, so

$$\text{Total area} = 20.75.$$

Because $f'(t)$ is in 100 million gallons per year, we conclude that the total U.S. consumption of gasoline over the given period was about 2,075 million gallons, or 2.075 billion gallons.

While counting rectangles might seem easy, it becomes awkward in cases involving large numbers of rectangles or partial rectangles whose area is not easy to determine. Rather than counting rectangles, we can get the area by averaging the left and right Riemann sums whose subdivisions are determined by the grid:

$$\text{Left sum} = (2.5 + 2.5 + 3 + 3 + 2.5 + 2.5 + 2.5 + 2.5)(1) = 21$$
$$\text{Right sum} = (2.5 + 3 + 3 + 2.5 + 2.5 + 2.5 + 2.5 + 2)(1) = 20.5$$
$$\text{Average} = \frac{21 + 20.5}{2} = 20.75.$$

To see why this works, look at the single interval $[1, 2]$. The left sum contributes $2.5 \times 1 = 2.5$ and the right sum contributes $3 \times 1 = 3$. The exact area is their average, 2.75 (Figure 18).

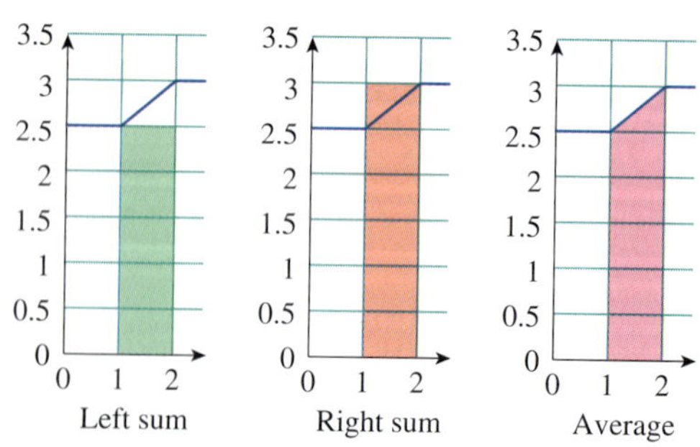

Figure 18

Before we go on... It is important to check that the units we are using in Example 4 match up correctly: t is given in *years* and $f'(t)$ is given in 100 million gallons per *year*. The integral is then given in

$$\text{Years} \times \frac{\text{100 million gallons}}{\text{Year}} = \text{100 million gallons}$$

If we had specified $f'(t)$ in, say, 100 million gallons per *day* but t in years, then we would have needed to convert either t or $f'(t)$ so that the units of time match. ■

The next example illustrates the use of technology in estimating definite integrals using Riemann sums.

EXAMPLE 5 Using Technology to Approximate the Definite Integral

Use technology to estimate the area under the graph of $f(x) = 1 - x^2$ over the interval [0, 1] using $n = 100$, $n = 200$, and $n = 500$ subdivisions.

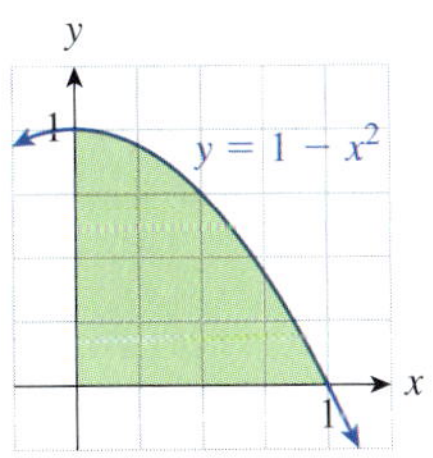

Figure 19

Solution We need to estimate the area under the parabola shown in Figure 19.

From the discussion above,

$$\text{Area} = \int_0^1 (1 - x^2)\,dx.$$

The Riemann sum with $n = 100$ has $\Delta x = (b - a)/n = (1 - 0)/100 = 0.01$ and is given by

$$\sum_{k=0}^{99} f(x_k)\Delta x = [f(0) + f(0.01) + \cdots + f(0.99)](0.01).$$

Similarly, the Riemann sum with $n = 200$ has $\Delta x = (b - a)/n = (1 - 0)/200 = 0.005$ and is given by

$$\sum_{k=0}^{199} f(x_k)\Delta x = [f(0) + f(0.005) + \cdots + f(0.995)](0.005).$$

For $n = 500$, $x = (b - a)/n = (1 - 0)/500 = 0.002$ and the Riemann sum is

$$\sum_{k=0}^{499} f(x_k)\Delta x = [f(0) + f(0.002) + \cdots + f(0.998)](0.002).$$

Using technology to evaluate these Riemann sums, we find:

$$n = 100: \sum_{k=0}^{99} f(x_k)\Delta x = 0.67165$$

$$n = 200: \sum_{k=0}^{199} f(x_k)\Delta x = 0.6691625$$

$$n = 500: \sum_{k=0}^{499} f(x_k)\Delta x = 0.667666$$

so we estimate that the area under the curve is about 0.667. (The exact answer is 2/3, as we will be able to verify using the techniques in the next section.)

using Technology

See the Technology Guides at the end of the chapter to find out how to compute these sums on a TI-83/84 Plus and Excel.

Web Site
www.AppliedCalc.org
Follow the path
Chapter 6
→ Math Tools for Chapter 6
→ Numerical Integration Utility
to obtain a utility that computes left and right Riemann sums. In the utilities, there is also a downloadable Excel spreadsheet that computes and also graphs Riemann sums (Riemann Sum Grapher).

EXAMPLE 6 Motion

A fast car has velocity $v(t) = 6t^2 + 10t$ ft/s after t seconds (as measured by a radar gun). Use several values of n to find the distance covered by the car from time $t = 3$ seconds to time $t = 4$ seconds.

Solution Because the velocity $v(t)$ is rate of change of position, the total change in position over the interval [3, 4] is

$$\text{Distance covered} = \text{Total change in position} = \int_3^4 v(t)\,dt = \int_3^4 (6t^2 + 10t)\,dt.$$

As in Examples 1 and 5, we can subdivide the one-second interval [3, 4] into smaller and smaller pieces to get more and more accurate approximations of the integral. By computing Riemann sums for various values of n, we get the following results.

$$n = 10: \sum_{k=0}^{9} v(t_k)\Delta t = 106.41 \qquad n = 100: \sum_{k=0}^{99} v(t_k)\Delta t \approx 108.740$$

$$n = 1{,}000: \sum_{k=0}^{999} v(t_k)\Delta t \approx 108.974 \qquad n = 10{,}000: \sum_{k=0}^{9999} v(t_k)\Delta t \approx 108.997$$

These calculations suggest that the total distance covered by the car, the value of the definite integral, is approximately 109 feet.

Web Site
www.AppliedCalc.org
At the Web site you will find the following optional online interactive section: Internet Topic: Numerical Integration.

➡ **Before we go on...** Do Example 6 using antiderivatives instead of Riemann sums, as in Section 6.1. Do you notice a relationship between antiderivatives and definite integrals? This will be explored in the next section. ■

6.3 EXERCISES

▼ more advanced ◆ challenging

T indicates exercises that should be solved using technology

In Exercises 1–8, use the given graph to estimate the left Riemann sum for the given interval with the stated number of subdivisions. HINT [See Example 2.]

1. [0, 5], $n = 5$

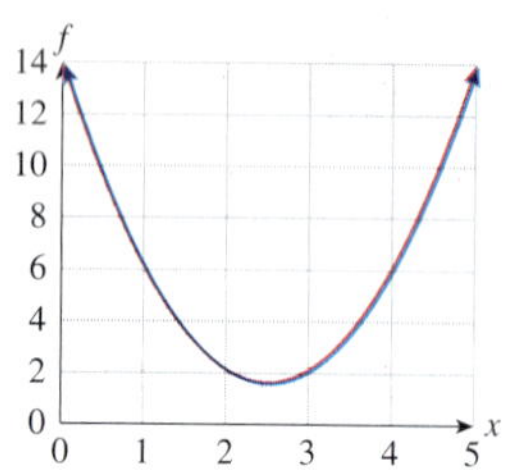

2. [0, 8], $n = 4$

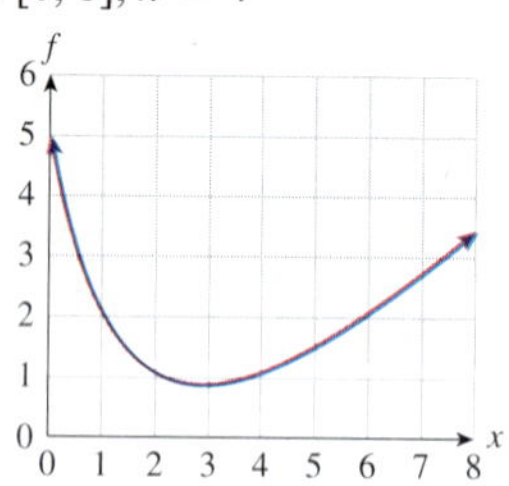

3. [1, 9], $n = 4$

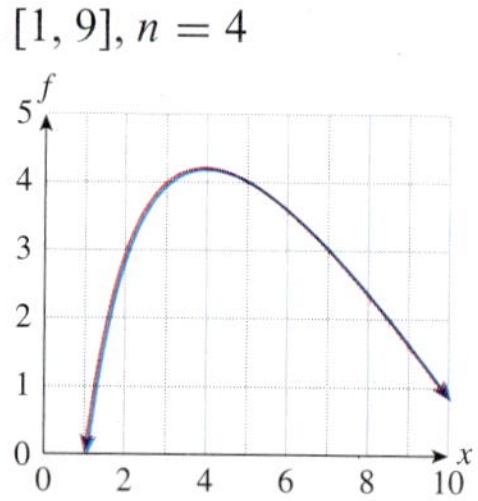

4. [0.5, 2.5], $n = 4$

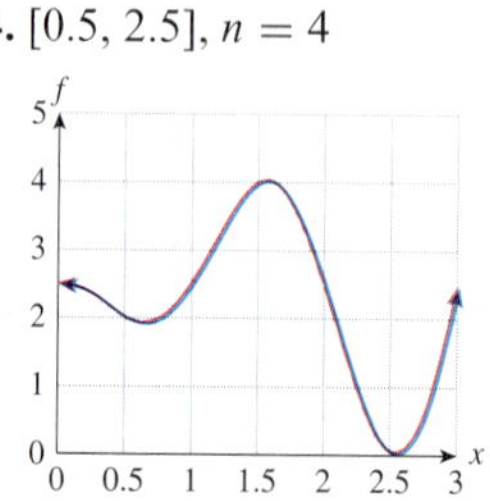

5. [1, 3.5], $n = 5$

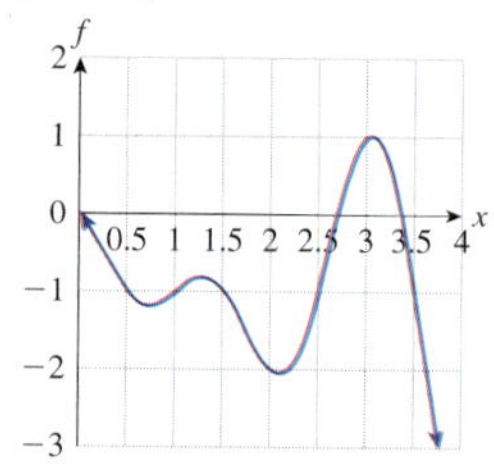

6. [0.5, 3.5], $n = 3$

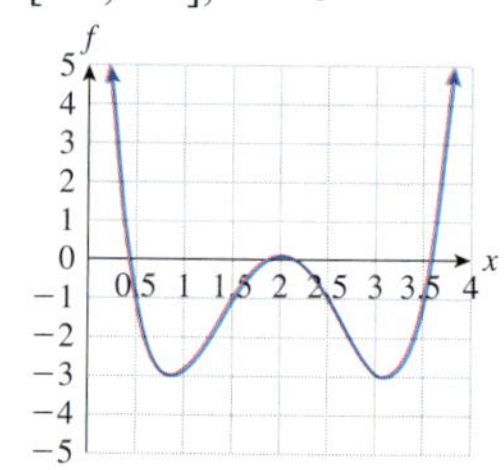

7. [0, 3]; $n = 3$

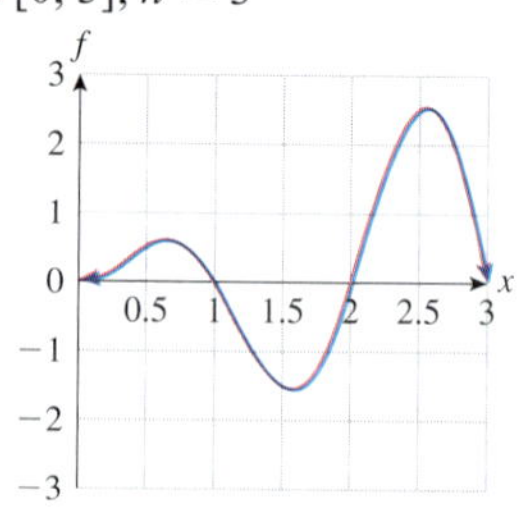

8. [0.5, 3]; $n = 5$

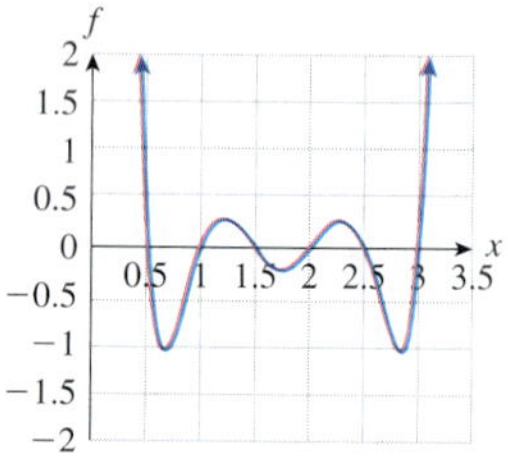

Calculate the left Riemann sums for the given functions over the given interval in Exercises 9–18, using the given values of n. (When rounding, round answers to four decimal places.) HINT [See Example 3.]

9. $f(x) = 4x - 1$ over $[0, 2]$, $n = 4$

10. $f(x) = 1 - 3x$ over $[-1, 1]$, $n = 4$

11. $f(x) = x^2$ over $[-2, 2]$, $n = 4$

12. $f(x) = x^2$ over $[1, 5]$, $n = 4$

13. $f(x) = \dfrac{1}{1+x}$ over $[0, 1]$, $n = 5$

14. $f(x) = \dfrac{x}{1+x^2}$ over $[0, 1]$, $n = 5$

15. $f(x) = e^{-x}$ over $[0, 10]$, $n = 5$

16. $f(x) = e^{-x}$ over $[-5, 5]$, $n = 5$

17. $f(x) = e^{-x^2}$ over $[0, 10]$, $n = 4$

18. $f(x) = e^{-x^2}$ over $[0, 100]$, $n = 4$

Use geometry (not Riemann sums) to compute the integrals in Exercises 19–28. HINT [See Quick Examples page 486.]

19. $\int_0^1 1\,dx$ **20.** $\int_0^2 5\,dx$

21. $\int_0^1 x\,dx$ **22.** $\int_1^2 x\,dx$

23. $\int_0^1 \frac{x}{2}\,dx$ **24.** $\int_1^2 \frac{x}{2}\,dx$

25. $\int_2^4 (x-2)\,dx$ **26.** $\int_3^6 (x-3)\,dx$

27. $\int_{-1}^1 x^3\,dx$ **28.** $\int_{-2}^2 \frac{x}{2}\,dx$

In Exercises 29–34, the graph of the derivative $f'(t)$ of $f(t)$ is shown. Compute the total change of $f(t)$ over the given interval. HINT [See Example 4.]

29. $[1, 5]$

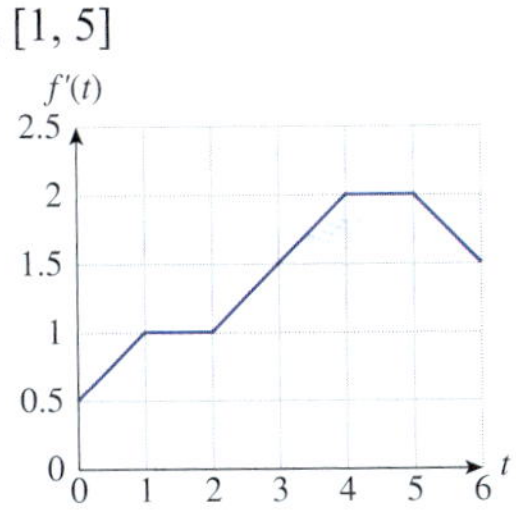

30. $[2, 6]$

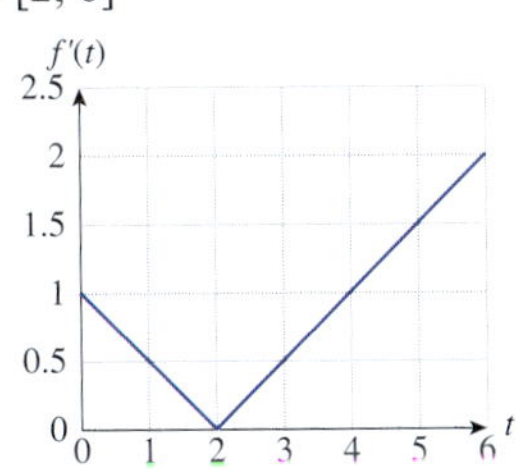

31. $[2, 6]$

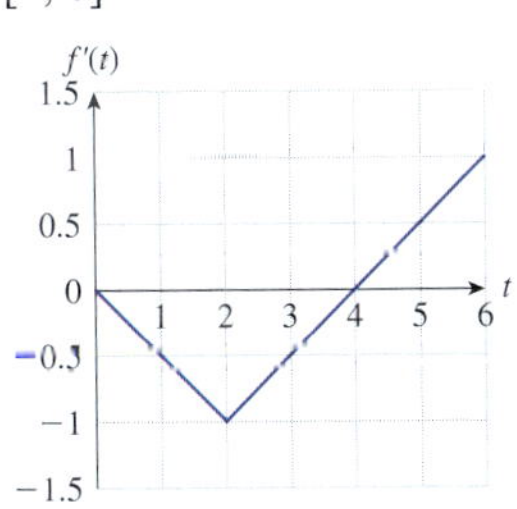

32. $[0, 5]$

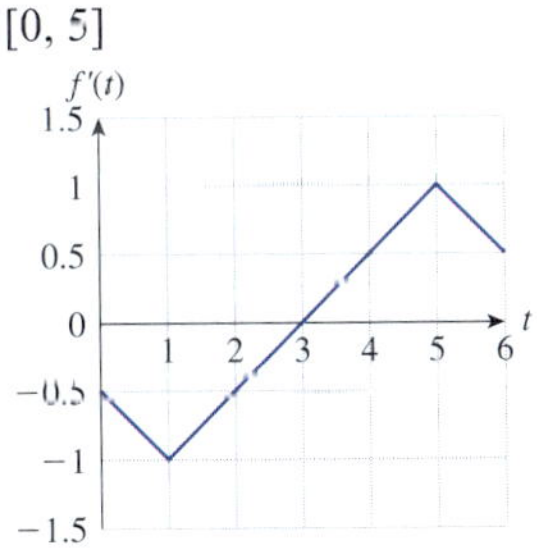

33. $[-1, 2]$

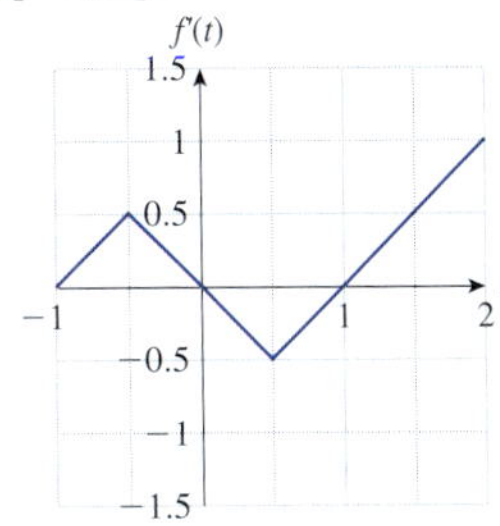

34. $[-1, 2]$

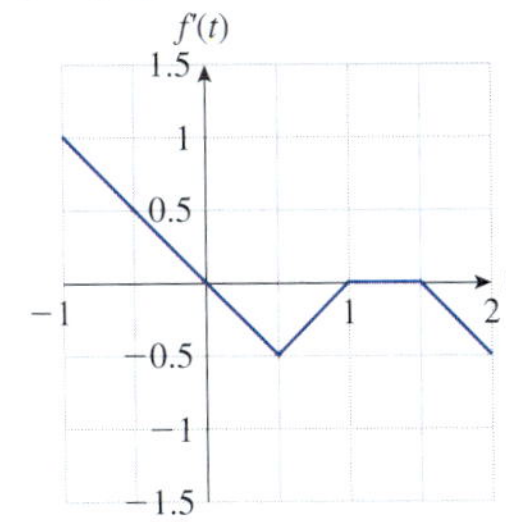

*In Exercises 35–38, use technology to approximate the given integrals with Riemann sums, using **(a)** $n = 10$, **(b)** $n = 100$, and **(c)** $n = 1{,}000$. Round all answers to four decimal places.* HINT [See Example 5.]

35. ▼ $\int_0^1 4\sqrt{1-x^2}\,dx$ **36.** ▼ $\int_0^1 \frac{4}{1+x^2}\,dx$

37. ▼ $\int_2^3 \frac{2x^{1.2}}{1+3.5x^{4.7}}\,dx$ **38.** ▼ $\int_3^4 3xe^{1.3x}\,dx$

APPLICATIONS

39. ***Cost*** The marginal cost function for the manufacture of portable MP3 players is given by

$$C'(x) = 20 - \frac{x}{200}$$

where x is the number of MP3 players manufactured. Use a Riemann sum with $n = 5$ to estimate the cost of producing the first 5 MP3 players. HINT [See Examples 1 and 3.]

40. ***Cost*** Repeat the preceding exercise using the marginal cost function

$$C'(x) = 25 - \frac{x}{50}.$$

HINT [See Examples 1 and 3.]

41. ***Bottled Water Sales*** The rate of U.S. sales of bottled water for the period 2000–2008 can be approximated by

$$s(t) = 12t^2 + 500t + 4{,}700 \text{ million gallons per year} \quad (0 \le t \le 8)$$

where t is time in years since the start of 2000.[19] Use a Riemann sum with $n = 5$ to estimate the total U.S. sales of bottled water from the start of 2000 to the start of 2005. (Round your answer to the nearest billion gallons.) HINT [See Example 3.]

[19] The 2008 figure is an estimate. Source: Beverage Marketing Corporation (www.bottledwater.org).

42. ***Bottled Water Sales*** The rate of U.S. per capita sales of bottled water for the period 2000–2008 can be approximated by

$$s(t) = 0.04t^2 + 1.5t + 17 \text{ gallons per year} \quad (0 \le t \le 8)$$

where t is the time in years since the start of 2000.[20] Use a Riemann sum with $n = 5$ to estimate the total U.S. per capita sales of bottled water from the start of 2003 to the start of 2008. (Round your answer to the nearest gallon.) HINT [See Example 3.]

43. ***Online Auctions: United States*** The following graph shows the approximate rate of change $n(t)$ of the number of items listed on **eBay**, in millions (t is time in quarters; $t = 1$ represents the first quarter in 2005).[21]

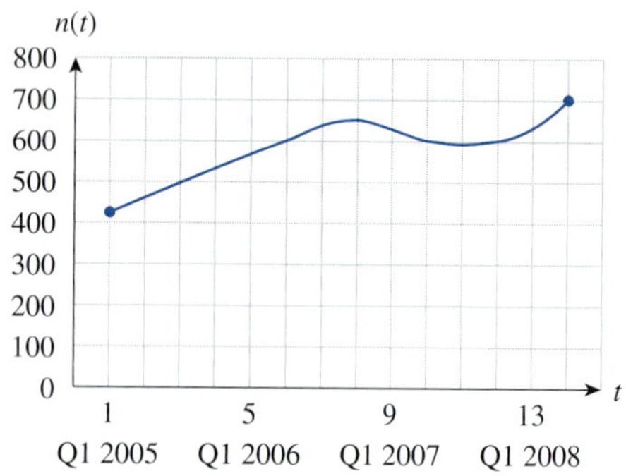

Use the graph to estimate the total number of items listed on **eBay** from Q2 2006 to Q2 2008 (the interval [6, 14]). Use a left Riemann sum with four subdivisions. HINT [See Example 2.]

44. ***Online Stores: United States*** The following graph shows the approximate rate of change $s(t)$ of the number of **eBay** stores (t is time in quarters; $t = 1$ represents the first quarter in 2005).[22]

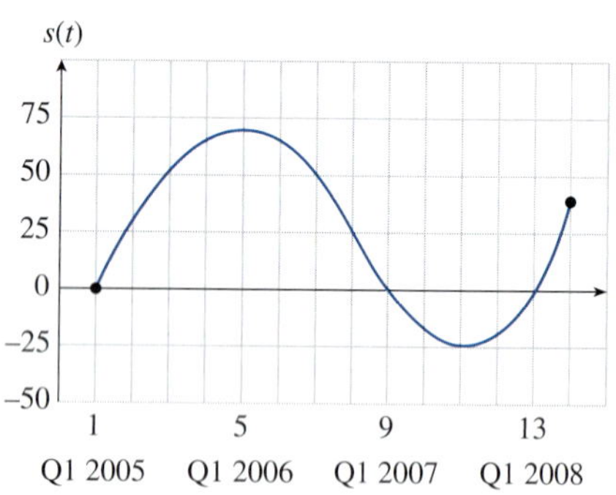

Use the graph to estimate the total net number of **eBay** stores opened between Q3 2006 and Q1 2008 (the interval [7, 13]). Use a left Riemann sum with three subdivisions. HINT [See Example 2.]

45. ***Scientific Research*** The rate of change $r(t)$ of the total number of research articles in the prominent journal *Physical Review* written by researchers in Europe is shown in the following graph:

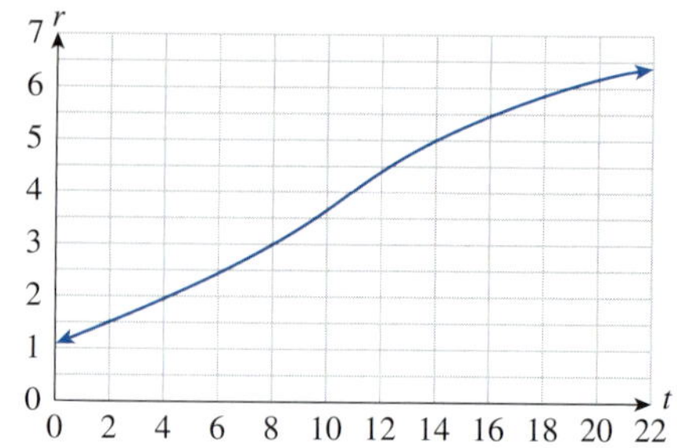

Here, t is time in years ($t = 0$ represents the start of 1983).[23]

a. Use both left and right Riemann sums with eight subdivisions to estimate the total number of articles in *Physical Review* written by researchers in Europe during the 16-year period beginning at the start of 1983. (Estimate each value of $r(t)$ to the nearest 0.5.) HINT [See Example 2.]

b. Use the answers from part (a) to obtain an estimate of $\int_0^{16} r(t)\, dt$. HINT [See Example 4.] Interpret the result.

46. ***Scientific Research*** The rate of change $r(t)$ of the total number of research articles in the prominent journal *Physical Review* written by researchers in the United States is shown in the following graph:

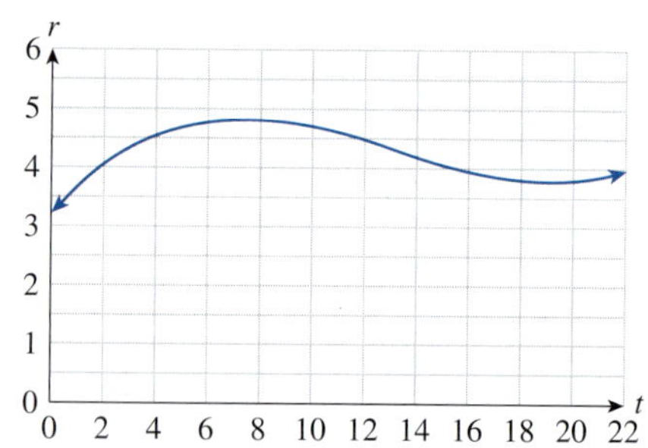

Here, t is time in years ($t = 0$ represents the start of 1983).[24]

a. Use both left and right Riemann sums with six subdivisions to estimate the total number of articles in *Physical Review* written by researchers in the United States during the 12-year period beginning at the start of 1993. (Estimate each value of $r(t)$ to the nearest 0.25.) HINT [See Example 2.]

b. Use the answers from part (a) to obtain an estimate of $\int_{10}^{22} r(t)\, dt$. HINT [See Example 4.] Interpret the result.

[20] The 2008 figure is an estimate. Source: Beverage Marketing Corporation (www.bottledwater.org).

[21] Source for data: eBay company report (www.investor.ebay.com).

[22] Ibid.

[23] Based on data from 1983 to 2003. Source: The American Physical Society/*New York Times*, May 3, 2003, p. A1.

[24] Ibid.

47. ***Graduate Degrees: Women*** The following graph shows the approximate number $n'(t)$ of doctoral degrees per year awarded to women in the United States during 2005–2014 ($t = 0$ represents 2005 and each unit of the y-axis represents 10,000 degrees).[25]

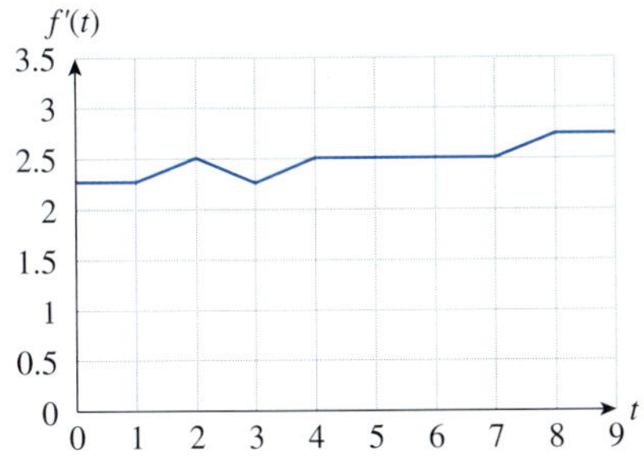

Use the graph to estimate the total number of doctoral degrees awarded to women from 2005 to 2010. HINT [See Example 4.]

48. ***Graduate Degrees: Men*** The following graph shows the approximate number $f'(t)$ of doctoral degrees per year awarded to men in the United States during 2005–2014 ($t = 0$ represents 2005 and each unit of the y-axis represents 10,000 degrees).[26]

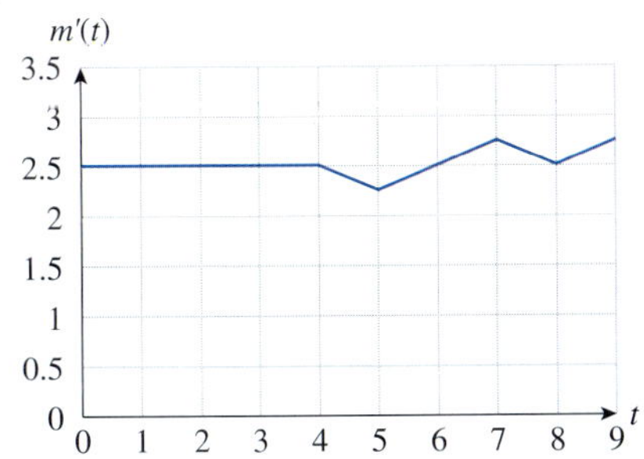

Use the graph to estimate the total number of doctoral degrees awarded to men from 2009 to 2014. HINT [See Example 4.]

Net Income: **General Electric.** *Exercises 49 and 50 are based on the following graph, which shows* **GE***'s approximate net income in billions of dollars each quarter from 2006 through the middle of 2008.*[27]

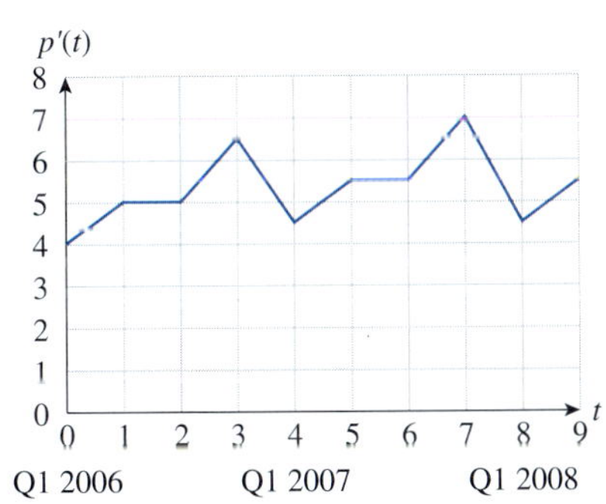

49. Compute the left- and right-Riemann sum estimates of $\int_0^4 p'(t)\,dt$ using $\Delta x = 1$. Which of these two sums gives the actual total net income earned by **GE** during 2006? Explain.

50. Compute the left- and right-Riemann sum estimates of $\int_3^7 p'(t)\,dt$ using $\Delta x = 1$. Which of these two sums gives the actual total net income earned by **GE** during 2007? Explain.

51. ***Visiting Students*** The following graph shows the approximate rate of change $c'(t)$ in the number of students from China who have taken the GRE exam required for admission to U.S. universities (t is time in years since 2000):[28]

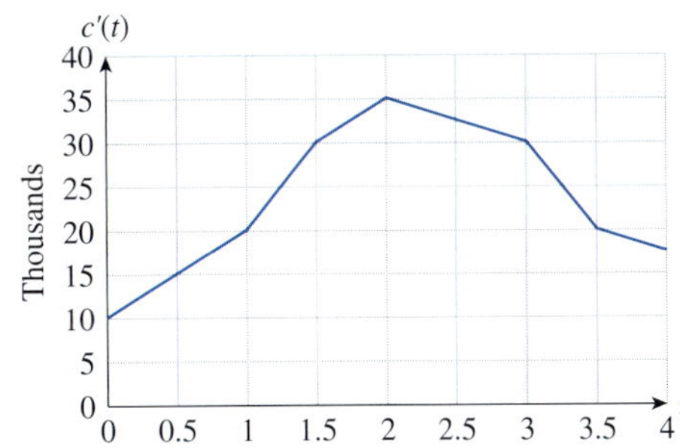

Use the graph to estimate, to the nearest 1,000, the total number of students from China who took the GRE exams during the period 2002–2004.

52. ***Visiting Students*** Repeat Exercise 51, using the following graph for students from India:[29]

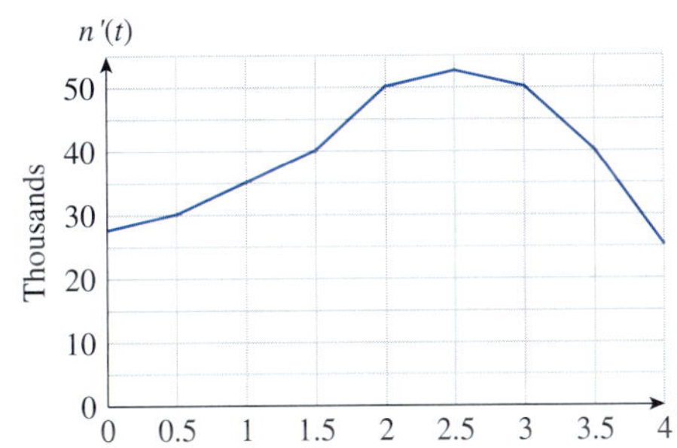

53. ***Motion under Gravity*** The velocity of a stone moving under gravity t seconds after being thrown up at 30 ft/s is given by $v(t) = -32t + 30$ ft/s. Use a Riemann sum with 5 subdivisions to estimate $\int_0^4 v(t)\,dt$. What does the answer represent? HINT [See Example 6.]

54. ***Motion under Gravity*** The velocity of a stone moving under gravity t seconds after being thrown up at 4 m/s is given by $v(t) = -9.8t + 4$ m/s. Use a Riemann sum with 5 subdivisions to estimate $\int_0^1 v(t)\,dt$. What does the answer represent? HINT [See Example 6.]

55. ***Motion*** A model rocket has upward velocity $v(t) = 40t^2$ ft/s, t seconds after launch. Use a Riemann sum with $n = 10$ to estimate how high the rocket is 2 seconds after launch. HINT [See Example 6.]

[25] Source for data and projections: National Center for Education Statistics (www.nces.ed.gov).

[26] Ibid.

[27] Source: Company reports (www.ge.com/investors).

[28] Source: Educational Testing Services/Shanghai and Jiao Tong University/*New York Times*, December 21, 2004, p. A25.

[29] Ibid.

56. ***Motion*** A race car has a velocity of $v(t) = 600(1 - e^{-0.5t})$ ft/s, t seconds after starting. Use a Riemann sum with $n = 10$ to estimate how far the car has traveled in the first 4 seconds. (Round your answer to the nearest whole number.) HINT [See Example 6.]

57. ***Oil Production in Mexico*** The rate of oil production by Pemex, Mexico's national oil company, can be approximated by

$$p(t) = -8.03t^2 + 73t + 1{,}060 \text{ million barrels per year} \quad (1 \le t \le 9)$$

where t is time in years since the start of 2000.[30] Estimate $\int_1^9 p(t)\,dt$ using a Riemann sum with $n = 150$. (Round your answer to three significant digits.) Interpret the answer. HINT [See Example 5.]

58. ***Oil Imports from Mexico*** The rate of oil imports to the U.S. from Mexico can be approximated by

$$r(t) = -5.48t^2 + 36.5t + 510 \text{ million barrels per year} \quad (0 \le t \le 8)$$

where t is time in years since the start of 2000.[31] Estimate $\int_0^8 r(t)\,dt$ using a Riemann sum with $n = 150$. (Round your answer to three significant digits.) Interpret the answer. HINT [See Example 5.]

59. ***Big Brother*** The total number of wiretaps authorized each year by U.S. state and federal courts from 1990 to 2010 can be approximated by

$$w(t) = 820e^{0.051t} \quad (0 \le t \le 20)$$

(t is time in years since the start of 1990).[32] Estimate $\int_0^{15} w(t)\,dt$ using a (left) Riemann sum with $n = 100$. (Round your answer to the nearest 10.) Interpret the answer. HINT [See Example 5.]

60. ***Big Brother*** The number of wiretaps authorized each year by U.S. state courts from 1990 to 2010 can be approximated by

$$w(t) = 440e^{0.06t} \quad (0 \le t \le 20)$$

(t is time in years since the start of 1990).[33] Estimate $\int_{10}^{15} w(t)\,dt$ using a (left) Riemann sum with $n = 100$. (Round your answer to the nearest 10.) Interpret the answer. HINT [See Example 5.]

61. ***Surveying*** My uncle intends to build a kidney-shaped swimming pool in his small yard, and the town zoning board will approve the project only if the total area of the pool does not exceed 500 square feet. The accompanying figure shows a diagram of the planned swimming pool, with measurements of its width at the indicated points. Will my uncle's plans be approved? Use a (left) Riemann sum to approximate the area.

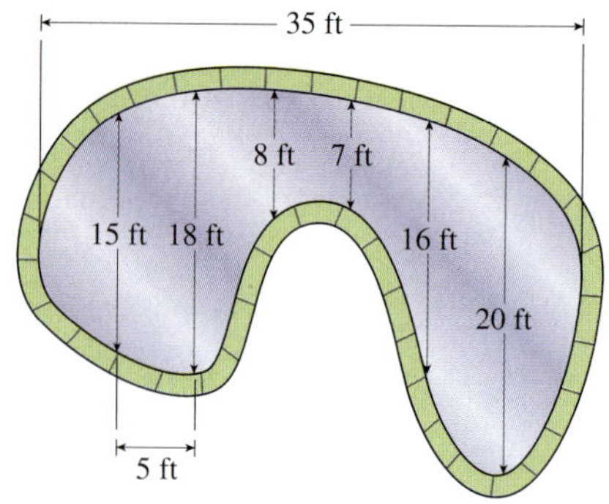

62. ***Pollution*** An aerial photograph of an ocean oil spill shows the pattern in the accompanying diagram. Assuming that the oil slick has a uniform depth of 0.01 m, how many cubic meters of oil would you estimate to be in the spill? (Volume = Area × Thickness. Use a (left) Riemann sum to approximate the area.)

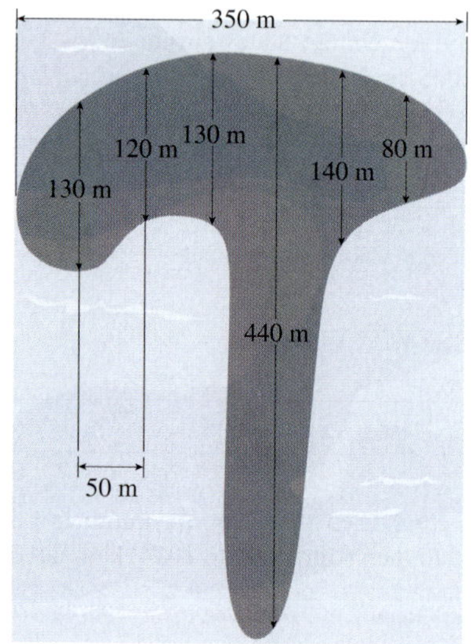

63. ***Oil Consumption: United States*** During the period 1980–2008 the United States was consuming oil at a rate of about

$$q(t) = 76t + 5{,}540 \text{ million barrels per year} \quad (0 \le t \le 28)$$

where t is time in years since the start of 1980.[34] During the same period, the price per barrel of crude oil in constant 2008 dollars was about

$$p(t) = 0.45t^2 - 12t + 105 \text{ dollars}^{35} \quad (0 \le t \le 28).$$

a. Graph the function $r(t) = p(t)q(t)$ for $0 \le t \le 28$, indicating the area that represents $\int_{10}^{20} r(t)\,dt$. What does this area signify?

[30] Source for data: Energy Information Administration/Pemex (http://www.eia.doe.gov.)

[31] Ibid.

[32] Source for data: 2007 Wiretap Report, Administrative Office of the United States Courts (www.uscourts.gov/wiretap07/2007WTText.pdf).

[33] Ibid.

[34] Source for data: BP Statistical Review of World Energy (www.bp.com/statisticalreview).

[35] Source for data: www.inflationdata.com.

b. Estimate the area in part (a) using a Riemann sum with $n = 200$. (Round the answer to 3 significant digits.) Interpret the answer.

64. ***Oil Consumption: China*** Repeat Exercise 63 using instead the rate of consumption of oil in China: During the period 1980–2008 China was consuming oil at a rate of about

$$q(t) = 82t + 221 \text{ million barrels per year}^{36} \quad (0 \le t \le 28).$$

The Normal Curve The normal distribution curve, which models the distributions of data in a wide range of applications, is given by the function

$$p(x) = \frac{1}{\sqrt{2\pi}\,\sigma} e^{-(x-\mu)^2/2\sigma^2}$$

where $\pi = 3.14159265\ldots$ and σ and μ are constants called the ***standard deviation*** *and the* ***mean****, respectively. Its graph (when $\sigma = 1$ and $\mu = 2$) is shown in the figure. Exercises 65 and 66 illustrate its use.*

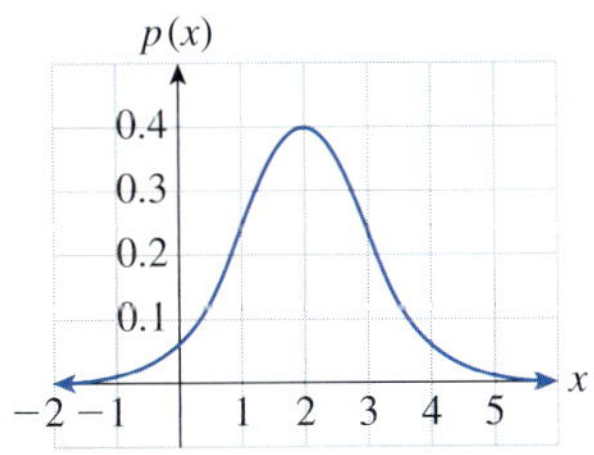

65. ***Test Scores*** Enormous State University's Calculus I test scores are modeled by a normal distribution with $\mu = 72.6$ and $\sigma = 5.2$. The percentage of students who obtained scores between a and b on the test is given by

$$\int_a^b p(x)\,dx.$$

a. Use a Riemann sum with $n = 40$ to estimate the percentage of students who obtained between 60 and 100 on the test.

b. What percentage of students scored less than 30?

66. ***Consumer Satisfaction*** In a survey, consumers were asked to rate a new toothpaste on a scale of 1–10. The resulting data are modeled by a normal distribution with $\mu = 4.5$ and $\sigma = 1.0$. The percentage of consumers who rated the toothpaste with a score between a and b on the test is given by

$$\int_a^b p(x)\,dx.$$

a. Use a Riemann sum with $n = 10$ to estimate the percentage of customers who rated the toothpaste 5 or above. (Use the range 4.5 to 10.5.)

b. What percentage of customers rated the toothpaste 0 or 1? (Use the range −0.5 to 1.5.)

COMMUNICATION AND REASONING EXERCISES

67. If $f(x) = 6$, then the left Riemann sum ________ (increases/decreases/stays the same) as n increases.

68. If $f(x) = -1$, then the left Riemann sum ________ (increases/decreases/stays the same) as n increases.

69. If f is an increasing function of x, then the left Riemann sum ________ (increases/decreases/stays the same) as n increases.

70. If f is a decreasing function of x, then the left Riemann sum ________ (increases/decreases/stays the same) as n increases.

71. If $\int_a^b f(x)\,dx = 0$, what can you say about the graph of f?

72. Sketch the graphs of two (different) functions $f(x)$ and $g(x)$ such that $\int_a^b f(x)\,dx = \int_a^b g(x)\,dx$.

73. The definite integral counts the area under the x-axis as negative. Give an example that shows how this can be useful in applications.

74. Sketch the graph of a nonconstant function whose Riemann sum with $n = 1$ gives the exact value of the definite integral.

75. Sketch the graph of a nonconstant function whose Riemann sums with $n = 1$, 5, and 10 are all zero.

76. Besides left and right Riemann sums, another approximation of the integral is the **mid-point** approximation, in which we compute the sum

$$\sum_{k=1}^{n} f(\bar{x}_k)\,\Delta x$$

where $\bar{x}_k = (x_{k-1} + x_k)/2$ is the point midway between the left and right endpoints of the interval $[x_{k-1}, x_k]$. Why is it true that the midpoint approximation is exact if f is linear? (Draw a picture.)

77. Your cell phone company charges you $c(t) = \dfrac{20}{t + 100}$ dollars for the tth minute. You make a 60-minute phone call. What kind of (left) Riemann sum represents the total cost of the call? Explain.

78. Your friend's cell phone company charges her $c(t) = \dfrac{20}{t + 100}$ dollars for the $(t + 1)$st minute. Your friend makes a 60-minute phone call. What kind of (left) Riemann sum represents the total cost of the call? Explain.

79. Give a formula for the **right Riemann Sum** with n equal subdivisions $a = x_0 < x_1 < \cdots < x_n = b$ for f over the interval $[a, b]$.

80. Refer to Exercise 79. If f is continuous, what happens to the difference between the left and right Riemann sums as $n \to \infty$? Explain.

81. When approximating a definite integral by computing Riemann sums, how might you judge whether you have chosen n large enough to get your answer accurate to, say, three decimal places?

[36] Source for data: BP Statistical Review of World Energy (www.bp.com/statisticalreview).

6.4 The Definite Integral: Algebraic Approach and the Fundamental Theorem of Calculus

In Section 6.3 we saw that the definite integral of the marginal cost function gives the total cost. However, in Section 6.1 we used antiderivatives to recover the cost function from the marginal cost function, so we *could* use antiderivatives to compute total cost. The following example, based on Example 5 in Section 6.1, compares these two approaches.

EXAMPLE 1 Finding Cost from Marginal Cost

The marginal cost of producing baseball caps at a production level of x caps is $4 - 0.001x$ dollars per cap. Find the total change of cost if production is increased from 100 to 200 caps.

Solution

Method 1: Using an Antiderivative (based on Example 5 in Section 6.1): Let $C(x)$ be the cost function. Because the marginal cost function is the derivative of the cost function, we have $C'(x) = 4 - 0.001x$ and so

$$\begin{aligned} C(x) &= \int (4 - 0.001x)\,dx \\ &= 4x - 0.001\frac{x^2}{2} + K \\ &= 4x - 0.0005x^2 + K. \end{aligned}$$

K is the constant of integration.

Although we do not know what to use for the value of the constant K, we can say:

$$\begin{aligned} \text{Cost at production level of 100 caps} &= C(100) \\ &= 4(100) - 0.0005(100)^2 + K \\ &= \$395 + K \\ \text{Cost at production level of 200 caps} &= C(200) \\ &= 4(200) - 0.0005(200)^2 + K \\ &= \$780 + K. \end{aligned}$$

Therefore,

$$\begin{aligned} \text{Total change in cost} &= C(200) - C(100) \\ &= (\$780 + K) - (\$395 + K) = \$385. \end{aligned}$$

Notice how the constant of integration simply canceled out! So, we could choose any value for K that we wanted (such as $K = 0$) and still come out with the correct total change. Put another way, we could use *any antiderivative* of $C'(x)$, such as

$$F(x) = 4x - 0.0005x^2$$

$F(x)$ is any antiderivative of $C'(x)$ whereas $C(x)$ is the actual cost function.

or

$$F(x) = 4x - 0.0005x^2 + 4$$

compute $F(200) - F(100)$, and obtain the total change, \$385.

Summarizing this method: To compute the total change of $C(x)$ over the interval $[100, 200]$, use any antiderivative $F(x)$ of $C'(x)$, and compute $F(200) - F(100)$.

Method 2: Using a Definite Integral (based on Example 1 in Section 6.3): Because the marginal cost $C'(x)$ is the rate of change of the total cost function $C(x)$, the total change in $C(x)$ over the interval $[100, 200]$ is given by

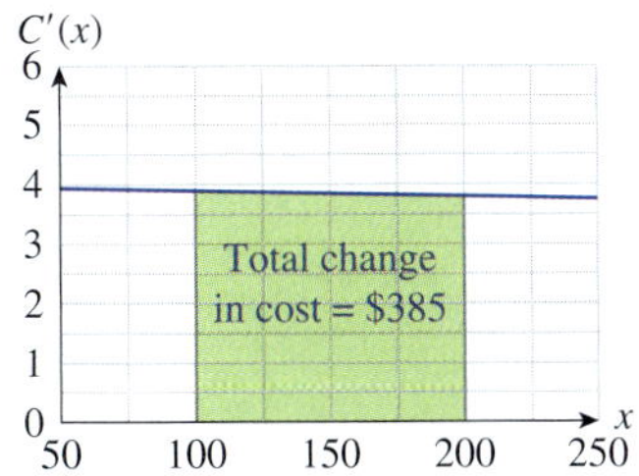

Figure 20

$$\begin{aligned}\text{Total change in cost} &= \text{Area under the marginal cost function curve}\\ &= \int_{100}^{200} C'(x)\,dx\\ &= \int_{100}^{200} (4 - 0.001x)\,dx \qquad \text{See Figure 20.}\\ &= \$385. \qquad \text{Using geometry or Riemann sums}\end{aligned}$$

Putting these two methods together gives us the following surprising result:

$$\int_{100}^{200} C'(x)\,dx = F(200) - F(100)$$

where $F(x)$ is any antiderivative of $C'(x)$.

Now, there is nothing special in Example 1 about the specific function $C'(x)$ or the choice of end-points of integration. So if we replace $C'(x)$ by a general continuous function $f(x)$, we can write

$$\int_a^b f(x)\,dx = F(b) - F(a)$$

where $F(x)$ is any antiderivative of $f(x)$. This result is known as the **Fundamental Theorem of Calculus**.

The Fundamental Theorem of Calculus (FTC)

Let f be a continuous function defined on the interval $[a, b]$ and let F be *any* antiderivative of f defined on $[a, b]$. Then

$$\int_a^b f(x)\,dx = F(b) - F(a).$$

Moreover, an antiderivative of f is guaranteed to exist.

In Words: Every continuous function has an antiderivative. To compute the definite integral of $f(x)$ over $[a, b]$, first find an antiderivative $F(x)$, then evaluate it at $x = b$, evaluate it at $x = a$, and subtract the two answers.

Quick Example

Because $F(x) = x^2$ is an antiderivative of $f(x) = 2x$,

$$\int_0^1 2x\,dx = F(1) - F(0) = 1^2 - 0^2 = 1.$$

Note The Fundamental Theorem of Calculus actually applies to some other functions besides the continuous ones. The function f is **piecewise continuous** on $[a, b]$ if it is defined and continuous at all but finitely many points in the interval, and at each point where the function is not defined or is discontinuous, the left and right limits of f exist and are finite. (See Figure 21.)

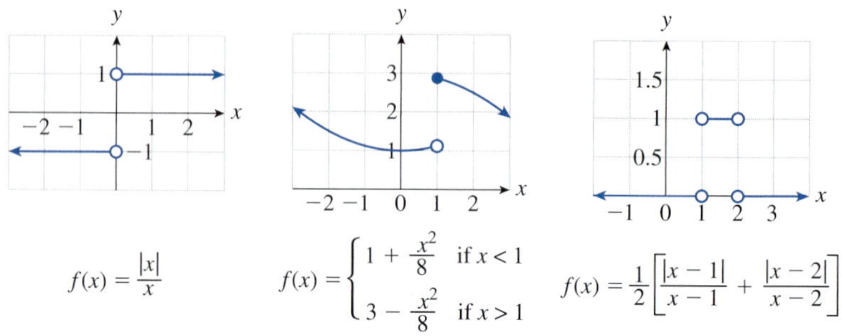

Figure 21

The FTC also applies to any piecewise continuous function f, as long as we specify that the antiderivative F that we choose be continuous. To be precise, to say that F is an antiderivative of f means here that $F'(x) = f(x)$ except at the points at which f is discontinuous or not defined, where $F'(x)$ may not exist. For example, if f is the step function $f(x) = |x|/x$, shown on the left in Figure 21, then we can use $F(x) = |x|$. (Note that F is continuous, and $F'(x) = |x|/x$ except when $x = 0$.) ■

EXAMPLE 2 Using the FTC to Calculate a Definite Integral

Calculate $\int_0^1 (1 - x^2)\,dx$.

Solution To use the FTC, we need to find an antiderivative of $1 - x^2$. But we know that

$$\int (1 - x^2)\,dx = x - \frac{x^3}{3} + C.$$

We need only one antiderivative, so let's take $F(x) = x - x^3/3$. The FTC tells us that

$$\int_0^1 (1 - x^2)\,dx = F(1) - F(0) = \left(1 - \frac{1}{3}\right) - (0) = \frac{2}{3}$$

which is the value we estimated in Section 6.3.

➡ **Before we go on...** A useful piece of notation is often used here. We write*

$$[F(x)]_a^b = F(b) - F(a).$$

* **NOTE** There seem to be several notations in use, actually. Another common notation is $F(x)\Big|_a^b$.

Thus, we can rewrite the computation in Example 2 as

$$\int_0^1 (1 - x^2)\,dx = \left[x - \frac{x^3}{3}\right]_0^1$$

Substitute $x = 1$. Substitute $x = 0$.

$$= \left(1 - \frac{1}{3}\right) - \left(0 - \frac{0}{3}\right)$$

$$= \left(1 - \frac{1}{3}\right) - (0) = \frac{2}{3}.$$ ■

EXAMPLE 3 More Use of the FTC

Compute the following definite integrals.

a. $\displaystyle\int_0^1 (2x^3 + 10x + 1)\,dx$ **b.** $\displaystyle\int_1^5 \left(\frac{1}{x^2} + \frac{1}{x}\right) dx$

Solution

$$\text{a. } \int_0^1 (2x^3 + 10x + 1)\,dx = \left[\frac{1}{2}x^4 + 5x^2 + x\right]_0^1$$

Substitute $x = 1$. Substitute $x = 0$.

$$= \left(\frac{1}{2} + 5 + 1\right) - \left(\frac{1}{2}(0) + 5(0) + 0\right)$$

$$= \left(\frac{1}{2} + 5 + 1\right) - (0) = \frac{13}{2}$$

$$\text{b. } \int_1^5 \left(\frac{1}{x^2} + \frac{1}{x}\right) dx = \int_1^5 (x^{-2} + x^{-1})\,dx$$

$$= [-x^{-1} + \ln|x|]_1^5$$

Substitute $x = 5$. Substitute $x = 1$.

$$= \left(-\frac{1}{5} + \ln 5\right) - (-1 + \ln 1)$$

$$= \frac{4}{5} + \ln 5$$

When calculating a definite integral, we may have to use substitution to find the necessary antiderivative. We could substitute, evaluate the indefinite integral with respect to u, express the answer in terms of x, and then evaluate at the limits of integration. However, there is a shortcut, as we shall see in the next example.

EXAMPLE 4 Using the FTC with Substitution

Evaluate $\displaystyle\int_1^2 (2x - 1)e^{2x^2-2x}\,dx$.

Solution The shortcut we promised is to put *everything* in terms of u, including the limits of integration.

$$u = 2x^2 - 2x$$

$$\frac{du}{dx} = 4x - 2$$

$$dx = \frac{1}{4x - 2}\,du$$

When $x = 1$, $u = 0$. Substitute $x = 1$ in the formula for u.

When $x = 2$, $u = 4$. Substitute $x = 2$ in the formula for u.

We get the value $u = 0$, for example, by substituting $x = 1$ in the equation $u = 2x^2 - 2x$. We can now rewrite the integral.

$$\int_1^2 (2x-1)e^{2x^2-2x}\,dx = \int_0^4 (2x-1)e^u \frac{1}{4x-2}\,du$$
$$= \int_0^4 \frac{1}{2} e^u\,du$$
$$= \left[\frac{1}{2}e^u\right]_0^4 = \frac{1}{2}e^4 - \frac{1}{2}$$

➡ **Before we go on...** The alternative, longer calculation in Example 4 is first to calculate the indefinite integral:

$$\int (2x-1)e^{2x^2-2x}\,dx = \int \frac{1}{2}e^u\,du$$
$$= \frac{1}{2}e^u + C = \frac{1}{2}e^{2x^2-2x} + C.$$

Then we can say that

$$\int_1^2 (2x-1)e^{2x^2-2x}\,dx = \left[\frac{1}{2}e^{2x^2-2x}\right]_1^2 = \frac{1}{2}e^4 - \frac{1}{2}.$$ ■

using Technology

The TI-83/84 Plus and the Web site use numerical methods that give good quite accurate approximations for definite integrals, and so we can use them to check our answers.

TI-83/84 Plus

Home Screen:
`fnInt(5/(10*x+1),x,0,2)`
(`fnInt` is MATH → 9)

Web Site

www.AppliedCalc.org
Follow the path
Chapter 6
→ Math Tools for Chapter 6
→ Numerical Integration Utility

Enter `5/(10*x+1)` for $f(x)$, 0 and 2 for the left and right end points, and press "Adaptive Quadrature."

APPLICATIONS

EXAMPLE 5 Total Cost

In Section 6.3 we considered the following example. Your cell phone company offers you an innovative pricing scheme. When you make a call, the marginal cost is

$$c(t) = \frac{5}{10t+1} \text{ dollars per hour.}$$

Compute the total cost of a 2-hour phone call.

Solution We calculate

$$\text{Total Cost} = \int_0^2 \frac{5}{10t+1}\,dt = 5\int_0^2 \frac{1}{10t+1}\,dt$$
$$= 5\left[\frac{1}{10}\ln(10t+1)\right]_0^2$$

See the shortcuts on page 474.

$$= \frac{5}{10}[\ln(21) - \ln(1)]$$
$$= \frac{1}{2}\ln 21 \approx \$1.52.$$

Compare this with Example 1 of Section 6.3, where we found the same answer by approximating with Riemann sums.

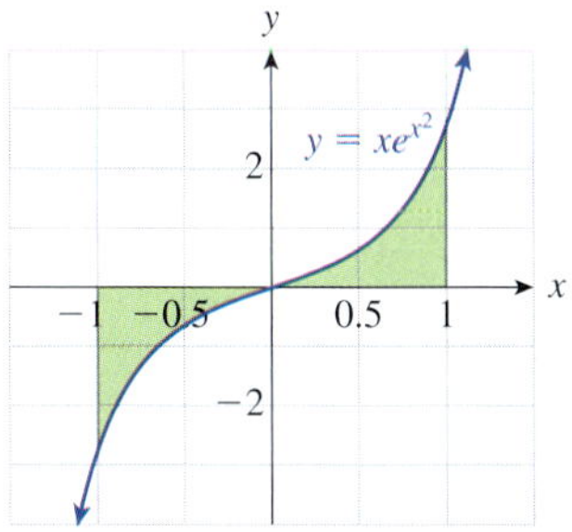

$\int_{-1}^{1} xe^{x^2}\,dx$

Figure 22

EXAMPLE 6 Computing Area

Find the total area of the region enclosed by the graph of $y = xe^{x^2}$, the x-axis, and the vertical lines $x = -1$ and $x = 1$.

Solution The region whose area we want is shown in Figure 22. Notice the symmetry of the graph. Also, half the region we are interested in is above the x-axis, while the other half is below. If we calculated the integral $\int_{-1}^{1} xe^{x^2}\,dx$, the result would be

$$\text{Area above } x\text{-axis} - \text{Area below } x\text{-axis} = 0,$$

which does not give us the total area. To prevent the area below the x-axis from being combined with the area above the axis, we do the calculation in two parts, as illustrated in Figure 23.

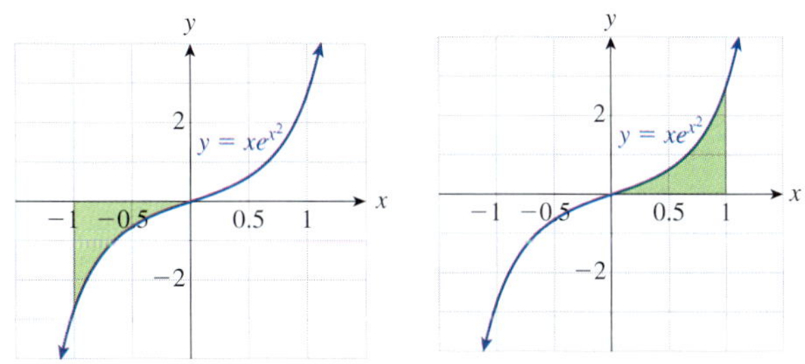

$\int_{-1}^{0} xe^{x^2}\,dx$ $\int_{0}^{1} xe^{x^2}\,dx$

Figure 23

(In Figure 23 we broke the integral at $x = 0$ because that is where the graph crosses the x-axis.) These integrals can be calculated using the substitution $u = x^2$:

$$\int_{-1}^{0} xe^{x^2}\,dx = \frac{1}{2}\left[e^{x^2}\right]_{-1}^{0} = \frac{1}{2}(1 - e) \approx -0.85914$$

Why is it negative?

$$\int_{0}^{1} xe^{x^2}\,dx = \frac{1}{2}\left[e^{x^2}\right]_{0}^{1} = \frac{1}{2}(e - 1) \approx 0.85914$$

To obtain the total area, we should add the *absolute* values of these answers because we don't wish to count any area as negative. Thus,

$$\text{Total area} \approx 0.85914 + 0.85914 = 1.71828.$$

6.4 EXERCISES

▼ more advanced ◆ challenging

▮ indicates exercises that should be solved using technology

Evaluate the integrals in Exercises 1–44. HINT [See Example 2.]

1. $\int_{-1}^{1} (x^2 + 2)\,dx$

2. $\int_{-2}^{1} (x - 2)\,dx$

3. $\int_{0}^{1} (12x^5 + 5x^4 - 6x^2 + 4)\,dx$

4. $\int_{0}^{1} (4x^3 - 3x^2 + 4x - 1)\,dx$

5. $\int_{-2}^{2} (x^3 - 2x)\,dx$

6. $\int_{-1}^{1} (2x^3 + x)\,dx$

7. $\int_{1}^{3} \left(\frac{2}{x^2} + 3x\right)dx$

8. $\int_{2}^{3} \left(x + \frac{1}{x}\right)dx$

9. $\int_{0}^{1} (2.1x - 4.3x^{1.2})\,dx$

10. $\int_{-1}^{0} (4.3x^2 - 1)\,dx$

11. $\int_0^1 2e^x\,dx$

12. $\int_{-1}^0 3e^x\,dx$

13. $\int_0^1 \sqrt{x}\,dx$

14. $\int_{-1}^1 \sqrt[3]{x}\,dx$

15. $\int_0^1 2^x\,dx$

16. $\int_0^1 3^x\,dx$

17. $\int_0^1 18(3x+1)^5\,dx$

18. $\int_0^1 8(-x+1)^7\,dx$

HINT [for Exercises 19–44: See Example 4 or use a shortcut (page 474) if applicable.]

19. $\int_{-1}^1 e^{2x-1}\,dx$

20. $\int_0^2 e^{-x+1}\,dx$

21. $\int_0^2 2^{-x+1}\,dx$

22. $\int_{-1}^1 3^{2x-1}\,dx$

23. $\int_0^{50} e^{-0.02x-1}\,dx$

24. $\int_{-20}^0 3e^{2.2x}\,dx$

25. $\int_0^4 |-3x+4|\,dx$

26. $\int_{-4}^4 |-x-2|\,dx$

27. $\int_{-1.1}^{1.1} e^{x+1}\,dx$

28. $\int_0^{\sqrt{2}} x\sqrt{2x^2+1}\,dx$

29. $\int_{-\sqrt{2}}^{\sqrt{2}} 3x\sqrt{2x^2+1}\,dx$

30. $\int_{-1.2}^{1.2} e^{-x-1}\,dx$

31. $\int_0^1 5xe^{x^2+2}\,dx$

32. $\int_0^2 \frac{3x}{x^2+2}\,dx$

33. $\int_2^3 \frac{x^2}{x^3-1}\,dx$

34. $\int_2^3 \frac{x}{2x^2-5}\,dx$

35. $\int_0^1 x(1.1)^{-x^2}\,dx$

36. $\int_0^1 x^2(2.1)^{x^3}\,dx$

37. $\int_1^2 \frac{e^{1/x}}{x^2}\,dx$

38. $\int_1^2 \frac{\sqrt{\ln x}}{x}\,dx$

39. ▼ $\int_0^2 \frac{x}{x+1}\,dx$

40. ▼ $\int_{-1}^1 \frac{2x}{x+2}\,dx$

41. ▼ $\int_1^2 x(x-2)^5\,dx$

42. ▼ $\int_1^2 x(x-2)^{1/3}\,dx$

43. ▼ $\int_0^1 x\sqrt{2x+1}\,dx$

44. ▼ $\int_{-1}^0 2x\sqrt{x+1}\,dx$

Calculate the total area of the regions described in Exercises 45–54. Do not count area beneath the x-axis as negative. HINT [See Example 6.]

45. Bounded by the line $y = x$, the x-axis, and the lines $x = 0$ and $x = 1$

46. Bounded by the line $y = 2x$, the x-axis, and the lines $x = 1$ and $x = 2$

47. Bounded by the curve $y = \sqrt{x}$, the x-axis, and the lines $x = 0$ and $x = 4$

48. Bounded by the curve $y = 2\sqrt{x}$, the x-axis, and the lines $x = 0$ and $x = 16$

49. Bounded by the graph of $y = |2x - 3|$, the x-axis, and the lines $x = 0$ and $x = 3$

50. Bounded by the graph of $y = |3x - 2|$, the x-axis, and the lines $x = 0$ and $x = 3$

51. ▼ Bounded by the curve $y = x^2 - 1$, the x-axis, and the lines $x = 0$ and $x = 4$

52. ▼ Bounded by the curve $y = 1 - x^2$, the x-axis, and the lines $x = -1$ and $x = 2$

53. ▼ Bounded by the x-axis, the curve $y = xe^{x^2}$, and the lines $x = 0$ and $x = (\ln 2)^{1/2}$

54. ▼ Bounded by the x-axis, the curve $y = xe^{x^2-1}$, and the lines $x = 0$ and $x = 1$

APPLICATIONS

55. ***Cost*** The marginal cost of producing the xth box of light bulbs is $5 + x^2/1{,}000$ dollars. Determine how much is added to the total cost by a change in production from $x = 10$ to $x = 100$ boxes. HINT [See Example 5.]

56. ***Revenue*** The marginal revenue of the xth box of flash cards sold is $100e^{-0.001x}$ dollars. Find the revenue generated by selling items 101 through 1,000. HINT [See Example 5.]

57. ***Motion*** A car traveling down a road has a velocity of $v(t) = 60 - e^{-t/10}$ mph at time t hours. Find the distance it has traveled from time $t = 1$ hour to time $t = 6$ hours. (Round your answer to the nearest mile.)

58. ***Motion*** A ball thrown in the air has a velocity of $v(t) = 100 - 32t$ ft/s at time t seconds. Find the total displacement of the ball between times $t = 1$ second and $t = 7$ seconds, and interpret your answer.

59. ***Motion*** A car slows to a stop at a stop sign, then starts up again, in such a way that its speed at time t seconds after it starts to slow is $v(t) = |-10t + 40|$ ft/s. How far does the car travel from time $t = 0$ to time $t = 10$ s?

60. ***Motion*** A truck slows, doesn't quite stop at a stop sign, and then speeds up again in such a way that its speed at time t seconds is $v(t) = 10 + |-5t + 30|$ ft/s. How far does the truck travel from time $t = 0$ to time $t = 10$?

61. ***Bottled Water Sales*** (Compare Exercise 41 in Section 6.3.) The rate of U.S. sales of bottled water for the period 2000–2008 can be approximated by

$$s(t) = 12t^2 + 500t + 4{,}700 \text{ million gallons per year} \quad (0 \le t \le 8),$$

where t is time in years since the start of 2000.[37] Use the FTC to estimate the total U.S. sales of bottled water from the start of 2000 to the start of 2005. (Round your answer to the nearest billion gallons.)

[37] The 2008 figure is an estimate. Source: Beverage Marketing Corporation (www.bottledwater.org).

62. ***Bottled Water Sales*** (Compare Exercise 42 in Section 6.3.) The rate of U.S. per capita sales of bottled water for the period 2000–2008 can be approximated by

$$s(t) = 0.04t^2 + 1.5t + 17 \text{ gallons per year} \quad (0 \le t < 8),$$

where t is the time in years since the start of 2000.[38] Use the FTC to estimate the total U.S. per capita sales of bottled water from the start of 2003 to the start of 2008. (Round your answer to the nearest gallon.)

63. ***Oil Production in Mexico*** (Compare Exercise 57 in Section 6.3.) The rate of oil production by **Pemex**, Mexico's national oil company, can be approximated by

$$p(t) = -8.03t^2 + 73t + 1{,}060 \text{ million barrels per year,} \quad (1 \le t \le 9)$$

where t is time in years since the start of 2000.[39] Use a definite integral to estimate total production of oil from the start of 2001 to the start of 2009.

64. ***Oil Imports from Mexico*** (Compare Exercise 58 in Section 6.3.) The rate of oil imports to the United States from Mexico can be approximated by

$$r(t) = -5.48t^2 + 36.5t + 510 \text{ million barrels per year} \quad (0 \le t \le 8)$$

where t is time in years since the start of 2000.[40] Use a definite integral to estimate total oil imports from the start of 2000 to the start of 2008.

65. ***Big Brother*** (Compare Exercise 59 in Section 6.3.) The total number of wiretaps authorized each year by U.S. state and federal courts from 1990 to 2010 can be approximated by

$$w(t) = 820e^{0.051t} \quad (0 \le t \le 20)$$

(t is time in years since the start of 1990).[41] Compute $\int_0^{15} w(t)\,dt$. (Round your answer to the nearest 10.) Interpret the answer.

66. ***Big Brother*** (Compare Exercise 60 in Section 6.3.) The number of wiretaps authorized each year by U.S. state courts from 1990 to 2010 can be approximated by

$$w(t) = 440e^{0.06t} \quad (0 \le t \le 20)$$

(t is time in years since the start of 1990).[42] Compute $\int_{10}^{15} w(t)\,dt$. (Round your answer to the nearest 10.) Interpret the answer.

67. ***Economic Growth*** The value of sold goods in Mexico can be approximated by

$$v(x) = 210 - 62e^{-0.05x} \text{ trillion pesos per month} \quad (x \ge 0)$$

where x is time in months since January 2005.[43] Find an expression for the total value $V(t)$ of sold goods in Mexico from January 2005 to time t. HINT [Use the shortcut on page 474.]

68. ***Economic Contraction*** The number of housing starts in the United States can be approximated by

$$n(t) = \frac{1}{12}(1.1 + 1.2e^{-0.08x}) \text{ million homes per month} \quad (x \ge 0)$$

where x is time in months since January 2006.[44] Find an expression for the total number $N(t)$ of housing starts in the United States from January 2006 to time t. HINT [Use the shortcut on page 474.]

69. ***Fuel Consumption*** The way Professor Waner drives, he burns gas at the rate of $1 - e^{-t}$ gallons each hour, t hours after a fill-up. Find the number of gallons of gas he burns in the first 10 hours after a fill-up.

70. ***Fuel Consumption*** The way Professor Costenoble drives, he burns gas at the rate of $1/(t+1)$ gallons each hour, t hours after a fill-up. Find the number of gallons of gas he burns in the first 10 hours after a fill-up.

71. ▼ ***Sales*** Weekly sales of your *Lord of the Rings* T-shirts have been falling by 5% per week. Assuming you are now selling 50 T-shirts per week, how many shirts will you sell during the coming year? (Round your answer to the nearest shirt.)

72. ▼ ***Sales*** Annual sales of fountain pens in Littleville are 4,000 per year and are increasing by 10% per year. How many fountain pens will be sold over the next five years?

73. T ***Embryo Development*** The oxygen consumption of a bird embryo increases from the time the egg is laid through the time the chick hatches. In a typical galliform bird, the oxygen consumption can be approximated by

$$c(t) = -0.065t^3 + 3.4t^2 - 22t + 3.6 \text{ milliliters per day} \quad (8 \le t \le 30)$$

where t is the time (in days) since the egg was laid.[45] (An egg will typically hatch at around $t = 28$.) Use technology to estimate the total amount of oxygen consumed during the ninth and tenth days ($t = 8$ to $t = 10$). Round your answer to the nearest milliliter. HINT [See the technology note in the margin on page 500.]

74. T ***Embryo Development*** The oxygen consumption of a turkey embryo increases from the time the egg is laid through the

[38] The 2008 figure is an estimate. Source: Beverage Marketing Corporation (www.bottledwater.org).

[39] Source for data: Energy Information Administration/Pemex (http://www.eia.doe.gov).

[40] Ibid.

[41] Source for data: 2007 Wiretap Report, Administrative Office of the United States Courts (www.uscourts.gov/wiretap07/2007WTText.pdf).

[42] Ibid.

[43] Source: Instituto Nacional de Estadística y Geografía (INEGI) (www.inegi.org.mx).

[44] Source for data: *New York Times*, February 17, 2007, p. C3.

[45] The model approximates graphical data published in the article "The Brush Turkey" by Roger S. Seymour, *Scientific American*, December 1991, pp. 108–114.

time the chick hatches. In a brush turkey, the oxygen consumption can be approximated by

$$c(t) = -0.028t^3 + 2.9t^2 - 44t + 95 \text{ milliliters per day} \quad (20 \le t \le 50)$$

where t is the time (in days) since the egg was laid.[46] (An egg will typically hatch at around $t = 50$.) Use technology to estimate the total amount of oxygen consumed during the 21st and 22nd days ($t = 20$ to $t = 22$). Round your answer to the nearest 10 milliliters. HINT [See the technology note in the margin on page 500.]

75. **T** ***Online Auctions: United States*** The rate of change $n(t)$ of the number of items listed on **eBay** can be approximated by

$$n(t) = 0.07t^4 - 1.7t^3 + 9.9t^2 + 17t + 390 \text{ million items/quarter} \quad (1 \le t \le 14)$$

(t is time in quarters; $t = 1$ represents the start of the first quarter in 2005.)[47] Use technology to compute $\int_1^9 n(t)\,dt$ correct to the nearest 100. Interpret your answer.

76. **T** ***Online Stores: United States*** The rate of change $s(t)$ of the number of **eBay** stores can be approximated by

$$s(t) = 0.035t^4 - 0.55t^3 - 1.45t^2 + 37t - 36 \text{ stores/quarter} \quad (1 \le t \le 14)$$

(t is time in quarters; $t = 1$ represents the start of the first quarter in 2005).[48] Use technology to compute $\int_5^{13} s(t)\,dt$ correct to the nearest whole number. Interpret your answer.

77. ▼ ***Total Cost*** Use the Fundamental Theorem of Calculus to show that if $m(x)$ is the marginal cost at a production level of x items, then the cost function $C(x)$ is given by

$$C(x) = C(0) + \int_0^x m(t)\,dt.$$

What do we call $C(0)$?

78. ▼ ***Total Sales*** The total cost of producing x items is given by

$$C(x) = 246.76 + \int_0^x 5t\,dt.$$

Find the fixed cost and the marginal cost of producing the 10th item.

79. ***Scientific Research*** (Compare Exercise 79 in Section 6.2.) The number of research articles in the prominent journal *Physical Review* written by researchers in Europe can be approximated by

$$E(t) = \frac{7e^{0.2t}}{5 + e^{0.2t}} \text{ thousand articles per year} \quad (t \ge 0)$$

where t is time in years ($t = 0$ represents 1983).[49] Use a definite integral to estimate the number of articles were written by researchers in Europe from 1983 to 2003. (Round your answer to the nearest 1,000 articles.) HINT [See Example 7 in Section 6.2.]

80. ***Scientific Research*** (Compare Exercise 80 in Section 6.2.) The number of research articles in the prominent journal *Physical Review* written by researchers in the United States can be approximated by

$$U(t) = \frac{4.6e^{0.6t}}{0.4 + e^{0.6t}} \text{ thousand articles per year} \quad (t \ge 0)$$

where t is time in years ($t = 0$ represents 1983).[50] Use a definite integral to estimate the total number of articles written by researchers in the United States from 1983 to 2003. HINT [See Example 7 in Section 6.2.]

81. ▼ ***The Logistic Function and High School Graduates***

a. Show that the logistic function $f(x) = \dfrac{N}{1 + Ab^{-x}}$ can be written in the form

$$f(x) = \frac{Nb^x}{A + b^x}.$$

HINT [See the note after Example 7 in Section 6.2.]

b. Use the result of part (a) and a suitable substitution to show that

$$\int \frac{N}{1 + Ab^{-x}}\,dx = \frac{N \ln(A + b^x)}{\ln b} + C.$$

c. The rate of graduation of private high school students in the United States for the period 1994–2008 was approximately

$$r(t) = 220 + \frac{110}{1 + 3.8(1.27)^{-t}} \text{ thousand students per year} \quad (0 \le t \le 14)$$

t years since 1994.[51] Use the result of part (b) to estimate the total number of private high school graduates over the period 2000–2008.

82. ▼ ***The Logistic Function and Grant Spending***

a. Show that the logistic function $f(x) = \dfrac{N}{1 + Ae^{-kx}}$ can be written in the form

$$f(x) = \frac{Ne^{kx}}{A + e^{kx}}.$$

HINT [See the note after Example 7 in Section 6.2.]

b. Use the result of part (a) and a suitable substitution to show that

$$\int \frac{N}{1 + Ae^{-kx}}\,dx = \frac{N \ln(A + e^{kx})}{k} + C.$$

---

[46] The model approximates graphical data published in the article "The Brush Turkey" by Roger S. Seymour, *Scientific American*, December 1991, pp. 108–114.

[47] Source for data: eBay company report (www.investor.ebay.com).

[48] Ibid.

[49] Based on data from 1983 to 2003. Source: The American Physical Society/*New York Times*, May 3, 2003, p. A1.

[50] Ibid.

[51] Based on a logistic regression. Source for data: National Center for Educational Statistics (www.nces.ed.gov/).

c. The rate of spending on grants by U.S. foundations in the period 1993–2003 was approximately

$$s(t) = 11 + \frac{20}{1 + 1{,}800e^{-0.9t}} \text{ billion dollars per year} \quad (3 \le t \le 13)$$

where t is the number of years since 1990.[52] Use the result of part (b) to estimate, to the nearest \$10 billion, the total spending on grants from 1998 to 2003.

83. ◆ ***Kinetic Energy*** The work done in accelerating an object from velocity v_0 to velocity v_1 is given by

$$W = \int_{v_0}^{v_1} v\frac{dp}{dv}\,dv$$

where p is its momentum, given by $p = mv$ (m = mass). Assuming that m is a constant, show that

$$W = \frac{1}{2}mv_1^2 - \frac{1}{2}mv_0^2.$$

The quantity $\frac{1}{2}mv^2$ is referred to as the **kinetic energy** of the object, so the work required to accelerate an object is given by its change in kinetic energy.

84. ◆ ***Einstein's Energy Equation*** According to the special theory of relativity, the apparent mass of an object depends on its velocity according to the formula

$$m = \frac{m_0}{\left(1 - \frac{v^2}{c^2}\right)^{1/2}}$$

where v is its velocity, m_0 is the "rest mass" of the object (that is, its mass when $v = 0$), and c is the velocity of light: approximately 3×10^8 meters per second.

a. Show that, if $p = mv$ is the momentum,

$$\frac{dp}{dv} = \frac{m_0}{\left(1 - \frac{v^2}{c^2}\right)^{3/2}}.$$

b. Use the integral formula for W in the preceding exercise, together with the result in part (a) to show that the work required to accelerate an object from a velocity of v_0 to v_1 is given by

$$W = \frac{m_0c^2}{\sqrt{1 - \frac{v_1^2}{c^2}}} - \frac{m_0c^2}{\sqrt{1 - \frac{v_0^2}{c^2}}}.$$

We call the quantity $\frac{m_0c^2}{\sqrt{1 - \frac{v^2}{c^2}}}$ the **total relativistic energy** of an object moving at velocity v. Thus, the work to accelerate an object from one velocity to another is given by the change in its total relativistic energy.

c. Deduce (as Albert Einstein did) that the total relativistic energy E of a body at rest with rest mass m is given by the famous equation

$$E = mc^2.$$

COMMUNICATION AND REASONING EXERCISES

85. Explain how the indefinite integral and the definite integral are related.

86. What is "definite" about the definite integral?

87. The total change of a quantity from time a to time b can be obtained from its rate of change by doing what?

88. Complete the following: The total sales from time a to time b are obtained from the marginal sales by taking its _____ _____ from _____ to _____.

89. What does the Fundamental Theorem of Calculus permit one to do?

90. If Felice and Philipe have different antiderivatives of f and each uses his or her own antiderivative to compute $\int_a^b f(x)\,dx$, they might get different answers, right?

91. ▼ Give an example of a nonzero velocity function that will produce a displacement of 0 from time $t = 0$ to time $t = 10$.

92. ▼ Give an example of a nonzero function whose definite integral over the interval [4, 6] is zero.

93. ▼ Give an example of a decreasing function $f(x)$ with the property that $\int_a^b f(x)\,dx$ is positive for every choice of a and $b > a$.

94. ▼ Explain why, in computing the total change of a quantity from its rate of change, it is useful to have the definite integral subtract area below the x-axis.

95. ◆ If $f(x)$ is a continuous function defined for $x \ge a$, define a new function $F(x)$ by the formula

$$F(x) = \int_a^x f(t)\,dt.$$

Use the Fundamental Theorem of Calculus to deduce that $F'(x) = f(x)$. What, if anything, is interesting about this result?

96. ◆ Use the result of Exercise 95 and technology to compute a table of values for $x = 1, 2, 3$ for an antiderivative $A(x)$ of e^{-x^2} with the property that $A(0) = 0$. (Round answers to two decimal places.)

[52] Based on a logistic regression. Source for data: The Foundation Center, *Foundation Growth and Giving Estimates*, 2004, downloaded from the Center's Web site (www.fdncenter.org).

CHAPTER 6 REVIEW

KEY CONCEPTS

Web Site www.AppliedCalc.org
Go to the student Web site at www.AppliedCalc.org to find a comprehensive and interactive Web-based summary of Chapter 6.

6.1 The Indefinite Integral

An antiderivative of a function f is a function F such that $F' = f$. *p. 452*

Indefinite integral $\int f(x)\,dx$ *p. 452*

Power rule for the indefinite integral:

$$\int x^n\,dx = \frac{x^{n+1}}{n+1} + C$$

(if $n \neq -1$) *p. 454*

$$\int x^{-1}\,dx = \ln|x| + C \quad p.\ 455$$

Indefinite Integral of e^x and b^x:

$$\int e^x\,dx = e^x + C$$

$$\int b^x\,dx = \frac{b^x}{\ln b} + C \quad p.\ 455$$

Sums, differences, and constant multiples:

$$\int [f(x) \pm g(x)]\,dx = \int f(x)\,dx \pm \int g(x)\,dx$$

$$\int kf(x)\,dx = k\int f(x)\,dx$$

(k constant) *p. 455*

Combining the rules *p. 458*

Position, velocity, and acceleration:

$$v = \frac{ds}{dt} \qquad s(t) = \int v(t)\,dt$$

$$a = \frac{dv}{dt} \qquad v(t) = \int a(t)\,dt \quad p.\ 461$$

Motion in a straight line *p. 461*

Vertical motion under gravity *p. 463*

6.2 Substitution

Substitution rule:

$$\int f\,dx = \int \left(\frac{f}{du/dx}\right)du \quad p.\ 468$$

Using the substitution rule *p. 468*

Shortcuts: integrals of expressions involving $(ax + b)$:

$$\int (ax+b)^n\,dx = \frac{(ax+b)^{n+1}}{a(n+1)} + C$$

(if $n \neq -1$)

$$\int (ax+b)^{-1}\,dx = \frac{1}{a}\ln|ax+b| + C$$

$$\int e^{ax+b}\,dx = \frac{1}{a}e^{ax+b} + C$$

$$\int c^{ax+b}\,dx = \frac{1}{a\ln c}c^{ax+b} + C$$

$$\int |ax+b|\,dx = \frac{1}{2a}(ax+b)|ax+b| + C \quad p.\ 474$$

6.3 The Definite Integral: Numerical and Graphical Approaches

Left Riemann sum:

$$\sum_{k=0}^{n-1} f(x_k)\Delta x = [f(x_0) + f(x_1) + \cdots + f(x_{n-1})]\Delta x$$

p. 482

Computing the Riemann sum from a graph *p. 484*

Computing the Riemann sum from a formula *p. 484*

Definite integral of f from a to b:

$$\int_a^b f(x)\,dx = \lim_{n\to\infty}\sum_{k=0}^{n-1} f(x_k)\Delta x. \quad p.\ 485$$

Estimating the definite integral from a graph *p. 487*

Estimating the definite integral using technology *p. 489*

Application to motion in a straight line *p. 490*

6.4 The Definite Integral: Algebraic Approach and the Fundamental Theorem of Calculus

The Fundamental Theorem of Calculus (FTC) *p. 497*

Using the FTC to compute definite integrals *p. 498*

Computing total cost from marginal cost *p. 500*

Computing area *p. 501*

REVIEW EXERCISES

Evaluate the indefinite integrals in Exercises 1–12.

1. $\int (x^2 - 10x + 2)\,dx$

2. $\int (e^x + \sqrt{x})\,dx$

3. $\int \left(\frac{4x^2}{5} - \frac{4}{5x^2}\right)dx$

4. $\int \left(\frac{3x}{5} - \frac{3}{5x}\right)dx$

5. $\int e^{-2x+11}\,dx$

6. $\int \frac{dx}{(4x-3)^2}$

7. $\int x(x^2+4)^{10}\,dx$

8. $\int \frac{x^2+1}{(x^3+3x+2)^2}\,dx$

9. $\int 5e^{-2x}\,dx$

10. $\int xe^{-x^2/2}\,dx$

11. $\int \frac{x+1}{x+2}\,dx$

12. $\int x\sqrt{x-1}\,dx$

In Exercises 13 and 14, use the given graph to estimate the left Riemann sum for the given interval with the stated number of subdivisions.

13. $[0, 3]$, $n = 6$

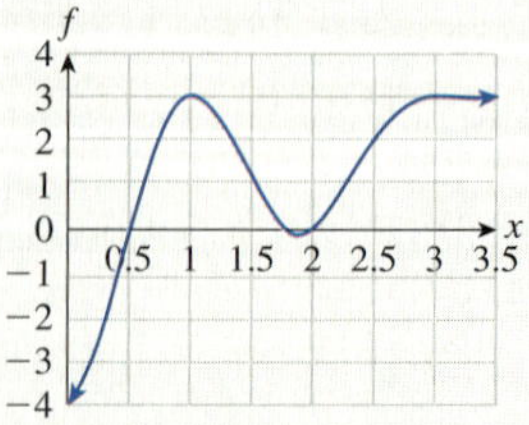

14. $[1, 3]$, $n = 4$

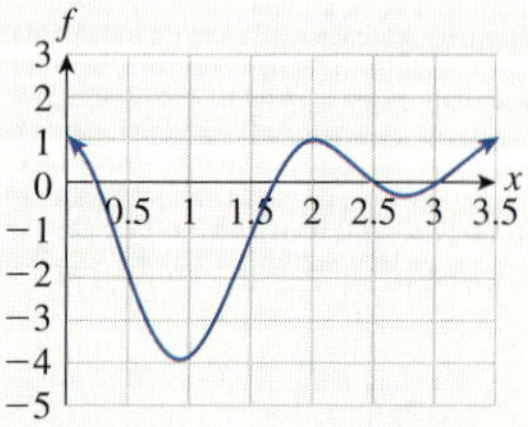

Calculate the left Riemann sums for the given functions over the given interval in Exercises 15–18, using the given values of n. (When rounding, round answers to four decimal places.)

15. $f(x) = x^2 + 1$ over $[-1, 1]$, $n = 4$

16. $f(x) = (x - 1)(x - 2) - 2$ over $[0, 4]$, $n = 4$

17. $f(x) = x(x^2 - 1)$ over $[0, 1]$, $n = 5$

18. $f(x) = \dfrac{x-1}{x-2}$ over $[0, 1.5]$, $n = 3$

T *In Exercises 19 and 20, use technology to approximate the given definite integrals using left Riemann sums with n = 10, 100, and 1,000. (Round answers to four decimal places.)*

19. $\displaystyle\int_0^1 e^{-x^2}\,dx$ **20.** $\displaystyle\int_1^3 x^{-x}\,dx$

In Exercises 21 and 22 the graph of the derivative $f'(x)$ of $f(x)$ is shown. Compute the total change of $f(x)$ over the given interval.

21. $[-1, 2]$ **22.** $[0, 2]$

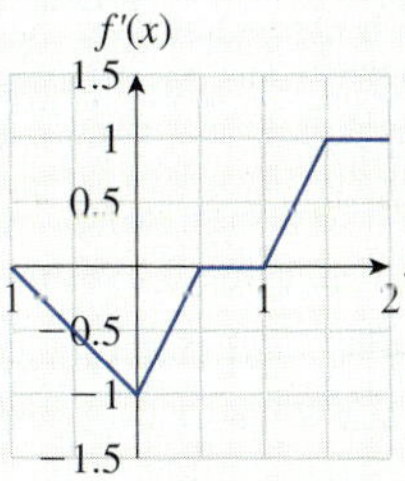

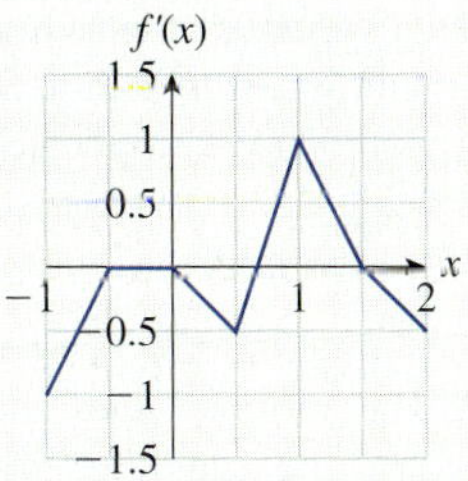

Evaluate the definite integrals in Exercises 23–30, using the Fundamental Theorem of Calculus.

23. $\displaystyle\int_0^1 (x - x^3)\,dx$ **24.** $\displaystyle\int_0^9 \frac{1}{x+1}\,dx$

25. $\displaystyle\int_{-1}^1 (1 + e^x)\,dx$ **26.** $\displaystyle\int_0^9 (x + \sqrt{x})\,dx$

27. $\displaystyle\int_0^2 x^2\sqrt{x^3+1}\,dx$ **28.** $\displaystyle\int_{-1}^1 3^{2x-2}\,dx$

29. $\displaystyle\int_0^{\ln 2} \frac{e^{-2x}}{1+4e^{-2x}}\,dx$ **30.** $\displaystyle\int_0^1 3xe^{-x^2}\,dx$

In Exercises 31–34, find the areas of the specified regions. (Do not count area below the x-axis as negative.)

31. The area bounded by $y = 4 - x^2$, the x-axis, and the lines $x = -2$ and $x = 2$

32. The area bounded by $y = 4 - x^2$, the x-axis, and the lines $x = 0$ and $x = 5$

33. The area bounded by $y = xe^{-x^2}$, the x-axis, and the lines $x = 0$ and $x = 5$

34. The area bounded by $y = |2x|$, the x-axis, and the lines $x = -1$ and $x = 1$

APPLICATIONS

35. ***Sales*** The rate of net sales (sales minus returns) of *The Secret Loves of John O*, a romance novel by Margó Dufón, can be approximated by

$$n(t) = 196 + t^2 - 0.16t^5 \text{ copies per week}$$

t weeks since its release.

a. Find the total net sales N as a function of time t.

b. How many books are still held by customers after 6 weeks? (Round your answer to the nearest book.)

36. ***Demand*** If OHaganBooks.com were to give away its latest bestseller, *A River Burns Through It*, the demand q would be 100,000 books. The marginal demand (dq/dp) for the book is $-20p$ at a price of p dollars.

a. What is the demand function for this book?

b. At what price does demand drop to zero?

37. ***Motion Under Gravity*** Billy-Sean O'Hagan's friend Juan (Billy-Sean is John O'Hagan's son, currently a senior in college) says he can throw a baseball vertically upward at 100 feet per second. Assuming Juan's claim is true,

a. Where would the baseball be at time t seconds?

b. How high would the ball go?

c. When would it return to Juan's hand?

38. ***Motion Under Gravity*** An overworked employee at OHaganBooks.com goes to the top of the company's 100 foot tall headquarters building and flings a book up into the air at a speed of 60 feet per second.

a. When will the book hit the ground 100 feet below? (Neglect air resistance.)

b. How fast will it be traveling when it hits the ground?

c. How high will the book go?

39. ***Sales*** Sales at the OHaganBooks.com Web site of *Larry Potter and the Riemann Sum* fluctuated rather wildly in the first 5 months of last year as the following graph shows:

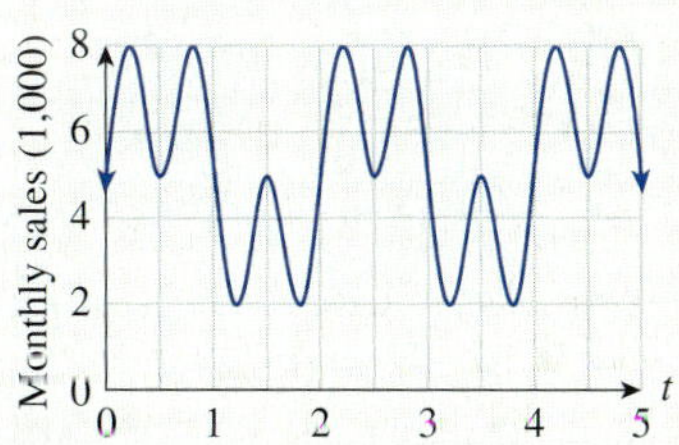

Puzzled by the graph, CEO John O'Hagan asks Jimmy Duffin[53] to estimate the total sales over the entire 5-month period shown. Jimmy decides to use a left Riemann sum with 10 partitions to estimate the total sales. What does he find?

40. ***Sales*** The following graph shows the approximate rate of change $s(t)$ of the total value, in thousands of dollars, of Spanish books sold online at OHaganBooks.com (t is the number of months since January 1):

[53] Marjory Duffin's nephew, currently at OHaganBooks.com on a summer internship.

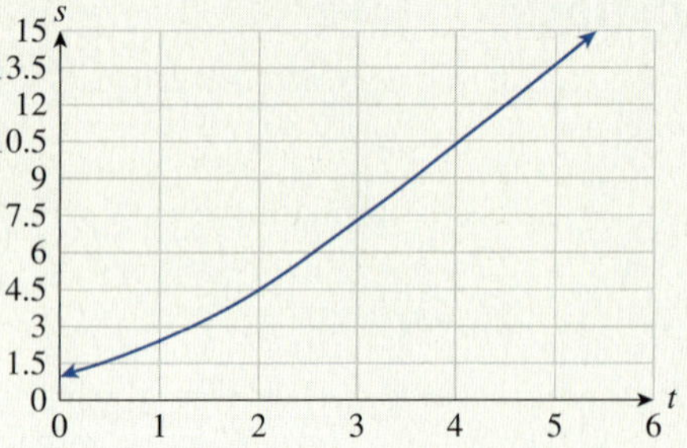

Use the graph to estimate the total value of Spanish books sold from March 1 through June 1. (Use a left Riemann sum with three subdivisions.)

41. ***Promotions*** Unlike sales of *Larry Potter and the Riemann Sum*, sales at OHaganBooks.com of the special leather-bound gift editions of *Lord of the Rings* have been suffering lately, as shown in the following graph (negative sales indicate returns by dissatisfied customers; t is time in months since January 1 of this year):

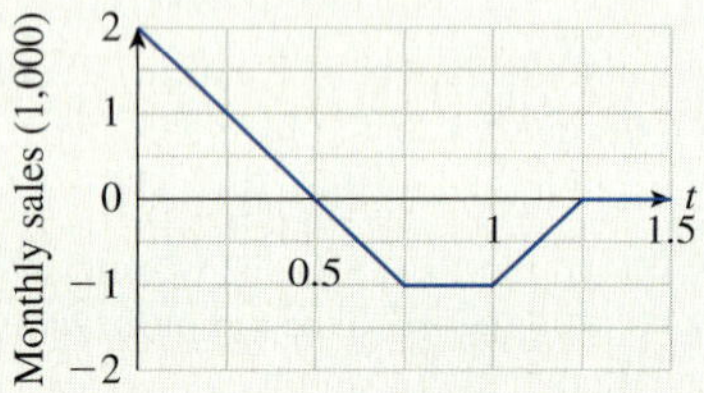

Use the graph to compute the total (net) sales over the period shown.

42. ***Sales*** Even worse than with the leather-bound *Lord of the Rings*, sales of *Investing in Real Estate* have been dismal, as shown in the following graph (negative sales indicate returns by dissatisfied customers; t is time in months since January 1 of this year):

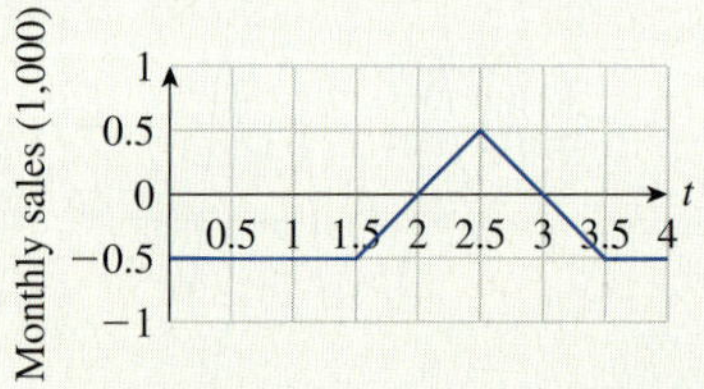

Use the graph to compute the total (net) sales over the period shown.

43. ***Web Site Activity*** The number of "hits" on the OHaganBooks.com Web site has been steadily increasing over the past month in response to recent publicity over a software glitch that caused the company to pay customers for buying books online. The activity can be modeled by

$$n(t) = 1{,}000t - 10t^2 + t^3 \text{ hits per day}$$

where t is time in days since news about the software glitch was first publicized on GrungeReport.com. Use a left Riemann sum with five partitions to estimate the total number of hits during the first 10 days of the period.

44. ***Web Site Crashes*** The latest DoorsXL servers OHaganBooks.com has been using for its Web site have been crashing with increasing frequency lately. One of the student summer interns has estimated the number of crashes to be

$$q(t) = 0.05t^2 + 0.4t + 9 \text{ crashes per week} \quad (0 \le t \le 10).$$

where t is the number of weeks since the DoorsXL system was first installed. Use a Riemann sum with five partitions to estimate the total number of crashes from the start of week 5 to the start of week 10. (Round your answer to the nearest crash.)

45. ***Student Intern Costs*** The marginal monthly cost of maintaining a group of summer student interns at OHaganBooks.com is calculated to be

$$c(x) = \frac{1{,}000(x+3)^2}{(8+(x+3)^3)^{3/2}} \text{ thousand dollars per additional student.}$$

Compute, to the nearest \$100, the total monthly cost if O'HaganBooks increases the size of the student intern program from five students to seven students.

46. ***Legal Costs*** The legal team maintained by OHaganBooks.com to handle the numerous lawsuits brought against the company by disgruntled clients may have to be expanded. The marginal monthly cost to maintain a team of x lawyers is estimated (by a method too complicated to explain) as

$$c(x) = (x-2)^2[8-(x-2)^3]^{3/2} \text{ thousand dollars per additional lawyer.}$$

Compute, to the nearest \$1,000, the total monthly cost if O'HaganBooks goes ahead with a proposal to increase the size of the legal team from two to four.

47. ***Projected Sales*** When OHaganBooks.com was about to go online, it estimated that its weekly sales would begin at about 6,400 books per week, with sales increasing at such a rate that weekly sales would double about every 2 weeks. If these estimates had been correct, how many books would the company have sold in the first 5 weeks? (Round your answer to the nearest 1,000 books.)

48. ***Projected Sales*** Once OHaganBooks.com actually went online, its weekly sales began at about 7,500 books per week, with weekly sales doubling every 3 weeks. How many books did the company actually sell in the first 5 weeks? (Round your answer to the nearest 1,000 books.)

49. ***Actual Sales*** OHaganBooks.com modeled its revised weekly sales over a period of time after it went online with the function

$$s(t) = 6{,}053 + \frac{4{,}474e^{0.55t}}{e^{0.55t} + 14.01}$$

where t is the time in weeks after it went online. According to this model, how many books did it actually sell in the first 5 weeks?

50. ***Computer Usage*** A consultant recently hired by OHaganBooks.com estimates total weekly computer usage as

$$w(t) = 620 + \frac{900e^{0.25t}}{3 + e^{0.25t}} \text{ hours} \quad (0 \le t \le 20)$$

where t is time in weeks since January 1 of this year. Use the model to estimate the total computer usage during the first 14 weeks of the year.

Case Study Spending on Housing Construction

Bill Varie/CORBIS

You are a summer intern at the U.S. Department of Housing and Urban Development, which has begun an ambitious public works program under the direction of the new administration. Yesterday you received the following memo from the senior aide to the Assistant Undersecretary:

TO: SW
FROM: SC
SUBJECT: Residential Construction Spending. Urgent!

Help! There is a subcommittee meeting in two hours and the office of the Undersecretary has asked me to immediately produce some mathematical formulas to (1) model the trend in residential construction spending since January 2006, when it was $618.7 billion, and (2) estimate the average spent per month on residential construction over a specified period of time. All I have on hand so far is data giving the month-over-month percentage changes (attached). Do you have any ideas?

ATTACHMENT*

Month	% Change	Month	% Change
1	1.16	16	−1.59
2	1.17	17	−1.62
3	1.33	18	−1.58
4	0.67	19	−1.67
5	0.42	20	−1.77
6	−0.04	21	−1.92
7	−0.24	22	−2.06
8	−0.44	23	−2.19
9	−0.70	24	−2.45
10	−0.94	25	−2.50
11	−1.31	26	−2.65
12	−1.34	27	−2.85
13	−1.49	28	−2.65
14	−1.75	29	−2.66
15	−1.74	30	−2.66

*Based on 12-month moving average. Source for data: U.S. Census Bureau: Manufacturing, Mining and Construction Statistics, Data 360 (www.data360.org/dataset.aspx?Data_Set_Id=3627).

Getting to work, you decide that the first thing to do is fit these data to a mathematical curve that you can use to project future changes in construction spending. You graph the data to get a sense of what mathematical models might be appropriate (Figure 24).

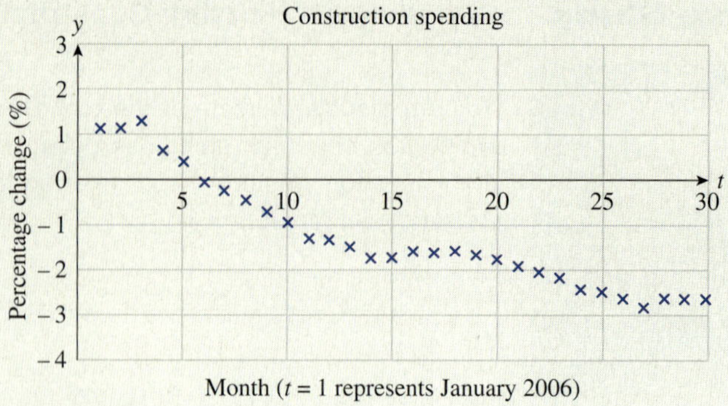

Figure 24

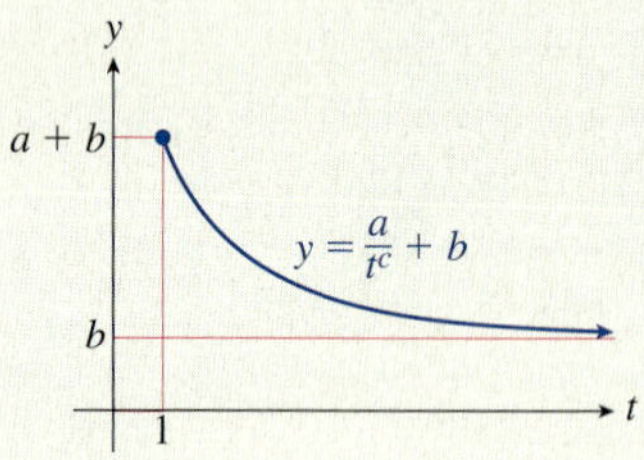

Figure 25

The graph suggests a decreasing trend, leveling off at about −3%. You recall that there are many curves that behave this way. One of the simplest is the curve

$$y = \frac{a}{t^c} + b \qquad (t \geq 1)$$

where a, b, and c are constants (Figure 25).[54]

You convert all the percentages to decimals, giving the following table of data:

t	y	t	y
1	0.0116	16	−0.0159
2	0.0117	17	−0.0162
3	0.0133	18	−0.0158
4	0.0067	19	−0.0167
5	0.0042	20	−0.0177
6	−0.0004	21	−0.0192
7	−0.0024	22	−0.0206
8	−0.0044	23	−0.0219
9	−0.0070	24	−0.0245
10	−0.0094	25	−0.0250
11	−0.0131	26	−0.0265
12	−0.0134	27	−0.0285
13	−0.0149	28	−0.0265
14	−0.0175	29	−0.0266
15	−0.0174	30	−0.0266

✱ NOTE To do this, you can use Excel's Solver, for example. However, the exact results may vary slightly, even when using the same program, depending on the settings and initial guesses for the coefficients.

You then find the values of a, b, and c that best fit the given data:*

$$a = 3.867405726, \quad b = -3.844754206, \quad c = 0.003685514$$

These values give you the following model for construction spending (with figures rounded to five significant digits):

$$y = \frac{3.8674}{t^{0.0036855}} - 3.8448.$$

[54] There is a good mathematical reason for choosing a curve of this form: It is a first approximation (for $t \geq 1$) to a general rational function that approaches a constant as $t \to +\infty$.

(It is interesting that the model predicts the change in construction spending leveling off to about −3.84%.) Figure 26 shows the graph of y superimposed on the data.

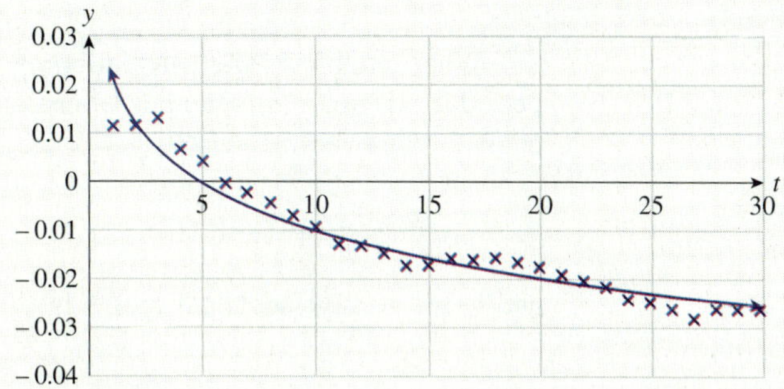

Figure 26

Now that you have a model for the month-over-month change in construction spending, you must use it to find the actual spending on construction. First, you realize that the model gives the *fractional rate of increase* of construction spending (because it is specified as a percentage, or fraction, of the total spending). In other words, if $p(t)$ represents the construction cost in month t, Then

$$y = \frac{dp/dt}{p} = \frac{d}{dt}(\ln p). \qquad \text{By the chain rule for derivatives}$$

You find an equation for actual monthly construction cost at time t by solving for p:

$$\begin{aligned}\ln p &= \int y\, dt \\ &= \int \left(\frac{a}{t^c} + b\right) dt \\ &= \frac{at^{1-c}}{1-c} + bt + K \\ &= dt^{1-c} + bt + K,\end{aligned}$$

where

$$d = \frac{a}{1-c} = \frac{3.867405726}{1 - 0.003685514} \approx 3.8817,$$

a and b are as above, and K is the constant of integration. So

$$p(t) = e^{dt^{1-c}+bt+K}.$$

To compute K, you substitute the initial data from the memo: $p(1) = 618.7$. Thus,

$$618.7 = e^{d+b+K} = e^{3.8817-3.8448+K} = e^{0.0369+K}.$$

Thus,

$$\ln(618.7) = 0.0369 + K,$$

which gives

$$K = \ln(618.7) - 0.0369 \approx 6.3907 \text{ (to five significant digits).}$$

Now you can write down the following formula for the monthly spending on residential construction as a function of t, the number of months since the beginning of 2006:

$$p(t) = e^{dt^{1-c}+bt+K} = e^{(3.8817t^{1-0.0036855}-3.8448t+6.3907)}.$$

What remains is the calculation of the average spent per month over a specified period $[r, s]$. Since p is the rate of change of the total spent, the total spent on housing construction over this period is

$$P = \int_r^s p(t)\,dt$$

and so the average spent per month is

$$\bar{P} = \frac{1}{s-r}\int_r^s p(t)\,dt. \qquad \frac{1}{\text{Number of months}} \times \text{Total spent}$$

Substituting the formula for $p(t)$ gives

$$\bar{P} = \frac{1}{s-r}\int_r^s e^{(3.8817t^{1-0.0036855}-3.8448t+6.3907)}\,dt.$$

You cannot find an explicit antiderivative for the integrand, so you decide that the only way to compute it is numerically. You send the following memo to SC.

> **TO:** SC
> **FROM:** SW
> **SUBJECT:** The formula you wanted
>
> Spending in the United States on housing construction in the tth month of 2006 can be modeled by
>
> $$p(t) = e^{(3.8817t^{1-0.0036855}-3.8448t+6.3907)} \text{ million dollars.}$$
>
> Further, the average spent per month from month r to month s (since the start of January 2006) can be computed as
>
> $$\bar{P} = \frac{1}{s-r}\int_r^s e^{(3.8817t^{1-0.0036855}-3.8448t+6.3907)}\,dt.$$
>
> To calculate it easily (and impress the subcommittee members), I suggest you have a graphing calculator on hand and enter the following on your graphing calculator (watch the parentheses!):
>
> ```
> Y1=1/(S-R)*fnInt(e^(3.8817T^(1-0.0036855)-3.8448T+6.3907),T,R,S)
> ```
>
> Then suppose, for example, you need to estimate the average for the period March 1, 2006 ($t = 3$) to February 1, 2007 ($t = 14$) All you do is enter
>
> ```
> 3→R
> 14→S
> Y1
> ```
>
> and your calculator will give you the result: The average spending was $624 million per month.
>
> Good luck with the meeting!

EXERCISES

1. Use the actual January 2006 spending figure of \$618.7 million and the percentage changes in the table to compute the actual spending in February, March, and April of that year. Also use the model of monthly spending to estimate those figures, and compare the predicted values with the actual figures. Is it unacceptable that the April figures agree to only one significant digit? Explain.

2. Use the model developed above to estimate the average monthly spending on residential construction over the 12-month period beginning June 1, 2006. (Round your answer to the nearest \$1 million.)

3. What (if any) advantages are there to using a model for residential construction spending when the actual residential construction spending figures are available?

4. The formulas for $p(t)$ and $\bar{P}$ were based on the January 2006 spending figure of \$618.7 million. Change the models to allow for a possibly revised January 2006 spending figure of $\$p_0$ million.

5. If we had used quadratic regression to model the construction spending data, we would have obtained

$$y = 0.00005t^2 - 0.0028t + 0.158.$$

(See the graph.)

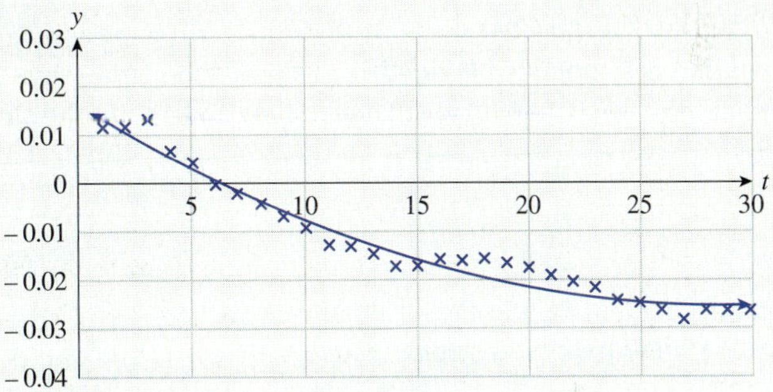

Use this formula and the given January 2006 spending figure to obtain corresponding models for $p(t)$ and $\bar{P}$.

6. Compare the model in the text with the quadratic model in Exercise 5 in terms of both short- and long-term predictions; in particular, when does the quadratic model predict construction spending will have reached its biggest monthly decrease? Are either of these models realistic in the near-term? in the long term?

TI-83/84 Plus Technology Guide

Section 6.3

Example 5 (page 489) Estimate the area under the graph of $f(x) = 1 - x^2$ over the interval [0, 1] using $n = 100$, $n = 200$, and $n = 500$ partitions.

Solution with Technology

There are several ways to compute Riemann sums with a graphing calculator. We illustrate one method. For $n = 100$, we need to compute the sum

$$\sum_{k=0}^{99} f(x_k)\Delta x = [f(0) + f(0.01) + \cdots + f(0.99)](0.01).$$

See discussion in Example 5.

Thus, we first need to calculate the numbers $f(0)$, $f(0.01)$, and so on, and add them up. The TI-83/84 Plus has a built-in `sum` function (available in the LIST MATH menu), which, like the SUM function in a spreadsheet, sums the entries in a list.

1. To generate a list that contains the numbers we want to add together, use the `seq` function (available in the LIST OPS menu). If we enter

   ```
   seq(1-X^2,X,0,0.99,0.01)
   ```

   seq: 2ND LIST OPS 5

   the calculator will calculate a list by evaluating `1-X^2` for values of `X` from 0 to 0.99 in steps of 0.01.
2. To take the sum of all these numbers, we wrap the `seq` function in a call to `sum`:

   ```
   sum(seq(1-X^2,X,0,0.99,0.01))
   ```

   sum: 2ND LIST MATH 5

   This gives the sum

   $$f(0) + f(0.01) + \cdots + f(0.99) = 67.165.$$
3. To obtain the Riemann sum, we need to multiply this sum by $\Delta x = 0.01$, and we obtain the estimate of $67.165 \times 0.01 = 0.67165$ for the Riemann sum:

We obtain the other Riemann sums similarly, as shown here:

$n = 200$

$n = 500$

One disadvantage of this method is that the TI-83/84 Plus can generate and sum a list of at most 999 entries. The LEFTSUM program below calculates left Riemann sums for any n. The TI-83/84 Plus also has a built-in function `fnInt`, which finds a very accurate approximation of a definite integral, using a more sophisticated technique than the one we are discussing here.

The LEFTSUM program for the TI-83/84 Plus

The following program calculates (left) Riemann sums for any n. The latest version of this program (and others) is available at the Web site.

```
PROGRAM: LEFTSUM
:Input "LEFT ENDPOINT? ",A        Prompts for the left end-point a
:Input "RIGHT ENDPOINT? ",B       Prompts for the right end-point b
:Input "N? ",N                    Prompts for the number of rectangles
:(B-A)/N→D                        D is Δx = (b − a)/n.
:Ø→L                              L will eventually be the left sum.
:A→X                              X is the current x-coordinate.
:For(I,1,N)                       Start of a loop—recall the sigma notation.
:L+Y1→L                           Add f(x_{i−1}) to L.
:A+I*D→X                          Uses formula x_i = a + iΔx
:End                              End of loop
:L*D→L                            Multiply by Δx.
:Disp "LEFT SUM IS ",L
:Stop
```

EXCEL Technology Guide

Section 6.3

Example 5 (page 489) Estimate the area under the graph of $f(x) = 1 - x^2$ over the interval [0, 1] using $n = 100$, $n = 200$, and $n = 500$ partitions.

Solution with Technology

We need to compute various sums:

$$\sum_{k=0}^{99} f(x_k)\Delta x = [f(0) + f(0.01) + \cdots + f(0.99)](0.01)$$

See discussion in Example 5.

$$\sum_{k=0}^{199} f(x_k)\Delta x = [f(0) + f(0.005) + \cdots + f(0.995)](0.005)$$

$$\sum_{k=0}^{499} f(x_k)\Delta x = [f(0) + f(0.002) + \cdots + f(0.998)](0.002).$$

Here is how you can compute them all on the same spreadsheet.

1. Enter the values for the end-points a and b, thc number of subdivisions n, and the formula $\Delta x = (b - a)/n$:

	A	B	C	D
1	x	f(x)	a	0
2			b	1
3			n	100
4			Delta x	`=(D2-D1)/D3`

2. Next, we compute all the x-values we might need in column A. Because the largest value of n that we will be using is 500, we will need a total of 501 values of x. Note that the value in each cell below A3 is obtained from the one above by adding Δx.

	A	B	C	D
1	x	f(x)	a	0
2	`=D1`		b	1
3	`=A2+$D$4`		n	100
4			Delta x	0.01
5				
501				
502				

	A	B	C	D
1	x	f(x)	a	0
2	0		b	1
3	0.01		n	100
4	0.02		Delta x	0.01
5	0.03			
501	4.99			
502	5			

(The fact that the values of x currently go too far will be corrected in the next step.)

3. We need to calculate the numbers $f(0)$, $f(0.01)$, and so on, but only those for which the corresponding x-value is less than b. To do this, we use a logical formula like we did with piecewise-defined functions in Chapter 1:

	A	B	C	D
1	x	f(x)	a	0
2	0	`=(1-A2^2)*(A2<$D$2)`	b	1
3	0.01		n	100
4	0.02		Delta x	0.01
5	0.03			
501	4.99			
502	5			

When the value of x is b or above, the function will evaluate to zero, because we do not want to count it.

4. Finally, we compute the Riemann sum by adding up everything in Column B and multiplying by Δx:

	A	B	C	D
1	x	f(x)	a	0
2	0	1	b	1
3	0.01	0.9999	n	100
4	0.02	0.9996	Delta x	0.01
5	0.03	0.9991	Left Sum	`=SUM(B:B)*D4`
6	0.04	0.9984		

↓

	A	B	C	D
1	x	f(x)	a	0
2	0	1	b	1
3	0.01	0.9999	n	100
4	0.02	0.9996	Delta x	0.01
5	0.03	0.9991	Left Sum	0.67165
6	0.04	0.9984		

Now it is easy to obtain the sums for $n = 200$ and $n = 500$: Simply change the value of n in cell D3:

	A	B	C	D
1	x	f(x)	a	0
2	0	1	b	1
3	0.005	0.999975	n	200
4	0.01	0.9999	Delta x	0.005
5	0.015	0.999775	Left Sum	0.6691625
6	0.02	0.9996		

	A	B	C	D
1	x	f(x)	a	0
2	0	1	b	1
3	0.002	0.999996	n	500
4	0.004	0.999984	Delta x	0.002
5	0.006	0.999964	Left Sum	0.667666
6	0.008	0.999936		

7 Further Integration Techniques and Applications of the Integral

Case Study Estimating Tax Revenues

You have just been hired by the incoming administration to coordinate national tax policy, and the so-called experts on your staff can't seem to agree on which of two tax proposals will result in more revenue for the government. The data you have are the two income tax proposals (graphs of tax vs. income) and the distribution of incomes in the country. **How do you use this information to decide which tax policy will result in more revenue?**

Dmitriy Shironosov, 2009/Used under license from Shutterstock.com

Web Site

www.AppliedCalc.org

At the Web site you will find:

- A detailed chapter summary
- A true/false quiz
- Additional review exercises
- A numerical integration utility
- Graphing calculator programs for numerical integration
- Graphers, Excel tutorials, and other resources

Introduction

In the preceding chapter, we learned how to compute many integrals and saw some of the applications of the integral. In this chapter, we look at some further techniques for computing integrals and then at more applications of the integral. We also see how to extend the definition of the definite integral to include integrals over infinite intervals, and show how such integrals can be used for long-term forecasting. Finally, we introduce the beautiful theory of differential equations and some of its numerous applications.

7.1 Integration by Parts

Integration by parts is an integration technique that comes from the product rule for derivatives. The tabular method we present here has been around for some time and makes integration by parts quite simple, particularly in problems where it has to be used several times.*

* **NOTE** The version of the tabular method we use was developed and taught to us by Dan Rosen at Hofstra University.

We start with a little notation to simplify things while we introduce integration by parts. (We use this notation only in the next few pages.) If u is a function, denote its derivative by $D(u)$ and an antiderivative by $I(u)$. Thus, for example, if $u = 2x^2$, then

$$D(u) = 4x$$

and

$$I(u) = \frac{2x^3}{3}.$$

[If we wished, we could instead take $I(u) = \dfrac{2x^3}{3} + 46$, but we usually opt to take the simplest antiderivative.]

Integration by Parts

If u and v are continuous functions of x, and u has a continuous derivative, then

$$\int u \cdot v\, dx = u \cdot I(v) - \int D(u) I(v)\, dx.$$

Quick Example

(Discussed more fully in Example 1 below)

$$\int x \cdot e^x dx = xI(e^x) - \int D(x)I(e^x)\, dx$$

$$= xe^x - \int 1 \cdot e^x\, dx \qquad I(e^x) = e^x;\ D(x) = 1$$

$$= xe^x - e^x + C. \qquad \int e^x\, dx = e^x + C$$

As the Quick Example shows, although we could not immediately integrate $u \cdot v = x \cdot e^x$, we could easily integrate $D(u)I(v) = 1 \cdot e^x = e^x$.

Derivation of Integration by Parts Formula

As we mentioned, the integration-by-parts formula comes from the product rule for derivatives. We apply the product rule to the function $uI(v)$

$$\begin{aligned} D[u \cdot I(v)] &= D(u)I(v) + uD(I(v)) \\ &= D(u)I(v) + uv \end{aligned}$$

because $D(I(v))$ is the derivative of an antiderivative of v, which is v. Integrating both sides gives

$$u \cdot I(v) = \int D(u)I(v)\,dx + \int uv\,dx.$$

A simple rearrangement of the terms now gives us the integration-by-parts formula.

The integration-by-parts formula is easiest to use via the tabular method illustrated in the following example, where we repeat the calculation we did in the Quick Example above.

EXAMPLE 1 Integration by Parts: Tabular Method

Calculate $\int xe^x\,dx$.

Solution First, the reason we *need* to use integration by parts to evaluate this integral is that none of the other techniques of integration that we've talked about up to now will help us. Furthermore, we cannot simply find antiderivatives of x and e^x and multiply them together. [You should check that $(x^2/2)e^x$ is *not* an antiderivative of xe^x.] However, as we saw above, this integral can be found by integration by parts. We want to find the integral of the *product* of x and e^x. We must make a decision: Which function will play the role of u and which will play the role of v in the integration-by-parts formula? Because the derivative of x is just 1, differentiating makes it simpler, so we try letting x be u and letting e^x be v. We need to calculate $D(u)$ and $I(v)$, which we record in the following table.

	D	I
$+$	x	e^x
$-\int$	1 →	e^x

The table is read as $+x \cdot e^x - \int 1 \cdot e^x\,dx$

Below x in the D column, we put $D(x) = 1$; below e^x in the I column, we put $I(e^x) = e^x$. The arrow at an angle connecting x and $I(e^x)$ reminds us that the product $xI(e^x)$ will appear in the answer; the plus sign on the left of the table reminds us that it is $+xI(e^x)$ that appears. The integral sign and the horizontal arrow connecting $D(x)$ and $I(e^x)$ remind us that the *integral* of the product $D(x)I(e^x)$ also appears in the answer; the minus sign on the left reminds us that we need to subtract this integral. Combining these two contributions, we get

$$\int xe^x\,dx = xe^x - \int e^x\,dx.$$

The integral that appears on the right is much easier than the one we began with, so we can complete the problem:

$$\int xe^x\,dx = xe^x - \int e^x\,dx = xe^x - e^x + C.$$

Before we go on... In Example 1, what if we had made the opposite decision and put e^x in the D column and x in the I column? Then we would have had the following table:

	D	I
$+$	e^x	x
$-\int$	e^x	$\to x^2/2$

This gives

$$\int xe^x\,dx = \frac{x^2}{2}e^x - \int \frac{x^2}{2}e^x\,dx.$$

The integral on the right is harder than the one we started with, not easier! How do we know beforehand which way to go? We don't. We have to be willing to do a little trial and error: We try it one way, and if it doesn't make things simpler, we try it another way. *Remember, though, that the function we put in the I column must be one that we can integrate.* ■

EXAMPLE 2 Repeated Integration by Parts

Calculate $\int x^2e^{-x}\,dx$.

Solution Again, we have a product—the integrand is the product of x^2 and e^{-x}. Because differentiating x^2 makes it simpler, we put it in the D column and get the following table:

	D	I
$+$	x^2	e^{-x}
$-\int$	$2x$	$\to -e^{-x}$

This table gives us

$$\int x^2e^{-x}\,dx = x^2(-e^{-x}) - \int 2x(-e^{-x})\,dx.$$

The last integral is simpler than the one we started with, but it still involves a product. It's a good candidate for another integration by parts. The table we would use would start with $2x$ in the D column and $-e^{-x}$ in the I column, which is exactly what we see in the last row of the table we've already made. Therefore, we *continue the process*, elongating the table above:

	D	I
$+$	x^2	e^{-x}
$-$	$2x$	$-e^{-x}$
$+\int$	2	$\to e^{-x}$

(Notice how the signs on the left alternate. Here's why: To compute $-\int 2x(-e^{-x})\,dx$ we use the negative of the following table:

	D	I
$+$	$2x$	$-e^{-x}$
$-\int$	2	e^{-x}

so we reverse all the signs.)

Now, we still have to compute an integral (the integral of the product of the functions in the bottom row) to complete the computation. But why stop here? Let's continue the process one more step:

	D	I
$+$	x^2	e^{-x}
$-$	$2x$	$-e^{-x}$
$+$	2	e^{-x}
$-\int$	0	$-e^{-x}$

In the bottom line we see that all that is left to integrate is $0(-e^{-x}) = 0$. Because the indefinite integral of 0 is C, we can read the answer from the table as

$$\begin{aligned}\int x^2e^{-x}\,dx &= x^2(-e^{-x}) - 2x(e^{-x}) + 2(-e^{-x}) + C\\ &= -x^2e^{-x} - 2xe^{-x} - 2e^{-x} + C\\ &= -e^{-x}(x^2 + 2x + 2) + C.\end{aligned}$$

In Example 2 we saw a technique that we can summarize as follows:

Integrating a Polynomial Times a Function

If one of the factors in the integrand is a polynomial and the other factor is a function that can be integrated repeatedly, put the polynomial in the D column and keep differentiating until you get zero. Then complete the I column to the same depth, and read off the answer.

For practice, redo Example 1 using this technique.

It is not always the case that the integrand is a polynomial times something easy to integrate, so we can't always expect to end up with a zero in the D column. In that case we hope that at some point we will be able to integrate the product of the functions in the last row. Here are some examples.

EXAMPLE 3 Polynomial Times a Logarithm

Calculate: **a.** $\int x \ln x\, dx$ **b.** $\int (x^2 - x) \ln x\, dx$ **c.** $\int \ln x\, dx$

Solution

a. This is a product and therefore a good candidate for integration by parts. Our first impulse is to differentiate x, but that would mean integrating $\ln x$, and we do not (yet) know how to do that. So we try it the other way around and hope for the best.

	D	I
$+$	$\ln x$	x
$-\int$	$\frac{1}{x}$	$\frac{x^2}{2}$

Why did we stop? If we continued the table, both columns would get more complicated. However, if we stop here we get

$$\int x \ln x\, dx = (\ln x)\left(\frac{x^2}{2}\right) - \int \left(\frac{1}{x}\right)\left(\frac{x^2}{2}\right) dx$$

$$= \frac{x^2}{2} \ln x - \frac{1}{2} \int x\, dx$$

$$= \frac{x^2}{2} \ln x - \frac{x^2}{4} + C.$$

b. We can use the same technique we used in part (a) to integrate any polynomial times the logarithm of x:

	D	I
$+$	$\ln x$	$x^2 - x$
$-\int$	$\frac{1}{x}$	$\frac{x^3}{3} - \frac{x^2}{2}$

$$\int (x^2 - x) \ln x\, dx = (\ln x)\left(\frac{x^3}{3} - \frac{x^2}{2}\right) - \int \left(\frac{1}{x}\right)\left(\frac{x^3}{3} - \frac{x^2}{2}\right) dx$$

$$= \left(\frac{x^3}{3} - \frac{x^2}{2}\right) \ln x - \int \left(\frac{x^2}{3} - \frac{x}{2}\right) dx$$

$$= \left(\frac{x^3}{3} - \frac{x^2}{2}\right) \ln x - \frac{x^3}{9} + \frac{x^2}{4} + C.$$

c. The integrand $\ln x$ is not a product. We can, however, *make* it into a product by thinking of it as $1 \cdot \ln x$. Because this is a polynomial times $\ln x$, we proceed as in parts (a) and (b):

	D	I
$+$	$\ln x$	1
$-\int$	$1/x$	x

We notice that the product of $1/x$ and x is just 1, which we know how to integrate, so we can stop here:

$$\begin{aligned}\int \ln x\,dx &= x\ln x - \int \left(\frac{1}{x}\right) x\,dx\\ &= x\ln x - \int 1\,dx\\ &= x\ln x - x + C.\end{aligned}$$

FAQs

Whether to Use Integration by Parts, and What Goes in the *D* and *I* Columns

Q: *Will integration by parts always work to integrate a product?*

A: No. Although integration by parts often works for products in which one factor is a polynomial, it will almost *never* work in the examples of products we saw when discussing substitution in Section 6.2. For example, although integration by parts can be used to compute $\int (x^2 - x)e^{2x-1}\,dx$ (put $x^2 - x$ in the *D* column and e^{2x-1} in the *I* column), it *cannot* be used to compute $\int (2x-1)e^{x^2-x}\,dx$ (put $u = x^2 - x$). Recognizing when to use integration by parts is best learned by experience.

Q: *When using integration by parts, which expression goes in the D column, and which in the I column?*

A: Although there is no general rule, the following guidelines are useful:

- To integrate a product in which one factor is a polynomial and the other can be integrated several times, put the polynomial in the *D* column and the other factor in the *I* column. Then differentiate the polynomial until you get zero.
- If one of the factors is a polynomial but the other factor cannot be integrated easily, put the polynomial in the *I* column and the other factor in the *D* column. Stop when the product of the functions in the bottom row can be integrated.
- If neither factor is a polynomial, put the factor that seems easier to integrate in the *I* column and the other factor in the *D* column. Again, stop the table as soon as the product of the functions in the bottom row can be integrated.
- If your method doesn't work, try switching the functions in the *D* and *I* columns or try breaking the integrand into a product in a different way. If none of this works, maybe integration by parts isn't the technique to use on this problem.

7.1 EXERCISES

▼ more advanced ◆ challenging

T indicates exercises that should be solved using technology

Evaluate the integrals in Exercises 1–40 using integration by parts where possible. HINT [See Examples 1–3.]

1. $\int 2xe^x\,dx$

2. $\int 3xe^{-x}\,dx$

3. $\int (3x-1)e^{-x}\,dx$

4. $\int (1-x)e^x\,dx$

5. $\int (x^2-1)e^{2x}\,dx$

6. $\int (x^2+1)e^{-2x}\,dx$

7. $\int (x^2+1)e^{-2x+4}\,dx$

8. $\int (x^2+1)e^{3x+1}\,dx$

9. $\int (2-x)2^x\,dx$

10. $\int (3x-2)4^x\,dx$

11. $\int (x^2-1)3^{-x}\,dx$

12. $\int (1-x^2)2^{-x}\,dx$

13. ▼ $\int \frac{x^2-x}{e^x}\,dx$

14. ▼ $\int \frac{2x+1}{e^{3x}}\,dx$

15. $\int x(x+2)^6\,dx$ (See note.[1])

16. $\int x^2(x-1)^6\,dx$ (See note.[1])

17. ▼ $\int \frac{x}{(x-2)^3}\,dx$

18. ▼ $\int \frac{x}{(x-1)^2}\,dx$

19. $\int x^3 \ln x\,dx$

20. $\int x^2 \ln x\,dx$

21. $\int (t^2+1)\ln(2t)\,dt$

22. $\int (t^2-t)\ln(-t)\,dt$

23. $\int t^{1/3}\ln t\,dt$

24. $\int t^{-1/2}\ln t\,dt$

25. $\int \log_3 x\,dx$

26. $\int x\log_2 x\,dx$

27. ▼ $\int (xe^{2x}-4e^{3x})\,dx$

28. ▼ $\int (x^2e^{-x}+2e^{-x+1})\,dx$

29. ▼ $\int \left(x^2e^x - xe^{x^2}\right)dx$

30. ▼ $\int \left[(2x+1)\,e^{x^2+x} - x^2e^{2x+1}\right]dx$

31. $\int_0^1 (x+1)e^x\,dx$

32. $\int_{-1}^1 (x^2+x)e^{-x}\,dx$

33. $\int_0^1 x^2(x+1)^{10}\,dx$

34. $\int_0^1 x^3(x+1)^{10}\,dx$

35. ▼ $\int (3x-4)\sqrt{2x-1}\,dx$ (See note.[1])

36. ▼ $\int \frac{2x+1}{\sqrt{3x-2}}\,dx$ (See note.[1])

37. $\int_1^2 x\ln(2x)\,dx$

38. $\int_1^2 x^2\ln(3x)\,dx$

39. $\int_0^1 x\ln(x+1)\,dx$

40. $\int_0^1 x^2\ln(x+1)\,dx$

41. Find the area bounded by the curve $y = xe^{-x}$, the x-axis, and the lines $x = 0$ and $x = 10$.

42. Find the area bounded by the curve $y = x\ln x$, the x-axis, and the lines $x = 1$ and $x = e$.

43. Find the area bounded by the curve $y = (x+1)\ln x$, the x-axis, and the lines $x = 1$ and $x = 2$.

44. Find the area bounded by the curve $y = (x-1)e^x$, the x-axis, and the lines $x = 0$ and $x = 2$.

[1] Several exercises, including these, can also be done by using substitution, although integration by parts is easier and should be used instead.

Integrals of Functions Involving Absolute Values *In Exercises 45–52, use integration by parts to evaluate the given integral using the following integral formulas where necessary. (You have seen some of these before; all can be checked by differentiating.)*

Integral Formula	Shortcut Version
$\int \frac{\lvert x\rvert}{x}\,dx = \lvert x\rvert + C$	$\int \frac{\lvert ax+b\rvert}{ax+b}\,dx = \frac{1}{a}\lvert ax+b\rvert + C$ Because $\frac{d}{dx}\lvert x\rvert = \frac{\lvert x\rvert}{x}$.
$\int \lvert x\rvert\,dx = \frac{1}{2}x\lvert x\rvert + C$	$\int \lvert ax+b\rvert\,dx = \frac{1}{2a}(ax+b)\lvert ax+b\rvert + C$
$\int x\lvert x\rvert\,dx = \frac{1}{3}x^2\lvert x\rvert + C$	$\int (ax+b)\lvert ax+b\rvert\,dx = \frac{1}{3a}(ax+b)^2\lvert ax+b\rvert + C$
$\int x^2\lvert x\rvert\,dx = \frac{1}{4}x^3\lvert x\rvert + C$	$\int (ax+b)^2\lvert ax+b\rvert\,dx = \frac{1}{4a}(ax+b)^3\lvert ax+b\rvert + C$

45. $\int x\lvert x-3\rvert\,dx$

46. $\int x\lvert x+4\rvert\,dx$

47. $\int 2x\frac{\lvert x-3\rvert}{x-3}\,dx$

48. $\int 3x\frac{\lvert x+4\rvert}{x+4}\,dx$

49. ▼ $\int 2x^2\lvert -x+4\rvert\,dx$

50. ▼ $\int 3x^2\lvert 2x-3\rvert\,dx$

51. ▼ $\int (x^2-2x+3)\lvert x-4\rvert\,dx$

52. ▼ $\int (x^2-x+1)\lvert 2x-4\rvert\,dx$

APPLICATIONS

53. ***Displacement*** A rocket rising from the ground has a velocity of $2{,}000te^{-t/120}$ ft/s, after t seconds. How far does it rise in the first two minutes?

54. ***Sales*** Weekly sales of graphing calculators can be modeled by the equation

$$s(t) = 10 - te^{-t/20}$$

where s is the number of calculators sold per week after t weeks. How many graphing calculators (to the nearest unit) will be sold in the first 20 weeks?

55. ***Total Cost*** The marginal cost of the xth box of light bulbs is $10 + [\ln(x+1)]/(x+1)^2$, and the fixed cost is \$5,000. Find the total cost to make x boxes of bulbs.

56. ***Total Revenue*** The marginal revenue for selling the xth box of light bulbs is $10 + 0.001x^2e^{-x/100}$. Find the total revenue generated by selling 200 boxes of bulbs.

57. ***Spending on Gasoline*** During 2000–2008, the United States consumed gasoline at a rate of about

$$q(t) = -3.5t + 280 \text{ billion gallons per year.} \quad (0 \le t \le 8)$$

(t is the number of years since 2000.)[2] During the same period, the price of gasoline was approximately

$$p(t) = 1.2e^{0.12t} \text{ dollars per gallon.}$$

Use an integral to estimate, to the nearest billion dollars, the total spent on gasoline during the given period. HINT [Rate of spending $= p(t)q(t)$.]

58. ***Spending on Gasoline*** During 1992–2000, the United States consumed gasoline at a rate of about

$$q(t) = 3.2t + 240 \text{ billion gallons per year.} \quad (0 \le t \le 8)$$

(t is the number of years since 1992.)[3] During the same period, the price of gasoline was approximately

$$p(t) = 1.0e^{0.02t} \text{ dollars per gallon.}$$

Use an integral to estimate, to the nearest billion dollars, the total spent on gasoline during the given period. HINT [Rate of spending $= p(t)q(t)$.]

59. ***Bottled Water Sales*** The rate of U.S. sales of bottled water for the period 2000–2008 can be approximated by

$$s(t) = 600t + 4{,}600 \text{ million gallons per year} \quad (0 \le t \le 8)$$

where t is time in years since the start of 2000.[4] After conducting a survey of sales in your town, you estimate that consumption in gyms accounts for a fraction

$$f(t) = \sqrt{0.1 + 0.02t}$$

of all bottled water sold. Assuming your model is correct, estimate, to the nearest million gallons, the total amount of bottled water consumed in gyms from the start of 2000 to the start of 2005. HINT [Rate of consumption $= s(t)f(t)$. Also see Exercise 35.]

60. ***Bottled Water Sales*** The rate of U.S. per capita sales of bottled water for the period 2000–2008 can be approximated by

$$s(t) = 1.8t + 16 \text{ gallons per year} \quad (0 \le t \le 8)$$

where t is the time in years since the start of 2000.[5] After conducting a survey of sales in your state, you estimate that consumption in gyms accounts for a fraction

$$f(t) = \sqrt{0.2 + 0.04t}$$

of all bottled water consumed. Assuming your model is correct, estimate, to the nearest gallon, the total amount of bottled water consumed per capita in gyms from the start of 2000 to the start of 2005. HINT [Rate of consumption $= s(t)f(t)$. Also see Exercise 36.]

61. ***Housing*** The following graph shows the annual number of housing starts in the United States during 2000–2008 together with a quadratic approximating model

$$s(t) = -30t^2 + 240t + 800 \text{ thousand homes per year.} \quad (0 \le t \le 8)$$

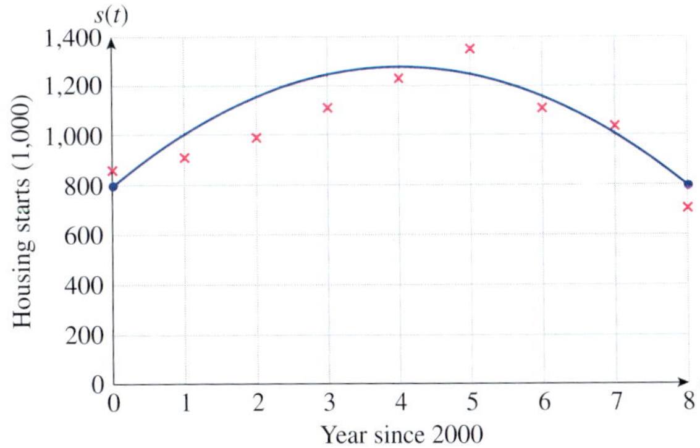

(t is the time in years since 2000.)[6] At the same time, the homes being built were getting larger: The average area per home was approximately

$$a(t) = 40t + 2{,}000 \text{ square feet.}$$

Use the given models to estimate the total housing area under construction over the given period. (Use integration by parts to evaluate the integral and round your answer to the nearest billion square feet.) HINT [Rate of change of area under construction $= s(t)a(t)$.]

62. ***Housing for Sale*** The following graph shows the number of housing starts for sale purposes in the United States during 2000–2008 together with a quadratic approximating model

$$s(t) = -33t^2 + 240t + 700 \text{ thousand homes per year.} \quad (0 \le t \le 8)$$

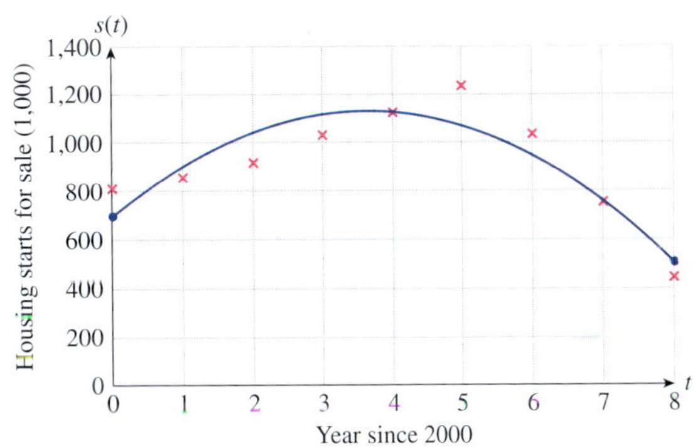

(t is the time in years since 2000.)[7] At the same time, the homes being built were getting larger: The average area per home was approximately

$$a(t) = 40t + 2{,}000 \text{ square feet.}$$

[2] Source for data: Energy Information Administration (Department of Energy) (http://www.eia.doe.gov).

[3] Ibid.

[4] Source for data: *Beverage Marketing Corporation* (www.bottledwater.org) (The 2008 figure is an estimate.)

[5] Ibid.

[6] Source for data: U.S. Census Bureau (www.census.gov).

[7] Ibid.

Use the given models to estimate the total housing area under construction for sale purposes over the given period. (Use integration by parts to evaluate the integral and round your answer to the nearest billion square feet.) HINT [Rate of change of area under construction $= s(t)a(t)$.]

63. ▼ ***Oil Production in Mexico:* Pemex** The rate of oil production by **Pemex**, Mexico's national oil company, can be approximated by

$$q(t) = -8t^2 + 70t + 1{,}000 \text{ million barrels per year} \quad (0 \le t \le 9)$$

where t is time in years since the start of 2000.[8] During that time, the price of oil was approximately[9]

$$p(t) = 25e^{0.1t} \text{ dollars per barrel.}$$

Obtain an expression for **Pemex**'s total oil revenue $R(x)$ since the start of 2000 to the start of year x as a function of x. (Do not simplify the answer.) HINT [Rate of revenue $= p(t)q(t)$.]

64. ▼ ***Oil Imports from Mexico*** The rate of oil imports to the United States from Mexico can be approximated by

$$r(t) = -5t^2 + 40t + 500 \text{ million barrels per year} \quad (0 \le t \le 8)$$

where t is time in years since the start of 2000.[10] During that time, the price of oil was approximately[11]

$$p(t) = 25e^{0.1t} \text{ dollars per barrel.}$$

Obtain an expression for the total oil revenue $R(t)$ Mexico earned from the United States since the start of 2000. HINT [Rate of revenue $= p(t)q(t)$.]

65. ▼ ***Revenue*** You have been raising the price of your *Lord of the Rings* T-shirts by 50¢ per week, and sales have been falling continuously at a rate of 2% per week. Assuming you are now selling 50 T-shirts per week and charging \$10 per T-shirt, how much revenue will you generate during the coming year? (Round your answer to the nearest dollar.) HINT [Weekly revenue = weekly sales × price per T-shirt.]

66. ▼ ***Revenue*** Luckily, sales of your *Star Wars* T-shirts are now 50 T-shirts per week and increasing continuously at a rate of 5% per week. You are now charging \$10 per T-shirt and are decreasing the price by 50¢ per week. How much revenue will you generate during the next six weeks?

Integrals of Piecewise Linear Functions *Exercises 67 and 68 are based on the following formula that can be used to represent a piecewise linear function as a closed form function:*

$$\begin{cases} p(x) & \text{if } x < a \\ q(x) & \text{if } x > a \end{cases} = p(x) + \frac{1}{2}[q(x) - p(x)]\left[1 + \frac{|x-a|}{x-a}\right]$$

Such functions can then be integrated using the technique of Exercises 45–52.

67. ◆ ***Population: Mexico*** The rate of change of population in Mexico over 1990–2010 was approximately

$$r(t) = \begin{cases} -0.1t + 3 & \text{if } 0 \le t \le 10 \\ -0.05t + 2.5 & \text{if } 10 \le t \le 20 \end{cases} \quad \text{million people per year}$$

where t is time in years since 1990.

a. Use the formula given before the exercise to represent $r(t)$ as a closed-form function. HINT [Use the formula with $a = 10$.]

b. Use the result of part (a) and a definite integral to estimate the total increase in population over the given 20-year period. HINT [Break up the integral into two, and use the technique of Exercises 45–52 to evaluate one of them.]

68. ◆ ***Population: Mexico*** The rate of change of population in Mexico over 1950–1990 was approximately

$$r(t) = \begin{cases} 0.05t + 2.5 & \text{if } 0 \le t \le 20 \\ -0.075t + 5 & \text{if } 20 \le t \le 40 \end{cases} \quad \text{million people per year}$$

where t is time in years since 1950.

a. Use the formula given before the exercise to represent $r(t)$ as a closed-form function. HINT [Use the formula with $a = 20$.]

b. Use the result of part (a) and a definite integral to estimate the total increase in population over the given 40-year period. HINT [Break up the integral into two, and use the technique of Exercises 45–52 to evaluate one of them.]

COMMUNICATION AND REASONING EXERCISES

69. Your friend Janice claims that integration by parts allows one to integrate any product of two functions. Prove her wrong by giving an example of a product of two functions that cannot be integrated using integration by parts.

70. Complete the following sentence in words: The integral of $u\,v$ is the first times the integral of the second minus the integral of ______.

71. Give an example of an integral that can be computed in two ways: by substitution or integration by parts.

72. Give an example of an integral that can be computed by substitution but not by integration by parts. (You need not compute the integral.)

In Exercises 73–80, indicate whether the given integral calls for integration by parts or substitution.

73. $\int (6x - 1)e^{3x^2 - x}\,dx$

74. $\int \frac{x^2 - 3x + 1}{e^{2x-3}}\,dx$

[8] Source for data: Energy Information Administration/Pemex (http://www.eia.doe.gov).

[9] Source for data: BP Statistical Review of World Energy (www.bp.com/statisticalreview).

[10] Source for data: Energy Information Administration/Pemex (http://www.eia.doe.gov).

[11] Source for data: BP Statistical Review of World Energy (www.bp.com/statisticalreview).

75. $\int (3x^2 - x)e^{6x-1}\,dx$

76. $\int \frac{2x-3}{e^{x^2-3x+1}}\,dx$

77. $\int \frac{1}{(x+1)\ln(x+1)}\,dx$

78. $\int \frac{\ln(x+1)}{x+1}\,dx$

79. $\int \ln(x^2)\,dx$

80. $\int (x+1)\ln(x+1)\,dx$

81. ▼ If $p(x)$ is a polynomial of degree n and $f(x)$ is some function of x, how many times do we generally have to integrate $f(x)$ to compute $\int p(x)f(x)\,dx$?

82. ▼ Use integration by parts to show that $\int(\ln x)^2\,dx = x(\ln x)^2 - 2x\ln x + 2x + C$.

83. ◆ ***Hermite's Identity*** If $f(x)$ is a polynomial of degree n, show that

$$\int_0^b f(x)e^{-x}\,dx = F(0) - F(b)e^{-b}$$

where $F(x) = f(x) + f'(x) + f''(x) + \cdots + f^{(n)}(x)$. (This is the sum of f and all of its derivatives.)

84. ◆ Write down a formula similar to Hermite's identity for $\int_0^b f(x)e^x\,dx$ when $f(x)$ is a polynomial of degree n.

7.2 Area Between Two Curves and Applications

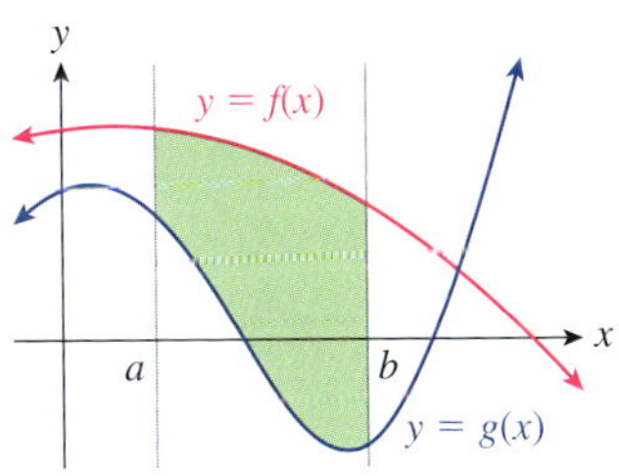

Figure 1

As we saw in the preceding chapter, we can use the definite integral to calculate the area between the graph of a function and the x-axis. With only a little more work, we can use it to calculate the area between two graphs. Figure 1 shows the graphs of two functions, $f(x)$ and $g(x)$, with $f(x) \geq g(x)$ for every x in the interval $[a, b]$.

To find the shaded area between the graphs of the two functions, we use the following formula.

Area Between Two Graphs

If $f(x) \geq g(x)$ for all x in $[a, b]$ (so that the graph of f does not move below that of g), then the area of the region between the graphs of f and g and between $x = a$ and $x = b$ is given by

$$A = \int_a^b [f(x) - g(x)]\,dx.$$

Caution If the graphs of f and g cross in the interval, the above formula does not hold; for instance, if $f(x) = x$ and $g(x) = -x$ then the total area shown in the figure is 2 square units, whereas $\int_{-1}^{1}[f(x) - g(x)]\,dx = 0$.

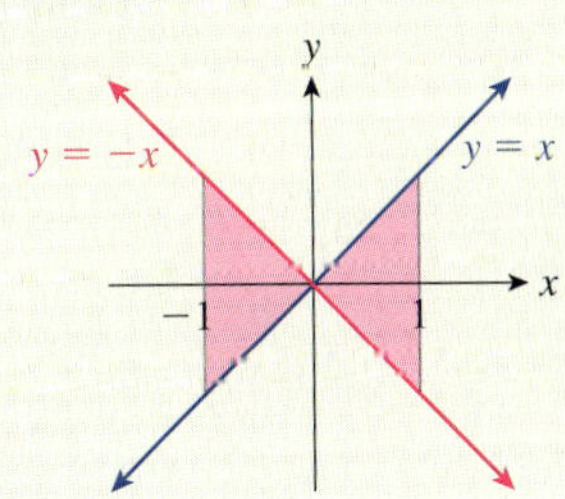

Let's look at an example and then discuss why the formula works.

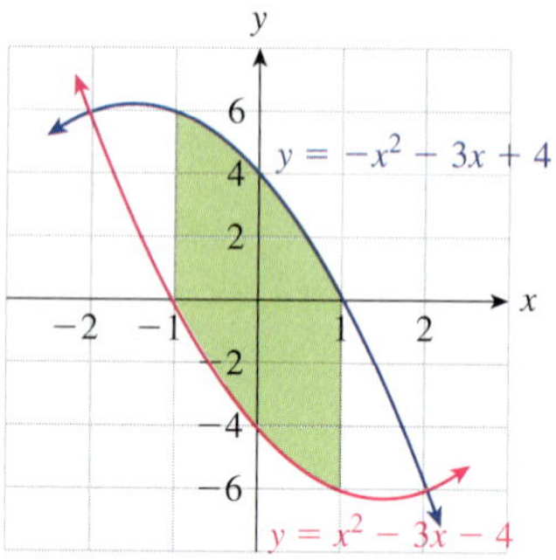

Figure 2

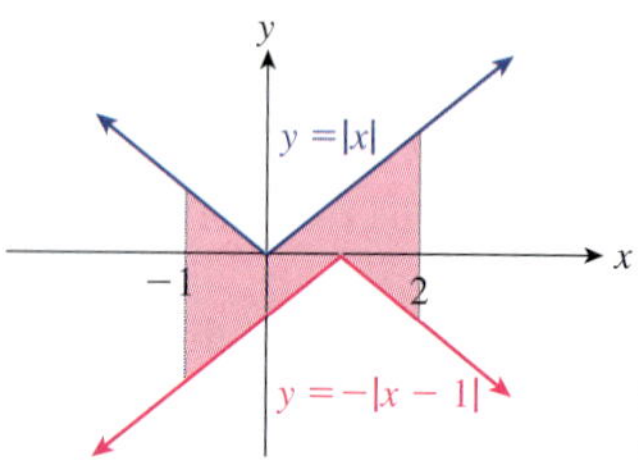

Figure 3

EXAMPLE 1 The Area Between Two Curves

Find the areas of the following regions:

a. Between $f(x) = -x^2 - 3x + 4$ and $g(x) = x^2 - 3x - 4$ and between $x = -1$ and $x = 1$

b. Between $f(x) = |x|$ and $g(x) = -|x - 1|$ over $[-1, 2]$

Solution

a. The area in question is shown in Figure 2. Because the graph of f lies above the graph of g in the interval $[-1, 1]$, we have $f(x) \geq g(x)$ for all x in $[-1, 1]$. Therefore, we can use the formula given above and calculate the area as follows:

$$\begin{aligned} A &= \int_{-1}^{1} [f(x) - g(x)]\, dx \\ &= \int_{-1}^{1} [(-x^2 - 3x + 4) - (x^2 - 3x - 4)]\, dx \\ &= \int_{-1}^{1} (8 - 2x^2)\, dx \\ &= \left[8x - \frac{2}{3}x^3\right]_{-1}^{1} \\ &= \frac{44}{3} \end{aligned}$$

b. The given area can be broken up into triangles and rectangles (see Figure 3), but we already know a formula for the antiderivative of $|ax + b|$ for constants a and b, so we can use calculus instead:

$$\begin{aligned} A &= \int_{-1}^{2} [f(x) - g(x)]\, dx \\ &= \int_{-1}^{2} [|x| - (-|x - 1|)]\, dx \\ &= \int_{-1}^{2} [|x| + |x - 1|]\, dx \\ &= \frac{1}{2}\left[x|x| + (x - 1)|x - 1|\right]_{-1}^{2} \\ &= \frac{1}{2}[(4 + 1) - (-1 - 4)] \\ &= \frac{1}{2}(10) = 5 \end{aligned}$$

$\int |ax + b|\, dx = \frac{1}{2a}(ax + b)|ax + b| + C$

Q: *Why does the formula for the area between two curves work?*

A: Let's go back once again to the general case illustrated in Figure 1, where we were given two functions f and g with $f(x) \geq g(x)$ for every x in the interval $[a, b]$. To avoid complicating the argument by the fact that the graph of g, or f, or both, may dip below the x-axis in the interval $[a, b]$ (as occurs in Figure 1 and also in Example 1), we shift both graphs vertically upward by adding a big enough constant M to lift them both above the x-axis in the interval $[a, b]$, as shown in Figure 4.

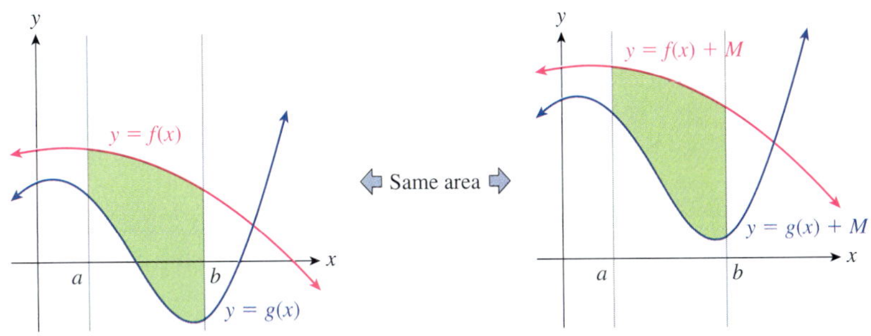

Figure 4

As the figure illustrates, the area of the region between the graphs is not affected, so we will calculate the area of the region shown on the right of Figure 4. That calculation is shown in Figure 5.

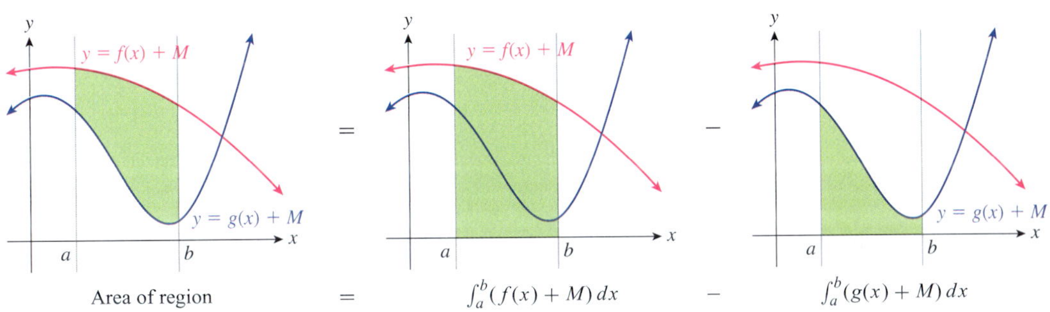

Figure 5

From the figure, the area we want is

$$\int_a^b (f(x) + M)\,dx - \int_a^b (g(x) + M)\,dx = \int_a^b [(f(x) + M) - (g(x) + M)]\,dx$$

$$= \int_a^b [f(x) - g(x)]\,dx,$$

whioh io tho formula wo gavo originally.

So far, we've been assuming that $f(x) \geq g(x)$, so that the graph of f never dips below the graph of g and so the graphs cannot cross (although they can touch). Example 2 shows how we compute the area between graphs that *do* cross.

EXAMPLE 2 Regions Enclosed by Crossing Curves

Find the area of the region between $y = 3x^2$ and $y = 1 - x^2$ and between $x = 0$ and $x = 1$.

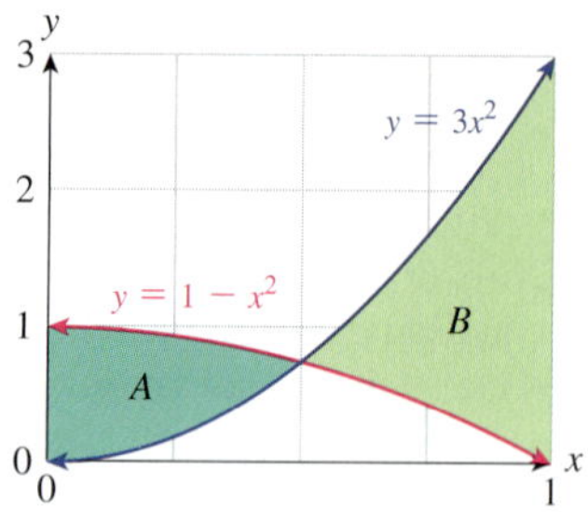

Figure 6

Solution The area we wish to calculate is shown in Figure 6. From the figure, we can see that neither graph lies above the other over the whole interval. To get around this, we break the area into the two pieces on either side of the point at which the graphs cross and then compute each area separately. To do this, we need to know exactly where that crossing point is. The crossing point is where $3x^2 = 1 - x^2$, so we solve for x:

$$\begin{aligned} 3x^2 &= 1 - x^2 \\ 4x^2 &= 1 \\ x^2 &= \frac{1}{4} \\ x &= \pm\frac{1}{2}. \end{aligned}$$

Because we are interested only in the interval [0, 1], the crossing point we're interested in is at $x = 1/2$.

Now, to compute the areas A and B, we need to know which graph is on top in each of these areas. We can see that from the figure, but what if the functions were more complicated and we could not easily draw the graphs? To be sure, we can test the values of the two functions at some point in each region. But we really need not worry. If we make the wrong choice for the top function, the integral will yield the negative of the area (why?), so we can simply take the absolute value of the integral to get the area of the region in question. For this example, we have

$$\begin{aligned} A &= \int_0^{1/2} [(1 - x^2) - 3x^2]\,dx = \int_0^{1/2} (1 - 4x^2)\,dx \\ &= \left[x - \frac{4x^3}{3}\right]_0^{1/2} \\ &= \left(\frac{1}{2} - \frac{1}{6}\right) - (0 - 0) = \frac{1}{3} \end{aligned}$$

and

$$\begin{aligned} B &= \int_{1/2}^{1} [3x^2 - (1 - x^2)]\,dx = \int_{1/2}^{1} (4x^2 - 1)\,dx \\ &= \left[\frac{4x^3}{3} - x\right]_{1/2}^{1} \\ &= \left(\frac{4}{3} - 1\right) - \left(\frac{1}{6} - \frac{1}{2}\right) = \frac{2}{3}. \end{aligned}$$

This gives a total area of $A + B = \frac{1}{3} + \frac{2}{3} = 1$.

Before we go on... What would have happened in Example 2 if we had not broken the area into two pieces but had just calculated the integral of the difference of the two functions? We would have calculated

$$\int_0^1 [(1 - x^2) - 3x^2]\,dx = \int_0^1 [1 - 4x^2]\,dx = \left[x - \frac{4x^3}{3}\right]_0^1 = -\frac{1}{3},$$

which is not even close to the right answer. What this integral calculated was actually $A - B$ rather than $A + B$. Why? ■

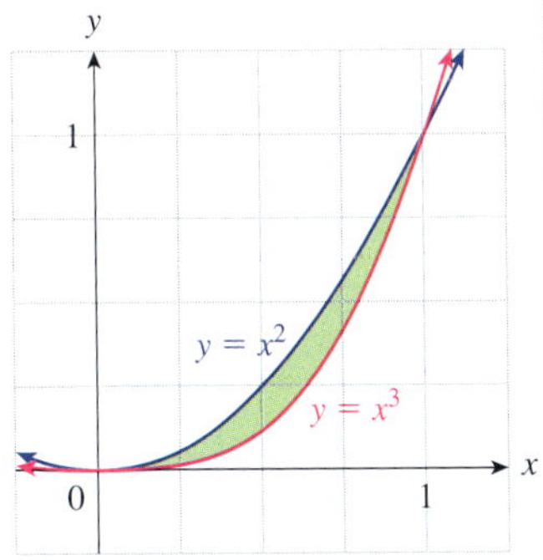

Figure 7

EXAMPLE 3 The Area Enclosed by Two Curves

Find the area enclosed by $y = x^2$ and $y = x^3$.

Solution This example has a new wrinkle: we are not told what interval to use for x. However, if we look at the graph in Figure 7, we see that the question can have only one meaning.

We are being asked to find the area of the shaded sliver, which is the only region that is actually *enclosed* by the two graphs. This sliver is bounded on either side by the two points where the graphs cross, so our first task is to find those points. They are the points where $x^2 = x^3$, so we solve for x:

$$\begin{aligned} x^2 &= x^3 \\ x^3 - x^2 &= 0 \\ x^2(x-1) &= 0 \\ x = 0 \quad &\text{or} \quad x = 1 \end{aligned}$$

Thus, we must integrate over the interval [0, 1]. Although we see from the diagram (or by substituting $x = 1/2$) that the graph of $y = x^2$ is above that of $y = x^3$, if we didn't notice that we might calculate

$$\int_0^1 (x^3 - x^2)\,dx = \left[\frac{x^4}{4} - \frac{x^3}{3}\right]_0^1 = -\frac{1}{12}.$$

This tells us that the required area is $1/12$ square units and also that we had our integral reversed. Had we calculated $\int_0^1 (x^2 - x^3)\,dx$ instead, we would have found the correct answer, $1/12$.

We can summarize the procedure we used in the preceding two examples.

Finding the Area Between the Graphs of $f(x)$ and $g(x)$

1. Find all points of intersection by solving $f(x) = g(x)$ for x. This either determines the interval over which you will integrate or breaks up a given interval into regions between the intersection points.
2. Determine the area of each region you found by integrating the difference of the larger and the smaller function. (If you accidentally take the smaller minus the larger, the integral will give the negative of the area, so just take the absolute value.)
3. Add together the areas you found in step 2 to get the total area.

Q: *Is there any quick and easy method to find the area between two graphs without having to find all points of intersection? What if it is hard or impossible to find out where the curves intersect?*

A: We can use technology to give the approximate area between two graphs. First recall that, if $f(x) \geq g(x)$ for all x in $[a, b]$, then the area between their graphs over $[a, b]$ is given by $\int_a^b [f(x) - g(x)]\,dx$, whereas if $g(x) \geq f(x)$, the area is given by $\int_a^b [g(x) - f(x)]\,dx$. Notice that both expressions are equal to

$$\int_a^b |f(x) - g(x)|\,dx$$

telling us that we can use this same formula in both cases.

Area Between Two Graphs: Approximation using Technology

The area of the region between the graphs of f and g and between $x = a$ and $x = b$ is given by

$$A = \int_a^b |f(x) - g(x)|\, dx.$$

Quick Example

To approximate the area of the region between $y = 3x^2$ and $y = 1 - x^2$ and between $x = 0$ and $x = 1$ we calculated in Example 2, use technology to compute

$$\int_0^1 |3x^2 - (1 - x^2)|\, dx = 1.$$

TI-83/84 Plus `fnInt(abs(3x^2-(1-x^2)),X,0,1)`

Web Site Online Utilities → Numerical Integration Utility

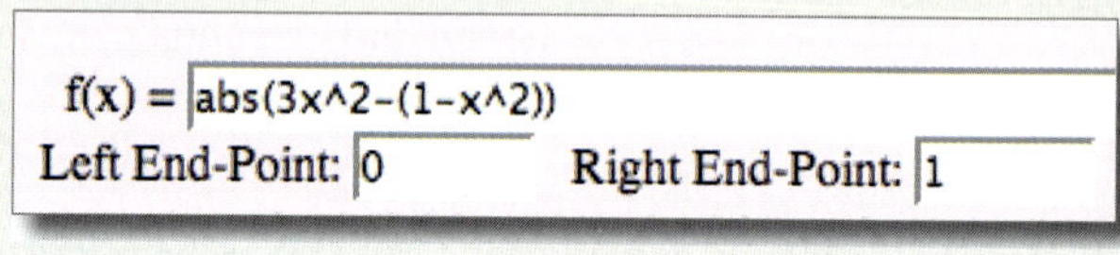

7.2 EXERCISES

▼ more advanced ◆ challenging

T indicates exercises that should be solved using technology

Find the area of the shaded region in Exercises 1–8. (We suggest you use technology to check your answers.)

1.

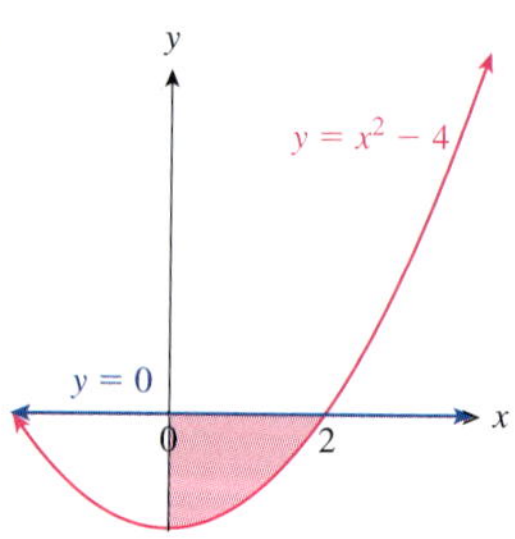

2.

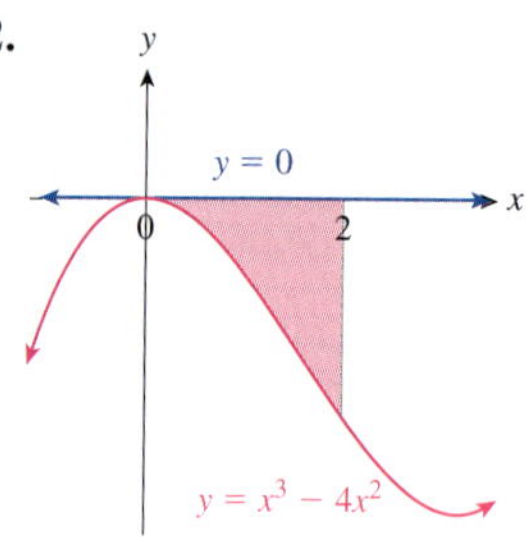

3.

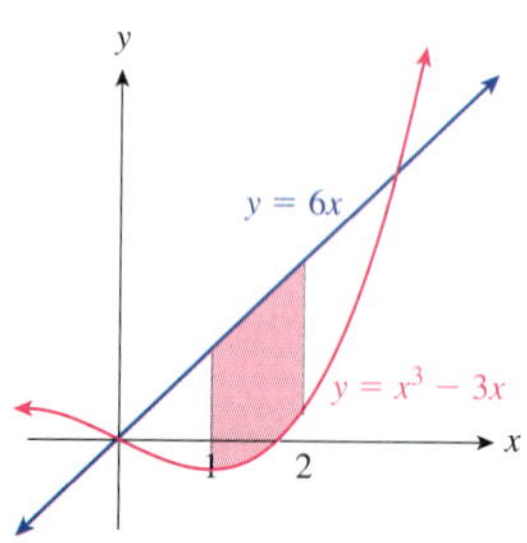

4.

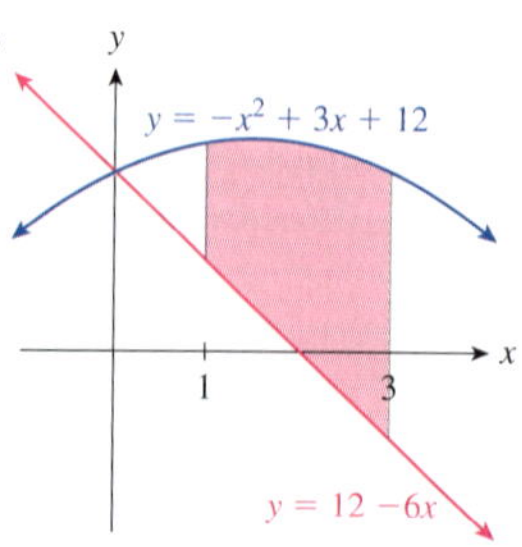

5.

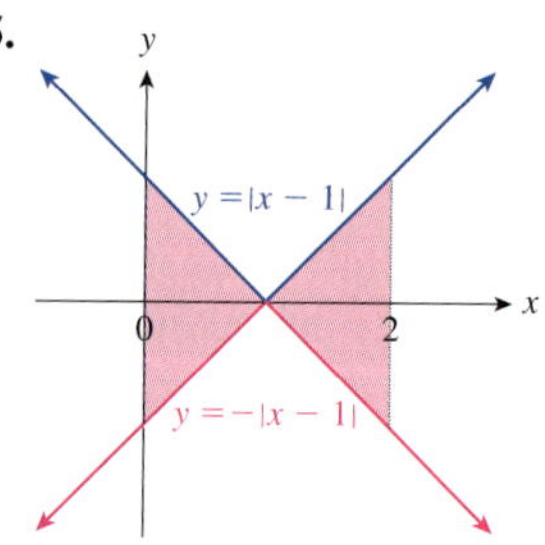

6.

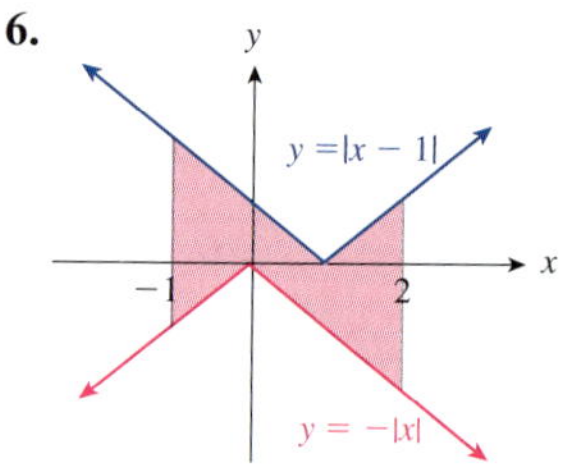

7. ▼

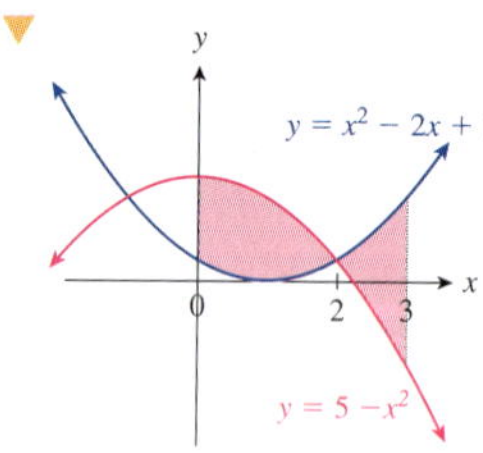

8. ▼

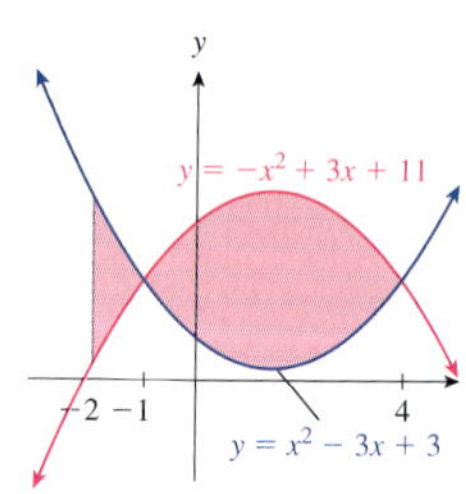

Find the area of the indicated region in Exercises 9–42. We suggest you graph the curves to check whether one is above the other or whether they cross, and that you use technology to check your answers.

9. Between $y = x^2$ and $y = -1$ for x in $[-1, 1]$ HINT [See Example 1.]

10. Between $y = x^3$ and $y = -1$ for x in $[-1, 1]$ HINT [See Example 1.]

11. Between $y = -x$ and $y = x$ for x in $[0, 2]$

12. Between $y = -x$ and $y = x/2$ for x in $[0, 2]$

13. Between $y = |x|$ and $y = x^2$ in $[-1, 1]$

14. Between $y = -|x|$ and $y = x^2 - 2$ in $[-1, 1]$

15. Between $y = x$ and $y = x^2$ for x in $[-1, 1]$ HINT [See Example 2.]

16. Between $y = x$ and $y = x^3$ for x in $[-1, 1]$ HINT [See Example 2.]

17. Between $y = x^2 - 2x$ and $y = -x^2 + 4x - 4$ for x in $[0, 2]$

18. Between $y = x^2 - 4x + 2$ and $y = -x^2 + 4x - 4$ for x in $[0, 3]$

19. Between $y = 2x^2 + 10x - 5$ and $y = -x^2 + 4x + 4$ for x in $[-3, 2]$

20. Between $y = 2x^2 + 7x - 2$ and $y = -x^2 + 4x + 4$ for x in $[-2, 2]$

21. Between $y = e^x$ and $y = x$ for x in $[0, 1]$

22. Between $y = e^{-x}$ and $y = -x$ for x in $[0, 1]$

23. Between $y = (x - 1)^2$ and $y = -(x - 1)^2$ for x in $[0, 1]$

24. Between $y = x^2(x^3 + 1)^{10}$ and $y = -x(x^2 + 1)^{10}$ for x in $[0, 1]$

25. Enclosed by $y = x$ and $y = x^4$ HINT [See Example 3.]

26. Enclosed by $y = x$ and $y = -x^4$ HINT [See Example 3.]

27. Enclosed by $y = x^3$ and $y = x^4$

28. Enclosed by $y = x$ and $y = x^3$

29. Enclosed by $y = x^2$ and $y = x^4$

30. Enclosed by $y = x^4 - x^2$ and $y = x^2 - x^4$

31. Enclosed by $y = x^2 - 2x$ and $y = -x^2 + 4x - 4$

32. Enclosed by $y = x^2 - 4x + 2$ and $y = -x^2 + 4x - 4$

33. Enclosed by $y = 2x^2 + 10x - 5$ and $y = -x^2 + 4x + 4$

34. Enclosed by $y = 2x^2 + 7x - 2$ and $y = -x^2 + 4x + 4$

35. Enclosed by $y = e^x$, $y = 2$, and the y-axis

36. Enclosed by $y = e^{-x}$, $y = 3$, and the y-axis

37. Enclosed by $y = \ln x$, $y = 2 - \ln x$, and $x = 4$

38. Enclosed by $y = \ln x$, $y = 1 - \ln x$, and $x = 4$

39. T Enclosed by $y = e^x$, $y = 2x + 1$, $x = -1$, and $x = 1$ (Round answer to four significant digits.) HINT [See Quick Example page 532.]

40. T Enclosed by $y = 2^x$, $y = x + 2$, $x = -2$, and $x = 2$ (Round answer to four significant digits.) HINT [See Quick Example page 532.]

41. T Enclosed by $y = \ln x$ and $y = \frac{x}{2} - \frac{1}{2}$ (Round answer to four significant digits.) [First use technology to determine approximately where the graphs cross.]

42. T Enclosed by $y = \ln x$ and $y = x - 2$ (Round answer to four significant digits.) [First use technology to determine approximately where the graphs cross.]

APPLICATIONS

43. ***Revenue and Cost*** Suppose your daily revenue from selling used DVDs is

$$R(t) = 100 + 10t \quad (0 \le t \le 5)$$

dollars per day, where t represents days from the beginning of the week, while your daily costs are

$$C(t) = 90 + 5t \quad (0 \le t \le 5)$$

dollars per day. Find the area between the graphs of $R(t)$ and $C(t)$ for $0 \le t \le 5$. What does your answer represent?

44. ***Income and Expenses*** Suppose your annual income is

$$I(t) = 50{,}000 + 2{,}000t \quad (0 \le t \le 3)$$

dollars per year, where t represents the number of years since you began your job, while your annual expenses are

$$E(t) = 45{,}000 + 1{,}500t \quad (0 \le t \le 3)$$

dollars per year. Find the area between the graphs of $I(t)$ and $E(t)$ for $0 \le t \le 3$. What does your answer represent?

45. ***Housing*** The total number of housing starts in the United States during 2000–2008 was approximately

$$h(t) = -30t^2 + 240t + 800 \text{ thousand homes per year} \quad (0 \le t \le 8)$$

where t is time in years since the start of 2000.[12] During that time, the number of housing starts for sale purposes in the United States was approximately

$$s(t) = -33t^2 + 240t + 700 \text{ thousand homes per year.} \quad (0 \le t \le 8)$$

Compute the area shown in the graph, and interpret the answer.

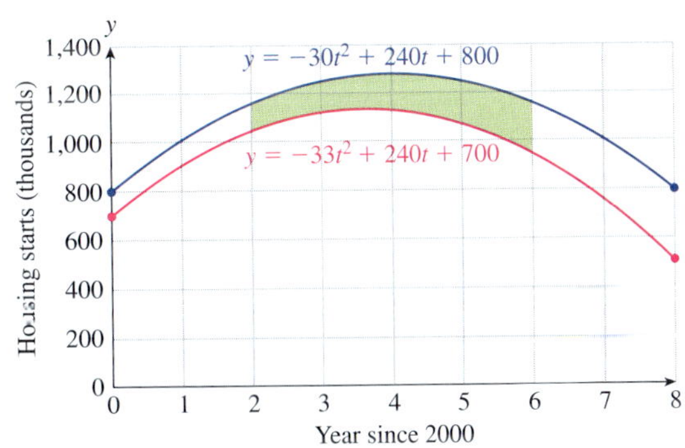

46. ***Oil Production in Mexico: Pemex*** The rate of oil production by **Pemex**, Mexico's national oil company, can be approximated by

$$q(t) = -8t^2 + 70t + 1{,}000 \text{ million barrels per year} \quad (0 \le t \le 9)$$

[12] Source for data: U.S. Census Bureau (www.census.gov).

where t is time in years since the start of 2000.[13] During that time, Mexico exported oil to the United States at a rate of

$$r(t) = -5t^2 + 40t + 500 \text{ million barrels per year.} \quad (0 \le t \le 8)$$

Compute the area shown in the graph, and interpret the answer.

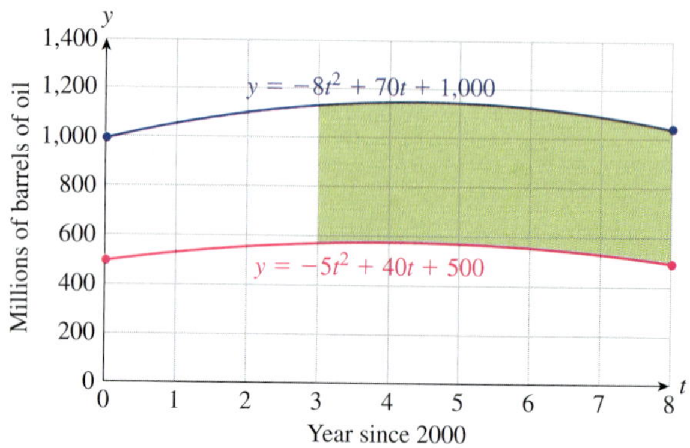

MySpace ***and*** **Facebook** *Exercises 47 and 48 are based on the following models, which show: the rate at which new members joined* **Facebook** *and* **MySpace**:

Facebook:[14] $f(t) = 12t^2 - 20t + 10$ *million members per year* $(0 < t < 3.5)$

MySpace:[15] $m(t) = 10.5t^2 + 25t + 18.5$ *million members per year* $(0 \le t \le 3.5)$

(t is time in years since the start of 2005.)

47. a. Use a graph to determine which of the two network sites was experiencing a larger influx of new members from June 2005 to the start of 2007. Use an integral to estimate how many more people joined that network site than joined its competitor during that period. (Round your answer to the nearest million.)

b. To what area does the integral used in part (a) correspond?

48. a. Use a graph to determine which of the two network sites was experiencing a larger influx of new members from the start of 2007 to June 2008. Use an integral to estimate how many more people joined that network site than joined its competitor during that period. (Round your answer to the nearest million.)

b. To what area does the integral used in part (a) correspond?

Big Brother *Exercises 49 and 50 are based on the following models, which show the total number of wiretaps authorized per year by all state and federal courts in the United States and the number authorized by state courts.*

State and federal courts: $a(t) = 820e^{0.051t} \quad (0 \le t \le 20)$

State courts: $s(t) = 440e^{0.06t} \quad (0 \le t \le 20)$

(t is time in years since the start of 1990.)[16]

49. ▼ Estimate the area between the graphs of the two functions over the interval $[0, t]$. Interpret your answer.

50. ▼ Estimate the area between the graphs of the two functions over the interval $[t, 20]$. Interpret your answer.

COMMUNICATION AND REASONING EXERCISES

51. If f and g are continuous functions with $\int_a^b [f(x) - g(x)]\,dx = 0$, it follows that the area between the graphs of f and g is zero, right?

52. You know that f and g are continuous and their graphs do not cross on the interval $[a, b]$, so you calculate $\int_a^b [f(x) - g(x)]\,dx$ and find that the answer is -40. Why is it negative, and what is the area between the curves?

53. The following graph shows annual U.S. exports and imports for the period 1960–2007.[17]

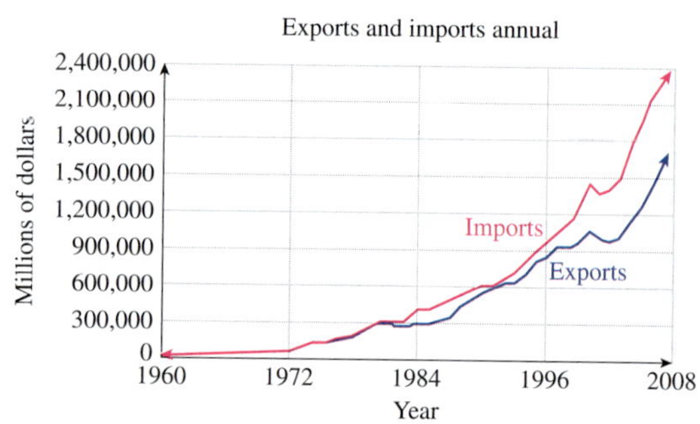

What does the area between the export and import curves represent?

54. The following graph shows a fictitious country's monthly exports and imports for the period 1997–2001.

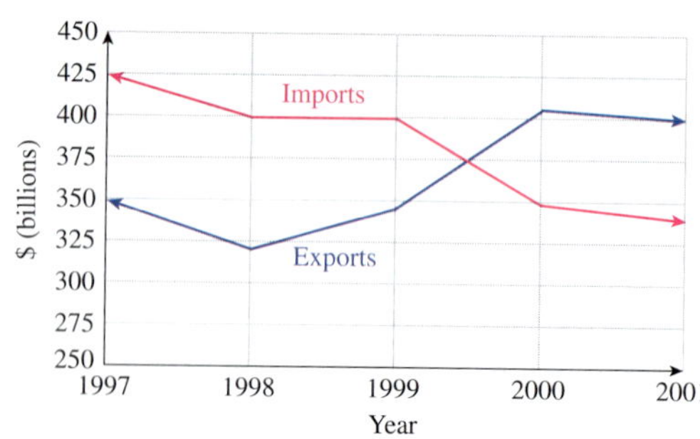

[13] Source for data: Energy Information Administration/Pemex (http://www.eia.doe.gov).

[14] Sources for data: Some data are interpolated (www.facebook.com/, www.insidehighered.com).

[15] Source for data: www.swivel.com/data_sets.

[16] Source for data: 2007 Wiretap Report, Administrative Office of the United States Courts (www.uscourts.gov/wiretap07/2007WTText.pdf).

[17] Source: Data 360 (www.data360.org).

What does the total area enclosed by the export and import curves represent, and what does the definite integral of the difference, Exports − Imports, represent?

55. ▼ The following graph shows the daily revenue and cost in your Adopt-a-Chia operation t days from its inception.

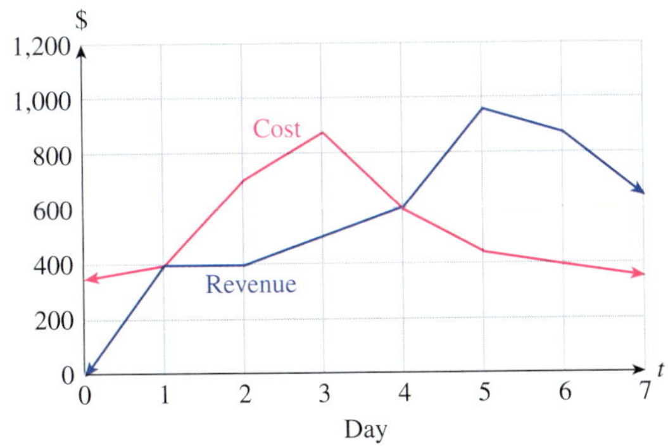

Multiple choice: The area between the cost and revenue curves represents:

(A) The accumulated loss through day 4 plus the accumulated profit for days 5 through 7
(B) The accumulated profit for the week
(C) The accumulated loss for the week
(D) The accumulated cost through day 4 plus the accumulated revenue for days 5 through 7

56. ▼ The following graph shows daily orders and inventory (stock on hand) for your Jackson Pollock paint-by-number sets t days into last week.

a. Multiple choice: Which is greatest?

(A) $\int_0^7 (\text{Orders} - \text{Inventory})\, dt$

(B) $\int_0^7 (\text{Inventory} - \text{Orders})\, dt$

(C) The area between the Orders and Inventory curves.

b. Multiple choice: The answer to part (a) measures

(A) The accumulated gap between orders and inventory
(B) The accumulated surplus through day 3 minus the accumulated shortage for days 3 through 5 plus the accumulated surplus through days 6 through 7
(C) The total net surplus
(D) The total net loss

57. ▼ What is wrong with the following claim: "I purchased Novartis AG shares for $50 at the beginning of March 2008. My total profit per share from this investment from March through August is represented by the area between the stock price curve and the purchase price curve as shown on the following graph, where t is time in months since March 1, 2008."[18]

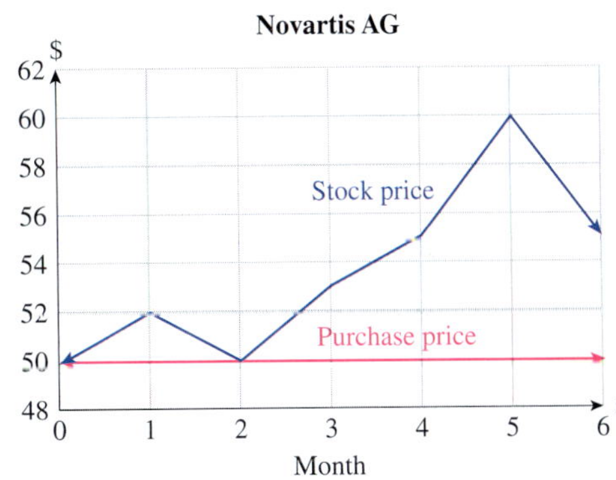

58. ▼ Your pharmaceutical company monitors the amount of medication in successive batches of 100 mg tetracycline capsules, and obtains the following graph.

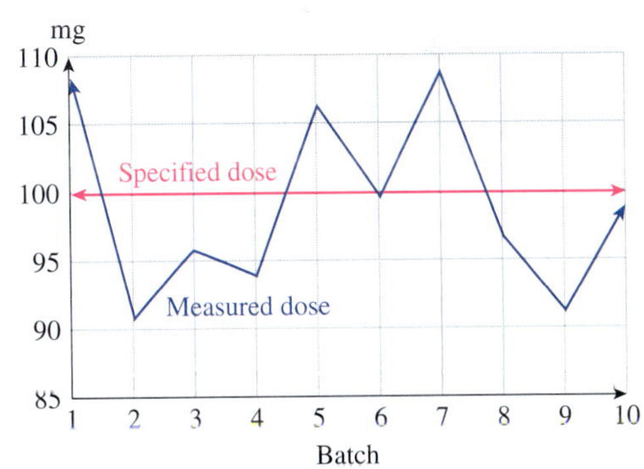

Your production manager claims that the batches of tetracycline conform to the exact dosage requirement because half of the area between the graphs is above the Specified dose line and half is below it. Comment on this reasoning.

[18] Source: www.finance.google.com.

7.3 Averages and Moving Averages

Averages

To find the average of, say, 20 numbers, we simply add them up and divide by 20. More generally, if we want to find the **average**, or **mean**, of the n numbers $y_1, y_2, y_3, \ldots y_n$, we add them up and divide by n. We write this average as $\bar{y}$ ("y-bar").

Average, or Mean, of a Collection of Values

$$\bar{y} = \frac{y_1 + y_2 + \cdots + y_n}{n}$$

Quick Example

The average of $\{0, 2, -1, 5\}$ is $\bar{y} = \dfrac{0+2-1+5}{4} = \dfrac{6}{4} = 1.5.$

But, we also use the word *average* in other senses. For example, we speak of the average speed of a car during a trip.

Vladimir Pcholkin/Getty Images

EXAMPLE 1 Average Speed

Over the course of 2 hours, my speed varied from 50 miles per hour to 60 miles per hour, following the function $v(t) = 50 + 2.5t^2$, $0 \le t \le 2$. What was my average speed over those two hours?

Solution Recall that average speed is simply the total distance traveled divided by the time it took. Recall, also, that we can find the distance traveled by integrating the speed:

$$\begin{aligned}\text{Distance traveled} &= \int_0^2 v(t)\,dt \\ &= \int_0^2 (50 + 2.5t^2)\,dt \\ &= \left[50t + \frac{2.5}{3}t^3\right]_0^2 \\ &= 100 + \frac{20}{3} \\ &\approx 106.67 \text{ miles}\end{aligned}$$

It took 2 hours to travel this distance, so the average speed was

$$\text{Average speed} \approx \frac{106.67}{2} \approx 53.3 \text{ mph}$$

In general, if we travel with velocity $v(t)$ from time $t = a$ to time $t = b$, we will travel a distance of $\int_a^b v(t)\,dt$ in time $b - a$, which gives an average velocity of

$$\text{Average velocity} = \frac{1}{b-a}\int_a^b v(t)\,dt.$$

Thinking of this calculation as finding the average value of the velocity function, we generalize and make the following definition.

Average Value of a Function

The **average**, or **mean**, of a function $f(x)$ on an interval $[a, b]$ is

$$\bar{f} = \frac{1}{b-a}\int_a^b f(x)\,dx.$$

Quick Example

The average of $f(x) = x$ on $[1, 5]$ is

$$\bar{f} = \frac{1}{b-a}\int_a^b f(x)\,dx$$

$$= \frac{1}{5-1}\int_1^5 x\,dx$$

$$= \frac{1}{4}\left[\frac{x^2}{2}\right]_1^5 = \frac{1}{4}\left(\frac{25}{2} - \frac{1}{2}\right) = 3.$$

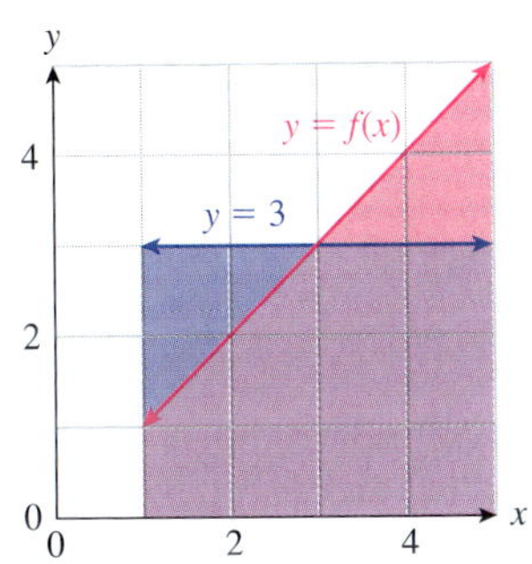

Figure 8

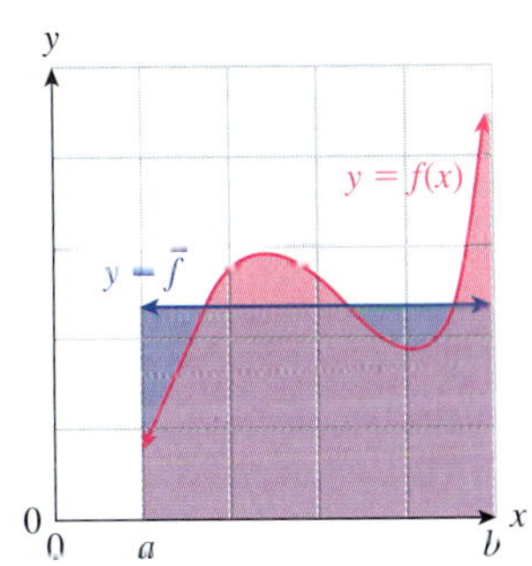

Figure 9

Interpreting the Average of a Function Geometrically

The average of a function has a geometric interpretation. Referring to the Quick Example above, we can compare the graph of $y = f(x)$ with the graph of $y = 3$, both over the interval $[1, 5]$ (Figure 8). We can find the area under the graph of $f(x) = x$ by geometry or by calculus; it is 12. The area in the rectangle under $y = 3$ is also 12.

In general, the average $\bar{f}$ of a positive function over the interval $[a, b]$ gives the height of the rectangle over the interval $[a, b]$ that has the same area as the area under the graph of $f(x)$ as illustrated in Figure 9. The equality of these areas follows from the equation

$$(b-a)\bar{f} = \int_a^b f(x)\,dx.$$

EXAMPLE 2 Average Balance

A savings account at the People's Credit Union pays 3% interest, compounded continuously, and at the end of the year you get a bonus of 1% of the average balance in the account during the year. If you deposit \$10,000 at the beginning of the year, how much interest and how large a bonus will you get?

Solution We can use the continuous compound interest formula to calculate the amount of money you have in the account at time t:

$$A(t) = 10{,}000e^{0.03t}$$

where t is measured in years. At the end of 1 year, the account will have

$$A(1) = \$10{,}304.55$$

so you will have earned \$304.55 interest. To compute the bonus, we need to find the average amount in the account, which is the average of $A(t)$ over the interval [0, 1]. Thus,

$$\begin{aligned}\bar{A} &= \frac{1}{b-a}\int_a^b A(t)\,dt \\ &= \frac{1}{1-0}\int_0^1 10{,}000e^{0.03t}\,dt = \frac{10{,}000}{0.03}\left[e^{0.03t}\right]_0^1 \\ &\approx \$10{,}151.51\end{aligned}$$

The bonus is 1% of this, or \$101.52.

Before we go on... The 1% bonus in Example 2 was one-third of the total interest. Why did this happen? What fraction of the total interest would the bonus be if the interest rate was 4%, 5%, or 10%? ■

Moving Averages

Suppose you follow the performance of a company's stock by recording the daily closing prices. The graph of these prices may seem jagged or "jittery" due to random day-to-day fluctuations. To see any trends, you would like a way to "smooth out" these data. The **moving average** is one common way to do that.

EXAMPLE 3 Stock Prices

The following table shows Colossal Conglomerate's closing stock prices for 20 consecutive trading days.

Day	1	2	3	4	5	6	7	8	9	10
Price	20	22	21	24	24	23	25	26	20	24
Day	11	12	13	14	15	16	17	18	19	20
Price	26	26	25	27	28	27	29	27	25	24

Plot these prices and the 5-day moving average.

Solution The 5-day moving average is the average of each day's price together with the prices of the preceding 4 days. We can compute the 5-day moving averages starting on the fifth day. We get these numbers:

Day	1	2	3	4	5	6	7	8	9	10
Moving Average					22.2	22.8	23.4	24.4	23.6	23.6
Day	11	12	13	14	15	16	17	18	19	20
Moving Average	24.2	24.4	24.2	25.6	26.4	26.6	27.2	27.6	27.2	26.4

using Technology

See the Technology Guides at the end of the chapter to find out how to tabulate and graph moving averages on a TI-83/84 Plus or Excel.

Here is an outline:

TI-83/84 Plus

STAT EDIT; enter the days in L_1, and the prices in L_2.

Home screen:

```
seq((L2(X)+L2(X-1)
+L2(X-2)+L2(X-3)
+L2(X-4))/5,X,5,20)
→L3
```

[More details on page 581.]

Excel

Day data in A2–A21

Price data in B2–B21

Enter =AVERAGE(B2:B6) in C6, and copy down to C21.

Graph: Highlight columns A–C and insert a Scatter Chart.

[More details on page 582.]

The closing stock prices and moving averages are plotted in Figure 10.

Figure 10

As you can see, the moving average is less volatile than the closing price. Because the moving average incorporates the stock's performance over 5 days at a time, a single day's fluctuation is smoothed out. Look at day 9 in particular. The moving average also tends to lag behind the actual performance because it takes past history into account. Look at the downturns at days 6 and 18 in particular.

The period of 5 days for a moving average, as used in Example 3, is arbitrary. Using a longer period of time would smooth the data more but increase the lag. For data used as economic indicators, such as housing prices or retail sales, it is common to compute the 4-quarter moving average to smooth out seasonal variations.

It is also sometimes useful to compute moving averages of continuous functions. We may want to do this if we use a mathematical model of a large collection of data. Also, some physical systems have the effect of converting an input function (an electrical signal, for example) into its moving average. By an ***n*-unit moving average** of a function $f(x)$ we mean the function $\bar{f}$ for which $\bar{f}(x)$ is the average of the value of $f(x)$ on $[x-n, x]$. Using the formula for the average of a function, we get the following formula.

n-Unit Moving Average of a Function

The n-unit moving average of a function f is

$$\bar{f}(x) = \frac{1}{n}\int_{x-n}^{x} f(t)\,dt.$$

Quick Example

The 2-unit moving average of $f(x) = x^2$ is

$$\bar{f}(x) = \frac{1}{2}\int_{x-2}^{x} t^2\,dt = \frac{1}{6}\left[t^3\right]_{x-2}^{x} = x^2 - 2x + \frac{4}{3}.$$

The graphs of $f(x)$ and $\bar{f}(x)$ are shown in Figure 11.

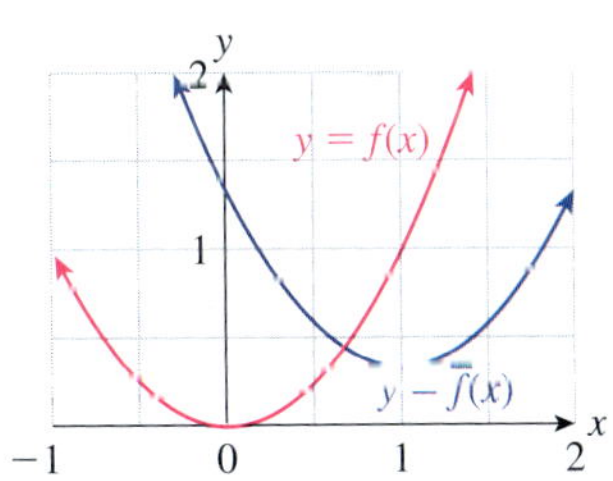

Figure 11

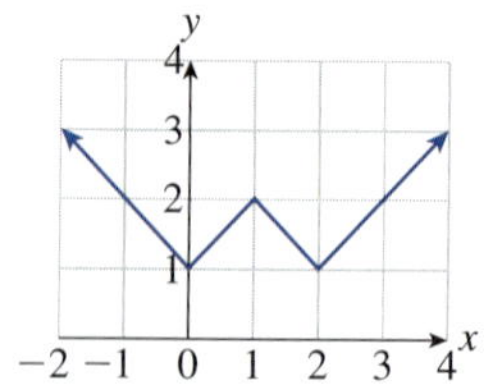

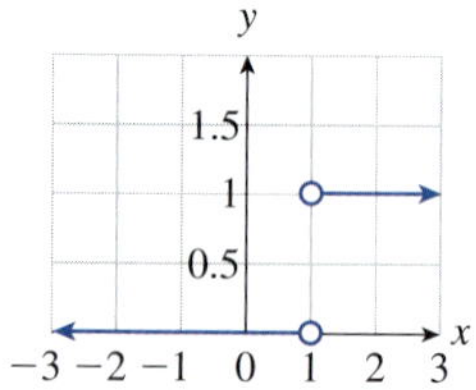

Figure 12

EXAMPLE 4 Moving Averages: Saw Tooth and Step Functions

Graph the following functions, and then compute and graph their 1-unit moving averages.

$$f(x) = |x| - |x-1| + |x-2| \qquad \text{Saw tooth}$$

$$g(x) = \frac{1}{2}\left[1 + \frac{|x-1|}{x-1}\right] \qquad \text{Unit step at } x = 1$$

Solution The graphs of f and g are shown in Figure 12. (Notice that the step function is not defined at $x = 1$. Most graphers will show the step function as an actual step by connecting the points (1, 0) and (1, 1) with a vertical line.)

The 1-step moving averages are:

$$\bar{f}(x) = \int_{x-1}^{x} f(t)\,dt = \int_{x-1}^{x} [|t| - |t-1| + |t-2|]\,dt$$

$$= \frac{1}{2}[t|t| - (t-1)|t-1| + (t-2)|t-2|]_{x-1}^{x}$$

$$= \frac{1}{2}([x|x| - (x-1)|x-1| + (x-2)|x-2|]$$
$$- [(x-1)|x-1| - (x-2)|x-2| + (x-3)|x-3|])$$

$$= \frac{1}{2}[x|x| - 2(x-1)|x-1| + 2(x-2)|x-2| - (x-3)|x-3|]$$

$$\bar{g}(x) = \int_{x-1}^{x} g(t)\,dt = \int_{x-1}^{x} \frac{1}{2}\left[1 + \frac{|t-1|}{t-1}\right]dt$$

$$= \frac{1}{2}[t + |t-1|]_{x-1}^{x}$$

$$= \frac{1}{2}[(x + |x-1|) - (x - 1 + |x-2|)]$$

$$= \frac{1}{2}[1 + |x-1| - |x-2|]$$

The graphs of $\bar{f}$ and $\bar{g}$ are shown in Figure 13.

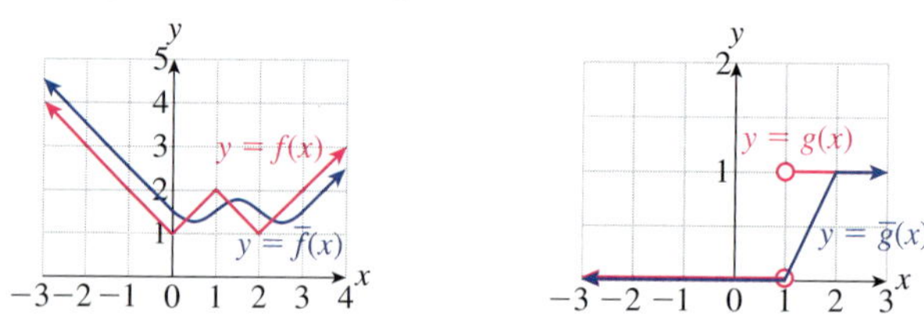

Figure 13

Notice how the graph of $\bar{f}$ smooths out the zig-zags of the sawtooth function.

using Technology

TI-83/84 Plus

We can graph the moving average of f in Example 4 on a TI-83/84 Plus as follows:

```
Y1=abs(X)-abs(X-1)
+abs(X-2)
Y2=fnInt(Y1(T),T,
X-1,X)
ZOOM 0
```

For g, change Y_1 to

```
Y1=.5*(1+abs(X-1)/
(X-1))
```

➡ **Before we go on...** Figure 14 shows the 2-unit moving average of f in Example 4,

$$\bar{f}(x) = \frac{1}{2}\int_{x-2}^{x} f(t)\,dt$$

$$= \frac{1}{4}(x|x| - (x-1)|x-1| + (x-3)|x-3| - (x-4)|x-4|)$$

Notice how the 2-point moving average has completely eliminated the zig-zags, illustrating how moving averages can be used to remove seasonal fluctuations in real-life situations. ■

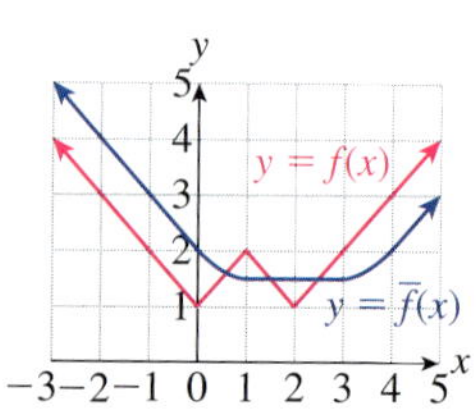

Figure 14

7.3 EXERCISES

▼ more advanced ◆ challenging

T indicates exercises that should be solved using technology

Find the averages of the functions in Exercises 1–8 over the given intervals. Plot each function and its average on the same graph (as in Figure 8). HINT [See Quick Example page 537.]

1. $f(x) = x^3$ over $[0, 2]$ **2.** $f(x) = x^3$ over $[-1, 1]$

3. $f(x) = x^3 - x$ over $[0, 2]$ **4.** $f(x) = x^3 - x$ over $[0, 1]$

5. $f(x) = e^{-x}$ over $[0, 2]$ **6.** $f(x) = e^x$ over $[-1, 1]$

7. $f(x) = |2x - 5|$ over $[0, 4]$

8. $f(x) = |-x + 2|$ over $[-1, 3]$

In Exercises 9 and 10, complete the given table with the values of the 3-unit moving average of the given function. HINT [See Example 3.]

9.

x	0	1	2	3	4	5	6	7
$r(x)$	3	5	10	3	2	5	6	7
$\bar{r}(x)$								

10.

x	0	1	2	3	4	5	6	7
$s(x)$	2	9	7	3	2	5	7	1
$\bar{s}(x)$								

In Exercises 11 and 12, some values of a function and its 3-unit moving average are given. Supply the missing information.

11.

x	0	1	2	3	4	5	6	7
$r(x)$	1	2			11		10	2
$\bar{r}(x)$			3	5		11		

12.

x	0	1	2	3	4	5	6	7
$s(x)$	1	5		1				
$\bar{s}(x)$			5		5	2	3	2

Calculate the 5-unit moving average of each function in Exercises 13–24. Plot each function and its moving average on the same graph, as in Example 4. (You may use graphing technology for these plots, but you should compute the moving averages analytically.) HINT [See Quick Example page 539, and Example 4.]

13. $f(x) = x^3$ **14.** $f(x) = x^3 - x$

15. $f(x) = x^{2/3}$ **16.** $f(x) = x^{2/3} + x$

17. $f(x) = e^{0.5x}$ **18.** $f(x) = e^{-0.02x}$

19. $f(x) = \sqrt{x}$ **20.** $f(x) = x^{1/3}$

21. $f(x) = 1 - \dfrac{|2x - 1|}{2x - 1}$ **22.** $f(x) = 2 + \dfrac{|3x + 1|}{3x + 1}$

23. ▼ $f(x) = 2 - |x + 1| + |x|$ [Do not simplify the answer.]

24. ▼ $f(x) = |2x + 1| - |2x| - 2$ [Do not simplify the answer.]

T *In Exercises 25–34, use graphing technology to plot the given functions together with their 3-unit moving averages.* HINT [See Technology Note for Example 4.]

25. $f(x) = \dfrac{10x}{1 + 5|x|}$ **26.** $f(x) = \dfrac{1}{1 + e^x}$

27. $f(x) = \ln(1 + x^2)$ **28.** $f(x) = e^{1-x^2}$

29. $f(x) = |x| - |x - 1| + |x - 2| - |x - 3| + |x - 4|$

30. $f(x) = |x| - 2|x - 1| + 2|x - 2| - 2|x - 3| + |x - 4|$

31. $f(x) = \dfrac{|x|}{x} - \dfrac{|x-1|}{x-1} + \dfrac{|x-2|}{x-2} - \dfrac{|x-3|}{x-3}$

32. $f(x) = \dfrac{|x|}{x} - 2\dfrac{|x-1|}{x-1} + 2\dfrac{|x-2|}{x-2} - \dfrac{|x-3|}{x-3}$

33. $f(x) = \dfrac{|x|}{x} + \dfrac{|x-1|}{x-1} + \dfrac{|x-2|}{x-2} + \dfrac{|x-3|}{x-3}$

34. $f(x) = 4 - \dfrac{|x|}{x} - \dfrac{|x-1|}{x-1} - \dfrac{|x-2|}{x-2} - \dfrac{|x-3|}{x-3}$

APPLICATIONS

35. ***Profits:* British Petroleum** The following table shows **BP**'s quarterly profits for a 5-quarter period beginning the second quarter of 2007:[19]

Quarter	2007 Q2	2007 Q3	2007 Q4	2008 Q1	2008 Q2
Profit ($ million)	7,441	4,478	4,504	7,583	9,588

What was the average quarterly profit earned by **BP** during the given period? HINT [See Quick Example page 536.]

36. ***Revenues:* British Petroleum** The following table shows **BP**'s quarterly revenues for a 5-quarter period beginning the second quarter of 2007:[20]

Quarter	2007 Q2	2007 Q3	2007 Q4	2008 Q1	2008 Q2
Revenue ($ million)	73,083	72,610	81,222	89,223	110,903

What was the average quarterly profit earned by **BP** during the given period? HINT [See Quick Example page 536.]

37. ***Television Advertising*** The cost, in millions of dollars, of a 30-second television ad during the Super Bowl in the years 2000 to 2007 can be approximated by

$$C(t) = 0.13t + 1.5 \text{ million dollars.} \quad (0 \le t \le 7)$$

[19] Source: British Petroleum (www.bp.com).

[20] Ibid.

(t is the number of years since 2000).[21] What was the average cost of a Super Bowl ad during the given period? HINT [See Example 1.]

38. ***Television Advertising*** The cost, in millions of dollars, of a 30-second television ad during the Super Bowl in the years 1990 to 1998 can be approximated by

$$C(t) = 0.08t + 0.6 \text{ million dollars.} \quad (0 \le t \le 8)$$

(t is the number of years since 1990).[22] What was the average cost of a Super Bowl ad during the given period? HINT [See Example 1.]

39. ***Membership:* Facebook** The number of new members joining **Facebook** each year can be modeled by

$$m(t) = 12t^2 - 20t + 10 \text{ million members per year} \quad (0 \le t \le 3.5)$$

where t is time in years since the start of 2005.[23] What was the average number of new members joining **Facebook** each year from the start of 2005 to the start of 2008?

40. ***Membership:* MySpace** The number of new members joining **Myspace** each year can be modeled by

$$m(t) = 10.5t^2 + 14t - 6 \text{ million members per year} \quad (0 \le t \le 3.5)$$

where t is time in years since the start of 2004.[24] What was the average number of new members joining **MySpace** each year from the start of 2004 to the start of 2007?

41. ***Freon Production*** Annual production of ozone-layer damaging Freon 22 (chlorodifluoromethane) in developing countries from 2000 to 2010 can be modeled by

$$F(t) = 97.2(1.20)^t \text{ tons.} \quad (0 \le t \le 10)$$

(t is the number of years since 2000).[25] What was the average annual production over the period shown? (Round your answer to the nearest ton.) HINT [See Example 2.]

42. ***Health Expenditures*** Annual expenditure on health in the United States from 1980 to 2010 can be modeled by

$$F(t) = 282(1.08)^t \text{ billion dollars.} \quad (0 \le t \le 30)$$

(t is the number of years since 1980).[26] What was the average annual expenditure over the period shown? (Round your answer to the nearest billion dollars.) HINT [See Example 2.]

43. ***Investments*** If you invest \$10,000 at 8% interest compounded continuously, what is the average amount in your account over one year?

44. ***Investments*** If you invest \$10,000 at 12% interest compounded continuously, what is the average amount in your account over one year?

45. ▼ ***Average Balance*** Suppose you have an account (paying no interest) into which you deposit \$3,000 at the beginning of each month. You withdraw money continuously so that the amount in the account decreases linearly to 0 by the end of the month. Find the average amount in the account over a period of several months. (Assume that the account starts at \$0 at $t = 0$ months.)

46. ▼ ***Average Balance*** Suppose you have an account (paying no interest) into which you deposit \$4,000 at the beginning of each month. You withdraw \$3,000 during the course of each month, in such a way that the amount decreases linearly. Find the average amount in the account in the first two months. (Assume that the account starts at \$0 at $t = 0$ months.)

47. T ***Online Auctions:* eBay** The number of items sold each quarter on **eBay** from the first quarter of 2005 through the second quarter of 2008 can be approximated by

$$n(t) = 0.07t^4 - 1.7t^3 + 9.9t^2 + 17t + 390 \text{ million items.} \quad (1 \le t \le 14)$$

(t is time in quarters; $t = 1$ represents the start of first quarter in 2005).[27] Use technology to estimate the average number of items sold per quarter on **eBay** during the given period. (Round your answer to the nearest million.)

48. T ***Online Stores:* eBay** The number of stores opened on **eBay** each quarter from the first quarter of 2005 through the second quarter of 2008 can be approximated by

$$s(t) = 0.035t^4 - 0.55t^3 - 1.45t^2 + 37t - 36 \text{ stores.} \quad (1 \le t \le 14)$$

(t is time in quarters; $t = 1$ represents the start of first quarter in 2005).[28] Use technology to estimate the average number of stores opened per quarter on **eBay** during the given period. (Round your answer to the nearest whole number.)

---

[21] Source: Advertising Age (www.adage.com/SuperBowlBuyers/superbowlhistory07.html).

[22] Ibid.

[23] Sources for data: Some data are interpolated. (www.facebook.com, www.insidehighered.com).

[24] Source for data: www.swivel.com/data_sets.

[25] Figures are approximate. Source: Lampert Kuijpers (Panel of the Montreal Protocol), National Bureau of Statistics in China, via CEIC DSata/*New York Times*, February 23, 2007, p. C1.

[26] Data are rounded. 2005 and 2010 figures are projections. Source: Centers for Medicare and Medicaid Services, "National Health Expenditures," 2002 version, released January 2004 (www.cms.hhs.gov/statistics/nhe/).

[27] Source for data: eBay company report (www.investor.ebay.com).

[28] Ibid.

49. ***Stock Prices:* Exxon Mobil** The following table shows the approximate price of **Exxon Mobil** stock in October of each year from 1999 through 2008. Complete the table by computing the 4-year moving averages. (Note the steep drop in October 2008.) Round each average to the nearest dollar.

Year t	1999	2000	2001	2002	2003	2004	2005	2006	2007	2008
Stock Price	36	44	41	36	38	49	56	71	92	62
Moving Average (rounded)										

The stock price spiked in 2007 and then dropped steeply in 2008. What happened to the corresponding moving average? HINT [See Example 3.]

50. ***Stock Prices:* Nokia** The following table shows the approximate price of **Nokia Corporation** stocks in October of each year from 1999 through 2008. Complete the table by computing the 4-year moving averages. (Note the steep drop in October 2008.) Round each average to the nearest dollar.

Year t	1999	2000	2001	2002	2003	2004	2005	2006	2007	2008
Stock Price	25	38	20	13	16	15	16	20	40	15
Moving Average (rounded)										

How does the average year-by-year change in the moving average compare with the average year-by-year change in the stock price? HINT [See Example 3.]

51. ▼ ***Cancun*** The *Playa Loca Hotel* in Cancun has an advertising brochure with the following chart, showing the year-round temperature.

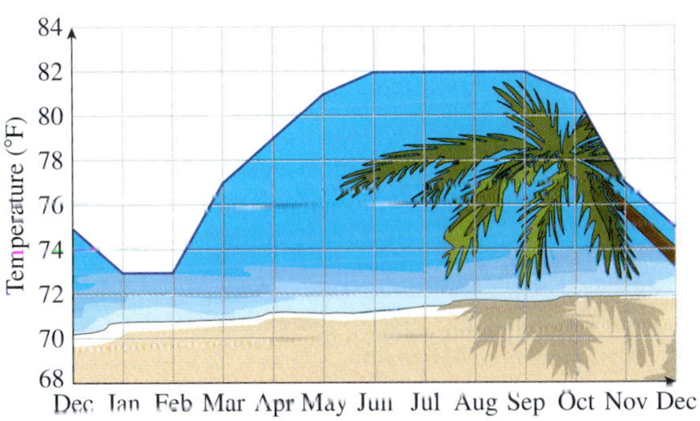

Source: http://www.holiday-weather.com (Temperatures are rounded.)

a. Estimate and plot the year-round 6-month moving average. (Use graphing technology, if available, to check your graph.)

b. What can you say about the 12-month moving average?

52. ▼ ***Reykjavik*** Repeat the preceding exercise, using the following data from the brochure of the *Tough Traveler Lodge* in Reykjavik.

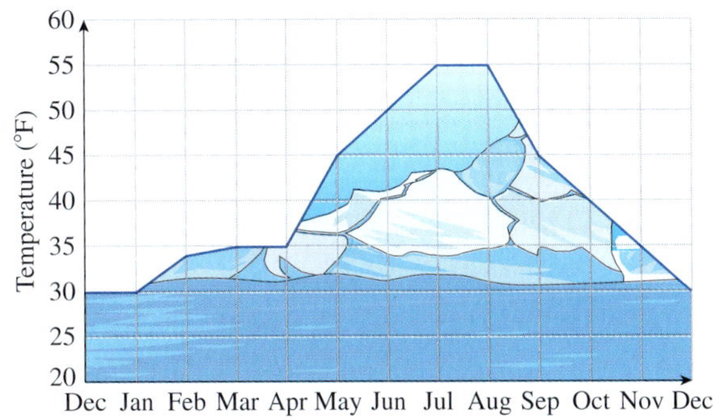

Source: www.holiday-weather.com (Temperatures are rounded.)

53. T ▼ ***Sales:* Apple** The following table shows approximate quarterly sales of iPods in millions of units, starting in the first quarter of 2004.[29]

Quarter	2004 Q1	2004 Q2	2004 Q3	2004 Q4	2005 Q1	2005 Q2	2005 Q3	2005 Q4	2006 Q1	2006 Q2
Housing Starts (1,000)	345	456	440	370	369	485	471	392	382	433
Quarter	2006 Q3	2006 Q4	2007 Q1	2007 Q2	2007 Q3	2007 Q4	2008 Q1	2008 Q2	2008 Q3	2008 Q4
Housing Starts (1,000)	372	278	260	333	265	188	162	193	162	116

a. Use technology to compute and plot the 4-quarter moving average of these data.

b. The graph of the moving average for the last four quarters will appear almost linear during 2008. Use the 2008 Q1 and 2008 Q4 figures of the moving average to give an estimate (to the nearest 0.1 million units) of the rate of change of iPod sales during 2008.

54. T ▼ ***Housing Starts*** The following table shows the number of housing starts for one-family units, in thousands of units, starting in the first quarter of 2004.[30]

Quarter	2004 Q1	2004 Q2	2004 Q3	2004 Q4	2005 Q1	2005 Q2	2005 Q3	2005 Q4	2006 Q1	2006 Q2
Sales (millions)	0.8	0.9	2	4.5	5.3	6.2	6.5	14	8.5	8.1
Quarter	2006 Q3	2006 Q4	2007 Q1	2007 Q2	2007 Q3	2007 Q4	2008 Q1	2008 Q2	2008 Q3	2008 Q4
Sales (millions)	8.7	21.1	10.5	9.7	10.3	22.1	10.6	11	11.3	23.9

---

[29] Source for data: Apple quarterly earnings reports (www.apple.com/investor/).

[30] Source for data: U.S. Census Bureau (www.census.gov).

a. Use technology to compute and plot the 4-quarter moving average of these data.
b. The graph of the moving average for the last four quarters will appear almost linear during 2008. Use the 2008 Q1 and 2008 Q4 figures of the moving average to give an estimate (to the nearest thousand units) of the rate of change of housing starts during 2008.

55. ***Bottled Water Sales*** The rate of U.S. sales of bottled water for the period 2000–2008 can be approximated by

$$s(t) = 12t^2 + 500t + 4{,}700 \text{ million gallons per year} \quad (0 \le t \le 8)$$

where t is time in years since the start of 2000.[31]

a. Estimate the average annual sales of bottled water over the period 2000–2008, to the nearest 100 million gallons per year. HINT [See Quick Example page 537.]
b. Compute the two-year moving average of s. (You need not simplify the answer.) HINT [See Quick Example page 539.]
c. Without simplifying the answer in part (b), say what kind of function the moving average is.

56. ***Bottled Water Sales*** The rate of U.S. per capita sales of bottled water for the period 2000–2008 can be approximated by

$$s(t) = 0.04t^2 + 1.5t + 17 \text{ gallons per year} \quad (0 \le t \le 8)$$

where t is the time in years since the start of 2000.[32] Repeat the preceding exercise as applied to per capita sales. (Give your answer to (a) to the nearest gallon per year.)

57. ***Medicare Spending*** Annual federal spending on Medicare (in constant 2000 dollars) was projected to increase from \$240 billion in 2000 to \$600 billion in 2025.[33]

a. Use this information to express s, the annual spending on Medicare (in billions of dollars), as a linear function of t, the number of years since 2000.
b. Find the 4-year moving average of your model.
c. What can you say about the slope of the moving average?

58. ***Pasta Imports in the 90s*** In 1990, the United States imported 290 million pounds of pasta. From 1990 to 2000 imports increased by an average of 40 million pounds per year.[34]

a. Use these data to express q, the annual U.S. imports of pasta (in millions of pounds), as a linear function of t, the number of years since 1990.
b. Find the 4-year moving average of your model.
c. What can you say about the slope of the moving average?

59. ▼ ***Moving Average of a Linear Function*** Find a formula for the a-unit moving average of a general linear function $f(x) = mx + b$.

60. ▼ ***Moving Average of an Exponential Function*** Find a formula for the a-unit moving average of a general exponential function $f(x) = Ae^{kx}$.

COMMUNICATION AND REASONING EXERCISES

61. Explain why it is sometimes more useful to consider the moving average of a stock price rather than the stock price itself.

62. Sales this month were sharply lower than they were last month, but the 12-unit moving average this month was higher than it was last month. How can that be?

63. Your company's six-month moving average of sales is constant. What does that say about the sales figures?

64. Your monthly salary has been increasing steadily for the past year, and your average monthly salary over the past year was x dollars. Would you have earned more money if you had been paid x dollars per month? Explain your answer.

65. ▼ What property does a (nonconstant) function have if its average value over an interval is zero? Sketch a graph of such a function.

66. ▼ Can the average value of a function f on an interval be greater than its value at every point in that interval? Explain.

67. ▼ Criticize the following claim: The average value of a function on an interval is midway between its highest and lowest value.

68. ▼ Your manager tells you that 12-month moving averages gives at least as much information as shorter-term moving averages and very often more. How would you argue that he is wrong?

69. ▼ Which of the following most closely approximates the original function, (A) its 10-unit moving average, (B) its 1-unit moving average, or (C) its 0.8-unit moving average? Explain your answer.

70. ▼ Is an increasing function larger or smaller than its 1-unit moving average? Explain.

---

[31] Source for data: *Beverage Marketing Corporation* (www.bottledwater.org) (The 2008 figure is an estimate.)

[32] Ibid.

[33] Data are rounded. Source: The Urban Institute's Analysis of the 1999 Trustee's Report (www.urban.org).

[34] Data are rounded. Sources: Department of Commerce/*New York Times*, September 5, 1995, p. D4, International Trade Administration (www.ita.doc.gov/) March 31, 2002.

7.4 Applications to Business and Economics: Consumers' and Producers' Surplus and Continuous Income Streams

Consumers' Surplus

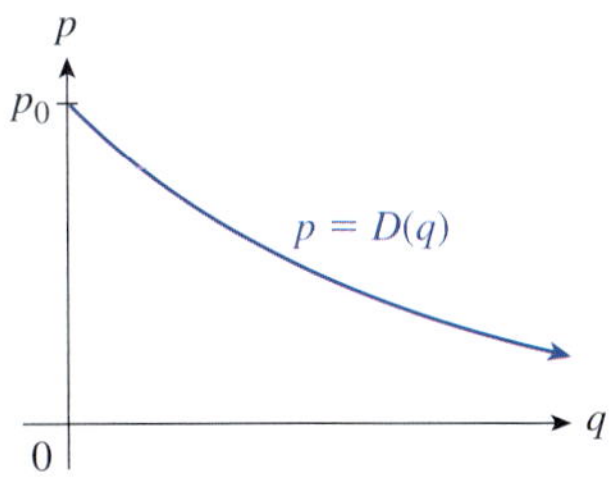

Figure 15

Consider a general demand curve presented, as is traditional in economics, as $p = D(q)$, where p is unit price and q is demand measured, say, in annual sales (Figure 15). Thus, $D(q)$ is the price at which the demand will be q units per year. The price p_0 shown on the graph is the highest price that customers are willing to pay.

Suppose, for example, that the graph in Figure 15 is the demand curve for a particular new model of computer. When the computer first comes out and supplies are low (q is small), "early adopters" will be willing to pay a high price. This is the part of the graph on the left, near the p-axis. As supplies increase and the price drops, more consumers will be willing to pay and more computers will be sold. We can ask the following question: How much are consumers willing to spend for the first $\bar{q}$ units?

Consumers' Willingness to Spend

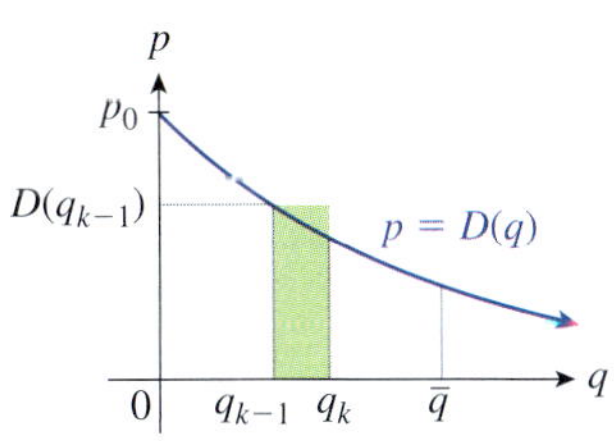

Figure 16

We can approximate consumers' willingness to spend on the first $\bar{q}$ units as follows. We partition the interval $[0, \bar{q}]$ into n subintervals of equal length, as we did when discussing Riemann sums. Figure 16 shows a typical subinterval, $[q_{k-1}, q_k]$.

The price consumers are willing to pay for each of units q_{k-1} through q_k is approximately $D(q_{k-1})$, so the total that consumers are willing to spend for these units is approximately $D(q_{k-1})(q_k - q_{k-1}) = D(q_{k-1})\Delta q$, the area of the shaded region in Figure 16. Thus, the total amount that consumers are willing to spend for items 0 through $\bar{q}$ is

$$W \approx D(q_0)\Delta q + D(q_1)\Delta q + \cdots + D(q_{n-1})\Delta q = \sum_{k=0}^{n-1} D(q_k)\Delta q,$$

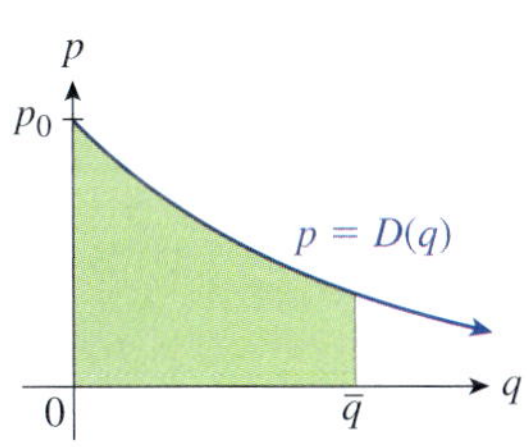

Figure 17

which is a Riemann sum. The approximation becomes better the larger n becomes, and in the limit the Riemann sums converge to the integral

$$W = \int_0^{\bar{q}} D(q)\, dq.$$

This quantity, the area shaded in Figure 17, is the total consumers' willingness to spend to buy the first $\bar{q}$ units.

Consumers' Expenditure

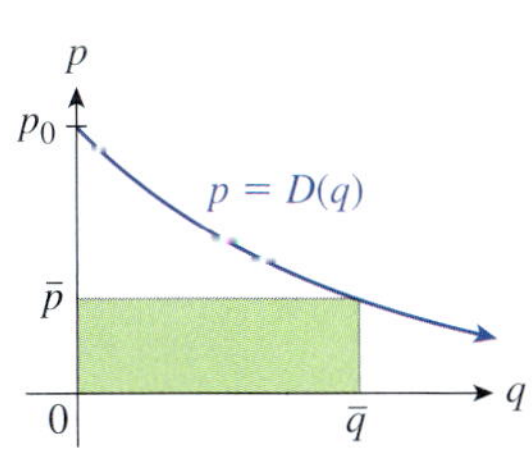

Figure 18

Now suppose that the manufacturer simply sets the price at $\bar{p}$, with a corresponding demand of $\bar{q}$, so $D(\bar{q}) = \bar{p}$. Then the amount that consumers will actually spend to buy these $\bar{q}$ is $\bar{p}\,\bar{q}$, the product of the unit price and the quantity sold. This is the area of the rectangle shown in Figure 18. Notice that we can write $\bar{p}\,\bar{q} = \int_0^{\bar{q}} \bar{p}\, dq$, as suggested by the figure.

The difference between what consumers are willing to pay and what they actually pay is money in their pockets and is called the **consumers' surplus**.

Consumers' Surplus

If demand for an item is given by $p = D(q)$, the selling price is $\bar{p}$, and $\bar{q}$ is the corresponding demand [so that $D(\bar{q}) = \bar{p}$], then the **consumers' surplus** is the difference between willingness to spend and actual expenditure:

$$CS = \int_0^{\bar{q}} D(q)\,dq - \bar{p}\bar{q} = \int_0^{\bar{q}} (D(q) - \bar{p})\,dq$$

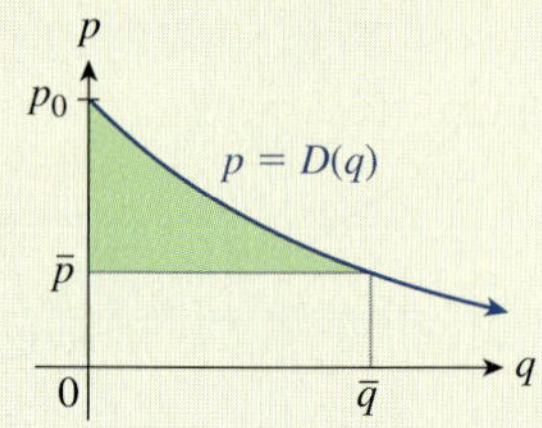

Graphically, it is the area between the graphs of $p = D(q)$ and $p = \bar{p}$, as shown in the figure.

EXAMPLE 1 Consumers' Surplus

Your used-CD store has an exponential demand equation of the form

$$p = 15e^{-0.01q}$$

where q represents daily sales of used CDs and p is the price you charge per CD. Calculate the daily consumers' surplus if you sell your used CDs at \$5 each.

Solution We are given $D(q) = 15e^{-0.01q}$ and $\bar{p} = 5$. We also need $\bar{q}$. By definition,

$$D(\bar{q}) = \bar{p}$$

or $\quad 15e^{-0.01\bar{q}} = 5$

which we must solve for $\bar{q}$:

$$e^{-0.01\bar{q}} = \frac{1}{3}$$

$$-0.01\bar{q} = \ln\left(\frac{1}{3}\right) = -\ln 3$$

$$\bar{q} = \frac{\ln 3}{0.01} \approx 109.8612$$

We now have

$$\begin{aligned} CS &= \int_0^{\bar{q}} (D(q) - \bar{p})\,dq \\ &= \int_0^{109.8612} (15e^{-0.01q} - 5)\,dq \\ &= \left[\frac{15}{-0.01}e^{-0.01q} - 5q\right]_0^{109.8612} \\ &\approx (-500 - 549.31) - (-1{,}500 - 0) \\ &= \$450.69 \text{ per day} \end{aligned}$$

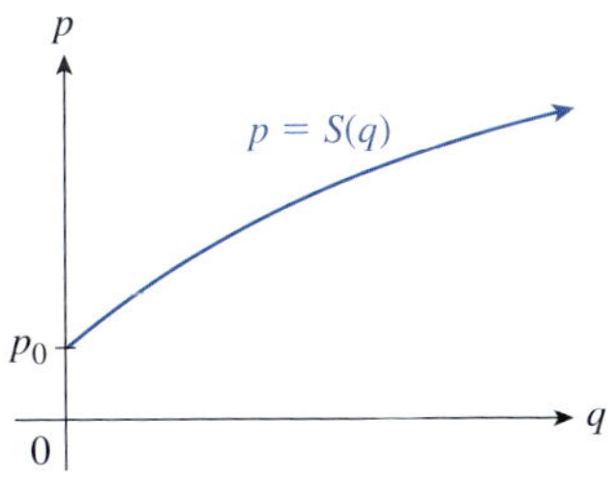

Figure 19

Producers' Surplus

We can also calculate extra income earned by producers. Consider a supply equation of the form $p = S(q)$, where $S(q)$ is the price at which a supplier is willing to supply q items (per time period). Because a producer is generally willing to supply more units at a higher price per unit, a supply curve usually has a positive slope, as shown in Figure 19. The price p_0 is the lowest price that a producer is willing to charge.

Arguing as before, we see that the minimum amount of money producers are willing to receive in exchange for $\bar{q}$ items is $\int_0^{\bar{q}} S(q)\,dq$. On the other hand, if the producers charge $\bar{p}$ per item for $\bar{q}$ items, their actual revenue is $\bar{p}\bar{q} = \int_0^{\bar{q}} \bar{p}\,dq$.

The difference between the producers' actual revenue and the minimum they would have been willing to receive is the **producers' surplus**.

Producers' Surplus

The **producers' surplus** is the extra amount earned by producers who were willing to charge less than the selling price of $\bar{p}$ per unit and is given by

$$PS = \int_0^{\bar{q}} [\bar{p} - S(q)]\,dq$$

where $S(\bar{q}) = \bar{p}$. Graphically, it is the area of the region between the graphs of $p = \bar{p}$ and $p = S(q)$ for $0 \le q \le \bar{q}$, as in the figure.

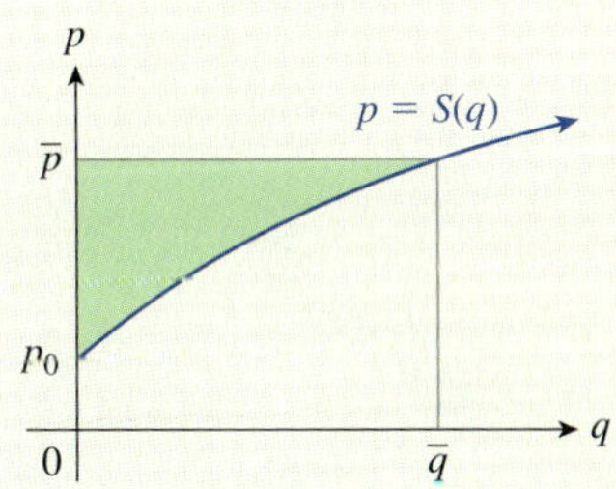

EXAMPLE 2 Producers' Surplus

My tie-dye T-shirt enterprise has grown to the extent that I am now able to produce T-shirts in bulk, and several campus groups have begun placing orders. I have informed one group that I am prepared to supply $20\sqrt{p-4}$ T-shirts at a price of p dollars per shirt. What is my total surplus if I sell T-shirts to the group at \$8 each?

Solution We need to calculate the producers' surplus when $\bar{p} = 8$. The supply equation is

$$q = 20\sqrt{p-4}$$

but in order to use the formula for producers' surplus, we need to express p as a function of q. First, we square both sides to remove the radical sign:

$$q^2 = 400(p-4)$$

so

$$p - 4 = \frac{q^2}{400}$$

giving

$$p = S(q) = \frac{q^2}{400} + 4.$$

We now need the value of $\bar{q}$ corresponding to $\bar{p} = 8$. Substituting $p = 8$ in the original equation, gives

$$\bar{q} = 20\sqrt{8-4} = 20\sqrt{4} = 40.$$

Thus,

$$\begin{aligned} PS &= \int_0^{\bar{q}} (\bar{p} - S(q))\,dq \\ &= \int_0^{40} \left[8 - \left(\frac{q^2}{400} + 4\right)\right] dq \\ &= \int_0^{40} \left(4 - \frac{q^2}{400}\right) dq \\ &= \left[4q - \frac{q^3}{1{,}200}\right]_0^{40} \approx \$106.67 \end{aligned}$$

Thus, I earn a surplus of $106.67 if I sell T-shirts to the group at $8 each.

EXAMPLE 3 Equilibrium

To continue the preceding example: A representative informs me that the campus group is prepared to order only $\sqrt{200(16-p)}$ T-shirts at p dollars each. I would like to produce as many T-shirts for them as possible but avoid being left with unsold T-shirts. Given the supply curve from the preceding example, what price should I charge per T-shirt, and what are the consumers' and producers' surpluses at that price?

Solution The price that guarantees neither a shortage nor a surplus of T-shirts is the **equilibrium price**, the price where supply equals demand. We have

$$\begin{aligned} \text{Supply:} \quad & q = 20\sqrt{p-4}. \\ \text{Demand:} \quad & q = \sqrt{200(16-p)}. \end{aligned}$$

Equating these gives

$$\begin{aligned} 20\sqrt{p-4} &= \sqrt{200(16-p)} \\ 400(p-4) &= 200(16-p), \\ 400p - 1{,}600 &= 3{,}200 - 200p \\ 600p &= 4{,}800 \\ p &= \$8 \text{ per T-shirt.} \end{aligned}$$

We therefore take $\bar{p} = 8$ (which happens to be the price we used in the preceding example). We get the corresponding value for q by substituting $p = 8$ into either the demand or supply equation:

$$\bar{q} = 20\sqrt{8-4} = 40.$$

Thus, $\bar{p} = 8$ and $\bar{q} = 40$.

We must now calculate the consumers' surplus and the producers' surplus. We calculated the producers' surplus for $\bar{p} = 8$ in the preceding example.

$$PS = \$106.67.$$

For the consumers' surplus, we must first express p as a function of q for the demand equation. Thus, we solve the demand equation for p as we did for the supply equation and we obtain

$$\text{Demand:} \quad D(q) = 16 - \frac{q^2}{200}.$$

Therefore,

$$\begin{aligned} CS &= \int_0^{\bar{q}} (D(q) - \bar{p})\, dq \\ &= \int_0^{40} \left[\left(16 - \frac{q^2}{200}\right) - 8\right] dq \\ &= \int_0^{40} \left(8 - \frac{q^2}{200}\right) dq \\ &= \left[8q - \frac{q^3}{600}\right]_0^{40} \approx \$213.33 \end{aligned}$$

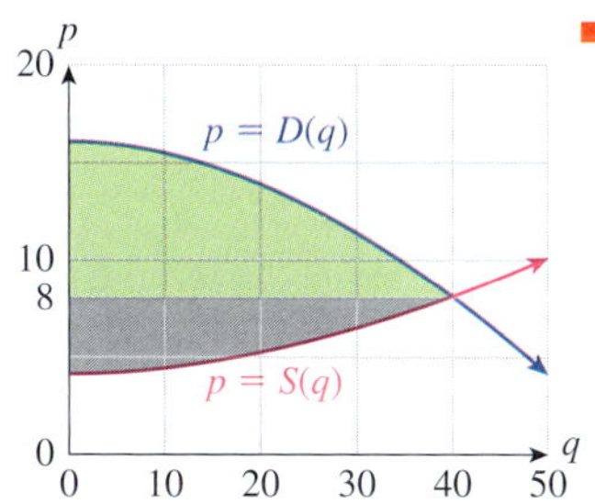

Figure 20

➡ **Before we go on...** Figure 20 shows both the consumers' surplus (top portion) and the producers' surplus (bottom portion) from Example 3. Because extra money in people's pockets is a good thing, the total of the consumers' and the producers' surpluses is called the **total social gain**. In this case it is

$$\text{Social gain} = CS + PS = 213.33 + 106.67 = \$320.00.$$

As you can see from the figure, the total social gain is also the area between two curves and equals

$$\int_0^{40} (D(q) - S(q))\, dq.$$ ■

Continuous Income Streams

For purposes of calculation, it is often convenient to assume that a company with a high sales volume receives money continuously. In such a case, we have a function $R(t)$ that represents the rate at which money is being received by the company at time t.

Andersen Ross/Digital Vision/Getty Images

EXAMPLE 4 Continuous Income

An ice cream store's business peaks in late summer; the store's summer revenue is approximated by

$$R(t) = 300 + 4.5t - 0.05t^2 \text{ dollars per day} \quad (0 \le t \le 92)$$

where t is measured in days after June 1. What is its total revenue for the months of June, July, and August?

Solution Let's approximate the total revenue by breaking up the interval [0, 92] representing the three months into n subintervals $[t_{k-1}, t_k]$, each with length Δt. In the interval $[t_{k-1}, t_k]$ the store receives money at a rate of approximately $R(t_{k-1})$ dollars per day for Δt days, so it will receive a total of $R(t_{k-1})\Delta t$ dollars. Over the whole summer, then, the store will receive approximately

$$R(t_0)\Delta t + R(t_1)\Delta t + \cdots + R(t_{n-1})\Delta t \text{ dollars.}$$

As we let n become large to better approximate the total revenue, this Riemann sum approaches the integral

$$\text{Total revenue} = \int_0^{92} R(t)\,dt.$$

Substituting the function we were given, we get

$$\begin{aligned}\text{Total revenue} &= \int_0^{92} (300 + 4.5t - 0.05t^2)\,dt \\ &= \left[300t + 2.25t^2 - \frac{0.05}{3}t^3\right]_0^{92} \\ &\approx \$33{,}666.\end{aligned}$$

Before we go on... We could approach the calculation in Example 4 another way: $R(t) = S'(t)$ where $S(t)$ is the total revenue earned up to day t. By the Fundamental Theorem of Calculus,

$$\text{Total revenue} = S(92) - S(0) = \int_0^{92} R(t)\,dt.$$

We did the calculation using Riemann sums mainly as practice for the next example. ■

Generalizing Example 4, we can say the following:

Total Value of a Continuous Income Stream

If the rate of receipt of income is $R(t)$ dollars per unit of time, then the total income received from time $t = a$ to $t = b$ is

$$\text{Total value} = TV = \int_a^b R(t)\,dt$$

EXAMPLE 5 Future Value

Suppose the ice cream store in the preceding example deposits its receipts in an account paying 5% interest per year compounded continuously. How much money will it have in its account at the end of August?

Solution Now we have to take into account not only the revenue but also the interest it earns in the account. Again, we break the interval [0, 92] into n subintervals. During the interval $[t_{k-1}, t_k]$, approximately $R(t_{k-1})\Delta t$ dollars are deposited in the account. That money will earn interest until the end of August, a period of $92 - t_{k-1}$ days, or

$(92 - t_{k-1})/365$ years. The formula for continuous compounding tells us that by the end of August, those $R(t_{k-1})\Delta t$ dollars will have turned into

$$R(t_{k-1})\Delta t e^{0.05(92-t_{k-1})/365} = R(t_{k-1})e^{0.05(92-t_{k-1})/365}\Delta t \text{ dollars.}$$

(Recall that 5% is the *annual* interest rate.) Adding up the contributions from each subinterval, we see that the total in the account at the end of August will be approximately

$$R(t_0)e^{0.05(92-t_0)/365}\Delta t + R(t_1)e^{0.05(92-t_1)/365}\Delta t + \cdots + R(t_{n-1})e^{0.05(92-t_{n-1})/365}\Delta t.$$

This is a Riemann sum; as n gets large the sum approaches the integral

$$\text{Future value} = FV = \int_0^{92} R(t)e^{0.05(92-t)/365}\,dt.$$

Substituting $R(t) = 300 + 4.5t - 0.05t^2$, we obtain

$$FV = \int_0^{92} (300 + 4.5t - 0.05t^2)e^{0.05(92-t)/365}\,dt$$

$$\approx \$33{,}880.$$ Using technology or integration by parts

Before we go on... The interest earned in the account in Example 5 was fairly small. (Compare this answer to that in Example 4.) Not only was the money in the account for only three months, but much of it was put in the account toward the end of that period, so had very little time to earn interest. ■

Generalizing again, we have the following:

Future Value of a Continuous Income Stream

If the rate of receipt of income from time $t = a$ to $t = b$ is $R(t)$ dollars per unit of time and the income is deposited as it is received in an account paying interest at rate r per unit of time, compounded continuously, then the amount of money in the account at time $t = b$ is

$$\text{Future value} = FV = \int_a^b R(t)e^{r(b-t)}\,dt.$$

EXAMPLE 6 Present Value

You are thinking of buying the ice cream store discussed in the preceding two examples. What is its income stream worth to you on June 1? Assume that you have access to the same account paying 5% per year compounded continuously.

Solution The value of the income stream on June 1 is the amount of money that, if deposited June 1, would give you the same future value as the income stream will. If we let PV denote this "present value," its value after 92 days will be

$$PVe^{0.05\times 92/365}.$$

We equate this with the future value of the income stream to get

$$PVe^{0.05\times 92/365} = \int_0^{92} R(t)e^{0.05(92-t)/365}\,dt$$

so

$$PV = \int_0^{92} R(t)e^{-0.05t/365}\,dt.$$

Substituting the formula for $R(t)$ and integrating using technology or integration by parts, we get

$$PV \approx \$33{,}455.$$

The general formula is the following:

Present Value of a Continuous Income Stream

If the rate of receipt of income from time $t = a$ to $t = b$ is $R(t)$ dollars per unit of time and the income is deposited as it is received in an account paying interest at rate r per unit of time, compounded continuously, then the value of the income stream at time $t = a$ is

$$\text{Present value} = PV = \int_a^b R(t)e^{r(a-t)}dt.$$

We can derive this formula from the relation

$$FV = PVe^{r(b-a)}$$

because the present value is the amount that would have to be deposited at time $t = a$ to give a future value of FV at time $t = b$.

Note These formulas are more general than we've said. They still work when $R(t) < 0$ if we interpret negative values as money flowing *out* rather than in. That is, we can use these formulas for income we receive or for payments that we make, or for situations where we sometimes receive money and sometimes pay it out. These formulas can also be used for flows of quantities other than money. For example, if we use an exponential model for population growth and we let $R(t)$ represent the rate of immigration $[R(t) > 0]$ or emigration $[R(t) < 0]$, then the future value formula gives the future population. ■

7.4 EXERCISES

▼ more advanced ◆ challenging

T indicates exercises that should be solved using technology

Calculate the consumers' surplus at the indicated unit price $\bar{p}$ for each of the demand equations in Exercises 1–12. HINT [See Example 1.]

1. $p = 10 - 2q$; $\bar{p} = 5$

2. $p = 100 - q$; $\bar{p} = 20$

3. $p = 100 - 3\sqrt{q}$; $\bar{p} = 76$

4. $p = 10 - 2q^{1/3}$; $\bar{p} = 6$

5. $p = 500e^{-2q}$; $\bar{p} = 100$

6. $p = 100 - e^{0.1q}$; $\bar{p} = 50$

7. $q = 100 - 2p$; $\bar{p} = 20$

8. $q = 50 - 3p$; $\bar{p} = 10$

9. $q = 100 - 0.25p^2$; $\bar{p} = 10$

10. $q = 20 - 0.05p^2$; $\bar{p} = 5$

11. $q = 500e^{-0.5p} - 50$; $\bar{p} = 1$

12. $q = 100 - e^{0.1p}$; $\bar{p} = 20$

Calculate the producers' surplus for each of the supply equations in Exercises 13–24 at the indicated unit price $\bar{p}$. HINT [See Example 2.]

13. $p = 10 + 2q$; $\bar{p} = 20$

14. $p = 100 + q$; $\bar{p} = 200$

15. $p = 10 + 2q^{1/3}$; $\bar{p} = 12$

16. $p = 100 + 3\sqrt{q}$; $\bar{p} = 124$

17. $p = 500e^{0.5q}$; $\bar{p} = 1{,}000$

18. $p = 100 + e^{0.01q}$; $\bar{p} = 120$

19. $q = 2p - 50$; $\bar{p} = 40$

20. $q = 4p - 1{,}000$; $\bar{p} = 1{,}000$

21. $q = 0.25p^2 - 10$; $\bar{p} = 10$

22. $q = 0.05p^2 - 20$; $\bar{p} = 50$

23. $q = 500e^{0.05p} - 50$; $\bar{p} = 10$

24. $q = 10(e^{0.1p} - 1)$; $\bar{p} = 5$

In Exercises 25–30, find the total value of the given income stream and also find its future value (at the end of the given interval) using the given interest rate. HINT [See Examples 4, 5.]

25. $R(t) = 30{,}000$, $0 \le t \le 10$, at 7%

26. $R(t) = 40{,}000$, $0 \le t \le 5$, at 10%

27. $R(t) = 30{,}000 + 1{,}000t$, $0 \le t \le 10$, at 7%

28. $R(t) = 40{,}000 + 2{,}000t$, $0 \le t \le 5$, at 10%

29. $R(t) = 30{,}000e^{0.05t}$, $0 \le t \le 10$, at 7%

30. $R(t) = 40{,}000e^{0.04t}$, $0 \le t \le 5$, at 10%

In Exercises 31–36, find the total value of the given income stream and also find its present value (at the beginning of the given interval) using the given interest rate. HINT [See Examples 4, 6.]

31. $R(t) = 20{,}000$, $0 \le t \le 5$, at 8%

32. $R(t) = 50{,}000$, $0 \le t \le 10$, at 5%

33. $R(t) = 20{,}000 + 1{,}000t$, $0 \le t \le 5$, at 8%

34. $R(t) = 50{,}000 + 2{,}000t$, $0 \le t \le 10$, at 5%

35. $R(t) = 20{,}000e^{0.03t}$, $0 \le t \le 5$, at 8%

36. $R(t) = 50{,}000e^{0.06t}$, $0 \le t \le 10$, at 5%

APPLICATIONS

37. ***College Tuition*** A study of U.S. colleges and universities resulted in the demand equation $q = 20{,}000 - 2p$, where q is the enrollment at a public college or university and p is the average annual tuition (plus fees) it charges.[35] Officials at Enormous State University have developed a policy whereby the number of students it will accept per year at a tuition level of p dollars is given by $q = 7{,}500 + 0.5p$. Find the equilibrium tuition price $\bar{p}$ and the consumers' and producers' surpluses at this tuition level. What is the total social gain at the equilibrium price? HINT [See Example 3.]

38. ***Fast Food*** A fast-food outlet finds that the demand equation for its new side dish, "Sweetdough Tidbit," is given by

$$p = \frac{128}{(q+1)^2}$$

where p is the price (in cents) per serving and q is the number of servings that can be sold per hour at this price. At the same time, the franchise is prepared to sell $q = 0.5p - 1$ servings per hour at a price of p cents. Find the equilibrium price p and the consumers' and producers' surpluses at this price level. What is the total social gain at the equilibrium price? HINT [See Example 3.]

39. ▼ ***Linear Demand*** Given a linear demand equation $q = -mp + b$ $(m > 0)$, find a formula for the consumers' surplus at a price level of $\bar{p}$ per unit.

40. ▼ ***Linear Supply*** Given a linear supply equation of the form $q = mp + b$ $(m > 0)$, find a formula for the producers' surplus at a price level of $\bar{p}$ per unit.

41. ***Revenue:*** **Nokia** The annual net sales (revenue) earned by **Nokia** in the years starting January 2000 to January 2008 can be approximated by

$$R(t) = 0.6t^2 - 2t + 30 \text{ billion euros per year} \quad (0 \le t \le 8)$$

where t is time in years ($t = 0$ represents January 2000).[36] Estimate, to the nearest €10 billion, **Nokia**'s total revenue from January 2002 to January 2008. HINT [See Example 4.]

42. ***Revenue:*** **Nintendo** The annual net sales (revenue) earned by **Nintendo Co., Ltd.**, in the fiscal years ending March 31, 2000 to March 31, 2008, can be approximated by

$$R(t) = 12t^3 - 100t^2 + 180t + 500 \text{ billion yen per year} \quad (0 \le t \le 8)$$

where t is time in years ($t = 0$ represents March 31, 2000).[37] Estimate, to the nearest ¥100 billion, **Nintendo**'s total revenue from March 31, 2002 to March 31, 2006. HINT [See Example 4.]

43. ***Revenue:*** **Wal-Mart** The annual revenue earned by **Wal-Mart** in the years starting January 2000 to January 2008 can be approximated by

$$R(t) = 162e^{0.11t} \text{ billion dollars per year} \quad (0 \le t \le 8)$$

where t is time in years ($t = 0$ represents January 2000).[38] Estimate, to the nearest \$10 billion, **Wal-Mart**'s total revenue from January 2000 to January 2006.

44. ***Revenue:*** **Target** The annual revenue earned by **Target** for fiscal years 1998 through 2008 can be approximated by

$$R(t) = 27e^{0.11t} \text{ billion dollars per year} \quad (0 \le t \le 8)$$

where t is time in years ($t = 0$ represents the beginning of fiscal year 2000).[39] Estimate, to the nearest \$10 billion, **Target**'s total revenue from the beginning of fiscal year 2004 to the beginning of fiscal year 2008.

[35] Idea based on a study by A. L. Ostrosky, Jr. and J. V. Koch, as cited in their book, *Introduction to Mathematical Economics* (Waveland Press, Illinois, 1979, p. 133). The data used here are fictitious, however.

[36] Source for data: Nokia financial statements (www.investors.nokia.com).

[37] Source for data: Nintendo annual reports (www.nintendo.com/corp).

[38] Source for data: Wal-Mart annual reports (www.walmartstores.com/Investors).

[39] The model is based on an exponential regression. Source for data: Target 2003 annual report (www.targetcorp.com).

45. ▼ ***Revenue*** Refer to Exercise 41. Suppose that, from January 2000 on, **Nokia** invested its revenue in an investment yielding 4% compounded continuously. What, to the nearest €10 billion, would the total value of **Nokia**'s revenue from January 2002 to January 2008 have been as of January 2008? HINT [See Example 5.]

46. ▼ ***Revenue*** Refer to Exercise 42. Suppose that, from March 2000 on, **Nintendo** invested its revenue in an investment yielding 5% compounded continuously. What, to the nearest ¥100 billion, would the total value of **Nintendo**'s revenue from March 31, 2002 to March 31, 2006 have been as of March 31, 2006? HINT [See Example 5.]

47. ▼ ***Revenue*** Refer to Exercise 43. Suppose that, from January 2000 on, **Wal-Mart** invested its revenue in an investment that depreciated continuously at a rate of 5% per year. What, to the nearest $10 billion, would the total value of **Wal-Mart**'s revenues from January 2000 to January 2006 have been?

48. ▼ ***Revenue*** Refer to Exercise 44. Suppose that, from the start of fiscal year 2004 on, **Target** invested its revenue in an investment that depreciated continuously at a rate of 3% per year. What, to the nearest $10 billion, would the total value of **Target**'s revenue from the beginning of fiscal year 2004 to the beginning of fiscal year 2008 have been?

49. ▼ ***Saving for Retirement*** You are saving for your retirement by investing $700 per month in an annuity with a guaranteed interest rate of 6% per year. With a continuous stream of investment and continuous compounding, how much will you have accumulated in the annuity by the time you retire in 45 years?

50. ▼ ***Saving for College*** When your first child is born, you begin to save for college by depositing $400 per month in an account paying 12% interest per year. With a continuous stream of investment and continuous compounding, how much will you have accumulated in the account by the time your child enters college 18 years later?

51. ▼ ***Saving for Retirement*** You begin saving for your retirement by investing $700 per month in an annuity with a guaranteed interest rate of 6% per year. You increase the amount you invest at the rate of 3% per year. With continuous investment and compounding, how much will you have accumulated in the annuity by the time you retire in 45 years?

52. ▼ ***Saving for College*** When your first child is born, you begin to save for college by depositing $400 per month in an account paying 12% interest per year. You increase the amount you save by 2% per year. With continuous investment and compounding, how much will have accumulated in the account by the time your child enters college 18 years later?

53. ▼ ***Bonds*** The U.S. Treasury issued a 30-year bond on October 15, 2001, paying 3.375% interest.[40] Thus, if you bought $100,000 worth of these bonds you would receive $3,375 per year in interest for 30 years. An investor wishes to buy the rights to receive the interest on $100,000 worth of these bonds. The amount the investor is willing to pay is the present value of the interest payments, assuming a 4% rate of return. Assuming (incorrectly, but approximately) that the interest payments are made continuously, what will the investor pay? HINT [See Example 6.]

54. ▼ ***Bonds*** The Megabucks Corporation is issuing a 20-year bond paying 7% interest. (See the preceding exercise.) An investor wishes to buy the rights to receive the interest on $50,000 worth of these bonds, and seeks a 6% rate of return. Assuming that the interest payments are made continuously, what will the investor pay? HINT [See Example 6.]

55. ▼ ***Valuing Future Income*** Inga was injured and can no longer work. As a result of a lawsuit, she is to be awarded the present value of the income she would have received over the next 20 years. Her income at the time she was injured was $100,000 per year, increasing by $5,000 per year. What will be the amount of her award, assuming continuous income and a 5% interest rate?

56. ▼ ***Valuing Future Income*** Max was injured and can no longer work. As a result of a lawsuit, he is to be awarded the present value of the income he would have received over the next 30 years. His income at the time he was injured was $30,000 per year, increasing by $1,500 per year. What will be the amount of his award, assuming continuous income and a 6% interest rate?

COMMUNICATION AND REASONING EXERCISES

57. Complete the following: The future value of a continuous income stream earning 0% interest is the same as the _____ value.

58. Complete the following: The present value of a continuous income stream earning 0% interest is the same as the _____ value.

59. ▼ Your study group friend says that the future value of a continuous stream of income is always greater than the total value, assuming a positive rate of return. Is she correct? Why?

60. ▼ Your other study group friend says that the present value of a continuous stream of income can sometimes be greater than the total value, depending on the (positive) interest rate. Is he correct? Explain.

61. ▼ Arrange from smallest to largest: Total Value, Future Value, Present Value of a continuous stream of income (assuming a positive income and positive rate of return).

62. ▼ **a.** Arrange the following functions from smallest to largest $R(t)$, $R(t)e^{r(b-t)}$, $R(t)e^{r(a-t)}$, where $a \le t \le b$, and r and $R(t)$ are positive.

b. Use the result from part (a) to justify your answers in Exercises 59–61.

[40]Note: The U.S. Treasury suspended selling Treasury Bonds after October 2001, but resumed issuing them in February, 2006. Source: The Bureau of the Public Debt (www.publicdebt.treas.gov).

7.5 Improper Integrals and Applications

All the definite integrals we have seen so far have had the form $\int_a^b f(x)\,dx$, with a and b finite and $f(x)$ piecewise continuous on the closed interval $[a, b]$. If we relax one or both of these requirements somewhat, we obtain what are called **improper integrals**. There are various types of improper integrals.

Integrals in Which a Limit of Integration is Infinite

Integrals in which one or more limits of integration are infinite can be written as

$$\int_a^{+\infty} f(x)\,dx, \int_{-\infty}^{b} f(x)\,dx, \quad \text{or} \quad \int_{-\infty}^{+\infty} f(x)\,dx.$$

Let's concentrate for a moment on the first form, $\int_a^{+\infty} f(x)\,dx$. What does the $+\infty$ mean here? As it often does, it means that we are to take a limit as something gets large. Specifically, it means the limit as the upper bound of integration gets large.

Improper Integral with an Infinite Limit of Integration

We define

$$\int_a^{+\infty} f(x)\,dx = \lim_{M\to+\infty} \int_a^{M} f(x)\,dx,$$

provided the limit exists. If the limit exists, we say that $\int_a^{+\infty} f(x)\,dx$ **converges**. Otherwise, we say that $\int_a^{+\infty} f(x)\,dx$ **diverges**. Similarly, we define

$$\int_{-\infty}^{b} f(x)\,dx = \lim_{M\to-\infty} \int_M^{b} f(x)\,dx,$$

provided the limit exists. Finally, we define

$$\int_{-\infty}^{+\infty} f(x)\,dx = \int_{-\infty}^{a} f(x)\,dx + \int_a^{+\infty} f(x)\,dx$$

for some convenient a, provided *both* integrals on the right converge.

Quick Examples

1. $\displaystyle\int_1^{+\infty} \frac{dx}{x^2} = \lim_{M\to+\infty} \int_1^{M} \frac{dx}{x^2} = \lim_{M\to+\infty} \left[-\frac{1}{x}\right]_1^M = \lim_{M\to+\infty} \left(-\frac{1}{M}+1\right) = 1$

   Converges

2. $\displaystyle\int_1^{+\infty} \frac{dx}{x} = \lim_{M\to+\infty} \int_1^{M} \frac{dx}{x} = \lim_{M\to+\infty} [\ln|x|]_1^M = \lim_{M\to+\infty} (\ln M - \ln 1) = +\infty$

   Diverges

3. $\displaystyle\int_{-\infty}^{-1} \frac{dx}{x^2} = \lim_{M\to-\infty} \int_M^{-1} \frac{dx}{x^2} = \lim_{M\to-\infty} \left[-\frac{1}{x}\right]_M^{-1} = \lim_{M\to-\infty} \left(1+\frac{1}{M}\right) = 1$

   Converges

4. $\displaystyle\int_{-\infty}^{+\infty} e^{-x}\,dx = \int_{-\infty}^{0} e^{-x}\,dx + \int_{0}^{+\infty} e^{-x}\,dx$

$$= \lim_{M\to-\infty}\int_{M}^{0} e^{-x}\,dx + \lim_{M\to+\infty}\int_{0}^{M} e^{-x}\,dx$$

$$= \lim_{M\to-\infty} -[e^{-x}]_{M}^{0} + \lim_{M\to+\infty} -[e^{-x}]_{0}^{M}$$

$$= \lim_{M\to-\infty}(e^{-M}-1) + \lim_{M\to+\infty}(1-e^{-M})$$

$$= +\infty + 1$$ Diverges

5. $\displaystyle\int_{-\infty}^{+\infty} xe^{-x^2}\,dx = \int_{-\infty}^{0} xe^{-x^2}\,dx + \int_{0}^{+\infty} xe^{-x^2}\,dx$

$$= \lim_{M\to-\infty}\int_{M}^{0} xe^{-x^2}\,dx + \lim_{M\to+\infty}\int_{0}^{M} xe^{-x^2}\,dx$$

$$= \lim_{M\to-\infty}\left[-\frac{1}{2}e^{-x^2}\right]_{M}^{0} + \lim_{M\to+\infty}\left[-\frac{1}{2}e^{-x^2}\right]_{0}^{M}$$

$$= \lim_{M\to-\infty}\left(-\frac{1}{2}+\frac{1}{2}e^{-M^2}\right) + \lim_{M\to+\infty}\left(-\frac{1}{2}e^{-M^2}+\frac{1}{2}\right)$$

$$= -\frac{1}{2}+\frac{1}{2} = 0$$ Converges

Q: *We learned that the integral can be interpreted as the area under the curve. Is this still true for improper integrals?*

A: Yes. Figure 21 illustrates how we can represent an improper integral as the area of an infinite region.

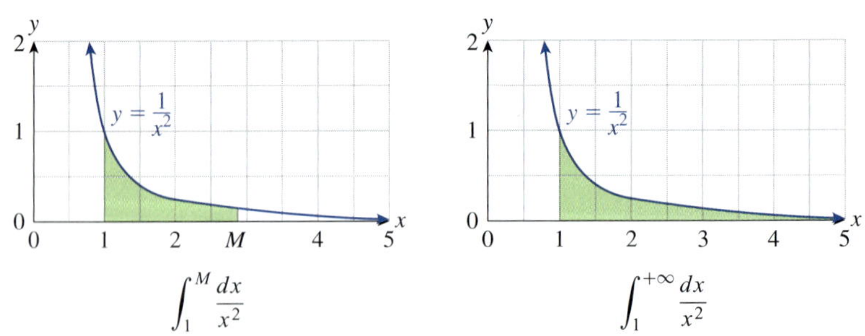

Figure 21

On the left we see the area represented by $\int_1^M dx/x^2$. As M gets larger, the integral approaches $\int_1^{+\infty} dx/x^2$. In the picture, think of M being moved farther and farther along the x-axis in the direction of increasing x, resulting in the region shown on the right.

Q: *Wait! We calculated $\int_1^{+\infty} dx/x^2 = 1$. Does this mean that the infinitely long area in Figure 21 has an area of only 1 square unit?*

A: That is exactly what it means. If you had enough paint to cover 1 square unit, you would never run out of paint while painting the region in Figure 21. This is one of the places where mathematics seems to contradict common sense. But common sense is notoriously unreliable when dealing with infinities.

Andy Crawford/Dorling Kindersley/ Getty Images

EXAMPLE 1 Future Sales of CDs

In 2006, music downloads were starting to make inroads into the sales of CDs.* Approximately 140 million CD albums were sold in the first quarter of 2006, and sales declined by about 3.5% per quarter for the following 2 years. Suppose that this rate of decrease were to continue indefinitely. How many CD albums, total, would be sold from the first quarter of 2006 on?

Solution Recall that the total sales between two dates can be computed as the definite integral of the rate of sales. So, if we wanted the sales between the first quarter of 2006 and a time far in the future, we would compute $\int_0^M s(t)\,dt$ with a large M, where $s(t)$ is the quarterly sales t quarters after the first quarter of 2006. Because we want to know the *total* number of CD albums sold from the first quarter of 2006 on, we let $M \to +\infty$; that is, we compute $\int_0^{+\infty} s(t)\,dt$.

Because sales of CD albums are decreasing by 3.5% per quarter, we can model $s(t)$ by

$$s(t) = 140(0.965)^t \text{ million CD albums}$$

where t is the number of quarters since the first quarter of 2006.

$$\begin{aligned}
\text{Total sales from the first quarter of 2006 on} &= \int_0^{+\infty} 140(0.965)^t\,dt \\
&= \lim_{M\to+\infty} \int_0^M 140(0.965)^t\,dt \\
&= \frac{140}{\ln 0.965} \lim_{M\to+\infty} [0.965^t]_0^M \\
&= \frac{140}{\ln 0.965} \lim_{M\to+\infty} (0.965^M - 0.965^0) \\
&= \frac{140}{\ln 0.965}(-1) \\
&\approx 3929.6 \text{ million CD albums.}
\end{aligned}$$

*Source: "Album sales decline, but is the slump slowing?" USA Today, (www.usatoday.com/life/music/news/2008-04-02-album-sales_N.htm) April 3, 2008.

using Technology

You can estimate the integral in Example 1 with technology by computing $\int_0^M 140(0.965)^t\,dt$ for $M = 10, 100, 1000, \ldots$. You will find that the resulting values appear to converge to about 3,929.6. (Stop when the effect of further increases of M has no effect at this level of accuracy.)

TI-83/84 Plus

```
Y1=140*0.965^X
```

Home screen:

```
fnInt(Y1,X,0,10)
fnInt(Y1,X,0,100)
fnInt(Y1,X,0,1000)
```

Web Site

www.AppliedCalc.org:

Online Utilities

→ Numerical Integration Utility

Enter

```
140*0.965^X
```

for $f(x)$. Enter 0 and 10 for the left and right end-points and press "Adaptive Quadrature" for the most accurate estimate of the integral. Repeat with the right end-point set to 100, 1000, and higher.

Integrals in Which the Integrand Becomes Infinite

We can sometimes compute integrals $\int_a^b f(x)\,dx$ in which $f(x)$ becomes infinite. As we'll see in Example 4, the Fundamental Theorem of Calculus does not work for such integrals. The first case to consider is when $f(x)$ approaches $\pm\infty$ at either a or b.

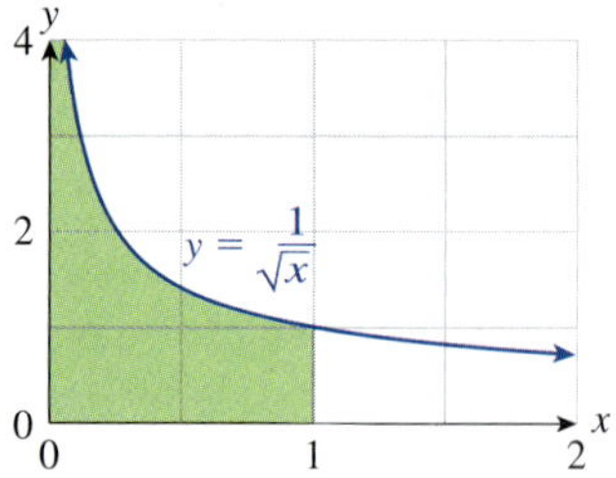

Figure 22

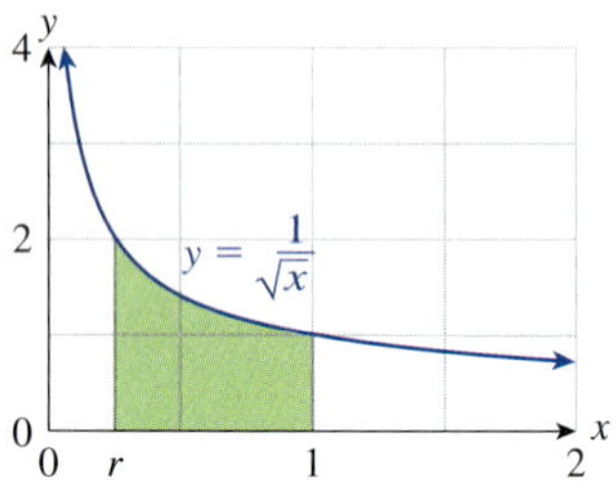

Figure 23

EXAMPLE 2 Integrand Infinite at One Endpoint

Calculate $\int_0^1 \frac{1}{\sqrt{x}}\,dx$.

Solution Notice that the integrand approaches $+\infty$ as x approaches 0 from the right and is not defined at 0. This makes the integral an improper integral. Figure 22 shows the region whose area we are trying to calculate; it extends infinitely vertically rather than horizontally.

Now, if $0 < r < 1$, the integral $\int_r^1 (1/\sqrt{x})\,dx$ is a proper integral because we avoid the bad behavior at 0. This integral gives the area shown in Figure 23. If we let r approach 0 from the right, the area in Figure 23 will approach the area in Figure 22. So, we calculate

$$\begin{aligned}\int_0^1 \frac{1}{\sqrt{x}}\,dx &= \lim_{r\to 0^+} \int_r^1 \frac{1}{\sqrt{x}}\,dx \\ &= \lim_{r\to 0^+} [2\sqrt{x}]_r^1 \\ &= \lim_{r\to 0^+} (2 - 2\sqrt{r}) \\ &= 2.\end{aligned}$$

Thus, we again have an infinitely long region with finite area.

Generalizing, we make the following definition.

Improper Integral in Which the Integrand Becomes Infinite

If $f(x)$ is defined for all x with $a < x \le b$ but approaches $\pm\infty$ as x approaches a, we define

$$\int_a^b f(x)\,dx = \lim_{r\to a^+} \int_r^b f(x)\,dx$$

provided the limit exists. Similarly, if $f(x)$ is defined for all x with $a \le x < b$ but approaches $\pm\infty$ as x approaches b, we define

$$\int_a^b f(x)\,dx = \lim_{r\to b^-} \int_a^r f(x)\,dx$$

provided the limit exists. In either case, if the limit exists, we say that $\int_a^b f(x)\,dx$ **converges**. Otherwise, we say that $\int_a^b f(x)\,dx$ **diverges**.

Note We saw in Chapter 6 that the Fundamental Theorem of Calculus applies to piecewise continuous functions as well as continuous ones. Examples are $f(x) = |x|/x$ and $(x^2 - 1)/(x - 1)$. The integrals of such functions are not improper, and we can use the Fundamental Theorem of Calculus to evaluate such integrals in the usual way. ■

EXAMPLE 3 Testing for Convergence

Does $\int_{-1}^{3} \frac{x}{x^2 - 9}\,dx$ converge? If so, to what?

Solution We first check to see where, if anywhere, the integrand approaches $\pm\infty$. That will happen where the denominator becomes 0, so we solve $x^2 - 9 = 0$.

$$\begin{aligned} x^2 - 9 &= 0 \\ x^2 &= 9 \\ x &= \pm 3 \end{aligned}$$

The solution $x = -3$ is outside of the range of integration, so we ignore it. The solution $x = 3$ is, however, the right endpoint of the range of integration, so the integral is improper. We need to investigate the following limit:

$$\int_{-1}^{3} \frac{x}{x^2 - 9}\,dx = \lim_{r \to 3^-} \int_{-1}^{r} \frac{x}{x^2 - 9}\,dx.$$

Now, to calculate the integral we use a substitution:

$$\begin{aligned} u &= x^2 - 9 \\ \frac{du}{dx} &= 2x \\ dx &= \frac{1}{2x}\,du \end{aligned}$$

when $x = r$, $u = r^2 - 9$

when $x = -1$, $u = (-1)^2 - 9 = -8$

Thus,

$$\begin{aligned} \int_{-1}^{r} \frac{x}{x^2 - 9}\,dx &= \int_{-8}^{r^2 - 9} \frac{1}{2u}\,du \\ &= \frac{1}{2}[\ln|u|]_{-8}^{r^2-9} \\ &= \frac{1}{2}(\ln|r^2 - 9| - \ln 8). \end{aligned}$$

Now we take the limit:

$$\begin{aligned} \int_{-1}^{3} \frac{x}{x^2 - 9}\,dx &= \lim_{r \to 3^-} \int_{-1}^{r} \frac{x}{x^2 - 9}\,dx \\ &= \lim_{r \to 3^-} \frac{1}{2}(\ln|r^2 - 9| - \ln 8) \\ &= -\infty \end{aligned}$$

because, as $r \to 3$, $r^2 - 9 \to 0$, and so $\ln|r^2 - 9| \to -\infty$. Thus, this integral diverges.

EXAMPLE 4 Integrand Infinite Between the Endpoints

Does $\int_{-2}^{3} \frac{1}{x^2}\,dx$ converge? If so, to what?

Solution Again we check to see if there are any points at which the integrand approaches $\pm\infty$. There is such a point, at $x = 0$. This is between the endpoints of the range of integration. To deal with this we break the integral into two integrals:

$$\int_{-2}^{3} \frac{1}{x^2}\,dx = \int_{-2}^{0} \frac{1}{x^2}\,dx + \int_{0}^{3} \frac{1}{x^2}\,dx.$$

Each integral on the right is an improper integral with the integrand approaching $\pm\infty$ at an endpoint. If both of the integrals on the right converge, we take the sum as the value of the integral on the left. So now we compute

$$\begin{aligned}\int_{-2}^{0} \frac{1}{x^2}\,dx &= \lim_{r\to 0^-} \int_{-2}^{r} \frac{1}{x^2}\,dx \\ &= \lim_{r\to 0^-} \left[-\frac{1}{x}\right]_{-2}^{r} \\ &= \lim_{r\to 0^-} \left(-\frac{1}{r} - \frac{1}{2}\right),\end{aligned}$$

which diverges to $+\infty$. There is no need now to check $\int_0^3 (1/x^2)\,dx$; because one of the two pieces of the integral diverges, we simply say that $\int_{-2}^3 (1/x^2)\,dx$ diverges.

Before we go on... What if we had been sloppy in Example 4 and had not checked first whether the integrand approached $\pm\infty$ somewhere? Then we probably would have applied the Fundamental Theorem of Calculus and done the following:

$$\int_{-2}^{3} \frac{1}{x^2}\,dx = \left(-\frac{1}{x}\right)_{-2}^{3} = \left(-\frac{1}{3} - \frac{1}{2}\right) = -\frac{5}{6} \qquad \text{✗ WRONG!}$$

Notice that the answer this "calculation" gives is patently ridiculous. Because $1/x^2 > 0$ for all x for which it is defined, any definite integral of $1/x^2$ must give a positive answer. *Moral:* Always check to see whether the integrand blows up anywhere in the range of integration. If it does, the FTC does not apply, and we must use the methods of this example. ■

We end with an example of what to do if an integral is improper for more than one reason.

EXAMPLE 5 An Integral Improper in Two Ways

Does $\int_0^{+\infty} \frac{1}{\sqrt{x}}\,dx$ converge? If so, to what?

Solution This integral is improper for two reasons. First, the range of integration is infinite. Second, the integrand blows up at the endpoint 0. In order to separate these two problems, we break up the integral at some convenient point:

$$\int_0^{+\infty} \frac{1}{\sqrt{x}}\,dx = \int_0^1 \frac{1}{\sqrt{x}}\,dx + \int_1^{+\infty} \frac{1}{\sqrt{x}}\,dx$$

We chose to break the integral at 1. Any positive number would have sufficed, but 1 is generally easier to use in calculations.

The first piece, $\int_0^1 (1/\sqrt{x})\,dx$, we discussed in Example 2; it converges to 2. For the second piece we have

$$\begin{aligned}\int_1^{+\infty} \frac{1}{\sqrt{x}}\,dx &= \lim_{M\to+\infty} \int_1^M \frac{1}{\sqrt{x}}\,dx \\ &= \lim_{M\to+\infty} [2\sqrt{x}]_1^M \\ &= \lim_{M\to+\infty} (2\sqrt{M} - 2),\end{aligned}$$

which diverges to $+\infty$. Because the second piece of the integral diverges, we conclude that $\int_0^{+\infty} (1/\sqrt{x})\,dx$ diverges.

7.5 EXERCISES

▼ more advanced ◆ challenging

T indicates exercises that should be solved using technology

Note: *For some of the exercises in this section you need to assume the fact that* $\lim_{M\to+\infty} M^n e^{-M} = 0$ *for all n.*

In Exercises 1–26, decide whether or not the given integral converges. If the integral converges, compute its value. HINT [See Quick Examples page 555.]

1. $\int_1^{+\infty} x\,dx$

2. $\int_0^{+\infty} e^{-x}\,dx$

3. $\int_{-2}^{+\infty} e^{-0.5x}\,dx$

4. $\int_1^{+\infty} \frac{1}{x^{1.5}}\,dx$

5. $\int_{-\infty}^{2} e^{x}\,dx$

6. $\int_{-\infty}^{-1} \frac{1}{x^{1/3}}\,dx$

7. $\int_{-\infty}^{-2} \frac{1}{x^2}\,dx$

8. $\int_{-\infty}^{0} e^{-x}\,dx$

9. $\int_0^{+\infty} x^2 e^{-6x}\,dx$

10. $\int_0^{+\infty} (2x-4)e^{-x}\,dx$

11. $\int_0^5 \frac{2}{x^{1/3}}\,dx$ HINT [See Example 2.]

12. $\int_0^2 \frac{1}{x^2}\,dx$

13. $\int_{-1}^2 \frac{3}{(x+1)^2}\,dx$ HINT [See Example 3.]

14. $\int_{-1}^2 \frac{3}{(x+1)^{1/2}}\,dx$

15. $\int_{-1}^2 \frac{3x}{x^2-1}\,dx$ HINT [See Example 4.]

16. $\int_{-1}^2 \frac{3}{x^{1/3}}\,dx$

17. $\int_{-2}^2 \frac{1}{(x+1)^{1/5}}\,dx$

18. $\int_{-2}^2 \frac{2x}{\sqrt{4-x^2}}\,dx$

19. $\int_{-1}^1 \frac{2x}{x^2-1}\,dx$

20. $\int_{-1}^2 \frac{2x}{x^2-1}\,dx$

21. $\int_{-\infty}^{+\infty} xe^{-x^2}\,dx$ **22.** $\int_{-\infty}^{\infty} xe^{1-x^2}\,dx$

23. $\int_{0}^{+\infty} \frac{1}{x \ln x}\,dx$ HINT [See Example 5.]

24. $\int_{0}^{+\infty} \ln x\,dx$

25. ▼ $\int_{0}^{+\infty} \frac{2x}{x^2-1}\,dx$ **26.** ▼ $\int_{-\infty}^{0} \frac{2x}{x^2-1}\,dx$

T *In Exercises 27–34 use technology to approximate the given integrals with $M = 10, 100, 1{,}000, \ldots$ and hence decide whether the associated improper integral converges and estimate its value to four significant digits if it does.* HINT [See the Technology Note for Example 1.]

27. $\int_{1}^{M} \frac{1}{x^2}\,dx$ **28.** $\int_{0}^{M} e^{-x^2}\,dx$

29. $\int_{0}^{M} \frac{x}{1+x}\,dx$ **30.** $\int_{1/M}^{1} \frac{1}{\sqrt{x}}\,dx$

31. $\int_{1+1/M}^{2} \frac{1}{\sqrt{x-1}}\,dx$ **32.** $\int_{1}^{M} \frac{1}{x}\,dx$

33. $\int_{0}^{1-1/M} \frac{1}{(1-x)^2}\,dx$ **34.** $\int_{0}^{2-1/M} \frac{1}{(2-x)^3}\,dx$

APPLICATIONS

35. ***New Home Sales*** Sales of new homes in the United States decreased dramatically from 2006 to 2008, as shown in the model

$$n(t) = 1.05e^{-0.376t} \text{ million homes per year} \quad (0 \le t \le 2)$$

where t is the year since 2006.[41] If this trend were to have continued into the indefinite future, estimate the total number of new homes that would have been sold in the United States from 2006 on. HINT [See Example 1.]

36. ***Revenue from New Home Sales*** Revenue from the sale of new homes in the United States decreased dramatically from 2006 to 2008 as shown in the model

$$r(t) = 321e^{-0.429t} \text{ billion dollars per year} \quad (0 \le t \le 2)$$

where t is the year since 2006.[42] If this trend were to have continued into the indefinite future, estimate the total revenue from the sale of new homes in the United States from 2006 on. HINT [See Example 1.]

37. ***Cigarette Sales*** According to data published by the Federal Trade Commission, the number of cigarettes sold domestically has been decreasing by about 3% per year from the 2000 total of about 415 billion.[43] Use an exponential model to forecast the total number of cigarettes sold from 2000 on. (Round your answer to the nearest 100 billion cigarettes.) HINT [Use a model of the form Ab^t.]

38. ***Sales*** Sales of the text *Calculus and You* have been declining continuously at a rate of 5% per year. Assuming that *Calculus and You* currently sells 5,000 copies per year and that sales will continue this pattern of decline, calculate total future sales of the text. HINT [Use a model of the form Ae^{rt}.]

39. ▼ ***Sales*** My financial adviser has predicted that annual sales of Frodo T-shirts will continue to decline by 10% each year. At the moment, I have 3,200 of the shirts in stock and am selling them at a rate of 200 per year. Will I ever sell them all?

40. ▼ ***Revenue*** Alarmed about the sales prospects for my Frodo T-shirts (see the preceding exercise), I will try to make up lost revenues by increasing the price by \$1 each year. I now charge \$10 per shirt. What is the total amount of revenue I can expect to earn from sales of my T-shirts, assuming the sales levels described in the previous exercise? (Give your answer to the nearest \$1,000.)

41. ▼ ***Education*** Let $N(t)$ be the number of high school students graduated in the United States in year t. This number is projected to change at a rate of about

$$N'(t) = 0.20t^{-0.93} \text{ million graduates per year} \quad (1 \le t \le 21)$$

where t is time in years since 2000.[44] In 2001, there were about 2.8 million high school students graduated. By extrapolating the model, what can you say about the number of high school students graduated in a year far in the future?

42. ▼ ***Education, Martian*** Let $M(t)$ be the number of high school students graduated in the Republic of Mars in year t. This number is projected to change at a rate of about

$$M'(t) = 0.321t^{-1.10} \text{ thousand graduates per year} \quad (1 \le t \le 50)$$

where t is time in years since 2020. In 2021, there were about 1,300 high school students graduated. By extrapolating the model, what can you say about the number of high school students graduated in a year far in the future?

[41] Based on new home sales data at www.census.gov.

[42] Ibid.

[43] Source for data: Federal Trade Commission Cigarette Report for 2004 and 2005 available at www.ftc.gov.

[44] Based on a regression model. Source for data: U.S. Department of Education, 2002 (www.nces.ed.gov/).

43. ▼ ***Cell Phone Revenues*** The number of cell phone subscribers in China in the early 2000s was projected to follow the equation,[45]

$$N(t) = 39t + 68 \text{ million subscribers}$$

in year t ($t = 0$ represents 2000). The average annual revenue per cell phone user was \$350 in 2000.

a. Assuming that, due to competition, the revenue per cell phone user decreases continuously at an annual rate of 10%, give a formula for the annual revenue in year t.

b. Using the model you obtained in part (a) as an estimate of the rate of change of total revenue, estimate the total revenue from 2000 into the indefinite future.

44. ▼ ***Vid Phone Revenues*** The number of vid phone subscribers in the Republic of Mars for the period 2200–2300 was projected to follow the equation

$$N(t) = 18t - 10 \text{ thousand subscribers}$$

in year t ($t = 0$ represents 2200). The average annual revenue per vid phone user was Ƶ40 in 2200.[46]

a. Assuming that, due to competition, the revenue per vid phone user decreases continuously at an annual rate of 20%, give a formula for the annual revenue in year t.

b. Using the model you obtained in part (a) as an estimate of the rate of change of total revenue, estimate the total revenue from 2200 into the indefinite future.

45. T ***Development Assistance*** According to data published by the World Bank, development assistance to low income countries from 2000 through 2008 was approximately

$$q(t) = 0.2t^2 + 3.5t + 60 \text{ billion dollars per year}$$

where t is time in years since 2000.[47] Assuming a world-wide inflation rate of 3% per year, and that the above model remains accurate into the indefinite future, find the value of all development assistance to low income countries from 2000 on in constant dollars. (The constant dollar value of $q(t)$ dollars t years from now is given by $q(t)e^{-rt}$, where r is the fractional rate of inflation. Give your answer to the nearest \$100 billion.) HINT [See Technology Note for Example 1.]

46. T ***Humanitarian Aid*** Repeat the preceding exercise, using the following model for humanitarian aid.[48]

$$q(t) = 0.07t^2 + 0.3t + 5 \text{ billion dollars per year}$$

47. ▼ ***Hair Mousse Sales*** The amount of extremely popular hair mousse sold online at your Web site can be approximated by

$$N(t) = \frac{80(7)^t}{20 + 7^t} \text{ million gallons per year.}$$

($t = 0$ represents the current year.) Investigate the integrals $\int_0^{+\infty} N(t)\,dt$ and $\int_{-\infty}^0 N(t)\,dt$ and interpret your answers.

48. ▼ ***Chocolate Mousse Sales*** The weekly demand for your company's Lo-Cal Mousse is modeled by the equation

$$q(t) = \frac{50e^{2t-1}}{1 + e^{2t-1}} \text{ gallons per week}$$

where t is time from now in weeks. Investigate the integrals $\int_0^{+\infty} q(t)\,dt$ and $\int_{-\infty}^0 q(t)\,dt$ and interpret your answers.

T ***The Normal Curve*** *Exercises 49–52 require the use of a graphing calculator or computer programmed to do numerical integration. The normal distribution curve, which models the distributions of data in a wide range of applications, is given by the function*

$$p(x) = \frac{1}{\sqrt{2\pi}\,\sigma} e^{-(x-\mu)^2/2\sigma^2}$$

where $\pi = 3.14159265\ldots$ *and* σ *and* μ *are constants called the standard deviation and the mean, respectively. Its graph (for* $\sigma = 1$ *and* $\mu = 2$*) is shown in the figure.*

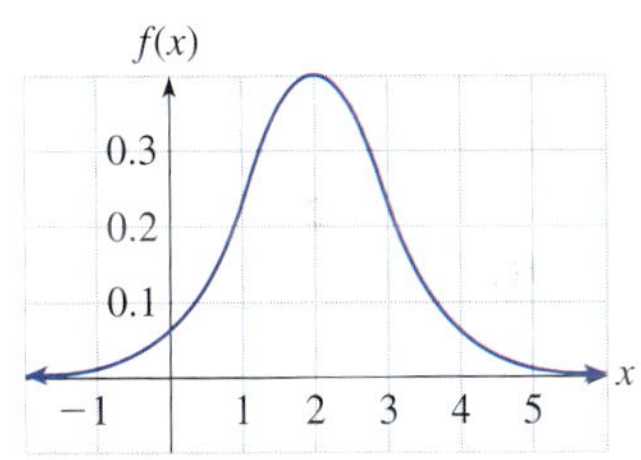

49. ▼ With $\sigma = 4$ and $\mu = 1$, approximate $\int_{-\infty}^{+\infty} p(x)\,dx$. HINT [See Example 5 and Technology Note for Example 1.]

50. ▼ With $\sigma = 1$ and $\mu = 0$, approximate $\int_0^{+\infty} p(x)\,dx$.

51. ▼ With $\sigma = 1$ and $\mu = 0$, approximate $\int_1^{+\infty} p(x)\,dx$.

52. ▼ With $\sigma = 1$ and $\mu = 0$, approximate $\int_{-\infty}^1 p(x)\,dx$.

53. ◆ ***Variable Sales*** The value of your Chateau Petit Mont Blanc 1963 vintage burgundy is increasing continuously at an annual rate of 40%, and you have a supply of 1,000 bottles worth \$85 each at today's prices. In order to ensure a steady income, you have decided to sell your wine at a diminishing rate—starting at 500 bottles per year, and then decreasing this figure continuously at a fractional rate of 100% per year. How much income (to the nearest dollar) can you expect to generate by this scheme? HINT [Use the formula for continuously compounded interest.]

[45] Based on a regression of projected figures (coefficients are rounded). Source: Intrinsic Technology/*New York Times*, Nov. 24, 2000, p. C1.

[46] Ƶ designates Zonars, the designated currency for the city-state of Ularek, Mars. Source: (www.marsnext.com/comm/zonars.html).

[47] The authors' approximation, based on data from the World Bank, obtained from www.worldbank.org.

[48] Ibid.

54. ◆ ***Panic Sales*** Unfortunately, your large supply of Chateau Petit Mont Blanc is continuously turning to vinegar at a fractional rate of 60% per year! You have thus decided to sell off your Petit Mont Blanc at \$50 per bottle, but the market is a little thin, and you can only sell 400 bottles per year. Because you have no way of knowing which bottles now contain vinegar until they are opened, you shall have to give refunds for all the bottles of vinegar. What will your net income be before all the wine turns to vinegar?

55. ◆ ***Meteor Impacts*** The frequency of meteor impacts on earth can be modeled by

$$n(k) = \frac{1}{5.6997k^{1.081}}$$

where $n(k) = N'(k)$, and $N(k)$ is the average number of meteors of energy less than or equal to k megatons that will hit the earth in one year.[49] (A small nuclear bomb releases on the order of one megaton of energy.)

a. How many meteors of energy at least $k = 0.2$ hit the earth each year?
b. Investigate and interpret the integral $\int_0^1 n(k)\,dk$.

56. ◆ ***Meteor Impacts*** (continuing the previous exercise)

a. Explain why the integral

$$\int_a^b kn(k)\,dk$$

computes the total energy released each year by meteors with energies between a and b megatons.

b. Compute and interpret

$$\int_0^1 kn(k)\,dk.$$

c. Compute and interpret

$$\int_1^{+\infty} kn(k)\,dk.$$

57. ◆ ***The Gamma Function*** The gamma function is defined by the formula

$$\Gamma(x) = \int_0^{+\infty} t^{x-1}e^{-t}\,dt.$$

a. Find $\Gamma(1)$ and $\Gamma(2)$.
b. Use integration by parts to show that for every positive integer n, $\Gamma(n+1) = n\Gamma(n)$.
c. Deduce that $\Gamma(n) = (n-1)!\ [= (n-1)(n-2)\cdots 2\cdot 1]$ for every positive integer n.

58. ◆ ***Laplace Transforms*** The Laplace Transform $F(x)$ of a function $f(t)$ is given by the formula

$$F(x) = \int_0^{+\infty} f(t)e^{-xt}\,dt \quad (x > 0).$$

a. Find $F(x)$ for $f(t) = 1$ and for $f(t) = t$.
b. Find a formula for $F(x)$ if $f(t) = t^n$ $(n = 1, 2, 3, \ldots)$.
c. Find a formula for $F(x)$ if $f(t) = e^{at}$ (a constant).

COMMUNICATION AND REASONING EXERCISES

59. Why can't the Fundamental Theorem of Calculus be used to evaluate $\displaystyle\int_{-1}^{1} \frac{1}{x}\,dx$?

60. Why can't the Fundamental Theorem of Calculus be used to evaluate $\displaystyle\int_{1}^{+\infty} \frac{1}{x^2}\,dx$?

61. It sometimes happens that the Fundamental Theorem of Calculus gives the correct answer for an improper integral. Does the FTC give the correct answer for improper integrals of the form

$$\int_{-a}^{a} \frac{1}{x^{1/r}}\,dx$$

if $r = 3, 5, 7, \ldots$?

62. Does the FTC give the correct answer for improper integrals of the form

$$\int_{-a}^{a} \frac{1}{x^r}\,dx$$

if $r = 3, 5, 7, \ldots$?

63. Which of the following integrals are improper, and why? (Do not evaluate any of them.)

a. $\displaystyle\int_{-1}^{1} \frac{|x|}{x}\,dx$ **b.** $\displaystyle\int_{-1}^{1} x^{-1/3}\,dx$ **c.** $\displaystyle\int_{0}^{2} \frac{x-2}{x^2-4x+4}\,dx$

64. Which of the following integrals are improper, and why? (Do not evaluate any of them.)

a. $\displaystyle\int_{-1}^{1} \frac{|x-1|}{x-1}\,dx$ **b.** $\displaystyle\int_{0}^{1} \frac{1}{x^{2/3}}\,dx$ **c.** $\displaystyle\int_{0}^{2} \frac{x^2-4x+4}{x-2}\,dx$

65. T ▼ How could you use technology to approximate improper integrals? (Your discussion should refer to each type of improper integral.)

66. T ▼ Use technology to approximate the integrals $\int_0^M e^{-(x-10)^2}\,dx$ for larger and larger values of M, using Riemann sums with 500 subdivisions. What do you find? Comment on the answer.

67. ▼ Make up an interesting application whose solution is $\int_{10}^{+\infty} 100te^{-0.2t}dt = \$1{,}015.01$.

68. ▼ Make up an interesting application whose solution is $\int_{100}^{+\infty} \frac{1}{r^2}\,dr = 0.01$.

[49] The authors' model, based on data published by NASA International Near-Earth-Object Detection Workshop (*The New York Times*, Jan. 25, 1994, p. C1).

7.6 Differential Equations and Applications

A **differential equation** is an equation that involves a derivative of an unknown function. A **first-order differential equation** involves only the first derivative of the unknown function. A **second-order differential equation** involves the second derivative of the unknown function (and possibly the first derivative). Higher-order differential equations are defined similarly. In this book, we will deal only with first-order differential equations.

To **solve** a differential equation means to find the unknown function. Many of the laws of science and other fields describe how things change. When expressed mathematically, these laws take the form of equations involving derivatives—that is, differential equations. The field of differential equations is a large and very active area of study in mathematics, and we shall see only a small part of it in this section.

Jon Ferrey/Allsport/Getty Images

EXAMPLE 1 Motion

A dragster accelerates from a stop so that its speed t seconds after starting is $40t$ ft/s. How far will the car go in 8 seconds?

Solution We wish to find the car's position function $s(t)$. We are told about its speed, which is ds/dt. Precisely, we are told that

$$\frac{ds}{dt} = 40t.$$

This is the differential equation we have to solve to find $s(t)$. But we already know how to solve this kind of differential equation; we integrate

$$s(t) = \int 40t\,dt = 20t^2 + C.$$

We now have the **general solution** to the differential equation. By letting C take on different values, we get all the possible solutions. We can specify the one **particular solution** that gives the answer to our problem by imposing the **initial condition** that $s(0) = 0$. Substituting into $s(t) = 20t^2 + C$, we get

$$0 = s(0) = 20(0)^2 + C = C$$

so $C = 0$ and $s(t) = 20t^2$. To answer the question, the car travels $20(8)^2 = 1{,}280$ feet in 8 seconds.

We did not have to work hard to solve the differential equation in Example 1. In fact, any differential equation of the form $dy/dx = f(x)$ can (in theory) be solved by integrating. (Whether we can actually carry out the integration is another matter!)

Simple Differential Equations

A **simple** differential equation has the form

$$\frac{dy}{dx} = f(x).$$

Its general solution is

$$y = \int f(x)\,dx.$$

Quick Example

The differential equation

$$\frac{dy}{dx} = 2x^2 - 4x^3$$

is simple and has general solution

$$y = \int f(x)\,dx = \frac{2x^3}{3} - x^4 + C.$$

Not all differential equations are simple, as the next example shows.

EXAMPLE 2 Separable Differential Equation

Consider the differential equation $\dfrac{dy}{dx} = \dfrac{x}{y^2}$.

a. Find the general solution.

b. Find the particular solution that satisfies the initial condition $y(0) = 2$.

Solution

a. This is not a simple differential equation because the right-hand side is a function of both x and y. We cannot solve this equation by just integrating; the solution to this problem is to "separate" the variables.

Step 1: *Separate the variables algebraically.* We rewrite the equation as

$$y^2\,dy = x\,dx.$$

Step 2: *Integrate both sides.*

$$\int y^2\,dy = \int x\,dx$$

giving

$$\frac{y^3}{3} = \frac{x^2}{2} + C.$$

Step 3: *Solve for the dependent variable.* We solve for y:

$$y^3 = \frac{3}{2}x^2 + 3C = \frac{3}{2}x^2 + D$$

(Rewriting $3C$ as D, an equally arbitrary constant), so

$$y = \left(\frac{3}{2}x^2 + D\right)^{1/3}.$$

This is the general solution of the differential equation.

b. We now need to find the value for D that will give us the solution satisfying the condition $y(0) = 2$. Substituting 0 for x and 2 for y in the general solution, we get

$$2 = \left(\frac{3}{2}(0)^2 + D\right)^{1/3} = D^{1/3}$$

so

$$D = 2^3 = 8.$$

Thus, the particular solution we are looking for is

$$y = \left(\frac{3}{2}x^2 + 8\right)^{1/3}.$$

Before we go on... We can check the general solution in Example 2 by calculating both sides of the differential equation and comparing.

$$\frac{dy}{dx} = \frac{d}{dx}\left(\frac{3}{2}x^2 + D\right)^{1/3} = x\left(\frac{3}{2}x^2 + D\right)^{-2/3}$$

$$\frac{x}{y^2} = \frac{x}{\left(\frac{3}{2}x^2 + 8\right)^{2/3}} = x\left(\frac{3}{2}x^2 + D\right)^{-2/3}$$ ✓ ■

Q: *In Example 2, we wrote $y^2\,dy$ and $x\,dx$. What do they mean?*

A: Although it is possible to give meaning to these symbols, for us they are just a notational convenience. We could have done the following instead:

$$y^2\frac{dy}{dx} = x$$

Now we integrate both sides with respect to x.

$$\int y^2\frac{dy}{dx}\,dx = \int x\,dx$$

We can use substitution to rewrite the left-hand side:

$$\int y^2\frac{dy}{dx}\,dx = \int y^2\,dy,$$

which brings us back to the equation

$$\int y^2\,dy = \int x\,dx.$$

We were able to separate the variables in the preceding example because the right-hand side, x/y^2, was a *product* of a function of x and a function of y—namely,

$$\frac{x}{y^2} = x\left(\frac{1}{y^2}\right).$$

In general, we can say the following:

Separable Differential Equation

A **separable** differential equation has the form

$$\frac{dy}{dx} = f(x)g(y).$$

We solve a separable differential equation by separating the xs and the ys algebraically, writing

$$\frac{1}{g(y)}\,dy = f(x)\,dx$$

and then integrating:

$$\int \frac{1}{g(y)}\,dy = \int f(x)\,dx.$$

EXAMPLE 3 Rising Medical Costs

Spending on Medicare from 2000 to 2025 was projected to rise continuously at an instantaneous rate of 3.7% per year.* Find a formula for Medicare spending y as a function of time t in years since 2000.

Solution When we say that Medicare spending y was going up continuously at an instantaneous rate of 3.7% per year, we mean that

the instantaneous rate of increase of y *was 3.7% of its value*

or $$\frac{dy}{dt} = 0.037y.$$

This is a separable differential equation. Separating the variables gives

$$\frac{1}{y}dy = 0.037\,dt.$$

Integrating both sides, we get

$$\int \frac{1}{y}dy = \int 0.037\,dt$$

so $$\ln y = 0.037t + C.$$

(We should write $\ln|y|$, but we know that the medical costs are positive.) We now solve for y.

$$y = e^{0.037t+C} = e^C e^{0.037t} = Ae^{0.037t}$$

where A is a positive constant. This is the formula we used before for continuous percentage growth.

*Spending is in constant 2000 dollars. Source for projected data: The Urban Institute's Analysis of the 1999 Trustee's Report (www.urban.org).

➡ **Before we go on...** To determine A in Example 3 we need to know, for example, Medicare spending at time $t = 0$ (the initial condition). The source cited estimates Medicare spending as \$239.6 billion in 2000. Substituting $t = 0$ in the equation above gives

$$239.6 = Ae^0 = A.$$

Thus, projected Medicare spending is

$$y = 239.6e^{0.037t} \text{ billion dollars}$$

t years after 2000. ■

EXAMPLE 4 Newton's Law of Cooling

Newton's Law of Cooling states that a hot object cools at a rate proportional to the difference between its temperature and the temperature of the surrounding environment (the **ambient temperature**). If a hot cup of coffee, at 170°F, is left to sit in a room at 70°F, how will the temperature of the coffee change over time?

Solution We let $H(t)$ denote the temperature of the coffee at time t. Newton's Law of Cooling tells us that $H(t)$ *decreases* at a rate proportional to the difference between $H(t)$ and 70°F, the ambient temperature. In other words,

$$\frac{dH}{dt} = -k(H - 70)$$

where k is some positive constant.* Note that $H \geq 70$: The coffee will never cool to less than the ambient temperature.

* **NOTE** When we say that a quantity Q is *proportional* to a quantity R, we mean that $Q = kR$ for some constant k. The constant k is referred to as the **constant of proportionality**.

The variables here are H and t, which we can separate as follows:

$$\frac{dH}{H - 70} = -k\,dt.$$

Integrating, we get

$$\int \frac{dH}{H - 70} = \int (-k)\,dt$$

so $\quad \ln(H - 70) = -kt + C.$

(Note that $H - 70$ is positive, so we don't need absolute values.) We now solve for H:

$$\begin{aligned} H - 70 &= e^{-kt+C} \\ &= e^C e^{-kt} \\ &= Ae^{-kt} \end{aligned}$$

so

$$H(t) = 70 + Ae^{-kt}$$

where A is some positive constant. We can determine the constant A using the initial condition $H(0) = 170$:

$$170 = 70 + Ae^0 = 70 + A$$

so $\quad A = 100.$

Therefore,

$$H(t) = 70 + 100e^{-kt}$$

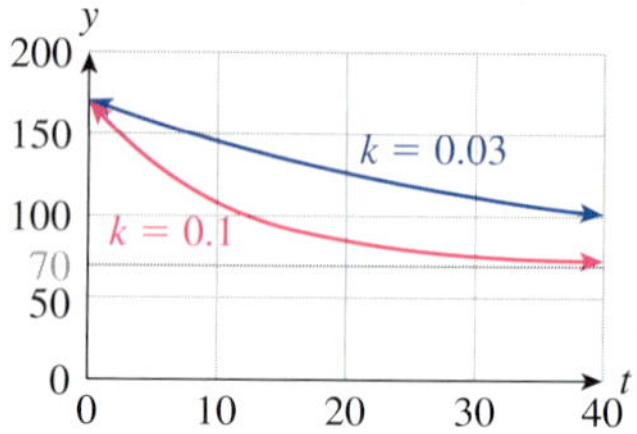

Figure 24

Q: *But what is k?*

A: The constant k determines the rate of cooling. Its value depends on the units of time we are using, on the substance cooling–in this case the coffee–and its container. Because k depends so heavily on the particular circumstances, it's usually easiest to determine it experimentally. Figure 24 shows two possible graphs, one with $k = 0.1$ and the other with $k = 0.03$ ($k \approx 0.03$ would be reasonable for a cup of coffee in a polystyrene container with t measured in minutes).

In any case, we can see from the graph or the formula for $H(t)$ that the temperature of the coffee will approach the ambient temperature exponentially.

Before we go on... Notice that the calculation in Example 4 shows that the temperature of an object cooling according to Newton's law is given in general by

$$H(t) = T_a + (T_0 - T_a)e^{-kt}$$

where T_a is the ambient temperature (70° in the example) and T_0 is the initial temperature (170° in the example). The formula also holds if the ambient temperature is higher than the initial temperature ("Newton's Law of Heating"). ■

7.6 EXERCISES

▼ more advanced ◆ challenging

T indicates exercises that should be solved using technology

Find the general solution of each differential equation in Exercises 1–10. Where possible, solve for y as a function of x.

1. $\frac{dy}{dx} = x^2 + \sqrt{x}$ HINT [See Quick Example page 576.]

2. $\frac{dy}{dx} = \frac{1}{x} + 3$ HINT [See Quick Example page 576.]

3. $\frac{dy}{dx} = \frac{x}{y}$ HINT [See Example 2a.]

4. $\frac{dy}{dx} = \frac{y}{x}$ HINT [See Example 2a.]

5. $\frac{dy}{dx} = xy$

6. $\frac{dy}{dx} = x^2y$

7. $\frac{dy}{dx} = (x+1)y^2$

8. $\frac{dy}{dx} = \frac{1}{(x+1)y^2}$

9. $x\frac{dy}{dx} = \frac{1}{y}\ln x$

10. $\frac{1}{x}\frac{dy}{dx} = \frac{1}{y}\ln x$

For each differential equation in Exercises 11–20, find the particular solution indicated. HINT [See Example 2b.]

11. $\frac{dy}{dx} = x^3 - 2x$; $y = 1$ when $x = 0$

12. $\frac{dy}{dx} = 2 - e^{-x}$; $y = 0$ when $x = 0$

13. $\frac{dy}{dx} = \frac{x^2}{y^2}$; $y = 2$ when $x = 0$

14. $\frac{dy}{dx} = \frac{y^2}{x^2}$; $y = \frac{1}{2}$ when $x = 1$

15. $x\frac{dy}{dx} = y$; $y(1) = 2$

16. $x^2\frac{dy}{dx} = y$; $y(1) = 1$

17. $\frac{dy}{dx} = x(y+1)$; $y(0) = 0$

18. $\frac{dy}{dx} = \frac{y+1}{x}$; $y(1) = 2$

19. $\frac{dy}{dx} = \frac{xy^2}{x^2+1}$; $y(0) = -1$

20. $\frac{dy}{dx} = \frac{xy}{(x^2+1)^2}$; $y(0) = 1$

APPLICATIONS

21. ***Sales*** Your monthly sales of Green Tea Ice Cream are falling at an instantaneous rate of 5% per month. If you currently sell 1,000 quarts per month, find the differential equation describing your change in sales, and then solve it to predict your monthly sales. HINT [See Example 3.]

22. ***Profit*** Your monthly profit on sales of Avocado Ice Cream is rising at an instantaneous rate of 10% per month. If you currently make a profit of \$15,000 per month, find the differential equation describing your change in profit, and solve it to predict your monthly profits. HINT [See Example 3.]

23. ***Newton's Law of Cooling*** For coffee in a ceramic cup, suppose $k \approx 0.05$ with time measured in minutes. **(a)** Use Newton's Law of Cooling to predict the temperature of the coffee, initially at a temperature of 200°F, that is left to sit in a room at 75°F. **(b)** When will the coffee have cooled to 80°F? HINT [See Example 4.]

24. ***Newton's Law of Cooling*** For coffee in a paper cup, suppose $k \approx 0.08$ with time measured in minutes. **(a)** Use Newton's Law of Cooling to predict the temperature of the coffee, initially at a temperature of 210°F, that is left to sit in a room at

60°F. **(b)** When will the coffee have cooled to 70°F? HINT [See Example 4.]

25. ***Cooling*** A bowl of clam chowder at 190°F is placed in a room whose air temperature is 75°F. After 10 minutes, the soup has cooled to 150°F. Find the value of k in Newton's Law of Cooling, and hence find the temperature of the chowder as a function of time.

26. ***Heating*** Suppose that a pie, at 20°F, is put in an oven at 350°F. After 15 minutes, its temperature has risen to 80°F. Find the value of k in Newton's Law of Heating (see the note after Example 4), and hence find the temperature of the pie as a function of time.

27. ***Market Saturation*** You have just introduced a new flat-screen monitor to the market. You predict that you will eventually sell 100,000 monitors and that your monthly rate of sales will be 10% of the difference between the saturation value of 100,000 and the total number you have sold up to that point. Find a differential equation for your total sales (as a function of the month) and solve. (What are your total sales at the moment when you first introduce the monitor?)

28. ***Market Saturation*** Repeat the preceding exercise, assuming that monthly sales will be 5% of the difference between the saturation value (of 100,000 monitors) and the total sales to that point, and assuming that you sell 5,000 monitors to corporate customers before placing the monitor on the open market.

29. ***Determining Demand*** Nancy's Chocolates estimates that the elasticity of demand for its dark chocolate truffles is $E = 0.05p - 1.5$ where p is the price per pound. Nancy's sells 20 pounds of truffles per week when the price is \$20 per pound. Find the formula expressing the demand q as a function of p. Recall that the elasticity of demand is given by

$$E = -\frac{dq}{dp} \times \frac{p}{q}.$$

30. ***Determining Demand*** Nancy's Chocolates estimates that the elasticity of demand for its chocolate strawberries is $E = 0.02p - 0.5$ where p is the price per pound. It sells 30 pounds of chocolate strawberries per week when the price is \$30 per pound. Find the formula expressing the demand q as a function of p. Recall that the elasticity of demand is given by

$$E = -\frac{dq}{dp} \times \frac{p}{q}$$

Linear Differential Equations *Exercises 31–36 are based on* ***first order linear differential equations with constant coefficients****. These have the form*

$$\frac{dy}{dt} + py = f(t) \quad (p \text{ constant})$$

and the general solution is

$$y = e^{-pt} \int f(t)e^{pt}\, dt. \quad \text{(Check this by substituting!)}$$

31. Solve the linear differential equation

$$\frac{dy}{dt} + y = e^{-t};\ y = 1 \text{ when } t = 0.$$

32. Solve the linear differential equation

$$\frac{dy}{dt} - y = e^{2t};\ y = 2 \text{ when } t = 0.$$

33. Solve the linear differential equation

$$2\frac{dy}{dt} - y = 2t;\ y = 1 \text{ when } t = 0.$$

HINT [First rewrite the differential equation in the form $\frac{dy}{dt} + py = f(t)$.]

34. Solve the linear differential equation

$$2\frac{dy}{dt} + y = -t. \quad y = 1 \text{ when } t = 0$$

HINT [First rewrite the differential equation in the form $\frac{dy}{dt} + py = f(t)$.]

35. ▼ ***Electric Circuits*** The flow of current $i(t)$ in an electric circuit without capacitance satisfies the linear differential equation

$$L\frac{di}{dt} + Ri = V(t)$$

where L and R are constants (the *inductance* and *resistance* respectively) and $V(t)$ is the applied voltage. (See figure.)

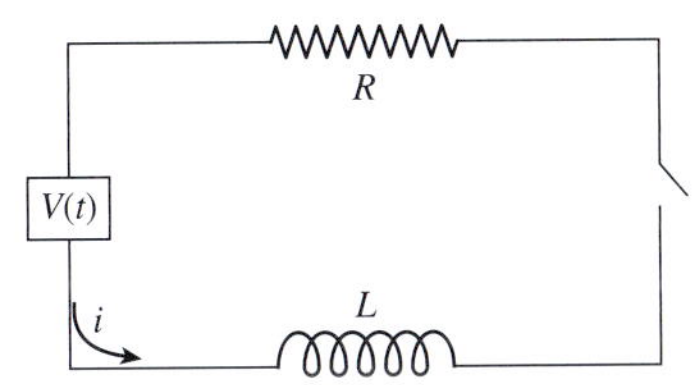

If the voltage is supplied by a 10-volt battery and the switch is turned on at time $t = 1$, then the voltage V is a step function that jumps from 0 to 10 at $t = 1$: $V(t) = 5\left[1 + \frac{|t-1|}{t-1}\right]$. Find the current as a function of time for $L = R = 1$. Use a grapher to plot the resulting current as a function of time. (Assume there is no current flowing at time $t = 0$.) [Use the following integral formula: $\int \left[1 + \frac{|t-1|}{t-1}\right] e^t\, dt = \left[1 + \frac{|t-1|}{t-1}\right](e^t - e) + C$.]

36. ▼ ***Electric Circuits*** Repeat Exercise 35 for $L = 1$, $R = 5$, and $V(t) = 5\left[1 + \frac{|t-2|}{t-2}\right]$. (The switch flipped on at time $t = 2$). [Use the following integral formula: $\int \left[1 + \frac{|t-2|}{t-2}\right] e^{5t}\, dt = \left(1 + \frac{|t-2|}{t-2}\right)\left(\frac{e^{5t} - e^{10}}{5}\right) + C$.]

37. ▼ ***Approach to Equilibrium*** The Extrasoft Toy Co. has just released its latest creation, a plush platypus named "Eggbert." The demand function for Eggbert dolls is $D(p) = 50{,}000 - 500p$ dolls per month when the price is p dollars. The supply function is $S(p) = 30{,}000 + 500p$ dolls per month when the price is p dollars. This makes the equilibrium price \$20. The Evans price adjustment model assumes that if the price is set at a value other than the equilibrium price, it will change over time in such a way that its rate of change is proportional to the shortage $D(p) - S(p)$.

a. Write the differential equation given by the Evans price adjustment model for the price p as a function of time.
b. Find the general solution of the differential equation you wrote in (a). (You will have two unknown constants, one being the constant of proportionality.)
c. Find the particular solution in which Eggbert dolls are initially priced at \$10 and the price rises to \$12 after one month.

38. ▼ ***Approach to Equilibrium*** Spacely Sprockets has just released its latest model, the Dominator. The demand function is $D(p) = 10{,}000 - 1{,}000p$ sprockets per year when the price is p dollars. The supply function is $S(p) = 8{,}000 + 1{,}000p$ sprockets per year when the price is p dollars.

a. Using the Evans price adjustment model described in the preceding exercise, write the differential equation for the price $p(t)$(as a function of time.
b. Find the general solution of the differential equation you wrote in (a).
c. Find the particular solution in which Dominator sprockets are initially priced at \$5 each but fall to \$3 each after one year.

39. ▼ ***Logistic Equation*** There are many examples of growth in which the rate of growth is slow at first, becomes faster, and then slows again as a limit is reached. This pattern can be described by the differential equation

$$\frac{dy}{dt} = ay(L - y)$$

where a is a constant and L is the limit of y. Show by substitution that

$$y = \frac{CL}{e^{-aLt} + C}$$

is a solution of this equation, where C is an arbitrary constant.

40. ▼ ***Logistic Equation*** Using separation of variables and integration with a table of integrals or a symbolic algebra program, solve the differential equation in the preceding exercise to derive the solution given there.

T *Exercises 41–44 require the use of technology.*

41. ▼ ***Market Saturation*** You have just introduced a new model of DVD player. You predict that the market will saturate at 2,000,000 DVD players and that your total sales will be governed by the equation

$$\frac{dS}{dt} = \frac{1}{4}S(2 - S)$$

where S is the total sales in millions of DVD players and t is measured in months. If you give away 1,000 DVD players when you first introduce them, what will S be? Sketch the graph of S as a function of t. About how long will it take to saturate the market? (See Exercise 39.)

42. ▼ ***Epidemics*** A certain epidemic of influenza is predicted to follow the function defined by

$$\frac{dA}{dt} = \frac{1}{10}A(20 - A)$$

where A is the number of people infected in millions and t is the number of months after the epidemic starts. If 20,000 cases are reported initially, find $A(t)$ and sketch its graph. When is A growing fastest? How many people will eventually be affected? (See Exercise 39.)

43. ▼ ***Growth of Tumors*** The growth of tumors in animals can be modeled by the Gompertz equation:

$$\frac{dy}{dt} = -ay\ln\left(\frac{y}{b}\right)$$

where y is the size of a tumor, t is time, and a and b are constants that depend on the type of tumor and the units of measurement.

a. Solve for y as a function of t.
b. If $a = 1$, $b = 10$, and $y(0) = 5\text{ cm}^3$ (with t measured in days), find the specific solution and graph it.

44. ▼ ***Growth of Tumors*** Refer back to the preceding exercise. Suppose that $a = 1$, $b = 10$, and $y(0) = 15\text{ cm}^3$. Find the specific solution and graph it. Comparing its graph to the one obtained in the preceding exercise, what can you say about tumor growth in these instances?

COMMUNICATION AND REASONING EXERCISES

45. What is the difference between a particular solution and the general solution of a differential equation? How do we get a particular solution from the general solution?

46. Why is there always an arbitrary constant in the general solution of a differential equation? Why are there not two or more arbitrary constants in a first-order differential equation?

47. ▼ Show by example that a **second-order** differential equation, one involving the second derivative y'', usually has two arbitrary constants in its general solution.

48. ▼ Find a differential equation that is not separable.

49. ▼ Find a differential equation whose general solution is $y = 4e^{-x} + 3x + C$.

50. ▼ Explain how, knowing the elasticity of demand as a function of either price or demand, you may find the demand equation. (See Exercise 29.)

CHAPTER 7 REVIEW

KEY CONCEPTS

Web Site www.AppliedCalc.org

Go to the student Web site at www.AppliedCalc.org to find a comprehensive and interactive Web-based summary of Chapter 7.

7.1 Integration by Parts

Integration by parts formula:

$$\int u \cdot v\,dx = u \cdot I(v) - \int D(u)I(v)\,dx$$ *p. 518*

Tabular method for integration by parts *p. 519*

Integrating a polynomial times a logarithm *p. 521*

7.2 Area Between Two Curves and Applications

If $f(x) \geq g(x)$ for all x in $[a, b]$, then the area of the region between the graphs of f and g and between $x = a$ and $x = b$ is given by

$$A = \int_a^b [f(x) - g(x)]\,dx$$ *p. 527*

Regions enclosed by crossing curves *p. 529*

Area enclosed by two curves *p. 531*

General instructions for finding the area between the graphs of $f(x)$ and $g(x)$ *p. 531*

Approximating the area between two curves using technology:

$$A = \int_a^b |f(x) - g(x)|\,dx$$ *p. 532*

7.3 Averages and Moving Averages

Average, or mean, of a collection of values

$$\bar{y} = \frac{y_1 + y_2 + \cdots + y_n}{n}$$ *p. 536*

The *average*, or *mean*, of a function $f(x)$ on an interval $[a, b]$ is

$$\bar{f} = \frac{1}{b-a}\int_a^b f(x)\,dx$$ *p. 537*

Average balance *p. 537*

Computing the moving average of a set of data *p. 538*

n-Unit moving average of a function:

$$\bar{f}(x) = \frac{1}{n}\int_{x-n}^{x} f(t)\,dt$$ *p. 539*

Computing moving averages of saw tooth and step functions *p. 540*

7.4 Applications to Business and Economics: Consumers' and Producers' Surplus and Continuous Income Streams

Consumers' surplus:

$$CS = \int_0^{\bar{q}} [D(q) - \bar{p}]\,dq$$ *p. 546*

Producers' surplus:

$$PS = \int_0^{\bar{q}} [\bar{p} - S(q)]\,dq$$ *p. 547*

Equilibrium price *p. 548*

Social gain $= CS + PS$ *p. 549*

Total value of a continuous income stream: $TV = \int_a^b R(t)\,dt$ *p. 550*

Future value of a continuous income stream: $FV = \int_a^b R(t)e^{r(b-t)}\,dt$ *p. 551*

Present value of a continuous income stream: $PV = \int_a^b R(t)e^{r(a-t)}\,dt$ *p. 552*

7.5 Improper Integrals and Applications

Improper integral with an infinite limit of integration:

$$\int_a^{+\infty} f(x)\,dx,\quad \int_{-\infty}^{b} f(x)\,dx,\quad \int_{-\infty}^{+\infty} f(x)\,dx$$ *p. 555*

Improper integral in which the integrand becomes infinite *p. 558*

Testing for convergence *p. 559*

Integrand infinite between the endpoints *p. 560*

Integral improper in two ways *p. 561*

7.6 Differential Equations and Applications

Simple differential equations:

$$\frac{dy}{dx} = f(x)$$ *p. 566*

Separable differential equations:

$$\frac{dy}{dx} = f(x)g(y)$$ *p. 568*

Newton's Law of Cooling *p. 569*

REVIEW EXERCISES

Evaluate the integrals in Exercises 1–10.

1. $\int (x^2 + 2)e^x\,dx$

2. $\int (x^2 - x)e^{-3x+1}\,dx$

3. $\int x^2 \ln(2x)\,dx$

4. $\int \log_5 x\,dx$

5. $\int 2x|2x + 1|\,dx$

6. $\int 3x|-x + 5|\,dx$

7. $\int 5x\frac{|-x + 3|}{-x + 3}\,dx$

8. $\int 2x\frac{|3x + 1|}{3x + 1}\,dx$

9. $\int_{-2}^{2} (x^3 + 1)e^{-x}\,dx$

10. $\int_1^e x^2 \ln x\,dx$

In Exercises 11–14, find the areas of the given regions.

11. Between $y = x^3$ and $y = 1 - x^3$ for x in $[0, 1]$

12. Between $y = e^x$ and $y = e^{-x}$ for x in $[0, 2]$

13. Enclosed by $y = 1 - x^2$ and $y = x^2$

14. Between $y = x$ and $y = xe^{-x}$ for x in $[0, 2]$

In Exercises 15–18, find the average value of the given function over the indicated interval

15. $f(x) = x^3 - 1$ over $[-2, 2]$

16. $f(x) = \dfrac{x}{x^2 + 1}$ over $[0, 1]$

17. $f(x) = x^2e^x$ over $[0, 1]$

18. $f(x) = (x + 1)\ln x$ over $[1, 2e]$

In Exercises 19–22, find the 2-unit moving averages of the given function.

19. $f(x) = 3x + 1$

20. $f(x) = 6x^2 + 12$

21. $f(x) = x^{4/3}$

22. $f(x) = \ln x$

In Exercises 23 and 24, calculate the consumers' surplus at the indicated unit price $\bar{p}$ for the given demand equation.

23. $p = 50 - \frac{1}{2}q;\ \bar{p} = 10$

24. $p = 10 - q^{1/2};\ \bar{p} = 4$

In Exercises 25 and 26, calculate the producers' surplus at the indicated unit price $\bar{p}$ for the given supply equation.

25. $p = 50 + \frac{1}{2}q;\ \bar{p} = 100$

26. $p = 10 + q^{1/2};\ \bar{p} = 40$

In Exercises 27–32 decide whether the given integral converges. If the integral converges, compute its value.

27. $\int_1^{\infty} \frac{1}{x^5}\,dx$

28. $\int_0^1 \frac{1}{x^5}\,dx$

29. $\int_{-1}^{1} \frac{x}{(x^2-1)^{5/3}}\,dx$

30. $\int_0^2 \frac{x}{(x^2-1)^{1/3}}\,dx$

31. $\int_0^{+\infty} 2xe^{-x^2}\,dx$

32. $\int_0^{+\infty} x^2 e^{-6x^3}\,dx$

Solve the differential equations in Exercises 33–36.

33. $\frac{dy}{dx} = x^2y^2$

34. $\frac{dy}{dx} = xy + 2x$

35. $xy\frac{dy}{dx} = 1;\ y(1) = 1$

36. $y(x^2+1)\frac{dy}{dx} = xy^2;\ y(0) = 2$

APPLICATIONS

37. ***Spending on Stationery*** Alarmed by the volume of pointless memos and reports being copied and circulated by management at OHaganBooks.com, John O'Hagan ordered a 5-month audit of paper usage at the company. He found that management consumed paper at a rate of

$$q(t) = 45t + 200 \text{ thousand sheets per month} \quad (0 \le t \le 5)$$

(t is the time in months since the audit began). During the same period, the price of paper was escalating; the company was charged approximately

$$p(t) = 9e^{0.09t} \text{ dollars per thousand sheets.}$$

Use an integral to estimate, to the nearest hundred dollars, the total spent on paper for management during the given period.

38. ***Spending on Shipping*** During the past 10 months, OHaganBooks.com shipped orders at a rate of about

$$q(t) = 25t + 3{,}200 \text{ packages per month} \quad (0 \le t \le 10)$$

(t is the time in months since the beginning of the year). During the same period, the cost of shipping a package averaged approximately

$$p(t) = 4e^{0.04t} \text{ dollars per package.}$$

Use an integral to estimate, to the nearest thousand dollars, the total spent on shipping orders during the given period.

39. ***Education Costs*** Billy-Sean O'Hagan, having graduated *summa cum laude* from college, has been accepted by the doctoral program in biophysics at Oxford. John O'Hagan estimates that the total cost (minus scholarships) he will need to pay is \$2,000 per month, but that this cost will escalate at a continuous compounding rate of 1% per month.

a. What, to the nearest dollar, is the average monthly cost over the course of two years?

b. Find the four-month moving average of the monthly cost.

40. ***Investments*** OHaganBooks.com keeps its cash reserves in a hedge fund paying 6% compounded continuously. It starts a year with \$1 million in reserves and does not withdraw or deposit any money.

a. What is the average amount it will have in the fund over the course of two years?

b. Find the one-month moving average of the amount it has in the fund.

41. ***Consumers' and Producers' Surplus*** Currently, the hottest selling item at OHaganBooks.com is *Mensa for Dummies*[50] with a demand curve of $q = 20{,}000(28 - p)^{1/3}$ books per week, and a supply curve of $q = 40{,}000(p - 19)^{1/3}$ books per week.

a. Find the equilibrium price and demand.

b. Find the consumers' and producers' surpluses at the equilibrium price.

42. ***Consumers' and Producers' Surplus*** OHaganBooks.com is about to start selling a new coffee table book, *Computer Designs of the Late Twentieth Century*. It estimates the demand curve to be $q = 1{,}000\sqrt{200 - 2p}$, and its willingness to order books from the publisher is given by the supply curve $q = 1{,}000\sqrt{10p - 400}$.

a. Find the equilibrium price and demand.

b. Find the consumers' and producers' surpluses at the equilibrium price.

43. ***Revenue*** Sales of the bestseller *A River Burns Through It* are dropping at OHaganBooks.com. To try to bolster sales, the company is dropping the price of the book, now \$40, at a rate of \$2 per week. As a result, this week OHaganBooks.com will sell 5,000 copies, and it estimates that sales will fall continuously at a rate of 10% per week. How much revenue will it earn on sales of this book over the next 8 weeks?

44. ***Foreign Investments*** Panicked by the performance of the U.S. stock market, Marjory Duffin is investing her 401(k) money in a Russian hedge fund at a rate of approximately

$$q(t) = 1.7t^2 - 0.5t + 8 \text{ thousand shares per month}$$

where t is time in months since the stock market began to plummet. At the time she started making the investments, the hedge fund was selling for \$1 per share, but subsequently

[50] The actual title is: *Let Us Just Have A Ball! Mensa for Dummies* by Wendu Mekbib (Author) Silhouette Publishing Corporation.

declined in value at a continuous rate of 5% per month. What was the total amount of money Marjory Duffin invested after one year? (Answer to the nearest \$1,000.)

45. ***Investments*** OHaganBooks.com CEO John O'Hagan has started a gift account for the Marjory Duffin Foundation. The account pays 6% compounded continuously and is initially empty. OHaganBooks .com deposits money continuously into it, starting at the rate of \$100,000 per month and increasing continuously by \$10,000 per month.
   a. How much money will the company have in the account at the end of two years?
   b. How much of the amount you found in part (a) was principal deposited and how much was interest earned? (Round answers to the nearest \$1,000.)

46. ***Savings*** John O'Hagan had been saving money for Billy-Sean's education since he had been a wee lad. O'Hagan began depositing money at the rate of \$1,000 per month and increased his deposits continuously by \$50 per month. If the account earned 5% compounded continuously and O'Hagan continued these deposits for 15 years,
   a. How much money did he accumulate?
   b. How much was money deposited and how much was interest?

47. ***Acquisitions*** The Megabucks Corporation is considering buying OHaganBooks.com. They estimate OHaganBooks.com's revenue stream at \$50 million per year, growing continuously at a 10% rate. Assuming interest rates of 6%, how much is OHaganBooks.com's revenue for the next year worth now?

48. ***More Acquisitions*** OHaganBooks.com is thinking of buying JungleBooks and would like to recoup its investment after three years. The estimated net profit for JungleBooks is \$40 million per year, growing linearly by \$5 million per year. Assuming interest rates of 4%, how much should OHaganBooks.com pay for JungleBooks?

49. ***Incompetence*** OHaganBooks.com is shopping around for a new bank. A junior executive at one bank offers them the following interesting deal: The bank will pay them interest continuously at a rate numerically equal to 0.01% of the square of the amount of money they have in the account at any time. By considering what would happen if \$10,000 was deposited in such an account, explain why the junior executive was fired shortly after this offer was made.

50. ***Shrewd Bankers*** The new junior officer at the bank (who replaced the one fired in the preceding exercise) offers OHaganBook.com the following deal for the \$800,000 they plan to deposit: While the amount in the account is less than \$1 million, the bank will pay interest continuously at a rate equal to 10% of the difference between \$1 million and the amount of money in the account. When it rises over \$1 million, the bank will pay interest of 20%. Why should OHaganBooks.com not take this offer?

Case Study

Estimating Tax Revenues

You have just been hired by the incoming administration of your country as chief consultant for national tax policy, and you have been getting conflicting advice from the finance experts on your staff. Several of them have come up with plausible suggestions for new tax structures, and your job is to choose the plan that results in more revenue for the government.

Before you can evaluate their plans, you realize that it is essential to know your country's income distribution—that is, how many people earn how much money.* One might think that the most useful way of specifying income distribution would be to use a function that gives the exact number $f(x)$ of people who earn a given salary x. This would necessarily be a discrete function—it makes sense only if x happens to be a whole number of cents. There is, after all, no one earning a salary of exactly \$22,000.142567! Furthermore, this function would behave rather erratically, because there are, for example, probably many more people making a salary of exactly \$30,000 than exactly \$30,000.01. Given these problems, it is far more convenient to start with the function defined by

$$N(x) = \textit{the total number of people earning between } 0 \textit{ and } x \textit{ dollars.}$$

Actually, you would want a "smoothed" version of this function. The graph of $N(x)$ might look like the one shown in Figure 25.

* **NOTE** To simplify our discussion, we are assuming that (1) all tax revenues are based on earned income and that (2) everyone in the population we consider earns some income.

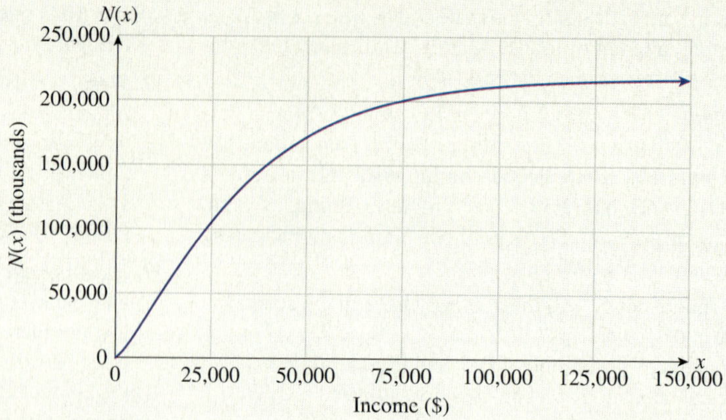

Figure 25

If we take the *derivative* of $N(x)$, we get an income distribution function. Its graph might look like the one shown in Figure 26.

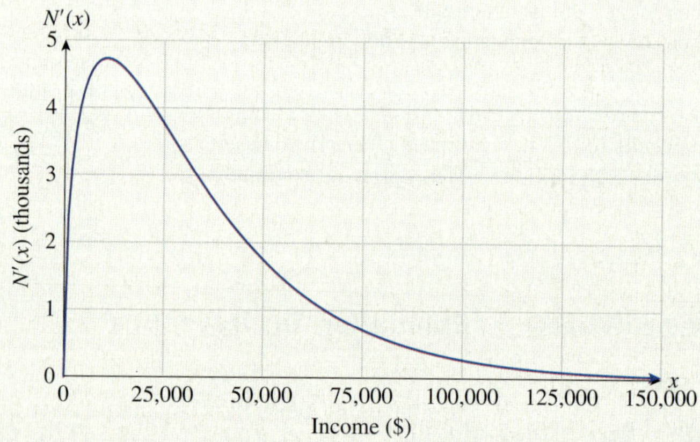

Figure 26

Because the derivative measures the rate of change, its value at x is the additional number of taxpayers per \$1 increase in salary. Thus, the fact that $N'(25{,}000) \approx 3{,}800$ tells us that approximately 3,800 people are earning a salary of between \$20,000 and \$20,001. In other words, N' shows the distribution of incomes among the population—hence, the name "distribution function."*

*** NOTE** A very similar idea is used in probability. See the optional chapter "Calculus Applied to Probability and Statistics" on the Web site.

You thus send a memo to your experts requesting the income distribution function for the nation. After much collection of data, they tell you that the income distribution function is

$$N'(x) = 100x^{0.466}e^{-x/23{,}000}.$$

This is in fact the function whose graph is shown in Figure 26 and is an example of a **gamma distribution**.* (You might find it odd that you weren't given the original function N, but it will turn out that you don't need it. How would you compute it?)

*** NOTE** Gamma distributions are often good models for income distributions. The one used in the text is the authors' approximation of the income distribution in the United States in 2007. Source for data: U.S. Census Bureau, Current Population Survey, 2008 Annual Social and Economic Supplement.

Given this income distribution, your financial experts have come up with the two possible tax policies illustrated in Figures 27 and 28.

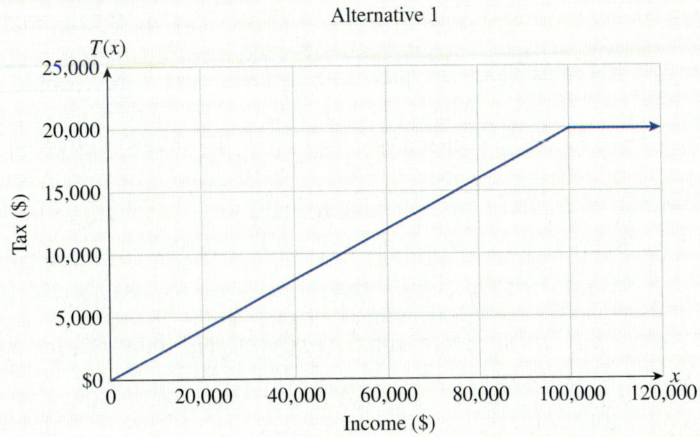

Figure 27

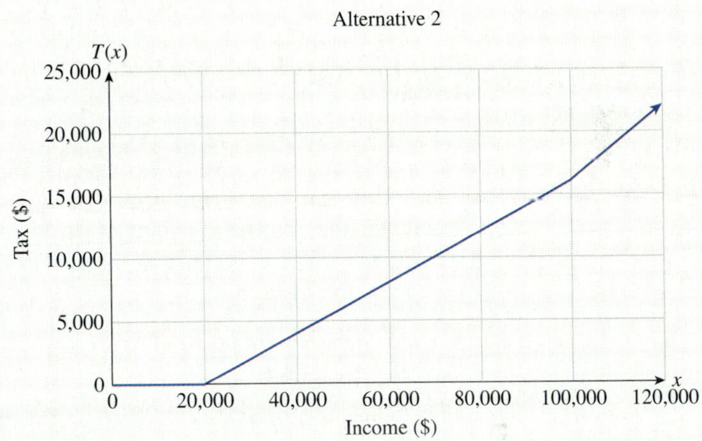

Figure 28

In the first alternative, all taxpayers pay 20% of their income in taxes, except that no one pays more than \$20,000 in taxes. In the second alternative, there are three tax brackets, described by the following table:

Income	*Marginal tax rate*
\$0–20,000	0%
\$20,000 100,000	20%
Above \$100,000	30%

Now you must determine which alternative will generate more annual tax revenue.

Each of Figures 27 and 28 is the graph of a function, T. Rather than using the formulas for these particular functions, you begin by working with the general situation. You have an income distribution function N' and a tax function T, both functions of annual income. You need to find a formula for total tax revenues. First you decide to use a cutoff so that you need to work only with incomes in some finite bracket $[0, M]$; you might use, for example, $M = \$10$ million. (Later you will let M approach $+\infty$.) Next, you subdivide the interval $[0, M]$ into a large number of intervals of small width, Δx. If $[x_{k-1}, x_k]$ is a typical such interval, you wish to calculate the approximate tax revenue from people whose total incomes lie between x_{k-1} and x_k. You will then sum over k to get the total revenue.

You need to know how many people are making incomes between x_{k-1} and x_k. Because $N(x_k)$ people are making incomes *up to* x_k and $N(x_{k-1})$ people are making incomes up to x_{k-1}, the number of people making incomes between x_{k-1} and x_k is $N(x_k) - N(x_{k-1})$. Because x_k is very close to x_{k-1}, the incomes of these people are all approximately equal to x_{k-1} dollars, so each of these taxpayers is paying an annual tax of about $T(x_{k-1})$. This gives a tax revenue of

$$[N(x_k) - N(x_{k-1})]T(x_{k-1}).$$

Now you do a clever thing. You write $x_k - x_{k-1} = \Delta x$ and replace $N(x_k) - N(x_{k-1})$ by

$$\frac{N(x_k) - N(x_{k-1})}{\Delta x}\Delta x.$$

This gives you a tax revenue of about

$$\frac{N(x_k) - N(x_{k-1})}{\Delta x}T(x_{k-1})\Delta x$$

from wage-earners in the bracket $[x_{k-1}, x_k]$. Summing over k gives an approximate total revenue of

$$\sum_{k=1}^{n}\frac{N(x_k) - N(x_{k-1})}{\Delta x}T(x_{k-1})\Delta x$$

where n is the number of subintervals. The larger n is, the more accurate your estimate will be, so you take the limit of the sum as $n \to \infty$. When you do this, two things happen. First, the quantity

$$\frac{N(x_k) - N(x_{k-1})}{\Delta x}$$

approaches the derivative, $N'(x_{k-1})$. Second, the sum, which you recognize as a Riemann sum, approaches the integral

$$\int_0^M N'(x)T(x)\,dx.$$

You now take the limit as $M \to +\infty$ to get

$$\text{Total tax revenue} = \int_0^{+\infty} N'(x)T(x)\,dx.$$

This improper integral is fine in theory, but the actual calculation will have to be done numerically, so you stick with the upper limit of \$10 million for now. You will have to check that it is reasonable at the end. (Notice that, by the graph of N', it appears that extremely few, if any, people earn that much.) Now you already have a formula for $N'(x)$, but you still need to write formulas for the tax functions $T(x)$ for both alternatives.

Alternative 1 The graph in Figure 27 rises linearly from 0 to 20,000 as x ranges from 0 to 100,000, and then stays constant at 20,000. The slope of the first part is $20{,}000/100{,}000 = 0.2$. The taxation function is therefore

$$T_1(x) = \begin{cases} 0.2x & \text{if } 0 \le x < 100{,}000 \\ 20{,}000 & \text{if } x \ge 100{,}000 \end{cases}.$$

For use of technology, it's convenient to express this in closed form using absolute values:

$$T_1(x) = 0.2x + \frac{1}{2}\left(1 + \frac{|x - 100{,}000|}{x - 100{,}000}\right)(20{,}000 - 0.2x)$$

The total revenue generated by this tax scheme is, therefore,

$$R_1 = \int_0^{10{,}000{,}000} (100x^{0.466}e^{-x/23{,}000}) \times \left[0.2x + \frac{1}{2}\left(1 + \frac{|x - 100{,}000|}{x - 100{,}000}\right)(20{,}000 - 0.2x)\right] dx$$

You decide not to attempt this by hand! You use numerical integration software to obtain a grand total of $R_1 = \$1{,}445{,}860{,}000{,}000$, or \$1.44586 trillion (rounded to six significant digits).

Alternative 2 The graph in Figure 28 rises with a slope of 0.2 from 0 to 16,000 as x ranges from 20,000 to 100,000, then rises from that point on with a slope of 0.3. (This is why we say that the *marginal* tax rates are 20% and 30%, respectively.) The taxation function is therefore

$$T_2(x) = \begin{cases} 0 & \text{if } 0 \le x < 20{,}000 \\ 0.2(x - 20{,}000) & \text{if } 20{,}000 \le x < 100{,}000. \\ 16{,}000 + 0.3(x - 100{,}000) & \text{if } x \ge 100{,}000 \end{cases}$$

Again, you express this in closed form using absolute values:

$$\begin{aligned} T_2(x) &= [0.2(x - 20{,}000)]\frac{1}{2}\left(\frac{|x - 20{,}000|}{x - 20{,}000} - \frac{|x - 100{,}000|}{x - 100{,}000}\right) \\ &\quad + [16{,}000 + 0.3(x - 100{,}000)]\frac{1}{2}\left(1 + \frac{|x - 100{,}000|}{x - 100{,}000}\right) \\ &= 0.1(x - 20{,}000)\left(\frac{|x - 20{,}000|}{x - 20{,}000} - \frac{|x - 100{,}000|}{x - 100{,}000}\right) \\ &\quad + [8{,}000 + 0.15(x - 100{,}000)]\left(1 + \frac{|x - 100{,}000|}{x - 100{,}000}\right) \end{aligned}$$

Values of x between 0 and 20,000 do not contribute to the integral, so

$$R_2 = \int_{20{,}000}^{10{,}000{,}000} 100x^{0.466}e^{-x/23{,}000}\,T_2(x)\,dx$$

with $T_2(x)$ as above. Numerical integration software gives $R_2 = \$0.777589$ trillion—considerably less than Alternative 1. Thus, even though Alternative 2 taxes the wealthy more heavily, it yields less total revenue.

Now what about the cutoff at \$10 million annual income? If you try either integral again with an upper limit of \$100 million, you will see no change in either result to six significant digits. There simply are not enough taxpayers earning an income above \$10,000,000 to make a difference. You conclude that your answers are sufficiently accurate and that the first alternative provides more tax revenue.

EXERCISES

In Exercises 1–4, calculate the total tax revenue for a country with the given income distribution and tax policies (all currency in dollars).

1. T $N'(x) = 100x^{0.466}e^{-x/23{,}000}$; 25% tax on all income

2. T $N'(x) = 100x^{0.4}e^{-x/30{,}000}$; 45% tax on all income

3. T $N'(x) = 100x^{0.466}e^{-x/23{,}000}$; tax brackets as in the following tax table:

Income	*Marginal tax rate*
\$0–30,000	0%
\$30,000–250,000	10%
Above \$250,000	80%

4. T $N'(x) = 100x^{0.4}e^{-x/30{,}000}$; no tax on any income below \$250,000, 100% marginal tax rate on any income above \$250,000

5. Let $N'(x)$ be an income distribution function.

a. If $0 \le a < b$, what does $\int_a^b N'(x)\,dx$ represent? **HINT** [Use the Fundamental Theorem of Calculus.]

b. What does $\int_0^{+\infty} N'(x)\,dx$ represent?

6. Let $N'(x)$ be an income distribution function. What does $\int_0^{+\infty} xN'(x)\,dx$ represent? **HINT** [Argue as in the text.]

7. Let $P(x)$ be the number of people earning more than x dollars.

a. What is $N(x) + P(x)$?

b. Show that $P'(x) = -N'(x)$.

c. Use integration by parts to show that, if $T(0) = 0$, then the total tax revenue is

$$\int_0^{+\infty} P(x)T'(x)\,dx.$$

[Note: You may assume that $T'(x)$ is continuous, but the result is still true if we assume only that $T(x)$ is continuous and piecewise continuously differentiable.]

8. Income tax functions T are most often described, as in the text, by tax brackets and marginal tax rates.

a. If one tax bracket is $a < x \le b$, show that $\int_a^b P(x)\,dx$ is the total income earned in the country that falls into that bracket (P as in the preceding exercise).

b. Use (a) to explain directly why $\int_0^{+\infty} P(x)T'(x)\,dx$ gives the total tax revenue in the case where T is described by tax brackets and constant marginal tax rates in each bracket.

TI-83/84 Plus Technology Guide

Section 7.3

Example 3 (page 538) The following table shows Colossal Conglomerate's closing stock prices for 20 consecutive trading days:

Day	1	2	3	4	5	6	7	8	9	10
Price	20	22	21	24	24	23	25	26	20	24
Day	11	12	13	14	15	16	17	18	19	20
Price	26	26	25	27	28	27	29	27	25	24

Plot these prices and the 5-day moving average.

Solution with Technology

Here is how to automate this calculation on a TI-83 Plus.

1. Use

```
seq(X,X,1,20)→L1
```

[2ND] [STAT] → OPS → 5
[STO] [2ND] [STAT] → L1

to enter the sequence of numbers 1 through 20 into the list L_1, representing the trading days.

2. Using the list editor accessible through the [STAT] menu, enter the daily stock prices in list L_2.

3. Calculate the list of 5-day moving averages by using the following command:

```
seq((L2(X)+L2(X-1)+L2(X-2)+L2(X-3)
  +L2(X-4))/5,X,5,20)→L3
```

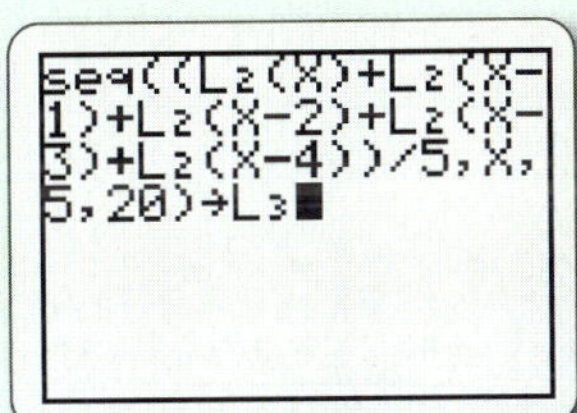

This has the effect of putting the moving averages into elements 1 through 15 of list L_3.

4. If you wish to plot the moving average on the same graph as the daily prices, you will want the averages in L_3 to match up with the prices in L_2. One way to do this is to put four more entries at the beginning of L_3—say, copies of the first four entries of L_2. The following command accomplishes this:

```
augment(seq(L2(X),X,1,4),L3)→L3
```

[2ND] [STAT] → OPS → 9

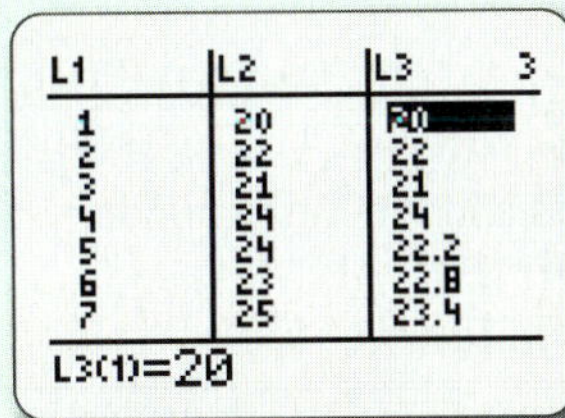

5. You can now graph the prices and moving averages by creating an xyLine scatter plot through the [STAT PLOT] menu, with L_1 being the Xlist and L_2 being the Ylist for Plot1, and L_1 being the Xlist and L_3 the Ylist for Plot2:

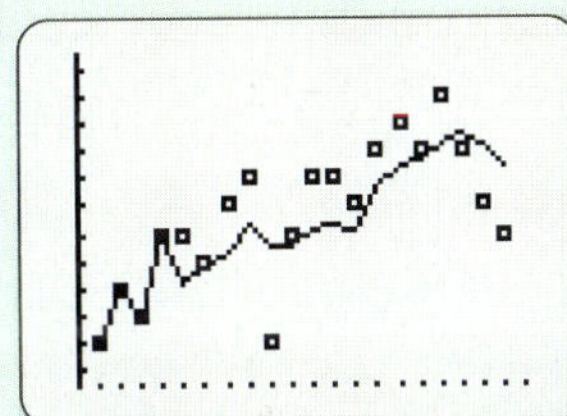

EXCEL Technology Guide

Section 7.3

Example 3 (page 538) The following table shows Colossal Conglomerate's closing stock prices for 20 consecutive trading days:

Day	1	2	3	4	5	6	7	8	9	10
Price	20	22	21	24	24	23	25	26	20	24
Day	11	12	13	14	15	16	17	18	19	20
Price	26	26	25	27	28	27	29	27	25	24

Plot these prices and the 5-day moving average.

Solution with Technology

1. Compute the moving averages in a column next to the daily prices, as shown here:

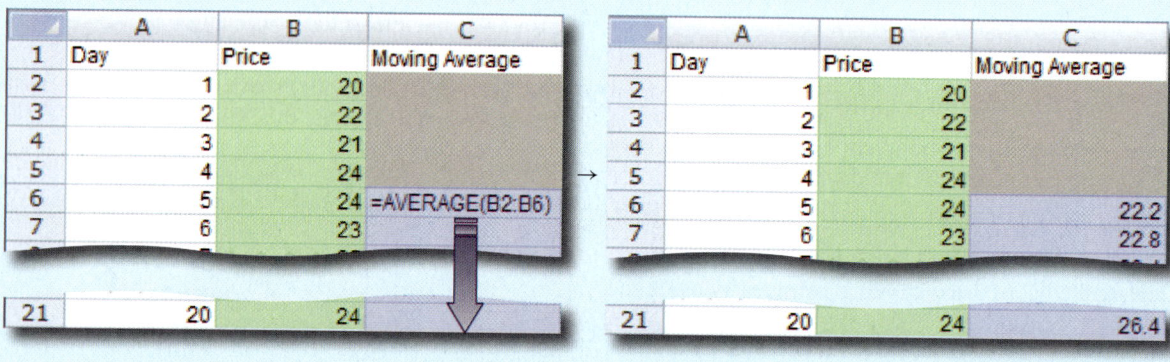

2. You can then graph the price and moving average using a scatter plot.

8 Functions of Several Variables

Web Site
www.AppliedCalc.org
At the Web site you will find:

- A detailed chapter summary
- A true/false quiz
- A surface grapher
- An Excel surface grapher
- A linear multiple regression utility
- The following optional extra sections:

  Maxima and Minima: Boundaries and the Extreme Value Theorem

  The Chain Rule for Functions of Several Variables

Case Study Modeling College Population

College Malls Inc. is planning to build a national chain of shopping malls in college neighborhoods. The company is planning to lease only to stores that target the specific age demographics of the national college student population. To decide which age brackets to target, the company has asked you, a paid consultant, for an analysis of the college population by student age, and of its trends over time. **How can you analyze the relevant data?**

David Pearson/Alamy

Introduction

We have studied functions of a single variable extensively. But not every useful function is a function of only one variable. In fact, most are not. For example, if you operate an online bookstore in competition with Amazon.com, BN.com, and Booksamillion.com, your sales may depend on those of your competitors. Your company's daily revenue might be modeled by a function such as

$$R(x, y, z) = 10{,}000 - 0.01x - 0.02y - 0.01z + 0.00001yz$$

where x, y, and z are the online daily revenues of Amazon.com, BN.com, and Booksamillion.com, respectively. Here, R is a function of three variables because it *depends on* x, y, and z. As we shall see, the techniques of calculus extend readily to such functions. Among the applications we shall look at is optimization: finding, where possible, the maximum or minimum of a function of two or more variables.

8.1 Functions of Several Variables from the Numerical, Algebraic, and Graphical Viewpoints

Numerical and Algebraic Viewpoints

Recall that a function of one variable is a rule for manufacturing a new number $f(x)$ from a single independent variable x. A function of two or more variables is similar, but the new number now depends on more than one independent variable.

Function of Several Variables

A **real-valued function,** f, **of** $x, y, z, \ldots$ is a rule for manufacturing a new number, written $f(x, y, z, \ldots)$, from the values of a sequence of independent variables $(x, y, z, \ldots)$. The function f is called a **real-valued function of two variables** if there are two independent variables, a **real-valued function of three variables** if there are three independent variables, and so on.

Quick Examples

1. $f(x, y) = x - y$	Function of two variables
$f(1, 2) = 1 - 2 = -1$	Substitute 1 for x and 2 for y.
$f(2, -1) = 2 - (-1) = 3$	Substitute 2 for x and -1 for y.
$f(y, x) = y - x$	Substitute y for x and x for y.
2. $g(x, y) = x^2 + y^2$	Function of two variables
$g(-1, 3) = (-1)^2 + 3^2 = 10$	Substitute -1 for x and 3 for y.
3. $h(x, y, z) = x + y + xz$	Function of three variables
$h(2, 2, -2) = 2 + 2 + 2(-2) = 0$	Substitute 2 for x, 2 for y, and -2 for z.

Note It is often convenient to use $x_1, x_2, x_3, \ldots$ for the independent variables, so that, for instance, the third example above would be $h(x_1, x_2, x_3) = x_1 + x_2 + x_1x_3$. ■

Figure 1 illustrates the concept of a function of two variables: In goes a pair of numbers and out comes a single number.

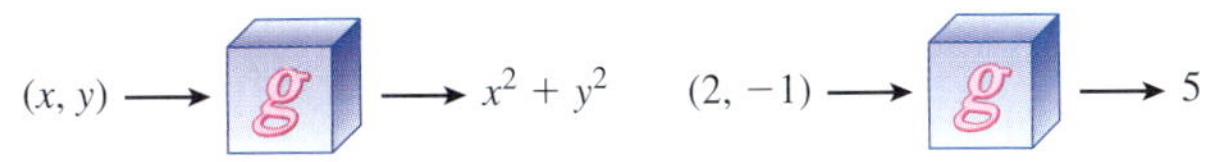

Figure 1

As with functions of one variable, functions of several variables can be represented numerically (using a table of values), algebraically (using a formula as in the above examples), and sometimes graphically (using a graph).

Let's now look at a number of examples of interesting functions of several variables.

Roy Mehta/Photonica/Getty Images

using Technology

See the Technology Guides at the end of the chapter to see how you can use a TI-83/84 Plus and Excel to display various values of $C(x, y)$ in Example 1. Here is an outline:

TI-83/84 Plus

```
Y1=10000+20X+40Y
```

To evaluate $C(10, 30)$:

```
10 → X
30 → Y
Y1
```

[More details on page 646.]

Excel

x-values down column A starting in A2

y-values down column B starting in B2

```
=10000+20*A2+40*B2
```

in C2; copy down column C. [More details on page 646.]

EXAMPLE 1 Cost Function

You own a company that makes two models of speakers: the Ultra Mini and the Big Stack. Your total monthly cost (in dollars) to make x Ultra Minis and y Big Stacks is given by

$$C(x, y) = 10{,}000 + 20x + 40y.$$

What is the significance of each term in this formula?

Solution The terms have meanings similar to those we saw for linear cost functions of a single variable. Let us look at the terms one at a time.

Constant Term Consider the monthly cost of making no speakers at all ($x = y = 0$). We find

$$C(0, 0) = 10{,}000. \qquad \text{Cost of making no speakers is \$10,000.}$$

Thus, the constant term 10,000 is the **fixed cost**, the amount you have to pay each month even if you make no speakers.

Coefficients of x and y Suppose you make a certain number of Ultra Minis and Big Stacks one month and the next month you increase production by one Ultra Mini. The costs are

$$\begin{aligned} C(x, y) &= 10{,}000 + 20x + 40y && \text{First month} \\ C(x+1, y) &= 10{,}000 + 20(x+1) + 40y && \text{Second month} \\ &= 10{,}000 + 20x + 20 + 40y \\ &= C(x, y) + 20. \end{aligned}$$

Thus, each Ultra Mini adds \$20 to the total cost. We say that \$20 is the **marginal cost** of each Ultra Mini. Similarly, because of the term $40y$, each Big Stack adds \$40 to the total cost. The marginal cost of each Big Stack is \$40.

This cost function is an example of a *linear function of two variables*. The coefficients of x and y play roles similar to that of the slope of a line. In particular, they give the rates of change of the function as each variable increases while the other stays constant (think about it). We shall say more about linear functions below.

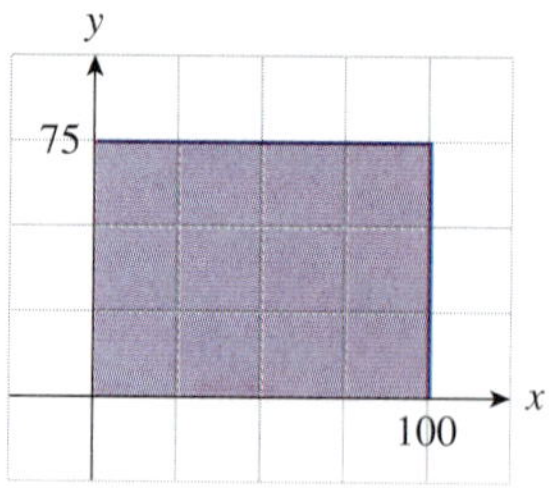

Figure 2

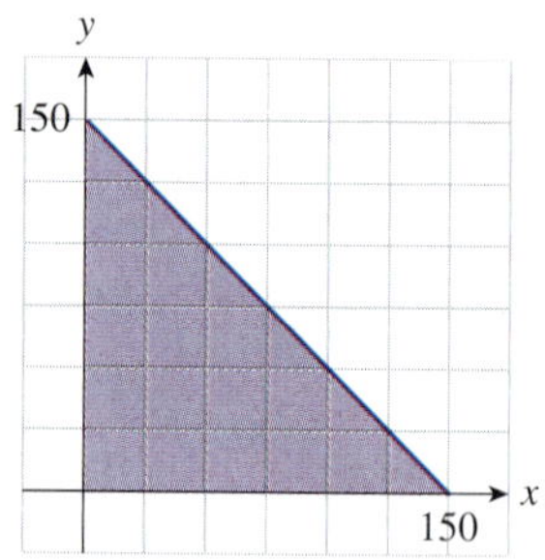

Figure 3

Before we go on... In Example 1 which values of x and y may we substitute into $C(x, y)$? Certainly we must have $x \geq 0$ and $y \geq 0$ because it makes no sense to speak of manufacturing a negative number of speakers. Also, there is certainly some upper bound to the number of speakers that can be made in a month. The bound might take one of several forms. The number of each model may be bounded—say $x \leq 100$ and $y \leq 75$. The inequalities $0 \leq x \leq 100$ and $0 \leq y \leq 75$ describe the region in the plane shaded in Figure 2.

Another possibility is that the *total* number of speakers is bounded—say, $x + y \leq 150$. This, together with $x \geq 0$ and $y \geq 0$, describes the region shaded in Figure 3.

In either case, the region shown represents the pairs (x, y) for which $C(x, y)$ is defined. Just as with a function of one variable, we call this region the **domain** of the function. As before, when the domain is not given explicitly, we agree to take the largest domain possible. ■

EXAMPLE 2 Faculty Salaries

David Katz came up with the following function for the salary of a professor with 10 years of teaching experience in a large university.

$$S(x, y, z) = 13{,}005 + 230x + 18y + 102z$$

Here, S is the salary in 1969–1970 in dollars per year, x is the number of books the professor has published, y is the number of articles published, and z is the number of "excellent" articles published.* What salary do you expect that a professor with 10 years' experience earned in 1969–1970 if she published 2 books, 20 articles, and 3 "excellent" articles?

Solution All we need to do is calculate

$$\begin{aligned} S(2, 20, 3) &= 13{,}005 + 230(2) + 18(20) + 102(3) \\ &= \$14{,}131. \end{aligned}$$

*David A. Katz, "Faculty Salaries, Promotions and Productivity at a Large University," *American Economic Review*, June 1973, pp. 469–477. Prof. Katz's equation actually included other variables, such as the number of dissertations supervised; our equation assumes that all of these are zero.

Before we go on... In Example 1, we gave a linear function of two variables. In Example 2 we have a linear function of three variables. Katz came up with his model by surveying a large number of faculty members and then finding the linear function "best" fitting the data. Such models are called **multiple linear regression** models. In the Case Study at the end of this chapter, we shall see a spreadsheet method of finding the coefficients of a multiple regression model from a set of observed data.

What does this model say about the value of a single book or a single article? If a book takes 15 times as long to write as an article, how would you recommend that a professor spend her writing time? ■

Here are two simple kinds of functions of several variables.

Linear Function

A function f of n variables is **linear** if f has the property that

$$f(x_1, x_2, \ldots, x_n) = a_0 + a_1x_1 + \cdots + a_nx_n. \qquad (a_0, a_1, a_2, \ldots, a_n \text{ constants})$$

Quick Examples

1. $f(x, y) = 3x - 5y$ — Linear function of x and y
2. $C(x, y) = 10{,}000 + 20x + 40y$ — Example 1
3. $S(x_1, x_2, x_3) = 13{,}005 + 230x_1 + 18x_2 + 102x_3$ — Example 2

Interaction Function

If we add to a linear function one or more terms of the form bx_ix_j (b a nonzero constant and $i \neq j$), we get a **second-order interaction function**.

Quick Examples

1. $C(x, y) = 10{,}000 + 20x + 40y + 0.1xy$
2. $R(x_1, x_2, x_3) = 10{,}000 - 0.01x_1 - 0.02x_2 - 0.01x_3 + 0.00001x_2x_3$

So far, we have been specifying functions of several variables **algebraically**—by using algebraic formulas. If you have ever studied statistics, you are probably familiar with statistical tables. These tables may also be viewed as representing functions **numerically**, as the next example shows.

EXAMPLE 3 Function Represented Numerically: Body Mass Index

The following table lists some values of the "body mass index," which gives a measure of the massiveness of your body, taking height into account.* The variable w represents your weight in pounds, and h represents your height in inches. An individual with a body mass index of 25 or above is generally considered overweight.

$h \downarrow$ \ $w \rightarrow$	130	140	150	160	170	180	190	200	210
60	25.2	27.1	29.1	31.0	32.9	34.9	36.8	38.8	40.7
61	24.4	26.2	28.1	30.0	31.9	33.7	35.6	37.5	39.4
62	23.6	25.4	27.2	29.0	30.8	32.7	34.5	36.3	38.1
63	22.8	24.6	26.4	28.1	29.9	31.6	33.4	35.1	36.9
64	22.1	23.8	25.5	27.2	28.9	30.7	32.4	34.1	35.8
65	21.5	23.1	24.8	26.4	28.1	29.7	31.4	33.0	34.7
66	20.8	22.4	24.0	25.6	27.2	28.8	30.4	32.0	33.6
67	20.2	21.8	23.3	24.9	26.4	28.0	29.5	31.1	32.6
68	19.6	21.1	22.6	24.1	25.6	27.2	28.7	30.2	31.7
69	19.0	20.5	22.0	23.4	24.9	26.4	27.8	29.3	30.8
70	18.5	19.9	21.4	22.8	24.2	25.6	27.0	28.5	29.9
71	18.0	19.4	20.8	22.1	23.5	24.9	26.3	27.7	29.1
72	17.5	18.8	20.2	21.5	22.9	24.2	25.6	26.9	28.3
73	17.0	18.3	19.6	20.9	22.3	23.6	24.9	26.2	27.5
74	16.6	17.8	19.1	20.4	21.7	22.9	24.2	25.5	26.7
75	16.1	17.4	18.6	19.8	21.1	22.3	23.6	24.8	26.0
76	15.7	16.9	18.1	19.3	20.5	21.7	22.9	24.2	25.4

* **NOTE** It is interesting that weight-lifting competitions are usually based on weight, rather than body mass index. As a consequence, taller people are at a significant disadvantage in these competitions because they must compete with shorter, stockier people of the same weight. (An extremely thin, very tall person can weigh as much as a muscular short person, although his body mass index would be significantly lower.)

As the table shows, the value of the body mass index depends on two quantities: w and h. Let us write $M(w, h)$ for the body mass index function. What are $M(140, 62)$ and $M(210, 63)$?

Solution We can read the answers from the table:

$$M(140, 62) = 25.4 \qquad w = 140 \text{ lb}, h = 62 \text{ in}$$

and

$$M(210, 63) = 36.9. \qquad w = 210 \text{ lb}, h = 63 \text{ in}$$

The function $M(w, h)$ is actually given by the formula

$$M(w, h) = \frac{0.45w}{(0.0254h)^2}.$$

[The factor 0.45 converts the weight to kilograms, and 0.0254 converts the height to meters. If w is in kilograms and h is in meters, the formula is simpler: $M(w, h) = w/h^2$.]

using Technology

See the Technology Guides at the end of the chapter to see how to use Excel to create the table in Example 3. Here is an outline:

Excel

w-values 130 to 210 in B1–J1

h-values 60 to 76 in A2–A18

```
=0.45*B$1/
(0.0254*$A2)^2
```

in B2; copy down and across through J18. [More details on page 646.]

Geometric Viewpoint: Three-Dimensional Space and the Graph of a Function of Two Variables

Just as functions of a single variable have graphs, so do functions of two or more variables. Recall that the graph of $f(x)$ consists of all points $(x, f(x))$ in the xy-plane. By analogy, we would like to say that the graph of a function of *two* variables, $f(x, y)$, consists of all points of the form $(x, y, f(x, y))$. Thus, we need three axes: the x-, y-, and z-axes. In other words, our graph will live in **three-dimensional space**, or **3-space**.*

*** NOTE** If we were dealing instead with a function of *three* variables, then we would need to go to *four-dimensional* space. Here we run into visualization problems (to say the least!) so we won't discuss the graphs of functions of three or more variables in this text.

Just as we had two mutually perpendicular axes in two-dimensional space (the xy-plane; see Figure 4(a)), so we have three mutually perpendicular axes in three-dimensional space (Figure 4(b)).

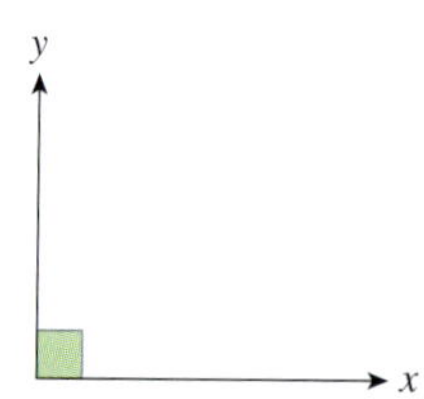

Two-dimensional space

Figure 4(a)

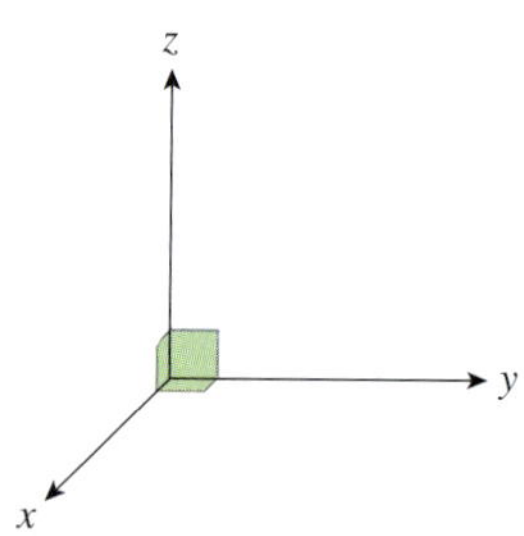

Three-dimensional space

Figure 4(b)

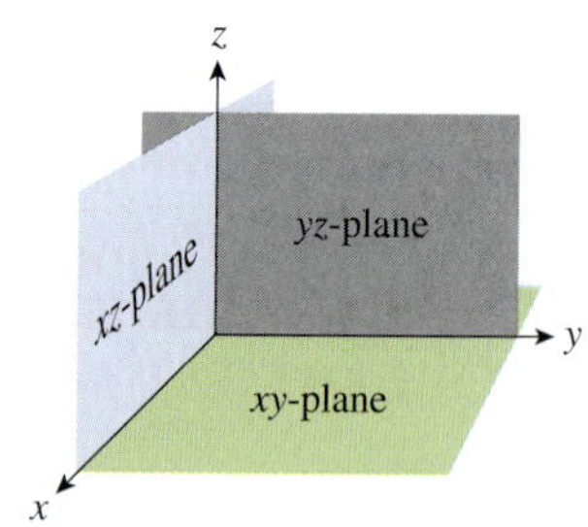

Figure 5

In both 2-space and 3-space, the axis labeled with the last letter goes up. Thus, the z-direction is the "up" direction in 3-space, rather than the y-direction.

Three important planes are associated with these axes: the xy-plane, the yz-plane, and the xz-plane. These planes are shown in Figure 5. Any two of these planes intersect in one of the axes (for example, the xy- and xz-planes intersect in the x-axis) and all three meet at the origin. Notice that the xy-plane consists of all points with z-coordinate zero, the xz-plane consists of all points with $y = 0$, and the yz-plane consists of all points with $x = 0$.

In 3-space, each point has *three* coordinates, as you might expect: the x-coordinate, the y-coordinate, and the z-coordinate. To see how this works, look at the following examples.

EXAMPLE 4 Plotting Points in Three Dimensions

Locate the points $P(1, 2, 3)$, $Q(-1, 2, 3)$, $R(1, -1, 0)$, and $S(1, 2, -2)$ in 3-space.

Solution To locate P, the procedure is similar to the one we used in 2-space: Start at the origin, proceed one unit in the x direction, then proceed two units in the y direction, and finally, proceed three units in the z direction. We wind up at the point P shown in Figures 6(a) and 6(b).

Here is another, extremely useful way of thinking about the location of P. First, look at the x- and y-coordinates, obtaining the point $(1, 2)$ in the xy-plane. The point we want is then three units vertically above the point $(1, 2)$ because the z-coordinate of a point is just its height above the xy-plane. This strategy is shown in Figure 6(c).

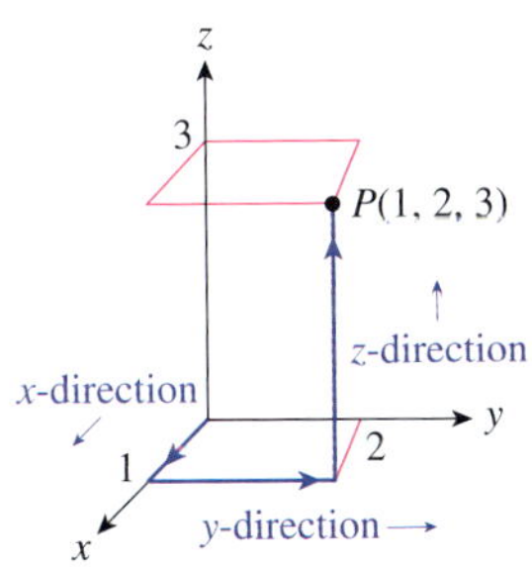

Figure 6(a)

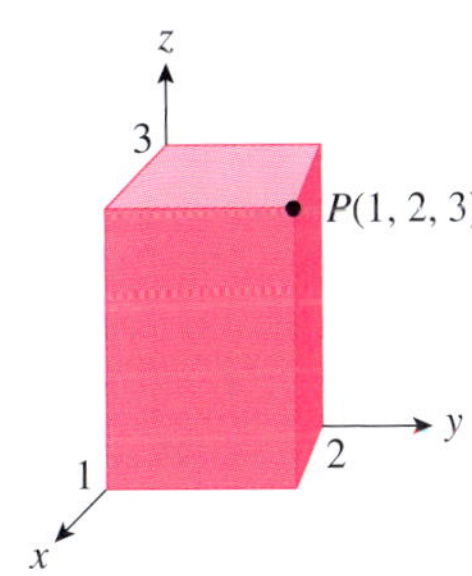

Figure 6(b)

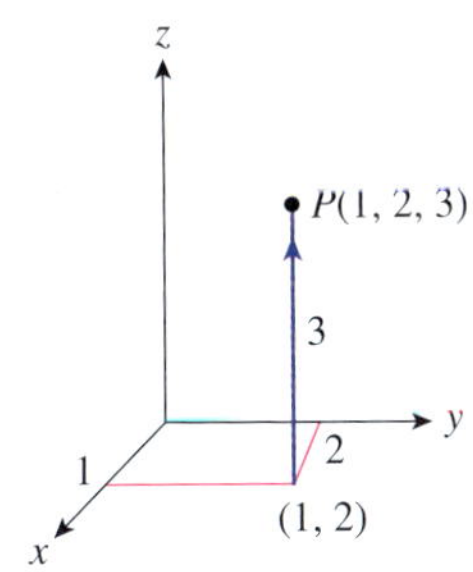

Figure 6(c)

Plotting the points Q, R, and S is similar, using the convention that negative coordinates correspond to moves back, left, or down. (See Figure 7.)

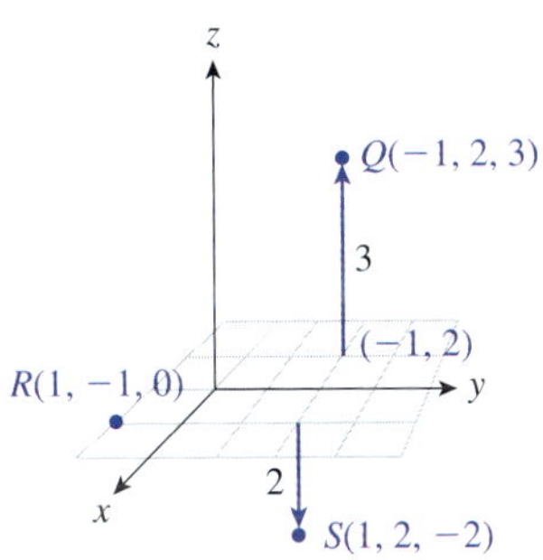

Figure 7

Our next task is to describe the graph of a function $f(x, y)$ of two variables.

Graph of a Function of Two Variables

The **graph of the function f of two variables** is the set of all points $(x, y, f(x, y))$ in three-dimensional space, where we restrict the values of (x, y) to lie in the domain of f. In other words, the graph is the set of all the points (x, y, z) with $z = f(x, y)$.

For *every* point (x, y) in the domain of f, the z-coordinate of the corresponding point on the graph is given by evaluating the function at (x, y). Thus, there will be a point on the graph above *every* point in the domain of f, so that the graph is usually a *surface* of some sort.

EXAMPLE 5 Graph of a Function of Two Variables

Describe the graph of $f(x, y) = x^2 + y^2$.

Solution Your first thought might be to make a table of values. You could choose some values for x and y and then, for each such pair, calculate $z = x^2 + y^2$. For example, you might get the following table:

$y \downarrow$ \ $x \rightarrow$	**−1**	**0**	**1**
−1	2	1	2
0	1	0	1
1	2	1	2

$f(x, y) = x^2 + y^2$

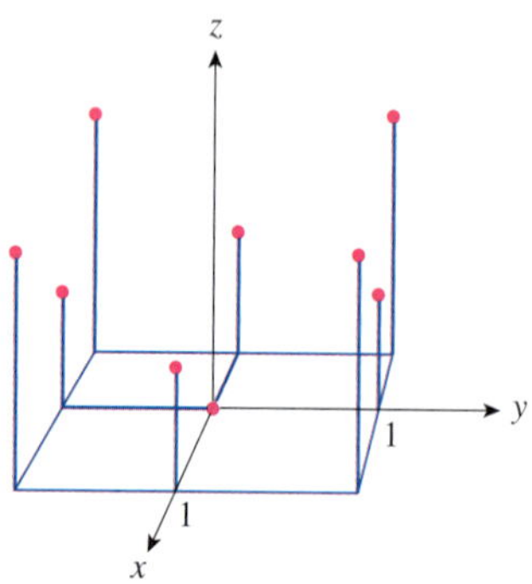

Figure 8

This gives the following nine points on the graph of f: $(-1, -1, 2)$, $(-1, 0, 1)$, $(-1, 1, 2)$, $(0, -1, 1)$, $(0, 0, 0)$, $(0, 1, 1)$, $(1, -1, 2)$, $(1, 0, 1)$, and $(1, 1, 2)$. These points are shown in Figure 8.

The points on the xy-plane we chose for our table are the grid points in the xy-plane, and the corresponding points on the graph are marked with solid dots. The problem is that this small number of points hardly tells us what the surface looks like, and even if we plotted more points, it is not clear that we would get anything more than a mass of dots on the page.

What can we do? There are several alternatives. One place to start is to use technology to draw the graph. (See the technology note on the next page.) We then obtain something like Figure 9. This particular surface is called a **paraboloid**.

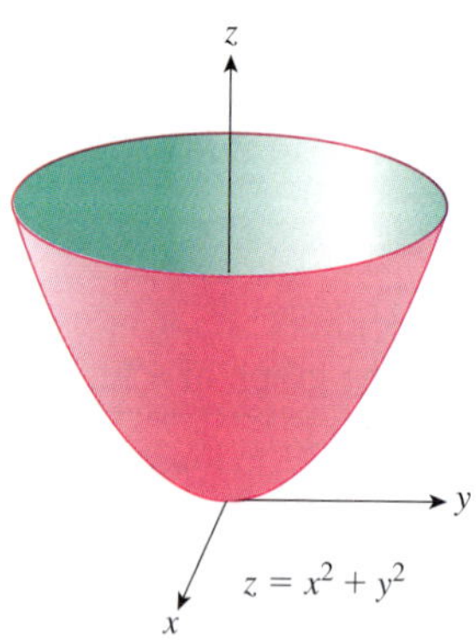

Figure 9

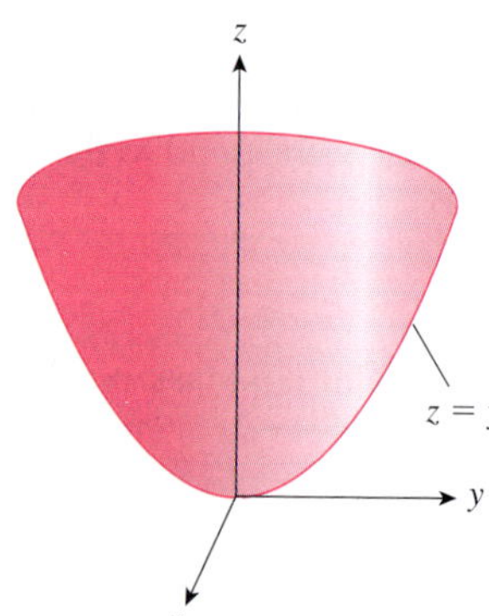

Figure 10

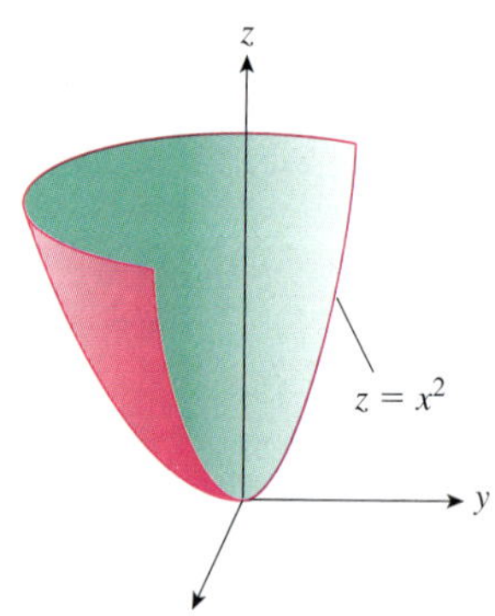

Figure 11

If we slice vertically through this surface along the yz-plane, we get the picture in Figure 10. The shape of the front edge, where we cut, is a parabola. To see why, note that the yz-plane is the set of points where $x = 0$. To get the intersection of $x = 0$ and $z = x^2 + y^2$, we substitute $x = 0$ in the second equation, getting $z = y^2$. This is the equation of a parabola in the yz-plane.

Similarly, we can slice through the surface with the xz-plane by setting $y = 0$. This gives the parabola $z = x^2$ in the xz-plane (Figure 11).

✱ NOTE See Section 0.7 for a discussion of equations of circles.

using Technology

We can use technology to obtain the graph of the function in Example 5:

Excel

Table of values:
x-values -3 to 3 in B1–H1
y-values -3 to 3 in A2–A8
`=B$1^2+$A2^2`
in B2; copy down and across through H8.
Graph: Highlight A1 through H8 and insert a Surface chart.
[More details on page 647.]

Web Site

www.AppliedCalc.org
Everything for Calculus
→ Chapter 8
→ Math Tools (Surface Graphing Utility)
Enter `x^2+y^2` for $f(x, y)$
Set xMin = −3, xMax = 3,
yMin = −3, yMax = 3
Press "Graph."

We can also look at horizontal slices through the surface, that is, slices by planes parallel to the xy-plane. These are given by setting $z = c$ for various numbers c. For example, if we set $z = 1$, we will see only the points with height 1. Substituting in the equation $z = x^2 + y^2$ gives the equation

$$1 = x^2 + y^2,$$

which is the equation of a circle of radius 1.* If we set $z = 4$, we get the equation of a circle of radius 2:

$$4 = x^2 + y^2.$$

In general, if we slice through the surface at height $z = c$, we get a circle (of radius $\sqrt{c}$). Figure 12 shows several of these circles.

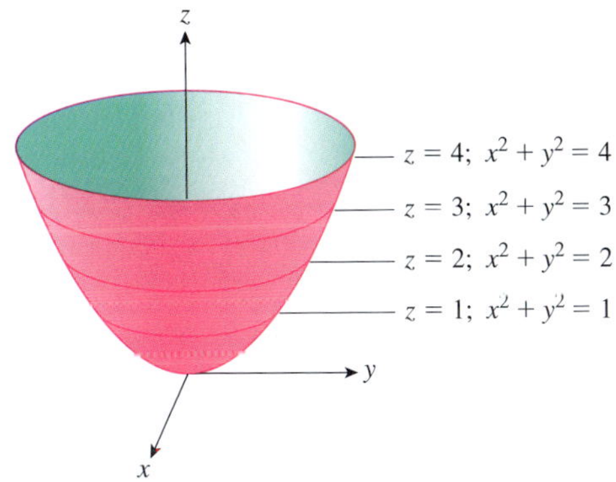

Figure 12

Looking at these circular slices, we see that this surface is the one we get by taking the parabola $z = x^2$ and spinning it around the z-axis. This is an example of what is known as a **surface of revolution**.

➡Before we go on... The graph of any function of the form $f(x, y) = Ax^2 + By^2 + Cxy + Dx + Ey + F$ ($A, B, \ldots, F$ constants), with $4AB - C^2$ positive, can be shown to be a paraboloid of the same general shape as that in Example 5 if A and B are positive, or upside-down if A and B are negative. If $A \neq B$, the horizontal slices will be ellipses rather than circles.

Notice that each horizontal slice through the surface in Example 5 was obtained by putting $z = constant$. This gave us an equation in x and y that described a curve. These curves are called the **level curves** of the surface $z = f(x, y)$. In Example 5, the equations are of the form $x^2 + y^2 = c$ (c constant), and so the level curves are circles. Figure 13 shows the level curves for $c = 0, 1, 2, 3,$ and 4.

The level curves give a contour map or topographical map of the surface. Each curve shows all of the points on the surface at a particular height c. You can use this contour map to visualize the shape of the surface. Imagine moving the contour at $c = 1$ to a height of 1 unit above the xy-plane, the contour at $c = 2$ to a height of 2 units above the xy-plane, and so on. You will end up with something like Figure 12. ■

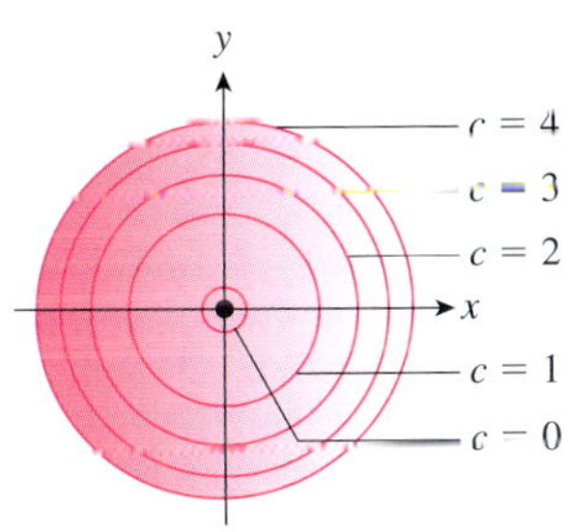

Level curves of the paraboloid $z = x^2 + y^2$

Figure 13

The following summary includes the techniques we have just used plus some additional ones:

Analyzing the Graph of a Function of Two Variables

If possible, use technology to render the graph of a given function $z = f(x, y)$. Given the function $z = f(x, y)$, you can analyze its graph as follows:

Step 1 Obtain the ***x*-, *y*-, and *z*-intercepts** (the places where the surface crosses the coordinate axes).

***x*-Intercept(s):** Set $y = 0$ and $z = 0$ and solve for x.

***y*-Intercept(s):** Set $x = 0$ and $z = 0$ and solve for y.

***z*-Intercept:** Set $x = 0$ and $y = 0$ and compute z.

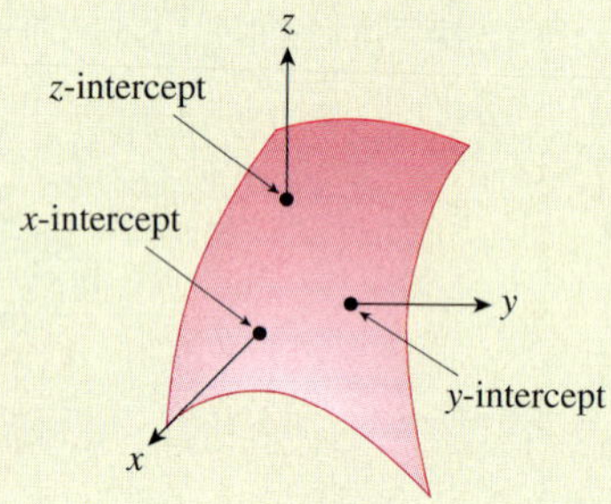

Step 2 Slice the surface along planes parallel to the *xy*-, *yz*-, and *xz*-planes.

z* = *constant **(level curves)** Set $z = constant$ and analyze the resulting curves. These are the curves resulting from horizontal slices.

x* = *constant Set $x = constant$ and analyze the resulting curves. These are the curves resulting from slices parallel to the *yz*-plane.

y* = *constant Set $y = constant$ and analyze the resulting curves. These are the curves resulting from slices parallel to the *xz*-plane.

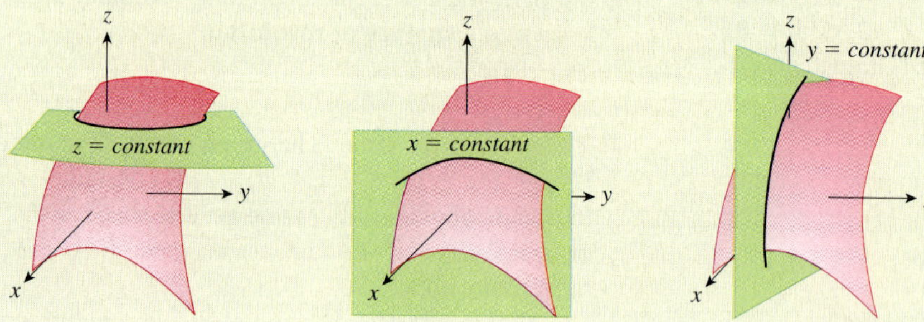

Spreadsheets often have built-in features to render surfaces such as the paraboloid in Example 5. In the following example, we use Excel to graph another surface and then analyze it as above.

EXAMPLE 6 **Analyzing a Surface**

Describe the graph of $f(x, y) = x^2 - y^2$.

Solution First we obtain a picture of the graph using technology. Figure 14 shows two graphs obtained using resources at the Web site.

Chapter 8 → Math Tools for Chapter 8 → Surface Graphing Utility

Chapter 8 → Math Tools for Chapter 8 → Excel Surface Graphing Utility

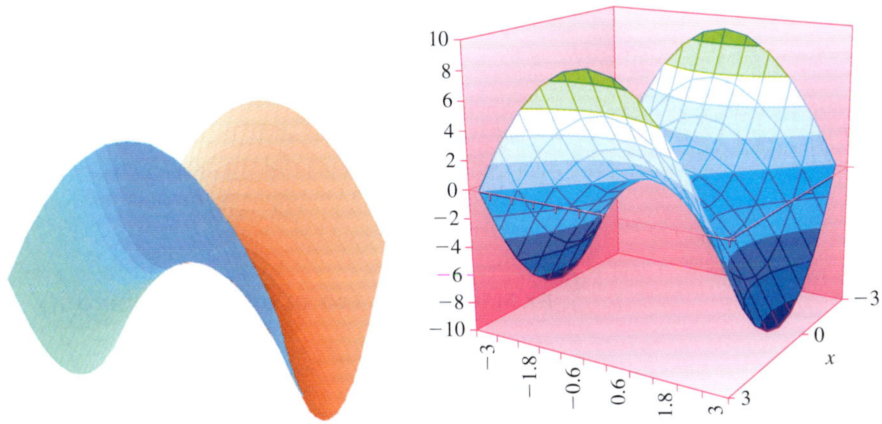

Figure 14

See the Technology Guides at the end of the chapter to find out how to obtain a similar graph from scratch on an ordinary Excel sheet.

The graph shows an example of a "saddle point" at the origin. (We return to this idea in Section 8.3.) To analyze the graph for the features shown in the box above, replace $f(x, y)$ by z to obtain

$$z = x^2 - y^2.$$

Step 1: *Intercepts* Setting any two of the variables x, y, and z equal to zero results in the third also being zero, so the x-, y-, and z-intercepts are all 0. In other words, the surface touches all three axes in exactly one point, the origin.

Step 1: *Slices* Slices in various directions show more interesting features.

Slice by $x = c$ This gives $z = c^2 - y^2$, which is the equation of a parabola that opens downward. You can see two of these slices ($c = -3$, $c = 3$) as the front and back edges of the surface in Figure 14. (More are shown in Figure 15(a).)

Slice by $y = c$ This gives $z = x^2 - c^2$, which is the equation of a parabola once again—this time, opening upward. You can see two of these slices ($c = -3$, $c = 3$) as the left and right edges of the surface in Figure 14. (More are shown in Figure 15(b).)

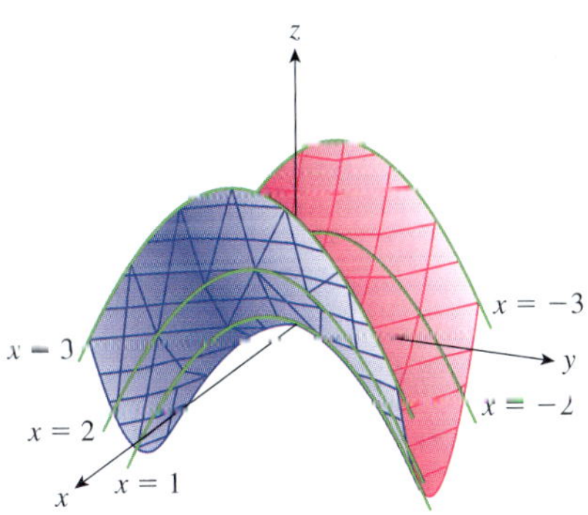

x = constant

Figure 15(a)

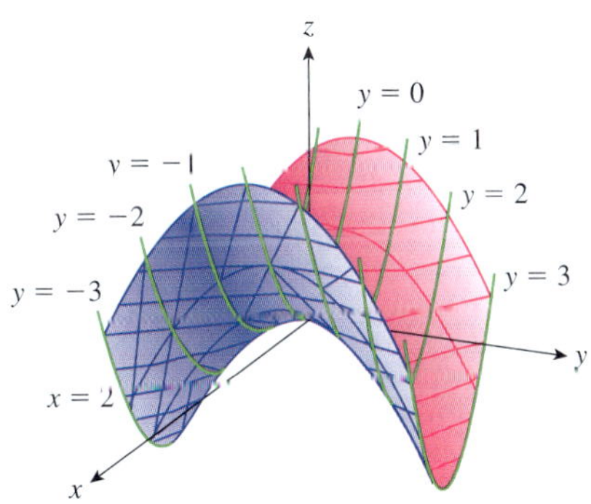

y = constant

Figure 15(b)

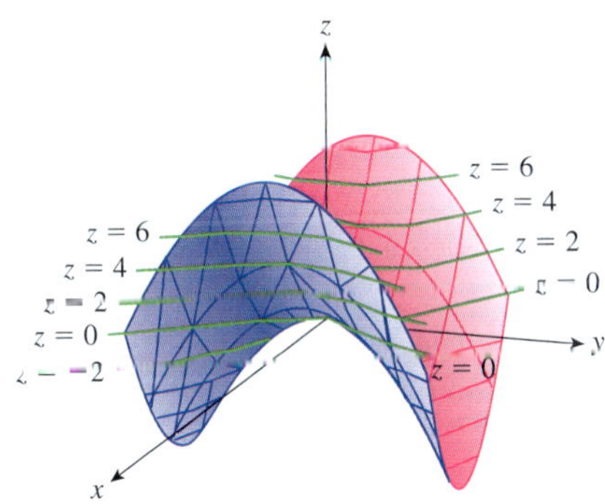

z = constant

Figure 15(c)

Slice by $z = c$ This gives $x^2 - y^2 = c$, which is a hyperbola. The level curves for various values of c are visible in Figure 14 as the horizontal slices. (See Figure 15(c).) The case $c = 0$ is interesting: The equation $x^2 - y^2 = 0$ can be rewritten as $x = \pm y$ (why?), which represents two lines at right-angles to each other.

To obtain really beautiful renderings of surfaces, you could use one of the commercial computer algebra software packages, such as Mathematica® or Maple®, or, if you use a Mac, the built-in grapher (grapher.app located in the Utilities folder).

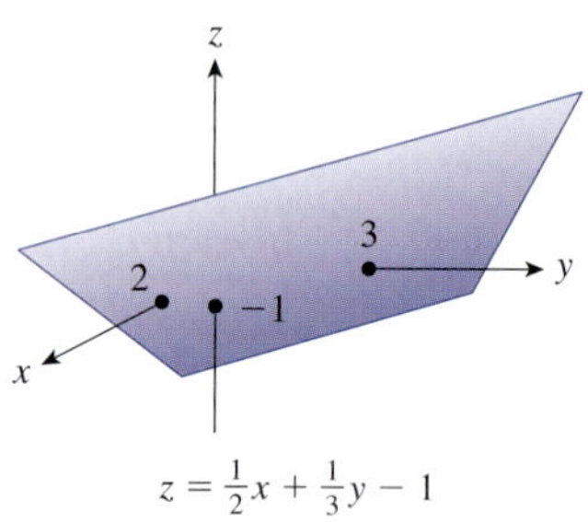

Figure 16

EXAMPLE 7 Graph of a Linear Function

Describe the graph of $g(x, y) = \frac{1}{2}x + \frac{1}{3}y - 1$.

Solution Notice first that g is a linear function of x and y. Figure 16 shows a portion of the graph, which is a plane.

We can get a good idea of what plane this is by looking at the x-, y-, and z-intercepts.

x-intercept Set $y = z = 0$, which gives $x = 2$.

y-intercept Set $x = z = 0$, which gives $y = 3$.

z-intercept Set $x = y = 0$, which gives $z = -1$.

Three points are enough to define a plane, so we can say that the plane is the one passing through the three points (2, 0, 0), (0, 3, 0), and (0, 0, −1).

Note It can be shown that the graph of every linear function of two variables is a plane. What do the level curves look like? ■

8.1 EXERCISES

▼ more advanced ◆ challenging

T indicates exercises that should be solved using technology

*For each function in Exercises 1–4, evaluate **(a)** $f(0, 0)$; **(b)** $f(1, 0)$; **(c)** $f(0, -1)$; **(d)** $f(a, 2)$; **(e)** $f(y, x)$; **(f)** $f(x + h, y + k)$* HINT [See Quick Examples page 584.]

1. $f(x, y) = x^2 + y^2 - x + 1$

2. $f(x, y) = x^2 - y - xy + 1$

3. $f(x, y) = 0.2x + 0.1y - 0.01xy$

4. $f(x, y) = 0.4x - 0.5y - 0.05xy$

*For each function in Exercises 5–8, evaluate **(a)** $g(0, 0, 0)$; **(b)** $g(1, 0, 0)$; **(c)** $g(0, 1, 0)$; **(d)** $g(z, x, y)$; **(e)** $g(x + h, y + k, z + l)$, provided such a value exists.*

5. $g(x, y, z) = e^{x+y+z}$

6. $g(x, y, z) = \ln(x + y + z)$

7. $g(x, y, z) = \frac{xyz}{x^2 + y^2 + z^2}$

8. $g(x, y, z) = \frac{e^{xyz}}{x + y + z}$

9. Let $f(x, y, z) = 1.5 + 2.3x - 1.4y - 2.5z$. Complete the following sentences. HINT [See Example 1.]

a. f ___ by ___ units for every 1 unit of increase in x.
b. f ___ by ___ units for every 1 unit of increase in y.
c. _______ by 2.5 units for every _______.

10. Let $g(x, y, z) = 0.01x + 0.02y - 0.03z - 0.05$. Complete the following sentences.

a. g ___ by ___ units for every 1 unit of increase in z.
b. g ___ by ___ units for every 1 unit of increase in x.
c. _______ by 0.02 units for every _______.

In Exercises 11–18, classify each function as linear, interaction, or neither. HINT [See Quick Examples page 587.]

11. $L(x, y) = 3x - 2y + 6xy - 4y^2$

12. $L(x, y, z) = 3x - 2y + 6xz$

13. $P(x_1, x_2, x_3) = 0.4 + 2x_1 - x_3$

14. $Q(x_1, x_2) = 4x_2 - 0.5x_1 - x_1^2$

15. $f(x, y, z) = \dfrac{x + y - z}{3}$

16. $g(x, y, z) = \dfrac{xz - 3yz + z^2}{4z} \quad (z \neq 0)$

17. $g(x, y, z) = \dfrac{xz - 3yz + z^2 y}{4z} \quad (z \neq 0)$

18. $f(x, y) = x + y + xy + x^2 y$

In Exercises 19 and 20, use the given tabular representation of the function f to compute the quantities asked for. HINT [See Example 3.]

19. $x \rightarrow$ (columns), $y \downarrow$ (rows)

	10	20	30	40
10	−1	107	162	−3
20	−6	194	294	−14
30	−11	281	426	−25
40	−16	368	558	−36

a. $f(20, 10)$ **b.** $f(40, 20)$ **c.** $f(10, 20) - f(20, 10)$

20. $x \rightarrow$ (columns), $y \downarrow$ (rows)

	10	20	30	40
10	162	107	−5	−7
20	294	194	−22	−30
30	426	281	−39	−53
40	558	368	−56	−76

a. $f(10, 30)$ **b.** $f(20, 10)$ **c.** $f(10, 40) + f(10, 20)$

T *In Exercises 21 and 22, use a spreadsheet or some other method to complete the given tables.*

21. $P(x, y) = x - 0.3y + 0.45xy$

$x \rightarrow$ (columns), $y \downarrow$ (rows)

	10	20	30	40
10				
20				
30				
40				

22. $Q(x, y) = 0.4x + 0.1y - 0.06xy$

$x \rightarrow$ (columns), $y \downarrow$ (rows)

	10	20	30	40
10				
20				
30				
40				

23. T ▼ The following statistical table lists some values of the "Inverse F distribution" ($\alpha = 0.5$):

$n \rightarrow$ (columns), $d \downarrow$ (rows)

	1	2	3	4	5	6	7	8	9	10
1	161.4	199.5	215.7	224.6	230.2	234.0	236.8	238.9	240.5	241.9
2	18.51	19.00	19.16	19.25	19.30	19.33	19.35	19.37	19.39	19.40
3	10.13	9.552	9.277	9.117	9.013	8.941	8.887	8.812	8.812	8.785
4	7.709	6.944	6.591	6.388	6.256	6.163	6.094	5.999	5.999	5.964
5	6.608	5.786	5.409	5.192	5.050	4.950	4.876	4.772	4.772	4.735
6	5.987	5.143	4.757	4.534	4.387	4.284	4.207	4.099	4.099	4.060
7	5.591	4.737	4.347	4.120	3.972	3.866	3.787	3.677	3.677	3.637
8	5.318	4.459	4.066	3.838	3.688	3.581	3.500	3.388	3.388	3.347
9	5.117	4.256	3.863	3.633	3.482	3.374	3.293	3.179	3.179	3.137
10	4.965	4.103	3.708	3.478	3.326	3.217	3.135	3.020	3.020	2.978

In Excel, you can compute the value of this function at (n, d) by the formula

`=FINV(0.05, n, d)` The 0.05 is the value of alpha (α).

Use Excel to re-create this table.

24. T ▼ The formula for body mass index $M(w, h)$, if w is given in kilograms and h is given in meters, is

$$M(w, h) = \frac{w}{h^2}$$ See Example 3.

Use this formula to complete the following table in Excel:

$w \rightarrow$ (columns), $h \downarrow$ (rows)

	70	80	90	100	110	120	130
1.8							
1.85							
1.9							
1.95							
2							
2.05							
2.1							
2.15							
2.2							
2.25							
2.3							

T *In Exercises 25–28, use either a graphing calculator or a spreadsheet to complete each table. Express all your answers as decimals rounded to four decimal places.*

25.

x	y	$f(x, y) = x^2\sqrt{1 + xy}$
3	1	
1	15	
0.3	0.5	
56	4	

26.

x	y	$f(x, y) = x^2e^y$
0	2	
−1	5	
1.4	2.5	
11	9	

27.

x	y	$f(x, y) = x\ln(x^2 + y^2)$
3	1	
1.4	−1	
e	0	
0	e	

28.

x	y	$f(x, y) = \dfrac{x}{x^2 - y^2}$
−1	2	
0	0.2	
0.4	2.5	
10	0	

29. ▼ Brand Z's annual sales are affected by the sales of related products X and Y as follows: Each \$1 million increase in sales of brand X causes a \$2.1 million decline in sales of brand Z, whereas each \$1 million increase in sales of brand Y results in an increase of \$0.4 million in sales of brand Z. Currently, brands X, Y, and Z are each selling \$6 million per year. Model the sales of brand Z using a linear function.

30. ▼ Let $f(x, y, z) = 43.2 - 2.3x + 11.3y - 4.5z$. Complete the following: An increase of 1 in the value of y causes the value of f to ___ by ___, whereas increasing the value of x by 1 and ___ the value of z by ___ causes a decrease of 11.3 in the value of f.

31. Sketch the cube with vertices (0, 0, 0), (1, 0, 0), (0, 1, 0), (0, 0, 1), (1, 1, 0), (1, 0, 1), (0, 1, 1), and (1, 1, 1). HINT [See Example 4.]

32. Sketch the cube with vertices (−1, −1, −1), (1, −1, −1), (−1, 1, −1), (−1, −1, 1), (1, 1, −1), (1, −1, 1), (−1, 1, 1), and (1, 1, 1). HINT [See Example 4.]

33. Sketch the pyramid with vertices (1, 1, 0), (1, −1, 0), (−1, 1, 0), (−1, −1, 0), and (0, 0, 2).

34. Sketch the solid with vertices (1, 1, 0), (1, −1, 0), (−1, 1, 0), (−1, −1, 0), (0, 0, −1), and (0, 0, 1).

Sketch the planes in Exercises 35–40.

35. $z = -2$ **36.** $z = 4$

37. $y = 2$ **38.** $y = -3$

39. $x = -3$ **40.** $x = 2$

Match each equation in Exercises 41–48 with one of the graphs below. (If necessary, use technology to render the surfaces.) HINT [See Examples 5, 6, and 7.]

41. $f(x, y) = 1 - 3x + 2y$ **42.** $f(x, y) = 1 - \sqrt{x^2 + y^2}$

43. $f(x, y) = 1 - (x^2 + y^2)$ **44.** $f(x, y) = y^2 - x^2$

45. $f(x, y) = -\sqrt{1 - (x^2 + y^2)}$

46. $f(x, y) = 1 + (x^2 + y^2)$

47. $f(x, y) = \dfrac{1}{x^2 + y^2}$ **48.** $f(x, y) = 3x - 2y + 1$

(A)

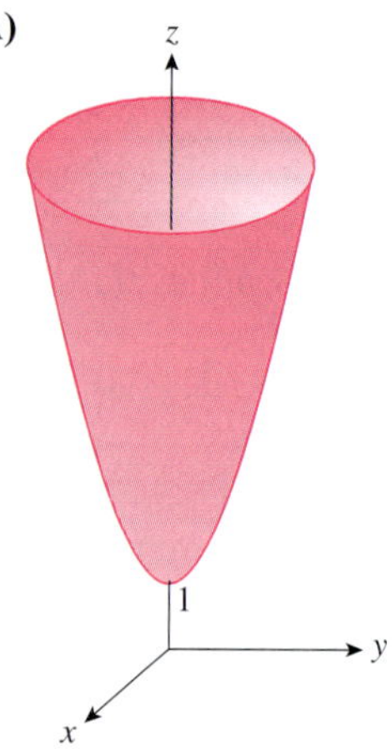

(B)

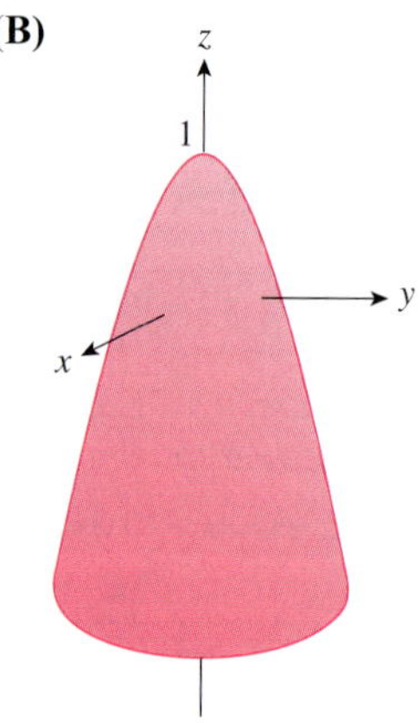

(C)

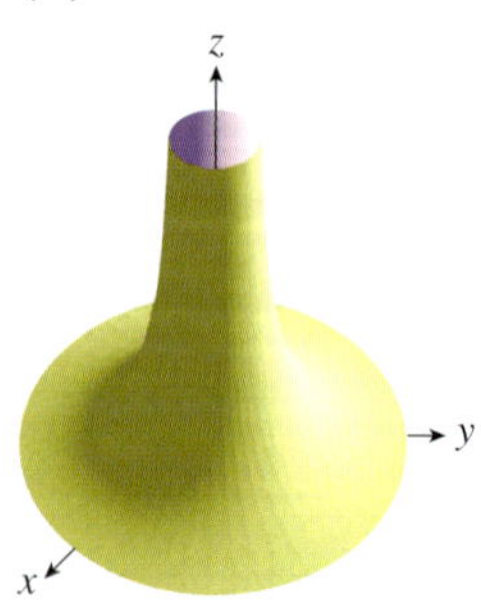

(D)

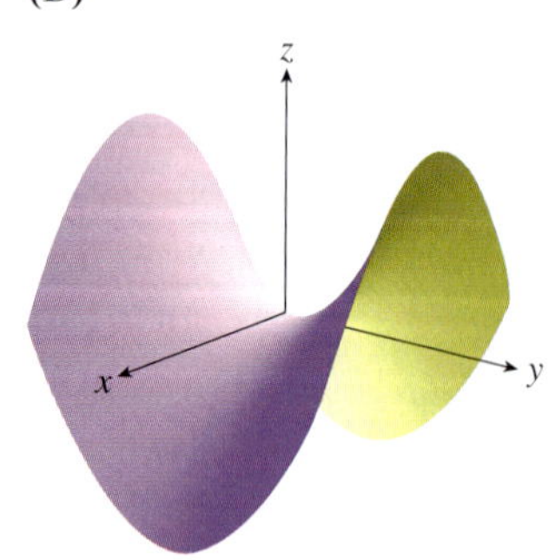

(E)

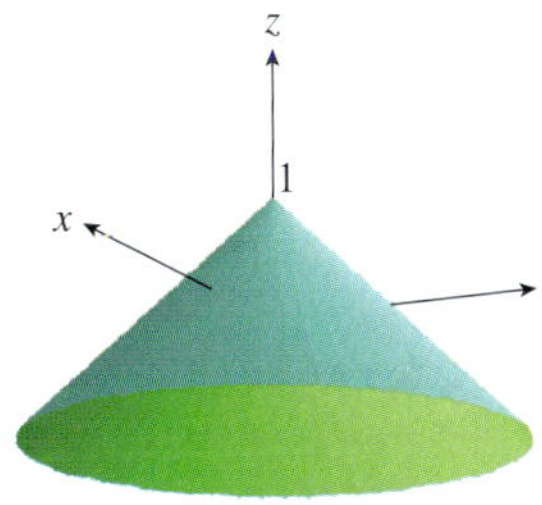

(F)

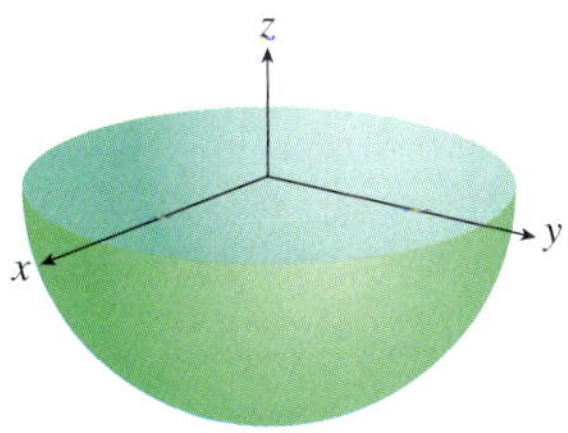

(G)

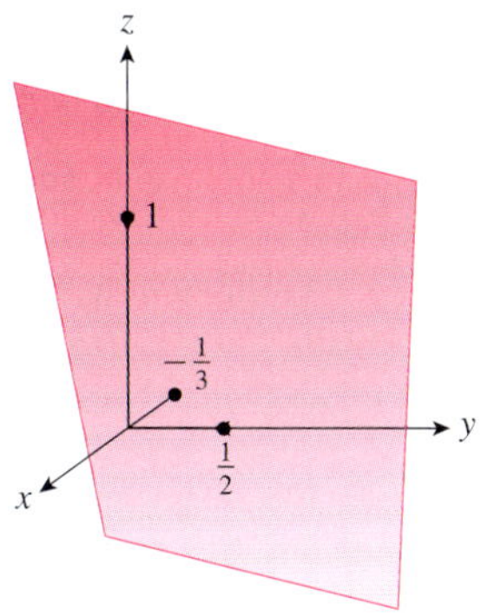

(H)

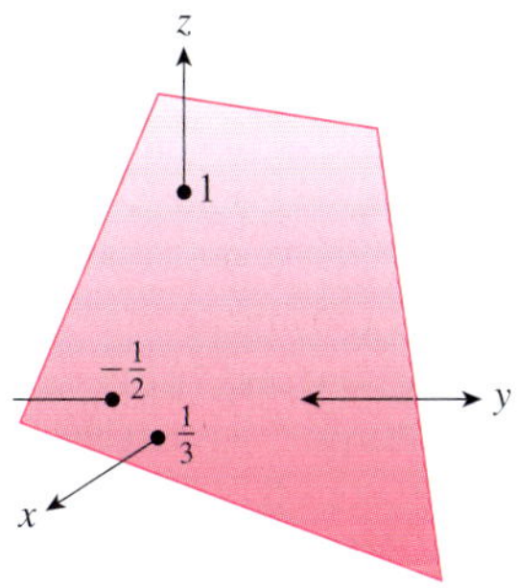

In Exercises 49–54 sketch the level curves $f(x, y) = c$ for the given function and values of c. HINT [See Example 5.]

49. $f(x, y) = 2x^2 + 2y^2$; $c = 0, 2, 18$

50. $f(x, y) = 3x^2 + 3y^2$; $c = 0, 3, 27$

51. $f(x, y) = y + 2x^2$; $c = -2, 0, 2$

52. $f(x, y) = 2y - x^2$; $c = -2, 0, 2$

53. $f(x, y) = 2xy - 1$; $c = -1, 0, 1$

54. $f(x, y) = 2 + xy$; $c = -2, 0, 2$

Sketch the graphs of the functions in Exercises 55–76. HINT [See Example 7.]

55. $f(x, y) = 1 - x - y$ **56.** $f(x, y) = x + y - 2$

57. $g(x, y) = 2x + y - 2$ **58.** $g(x, y) = 3 - x + 2y$

59. $h(x, y) = x + 2$ **60.** $h(x, y) = 3 - y$

61. $r(x, y) = x + y$ **62.** $r(x, y) = x - y$

Use of technology is suggested in Exercises 63–76. HINT [See Example 6.]

63. $s(x, y) = 2x^2 + 2y^2$. Show cross sections at $z = 1$ and $z = 2$.

64. $s(x, y) = -(x^2 + y^2)$. Show cross sections at $z = -1$ and $z = -2$.

65. $t(x, y) = x^2 + 2y^2$. Show cross sections at $x = 0$ and $z = 1$.

66. $t(x, y) = \frac{1}{2}x^2 + y^2$. Show cross sections at $x = 0$ and $z = 1$.

67. $f(x, y) = 2 + \sqrt{x^2 + y^2}$. Show cross sections at $z = 3$ and $y = 0$.

68. $f(x, y) = 2 - \sqrt{x^2 + y^2}$. Show cross sections at $z = 0$ and $y = 0$.

69. $f(x, y) = -2\sqrt{x^2 + y^2}$. Show cross sections at $z = -4$ and $y = 1$.

70. $f(x, y) = 2 + 2\sqrt{x^2 + y^2}$. Show cross sections at $z = 4$ and $y = 1$.

71. $f(x, y) = y^2$ **72.** $g(x, y) = x^2$

73. $h(x, y) = \frac{1}{y}$ **74.** $k(x, y) = e^y$

75. $f(x, y) = e^{-(x^2+y^2)}$ **76.** $g(x, y) = \frac{1}{\sqrt{x^2 + y^2}}$

APPLICATIONS

77. ***Marginal Cost*** Your weekly cost (in dollars) to manufacture x cars and y trucks is

$$C(x, y) = 240{,}000 + 6{,}000x + 4{,}000y.$$

a. What is the marginal cost of a car? Of a truck? HINT [See Example 1.]

b. Describe the graph of the cost function C. HINT [See Example 7]

c. Describe the slice $x = 10$. What cost function does this slice describe?

d. Describe the level curve $z = 480{,}000$. What does this curve tell you about costs?

78. ***Marginal Cost*** Your weekly cost (in dollars) to manufacture x bicycles and y tricycles is

$$C(x, y) = 24{,}000 + 60x + 20y.$$

a. What is the marginal cost of a bicycle? Of a tricycle? HINT [See Example 1.]

b. Describe the graph of the cost function C. HINT [See Example 7.]

c. Describe the slice by $y = 100$. What cost function does this slice describe?

d. Describe the level curve $z = 72{,}000$. What does this curve tell you about costs?

79. ***Marginal Cost*** Your sales of online video and audio clips are booming. Your Internet provider, Moneydrain.com, wants to get in on the action and has offered you unlimited technical assistance and consulting if you agree to pay Moneydrain 3¢ for every video clip and 4¢ for every audio clip you sell on the site. Further, Moneydrain agrees to charge you only \$10 per month to host your site. Set up a (monthly) cost function for the scenario, and describe each variable.

80. ***Marginal Cost*** Your Cabaret nightspot "Jazz on Jupiter" has become an expensive proposition: You are paying monthly costs of \$50,000 just to keep the place running. On top of that, your regular cabaret artist is charging you \$3,000 per performance, and your jazz ensemble is charging \$1,000 per hour. Set up a (monthly) cost function for the scenario, and describe each variable.

81. ***Scientific Research*** In each year from 1983 to 2003, the percentage y of research articles in *Physical Review* written by researchers in the United States can be approximated by

$$y = 82 - 0.78t - 1.02x \text{ percentage points} \quad (0 \le t \le 20)$$

where t is the year since 1983 and x is the percentage of articles written by researchers in Europe.[1]

a. In 2003, researchers in Europe wrote 38% of the articles published by the journal that year. What percentage was written by researchers in the United States?
b. In 1983, researchers in the United States wrote 61% of the articles published that year. What percentage was written by researchers in Europe?
c. What are the units of measurement of the coefficient of t?

82. ***Scientific Research*** The number z of research articles in *Physical Review* that were written by researchers in the United States from 1993 through 2003 can be approximated by

$$z = 5{,}960 - 0.71x + 0.50y \quad (3{,}000 \le x, y \le 6{,}000)$$

articles each year, where x is the number of articles written by researchers in Europe and y is the number written by researchers in other countries (excluding Europe and the United States).[2]

a. In the year 2000, approximately 5,500 articles were written by researchers in Europe, and 4,500 by researchers in other countries. How many (to the nearest 100) were written by researchers in the United States?
b. According to the model, if 5,000 articles were written in Europe and an equal number by researchers in the United States and other countries, what would that number be?
c. What is the significance of the fact that the coefficient of x is negative?

83. ***Market Share in the 1900s:* Chrysler, Ford, General Motors** In the late 1900s, the relationship between the domestic market shares of three major U.S. manufacturers of cars and light trucks could be modeled by

$$x_3 = 0.66 - 2.2x_1 - 0.02x_2$$

where x_1, x_2, and x_3 are, respectively, the fractions of the market held by **Chrysler**, **Ford**, and **General Motors**.[3] Thinking of **General Motors**' market share as a function of the shares of the other two manufacturers, describe the graph of the resulting function. How are the different slices by $x_1 = constant$ related to one another? What does this say about market share?

84. ***Market Share in the 1900s:* Kellogg, General Mills, General Foods** In the late 1900s, the relationship among the domestic market shares of three major manufacturers of breakfast cereal was

$$x_1 = -0.4 + 1.2x_2 + 2x_3$$

where x_1, x_2, and x_3 are, respectively, the fractions of the market held by **Kellogg**, **General Mills**, and **General Foods**.[4] Thinking of **Kellogg**'s market share as a function of shares of the other two manufacturers, describe the graph of the resulting function. How are the different slices by $x_2 = constant$ related to one another? What does this say about market share?

85. ***Prison Population*** The number of prisoners in federal prisons in the United States can be approximated by

$$N(x, y) = 27 - 0.08x + 0.08y + 0.0002xy \text{ thousand inmates}$$

where x is the number, in thousands, in state prisons, and y is the number, in thousands, in local jails.[5]

a. In 2007 there were approximately 1.3 million in state prisons and 781 thousand in local jails. Estimate, to the nearest thousand, the number of prisoners in federal prisons that year.
b. Obtain N as a function of x for $y = 300$, and again for $y = 500$. Interpret the slopes of the resulting linear functions.

86. ***Prison Population*** The number of prisoners in state prisons in the United States can be approximated by

$$N(x, y) = -260 + 7x + 2y - 0.009xy \text{ thousand inmates}$$

where x is the number, in thousands, in federal prisons, and y is the number, in thousands, in local jails.[6]

a. In 2007 there were approximately 189 thousand in federal prisons and 781 thousand in local jails. Estimate, to the nearest 0.1 million, the number of prisoners in state prisons that year.
b. Obtain N as a function of y for $x = 80$, and again for $x = 100$. Interpret the slopes of the resulting linear functions.

87. ***Marginal Cost (Interaction Model)*** Your weekly cost (in dollars) to manufacture x cars and y trucks is

$$C(x, y) = 240{,}000 + 6{,}000x + 4{,}000y - 20xy.$$

(Compare with Exercise 77.)

a. Describe the slices $x =$ constant and $y =$ constant.
b. Is the graph of the cost function a plane? How does your answer relate to part (a)?
c. What are the slopes of the slices $x = 10$ and $x = 20$? What does this say about cost?

---

[1] Source: The American Physical Society/*New York Times*, May 3, 2003, p. A1.

[2] Ibid.

[3] The model is based on a linear regression. Source of data: Ward's AutoInfoBank/*The New York Times*, July 29, 1998, p. D6.

[4] The models are based on a linear regression. Source of data: Bloomberg Financial Markets/*The New York Times*, November 28, 1998, p. C1.

[5] Source for data: Sourcebook of Criminal Justice Statistics Online (www.albany.edu/sourcebook/wk1/t6132007.wk1).

[6] Ibid.

88. ***Marginal Cost (Interaction Model)*** Repeat the preceding exercise using the weekly cost to manufacture x bicycles and y tricycles given by

$$C(x, y) = 24{,}000 + 60x + 20y + 0.3xy.$$

(Compare with Exercise 78.)

89. ▼ ***Online Revenue*** Let us look once again at the example we used to introduce the chapter. Your major online bookstore is in direct competition with **Amazon.com**, **BN.com**, and **Borders.com**. Your company's daily revenue in dollars is given by

$$R(x, y, z) = 10{,}000 - 0.01x - 0.02y - 0.01z + 0.00001yz,$$

where x, y, and z are the online daily revenues of **Amazon.com**, **BN.com**, and **Borders.com** respectively.

a. If, on a certain day, **Amazon.com** shows revenue of \$12,000, while **BN.com** and **Borders.com** each show \$5,000, what does the model predict for your company's revenue that day?

b. If **Amazon.com** and **BN.com** each show daily revenue of \$5,000, give an equation showing how your daily revenue depends on that of **Borders.com**.

90. ▼ ***Online Revenue*** Repeat the preceding exercise, using the revised revenue function

$$R(x, y, z) = 20{,}000 - 0.02x - 0.04y - 0.01z + 0.00001yz.$$

91. ▼ ***Profits:* Walmart, Target** The following table shows the approximate net earnings, in billions of dollars, of **Walmart** and **Target** in 2000, 2004, and 2008.[7]

	2000	2004	2008
Walmart	160	250	370
Target	27	42	62

Model **Walmart**'s net earnings as a function of **Target**'s net earnings and time, using a linear function of the form

$$f(x, t) = Ax + Bt + C \quad (A, B, C \text{ constants})$$

where f is **Walmart**'s net earnings (in billions of dollars), x is **Target**'s net earnings (in billions of dollars), and t is time in years since 2000. In 2006 **Target**'s net earnings were about \$52.5 billion. What does your model estimate as **Walmart**'s net earnings that year?

92. ▼ ***Profits:* Nintendo, Nokia** The following table shows the approximate net earnings of **Nintendo** (in billions of yen) and **Nokia** (in billions of euro) in 2000, 2004, and 2008.[8]

	2000	2004	2008
Nintendo	530	510	1700
Nokia	30	30	52

Model **Nintendo**'s net earnings as a function of **Nokia**'s net earnings and time, using a linear function of the form

$$f(x, t) = Ax + Bt + C \quad (A, B, C \text{ constants})$$

where f is **Nintendo**'s net earnings (in billions of yen), x is **Nokia**'s net earnings (in billions of euro), and t is time in years since 2000. In 2007 **Nokia**'s net earnings were about €50 billion. What does your model estimate as **Nokia**'s net earnings that year?

93. ▼ ***Utility*** Suppose your newspaper is trying to decide between two competing desktop publishing software packages, Macro Publish and Turbo Publish. You estimate that if you purchase x copies of Macro Publish and y copies of Turbo Publish, your company's daily productivity will be

$$U(x, y) = 6x^{0.8}y^{0.2} + x$$

where $U(x, y)$ is measured in pages per day (U is called a *utility function*). If $x = y = 10$, calculate the effect of increasing x by one unit, and interpret the result.

94. ▼ ***Housing Costs***[9] The cost C (in dollars) of building a house is related to the number k of carpenters used and the number e of electricians used by

$$C(k, e) = 15{,}000 + 50k^2 + 60e^2.$$

If $k = e = 10$, compare the effects of increasing k by one unit and of increasing e by one unit. Interpret the result.

95. ▼ ***Volume*** The volume of an ellipsoid with cross-sectional radii a, b, and c is $V(a, b, c) = \frac{4}{3}\pi abc$.

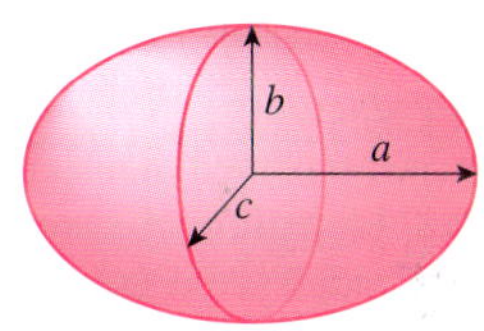

a. Find at least two sets of values for a, b, and c such that $V(a, b, c) = 1$.

b. Find the value of a such that $V(a, a, a) = 1$, and describe the resulting ellipsoid.

96. ▼ ***Volume*** The volume of a right elliptical cone with height h and radii a and b of its base is $V(a, b, h) = \frac{1}{3}\pi abh$.

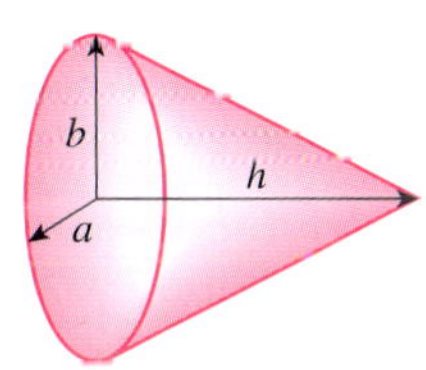

a. Find at least two sets of values for a, b, and h such that $V(a, b, h) = 1$.

b. Find the value of a such that $V(a, a, a) = 1$, and describe the resulting cone.

[7] Sources: www.walmartstores.com/Investors, www.investors.target.com

[8] Sources: www.nintendo.com/corp/, www.investors.nokia.com

[9] Based on an Exercise in *Introduction to Mathematical Economics* by A. L. Ostrosky Jr. and J. V. Koch (Waveland Press, Illinois, 1979).

Exercises 97–100 involve "Cobb-Douglas" productivity functions. These functions have the form

$$P(x, y) = Kx^a y^{1-a}$$

where P stands for the number of items produced per year, x is the number of employees, and y is the annual operating budget. (The numbers K and a are constants that depend on the situation we are looking at, with $0 \le a \le 1$.)

97. *Productivity* How many items will be produced per year by a company with 100 employees and an annual operating budget of \$500,000 if $K = 1{,}000$ and $a = 0.5$? (Round your answer to one significant digit.)

98. *Productivity* How many items will be produced per year by a company with 50 employees and an annual operating budget of \$1,000,000 if $K = 1{,}000$ and $a = 0.5$? (Round your answer to one significant digit.)

99. ▼ *Modeling Production with Cobb-Douglas* Two years ago my piano manufacturing plant employed 1,000 workers, had an operating budget of \$1 million, and turned out 100 pianos. Last year, I slashed the operating budget to \$10,000, and production dropped to 10 pianos.

a. Use the data for each of the two years and the Cobb-Douglas formula to obtain two equations in K and a.
b. Take logs of both sides in each equation and obtain two linear equations in a and $\log K$.
c. Solve these equations to obtain values for a and K.
d. Use these values in the Cobb-Douglas formula to predict production if I increase the operating budget back to \$1 million but lay off half the work force.

100. ▼ *Modeling Production with Cobb-Douglas* Repeat the preceding exercise using the following data: Two years ago—1,000 employees, \$1 million operating budget, 100 pianos; last year—1,000 employees, \$100,000 operating budget, 10 pianos.

101. ▼ *Pollution* The burden of man-made aerosol sulfate in the earth's atmosphere, in grams per square meter, is

$$B(x, n) = \frac{xn}{A}$$

where x is the total weight of aerosol sulfate emitted into the atmosphere per year and n is the number of years it remains in the atmosphere. A is the surface area of the earth, approximately 5.1×10^{14} square meters.[10]

a. Calculate the burden, given the 1995 estimated values of $x = 1.5 \times 10^{14}$ grams per year, and $n = 5$ days.
b. What does the function $W(x, n) = xn$ measure?

102. ▼ *Pollution* The amount of aerosol sulfate (in grams) was approximately 45×10^{12} grams in 1940 and has been increasing exponentially ever since, with a doubling time of approximately 20 years.[11] Use the model from the preceding exercise to give a formula for the atmospheric burden of aerosol sulfate as a function of the time t in years since 1940 and the number of years n it remains in the atmosphere.

103. ▼ *Alien Intelligence* Frank Drake, an astronomer at the University of California at Santa Cruz, devised the following equation to estimate the number of planet-based civilizations in our Milky Way galaxy willing and able to communicate with Earth:[12]

$$N(R, f_p, n_e, f_l, f_i, f_c, L) = Rf_p n_e f_l f_i f_c L$$

R = the number of new stars formed in our galaxy each year
f_p = the fraction of those stars that have planetary systems
n_e = the average number of planets in each such system that can support life
f_l = the fraction of such planets on which life actually evolves
f_i = the fraction of life-sustaining planets on which intelligent life evolves
f_c = the fraction of intelligent-life-bearing planets on which the intelligent beings develop the means and the will to communicate over interstellar distances
L = the average lifetime of such technological civilizations (in years)

a. What would be the effect on N if any one of the variables were doubled?
b. How would you modify the formula if you were interested only in the number of intelligent-life-bearing planets in the galaxy?
c. How could one convert this function into a linear function?
d. (For discussion) Try to come up with an estimate of N.

104. ▼ *More Alien Intelligence* The formula given in the preceding exercise restricts attention to planet-based civilizations in our galaxy. Give a formula that includes intelligent planet-based aliens from the galaxy Andromeda. (Assume that all the variables used in the formula for the Milky Way have the same values for Andromeda.)

COMMUNICATION AND REASONING EXERCISES

105. Let $f(x, y) = \dfrac{x}{y}$. How are $f(x, y)$ and $f(y, x)$ related?

106. Let $f(x, y) = x^2y^3$. How are $f(x, y)$ and $f(-x, -y)$ related?

[10] Source: Robert J. Charlson and Tom M. L. Wigley, "Sulfate Aerosol and Climatic Change," *Scientific American*, February, 1994, pp. 48–57.

[11] Ibid.

[12] Source: "First Contact" (Plume Books/Penguin Group)/*The New York Times*, October 6, 1992, p. C1.

107. Give an example of a function of the two variables x and y with the property that interchanging x and y has no effect.

108. Give an example of a function f of the two variables x and y with the property that $f(x, y) = -f(y, x)$.

109. Give an example of a function f of the three variables x, y, and z with the property that $f(x, y, z) = f(y, x, z)$ and $f(-x, -y, -z) = -f(x, y, z)$.

110. Give an example of a function f of the three variables x, y, and z with the property that $f(x, y, z) = f(y, x, z)$ and $f(-x, -y, -z) = f(x, y, z)$.

111. Illustrate by means of an example how a real-valued function of the two variables x and y gives different real-valued functions of one variable when we restrict y to be different constants.

112. Illustrate by means of an example how a real-valued function of one variable x gives different real-valued functions of the two variables y and z when we substitute for x suitable functions of y and z.

113. ▼ If f is a linear function of x and y, show that if we restrict y to be a fixed constant, then the resulting function of x is linear. Does the slope of this linear function depend on the choice of y?

114. ▼ If f is an interaction function of x and y, show that if we restrict y to be a fixed constant, then the resulting function of x is linear. Does the slope of this linear function depend on the choice of y?

115. ▼ Suppose that $C(x, y)$ represents the cost of x CDs and y cassettes. If $C(x, y+1) < C(x+1, y)$ for every $x \geq 0$ and $y \geq 0$, what does this tell you about the cost of CDs and cassettes?

116. ▼ Suppose that $C(x, y)$ represents the cost of renting x DVDs and y video games. If $C(x+2, y) < C(x, y+1)$ for every $x \geq 0$ and $y \geq 0$, what does this tell you about the cost of renting DVDs and video games?

117. Complete the following: The graph of a linear function of two variables is a ____ .

118. Complete the following: The level curves of a linear function of two variables are ___ .

119. ▼ ***Heat-Seeking Missiles*** The following diagram shows some level curves of the temperature, in degrees Fahrenheit, of a region in space, as well as the location, on the 100-degree curve, of a heat-seeking missile moving through the region. (These level curves are called **isotherms**.) In which of the three directions shown should the missile be traveling so as to experience the fastest rate of increase in temperature at the given point? Explain your answer.

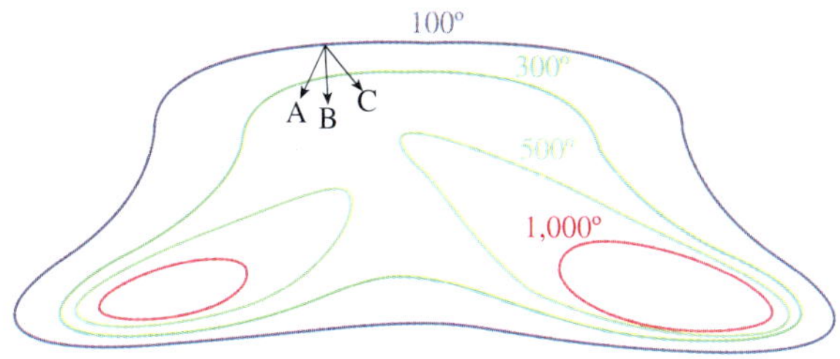

120. ▼ ***Hiking*** The following diagram shows some level curves of the altitude of a mountain valley, as well as the location, on the 2,000-ft curve, of a hiker. The hiker is currently moving at the greatest possible rate of descent. In which of the three directions shown is he moving? Explain your answer.

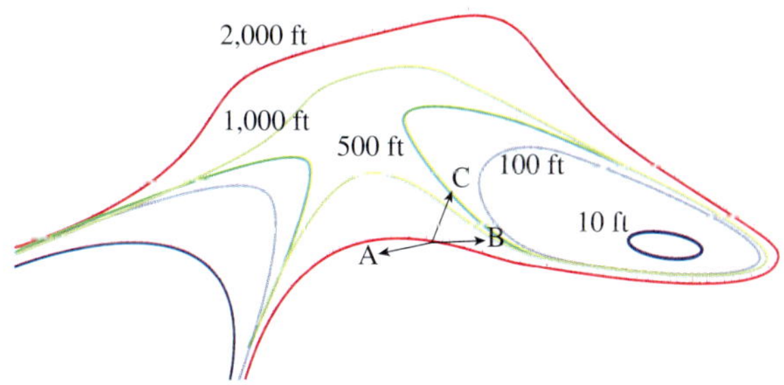

121. Your study partner Slim claims that because the surface $z = f(x, y)$ you have been studying is a plane, it follows that all the slices $x = constant$ and $y = constant$ are straight lines. Do you agree or disagree? Explain.

122. Your other study partner Shady just told you that the surface $z = xy$ you have been trying to graph must be a plane because you've already found that the slices $x = constant$ and $y = constant$ are all straight lines. Do you agree or disagree? Explain.

123. Why do we not sketch the graphs of functions of three or more variables?

124. The surface of a mountain can be thought of as the graph of what function?

125. Why is three-dimensional space used to represent the graph of a function of two variables?

126. Why is it that we can sketch the graphs of functions of two variables on the two-dimensional flat surfaces of these pages?

8.2 Partial Derivatives

Recall that if f is a function of x, then the derivative df/dx measures how fast f changes as x increases. If f is a function of two or more variables, we can ask how fast f changes as each variable increases while the others remain fixed. These rates of change are called the "partial derivatives of f," and they measure how each variable contributes to the change in f. Here is a more precise definition.

Partial Derivatives

The **partial derivative of f with respect to x** is the derivative of f with respect to x, when all other variables are treated as constant. Similarly, the **partial derivative of f with respect to y** is the derivative of f with respect to y, with all other variables treated as constant, and so on for other variables. The partial derivatives are written as $\frac{\partial f}{\partial x}, \frac{\partial f}{\partial y}$, and so on. The symbol ∂ is used (instead of d) to remind us that there is more than one variable and that we are holding the other variables fixed.

Quick Examples

1. Let $f(x, y) = x^2 + y^2$.

$$\frac{\partial f}{\partial x} = 2x + 0 = 2x$$ Because y^2 is treated as a constant

$$\frac{\partial f}{\partial y} = 0 + 2y = 2y$$ Because x^2 is treated as a constant

2. Let $z = x^2 + xy$.

$$\frac{\partial z}{\partial x} = 2x + y$$ $\frac{\partial}{\partial x}(xy) = \frac{\partial}{\partial x}(x \cdot \text{constant}) = \text{constant} = y$

$$\frac{\partial z}{\partial y} = 0 + x$$ $\frac{\partial}{\partial y}(xy) = \frac{\partial}{\partial x}(\text{constant} \cdot y) = \text{constant} = x$

3. Let $f(x, y) = x^2y + y^2x - xy + y$.

$$\frac{\partial f}{\partial x} = 2xy + y^2 - y$$ y is treated as a constant.

$$\frac{\partial f}{\partial y} = x^2 + 2xy - x + 1$$ x is treated as a constant.

Interpretation

$\frac{\partial f}{\partial x}$ is the rate at which f changes as x changes, for a fixed (constant) y.

$\frac{\partial f}{\partial y}$ is the rate at which f changes as y changes, for a fixed (constant) x.

EXAMPLE 1 Marginal Cost: Linear Model

We return to Example 1 from Section 8.1. Suppose that you own a company that makes two models of speakers, the Ultra Mini and the Big Stack. Your total monthly cost (in dollars) to make x Ultra Minis and y Big Stacks is given by

$$C(x, y) = 10{,}000 + 20x + 40y.$$

What is the significance of $\frac{\partial C}{\partial x}$ and of $\frac{\partial C}{\partial y}$?

Solution First we compute these partial derivatives:

$$\frac{\partial C}{\partial x} = 20$$

$$\frac{\partial C}{\partial y} = 40.$$

We interpret the results as follows: $\frac{\partial C}{\partial x} = 20$ means that the cost is increasing at a rate of \$20 per additional Ultra Mini (if production of Big Stacks is held constant); $\frac{\partial C}{\partial y} = 40$ means that the cost is increasing at a rate of \$40 per additional Big Stack (if production of Ultra Minis is held constant). In other words, these are the **marginal costs** of each model of speaker.

Before we go on... How much does the cost rise if you increase x by Δx and y by Δy? In Example 1, the change in cost is given by

$$\Delta C = 20\Delta x + 40\Delta y = \frac{\partial C}{\partial x}\Delta x + \frac{\partial C}{\partial y}\Delta y.$$

This suggests the **chain rule for several variables**. Part of this rule says that if x and y are both functions of t, then C is a function of t through them, and the rate of change of C with respect to t can be calculated as

$$\frac{dC}{dt} = \frac{\partial C}{\partial x} \cdot \frac{dx}{dt} + \frac{\partial C}{\partial y} \cdot \frac{dy}{dt}.$$

See the optional section on the chain rule for several variables for further discussion and applications of this interesting result. ■

EXAMPLE 2 Marginal Cost: Interaction Model

Another possibility for the cost function in the preceding example is the interaction model

$$C(x, y) = 10{,}000 + 20x + 40y + 0.1xy.$$

a. *Now* what are the marginal costs of the two models of speakers?

b. What is the marginal cost of manufacturing Big Stacks at a production level of 100 Ultra Minis and 50 Big Stacks per month?

Solution

a. We compute the partial derivatives:

$$\frac{\partial C}{\partial x} = 20 + 0.1y$$

$$\frac{\partial C}{\partial y} = 40 + 0.1x.$$

Thus, the marginal cost of manufacturing Ultra Minis increases by \$0.1 or 10¢ for each Big Stack that is manufactured. Similarly, the marginal cost of manufacturing Big Stacks increases by 10¢ for each Ultra Mini that is manufactured.

b. From part (a), the marginal cost of manufacturing Big Stacks is

$$\frac{\partial C}{\partial y} = 40 + 0.1x.$$

At a production level of 100 Ultra Minis and 50 Big Stacks per month, we have $x = 100$ and $y = 50$. Thus, the marginal cost of manufacturing Big Stacks at these production levels is

$$\left.\frac{\partial C}{\partial y}\right|_{(100,50)} = 40 + 0.1(100) = \$50 \text{ per Big Stack.}$$

Partial derivatives of functions of three variables are obtained in the same way as those for functions of two variables, as the following example shows.

EXAMPLE 3 Function of Three Variables

Calculate $\frac{\partial f}{\partial x}$, $\frac{\partial f}{\partial y}$, and $\frac{\partial f}{\partial z}$ if $f(x, y, z) = xy^2z^3 - xy$.

Solution Although we now have three variables, the calculation remains the same: $\partial f/\partial x$ is the derivative of f with respect to x, with *both* other variables, y and z, held constant:

$$\frac{\partial f}{\partial x} = y^2z^3 - y.$$

Similarly, $\partial f/\partial y$ is the derivative of f with respect to y, with both x and z held constant:

$$\frac{\partial f}{\partial y} = 2xyz^3 - x.$$

Finally, to find $\partial f/\partial z$, we hold both x and y constant and take the derivative with respect to z.

$$\frac{\partial f}{\partial z} = 3xy^2z^2.$$

Note The procedure for finding a partial derivative is the same for any number of variables: To get the partial derivative with respect to any one variable, we treat all the others as constants. ■

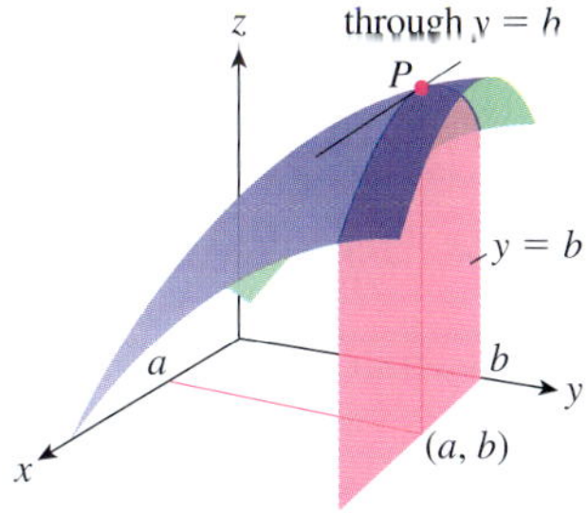

$\left.\frac{\partial f}{\partial x}\right|_{(a,b)}$ is the slope of the tangent line at the point $P(a, b, f(a, b))$ along the slice through $y = b$.

Figure 17

Geometric Interpretation of Partial Derivatives

Recall that if f is a function of one variable x, then the derivative df/dx gives the slopes of the tangent lines to its graph. Now, suppose that f is a function of x and y. By definition, $\partial f/\partial x$ is the derivative of the function of x we get by holding y fixed. If we evaluate this derivative at the point (a, b), we are holding y fixed at the value b, taking the ordinary derivative of the resulting function of x, and evaluating this at $x = a$. Now, holding y fixed at b amounts to slicing through the graph of f along the plane $y = b$, resulting in a curve. Thus, the partial derivative is the slope of the tangent line to this curve at the point where $x = a$ and $y = b$, along the plane $y = b$ (Figure 17). This fits with our interpretation of $\partial f/\partial x$ as the rate of increase of f with increasing x when y is held fixed at b.

The other partial derivative, $\partial f/\partial y|_{(a,b)}$ is, similarly, the slope of the tangent line at the same point $P(a, b, f(a, b))$ but along the slice by the plane $x = a$. You should draw the corresponding picture for this on your own.

Second-Order Partial Derivatives

Just as for functions of a single variable, we can calculate second derivatives. Suppose, for example, that we have a function of x and y, say, $f(x, y) = x^2 - x^2y^2$. We know that

$$\frac{\partial f}{\partial x} = 2x - 2xy^2.$$

If we take the partial derivative with respect to x once again, we obtain

$$\frac{\partial}{\partial x}\left(\frac{\partial f}{\partial x}\right) = 2 - 2y^2. \qquad \text{Take } \frac{\partial}{\partial x} \text{ of } \frac{\partial f}{\partial x}.$$

(The symbol $\partial/\partial x$ means "the partial derivative with respect to x," just as d/dx stands for "the derivative with respect to x.") This is called the **second-order partial derivative** and is written $\frac{\partial^2 f}{\partial x^2}$. We get the following derivatives similarly:

$$\frac{\partial f}{\partial y} = -2x^2y$$

$$\frac{\partial^2 f}{\partial y^2} = -2x^2. \qquad \text{Take } \frac{\partial}{\partial y} \text{ of } \frac{\partial f}{\partial y}.$$

Now what if we instead take the partial derivative with respect to y of $\partial f/\partial x$?

$$\frac{\partial^2 f}{\partial y\partial x} = \frac{\partial}{\partial y}\left(\frac{\partial f}{\partial x}\right) \qquad \text{Take } \frac{\partial}{\partial y} \text{ of } \frac{\partial f}{\partial x}.$$

$$= \frac{\partial}{\partial y}[2x - 2xy^2] = -4xy.$$

Here, $\frac{\partial^2 f}{\partial y\partial x}$ means "first take the partial derivative with respect to x and then with respect to y," and is called a **mixed partial derivative**. If we differentiate in the opposite order, we get

$$\frac{\partial^2 f}{\partial x\partial y} = \frac{\partial}{\partial x}\left(\frac{\partial f}{\partial y}\right) = \frac{\partial}{\partial x}[-2x^2y] = -4xy,$$

the same expression as $\dfrac{\partial^2 f}{\partial y \partial x}$. This is no coincidence: The mixed partial derivatives $\dfrac{\partial^2 f}{\partial x \partial y}$ and $\dfrac{\partial^2 f}{\partial y \partial x}$ are always the same as long as the first partial derivatives are both differentiable functions of x and y and the mixed partial derivatives are continuous. Because all the functions we shall use are of this type, we can take the derivatives in any order we like when calculating mixed derivatives.

Here is another notation for partial derivatives that is especially convenient for second-order partial derivatives:

$$f_x \text{ means } \frac{\partial f}{\partial x}$$

$$f_y \text{ means } \frac{\partial f}{\partial y}$$

$$f_{xy} \text{ means } (f_x)_y = \frac{\partial^2 f}{\partial y \partial x} \quad \text{(Note the order in which the derivatives are taken.)}$$

$$f_{yx} \text{ means } (f_y)_x = \frac{\partial^2 f}{\partial x \partial y}.$$

8.2 EXERCISES

▼ more advanced ◆ challenging

T indicates exercises that should be solved using technology

In Exercises 1–18, calculate $\dfrac{\partial f}{\partial x}, \dfrac{\partial f}{\partial y}, \left.\dfrac{\partial f}{\partial x}\right|_{(1,-1)}$, *and* $\left.\dfrac{\partial f}{\partial y}\right|_{(1,-1)}$ *when defined.* HINT [See Quick Examples page 602.]

1. $f(x, y) = 10{,}000 - 40x + 20y$
2. $f(x, y) = 1{,}000 + 5x - 4y$
3. $f(x, y) = 3x^2 - y^3 + x - 1$
4. $f(x, y) = x^{1/2} - 2y^4 + y + 6$
5. $f(x, y) = 10{,}000 - 40x + 20y + 10xy$
6. $f(x, y) = 1{,}000 + 5x - 4y - 3xy$
7. $f(x, y) = 3x^2y$
8. $f(x, y) = x^4y^2 - x$
9. $f(x, y) = x^2y^3 - x^3y^2 - xy$
10. $f(x, y) = x^{-1}y^2 + xy^2 + xy$
11. $f(x, y) = (2xy + 1)^3$
12. $f(x, y) = \dfrac{1}{(xy + 1)^2}$
13. ▼ $f(x, y) = e^{x+y}$
14. ▼ $f(x, y) = e^{2x+y}$
15. ▼ $f(x, y) = 5x^{0.6}y^{0.4}$
16. ▼ $f(x, y) = -2x^{0.1}y^{0.9}$
17. ▼ $f(x, y) = e^{0.2xy}$
18. ▼ $f(x, y) = xe^{xy}$

In Exercises 19–28, find $\dfrac{\partial^2 f}{\partial x^2}, \dfrac{\partial^2 f}{\partial y^2}, \dfrac{\partial^2 f}{\partial x \partial y}$, *and* $\dfrac{\partial^2 f}{\partial y \partial x}$, *and evaluate them all at* $(1, -1)$ *if possible.* HINT [See Discussion on page 605.]

19. $f(x, y) = 10{,}000 - 40x + 20y$
20. $f(x, y) = 1{,}000 + 5x - 4y$
21. $f(x, y) = 10{,}000 - 40x + 20y + 10xy$
22. $f(x, y) = 1{,}000 + 5x - 4y - 3xy$
23. $f(x, y) = 3x^2y$
24. $f(x, y) = x^4y^2 - x$
25. ▼ $f(x, y) = e^{x+y}$
26. ▼ $f(x, y) = e^{2x+y}$
27. ▼ $f(x, y) = 5x^{0.6}y^{0.4}$
28. ▼ $f(x, y) = -2x^{0.1}y^{0.9}$

In Exercises 29–40, find $\dfrac{\partial f}{\partial x}, \dfrac{\partial f}{\partial y}, \dfrac{\partial f}{\partial z}$, *and their values at* $(0, -1, 1)$ *if possible.* HINT [See Example 3.]

29. $f(x, y, z) = xyz$
30. $f(x, y, z) = xy + xz - yz$
31. ▼ $f(x, y, z) = -\dfrac{4}{x + y + z^2}$
32. ▼ $f(x, y, z) = \dfrac{6}{x^2 + y^2 + z^2}$
33. ▼ $f(x, y, z) = xe^{yz} + ye^{xz}$
34. ▼ $f(x, y, z) = xye^z + xe^{yz} + e^{xyz}$
35. ▼ $f(x, y, z) = x^{0.1}y^{0.4}z^{0.5}$
36. ▼ $f(x, y, z) = 2x^{0.2}y^{0.8} + z^2$
37. ▼ $f(x, y, z) = e^{xyz}$
38. ▼ $f(x, y, z) = \ln(x + y + z)$
39. ▼ $f(x, y, z) = \dfrac{2{,}000z}{1 + y^{0.3}}$
40. ▼ $f(x, y, z) = \dfrac{e^{0.2x}}{1 + e^{-0.1y}}$

APPLICATIONS

41. ***Marginal Cost (Linear Model)*** Your weekly cost (in dollars) to manufacture x cars and y trucks is

$$C(x, y) = 240{,}000 + 6{,}000x + 4{,}000y.$$

Calculate and interpret $\frac{\partial C}{\partial x}$ and $\frac{\partial C}{\partial y}$. HINT [See Example 1.]

42. ***Marginal Cost (Linear Model)*** Your weekly cost (in dollars) to manufacture x bicycles and y tricycles is

$$C(x, y) = 24{,}000 + 60x + 20y.$$

Calculate and interpret $\frac{\partial C}{\partial x}$ and $\frac{\partial C}{\partial y}$.

43. ***Scientific Research*** In each year from 1983 to 2003, the percentage y of research articles in *Physical Review* written by researchers in the United States can be approximated by

$$y = 82 - 0.78t - 1.02x \text{ percentage points} \quad (0 \le t \le 20)$$

where t is the year since 1983 and x is the percentage of articles written by researchers in Europe.[13] Calculate and interpret $\frac{\partial y}{\partial t}$ and $\frac{\partial y}{\partial x}$.

44. ***Scientific Research*** The number z of research articles in *Physical Review* that were written by researchers in the United States from 1993 through 2003 can be approximated by

$$z = 5{,}960 - 0.71x + 0.50y \quad (3{,}000 \le x, y \le 6{,}000)$$

articles each year, where x is the number of articles written by researchers in Europe and y is the number written by researchers in other countries (excluding Europe and the United States).[14] Calculate and interpret $\frac{\partial z}{\partial x}$ and $\frac{\partial z}{\partial y}$.

45. ***Marginal Cost (Interaction Model)*** Your weekly cost (in dollars) to manufacture x cars and y trucks is

$$C(x, y) = 240{,}000 + 6{,}000x + 4{,}000y - 20xy.$$

(Compare with Exercise 77.) Compute the marginal cost of manufacturing cars at a production level of 10 cars and 20 trucks. HINT [See Example 2.]

46. ***Marginal Cost (Interaction Model)*** Your weekly cost (in dollars) to manufacture x bicycles and y tricycles is

$$C(x, y) = 24{,}000 + 60x + 20y + 0.3xy.$$

(Compare with Exercise 78.) Compute the marginal cost of manufacturing tricycles at a production level of 10 bicycles and 20 tricycles. HINT [See Example 2.]

[13] Source: The American Physical Society/*New York Times*, May 3, 2003, p. A1.

[14] Ibid.

47. ***Brand Loyalty:* Mazda** The fraction of **Mazda** car owners who chose another new **Mazda** can be modeled by the following function:[15]

$$M(c, f, g, h, t) = 1.1 - 3.8c + 2.2f + 1.9g - 1.7h - 1.3t.$$

Here, c is the fraction of **Chrysler** car owners who remained loyal to **Chrysler**, f is the fraction of **Ford** car owners remaining loyal to **Ford**, g the corresponding figure for **General Motors**, h the corresponding figure for **Honda**, and t for **Toyota**.

a. Calculate $\frac{\partial M}{\partial c}$ and $\frac{\partial M}{\partial f}$ and interpret the answers.

b. One year it was observed that $c = 0.56$, $f = 0.56$, $g = 0.72$, $h = 0.50$, and $t = 0.43$. According to the model, what percentage of **Mazda** owners remained loyal to **Mazda**? (Round your answer to the nearest percentage point.)

48. ***Brand Loyalty*** The fraction of **Mazda** car owners who chose another new **Mazda** can be modeled by the following function:[16]

$$M(c, f) = 9.4 + 7.8c + 3.6c^2 - 38f - 22cf + 43f^2$$

where c is the fraction of **Chrysler** car owners who remained loyal to **Chrysler** and f is the fraction of **Ford** car owners remaining loyal to **Ford**.

a. Calculate $\frac{\partial M}{\partial c}$ and $\frac{\partial M}{\partial f}$ evaluated at the point (0.7, 0.7), and interpret the answers.

b. One year it was observed that $c = 0.56$, and $f = 0.56$. According to the model, what percentage of **Mazda** owners remained loyal to **Mazda**? (Round your answer to the nearest percentage point.)

49. ▼ ***Income Gap*** The following model is based on data on the median family incomes of Hispanic and white families in the United States for the period 1980–2008:[17]

$$z(t, x) = 31{,}200 + 270t + 13{,}500x + 140xt$$

where

$z(t, x)$ = median family income

t = year ($t = 0$ represents 1980)

$$x = \begin{cases} 0 & \text{if the income was for a Hispanic family} \\ 1 & \text{if the income was for a white family.} \end{cases}$$

[15] The model is an approximation of a linear regression based on data from the period 1988–1995. Source for data: Chrysler, Maritz Market Research, Consumer Attitude Research, and Strategic Vision/*The New York Times*, November 3, 1995, p. D2.

[16] The model is an approximation of a second order regression based on data from the period 1988–1995. Source for data: Chrysler, Maritz Market Research, Consumer Attitude Research, and Strategic Vision/*The New York Times*, November 3, 1995, p. D2.

[17] Incomes are in 2007 dollars. Source for data: U.S. Census Bureau (www.census.gov).

a. Use the model to estimate the median income of a Hispanic family and of a white family in 2000.
b. According to the model, how fast was the median income for a Hispanic family increasing in 2000? How fast was the median income for a white family increasing in 2000?
c. Do the answers in part (b) suggest that the income gap between white and Hispanic families was widening or narrowing during the given period?
d. What does the coefficient of xt in the formula for $z(t, x)$ represent in terms of the income gap?

50. ▼ ***Income Gap*** The following model is based on data on the median family incomes of black and white families in the United States for the period 1980–2008:[18]

$$z(t, x) = 24,500 + 390t + 20,200x + 20xt$$

where

$$z(t, x) = \text{median family income}$$
$$t = \text{year } (t = 0 \text{ represents } 1980)$$
$$x = \begin{cases} 0 & \text{if the income was for a black family} \\ 1 & \text{if the income was for a white family.} \end{cases}$$

a. Use the model to estimate the median income of a black family and of a white family in 2000.
b. According to the model, how fast was the median income for a black family increasing in 2000? How fast was the median income for a white family increasing in 2000?
c. Do the answers in part (b) suggest that the income gap between white and black families was widening or narrowing during the given period?
d. What does the coefficient of xt in the formula for $z(t, x)$ represent in terms of the income gap?

51. ▼ ***Marginal Cost*** Your weekly cost (in dollars) to manufacture x cars and y trucks is

$$C(x, y) = 200,000 + 6,000x + 4,000y - 100,000e^{-0.01(x+y)}.$$

What is the marginal cost of a car? Of a truck? How do these marginal costs behave as total production increases?

52. ▼ ***Marginal Cost*** Your weekly cost (in dollars) to manufacture x bicycles and y tricycles is

$$C(x, y) = 20,000 + 60x + 20y + 50\sqrt{xy}.$$

What is the marginal cost of a bicycle? Of a tricycle? How do these marginal costs behave as x and y increase?

53. ▼ ***Average Cost*** If you average your costs over your total production, you get the **average cost**, written $\bar{C}$:

$$\bar{C}(x, y) = \frac{C(x, y)}{x + y}.$$

Find the average cost for the cost function in Exercise 51. Then find the marginal average cost of a car and the marginal average cost of a truck at a production level of 50 cars and 50 trucks. Interpret your answers.

54. ▼ ***Average Cost*** Find the average cost for the cost function in Exercise 52. (See the preceding exercise.) Then find the marginal average cost of a bicycle and the marginal average cost of a tricycle at a production level of five bicycles and five tricycles. Interpret your answers.

55. ▼ ***Marginal Revenue*** As manager of an auto dealership, you offer a car rental company the following deal: You will charge \$15,000 per car and \$10,000 per truck, but you will then give the company a discount of \$5,000 times the square root of the total number of vehicles it buys from you. Looking at your marginal revenue, is this a good deal for the rental company?

56. ▼ ***Marginal Revenue*** As marketing director for a bicycle manufacturer, you come up with the following scheme: You will offer to sell a dealer x bicycles and y tricycles for

$$R(x, y) = 3,500 - 3,500e^{-0.02x-0.01y} \text{ dollars.}$$

Find your marginal revenue for bicycles and for tricycles. Are you likely to be fired for your suggestion?

57. ▼ ***Research Productivity*** Here we apply a variant of the Cobb-Douglas function to the modeling of research productivity. A mathematical model of research productivity at a particular physics laboratory is

$$P = 0.04x^{0.4}y^{0.2}z^{0.4}$$

where P is the annual number of groundbreaking research papers produced by the staff, x is the number of physicists on the research team, y is the laboratory's annual research budget, and z is the annual National Science Foundation subsidy to the laboratory. Find the rate of increase of research papers per government-subsidy dollar at a subsidy level of \$1,000,000 per year and a staff level of 10 physicists if the annual budget is \$100,000.

58. ▼ ***Research Productivity*** A major drug company estimates that the annual number P of patents for new drugs developed by its research team is best modeled by the formula

$$P = 0.3x^{0.3}y^{0.4}z^{0.3}$$

where x is the number of research biochemists on the payroll, y is the annual research budget, and z is the size of the bonus awarded to discoverers of new drugs. Assuming that the company has 12 biochemists on the staff, has an annual research budget of \$500,000 and pays \$40,000 bonuses to developers of new drugs, calculate the rate of growth in the annual number of patents per new research staff member.

59. ▼ ***Utility*** Your newspaper is trying to decide between two competing desktop publishing software packages, Macro Publish and Turbo Publish. You estimate that if you purchase x copies of Macro Publish and y copies of Turbo Publish, your company's daily productivity will be

$$U(x, y) = 6x^{0.8}y^{0.2} + x.$$

$U(x, y)$ is measured in pages per day.

a. Calculate $\left.\frac{\partial U}{\partial x}\right|_{(10,5)}$ and $\left.\frac{\partial U}{\partial y}\right|_{(10,5)}$ to two decimal places, and interpret the results.

[18] Incomes are in 2007 dollars. Source for data: U.S. Census Bureau (www.census.gov).

b. What does the ratio $\left.\frac{\partial U}{\partial x}\right|_{(10,5)} \Big/ \left.\frac{\partial U}{\partial y}\right|_{(10,5)}$ tell about the usefulness of these products?

60. ▼ ***Grades***[19] A production formula for a student's performance on a difficult English examination is given by

$$g(t, x) = 4tx - 0.2t^2 - x^2$$

where g is the grade the student can expect to get, t is the number of hours of study for the examination, and x is the student's grade point average.

a. Calculate $\left.\frac{\partial g}{\partial t}\right|_{(10,3)}$ and $\left.\frac{\partial g}{\partial x}\right|_{(10,3)}$ and interpret the results.

b. What does the ratio $\left.\frac{\partial g}{\partial t}\right|_{(10,3)} \Big/ \left.\frac{\partial g}{\partial x}\right|_{(10,3)}$ tell about the relative merits of study and grade point average?

61. ▼ ***Electrostatic Repulsion*** If positive electric charges of Q and q coulombs are situated at positions (a, b, c) and (x, y, z) respectively, then the force of repulsion they experience is given by

$$F = K\frac{Qq}{(x-a)^2 + (y-b)^2 + (z-c)^2}$$

where $K \approx 9 \times 10^9$, F is given in newtons, and all positions are measured in meters. Assume that a charge of 10 coulombs is situated at the origin, and that a second charge of 5 coulombs is situated at (2, 3, 3) and moving in the y-direction at one meter per second. How fast is the electrostatic force it experiences decreasing? (Round the answer to one significant digit.)

62. ▼ ***Electrostatic Repulsion*** Repeat the preceding exercise, assuming that a charge of 10 coulombs is situated at the origin and that a second charge of 5 coulombs is situated at (2, 3, 3) and moving in the negative z direction at one meter per second. (Round the answer to one significant digit.)

63. ▼ ***Investments*** Recall that the compound interest formula for annual compounding is

$$A(P, r, t) = P(1 + r)^t$$

where A is the future value of an investment of P dollars after t years at an interest rate of r.

a. Calculate $\frac{\partial A}{\partial P}$, $\frac{\partial A}{\partial r}$, and $\frac{\partial A}{\partial t}$, all evaluated at (100, 0.10, 10). (Round your answers to two decimal places.) Interpret your answers.

b. What does the function $\left.\frac{\partial A}{\partial P}\right|_{(100,0.10,t)}$ of t tell about your investment?

64. ▼ ***Investments*** Repeat the preceding exercise, using the formula for continuous compounding:

$$A(P, r, t) = Pe^{rt}$$

65. ▼ ***Modeling with the Cobb-Douglas Production Formula*** Assume you are given a production formula of the form

$$P(x, y) = Kx^a y^b \quad (a + b = 1).$$

a. Obtain formulas for $\frac{\partial P}{\partial x}$ and $\frac{\partial P}{\partial y}$, and show that $\frac{\partial P}{\partial x} = \frac{\partial P}{\partial y}$ precisely when $x/y = a/b$.

b. Let x be the number of workers a firm employs and let y be its monthly operating budget in thousands of dollars. Assume that the firm currently employs 100 workers and has a monthly operating budget of \$200,000. If each additional worker contributes as much to productivity as each additional \$1,000 per month, find values of a and b that model the firm's productivity.

66. ▼ ***Housing Costs***[20] The cost C of building a house is related to the number k of carpenters used and the number e of electricians used by

$$C(k, e) = 15{,}000 + 50k^2 + 60e^2.$$

If three electricians are currently employed in building your new house and the marginal cost per additional electrician is the same as the marginal cost per additional carpenter, how many carpenters are being used? (Round your answer to the nearest carpenter.)

67. ▼ ***Nutrient Diffusion*** Suppose that one cubic centimeter of nutrient is placed at the center of a circular petri dish filled with water. We might wonder how the nutrient is distributed after a time of t seconds. According to the classical theory of diffusion, the concentration of nutrient (in parts of nutrient per part of water) after a time t is given by

$$u(r, t) = \frac{1}{4\pi Dt}e^{-\frac{r^2}{4Dt}}.$$

Here D is the *diffusivity*, which we will take to be 1, and r is the distance from the center in centimeters. How fast is the concentration increasing at a distance of 1 cm from the center 3 seconds after the nutrient is introduced?

68. ▼ ***Nutrient Diffusion*** Refer back to the preceding exercise. How fast is the concentration increasing at a distance of 4 cm from the center 4 seconds after the nutrient is introduced?

COMMUNICATION AND REASONING EXERCISES

69. Given that $f(a, b) = r$, $f_x(a, b) = s$, and $f_y(a, b) = t$, complete the following: _____ is increasing at a rate of _____ units per unit of x, _____ is increasing at a rate of _____ units per unit of y, and the value of _____ is _____ when $x =$ _____ and $y =$ _____.

70. A firm's productivity depends on two variables, x and y. Currently, $x = a$ and $y = b$, and the firm's productivity is 4,000 units. Productivity is increasing at a rate of 400 units per unit *decrease* in x, and is decreasing at a rate of 300 units per unit increase in y. What does all of this information tell you about the firm's productivity function $g(x, y)$?

71. Complete the following: Let $f(x, y, z)$ be the cost to build a development of x cypods (one-bedroom units) in the city-state of Utarek, Mars, y argaats (two-bedroom units) and z orbici

[19] Based on an Exercise in *Introduction to Mathematical Economics* by A. L. Ostrosky Jr. and J. V. Koch (Waveland Press, Illinois, 1979).

[20] Ibid.

(singular: orbicus; three-bedroom units) in Ƶ (zonars, the designated currency in Utarek).[21] Then $\dfrac{\partial f}{\partial z}$ measures _____ and has units _____ .

72. Complete the following: Let $f(t, x, y)$ be the projected number of citizens of the Principality State of Voodice, Luna[22] in year t since its founding, assuming the presence of x lunar vehicle factories and y domed settlements. Then $\dfrac{\partial f}{\partial x}$ measures _____ and has units _____ .

73. Give an example of a function $f(x, y)$ with $f_x(1, 1) = -2$ and $f_y(1, 1) = 3$.

74. Give an example of a function $f(x, y, z)$ that has all of its partial derivatives equal to nonzero constants.

[21] Source: www.marsnext.com/comm/zonars.html

[22] Source: www.voodice.info

75. ▼ The graph of $z = b + mx + ny$ (b, m, and n constants) is a plane.

   a. Explain the geometric significance of the numbers b, m, and n.

   b. Show that the equation of the plane passing through (h, k, l) with slope m in the x direction (in the sense of $\partial/\partial x$) and slope n in the y direction is

$$z = l + m(x - h) + n(y - k).$$

76. ▼ The **tangent plane** to the graph of $f(x, y)$ at $P(a, b, f(a, b))$ is the plane containing the lines tangent to the slice through the graph by $y = b$ (as in Figure 17) and the slice through the graph by $x = a$. Use the result of the preceding exercise to show that the equation of the tangent plane is

$$z = f(a, b) + f_x(a, b)(x - a) + f_y(a, b)(y - b).$$

8.3 Maxima and Minima

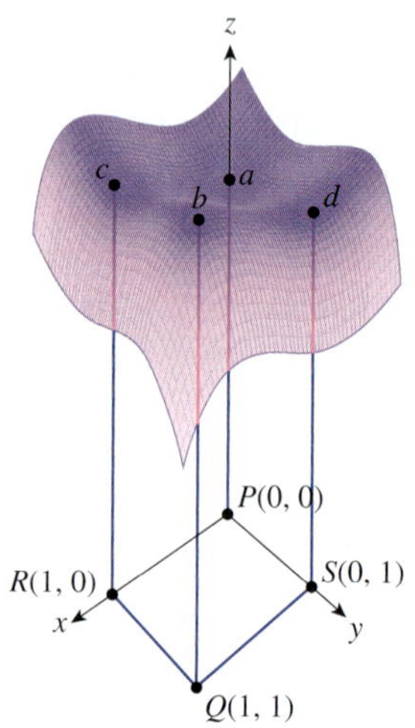

Figure 18

In Chapter 5, on applications of the derivative, we saw how to locate relative extrema of a function of a single variable. In this section we extend our methods to functions of two variables. Similar techniques work for functions of three or more variables.

Figure 18 shows a portion of the graph of the function $f(x, y) = 2(x^2 + y^2) - (x^4 + y^4) + 1$. The graph in Figure 18 resembles a "flying carpet," and several interesting points, marked a, b, c, and d are shown.

1. The point a has coordinates $(0, 0, f(0, 0))$, is directly above the origin $(0, 0)$, and is the lowest point in its vicinity; water would puddle there. We say that f has a **relative minimum** at $(0, 0)$ because $f(0, 0)$ is smaller than $f(x, y)$ for any (x, y) near $(0, 0)$.
2. Similarly, the point b is higher than any point in its vicinity. Thus, we say that f has a **relative maximum** at $(1, 1)$.
3. The points c and d represent a new phenomenon and are called **saddle points**. They are neither relative maxima nor relative minima but seem to be a little of both.

To see more clearly what features a saddle point has, look at Figure 19, which shows a portion of the graph near the point d.

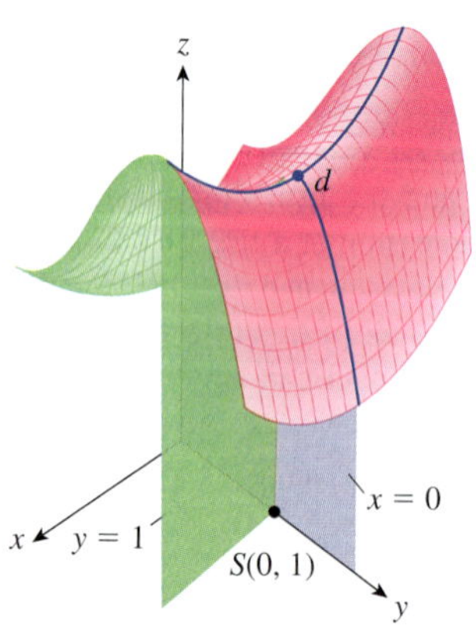

Figure 19

If we slice through the graph along $y = 1$, we get a curve on which d is the *lowest* point. Thus, d looks like a relative minimum along this slice. On the other hand, if we slice through the graph along $x = 0$, we get another curve, on which d is the *highest* point, so d looks like a relative maximum along this slice. This kind of behavior characterizes a saddle point: f has a **saddle point** at (r, s) if f has a relative minimum at (r, s) along some slice through that point and a relative maximum along another slice through that point. If you look at the other saddle point, c, in Figure 18, you see the same characteristics.

While numerical information can help us locate the approximate positions of relative extrema and saddle points, calculus permits us to locate these points accurately, as we did for functions of a single variable. Look once again at Figure 18, and notice the following:

- The points P, Q, R, and S are all in the **interior** of the domain of f; that is, none lie on the boundary of the domain. Said another way, we can move some distance in any direction from any of these points without leaving the domain of f.

- The tangent lines along the slices through these points parallel to the x- and y-axes are *horizontal*. Thus, the partial derivatives $\partial f/\partial x$ and $\partial f/\partial y$ are zero when evaluated at any of the points P, Q, R, and S. This gives us a way of locating candidates for relative extrema and saddle points.

The following summary generalizes and also expands on some of what we have just said:

Relative and Absolute Maxima and Minima

The function f of n variables has a **relative maximum** at $(x_1, x_2, \ldots, x_n) = (r_1, r_2, \ldots, r_n)$ if $f(r_1, r_2, \ldots, r_n) \geq f(x_1, x_2, \ldots, x_n)$ for every point $(x_1, x_2, \ldots, x_n)$ near* $(r_1, r_2, \ldots, r_n)$ in the domain of f. We say that f has an **absolute maximum** at $(r_1, r_2, \ldots, r_n)$ if $f(r_1, r_2, \ldots, r_n) \geq f(x_1, x_2, \ldots, x_n)$ for every point $(x_1, x_2, \ldots, x_n)$ in the domain of f. The terms **relative minimum** and **absolute minimum** are defined in a similar way.

* **NOTE** For $(x, y, \ldots)$ to be near $(r, s, \ldots)$ we mean that x is in some open interval $(r - h_1, r + h_1)$ centered at r, y is in some open interval $(s - h_2, s + h_2)$ centered at s, and so on.

Locating Candidates for Relative Extrema and Saddle Points in the Interior of the Domain of *f*:

- Set $\dfrac{\partial f}{\partial x_1} = 0, \dfrac{\partial f}{\partial x_2} = 0, \ldots, \dfrac{\partial f}{\partial x_n} = 0$, simultaneously, and solve for $x_1, x_2, \ldots, x_n$.
- Check that the resulting points $(x_1, x_2, \ldots, x_n)$ are in the interior of the domain of f.

Points at which all the partial derivatives of f are zero are called **critical points**. The critical points are the only candidates for relative extrema and saddle points in the interior of the domain of f, assuming that its partial derivatives are defined at every point.*

* **NOTE** One can use the techniques of the next section to find extrema on the *boundary* of the domain of a function; for a complete discussion, see the optional extra section: *Maxima and Minima: Boundaries and the Extreme Value Theorem.* (We shall not consider the analogs of the singular points.)

Quick Examples

In each of the following Quick Examples, the domain is the whole Cartesian plane, and the partial derivatives are defined at every point, so the critical points give us the only candidates for relative extrema and saddle points:

1. Let $f(x, y) = x^3 + (y - 1)^2$. Then $\dfrac{\partial f}{\partial x} = 3x^2$ and $\dfrac{\partial f}{\partial y} = 2(y - 1)$. Thus, we solve the system

   $$3x^2 = 0 \quad \text{and} \quad 2(y - 1) = 0.$$

   The first equation gives $x = 0$, and the second gives $y = 1$. Thus, the only critical point is (0, 1).
2. Let $f(x, y) = x^2 - 4xy + 8y$. Then $\dfrac{\partial f}{\partial x} = 2x - 4y$ and $\dfrac{\partial f}{\partial y} = -4x + 8$. Thus, we solve

   $$2x - 4y = 0 \quad \text{and} \quad -4x + 8 = 0.$$

   The second equation gives $x = 2$, and the first then gives $y = 1$. Thus, the only critical point is (2, 1).

3. Let $f(x, y) = e^{-(x^2+y^2)}$. Taking partial derivatives and setting them equal to zero gives

$$-2xe^{-(x^2+y^2)} = 0 \qquad \text{We set } \frac{\partial f}{\partial x} = 0.$$

$$-2ye^{-(x^2+y^2)} = 0. \qquad \text{We set } \frac{\partial f}{\partial y} = 0.$$

The first equation implies that $x = 0$,* and the second implies that $y = 0$. Thus, the only critical point is $(0, 0)$.

* **NOTE** Recall that if a product of two numbers is zero, then one or the other must be zero. In this case the number $e^{-(x^2+y^2)}$ can't be zero (because e^u is never zero), which gives the result claimed.

In the next example we first locate all critical points, and then classify each one as a relative maximum, minimum, saddle point, or none of these.

EXAMPLE 1 Locating and Classifying Critical Points

Locate all critical points of $f(x, y) = x^2y - x^2 - 2y^2$. Graph the function to classify the critical points as relative maxima, minima, saddle points, or none of these.

Solution The partial derivatives are

$$f_x = 2xy - 2x = 2x(y - 1)$$
$$f_y = x^2 - 4y.$$

Setting these equal to zero gives

$$x = 0 \text{ or } y = 1$$
$$x^2 = 4y.$$

We get a solution by choosing either $x = 0$ or $y = 1$ and substituting into $x^2 = 4y$.

Case 1: $x = 0$ Substituting into $x^2 = 4y$ gives $0 = 4y$ and hence $y = 0$. Thus, the critical point for this case is $(x, y) = (0, 0)$.

Case 2: $y = 1$ Substituting into $x^2 = 4y$ gives $x^2 = 4$ and hence $x = \pm 2$. Thus, we get two critical points for this case: $(2, 1)$ and $(-2, 1)$.

We now have three critical points altogether: $(0, 0)$, $(2, 1)$, and $(-2, 1)$. Because the domain of f is the whole Cartesian plane and the partial derivatives are defined at every point, these critical points are the only candidates for relative extrema and saddle points. We get the corresponding points on the graph by substituting for x and y in the equation for f to get the z-coordinates. The points are $(0, 0, 0)$, $(2, 1, -2)$, and $(-2, 1, -2)$.

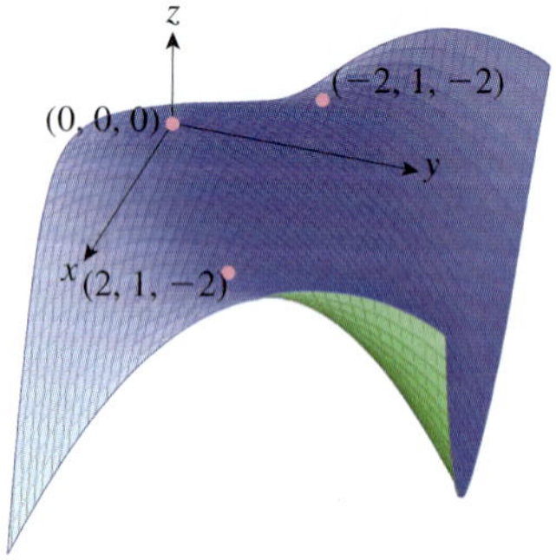

Figure 20

Classifying the Critical Points Graphically To classify the critical points graphically, we look at the graph of f shown in Figure 20.

Examining the graph carefully, we see that the point $(0, 0, 0)$ is a relative maximum. As for the other two critical points, are they saddle points or are they relative maxima? They are relative maxima along the y-direction, but they are relative minima along the lines $y = \pm x$ (see the top edge of the picture, which shows a dip at $(-2, 1, 2)$) and so they are saddle points. If you don't believe this, we will get more evidence following and in a later example.

T ***Classifying the Critical Points Numerically*** We can use a tabular representation of the function to classify the critical points numerically. The following tabular representation of the function can be obtained using Excel. (See the Excel Technology Guide discussion of Section 8.1 Example 3 at the end of the chapter for information on using Excel to generate such a table.)

$x \rightarrow$ / $y \downarrow$

	−3	**−2**	**−1**	**0**	**1**	**2**	**3**
−3	−54	−34	−22	−18	−22	−34	−54
−2	−35	−20	−11	−8	−11	−20	−35
−1	−20	−10	−4	−2	−4	−10	−20
0	−9	−4	−1	**0**	−1	−4	−9
1	−2	**−2**	−2	−2	−2	**−2**	−2
2	1	−4	−7	−8	−7	−4	1
3	0	−10	−16	−18	−16	−10	0

The shaded and colored cells show rectangular neighborhoods of the three critical points (0, 0), (2, 1), and (−2, 1). (Notice that they overlap.) The values of f at the points are at the centers of these rectangles. Looking at the gray neighborhood of $(x, y) = (0, 0)$, we see that $f(0, 0) = 0$ is the largest value of f in the shaded cells, suggesting that f has a maximum at (0, 0). The shaded neighborhood of (2, 1) on the right shows $f(2, 1) = -2$ as the maximum along some slices (e.g., the vertical slice), and a minimum along the diagonal slice from top left to bottom right. This is what results in a saddle point on the graph. The point (−2, 1) is similar, and thus f also has a saddle point at (−2, 1).

Q: *Is there an algebraic way of deciding whether a given point is a relative maximum, relative minimum, or saddle point?*

A: There is a "second derivative test" for functions of two variables, stated as follows.

Second Derivative Test for Functions of Two Variables

Suppose (a, b) is a critical point in the interior of the domain of the function f of two variables. Let H be the quantity

$$H = f_{xx}(a, b)f_{yy}(a, b) - [f_{xy}(a, b)]^2.$$

H is called the Hessian

Then, if H is *positive*,

- f has a relative minimum at (a, b) if $f_{xx}(a, b) > 0$.
- f has a relative maximum at (a, b) if $f_{xx}(a, b) < 0$.

If H is *negative*,

- f has a saddle point at (a, b).

If $H = 0$, the test tells us nothing, so we need to look at the graph or a numerical table to see what is going on.

Quick Examples

1. Let $f(x, y) = x^2 - y^2$. Then

$$f_x = 2x \quad \text{and} \quad f_y = -2y,$$

which gives (0, 0) as the only critical point. Also,

$$f_{xx} = 2,\ f_{xy} = 0, \quad \text{and} \quad f_{yy} = -2,$$

Note that these are constant.

which gives $H = (2)(-2) - 0^2 = -2$. Because H is negative, we have a saddle point at (0, 0).

2. Let $f(x, y) = x^2 + 2y^2 + 2xy + 4x$. Then

$$f_x = 2x + 2y + 4 \quad \text{and} \quad f_y = 2x + 4y.$$

Setting these equal to zero gives a system of two linear equations in two unknowns:

$$x + y = -2$$
$$x + 2y = 0.$$

This system has solution $(-4, 2)$, so this is our only critical point. The second partial derivatives are $f_{xx} = 2$, $f_{xy} = 2$, and $f_{yy} = 4$, so $H = (2)(4) - 2^2 = 4$. Because $H > 0$ and $f_{xx} > 0$, we have a relative minimum at $(-4, 2)$.

Note There is a second derivative test for functions of three or more variables, but it is considerably more complicated. We stick with functions of two variables for the most part in this book. The justification of the second derivative test is beyond the scope of this book. ■

EXAMPLE 2 Using the Second Derivative Test

Use the second derivative test to analyze the function $f(x, y) = x^2y - x^2 - 2y^2$ discussed in Example 1, and confirm the results we got there.

Solution We saw in Example 1 that the first-order derivatives are

$$f_x = 2xy - 2x = 2x(y - 1)$$
$$f_y = x^2 - 4y$$

and the critical points are (0, 0), (2, 1), and $(-2, 1)$. We also need the second derivatives:

$$f_{xx} = 2y - 2$$
$$f_{xy} = 2x$$
$$f_{yy} = -4.$$

The point (0, 0): $f_{xx}(0, 0) = -2$, $f_{xy}(0, 0) = 0$, $f_{yy}(0, 0) = -4$, so $H = 8$. Because $H > 0$ and $f_{xx}(0, 0) < 0$, the second derivative test tells us that f has a relative maximum at (0, 0).

The point (2, 1): $f_{xx}(2, 1) = 0$, $f_{xy}(2, 1) = 4$ and $f_{yy}(2, 1) = -4$, so $H = -16$. Because $H < 0$, we know that f has a saddle point at (2, 1).

The point (−2, 1): $f_{xx}(-2, 1) = 0$, $f_{xy}(-2, 1) = -4$ and $f_{yy}(-2, 1) = -4$, so once again $H = -16$, and f has a saddle point at $(-2, 1)$.

Deriving the Regression Formulas

Back in Section 1.4, we presented the following set of formulas for the **regression** or **best-fit** line associated with a given set of data points $(x_1, y_1), (x_2, y_2), \dots, (x_n, y_n)$.

Regression Line

The line that best fits the n data points $(x_1, y_1), (x_2, y_2), \dots, (x_n, y_n)$ has the form

$$y = mx + b$$

where

$$m = \frac{n\left(\sum xy\right) - \left(\sum x\right)\left(\sum y\right)}{n\left(\sum x^2\right) - \left(\sum x\right)^2}$$

$$b = \frac{\sum y - m\left(\sum x\right)}{n}$$

$n =$ number of data points.

Derivation of the Regression Line Formulas

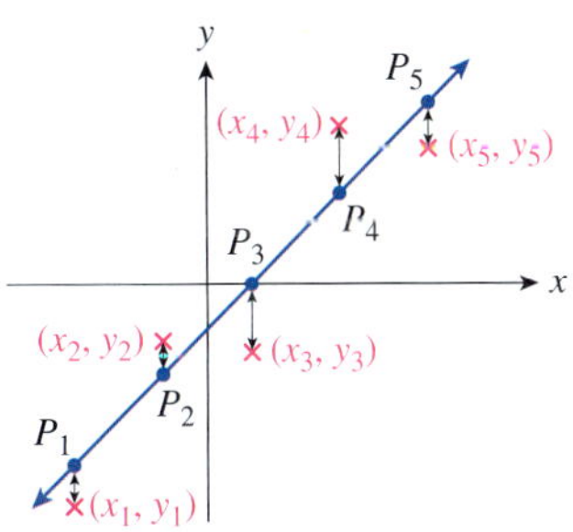

Figure 21

Recall that the regression line is defined to be the line that minimizes the sum of the squares of the **residuals**, measured by the vertical distances shown in Figure 21, which shows a regression line associated with $n = 5$ data points. In the figure, the points $P_1, \dots, P_n$ on the regression line have coordinates $(x_1, mx_1 + b), (x_2, mx_2 + b), \dots, (x_n, mx_n + b)$. The residuals are the quantities $y_{\text{Observed}} - y_{\text{Predicted}}$:

$$y_1 - (mx_1 + b), y_2 - (mx_2 + b), \dots, y_n - (mx_n + b).$$

The sum of the squares of the residuals is therefore

$$S(m, b) = [y_1 - (mx_1 + b)]^2 + [y_2 - (mx_2 + b)]^2 + \cdots + [y_n - (mx_n + b)]^2$$

and this is the quantity we must minimize by choosing m and b. Because we reason that there is a line that minimizes this quantity, there must be a relative minimum at that point. We shall see in a moment that the function S has at most one critical point, which must therefore be the desired absolute minimum. To obtain the critical points of S, we set the partial derivatives equal to zero and solve:

$$\begin{aligned} S_m = 0: &\quad -2x_1[y_1 - (mx_1 + b)] - \cdots - 2x_n[y_n - (mx_n + b)] = 0 \\ S_b = 0: &\quad -2[y_1 - (mx_1 + b)] - \cdots - 2[y_n - (mx_n + b)] = 0. \end{aligned}$$

Dividing by -2 and gathering terms allows us to rewrite the equations as

$$\begin{aligned} m\left(x_1^2 + \cdots + x_n^2\right) + b(x_1 + \cdots + x_n) &= x_1y_1 + \cdots + x_ny_n \\ m(x_1 + \cdots + x_n) + nb &= y_1 + \cdots + y_n. \end{aligned}$$

We can rewrite these equations more neatly using $\sum$-notation:

$$\begin{aligned} m\left(\sum x^2\right) + b\left(\sum x\right) &= \sum xy \\ m\left(\sum x\right) + nb &= \sum y. \end{aligned}$$

This is a system of two linear equations in the two unknowns m and b. It may or may not have a unique solution. When there is a unique solution, we can conclude that the best fit line is given by solving these two equations for m and b. Alternatively, there is a general formula for the solution of any system of two equations in two unknowns, and if we apply this formula to our two equations, we get the regression formulas above.

8.3 EXERCISES

▼ more advanced ◆ challenging

T indicates exercises that should be solved using technology

In Exercises 1–4, classify each labeled point on the graph as one of the following:

a. a relative maximum
b. a relative minimum
c. a saddle point
d. a critical point but neither a relative extremum nor a saddle point
e. none of the above HINT [See Example 1.]

1.

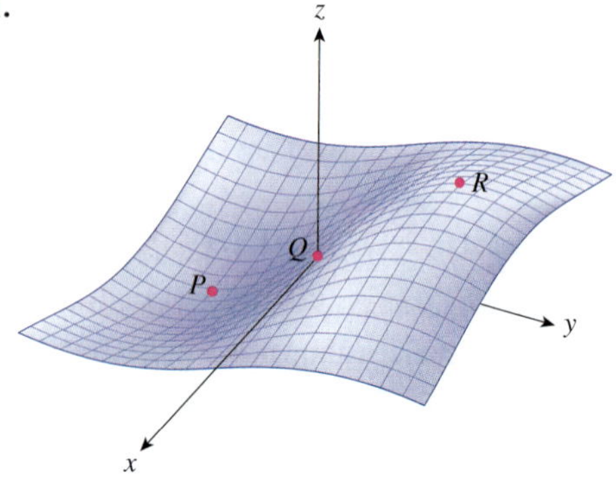

2.

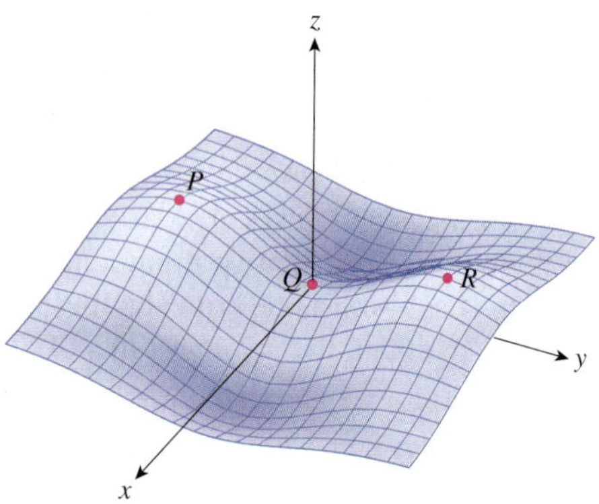

3.

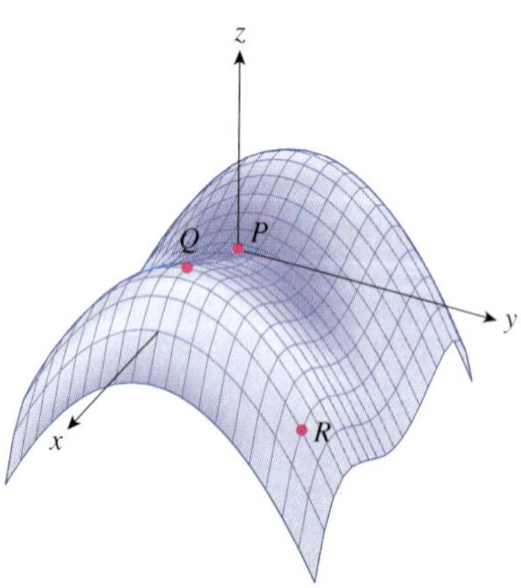

4.

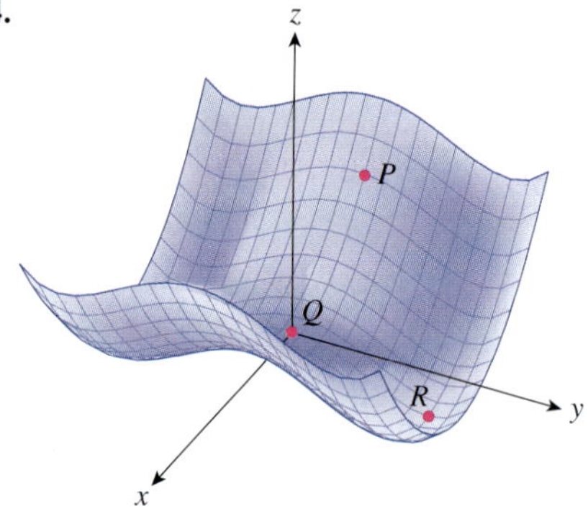

In Exercises 5–10, classify the shaded value in each table as one of the following:

a. a relative maximum
b. a relative minimum
c. a saddle point
d. neither a relative extremum nor a saddle point

5. $x \rightarrow$, $y \downarrow$

	−3	−2	−1	0	1	2
−3	10	5	2	1	2	5
−2	9	4	1	0	1	4
−1	10	5	2	1	2	5
0	13	8	5	4	5	8
1	18	13	10	9	10	13
2	25	20	17	16	17	20
3	34	29	26	25	26	29

6. $x \rightarrow$, $y \downarrow$

	−3	−2	−1	0	1	2
−3	5	0	−3	−4	−3	0
−2	8	3	0	−1	0	3
−1	9	4	1	0	1	4
0	8	3	0	−1	0	3
1	5	0	−3	−4	−3	0
2	0	−5	−8	−9	−8	−5
3	−7	−12	−15	−16	−15	−12

7.

y↓ \ x→	−3	−2	−1	0	1	2
−3	5	0	−3	−4	−3	0
−2	8	3	0	−1	0	3
−1	9	4	1	0	1	4
0	8	3	0	−1	0	3
1	5	0	−3	−4	−3	0
2	0	−5	−8	−9	−8	−5
3	−7	−12	−15	−16	−15	−12

8.

y↓ \ x→	−3	−2	−1	0	1	2
−3	2	3	2	−1	−6	−13
−2	3	4	3	0	−5	−12
−1	2	3	2	−1	−6	−13
0	−1	0	−1	−4	−9	−16
1	−6	−5	−6	−9	−14	−21
2	−13	−12	−13	−16	−21	−28
3	−22	−21	−22	−25	−30	−37

9.

y↓ \ x→	−3	−2	−1	0	1	2
−3	4	5	4	1	−4	−11
−2	3	4	3	0	−5	−12
−1	4	5	4	1	−4	−11
0	7	8	7	4	−1	−8
1	12	13	12	9	4	−3
2	19	20	19	16	11	4
3	28	29	28	25	20	13

10.

y↓ \ x→	−3	−2	−1	0	1	2
−3	100	101	100	97	92	85
−2	99	100	99	96	91	84
−1	98	99	98	95	90	83
0	91	92	91	88	83	76
1	72	73	72	69	64	57
2	35	36	35	32	27	20
3	−26	−25	−26	−29	−34	−41

Locate and classify all the critical points of the functions in Exercises 11–36. HINT [See Example 2.]

11. $f(x, y) = x^2 + y^2 + 1$

12. $f(x, y) = 4 - (x^2 + y^2)$

13. $g(x, y) = 1 - x^2 - x - y^2 + y$

14. $g(x, y) = x^2 + x + y^2 - y - 1$

15. $k(x, y) = x^2 - 3xy + y^2$

16. $k(x, y) = x^2 - xy + 2y^2$

17. $f(x, y) = x^2 + 2xy + 2y^2 - 2x + 4y$

18. $f(x, y) = x^2 + xy - y^2 + 3x - y$

19. $g(x, y) = -x^2 - 2xy - 3y^2 - 3x - 2y$

20. $g(x, y) = -x^2 - 2xy + y^2 + x - 4y$

21. $h(x, y) = x^2y - 2x^2 - 4y^2$

22. $h(x, y) = x^2 + y^2 - y^2x - 4$

23. $f(x, y) = x^2 + 2xy^2 + 2y^2$

24. $f(x, y) = x^2 + x^2y + y^2$

25. $s(x, y) = e^{x^2+y^2}$

26. $s(x, y) = e^{-(x^2+y^2)}$

27. $t(x, y) = x^4 + 8xy^2 + 2y^4$

28. $t(x, y) = x^3 - 3xy + y^3$

29. $f(x, y) = x^2 + y - e^y$

30. $f(x, y) = xe^y$

31. $f(x, y) = e^{-(x^2+y^2+2x)}$

32. $f(x, y) = e^{-(x^2+y^2-2x)}$

33. ▼ $f(x, y) = xy + \frac{2}{x} + \frac{2}{y}$

34. ▼ $f(x, y) = xy + \frac{4}{x} + \frac{2}{y}$

35. ▼ $g(x, y) = x^2 + y^2 + \frac{2}{xy}$

36. ▼ $g(x, y) = x^3 + y^3 + \frac{3}{xy}$

37. ▼ Refer back to Exercise 11. Which (if any) of the critical points of $f(x, y) = x^2 + y^2 + 1$ are absolute extrema?

38. ▼ Refer back to Exercise 12. Which (if any) of the critical points of $f(x, y) = 4 - (x^2 + y^2)$ are absolute extrema?

39. T ▼ Refer back to Exercise 21. Which (if any) of the critical points of $h(x, y) = x^2y - 2x^2 - 4y^2$ are absolute extrema?

40. T ▼ Refer back to Exercise 22. Which (if any) of the critical points of $h(x, y) = x^2 + y^2 - y^2x - 4$ are absolute extrema?

APPLICATIONS

41. ***Brand Loyalty*** Suppose the fraction of **Mazda** car owners who chose another new **Mazda** can be modeled by the following function:[23]

$$M(c, f) = 11 + 8c + 4c^2 - 40f - 20cf + 40f^2$$

where c is the fraction of **Chrysler** car owners who remained loyal to **Chrysler** and f is the fraction of **Ford** car owners

[23] This model is not accurate, although it was inspired by an approximation of a second order regression based on data from the period 1988–1995. Source for original data: Chrysler, Maritz Market Research, Consumer Attitude Research, and Strategic Vision/*The New York Times*, November 3, 1995, p. D2.

remaining loyal to Ford. Locate and classify all the critical points and interpret your answer. HINT [See Example 2.]

42. ***Brand Loyalty*** Repeat the preceding exercise using the function:

$$M(c, f) = -10 - 8f - 4f^2 + 40c + 20fc - 40c^2$$

HINT [See Example 2.]

43. ▼ ***Pollution Control*** The cost of controlling emissions at a firm goes up rapidly as the amount of emissions reduced goes up. Here is a possible model:

$$C(x, y) = 4{,}000 + 100x^2 + 50y^2$$

where x is the reduction in sulfur emissions, y is the reduction in lead emissions (in pounds of pollutant per day), and C is the daily cost to the firm (in dollars) of this reduction. Government clean-air subsidies amount to \$500 per pound of sulfur and \$100 per pound of lead removed. How many pounds of pollutant should the firm remove each day in order to minimize *net* cost (cost minus subsidy)?

44. ▼ ***Pollution Control*** Repeat the preceding exercise using the following information:

$$C(x, y) = 2{,}000 + 200x^2 + 100y^2$$

with government subsidies amounting to \$100 per pound of sulfur and \$500 per pound of lead removed per day.

45. ▼ ***Revenue*** Your company manufactures two models of speakers, the Ultra Mini and the Big Stack. Demand for each depends partly on the price of the other. If one is expensive, then more people will buy the other. If p_1 is the price of the Ultra Mini, and p_2 is the price of the Big Stack, demand for the Ultra Mini is given by

$$q_1(p_1, p_2) = 100{,}000 - 100p_1 + 10p_2$$

where q_1 represents the number of Ultra Minis that will be sold in a year. The demand for the Big Stack is given by

$$q_2(p_1, p_2) = 150{,}000 + 10p_1 - 100p_2.$$

Find the prices for the Ultra Mini and the Big Stack that will maximize your total revenue.

46. ▼ ***Revenue*** Repeat the preceding exercise, using the following demand functions:

$$q_1(p_1, p_2) = 100{,}000 - 100p_1 + p_2$$
$$q_2(p_1, p_2) = 150{,}000 + p_1 - 100p_2.$$

47. ▼ ***Luggage Dimensions: American Airlines*** American Airlines requires that the total outside dimensions (length + width + height) of a checked bag not exceed 62 inches.[24] What are the dimensions of the largest volume bag that you can check on an American flight?

[24] According to information on its Web site (www.aa.com).

48. ▼ ***Carry-on Bag Dimensions: American Airlines*** American Airlines requires that the total outside dimensions (length + width + height) of a carry-on bag not exceed 45 inches.[25] What are the dimensions of the largest volume bag that you can carry on an American flight?

49. ▼ ***Package Dimensions: USPS*** The U.S. Postal Service (USPS) will accept only packages with a length plus girth no more than 108 inches.[26] (See the figure.)

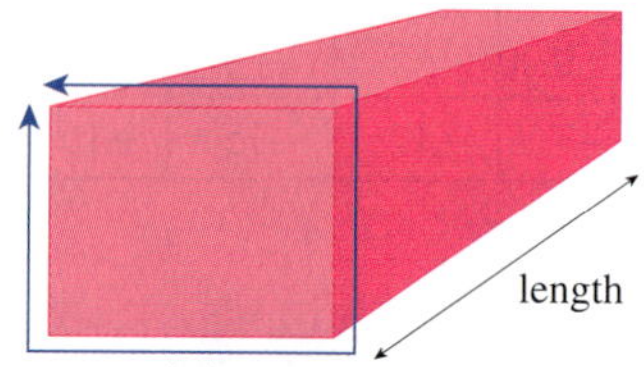

What are the dimensions of the largest volume package that the USPS will accept? What is its volume?

50. ▼ ***Package Dimensions: UPS*** United Parcel Service (UPS) will accept only packages with length no more than 108 inches and length plus girth no more than 165 inches.[27] (See figure for the preceding exercise.) What are the dimensions of the largest volume package that UPS will accept? What is its volume?

COMMUNICATION AND REASONING EXERCISES

51. Sketch the graph of a function that has one extremum and no saddle points.

52. Sketch the graph of a function that has one saddle point and one extremum.

53. ▼ Sketch the graph of a function that has one relative extremum, no absolute extrema, and no saddle points.

54. ▼ Sketch the graph of a function that has infinitely many absolute maxima.

55. Let $H = f_{xx}(a, b)f_{yy}(a, b) - f_{xy}(a, b)^2$. What condition on H guarantees that f has a relative extremum at the point (a, b)?

56. Let H be as in the preceding exercise. Give an example to show that it is possible to have $H = 0$ and a relative minimum at (a, b).

57. ▼ Suppose that when the graph of $f(x, y)$ is sliced by a vertical plane through (a, b) parallel to either the xz-plane or the yz-plane, the resulting curve has a relative maximum at (a, b). Does this mean that f has a relative maximum at (a, b)? Explain your answer.

[25] Ibid.

[26] The requirement for packages sent other than Parcel Post, as of September 2008 (www.usps.com).

[27] The requirement as of September 2008 (www.ups.com).

58. ▼Suppose that f has a relative maximum at (a, b). Does it follow that, if the graph of f is sliced by a vertical plane parallel to either the xz-plane or the yz-plane, the resulting curve has a relative maximum at (a, b)? Explain your answer.

59. ▼***Average Cost*** Let $C(x, y)$ be any cost function. Show that when the average cost is minimized, the marginal costs C_x and C_y both equal the average cost. Explain why this is reasonable.

60. ▼***Average Profit*** Let $P(x, y)$ be any profit function. Show that when the average profit is maximized, the marginal profits P_x and P_y both equal the average profit. Explain why this is reasonable.

61. ◆ The tangent plane to a graph was introduced in Exercise 76 in the preceding section. Use the equation of the tangent plane given there to explain why the tangent plane is parallel to the xy-plane at a relative maximum or minimum of $f(x, y)$.

62. ◆ Use the equation of the tangent plane given in Exercise 76 in the preceding section to explain why the tangent plane is parallel to the xy-plane at a saddle point of $f(x, y)$.

8.4 Constrained Maxima and Minima and Applications

So far we have looked only at the relative extrema of functions with no constraints. However, in Section 5.2 we saw examples in which we needed to find the maximum or minimum of an objective function subject to one or more constraints on the independent variables. For instance, consider the following problem:

$$\text{Minimize } S = xy + 2xz + 2yz \quad \text{subject to } xyz = 4 \text{ with } x > 0,\ y > 0,\ z > 0.$$

One strategy for solving such problems is essentially the same as the strategy we used earlier: Solve the constraint equation for one of the variables, substitute into the objective function, and then optimize the resulting function using the methods of the preceding section. We will call this the *substitution method.* An alternative method, called the *method of Lagrange multipliers*, is useful when it is difficult or impossible to solve the constraint equation for one of the variables, and even when it is possible to do so.

Substitution Method

EXAMPLE 1 Using Substitution

Minimize $S = xy + 2xz + 2yz$ subject to $xyz = 4$ with $x > 0$, $y > 0$, $z > 0$.

Solution As suggested in the above discussion, we proceed as follows:

Solve the constraint equation for one of the variables and then substitute in the objective function. The constraint equation is $xyz = 4$. Solving for z gives

$$z = \frac{4}{xy}.$$

The objective function is $S = xy + 2xz + 2yz$, so substituting $z = 4/xy$ gives

$$S = xy + 2x\frac{4}{xy} + 2y\frac{4}{xy}$$

$$= xy + \frac{8}{y} + \frac{8}{x}.$$

Minimize the resulting function of two variables. We use the method in Section 8.4 to find the minimum of $S = xy + \frac{8}{y} + \frac{8}{x}$ for $x > 0$ and $y > 0$. We look for critical points:

$$S_x = y - \frac{8}{x^2} \qquad S_y = x - \frac{8}{y^2}$$

$$S_{xx} = \frac{16}{x^3} \qquad S_{xy} = 1 \qquad S_{yy} = \frac{16}{y^3}.$$

We now equate the first partial derivatives to zero:

$$y = \frac{8}{x^2} \quad \text{and} \quad x = \frac{8}{y^2}.$$

To solve for x and y, we substitute the first of these equations in the second, getting

$$x = \frac{x^4}{8}$$

$$x^4 - 8x = 0$$

$$x(x^3 - 8) = 0.$$

The two solutions are $x = 0$, which we reject because x cannot be zero, and $x = 2$. Substituting $x = 2$ in $y = 8/x^2$ gives $y = 2$ also. Thus, the only critical point is (2, 2). To apply the second derivative test, we compute

$$S_{xx}(2, 2) = 2 \qquad S_{xy}(2, 2) = 1 \qquad S_{yy}(2, 2) = 2$$

and find that $H = 3 > 0$, so we have a relative minimum at (2, 2).

The corresponding value of z is given by the constraint equation:

$$z = \frac{4}{xy} = \frac{4}{4} = 1.$$

The corresponding value of the objective function is

$$S = xy + \frac{8}{y} + \frac{8}{x} = 4 + \frac{8}{2} + \frac{8}{2} = 12.$$

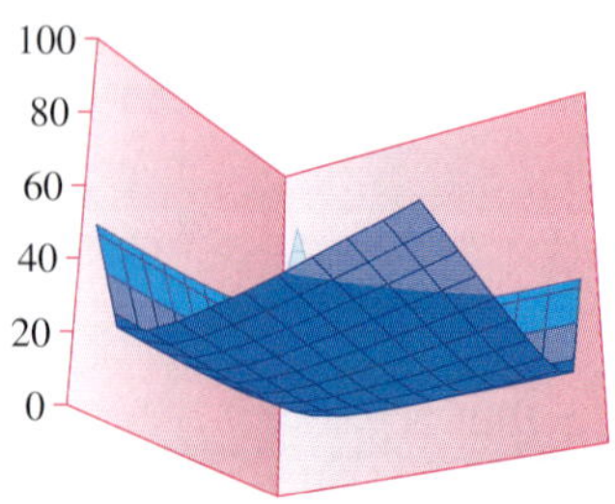

Graph of $S = xy + \frac{8}{y} + \frac{8}{x}$ $(0.2 \le x \le 5, 0.2 \le y \le 5)$

Figure 22

Figure 22 shows a portion of the graph of $S = xy + \frac{8}{y} + \frac{8}{x}$ for positive x and y (drawn using the Excel Surface Grapher in the Chapter 8 utilities at the Web site), and suggests that there is a single absolute minimum, which must be at our only candidate point (2, 2).

We conclude that the minimum of S is 12 and occurs at (2, 2, 1).

The Method of Lagrange Multipliers

As we mentioned above, the method of Lagrange multipliers has the advantage that it can be used in constrained optimization problems when it is difficult or impossible to solve a constraint equation for one of the variables. We restrict attention to the case of a single constraint equation, although the method also generalizes to any number of constraint equations.

Locating Relative Extrema Using the Method of Lagrange Multipliers

To locate the candidates for relative extrema of a function $f(x, y, \ldots)$ subject to the constraint $g(x, y, \ldots) = 0$:

1. Construct the **Lagrangian function**

$$L(x, y, \ldots) = f(x, y, \ldots) - \lambda g(x, y, \ldots)$$

   where λ is a new unknown called a **Lagrange multiplier.**

2. The candidates for the relative extrema occur at the critical points of $L(x, y, \ldots)$. To find them, set all the partial derivatives of $L(x, y, \ldots)$ equal to zero and solve the resulting system, together with the constraint equation $g(x, y, \ldots) = 0$, for the unknowns $x, y, \ldots$ and λ.

The points $(x, y, \ldots)$ that occur in solutions are then the candidates for the relative extrema of f subject to $g = 0$.

Although the justification for the method of Lagrange multipliers is beyond the scope of this text (a derivation can be found in many vector calculus textbooks), we will demonstrate by example how it is used.

EXAMPLE 2 Using Lagrange Multipliers

Use the method of Lagrange multipliers to find the maximum value of $f(x, y) = 2xy$ subject to $x^2 + 4y^2 = 32$.

Solution We start by rewriting the problem with the constraint in the form $g(x, y) = 0$:

$$\text{Maximize } f(x, y) = 2xy \text{ subject to } x^2 + 4y^2 - 32 = 0.$$

Here, $g(x, y) = x^2 + 4y^2 - 32$, and the Lagrangian function is

$$\begin{aligned} L(x, y) &= f(x, y) - \lambda g(x, y) \\ &= 2xy - \lambda(x^2 + 4y^2 - 32). \end{aligned}$$

The system of equations we need to solve is thus

$$\begin{aligned} L_x &= 0: \quad & 2y - 2\lambda x &= 0 \\ L_y &= 0: \quad & 2x - 8\lambda y &= 0 \\ g &= 0. \quad & x^2 + 4y^2 - 32 &= 0. \end{aligned}$$

It is often convenient to solve such a system by first solving one of the equations for λ and then substituting in the remaining equations. Thus, we start by solving the first equation to obtain

$$\lambda = \frac{y}{x}.$$

(A word of caution: Because we divided by x, we made the implicit assumption that $x \neq 0$, so before continuing we should check what happens if $x = 0$. But if $x = 0$, then the first equation, $2y = 2\lambda x$, tells us that $y = 0$ as well, and this contradicts the third equation: $x^2 + 4y^2 - 32 = 0$. Thus, we can rule out the possibility that $x = 0$.) Substituting the value of λ in the second equation gives

$$2x - 8\left(\frac{y}{x}\right)y = 0 \quad \text{or} \quad x^2 = 4y^2.$$

We can now substitute $x^2 = 4y^2$ in the constraint equation, obtaining

$$\begin{aligned} 4y^2 + 4y^2 - 32 &= 0 \\ 8y^2 &= 32 \\ y &= \pm 2. \end{aligned}$$

We now substitute back to obtain

$$x^2 = 4y^2 = 16,$$

or $\qquad x = \pm 4.$

We don't need the value of λ, so we won't solve for it. Thus, the candidates for relative extrema are given by $x = \pm 4$ and $y = \pm 2$, that is, the four points $(-4, -2)$, $(-4, 2)$, $(4, -2)$, and $(4, 2)$. Recall that we are seeking the values of x and y that give the maximum value for $f(x, y) = 2xy$. Because we now have only four points to choose from, we compare the values of f at these four points and conclude that the maximum value of f occurs when $(x, y) = (-4, -2)$ or $(4, 2)$.

Something is suspicious in Example 2. We didn't check to see whether these candidates were relative extrema to begin with, let alone absolute extrema! How do we justify this omission? One of the difficulties with using the method of Lagrange multipliers is that it does not provide us with a test analogous to the second derivative test for functions of several variables. However, if you grant that the function in question does have an absolute maximum, then we require no test, because one of the candidates must give this maximum.

Q: *But how do we know that the given function has an absolute maximum?*

A: The best way to see this is by giving a geometric interpretation. The constraint $x^2 + 4y^2 = 32$ tells us that the point (x, y) must lie on the ellipse shown in Figure 23. The function $f(x, y) = 2xy$ gives the area of the rectangle shaded in Figure 23. There must be a largest such rectangle, because the area varies continuously from 0 when (x, y) is on the x-axis, to positive when (x, y) is in the first quadrant, to 0 again when (x, y) is on the y-axis, so f must have an absolute maximum for at least one pair of coordinates (x, y).

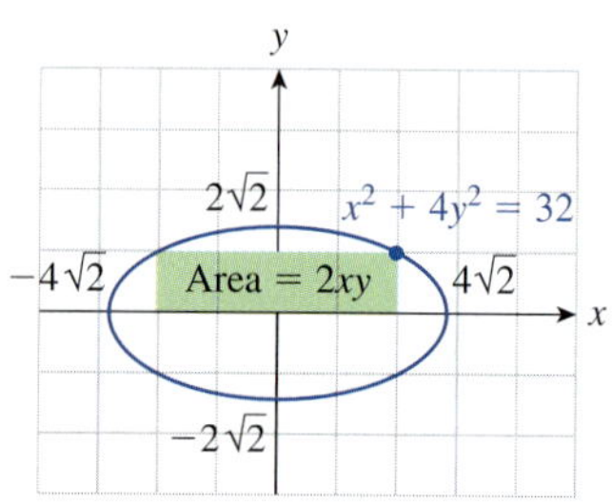

Figure 23

We now show how to use Lagrange multipliers to solve the minimization problem in Example 1:

EXAMPLE 3 Using Lagrange Multipliers: Function of Three Variables

Use the method of Lagrange multipliers to find the minimum value of $S = xy + 2xz + 2yz$ subject to $xyz = 4$ with $x > 0$, $y > 0$, $z > 0$.

Solution We start by rewriting the problem in standard form:

$$\begin{aligned} &\text{Maximize } f(x, y, z) = xy + 2xz + 2yz \\ &\quad \text{subject to } xyz - 4 = 0 \text{ (with } x > 0, y > 0, z > 0). \end{aligned}$$

Here, $g(x, y, z) = xyz - 4$, and the Lagrangian function is

$$L(x, y, z) = f(x, y, z) - \lambda g(x, y, z)$$
$$= xy + 2xz + 2yz - \lambda(xyz - 4).$$

The system of equations we need to solve is thus

$$\begin{aligned} L_x = 0{:}\quad & y + 2z - \lambda yz = 0 \\ L_y = 0{:}\quad & x + 2z - \lambda xz = 0 \\ L_z = 0{:}\quad & 2x + 2y - \lambda xy = 0 \\ g = 0{:}\quad & xyz - 4 = 0. \end{aligned}$$

As in the last example, we solve one of the equations for λ and substitute in the others. The first equation gives

$$\lambda = \frac{1}{z} + \frac{2}{y}.$$

Substituting this into the second equation gives

$$x + 2z = x + \frac{2xz}{y}$$

or $$2 = \frac{2x}{y},$$ Subtract x from both sides and then divide by z.

giving $$y = x.$$

Substituting the expression for λ into the third equation gives

$$2x + 2y = \frac{xy}{z} + 2x$$

or $$2 = \frac{x}{z},$$ Subtract $2x$ from both sides and then divide by y.

giving $$z = \frac{x}{2}.$$

Now we have both y and z in terms of x. We substitute these values in the last (constraint) equation:

$$x(x)\left(\frac{x}{2}\right) - 4 = 0$$
$$x^3 = 8$$
$$x = 2.$$

Thus, $y = x = 2$, and $z = \frac{x}{2} = 1$. Therefore, the only critical point occurs at (2, 2, 1) as we found in Example 1, and the corresponding value of S is

$$S = xy + 2xz + 2yz = (2)(2) + 2(2)(1) + 2(2)(1) = 12.$$

Before we go on... Again, the method of Lagrange multipliers does not tell us whether the critical point in Example 3 is a maximum, minimum, or neither. However, if you grant that the function in question does have an absolute minimum, then the values we found must give this minimum value. ■

APPLICATIONS

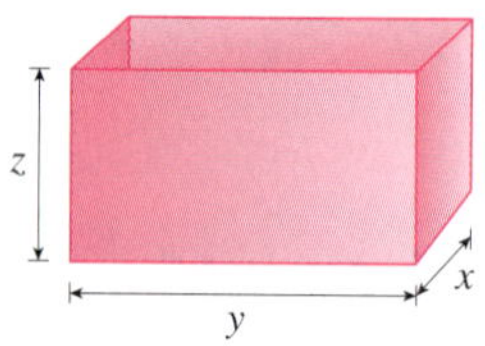

Figure 24

EXAMPLE 4 Minimizing Area

Find the dimensions of an open-top rectangular box that has a volume of 4 cubic feet and the smallest possible surface area.

Solution Our first task is to rephrase this request as a mathematical optimization problem. Figure 24 shows a picture of the box with dimensions x, y, and z. We want to minimize the total surface area, which is given by

$$S = xy + 2xz + 2yz. \qquad \text{Base + Sides + Front and Back}$$

This is our objective function. We can't simply choose x, y, and z to all be zero, however, because the enclosed volume must be 4 cubic feet. So,

$$xyz = 4. \qquad \text{Constraint}$$

This is our constraint equation. Other unstated constraints are $x > 0$, $y > 0$, and $z > 0$, because the dimensions of the box must be positive. We now restate the problem as follows:

$$\text{Minimize } S = xy + 2xz + 2yz \quad \text{subject to } xyz = 4, x > 0, y > 0, z > 0.$$

But this is exactly the problem in Examples 1 and 3, and has a solution $x = 2$, $y = 2$, $z = 1$, $S = 12$. Thus, the required dimensions of the box are

$$x = 2 \text{ ft}, y = 2 \text{ ft}, z = 1 \text{ ft},$$

requiring a total surface area of 12 ft^2.

Q: *In Example 1 we checked that we had a relative minimum at $(x, y) = (2, 2)$ and we were persuaded graphically that this was probably an absolute minimum. Can we be* sure *that this relative minimum is an absolute minimum?*

A: Yes. There must be a least surface area among all boxes that hold 4 cubic feet. (Why?) Because this would give a relative minimum of S and because the only possible relative minimum of S occurs at (2, 2), this is the absolute minimum.

EXAMPLE 5 Maximizing productivity

An electric motor manufacturer uses workers and robots on its assembly line, and has a Cobb-Douglas productivity function* of the form

$$P(x, y) = 10x^{0.2}y^{0.8} \text{ motors manufactured per day}$$

where x is the number of assembly-line workers and y is the number of robots. Daily operating costs amount to \$100 per worker and \$16 per robot. How many workers and robots should be used to maximize productivity if the manufacturer has a daily budget of \$4,000?

Solution Our objective function is the productivity $P(x, y)$, and the constraint is

$$100x + 16y = 4{,}000.$$

*** NOTE** Cobb-Douglas production formulas were discussed in Section 4.6.

Fredrik Persson/AP Photo

So, the optimization problem is:

Maximize $P(x, y) = 10x^{0.2}y^{0.8}$ subject to $100x + 16y = 4{,}000$ $(x \geq 0, y \geq 0)$.

Here, $g(x, y) = 100x + 16y - 4{,}000$, and the Lagrangian function is

$$\begin{aligned} L(x, y) &= P(x, y) - \lambda g(x, y) \\ &= 10x^{0.2}y^{0.8} - \lambda(100x + 16y - 4{,}000). \end{aligned}$$

The system of equations we need to solve is thus

$$\begin{aligned} L_x = 0: &\quad 2x^{-0.8}y^{0.8} - 100\lambda = 0 \\ L_y = 0: &\quad 8x^{0.2}y^{-0.2} - 16\lambda = 0 \\ g = 0: &\quad 100x + 16y = 4{,}000. \end{aligned}$$

We can rewrite the first two equations as:

$$2\left(\frac{y}{x}\right)^{0.8} = 100\lambda \qquad 8\left(\frac{x}{y}\right)^{0.2} = 16\lambda.$$

Dividing the first by the second to eliminate λ gives

$$\frac{1}{4}\left(\frac{y}{x}\right)^{0.8}\left(\frac{y}{x}\right)^{0.2} = \frac{100}{16}$$

that is, $$\frac{1}{4}\frac{y}{x} = \frac{25}{4},$$

giving $$y = 25x.$$

Substituting this result into the constraint equation gives

$$\begin{aligned} 100x + 16(25x) &= 4{,}000 \\ 500x &= 4{,}000 \end{aligned}$$

so $\quad x = 8$ workers, $\quad y = 25x = 200$ robots

for a productivity of

$$P(8, 200) = 10(8)^{0.2}(200)^{0.8} \approx 1{,}051 \text{ motors manufactured per day.}$$

FAQ

When to Use Lagrange Multipliers

Q: *When can I use the method of Lagrange multipliers? When should I use it?*

A: We have discussed the method only when there is a single equality constraint. There is a generalization, which we have not discussed, that works when there are more equality constraints (we need to introduce one multiplier for each constraint). So, if you have a problem with more than one equality constraint, or with any inequality constraints, you must use the substitution method. On the other hand, if you have one equality constraint, and it would be difficult to solve it for one of the variables, then you should use Lagrange multipliers.

8.4 EXERCISES

▼ more advanced ◆ challenging

T indicates exercises that should be solved using technology

In Exercises 1–6, solve the given optimization problem by using substitution. HINT [See Example 1.]

1. Find the maximum value of $f(x, y, z) = 1 - x^2 - y^2 - z^2$ subject to $z = 2y$. Also find the corresponding point(s) (x, y, z).

2. Find the minimum value of $f(x, y, z) = x^2 + y^2 + z^2 - 2$ subject to $x = y$. Also find the corresponding point(s) (x, y, z).

3. Find the maximum value of $f(x, y, z) = 1 - x^2 - x - y^2 + y - z^2 + z$ subject to $3x = y$. Also find the corresponding point(s) (x, y, z).

4. Find the maximum value of $f(x, y, z) = 2x^2 + 2x + y^2 - y + z^2 - z - 1$ subject to $z = 2y$. Also find the corresponding point(s) (x, y, z).

5. Minimize $S = xy + 4xz + 2yz$ subject to $xyz = 1$ with $x > 0$, $y > 0, z > 0$.

6. Minimize $S = xy + xz + yz$ subject to $xyz = 2$ with $x > 0$, $y > 0, z > 0$.

In Exercises 7–18, use Lagrange multipliers to solve the given optimization problem. HINT [See Example 2.]

7. Find the maximum value of $f(x, y) = xy$ subject to $x + 2y = 40$. Also find the corresponding point(s) (x, y).

8. Find the maximum value of $f(x, y) = xy$ subject to $3x + y = 60$. Also find the corresponding point(s) (x, y).

9. Find the maximum value of $f(x, y) = 4xy$ subject to $x^2 + y^2 = 8$. Also find the corresponding point(s) (x, y).

10. Find the maximum value of $f(x, y) = xy$ subject to $y = 3 - x^2$. Also find the corresponding point(s) (x, y).

11. Find the minimum value of $f(x, y) = x^2 + y^2$ subject to $x + 2y = 10$. Also find the corresponding point(s) (x, y).

12. Find the minimum value of $f(x, y) = x^2 + y^2$ subject to $xy^2 = 16$. Also find the corresponding point(s) (x, y).

13. The problem in Exercise 1. HINT [See Example 3.]

14. The problem in Exercise 2. HINT [See Example 3.]

15. The problem in Exercise 3.

16. The problem in Exercise 4.

17. The problem in Exercise 5.

18. The problem in Exercise 6.

APPLICATIONS

Exercises 19–22 were solved in Section 5.2. This time, use the method of Lagrange multipliers to solve them.

19. ***Fences*** I want to fence in a rectangular vegetable patch. The fencing for the east and west sides costs \$4 per foot, and the fencing for the north and south sides costs only \$2 per foot. I have a budget of \$80 for the project. What is the largest area I can enclose?

20. ***Fences*** My orchid garden abuts my house so that the house itself forms the northern boundary. The fencing for the southern boundary costs \$4 per foot, and the fencing for the east and west sides costs \$2 per foot. If I have a budget of \$80 for the project, what is the largest area I can enclose?

21. ***Revenue*** Hercules Films is deciding on the price of the video release of its film *Son of Frankenstein.* Its marketing people estimate that at a price of p dollars, it can sell a total of $q = 200{,}000 - 10{,}000p$ copies. What price will bring in the greatest revenue?

22. ***Profit*** Hercules Films is also deciding on the price of the video release of its film *Bride of the Son of Frankenstein.* Again, marketing estimates that at a price of p dollars it can sell $q = 200{,}000 - 10{,}000p$ copies, but each copy costs \$4 to make. What price will give the greatest *profit*?

23. ***Geometry*** At what points on the sphere $x^2 + y^2 + z^2 = 1$ is the product xyz a maximum? (The method of Lagrange multipliers can be used.)

24. ***Geometry*** At what point on the surface $z = (x^2 + x + y^2 + 4)^{1/2}$ is the quantity $x^2 + y^2 + z^2$ a minimum? (The method of Lagrange multipliers can be used.)

25. ▼ ***Geometry*** What point on the surface $z = x^2 + y - 1$ is closest to the origin? HINT [Minimize the square of the distance from (x, y, z) to the origin.]

26. ▼ ***Geometry*** What point on the surface $z = x + y^2 - 3$ is closest to the origin? HINT [Minimize the square of the distance from (x, y, z) to the origin.]

27. ▼ ***Geometry*** Find the point on the plane $-2x + 2y + z - 5 = 0$ closest to $(-1, 1, 3)$. HINT [Minimize the square of the distance from the given point to a general point on the plane.]

28. ▼ ***Geometry*** Find the point on the plane $2x - 2y - z + 1 = 0$ closest to $(1, 1, 0)$.

29. ***Construction Cost*** A closed rectangular box is made with two kinds of materials. The top and bottom are made with heavy-duty cardboard costing 20¢ per square foot, and the sides are made with lightweight cardboard costing 10¢ per square foot. Given that the box is to have a capacity of 2 cubic feet, what should its dimensions be if the cost is to be minimized? HINT [See Example 4.]

30. ***Construction Cost*** Repeat the preceding exercise assuming that the heavy-duty cardboard costs 30¢ per square foot, the lightweight cardboard costs 5¢ per square foot, and the box is to have a capacity of 6 cubic feet. HINT [See Example 4.]

31. ***Package Dimensions: USPS*** The **U.S. Postal Service (USPS)** will accept only packages with a length plus girth no more than 108 inches.[28] (See the figure.)

[28] The requirement for packages sent other than Parcel Post, as of September 2008 (www.usps.com).

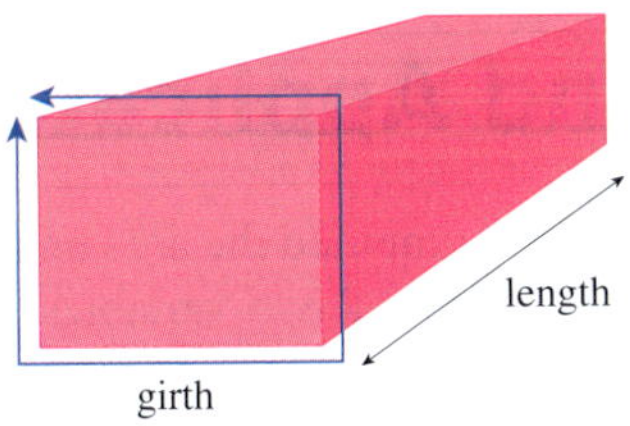

What are the dimensions of the largest volume package that the USPS will accept? What is its volume? (This exercise is the same as Exercise 49 in the preceding section. This time, solve it using Lagrange multipliers.)

32. ***Package Dimensions: UPS*** United Parcel Service (UPS) will accept only packages with length no more than 108 inches and length plus girth no more than 165 inches.[29] (See figure for the preceding exercise.) What are the dimensions of the largest volume package that UPS will accept? What is its volume? (This exercise is the same as Exercise 50 in the preceding section. This time, solve it using Lagrange multipliers.)

33. ▼ ***Construction Cost*** My company wishes to manufacture boxes similar to those described in Exercise 29 as cheaply as possible, but unfortunately the company that manufactures the cardboard is unable to give me price quotes for the heavy-duty and lightweight cardboard. Find formulas for the dimensions of the box in terms of the price per square foot of heavy-duty and lightweight cardboard.

34. ▼ ***Construction Cost*** Repeat the preceding exercise, assuming that only the bottoms of the boxes are to be made using heavy-duty cardboard.

35. ▼ ***Geometry*** Find the dimensions of the rectangular box with largest volume that can be inscribed above the xy-plane and under the paraboloid $z = 1 - (x^2 + y^2)$.

36. ▼ ***Geometry*** Find the dimensions of the rectangular box with largest volume that can be inscribed above the xy-plane and under the paraboloid $z = 2 - (2x^2 + y^2)$.

37. ***Productivity*** The Gym Shirt Company manufactures cotton socks. Production is partially automated through the use of robots. Daily operating costs amount to \$150 per laborer and \$60 per robot. The number of pairs of socks the company can manufacture in a day is given by a Cobb-Douglas production formula

$$q = 50n^{0.6}r^{0.4}$$

where q is the number of pairs of socks that can be manufactured by n laborers and r robots. Assuming that the company has a daily operating budget of \$1,500 and wishes to maximize productivity, how many laborers and how many robots should it use? What is the productivity at these levels? HINT [See Example 5.]

38. ***Productivity*** Your automobile assembly plant has a Cobb-Douglas production function given by

$$q = 100x^{0.3}y^{0.7}$$

where q is the number of automobiles it produces per year, x is the number of employees, and y is the monthly assembly-line budget (in thousands of dollars). Annual operating costs amount to an average of \$60 thousand per employee plus the operating budget of \$12y thousand. Your annual budget is \$1,200,000. How many employees should you hire and what should your assembly-line budget be to maximize productivity? What is the productivity at these levels?

COMMUNICATION AND REASONING EXERCISES

39. Outline two methods of solution of the problem "*Maximize* $f(x, y, z)$ *subject to* $g(x, y, z) = 0$" and give an advantage and disadvantage of each.

40. Suppose we know that $f(x, y)$ has both partial derivatives in its domain D: $x > 0$, $y > 0$, and that (a, b) is the only point in D such that $f_x(a, b) = f_y(a, b) = 0$. Must it be the case that, if f has an absolute maximum, it occurs at (a, b)? Explain.

41. Under what circumstances would it be necessary to use the method of Lagrange multipliers?

42. Under what circumstances would the method of Lagrange multipliers not apply?

43. Restate the following problem as a maximization problem of the form *Maximize* $f(x, y)$ *subject to* $g(x, y) = 0$":

Find the maximum value of $h(x) = 1 - 2x^2$.

44. Restate the following problem as a maximization problem of the form *Maximize* $f(x, y, z)$ *subject to* $g(x, y, z) = 0$":

Find the maximum value of $h(x, y) = 1 - 2(x^2 + y^2)$.

45. ▼ If the partial derivatives of a function of several variables are never 0, is it possible for the function to have relative extrema on some domain? Explain your answer.

46. ◆ A **linear programming problem in two variables** is a problem of the form: *Maximize (or minimize)* $f(x, y)$ *subject to constraints of the form* $C(x, y) \geq 0$ *or* $C(x, y) \leq 0$. Here, the objective function f and the constraints C are linear functions. There may be several linear constraints in one problem. Explain why the solution cannot occur in the interior of the domain of f.

47. ◆ Refer back to Exercise 46. Explain why the solution will actually be at a corner of the domain of f (where two or more of the line segments that make up the boundary meet). This result—or rather a slight generalization of it—is known as the Fundamental Theorem of Linear Programming.

[29] The requirement as of September 2008 (www.ups.com).

8.5 Double Integrals and Applications

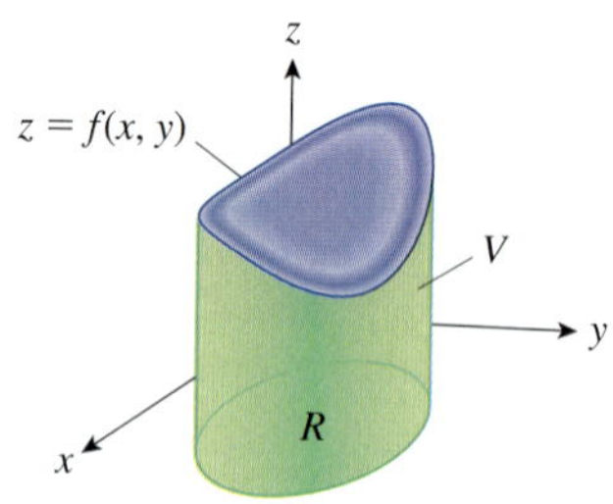

Figure 25

When discussing functions of one variable, we computed the area under a graph by integration. The analog for the graph of a function of two variables is the *volume V* under the graph, as in Figure 25. Think of the region R in the xy-plane as the "shadow" under the portion of the surface $z = f(x, y)$ shown.

By analogy with the definite integral of a function of one variable, we make the following definition:

Geometric Definition of the Double Integral

The **double integral of *f*(*x*, *y*) over the region *R* in the *xy*-plane** is defined as

(Volume *above* the region R and under the graph of f)

− (Volume *below* the region R and above the graph of f)

We denote the double integral of $f(x, y)$ over the region R by $\iint_R f(x, y)\,dx\,dy$.

Quick Example

Take $f(x, y) = 2$ and take R to be the rectangle $0 \le x \le 1, 0 \le y \le 1$. Then the graph of f is a flat horizontal surface $z = 2$, and

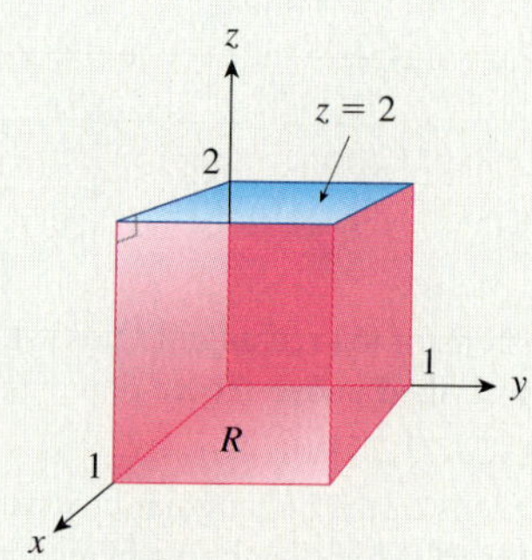

$$\iint_R f(x, y)\,dx\,dy = \text{Volume of box}$$
$$= \text{Width} \times \text{Length} \times \text{Height} = 1 \times 1 \times 2 = 2.$$

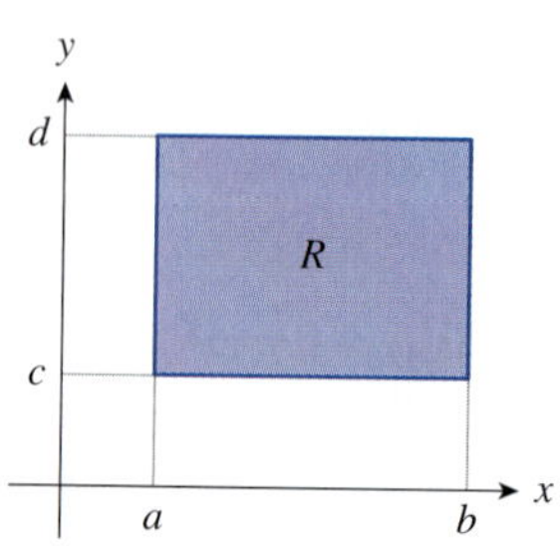

Figure 26

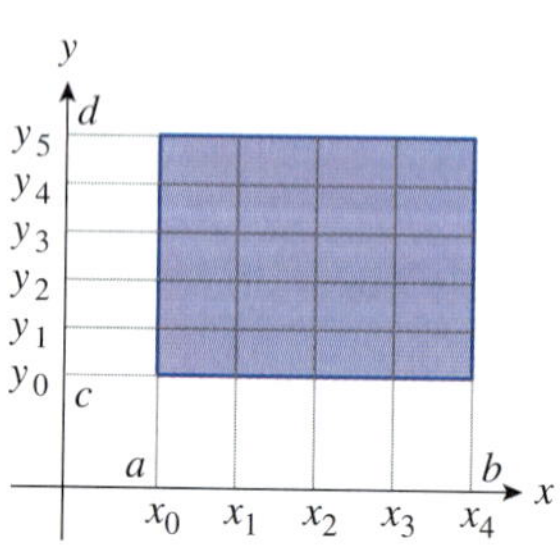

Figure 27

As we saw in the case of the definite integral of a function of one variable, we also desire *numerical* and *algebraic* definitions for two reasons: (1) to make the mathematical definition more precise, so as not to rely on the notion of "volume," and (2) for direct computation of the integral using technology or analytical tools.

We start with the simplest case, when the region R is a rectangle $a \le x \le b$ and $c \le y \le d$. (See Figure 26.) To compute the volume over R, we mimic what we did to find the area under the graph of a function of one variable. We break up the interval $[a, b]$ into m intervals all of width $\Delta x = (b - a)/m$, and we break up $[c, d]$ into n intervals all of width $\Delta y = (d - c)/n$. Figure 27 shows an example with $m = 4$ and $n = 5$.

This gives us mn rectangles defined by $x_{i-1} \le x \le x_i$ and $y_{j-1} \le y \le y_j$. Over one of these rectangles, f is approximately equal to its value at one corner—say $f(x_i, y_j)$. The volume under f over this small rectangle is then approximately the volume of the

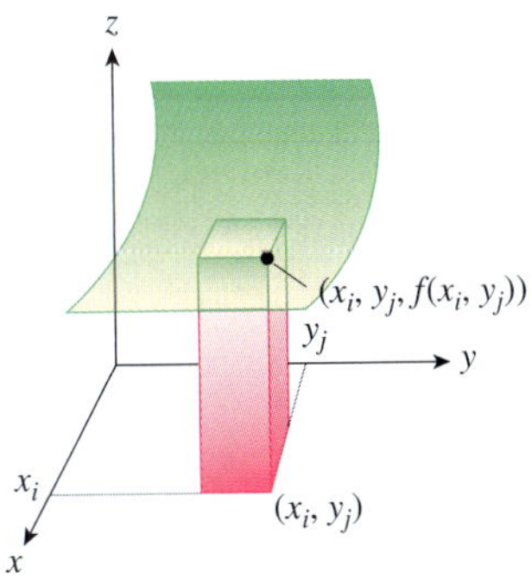

Figure 28

rectangular brick (size exaggerated) shown in Figure 28. This brick has height $f(x_i, y_j)$, and its base is Δx by Δy. Its volume is therefore $f(x_i, y_j)\Delta x\,\Delta y$. Adding together the volumes of all of the bricks over the small rectangles in R, we get

$$\iint_R f(x, y)\,dx\,dy \approx \sum_{j=1}^{n}\sum_{i=1}^{m} f(x_i, y_j)\Delta x\,\Delta y.$$

This double sum is called a **double Riemann sum**. We define the double integral to be the limit of the Riemann sums as m and n go to infinity.

Algebraic Definition of the Double Integral

$$\iint_R f(x, y)\,dx\,dy = \lim_{n\to\infty}\lim_{m\to\infty}\sum_{j=1}^{n}\sum_{i=1}^{m} f(x_i, y_j)\Delta x\,\Delta y$$

Note This definition is adequate (the limit exists) when f is continuous. More elaborate definitions are needed for general functions. ■

This definition also gives us a clue about how to compute a double integral. The innermost sum is $\sum_{i=1}^{m} f(x_i, y_j)\Delta x$, which is a Riemann sum for $\int_a^b f(x, y_j)\,dx$. The innermost limit is therefore

$$\lim_{m\to\infty}\sum_{i=1}^{m} f(x_i, y_j)\Delta x = \int_a^b f(x, y_j)\,dx.$$

The outermost limit is then also a Riemann sum, and we get the following way of calculating double integrals:

Computing the Double Integral over a Rectangle

If R is the rectangle $a \le x \le b$ and $c \le y \le d$, then

$$\iint_R f(x, y)\,dx\,dy = \int_c^d\left(\int_a^b f(x, y)\,dx\right)dy = \int_a^b\left(\int_c^d f(x, y)\,dy\right)dx.$$

The second formula comes from switching the order of summation in the double sum.

Quick Example

If R is the rectangle $1 \le x \le 2$ and $1 \le y \le 3$, then

$$\iint_R 1\,dx\,dy = \int_1^3\left(\int_1^2 1\,dx\right)dy$$

$$= \int_1^3 [x]_{x=1}^2\,dy \qquad \text{Evaluate the inner integral.}$$

$$= \int_1^3 1\,dy \qquad [x]_{x=1}^2 = 2 - 1 = 1.$$

$$= [y]_{y=1}^3 = 3 - 1 = 2.$$

The Quick Example used a constant function for the integrand. Here is an example in which the integrand is not constant.

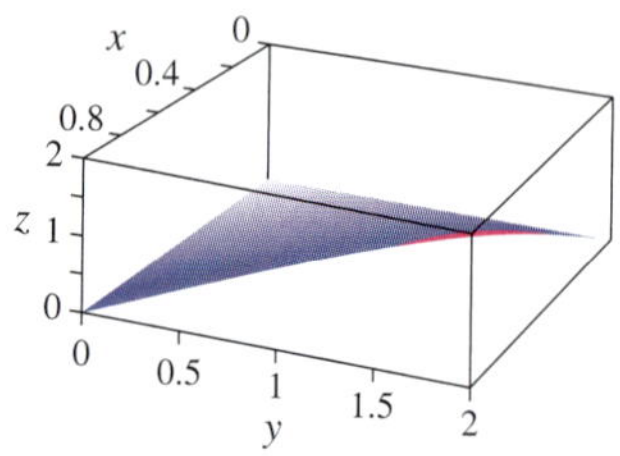

$z = xy$, $0 \le x \le 1$ and $0 \le y \le 2$

Figure 29

EXAMPLE 1 Double Integral over a Rectangle

Let R be the rectangle $0 \le x \le 1$ and $0 \le y \le 2$. Compute $\iint_R xy\,dx\,dy$. This integral gives the volume of the part of the boxed region under the surface $z = xy$ shown in Figure 29.

Solution

$$\iint_R xy\,dx\,dy = \int_0^2 \int_0^1 xy\,dx\,dy$$

(We usually drop the parentheses around the inner integral like this.) As in the Quick Example, we compute this **iterated integral** from the inside out. First we compute

$$\int_0^1 xy\,dx.$$

To do this computation, we do as we did when finding partial derivatives: We treat y as a constant. This gives

$$\int_0^1 xy\,dx = \left[\frac{x^2}{2} \cdot y\right]_{x=0}^1 = \frac{1}{2}y - 0 = \frac{y}{2}.$$

We can now calculate the outer integral.

$$\int_0^2 \int_0^1 xy\,dx\,dy = \int_0^2 \frac{y}{2}\,dy = \left[\frac{y^2}{4}\right]_0^2 = 1$$

Before we go on... We could also reverse the order of integration in Example 1.

$$\int_0^1 \int_0^2 xy\,dy\,dx = \int_0^1 \left(\left[x \cdot \frac{y^2}{2}\right]_{y=0}^2\right) dx = \int_0^1 2x\,dx = \left[x^2\right]_0^1 = 1$$

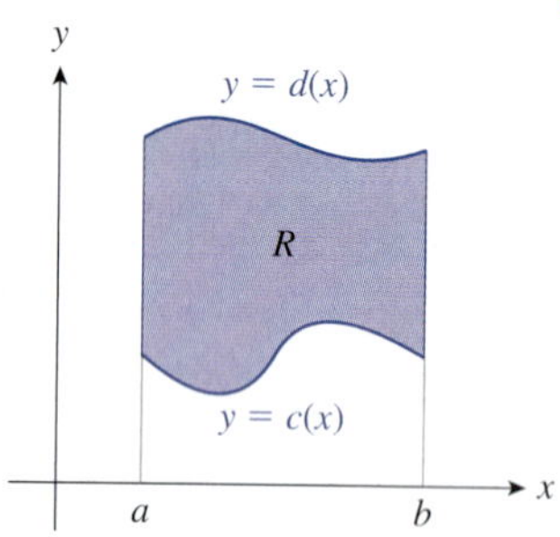

Figure 30

Often we need to integrate over regions R that are not rectangular. There are two cases that come up. The first is a region like the one shown in Figure 30. In this region, the bottom and top sides are defined by functions $y = c(x)$ and $y = d(x)$, respectively, so that the whole region can be described by the inequalities $a \le x \le b$ and $c(x) \le y \le d(x)$. To evaluate a double integral over such a region, we have the following formula:

Computing the Double Integral over a Nonrectangular Region

If R is the region $a \le x \le b$ and $c(x) \le y \le d(x)$ (Figure 30), then we integrate over R according to the following equation:

$$\iint_R f(x, y)\,dx\,dy = \int_a^b \int_{c(x)}^{d(x)} f(x, y)\,dy\,dx.$$

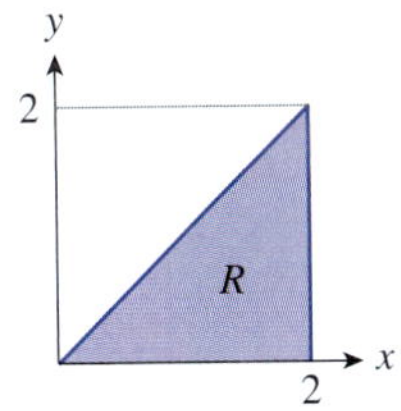

Figure 31

EXAMPLE 2 Double Integral over a Nonrectangular Region

R is the triangle shown in Figure 31. Compute $\iint_R x\,dx\,dy$.

Solution R is the region described by $0 \le x \le 2$, $0 \le y \le x$. We have

$$\begin{aligned}\iint_R x\,dx\,dy &= \int_0^2 \int_0^x x\,dy\,dx \\ &= \int_0^2 [xy]_{y=0}^x\,dx \\ &= \int_0^2 x^2\,dx \\ &= \left[\frac{x^3}{3}\right]_0^2 = \frac{8}{3}.\end{aligned}$$

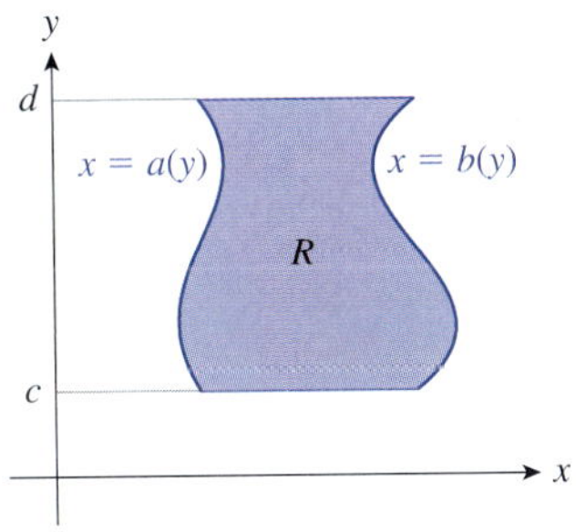

Figure 32

The second type of region is shown in Figure 32. This is the region described by $c \le y \le d$ and $a(y) \le x \le b(y)$. To evaluate a double integral over such a region, we have the following formula:

Double Integral over a Nonrectangular Region (continued)

If R is the region $c \le y \le d$ and $a(y) \le x \le b(y)$ (Figure 32), then we integrate over R according to the following equation:

$$\iint_R f(x, y)\,dx\,dy = \int_c^d \int_{a(y)}^{b(y)} f(x, y)\,dx\,dy.$$

EXAMPLE 3 Double Integral over a Nonrectangular Region

Redo Example 2, integrating in the opposite order.

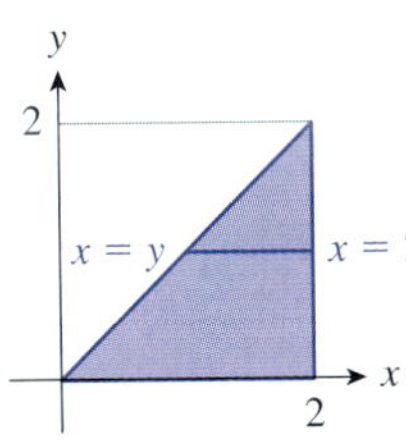

Figure 33

Solution We can integrate in the opposite order if we can describe the region in Figure 31 in the way shown in Figure 32. In fact, it is the region $0 \le y \le 2$ and $y \le x \le 2$. To see this, we draw a horizontal line through the region, as in Figure 33. The line extends from $x = y$ on the left to $x = 2$ on the right, so $y \le x \le 2$. The possible heights for such a line are $0 \le y \le 2$. We can now compute the integral:

$$\begin{aligned}\iint_R x\,dx\,dy &= \int_0^2 \int_y^2 x\,dx\,dy \\ &= \int_0^2 \left[\frac{x^2}{2}\right]_{x=y}^2 dy \\ &= \int_0^2 \left(2 - \frac{y^2}{2}\right) dy \\ &= \left[2y - \frac{y^3}{6}\right]_0^2 = \frac{8}{3}.\end{aligned}$$

Note Many regions can be described in two different ways, as we saw in Examples 2 and 3. Sometimes one description will be much easier to work with than the other, so it pays to consider both. ■

APPLICATIONS

There are many applications of double integrals besides finding volumes. For example, we can use them to find *averages*. Remember that the average of $f(x)$ on $[a, b]$ is given by $\int_a^b f(x)\,dx$ divided by $(b - a)$, the length of the interval.

Average of a Function of Two Variables

The average of $f(x, y)$ on the region R is

$$\bar{f} = \frac{1}{A}\iint_R f(x, y)\,dx\,dy.$$

Here, A is the area of R. We can compute the area A geometrically, or by using the techniques from the chapter on applications of the integral, or by computing

$$A = \iint_R 1\,dx\,dy.$$

Quick Example

The average value of $f(x, y) = xy$ on the rectangle given by $0 \le x \le 1$ and $0 \le y \le 2$ is

$$\bar{f} = \frac{1}{2}\iint_R xy\,dx\,dy$$ The area of the rectangle is 2.

$$= \frac{1}{2}\int_0^2\int_0^1 xy\,dx\,dy$$

$$= \frac{1}{2} \cdot 1 = \frac{1}{2}.$$ We calculated the integral in Example 1.

EXAMPLE 4 Average Revenue

Your company is planning to price its new line of subcompact cars at between \$10,000 and \$15,000. The marketing department reports that if the company prices the cars at p dollars per car, the demand will be between $q = 20{,}000 - p$ and $q = 25{,}000 - p$ cars sold in the first year. What is the average of all the possible revenues your company could expect in the first year?

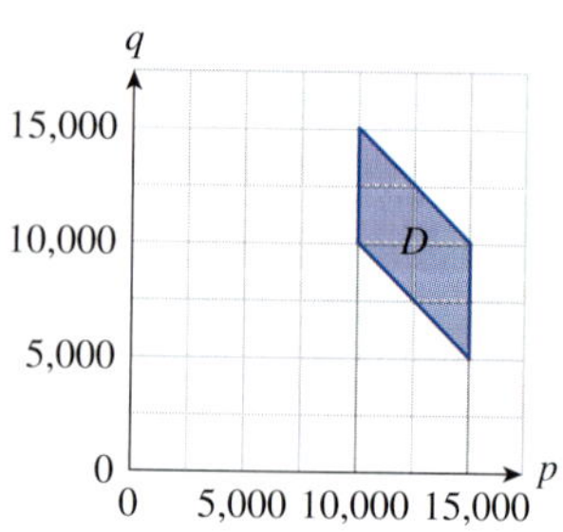

Figure 34

Solution Revenue is given by $R = pq$ as usual, and we are told that

$$10{,}000 \le p \le 15{,}000$$

and $$20{,}000 - p \le q \le 25{,}000 - p.$$

This domain D of prices and demands is shown in Figure 34.

To average the revenue R over the domain D, we need to compute the area A of D. Using either calculus or geometry, we get $A = 25{,}000{,}000$. We then need to integrate R over D:

$$\begin{aligned}\iint_D pq\,dp\,dq &= \int_{10{,}000}^{15{,}000}\int_{20{,}000-p}^{25{,}000-p} pq\,dq\,dp\\ &= \int_{10{,}000}^{15{,}000}\left[\frac{pq^2}{2}\right]_{q=20{,}000-p}^{25{,}000-p} dp\\ &= \frac{1}{2}\int_{10{,}000}^{15{,}000}[p(25{,}000-p)^2 - p(20{,}000-p)^2]\,dp\\ &= \frac{1}{2}\int_{10{,}000}^{15{,}000}[225{,}000{,}000p - 10{,}000p^2]\,dp\\ &\approx 3{,}072{,}900{,}000{,}000{,}000.\end{aligned}$$

The average of all the possible revenues your company could expect in the first year is therefore

$$\bar{R} = \frac{3{,}072{,}900{,}000{,}000{,}000}{25{,}000{,}000} \approx \$122{,}900{,}000.$$

Before we go on... To check that the answer obtained in Example 4 is reasonable, notice that the revenues at the corners of the domain are \$100,000,000 per year, \$150,000,000 per year (at two corners), and \$75,000,000 per year. Some of these are smaller than the average and some larger, as we would expect. ■

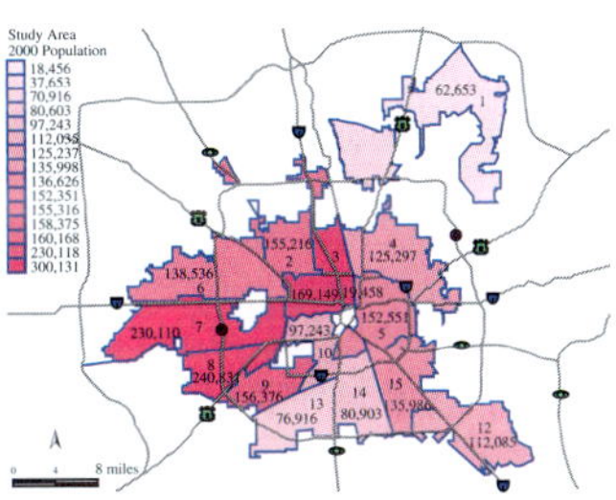

Darker regions have higher population density

Figure 35

Another useful application of the double integral comes about when we consider density. For example, suppose that $P(x, y)$ represents the population density (in people per square mile, say) in the city of Houston, shown in Figure 35.

If we break the city up into small rectangles (for example, city blocks), then the population in the small rectangle $x_{i-1} \le x \le x_i$ and $y_{j-1} \le y \le y_j$ is approximately $P(x_i, y_j)\Delta x \Delta y$. Adding up all of these population estimates, we get

$$\text{Total population} \approx \sum_{j=1}^{n}\sum_{i=1}^{m} P(x_i, y_j)\,\Delta x\,\Delta y.$$

Because this is a double Riemann sum, when we take the limit as m and n go to infinity, we get the following calculation of the population of the city:

$$\text{Total population} = \iint_{\text{City}} P(x, y)\,dx\,dy.$$

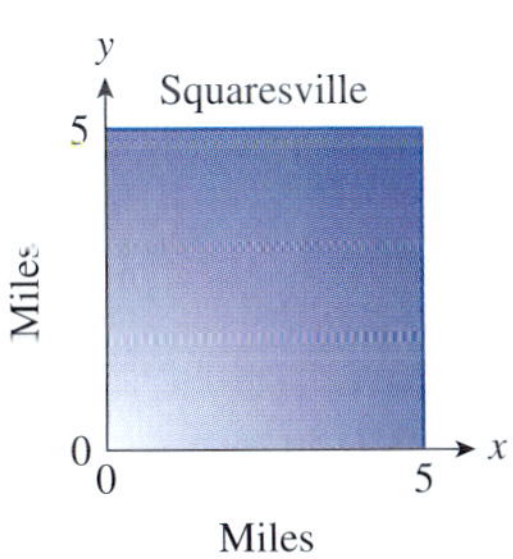

Figure 36

EXAMPLE 5 Population

Squaresville is a city in the shape of a square 5 miles on a side. The population density at a distance of x miles east and y miles north of the southwest corner is $P(x, y) = x^2 + y^2$ thousand people per square mile. Find the total population of Squaresville.

Solution Squaresville is pictured in Figure 36, in which we put the origin in the southwest corner of the city.

To compute the total population, we integrate the population density over the city S.

$$\text{Total population} = \iint_{\text{Squaresville}} P(x, y)\, dx\, dy$$
$$= \int_0^5 \int_0^5 (x^2 + y^2)\, dx\, dy$$
$$= \int_0^5 \left[\frac{x^3}{3} + xy^2 \right]_{x=0}^{5} dy$$
$$= \int_0^5 \left[\frac{125}{3} + 5y^2 \right] dy$$
$$= \frac{1{,}250}{3} \approx 417 \text{ thousand people}$$

Before we go on... Note that the average population density is the total population divided by the area of the city, which is about 17,000 people per square mile. Compare this calculation with the calculations of averages in the previous two examples. ■

8.5 EXERCISES

▼ more advanced ◆ challenging

T indicates exercises that should be solved using technology

Compute the integrals in Exercises 1–16. HINT [See Example 1.]

1. $\int_0^1 \int_0^1 (x - 2y)\, dx\, dy$

2. $\int_{-1}^1 \int_0^2 (2x + 3y)\, dx\, dy$

3. $\int_0^1 \int_0^2 (ye^x - x - y)\, dx\, dy$

4. $\int_1^2 \int_2^3 \left(\frac{1}{x} + \frac{1}{y} \right) dx\, dy$

5. $\int_0^2 \int_0^3 e^{x+y}\, dx\, dy$

6. $\int_0^1 \int_0^1 e^{x-y}\, dx\, dy$

7. $\int_0^1 \int_0^{2-y} x\, dx\, dy$

8. $\int_0^1 \int_0^{2-y} y\, dx\, dy$

9. $\int_{-1}^1 \int_{y-1}^{y+1} e^{x+y}\, dx\, dy$ HINT [See Example 2.]

10. $\int_0^1 \int_y^{y+2} \frac{1}{\sqrt{x+y}}\, dx\, dy$ HINT [See Example 2.]

11. $\int_0^1 \int_{-x^2}^{x^2} x\, dy\, dx$

12. $\int_1^4 \int_{-\sqrt{x}}^{\sqrt{x}} \frac{1}{x}\, dy\, dx$

13. $\int_0^1 \int_0^x e^{x^2}\, dy\, dx$

14. $\int_0^1 \int_0^{x^2} e^{x^3+1}\, dy\, dx$

15. $\int_0^2 \int_{1-x}^{8-x} (x + y)^{1/3}\, dy\, dx$

16. $\int_1^2 \int_{1-2x}^{x^2} \frac{x+1}{(2x+y)^3}\, dy\, dx$

In Exercises 17–24, find $\iint_R f(x, y)\, dx\, dy$*, where R is the indicated domain. (Remember that you often have a choice as to the order of integration.)* HINT [See Example 2.]

17. $f(x, y) = 2$

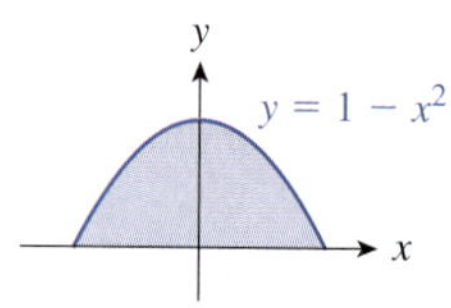

18. $f(x, y) = x$

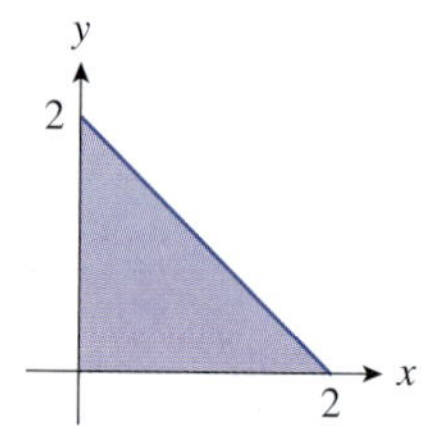

19. $f(x, y) = 1 + y$ HINT [See Example 3.]

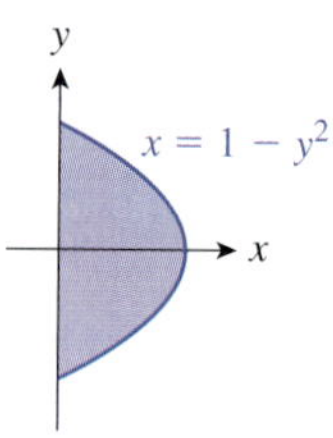

20. $f(x, y) = e^{x+y}$ HINT [See Example 3.]

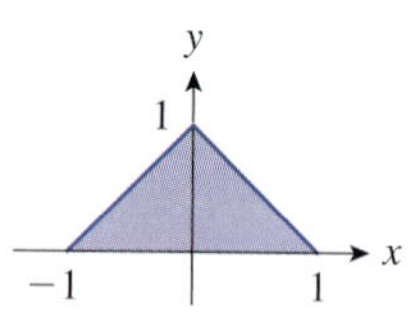

21. $f(x, y) = xy^2$

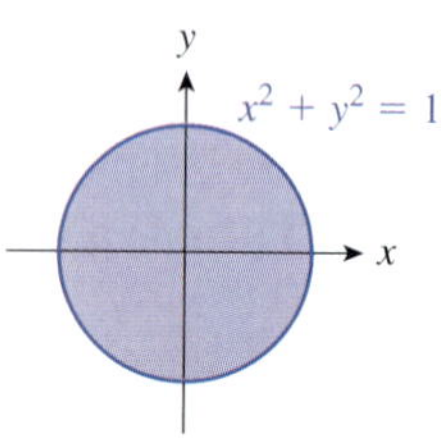

22. $f(x, y) = xy^2$

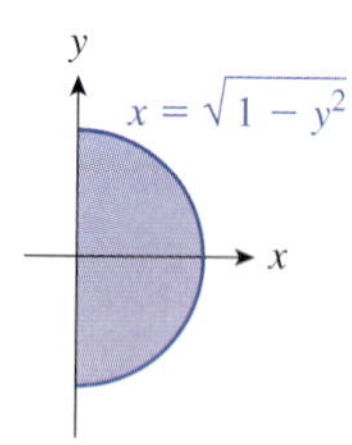

23. $f(x, y) = x^2 + y^2$

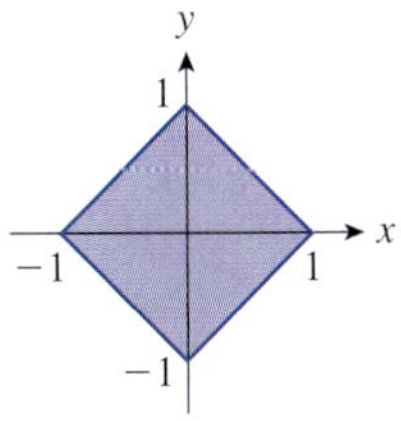

24. $f(x, y) = x^2$

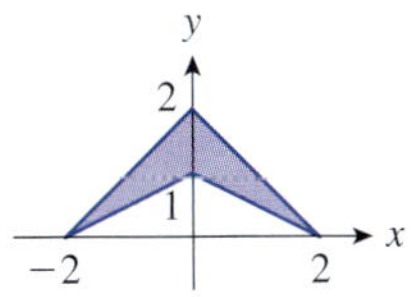

In Exercises 25–30, find the average value of the given function over the indicated domain. HINT [See Quick Examples page 632.]

25. $f(x, y) = y$

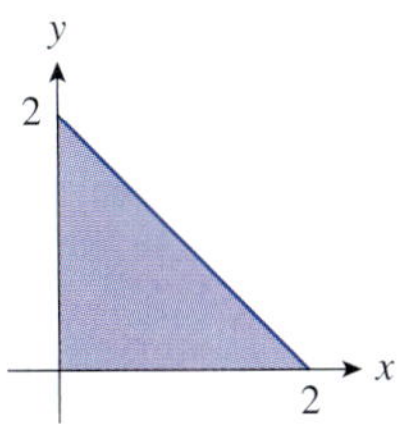

26. $f(x, y) = 2 + x$

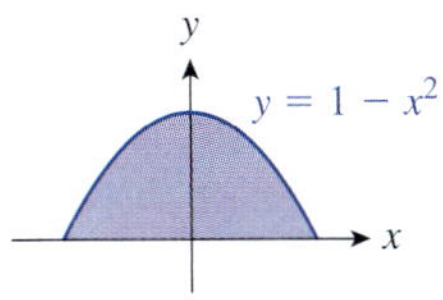

27. $f(x, y) = e^y$

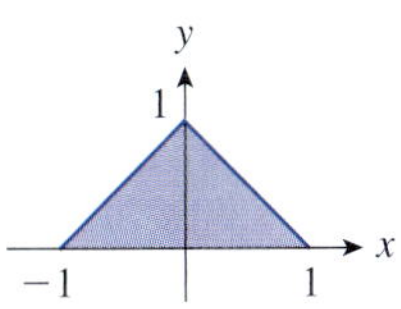

28. $f(x, y) = y$

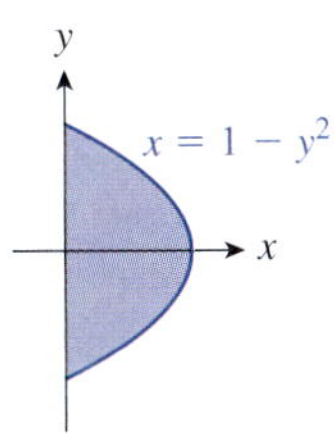

29. $f(x, y) = x^2$

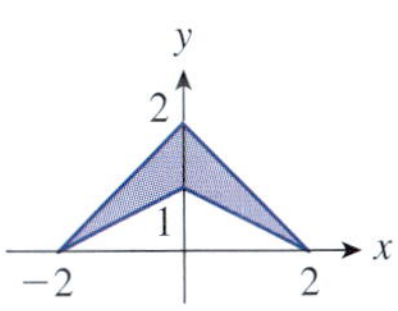

30. $f(x, y) = x^2 + y^2$

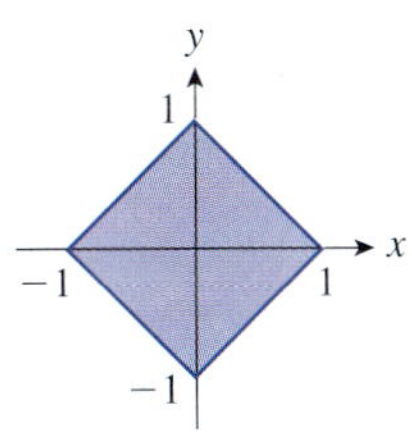

In Exercises 31–36, sketch the region over which you are integrating, and then write down the integral with the order of integration reversed (changing the limits of integration as necessary).

31. ▼ $\int_0^1 \int_0^y f(x, y)\,dx\,dy$

32. ▼ $\int_{-1}^1 \int_0^{1+y} f(x, y)\,dx\,dy$

33. ▼ $\int_{-1}^1 \int_0^{\sqrt{1+y}} f(x, y)\,dx\,dy$

34. ▼ $\int_{-1}^1 \int_0^{\sqrt{1-y}} f(x, y)\,dx\,dy$

35. ▼ $\int_1^2 \int_1^{4/x^2} f(x, y)\,dy\,dx$

36. ▼ $\int_1^{e^2} \int_0^{\ln x} f(x, y)\,dy\,dx$

37. Find the volume under the graph of $z = 1 - x^2$ over the region $0 \le x \le 1$ and $0 \le y \le 2$.

38. Find the volume under the graph of $z = 1 - x^2$ over the triangle $0 \le x \le 1$ and $0 \le y \le 1 - x$.

39. ▼ Find the volume of the tetrahedron shown in the figure. Its corners are (0, 0, 0), (1, 0, 0), (0, 1, 0), and (0, 0, 1).

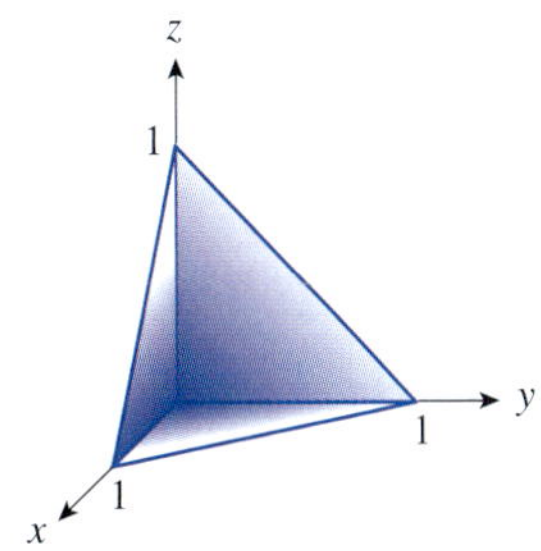

40. ▼ Find the volume of the tetrahedron with corners at (0, 0, 0), $(a, 0, 0)$, $(0, b, 0)$, and $(0, 0, c)$.

APPLICATIONS

41. ***Productivity*** A productivity model at the Handy Gadget Company is

$$P = 10{,}000x^{0.3}y^{0.7}$$

where P is the number of gadgets the company turns out per month, x is the number of employees at the company, and y is the monthly operating budget in thousands of dollars. Because the company hires part-time workers, it uses anywhere between 45 and 55 workers each month, and its operating budget varies from \$8,000 to \$12,000 per month. What is the average of the possible numbers of gadgets it can turn out per month? (Round the answer to the nearest 1,000 gadgets.) HINT [See Quick Examples page 632.]

42. ***Productivity*** Repeat the preceding exercise using the productivity model

$$P = 10{,}000x^{0.7}y^{0.3}.$$

43. ***Revenue*** Your latest CD-ROM of clip art is expected to sell between $q = 8{,}000 - p^2$ and $q = 10{,}000 - p^2$ copies if priced at p dollars. You plan to set the price between \$40 and \$50. What is the average of all the possible revenues you can make? HINT [See Example 4.]

44. ***Revenue*** Your latest DVD drive is expected to sell between $q = 180{,}000 - p^2$ and $q = 200{,}000 - p^2$ units if priced at p dollars. You plan to set the price between \$300 and \$400. What is the average of all the possible revenues you can make? HINT [See Example 4.]

45. ***Revenue*** Your self-published novel has demand curves between $p = 15{,}000/q$ and $p = 20{,}000/q$. You expect to sell between 500 and 1,000 copies. What is the average of all the possible revenues you can make?

46. ***Revenue*** Your self-published book of poetry has demand curves between $p = 80{,}000/q^2$ and $p = 100{,}000/q^2$. You expect to sell between 50 and 100 copies. What is the average of all the possible revenues you can make?

47. ***Population Density*** The town of West Podunk is shaped like a rectangle 20 miles from west to east and 30 miles from north to south. (See the figure.) It has a population density of $P(x, y) = e^{-0.1(x+y)}$ hundred people per square mile x miles east and y miles north of the southwest corner of town. What is the total population of the town? HINT [See Example 5.]

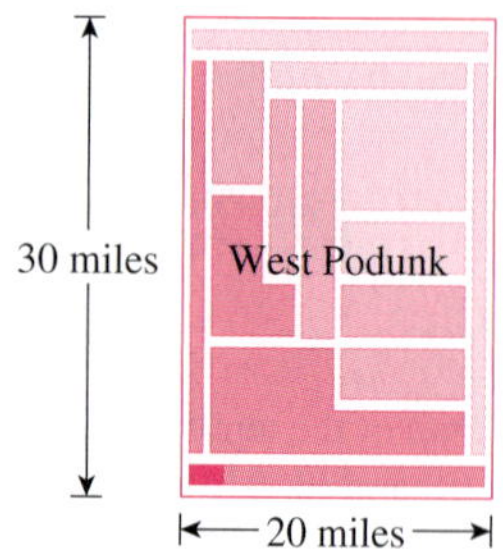

48. ***Population Density*** The town of East Podunk is shaped like a triangle with an east-west base of 20 miles and a north-south height of 30 miles. (See the figure.) It has a population density of $P(x, y) = e^{-0.1(x+y)}$ hundred people per square mile x miles east and y miles north of the southwest corner of town. What is the total population of the town? HINT [See Example 5.]

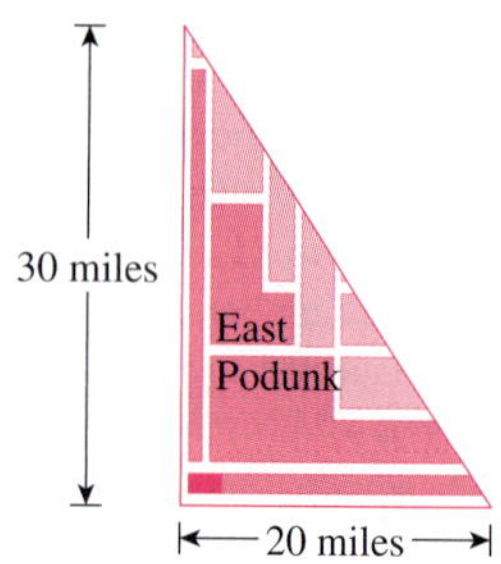

49. ***Temperature*** The temperature at the point (x, y) on the square with vertices (0, 0), (0, 1), (1, 0), and (1, 1) is given by $T(x, y) = x^2 + 2y^2$. Find the average temperature on the square.

50. ***Temperature*** The temperature at the point (x, y) on the square with vertices (0, 0), (0, 1), (1, 0), and (1, 1) is given by $T(x, y) = x^2 + 2y^2 - x$. Find the average temperature on the square.

COMMUNICATION AND REASONING EXERCISES

51. Explain how double integrals can be used to compute the area between two curves in the xy plane.

52. Explain how double integrals can be used to compute the volume of solids in 3-space.

53. Complete the following: The first step in calculating an integral of the form $\int_a^b \int_{r(x)}^{s(x)} f(x, y)\, dy\, dx$ is to evaluate the integral _____, obtained by holding ___ constant and integrating with respect to ___ .

54. If the units of $f(x, y)$ are zonars per square meter, and x and y are given in meters, what are the units of $\int_a^b \int_{r(x)}^{s(x)} f(x, y)\, dy\, dx$?

55. If the units of $\int_a^b \int_{r(x)}^{s(x)} f(x, y)\, dy\, dx$ are paintings, the units of x are picassos, and the units of y are dalis, what are the units of $f(x, y)$?

56. Complete the following: If the region R is bounded on the left and right by vertical lines and on the top and bottom by the graphs of functions of x, then we integrate over R by first integrating with respect to ____ and then with respect to ___.

57. ▼ Show that if a, b, c, and d are constant, then $\int_a^b \int_c^d f(x) g(y)\, dx\, dy = \int_c^d f(x)\, dx \int_a^b g(y)\, dy$. Test this result on the integral $\int_0^1 \int_1^2 ye^x\, dx\, dy$.

58. ▼ Refer to Exercise 57. If a, b, c, and d are constants, can $\displaystyle\int_a^b \int_c^d \frac{f(x)}{g(y)}\, dx\, dy$ be expressed as a product of two integrals? Explain.

KEY CONCEPTS

Web Site www.AppliedCalc.org

Go to the student Web site at www.AppliedCalc.org to find a comprehensive and interactive Web-based summary of Chapter 8.

8.1 Functions of Several Variables from the Numerical, Algebraic, and Graphical Viewpoints

A real-valued function, f, of $x, y, z, \ldots$ *p. 584*

Cost functions *p. 585*

A linear function of the variables $x_1, x_2, \ldots, x_n$ is a function of the form $f(x_1, x_2, \ldots, x_n) = a_0 + a_1x_1 + \ldots + a_nx_n$ $(a_0, a_1, \ldots, a_n$ constants) *p. 586*

Representing functions of two variables numerically *p. 587*

Using Excel to represent a function of two variables *p. 588*

Plotting points in three dimensions *p. 589*

Graph of a function of two variables *p. 590*

Analyzing the graph of a function of two variables *p. 592*

Graph of a linear function *p. 594*

8.2 Partial Derivatives

Definition of partial derivatives *p. 602*

Application to marginal cost: linear cost function *p. 603*

Application to marginal cost: interaction cost function *p. 603*

Geometric interpretation of partial derivatives *p. 605*

Second order partial derivatives *p. 605*

8.3 Maxima and Minima

Definition of relative maximum and minimum *p. 611*

Locating candidates for relative maxima and minima *p. 611*

Classifying critical points graphically *p. 612*

Classifying critical points numerically *p. 613*

Second derivative test for a function of two variables *p. 613*

Using the second derivative test *p. 614*

Deriving the regression formulas:

$$m = \frac{n\left(\sum xy\right) - \left(\sum x\right)\left(\sum y\right)}{n\left(\sum x^2\right) - \left(\sum x\right)^2}$$

$$b = \frac{\sum y - m\left(\sum x\right)}{n}$$

$n =$ number of data points *p. 615*

8.4 Constrained Maxima and Minima and Applications

Constrained maximum and minimum problem *p. 619*

Solving constrained maxima and minima problems using substitution *p. 619*

The method of Lagrange multipliers *p. 620*

Using Lagrange multipliers *p. 621*

8.5 Double Integrals and Applications

Geometric definition of the double integral *p. 628*

Algebraic definition of the double integral

$$\iint_R f(x, y)\,dx\,dy = \lim_{n\to\infty}\lim_{m\to\infty}\sum_{j=1}^{n}\sum_{i=1}^{m} f(x_i, y_j)\Delta x\,\Delta y$$

p. 629

Computing the double integral over a rectangle *p. 629*

Computing the double integral over nonrectangular regions *p. 630*

Average of $f(x, y)$ on the region R:

$$\bar{f} = \frac{1}{A}\iint_R f(x, y)\,dx\,dy \quad p.\ 632$$

REVIEW EXERCISES

1. Let $f(x, y, z) = \dfrac{x}{y + xz} + x^2y$. Evaluate $f(0, 1, 1)$, $f(2, 1, 1)$, $f(-1, 1, -1)$, $f(z, z, z)$, and $f(x + h, y + k, z + l)$.

2. Let $g(x, y, z) = xy(x + y - z) + x^2$. Evaluate $g(0, 0, 0)$, $g(1, 0, 0)$, $g(0, 1, 0)$, $g(x, x, x)$, and $g(x, y + k, z)$.

3. Let $f(x, y, z) = 2.72 - 0.32x - 3.21y + 12.5z$. Complete the following: f ___ by ___ units for every 1 unit of increase in x, and ___ by ___ units for every unit of increase in z.

4. Let $g(x, y, z) = 2.16x + 11y - 1.53z + 31.4$. Complete the following: g ___ by ___ units for every 1 unit of increase in y, and ___ by ___ units for every unit of increase in z.

In Exercises 5–6 complete the given table for values for $h(x, y) = 2x^2 + xy - x$.

5.

$y \downarrow$ \ $x \rightarrow$	−1	0	1
−1			
0			
1			

6.

$y \downarrow$ \ $x \rightarrow$	−2	2	3
−2			
2			
3			

7. Give a formula for a (single) function f with the property that $f(x, y) = -f(y, x)$ and $f(1, -1) = 3$.

8. Let $f(x, y) = x^2 + (y + 1)^2$. Show that $f(y, x) = f(x + 1, y - 1)$.

In Exercises 9–14, compute the partial derivatives shown for the given function.

9. $f(x, y) = x^2 + xy$; find f_x, f_y, and f_{yy}

10. $f(x, y) = \dfrac{6}{xy} + \dfrac{xy}{6}$; find f_x, f_y, and f_{yy}

11. $f(x, y) = 4x + 5y - 6xy$; find $f_{xx}(1, 0) - f_{xx}(3, 2)$

12. $f(x, y) = e^{xy} + e^{3x^2 - y^2}$; find $\dfrac{\partial f}{\partial x}$ and $\dfrac{\partial^2 f}{\partial x \partial y}$

13. $f(x, y, z) = \dfrac{x}{x^2 + y^2 + z^2}$; find $\dfrac{\partial f}{\partial x}, \dfrac{\partial f}{\partial y}, \dfrac{\partial f}{\partial z}$ and $\left.\dfrac{\partial f}{\partial x}\right|_{(0,1,0)}$

14. $f(x, y, z) = x^2 + y^2 + z^2 + xyz$; find $f_{xx} + f_{yy} + f_{zz}$

In Exercises 15–19, locate and classify all critical points.

15. $f(x, y) = (x - 1)^2 + (2y - 3)^2$

16. $g(x, y) = (x - 1)^2 - 3y^2 + 9$

17. $h(x, y) = e^{xy}$

18. $j(x, y) = xy + x^2$

19. $f(x, y) = \ln(x^2 + y^2) - (x^2 + y^2)$

In Exercises 20–23, solve the following constrained optimization problems by using substitution to eliminate a variable. (Do not use Lagrange multipliers.)

20. Find the minimum value of $f(x, y, z) = x^2 + y^2 + z^2 - 1$ subject to $x = y + z$. Also find the corresponding point(s) (x, y, z).

21. Find the largest value of xyz subject to $x + y + z = 1$ with $x > 0, y > 0, z > 0$. Also find the corresponding point(s) (x, y, z).

22. Minimize $S = xy + x^2z^2 + 4yz$ subject to $xyz = 1$ with $x > 0, y > 0, z > 0$.

23. Find the point on the surface $z = \sqrt{x^2 + 2(y - 3)^2}$ closest to the origin.

In Exercises 24–28, use Lagrange multipliers to solve the given optimization problem.

24. Find the maximum value of $f(x, y) = xy$ subject to $y = e^{-x}$. Also find the corresponding point(s) (x, y).

25. Find the minimum value of $f(x, y) = x^2 + y^2$ subject to $xy = 2$. Also find the corresponding point(s) (x, y).

26. The problem in Exercise 20.

27. The problem in Exercise 22.

28. The problem in Exercise 23.

In Exercises 29–34, compute the given quantities.

29. $\displaystyle\int_0^1 \int_0^2 (2xy)\,dx\,dy$

30. $\displaystyle\int_1^2 \int_0^1 xye^{x+y}\,dx\,dy$

31. $\displaystyle\int_0^2 \int_0^{2x} \frac{1}{x^2 + 1}\,dy\,dx$

32. The average value of xye^{x+y} over the rectangle $0 \le x \le 1$, $1 \le y \le 2$

33. $\iint_R (x^2 - y^2)\,dx\,dy$, where R is the region shown in the figure

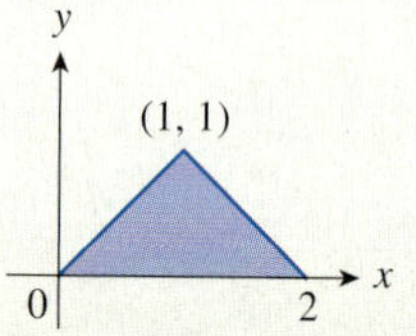

34. The volume under the graph of $z = 1 - y$ over the region in the xy plane between the parabola $y = 1 - x^2$ and the x-axis.

APPLICATIONS

35. ***Web Site Traffic*** OHaganBooks.com has two principal competitors: JungleBooks.com and FarmerBooks.com. Current Web site traffic at OHaganBooks.com is estimated at 5,000 hits per day. This number is predicted to decrease by 0.8 for every new customer of JungleBooks.com and by 0.6 for every new customer of FarmerBooks.com.

a. Use this information to model the daily Web site traffic at OHaganBooks.com as a linear function of the new customers of its two competitors.

b. According to the model, if Junglebooks.com gets 100 new customers and OHaganBooks.com traffic drops to 4,770 hits per day, how many new customers has FarmerBooks .com obtained?

c. The model in part (a) did not take into account the growth of the total online consumer base. OHaganBooks .com expects to get approximately one additional hit per day for every 10,000 new Internet shoppers. Modify your model in part (a) so as to include this information using a new independent variable.

d. How many new Internet shoppers would it take to offset the effects on traffic at OHaganBooks.com of 100 new customers at each of its competitor sites?

36. ***Productivity*** Billy-Sean O'Hagan is writing up his Ph.D. thesis in biophysics but finds that his productivity is affected by the temperature and the number of text messages he receives per hour. On a brisk winter's day when the temperature is 0°C and there are no text messages, Billy-Sean can produce 15 pages of his thesis. His productivity goes down by 0.3 pages per degree Celsius increase in the temperature and by 1.2 pages for each additional text message per hour.

a. Use this information to model Billy-Sean's productivity p as a function of the temperature and the hourly rate of text messages.

b. The other day the temperature was 20°C and Billy-Sean managed to produce only three pages of his thesis. What was the hourly rate of incoming text messages?

c. Billy reasons that each cup of coffee he drinks per hour can counter the effect on his productivity of two text messages per hour. Modify the model in part (a) to take consumption of coffee into account.

d. What would the domain of your function look like to ensure that p is never negative?

37. ***Internet Advertising*** To increase business at OHaganBooks .com, you have purchased banner ads at well-known Internet portals and have advertised on television. The following interaction model shows the average number h of hits per day as a function of monthly expenditures x on banner ads and y on television advertising (x and y are in dollars).

$$h(x, y) = 1{,}800 + 0.05x + 0.08y + 0.00003xy$$

a. Based on your model, how much traffic can you anticipate if you spend \$2,000 per month for banner ads and \$3,000 per month on television advertising?

b. Evaluate $\frac{\partial h}{\partial y}$, specify its units of measurement, and indicate whether it increases or decreases with increasing x.

c. How much should the company spend on banner ads to obtain 1 hit per day for each \$5 spent per month on television advertising?

38. ***Company Retreats*** Their companies having recently been bailed out by the government at taxpayer expense, Marjory Duffin and John O'Hagan are planning a joint winter business retreat in Cancun, but they are not sure how many sales reps to take along. The following interaction model shows the estimated cost C to their companies (in dollars) as a function of the number of sales reps x and the length of time t in days.

$$C(x, t) = 20{,}000 - 100x + 600t + 300xt$$

a. Based on the model, how much would it cost to take five sales reps along for a 10-day retreat?

b. Evaluate $\frac{\partial C}{\partial t}$, specify its units of measurement, and indicate whether it increases or decreases with increasing x.

c. How many reps should they take along if they wish to limit the marginal daily cost to \$1,200?

39. ***Internet Advertising*** Refer to the model in Exercise 37. One or more of the following statements is correct. Identify which one(s).

(A) If nothing is spent on television advertising, one more dollar spent per month in banner ads will buy approximately 0.05 hits per day at OHaganBooks.com.

(B) If nothing is spent on television advertising, one more hit per day at OHaganBooks.com will cost the company about 5¢ per month in banner ads.

(C) If nothing is spent on banner ads, one more hit per day at OHaganBooks.com will cost the company about 5¢ per month in banner ads.

(D) If nothing is spent on banner ads, one more dollar spent per month in banner ads will buy approximately 0.05 hits per day at OHaganBooks.com.

(E) Hits at OHaganBooks.com cost approximately 5¢ per month spent on banner ads, and this cost increases at a rate of 0.003¢ per month, per hit.

40. ***Company Retreats*** Refer to the model in Exercise 38. One or more of the following statements is correct. Identify which one(s).

(A) If the retreat lasts for 10 days, the daily cost per sales rep is \$400.

(B) If the retreat lasts for 10 days, each additional day will cost the company \$2,900.

(C) If the retreat lasts for 10 days, each additional sales rep will cost the company \$800.

(D) If the retreat lasts for 10 days, the daily cost per sales rep is \$2,900.

(E) If the retreat lasts for 10 days, each additional sales rep will cost the company \$2,900.

41. ***Productivity*** The holiday season is now at its peak and OHaganBooks.com has been understaffed and swamped with orders. The current backlog (orders unshipped for two or more days) has grown to a staggering 50,000, and new orders are coming in at a rate of 5,000 per day. Research based on productivity data at OHaganBooks.com results in the following model:

$$P(x, y) = 1{,}000x^{0.9}y^{0.1} \text{ additional orders filled per day}$$

where x is the number of additional personnel hired and y is the daily budget (excluding salaries) allocated to eliminating the backlog.

a. How many additional orders will be filled per day if the company hires 10 additional employees and budgets an additional \$1,000 per day? (Round the answer to the nearest 100.)

b. In addition to the daily budget, extra staffing costs the company \$150 per day for every new staff member hired. In order to fill at least 15,000 additional orders per day at a minimum total daily cost, how many new staff members should the company hire? (Use the method of Lagrange multipliers.)

42. ***Productivity*** The holiday season has now ended, and orders at OHaganBooks.com have plummeted, leaving staff members in the shipping department with little to do besides spend their time on their Facebook pages, so the company is considering laying off a number of personnel and slashing the shipping budget. Research based on productivity data at OHaganBooks.com results in the following model:

$$C(x, y) = 1{,}000x^{0.8}y^{0.2} \text{ fewer orders filled per day}$$

where x is the number of personnel laid off and y is the cut in the shipping budget (excluding salaries).

a. How many fewer orders will be filled per day if the company lays off 15 additional employees and cuts the budget by an additional \$2,000 per day? (Round the answer to the nearest 100.)

b. In addition to the cut in the shipping budget, the layoffs will save the company \$200 per day for every new staff member laid off. The company needs to meet a target of 20,000 fewer orders per day but, for tax reasons, it must minimize the total resulting savings. How many new staff members should the company lay off? (Use the method of Lagrange multipliers.)

43. ***Profit*** If OHaganBooks.com sells x paperback books and y hardcover books per week, it will make an average weekly profit of

$$P(x, y) = 3x + 10y \text{ dollars.}$$

If it sells between 1,200 and 1,500 paperback books and between 1,800 and 2,000 hardcover books per week, what is the average of all its possible weekly profits?

44. ***Cost*** It costs Duffin publishers

$$C(x, y) = x^2 + 2y \text{ dollars}$$

to produce x coffee table art books and y paperback books per week. If it produces between 100 and 120 art books and between 800 and 1,000 paperbacks per week, what is the average of all its possible weekly costs?

Case Study — Modeling College Population

David Pearson/Alamy

College Malls, Inc. is planning to build a national chain of shopping malls in college neighborhoods. However, malls in general have been experiencing large numbers of store closings due to, among other things, misjudgments of the shopper demographics. As a result, the company is planning to lease only to stores that target the specific age demographics of the national college student population.

As a marketing consultant to College Malls, you will be providing the company with a report that addresses the following specific issues:

- A quick way of estimating the number of students of any specified age and in any particular year, and the effect of increasing age on the college population
- The ages that correspond to relatively high and low college populations
- How fast the 20-year-old and 25-year-old student populations are increasing
- Some near-term projections of the student population trend

You decide that a good place to start would be with a visit to the Census Bureau's Web site at www.census.gov. After some time battling with search engines, all you can find is some data on college enrollment for three age brackets for the period 1980–2005, as shown in the following table:[30]

College Enrollment (thousands)

Year	1980	1985	1990	1995	1999	2000	2001	2002	2003	2004	2005
18–24	7,229	7,537	7,964	8,541	9,260	9,451	9,629	10,033	10,364	10,608	10,831
25–34	2,703	3,063	3,161	3,349	3,095	3,207	3,422	3,401	3,494	3,695	3,601
35–41	1,207	1,661	2,318	2,669	2,699	2,507	2,684	2,868	2,630	2,883	2,857

The data are inadequate for several reasons: The data are given only for certain years, and in age brackets rather than year-by-year; nor is it obvious as to how you would project the figures. However, you notice that the table is actually a numerical representation of a function of two variables: year and age. Since the age brackets are of different sizes, you "normalize" the data by dividing each figure by the number of years represented in the corresponding age bracket; for instance, you divide the 1980 figure for the first age group by 7 in order to obtain the average enrollment for each year of age in that group.

[30] Source: Bureau of the Census, 2008 (www.census.gov/population/www/socdemo/school.html).

You then rewrite the resulting table representing the years by values of t and each age bracket by the age x at its center (enrollment values are rounded):

$x \downarrow$ \ $t \rightarrow$	0	5	10	15	19	20	21	22	23	24	25
21	1,033	1,077	1,138	1,220	1,323	1,350	1,376	1,433	1,481	1,515	1,547
29.5	270	306	316	335	310	321	342	340	349	370	360
38	172	237	331	381	386	358	383	410	376	412	408

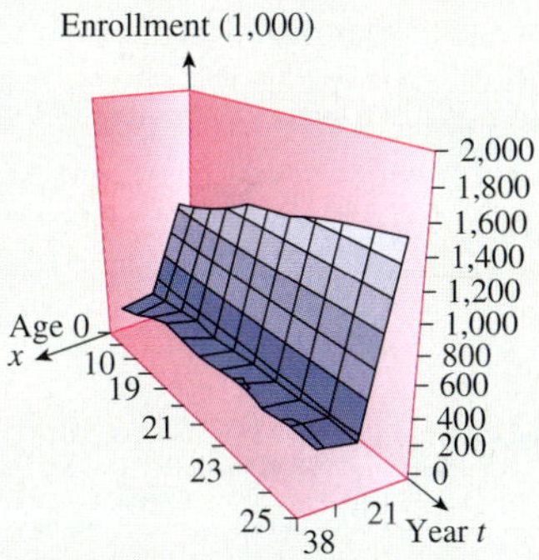

Figure 37

In order to see a visual representation of what the data are saying, you use Excel to graph the data as a surface (Figure 37). It is important to notice that Excel does not scale the t-axis as you would expect: It uses one subdivision for each year shown in the chart, and the result is an uneven scaling of the t-axis. Despite this drawback, you do see two trends after looking at views of the graph from various angles. First, enrollment of 38-year-olds (the front edge of the graph) seems to be increasing faster than enrollment of 21-year-olds. Second, the enrollment for all ages seem to be increasing approximately linearly with time since 1999, although at different rates for different age groups; for instance, the front and rear edges rise more-or-less linearly, but do not seem to be parallel.

At this point you realize that a mathematical model of these data would be useful; not only would it "smooth out the bumps" but it would give you a way to estimate enrollment N at each specific age, project the enrollments, and thereby complete the project for College Malls. Although technology can give you a regression model for data such as this, it is up to you to decide on the form of the model. It is in choosing an appropriate model that your analysis of the graph comes in handy. Because N should vary linearly with time t for each value of x, you would like

$$N = mt + k$$

for each value of x. Also, because there are three values of x for every value of time, you try a quadratic model for N as a function of x:

$$N = a + bx + cx^2$$

Putting these together, you get the following candidate model

$$N(t, x) = a_1 + a_2t + a_3x + a_4x^2$$

where a_1, a_2, a_3, and a_4 are constants. However, for each specific age $x = k$, you get

$$N(t, k) = a_1 + a_2t + a_3k + a_4k^2 = \text{Constant} + a_2t$$

with the same slope a_2 for every choice of the age k, contrary to your observation that enrollment for different age groups is rising at different rates, so you will need a more elaborate model. You recall from your applied calculus course that interaction functions give a way to model the effect of one variable on the rate of change of another, so, as an experiment, you try adding interaction terms to your model:

$$\text{Model 1: } N(t, x) = a_1 + a_2t + a_3x + a_4x^2 + a_5xt \qquad \text{Second order model}$$

$$\text{Model 2: } N(t, x) = a_1 + a_2t + a_3x + a_4x^2 + a_5xt + a_6x^2t. \qquad \text{Third order model}$$

(Model 1 is referred to as a second order model because it contains no products of more than two independent variables, whereas Model 2 contains the third order term $x^2t = x \cdot x \cdot t$.) If you study these two models for specific values k of x you get:

$$\text{Model 1: } N = \text{Constant} + (a_2 + a_5k)t \qquad \text{Slope depends linearly on age}$$

$$\text{Model 2: } N = \text{Constant} + (a_2 + a_5k + a_6k^2)t. \qquad \text{Slope depends quadratically on age}$$

	A	B	C
1	N	t	x
2	1033	0	21
3	1077	5	21
4	1138	10	21
5	1220	15	21
6	1323	19	21
7	1350	20	21
8	1376	21	21
9	1433	22	21
10	1481	23	21
11	1515	24	21
12	1547	25	21
13	270	0	29.5
14	306	5	29.5
15	316	10	29.5
16	335	15	29.5
17	310	19	29.5
18	321	20	29.5
19	342	21	29.5
20	340	22	29.5
21	349	23	29.5
22	370	24	29.5
23	360	25	29.5
24	172	0	38
25	237	5	38
26	331	10	38
27	381	15	38
28	386	19	38
29	358	20	38
30	383	21	38
31	410	22	38
32	376	23	38
33	412	24	38
34	408	25	38

Figure 38

This is encouraging: Both models show different slopes for different ages. Model 1 would predict that the slope either increases with increasing age (a_5 positive) or decreases with increasing age (a_5 negative). However, the graph suggests that the slope is larger for both younger and older students, but smaller for students of intermediate age, contrary to what Model 1 predicts. So you decide to go with the more flexible Model 2, which permits the slope to decrease and then increase with increasing age. This is exactly what you observe on the graph.

You decide to use Excel to generate your model. However, the data as shown in the table are not in a form Excel can use for regression. The data need to be organized into columns: Column A for the dependent variable N and Columns B–C for the independent variables, as shown in Figure 38. You then add columns for the higher order terms x^2, xt, and x^2t as shown below:

	A	B	C	D	E	F
1	N	t	x	x^2	x*t	x^2*t
2	1033	0	21	=C2^2	=C2*B2	=C2^2*B2
3	1077	5	21			
4	1138	10	21			
5	1220	15	21			
6	1323	19	21			
7	1350	20	21			
8	1376	21	21			

↓

	A	B	C	D	E	F
1	N	t	x	x^2	x*t	x^2*t
2	1033	0	21	441	0	0
3	1077	5	21	441	105	2205
4	1138	10	21	441	210	4410
5	1220	15	21	441	315	6615
6	1323	19	21	441	399	8379
7	1350	20	21	441	420	8820
8	1376	21	21	441	441	9261

Next, in the "Analysis" section of the Data tab, choose "Data analysis." (If this command is not available, you will need to load the Analysis ToolPak add-in.) Choose "Regression" from the list that appears, and in the resulting dialogue box enter the location of the data and where you want to put the results as shown in Figure 39; identify where the dependent and independent variables are (A1–A34 for the Y range, and B1–F34 for the X range), check "Labels," and click "OK."

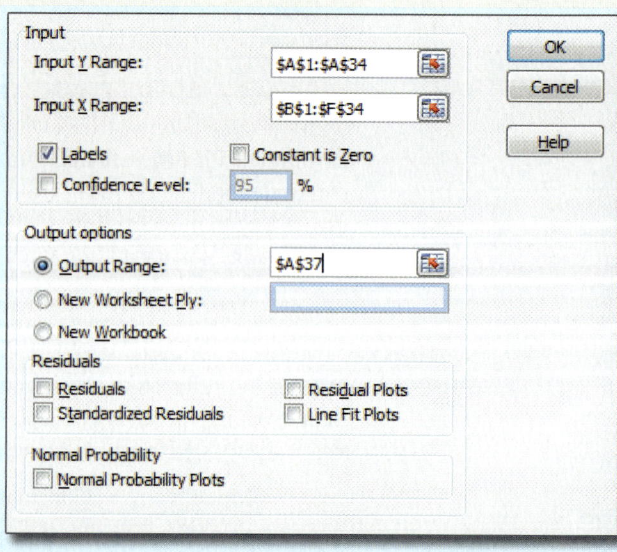

Figure 39

A portion of the output is shown below, with some of the important statistics highlighted.

	A	B	C	D	E	F
37	SUMMARY OUTPUT					
38						
39	*Regression Statistics*					
40	Multiple R	0.99807463				
41	R Square	0.99615297				
42	Adjusted R Square	0.99544056				
43	Standard Error	32.4693397				
44	Observations	33				
45						
46	ANOVA					
47		*df*	*SS*	*MS*	*F*	*Significance F*
48	Regression	5	7370748.67	1474149.73	1398.28173	1.0623E-31
49	Residual	27	28464.96658	1054.25802		
50	Total	32	7399213.636			
51						
52		*Coefficients*	*Standard Error*	*t Stat*	*P-value*	*Lower 95%*
53	Intercept	5317.742	324.3264401	16.3962642	1.4646E-15	4652.279561
54	t	165.703742	17.52824924	9.45352497	4.6746E-10	129.7387692
55	x	-296.5247	22.98384198	-12.901442	4.6484E-13	-343.6836131
56	x^2	4.26166483	0.38821602	10.9775605	1.8505E-11	3.465111887
57	x*t	-10.325579	1.242163638	-8.3125755	6.3811E-09	-12.87428658
58	x^2*t	0.16304663	0.020981167	7.77109439	2.3426E-08	0.119996859

The desired constants a_1, a_2, a_3, a_4, and a_5 appear in the coefficients column at the bottom left, in the correct order: a_1 is the "intercept," a_2 is the coefficient of t, and so on. Thus, if we round to five significant digits, we have

$$a_1 = 5{,}317.7 \quad a_2 = 165.70 \quad a_3 = -296.52$$
$$a_4 = 4.2617 \quad a_5 = -10.326 \quad a_6 = 0.16305$$

which gives our regression model:

$$N(t, x) = 5{,}317.7 + 165.70t - 296.52x + 4.2617x^2 - 10.326xt + 0.16305x^2t.$$

Fine, you say to yourself, now you have the model, but how good a fit is it to the data? That is where the "Multiple R" at the top of the data analysis comes in. R is called the **multiple coefficient of correlation,** and generalizes the coefficient of correlation discussed in the section on regression in Chapter 1: The closer R is to 1, the better the fit. We can interpret its square, given in the table as "R Square" with value 0.996, as indicating that approximately 99.6% of the variation in college enrollment is explained by the regression model, indicating an excellent fit. The "P-values" at bottom right are also indicators of the appropriateness of the model; a P-value close to zero indicates a high degree of confidence that the corresponding independent variable contributes significantly to the value of the college enrollment (a P-value less than 0.05 would indicate, for example, that the corresponding coefficient is statistically significant at a level of more than 95% confidence) whereas a P-value close to 1 indicates low confidence; there is a P-value for each coefficient. Because all the values are extremely tiny, you are confident indeed that the model is an appropriate one. Another statistical indicator is the value of "F" on the right—an indicator of confidence in the model as a whole. The fact that it too is large and its "Significance F" is tiny is yet another good sign.*

*** NOTE** We are being deliberately vague about the exact meaning of these statistics, which are discussed fully in many applied statistics texts.

As comforting as these statistics are, nothing can be quite as persuasive as a graph. You turn to the graphing software of your choice and notice that the graph of the model appears to be a faithful representation of the data. (See Figure 40.)

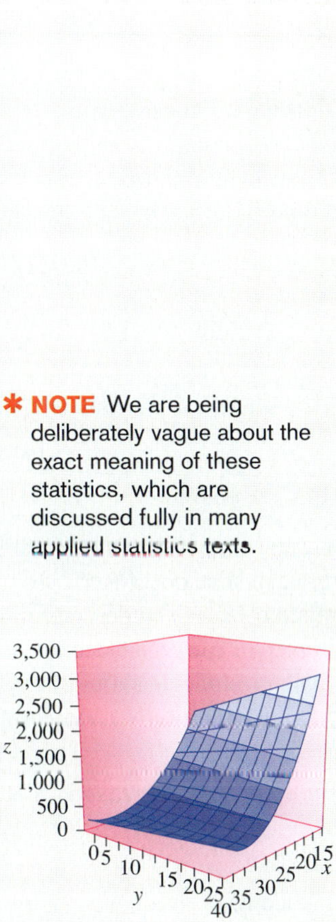

Figure 40

Now you get to work, using the model to address the questions posed by College Malls.

1. *A quick way of estimating the number of students of any specified age and in any particular year, and the effect of increasing age on the college population.* You already have a quantitative relationship in the form of the regression model. As for the second part of the question, the rate of change of college enrollment with respect to age is given by the partial derivative:

$$\frac{\partial N}{\partial x} = -296.52 + 8.5234x - 10.326t + 0.32610xt \text{ thousand students per additional year of age.}$$

Thus, for example, with $x = 20$ in 2004 ($t = 24$), we have

$$\begin{aligned}\frac{\partial N}{\partial x} &= -296.52 + 8.5234(20) - 10.326(24) + 0.32610(20)(24) \\ &\approx -217 \text{ thousand students per additional year of age,}\end{aligned}$$

so there were about 217,000 fewer students of age 21 than age 20 in 2004.
On the other hand, when $x = 38$ in the same year, one has

$$\begin{aligned}\frac{\partial N}{\partial x} &= -296.52 + 8.5234(38) - 10.326(24) + 0.32610(38)(24) \\ &\approx 77 \text{ thousand students per additional year of age,}\end{aligned}$$

so there were about 77,000 more students of age 39 than age 38 that year.

2. *The ages that correspond to relatively high and low college populations.* Although a glance at the graph shows you that there are no relative maxima, holding t constant (that is, on any given year) gives a parabola along the corresponding slice and hence a minimum somewhere along the slice.

$$\frac{\partial N}{\partial x} = 0$$

when $$-296.52 + 8.5234x - 10.326t + 0.32610xt = 0,$$

which gives $$x = \frac{296.52 + 10.326t}{8.5234 + 0.32610t} \text{ years of age.}$$

For instance, in 2000 ($t = 20$), the age at which there are the fewest students (in the given range) is 33.4 years of age. The relative maxima for each slice occur at the front and back edges of the surface, meaning that there are relatively more students of the lowest and highest ages represented. The absolute maximum for each slice occurs, as expected, at the lowest age. In short, a mall catering to college students should focus mostly on freshman-age students, least on 33-year-olds, and somewhat more on people around age 40.

3. *How fast the 20-year-old and 25-year-old student populations are increasing.* The rate of change of student population with respect to time is

$$\frac{\partial N}{\partial t} = 165.70 - 10.326x + 0.16305x^2 \text{ thousand students per year.}$$

For the three age groups in question, we obtain

$$x = 20:\ 165.70 - 10.326(20) + 0.16305(20)^2 = 24.4 \text{ thousand students/year}$$

$$x = 25:\ 165.70 - 10.326(25) + 0.16305(25)^2 \approx 9.5 \text{ thousand students/year.}$$

(Note that these rates of change are independent of time as we chose a model that is linear in time.)

4. *Some near-term projections of the student population trend.* As we have seen throughout the book, extrapolation can be a risky venture; however, *near-term* extrapolation from a good model can be reasonable. You enter the model in an Excel spreadsheet to obtain the following predicted college enrollments (in thousands) for the years 2006–2010:

$x \downarrow$ \ $t \rightarrow$	**26**	**27**	**28**	**29**	**30**
21	1,510	1,531	1,551	1,572	1,593
29.5	357	359	362	365	368
38	431	440	449	458	467

EXERCISES

1. Use Excel to obtain Model 1:

$$N(t, x) = a_1 + a_2 t + a_3 x + a_4 x^2 + a_5 x t.$$

Compare the fit of this model with that of the quadratic model above. Comment on the result.

2. How much is there to be gained by adding an additional term of the form $a_7 t^2$ to the original model? (Perform the regression and analyze the result by referring to the P-value for the resulting coefficient of t^2.) How does the addition of this term affect the slices $x =$ constant?

3. Compute and interpret $\left.\frac{\partial N}{\partial t}\right|_{(10,18)}$ and $\left.\frac{\partial^2 N}{\partial t \partial x}\right|_{(10,18)}$ for the model in the text. What are their units of measurement?

4. Notice that the derivatives in the preceding exercise do not depend on time. What additional polynomial term(s) would make both $\partial N/\partial t$ and $\partial^2 N/\partial t \partial x$ depend on time? (Write down the entire model.) Of what order is your model?

5. Test the model you constructed in the preceding question by discussing the P-values of the additional coefficients.

TI-83/84 Plus Technology Guide

Section 8.1

Example 1 (page 585) You own a company that makes two models of speakers: the Ultra Mini and the Big Stack. Your total monthly cost (in dollars) to make x Ultra Minis and y Big Stacks is given by

$$C(x, y) = 10{,}000 + 20x + 40y.$$

Compute several values of this function.

Solution with Technology

You can have a TI-83/84 Plus compute $C(x, y)$ numerically as follows:

1. In the "Y=" screen, enter

```
Y1=10000+20X+40Y
```

2. To evaluate, say, $C(10, 30)$ (the cost to make 10 Ultra Minis and 30 Big Stacks), enter

```
10→X
30→Y
Y1
```

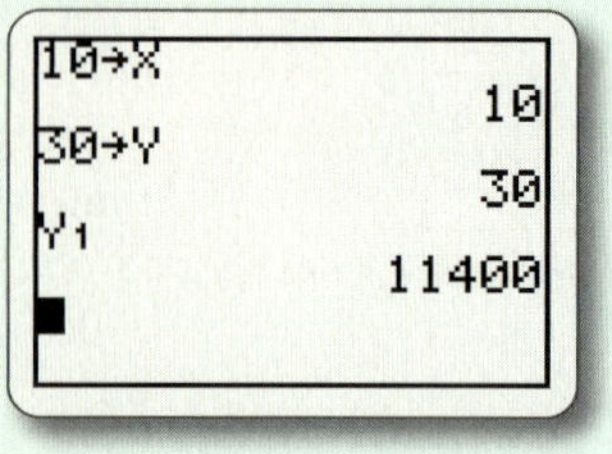

and the calculator will evaluate the function and give the answer $C(10, 30) = 11{,}400$.

This procedure is too laborious if you want to calculate $f(x, y)$ for a large number of different values of x and y.

EXCEL Technology Guide

Section 8.1

Example 1 (page 585) You own a company that makes two models of speakers: the Ultra Mini and the Big Stack. Your total monthly cost (in dollars) to make x Ultra Minis and y Big Stacks is given by

$$C(x, y) = 10{,}000 + 20x + 40y.$$

Compute several values of this function.

Solution with Technology

Excel handles functions of several variables easily. The following setup shows how a table of values of C can be created, using values of x and y you enter:

	A	B	C
1	x	y	C(x, y)
2	10	30	=10000+20*A2+40*B2
3	20	30	
4	15	0	
5	0	30	
6	30	30	

↓

	A	B	C
1	x	y	C(x, y)
2	10	30	11400
3	20	30	11600
4	15	0	10300
5	0	30	11200
6	30	30	11800

A disadvantage of this layout is that it's not easy to enter values of x and y systematically in two columns. Can you find a way to remedy this? (See Example 3 for one method.)

Example 3 (page 587) Use technology to create a table of values of the body mass index

$$M(w, h) = \frac{0.45w}{(0.0254h)^2}.$$

Solution with Technology

We can use this formula to recreate a table in Excel, as follows:

	A	B	C	D
1		130	140	150
2	60	=0.45*B$1/(0.0254*$A2)^2		
3	61			
4	62			
5	63			
6	64			
7	65			
8	66			
9	67			

In the formula in cell B2 we have used B$1 instead of B1 for the w-coordinate because we want all references to w to use the same row (1). Similarly, we want all references to h to refer to the same column (A), so we used $A2 instead of A2.

We copy the formula in cell B2 to all of the red shaded area to obtain the desired table:

	A	B	C	D	
1		130	140	150	3
2	60	25.18755038	27.12505425	29.06255813	3
3	61	24.36849808	26.24299793	28.11749778	29.9
4	62	23.58875685	25.40327661	27.21779637	29.0
5	63	22.84585068	24.60322381	26.36059694	28.1
6	64	22.13749545	23.84037971	25.54326398	27.2
7	65	21.46158138	23.11247226	24.76336314	26.4
8	66	20.81615733	22.41740021	24.01864308	25.6
9	67	20.19941665	21.75321793	23.30701921	24.8

Example 5 (page 590) Obtain the graph of $f(x, y) = x^2 + y^2$.

Solution with Technology

1. Set up a table showing a range of values of x and y and the corresponding values of the function (see Example 3):

	A	B	C	D	E	F	G	H
1		-3	-2	-1	0	1	2	3
2	-3	=B1^2+A2^2						
3	-2							
4	-1							
5	0							
6	1							
7	2							
8	3							

↓

	A	B	C	D	E	F	G	H
1		-3	-2	-1	0	1	2	3
2	-3	18	13	10	9	10	13	18
3	-2	13	8	5	4	5	8	13
4	-1	10	5	2	1	2	5	10
5	0	9	4	1	0	1	4	9
6	1	10	5	2	1	2	5	10
7	2	13	8	5	4	5	8	13
8	3	18	13	10	9	10	13	18

2. Select the cells with the values (B2: H8) and insert a chart, with the "Surface" option selected and "Series in Columns" selected as the data option, to obtain a graph like the following:

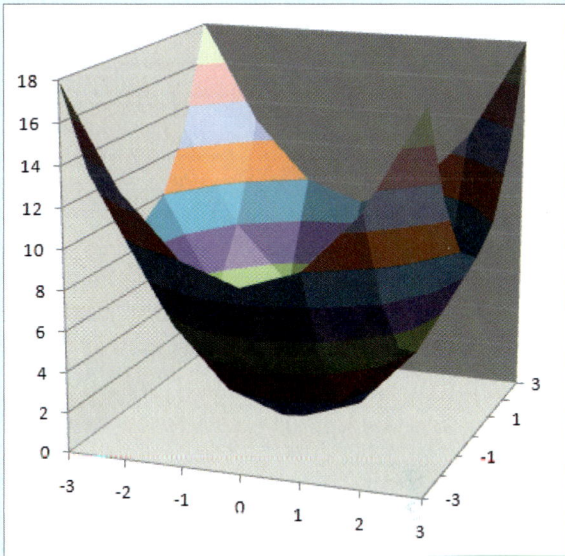

Business Finance Section

5 The Mathematics of Finance

Case Study Adjustable Rate and Subprime Mortgages

Mr. and Mrs. Wong have an appointment tomorrow with you, their investment counselor, to discuss their plan to purchase a $400,000 house in Orlando, Florida. Their combined annual income is $80,000 per year, which they estimate will increase by 4% annually over the foreseeable future, and they are considering three different specialty 30-year mortgages:

Hybrid: The interest is fixed at a low introductory rate of 4% for 5 years.

Interest-Only: During the first 5 years, the rate is set at 4.2% and no principal is paid.

Negative Amortization: During the first 5 years, the rate is set at 4.7% based on a principal of 60% of the purchase price of the home.

How would you advise them?

Gary Faber/Photodisc/Getty Images

Web Site
www.FiniteMath.org
At the Web site you will find:

- A detailed chapter summary
- A true/false quiz
- A time value of money utility
- Excel tutorials for every section

Introduction

A knowledge of the mathematics of investments and loans is important not only for business majors but also for everyone who deals with money, which is all of us. This chapter is largely about *interest*: interest paid by an investment, interest paid on a loan, and variations on these.

We focus on three forms of investment: investments that pay simple interest, investments in which interest is compounded, and annuities. An investment that pays *simple interest* periodically gives interest directly to the investor, perhaps in the form of a monthly check. If instead, the interest is reinvested, the interest is *compounded*, and the value of the account grows as the interest is added. An *annuity* is an investment earning compound interest into which periodic payments are made or from which periodic withdrawals are made; in the case of periodic payments, such an investment is more commonly called a *sinking fund*. From the point of view of the lender, a loan is a kind of annuity.

We also look at bonds, the primary financial instrument used by companies and governments to raise money. Although bonds nominally pay simple interest, determining their worth, particularly in the secondary market, requires an annuity calculation.

5.1 Simple Interest

You deposit \$1,000, called the **principal** or **present value**, into a savings account. The bank pays you 5% interest, in the form of a check, each year. How much interest will you earn each year? Because the bank pays you 5% interest each year, your annual (or yearly) interest will be 5% of \$1,000, or $0.05 \times 1{,}000 = \$50$.

Generalizing this calculation, call the present value PV and the interest rate (expressed as a decimal) r. Then INT, the annual interest paid to you, is given by

$$INT = PVr.$$

If the investment is made for a period of t years, then the total interest accumulated is t times this amount, which gives us the following:

Simple Interest

The **simple interest** on an investment (or loan) of PV dollars at an annual interest rate of r for a period of t years is

$$INT = PVrt.$$

Quick Example

The simple interest over a period of 4 years on a \$5,000 investment earning 8% per year is

$$\begin{aligned} INT &= PVrt \\ &= (5{,}000)(0.08)(4) = \$1{,}600. \end{aligned}$$

Note on Multiletter Variables

Multiletter variables like PV and INT are traditional in finance textbooks, calculators (like the TI-83/84 Plus), and spreadsheets (like Excel). Just watch out for expressions like PVr, which is the product of two things, PV and r, not three.

Given your \$1,000 investment at 5% simple interest, how much money will you have after 2 years? To find the answer, we need to add the accumulated interest to the principal to get the **future value** (FV) of your deposit.

$$FV = PV + INT = \$1{,}000 + (1{,}000)(0.05)(2) = \$1{,}100$$

In general, we can compute the future value as follows:

$$FV = PV + INT = PV + PVrt = PV(1 + rt).$$

Future Value for Simple Interest

The **future value** of an investment of PV dollars at an annual simple interest rate of r for a period of t years is

$$FV = PV(1 + rt).$$

Quick Examples

1. The value, at the end of 4 years, of a \$5,000 investment earning 8% simple interest per year is

$$FV = PV(1 + rt)$$
$$= 5{,}000[1 + (0.08)(4)] = \$6{,}600.$$

2. Writing the future value in Quick Example 1 as a function of time, we get

$$FV = 5{,}000(1 + 0.08t)$$
$$= 5{,}000 + 400t$$

which is a linear function of time t. The intercept is $PV = \$5{,}000$, and the slope is the annual interest, \$400 per year.

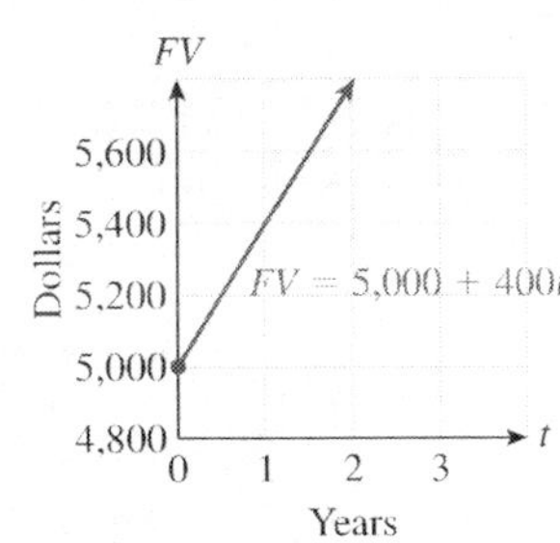

In general: *Simple interest growth is a linear function of time, with intercept given by the present value and slope given by annual interest.*

EXAMPLE 1 Savings Accounts

In December 2008, the Amex Bank of Canada was paying 2.40% interest on savings accounts.* If the interest is paid as simple interest, find the future value of a \$2,000 deposit in 6 years. What is the total interest paid over the period?

*Source: Canoe Money (http://money.canoe.ca/rates/savings.html).

Solution We use the future value formula:

$$\begin{aligned} FV &= PV(1 + rt) \\ &= 2{,}000[1 + (0.024)(6)] = 2{,}000[1.144] = \$2{,}288. \end{aligned}$$

The total interest paid is given by the simple interest formula.

$$\begin{aligned} INT &= PVrt \\ &= (2{,}000)(0.024)(6) = \$288 \end{aligned}$$

Note To find the interest paid, we could also have computed

$$INT = FV - PV = 2{,}288 - 2{,}000 = \$288.$$

■

➡ **Before we go on...** In the preceding example, we could look at the future value as a function of time:

$$FV = 2{,}000(1 + 0.024t) = 2{,}000 + 48t.$$

Thus, the future value is growing linearly at a rate of \$48 per year. ■

Fisher/Thatcher/Stone/Getty Images

EXAMPLE 2 Bridge Loans

When "trading up," homeowners sometimes have to buy a new house before they sell their old house. One way to cover the costs of the new house until they get the proceeds from selling the old house is to take out a short-term *bridge loan*. Suppose a bank charges 12% simple annual interest on such a loan. How much will be owed at the maturation (the end) of a 90-day bridge loan of \$90,000?

Solution We use the future value formula

$$FV = PV(1 + rt)$$

with $t = 90/365$, the fraction of a year represented by 90 days:

$$\begin{aligned} FV &= 90{,}000[1 + (0.12)(90/365)] \\ &= \$92{,}663.01. \end{aligned}$$

(We will always round our answers to the nearest cent after calculation. Be careful not to round intermediate results.)

➡ **Before we go on...** Many banks use 360 days for this calculation rather than 365, which makes a "year" for the purposes of the loan slightly shorter than a calendar year. The effect is to increase the amount owed:

$$FV = 90{,}000[1 + (0.12)(90/360)] = \$92{,}700.$$

■

One of the primary ways companies and governments raise money is by selling **bonds**. At its most straightforward, a corporate bond promises to pay simple interest, usually twice a year, for a length of time until it **matures**, at which point it returns the original investment to the investor (U.S. Treasury notes and bonds are similar). Things get more complicated when the selling price is negotiable, as we will see in the following sections.

EXAMPLE 3 Corporate Bonds

The Megabucks Corporation is issuing 10-year bonds paying an annual rate of 6.5%. If you buy \$10,000 worth of bonds, how much interest will you earn every 6 months, and how much interest will you earn over the life of the bonds?

Solution Using the simple interest formula, every 6 months you will receive

$$\begin{aligned} INT &= PVrt \\ &= (10{,}000)(0.065)\left(\frac{1}{2}\right) = \$325. \end{aligned}$$

Over the 10-year life of the bonds, you will earn

$$\begin{aligned} INT &= PVrt \\ &= (10{,}000)(0.065)(10) = \$6{,}500 \end{aligned}$$

in interest. So, at the end of 10 years, when your original investment is returned to you, your \$10,000 will have turned into \$16,500.

We often want to turn an interest calculation around: Rather than starting with the present value and finding the future value, there are times when we know the future value and need to determine the present value. Solving the future value formula for PV gives us the following.

Present Value for Simple Interest

The present value of an investment at an annual simple interest rate of r for a period of t years, with future value FV, is

$$PV = \frac{FV}{1+rt}.$$

Quick Example

If an investment earns 5% simple interest and will be worth \$1,000 in 4 years, then its present value (its initial value) is

$$\begin{aligned} PV &= \frac{FV}{1+rt} \\ &= \frac{1{,}000}{1+(0.05)(4)} = \$833.33. \end{aligned}$$

Here is a typical example. U.S. Treasury bills (T-bills) are short-term investments (up to 1 year) that pay you a set amount after a period of time.

EXAMPLE 4 Treasury Bills

A U.S. Treasury bill paying \$10,000 after 6 months earns 3.67% simple annual interest. How much would it cost to buy?

Solution The future value of the T-bill is \$10,000; the price we pay is its present value. We know that

$$FV = \$10{,}000$$
$$r = 0.0367$$

and

$$t = 0.5.$$

Substituting into the present value formula, we have

$$PV = \frac{10{,}000}{1 + (0.0367)(0.5)} = 9{,}819.81$$

so we will pay \$9,819.81 for the T-bill.

➡ **Before we go on...** The simplest way to find the interest earned on the T-bill is by subtraction:

$$INT = FV - PV = 10{,}000 - 9{,}819.81 = \$180.19.$$ ■

Here is some additional terminology on treasury bills:

Treasury Bills (T-Bills): Maturity Value, Discount Rate, and Yield

The **maturity value** of a T-bill is the amount of money it will pay at the end of its life, that is, upon **maturity**.

Quick Example

A 1-year \$10,000 T-bill has a maturity value of \$10,000, and so will pay you \$10,000 after one year.

The cost of a T-bill is generally less than its maturity value.* In other words, a T-bill will generally sell at a *discount*, and the **discount rate** is the *annualized* percentage of this discount; that is, the percentage is adjusted to give an annual percentage. (See Quick Examples 2 and 3.)

* **NOTE** An exception occurred during the financial meltdown of 2008, when T-bills were heavily in demand as "safe haven" investments and were sometimes selling at—or even above—their maturity values.

Quick Examples

1. A 1-year \$10,000 T-bill with a discount rate of 5% will sell for 5% less than its maturity value of \$10,000, that is, for \$9,500.
2. A 6-month \$10,000 T-bill with a discount rate of 5% will sell at an actual discount of half of that—2.5% less than its maturity value, or \$9,750—because 6 months is half of a year.
3. A 3-month \$10,000 T-bill with a discount rate of 5% will sell at an actual discount of a fourth of that: 1.25% less than its maturity value, or \$9,875.

The annual **yield** of a T-bill is the simple annual interest rate an investor earns when the T-bill matures, as calculated in the next example.

EXAMPLE 5 Treasury Bills

A T-bill paying \$10,000 after 6 months sells at a discount rate of 3.6%. What does it sell for? What is the annual yield?

Solution The (annualized) discount rate is 3.6%; so, for a 6-month bill, the actual discount will be half of that: $3.6\%/2 = 1.8\%$ below its maturity value. This makes the selling price

$$10{,}000 - (0.018)(10{,}000) = \$9{,}820. \qquad \text{Maturity value} - \text{Discount}$$

To find the annual interest rate, note that the present value of the investment is the price the investor pays, \$9,820, and the future value is its maturity value, \$10,000 six months later. So,

$$PV = \$9{,}820 \qquad FV = \$10{,}000 \qquad t = 0.5$$

and we wish to find r. Substituting in the future value formula, we get

$$FV = PV(1 + rt)$$
$$10{,}000 = 9{,}820(1 + 0.5r)$$

so

$$1 + 0.5r = 10{,}000/9{,}820$$

and

$$r = (10{,}000/9{,}820 - 1)/0.5 \approx 0.0367.$$

Thus, the T-bill is paying 3.67% simple annual interest.

➡ **Before we go on...** The T-bill in Example 5 is the same one as in Example 4 (with a bit of rounding). The interest rate and the discount rate are two different ways of telling what the investment pays. One of the Communication and Reasoning Exercises asks you to find a formula for the interest rate in terms of the discount rate. ■

Fees on loans can also be thought of as a form of interest.

EXAMPLE 6 Tax Refunds

You are expecting a tax refund of \$800. Because it may take up to 6 weeks to get the refund, your tax preparation firm offers, for a fee of \$40, to give you an "interest-free" loan of \$800 to be paid back with the refund check. If we think of the fee as interest, what simple annual interest rate is the firm actually charging?

Solution If we view the \$40 as interest, then the future value of the loan (the value of the loan to the firm, or the total you will pay the firm) is \$840. Thus, we have

$$FV = 840$$
$$PV = 800$$
$$t = 6/52 \qquad \text{Using 52 weeks in a year}$$

and we wish to find r. Substituting, we get

$$FV = PV(1 + rt)$$

$$840 = 800(1 + 6r/52) = 800 + \frac{4{,}800r}{52}$$

so

$$\frac{4{,}800r}{52} = 840 - 800 = 40$$

$$r = \frac{40 \times 52}{4{,}800} \approx 0.43.$$

In other words, the firm is charging you 43% annual interest! Save your money and wait 6 weeks for your refund.

5.1 EXERCISES

▼ more advanced ◆ challenging

T indicates exercises that should be solved using technology

In Exercises 1–6, compute the simple interest for the specified period and the future value at the end of the period. Round all answers to the nearest cent. HINT [See Example 1.]

1. \$2,000 is invested for 1 year at 6% per year.
2. \$1,000 is invested for 10 years at 4% per year.
3. \$20,200 is invested for 6 months at 5% per year.
4. \$10,100 is invested for 3 months at 11% per year.
5. You borrow \$10,000 for 10 months at 3% per year.
6. You borrow \$6,000 for 5 months at 9% per year.

In Exercises 7–12, find the present value of the given investment. HINT [See the Quick Example on page 351.]

7. An investment earns 2% per year and is worth \$10,000 after 5 years.
8. An investment earns 5% per year and is worth \$20,000 after 2 years.
9. An investment earns 7% per year and is worth \$1,000 after 6 months.
10. An investment earns 10% per year and is worth \$5,000 after 3 months.
11. An investment earns 3% per year and is worth \$15,000 after 15 months.
12. An investment earns 6% per year and is worth \$30,000 after 20 months.

APPLICATIONS

In Exercises 13–28, compute the specified quantity. Round all answers to the nearest month, the nearest cent, or the nearest 0.001%, as appropriate.

13. ***Simple Loans*** You take out a 6-month, \$5,000 loan at 8% simple interest. How much would you owe at the end of the 6 months? HINT [See Example 2.]
14. ***Simple Loans*** You take out a 15-month, \$10,000 loan at 11% simple interest. How much would you owe at the end of the 15 months? HINT [See Example 2.]
15. ***Savings*** How much would you have to deposit in an account earning 4.5% simple interest if you wanted to have \$1,000 after 6 years? HINT [See Example 4.]
16. ***Simple Loans*** Your total payment on a 4-year loan, which charged 9.5% simple interest, amounted to \$30,360. How much did you originally borrow? HINT [See Example 4.]
17. ***Bonds*** A 5-year bond costs \$1,000 and will pay a total of \$250 interest over its lifetime. What is its annual interest rate? HINT [See Example 3.]
18. ***Bonds*** A 4-year bond costs \$10,000 and will pay a total of \$2,800 in interest over its lifetime. What is its annual interest rate? HINT [See Example 3.]
19. ***Treasury Bills*** In December 2008 (in the midst of the financial crisis) a \$5,000 6-month T-bill was selling at a discount rate of only 0.25%.[1] What was its simple annual yield? HINT [See Example 5.]
20. ***Treasury Bills*** In December 2008 (in the midst of the financial crisis) a \$15,000 3-month T-bill was selling at a discount rate of only 0.06%.[2] What was its simple annual yield? HINT [See Example 5.]

[1] Discount rate on December 29, 2008. Source: U.S. Treasury (www.ustreas.gov/offices/domestic-finance/debt-management/interest-rate/daily_treas_bill_rates.shtml).

[2] Ibid.

21. Simple Loans A $4,000 loan, taken now, with a simple interest rate of 8% per year, will require a total repayment of $4,640. When will the loan mature?

22. Simple Loans The simple interest on a $1,000 loan at 8% per year amounted to $640. When did the loan mature?

23. Treasury Bills At auction on August 18, 2005, 6-month T-bills were sold at a discount of 3.705%.[3] What was the simple annual yield? HINT [See Example 5.]

24. Treasury Bills At auction on August 18, 2005, 3-month T-bills were sold at a discount of 3.470%.[4] What was the simple annual yield? HINT [See Example 5.]

25. Fees You are expecting a tax refund of $1,000 in 4 weeks. A tax preparer offers you a $1,000 loan for a fee of $50 to be repaid by your refund check when it arrives in 4 weeks. Thinking of the fee as interest, what simple interest rate would you be paying on this loan? HINT [See Example 6.]

26. Fees You are expecting a tax refund of $1,500 in 3 weeks. A tax preparer offers you a $1,500 loan for a fee of $60 to be repaid by your refund check when it arrives in 3 weeks. Thinking of the fee as interest, what simple interest rate would you be paying on this loan?

27. Fees You take out a 2-year, $5,000 loan at 9% simple annual interest. The lender charges you a $100 fee. Thinking of the fee as additional interest, what is the actual annual interest rate you will pay?

28. Fees You take out a 3-year, $7,000 loan at 8% simple annual interest. The lender charges you a $100 fee. Thinking of the fee as additional interest, what is the actual annual interest rate you will pay?

Stock Investments *Exercises 29–34 are based on the following chart, which shows monthly figures for* Apple Inc. *stock in 2008.*[5]

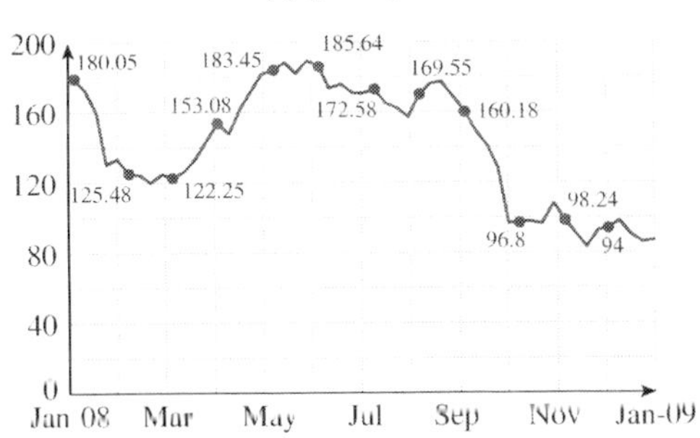

Marked are the following points on the chart:

Jan. 2008	Feb. 2008	Mar. 2008	Apr. 2008	May 2008	June 2008
180.05	125.48	122.25	153.08	183.45	185.64
July 2008	Aug. 2008	Sep. 2008	Oct. 2008	Nov. 2008	Dec. 2008
172.58	169.55	160.18	96.80	98.24	94.00

[3] Source: The Bureau of the Public Debt's Web site: www.publicdebt.treas.gov.

[4] Ibid.

[5] Source: Yahoo! Finance (http://finance.yahoo.com).

29. Calculate to the nearest 0.01% your annual percentage loss (on a simple interest basis) if you had bought Apple stock in June and sold in December.

30. Calculate to the nearest 0.01% your annual percentage return (on a simple interest basis) if you had bought Apple stock in February and sold in June.

31. Suppose you bought Apple stock in January 2008. If you later sold at one of the marked dates on the chart, which of those dates would have given you the largest annual return (on a simple interest basis), and what would that return have been?

32. Suppose you bought Apple stock in April. If you later sold at one of the marked dates on the chart, which of those dates would have given you the largest annual loss (on a simple interest basis), and what would that loss have been?

33. Did Apple's stock undergo simple interest change in the period January through April? (Give a reason for your answer.)

34. If Apple's stock had undergone simple interest change from April to June at the same simple rate as from March to April, what would the price have been in June?

Population *Exercises 35–40 are based on the following graph, which shows the population of San Diego County from 1950 to 2000.*[6]

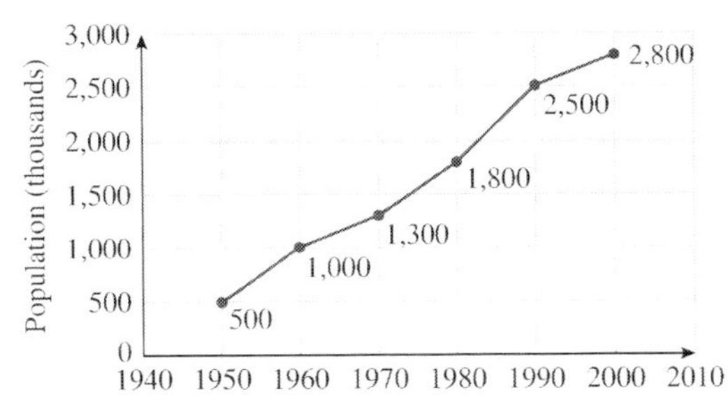

35. At what annual (simple interest) rate did the population of San Diego County increase from 1950 to 2000?

36. At what annual (simple interest) rate did the population of San Diego County increase from 1950 to 1990?

37. If you used your answer to Exercise 35 as the annual (simple interest) rate at which the population was growing since 1950, what would you predict the San Diego County population to be in 2010?

38. If you used your answer to Exercise 36 as the annual (simple interest) rate at which the population was growing since 1950, what would you predict the San Diego County population to be in 2010?

39. Use your answer to Exercise 35 to give a linear model for the population of San Diego County from 1950 to 2000. Draw the graph of your model over that period of time.

40. Use your answer to Exercise 36 to give a linear model for the population of San Diego County from 1950 to 2000. Draw the graph of your model over that period of time.

[6] Source: Census Bureau/*New York Times*, April 2, 2002, p. F3.

COMMUNICATION AND REASONING EXERCISES

41. One or more of the following graphs represent the future value of an investment earning simple interest. Which one(s)? Give the reason for your choice(s).

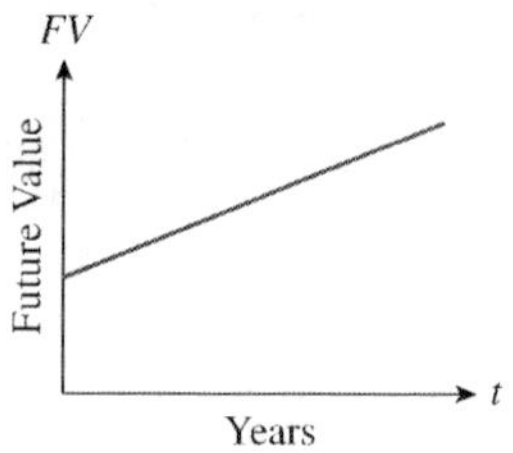

(A)

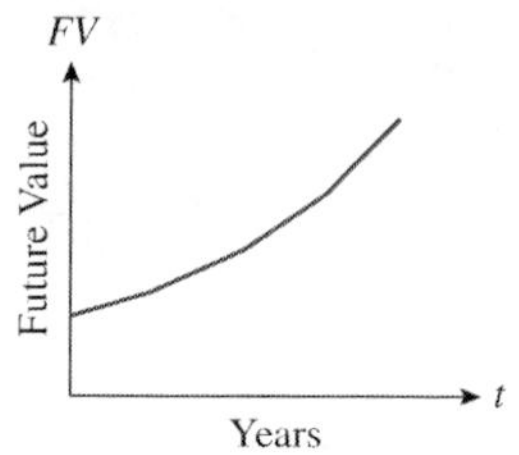

(B)

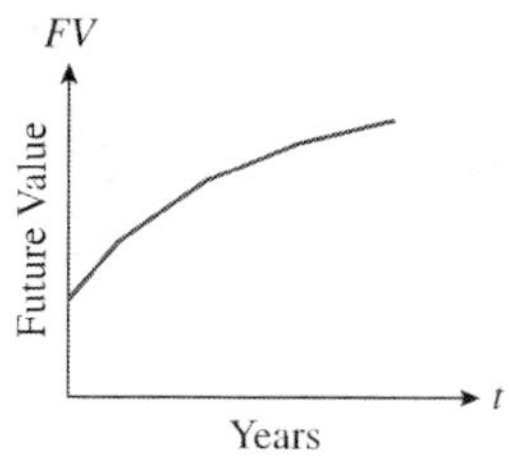

(C)

42. Given that $FV = 5t + 400$, for what interest rate is this the equation of future value (in dollars) as a function of time t (in years)?

43. ***Interpreting the News*** You hear the following on your local radio station's business news: "The economy last year grew by 1%. This was the second year in a row in which the economy showed a 1% growth." Because the rate of growth was the same two years in a row, this represents simple interest growth, right? Explain your answer.

44. ***Interpreting the News*** You hear the following on your local radio station's business news: "The economy last year grew by 1%. This was the second year in a row in which the economy showed a 1% growth." This means that, in dollar terms, the economy grew more last year than the year before. Why?

45. Explain why simple interest is not the appropriate way to measure interest on a savings account that pays interest directly into your account.

46. Suppose that a one-year T-bill sells at a discount rate d. Find a formula, in terms of d, for the simple annual interest rate r the bill will pay.

5.2 Compound Interest

You deposit \$1,000 into a savings account. The bank pays you 5% interest, which it deposits into your account, or **reinvests**, at the end of each year. At the end of 5 years, how much money will you have accumulated? Let us compute the amount you have at the end of each year. At the end of the first year, the bank will pay you simple interest of 5% on your \$1,000, which gives you

$$\begin{aligned} PV(1+rt) &= 1{,}000(1+0.05) \\ &= \$1{,}050. \end{aligned}$$

At the end of the second year, the bank will pay you another 5% interest, but this time computed on the total in your account, which is \$1,050. Thus, you will have a total of

$$1{,}050(1+0.05) = \$1{,}102.50.$$

If you were being paid simple interest on your original \$1,000, you would have only \$1,100 at the end of the second year. The extra \$2.50 is the interest earned on the \$50 interest added to your account at the end of the first year. Having interest earn interest is called **compounding** the interest. We could continue like this until the end of the fifth year, but notice what we are doing: Each year we are multiplying by $1 + 0.05$. So, at the end of 5 years, you will have

$$1{,}000(1+0.05)^5 \approx \$1{,}276.28.$$

It is interesting to compare this to the amount you would have if the bank paid you simple interest:

$$1{,}000(1 + 0.05 \times 5) = \$1{,}250.00.$$

The extra \$26.28 is again the effect of compounding the interest.

Banks often pay interest more often than once a year. Paying interest quarterly (four times per year) or monthly is common. If your bank pays interest monthly, how much will your \$1,000 deposit be worth after 5 years? The bank will not pay you 5% interest every month, but will give you 1/12 of that,[7] or 5/12% interest each month. Thus, instead of multiplying by $1 + 0.05$ every year, we should multiply by $1 + 0.05/12$ each month. Because there are $5 \times 12 = 60$ months in 5 years, the total amount you will have at the end of 5 years is

$$1{,}000\left(1 + \frac{0.05}{12}\right)^{60} \approx \$1{,}283.36.$$

Compare this to the \$1,276.28 you would get if the bank paid the interest every year. You earn an extra \$7.08 if the interest is paid monthly because interest gets into your account and starts earning interest earlier. The amount of time between interest payments is called the **compounding period**.

The following table summarizes the results above.

Time in Years	*Amount with Simple Interest*	*Amount with Annual Compounding*	*Amount with Monthly Compounding*
0	\$1,000	\$1,000	\$1,000
1	$1{,}000(1 + 0.05)$ $= \$1{,}050$	$1{,}000(1 + 0.05)$ $= \$1{,}050$	$1{,}000(1 + 0.05/12)^{12}$ $= \$1{,}051.16$
2	$1{,}000(1 + 0.05 \times 2)$ $= \$1{,}100$	$1{,}000(1 + 0.05)^2$ $= \$1{,}102.50$	$1{,}000(1 + 0.05/12)^{24}$ $= \$1{,}104.94$
5	$1{,}000(1 + 0.05 \times 5)$ $= \$1{,}250$	$1{,}000(1 + 0.05)^5$ $= \$1{,}276.28$	$1{,}000(1 + 0.05/12)^{60}$ $= \$1{,}283.36$

The preceding calculations generalize easily to give the general formula for future value when interest is compounded.

Future Value for Compound Interest

The future value of an investment of PV dollars earning interest at an annual rate of r compounded (reinvested) m times per year for a period of t years is

$$FV = PV\left(1 + \frac{r}{m}\right)^{mt}$$

or

$$FV = PV(1 + i)^n$$

where $i = r/m$ is the interest paid each compounding period and $n = mt$ is the total number of compounding periods.

[7]This is approximate. They will actually give you 31/365 of the 5% at the end of January and so on, but it's simpler and reasonably accurate to call it 1/12.

Quick Examples

1. To find the future value after 5 years of a \$10,000 investment earning 6% interest, with interest reinvested every month, we set $PV = 10{,}000$, $r = 0.06$, $m = 12$, and $t = 5$. Thus,

$$FV = PV\left(1 + \frac{r}{m}\right)^{mt} = 10{,}000\left(1 + \frac{0.06}{12}\right)^{60} \approx \$13{,}488.50.$$

2. Writing the future value in Quick Example 1 as a function of time, we get

$$\begin{aligned} FV &= 10{,}000\left(1 + \frac{0.06}{12}\right)^{12t} \\ &= 10{,}000(1.005)^{12t} \end{aligned}$$

which is an **exponential** function of time t.

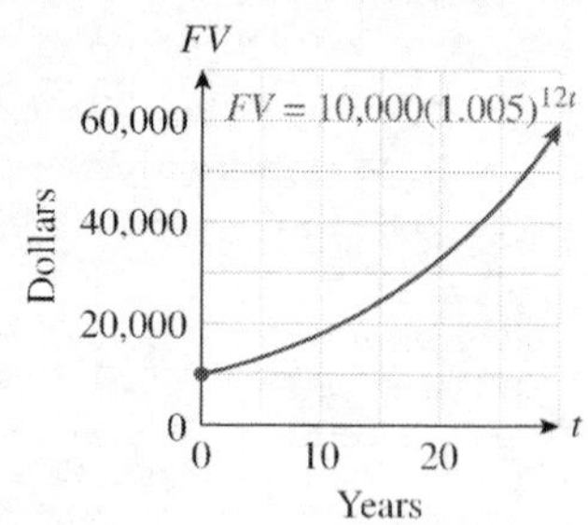

In general: *Compound interest growth is an exponential function of time.*

using Technology

While technology can be used for compound interest calculations using the formulas, all three technologies discussed in this book have built-in mathematics of finance capabilities. See the Technology Guides at the end of the chapter for details on using a TI-83/84 Plus or Excel to do the calculations in Example 1. [Details: TI-83/84 Plus: page 388, Excel: page 391]

Web Site
www.FiniteMath.org
→ Online Utilities
→ Time Value of Money Utility

This utility is similar to the TVM Solver on the TI-83/84 Plus. To compute the future value, enter the values shown, and click "Compute" next to FV.

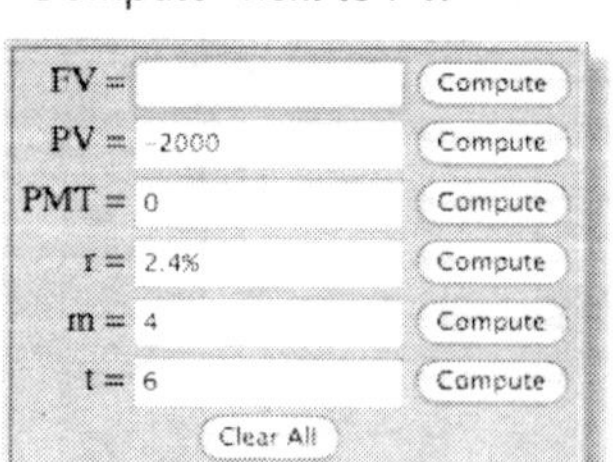

(For an explanation of the terms, see the Technology Guides for the TI-83/84 Plus.)

EXAMPLE 1 Savings Accounts

In December 2008, the Amex Bank of Canada was paying 2.40% interest on savings accounts.* If the interest is compounded quarterly, find the future value of a \$2,000 deposit in 6 years. What is the total interest paid over the period?

Solution We use the future value formula with $m = 4$:

$$\begin{aligned} FV &= PV\left(1 + \frac{r}{m}\right)^{mt} \\ &= 2{,}000\left(1 + \frac{0.0240}{4}\right)^{(4)(6)} \approx \$2{,}308.77 \end{aligned}$$

The total interest paid is

$$INT = FV - PV = 2{,}308.77 - 2{,}000 = \$308.77.$$

* Source: Canoe Money (http://money.canoe.ca/rates/savings.html).

Example 1 illustrates the concept of the **time value of money**: A given amount of money received now will usually be worth a different amount to us than the same amount received some time in the future. In the example above, we can say that \$2,000 received now is worth the same as \$2,308.77 received 6 years from now, because if we receive \$2,000 now, we can turn it into \$2,308.77 by the end of 6 years.

We often want to know, for some amount of money in the future, what is the equivalent value at present. As we did for simple interest, we can solve the future value formula for the present value and obtain the following formula.

Present Value for Compound Interest

The present value of an investment earning interest at an annual rate of r compounded m times per year for a period of t years, with future value FV, is

$$PV = \frac{FV}{\left(1+\frac{r}{m}\right)^{mt}}$$

or

$$PV = \frac{FV}{(1+i)^n} = FV(1+i)^{-n}$$

where $i = r/m$ is the interest paid each compounding period and $n = mt$ is the total number of compounding periods.

Quick Examples

To find the amount we need to invest in an investment earning 12% per year, compounded annually, so that we will have \$1 million in 20 years, use $FV = \$1{,}000{,}000$, $r = 0.12$, $m = 1$, and $t = 20$:

$$PV = \frac{FV}{\left(1+\frac{r}{m}\right)^{mt}} = \frac{1{,}000{,}000}{(1+0.12)^{20}} \approx \$103{,}666.77.$$

Put another way, \$1,000,000 20 years from now is worth only \$103,666.77 to us now, if we have a 12% investment available.

In the preceding section, we mentioned that a bond pays interest until it reaches maturity, at which point it pays you back an amount called its **maturity value** or **par value**. The two parts, the interest and the maturity value, can be separated and sold and traded by themselves. A **zero coupon bond** is a form of corporate bond that pays no interest during its life but, like U.S. Treasury bills, promises to pay you the maturity value when it reaches maturity. Zero coupon bonds are often created by removing or *stripping* the interest coupons from an ordinary bond, so are also known as **strips**. Zero coupon bonds sell for less than their maturity value, and the return on the investment is the difference between what the investor pays and the maturity value. Although no interest is actually paid, we measure the return on investment by thinking of the interest rate that would make the selling price (the present value) grow to become the maturity value (the future value).[8]

[8] The IRS refers to this kind of interest as **original issue discount (OID)** and taxes it as if it were interest actually paid to you each year.

EXAMPLE 2 Zero Coupon Bonds

Megabucks Corporation is issuing 10-year zero coupon bonds. How much would you pay for bonds with a maturity value of \$10,000 if you wish to get a return of 6.5% compounded annually?*

Solution As we said earlier, we think of a zero coupon bond as if it were an account earning (compound) interest. We are asked to calculate the amount you will pay for the bond—the present value PV. We have

$$\begin{aligned} FV &= \$10{,}000 \\ r &= 0.065 \\ t &= 10 \\ m &= 1. \end{aligned}$$

We can now use the present value formula:

$$PV = \frac{FV}{\left(1+\frac{r}{m}\right)^{mt}}$$

$$PV = \frac{10{,}000}{\left(1+\frac{0.065}{1}\right)^{10\times 1}} \approx \$5{,}327.26.$$

Thus, you should pay \$5,327.26 to get a return of 6.5% annually.

* **NOTE** The return investors look for depends on a number of factors, including risk (the chance that the company will go bankrupt and you will lose your investment); the higher the risk, the higher the return. U.S. Treasuries are considered risk free because the federal government has never defaulted on its debts. On the other hand, so-called junk bonds are high-risk investments (below investment grade) and have correspondingly high yields.

using Technology

See the Technology Guides at the end of the chapter for details on using TVM Solver on the TI-83/84 Plus or the built-in finance functions in Excel to do the calculations in Example 2. [Details: TI-83/84 Plus: page 388, Excel: page 392]

Web Site

www.FiniteMath.org

→ Online Utilities

→ Time Value of Money Utility

This utility is similar to the TVM Solver on the TI-83/84 Plus. To compute the present value, enter the values shown, and click "Compute" next to PV.

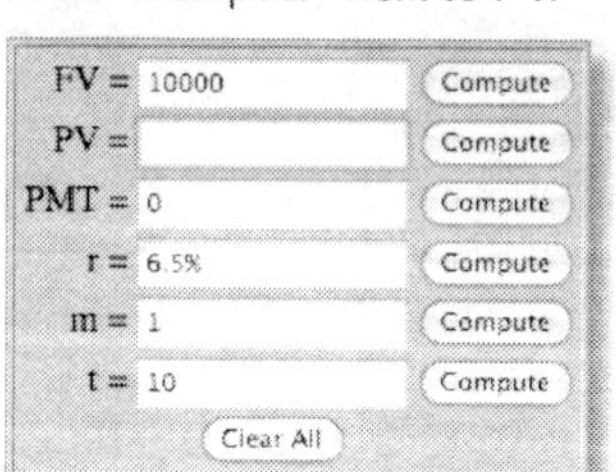

➡ Before we go on... Particularly in financial applications, you will hear the word **"discounted"** in place of "compounded" when discussing present value. Thus, the result of Example 2 might be phrased, "The present value of \$10,000 to be received 10 years from now, with an interest rate of 6.5% discounted annually, is \$5,327.26." ■

Time value of money calculations are often done to take into account inflation, which behaves like compound interest. Suppose, for example, that inflation is running at 5% per year. Then prices will increase by 5% each year, so if PV represents the price now, the price one year from now will be 5% higher, or $PV(1+0.05)$. The price a year from then will be 5% higher still, or $PV(1+0.05)^2$. Thus, the effects of inflation are compounded just as reinvested interest is.

EXAMPLE 3 Inflation

Inflation in East Avalon is 5% per year. TruVision television sets cost \$200 today. How much will a comparable set cost 2 years from now?

Solution To find the price of a television set 2 years from now, we compute the future value of \$200 at an inflation rate of 5% compounded yearly:

$$FV = 200(1+0.05)^2 = \$220.50.$$

EXAMPLE 4 Constant Dollars

Inflation in North Avalon is 6% per year. Which is really more expensive, a car costing \$20,000 today or one costing \$22,000 in 3 years?

Solution We cannot compare the two costs directly because inflation makes \$1 today worth more (it buys more) than a dollar 3 years from now. We need the two prices expressed in comparable terms, so we convert to **constant dollars**. We take the car costing \$22,000 three years from now and ask what it would cost in today's dollars. In other words, we convert the future value of \$22,000 to its present value:

$$\begin{aligned} PV &= FV(1+i)^{-n} \\ &= 22{,}000(1+0.06)^{-3} \\ &\approx \$18{,}471.62. \end{aligned}$$

Thus, the car costing \$22,000 in 3 years actually costs less than the one costing \$20,000 now.

Before we go on... In the presence of inflation, the only way to compare prices at different times is to convert all prices to constant dollars. We pick some fixed time and compute future or present values as appropriate to determine what things would have cost at that time. ■

There are some other interesting calculations related to compound interest, besides present and future values.

EXAMPLE 5 Effective Interest Rate

You have just won \$1 million in the lottery and are deciding what to do with it during the next year before you move to the South Pacific. Bank Ten offers 10% interest, compounded annually, while Bank Nine offers 9.8% compounded monthly. In which should you deposit your money?

Solution Let's calculate the future value of your \$1 million after one year in each of the banks:

$$\text{Bank Ten:} \quad FV = 1(1+0.10)^1 = \$1.1 \text{ million}$$

$$\text{Bank Nine:} \quad FV = 1\left(1+\frac{0.098}{12}\right)^{12} = \$1.1025 \text{ million.}$$

Bank Nine turns out to be better: It will pay you a total of \$102,500 in interest over the year, whereas Bank Ten will pay only \$100,000 in interest.

Another way of looking at the calculation in Example 5 is that Bank Nine gave you a total of 10.25% interest on your investment over the year. We call 10.25% the **effective interest rate** of the investment (also referred to as the **annual percentage yield**, or **APY** in the banking industry); the stated 9.8% is called the **nominal** interest rate. In general, to best compare two different investments, it is wisest to compare their *effective*—rather than nominal—interest rates.

Notice that we got 10.25% by computing

$$\left(1+\frac{0.098}{12}\right)^{12} = 1.1025$$

and then subtracting 1 to get 0.1025, or 10.25%. Generalizing, we get the following formula.

Effective Interest Rate

The effective interest rate r_{eff} of an investment paying a nominal interest rate of r_{nom} compounded m times per year is

$$r_{\text{eff}} = \left(1+\frac{r_{\text{nom}}}{m}\right)^{m} - 1.$$

To compare rates of investments with different compounding periods, always compare the effective interest rates rather than the nominal rates.

Quick Example

To calculate the effective interest rate of an investment that pays 8% per year, with interest reinvested monthly, set $r_{\text{nom}} = 0.08$ and $m = 12$, to obtain

$$r_{\text{eff}} = \left(1+\frac{0.08}{12}\right)^{12} - 1 \approx 0.0830, \quad \text{or} \quad 8.30\%.$$

EXAMPLE 6 How Long to Invest

You have \$5,000 to invest at 6% interest compounded monthly. How long will it take for your investment to grow to \$6,000?

Solution If we use the future value formula, we already have the values

$$\begin{aligned} FV &= 6{,}000 \\ PV &= 5{,}000 \\ r &= 0.06 \\ m &= 12. \end{aligned}$$

Substituting, we get

$$6{,}000 = 5{,}000\left(1+\frac{0.06}{12}\right)^{12t}.$$

If you are familiar with logarithms, you can solve explicitly for t as follows:

$$\left(1+\frac{0.06}{12}\right)^{12t} = \frac{6{,}000}{5{,}000} = 1.2$$

$$\log\left(1+\frac{0.06}{12}\right)^{12t} = \log 1.2$$

$$12t\log\left(1+\frac{0.06}{12}\right) = \log 1.2$$

$$t = \frac{\log 1.2}{12\log\left(1+\frac{0.06}{12}\right)} \approx 3.046 \approx 3 \text{ years.}$$

Another approach is to use a bit of trial and error to find the answer. Let's see what the future value is after 2, 3, and 4 years:

$$5{,}000\left(1+\frac{0.06}{12}\right)^{12\times 2} = 5{,}635.80 \qquad \text{Future value after 2 years}$$

$$5{,}000\left(1+\frac{0.06}{12}\right)^{12\times 3} = 5{,}983.40 \qquad \text{Future value after 3 years}$$

$$5{,}000\left(1+\frac{0.06}{12}\right)^{12\times 4} = 6{,}352.45. \qquad \text{Future value after 4 years}$$

From these calculations, it looks as if the answer should be a bit more than 3 years, but is certainly between 3 and 4 years. We could try 3.5 years next and then narrow it down systematically until we have a pretty good approximation of the correct answer.

Graphing calculators and spreadsheets give us alternative methods of solution.

using Technology

See the Technology Guides at the end of the chapter for details on using TVM Solver on the TI-83/84 Plus or the built-in finance functions in Excel to do the calculations in Example 6. [Details: TI-83/84 Plus: page 389, Excel: page 392]

Web Site

www.FiniteMath.org

→ Online Utilities

→ Time Value of Money Utility

This utility is similar to the TVM Solver on the TI-83/84 Plus. To compute the time needed, enter the values shown, and click "Compute" next to t.

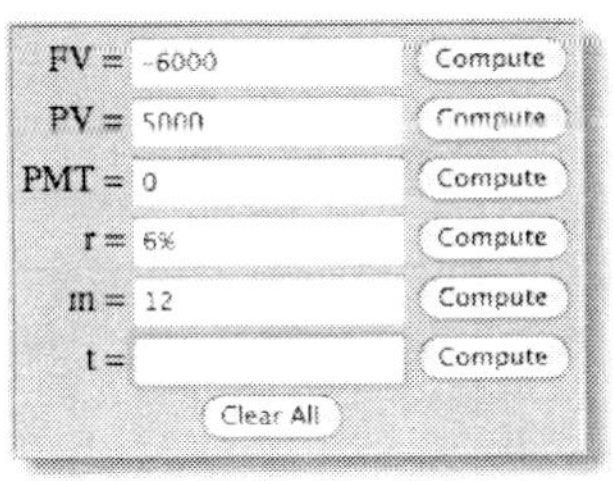

FAQs

Recognizing When to Use Compound Interest and the Meaning of Present Value

Q: *How do I distinguish a problem that calls for compound interest from one that calls for simple interest?*

A: Study the scenario to ascertain whether the interest is being withdrawn as it is earned or reinvested (deposited back into the account). If the interest is being withdrawn, the problem is calling for simple interest because the interest is not itself earning interest. If it is being reinvested, the problem is calling for compound interest.

Q: *How do I distinguish present value from future value in a problem?*

A: The present value always refers to the value of an investment before any interest is included (or, in the case of a depreciating investment, before any depreciation takes place). As an example, the future value of a bond is its maturity value. The value of $1 today in constant 1990 dollars is its present value (even though 1990 is in the past).

5.2 EXERCISES

▼ more advanced ◆ challenging

T indicates exercises that should be solved using technology

In Exercises 1–8, calculate, to the nearest cent, the future value of an investment of $10,000 at the stated interest rate after the stated amount of time. HINT [See Example 1.]

1. 3% per year, compounded annually, after 10 years

2. 4% per year, compounded annually, after 8 years

3. 2.5% per year, compounded quarterly (4 times/year), after 5 years

4. 1.5% per year, compounded weekly (52 times/year), after 5 years

5. 6.5% per year, compounded daily (assume 365 days/year), after 10 years

6. 11.2% per year, compounded monthly, after 12 years

7. 0.2% per month, compounded monthly, after 10 years

8. 0.45% per month, compounded monthly, after 20 years

In Exercises 9–14, calculate the present value of an investment that will be worth $1,000 at the stated interest rate after the stated amount of time. HINT [See Example 2.]

9. 10 years, at 5% per year, compounded annually

10. 5 years, at 6% per year, compounded annually

11. 5 years, at 4.2% per year, compounded weekly (assume 52 weeks per year)

12. 10 years, at 5.3% per year, compounded quarterly

13. 4 years, depreciating 5% each year

14. 5 years, depreciating 4% each year

In Exercises 15–20, find the effective annual interest rates of the given annual interest rates. Round your answers to the nearest 0.01%. HINT [See Quick Example on page 362.]

15. 5% compounded quarterly

16. 5% compounded monthly

17. 10% compounded monthly

18. 10% compounded daily (assume 365 days per year)

19. 10% compounded hourly (assume 365 days per year)

20. 10% compounded every minute (assume 365 days per year)

APPLICATIONS

21. ***Savings*** You deposit $1,000 in an account at the Lifelong Trust Savings and Loan that pays 6% interest compounded quarterly. By how much will your deposit have grown after 4 years?

22. ***Investments*** You invest $10,000 in Rapid Growth Funds, which appreciate by 2% per year, with yields reinvested quarterly. By how much will your investment have grown after 5 years?

23. ***Depreciation*** During 2008, the S&P 500 index depreciated by 37.6%.[9] Assuming that this trend were to continue, how much would a $3,000 investment in an S&P index fund be worth in 3 years?

24. ***Depreciation*** During 2008, Fidelity's Nasdaq Composite Fund depreciated by 42%.[10] Assuming that this trend were to continue, how much would a $5,000 investment in the fund be worth in 4 years?

25. ***Bonds*** You want to buy a 10-year zero coupon bond with a maturity value of $5,000 and a yield of 5.5% annually. How much will you pay?

26. ***Bonds*** You want to buy a 15-year zero coupon bond with a maturity value of $10,000 and a yield of 6.25% annually. How much will you pay?

27. ▼ ***Investments*** When I was considering what to do with my $10,000 Lottery winnings, my broker suggested I invest half of it in gold, the value of which was growing by 10% per year, and the other half in CDs, which were yielding 5% per year, compounded every 6 months. Assuming that these rates are sustained, how much will my investment be worth in 10 years?

28. ▼ ***Investments*** When I was considering what to do with the $10,000 proceeds from my sale of technology stock, my broker suggested I invest half of it in municipal bonds, whose value was growing by 6% per year, and the other half in CDs, which were yielding 3% per year, compounded every 2 months. Assuming that these interest rates are sustained, how much will my investment be worth in 10 years?

29. ▼ ***Depreciation*** During a prolonged recession, property values on Long Island depreciated by 2% every 6 months. If my house cost $200,000 originally, how much was it worth 5 years later?

30. ▼ ***Depreciation*** Stocks in the health industry depreciated by 5.1% in the first 9 months of 1993.[11] Assuming that this trend were to continue, how much would a $40,000 investment be worth in 9 years? HINT [Nine years corresponds to 12 nine-month periods.]

31. ▼ ***Retirement Planning*** I want to be earning an annual salary of $100,000 when I retire in 15 years. I have been offered a job that guarantees an annual salary increase of 4% per year, and the starting salary is negotiable. What salary should I request in order to meet my goal?

32. ▼ ***Retirement Planning*** I want to be earning an annual salary of $80,000 when I retire in 10 years. I have been offered a job that guarantees an annual salary increase of 5% per year, and

[9] Source: http://finance.google.com.

[10] Ibid.

[11] Source: *The New York Times*, October 9, 1993, p. 37.

the starting salary is negotiable. What salary should I request in order to meet my goal?

33. **Present Value** Determine the amount of money, to the nearest dollar, you must invest at 6% per year, compounded annually, so that you will be a millionaire in 30 years.

34. **Present Value** Determine the amount of money, to the nearest dollar, you must invest now at 7% per year, compounded annually, so that you will be a millionaire in 40 years.

35. **Stocks** Six years ago, I invested some money in Dracubunny Toy Co. stock, acting on the advice of a "friend." As things turned out, the value of the stock decreased by 5% every 4 months, and I discovered yesterday (to my horror) that my investment was worth only $297.91. How much did I originally invest?

36. **Sales** My recent marketing idea, the *Miracle Algae Growing Kit*, has been remarkably successful, with monthly sales growing by 6% every 6 months over the past 5 years. Assuming that I sold 100 kits the first month, how many kits did I sell in the first month of this year?

37. **Inflation** Inflation has been running 2% per year. A car now costs $30,000. How much would it have cost 5 years ago? HINT [See Example 3.]

38. **Inflation** (Compare Exercise 37.) Inflation has been running 1% every 6 months. A car now costs $30,000. How much would it have cost 5 years ago?

39. **Inflation** Housing prices have been rising 6% per year. A house now costs $200,000. What would it have cost 10 years ago?

40. **Inflation** (Compare Exercise 39.) Housing prices have been rising 0.5% each month. A house now costs $200,000. What would it have cost 10 years ago? HINT [See Example 4.]

41. **Constant Dollars** Inflation is running 3% per year when you deposit $1,000 in an account earning interest of 5% per year compounded annually. In *constant dollars*, how much money will you have two years from now? HINT [First calculate the value of your account in two years' time, and then find its present value based on the inflation rate.]

42. **Constant Dollars** Inflation is running 1% per month when you deposit $10,000 in an account earning 8% compounded monthly. In *constant dollars*, how much money will you have two years from now? HINT [See Exercise 41.]

43. **Investments** You are offered two investments. One promises to earn 12% compounded annually. The other will earn 11.9% compounded monthly. Which is the better investment? HINT [See Example 5.]

44. **Investments** You are offered three investments. The first promises to earn 15% compounded annually, the second will earn 14.5% compounded quarterly, and the third will earn 14% compounded monthly. Which is the best investment? HINT [See Example 5.]

45. **History** Legend has it that the Manhattan Indians sold Manhattan Island to the Dutch in 1626 for $24. In 2001, the total value of Manhattan real estate was estimated to be $136,106 million.[12] Suppose that the Manhattan Indians had taken that $24 and invested it at 6.2% compounded annually (a relatively conservative investment goal). Could the Manhattan Indians have bought back the island in 2001?

46. **History** Repeat Exercise 45, assuming that the Manhattan Indians had invested the $24 at 6.1% compounded annually.

Inflation *Exercises 47–54 are based on the following table, which shows the 2008 annual inflation rates in several Latin American countries.*[13] *Assume that the rates shown continue indefinitely.*

Country	Argentina	Brazil	Bolivia	Nicaragua	Venezuela	Mexico	Uruguay
Currency	Peso	Real	Boliviano	Gold cordoba	Bolivar	Peso	Peso
Inflation Rate (%)	9.2	6.3	15.1	13.8	25.7	5.0	8.5

47. If an item in Brazil now costs 100 reals, what do you expect it to cost 5 years from now? (Answer to the nearest real.)

48. If an item in Argentina now costs 1,000 pesos, what do you expect it to cost 5 years from now? (Answer to the nearest peso.)

49. If an item in Bolivia will cost 1,000 bolivianos in 10 years, what does it cost now? (Answer to the nearest boliviano.)

50. If an item in Mexico will cost 20,000 pesos in 10 years, what does it cost now? (Answer to the nearest peso.)

51. You wish to invest 1,000 bolivars in Venezuela at 8% annually, compounded twice a year. Find the value of your investment in 10 years, expressing the answer in constant bolivars. (Answer to the nearest bolivar.)

52. You wish to invest 1,000 pesos in Uruguay at 8% annually, compounded twice a year. Find the value of your investment in 10 years, expressing the answer in constant pesos. (Answer to the nearest peso.)

53. Which is the better investment: an investment in Mexico yielding 5.3% per year, compounded annually, or an investment in Nicaragua yielding 14% per year, compounded every 6 months? Support your answer with figures that show the future value of an investment of one unit of currency in constant units.

54. Which is the better investment: an investment in Argentina yielding 10% per year, compounded annually, or an investment in Uruguay, yielding 9% per year, compounded every 6 months? Support your answer with figures that show the future value of an investment of one unit of currency in constant units.

[12] Source: Queens County Overall Economic Development Corporation, 2001 (www.queensny.org/outlook/Outlook_Sp_7.pdf).

[13] Sources: www.bloomberg.com, www.csmonitor.com, www.indexmundi.com.

Stock Investments *Exercises 55–60 are based on the following chart, which shows monthly figures for* **Apple Inc.** *stock in 2008.*[14]

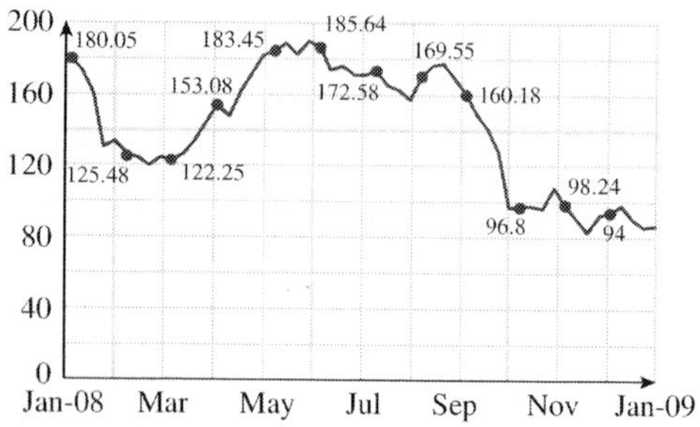

Marked are the following points on the chart:

Jan. 2008	Feb. 2008	Mar. 2008	Apr. 2008	May 2008	June 2008
180.05	125.48	122.25	153.08	183.45	185.64
July 2008	**Aug. 2008**	**Sep. 2008**	**Oct. 2008**	**Nov. 2008**	**Dec. 2008**
172.58	169.55	160.18	96.80	98.24	94.00

55. Calculate to the nearest 0.01% your annual percentage loss (assuming annual compounding) if you had bought Apple stock in June and sold in December.

56. Calculate to the nearest 0.01% your annual percentage return (assuming annual compounding) if you had bought Apple stock in February and sold in June.

57. Suppose you bought Apple stock in January 2008. If you later sold at one of the marked dates on the chart, which of those dates would have given you the largest annual return (assuming annual compounding), and what would that return have been?

58. Suppose you bought Apple stock in April. If you later sold at one of the marked dates on the chart, which of those dates would have given you the largest annual loss (assuming annual compounding), and what would that loss have been?

59. Did Apple's stock undergo compound interest change in the period January through April? (Give a reason for your answer.)

60. If Apple's stock had undergone compound interest change from April to June at the same monthly rate as from March to April, what would the price have been in June?

61. ***Competing Investments*** I just purchased $5,000 worth of municipal funds which are expected to yield 5.4% per year, compounded every 6 months. My friend has just purchased $6,000 worth of CDs which will earn 4.8% per year, compounded every 6 months. Determine when, to the nearest year, the value of my investment will be the same as hers, and what this value will be. HINT [You can either graph the values of both investments or make tables of the values of both investments.]

[14] Source: Yahoo! Finance (http://finance.yahoo.com).

62. ***Investments*** Determine when, to the nearest year, $3,000 invested at 5% per year, compounded daily, will be worth $10,000.

63. ***Epidemics*** At the start of 1985, the incidence of AIDS was doubling every 6 months and 40,000 cases had been reported in the United States. Assuming this trend would have continued, determine when, to the nearest tenth of a year, the number of cases would have reached 1 million.

64. ***Depreciation*** My investment in Genetic Splicing, Inc., is now worth $4,354 and is depreciating by 5% every 6 months. For some reason, I am reluctant to sell the stock and swallow my losses. Determine when, to the nearest year, my investment will drop below $50.

65. ◆ ***Bonds*** Once purchased, bonds can be sold in the secondary market. The value of a bond depends on the prevailing interest rates, which vary over time. Suppose that, in January, 1982, you bought a 30-year zero coupon U.S. Treasury bond with a maturity value of $100,000 and a yield of 15% annually.
a. How much did you pay for the bond?
b. In January 1999, your bond had 13 years remaining until maturity. Rates on U.S. Treasury bonds of comparable length were about 4.75%. If you sold your bond to an investor looking for a return of 4.75% annually, how much money would you have received?
c. Using your answers to parts (a) and (b), what was the annual yield on your 17-year investment?

66. ◆ ***Bonds*** Suppose that, in January 1999, you bought a 30-year zero coupon U.S. Treasury bond with a maturity value of $100,000 and a yield of 5% annually.
a. How much did you pay for the bond?
b. Suppose that, 15 years later, interest rates have risen again, to 12%. If you sell your bond to an investor looking for a return of 12%, how much money will you receive?
c. Using your answers to parts (a) and (b), what will be the annual yield on your 15 year investment?

COMMUNICATION AND REASONING EXERCISES

67. Why is the graph of the future value of a compound interest investment as a function of time not a straight line (assuming a nonzero rate of interest)?

68. An investment that earns 10% (compound interest) every year is the same as an investment that earns 5% (compound interest) every 6 months, right?

69. If a bacteria culture is currently 0.01g and increases in size by 10% each day, then its growth is linear, right?

70. At what point is the future value of a compound interest investment the same as the future value of a simple interest investment at the same annual rate of interest?

71. If two equal investments have the same effective interest rate and you graph the future value as a function of time for each of them, are the graphs necessarily the same? Explain your answer.

72. For what kind of compound interest investments is the effective rate the same as the nominal rate? Explain your answer.

73. For what kind of compound interest investments is the effective rate greater than the nominal rate? When is it smaller? Explain your answer.

74. If an investment appreciates by 10% per year for 5 years (compounded annually) and then depreciates by 10% per year (compounded annually) for 5 more years, will it have the same value as it had originally? Explain your answer.

75. You can choose between two investments that mature at different times in the future. If you knew the rate of inflation, how would you decide which is the better investment?

76. If you knew the various inflation rates for the years 1995 through 2001, how would you convert \$100 in 2002 dollars to 1995 dollars?

77. On the same set of axes, graph the future value of a \$100 investment earning 10% per year as a function of time over a 20-year period, compounded once a year, 10 times a year, 100 times a year, 1,000 times a year, and 10,000 times a year. What do you notice?

78. By graphing the future value of a \$100 investment that is depreciating by 1% each year, convince yourself that, eventually, the future value will be less than \$1.

5.3 Annuities, Loans, and Bonds

A typical defined-contribution pension fund works as follows*: Every month while you work, you and your employer deposit a certain amount of money in an account. This money earns (compound) interest from the time it is deposited. When you retire, the account continues to earn interest, but you may then start withdrawing money at a rate calculated to reduce the account to zero after some number of years. This account is an example of an **annuity**, an account earning interest into which you make periodic deposits or from which you make periodic withdrawals. In common usage, the term "annuity" is used for an account from which you make withdrawals. There are various terms used for accounts into which you make payments, based on their purpose. Examples include **savings account**, **pension fund**, or **sinking fund**. A sinking fund is generally used by businesses or governments to accumulate money to pay off an anticipated debt, but we'll use the term to refer to any account into which you make periodic payments.

* **NOTE** Defined-contribution pensions are increasingly common, replacing the defined-benefit pensions that were once the norm. In a defined-benefit pension the size of your pension is guaranteed; it is typically a percentage of your final working salary. In a defined-contribution plan the size of your pension depends on how well your investments do.

Sinking Funds

Suppose you make a payment of \$100 at the end of every month into an account earning 3.6% interest per year, compounded monthly. This means that your investment is earning $3.6\%/12 = 0.3\%$ per month. We write $i = 0.036/12 = 0.003$. What will be the value of the investment at the end of 2 years (24 months)?

Think of the deposits separately. Each earns interest from the time it is deposited, and the total accumulated after 2 years is the sum of these deposits and the interest they earn. In other words, the accumulated value is the sum of the future values of the deposits, taking into account how long each deposit sits in the account. Figure 1 shows a timeline with the deposits and the contribution of each to the final value. For example, the very last deposit (at the end of month 24) has no time to earn interest, so it contributes only \$100. The very first deposit, which earns interest for 23 months, by the future value formula for compound interest contributes $\$100(1 + 0.003)^{23}$ to the total. Adding together all of the future values gives us the total future value:

$$\begin{aligned} FV &= 100 + 100(1 + 0.003) + 100(1 + 0.003)^2 + \cdots + 100(1 + 0.003)^{23} \\ &= 100[1 + (1 + 0.003) + (1 + 0.003)^2 + \cdots + (1 + 0.003)^{23}] \end{aligned}$$

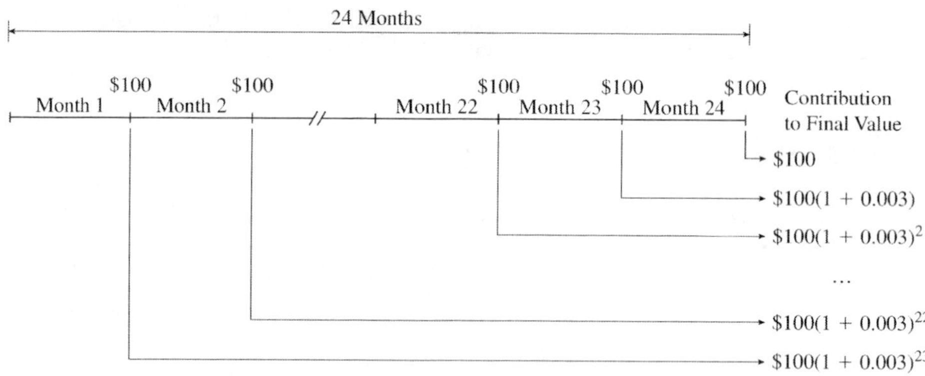

Figure 1

Fortunately, this sort of sum is well-known (to mathematicians, anyway*) and there is a convenient formula for its value†:

$$1 + x + x^2 + \cdots + x^{n-1} = \frac{x^n - 1}{x - 1}.$$

In our case, with $x = 1 + 0.003$, this formula allows us to calculate the future value:

$$FV = 100\frac{(1 + 0.003)^{24} - 1}{(1 + 0.003) - 1} = 100\frac{(1.003)^{24} - 1}{0.003} \approx \$2{,}484.65.$$

It is now easy to generalize this calculation.

* **NOTE** It is called a **geometric series**.

† **NOTE** The quickest way to convince yourself that this formula is correct is to multiply out $(x - 1)(1 + x + x^2 + \cdots + x^{n-1})$ and see that you get $x^n - 1$. You should also try substituting some numbers. For example, $1 + 3 + 3^2 = 13 = (3^3 - 1)/(3 - 1)$.

Future Value of a Sinking Fund

A **sinking fund** is an account earning compound interest into which you make periodic deposits. Suppose that the account has an annual rate of r compounded m times per year, so that $i = r/m$ is the interest rate per compounding period. If you make a payment of PMT at the end of each period, then the future value after t years, or $n = mt$ periods, will be

$$FV = PMT\frac{(1 + i)^n - 1}{i}.$$

Quick Example

At the end of each month you deposit $50 into an account earning 2% annual interest compounded monthly. To find the future value after 5 years, we use $i = 0.02/12$ and $n = 12 \times 5 = 60$ compounding periods, so

$$FV = 50\frac{(1 + 0.02/12)^{60} - 1}{0.02/12} = \$3{,}152.37.$$

using Technology

See the Technology Guides at the end of the chapter for a discussion of using a TI-83/84 Plus or Excel to compute the future value in Example 1.

EXAMPLE 1 Retirement Account

Your retirement account has $5,000 in it and earns 5% interest per year compounded monthly. Every month for the next 10 years you will deposit $100 into the account. How much money will there be in the account at the end of those 10 years?

TI-83/84 Plus

Press [FINANCE], choose `1:TVM Solver`.

Enter N = 120, I% = 5, PV = −5000, PMT = −100, P/Y = 12, C/Y = 12

Put cursor on FV line.

[ALPHA] [SOLVE]

[More details on page 389.]

Excel

Use the formula

`=FV(5/12,10*12,-100,-5000)`

[More details on page 392.]

Web Site

www.FiniteMath.org

Student Home

→ Online Utilities

→ Time Value of Money Utility

Enter the values shown and click "Compute" next to FV.

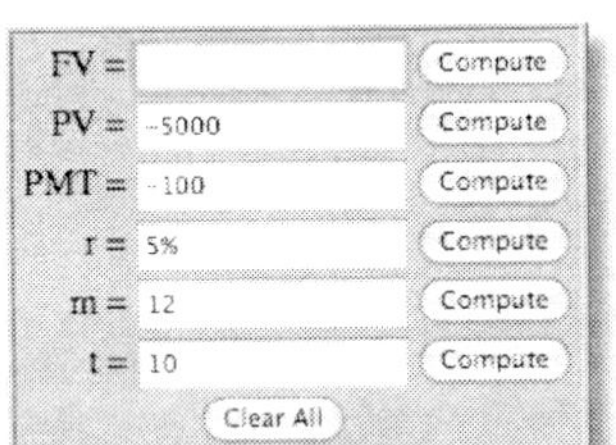

Solution This is a sinking fund with $PMT = \$100$, $r = 0.05$, $m = 12$, so $i = 0.05/12$, and $n = 12 \times 10 = 120$. Ignoring for the moment the \$5,000 already in the account, your payments have the following future value:

$$\begin{aligned} FV &= PMT\frac{(1+i)^n - 1}{i} \\ &= 100\frac{\left(1+\frac{0.05}{12}\right)^{120} - 1}{\frac{0.05}{12}} \\ &\approx \$15{,}528.23. \end{aligned}$$

What about the \$5,000 that was already in the account? That sits there and earns interest, so we need to find its future value as well, using the compound interest formula:

$$\begin{aligned} FV &= PV(1+i)^n \\ &= 5{,}000(1 + 0.05/12)^{120} \\ &= \$8{,}235.05. \end{aligned}$$

Hence, the total amount in the account at the end of 10 years will be

$$\$15{,}528.23 + 8{,}235.05 = \$23{,}763.28.$$

Sometimes we know what we want the future value to be and need to determine the payments necessary to achieve that goal. We can simply solve the future value formula for the payment.

Payment Formula for a Sinking Fund

Suppose that an account has an annual rate of r compounded m times per year, so that $i = r/m$ is the interest rate per compounding period. If you want to accumulate a total of FV in the account after t years, or $n = mt$ periods, by making payments of PMT at the end of each period, then each payment must be

$$PMT = FV\frac{i}{(1+i)^n - 1}.$$

using Technology

See the Technology Guides at the end of the chapter for a discussion of using a TI 83/84 Plus or Excel to compute the payments in Example 2.

TI-83/84 Plus

Press [FINANCE], choose `1:TVM Solver`.

Enter N = 68, I% = 4, PV = 0, FV = 100000, P/Y = 4, C/Y = 4

Put cursor on PMT line.

[ALPHA] [SOLVE]

[More details on page 389.]

EXAMPLE 2 Education Fund

Tony and Maria have just had a son, José Phillipe. They establish an account to accumulate money for his college education, in which they would like to have \$100,000 after 17 years. If the account pays 4% interest per year compounded quarterly, and they make deposits at the end of every quarter, how large must each deposit be for them to reach their goal?

Solution This is a sinking fund with $FV = \$100{,}000$, $m = 4$, $n = 4 \times 17 = 68$, and $r = 0.04$, so $i = 0.04/4 = 0.01$. From the payment formula, we get

$$PMT = 100{,}000\frac{0.01}{(1 + 0.01)^{68} - 1} \approx \$1{,}033.89.$$

So, Tony and Maria must deposit \$1,033.89 every quarter in order to meet their goal.

Excel

Use the formula

`=PMT(4/4,17*4,0,100000)`

[More details on page 392.]

Web Site

www.FiniteMath.org

Student Home

→ Online Utilities

→ Time Value of Money Utility

Enter the values shown and click "Compute" next to PMT.

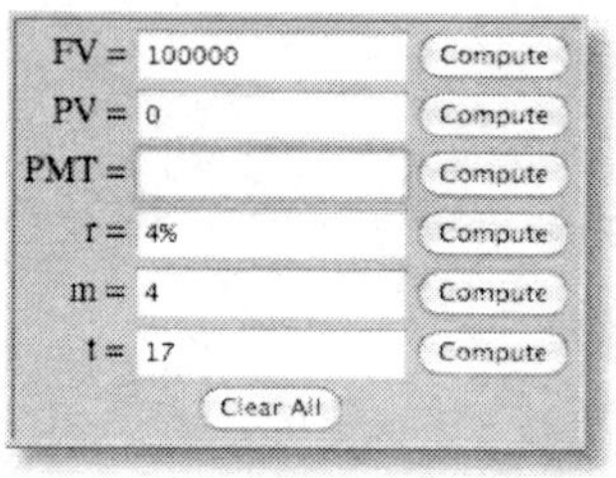

Annuities

Suppose we deposit an amount PV now in an account earning 3.6% interest per year, compounded monthly. Starting 1 month from now, the bank will send us monthly payments of \$100. What must PV be so that the account will be drawn down to \$0 in exactly 2 years?

As before, we write $i = r/m = 0.036/12 = 0.003$, and we have $PMT = 100$. The first payment of \$100 will be made 1 month from now, so its present value is

$$\frac{PMT}{(1+i)^n} = \frac{100}{1+0.003} = 100(1+0.003)^{-1} \approx \$99.70.$$

In other words, that much of the original PV goes toward funding the first payment. The second payment, 2 months from now, has a present value of

$$\frac{PMT}{(1+i)^n} = \frac{100}{(1+0.003)^2} = 100(1+0.003)^{-2} \approx \$99.40.$$

That much of the original PV funds the second payment. This continues for 2 years, at which point we receive the last payment, which has a present value of

$$\frac{PMT}{(1+i)^n} = \frac{100}{(1+0.003)^{24}} = 100(1+0.003)^{-24} \approx \$93.06$$

and that exhausts the account. Figure 2 shows a timeline with the payments and the present value of each.

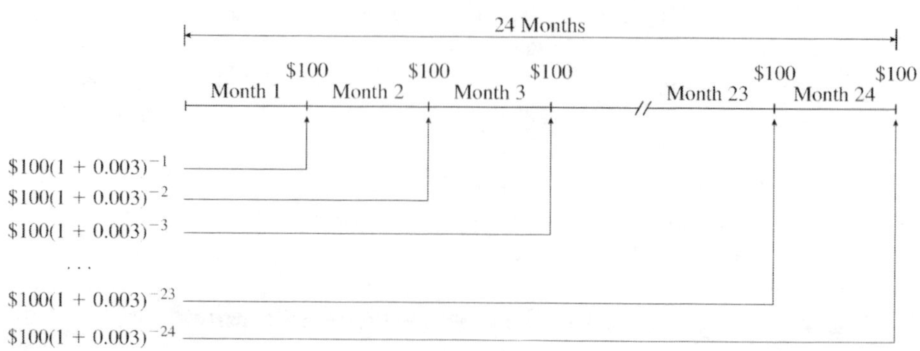

Figure 2

Because PV must be the sum of these present values, we get

$$\begin{aligned} PV &= 100(1+0.003)^{-1} + 100(1+0.003)^{-2} + \cdots + 100(1+0.003)^{-24} \\ &= 100[(1+0.003)^{-1} + (1+0.003)^{-2} + \cdots + (1+0.003)^{-24}]. \end{aligned}$$

We can again find a simpler formula for this sum:

$$\begin{aligned} x^{-1} + x^{-2} + \cdots + x^{-n} &= \frac{1}{x^n}(x^{n-1} + x^{n-2} + \cdots + 1) \\ &= \frac{1}{x^n} \cdot \frac{x^n - 1}{x - 1} = \frac{1 - x^{-n}}{x - 1}. \end{aligned}$$

So, in our case,

$$PV = 100\frac{1-(1+0.003)^{-24}}{(1+0.003)-1}$$

or

$$PV = 100\frac{1-(1.003)^{-24}}{0.003} \approx \$2,312.29.$$

If we deposit \$2,312.29 initially and the bank sends us \$100 per month for 2 years, our account will be exhausted at the end of that time.

Generalizing, we get the following formula:

Present Value of an Annuity

An **annuity** is an account earning compound interest from which periodic withdrawals are made. Suppose that the account has an annual rate of r compounded m times per year, so that $i = r/m$ is the interest rate per compounding period. Suppose also that the account starts with a balance of PV. If you receive a payment of PMT at the end of each compounding period, and the account is down to \$0 after t years, or $n = mt$ periods, then

$$PV = PMT\frac{1-(1+i)^{-n}}{i}.$$

Quick Example

At the end of each month you want to withdraw \$50 from an account earning 2% annual interest compounded monthly. If you want the account to last for 5 years (60 compounding periods), it must have the following amount to begin with:

$$PV = 50\frac{1-(1+0.02/12)^{-60}}{0.02/12} = \$2,852.62.$$

Note If you make your withdrawals at the end of each compounding period, as we've discussed so far, you have an **ordinary annuity**. If, instead, you make withdrawals at the beginning of each compounding period, you have an **annuity due**. Because each payment occurs one period earlier, there is one less period in which to earn interest, hence the present value must be larger by a factor of $(1+i)$ to fund each payment. So, the present value formula for an annuity due is

$$PV = PMT(1+i)\frac{1-(1+i)^{-n}}{i}.$$

In this book, we will concentrate on ordinary annuities. ■

using Technology

See the Technology Guides at the end of the chapter for a discussion of using a TI-83/84 Plus or Excel to compute the present value of the annuity in Example 3.

EXAMPLE 3 Trust Fund

You wish to establish a trust fund from which your niece can withdraw \$2,000 every six months for 15 years, at the end of which time she will receive the remaining money in the trust, which you would like to be \$10,000. The trust will be invested at 7% per year compounded every six months. How large should the trust be?

TI-83/84 Plus

Press FINANCE, choose `1:TVM Solver.`

Enter N = 30, I% = 7, PMT = 2000, FV = 10000, P/Y = 2, C/Y = 2

Put cursor on PV line.

ALPHA SOLVE

[More details on page 389.]

Excel

Use the formula

`=PV(7/2,15*2,2000, 10000)`

[More details on page 393.]

Web Site

www.FiniteMath.org

Student Home

→ Online Utilities

→ Time Value of Money Utility

Enter the values shown and click "Compute" next to PV.

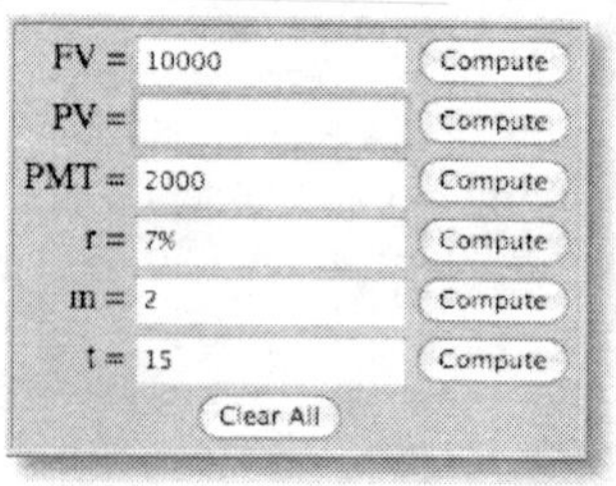

Solution We view this account as having two parts, one funding the semiannual payments and the other funding the \$10,000 lump sum at the end. The amount of money necessary to fund the semiannual payments is the present value of an annuity, with $PMT = 2{,}000$, $r = 0.07$ and $m = 2$, so $i = 0.07/2 = 0.035$, and $n = 2 \times 15 = 30$. Substituting gives

$$PV = 2{,}000\frac{1-(1+0.035)^{-30}}{0.035}$$
$$= \$36{,}784.09.$$

To fund the lump sum of \$10,000 after 15 years, we need the present value of \$10,000 under compound interest:

$$PV = 10{,}000(1+0.035)^{-30}$$
$$= \$3{,}562.78.$$

Thus the trust should start with \$36,784.09 + 3,562.78 = \$40,346.87.

Sometimes we know how much money we begin with and for how long we want to make withdrawals. We then want to determine the amount of money we can withdraw each period. For this, we simply solve the present value formula for the payment.

Payment Formula for an Ordinary Annuity

Suppose that an account has an annual rate of r compounded m times per year, so that $i = r/m$ is the interest rate per compounding period. Suppose also that the account starts with a balance of PV. If you want to receive a payment of PMT at the end of each compounding period, and the account is down to \$0 after t years, or $n = mt$ periods, then

$$PMT = PV\frac{i}{1-(1+i)^{-n}}.$$

EXAMPLE 4 Education Fund

Tony and Maria (see Example 2), having accumulated \$100,000 for José Phillipe's college education, would now like to make quarterly withdrawals over the next 4 years. How much money can they withdraw each quarter in order to draw down the account to zero at the end of the 4 years? (Recall that the account pays 4% interest compounded quarterly.)

Solution Now Tony and Maria's account is acting as an annuity with a present value of \$100,000. So, $PV = \$100{,}000$, $r = 0.04$ and $m = 4$, giving $i = 0.04/4 = 0.01$, and $n = 4 \times 4 = 16$. We use the payment formula to get

$$PMT = 100{,}000\frac{0.01}{1-(1+0.01)^{-16}} \approx \$6{,}794.46.$$

So, they can withdraw \$6,794.46 each quarter for 4 years, at the end of which time their account balance will be 0.

Paul Seheult/Eye Ubiquitous/CORBIS

using Technology

See the Technology Guides at the end of the chapter for a discussion of using a TI-83/84 Plus or Excel to compute the payment for the annuity in Example 4.

TI-83/84 Plus

Press [FINANCE], choose `1:TVM Solver.`

Enter N = 16, I% = 4,
PV = −100000, FV = 0,
P/Y = 4, C/Y = 4

Put cursor on PMT line.

[ALPHA] [SOLVE]

[More details on page 390.]

Excel

Use the formula

`=PMT(4/4,4*4,-100000,0)`

[More details on page 393.]

Web Site

www.FiniteMath.org

Student Home
→ Online Utilities
→ Time Value of Money Utility

Enter the values shown and click "Compute" next to PMT.

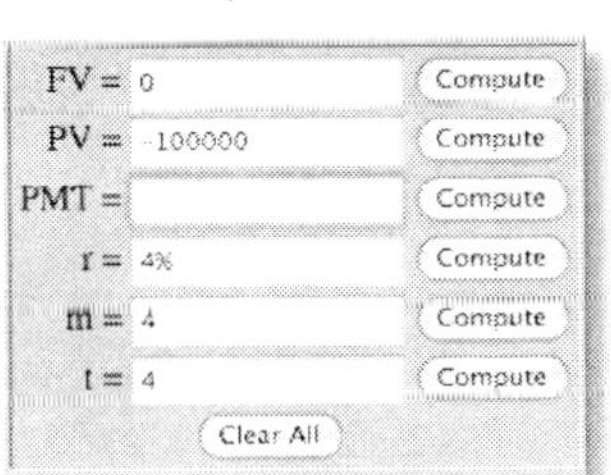

EXAMPLE 5 Saving for Retirement

Jane Q. Employee has just started her new job with Big Conglomerate, Inc., and is already looking forward to retirement. BCI offers her as a pension plan an annuity that is guaranteed to earn 6% annual interest compounded monthly. She plans to work for 40 years before retiring and would then like to be able to draw an income of \$7,000 per month for 20 years. How much do she and BCI together have to deposit per month into the fund to accomplish this?

Solution Here we have the situation we described at the beginning of the section: a sinking fund accumulating money to be used later as an annuity. We know the desired payment out of the annuity, so we work backward. The first thing we need to do is calculate the present value of the annuity required to make the pension payments. We use the annuity present value formula with $PMT = 7{,}000$, $i = r/m = 0.06/12 = 0.005$ and $n = 12 \times 20 = 240$.

$$\begin{aligned} PV &= PMT\frac{1-(1+i)^{-n}}{i} \\ &= 7{,}000\frac{1-(1+0.005)^{-240}}{0.005} \approx \$977{,}065.40 \end{aligned}$$

This is the total that must be accumulated in the sinking fund during the 40 years she plans to work. In other words, this is the *future* value, FV, of the sinking fund. (Thus, the present value in the first step of our calculation is the future value in the second step.) To determine the payments necessary to accumulate this amount, we use the sinking fund payment formula with $FV = 977{,}065.40$, $i = 0.005$, and $n = 12 \times 40 = 480$.

$$\begin{aligned} PMT &= FV\frac{i}{(1+i)^n - 1} \\ &= 977{,}065.40\frac{0.005}{1.005^{480} - 1} \\ &\approx \$490.62 \end{aligned}$$

So, if she and BCI collectively deposit \$490.62 per month into her retirement fund, she can retire with the income she desires.

Installment Loans

In a typical installment loan, such as a car loan or a home mortgage, we borrow an amount of money and then pay it back with interest by making fixed payments (usually every month) over some number of years. From the point of view of the lender, this is an annuity. Thus, loan calculations are identical to annuity calculations.

EXAMPLE 6 Home Mortgages

Marc and Mira are buying a house, and have taken out a 30-year, \$90,000 mortgage at 8% interest per year. What will their monthly payments be?

Solution From the bank's point of view, a mortgage is an annuity. In this case, the present value is $PV = \$90{,}000$, $r = 0.08$, $m = 12$, and $n = 12 \times 30 = 360$. To find the payments, we use the payment formula:

$$PMT = 90{,}000\frac{0.08/12}{1-(1+0.08/12)^{-360}} \approx \$660.39.$$

The word "mortgage" comes from the French for "dead pledge." The process of paying off a loan is called **amortizing** the loan, meaning to kill the debt owed.

EXAMPLE 7 Amortization Schedule

Continuing Example 6: Mortgage interest is tax-deductible, so it is important to know how much of a year's mortgage payments represents interest. How much interest will Marc and Mira pay in the first year of their mortgage?

Solution Let us calculate how much of each month's payment is interest and how much goes to reducing the outstanding principal. At the end of the first month Marc and Mira must pay one month's interest on $90,000, which is

$$\$90{,}000 \times \frac{0.08}{12} = \$600.$$

The remainder of their first monthly payment, $660.39 – 600 = $60.39 goes to reducing the principal. Thus, in the second month the outstanding principal is $90,000 – 60.39 = $89,939.61, and part of their second monthly payment will be for the interest on this amount, which is

$$\$89{,}939.61 \times \frac{0.08}{12} = \$599.60.$$

The remaining $660.39 – $599.60 = $60.79 goes to further reduce the principal. If we continue this calculation for the 12 months of the first year, we get the beginning of the mortgage's **amortization schedule**.

using Technology

See the Technology Guides at the end of the chapter for a discussion of using a TI-83/84 Plus or Excel to construct the amortization schedule in Example 7. [Details: TI-83/84 Plus: page 390, Excel: page 393.]

Month	*Interest Payment*	*Payment on Principal*	*Outstanding Principal*
0			$90,000.00
1	$600.00	$60.39	89,939.61
2	599.60	60.79	89,878.82
3	599.19	61.20	89,817.62
4	598.78	61.61	89,756.01
5	598.37	62.02	89,693.99
6	597.96	62.43	89,631.56
7	597.54	62.85	89,568.71
8	597.12	63.27	89,505.44
9	596.70	63.69	89,441.75
10	596.28	64.11	89,377.64
11	595.85	64.54	89,313.10
12	595.42	64.97	89,248.13
Total	$7,172.81	$751.87	

As we can see from the totals at the bottom of the columns, Marc and Mira will pay a total of $7,172.81 in interest in the first year.

Bonds

Suppose that a corporation offers a 10-year bond paying 6.5% with payments every 6 months. As we saw in Example 3 of Section 5.1, this means that if we pay \$10,000 for bonds with a maturity value of \$10,000, we will receive $6.5/2 = 3.25\%$ of \$10,000, or \$325, every 6 months for 10 years, at the end of which time the corporation will give us the original \$10,000 back. But bonds are rarely sold at their maturity value. Rather, they are auctioned off and sold at a price the bond market determines they are worth.

For example, suppose that bond traders are looking for an investment that has a **rate of return** or **yield** of 7% rather than the stated 6.5% (sometimes called the **coupon interest rate** to distinguish it from the rate of return). How much would they be willing to pay for the bonds above with a maturity value of \$10,000? Think of the bonds as an investment that will pay the owner \$325 every 6 months for 10 years, at the end of which time it will pay \$10,000. We can treat the \$325 payments as if they come from an annuity and determine how much an investor would pay for such an annuity if it earned 7% compounded semiannually. Separately, we determine the present value of an investment worth \$10,000 ten years from now, if it earned 7% compounded semiannually. For the first calculation, we use the annuity present value formula, with $i = 0.07/2$ and $n = 2 \times 10 = 20$.

$$\begin{aligned} PV &= PMT\frac{1-(1+i)^{-n}}{i} \\ &= 325\frac{1-(1+0.07/2)^{-20}}{0.07/2} \\ &= \$4{,}619.03 \end{aligned}$$

For the second calculation, we use the present value formula for compound interest:

$$\begin{aligned} PV &= 10{,}000(1+0.07/2)^{-20} \\ &= \$5{,}025.66. \end{aligned}$$

Thus, an investor looking for a 7% return will be willing to pay \$4,619.03 for the semiannual payments of \$325 and \$5,025.66 for the \$10,000 payment at the end of 10 years, for a total of $\$4{,}619.03 + 5{,}025.66 = \$9{,}644.69$ for the \$10,000 bond.

EXAMPLE 8 Bonds

Suppose that bond traders are looking for only a 6% yield on their investment. How much would they pay per \$10,000 for the 10-year bonds above, which have a coupon interest rate of 6.5% and pay interest every six months?

Solution We redo the calculation with $r = 0.06$. For the annuity calculation we now get

$$PV = 325\frac{1-(1+0.06/2)^{-20}}{0.06/2} = \$4{,}835.18.$$

For the compound interest calculation we get

$$PV = 10{,}000(1+0.06/2)^{-20} = \$5{,}536.76.$$

Thus, traders would be willing to pay a total of $\$4{,}835.18 + \$5{,}536.76 = \$10{,}371.94$ for bonds with a maturity value of \$10,000.

➡ Before we go on... Notice how the selling price of the bonds behaves as the desired yield changes. As desired yield goes up, the price of the bonds goes down, and as desired yield goes down, the price of the bonds goes up. When the desired yield equals the coupon interest rate, the selling price will equal the maturity value. Therefore, when the yield is higher than the coupon interest rate, the price of the bond will be below its maturity value, and when the yield is lower than the coupon interest rate, the price will be above the maturity value.

As we've mentioned before, the desired yield depends on many factors, but it generally moves up and down with prevailing interest rates. And interest rates have historically gone up and down cyclically. The effect on the value of bonds can be quite dramatic (see Exercises 65 and 66 in the preceding section). Because bonds can be sold again once bought, someone who buys bonds while interest rates are high and then resells them when interest rates decline can make a healthy profit. ■

using Technology

See the Technology Guides at the end of the chapter for a discussion of using a TI-83/84 Plus or Excel to compute the future value in Example 9.

TI-83/84 Plus

Press FINANCE, choose `1:TVM Solver.`
Enter N = 40, PV = −9800, PMT = 250, FV = 10000, P/Y = 2, C/Y = 2
Put cursor on I% line.
ALPHA SOLVE
[More details on page 391.]

Excel

Use the formula
`=RATE(20*2,250,-9800,10000)*2`
[More details on page 394.]

Web Site

www.FiniteMath.org
Student Home
→ Online Utilities
→ Time Value of Money Utility

Enter the values shown and click "Compute" next to r.

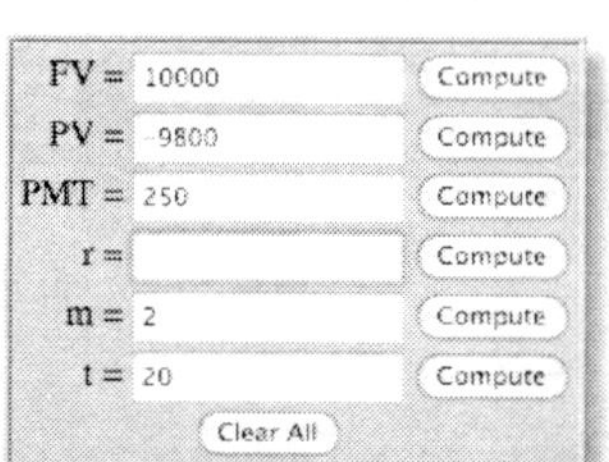

EXAMPLE 9 Rate of Return on a Bond

Suppose that a 5%, 20-year bond sells for \$9,800 per \$10,000 maturity value. What rate of return will investors get?

Solution Assuming the usual semiannual payments, we know the following about the annuity calculation:

$$PMT = 0.05 \times 10{,}000/2 = 250$$
$$n = 20 \times 2 = 40.$$

What we do not know is r or i, the annual or semiannual interest rate. So, we write

$$PV = 250\frac{1-(1+i)^{-40}}{i}.$$

For the compound interest calculation, we know $FV = 10{,}000$ and $n = 40$ again, so we write

$$PV = 10{,}000(1+i)^{-40}.$$

Adding these together should give the selling price of \$9,800:

$$250\frac{1-(1+i)^{-40}}{i} + 10{,}000(1+i)^{-40} = 9{,}800.$$

This equation cannot be solved for i directly. The best we can do by hand is a sort of trial and error approach, substituting a few values for i in the left-hand side of the above equation to get an estimate:

i	0.01	0.02	0.03
$250\frac{1-(1+i)^{-40}}{i} + 10{,}000(1+i)^{-40}$	14,925	11,368	8,844

Because we want the value to be 9,800, we see that the correct answer is somewhere between $i = 0.02$ and $i = 0.03$. Let us try the value midway between 0.02 and 0.03; namely $i = 0.025$.

i	0.02	0.025	0.03
$250\frac{1-(1+i)^{-40}}{i} + 10{,}000(1+i)^{-40}$	11,368	10,000	8,844

Now we know that the correct value of i is somewhere between 0.025 and 0.03, so we choose for our next estimate of i the number midway between them: 0.0275. We could continue in this fashion to obtain i as accurately as we like. In fact, $i \approx 0.02581$, corresponding to an annual rate of return of approximately 5.162%.

FAQ

Which Formula to Use

Q: *We have retirement accounts, trust funds, loans, bonds, and so on. Some are sinking funds, others are annuities. How do we distinguish among them, so we can tell which formula to use?*

A: In general, remember that a sinking fund is an interest-bearing fund into which payments are made, while an annuity is an interest-bearing fund from which money is withdrawn. Here is a list of some of the accounts we have discussed in this section:

- *Retirement Accounts* A retirement account is a sinking fund while payments are being made into the account (prior to retirement) and an annuity while a pension is being withdrawn (after retirement).
- *Education Funds* These are similar to retirement accounts.
- *Trust Funds* A trust fund is an annuity if periodic withdrawals are made.
- *Installment Loans* We think of an installment loan as an investment a bank makes in the lender. In this way, the lender's payments can be viewed as the bank's withdrawals, and so a loan is an annuity.
- *Bonds* A bond pays regular fixed amounts until it matures, at which time it pays its maturity value. We think of the bond as an annuity coupled with a compound interest investment funding the payment of the maturity value. We can then determine its present value based on the current market interest rate.

From a mathematical point of view, sinking funds and annuities are really the same thing. See the Communication and Reasoning Exercises for more about this.

5.3 EXERCISES

▼ more advanced ◆ challenging

T indicates exercises that should be solved using technology

Find the amount accumulated in the sinking funds in Exercises 1–6. (Assume end-of-period deposits and compounding at the same intervals as deposits.) HINT [See Example 1.]

1. \$100 deposited monthly for 10 years at 5% per year
2. \$150 deposited monthly for 20 years at 3% per year
3. \$1,000 deposited quarterly for 20 years at 7% per year
4. \$2,000 deposited quarterly for 10 years at 7% per year
5. ▼ \$100 deposited monthly for 10 years at 5% per year in an account containing \$5,000 at the start
6. ▼ \$150 deposited monthly for 20 years at 3% per year in an account containing \$10,000 at the start

Find the periodic payments necessary to accumulate the amounts given in Exercises 7–12 in a sinking fund. (Assume end-of-period deposits and compounding at the same intervals as deposits.) HINT [See Example 2.]

7. \$10,000 in a fund paying 5% per year, with monthly payments for 5 years
8. \$20,000 in a fund paying 3% per year, with monthly payments for 10 years
9. \$75,000 in a fund paying 6% per year, with quarterly payments for 20 years
10. \$100,000 in a fund paying 7% per year, with quarterly payments for 20 years
11. ▼ \$20,000 in a fund paying 5% per year, with monthly payments for 5 years, if the fund contains \$10,000 at the start

12. $30,000 in a fund paying 3% per year, with monthly payments for 10 years, if the fund contains $10,000 at the start

Find the present value of the annuity necessary to fund the withdrawals given in Exercises 13–18. (Assume end-of-period withdrawals and compounding at the same intervals as withdrawals.) HINT [See Example 3.]

13. $500 per month for 20 years, if the annuity earns 3% per year
14. $1,000 per month for 15 years, if the annuity earns 5% per year
15. $1,500 per quarter for 20 years, if the annuity earns 6% per year
16. $2,000 per quarter for 20 years, if the annuity earns 4% per year
17. $500 per month for 20 years, if the annuity earns 3% per year and if there is to be $10,000 left in the annuity at the end of the 20 years
18. $1,000 per month for 15 years, if the annuity earns 5% per year and if there is to be $20,000 left in the annuity at the end of the 15 years

Find the periodic withdrawals for the annuities given in Exercises 19–24. (Assume end-of-period withdrawals and compounding at the same intervals as withdrawals.) HINT [See Example 4.]

19. $100,000 at 3%, paid out monthly for 20 years
20. $150,000 at 5%, paid out monthly for 15 years
21. $75,000 at 4%, paid out quarterly for 20 years
22. $200,000 at 6%, paid out quarterly for 15 years
23. $100,000 at 3%, paid out monthly for 20 years, leaving $10,000 in the account at the end of the 20 years
24. $150,000 at 5%, paid out monthly for 15 years, leaving $20,000 in the account at the end of the 15 years

Determine the periodic payments on the loans given in Exercises 25–28. HINT [See Example 6.]

25. $10,000 borrowed at 9% for 4 years, with monthly payments
26. $20,000 borrowed at 8% for 5 years, with monthly payments
27. $100,000 borrowed at 5% for 20 years, with quarterly payments
28. $1,000,000 borrowed at 4% for 10 years, with quarterly payments

Determine the selling price, per $1,000 maturity value, of the bonds[15] in Exercises 29–32. (Assume twice-yearly interest payments.) HINT [See Example 8.]

29. 10 year, 4.875% bond, with a yield of 4.880%
30. 30 year, 5.375% bond, with a yield of 5.460%
31. 2 year, 3.625% bond, with a yield of 3.705%
32. 5 year, 4.375% bond, with a yield of 4.475%

Determine the yield on the bonds[16] in Exercises 33–36. (Assume twice-yearly interest payments.) HINT [See Example 9.]

33. 5 year, 3.5% bond, selling for $994.69 per $1,000 maturity value
34. 10 year, 3.375% bond, selling for $991.20 per $1,000 maturity value
35. 2 year, 3% bond, selling for $998.86 per $1,000 maturity value
36. 5 year, 4.625% bond, selling for $998.45 per $1,000 maturity value

APPLICATIONS

37. ***Pensions*** Your pension plan is an annuity with a guaranteed return of 3% per year (compounded monthly). You would like to retire with a pension of $5,000 per month for 20 years. If you work 40 years before retiring, how much must you and your employer deposit each month into the fund? HINT [See Example 5.]
38. ***Pensions*** Meg's pension plan is an annuity with a guaranteed return of 5% per year (compounded quarterly). She would like to retire with a pension of $12,000 per quarter for 25 years. If she works 45 years before retiring, how much money must she and her employer deposit each quarter? HINT [See Example 5.]
39. ***Pensions*** Your pension plan is an annuity with a guaranteed return of 4% per year (compounded quarterly). You can afford to put $1,200 per quarter into the fund, and you will work for 40 years before retiring. After you retire, you will be paid a quarterly pension based on a 25-year payout. How much will you receive each quarter?
40. ***Pensions*** Jennifer's pension plan is an annuity with a guaranteed return of 5% per year (compounded monthly). She can afford to put $300 per month into the fund, and she will work for 45 years before retiring. If her pension is then paid out monthly based on a 20-year payout, how much will she receive per month?
41. ***Car Loans*** While shopping for a car loan, you get the following offers: Solid Savings & Loan is willing to loan you $10,000 at 9% interest for 4 years. Fifth Federal Bank & Trust will loan you the $10,000 at 7% interest for 3 years. Both require monthly payments. You can afford to pay $250 per month. Which loan, if either, can you take?
42. ***Business Loans*** You need to take out a loan of $20,000 to start up your T-shirt business. You have two possibilities: One bank is offering a 10% loan for 5 years, and another is offering a 9% loan for 4 years. Which will have the lower monthly payments? On which will you end up paying more interest total?

[15] These are actual U.S. Treasury notes and bonds auctioned in 2001 and 2002. Source: The Bureau of the Public Debt's Web site (www.publicdebt.treas.gov/).

[16] Ibid.

Note: In Exercises 43–48 we suggest the use of the finance functions in the TI-83/84 Plus or amortization tables in Excel.

43. ***Mortgages*** You take out a 15-year mortgage for \$50,000, at 8%, to be paid off monthly. Construct an amortization table showing how much you will pay in interest each year, and how much goes toward paying off the principal. HINT [See Example 7.]

44. ***Mortgages*** You take out a 30-year mortgage for \$95,000 at 9.75%, to be paid off monthly. If you sell your house after 15 years, how much will you still owe on the mortgage? HINT [See Example 7.]

45. ***Mortgages*** This exercise describes a popular kind of mortgage. You take out a \$75,000, 30-year mortgage. For the first 5 years the interest rate is held at 5%, but for the remaining 25 years it rises to 9.5%. The payments for the first 5 years are calculated as if the 5% rate were going to remain in effect for all 30 years, and then the payments for the last 25 years are calculated to amortize the debt remaining at the end of the fifth year. What are your monthly payments for the first 5 years, and what are they for the last 25 years?

46. ***Adjustable Rate Mortgages*** You take out an adjustable rate mortgage for \$100,000 for 20 years. For the first 5 years, the rate is 4%. It then rises to 7% for the next 10 years, and then 9% for the last 5 years. What are your monthly payments in the first 5 years, the next 10 years, and the last 5 years? (Assume that each time the rate changes, the payments are recalculated to amortize the remaining debt if the interest rate were to remain constant for the remaining life of the mortgage.)

47. ***Refinancing*** Your original mortgage was a \$96,000, 30-year 9.75% mortgage. After 4 years you refinance the remaining principal for 30 years at 6.875%. What was your original monthly payment? What is your new monthly payment? How much will you save in interest over the course of the loan by refinancing?

48. ***Refinancing*** Kara and Michael take out a \$120,000, 30-year, 10% mortgage. After 3 years they refinance the remaining principal with a 15 year, 6.5% loan. What were their original monthly payments? What is their new monthly payment? How much did they save in interest over the course of the loan by refinancing?

49. ***Fees*** You take out a 2-year, \$5,000 loan at 9% interest with monthly payments. The lender charges you a \$100 fee that can be paid off, interest free, in equal monthly installments over the life of the loan. Thinking of the fee as additional interest, what is the actual annual interest rate you will pay?

50. ***Fees*** You take out a 3-year, \$7,000 loan at 8% interest with monthly payments. The lender charges you a \$100 fee that can be paid off, interest free, in equal monthly installments over the life of the loan. Thinking of the fee as additional interest, what is the actual annual interest rate you will pay?

51. ***Savings*** You wish to accumulate \$100,000 through monthly payments of \$500. If you can earn interest at an annual rate of 4% compounded monthly, how long (to the nearest year) will it take to accomplish your goal?

52. ***Retirement*** Alonzo plans to retire as soon as he has accumulated \$250,000 through quarterly payments of \$2,500. If Alonzo invests this money at 5.4% interest, compounded quarterly, when (to the nearest year) can he retire?

53. ***Loans*** You have a \$2,000 credit card debt, and you plan to pay it off through monthly payments of \$50. If you are being charged 15% interest per year, how long (to the nearest 0.5 years) will it take you to repay your debt?

54. ***Loans*** You owe \$2,000 on your credit card, which charges you 15% interest. Determine, to the nearest 1¢, the minimum monthly payment that will allow you to eventually repay your debt.

55. ***Savings*** You are depositing \$100 per month in an account that pays 4.5% interest per year (compounded monthly), while your friend Lucinda is depositing \$75 per month in an account that earns 6.5% interest per year (compounded monthly). When, to the nearest year, will her balance exceed yours?

56. ***Car Leasing*** You can lease a \$15,000 car for \$300 per month. For how long (to the nearest year) should you lease the car so that your monthly payments are lower than if you were purchasing it with an 8%-per-year loan?

COMMUNICATION AND REASONING EXERCISES

57. Your cousin Simon claims that you have wasted your time studying annuities: If you wish to retire on an income of \$1,000 per month for 20 years, you need to save \$1,000 per month for 20 years. Explain why he is wrong.

58. Your other cousin Trevor claims that you will earn more interest by accumulating \$10,000 through smaller payments than through larger payments made over a shorter period. Is he correct? Give a reason for your answer.

59. A real estate broker tells you that doubling the period of a mortgage halves the monthly payments. Is he correct? Support your answer by means of an example.

60. Another real estate broker tells you that doubling the size of a mortgage doubles the monthly payments. Is he correct? Support your answer by means of an example.

61. ◆ Consider the formula for the future value of a sinking fund with given payments. Show algebraically that the present value of that future value is the same as the present value of the annuity required to fund the same payments.

62. ◆ Give a non-algebraic justification for the result from the preceding exercise.

CHAPTER 5 REVIEW

KEY CONCEPTS

Web Site www.FiniteMath.org
Go to the student Web site at www.FiniteMath.org to find a comprehensive and interactive Web-based summary of Chapter 5.

5.1 Simple Interest

Simple interest *p. 348*
Future value *p. 349*
Bond, maturity value *p. 351*
Present value *p. 351*
Treasury bills (T-bills); maturity, discount rate, yield *pp. 351–352*

5.2 Compound Interest

Compound interest *p. 356*
Future value for compound interest *p. 357*
Present value for compound interest *p. 359*
Zero coupon bond or strip *p. 359*
Inflation, constant dollars *pp. 360–361*
Effective interest rate, annual percentage yield (APY) *p. 361*

5.3 Annuities, Loans, and Bonds

Annuity, sinking fund *p. 367*
Future value of a sinking fund *p. 368*
Payment formula for a sinking fund *p. 369*
Present value of an annuity *p. 371*
Ordinary annuity, annuity due *p. 371*
Payment formula for an annuity *p. 372*
Installment loan *p. 373*
Amortization schedule *p. 374*
Bond *p. 375*

REVIEW EXERCISES

In each of Exercises 1–6, find the future value of the investment.

1. \$6,000 for 5 years at 4.75% simple annual interest

2. \$10,000 for 2.5 years at 5.25% simple annual interest

3. \$6,000 for 5 years at 4.75% compounded monthly

4. \$10,000 for 2.5 years at 5.25% compounded semiannually

5. \$100 deposited at the end of each month for 5 years, at 4.75% interest compounded monthly

6. \$2,000 deposited at the end of each half-year for 2.5 years, at 5.25% interest compounded semiannually

In each of Exercises 7–12, find the present value of the investment.

7. Worth \$6,000 after 5 years at 4.75% simple annual interest

8. Worth \$10,000 after 2.5 years at 5.25% simple annual interest

9. Worth \$6,000 after 5 years at 4.75% compounded monthly

10. Worth \$10,000 after 2.5 years at 5.25% compounded semiannually

11. Funding \$100 withdrawals at the end of each month for 5 years, at 4.75% interest compounded monthly

12. Funding \$2,000 withdrawals at the end of each half-year for 2.5 years, at 5.25% interest compounded semiannually

In each of Exercises 13–18, find the amounts indicated.

13. The monthly deposits necessary to accumulate \$12,000 after 5 years in an account earning 4.75% compounded monthly

14. The semiannual deposits necessary to accumulate \$20,000 after 2.5 years in an account earning 5.25% compounded semiannually

15. The monthly withdrawals possible over 5 years from an account earning 4.75% compounded monthly and starting with \$6,000

16. The semiannual withdrawals possible over 2.5 years from an account earning 5.25% compounded semiannually and starting with \$10,000

17. The monthly payments necessary on a 5-year loan of \$10,000 at 4.75%

18. The semiannual payments necessary on a 2.5-year loan of \$15,000 at 5.25%

19. How much would you pay for a \$10,000, 5-year, 6% bond if you want a return of 7%? (Assume that the bond pays interest every 6 months.)

20. How much would you pay for a \$10,000, 5-year, 6% bond if you want a return of 5%? (Assume that the bond pays interest every 6 months.)

21. A \$10,000, 7-year, 5% bond sells for \$9,800. What return does it give you? (Assume that the bond pays interest every 6 months.)

22. A \$10,000, 7-year, 5% bond sells for \$10,200. What return does it give you? (Assume that the bond pays interest every 6 months.)

In each of Exercises 23–28, find the time requested, to the nearest 0.1 year.

23. The time it would take \$6,000 to grow to \$10,000 at 4.75% simple annual interest

24. The time it would take \$10,000 to grow to \$15,000 at 5.25% simple annual interest

25. The time it would take \$6,000 to grow to \$10,000 at 4.75% interest compounded monthly

26. The time it would take \$10,000 to grow to \$15,000 at 5.25% interest compounded semiannually

27. The time it would take to accumulate \$10,000 by depositing \$100 at the end of each month in an account earning 4.75% interest compounded monthly

28. The time it would take to accumulate \$15,000 by depositing \$2,000 at the end of each half-year in an account earning 5.25% compounded semiannually

APPLICATIONS

Stock Investments *Exercises 29–34 are based on the following table, which shows some values of ABCromD (ABCD) stock:*

Dec. 2002	Aug. 2004	Mar. 2005	May 2005	Aug. 2005	Dec. 2005
3.28	16.31	33.95	21.00	30.47	7.44
Jan. 2007	Mar. 2008	Oct. 2008	Nov. 2009	Feb. 2010	Aug. 2010
12.36	7.07	11.44	33.53	44.86	45.74

29. Marjory Duffin purchased ABCD stock in December 2002 and sold it in August 2010. Calculate her annual return on a simple interest basis to the nearest 0.01%.

30. John O'Hagan purchased ABCD stock in March 2005 and sold it in January 2007. Calculate his annual return on a simple interest basis to the nearest 0.01%.

31. Suppose Marjory Duffin had bought ABCD stock in January 2007. If she had later sold at one of the dates in the table, which of those dates would have given her the largest annual return on a simple interest basis, and what would that return have been?

32. Suppose John O'Hagan had purchased ABCD stock in August 2004. If he had later sold at one of the dates in the table, which of those dates would have given him the largest annual loss on a simple interest basis, and what would that loss have been?

33. Did ABCD stock undergo simple interest increase in the period December 2002 through March 2005? (Give a reason for your answer.)

34. If ABCD stock underwent simple interest increase from February 2010 through August 2010 and into 2011, what would the price be in December 2011?

35. ***Revenue*** Total Online revenues at OHaganBooks.com during 1999, its first year of operation, amounted to $150,000. After December, 1999, revenues increased by a steady 20% each year. Track OHaganBooks.com's revenues for the subsequent 5 years, assuming that this rate of growth continued. During which year did the revenue surpass $300,000?

36. ***Net Income*** Unfortunately, the picture for net income was not so bright: The company lost $20,000 in the fourth quarter of 1999. However, the quarterly loss decreased at an annual rate of 15%. How much did the company lose during the third quarter of 2001?

37. ***Stocks*** In order to finance anticipated expansion, CEO John O'Hagan is considering making a public offering of OHaganBooks.com shares at $3.00 a share. O'Hagan is not sure how many shares to offer, but would like the total value of the shares to reach a market value of at least $500,000 6 months after the initial offering. He estimates that the value of the stock will double in the first day of trading, and then appreciate at around 8% per month for the first 6 months. How many shares of stock should the company offer?

38. ***Stocks*** Unfortunately, the stock market took a dive on the very first day of OHaganBooks.com's IPO of 600,000 shares at $3.00 per share, and the shares ended the trading day 60% lower. Subsequently, as more and more bad news about the housing market emerged, the stock depreciated by 10% per week during the subsequent 5 weeks. What was the total market value of the stock at the end of 5 weeks?

39. ***Loans*** OHaganBooks.com is seeking a $250,000 loan to finance its continuing losses. One of the best deals available is offered by Industrial Bank, which offers a 10-year 9.5% loan. What would the monthly payments be for this loan?

40. ***Loans*** (See Exercise 39.) Expansion Loans offers an 8-year 6.5% loan. What would the monthly payments be for the $250,000 loan from Expansion?

41. ***Loans*** (See Exercise 39.) OHaganBooks.com can afford to pay only $3,000 per month to service its debt. What, to the nearest dollar, is the largest amount the company can borrow from Industrial Bank?

42. ***Loans*** (See Exercise 40.) OHaganBooks.com can afford to pay only $3,000 per month to service its debt. What, to the nearest dollar, is the largest amount the company can borrow from Expansion Loans?

43. ***Loans*** (See Exercise 39.) What interest rate would Industrial Bank have to offer in order to meet the company's loan requirements at a price (no more than $3,000 per month) it can afford?

44. ***Loans*** (See Exercise 40.) What interest rate would Expansion Loans have to offer in order to meet the company's loan requirements at a price (no more than $3,000 per month) it can afford?

Retirement Planning *OHaganBooks.com has just introduced a retirement package for the employees. Under the annuity plan operated by Sleepy Hollow, the monthly contribution by the company on behalf of each employee is $800. Each employee can then supplement that amount through payroll deductions. The current rate of return of Sleepy Hollow's retirement fund is 7.3%. Use this information in Exercises 45–52.*

45. Jane Callahan, the Web site developer at OHaganBooks.com, plans to retire in 10 years. She contributes $1,000 per month to the plan (in addition to the company contribution of $800). Currently, there is $50,000 in her retirement annuity. How much (to the nearest dollar) will it be worth when she retires?

46. Percy Egan, the assistant Web site developer at OHaganBooks.com, plans to retire in 8 years. He contributes $950 per month to the plan (in addition to the company contribution of $800). Currently, there is $60,000 in his retirement annuity. How much (to the nearest dollar) will it be worth when he retires?

47. When she retires, how much of Jane Callahan's retirement fund will have resulted from the company contribution? (See Exercise 45. The company did not contribute toward the $50,000 Callahan now has.)

48. When he retires, how much of Percy Egan's retirement fund will have resulted from the company contribution? (See Exercise 46. The company began contributing $800 per month to his retirement fund when he was hired 5 years ago. Assume the rate of return from Sleepy Hollow has been unchanged.)

49. (See Exercise 45.) Jane Callahan actually wants to retire with $500,000. How much should she contribute each month to the annuity?

50. (See Exercise 46.) Percy Egan actually wants to retire with $600,000. How much should he contribute each month to the annuity?

51. (See Exercise 45.) On second thought, Callahan wants to be in a position to draw at least $5,000 per month for 30 years after her retirement. She feels she can invest the proceeds of her retirement annuity at 8.7% per year in perpetuity. Given the information in Exercise 45, how much will she need to contribute to the plan starting now?

52. (See Exercise 46.) On second thought, Egan wants to be in a position to draw at least $6,000 per month for 25 years after his retirement. He feels he can invest the proceeds of his retirement annuity at 7.8% per year in perpetuity. Given the information in Exercise 46, how much will he need to contribute to the plan starting now?

Actually, Jane Callahan is quite pleased with herself; 1 year ago she purchased a $50,000 government bond paying 7.2% per year (with interest paid every 6 months) and maturing in 10 years, and interest rates have come down since then.

53. The current interest rate on 10-year government bonds is 6.3%. If she were to auction the bond at the current interest rate, how much would she get?

54. If she holds on to the bond for 6 more months and the interest rate drops to 6%, how much will the bond be worth then?

55. T Jane suspects that interest rates will come down further during the next 6 months. If she hopes to auction the bond for $54,000 in 6 months' time, what will the interest rate need to be at that time?

56. T If, in 6 months' time, the bond is auctioned for only $52,000, what will the interest rate be at that time?

Case Study Adjustable Rate and Subprime Mortgages

Gary Faber/Photodisc/Getty Images

The term **subprime mortgage** refers to mortgages given to home-buyers with a heightened perceived risk of default, as when, for instance, the price of the home being purchased is higher than the borrower can reasonably afford. Such loans are typically **adjustable rate** loans, meaning that the lending rate varies through the duration of the loan.* Subprime adjustable rate loans typically start at artificially low "teaser rates" that the borrower can afford, but then increase significantly over the life of the mortgage. The U.S. real estate bubble of 2000–2005 led to a frenzy of subprime lending, the rationale being that a borrower having trouble meeting mortgage payments could either sell the property at a profit or re-finance the loan, or the lending institution could earn a hefty profit by repossessing the property in the event of foreclosure.

Mr. and Mrs. Wong have an appointment tomorrow with you, their investment counselor, to discuss their plan to purchase a $400,000 house in Orlando, Florida. They have saved $20,000 for a down payment, so want to take out a $380,000 mortgage. Their combined annual income is $80,000 per year, which they estimate will increase by 4% annually over the foreseeable future, and they are considering three different specialty 30-year mortgages:

Hybrid: The interest is fixed at a low introductory rate of 4% for 5 years, and then adjusts annually to 5% over the U.S. federal funds rate.†

Interest-Only: During the first 5 years, the rate is set at 4.2% and no principal is paid. After that time, the mortgage adjusts annually to 5% over the U.S. federal funds rate.

Negative Amortization: During the first 5 years, the rate is set at 4.7% based on a principal of 60% of the purchase price of the home, with the result that the balance on the principal actually grows during this period. After that time, the mortgage adjusts annually to 5% over the U.S. federal funds rate.

* **NOTE** In an adjustable rate mortgage, the payments are recalculated each time the interest rate changes, based on the assumption that the new interest rate will be unchanged for the remaining life of the loan. We say that the loan is **re-amortized** at the new rate.

† **NOTE** The U.S. federal funds rate is the rate banks charge each other for loans and is often used to set rates for other loans. Manipulating this rate is one way the U.S. Federal Reserve regulates the money supply.

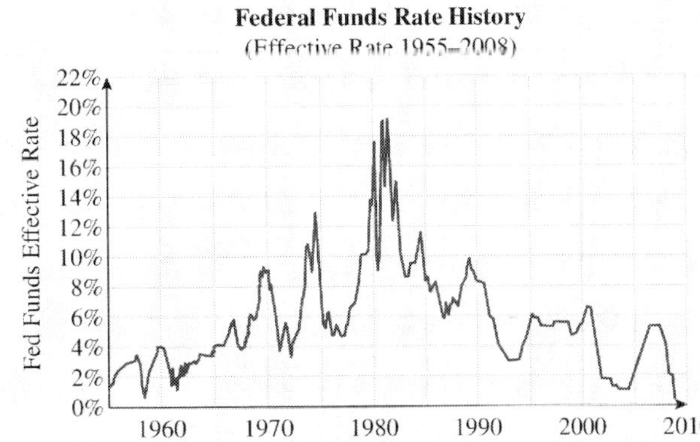

Source: MoneyCafe.com (www.moneycafe.com/library/fedfundsratehistory.htm).

Figure 3

You decide that you should create an Excel worksheet that will compute the monthly payments for the three types of loan. Of course, you have no way of predicting what the U.S. federal funds rate will be over the next 30 years (see Figure 3 for historical values), so you decide to include three scenarios for the federal funds rate in each case:

Scenario 1: Federal funds rate is 4.25% in year 6 and then increases by 0.25% per year.

Scenario 2: Federal funds rate is steady at 4% during the term of the loan.

Scenario 3: Federal funds rate is 15% in year 6 and then decreases by 0.25% per year.

Each worksheet will show month-by-month payments for the specific type of loan. Typically, to be affordable, payments should not exceed 28% of gross monthly income, so you will tabulate that quantity as well.

Hybrid Loan: You begin to create your worksheet by estimating 28% of the Wongs' monthly income, assuming a 4% increase each year (the income is computed using the compound interest formula for annual compounding):

	A	B	C	D	E	F	G	H
1	Year	20% of Monthly Income		Interest Rate			Monthly Payment	
2			Scenario 1	Scenario 2	Scenario 3	Scenario 1	Scenario 2	Scenario 3
3	1	=0.28*80000*(1.04)^(A3-1)/12						
4	2							
5	3							
30	28							
31	29							
32	30							

The next sheet shows the result, as well as the formulas for computing the interest rate in each scenario.

	A	B	C	D	E	F	G	H
1	Year	28% of Monthly Income		Interest Rate			Monthly Payment	
2			Scenario 1	Scenario 2	Scenario 3	Scenario 1	Scenario 2	Scenario 3
3	1	$1,866.67	4	4	4			
4	2	$1,941.33	4	4	4			
5	3	$2,018.99	4	4	4			
6	4	$2,099.75	4	4	4			
7	5	$2,183.74	4	4	4			
8	6	$2,271.09	9.25	9	15			
9	7	$2,361.93	=C8+0.25	=D8	=E8-0.25			
10	8	$2,456.41						
30	28	$5,382.29						
31	29	$5,597.58						
32	30	$5,821.48						

To compute the monthly payment, you decide to use the built-in function PMT, which has the format

$$PMT(i, n, PV, [FV], [type]),$$

where i = interest per period, n = total number of periods of the loan, PV = present value, FV = future value (optional); the *type*, also optional, is 0 or omitted if payments are at the end of each period, and 1 if at the start of each period. The present value will be the outstanding principal owed on the home each time the rate is changed and so that too will need to be known. During the first 5 years we can use as the present value the original cost of the home, but each year thereafter, the loan is re-amortized at the new interest rate, and so the outstanding principal will need to be computed. Although Excel has a built-in function that calculates payment on the principal, it returns only the payment for a single period (month), so without creating a month-by-month amortization table it would be difficult to use this function to track the outstanding principal. On the other hand, the total outstanding principal at any point in time can be computed using the future value formula FV. You decide to add three more columns to your Excel worksheet to show the principal outstanding at the start of each year. Here is the spreadsheet with the formulas for the payments and outstanding principal for the first 5 years.

	A	B	C	D	E	F	G	H	I	J	K
1	Year	28% of Monthly Income		Interest Rate			Monthly Payment			Balance on Principal	
2			Scenario 1	Scenario 2	Scenario 3	Scenario 1	Scenario 2	Scenario 3	Scenario 1	Scenario 2	Scenario 3
3	1	$1,866.67	4	4	4	=-PMT(C3/1200,360,I$3)			$380,000.00	$380,000.00	$380,000.00
4	2	$1,941.33	4	4	4				=FV(C3/1200,12*$A3,F3,-I$3)		
5	3	$2,018.99	4	4	4						
6	4	$2,099.75	4	4	4						
7	5	$2,183.74	4	4	4						
8	6	$2,271.09	9.25	9	15						
9	7	$2,361.93	9.5	9	14.75						
10	8	$2,456.41	9.75	9	14.5						

The two formulas will each be copied across to the adjacent two cells for the other scenarios. A few things to notice: The negative sign before *PMT* converts the negative quantity returned by *PMT* to a positive amount. The dollar sign in I$3 in the *PMT* formula fixes the present value for each year at the original cost of the home for the first 5 years, during which payments are computed as for a fixed rate loan. In the formula for the balance on principal at the start of each year, the number of periods is the total number of months up through the preceding year, and the present value is the same initial price of the home each year during the 5-year fixed-rate period.

The next sheet shows the calculated results for the fixed-rate period, and the new formulas to be added for the adjustable-rate period starting with the sixth year.

	C	D	E	F	G	H	I	J	K
1		Interest Rate			Monthly Payment			Balance on Principal	
2	Scenario 1	Scenario 2	Scenario 3	Scenario 1	Scenario 2	Scenario 3	Scenario 1	Scenario 2	Scenario 3
3	4	4	4	$1,814.18	$1,814.18	$1,814.18	$380,000.00	$380,000.00	$380,000.00
4	4	4	4	$1,814.18	$1,814.18	$1,814.18	$373,308.06	$373,308.06	$373,308.06
5	4	4	4	$1,814.18	$1,814.18	$1,814.18	$366,343.48	$366,343.48	$366,343.48
6	4	4	4	$1,814.18	$1,814.18	$1,814.18	$359,095.16	$359,095.16	$359,095.16
7	4	4	4	$1,814.18	$1,814.18	$1,814.18	$351,551.52	$351,551.52	$351,551.52
8	9.25	9	15	=-PMT(C8/1200,360-12*$A7,I8)			=FV(C7/1200,12,F7,-I7)		
9	9.5	9	14.75						
10	9.75	9	14.5						

Notice the changes: The loan is re-amortized each year starting with year 6, and the payment calculation needs to take into account the reduced, remaining lifetime of the loan each time. You now copy these formulas across for the remaining two scenarios, and then copy all six formulas down the remaining rows to complete the calculation. Following is a portion of the complete worksheet showing, in red, those years during which the monthly payment will exceed 28% of the gross monthly income.

	A	B	C	D	E	F	G	H	I	J	K
1	Year	28% of Monthly Income		Interest Rate			Monthly Payment			Balance on Principal	
2			Scenario 1	Scenario 2	Scenario 3	Scenario 1	Scenario 2	Scenario 3	Scenario 1	Scenario 2	Scenario 3
3	1	$1,866.67	4	4	4	$1,814.18	$1,814.18	$1,814.18	$380,000.00	$380,000.00	$380,000.00
4	2	$1,941.33	4	4	4	$1,814.18	$1,814.18	$1,814.18	$373,308.06	$373,308.06	$373,308.06
5	3	$2,018.99	4	4	4	$1,814.18	$1,814.18	$1,814.18	$366,343.48	$366,343.48	$366,343.48
6	4	$2,099.75	4	4	4	$1,814.18	$1,814.18	$1,814.18	$359,095.16	$359,095.16	$359,095.16
7	5	$2,183.74	4	4	4	$1,814.18	$1,814.18	$1,814.18	$351,551.52	$351,551.52	$351,551.52
8	6	$2,271.09	9.25	9	15	$2,943.39	$2,884.32	$4,402.22	$343,700.55	$343,700.55	$343,700.55
9	7	$2,361.93	9.5	9	14.75	$3,001.60	$2,884.32	$4,336.46	$340,018.68	$339,866.12	$342,337.80
10	8	$2,456.41	9.75	9	14.5	$3,058.89	$2,884.32	$4,271.84	$336,135.05	$335,671.99	$340,686.40
11	9	$2,554.66	10	9	14.25	$3,115.17	$2,884.32	$4,208.48	$332,020.97	$331,084.43	$338,694.96
12	10	$2,656.85	10.25	9	14	$3,170.39	$2,884.32	$4,146.53	$327,643.98	$326,066.52	$336,305.10
13	11	$2,763.12	10.5	9	13.75	$3,224.44	$2,884.32	$4,086.13	$322,967.19	$320,577.89	$333,450.89
14	12	$2,873.65	10.75	9	13.5	$3,277.24	$2,884.32	$4,027.41	$317,948.51	$314,574.40	$330,058.36
15	13	$2,988.59	11	9	13.25	$3,328.70	$2,884.32	$3,970.53	$312,539.70	$308,007.73	$326,016.01
16	14	$3,108.14	11.25	9	13	$3,378.71	$2,884.32	$3,915.64	$306,685.30	$300,825.07	$321,319.42
17	15	$3,232.46	11.5	9	12.75	$3,427.16	$2,884.32	$3,862.89	$300,321.34	$292,968.62	$315,780.91
18	16	$3,361.76	11.75	9	12.5	$3,473.93	$2,884.32	$3,812.43	$293,373.72	$284,375.19	$309,319.29
19	17	$3,496.23	12	9	12.25	$3,518.89	$2,884.32	$3,764.40	$285,756.40	$274,975.63	$301,814.76
20	18	$3,636.08	12.25	9	12	$3,561.90	$2,884.32	$3,718.94	$277,369.15	$264,694.33	$293,137.86
21	19	$3,781.52	12.5	9	11.75	$3,602.81	$2,884.32	$3,676.20	$268,094.83	$253,448.57	$283,149.60
22	20	$3,932.79	12.75	9	11.5	$3,641.47	$2,884.32	$3,636.32	$257,796.15	$241,147.88	$271,701.73

In the third scenario, the Wongs' payments would more than double at the start of the sixth year, and remain above what they can reasonably afford for 13 more years. Even if the Fed rate were to remain at the low rate of 4%, the monthly payments would still jump to above what the Wongs can afford at the start of the sixth year.

Interest-Only Loan: Here, the only change in the worksheet constructed previously is the computation of the payments for the first five years; because the loan is interest-only during this period, the monthly payment is computed as simple interest at 4.2% on the \$380,000 loan for a 30-year period:

$$INT = PVr = 380{,}000 \times .042/12 = \$1{,}330.00.$$

The formula you could use in the spreadsheet in cell F3 is `=C3/1200*I$3`, and then copy this across and down the entire block of payments for the first 5 years. The rest of the spreadsheet (including the balance on principal) will adjust itself accordingly with the formulas you had for the hybrid load. Below is a portion of the result, with a lot more red than in the hybrid loan case!

	A	B	C	D	E	F	G	H	I	J	K
1	Year	28% of Monthly Income		Interest Rate			Monthly Payment			Balance on Principal	
2			Scenario 1	Scenario 2	Scenario 3	Scenario 1	Scenario 2	Scenario 3	Scenario 1	Scenario 2	Scenario 3
3	1	$1,866.67	4.2	4.2	4.2	$1,330.00	$1,330.00	$1,330.00	$380,000.00	$380,000.00	$380,000.00
4	2	$1,941.33	4.2	4.2	4.2	$1,330.00	$1,330.00	$1,330.00	$380,000.00	$380,000.00	$380,000.00
5	3	$2,018.99	4.2	4.2	4.2	$1,330.00	$1,330.00	$1,330.00	$380,000.00	$380,000.00	$380,000.00
6	4	$2,099.75	4.2	4.2	4.2	$1,330.00	$1,330.00	$1,330.00	$380,000.00	$380,000.00	$380,000.00
7	5	$2,183.74	4.2	4.2	4.2	$1,330.00	$1,330.00	$1,330.00	$380,000.00	$380,000.00	$380,000.00
8	6	$2,271.09	9.25	9	15	$3,254.25	$3,188.95	$4,867.16	$380,000.00	$380,000.00	$380,000.00
9	7	$2,361.93	9.5	9	14.75	$3,318.61	$3,188.95	$4,794.45	$375,929.28	$375,760.60	$378,493.33
10	8	$2,456.41	9.75	9	14.5	$3,381.95	$3,188.95	$4,723.00	$371,635.48	$371,123.52	$376,667.51
11	9	$2,554.66	10	9	14.25	$3,444.18	$3,188.95	$4,652.96	$367,086.90	$366,051.44	$374,465.75
12	10	$2,656.85	10.25	9	14	$3,505.22	$3,188.95	$4,584.46	$362,247.64	$360,503.57	$371,823.49
13	11	$2,763.12	10.5	9	13.75	$3,564.98	$3,188.95	$4,517.68	$357,076.92	$354,435.28	$368,667.84
14	12	$2,873.65	10.75	9	13.5	$3,623.37	$3,188.95	$4,452.76	$351,528.19	$347,797.73	$364,917.01
15	13	$2,988.59	11	9	13.25	$3,680.26	$3,188.95	$4,389.88	$345,548.14	$340,537.54	$360,479.80
16	14	$3,108.14	11.25	9	13	$3,735.55	$3,188.95	$4,329.19	$339,075.44	$332,596.29	$355,255.12
17	15	$3,232.46	11.5	9	12.75	$3,789.12	$3,188.95	$4,270.87	$332,039.35	$323,910.09	$349,131.66
18	16	$3,361.76	11.75	9	12.5	$3,840.82	$3,188.95	$4,215.07	$324,357.97	$314,409.07	$341,987.61
19	17	$3,496.23	12	9	12.25	$3,890.53	$3,188.95	$4,161.97	$315,936.16	$304,016.79	$333,690.51
20	18	$3,636.08	12.25	9	12	$3,938.08	$3,188.95	$4,111.71	$306,663.10	$292,649.65	$324,097.21
21	19	$3,781.52	12.5	9	11.75	$3,983.32	$3,188.95	$4,064.46	$296,409.28	$280,216.18	$313,054.05
22	20	$3,932.79	12.75	9	11.5	$4,026.06	$3,188.95	$4,020.37	$285,022.93	$266,616.37	$300,397.13
23	21	$4,090.10	13	9	11.25	$4,066.11	$3,188.95	$3,979.57	$272,325.66	$251,740.81	$285,952.80
24	22	$4,253.70	13.25	9	11	$4,103.29	$3,188.95	$3,942.23	$258,107.21	$235,469.81	$269,538.33

In all three scenarios, this type of mortgage is worse for the Wongs than the hybrid loan; in particular, their payments in Scenario 3 would jump to more than double what they can afford at the start of the sixth year.

Negative Amortization Loan: Again, the only change in the worksheet is the computation of the payments for the first 5 years. This time, the loan amortizes negatively during the initial 5-year period, so the payment formula in this period is adjusted to reflect this

```
=-PMT(C3/1200,360,I$3*0.6).
```

	A	B	C	D	E	F	G	H	I	J	K
1	Year	28% of Monthly Income		Interest Rate			Monthly Payment			Balance on Principal	
2			Scenario 1	Scenario 2	Scenario 3	Scenario 1	Scenario 2	Scenario 3	Scenario 1	Scenario 2	Scenario 3
3	1	$1,866.67	4.7	4.7	4.7	$1,182.49	$1,182.49	$1,182.49	$380,000.00	$380,000.00	$380,000.00
4	2	$1,941.33	4.7	4.7	4.7	$1,182.49	$1,182.49	$1,182.49	$383,750.17	$383,750.17	$383,750.17
5	3	$2,018.99	4.7	4.7	4.7	$1,182.49	$1,182.49	$1,182.49	$387,680.45	$387,680.45	$387,680.45
6	4	$2,099.75	4.7	4.7	4.7	$1,182.49	$1,182.49	$1,182.49	$391,799.48	$391,799.48	$391,799.48
7	5	$2,183.74	4.7	4.7	4.7	$1,182.49	$1,182.49	$1,182.49	$396,116.32	$396,116.32	$396,116.32
8	6	$2,271.09	9.25	9	15	$3,431.01	$3,362.16	$5,131.53	$400,640.49	$400,640.49	$400,640.49
9	7	$2,361.93	9.5	9	14.75	$3,498.87	$3,362.16	$5,054.87	$396,348.66	$396,170.82	$399,051.98
10	8	$2,456.41	9.75	9	14.5	$3,565.64	$3,362.16	$4,979.54	$391,821.64	$391,281.87	$397,127.00
11	9	$2,554.66	10	9	14.25	$3,631.26	$3,362.16	$4,905.69	$387,025.99	$385,934.29	$394,805.64
12	10	$2,656.85	10.25	9	14	$3,695.62	$3,362.16	$4,833.48	$381,923.88	$380,085.08	$392,019.86
13	11	$2,763.12	10.5	9	13.75	$3,758.62	$3,362.16	$4,763.06	$376,472.30	$373,687.17	$388,692.80
14	12	$2,873.65	10.75	9	13.5	$3,820.18	$3,362.16	$4,694.62	$370,622.18	$366,689.09	$384,738.24
15	13	$2,988.59	11	9	13.25	$3,880.16	$3,362.16	$4,628.32	$364,317.31	$359,034.55	$380,060.01
16	14	$3,108.14	11.25	9	13	$3,938.46	$3,362.16	$4,564.34	$357,493.03	$350,661.95	$374,551.54
17	15	$3,232.46	11.5	9	12.75	$3,994.93	$3,362.16	$4,502.85	$350,074.77	$341,503.95	$368,095.48
18	16	$3,361.76	11.75	9	12.5	$4,049.45	$3,362.16	$4,444.02	$341,976.15	$331,486.86	$360,563.39
19	17	$3,496.23	12	9	12.25	$4,101.85	$3,362.16	$4,388.03	$333,096.90	$320,530.10	$351,815.60
20	18	$3,636.08	12.25	9	12	$4,151.99	$3,362.16	$4,335.05	$323,320.15	$308,545.52	$341,701.22
21	19	$3,781.52	12.5	9	11.75	$4,199.68	$3,362.16	$4,285.23	$312,509.37	$295,436.71	$330,058.23
22	20	$3,932.79	12.75	9	11.5	$4,244.74	$3,362.16	$4,238.74	$300,504.55	$281,098.20	$316,713.83
23	21	$4,090.10	13	9	11.25	$4,286.97	$3,362.16	$4,195.73	$287,117.60	$265,414.64	$301,484.92
24	22	$4,253.70	13.25	9	11	$4,326.17	$3,362.16	$4,156.36	$272,126.84	$248,259.85	$284,178.86
25	23	$4,423.85	13.5	9	10.75	$4,362.10	$3,362.16	$4,120.77	$255,270.30	$229,495.82	$264,594.34
26	24	$4,600.80	13.75	9	10.5	$4,394.53	$3,362.16	$4,089.09	$236,237.47	$208,971.60	$242,522.55

Clearly the Wongs should steer clear of this type of loan in order to be able to continue to afford making payments!

In short, it seems unlikely that the Wongs will be able to afford payments on any of the three mortgages in question, and you decide to advise them to either seek a less expensive home or wait until their income has appreciated to enable them to afford a home of this price.

EXERCISES

1. In the case of a hybrid loan, what would the federal funds rate have to be in Scenario 2 to ensure that the Wongs can afford to make all payments?
2. Repeat the preceding exercise in the case of an interest-only loan.
3. Repeat the preceding exercise in the case of a negative-amortization loan.
4. What home price, to the nearest $5,000, could the Wongs afford if they took out a hybrid loan, regardless of scenario? HINT [Adjust the original value of the loan on your spreadsheet to obtain the desired result.]
5. What home price, to the nearest $5,000, could the Wongs afford if they took out a negative-amortization loan, regardless of scenario? HINT [Adjust the original value of the loan on your spreadsheet to obtain the desired result.]
6. How long would the Wongs need to wait before they could afford to purchase a $400,000 home, assuming that their income continues to increase as above, they still have $20,000 for a down payment, and the mortgage offers remain the same.

TI-83/84 Plus Technology Guide

Section 5.2

Example 1 (page 358) In August, 2005, the Amex Bank of Canada was paying 2.40% interest on savings accounts. If the interest is compounded quarterly, find the future value of a \$2,000 deposit in 6 years. What is the total interest paid over the period?

Solution with Technology

We could calculate the future value using the TI-83/84 Plus by entering

```
2000(1+0.0240/4)^(4*6)
```

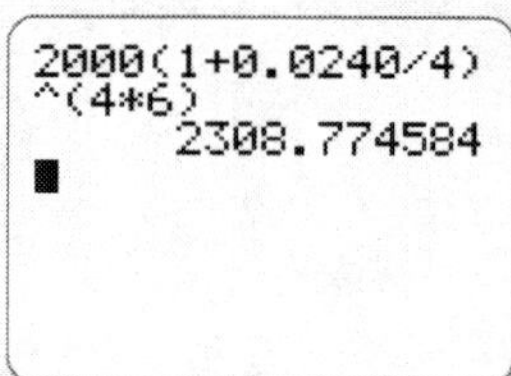

However, the TI-83/84 Plus has this and other useful calculations built into its TVM (Time Value of Money) Solver.

1. Press APPS then choose item `1:Finance` and then choose item `1:TVM Solver`. This brings up the TVM Solver window.

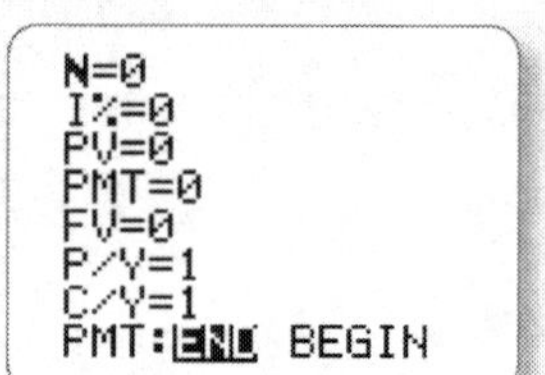

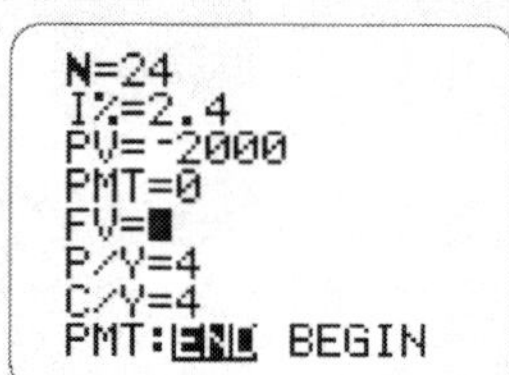

The second screen shows the values you should enter for this example. The various variables are:

N	Number of compounding periods
I%	Annual interest rate, as percent, not decimal
PV	Negative of present value
PMT	Payment per period (0 in this section)
FV	Future value
P/Y	Payments per year
C/Y	Compounding periods per year
PMT:	Not used in this section

Several things to notice:

- *I%* is the *annual* interest rate, corresponding to *r*, not *i*, in the compound interest formula.
- The present value, *PV*, is entered as a negative number. In general, when using the TVM Solver, any amount of money you give to someone else (such as the \$2,000 you deposit in the bank) will be a negative number, whereas any amount of money someone gives to you (such as the future value of your deposit, which the bank will give back to you) will be a positive number.
- *PMT* is not used in this example (it will be used in the next section) and should be 0.
- *FV* is the future value, which we shall compute in a moment; it doesn't matter what you enter now.
- *P/Y* and *C/Y* stand for payments per year and compounding periods per year, respectively: They should both be set to the number of compounding periods per year for compound interest problems (setting *P/Y* automatically sets *C/Y* to the same value).
- *PMT*: *END* or *BEGIN* is not used in this example and it doesn't matter which you select.

2. To compute the future value, use the up or down arrow to put the cursor on the *FV* line, then press ALPHA SOLVE.

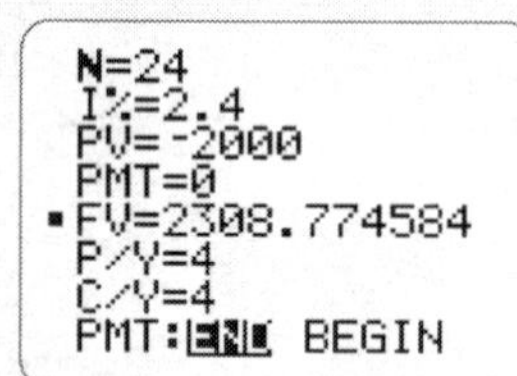

Example 2 (page 360) Megabucks Corporation is issuing 10-year zero coupon bonds. How much would you pay for bonds with a maturity value of \$10,000 if you wish to get a return of 6.5% compounded annually?

Solution with Technology

To compute the present value using a TI-83/84 Plus:

1. Enter the numbers shown on the left in the TVM Solver window.
2. Put the cursor on the PV line, and press ALPHA SOLVE.

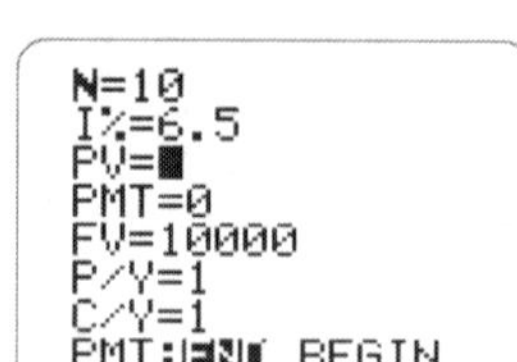

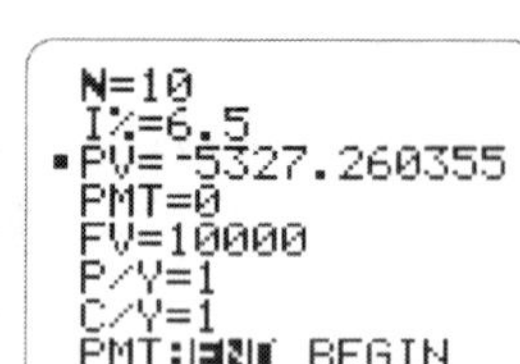

Why is the present value given as negative?

Example 6 (page 362) You have \$5,000 to invest at 6% interest compounded monthly. How long will it take for your investment to grow to \$6,000?

Solution with Technology

1. Enter the numbers shown on the left in the TVM Solver window.
2. Put the cursor on the N line, and press [ALPHA] [SOLVE].

```
N=■
I%=6
PV=-5000
PMT=0
FV=6000
P/Y=12
C/Y=12
PMT:END BEGIN
```

```
▪N=36.55539636
 I%=6
 PV=-5000
 PMT=0
 FV=6000
 P/Y=12
 C/Y=12
 PMT:END BEGIN
```

Recall that I% is the annual interest rate, corresponding to r in the formula, but N is the number of compounding periods, so number of months in this example. Thus, you will need to invest your money for about 36.5 months, or just over three years, before it grows to \$6,000.

Section 5.3

Example 1 (page 368) Your retirement account has \$5,000 in it and earns 5% interest per year compounded monthly. Every month for the next 10 years, you will deposit \$100 into the account. How much money will there be in the account at the end of those 10 years?

Solution with Technology

We can use the TVM Solver in the TI-83/84 Plus to calculate future values like these:

1. Enter the values shown on the left.
2. With the cursor on the FV line, press [ALPHA] [SOLVE] to find the future value.

```
N=120
I%=5
PV=-5000
PMT=-100
FV=■
P/Y=12
C/Y=12
PMT:END BEGIN
```

```
 N=120
 I%=5
 PV=-5000
 PMT=-100
▪FV=23763.27543
 P/Y=12
 C/Y=12
 PMT:END BEGIN
```

Note that the TVM Solver allows you to put the \$5,000 already in the account as the present value of the account.

3. Following the TI-83/84 Plus's usual convention, set *PV* to the *negative* of the present value because this is money you paid into the account.
4. Likewise, set *PMT* to −100 because you are paying \$100 each month.
5. Set the number of payment and compounding periods to 12 per year.
6. Finally, set the payments to be made at the end of each period.

Example 2 (page 369) Tony and Maria have just had a son, José Phillipe. They establish an account to accumulate money for his college education. They would like to have \$100,000 in this account after 17 years. If the account pays 4% interest per year compounded quarterly, and they make deposits at the end of every quarter, how large must each deposit be for them to reach their goal?

Solution with Technology

1. In the TVM Solver in the TI-83/84 Plus, enter the values shown below.
2. Solve for *PMT*.

```
N=68
I%=4
PV=0
PMT=■
FV=100000
P/Y=4
C/Y=4
PMT:END BEGIN
```

```
 N=68
 I%=4
 PV=0
▪PMT=-1033.8885…
 FV=100000
 P/Y=4
 C/Y=4
 PMT:END BEGIN
```

Why is *PMT* negative?

Example 3 (page 371) You wish to establish a trust fund from which your niece can withdraw \$2,000 every 6 months for 15 years, at which time she will receive the remaining money in the trust, which you would like to be \$10,000. The trust will be invested at 7% per year compounded every 6 months. How large should the trust be?

Solution with Technology

1. In the TVM Solver in the TI-83/84 Plus, enter the values shown below.
2. Solve for *PV*.

```
N=30
I%=7
PV=■
PMT=2000
FV=10000
P/Y=2
C/Y=2
PMT:END BEGIN
```

```
 N=30
 I%=7
▪PV=-40346.87493
 PMT=2000
 FV=10000
 P/Y=2
 C/Y=2
 PMT:END BEGIN
```

The payment and future value are positive because you (or your niece) will be receiving these amounts from the investment.

Note We have assumed that your niece receives the withdrawals at the end of each compounding period, so that the trust fund is an ordinary annuity. If, instead, she receives the payments at the beginning of each compounding period, it is an annuity due. You switch between the two types of annuity by changing PMT: END at the bottom to PMT: BEGIN. ■

```
N=30
I%=7
PV=■
PMT=2000
FV=10000
P/Y=2
C/Y=2
PMT:END BEGIN
```

```
N=30
I%=7
▪PV=-41634.31811
PMT=2000
FV=10000
P/Y=2
C/Y=2
PMT:END BEGIN
```

As mentioned in the text, the present value must be higher to fund payments at the beginning of each period, because the money in the account has less time to earn interest.

Example 4 (page 372) Tony and Maria (see Example 2), having accumulated $100,000 for José Phillipe's college education, would now like to make quarterly withdrawals over the next 4 years. How much money can they withdraw each quarter in order to draw down the account to zero at the end of the 4 years? (Recall that the account pays 4% interest compounded quarterly.)

Solution with Technology

1. Enter the values shown below in the TI-83/84 Plus TVM Solver.
2. Solve for *PMT*.

```
N=16
I%=4
PV=-100000
PMT=■
FV=0
P/Y=4
C/Y=4
PMT:END BEGIN
```

```
N=16
I%=4
PV=-100000
▪PMT=6794.459682
FV=0
P/Y=4
C/Y=4
PMT:END BEGIN
```

The present value is negative because Tony and Maria do not possess it; the bank does.

Example 7 (page 374) Marc and Mira are buying a house, and have taken out a 30-year, $90,000 mortgage at 8% interest per year. Mortgage interest is tax-deductible, so it is important to know how much of a year's mortgage payments represents interest. How much interest will Marc and Mira pay in the first year of their mortgage?

Solution with Technology

The TI-83/84 Plus has built-in functions to compute the values in an amortization schedule.

1. First, use the TVM Solver to find the monthly payment.

```
N=360
I%=8
PV=90000
▪PMT=-660.38811...
FV=0
P/Y=12
C/Y=12
PMT:END BEGIN
```

Three functions correspond to the last three columns of the amortization schedule given in the text: `ΣInt`, `ΣPrn`, and `bal` (found in the `Finance` menu accessed through [APPS]). They all require that the values of *I%*, *PV*, and *PMT* be entered or calculated ahead of time; calculating the payment in the TVM Solver in Step 1 accomplishes this.

2. Use `ΣInt(m,n,2)` to compute the sum of the interest payments from payment *m* through payment *n*. For example,

   `ΣInt(1,12,2)`

   will return −7,172.81, the total paid in interest in the first year, which answers the question asked in this example. (The last argument, 2, tells the calculator to round all intermediate calculations to two decimal places—that is, the nearest cent—as would the mortgage lender.)
3. Use `ΣPrn(m,n,2)` to compute the sum of the payments on principal from payment *m* through payment *n*. For example,

   `ΣPrn(1,12,2)`

   will return −751.87, the total paid on the principal in the first year.
4. Finally, `bal(n,2)` finds the balance of the principal outstanding after *n* payments. For example,

   `bal(12,2)`

   will return the value 89,248.13, the balance remaining at the end of one year.
5. To construct an amortization schedule as in the text, make sure that FUNC is selected in the [MODE] window; then enter the functions in the [Y=] window as shown below.

```
Plot1 Plot2 Plot3
\Y1■ΣInt(X,X,2)
\Y2■ΣPrn(X,X,2)
\Y3■bal(X,2)
\Y4=
\Y5=
\Y6=
\Y7=
```

6. Press [2ND] [TBLSET] and enter the values shown here.

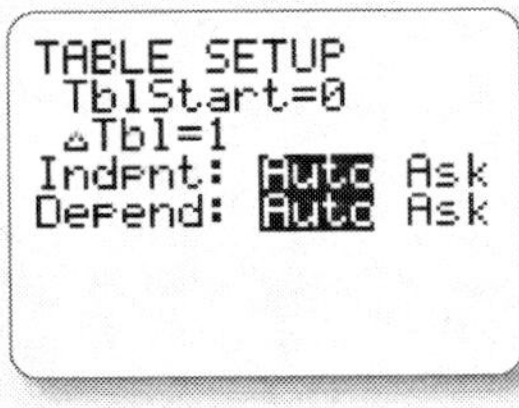

7. Press [2ND] [TABLE], to get the table shown here.

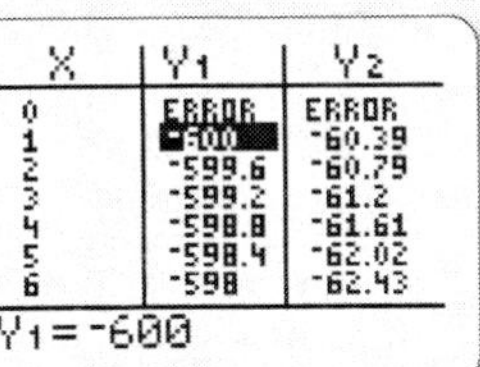

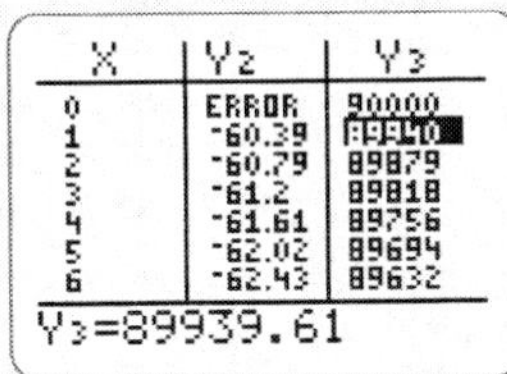

The column labeled X gives the month, the column labeled Y_1 gives the interest payment for each month, the column labeled Y_2 gives the payment on principal for each month, and the column labeled Y_3 (use the right arrow button to make it visible) gives the outstanding principal.

8. To see later months, use the down arrow. As you can see, some of the values will be rounded in the table, but by selecting a value (as the outstanding principal at the end of the first month is selected in the second screen) you can see its exact value at the bottom of the screen.

Example 9 (page 376) Suppose that a 5%, 20-year bond sells for \$9,800 per \$10,000 maturity value. What rate of return will investors get?

Solution with Technology

We can use the TVM Solver in the TI-83/84 Plus to find the interest rate just as we use it to find any other one of the variables.

1. Enter the values shown in the TVM Solver window.
2. Solve for *I*%. (Recall that *I*% is the annual interest rate, corresponding to *r* in the formula.)

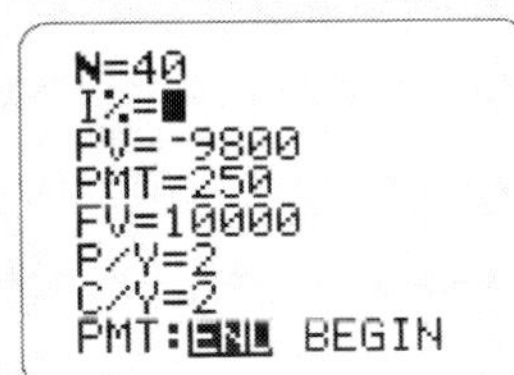

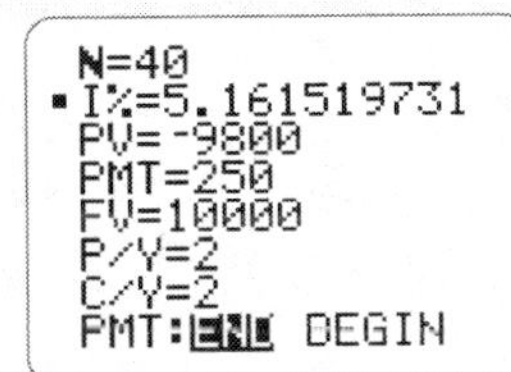

Thus, at \$9,800 per \$10,000 maturity value, these bonds yield 5.162% compounded semiannually.

EXCEL Technology Guide

Section 5.2

Example 1 (page 358) In August 2005, the Amex Bank of Canada was paying 2.40% interest on savings accounts. If the interest is compounded quarterly, find the future value of a \$2,000 deposit in 6 years. What is the total interest paid over the period?

Solution with Technology

You can either compute compound interest directly or use financial functions built into Excel. The following worksheet has more than we need for this example, but will be useful for other examples in this and the next section.

	A	B	C	D
1		Entered	Calculated	
2	Rate	2.40%		
3	Years	6		
4	Payment	\$0.00		
5	Present Value	-\$2,000.00		
6	Future Value		=FV(B2/B7,B3*B7,B4,B5)	
7	Periods per year	4		

For this example, the payment amount in B4 should be 0 (we shall use it in the next section).

1. Enter the other numbers as shown. As with other technologies, like the TVM Solver in the TI-83/84 Plus calculator, money that you pay to others (such as the \$2,000 you deposit in the bank) should be entered as negative, whereas money that is paid to you is positive.
2. The formula entered in C6 uses the built-in FV function to calculate the future value based on the entries in column B. This formula has the following format:

$$\text{FV}(i, n, PMT, PV).$$

i = interest per period — We use B2/B7 for the interest.

n = number of periods — We use B3*B7 for the number of periods.

PMT = payment per period — The payment is 0 (cell B4).

PV = present value — The present value is in cell B5.

Instead of using the built-in FV function, we could use

```
=-B5*(1+B2/B7)^(B3*B7)
```

based on the future value formula for compound interest. After calculation the result will appear in cell C6.

	A	B	C	D
1		Entered	Calculated	
2	Rate	2.40%		
3	Years	6		
4	Payment	$0.00		
5	Present Value	-$2,000.00		
6	Future Value		$2,308.77	
7	Periods per year	4		

Note that we have formatted the cells B4:C6 as currency with two decimal places. If you change the values in column B, the future value in column C will be automatically recalculated.

Example 2 (page 360) Megabucks Corporation is issuing 10-year zero coupon bonds. How much would you pay for bonds with a maturity value of $10,000 if you wish to get a return of 6.5% compounded annually?

Solution with Technology

You can compute present value in Excel using the PV worksheet function. The following worksheet is similar to the one in the preceding example, except that we have entered a formula for computing the present value from the entered values.

	A	B	C	D
1		Entered	Calculated	
2	Rate	6.50%		
3	Years	10		
4	Payment	$0.00		
5	Present Value		=PV(B2/B7,B3*B7,B4,B6)	
6	Future Value	$10,000.00		
7	Periods per year	1		

The next worksheet shows the calculated value.

	A	B	C	D
1		Entered	Calculated	
2	Rate	6.50%		
3	Years	10		
4	Payment	$0.00		
5	Present Value		-$5,327.26	
6	Future Value	$10,000.00		
7	Periods per year	1		

Why is the present value negative?

Example 6 (page 362) You have $5,000 to invest at 6% interest compounded monthly. How long will it take for your investment to grow to $6,000?

Solution with Technology

You can compute the requisite length of an investment in Excel using the NPER worksheet function. The following worksheets show the calculation.

	A	B	C	D
1		Entered	Calculated	
2	Rate	6.00%		
3	Years		=NPER(B2/B7,B4,B5,B6)/B7	
4	Payment	$0.00		
5	Present Value	-$5,000.00		
6	Future Value	$6,000.00		
7	Periods per year	12		

	A	B	C	D
1		Entered	Calculated	
2	Rate	6.00%		
3	Years		3.04628303	
4	Payment	$0.00		
5	Present Value	-$5,000.00		
6	Future Value	$6,000.00		
7	Periods per year	12		

The NPER function computes the number of compounding periods, months in this case, so we divide by B7, the number of periods per year, to calculate the number of years, which appears as 3.046. So, you need to invest your money for just over 3 years for it to grow to $6,000.

Section 5.3

Example 1 (page 368) Your retirement account has $5,000 in it and earns 5% interest per year compounded monthly. Every month for the next 10 years you will deposit $100 into the account. How much money will there be in the account at the end of those 10 years?

Solution with Technology

We can use exactly the same worksheet that we used in Example 1 in the preceding section. In fact, we included the "Payment" row in that worksheet just for this purpose.

	A	B	C	D
1		Entered	Calculated	
2	Rate	5.00%		
3	Years	10		
4	Payment	-$100.00		
5	Present Value	-$5,000.00		
6	Future Value		=FV(B2/B7,B3*B7,B4,B5)	
7	Periods per year	12		

	A	B	C	D
1		Entered	Calculated	
2	Rate	5.00%		
3	Years	10		
4	Payment	-$100.00		
5	Present Value	-$5,000.00		
6	Future Value		$23,763.28	
7	Periods per year	12		

Note that the FV function allows us to enter, as the last argument, the amount of money already in the account. Following the usual convention, we enter the present value and the payment as *negative*, because these are amounts you pay into the account.

Example 2 (page 369) Tony and Maria have just had a son, José Phillipe. They establish an account to accumulate money for his college education, in which they would like to have $100,000 after 17 years. If the account pays 4% interest per year compounded quarterly, and they make deposits at the end of every quarter, how large must each deposit be for them to reach their goal?

Solution with Technology

Use the following worksheet, in which the PMT worksheet function is used to calculate the required payments.

	A	B	C	D
1		Entered	Calculated	
2	Rate	4.00%		
3	Years	17		
4	Payment		=PMT(B2/B7,B3*B7,B5,B6)	
5	Present Value	$0.00		
6	Future Value	$100,000.00		
7	Periods per year	4		

	A	B	C	D
1		Entered	Calculated	
2	Rate	4.00%		
3	Years	17		
4	Payment		-$1,033.89	
5	Present Value	$0.00		
6	Future Value	$100,000.00		
7	Periods per year	4		

Why is the payment negative?

Example 3 (page 371) You wish to establish a trust fund from which your niece can withdraw $2,000 every 6 months for 15 years, at which time she will receive the remaining money in the trust, which you would like to be $10,000. The trust will be invested at 7% per year compounded every 6 months. How large should the trust be?

Solution with Technology

You can use the same worksheet as in Example 2 in Section 5.2.

	A	B	C	D
1		Entered	Calculated	
2	Rate	7.00%		
3	Years	15		
4	Payment	$2,000.00		
5	Present Value		=PV(B2/B7,B3*B7,B4,B6)	
6	Future Value	$10,000.00		
7	Periods per year	2		

	A	B	C	D
1		Entered	Calculated	
2	Rate	7.00%		
3	Years	15		
4	Payment	$2,000.00		
5	Present Value		-$40,346.87	
6	Future Value	$10,000.00		
7	Periods per year	2		

The payment and future value are positive because you (or your niece) will be receiving these amounts from the investment.

Note We have assumed that your niece receives the withdrawals at the end of each compounding period, so that the trust fund is an ordinary annuity. If, instead, she receives the payments at the beginning of each compounding period, it is an annuity due. You switch to an annuity due by adding an optional last argument of 1 to the PV function (and similarly for the other finance functions in Excel).

	A	B	C	D
1		Entered	Calculated	
2	Rate	7.00%		
3	Years	15		
4	Payment	$2,000.00		
5	Present Value		=PV(B2/B7,B3*B7,B4,B6,1)	
6	Future Value	$10,000.00		
7	Periods per year	2		

	A	B	C	D
1		Entered	Calculated	
2	Rate	7.00%		
3	Years	15		
4	Payment	$2,000.00		
5	Present Value		-$41,634.32	
6	Future Value	$10,000.00		
7	Periods per year	2		

As mentioned in the text, the present value must be higher to fund payments at the beginning of each period, because the money in the account has less time to earn interest. ■

Example 4 (page 372) Tony and Maria (see Example 2), having accumulated $100,000 for José Phillipe's college education, would now like to make quarterly withdrawals over the next 4 years. How much money can they withdraw each quarter in order to draw down the account to zero at the end of the four years? (Recall that the account pays 4% interest compounded quarterly.)

Solution with Technology

You can use the same worksheet as in Example 2.

	A	B	C	D
1		Entered	Calculated	
2	Rate	4.00%		
3	Years	4		
4	Payment		=PMT(B2/B7,B3*B7,B5,B6)	
5	Present Value	-$100,000.00		
6	Future Value	$0.00		
7	Periods per year	4		

	A	B	C	D
1		Entered	Calculated	
2	Rate	4.00%		
3	Years	4		
4	Payment		$6,794.46	
5	Present Value	-$100,000.00		
6	Future Value	$0.00		
7	Periods per year	4		

The present value is negative since Tony and Maria do not possess it; the bank does.

Example 7 (page 374) Marc and Mira are buying a house, and have taken out a 30-year, $90,000 mortgage at 8% interest per year. Mortgage interest is tax-deductible, so it is important to know how much of a year's mortgage payments represents interest. How much interest will Marc and Mira pay in the first year of their mortgage?

Solution with Technology

We construct an amortization schedule with which we can answer the question.

1. Begin with the worksheet below.

	A	B	C	D	E	F	G	H
1	Month	Interest Payment	Payment on Principal	Outstanding Principal				
2	0			$90,000.00	Rate	8%		
3					Years	30		
4					Payment	=DOLLAR(-PMT(F2/12,F3*12,D2))		

	A	B	C	D	E	F
1	Month	Interest Payment	Payment on Principal	Outstanding Principal		
2	0			$90,000.00	Rate	8%
3					Years	30
4					Payment	$660.39

Note the formula for the monthly payment:

```
=DOLLAR(-PMT(F2/12,F3*12,D2))
```

The function DOLLAR rounds the payment to the nearest cent, as the bank would.

2. Calculate the interest owed at the end of the first month using the formula

```
=DOLLAR(D2*F$2/12)
```

in cell B3.

3. The payment on the principal is the remaining part of the payment, so enter

```
=F$4-B3
```

in cell C3.

4. Calculate the outstanding principal by subtracting the payment on the principal from the previous outstanding principal, by entering

```
=D2-C3
```

in cell D3.

5. Copy the formulas in cells B3, C3, and D3 into the cells below them to continue the table.

	A	B	C	D	E	F
1	Month	Interest Payment	Payment on Principal	Outstanding Principal		
2	0			$90,000.00	Rate	8%
3	1	$600.00	$60.39	$89,939.61	Years	30
4	2				Payment	$660.39
5	3					
6	4					
7	5					
8	6					
9	7					
10	8					
11	9					
12	10					
13	11					
14	12					

The result should be something like the following:

	A	B	C	D	E	F
1	Month	Interest Payment	Payment on Principal	Outstanding Principal		
2	0			$90,000.00	Rate	8%
3	1	$600.00	$60.39	$89,939.61	Years	30
4	2	$599.60	$60.79	$89,878.82	Payment	$660.39
5	3	$599.19	$61.20	$89,817.62		
6	4	$598.78	$61.61	$89,756.01		
7	5	$598.37	$62.02	$89,693.99		
8	6	$597.96	$62.43	$89,631.56		
9	7	$597.54	$62.85	$89,568.71		
10	8	$597.12	$63.27	$89,505.44		
11	9	$596.70	$63.69	$89,441.75		
12	10	$596.28	$64.11	$89,377.64		
13	11	$595.85	$64.54	$89,313.10		
14	12	$595.42	$64.97	$89,248.13		

6. Adding the calculated interest payments gives us the total interest paid in the first year: $7,172.81.

Note Excel has built-in functions that compute the interest payment (IPMT) or the payment on the principle (PPMT) in a given period. We could also have used the built-in future value function (FV) to calculate the outstanding principal each month. The main problem with using these functions is that, in a sense, they are too accurate. They do not take into account the fact that payments and interest are rounded to the nearest cent. Over time, this rounding causes the actual value of the outstanding principal to differ from what the FV function would tell us. In fact, because the actual payment is rounded slightly upwards (to $660.39 from 660.38811. . .), the principal is reduced slightly faster than necessary and a last payment of $660.39 would be $2.95 larger than needed to clear out the debt. The lender would reduce the last payment by $2.95 for this reason; Marc and Mira will pay only $657.44 for their final payment. This is common: The last payment on an installment loan is usually slightly larger or smaller than the others, to compensate for the rounding of the monthly payment amount. ■

Example 9 (page 376) Suppose that a 5%, 20-year bond sells for $9,800 per $10,000 maturity value. What rate of return will investors get?

Solution with Technology Use the following worksheet, in which the RATE worksheet function is used to calculate the interest rate.

	A	B	C	D
1		Entered	Calculated	
2	Rate		=RATE(B3*B7,B4,B5,B6)*B7	
3	Years	20		
4	Payment	$250.00		
5	Present Value	-$9,800.00		
6	Future Value	$10,000.00		
7	Periods per year	2		

	A	B	C	D
1		Entered	Calculated	
2	Rate		5.162%	
3	Years	20		
4	Payment	$250.00		
5	Present Value	-$9,800.00		
6	Future Value	$10,000.00		
7	Periods per year	2		

5. $p = 21; x = 9, y = 3$ **7.** $c = 22; x = 8, y = 6$
9. $p = 45;\ x = 0, y = 15, z = 15$
11. $p = 220; x = 20, y = 20, z = 60$
13. $c = 30; x = 30, y = 0, z = 0$
15. $c = 50; x = 20, y = 10, z = 0, w = 20$
17. $c = 60; x = 24, y = 12$ **19.** $c = 20;\ x = 0,\ y = 20$
21. $R = [1/2 \quad 1/2 \quad 0], C = [0 \quad 1/3 \quad 2/3]^T, e = 0$
23. $R = [1/27 \quad 7/9 \quad 5/27], C = [8/27 \quad 5/27 \quad 14/27]^T$, $e = 8/27$ **25.** (A) **27.** 35 **29.** 400 packages from each for a minimum cost of \$52,000 **31. a.** (B), (D) **b.** 450 packages from Duffin House, 375 from Higgins Press for a minimum cost of \$52,500 **33.** 220 shares of EEE and 20 shares of RRR. The minimum total risk index is 500. **35.** 240 Sprinkles, 120 Storms, and no Hurricanes **37.** Order 600 packages from Higgins and none from the others, for a total cost of \$90,000. **39. a.** Let x = # science credits, y = # fine arts credits, z = # liberal arts credits, and w = # math credits. Minimize $C = 300x + 300y + 200z + 200w$ subject to: $x + y + z + w \geq 120$; $x - y \geq 0$; $-2x + w \leq 0$; $-y + 3z - 3w \leq 0$; $x \geq 0, y \geq 0, z \geq 0, w \geq 0$. **b.** Billy-Sean should take the following combination: Sciences—24 credits, Fine Arts—no credits, Liberal Arts—48 credits, Mathematics—48 credits, for a total cost of \$26,400. **41.** Smallest cost is \$20,000; New York to OHaganBooks.com: 600 packages, New York to FantasyBooks.com: 0 packages, Illinois to OHaganbooks.com: 0 packages, Illinois to FantasyBooks.com: 200 packages. **43.** FantasyBooks.com should choose between "2 for 1" and "3 for 2" with probabilities 20% and 80%, respectively. OHaganBooks.com should choose between "3 for 1" and "Finite Math" with probabilities 60% and 40%, respectively. OHaganBooks.com expects to gain 12,000 customers from FantasyBooks.com.

Chapter 5

Section 5.1

1. $INT = \$120, FV = \$2,120$ **3.** $INT = \$505, FV = \$20,705$ **5.** $INT = \$250, FV = \$10,250$ **7.** $PV = \$9,090.91$ **9.** $PV = \$966.18$ **11.** $PV = \$14,457.83$ **13.** \$5,200 **15.** \$787.40 **17.** 5% **19.** About 0.2503% **21.** In 2 years **23.** 3.775% **25.** 65% **27.** 10% **29.** 98.73% **31.** 7.45% if you had sold in June 2008 **33.** No. Simple interest increase is linear. The graph is visibly not linear in that time period. Further, the slopes of the lines through the successive pairs of marked points are quite different. **35.** 9.2% **37.** 3,260,000 **39.** $P = 500 + 46t$ (t = time in years since 1950) Graph:

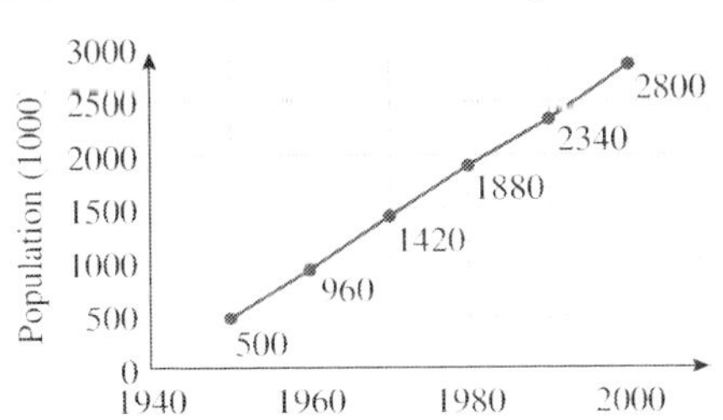

41. Graph (A) is the only possible choice, because the equation $FV = PV(1 + rt) = PV + PVrt$ gives the future value as a linear function of time. **43.** Wrong. In simple interest growth, the change each year is a fixed percentage of the *starting* value, and not the preceding year's value. (Also see Exercise 44.) **45.** Simple interest is always calculated on a constant amount, PV. If interest is paid into your account, then the amount on which interest is calculated does not remain constant.

Section 5.2

1. \$13,439.16 **3.** \$11,327.08 **5.** \$19,154.30 **7.** \$12,709.44 **9.** \$613.91 **11.** \$810.65 **13.** \$1,227.74 **15.** 5.09% **17.** 10.47% **19.** 10.52% **21.** \$268.99 **23.** \$728.91 **25.** \$2,927.15 **27.** \$21,161.79 **29.** \$163,414.56 **31.** \$55,526.45 per year **33.** \$174,110 **35.** \$750.00 **37.** \$27,171.92 **39.** \$111,678.96 **41.** \$1,039.21 **43.** The one earning 11.9% compounded monthly **45.** Yes. The investment will have grown to about \$150,281 million **47.** 136 reals **49.** 245 bolivianos **51.** 223 bolivars **53.** The Nicaragua investment is better: it is worth about 1.006 units of currency (in constant units) per unit invested as opposed to about 1.003 units for Mexico. **55.** 74.36% **57.** 7.61% if you had sold in June 2008 **59.** No. Compound interest increase is exponential, and exponential curves either increase continually (in the case of appreciation) or decrease continually (in the case of depreciation). The graph of the stock price first decreases and then increases during the given period, so the curve cannot be compound interest change. **61.** 31 years; about \$26,100 **63.** 2.3 years **65. a.** \$1,510.31 **b.** \$54,701.29 **c.** 23.51% **67.** The function $y = P(1 + r/m)^{mx}$ is not a linear function of x, but an exponential function. Thus, its graph is not a straight line. **69.** Wrong. Its growth is exponential and can be modeled by $0.01(1.10)^t$. **71.** The graphs are the same because the formulas give the same function of x; a compound-interest investment behaves as though it were being compounded once a year at the effective rate. **73.** The effective rate exceeds the nominal rate when the interest is compounded more than once a year because then interest is being paid on interest accumulated during each year, resulting in a larger effective rate. Conversely, if the interest is compounded less frequently than once a year, the effective rate is less than the nominal rate. **75.** Compare their future values in constant dollars. The investment with the larger future value is the better investment. **77.** The graphs are approaching a particular curve as m gets larger, approximately the curve given by the largest two values of m.

Section 5.3

1. \$15,528.23 **3.** \$171,793.82 **5.** \$23,763.28 **7.** \$147.05 **9.** \$491.12 **11.** \$105.38 **13.** \$90,155.46 **15.** \$69,610.99 **17.** \$95,647.68 **19.** \$554.60 **21.** \$1,366.41 **23.** \$524.14 **25.** \$248.85 **27.** \$1,984.65 **29.** \$999.61 **31.** \$998.47 **33.** 3.617% **35.** 3.059% **37.** \$973.54 **39.** \$7,451.49 **41.** You should take the loan from Solid Savings & Loan: It will have payments of \$248.85 per month. The payments on the other loan would be more than \$300 per month.

43. Answers using correctly rounded intermediate results:

Year	Interest	Payment on Principal
1	\$3,934.98	\$1,798.98
2	\$3,785.69	\$1,948.27
3	\$3,623.97	\$2,109.99
4	\$3,448.84	\$2,285.12
5	\$3,259.19	\$2,474.77
6	\$3,053.77	\$2,680.19
7	\$2,831.32	\$2,902.64
8	\$2,590.39	\$3,143.57
9	\$2,329.48	\$3,404.48
10	\$2,046.91	\$3,687.05
11	\$1,740.88	\$3,993.08
12	\$1,409.47	\$4,324.49
13	\$1,050.54	\$4,683.42
14	\$661.81	\$5,072.15
15	\$240.84	\$5,491.80

45. First five years: \$402.62/month; last 25 years: \$601.73 **47.** Original monthly payments were \$824.79. The new monthly payments will be \$613.46. You will save \$36,481.77 in interest. **49.** 10.81% **51.** 13 years **53.** 4.5 years **55.** 24 years **57.** He is wrong because his estimate ignores the interest that will be earned by your annuity—both while it is increasing and while it is decreasing. Your payments will be considerably smaller (depending on the interest earned). **59.** He is not correct. For instance, the payments on a \$100,000 10-year mortgage at 12% are \$1,434.71, while for a 20-year mortgage at the same rate, they are \$1,101.09, which is a lot more than half the 10-year mortgage payment. **61.** $PV = FV(1+i)^{-n} = PMT\frac{(1+i)^n - 1}{i}(1+i)^{-n} = PMT\frac{1-(1+i)^{-n}}{i}$

Chapter 5 Review

1. \$7,425.00 **3.** \$7,604.88 **5.** \$6,757.41 **7.** \$4,848.48 **9.** \$4,733.80 **11.** \$5,331.37 **13.** \$177.58 **15.** \$112.54 **17.** \$187.57 **19.** \$9,584.17 **21.** 5.346% **23.** 14.0 years **25.** 10.8 years **27.** 7.0 years **29.** 168.85% **31.** 85.28% if she sold in February 2010. **33.** No. Simple interest increase is linear. We can compare slopes between successive points to see if the slope remained roughly constant: From December 2002 to August 2004 the slope was $(16.31 - 3.28)/(20/12) = 7.818$, while from August 2004 to March 2005 the slope was $(33.95 - 16.31)/(7/12) = 30.24$. These slopes are quite different. **35.** 2003

Year	2000	2001	2002	2003	2004
Revenue	\$180,000	\$216,000	\$259,200	\$311,040	\$373,248

37. At least 52,515 shares **39.** \$3,234.94 **41.** \$231,844 **43.** 7.75% **45.** \$420,275 **47.** \$140,778 **49.** \$1,453.06 **51.** \$2,239.90 per month **53.** \$53,055.66 **55.** 5.99%

Chapter 6

Section 6.1

1. $F = \{$spring, summer, fall, winter$\}$ **3.** $I = \{1, 2, 3, 4, 5, 6\}$ **5.** $A = \{1, 2, 3\}$ **7.** $B = \{2, 4, 6, 8\}$ **9. a.** $S = \{(H, H), (H, T), (T, H), (T, T)\}$ **b.** $S = \{(H, H), (H, T), (T, T)\}$ **11.** $S = \{(1, 5), (2, 4), (3, 3), (4, 2), (5, 1)\}$ **13.** $S = \{(1, 5), (2, 4), (3, 3)\}$ **15.** $S = \emptyset$

17.

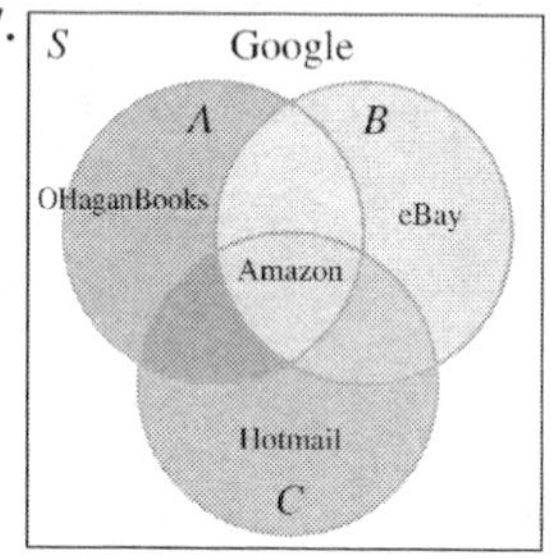

19.

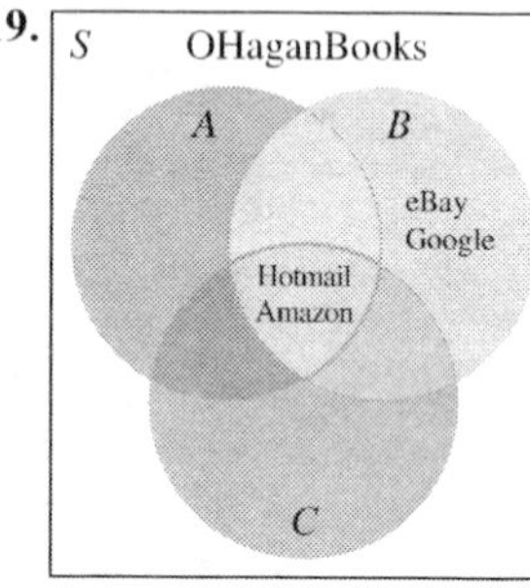

21. A **23.** A **25.** {June, Janet, Jill, Justin, Jeffrey, Jello, Sally, Solly, Molly, Jolly} **27.** {Jello} **29.** $\emptyset$ **31.** {Jello} **33.** {Janet, Justin, Jello, Sally, Solly, Molly, Jolly} **35.** {(small, triangle), (small, square), (medium, triangle), (medium, square), (large, triangle), (large, square)} **37.** {(small, blue), (small, green), (medium, blue), (medium, green), (large, blue), (large, green)}

39.

	A	B	C
1		Triangle	Square
2	Blue	Blue Triangle	Blue Square
3	Green	Green Triangle	Green Square

41.

	A	B	C
1		Blue	Green
2	Small	Small Blue	Small Green
3	Medium	Medium Blue	Medium Green
4	Large	Large Blue	Large Green

43. $B \times A = \{1H, 1T, 2H, 2T, 3H, 3T, 4H, 4T, 5H, 5T, 6H, 6T\}$ **45.** $A \times A \times A = \{HHH, HHT, HTH, HTT, THH, THT, TTH, TTT\}$ **47.** $\{(1,1), (1,3), (1,5), (3,1), (3,3), (3,5), (5,1), (5,3), (5,5)\}$ **49.** $\emptyset$ **51.** $\{(1,1), (1,3), (1,5), (3,1), (3,3), (3,5), (5,1), (5,3), (5,5), (2,2), (2,4), (2,6), (4,2), (4,4), (4,6), (6,2), (6,4), (6,6)\}$ **61.** $A \cap B = \{$Acme, Crafts$\}$ **63.** $B \cup C = \{$Acme, Brothers, Crafts, Dion, Effigy, Global, Hilbert$\}$ **65.** $A' \cap C = \{$Dion, Hilbert$\}$ **67.** $A \cap B' \cap C' = \emptyset$

International Finance Supplement

CHAPTER 1

An Introduction to International Financial Management

Required Investment
Sales Revenues
Operating Costs
Taxes

Cash Flows (CF)

$$\text{VALUE} = \sum_{t=0}^{N} \frac{CF_t}{(1+r)^t}$$

Capital Structure
Interest Rates
Asset Risk
Market Risk

Required Return (r)

e-resource

Visit ***http://crum.swlearning.com*** *to see the Web site accompanying this text. This ever-evolving site, for both students and instructors, is a tool for teaching and learning, and contains Excel chapter models and PowerPoint slides.*

In today's global economy, raw materials, finished goods, services, and money flow freely across most national boundaries, as do innovative ideas and new technologies. For example, world-class American companies and others around the world are making breakthroughs in foreign laboratories, obtaining capital from foreign investors, and putting foreign employees on the fast track to top management. Dozens of top U.S. manufacturers, including Dow Chemical, Colgate-Palmolive, Gillette, Hewlett-Packard, and Xerox, sell more of their products outside the United States than they do at home. Service firms are not far behind as Citigroup, Disney, McDonald's, and Time Warner all receive more than 20 percent of their revenues from foreign sales. Other multinational enterprises, or MNEs, from around the world are also experiencing similar sales and revenue growth in foreign markets, including the United States.

The trend is even more pronounced regarding profits. In recent years, Coca-Cola and many other companies have made more money in the Pacific Rim and western Europe than in the United States. However, like other companies, Coke has found that global investing also presents unique challenges and risks. Managers of multinational companies must deal with a wide range of issues that are not present when a company operates in a single country. Successful global companies such as Coca-Cola must conduct business in different countries, and they must be sensitive to the many subtleties of different cultures as well as political, legal, and economic systems.

This text is designed to help managers understand the global financial environment. We explain what "international financial management" is, and we explore the implications of globalization for firms' values. Why is it important for people to understand international financial management? Because it explains both how managers can increase their firm's value and why it is essential for them to do so. Today more than ever before, investors are forcing managers to focus on value maximization. Similarly, that is our focus in this text. Shown on the previous page is the basic valuation equation that we will use throughout the book.

In Part One of this text, we show that both the numerator and denominator of the valuation equation can be influenced by fluctuating economic conditions and associated changes in exchange rates. In Part Two, we use this equation to find the values of bonds, stocks, and other securities. We will also see how, given a security's price, we can use the equation to find its required rate of return. In Part Three, we discuss how managers use the valuation equation to make investment decisions within the firm. Finally, in Part Four, we show how the valuation equation is influenced by international corporate strategic decisions, including those involving risk management and multinational tax management.

In this first chapter, we introduce you to the international financial environment. We discuss some basic international concepts, reasons firms expand globally, unique challenges faced by multinationals, and the logic behind shareholder value maximization.

OVERVIEW OF INTERNATIONAL FINANCIAL MANAGEMENT

Multinational Enterprise (MNE) A firm that operates in an integrated fashion in a number of countries.

Multinational enterprises, or **MNEs**, control a large and growing share of the world's technological, marketing, and productive resources. Rather than merely buying resources from and selling goods to foreign nations, MNEs make direct investments in fully integrated operations—from the extraction of raw materials, through the manufacturing process, and on to distribution to consumers throughout the world. The traditional aspects of international commerce—namely, sourcing inputs from and selling goods and services to other countries (*importing* and *exporting*)—are also vitally important to MNEs, but their international involvement goes far beyond these traditional functions.

The emergence of "world companies" raises a host of questions for governments. Should domestic firms be favored, or does it make no difference what a company's nationality is as long as it provides domestic jobs? Should foreign companies be permitted to acquire domestic firms (*cross-border acquisitions*), or should the independence of local companies from foreign control be maintained? Should a company make an effort to keep jobs in its home country, or should it produce where the total costs of serving its markets are lowest? Clearly, even businesses that operate exclusively in the United States or in other countries are not immune to the effects of **globalization**. Increasing globalization creates additional competition from around the world, which then affects both local jobs and company profits. Globalization also has the potential to raise our standard of living by allowing us greater access to a wider range of products and services at more competitive prices. At the same time, though, it can challenge long-held cultural values about the role of businesses in society and the regulation of company practices, and it can even threaten local control of the country's economic destiny.

Globalization The increasing interconnectedness of economies, markets, and people across countries.

As you go through this book, you should gain a better appreciation of the problems facing governments and local societies as globalization trends advance and of the difficult, but profitable, challenges and opportunities facing managers of companies that seek to capitalize on this trend. Some of these firms will fit the definition of an MNE given earlier, but others will be more difficult to categorize. Even companies that consider themselves to be "domestic" are faced with many of the same issues as MNEs for at least three reasons:

1. Foreign competitors may enter these firms' traditional markets and use the benefits of their global scope to out-compete the domestic firms.
2. Some of the inputs used by domestic firms may be imported and/or some of their outputs may be sold to foreign customers, so changes in exchange rates can affect their costs and revenues, hence profits.
3. Even if all of a firm's inputs are domestic and all of its sales are to local customers, the globalization of capital markets will still affect the availability and cost of capital to both it and its customers.

Managers who ignore the implications of globalization for their companies do so at their own peril, so at a minimum, they need to understand how to protect themselves from adverse impacts of foreign competition and capital cost shifts, and then to take advantage of new opportunities that are created by such trends.

SELF-TEST QUESTIONS

Briefly explain whether the following statement is true or false: "MNEs' international involvement doesn't extend past importing and exporting goods and services."

What is a cross-border acquisition?

How does increased globalization raise our standard of living?

How might the trend toward globalization affect even a small, local homebuilder?

MAJOR DEVELOPMENTS IN INTERNATIONAL FINANCE

The world economy is quite fluid, so changes in one sector often have dramatic effects on economies throughout the world. Here we list a few of the major events that have shaped the current international financial environment:

1. The collapse of communism in eastern Europe and the former Soviet Union and the movement toward market economies (i.e., the privatization of state-run enterprises) throughout the world has created vast new markets for international commerce and foreign investment.
2. The European Community and the European Free Trade Association have created a "borderless" region where people, capital, goods, and services move freely among the European Union (EU) nations without the burden of tariffs. This consolidation has led to the creation of a single European currency called the "euro," and an increased militancy and clout regarding issues such as antitrust policy, product standards, and the protection of information.
3. The North American Free Trade Agreement (NAFTA) has moved the economies of the United States, Canada, and Mexico much closer together and made them more interdependent.
4. U.S., foreign bank, and capital market regulations have been loosened dramatically. One key deregulatory feature in the United States was the removal of interest rate ceilings, allowing banks to attract foreign deposits by raising rates. Another key feature was the removal of barriers to entry by foreign banks, which resulted in more cross-border banking transactions. At the same time, several foreign countries have significantly deregulated and liberalized

their foreign exchange and capital markets. All this has increased global competition in the financial services industry.

5. With the continuing reduction in trade and financial barriers, as well as advances in communication and computer technology, financial markets have become increasingly integrated (or at least less segmented), resulting in the increased globalization of these markets.
6. The emergence of China as an economic powerhouse in Asia has supplanted Japan somewhat in terms of influence and market interest. In addition to Hong Kong, with its separate economic system, new trading zones along the southeast China coast are hotbeds of investment growth and the focus of many foreign direct investments.
7. The rise of cross-border mergers and acquisitions are challenging domestic practices in many countries and causing governments to consider restrictions on the traditional flexibility of MNEs to integrate their global operations to their own benefit.
8. The ease with which imports can be brought into the United States, combined with low overseas production costs, especially in China, has led to a dramatic increase in the U.S. balance of trade deficit. This means that foreigners hold a large and increasing amount of U.S. debt securities. That, in turn, has implications for Federal Reserve monetary policy and thus the level of interest rates in the United States, which of course affects the U.S. economy.

SELF-TEST QUESTION

What are some major events that have affected the international financial environment?

WHY FIRMS "GO GLOBAL"

In response to the events just described, U.S. corporations have significantly increased their direct investments in other nations over the past two decades, and foreign corporations have mirrored the pattern and invested more in the United States. This trend is shown in Figure 1-1. These developments suggest an increasing degree of mutual interdependence among business enterprises and nations, what we have called "globalization." Table 1-1 demonstrates this trend on a country-by-country basis for 25 nations. The general trend from 1998 to 2000 shows an increase in foreign direct investment (FDI), followed by a steep drop-off that coincided with the bursting of the Internet bubble and the 2001 recession. However, this is not true of all countries, as China, France, Italy, and Mexico did not have huge declines and show almost no signs of the economic events that transpired. Notice that many nations maintained a relatively stable flow of FDI, and in some cases negative FDI even existed. An important question underlying this trend is, why do firms invest in foreign production facilities and related ventures? Various reasons have been put forth to explain why U.S. and foreign firms "go global." Some of the more important ones include the following:

1. *To seek production efficiency.* As competition increases in their domestic marketplace, and as demand increases in other markets, companies often reassess where it is best to produce their products. Depending on the nature of the production process, the availability of labor with the requisite skills, and the adequacy of transportation infrastructure, companies that operate in high-cost countries have strong incentives to shift production to lower-cost regions. For example, GE has production and assembly plants in Mexico, South Korea, and Singapore, and even Japanese manufacturers have started to shift some of

Figure 1-1 Direct Investment Positions on a Current-Cost Basis, 1982–2002

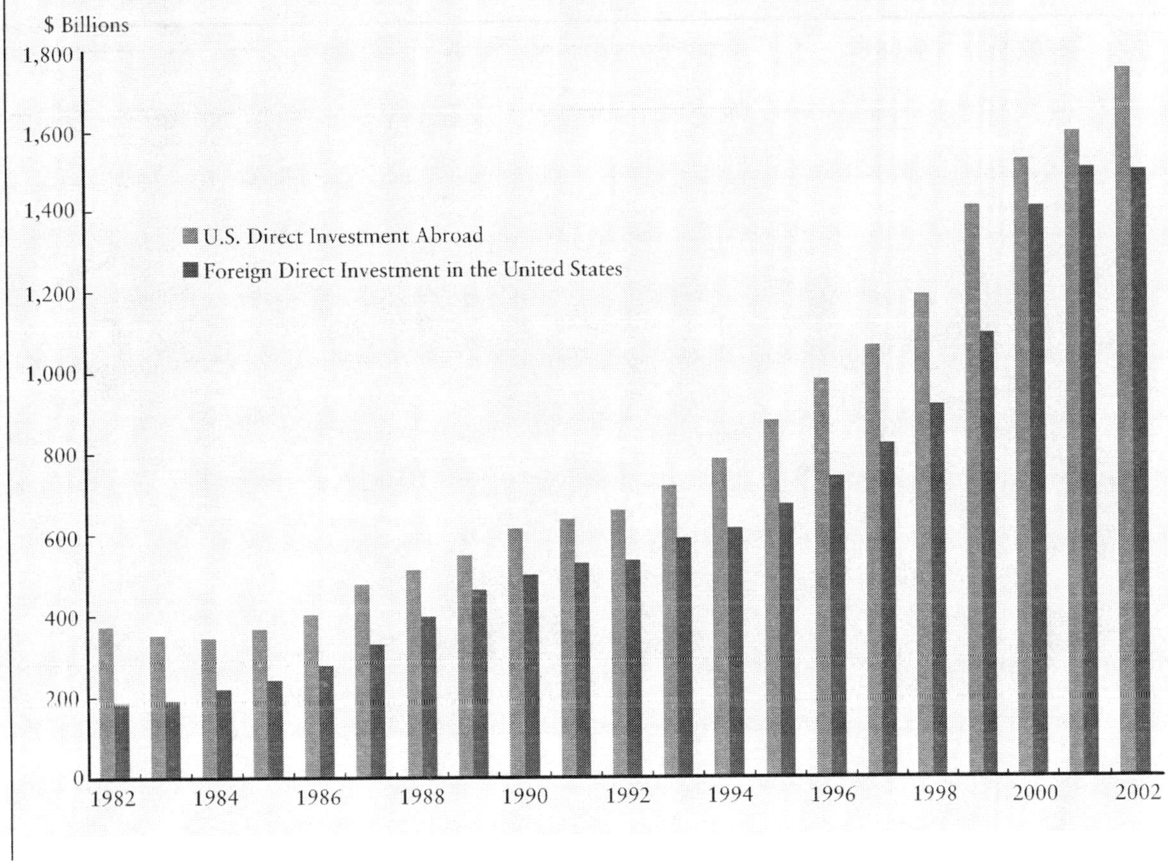

Source: Elena L. Nguyen, "The International Investment Position of the United States at Yearend 2002," *Survey of Current Business*, July 2003, 12–21.

their production to lower-cost countries in the Pacific Rim and the Americas. BMW, in response to high production costs in Germany, built assembly plants in the United States, among other countries. These examples illustrate how companies strive to remain competitive by locating manufacturing facilities wherever in the world they can produce and transport their products to meet the demand in their major markets at the lowest total unit landed costs.

Labor costs are one of the most important components in location decisions. Looking at Table 1-2, we see that labor costs vary considerably across countries. For instance, they are as high as $26.18 in Germany, $21.33 in the United States, and $18.83 in Japan, but only $2.38 in Mexico. Because labor is often immobile, particularly across countries, labor services in a country can be underpriced relative to its productivity for extended periods. This labor market imperfection makes it advantageous for firms to locate production abroad to benefit from lower cost labor services. The ability to shift production from country to country also has important implications for labor costs in other countries. For example, when Xerox threatened to move its copier rebuilding work to Mexico, its union in Rochester agreed to work-rule changes and productivity improvements that kept the operation in the United States. In addition, transportation costs often play a vital role in the location decision.

TABLE 1-1 **Foreign Direct Investment (Net Inflows) Measured in Billions of 2001 U.S. Dollars**

Country	1998	1999	2000	2001
Argentina	$ 7.29	$ 23.99	$ 11.66	$ 3.21
Australia	6.05	5.70	11.51	4.39
Belgium	22.69	142.70	214.43	73.63
Brazil	31.91	28.58	32.78	22.64
Canada	22.74	24.49	66.02	27.44
Chile	4.81	8.99	3.64	4.48
China	43.75	38.75	38.40	44.24
Costa Rica	0.61	0.62	0.41	0.45
France	29.52	46.63	43.17	52.50
Germany	23.64	54.58	207.72	31.53
Hong Kong, China	14.77	24.58	61.92	22.83
Indonesia	−0.36	−2.75	−4.55	−3.28
Israel	1.74	2.87	4.52	3.22
Italy	2.63	6.94	13.18	14.87
Japan	3.27	12.31	8.23	6.19
Kazakhstan	1.15	1.59	1.28	2.76
Kuwait	0.06	0.07	0.02	−0.04
Mexico	11.90	12.48	14.19	24.73
Morocco	0.33	0.85	0.20	2.66
Russian Federation	2.76	3.31	2.71	2.47
Singapore	6.39	11.80	5.41	8.61
South Africa	0.55	1.50	0.97	7.16
Sweden	19.413	59.386	22.125	13.085
United Kingdom	74.65	89.53	119.93	63.11
United States	179.03	289.44	307.74	130.80
Totals	$2,509.29	$2,887.94	$3,187.62	$2,564.69

Note: Net inflows represent new investments less withdrawals during the year.

Source: "Balance of Payments Statistics," *International Monetary Fund, Yearbook 2002.*

2. *To avoid political, trade, and regulatory hurdles.* Governments sometimes impose tariffs, quotas, and other restrictions on imported goods and services. They often do so to raise revenue, protect domestic industries, and pursue various political and economic policy objectives. To circumvent these government hurdles, firms often develop production facilities abroad. For instance, the primary reason Japanese auto companies moved production to the United States was to get around U.S. import quotas. Now, Honda, Nissan, Toyota, Mazda, and Mitsubishi are all assembling vehicles in the United States. This was also the situation with India in the 1970s when it was following a development strategy to compete domestically with imported products. One of the factors that prompted U.S. pharmaceutical maker SmithKline and Britain's Beecham to merge was that they wanted to avoid licensing and regulatory delays in their largest markets, western Europe and the United States. GlaxoSmithKline (the result of a 2000 merger between Glaxo Wellcome and SmithKline

TABLE 1-2 Hourly Compensation Costs for Manufacturing Production Workers Measured in U.S. Dollars

	1982	1987	1992	1997	2002
Canada	$10.45	$ 12.02	$17.17	$16.47	$16.02
France	7.85	12.29	17.47	17.24	17.42
Germany (former West Germany)	10.25	16.85	25.30	27.52	26.18
Japan	5.60	10.79	16.35	19.40	18.83
Mexico	1.97	1.04	2.17	1.70	2.38
Spain	5.28	7.63	13.50	12.16	12.04
Sweden	10.07	15.12	24.59	22.22	20.18
Taiwan	1.24	2.25	5.05	5.80	5.41
United Kingdom	6.92	9.09	14.37	15.57	17.47
United States	11.68	13.52	16.09	18.79	21.33

Source: "International Comparisons of Hourly Compensation Costs for Production Workers in Manufacturing, 1975–2002," U.S. Department of Labor, Bureau of Labor Statistics, Office of Productivity and Technology, September 2003.

Beecham) now identifies itself as an inside player in both Europe and the United States.

3. *To broaden their markets.* After a company's home market matures and competition becomes more intense, growth opportunities are often better in foreign markets. According to Vernon's product life-cycle theory, a firm first produces in its home market, where it can better develop its product and satisfy local customers.[1] This attracts competitors, but when the home market is expanding rapidly, new customers provide the sales growth desired. However, as the home market matures and the growth of total demand slows, competition becomes more intense. At the same time, demand for the product develops abroad, and this creates conditions favoring production in foreign countries both to satisfy foreign demand and to cut production and transportation costs so that the company can remain competitive. Thus, such homegrown firms as IBM, Coca-Cola, and McDonald's are aggressively expanding into overseas markets, and foreign firms such as Sony and Toshiba now dominate the U.S. consumer electronics market. Also, as products become more complex and development becomes more expensive, it is necessary to sell more units to cover overhead costs, so larger markets are critical. Thus, movie companies have "gone global" to get the volume necessary to support pictures such as *Titanic.*
4. *To seek raw materials and new technology.* Supplies of many raw materials that are important for industrial societies are geographically dispersed, so companies must go where the materials are found, no matter how challenging it may be to operate in some of the locations. For example, major deposits of oil are located on the northern coast of Alaska, in Siberia, and in the deserts

[1]Raymond Vernon, "International Investment and International Trade in the Product Cycle," *Quarterly Journal of Economics* 80, 1966, 190–207.

GOING GLOBAL AND YOUR JOB

We have all heard of the blue-collar American jobs lost because manufacturers have shifted production to low-cost countries such as Mexico, Taiwan, and Singapore. However, job losses are now moving up the food chain to relatively high-paying white-collar jobs in tech support, back-office operations, architecture, and even financial analysis. Why has this change occurred? In one word, *telecommunications*.

Most of us have bought a computer, printer, or software and then called an 800 number to ask a tech support person how to do something. Or, we have called a credit card company or a bank to find out about a charge or a check, or a catalog company to order something, or an airline to purchase a ticket. We or our parents may have hired an architect to design a house using architectural software, and we all use software that has millions of lines of code telling the computer how to perform different functions. In all of these situations, we are dealing with an American company, and we assume that Americans are on the other end of the line. But increasingly, the man or woman on the other end is Indian, Irish, Chinese, or some other foreign national.

Today, virtually the entire developed world is wired (or "wireless"), and the developing world is rapidly moving toward that goal. Moreover, information such as bank account and credit card records, catalog information, and airline flight schedules, along with design details on software or products under development, are kept in huge electronic databases. Those databases can be accessed from anywhere on earth—it's just as easy to access a database from Bombay as it is from Boston via satellite links. So, the people handling your catalog order, or flight reservation, or insurance claim, or who are programming the software for your next cell phone, can be physically located anywhere. Provided they are well educated and competent, you can be served from anywhere.†

In the past, men and women from India, China, and elsewhere would come to the United States for an education, then stay on and work in Silicon Valley, Wall Street, and the like. Now, though, many of those people are receiving an American education, returning home, and working for American, European, Japanese, and other multinational companies in their home countries. They may be employed by the multinational, but they often establish independent contracting companies, train their countrymen in the relevant technology, and then work under contract with the multinational.

Telecommunications makes this possible, and costs drive companies to use offshore service providers. For example, Microsoft managers were urged to "pick something to move offshore today . . . [because you can get] quality work at 50 percent to 60 percent of the cost. That's two heads for the price of one." Even Wall Street firms such as Lehman and Bear Stearns are eliminating $80,000-and-up financial analyst jobs and outsourcing them to Indian analysts, who can get constantly updated data from Thomson Analytics and other sources and analyze it just as well in Bombay as in New York, but at a fraction of the cost.

Globalization will produce winners and losers. Developing nations will certainly be winners, and this trend will lower costs for companies in some developed countries and also lower inflation rates. Moreover, worldwide efficiency and productivity will improve; developing nations will have higher incomes and thus can import more from developed nations—leading to gains for developed nations. However, at least some individual Americans who lose their jobs will be losers. Regardless of the final outcome, globalization of services and technology will affect all of us, and people, companies, and nations will have to deal with it.

†If you need to talk to someone, then spoken English is a requirement if you are American or British, or French for a French national, and so on. Note, though, that companies in Calcutta, Shanghai, and elsewhere literally have language schools that teach foreign nationals how to speak English with an American accent!

Source: Pete Engardio, Aaron Bernstein, and Manjeet Kripalani with Frederik Balfour, Brian Grow, and Jay Greene, "The New Global Job Shift," *BusinessWeek*, February 3, 2003, 50–60.

Vertically Integrated Investment
Occurs when a firm undertakes investment to secure its input supply at stable prices.

of the Middle East, all of which present unique challenges. This is why many U.S. oil companies, such as ExxonMobil, have major production facilities around the world to ensure access to the basic input resources needed to sustain the companies' primary business line. Because ExxonMobil has refineries, distribution facilities, and oil production fields, this type of investment is referred to as a **vertically integrated investment**, whereby the firm undertakes an investment to secure its supply of inputs at stable prices.

Natural Trade Barriers
Trade barriers that are not governmentally imposed.

Additionally, high transportation costs for raw materials can create **"natural" trade barriers** that are not governmentally imposed. To reduce transportation costs, firms often invest abroad. For example, cranberry grower Cran Chile invested in concentrate facilities because it is much cheaper to ship cranberry concentrate to the United States, its primary market, than it is to ship the raw berries. Furthermore, because no single nation holds a commanding advantage in all technologies, companies scour the globe for leading scientific and design ideas. By diversifying their production facilities around the world, firms gain better access to new technologies. For example, Xerox has introduced more than 80 different office copiers in the United States that were engineered and built by its Japanese joint venture, Fuji Xerox. Similarly, versions of the super concentrated detergent that Procter & Gamble first formulated in Japan in response to a rival's product are now being marketed in Europe and the United States.

5. *To protect the secrecy of their processes and products.* Firms often possess special intangible assets such as brand names, technological and marketing know-how, managerial expertise, and superior research and development (R&D) capabilities among others. Unfortunately, property rights in intangible assets are often difficult to protect, particularly in foreign markets. Firms sometimes invest abroad rather than license local foreign firms in order to protect the secrecy of their production process, distribution system, or the product itself. Once a firm's formula or production process is revealed to other local firms, they may then more easily develop similar products or processes, which will hurt firm sales.[2] For example, to protect their formula, Coke builds bottling plants and distribution networks in foreign markets but imports the concentrate or syrup required to make the product from the United States. In the 1960s, Coke faced strong pressure from the Indian government to reveal its formula in order to continue its operations in India. Rather than reveal its formula, Coke withdrew its operations from India until the foreign investment climate improved.
6. *To diversify.* By establishing worldwide production facilities and markets, firms can cushion the impact of adverse economic trends in any single country. For example, General Motors softened the blow of poor sales in the United States during a recent recession with strong sales by its European subsidiaries. Also, oil companies were able to weather the recent disruption in Venezuelan oil production by increasing production in Mexico and elsewhere in the world. In general, geographic diversification of inputs and outputs works because the economic fluctuations or political vagaries of different countries are not perfectly correlated. Therefore, companies investing overseas can benefit from diversification in the same way that individuals benefit from investing in a broad portfolio of stocks. However, because individual shareholders can diversify their investments internationally on their own, it makes less sense for firms to undertake foreign investments solely for diversification purposes. Note, though, that in countries that place constraints on foreign stock ownership or that do not have internationally traded companies, corporate diversification might make sense because then companies can do something that shareholders cannot duplicate easily in their individual portfolios.
7. *To retain customers.* If a company goes abroad and establishes production or distribution operations, it will need inputs and services at these new locations. If it can obtain what it needs from a single supplier that also operates

[2]According to the internalization theory of foreign direct investment, firms that have intangible assets often invest directly in foreign countries to avoid potential usage problems that may occur with their intangible assets and to also take advantage of potential economies of scale. We discuss FDI in more detail in Chapter 4.

in the same set of countries, then managing the relationship is much easier, and it is likely that economies of scale and other synergies will be obtained. Therefore, from the perspective of the supplier of inputs or services, it makes good business sense to follow customers abroad to retain the business. Large U.S. banks, such as Citibank and Chase, initially expanded abroad to supply banking services to their long-time customers, although they quickly capitalized on their global network to develop new customer relationships. The same history is also true for accounting, law, and advertising firms and other similar service providers.

SELF-TEST QUESTION Why do companies "go global"?

MULTINATIONAL VERSUS DOMESTIC FINANCIAL MANAGEMENT

Managers of multinational companies must deal with a wide range of issues that either are not present or are not as complex when a company operates in a single country. In theory, the concepts and procedures covered in your previous financial management courses are valid for both domestic and multinational operations. However, the following five major factors distinguish financial management in firms operating entirely within a single country from firms that operate globally:

1. *Different currency denominations.* Cash flows in various parts of a multinational corporate system will be denominated in different currencies. Hence, an analysis of exchange rates must be included in all financial analyses.
2. *Political risk.* Nations are free to place constraints on the transfer or use of corporate resources, and they can change regulations and tax rules at any time. At one extreme, they can even expropriate assets within their boundaries. Therefore, political risk can take on many subtle to more extreme forms. Of course, political risk is also present for companies operating in a single country, but the important reality for a multinational enterprise is that political risk not only exists but also varies from country to country, and it must be addressed explicitly in any financial analysis.
3. *Economic and legal ramifications.* Each country has its own unique economic and legal systems, and these differences can cause significant problems when a corporation tries to coordinate its worldwide operations. For example, differences in tax laws among countries cause a given economic transaction to have strikingly different after-tax consequences, depending on where the transaction occurs. Similarly, differences in legal systems of host nations, such as the Common Law of Great Britain versus the French Civil Law, complicate matters ranging from the simple recording of business transactions to the role played by the judiciary in resolving conflicts. Such differences can restrict multinational corporations' flexibility in deploying resources and make procedures that are required in one part of the company illegal in others. These differences also make it difficult for executives trained in one country to move easily to another.
4. *Role of governments.* Most financial models developed in the United States assume the existence of a competitive marketplace in which the participants determine the terms of trade. The government, through its power to establish basic ground rules, is involved in the process, but other than taxes, its role is minimal. Thus, the market provides the primary barometer of success, and it

gives the best clues about what must be done to remain competitive. This view of the process is reasonably correct for the United States and western Europe, but it does not accurately describe the situation in the rest of the world. Although market imperfections can complicate the decision process, they can also be valuable to the extent that they can be overcome by one firm but still serve as barriers to entry by competitors. Frequently, the terms under which companies compete, the actions that must be taken or avoided, and the terms of trade on various transactions are not determined in the marketplace but by direct negotiation between host governments and multinational enterprises. This is essentially a political process, and it must be treated as such. Thus, our traditional financial models have to be recast to include political and other noneconomic aspects of the decision. The ultimate outcome of such negotiations can provide access to additional profitable opportunities for the firm.

5. *Language and cultural differences.* The ability to communicate is critical in all business transactions. In this regard, U.S. citizens are often at a disadvantage because they are generally fluent only in English, while European and Japanese businesspeople are usually fluent in several languages, including English, hence they can operate in U.S. markets more easily than Americans can operate in their countries. At the same time, even within geographic regions that are considered relatively homogeneous, different countries have unique cultural heritages that shape values and influence the conduct of business. Multinational corporations find that matters such as defining the appropriate goals of the firm, attitudes toward risk, decision processes, performance evaluation and compensation system design, interactions with employees, and the ability to curtail unprofitable operations vary dramatically from one country to the next.

These five factors complicate financial management, and they increase the risks faced by multinational firms. However, the prospects for high returns and other factors make it worthwhile for firms to accept these risks and learn how to manage them.

SELF-TEST QUESTION Identify and briefly discuss five major factors that complicate financial management in multinational firms.

THE BASIC GOAL: CREATING STOCKHOLDER VALUE

Throughout this book we assume that management's basic overriding goal is to create value for stockholders. Stockholders own the firm—it legally belongs to them. As a result, they elect the directors, who then hire the managing executives. The directors, as representatives of the stockholders, monitor the actions of the managers, set the strategic direction of the company, and determine managers' compensation, rewarding them if performance is superior or replacing them if performance is poor.

To better understand the primary goal of creating stockholder value, we use the basic valuation equation given earlier as a key tool throughout this text. The basic valuation equation is

$$\text{Value} = \sum_{t=0}^{N} \frac{CF_t}{(1+r)^t},$$

where CF_t represents the expected cash flows at time t and r is the corresponding discount rate for the cash flows. In valuing a capital budgeting project, the valuation equation becomes

$$\text{Value} = \text{NPV} = \frac{CF_0}{(1+r)^0} + \frac{CF_1}{(1+r)^1} + \frac{CF_2}{(1+r)^2} + \cdots + \frac{CF_n}{(1+r)^n} = \sum_{t=0}^{n} \frac{CF_t}{(1+\text{WACC})^t}. \tag{1-1}$$

Note that Equation 1-1 is a special application of the basic valuation equation, where WACC is the project's risk-adjusted cost of capital and n is the project's life. To find a project's net present value (NPV), we estimate its relevant after-tax cash flows (CFs), its WACC, and n, after which we enter the data and solve the equation with a financial calculator or a computer spreadsheet. If a project has a positive NPV, it is generating more cash flow than is needed to cover its cost of capital, and the excess cash flow accrues to the firm's stockholders and thus raises company value. Therefore, if a firm takes on a project with a positive NPV, stockholder wealth increases by the amount of the NPV.

Societal Considerations

People with limited experience in business and economics, and sometimes even those with extensive experience, often argue that stock price maximization results in short-sighted decisions that are bad for employees, bad for consumers, and ultimately bad for society. They argue that firms should pursue nobler goals, such as the maximization of social well-being. Although such criticism makes for good press, and the lofty goals sound good, in practice the criticism is not warranted and the goals are not operational—people can't even agree how to define, much less measure, social well-being. Also, many of the activities that are criticized most often as being detrimental to social well-being (such as worker layoffs, plant closings, and similar restructuring activities) are based on the appropriate company objective of maintaining (or regaining) competitiveness. Companies that are not competitive in the global marketplace will not survive, and social well-being, however measured, will not be well served if companies go out of business.

There is another way to look at the issue. Of all the groups that have a direct stake in the continued health and prosperity of a company, only stockholders lack a contract that specifies the returns they are to receive. Virtually every other group has such a contractual relationship, and if the company does not honor the contract, they have recourse to the legal system to enforce it. This is why common stock is known as a "residual claimant"—only after all other claimants are fully satisfied do stockholders receive anything. Thus, maximizing the returns accruing to stockholders implies that all other groups with a direct stake in the fortunes of the firm have already received full and fair compensation. Also, we should always remember that in the real world, stock price maximization is constrained—we have laws that prevent managers from forming monopolies, from operating in an unsafe manner, from polluting the environment, and so forth. Stock prices respond positively to managerial actions that enhance the ability of firms to prosper in the long run, not to short-sighted actions. This means that the actions that managers legally take to maximize stock price, including restructuring when required to remain competitive, tend to provide long-run benefits to consumers as well as to stockholders.

Business students should understand why the goal of stock price maximization is indeed the proper foundation for our economic system, and why the same actions that maximize stock prices also benefit society. The economic logic behind this goal is spelled out here:

1. *Benefits to consumers.* Stock price maximization requires that corporations be efficient and produce goods and services with the required level of quality demanded by their target customer group at the lowest possible cost. They must also do this better than competitors in the minds of their customers. Moreover, the world is dynamic, so companies must continuously focus attention on producing products and services that consumers want and need, which leads to new and improved products in order to stay ahead of the competition. A competitive global economy also keeps their prices in check. For companies to maximize their stock prices, they must generate growth in sales by creating value for customers in the form of products that meet the needs of the customer, competitive prices, efficient and courteous service, adequate stocks of merchandise, and well-located business establishments.
2. *Benefits to employees.* There are cases in which a company's stock price increases when it announces plans to lay off employees. Properly interpreted, this is a signal from the market that some employees are redundant because of either bad management decisions in the past or changes in the business environment and that competitive advantage has shifted to other firms. In other words, it is a restructuring operation to regain competitive advantage, but viewed over time it is the exception rather than the rule for well-managed companies.[3] In general, companies that successfully increase stock prices also grow and add more employees, thus benefiting society. Note too that many governments across the world—including U.S. federal, state, and local governments—are privatizing some government-owned activities by selling these operations to investors. Perhaps not surprisingly, the sales and cash flows of recently privatized companies generally improve. Moreover, studies show that these newly privatized companies tend to grow at a higher rate and are more profitable than they were as state-owned enterprises, and their goods and/or services do not deteriorate in quality. When they are managed properly with the goal of stock price maximization, they require more employees, and society benefits.
3. *Other benefits.* Stockholders obviously benefit if the prices of their shares increase—it is better to be wealthier than poorer. Moreover, today many individuals around the world are stockholders, either directly or indirectly through retirement plans. Indeed, 45 percent of U.S. adults own stocks directly, and 80 percent own stocks through retirement programs. Thus, many members of society have an important stake in the stock market. Finally, strong stock prices benefit even those who have no direct or indirect ownership interest in the market, because strong markets stimulate both individual spending (through the "wealth effect") and corporate investment (because of lower capital costs and higher consumer spending). Obviously, all of this stimulates employment and economic growth.

Stockholder versus Stakeholder Wealth Maximization

Stockholder wealth maximization is generally accepted as the ultimate goal of financial management in many countries such as Australia, Canada, the United Kingdom, and the United States. However, it is not as widely accepted in many other parts of the world such as Asia and Europe. For instance, French and German firms often

[3]Maintaining employment is a public good and is socially desirable. However, if companies are forced to retain redundant workers, the cost of maintaining the public good is borne by the shareholders, not the public. When groups such as employees are hurt by company actions such as layoffs, government mechanisms should be available to mitigate the negative effects on the ex-employees—for example, unemployment insurance payments, job search assistance, and even retraining when necessary. These mechanisms transfer the cost of the public good to the general population so that it is not borne disproportionably by the shareholders.

Stakeholders Include a firm's employees, customers, banks, suppliers, and community, as well as its stockholders.

focus on maximizing the overall welfare of firms' **stakeholders**, with stockholders a subset of this group. Other stakeholders include the firm's employees, customers, banks, suppliers, and community. In Japan and Korea, managers often focus on increasing the prosperity of their interlocking business groups (called *keiretsu* and *chaebol*) by focusing on increasing their market share.

While stakeholder or business group maximization goals may sound noble, they often create long-run competitiveness problems and/or monopolistic behavior on the part of the business groups. In practice, these other goals are also very difficult to implement consistently and fairly. These problems, along with the increasing integration of world markets, have caused firms throughout the world to shift their focus more and more toward stockholder wealth maximization. Businesspeople throughout the world now realize that stockholder wealth maximization is a long-run goal and that firms cannot stay in business if they do not treat appropriately their employees, customers, and the environment in which they operate. When managers take actions to maximize the prices of their companies' stock, these same actions improve the quality of life for millions of ordinary citizens.

Agency Relationship Considerations

While stockholder wealth maximization provides strong business and societal benefits, in certain situations it can go awry. Managers are empowered by the owners of the firm—the shareholders—to make decisions. However, managers may have personal goals that compete with shareholder wealth maximization, and such potential conflicts of interest are addressed by *agency theory*.

An *agency relationship* arises whenever one individual (or more), called a *principal*, (1) hires another individual or organization, called an *agent*, to perform some service and then (2) delegates decision-making authority to that agent. Within the financial management context, the primary agency relationships are those (1) between stockholders and managers; (2) between managers and debtholders; and (3) among managers, stockholders, and debtholders in times of financial distress.[4]

Agency Problem A potential conflict of interest between the agent (manager) and (1) the outside stockholders, (2) the creditors (debtholders), and (3) stockholders and debtholders in times of financial distress.

A potential **agency problem** can arise whenever the manager of a firm owns less than 100 percent of the firm's common stock. If the firm is a proprietorship managed by its owner, the owner-manager will presumably operate so as to maximize his or her own welfare, with welfare measured in the form of increased personal wealth, more leisure, or perquisites.[5] However, if the owner-manager incorporates and then sells some of the stock to outsiders, a potential conflict of interest immediately arises. Now the owner-manager may decide to lead a more relaxed lifestyle and to work less strenuously to maximize shareholder wealth, because less of this wealth will accrue to him or her. Also, the owner-manager may decide to consume more perquisites, because some of these costs will be borne by the outside shareholders. In essence, the fact that the owner-manager will neither gain all the benefits of the wealth created by his or her efforts nor bear all of the costs of perquisites will increase the incentive to take actions that are not in the best interests of other shareholders. Agency conflicts are particularly important in large corporations because large firms' managers generally own only a small percentage of the stock. In this situation, shareholder wealth maximization could take a back seat to any number of conflicting managerial goals.

There are two extreme positions regarding how to deal with shareholder-manager agency conflicts. At one extreme, if a firm's managers were compensated solely on

[4]The classic work on the application of agency theory to financial management is Michael C. Jensen and William H. Meckling, "Theory of the Firm, Managerial Behavior, Agency Costs, and Ownership Structure," *Journal of Financial Economics*, October 1976, 305–360.

[5]*Perquisites* are executive fringe benefits such as luxurious offices, executive assistants, expense accounts, limousines, corporate jets, generous retirement plans, and the like.

the basis of stock price changes, agency costs would be low because managers would have a great deal of incentive to maximize shareholder wealth. However, it would be difficult if not impossible to hire competent managers under these terms, because the firm's earnings stream, and hence managers' compensation, would be affected by economic events that were not under managerial control. At the other extreme, stockholders could monitor every managerial action, but this would be costly and inefficient. The optimal solution lies somewhere in the middle, where executive compensation is tied to performance but some monitoring is also done.

In addition to conflicts between stockholders and managers, there can also be conflicts between stockholders (through managers) and creditors. Creditors have a contractual claim on part of the firm's earnings stream for payment of interest and principal on the debt, and they have a claim on the firm's assets in the event of bankruptcy. However, stockholders have control (through the managers) of decisions that affect the riskiness of the firm. Creditors lend funds at rates that are based on the firm's risk, which in turn is based on (1) the riskiness of the firm's existing assets, (2) expectations concerning the riskiness of future asset additions, (3) the existing capital structure, and (4) expectations concerning future capital structure decisions. These are the primary determinants of the riskiness of a firm's cash flows, hence the safety of its debt.

Now suppose stockholders, acting through management, cause a firm to sell some relatively safe assets and invest the proceeds in a large new project that is far riskier than the firm's old assets. This increased risk will cause the required rate of return on the firm's debt to increase, and that will cause the value of the outstanding debt to fall. If the risky project is successful, most of the benefits go to the stockholders, because creditors' returns are fixed at the old, low-risk rate. However, if the project is unsuccessful, the bondholders may have to share in the losses.

Can and should stockholders, through their managers/agents, try to expropriate wealth from creditors? The answer is no. First, creditors attempt to protect themselves from these types of actions by including restrictive covenants in debt agreements; so many times these actions are specifically prohibited. Second, even in the absence of covenants it does not make sense for a firm to deal unfairly with its creditors. Unethical behavior has no place in business, and if creditors perceive that a firm's managers are trying to take advantage of them, they will either refuse to deal further with the firm or will charge higher interest rates to compensate for the risk of possible exploitation. These higher interest rates, or the potential loss of access to capital markets, are detrimental to shareholders.

Corporate Governance System
Provides the financial and legal framework for regulating the relationship between a company's management and its shareholders.

Ultimately, it is important that firms have a strong **corporate governance system** in place to better monitor and address the various agency conflicts that arise within firms. Corporate governance mechanisms provide the financial and legal framework for regulating the relationship between a company's management and its shareholders. These mechanisms are particularly important when operating across countries where there is a wide variation in the legal protection of shareholders.[6] Some specific mechanisms used in the United States to motivate managers to act in shareholders' best interests include (1) managerial compensation plans, (2) direct intervention by shareholders, (3) the threat of firing, and (4) the threat of takeover. Promoting good

[6]There are two different philosophies of corporate governance corresponding roughly to the same division that we discussed in reference to the proper goals of management. The corporate governance system used in the United States is "outsider" or stockholder-based. Most of the literature available in the United States on corporate governance focuses on mechanisms that outsiders—stockholders—can use to make managers' welfare congruent with that of the stockholders. The other philosophy of corporate governance is "insider" or bank-based. In Europe, Japan, and other countries, banks are both creditors and stockholders. They have access to inside information and use their position to ensure that managers operate in the banks' best interest. Often this means that the companies take positions that protect the creditors at the expense of the stockholders.

Does Social Capital Add to a Nation's Wealth?

What factors cause one nation to be wealthy and another poor? Obviously, an abundance of natural resources creates wealth. In addition, economists have documented that geography, institutions, and a market economy matter. Other things equal, countries closer to the equator tend to be poorer than those located in moderate climates. Institutions such as law and democracy contribute to wealth, but economists believe that "the market guides selfish human actions to serve the common good." According to sociologists, there may be a factor missing from this list.

For the past decade, sociologists have considered one additional factor: social capital. What is social capital? Researchers do not always agree on the definition, much less on its effect on a nation's wealth.

One definition of social capital is trust or community. What is its relationship to a nation's wealth? The general concept is that the more people trust one another, the better off their society. Trust causes people to work together more efficiently. Consequently, we might conclude that the United States, with a successful economy, has an abundance of social capital. Studies have shown that this may or may not be the case.

A significant paper in this field, "Bowling Alone," by Robert Putnam of Harvard University, points out that Americans were far less likely to join organizations in the 1990s than they were in the 1950s. He proved this by showing the decline of bowling leagues. We might conclude from this paper that social capital (defined in these terms) is declining in the United States.

The idea that trust or community matters in determining a nation's wealth is not consistent with a basic assumption of traditional economic theory: Humans are self-interested beings. Some behavioral economists believe that this assumption has been accepted all too readily. Samuel Bowles and Herbert Gintis, both of the University of Massachusetts, argue that this assumption is flawed because people donate their time to all sorts of activities whose costs outweigh the benefits. They argue humans are social animals. On the other hand, others argue that these activities can be explained because people derive utility from helping others. However, the authors disagree. Through experiments they demonstrate how groups encourage individuals to act in the interests of the group as a whole. Many subjects punished free-riders, who then responded to the shame by cooperating. Thus, selfish preferences cannot explain the incentives of people in a village, school, or parish. They conclude that communities are the missing element, alongside markets and government, in understanding an economy.

Edward Glaeser and David Laibson of Harvard and Bruce Sacerdote of Dartmouth define social capital as an individual's social skills, including a long list of contacts, charisma, or popularity. They measure social capital by the number of organizations an individual belongs to and believe that social capital, like financial wealth, can be built by individuals. They argue that the more people invest in social capital the more likely they are to remain living in the same place. So, they attribute part of the "bowling alone" phenomenon to our society's rising mobility. In addition, they argue that as individuals invest more in social capital they are more likely to gain from that investment. If there is nothing in it for them, they will neglect their neighbors. From their analysis, the assumption of traditional economic theory is still alive—humans are self-interested beings.

How does this definition of social capital affect a nation's wealth? Decisions to invest in social capital affect everyone, not just the individuals making these decisions. If social capital is defined in terms of community or trust, it is tempting to say that the impact is only positive—but this may not be so. People can act in the interests of a group to the exclusion of others. (Entry barriers to the group may be kept high—excluding outsiders.)

So, what are we to conclude? Steven Durlauf of Wisconsin writes that conclusions about social capital are too far reaching. He argues that social capital cannot be precisely defined and that too much has been invested in its explanation of why nations become wealthy. It looks as if the verdict is out on social capital and its impact on the wealth of nations. Surely, though, more will be written about the subject on both sides.

Source: "A Question of Trust," *The Economist*, February 20, 2003.

Business Ethics A company's attitude and conduct toward its employees, customers, community, and stockholders.

ethical behavior in the firm can also help induce managers to act appropriately with regard to shareholders. **Business ethics** can be thought of as a company's attitude and conduct toward its employees, customers, community, and shareholders. It is important to remember that the shareholders own the firm and hence their capital is at risk. Therefore, it is only equitable that they receive a fair return on their investments.

SELF-TEST QUESTIONS

What is management's primary goal?
Why is common stock said to have a residual claim on assets and income?
How does the goal of stock price maximization benefit society at large?
What are agency costs, and who bears them?
What are some mechanisms that encourage managers to act in the best interests of stockholders? To not take advantage of bondholders?
Why should managers avoid taking actions that are unfair to any of the firm's stakeholders?
How might agency considerations arise in foreign operations?

ORGANIZATION OF THE BOOK

An important aspect of operating in an international environment is the ability to understand and work with foreign exchange rates, markets, and international institutions. Part One of this text covers these important elements. Following this introductory chapter, we discuss exchange rates, currency markets, and related institutions in Chapter 2, while Chapter 3 covers exchange rate analysis. Chapter 4 concludes Part One by examining the decision to invest abroad and the process of international expansion.

In Part Two, we cover fundamental financial concepts and analysis as well as international fixed-income and equity securities. This section provides students with financial tools and concepts that can be used to make financial decisions. Chapter 5 covers the fundamentals of risk and return in an international environment, while Chapters 6 and 7 discuss financial statement analysis from an international perspective. The remaining two chapters in this section, Chapters 8 and 9, cover debt and equity markets and provide students with an understanding of the various bond and equity-related instruments that are available in international securities markets, where debt and equity capital can be raised.

Part Three deals with corporate investment decisions. Firms normally finance using a mix of capital sources—short-term bank debt, long-term bonds, preferred stock, convertible bonds, retained earnings, and common stock. If, and only if, the firm invests in assets that return more than the cost of the capital used to acquire them will investments enhance stockholders' value. Therefore, for effective capital budgeting, it is essential to have a reasonably good estimate of projects' costs of capital, which is covered in Chapter 10. Chapter 11 goes on to discuss capital budgeting methodology, using projected cash flows and the cost of capital to determine projects' values. It examines the primary methods used to evaluate projects, and it brings in important international variations. Chapter 12 addresses additional risk considerations in capital budgeting and "real options" that might be embedded in a capital project. Real options are opportunities that arise as the result of making particular investments, and such an option can turn what appears at first glance to be a negative NPV project into a home run. In Chapter 13 we take up some strategic corporate financial decisions. We discuss target capital structures, international variations in those structures, and policies for distributing corporate income to shareholders. The decisions about how much debt or equity to use to finance a firm's growth, and the level of dividends to pay, are all interrelated and can substantially affect the firm's riskiness and value. In Chapter 14, we cover working capital, defined as current assets minus current liabilities. Here, we consider the management and financing of current assets and mechanisms for globally repositioning funds among operating units of the MNE.

Finally, Part Four covers international corporate financial strategy. Chapter 15 reviews risk management issues, including the use of derivatives in risk management, along with a discussion of foreign exchange and interest rate risk management. We conclude, in Chapter 16, with a discussion of international tax systems and multinational tax management.

It is worth noting that some instructors may choose to cover the chapters in a different sequence from their order in the book. The chapters are written to a large extent in a modular, self-contained manner, so such reordering should present no major difficulties.

SUMMARY

Over the past few decades, the global economy has become increasingly integrated, and a growing number of companies generate an increasing share of their profits from overseas operations. While multinational companies have more opportunities, they also face more risks and obstacles than do companies that operate only in their home country. However, even firms that do not expand beyond their home market are affected by the globalization of world markets. This chapter introduced the international financial environment. We discussed basic international concepts, key trends affecting global financial markets today, reasons firms expand globally, unique challenges multinationals face along with important differences between multinational and domestic financial management, and shareholder value maximization. The key concepts discussed in this chapter are listed below.

- International operations are becoming increasingly important to individual firms and to the national economy. A **multinational enterprise (MNE)** is a business organization that operates in an integrated fashion in a number of countries.
- **Globalization** refers to the increasing interconnectedness of economies, markets, and people across countries.
- The movement toward **market economies** throughout the world (i.e., privatization of state-run enterprises), the **creation of currency and free trade areas**, **deregulation of financial markets**, and the **increasing globalization of markets** are some major developments in the international financial environment.
- Companies "go global" for seven primary reasons: (1) **to seek production efficiency**; (2) **to avoid political, trade, and regulatory hurdles**; (3) **to broaden markets**; (4) **to seek raw materials and new technology**; (5) **to protect the secrecy of their processes and products**; (6) **to diversify**; and (7) **to retain customers.**
- Five major factors distinguish financial management as practiced by domestic firms relative to multinational firms: (1) **different currency denominations**, (2) **political risk**, (3) **economic and legal ramifications**, (4) **role of governments**, and (5) **language and cultural differences.**
- The goal of financial management is to **maximize shareholder value** by maximizing the value of a firm's stock.
- The value of a firm's stock is determined by the **cash flows** a firm produces, the **riskiness** of those cash flows, and the **timing** of the cash flows.

- The economic logic behind stock price maximization can be shown in three ways: (1) **benefits to consumers**, (2) **benefits to employees**, and (3) other benefits such as the **stimulation of employment and economic growth.**
- Efforts to maximize stock prices benefit society in several ways. These efforts help to **make business operations more efficient.** To maximize stock prices, managers offer goods and services that consumers desire; they price those goods and services as low as possible; and low prices require efficient, low-cost operations. The quest for stock price maximization also leads to **innovation, new products and services,** and **improved productivity.**
- Firms in some European and Asian countries seek to maximize the value contributed to the firm's **stakeholders**, including its employees, customers, suppliers, community, and business groups, in addition to stockholders. However, these approaches are not always operational and can be detrimental to competition. Recently, more firms in these countries are emphasizing shareholder wealth maximization due to the realization that this is a better long-run goal.
- An **agency relationship** arises whenever an individual or group, called a **principal**, hires someone called an **agent** to perform some service, and then the principal delegates decision-making power to the agent.
- Important agency relationships exist between **shareholders and managers;** between **managers and debtholders;** and among **managers, shareholders, and debtholders** during times of financial distress.
- An **agency problem** refers to a conflict between principals and agents. For example, managers, as agents, may pay themselves excessive salaries, obtain unreasonably large stock options, and the like, at the expense of the principals, the stockholders.
- **Corporate governance systems** provide the financial and legal framework for regulating the relationship between a company's management and its shareholders. These mechanisms are particularly important when operating across countries where there is a wide variation in the legal protection of shareholders.
- To mitigate agency problems, firms should have strong **corporate governance systems** in place, better **align managers' incentives** with corporate performance, and **promote good ethical behavior** in the firm.
- **Business ethics** refers to a company's attitudes and conduct toward its employees, customers, community, and shareholders.

QUESTIONS

(1-1) What is a multinational enterprise?

(1-2) Briefly explain the following statement, "MNEs' international involvement goes far beyond traditional mechanisms of international commerce."

(1-3) What are some of the questions that arise for governments due to the emergence of "world companies"?

(1-4) Briefly identify some of the effects of globalization for businesses.

(1-5) List three reasons even companies that consider themselves to be domestic are faced with many of the same issues as MNEs.

(1-6) What must managers know with respect to globalization?

(1-7) Is stock price maximization an appropriate goal for financial managers even those whose companies operate worldwide?

(1-8) Briefly explain the following statement, "In the real world, stock price maximization is constrained."

(1-9) Is stockholder wealth maximization generally accepted as the ultimate goal of financial management throughout the world? Explain your answer.

(1-10) Is globalization a good or bad trend? Explain your answer.

(1-11) If a firm is considering overseas operations, what benefits would a firm expect if it operated in Mexico? South Korea? Spain? Brazil? Latvia?

(1-12) How is multinational financial management different from domestic financial management?

(1-13) Explain the basic valuation equation with respect to maximizing a firm's value.

(1-14) How can the basic valuation equation be applied to financial assets (such as stocks and bonds) and capital budgeting projects?

(1-15) How can information asymmetries (through agency problems) affect individual perceptions of the basic value equation?

(1-16) Explain whether the goal of shareholder wealth maximization is a short-term or long-term goal. Can shareholder wealth maximization be distorted if considered along the wrong time dimension?

SELF-TEST PROBLEM

Solutions Appear in Appendix B

(ST-1) Define each of the following terms:

Key Terms

a. Multinational enterprise; globalization; cross-border acquisition
b. Vertically integrated investment; natural trade barriers
c. Stakeholders
d. Agency theory; agency relationship; agency problem
e. Corporate governance system; business ethics

THOMSON ONE Business School Edition QUESTIONS

EXAMINING COMPENSATION STRUCTURE

Overview

The Securities and Exchange Commission (SEC) requires public corporations to regularly provide annual reports, key financial statements, and other important information to their shareholders. For a description of the SEC and the various reports it requires, you can visit the SEC's web site: **http://www.sec.gov**. The SEC also has an online service entitled Edgar, which provides the public with easy access to the various SEC filings. You can find this information at **http://www.edgar-online.com**.

Thomson Financial has developed a tool that puts together a wide range of useful information from these various filings. This database, entitled *Thomson One: Business School Edition,* is particularly useful because it is easy to navigate, and it provides a way for investors to download relevant information to an *Excel* spreadsheet. Thomson Financial has made it possible for users of this textbook to have full access to the full range of this database for many leading companies. Throughout the textbook, we will highlight areas where students can use the database to learn more about illustrative companies. We begin below by considering the compensation plan ExxonMobil has established for its key executives.

ExxonMobil's Executive Compensation Plan

While managers run companies on a day-to-day basis, several key issues require shareholder approval. These issues include the election of board members, changes in the corporate charter, and approval of significant corporate events such as mergers. With this in mind, corporations are required to provide an annual proxy statement, which summarizes the information that shareholders need to consider when voting. These proxy statements also provide a great deal of useful information about the board of directors, the amount of stock held by officers and directors of the corporation, and details regarding executive compensation.

ExxonMobil has a performance share plan under which executive compensation is tied directly to certain thresholds, or hurdles, established by the Board's Compensation Committee. You can use *Thomson One: Business School Edition* to obtain this information. From the *Thomson One: Business School Edition* home page, enter ExxonMobil's ticker symbol (XOM) and click on "GO."

- Click on the tab labeled "Filings," and notice that a second line of tabs appears.
- On your screen you should see a category heading titled "Filings" with a list of filings and date filed. Click on "Proxy."
- At this point you can find information regarding ExxonMobil's election of directors, their compensation, its principal shareholders, and the details regarding its executive compensation.
- To see details of its executive compensation plan, click on "Executive Compensation."
- To see the details regarding the long-term incentive plan, select "Other Benefit Plans."

Discussion Questions

1. Briefly describe the details regarding ExxonMobil's compensation plan, including short-term and long-term incentives.
2. What other types of compensation does ExxonMobil provide to its executives?
3. Identify some of ExxonMobil's key executives, and outline their compensation.

Integrated Case

After Amanda (Mandy) Wu graduated from USC with a degree in finance two years ago, she took a job as assistant to Rick Schmidt, Medical Reports International's chief financial officer. MRI produces electronic diagnostic equipment that is used in hospitals, nursing homes, acute care facilities, and, increasingly, in private homes. The company began in 1981 as Medical Reports Inc., a manufacturer of heat-sensing devices that used a digital transmission system to report on acute care patients in large hospitals. It was based in southern California, where its founders resided, but over time the company broadened its line of diagnostic equipment and developed into an international leader in the medical equipment field. Most of its design work—much of which involves developing software and computer chips—is done in California, but most of its manufacturing is done in the Far East and Mexico, and it has sales offices throughout the world. To reflect its international presence, the company changed its name to Medical Reports International in 1995.

Table IC1-1 gives a list of the people involved in the case. One of MRI's founders, Mike Diaz, was an engineer, and the other, Bob Conti, was a sales representative for a medical supply distributor. Diaz developed a heat-sensing device that showed "hot spots" on patients, transmitted this information to hospital control stations, and thus alerted the medical staff to developing problems. Conti used his contacts with hospitals to sell the system. They were successful from the start, and the company was profitable enough to finance its early growth with internal funds plus bank loans supported by a strong cash flow. However, in 1985 Diaz and Conti realized that additional capital was needed if the company was to realize its full potential, so they sold a minority interest to Southern California Ventures (SCV), a leading venture capital company. By 1990, MRI was highly profitable. The firm clearly had a high value even though it was not publicly traded, hence there was no way to actually measure that value. Also, it had become clear that the company should expand internationally, and that would require additional capital.

At that point, the principals decided that MRI should go public and raise the required capital by issuing new stock. Also, SCV wanted to "harvest" some of its gains; Diaz and Conti wanted to diversify their personal holdings; and a number of key employees wanted to cash in on their stock options and

TABLE IC1-1 People Involved in the Case

Amanda (Mandy) Wu	Assistant to the CFO
Rick Schmidt	Vice president, CFO
Mike Diaz	Director, founder, chairman; engineering background
Bob Conti	Director, founder, president; medical products background
Vince Fuller	Director, officer of venture capital firm
Eduardo Suarez	Director, international marketing expert
Sally Williams	Director, college president; expert on adult education
Ellen Warren	Director, scientist; expert on medical equipment
John Costello	Consultant, finance professor

buy long-delayed homes and other things. So, MRI had a stock offering in which the company sold new stock to bring in additional capital, and Diaz, Conti, and SCV sold some of their shares to diversify their holdings. As a result of this offering, and of additional sales in the secondary market over the years, Diaz, Conti, and SVC each own 15 percent of the outstanding stock and public stockholders own the remaining 55 percent. Key employees hold options on 10 percent of the stock, which, when exercised, will reduce those percentages slightly.

Neither Diaz nor Conti, nor Vince Fuller, the SCV representative on the board, has had formal finance training. Fuller has an MBA, but he majored in marketing and avoided finance and accounting courses as much as possible. The other three members of the board, Sally Williams, Ellen Warren, and Eduardo Suarez, also lack financial training. Williams is the president of Foothills Community College, Warren is a respected scientist and medical equipment expert, and Suarez is an international marketing expert. After passage of the Sarbanes-Oxley Act in 2003, under which the boards of public companies can be held accountable for improper reporting, board members expressed a concern about their lack of financial expertise. Also, Schmidt thought that the board might be able to make better policy decisions if the members were better grounded in the fundamentals of financial management.

At a recent board meeting, Suarez suggested that Schmidt organize an educational program. One of his friends, a director of an aerospace company, told Suarez about such a program his company had set up and how valuable the board felt it had been. Suarez's suggestion was then discussed, and Schmidt was asked to establish an educational program for MRI's directors. The discussion then turned to the program's format—what should be its content, and who should conduct it? Williams, the college president, had been involved with various corporate training programs and suggested the best results would be obtained if the program was organized and conducted by a team consisting of a college finance professor, a senior MRI financial executive, and a sharp junior member of the company's finance staff. The team would construct a program, with *PowerPoint* slides and other exhibits, and practice their presentation on a group of company representatives who would (1) benefit from knowing more finance and (2) be able to provide feedback on what worked and did not work in terms of getting important concepts across. After that, there would be a 2-hour session before each monthly board meeting devoted to finance topics. Williams laughingly said, "At the end, we'll be awarded with a degree from MRI University." Diaz liked the idea but suggested that the six operating VPs be included, because he sensed that they too needed a stronger background in financial management. Suarez suggested that international finance topics should be covered, because international markets have the best growth potential and provide unique challenges. The remaining board members agreed with these suggestions, and Schmidt was asked to implement the idea, starting with the next monthly board meeting.

Schmidt has long felt uncomfortable because few board members had a good understanding of fundamentally important concepts like the cost of capital and capital budgeting. He thinks board discussions of key policy matters will produce better results if members know more about financial management. For several years, Schmidt has used John Costello, a brilliant finance professor at the local university, as a consultant and advisor on various financial issues. Costello is also an award-winning teacher, so he is a natural choice for the academic on the team. Schmidt himself is the logical choice for the MRI executive, and he quickly decided to ask Mandy Wu to serve as the junior MRI person. Wu is on the fast track for promotion within the company, and work on the project will give her valuable insights into the thoughts of senior executives and board members.

Immediately after the board meeting, Schmidt called Costello and Wu, explained the situation, and secured their participation in the program. At a meeting the next morning, they decided to divide the course into 16 sessions, each covering an important issue, following the organization of a leading textbook on international financial management. Costello suggested that each session be organized around a series of questions that would take participants through the fundamentals of the topic. Facts and issues relating to MRI itself would be used, so each session will teach fundamental analysis and apply the analysis to current issues facing MRI. Thus, the sessions will serve both educational and policy-making functions. The board members and executives are all quite intelligent, so they should pick things up quickly, but they are not conversant with many finance terms and concepts. Therefore, even elementary topics should at least be mentioned. Schmidt and Costello, with Wu looking on, prepared the following questions for the introductory session. Wu's task is to prepare answers for use during the session. Put yourself in Wu's position and prepare answers for the following questions.

Questions

a. What should be the primary goal of the firm? How is this goal achieved? Is this the firm's only goal? How does this goal relate to those who interact with the company (customers, employees, etc.)? How does this goal relate to other company goals? In your responses, consider the implications of international operations.

b. How will increased globalization affect MRI's financial decision making?

c. What major factors have helped shape the current international financial environment, and what business opportunities have they created?

d. MRI's first foray into international operations was selling to overseas customers, and later this expanded to overseas production. List and explain some examples of why firms "go global."

e. MRI's transition into becoming a multinational enterprise (MNE) was a rough one. Specifically, MRI's managers had trouble identifying the unique challenges that operating in an international environment presents. Identify some factors that distinguish multinational financial management from domestic financial management.

f. Define agency relationships. What agency problems can develop as a firm expands and decides to operate internationally?

g. Explain the importance of corporate governance and the need for managers to behave ethically.

CHAPTER 2

The Global Financial Environment: Markets, Institutions, Interest Rates, and Exchange Rates

Required Investment
Sales Revenues
Operating Costs
Taxes

Cash Flows (CF)

$$\text{VALUE} = \sum_{t=0}^{N} \frac{CF_t}{(1+r)^t}$$

Capital Structure
Interest Rates
Asset Risk
Market Risk

Required Return (r)

Managers and investors do not operate in a vacuum; rather, they make decisions within a global financial environment. Even if their decisions deal directly with only a small part of the global environment, such as a single country, the larger marketplace still influences its individual parts. The global environment includes local financial markets and institutions within specific countries as well as truly international markets and institutions that span many countries. It also includes the diverse spectrum of national tax and regulatory policies plus macroeconomic conditions within each country. This entire set of factors determines the financial alternatives, and it affects the outcomes of decisions made by both managers and investors. For all these reasons, it is important to understand the global environment and to anticipate the consequences of decisions in this context.

FIF

e-resource

*Visit **http://crum.swlearning.com** to see the Web site accompanying this text. This ever-evolving site, for both students and instructors, is a tool for teaching and learning, and contains Excel chapter models and PowerPoint slides.*

Sound financial decisions require an understanding of the economies, interest rates, exchange rates, and capital markets in all the countries in which a company operates. This makes sound financial decision making a difficult task. However, in recent years advances in information and transportation technologies, along with the global integration of banking and other organizational infrastructure, have helped ease this challenge. The Internet is a key factor here—indeed, the way people communicate and firms conduct business have literally been redefined by the Internet. Electronic commerce represents a paradigm shift from *the traditional business model—not only in finance but also in other business fields, as the long-promised benefits of electronic transfers of funds and data are being realized. Suddenly, capital markets in other countries are opening to foreign firms, and many companies are beginning to understand the value of dealing in the so-called euromarkets.*

In this chapter, we introduce the basic characteristics of U.S. and foreign financial markets. We show that different national economies are really global networks tied together by exchange rates, and that global investment and financing opportunities, while more complicated, are much less constrained and thus potentially much more profitable than strictly local operations. In terms of the valuation equation, this chapter's central focus is on the required return in the denominator—specifically, on how it is determined and how changes in risk and inflation lead to changes in the required return, r, over time.

FINANCIAL MARKETS

Physical Asset Markets
Also known as tangible, or real, asset markets. Capital budgeting decisions are important to these markets.

Financial Asset Markets
Marketplaces that deal with financial instruments such as stocks, bonds, notes, mortgages, currencies, and so on.

Financial Instruments
Pieces of paper or electronic entries in account ledgers with contractual provisions that spell out their owners' claims on specific real assets. Their value is defined in terms of a fixed number of money units.

Monetary
Relating to a unit of currency; value defined in terms of a fixed number of money units.

Businesses, individuals, and governments often need to raise capital. For example, suppose ChevronTexaco wants to develop a new oil field in western Kazakhstan. Because development costs are estimated at more than $3 billion, which is more cash than the company presently has available, ChevronTexaco will have to go to the financial markets to raise the money. Or suppose that Gilberto Sanchez, a Mexico City restaurateur, wants to add an outdoor patio to his restaurant. Where will he get the money to finance the addition and installation of the equipment? At the same time, some individuals and companies have incomes that are greater than their current expenditures, so they have funds available to invest. For example, Microsoft had roughly $52.8 billion of cash and marketable securities in December 2003 available for future investments.

Thus, there are many individuals, companies, and governments who need capital for investments, and another set of people, businesses, and governments that have excess funds to invest. Financial marketplaces bring these two groups together, and the more efficient financial markets are, the better off the world is.

Types of Markets

When we think of a "market," many different images may come to mind. Perhaps it conjures up a picture of a "farmers market," where you go to buy fresh vegetables, or a used car market, where you go to buy a car. You talk about "being in the market for a new computer system," or even a new job. These are all examples of **physical asset markets** (also called "tangible" or "real" asset markets), including the market for human capital. These markets are important, but so are financial markets, because borrowers often need capital to make investments in the physical asset markets. We will look closely at the decision process for acquiring physical assets later, in Chapters 11 and 12, where we discuss capital budgeting, but here we concentrate on **financial asset markets**, which deal with stocks, bonds, notes, mortgages, currencies, and other **financial instruments**. All of these assets (or liabilities if you are the borrower) are **monetary** in nature, and their values are defined in terms of money units. They are simply pieces of paper or, increasingly, electronic entries in account ledgers, and their contractual provisions spell out their owners' claims on specific real assets. For example, a corporate bond issued by Nestlé entitles its owner to a specific claim on the cash flows produced by Nestlé's physical assets, while a share of Nestlé stock entitles its owner to a different set of claims on

Derivative Security
Any asset whose value is derived from the value of some other "underlying" real or financial asset.

Spot Markets
The markets in which assets are bought or sold for "on-the-spot" delivery.

Futures Markets
The markets in which participants agree today to buy or sell an asset at some future date.

Money Markets
The financial markets in which funds are borrowed or loaned for short periods (less than one year)

Capital Markets
The financial markets for equity and for intermediate- or long-term debt (one year or longer).

Mortgage Markets
Markets that deal with loans on residential, commercial, and industrial real estate.

Consumer Credit Markets
Markets that involve loans on autos, appliances, education, vacations, and the like.

Currency Markets
Markets in which transactions for foreign exchange occur.

Physical Markets
Formal organizations having tangible physical locations that conduct auction markets in designated ("listed") securities.

those cash flows. Unlike these conventional financial instruments, **derivative securities** are not direct claims on either real assets or their cash flows. Instead, a derivative is a contractual claim whose value depends on the value of some other "underlying" asset. *Futures* and *options* are two important types of financial derivatives, and their values depend on the values of other assets, such as Japanese yen, pork bellies, or IBM stock. Therefore, the value of a derivative security is *derived* from the value of an underlying real or financial asset. The following list identifies the major types of financial markets:

1. **Spot markets** and **futures markets** are terms that refer to whether assets are being bought or sold for "on-the-spot" delivery (literally, within a few business days) or for delivery at some more distant future date.
2. **Money markets** are the markets for short-term, highly liquid debt securities. The New York and London money markets have long been the world's largest, but Tokyo and Singapore are growing rapidly. **Capital markets** are the markets for intermediate- or long-term corporate debt, and for equity securities. The New York Stock Exchange (NYSE), where stocks of the largest U.S. corporations are traded, is a prime example of a capital market. There is no hard and fast rule about this, but when describing debt markets, "short-term" generally means less than one year, "intermediate-term" means one to five years, and "long-term" means more than five years. Most equity securities never mature, so by definition equity markets are "long term."
3. **Mortgage markets** deal with loans on residential, commercial, and industrial real estate, while **consumer credit markets** involve loans on autos, appliances, education, vacations, and so on.
4. *Global, national, regional,* and *local markets* also exist. For example, the Eurobond market (headquartered mainly in London but with participants throughout the world) is a global market for long-term debt, while the Tokyo Stock Exchange is a national market for Japanese equities. Euronext is a regional market embracing Amsterdam, Brussels, Lisbon, and Paris, while the Jamaica Stock Exchange illustrates a strictly local market. Depending on an organization's size and operational scope, it may be able to acquire capital from sources all around the world, or it may be confined to a local market.
5. **Currency markets** deal with buying and selling different nations' currencies, or *foreign exchange*. There is a retail component of this market in which individuals interact globally and exchange one currency for another, but there is also a much larger commercial side in which companies engage in foreign exchange transactions as a part of their business operations.
6. **Physical markets** are those that have a trading floor and a physical presence, such as the New York Stock Exchange. **Electronic markets** do not have a physical location but rather exist as a computerized trading system such as the Nasdaq. As technology advances, electronic markets are replacing more and more physical markets. *The Wall Street Journal* and other financial publications have speculated for years that the NYSE will eventually be forced by competitive pressures to evolve into a mostly electronic market.
7. **Primary markets** are those in which corporations raise new capital. If Intel issued newly created common stock, this would be a primary market transaction because the company receives the cash proceeds from the sale.
8. **Secondary markets** are markets in which existing securities are traded among investors. Thus, if an investor bought 100 shares of Intel stock from another person, the purchase would occur in the secondary market. The NYSE is a secondary market since it deals in outstanding shares as opposed to newly issued shares. Secondary markets also exist for bonds, mortgages, various

Electronic Markets
Markets with no physical location, but rather exist as a computerized trading system.

Primary Markets
Markets in which corporations raise capital by issuing new securities.

Secondary Markets
Markets in which existing securities and other financial assets are traded among investors.

Initial Public Offering (IPO) Market
The market in which firms "go public" by offering shares to the public for the first time.

Private Markets
Markets in which transactions are negotiated directly between two parties.

Public Markets
Markets in which standardized contracts are traded on organized exchanges.

other types of loans, and other financial assets. The corporation whose securities are being traded in a secondary market is not involved in the transaction and, thus, does not receive any funds from the sale.

9. The **initial public offering (IPO) market** is a subset of the primary market. Here firms "go public" by offering shares to the general public for the first time. Microsoft had its IPO in 1986. Prior to that time, Bill Gates and other insiders owned all the firm's shares. In many IPOs, the insiders sell some of their shares to diversify, and the company sells newly created shares to raise additional capital.
10. **Private markets**, where transactions are negotiated between two parties, are differentiated from **public markets**, where standardized contracts are traded on organized exchanges. Bank loans and private debt placed with insurance companies are examples of private market transactions. Because these transactions are private, they may be structured in any manner that is legal and agreeable to both parties. By contrast, securities issued in public markets (e.g., common stock and corporate bonds) are ultimately held by a large number of individuals. Public securities must have fairly standardized contractual features that appeal to a broad range of investors. Their standardized features and diverse ownership also make public securities relatively liquid. Private market securities are, in contrast, more tailor-made but less liquid.

An almost unlimited number of market classifications could be developed, but this breakdown is sufficient to illustrate that there are many types of financial markets and that a diverse mix of financial assets is traded. Note also that precise classifications are more useful for pedagogic than practical purposes, because people dealing in these markets focus on the nature of the contract rather than its market classification.

A healthy, free-market economy is dependent on the efficient transfer of funds from net savers to individuals and companies who need capital. Without efficient financial markets, the broader economy will not be efficient, and it cannot reach its potential. For example, in the former Soviet Union, many promising businesses are held back because capital is not readily available. Without business expansion, employment, productivity, and the standard of living will be lower than the achievable level. Similarly, financial markets in developing nations must be nurtured by governments if those markets are to function efficiently. Indeed, economic efficiency is simply impossible without a good system for allocating capital within the economy.

Recent Trends

Financial markets have been evolving at an increasing rate during the past two decades. Advances in information technology have made the globalization of banking and commerce more feasible, and that, in turn, has led to deregulation and increased global competition. The result is a much more efficient, internationally integrated market, but one that is far more complex than existed a few years ago. While globalization has been largely positive, it has also created problems for policy makers. U.S. Federal Reserve Board Chairman Alan Greenspan has stated that modern financial markets "expose national economies to shocks from new and unexpected sources, and with little if any lag." He has also stated that central banks must develop new ways to evaluate and limit risks to the financial system. Large amounts of capital can move quickly around the world in response to interest and exchange rate changes, and those capital movements can disrupt local institutions and economies, much like your personal economy would be disrupted if someone suddenly made a large withdrawal from your bank account.

Globalization has accented the need for greater cooperation among international regulators. Various committees are currently addressing the following issues: (1) the integration of different national banking and securities industry structures, (2) the trend in Europe and North America toward financial service conglomerates, and (3) the reluctance of individual countries to relinquish national monetary policy control. Still, regulators agree that gaps in global market supervision need to be closed.

Derivatives markets have played a vital role in global risk management. As noted earlier, a derivative is any security whose value is *derived* from the value of some other "underlying" asset. An option to buy IBM stock at a fixed price is a derivative as is a contract to buy Japanese yen six months from now at a fixed exchange rate. The derivatives market has grown faster than any other market in recent years, providing corporations with new opportunities but also exposing them to new risks, some of which are not visible to investors because of the complexity of the transactions and accounting rules that do not require their disclosure.

Derivatives can be used either for risk reduction or for speculation. Suppose an importer's net income falls whenever the dollar weakens against the yen, as it did throughout much of 2003. That company could reduce its risk by purchasing derivatives that increase in value whenever the dollar declines. This is called a *hedging operation*, and its purpose is to reduce risk exposure. We will look at this in detail in Chapter 15. Speculation, on the other hand, is done in the hope of earning a high return, but it raises a firm's risk exposure. For example, in the late 1990s, Procter & Gamble disclosed that it lost $150 million on derivative investments, and Nick Leeson, a currency trader working for Barings Bank in Singapore, lost more than $4.5 billion in yen-dollar speculation activities that caused the bank to go bankrupt.[1] The size and complexity of derivatives transactions concern regulators, academics, and lawmakers. U.S. Fed Chairman Greenspan noted that, in theory, derivatives should allow companies to manage risk better but that it is not clear whether recent innovations have increased or decreased the inherent stability of the financial system.

Cross-Listing
Occurs when a company lists shares of stock on multiple exchanges to increase its global recognition.

Around-the-Clock Trading
Global financial institutions maintain offices in different time zones around the world and thus offer anytime trading.

Another recent trend has been the development of new financial instruments that allow a firm's shares to be traded in prices based on local currencies in many national markets, or in global equity markets. As companies become increasingly global in perspective, they are listing shares on multiple exchanges. **Cross-listing** and **around-the-clock trading** have resulted in greater global market integration, which can mean lower shareholder risk through globally diversified portfolios and lower financing costs for companies. However, it also challenges traditional notions of nationality as companies move their headquarters from one country to another to obtain more favorable taxation and other regulations.

For example, there has been a recent trend for U.S. companies to incorporate in Bermuda to lower their corporate taxes. Ingersoll Rand, Global Crossing, and Tyco International are a few of the American companies with their "official" headquarters in Bermuda.[2] Global Crossing and Tyco have been there since 1997, while Ingersoll Rand (a manufacturer of locks, golf carts and utility vehicles, refrigeration equipment, and industrial and construction equipment) recently reincorporated in Bermuda. Similarly, in 2002 Stanley Works, makers of Stanley tools, won a shareholder vote to reincorporate in Bermuda, but after a fight with the state of Connecticut and the

[1]Barings was one of the world's oldest banks, and its clients included the Queen of England. There are several published books detailing the events that led up to the Barings collapse, including Nicholas William Leeson and Edward Whitley, *Rogue Trader: How I Brought Down Barings Bank and Shook the Financial World* (Little Brown & Company, New York, NY, 1996); and Judith H. Rawnsley and Nicholas William Leeson, *Total Risk: Nick Leeson and the Fall of Barings Bank* (HarperCollins, London, U.K., 1995). In addition, a movie titled *Rogue Trader* starring Ewan McGregor was released in 1999.

[2]Generally, the "real" headquarters, along with most actual operations, remains wherever it was; the main thing that changes is the location stated on some legal papers. Companies with headquarters in Bermuda or the Bahamas may have only a postal box in a lawyer's office, with no real tie to the "headquarters country." Obviously, this is considered by many to be an abuse, but it is often difficult to draw the line between a "legitimate" and an "illegitimate" decision to change the country of incorporation.

AFL-CIO, Stanley abandoned its plan to move to Bermuda. Still, the trend for U.S. companies to reincorporate in countries with more favorable tax treatment is likely to continue, given increasing pressures from competition in the global marketplace.

SELF-TEST QUESTIONS

What is the purpose of financial marketplaces?
What are some major types of financial markets?
What is a financial instrument? Give some examples.
What is a derivative security?
What recent trends have affected the global financial environment?

FINANCIAL INSTITUTIONS

Transfers of capital between savers and borrowers take place in the three different ways diagrammed in Figure 2-1:

Investment Banking House
An organization that underwrites and distributes new investment securities and helps businesses obtain financing.

1. *Direct transfers* of money and securities, as shown in the top section, occur when a business sells its shares or bonds directly to savers, without using a financial institution. The business delivers its securities to savers, who in turn give the firm the money it needs.
2. As shown in the middle section, transfers may also go through an **investment banking house** such as Merrill Lynch, which *underwrites* the issue. An underwriter serves as a middleman and facilitates the issuance of securities. If more than one underwriter is involved, there is an *underwriting consortium,* or *syndicate,* of several institutions, and one serves as the *lead underwriter.* The company sells its shares or bonds to the investment bank, which in turn sells these same securities to savers. The firms' securities and the savers' money merely "pass through" the investment banking house. However, the investment bank does hold the securities for a period of time, so it is taking a risk—it may not be able to resell them to savers for as much as it paid. Because new

Figure 2-1 Diagram of the Capital Formation Process

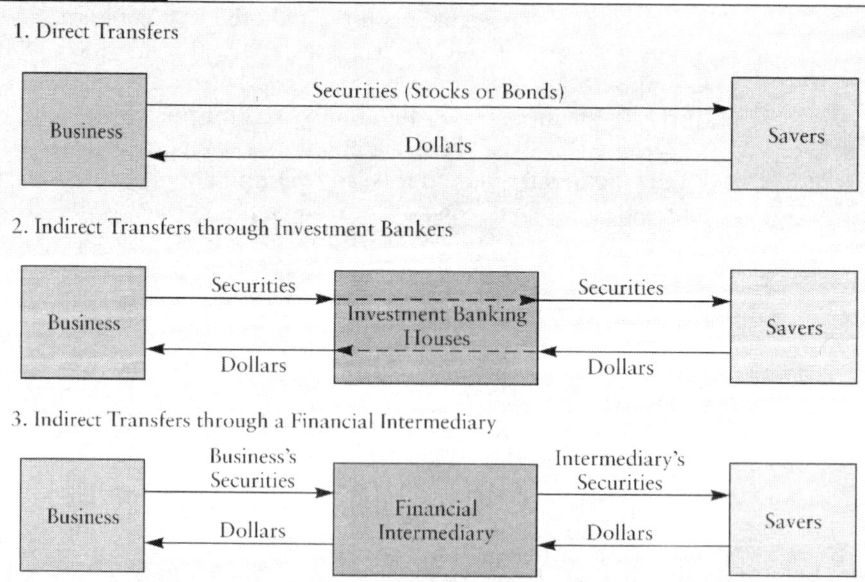

securities are involved and the company receives the proceeds of the sale, this is a primary market transaction.

3. Transfers can also be made through a **financial intermediary**, such as a bank or mutual fund. Here the intermediary first obtains funds from savers in exchange for its own securities, and then it uses this money to purchase and hold other assets (stocks, bonds, mortgages). For example, a saver might deposit funds in a bank, receiving from it a certificate of deposit, and the bank then lends the money to a small business to help it acquire working capital. Thus, intermediaries literally create new forms of capital—in this case, certificates of deposit, which are both safer and more liquid than small business loans and thus are better securities for most savers to hold. Financial intermediaries greatly increase the efficiency of money and capital markets.

Financial Intermediary
A specialized financial firm that facilitates the transfer of funds from savers to demanders of capital.

In our examples, we have assumed that the borrower is a business, but it is easy to visualize the demander of capital as a home purchaser, a government unit, or a student wanting to borrow funds for college. Of the three capital transfer mechanisms, direct transfers are the least important in developed nations, and larger firms generally employ investment banking houses to directly place securities with investors rather than acquiring capital through financial intermediaries.

Panel A of Table 2-1 lists the 10 largest U.S. bank and thrift holding companies, Panel B shows the leading world banking companies, and Panel C lists the ten leading global debt underwriters in terms of dollar volume of new issues (although many of the issues are in currencies other than dollars). Among the ten largest banks, only two (Citigroup and J. P. Morgan Chase) are based in the United States. Thus, while mergers have dramatically increased the size of U.S. banks, they are still small by

TABLE 2-1 10 Largest U.S. Bank and Thrift Holding Compaies, World Banking Companies, and Top 10 Leading Underwriters

Panel A U.S. Bank and Thrift Holding Companies[a]	*Panel B* World Banking Companies[b]	*Panel C* Leading Global Debt Underwriters[c]
Citigroup Inc.	Mizuho Holdings (Tokyo)	Citigroup
J. P. Morgan Chase & Co.	Citigroup (New York)	Morgan Stanley
Bank of America Corp.[d]	Sumitomo Mitsui Financial Group (Tokyo)	Merrill Lynch
Wells Fargo & Co.	UBS AG (Zurich)	Lehman Brothers
Wachovia Corp.	Allianz AG (Munich)	J. P. Morgan
MetLife Inc.	Deutsche Bank (Frankfurt)	Credit Suisse First Boston
Bank One Corp.	J. P. Morgan Chase & Co. (New York)	Deutsche Bank
Washington Mutual Inc.	HSBC Holdings PLC (London)	UBS
FleetBoston Financial Corp.[d]	ING Group NV (Amsterdam)	Goldman Sachs
U.S. Bancorp	BNP Paribas (Paris)	Banc of America

Notes:
[a]Ranked by total assets as of September 30, 2003. *Source:* "Top Bank and Thrift Holding Companies by Assets," *AmericanBanker.com*, September 30, 2003.
[b]Ranked by total assets as of December 31, 2002. *Source:* "World's Top Banking Companies by Assets," *AmericanBanker.com*, September 17, 2003.
[c]Ranked by dollar amount raised through new issues in 2003. For this ranking, the lead underwriter (manager) is given credit for the entire issue. *Source:* Adapted from *The Wall Street Journal*, January 2, 2004, R17.
[d]On March 17, 2004, Bank of America's acquisition of FleetBoston was approved by both firms' shareholders.

global standards. Note too that 6 of the top 10 underwriters are also major commercial banks or are part of bank holding companies, which demonstrates the blurring of distinctions among types of financial institutions.[3]

SELF-TEST QUESTIONS

In what three ways do transfers of capital between savers and borrowers take place?

Differentiate between an investment banking house and a financial intermediary.

THE COST OF MONEY

Interest Rate
The price paid to borrow debt capital.

Required Return
The minimum rate of return on a common stock that will induce a stockholder to purchase the stock.

Exchange Rate
The number of units of a given currency that can be purchased for one unit of another currency.

Time Preferences for Consumption
Consumer preferences for current consumption as opposed to saving for future consumption.

Expected Rate of Inflation
The amount by which prices are expected to increase over time.

Risk
In a financial market context, the chance that an investment will provide a low or negative return.

Production Opportunities
The returns available within an economy from investments in productive real assets.

Now that we have introduced the various types of financial markets and institutions, we discuss the market mechanics used to allocate capital between lenders and borrowers. In a market economy, capital is allocated through the price system. For debt securities, the cost of borrowed money is called the **interest rate.** Equity funds also have a cost—the investors must expect a return in the form of dividends and/or capital gains, and the minimum expectation that will get the investor to provide equity capital is called the investor's **required return.** Finally, if we want to purchase another country's currency, the price we pay is called an **exchange rate.**

Supply and demand determine the price (or cost) of capital, and four fundamental factors interact to determine this cost. On the supply side are three factors: **time preferences for consumption**, the **expected rate of inflation**, and **risk.** On the demand side there is one primary factor, the amount of and returns on **production opportunities.** To see how these factors determine the cost of money, visualize the preferences of business students as they go through school, get a job, and finally retire. While they are in school, most students' expenses exceed their incomes, so they have a strong preference to consume more than they are currently earning. By borrowing today and promising to repay loans out of future earnings, students can consume more than their incomes. After graduation and entering the workforce, they hold their consumption to a level below their incomes while they pay off their student loans, and perhaps save for a down payment on a house. If they marry and have children, they may begin saving for the children's education, and when the kids go off to college, they may borrow again to cover tuition. Then, they must pay off any loans and save in earnest for their eventual retirement. Thus, as their careers progress, they shift between consumption and saving, perhaps several times. There are many students (and parents) who need capital, and also people who are saving for retirement and thus supplying capital. The market system serves to allocate capital, and in the process it determines an equilibrium price for money.

The erosion of purchasing power is called *inflation*, and it is the second factor that influences the price of money. If inflation is expected, borrowers must pay a premium to enable savers to maintain the purchasing power of the money they lend.

Savers also worry about the *risk factor* inherent in their investments. Borrowers may, of course, not repay loans as specified in the contracts, but if savings are invested in a foreign-currency investment, there is additional risk because the exchange rate between the foreign currency and the home currency may change. Even if a saver earns a high rate of return on the foreign currency, wealth can be lost

[3]As a result of experience during the Great Depression of the 1930s, the federal government tried to make U.S. banks safer by restricting them to commercial banking and preventing them from engaging in investment banking. However, Japanese and European banks could engage in both commercial and investment banking. Thus, they could bid on a stock or bond underwriting and also offer to make short-term credit available, which gave them an advantage over U.S. banks. The banks then lobbied successfully to have their restrictions lifted, and today giant financial service firms own both huge banks and major investment banking businesses. For example, Citigroup owns the world's second largest bank, Citibank, and one of the largest investment banking firms, Smith Barney, along with many other financial businesses. The trend toward financial consolidations is going strong in 2004, with large global mergers occurring weekly.

when proceeds from the investment are converted back into the home currency. Most people prefer to minimize their risk, whether from default or exchange rates, so to overcome savers' reluctance to accept risks, borrowers must offer a rate sufficiently high to mitigate lenders' anxieties.

On the demand side, we must look at what borrowers intend to do with the money. Some money is borrowed to finance consumption (e.g., vacations), but the majority of all loans taken out are based on expectations regarding how profitably the borrowed funds can be invested. If many people conclude that it will be profitable to return to school for a graduate degree, or if many companies decide that new plants will provide high returns, then there will be a strong demand for money, and that will lead to higher interest rates. On the other hand, if prospective students conclude that they do not need graduate degrees, or companies have sufficient capacity, the demand for money will be weak and rates will be low.

Borrowers' expected rates of return on investments set the upper limit on what they will pay to acquire capital, while savers' time preferences for consumption establish how much consumption they are willing to defer and thus the lower limit on the cost for capital. Those two factors, along with inflation and risk, determine the market-clearing rate on different types of capital.

SELF-TEST QUESTIONS

What is the price paid to borrow money called?

What do investors expect companies to provide on equity securities as the price of that capital?

Name some factors that have a substantial impact on the cost of money. Which affect the supply side? The demand side?

What determines the upper and lower limits on the price of capital used for investment purposes? How is the market-clearing interest rate determined?

INTEREST AND EXCHANGE RATE DETERMINATION

Assume initially that there is no inflation, no risk, all transactions are in dollars, and you have \$1,000 that you can invest or spend. If you decide not to spend the money now, you will lend it to the federal government and receive \$1,030 at the end of the year. Your choices are diagrammed here:

Now	Next Year
Spend the \$1,000 today	Have none of the money next year
Invest the \$1,000 at 3% today	Receive and spend \$1,030 next year

Marginal Investor
The investor who operates at the margin and whose decisions determine the market interest rate.

Real Risk-Free Rate of Return (r*)
The interest rate that would exist on default-free U.S. Treasury securities if no inflation were expected.

Now assume that you are typical of all potential savers, so your preferences reflect those of the **marginal investor**, whose actions determine interest rates in the economy. Assume also that you are indifferent between these alternatives. If you were offered slightly more than 3 percent, then you would defer consumption and invest the \$1,000, while if you were offered slightly less you would spend the \$1,000 now. Thus, 3 percent is the indifference, or breakeven, rate, and it is defined as your *required rate of return*. Furthermore, because you are the marginal investor, then 3 percent will represent the cost of funds for investments of this type. Also, because the investment is riskless and no inflation is expected, the 3 percent return is also the **real, risk-free rate of return (r*)**. The term "real" means that this is the return if no inflation were expected. (Inflation is zero in this example.)

Potential borrowers make similar calculations. Depending on how profitably they think they can invest funds, they will offer more or less than 3 percent. In our example, we are implicitly assuming that borrowers at the margin have concluded

The Dangers of a Global Economy

An integrated global economy conjures images of new product markets and resources, but there is a dark side to this development. The United States has long been the driving force in the world economy, resulting in a highly correlated global economy. As of 2003, U.S. savings rates were at all-time lows, and government deficits (budget and current account deficits) were reaching record highs. Meanwhile, the rest of the world seemed to be sitting idly by, despite standing on the brink of a downward spiral. With no measures taken, the stage may have been set for a massive global recession. Similar conditions have existed before, but a fortunate string of events staved off disaster. Can luck hit twice?

In the last quarter of 2003, leading economic indicators suggested that the U.S. economy may be climbing out of its slump, prompting the world to sigh in relief. For decades, the United States has been the driving economic force for the rest of the world. However, overreliance on America is a serious problem facing the global economy, prompting former Treasury secretary Lawrence Summers to say, "The world economy is flying on one engine."

Since 1995 nearly 60 percent of the cumulative growth in world output has come from the United States, compared to the United States's 30 percent share of world gross domestic product (GDP). An extraordinary rise in U.S. spending is partly responsible, as U.S. domestic demand has risen 3.7 percent per year since 1995, nearly twice the rate of the rest of the world. And just as a one-engine plane is more likely to crash, a highly correlated global economy is more likely to collapse if only one of its engines is working properly. A significant drop in U.S. spending would have disastrous consequences for the rest of the world. Not an encouraging thought, as Americans have been spending beyond their means for many years and the American saving rate is at an all-time record low. Meanwhile, the U.S. current account deficit (the amount it must borrow from foreigners to spend more than it produces per year) has raced to more than 5 percent of national GDP. In the 1980s, the United States was the biggest creditor country in the world, but now it stands as the biggest debtor country in history.

The 1990s witnessed American firms using debt-financed investments to power the world economy. A decade and a stock market crash later, investment spending has collapsed as firms try to strengthen their balance sheets. Meanwhile, U.S. households have been slow to do the same. Sharp cuts in interest rates allowed households to borrow against their homes, and as of 2003 U.S. consumers' indebtedness was growing twice as fast as their income. The U.S. government has also exhibited increased spending—a large budget surplus in 2000 had turned into an even larger deficit by 2003.

The debt escalation can only go on for so long, because at some point creditors will refuse to continue lending money to the United States. Japan and China have been assuming a greater and greater share of U.S. government debt, which further ties the United States to the global economy. Their motives are to weaken their own currencies, thus buoying their exports to the United States.

There are some ways to ease the world off U.S. spending without causing it to tumble into a worldwide recession. Economic theory provides the most basic answer: Americans should spend less, foreigners should spend more, and the dollar should be weakened to allow for a U.S. shift away from importing and toward exporting. Such actions corrected a similar situation in the late 1980s, when the current account deficit was 3 percent. Then, the booming economies of Germany and Japan combined with a large drop in the value of the dollar to stave off recession. The odds of a repeat performance are slim.

With the Japanese and European economies in turmoil and their populations aging, dramatic improvement from somewhere is needed to provide support for the United States. The largest factor in easing this situation is likely to be the U.S. dollar. In the late 1980s, it fell 55 percent against the deutsche mark and 56 percent against the yen. By late 2003, the dollar had fallen 8 percent off its peak against the euro and yen. Without a dramatic shift in global demand, economists predict that the dollar will need to fall by at least 40 percent to make a significant impact on the U.S. current account deficit. However, if the dollar falls too fast the global economy is likely to roll back into a recession. Roadblocks exist. Japan has been actively keeping its currency weak, and China's economy would likely collapse if its currency (the yuan) strengthened too quickly.

Thus, the global economy is facing some serious problems. While the risk of a crash in the U.S. dollar or a serious global recession is real, sluggish growth and currency volatility seem more likely. The first step is recognizing how big the problem is.

Source: "Flying on One Engine," *The Economist*, September 18, 2003, **http://www.economist.com**.

that they will borrow if they can obtain funds for 3 percent. Millions of potential lenders and borrowers in the economy are making such decisions all the time, for thousands of different types of loans and investments, and this is the process that determines interest rates on different types of securities. Moreover, as conditions change—and they change daily, sometimes by large amounts—preferences, and thus the level of interest rates, will also change.

Further Analysis of Inflation

Inflation
The amount by which prices increase over time.

Now we can bring in inflation. Suppose a basket of goods and services that costs \$1,000 today is expected to cost \$1,035 in one year. This increase, expressed in percentage terms, is **inflation**, and it is given the symbol I and calculated as follows:

$$\begin{aligned} I &= \text{Change in price/Initial price} \\ &= (\$1{,}035 - \$1{,}000)/\$1{,}000 = 0.035 = 3.5\%. \end{aligned}$$

Nominal, Risk-Free Rate (r_{RF})
The rate of interest that both offsets inflation and provides the required real return on a riskless investment.

Just to preserve our real purchasing power—that is, to break even—we must earn enough to increase the initial investment by I, or 3.5 percent, and to the 3.0 percent required real rate, we must earn still more. The rate of interest that both offsets inflation and provides the required real return on a riskless investment is defined as the **nominal, risk-free rate, r_{RF}.**

It is tempting to estimate r_{RF} by simply adding the expected inflation rate to the required real rate, getting a rate of 6.5 percent, but that underestimates the correct r_{RF}:[4]

$$\text{Incorrect: } r_{RF} \neq r^* + I.$$

$$\text{Correct: } r_{RF} = r^* + I + r^*I. \quad (2\text{-}1)$$

The nominal, risk-free required rate of return in our example is thus

$$r_{RF} = 0.03000 + 0.03050 + 0.00105 = 0.06605 = 6.605\%.$$

Inflation Premium (IP)
A premium that must be added to the real risk-free rate to adjust for inflation; calculated as $I + r^*I$.

The value that must be added to the real risk-free rate is called the **inflation premium (IP)**, and its formula is given here:

$$IP = I + r^*I. \quad (2\text{-}2)$$

Based on our example, IP is 3.605 percent, which is higher than the expected inflation rate, 3.5 percent, because of the interaction term r^*I.

If both the real rate and expected inflation are low, as in our example, the interaction term included in the inflation premium will be very low, and the error that results from omitting it will be small. However, if inflation is high, as it often is in

---

[4] In our example with $r^* = 3\%$, $I = 3.5\%$, and \$1,000 to spend or invest, we need to end up with enough money to buy goods with an initial cost of \$1,000(1.03) = \$1,030 if we are to earn the 3 percent real return. However, \$1,000 of Year 0 goods will have a cost of \$1,035 given the 3.5 percent inflation rate, so we need \$1,035 just to break even in real terms. Moreover, that \$1,035 must be increased by 3 percent to provide the 3 percent required real return; that is, we must end up with \$1,035(1.03) = \$1,066.05 if we are to earn a 3 percent real rate of return. Putting all this together, we obtain this formula for r_{RF}:

$$\begin{aligned} (1 + r_{RF}) &= (1 + r^*)(1 + I) - 1 \\ r_{RF} &= r^* + I + r^*I. \end{aligned}$$

Substituting our illustrative numbers into this equation, we obtain the following result:

$$r_{RF} = 0.0300 + 0.0350 + 0.00105 = 0.06605.$$

If we earn this nominal return, we will end up with \$1,000(1.06605) = \$1,066.05, which will provide us with the required 3 percent real rate of return and protect the purchasing power of the initial \$1,000.

developing nations, then it is very important to use the correct formula, which means including the term r^*I. Also, note that if the term "risk-free rate" is used without the modifiers "real" or "nominal," you should assume that it refers to the *quoted, or nominal, rate*, which includes the inflation adjustment. We approximate the risk-free rate by the quoted return on U.S. Treasury securities or equivalent foreign government securities with similar maturities.

Risk Premium (RP) A premium that reflects the difference between the expected rate of return on a given risky asset and that on a less risky asset.

Risk

Investors dislike risk, so they add a **risk premium (RP)** when establishing the required rate on a risky security. To illustrate, assume that a student needs $1,000 for tuition, and because student loans are more risky than government loans, the corresponding rate is 10 percent. The difference between this risky loan rate and the nominal, risk-free rate calculated earlier is 10.000% − 6.605% = 3.395%, which compensates investors for bearing the risk inherent in the student loan.

Nominal, Risk-Adjusted Rate of Return (r) The actual rate charged on a loan; it compensates investors for postponing consumption, inflation, and risk.

The Nominal, Risk-Adjusted Rate (r)

The 10 percent student loan rate is a **nominal, risk-adjusted rate of return (r)**, and it compensates investors for postponing consumption (r^*), inflation (I), and risk (RP). It is also known as a **quoted**, or **stated**, **rate**, and it is the rate that would be given as the rate on a student loan. Quoted rates on many widely traded securities are reported in *The Wall Street Journal* or other financial publications, and they can be obtained from online sources.

Quoted (Stated) Rate The nominal, risk-adjusted rate of return, which is actually published in financial publications.

General Formula

The general formula used for the nominal risk-adjusted required rate of return is shown here:

$$r = r^* + IP + RP = r_{RF} + RP. \quad (2\text{-}3)$$

Here, r is the nominal risk-adjusted rate, r^* is the real risk-free rate, IP and RP are the inflation and risk premiums, respectively, and $r_{RF} = r^* + IP$.

The risk premium, RP, actually consists of at least five different components:[5]

Default Risk Premium (DRP) A premium that reflects the difference between the interest rate on a U.S. Treasury bond and a corporate bond of equal maturity and marketability.

1. *Default risk premium.* Default risk reflects the probability that a borrower will not pay the interest and principal on a loan when it comes due. The higher the probability of nonpayment, the greater the default risk and thus the higher the **default risk premium (DRP)**. U.S. Treasury securities have no

[5] We adjusted the inflation premium, IP, to account for the interaction between inflation and the real rate, r^*. It would be possible to make similar adjustments for the risk premium and identify interactions among r^*, I, and RP. However, this is not normally done. An exception is when the Capital Asset Pricing Model (CAPM), which we discuss in Chapter 5, is used to find local costs of capital in a country that is experiencing high inflation and the U.S. market risk premium (RP_M) is not an adequate proxy for the required risk premium in the local market. The reason for this is that the RP_M consists of a real required return to compensate for risk in the market plus an inflation adjustment to preserve the value of the real risk premium. When local inflation rates are significantly higher than U.S. inflation, this second adjustment must be made to the local RP_M. To see how this should be done, assume that $r^* = 3\%$ in both countries but $IP_{US} = 3\%$, $IP_{Indonesia} = 30\%$, and $RP_{M, US} = 7\%$. $r_{RF} = 3\% + 3\% = 6\%$ in the United States and $3\% + 30\% = 33\%$ in Indonesia. The return on the market $(r_M) = 6\% + 7\% = 13\%$ in the United States, but it is wrong to say that $r_M = 33\% + 7\% = 40\%$ in Indonesia (as some books do) because of the difference in inflation in the two countries. To find the inflation-adjusted RP_M for Indonesia, we need to find the real risk rate, r_{RSK}, for the United States and then apply this to the Indonesia case: $(1 + r^* + IP_{US})(1 + r_{RSK}) = (1 + r_{US})$; $(1 + 0.03 + 0.03)(1 + r_{RSK}) = (1 + 0.13)$; $r_{RSK} = 0.066038$ or 6.6038%. Then, the RP_M for Indonesia $= r_{RSK}(1 + r^* + IP_{Indonesia}) = 0.066038(1 + 0.03 + 0.30) = 0.08783$ or 8.783%. Thus, r_M for Indonesia $= 33\% + 8.783\% = 41.783\%$ instead of the 40 percent calculated (incorrectly) earlier. Some may wish to suggest that the risk aversion of Indonesian investors is different from that of U.S. investors so that the real risk rate should also differ for the two markets. We find merit in this suggestion, but if a person wants to assume that the risk rates are the same (an assumption that is often made), then it is necessary to make the inflation adjustment to risk premiums and avoid biasing the CAPM estimates.

default risk for American investors, hence they offer the lowest interest rates on taxable securities. Similarly, government securities denominated in the local currency in most other countries are assumed to be free of risk in their local capital markets, but investors outside the country may still build in a default premium due to exchange rate concerns and possible expropriation.

The difference between the quoted interest rate on a U.S. Treasury bond and that on a corporate bond with similar maturity, liquidity, and other features is the bond's default risk premium. The riskier the security, the higher the DRP. One indicator of the level of default risk on a corporate bond is its rating—the default premium rises as the credit rating declines.[6] Table 2-2 provides some representative interest rates and default premiums on long-term bonds during 2003. While the table does not show it, default risk premiums vary somewhat over time, but the figures in the table are representative of recent U.S. levels. DRPs in other countries will differ from those in the United States, so be careful when dealing with foreign bonds. It is important to estimate the DRP on the basis of data for the particular country. In other words, do not assume that the U.S. premium is a reasonable proxy for those in other countries. However, eurobonds denominated in U.S. dollars trade competitively with U.S. bonds, and thus do require risk premiums similar to those on U.S. bonds.

Liquidity Premium (LP)
A premium added to the equilibrium interest rate on a security if that security cannot be converted to cash on short notice and at close to "fair market value."

2. *Liquidity premium.* A "liquid" asset is one that can be converted into cash quickly and at a "fair market value." By definition, cash is the most liquid asset, and financial assets are usually more liquid than real assets. The harder it is to convert an asset into cash quickly and at near its fair market value, the lower its liquidity and thus the higher the required **liquidity premium (LP)**. U.S. Treasury securities are traded in a very broad, deep, and active market, so their liquidity premium is essentially zero. Although it is difficult to measure liquidity premiums accurately, the difference in premiums between the most and the least liquid financial assets of similar default risk and maturity in the U.S. is roughly 2 to 5 percent.

TABLE 2-2 Representative 2003 Default Risk Premiums

	Rate	DRP
U.S. Treasury	5.0%	—
AAA	5.7	0.7%
AA	6.1	1.1
A	6.4	1.4
BBB	6.8	1.8

Sources: U.S. Treasury data represent annual average of constant 20-year maturity, *Federal Reserve Statistical Release*, selected interest rates. Corporate yields represent long-term annual averages for maturities of 20 years or more obtained from "Moody's Corporate Bond Yield Averages," *Moody's Investors Service*.

[6]Sometimes the actual rate on a corporate bond is a better indicator of its risk than its rating. Institutional investors have excellent credit analysts; they may pick up changes in a bond's situation before the rating agencies change their classifications and through their buying or selling change the interest rate before the rating changes. For example, if credit analysts think that an AA-rated company's situation has deteriorated to the point where it should only have an A rating, then they will sell the bonds and drive their prices down until their yields rise to the same level as A-rated bonds.

Interest Rate Risk
The risk of capital losses to which investors are exposed due to rising interest rates.

Maturity Risk Premium (MRP)
A premium that reflects interest rate risk.

3. *Maturity risk premium.* As we shall demonstrate in Chapter 8, the prices of fixed-rate bonds fall when interest rates rise and rise when rates fall. The resulting risk of capital losses due to rising interest rates is known as **interest rate risk.** The price change associated with a given change in the interest rate is much greater for a long-term than a short-term bond. Therefore, the longer the time until maturity, the greater the price risk, hence the higher the **maturity risk premium (MRP)** that investors add when determining the required nominal interest rate. Even though U.S. Treasury bonds have no default risk and virtually no liquidity risk, long-term T-bonds are exposed to a significant maturity risk. Indeed, some long-term T-bonds lost fully half of their values during the 1980s, when interest rates rose sharply. The MRP for T-bonds has recently been in the range of 1 to 2 percent, but it varies somewhat over time depending on expectations regarding future interest rate volatility. In other countries, the same process determines their MRPs, and if interest rates in a given country have been more or less volatile than in the United States, their MRPs will reflect this difference.

Business Climate
Refers to a country's social, political, and economic environment.

Country Risk Premium (CRP)
A premium that reflects the risk that arises from investing or doing business in a particular country.

4. *Country risk premium. Country risk* refers to the risk associated with investing in a particular country, and it depends on the country's social, political, and economic environment, or its **business climate.** The more uncertain the business climate, the higher the **country risk premium (CRP).** See Table 2-3 for some recent rankings from *Euromoney* magazine. Perhaps surprisingly, this study, like most, concludes that the United States has more country risk than several other nations. This is particularly significant because even though people in the United States usually assume that our bonds have no country risk, others do not agree. Foreign investors are concerned about how changes in U.S. policies (say, tax and Federal Reserve policies) might affect their positions, so they build a country risk premium into their required rate of return on both corporate and Treasury debt. Because foreign investors are major purchasers of U.S. debt, and because the CRP is reflected in rates on all U.S. dollar-denominated bonds, perceptions about U.S. country risk affect the level of rates in this country.

Exchange Rate Risk Premium (ERP)
A premium that results from the possibility that an exchange rate change will lead to a loss in a bond's value.

5. *Exchange rate risk premium.* When a bond is *denominated* in a currency other than the investor's home-country currency, a change in the exchange rate will lead to a change in the bond's value to that investor. For example, suppose a Japanese investor buys a bond denominated in dollars, and the dollar then weakens. The dollars the investor receives will then buy fewer yen, and this will lower the bond's value to the Japanese investor. As a result, **exchange rate risk premiums (ERPs)** are built into the nominal rates of return on bonds. The size of the ERP depends on the stability of exchange rates between the two countries, and even U.S. Treasury securities have an ERP because so many of them are held by foreign investors.

We can now rewrite our equation for the nominal risk-adjusted required rate of return as shown in Equation 2-4:

$$r = r^* + IP + RP = r_{RF} + DRP + LP + MRP + CRP + ERP. \quad (2\text{-}4)$$

Exchange Rates

Until now we have dealt with interest rates in only one country. Interest rates are generally determined in the same way in all countries, but parameters such as the

TABLE 2-3 Selected Countries Ranked by Composite (Total) Risk

Rank	Country	Political Risk (25)	Economic Performance (25)	Debt Indicators (10)	Debt in Default or Rescheduled (10)	Credit Ratings (10)	Access to Bank Financing (5)	Access to Short-Term Financing (5)	Access to Capital Markets (5)	Discount on Forfeiting (5)	Total Risk (100)
1	Luxembourg	24.67	25.00	10.00	10.00	10.00	5.00	5.00	5.00	4.69	99.36
2	Norway	24.51	23.55	10.00	10.00	10.00	5.00	5.00	5.00	4.69	97.75
3	Switzerland	24.97	22.81	10.00	10.00	10.00	5.00	5.00	5.00	4.69	97.48
4	United States	24.95	21.69	10.00	10.00	10.00	5.00	5.00	5.00	5.00	96.64
12	Canada	24.64	17.98	10.00	10.00	9.79	5.00	5.00	5.00	4.69	92.11
49	China	16.84	11.22	9.66	10.00	5.83	0.12	2.50	2.00	3.34	61.52
51	Mexico	16.87	9.68	9.05	10.00	4.69	0.77	3.33	3.00	3.74	61.14
72	Russia	11.77	8.82	9.21	9.66	3.33	0.14	1.96	1.70	2.42	49.02
89	Indonesia	8.98	6.74	8.39	9.52	0.83	0.30	1.83	1.25	2.21	40.05
162	Argentina	3.26	7.39	7.75	4.15	0.21	0.02	3.00	0.00	0.00	25.78
185	North Korea	0.00	2.09	0.00	0.00	0.00	0.00	1.17	0.00	0.00	3.26

Source: "Country Risk Rankings," *Euromoney*, September 2003, **http://www.euromoney.com**.

inflation rate and the various premiums differ from country to country. These differences are reflected in local interest rates and, through them, in the exchange rate between the two currencies. We can show these linkages by extending our earlier example. The original data apply to Country A, whose currency is the dollar, and we now bring in a second country, Country B, whose currency is the peso. Our task is to derive the equilibrium exchange rate, given data on the other parameters. We start by assuming that r* is 3 percent in both countries, and we also assume that all the premiums are similar except for inflation. The inflation rate in Country A is 3.5 percent as given in the previous example, and it is 8 percent in Country B. With this information we can calculate r_{RF} in each country:

$$r_{RF, A} = [(1.03)(1.035) - 1] = 0.06605 \text{ or } 6.605\%.$$
$$r_{RF, B} = [(1.03)(1.08) - 1] = 0.11240 \text{ or } 11.240\%.$$

For simplicity, we assume the spot exchange rate is P1.0000/$; that is, the dollar and the peso currently have the same value.

Given this information, what should the exchange rate be in one year? Figure 2-2 shows the linkages that describe the situation. The top and bottom rows show the results from investing in each country. Given that risks are the same in both countries, an investor's position should end up the same regardless of which country he or she invests in, assuming that funds can be moved without material costs from country to country (which is true). In equilibrium, the exchange rate one year from now should be such that $1,066.05 is equivalent to P1,112.40 (i.e., we can exchange P1,112.40 for $1,066.05). If this situation does not hold, then effective rates in the two markets would be different, and investors would flock to the market with the higher effective rates. Put another way, an arbitrage opportunity would exist, and thus the market would not be in equilibrium.

We can solve for the equilibrium exchange rate one year hence as follows:

$$\text{P}1{,}112.40/X = \$1{,}066.05$$
$$X = \text{P}1.043478/\$.$$

If our expectations are borne out, a year from now it will require 1.043478 pesos to buy one dollar, meaning the dollar will have strengthened relative to the peso. The exchange rate would change because the two countries' inflation rates differed.

This example illustrates the linkages between inflation, interest rates, and exchange rates. We will discuss all this in greater depth in the next chapter where we discuss exchange rates, but for now you should simply recognize that interest rates, inflation rates, and exchange rates interact in the global marketplace, and if specific relationships do not hold, then arbitrage will occur until those relationships do hold.

Figure 2-2 Investing Globally in an Efficient Market

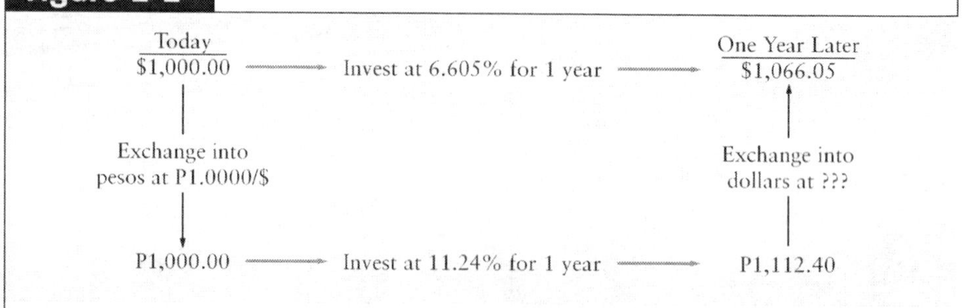

SELF-TEST QUESTIONS

Differentiate between the real risk-free rate and the nominal risk-free rate.
What is an inflation premium?
What is a default risk premium?
What is a liquidity premium?
What is a maturity risk premium?
What is a country risk premium?
What is an exchange rate risk premium?

INTERNATIONAL MONETARY ARRANGEMENTS

International Monetary System
The framework within which exchange rates are determined. It is the blueprint for international trade and capital flows.

In a competitive and efficient marketplace, which is characterized by an absence of government intervention to influence the exchange rate, the value of a country's currency will conform to the relationships discussed earlier. However, governments often intervene in the marketplace in an attempt to force the value of their currency to a level dictated by internal politics, so we need to look at some of the exchange rate policies and regimes in use around the world today. The **international monetary system** is the framework within which exchange rates are determined. Because exchange rates are a function of the supply and demand for various national currencies, the international monetary system is also the blueprint for international trade and capital flows. Thus, the international monetary system ties together global currency, money, capital, real estate, commodity, and real asset markets into a network of institutions and instruments, regulated by intergovernmental agreements, and driven by each country's unique political and economic objectives.[7]

International Monetary Terminology

While we will cover foreign exchange in greater detail in Chapter 3, it is useful to introduce some important concepts that will make it easier to understand the international monetary system:

1. An *exchange rate* is the price of one country's currency in terms of another currency. For example, on Wednesday, February 25, 2004, one U.S. dollar would buy 0.5348 British pound, 0.8003 euro, 1.3353 Canadian dollars, or 8.2781 Chinese yuan.[8]
2. A *spot exchange rate* is the quoted price for a unit of foreign currency to be delivered "on the spot," or within a very short period of time. The rate quoted above, £0.5348/$, is a spot rate as of the close of business on February 25, 2004.
3. A *forward exchange rate* is the quoted price for a unit of foreign currency to be delivered at a specified date in the future. If today were February 25, 2004, and we wanted to know how many pounds we could expect to receive for our dollars on August 25, 2004, we would look at the six-month forward rate,

[7]For a comprehensive history of the international monetary system and details of how it has evolved, consult one of the many economics books on the subject, including Robert Carbaugh, *International Economics* (Mason, OH: South-Western, 2004); Mordechai Kreinin, *International Economics: A Policy Approach*, 9th edition (Mason, OH: South-Western, 2002); and Joseph P. Daniels and David D. Van Hoose, *International Monetary and Financial Economics*, 2nd edition (Mason, OH: South-Western, 2002).

[8]The complete name for the Chinese currency is the *yuan renminbi*, but it is generally called either the yuan or the renminbi. In our experience, yuan is used more often, so we generally use that term. However, *The Wall Street Journal*, when it reports exchange rates, uses the term renminbi. Because China is becoming increasingly important in world commerce and as a U.S. trading partner, you should be familiar with both terms. Also, note that exchange rates could be, and in the past often were, defined in terms of a commodity, especially gold. In 1934, the United States defined the value of the dollar such that one ounce of gold equaled $35, but the "gold standard" was abandoned about 35 years later. If runaway inflation ever developed across the globe, countries would perhaps return to the gold standard, but if economic conditions continue to be relatively stable, that will not happen.

INSTABILITY IN THE EU

Despite early insistence on fiscal safeguards, France and Germany have become the habitual offenders of the European Union's "stability and growth" pact. The pact requires that member nations not run deficits greater than 3 percent of GDP. Successive breaches of the pact require disciplinary action, if the other member nations agree that enforcement is needed. However, there is a strong possibility that the other finance ministers would vote to let France and Germany slide on the infractions. Smaller nations are wary of making similar breaches, as there is no indication they would be given such favorable treatment if they violated the pact, and similar concerns face possible new members. Ultimately, the EU faces a tough task of coordinating the fiscal policies of its individual member nations, while having control only over the collective monetary policy of the EU.

France and Germany have been the cornerstones of the European Union since its inception. However, it is these two countries that have thrown the stability of the EU into question. Both Germany and France breached the 3 percent deficit limit in 2002 and 2003. The German deficit was 3.5 percent in 2002 and 3.8 percent in 2003, while France's deficits were 3.1 percent and 4.0 percent, respectively. Ironically, it was due to German insistence that the EU adopted the pact. In 2003, German finance minister Hans Eichel insisted that Germany would aim to comply with the pact in 2004, while French prime minister Jean-Pierre Raffarin had already given up on that goal. However, financial economists had little faith that either nation would reach the target, especially when Eichel was campaigning for a €16 billion ($17.4 billion) tax cut and Raffarin was adamant that finding jobs for the unemployed was his first priority.

If France and Germany fail to comply in 2004, protocol requires the EU to collect an interest-free deposit of between 0.2 and 0.5 percent of national GDP. Further breaches in 2005 would result in the loss of those deposits, amounting to more than €4 billion for Germany and €3 billion for France. Strict adherence to the measure would surely result in penalties for France and Germany, but if the countries are making an effort and deficits don't run out of control, fellow EU finance ministers are unlikely to vote to impose the sanctions.

French president Jacques Chirac has asked for a "temporary softening" of the stability pact, and German chancellor Gerhard Schröder has asked for leniency and more emphasis on the "growth" part of the "stability and growth" pact. Both leaders seem to suggest that the rules be applied only when no one is breaking them, and ignored when there are infractions. Together with Italy and Britain, the two offenders originally spearheaded the reform effort, much to the dismay of smaller states like Austria, Ireland, and the Netherlands, which wanted to run deficits. The smaller countries got their budgets in order, and they don't feel that the same leniency would be extended to them if they felt a need to default. Similar concerns surely plague countries that have yet to join the EU. Besides, as populations age, there will be more of a strain on annual budgetary burdens (an estimated shortfall of 2.2 percent of GDP for Germany by 2050). Turning a blind eye now could prove disastrous over the long run.

The current budgetary crises are primarily the result of economic downturns that should correct themselves in time. However, the pension problems are the result of long-term demographic trends that will take many years to play out. Proposals to quell the pension problems include postponing the retirement age and reducing benefits. These perceptions highlight the need to distinguish between cyclical and structural. The best solution to the pension problem would be a growing, vibrant economy and a workforce whose productivity is increasing; conditions that would also help solve the budget deficit problem.

The intent of the "stability and growth" pact was not to enforce the fiscal consolidation that Europe needs, but rather to safeguard the credibility of the young euro as a currency. With the euro's credibility no longer in doubt, the role of the "stability and growth" pact comes into question. French foreign minister Dominique de Villepin called for the coordination of EU member nations' fiscal policies. According to his argument, Europe needs coordination of the different countries' fiscal policies with the single monetary policy set by the European Central Bank (ECB). Obvious problems arise when the ECB sets monetary policy for all EU member nations, but has no ability to lower interest rates in response to a single country's efforts to repair its finances.

Source: "Deficits and Defiance," *The Economist*, September 2, 2003, **http://www.economist.com**.

which was £0.5434/$. Note that a *forward exchange contract* on February 25 would lock in this exchange rate, but no currency would change hands until August 25. The spot rate on August 25 might be quite different from £0.5434, in which case we would have a profit or a loss on the forward purchase.

4. A *fixed exchange rate* for a currency is set by the government and allowed to fluctuate only slightly (if at all) around the desired rate, called the *par value.* For example, China has fixed the exchange rate for the yuan at CNY 8.2781/$, and it has maintained this fixed rate for the past few years. Because the rate is fixed, only the spot rate is quoted in *The Wall Street Journal* and other publications.
5. A *floating* or *flexible exchange rate* is one that is not regulated by the government, so supply and demand in the market determine the currency's value. The U.S. dollar and the euro are examples of free-floating currencies. Note, though, that central banks do from time to time intervene in the market to nudge exchange rates up or down, even though they basically float.[9]
6. *Devaluation* or *revaluation of a currency* is the technical term referring to the decrease or increase in the par value of a currency whose value is fixed. This decision is made by the government, usually without warning.
7. *Depreciation* or *appreciation of a currency* refers to a decrease or increase in the foreign exchange value of a floating currency. These changes are caused by market forces rather than by governments.
8. A *soft* or *weak currency* is one that is expected to depreciate against most other currencies or else is being artificially maintained at an unrealistically high fixed rate by the government through open market purchases. A *hard* or *strong currency* is expected to appreciate against most other currencies or else is being artificially maintained by the government at an unrealistically low fixed rate. Many economists suggest the Chinese yuan is significantly undervalued, so it is a strong currency.[10]

The "Perfect" Currency

Before looking at currency arrangements, it might be instructive to summarize the characteristics of a "perfect," or ideal, currency. Most economists list three characteristics that are highly desirable in a money unit:

1. The currency's value should be fixed with respect to the currency used by the country's trading partners and other important nations; that is, its exchange rate should be stable. This would allow investors and businesses to plan with greater certainty regarding the future value of receipts and payments.
2. Each country would set its own domestic monetary and fiscal policy solely on the basis of its own economic situation. This would include policies regarding interest rates, managing inflation, fostering full employment, and ensuring economic prosperity for all.
3. Complete and unrestricted monetary flows should be permitted, allowing investors and businesses to move their wealth as they choose.

Many economists call these three characteristics "The Impossible Trinity" because it is impossible to achieve all three objectives together. For example, the United States has a floating-exchange-rate regime, no capital or exchange controls, and its monetary and fiscal policies are driven by domestic considerations. However, the

[9] Governments quite often buy or sell currencies in the open market to influence the floating rate, but they are simply changing supply or demand to influence the rate rather than using some nonmarket mandated adjustment.

[10] If the value of the yuan were allowed to rise, then a dollar would buy fewer than 8.2781 yuan, which would increase the dollar cost of Chinese goods imported into the United States; in turn it would reduce Chinese exports to the United States. The Chinese government wants to maintain export levels, so it has resisted revaluing the yuan.

U.S. dollar has not been stable against the currencies of such major trading partners as Canada, Europe, and Japan. In 2003, the dollar depreciated significantly against both the euro and the Japanese yen. China, on the other hand, maintains a fixed exchange rate with the U.S. dollar and independent monetary and fiscal policies, but the only way it can keep these policies intact is to impose strict capital and exchange controls. We see, then, that a country can maintain two of the three characteristics, but allowing all three to operate is impossible.

If we return to the problem diagrammed in Figure 2-2 we can illustrate this phenomenon. In that scenario, which mirrors the situation in the United States today, we assumed monetary independence and free capital movement, but the result was a change in the equilibrium exchange rate between the two countries. If we wanted to keep the exchange rate constant at P1.00/$, we could do so in the short run by preventing arbitrageurs from acquiring the foreign exchange needed to complete transactions. This is what China and some other countries do to maintain their fixed exchange rates. We could also maintain the exchange rate by keeping interest rates artificially low, but that would eventually overstimulate the economy and lead to inflation. If inflation in Country B was 8 percent and we wanted to keep the nominal rate at 6.605 percent, then the real rate would have to be negative! Thus, we see that countries have to make decisions about which leg of the trinity they want to abandon in order to maintain the other two. This decision is at the heart of a country's decision as to which currency arrangement to use.

Current Monetary Arrangements

At the most basic level, we can divide currency regimes into two broad groups: floating rates and fixed rates. Within the two regimes, there are gradations among sub-regimes in terms of how rigidly they adhere to the basic positions. Looking first at the floating-rate category, the two main subgroups are

Freely Floating
Occurs when the exchange rate is determined by supply and demand for the currency.

1. *Freely floating.* Here the exchange rate is determined by the supply and demand for the currency. Under a **freely-floating** regime, governments may occasionally intervene in the market to buy or sell their currency in order to stabilize fluctuations, but they do not attempt to alter the absolute level of the rate. This policy exists at one end of the continuum of exchange rate regimes. For example, the currencies of Australia, Brazil, and the Philippines are allowed to float.

Managed Float
Occurs when there is significant government intervention to control the exchange rate via manipulation of the currency's supply and demand.

2. *Managed floating.* Here there is significant government intervention to manage the exchange rate by manipulating the currency's supply and demand. The government rarely reveals its target exchange rate levels if it uses a **managed float** because this would make it too easy for currency speculators to profit. For example, the governments of Colombia, Israel, and Poland manage their respective currency's float.

Most developed countries follow either a freely floating or a managed floating regime. A few developing countries do as well, often reluctantly and as a result of a market that forces them to abandon a fixed-rate regime.

Types of fixed-exchange-rate regimes include the following:

1. *No local currency.* The most extreme position is for the country to have no local currency of its own. The country either uses another country's currency as its legal tender (such as the U.S. dollar in Panama and Ecuador) or else belongs to a group of countries that share a common currency (such as the euro). With this arrangement, the local government surrenders economic regulation.
2. *Currency board arrangement.* Under a variation of the first subregime, a country technically has its own currency but commits to exchange it for a specified foreign money unit at a fixed exchange rate. This requires it to impose domes-

tic currency restrictions unless it has the foreign currency reserves to cover requested exchanges. This is called a **currency board arrangement.** Argentina had a currency board arrangement before its crisis of January 2002, when it was forced to devalue the peso and default on its debt.

3. *Fixed peg arrangement.* In a **fixed peg arrangement** the country locks, or "pegs," its currency to another currency or basket of currencies at a fixed exchange rate. It allows the currency to vary only slightly from its desired rate, and if the currency moves outside specified limits (often set at ±1 percent of the target rate), it intervenes to force the currency back to within the limits. An example is China, where the yuan is pegged to the U.S. dollar at a rate of CNY 8.2781/$, and the Chinese central bank intervenes as necessary to maintain this target. Additional examples include Bhutan's ngultrum, which is pegged to the Indian rupee; the Falkland Islands' pound, which is pegged to the British pound; and Barbados's dollar, which is pegged to the U.S. dollar.

Currency Board Arrangement Occurs when a country has its own currency but commits to exchange it for a specified foreign money unit at a fixed exchange rate and legislates domestic currency restrictions, unless it has the foreign currency reserves to cover requested exchanges.

Fixed Peg Arrangement Occurs when a country locks its currency to a specific currency or basket of currencies at a fixed exchange rate. The exchange rate is allowed to vary only within ±1% of the target rate.

Other variations have been used, and new ones are developed from time to time. A majority of the world's countries employ some sort of fixed-exchange-rate arrangement. So, while the most important currencies (as measured by volume of transactions) are allowed to float, and the international monetary system is often called a "floating regime," most currencies are actually fixed in some manner.

Fixed or Floating: How Does a Country Choose?

Numerous arguments have been advanced over the years favoring one regime over others, and some researchers have suggested that countries tend to move to the polar extremes of freely floating or rigidly pegged rates. This is the so-called bipolar view. However, there seems to be no single answer to the question of the optimal exchange-rate regime. The exchange-rate regime should reflect national priorities regarding the relative importance of fighting inflation, holding down interest rates, combating unemployment, producing a favorable balance of payments, promoting trade flows, and so forth. Research suggests that priorities change as a nation goes through different stages of development, so it may be wise to follow one strategy in the early stages of development and then shift to another as the economy matures. In Table 2-4 we list some of the reasons advanced in favor of fixed and floating rates.

Some recent research by the International Monetary Fund suggests that for countries in a relatively early stage of economic development, the existence of a fixed regime appears to be a valuable tool for fighting inflation. Under such a regime, countries must intervene domestically to take counterinflationary actions, and knowledge of this requirement makes anti-inflation actions more credible. Of course, if people believe that inflation will be constrained, then this in and of itself helps hold down the inflation rate. Moreover, studies suggest that such actions do not constrain growth. However, as development progresses, and as a country becomes more closely integrated with its trading partners, there appear to be benefits from shifting to a more flexible regime. Thus, for developed countries, flexible or floating rates appear to offer higher growth potential without loss of credibility regarding an anti-inflationary stance.[11]

Emerging Markets and the Choice of Monetary Regime

As indicated earlier, empirical research suggests that developing countries benefit from adopting a fixed-rate regime until their economies become more integrated into the global marketplace. Also, the "bipolar view" seems to indicate that one of

[11] "Evolution and Performance of Exchange Rate Regimes," Working Paper no. 03/243, The International Monetary Fund, December 2003.

TABLE 2-4 Pros and Cons for Fixed- versus Floating-Exchange Rates

Factors Favoring Fixed Rates

1. More stable and predictable exchange rates, which reduces risks in international commerce.
2. Helps in the fight against inflation because countries must intervene domestically to take counterinflationary actions to maintain the exchange rate at a fixed level.

Factors Against Fixed Rates

1. Significant international reserves must be available for use in defending the currency.
2. If the fixed rate is inconsistent with the country's fundamental economic condition, either painful adjustments must be made in the domestic economy or the target exchange rate must be changed.

Factors Favoring Floating Rates

1. The market adjusts automatically for changing economic fundamentals, and no government intervention is needed.
2. Governments can follow domestic policies to reduce unemployment or to stimulate growth without having to explicitly consider international implications of the exchange rate.
3. International reserves do not have to be used to preserve the exchange rate because it is allowed to find its own equilibrium level.

Factors Against Floating Rates

1. Future exchange rates are less stable and less predictable, which increases the risks inherent in international transactions.
2. Distortions in the domestic economy can persist for longer periods if the government is not bound to intervene to defend its currency.

the extremes of the fixed-rate regime—either using another country's currency or tying the currency closely to another currency—is preferable to a more relaxed variant. In this section, we look at the two fixed-rate extremes and explain their characteristics in more detail.

Dollarization
Occurs when a country abandons its own currency and adopts the U.S. dollar as its legal tender.

DOLLARIZATION The most extreme fixed-rate position is for a country to abandon its own currency and adopt that of another country as its legal tender. Ecuador did this in 2000, when it made the U.S. dollar its official currency. **Dollarization**, as this phenomenon is called, is touted as having several benefits to the adopting country.

1. There is complete stability with the dollar, and, at least theoretically, the possibility of future currency crisis is eliminated.
2. It is easier to become more economically integrated with the United States when the currencies are the same.
3. Monetary policy is taken out of the hands of local politicians, hence it cannot be used for political purposes that might destabilize the economy (e.g., printing money or giving public employees outsized pay raises).

However, there are several counterarguments to dollarization.

1. The country loses sovereignty over its monetary policy. This could be significant in countries that are ultranationalistic.
2. The local central bank cannot create money, so it cannot serve effectively as a lender of last resort during a financial crisis.
3. The country loses the ability to profit from printing money. Thus, if it costs \$1 million to print \$1 billion worth of bills, and if \$1 billion of paper money is outstanding, then the government has effectively earned a profit of \$999 million. Similarly, a country can coin its own money, which is called *seigniorage*, and if it costs \$0.01 to mint a quarter but coin collectors will pay \$0.25 for the coin to add to their collection, the government profits from seigniorage.

Ecuador abandoned its national currency, the sucre, on September 9, 2000, and replaced it with the U.S. dollar. This decision to "dollarize" came after several years of economic crises and high inflation. Interestingly, the decision was not supported by either the U.S. government or the International Monetary Fund (IMF), but the Ecuadorian government proceeded with the program anyway. Inflation has been controlled since dollarization, and residents immediately returned significant amounts of money into the banking system and put those funds to work rather than holding them in mattresses. It may be too soon to judge the success of dollarization in Ecuador, but thus far it appears to be working.

Some economists suggest that regional trading groups would benefit from a single currency, as this would facilitate trade, make it easier for countries to capitalize on their comparative advantages, and thus improve economic efficiency. (This was the thinking that led to the adoption of the euro by the European Union.) For this reason, some economists believe that alliances such as NAFTA would benefit if Canada, Mexico, and the United States all had the same currency, and that similar benefits would accrue to all members of the Free Trade Area of the Americas if that regional alliance becomes a reality. However, it is difficult to get the countries involved to give up their separate national currencies, as demonstrated by the fact that Great Britain, Switzerland, and the Scandinavian countries (except for Finland) have refused to switch to the euro.

Currency Boards A less severe alternative is for the countries in a regional alliance to adopt a currency board. Under this alternative, each country maintains its own currency, but the value of that currency is locked to another currency (such as the dollar or the British pound). For this to work, the country's central bank must agree to back its money supply entirely with foreign currency reserves, and it cannot issue more of the local currency unless it has the reserves to support it. This 100 percent reserve requirement means that the only way the money supply can expand is if the country acquires more foreign currency reserves through trade, which effectively eliminates the possibility of fueling inflation by printing too much money.

Of course, the experience in Argentina tells us that while a currency board may solve one problem, if the government does not make sure that other economic variables are consistent with the currency value, problems will probably arise and destroy the arrangement's effectiveness. Argentina adopted a currency board and tied its peso to the U.S. dollar at a rate of P1.00/$ in 1991. They also allowed local investors to hold bank accounts in either pesos or dollars. Most Argentines were not convinced that the board would work, hence the peso-denominated accounts had to pay a risk premium—that is, offer a higher interest rate—to induce people to hold pesos. This skepticism was warranted, because in January 2002, the currency board was discontinued in the face of economic recession and political turmoil, and peso holders suffered large losses.

Self-Test Questions

What is an international monetary system?
What is the difference between spot and forward exchange rates?
What is the difference between floating- and fixed-exchange rates?
What is meant by a "perfect" currency?
Differentiate between devaluation/revaluation of a currency and depreciation/appreciation of a currency.
What is meant by a soft or weak currency? A hard or strong currency?
What are the two broad categories of the various currency regimes? What are the subgroups of these two broad categories?
What is dollarization?
What are currency boards?

SUMMARY

Managers and investors make decisions within a global financial environment. This includes the local financial markets and institutions in each country, as well as international markets and institutions across the globe. In addition, it includes national tax and regulatory policies across countries and the macroeconomic conditions of each country's economy. These factors impact the outcomes of financial decisions made by managers and investors. Consequently, it is important to understand the global environment in which firms operate. The key concepts discussed in this chapter are listed below.

- **Physical asset markets** are also known as tangible, or real, asset markets. Capital budgeting decisions are important to these markets.
- **Financial asset markets** are marketplaces that deal with financial instruments such as stocks, bonds, notes, mortgages, currencies, and the like.
- **Financial instruments** are pieces of paper or electronic entries in account ledgers with contractual provisions that spell out their owners' claims on specific real assets. Their value is defined in terms of a fixed number of money units.
- A **derivative security** is any asset whose value is derived from the value of some other "underlying" asset.
- **Spot markets** are the markets in which assets are bought or sold for "on-the-spot" delivery, while **futures markets** are the markets in which participants agree today to purchase or sell an asset at some future date.
- **Money markets** are the financial markets in which funds are borrowed or loaned for less than a year, while **capital markets** are the financial markets for equity and for intermediate- or long-term debt (one year or longer).
- **Mortgage markets** are markets that deal with loans on residential, commercial, and industrial real estate. **Consumer credit markets** are markets that involve loans on autos, appliances, education, vacations, and so on. **Currency markets** are the markets that transact in foreign exchange.
- **Physical markets** are formal organizations having tangible physical locations that conduct auction markets in designated ("listed") securities, while **electronic markets** are those with no physical location but that rather exist as a computerized trading system.
- **Primary markets** are markets in which corporations raise capital by issuing new securities, while **secondary markets** are those in which existing securities and other financial assets are traded among investors.
- **Initial public offering (IPO) market** is the market in which firms "go public" by offering shares to the public.
- **Private markets** are markets in which transactions are negotiated directly between two parties, while **public markets** are those in which standardized contracts are traded on organized exchanges.
- **Cross-listing** occurs when a company lists shares of stock on multiple exchanges to increase its global recognition.
- **Around-the-clock trading** occurs when global financial institutions maintain offices in different time zones around the world and thus offer anytime trading.
- **Transfers of capital** between savers and borrowers take place in three different ways: (1) **Direct transfers** occur when a business sells its shares or bonds directly to savers. (2) Transfers may go through an **investment banking house**, which *underwrites* the issue. An underwriter serves as a middleman and facilitates the issuance of securities. (3) Transfers may be made through a **financial intermediary.** Here the intermediary obtains funds from savers in exchange for its own securities and then lends the money to borrowers.
- The **interest rate** is the price paid to borrow debt capital, while the **required return** is the minimum rate of return on a common stock that will induce a stockholder to purchase the stock.
- An **exchange rate** is the number of units of a given currency that can be purchased for one unit of another currency.

- Four fundamental factors interact to determine the cost of money: (1) **time preferences for consumption**, (2) **expected inflation**, (3) **risk**, and (4) **production opportunities.**
- The **marginal investor** is the investor who operates at the margin and whose decisions determine the market interest rate.
- The **real risk-free rate of return (r*)** is the interest rate that would exist on default-free U.S. Treasury securities if no inflation were expected, while the **nominal, risk-free rate (r_{RF})** is the interest rate that both offsets inflation and provides the required real return on a riskless investment.
- An **inflation premium** is a premium that must be added to the real risk-free rate to adjust for inflation. It is calculated as $I + r^*I$, where I is the inflation rate.
- A **risk premium** is the premium that reflects the difference between the expected rate of return on a given risky asset and that on a less risky asset.
- The **nominal, risk-adjusted rate of return (r)** is the actual rate charged on a loan; it compensates investors for postponing consumption, inflation, and risk.
- The **quoted (stated) rate** is the nominal, risk-adjusted rate of return, which is actually published in financial publications.
- The **default risk premium** is the premium that reflects the difference between the interest rate on a U.S. Treasury bond and a corporate bond of equal maturity and marketability.
- The **liquidity premium** is a premium added to the equilibrium interest rate on a security if that security cannot be converted to cash on short notice and at close to "fair market value."
- **Interest rate risk** is the risk of capital losses to which investors are exposed due to rising interest rates.
- The **maturity risk premium** is a premium that reflects interest rate risk.
- The **country risk premium** is a premium that reflects the risk that arises from investing or doing business in a particular country.
- The **exchange rate risk premium** is a premium that results from the possibility that an exchange rate change will lead to a loss in a bond's value.
- **Business climate** refers to a country's social, political, and economic environment.
- The **international monetary system** is the framework in which exchange rates are determined; it is the blueprint for international trade and capital flows.
- A **spot exchange rate** is the quoted price for a unit of foreign currency to be delivered on the spot or within a very short period of time, while a **forward exchange rate** is the quoted price for a unit of foreign currency to be delivered at a specified date in the future.
- A **fixed exchange rate** is set by the government and regulated to fluctuate only slightly (if at all) around the desired rate.
- A **floating** or **flexible exchange rate** is not regulated by the government and allows supply and demand to determine the currency's value.
- **Devaluation** or **revaluation of a currency** is the technical term referring to the decrease or increase in the par value of a currency whose value is fixed.
- **Depreciation** or **appreciation of a currency** refers to a decrease or increase in the foreign exchange value of a floating currency.
- A **soft** or **weak currency** is expected to depreciate against major currencies or is being artificially maintained by the government at an unrealistically high fixed rate, while a **hard** or **strong currency** is expected to appreciate against major currencies or is being artificially maintained by the government at an unrealistically low fixed rate.
- **Freely floating** occurs when an exchange rate is determined by supply and demand for the currency, while **managed float** occurs when there is significant government intervention to control the exchange rate via manipulation of the currency's supply and demand.
- A **currency board arrangement** occurs when a country has its own currency but commits to exchange it for a specified foreign money unit at a fixed exchange rate and legislates domestic currency restrictions, unless it has the foreign currency reserves to back it up.

- A **fixed peg arrangement** occurs when a country locks its currency to a specific currency or basket of currencies at a fixed exchange rate. The exchange rate is allowed to vary only within ±1 percent of the central rate.
- **Dollarization** occurs when a country abandons its own currency and adopts the U.S. dollar as its legal tender.

QUESTIONS

(2-1) Differentiate between the following markets: (1) Spot versus futures markets, (2) money versus capital markets, (3) mortgage versus consumer credit markets, (4) physical versus electronic markets, (5) primary versus secondary markets, and (6) private versus public markets.

(2-2) Identify and briefly explain the four fundamental factors that interact to determine the cost of money.

(2-3) Identify and briefly explain the five different components that comprise the risk premium that make up the nominal, risk-adjusted required rate of return.

(2-4) How does inflation affect a country's exchange rate?

(2-5) What is "the impossible trinity," and why is it so named?

(2-6) Briefly identify and differentiate between the two subgroups of the floating-rate category.

(2-7) Briefly identify and differentiate among the different fixed-exchange-rate regimes.

(2-8) Briefly explain whether the following statement is true or false: "The international monetary system is often called a 'floating regime,' and that designation is generally descriptive for most currencies."

(2-9) What is meant by the "bipolar view" with respect to exchange-rate regimes?

(2-10) What must be considered in selecting an exchange-rate regime?

(2-11) What are the pros and cons for fixed-exchange rates? Floating-exchange rates?

(2-12) What are the advantages and disadvantages of dollarization? Give an example. Has it worked?

SELF-TEST PROBLEMS

Solutions Appear in Appendix B

(ST-1) Define each of the following terms:

Key Terms

a. Physical asset markets; financial asset markets
b. Financial instruments; derivative security; monetary
c. Spot markets; futures markets; money markets; capital markets
d. Mortgage markets; consumer credit markets; currency markets
e. Physical markets; electronic markets; primary markets; secondary markets
f. Initial public offering (IPO) market; private markets; public markets
g. Cross-listing; around-the-clock trading
h. Investment banking house; financial intermediary
i. Interest rate; required return; exchange rate
j. Time preferences for consumption; expected inflation; risk; production opportunities
k. Real risk-free rate of return (r^*); nominal, risk-free required rate of return (r_{RF})
l. Inflation premium; risk premium; nominal, risk-adjusted rate of return (r)
m. Quoted (stated) rate; marginal investor; business climate
n. Default risk premium (DRP); liquidity premium (LP); interest rate risk

o. Maturity risk premium (MRP); country risk premium (CRP); exchange rate risk premium (ERP)
p. International monetary system; freely floating; managed float
q. Currency board arrangement; fixed peg arrangement
r. Dollarization

(ST-2) Inflation and Real Risk-Free Rate
Suppose a basket of goods and services that costs \$2,500 today is expected to cost \$2,600 in one year. In addition, assume that the nominal, risk-free rate as approximated by a 1-year Treasury security is 7.64 percent.
a. What is the expected inflation rate, I?
b. What is the real risk-free rate of return, r*?
c. What is the inflation premium, IP?

(ST-3) Inflation Premium
If the real risk-free rate of return is 2.5 percent and the expected inflation rate is 3 percent, what is the calculated inflation premium?

(ST-4) Nominal, Risk-Adjusted Rate
A Japanese investor would like to purchase a long-term (20-year maturity) U.S. corporate, BBB-rated bond. You are given the following data:

- Default risk premium: 0.4 percent.
- Liquidity premium: 0.5 percent.
- Maturity risk premium: 1.2 percent.
- Country risk premium: 0.1 percent.
- Exchange rate risk premium: 0.2 percent.
- Nominal, risk-free rate of return (20-year Treasury): 5.7 percent.
- Real risk-free rate of return: 2.5 percent.
- Inflation premium (calculated over 20 years): 3.2 percent.

The corporation is not considered to be a "blue-chip" company, so small default risk and liquidity premiums should be added when calculating the estimated yield for the BBB-rated bond. What would be your estimate of this bond's yield?

(ST-5) Exchange Rate
You are given the following data:

	United States	Country X
r*	3.2%	3.2%
I	4.0	10.0

Country X's currency is the LC and today's spot exchange rate is LC 2.000/\$.
a. In equilibrium, what should be the exchange rate 1 year from now?
b. Has the dollar strengthened or weakened against LC? Explain.

STARTER PROBLEMS

(2-1) Inflation and Nominal, Risk-Free Rate
You are given the following data:

- Inflation premium, IP: 3.75 percent.
- Real risk-free rate, r*: 2.25 percent.

a. What is the expected inflation rate, I?
b. What is the nominal, risk-free rate, r_{RF}?

(2-2) Nominal, Risk-Free Rate and Risk Premium
You are given the following data:

- Inflation premium, IP: 3.5 percent.
- Real risk-free rate, r*: 2.2 percent.
- Nominal, risk-adjusted rate of return, r: 8.5 percent.

a. What is the nominal, risk-free rate, r_{RF}?
b. What is the risk premium, RP?

(2-3) Exchange Rate Risk Premium
A British investor recently purchased a U.S. blue-chip, AA-rated corporate bond with a 20-year maturity. The yield on the bond is 7.5 percent, and the default risk and liquidity premiums on this bond are zero. The real risk-free rate of return is

2.4 percent and the inflation premium (calculated over 20 years) is 3 percent. The maturity risk premium is 1.6 percent, and the country risk premium is 0.2 percent. The nominal risk-free rate of return is 5.4 percent. Given these data, what was the exchange rate risk premium on the bond?

(2-4) Exchange Rate — You are given the following data:

	United States	Foreign Country C
r*	2.3%	2.3%
I	2.5	12.5

Foreign Country C's currency is the LC and today's spot exchange rate is LC 3.75/$.

a. In equilibrium, what should be the exchange rate 1 year from now?
b. Has the dollar strengthened or weakened against LC? Explain.

EXAM-TYPE PROBLEMS

(2-5) Nominal Interest Rate — A U.S. investor recently purchased a U.S. Treasury bond with a 20-year maturity to earn a yield of 5.5 percent. A Mexican investor just purchased a U.S. AAA-rated, "blue-chip" corporate bond with a 20-year maturity for 6.2 percent. This bond can be converted to cash very quickly, so its liquidity premium is small, 0.1 percent. The spread between U.S. Treasury bonds and AAA-rated bonds with similar maturity and liquidity is 0.2 percent. Another Mexican investor purchased a U.S. BBB-rated, corporate bond with a 20-year maturity. The liquidity premium for this BBB-rated bond is 0.8 percent, and the spread between U.S. Treasury bonds and BBB-rated bonds with similar maturity and liquidity is 1.8 percent.

a. When comparing the corporate bonds to the U.S. Treasury bond, what are the implications about country and exchange rate risks faced by Mexicans when investing in U.S. Treasury securities?
b. What yield will the Mexican investor earn on the U.S. BBB-rated, corporate bond?

(2-6) Exchange Rate — You are given the following data:

	United States	Foreign Country D
r*	2.75%	2.75%
I	3.50	7.00

Country D's currency is the LC and the 1-year forward exchange rate is LC 2.75/$.

a. In equilibrium, what should be the spot exchange rate today?
b. During the year, has the dollar strengthened or weakened against LC? Explain.

(2-7) Nominal, Risk-Free Rate — If the real risk-free rate is 2.4 percent and the expected inflation rate is 2.8 percent, what is the nominal, risk-free rate, r_{RF}?

(2-8) Inflation Premium and Nominal, Risk-Free Rate — You have the following information: the expected inflation rate, I, is 3.2 percent, and the real risk-free rate, r*, is 2.1 percent.

a. What is the inflation premium, IP?
b. What is the nominal, risk-free rate, r_{RF}?

(2-9) Nominal, Risk-Free and Risk-Adjusted Rates — Assume that the real risk-free rate is 2.75 percent, the expected inflation rate is 2.433 percent, and the risk premium is 3.75 percent.

a. What is the inflation premium, IP?
b. What is the nominal, risk-free rate, r_{RF}?
c. What is the nominal, risk-adjusted rate, r?

Integrated Case

Rick Schmidt, MRI's CFO, along with Mandy Wu, Schmidt's assistant, and John Costello, a consultant and college finance professor, comprise a team established to teach the firm's directors and senior managers the basics of international financial management. The program includes 16 sessions, each lasting about 2 hours, with the sessions scheduled immediately before the monthly board meetings. One session has been completed, and the team is now preparing Session 2, which will deal with global financial markets and institutions, interest rates, and exchange rates. The prior session provided an overview of international finance.

For the financial environment session, the team will identify the major types of financial markets, discuss how capital flows through the economy from savers to borrowers, and identify the factors that determine interest rates. MRI is a multinational medical supply company, headquartered in southern California. MRI has customers in Europe and South America, and is beginning to branch out to Asia. As MRI has grown, it has looked for operating opportunities in these new foreign markets. MRI's most dynamic market continues to be Europe. While a majority of MRI's development and production has remained in the southwestern United States, there has been a growing trend to move these operations abroad. The foreign unit management teams, once dominated by MRI-USA executives, have also taken on a new look as local employees have risen to the managerial ranks.

Schmidt has invited a number of these executives from different units to the program sessions. There are several benefits Schmidt expects to generate by involving them in the education process. First, Wu has met with the foreign executives in the process of organizing the session, as their input and perspective on the global financial environment is invaluable. Wu has incorporated a number of their suggestions into her presentation and has encouraged these executives to be vocal at the sessions. Second, involving foreign executives forces MRI-USA directors to be aware of the implications of their actions on global operations. On firm-wide strategic matters, they often have the strongest voice and have been accused in the past of ignoring the effect on foreign operations. Third, it allows foreign executives to learn about the perspective of other foreign executives and the MRI-USA directors. A common complaint by the board of directors has been that foreign executives suffer from myopia in only seeing how actions affect themselves. Schmidt recognizes that everyone can benefit from stepping back and seeing the picture from all perspectives.

With those thoughts in mind, the three compiled the following set of questions, which they plan to discuss during the session. In other sessions, they plan to use an *Excel* model to answer quantitative questions, but because this session is entirely qualitative, no model has been developed. Assume that you are Mandy Wu, and that you must prepare answers for use during the session.

Questions

a. Differentiate between physical and financial asset markets and identify some of the assets that trade in each. Identify some major types of financial markets.

b. What trends have shaped the landscape of the current financial environment?

c. Describe the different ways in which capital is transferred.

d. What is the cost of money? What factors affect it?

e. Differentiate between the real risk-free rate and the nominal risk-free rate.

f. What is the difference between the inflation rate and the inflation premium? What causes them to be different?

g. Define the terms default risk premium (DRP), liquidity premium (LP), maturity risk premium (MRP), country risk premium (CRP), and exchange rate risk premium (ERP). Identify which kinds of securities carry these premiums (short-term versus long-term securities and corporate versus Treasury securities).

h. How does inflation affect the exchange rate between two countries? Provide an example.

i. Describe the international monetary system.

j. What is meant by the "perfect currency"?

k. What is the difference between a depreciation/appreciation of a currency and a devaluation/revaluation of a currency?

l. Identify the major floating- and fixed-exchange-rate systems. How does a country choose a system?

m. What is dollarization? Identify some benefits and drawbacks of dollarization.

APPENDIX C

Answers to End-of-Chapter Problems

We present here some intermediate steps and final answers to selected end-of-chapter problems. Please note that your answer may differ slightly from ours due to rounding differences. Also, although we hope not, some of the problems may have more than one correct solution, depending on what assumptions are made in working the problem. Finally, many of the problems involve some verbal discussion as well as numerical calculations; this verbal material is not presented here.

2-1 a. I = 3.6675%.
b. r_{RF} = 6%.

2-2 a. r_{RF} = 5.7%.
b. RP = 2.8%.

2-3 ERP = 0.3%.

2-4 a. LC4.115854/$.

2-5 a. CRP + ERP = 0.4%.
b. r_{BBB} = 8.5%.

2-6 a. LC2.66/$.

2-7 r_{RF} = 5.2672%.

2-8 a. IP = 3.2672%.
b. r_{RF} = 5.3672%.

2-9 a. IP = 2.5%.
b. r_{RF} = 5.25%.
c. r = 9.00%.

3-1 $0.1605/Dkr.

3-2 a. kr5.9049/SwF.
b. SwF0.1694/kr.

3-3 a. JD0.7093/$.
b. $1.4098/JD.

3-4 $r_\$$ = 2.5%.

3-5 a. R21.6145/A$.
b. A$0.0463/R.

3-6 a. %ΔFX = −2.92%.
b. 1.264490; 105.711495.
c. −0.93%; +1.17%.
d. $2,186.40.

3-7 a. NZ$129.99.
b. R556.28.
c. Mex$956.45.

3-8 SwF1.2997/$.

3-9 a. +3.30%.
b. 6.0875%.
c. 9.71%.
d. 3.30%.
e. −3.30%.
f. −3.30%.

4-1 b. Country C; return ranking = 22.
c. Country B; risk ranking = 10.
d. Overall country ranking: B, C, D, and A.
e. Overall country ranking: C, D, and A.

4-2 a. Yes, Country X.
b. Country Y.
c. Country Y: wheat; Country X: widgets.
d. P_{Wheat} = 1.5 crates of widgets.
e. Country Y; +1,200 bushels of wheat and +34,600 crates of widgets; ΔGWP = +1,200 bushels of wheat and +34,600 crates of widgets.
f. P_{Wheat} = 0.72 crate of widgets.
g. Country X; +1,200 bushels of wheat and +34,600 crates of widgets; ΔGWP = +1,200 bushels of wheat and +34,600 crates of widgets.
h. Both; Country X: +900 bushels of wheat and +24,225 crates of widgets; Country Y: +300 bushels of wheat and +10,375 crates of widgets; ΔGWP = +1,200 bushels of wheat and +34,600 crates of widgets.

4-3 b. Country P; return ranking = 16.
c. Country M; risk ranking = 7.
d. Overall country ranking: N, M, P, and O.

Answers to Selected Exercises

Chapter 0

Section 0.1

1. -48 **3.** $2/3$ **5.** -1 **7.** 9 **9.** 1 **11.** 33 **13.** 14 **15.** $5/18$ **17.** 13.31 **19.** 6 **21.** $43/16$ **23.** 0 **25.** `3*(2-5)` **27.** `3/(2-5)` **29.** `(3-1)/(8+6)` **31.** `3-(4+7)/8` **33.** `2/(3+x)-x*y^2` **35.** `3.1x^3-4x^(-2)-60/(x^2-1)` **37.** `(2/3)/5` **39.** `3^(4-5)*6` **41.** `3*(1+4/100)^(-3)` **43.** `3^(2*x-1)+4^x-1` **45.** `2^(2x^2-x+1)` **47.** `4*e^(-2*x)/(2-3e^(-2*x))` or `4(*e^(-2*x))/(2-3e^(-2*x))` **49.** `3(1-(-1/2)^2)^2+1`

Section 0.2

1. 27 **3.** -36 **5.** $4/9$ **7.** $-1/8$ **9.** 16 **11.** 2 **13.** 32 **15.** 2 **17.** x^5 **19.** $-\frac{y}{x}$ **21.** $\frac{1}{x}$ **23.** x^3y **25.** $\frac{z^4}{y^3}$ **27.** $\frac{x^6}{y^6}$ **29.** $\frac{x^4y^6}{z^4}$ **31.** $\frac{3}{x^4}$ **33.** $\frac{3}{4x^{2/3}}$ **35.** $1-0.3x^2-\frac{6}{5x}$ **37.** 2 **39.** $1/2$ **41.** $4/3$ **43.** $2/5$ **45.** 7 **47.** 5 **49.** -2.668 **51.** $3/2$ **53.** 2 **55.** 2 **57.** ab **59.** $x+9$ **61.** $x\sqrt[3]{a^3+b^3}$ **63.** $\frac{2y}{\sqrt{x}}$ **65.** $3^{1/2}$ **67.** $x^{3/2}$ **69.** $(xy^2)^{1/3}$ **71.** $x^{3/2}$ **73.** $\frac{3}{5}x^{-2}$ **75.** $\frac{3}{2}x^{-1.2}-\frac{1}{3}x^{-2.1}$ **77.** $\frac{2}{3}x-\frac{1}{2}x^{0.1}+\frac{4}{3}x^{-1.1}$ **79.** $\frac{3}{4}x^{1/2}-\frac{5}{3}x^{-1/2}+\frac{4}{3}x^{-3/2}$ **81.** $\frac{3}{4}x^{2/5}-\frac{7}{2}x^{-3/2}$ **83.** $(x^2+1)^{-3}-\frac{3}{4}(x^2+1)^{-1/3}$ **85.** $\sqrt[3]{2^2}$ **87.** $\sqrt[3]{x^4}$ **89.** $\sqrt[5]{\sqrt{x}\sqrt[3]{y}}$ **91.** $-\frac{3}{2\sqrt[4]{x}}$ **93.** $\frac{0.2}{\sqrt[3]{x^2}}+\frac{3\sqrt{x}}{7}$ **95.** $\frac{3}{4\sqrt{(1-x)^5}}$ **97.** 64 **99.** $\sqrt{3}$ **101.** $1/x$ **103.** xy **105.** $\left(\frac{y}{x}\right)^{1/3}$ **107.** ± 4 **109.** $\pm 2/3$ **111.** $-1, -1/3$ **113.** -2 **115.** 16 **117.** ± 1 **119.** $33/8$

Section 0.3

1. $4x^2+6x$ **3.** $2xy-y^2$ **5.** x^2-2x-3 **7.** $2y^2+13y+15$ **9.** $4x^2-12x+9$ **11.** x^2+2+1/x^2 **13.** $4x^2-9$ **15.** y^2-1/y^2 **17.** $2x^3+6x^2+2x-4$ **19.** $x^4-4x^3+6x^2-4x+1$ **21.** $y^5+4y^4+4y^3-y$ **23.** $(x+1)(2x+5)$ **25.** $(x^2+1)^5(x+3)^3(x^2+x+4)$ **27.** $-x^3(x^3+1)\sqrt{x+1}$ **29.** $(x+2)\sqrt{(x+1)^3}$ **31. a.** $x(2+3x)$ **b.** $x=0, -2/3$ **33. a.** $2x^2(3x-1)$ **b.** $x=0, 1/3$ **35. a.** $(x-1)(x-7)$ **b.** $x=1, 7$ **37. a.** $(x-3)(x+4)$ **b.** $x=3, -4$ **39. a.** $(2x+1)(x-2)$ **b.** $x=-1/2, 2$ **41. a.** $(2x+3)(3x+2)$ **b.** $x=-3/2, -2/3$ **43. a.** $(3x-2)(4x+3)$ **b.** $x=2/3, -3/4$ **45. a.** $(x+2y)^2$ **b.** $x=-2y$ **47. a.** $(x^2-1)(x^2-4)$ **b.** $x=\pm 1, \pm 2$

Section 0.4

1. $\frac{2x^2-7x-4}{x^2-1}$ **3.** $\frac{3x^2-2x+5}{x^2-1}$ **5.** $\frac{x^2-x+1}{x+1}$ **7.** $\frac{x^2-1}{x}$ **9.** $\frac{2x-3}{x^2y}$ **11.** $\frac{(x+1)^2}{(x+2)^4}$ **13.** $\frac{-1}{\sqrt{(x^2+1)^3}}$ **15.** $\frac{-(2x+y)}{x^2(x+y)^2}$

Section 0.5

1. -1 **3.** 5 **5.** $13/4$ **7.** $43/7$ **9.** -1 **11.** $(c-b)/a$ **13.** $x=-4, 1/2$ **15.** No solutions **17.** $\pm\sqrt{\frac{5}{2}}$ **19.** -1 **21.** $-1, 3$ **23.** $\frac{1\pm\sqrt{5}}{2}$ **25.** 1 **27.** $\pm 1, \pm 3$ **29.** $\pm\sqrt{\frac{-1\pm\sqrt{5}}{2}}$ **31.** $-1, -2, -3$ **33.** -3 **35.** 1 **37.** -2 **39.** $1, \pm\sqrt{5}$ **41.** $\pm 1, \pm\frac{1}{\sqrt{2}}$ **43.** $-2, -1, 2, 3$

Section 0.6

1. 0, 3 **3.** $\pm\sqrt{2}$ **5.** $-1, -5/2$ **7.** -3 **9.** $0, -1, 1$ **11.** $x=-1$ ($x=-2$ is not a solution.) **13.** $-2, -3/2, -1$ **15.** -1 **17.** $\pm\sqrt[4]{2}$ **19.** ± 1 **21.** ± 3 **23.** $2/3$ **25.** $-4, -1/4$

Section 0.7

1. $P(0, 2)$, $Q(4, -2)$, $R(-2, 3)$, $S(-3.5, -1.5)$, $T(-2.5, 0)$, $U(2, 2.5)$

3.

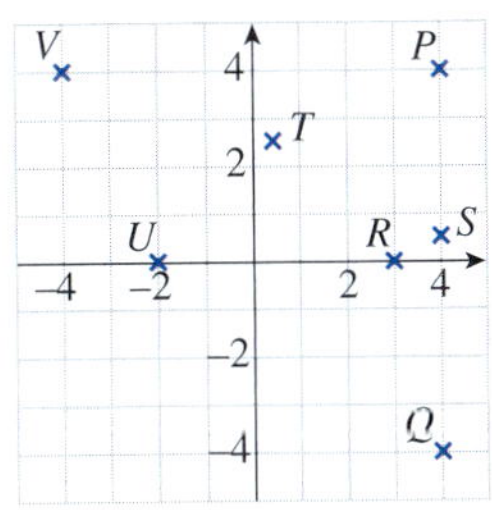

5.

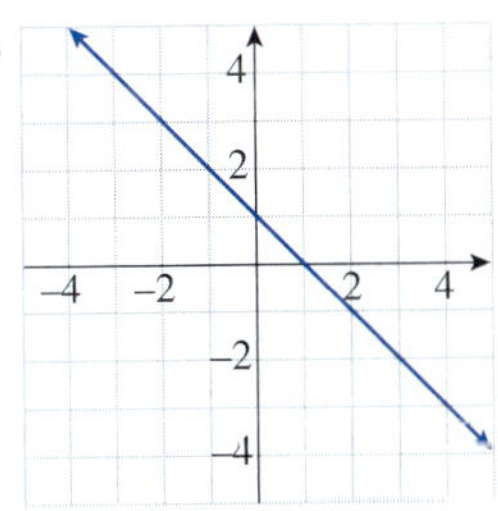

7.

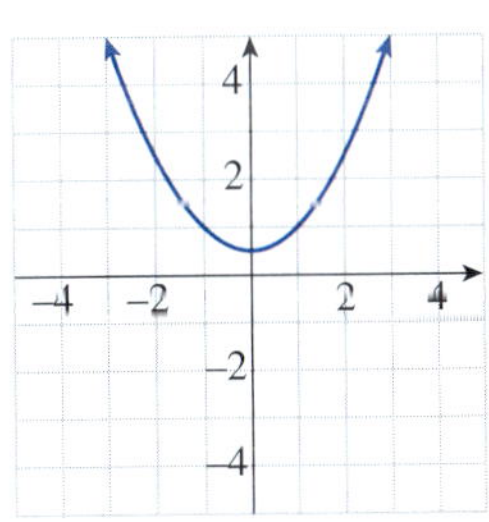

9.

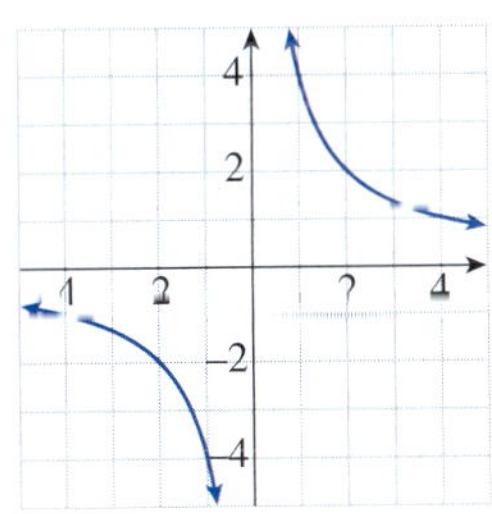

11.

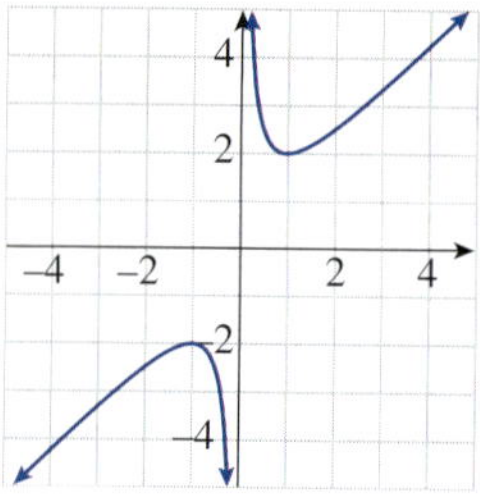

13. $\sqrt{2}$

15. $\sqrt{a^2+b^2}$

17. 1/2

19. Circle with center (0, 0) and radius 3

Chapter 1

Section 1.1

1. a. 2 **b.** 0.5 **3. a.** −1.5 **b.** 8 **c.** −8 **5. a.** 20 **b.** 30 **c.** 30 **d.** 20 **e.** 0 **7. a.** −1 **b.** 1.25 **c.** 0 **d.** 1 **e.** 0 **9. a.** Yes; $f(4) = 63/16$ **b.** Not defined **c.** Not defined **11. a.** Not defined **b.** Not defined **c.** Yes, $f(-10) = 0$ **13. a.** −7 **b.** −3 **c.** 1 **d.** $4y - 3$ **e.** $4(a+b) - 3$ **15. a.** 3 **b.** 6 **c.** 2 **d.** 6 **e.** $a^2 + 2a + 3$ **f.** $(x+h)^2 + 2(x+h) + 3$ **17. a.** 2 **b.** 0 **c.** 65/4 **d.** $x^2 + 1/x$ **e.** $(s+h)^2 + 1/(s+h)$ **f.** $(s+h)^2 + 1/(s+h) - (s^2 + 1/s)$

19.

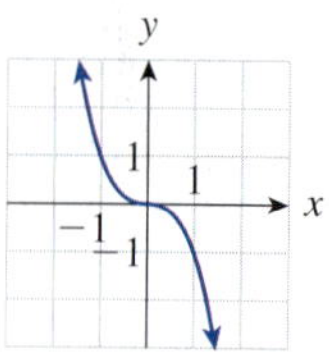

`-(x^3)`

21.

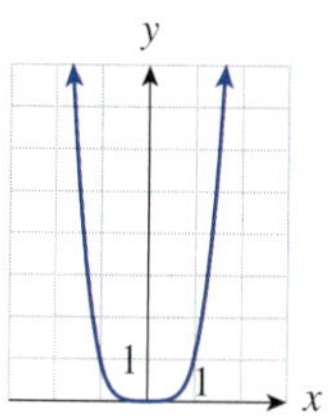

`x^4`

23.

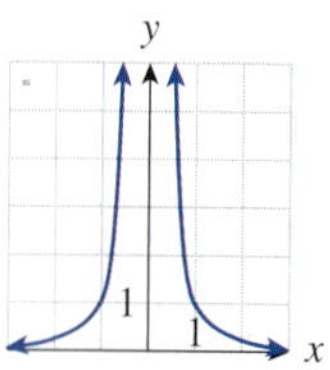

`1/x^2`

25. a. (I) **b.** (IV) **c.** (V) **d.** (VI) **e.** (III) **f.** (II)

27. `0.1*x^2-4*x+5`

x	0	1	2	3
f(*x*)	5	1.1	−2.6	−6.1
x	4	5	6	7
f(*x*)	−9.4	−12.5	−15.4	−18.1
x	8	9	10	
f(*x*)	−20.6	−22.9	−25	

29. `(x^2-1)/(x^2+1)`

x	0.5	1.5	2.5	3.5
h(*x*)	−0.6000	0.3846	0.7241	0.8491
x	4.5	5.5	6.5	7.5
h(*x*)	0.9059	0.9360	0.9538	0.9651
x	8.5	9.5	10.5	
h(*x*)	0.9727	0.9781	0.9820	

31. a. −1 **b.** 2 **c.** 2

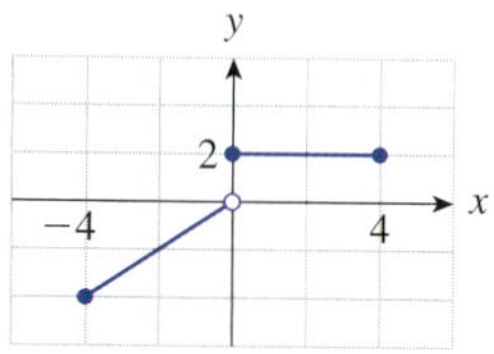

`x*(x<0)+2*(x>=0)`

33. a. 1 **b.** 0 **c.** 1

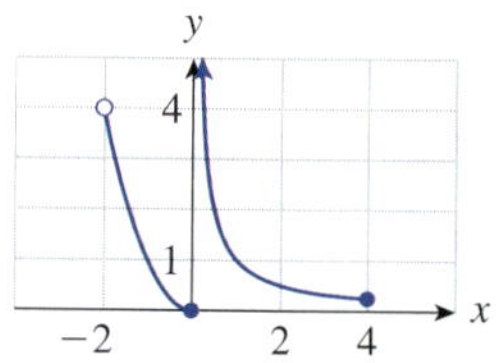

`(x^2)*(x<=0)+(1/x)*(0<x)`

35. a. 0 **b.** 2 **c.** 3 **d.** 3

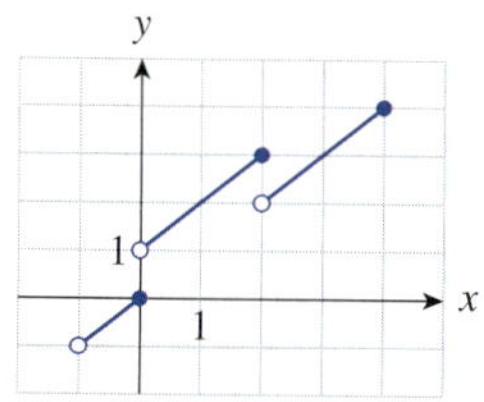

`x*(x<=0)+(x+1)*(0<x)*(x<=2)+x*(2<x)`

37. a. $h(2x+h)$ **b.** $2x+h$ **39. a.** $-h(2x+h)$ **b.** $-(2x+h)$ **41. a.** $I(3) = 1.55$. In 2003 the United States imported 1.55 million gallons/day. $I(5) = 1.5$. In 2005 the United States imported 1.5 million gallons/day. $I(6) = 1.5$. In 2006 the United States imported 1.5 million gallons/day. **b.** [1, 6] **c.** Graph:

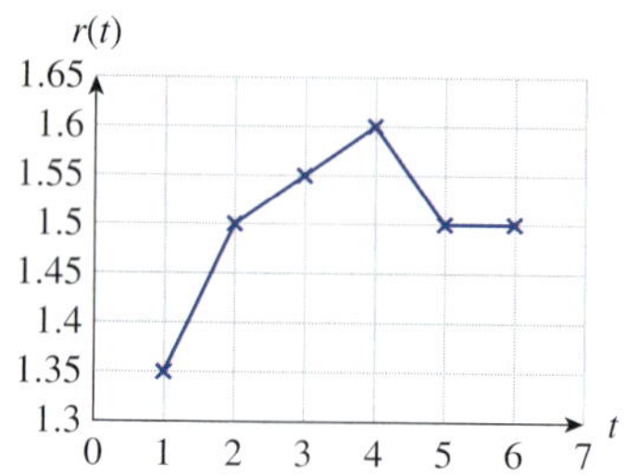

$I(4.5) \approx 1.55$. Thus the United States imported 1.55 million gallons/day in the year ending June 30, 2005 (or the year starting July 1, 2004). **43.** $f(4) \approx 1{,}600$, $f(5) \approx 1{,}700$, $f(6.5) \approx 1{,}300$. There were 1.6 million housing starts in 2004, 1.7 million housing starts in 2005, and 1.3 million housing starts in the year beginning July 2006. **45.** $f(7) - f(5)$ The change in the number of housing starts from 2005 to 2007 was larger in magnitude than the change from 2000 to 2005.

47. a. [0, 10]. $t \geq 0$ is not an appropriate domain because it would predict U.S. trade with China into the indefinite future with no basis. **b.** \$280 billion; U.S. trade with China in 2004 was valued at approximately \$280 billion. **49. a.** [1, 6] **b.** $L(2) \approx 400$; $L(5) \approx 900$, $L(6) \approx 1{,}100$. Sirius lost around \$400 million in 2002, \$900 million in 2005, and \$1,100 million (or \$1.1 billion) in 2006. **c.** $t \approx 4.5$; Sirius' losses were increasing fastest approximately midway through 2004.
51. a. $P(0) = 200$: At the start of 1995 the processor speed was 200 megahertz. $P(4) = 500$: At the start of 1999 the processor speed was 500 megahertz. $P(5) = 1{,}100$: At the start of 2000 the processor speed was 1,100 megahertz.
b. Graph:

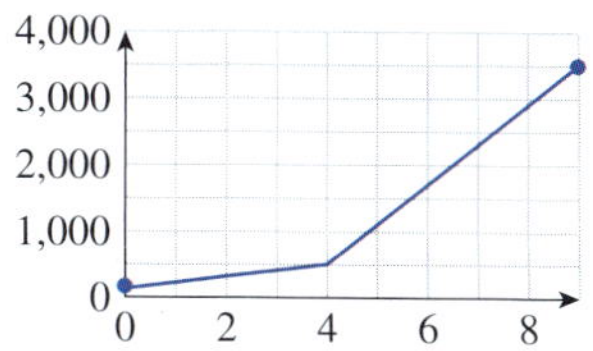

Midway through 2001 **c.**:

t	0	1	2	3	4	5	6	7	8	9
$P(t)$	200	275	350	425	500	1,100	1,700	2,300	2,900	3,500

53. $T(26{,}000) = \$3{,}508.75$; $T(65{,}000) = \$12{,}673.75$
55. a. `100*(1-12200/t^4.48)`
b. Graph:

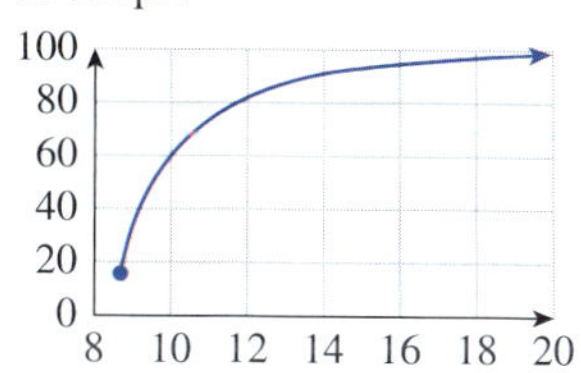

c.

t	9	10	11	12	13	14
$p(t)$	35.2	59.6	73.6	82.2	87.5	91.1
t	15	16	17	18	19	20
$p(t)$	93.4	95.1	96.3	97.1	97.7	98.2

d. 82.2% **e.** 14 months **57.** t; m **59.** $y(x) = 4x^2 - 2$ (or $f(x) = 4x^2 - 2$) **61.** True. We can construct a table of values from any graph by reading off a set of values. **63.** False. In a numerically specified function, only certain values of the function are specified so we cannot know its value on every real number in [0, 10], whereas an algebraically specified function would give values for every real number in [0, 10]. **65.** False: Functions with infinitely many points in their domain (such as $f(x) = x^2$) cannot be specified numerically. **67.** As the text reminds us: to evaluate f of a quantity (such as $x + h$) replace x everywhere by the *whole quantity* $x + h$, getting $f(x + h) = (x + h)^2 - 1$.
69. They are different portions of the graph of the associated equation $y = f(x)$. **71.** The graph of $g(x)$ is the same as the graph of $f(x)$, but shifted 5 units to the right.

Section 1.2

1. $N(t) = 200 + 10t$ (N = number of sound files, t = time in days) **3.** $A(x) = x^2/2$ **5.** $C(x) = 12x$

7. $C(x) = 1{,}500x + 1{,}200$ per day **a.** \$5,700 **b.** \$1,500 **c.** \$1,500 **d.** Variable cost = \$1,500x; Fixed cost = \$1,200; Marginal cost = \$1,500 per piano
e. Graph:

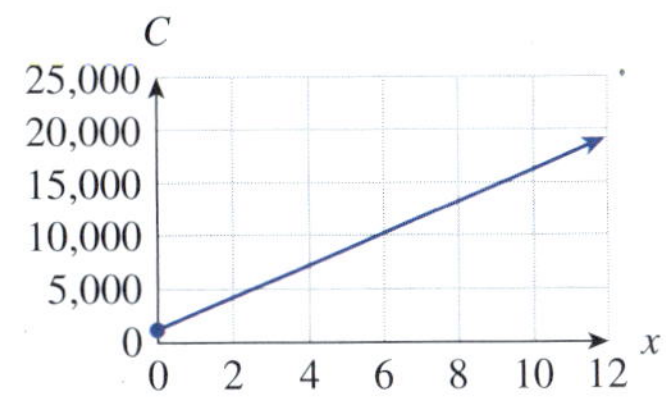

9. a. $C(x) = 0.4x + 70$, $R(x) = 0.5x$, $P(x) = 0.1x - 70$ **b.** $P(500) = -20$; a loss of \$20 **c.** 700 copies
11. $R(x) = 100x$, $P(x) = -2{,}000 + 90x - 0.2x^2$; at least 24 jerseys. **13.** $P(x) = -1.7 - 0.02x + 0.0001x^2$; approximately 264 thousand square feet
15. $P(x) = 100x - 5{,}132$, with domain [0, 405]. For profit, $x \geq 52$ **17.** 5,000 units **19.** $FC/(SP - VC)$
21. $P(x) = 579.7x - 20{,}000$, with domain $x \geq 0$; $x = 34.50$ g per day for break even
23. a. Graph:

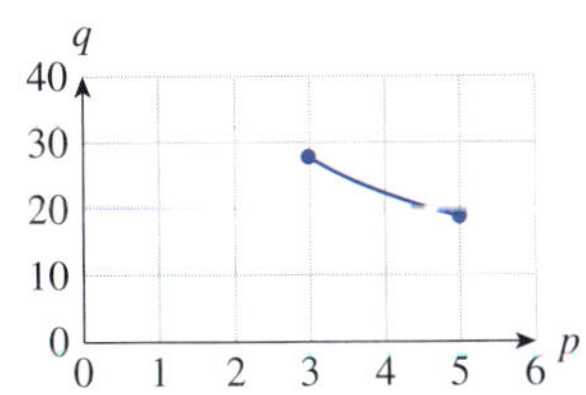

b. Ridership drops by about 3,070 rides per day.
25. a. 358,600 **b.** 361,200 **c.** \$6.00 **27.** \$240 per skateboard. **29. a.** \$110 per phone. **b.** Shortage of 25 million phones **31. a.** \$3.50 per ride.
Graph:

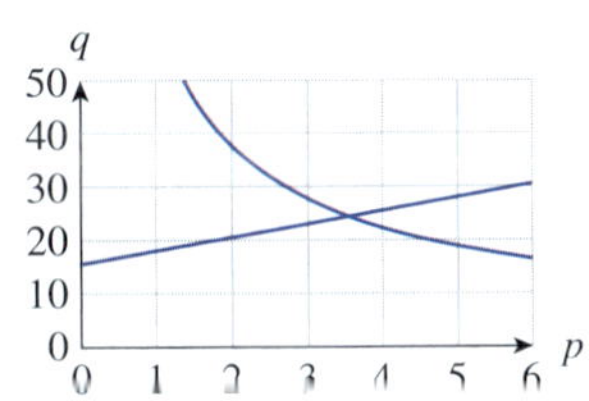

b. A surplus of around 9,170 rides.
33. a. \$12,000 **b.** $N(q) = 2{,}000 + 100q^2 - 500q$; $N(20) = \$32{,}000$ **35. a.** **(B)** **b.** \$36.8 billion
37. a. Models **(A)** and **(B)** **b.** Model **(A)**, which predicts about 3,757 tons of Freon. **39. a.** **(C)** **b.** \$20.80 per shirt if the team buys 70 shirts
Graph:

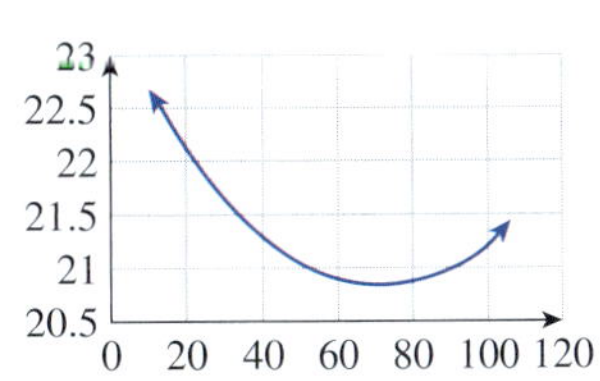

41. A quadratic model (B) is the best choice; the other models either predict perpetually increasing value of the euro or perpetually decreasing value of the euro. **43.** $A(t) = 5{,}000(1 + 0.0494/12)^{12t}$; \$7,061 **45.** At the beginning of 2016 **47.** 31.0 grams, 9.25 grams, 2.76 grams **49.** 20,000 years **51. a.** 1,000 years: 65%, 2,000 years: 42%, 3,000 years: 27% **b.** 1,600 years **53.** 30 **55.** Curve-fitting. The model is based on fitting a curve to a given set of observed data. **57.** The cost of downloading a movie was \$4 in January and is decreasing by 20¢ per month. **59.** Variable; marginal. **61.** Yes, as long as the supply is going up at a faster rate, as illustrated by the following graph:

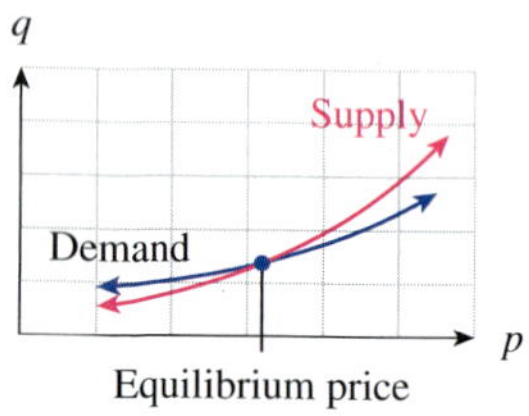

63. Extrapolate both models and choose the one that gives the most reasonable predictions.

Section 1.3

1. $m = 3$ **3.** $m = -1$ **5.** $m = 3/2$ **7.** $f(x) = -x/2 - 2$ **9.** $f(0) = -5$, $f(x) = -x - 5$ **11.** f is linear: $f(x) = 4x + 6$ **13.** g is linear: $g(x) = 2x - 1$ **15.** $-3/2$ **17.** $1/6$ **19.** Undefined **21.** 0 **23.** $-4/3$

25.

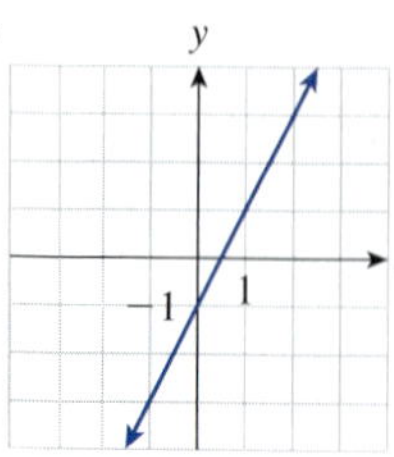

27.

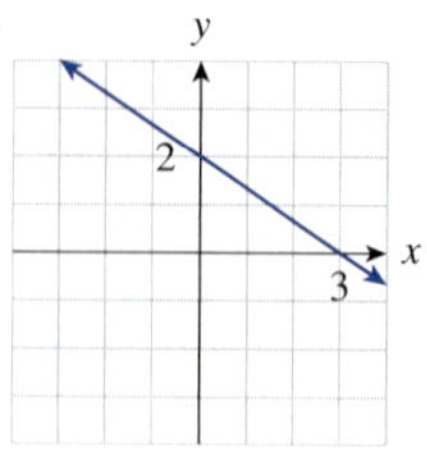

29.

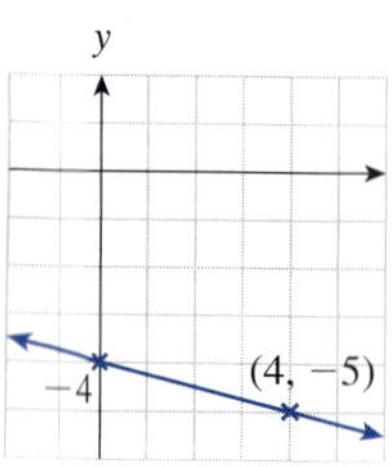

31.

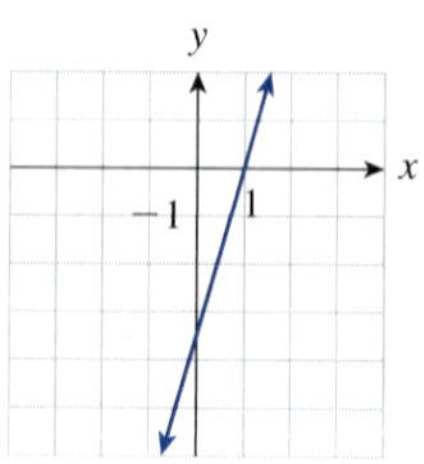

33.

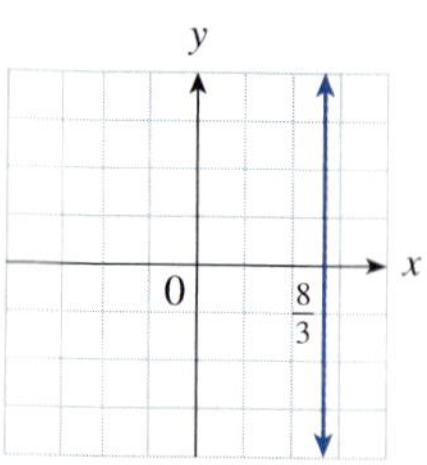

35.

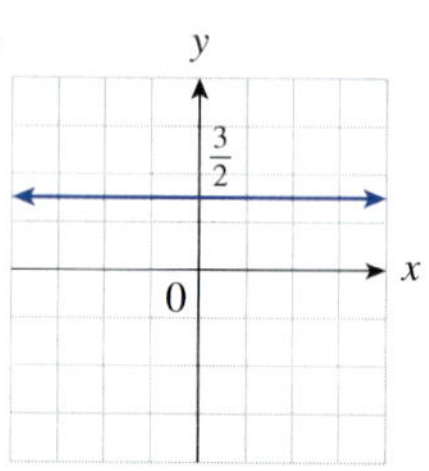

37.

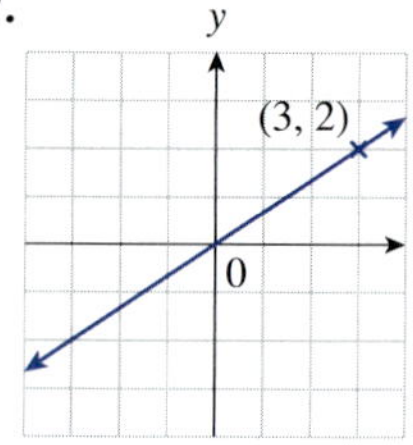

39. 2 **41.** 2 **43.** −2 **45.** Undefined **47.** 1.5 **49.** −0.09 **51.** 1/2 **53.** $(d - b)/(c - a)$ **55. a.** 1 **b.** 1/2 **c.** 0 **d.** 3 **e.** −1/3 **f.** −1 **g.** Undefined **h.** −1/4 **i.** −2

57. $y = 3x$ **59.** $y = \frac{1}{4}x - 1$

61. $y = 10x - 203.5$ **63.** $y = -5x + 6$ **65.** $y = -3x + 2.25$ **67.** $y = -x + 12$ **69.** $y = 2x + 4$ **71.** $y = \frac{q}{p}x$ **73.** $y = q$ **75.** Fixed cost = \$8,000, marginal cost = \$25 per bicycle **77.** $C = 145x + 75$; \$145 per iPod; \$14,575 **79.** $q = -40p + 2000$ **81. a.** $q = -p + 156.4$; 53.4 million phones; **b.** \$1, 1 million **83. a.** $q = -4{,}500p + 41{,}500$ **b.** Rides/day per \$1 increase in the fare; ridership decreases by 4,500 rides per day for every \$1 increase in the fare. **c.** 14,500 rides/day **85. a.** Demand: $q = -60p + 150$; supply: $q = 80p - 60$ **b.** \$1.50 each **87. a.** $q = 40t + 290$ **b.** $q(15) = 890$ million pounds **89. a.** $N = 1.45t - 4.05$ **b.** Million subscribers per year; the number of Sirius Satellite Radio subscribers grew at a rate of 1.45 million subscribers per year. **c.** 4.65 million subscribers, considerably less than the actual number. **91. a.** 2.5 ft/s **b.** 20 feet along the track **c.** after 6 seconds **93. a.** 130 miles per hour **b.** $s = 130t - 1{,}300$ **95.** $F = 1.8C + 32$; 86°F; 72°F; 14°F; 7°F **97.** $I(N) = 0.05N + 50{,}000$; $N = \$1{,}000{,}000$; marginal income is $m = 5¢$ per dollar of net profit **99.** $T(r) = (1/4)r + 45$; $T(100) = 70°\text{F}$ **101.** Increasing by \$130,000 per year. **103. a.** $y = -30t + 200$ **b.** $y = 60t - 250$

c. $y = \begin{cases} -30t + 200 & \text{if } 0 \le t \le 5 \\ 60t - 250 & \text{if } 5 < t \le 12 \end{cases}$ **d.** 170

105. $N = \begin{cases} 0.22t + 3 & \text{if } 0 \le t \le 5 \\ -0.15t + 4.85 & \text{if } 5 < t \le 9 \end{cases}$

3.8 million jobs

107. Compute the corresponding successive changes Δx in x and Δy in y, and compute the ratios $\Delta y/\Delta x$. If the answer is always the same number, then the values in the table come from a linear function. **109.** $f(x) = -\frac{a}{b}x + \frac{c}{b}$. If $b = 0$, then $\frac{a}{b}$ is undefined, and y cannot be specified as a function of x. (The graph of the resulting equation would be a vertical line.) **111.** slope, 3. **113.** If m is positive, then y will increase as x increases; if m is negative then y will decrease as x increases; if m is zero then y will not change as x changes. **115.** The slope increases, because an increase in the y-coordinate of the second point increases Δy while leaving Δx fixed. **117.** The units of the slope m are units of y (bootlags) per unit of x (zonars). The intercept b is on the y-axis, and is thus measured in units of y (bootlags). Thus, m is measured in bootlags per zonar and b is measured in bootlags. **119.** It must increase by 10 units each day, including the third. **121.** (B) **123.** Increasing the number of items from the break even results in a profit: Because the slope of the revenue graph is larger than the slope of the cost graph, it is higher than the cost graph to the right of the point of intersection, and hence corresponds to a profit.

Section 1.4

1. 6 **3.** 86 **5. a.** 0.5 (better fit) **b.** 0.75
7. a. 27.42 **b.** 27.16 (better fit)

9. $y = 1.5x - 0.6667$

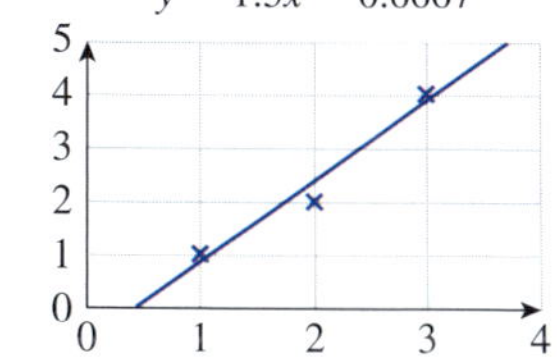

11. $y = 0.4118x + 0.9706$

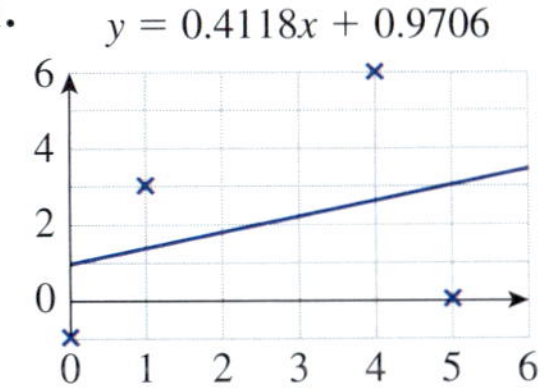

13. a. $r = 0.9959$ (best, not perfect) **b.** $r = 0.9538$
c. $r = 0.3273$ (worst)

15.

	x	y	xy	x^2
	3	500	1,500	9
	5	600	3,000	25
	7	800	5,600	49
Totals	15	1,900	10,100	83

$y = 75x + 258.33$; 858.33 million

17. $y = 3.65t + 4.62$; \$26.52 billion **19.** $y = 0.135x + 0.15$; 6.9 million jobs **21. a.** $y = 1.62x - 23.87$.
Graph:

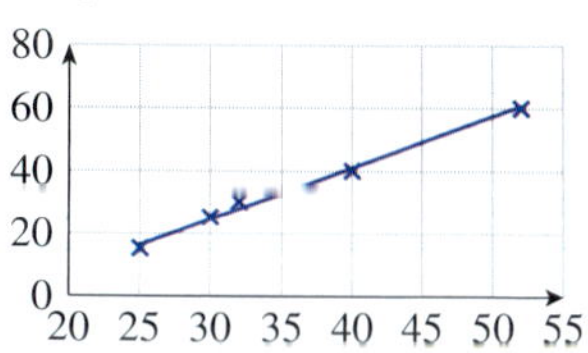

b. Each acre of cultivated land produces about 1.62 tons of soybeans. **23. a.** Regression line: $y = 0.032x + 0.042$
Graph:

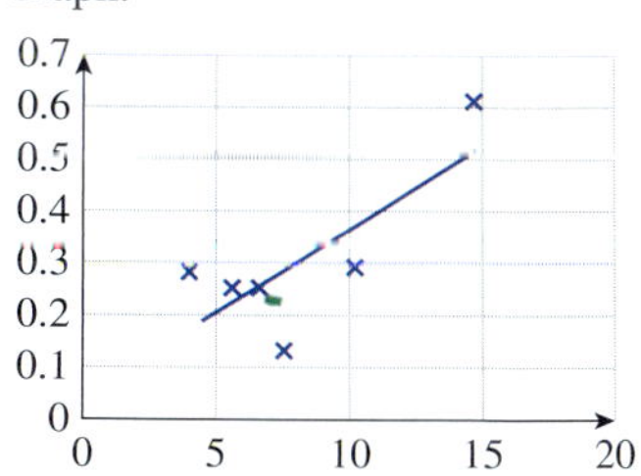

b. There are about 32 motorcycles per 1,000 automobiles.
c. $r \approx 0.7739$; not a strong correlation.

25. a. $p = 0.13t + 0.22$
Graph:

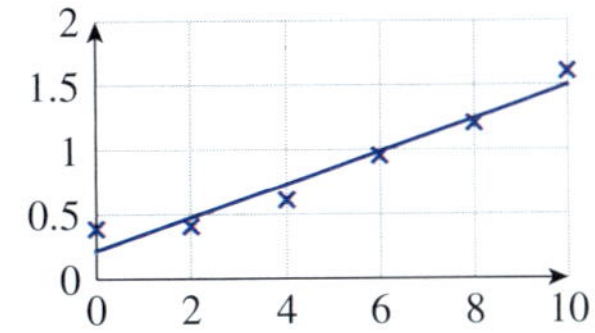

b. Yes; the first and last points lie above the regression line, while the central points lie below it, suggesting a curve.

c.

	A	B	C	D
1	t	p (Observed)	p (predicted)	Residual
2	0	0.38	0.22	0.16
3	2	0.4	0.48	-0.08
4	4	0.6	0.74	-0.14
5	6	0.95	1	-0.05
6	8	1.2	1.26	-0.06
7	10	1.6	1.52	0.08
8				

Notice that the residuals are positive at first, become negative, and then become positive, confirming the impression from the graph. **27.** The line that passes through (a, b) and (c, d) gives a sum-of-squares error SSE = 0, which is the smallest value possible. **29.** The regression line is the line passing through the given points. **31.** 0 **33.** No. The regression line through $(-1, 1)$, $(0, 0)$, and $(1, 1)$ passes through none of these points.

Chapter 1 Review

1. a. 1 **b.** −2 **c.** 0 **d.** −1 **3. a.** 1 **b.** 0 **c.** 0 **d.** −1

5.

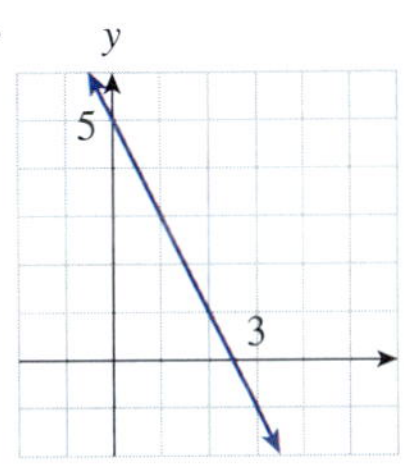

7.

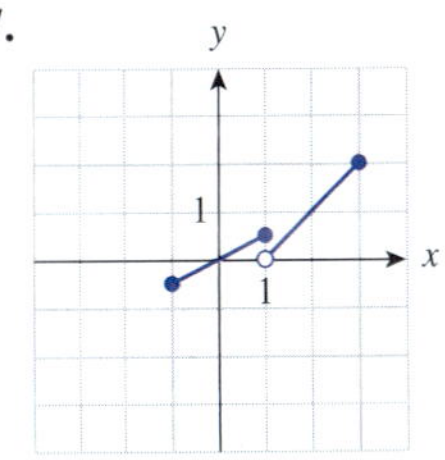

9. Absolute value **11.** Linear **13.** Quadratic
15. $y = -3x + 11$ **17.** $y = (1/2)x + 3/2$ **19.** The second line, $y = -x/4 + 1$, is a better fit **21.** $y = -0.214x + 1.14$, $r \approx -0.33$ **23. a.** Exponential.
Graph:

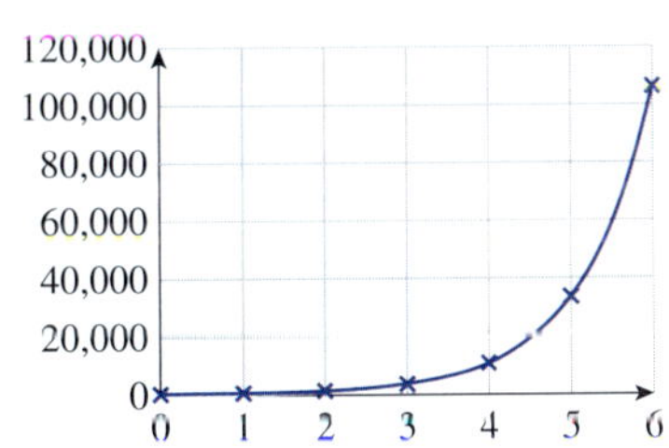

b. The ratios (rounded to 1 decimal place) are:

$V(1)/V(0)$	$V(2)/V(1)$	$V(3)/V(2)$	$V(4)/V(3)$	$V(5)/V(4)$	$V(6)/V(5)$
3	3.3	3.3	3.2	3.2	3.2

They are close to 3.2.

c. About 343,700 visits/day **25. a.** (A) **b.** (A) Leveling off (B) Rising (C) Rising; they begin to fall after 7 months (D) Rising **27. a.** 2,080 hits per day **b.** Probably not. This model predicts that Web site traffic will start to decrease as advertising increases beyond \$8,500 per month, and then drop toward zero. **29. a.** $q = -60p + 950$ **b.** 50 novels per month **c.** \$10, for a profit of \$1,200.

Chapter 2

Section 2.1

1. Vertex: $(-3/2, -1/4)$, y-intercept: 2, x-intercepts: $-2, -1$

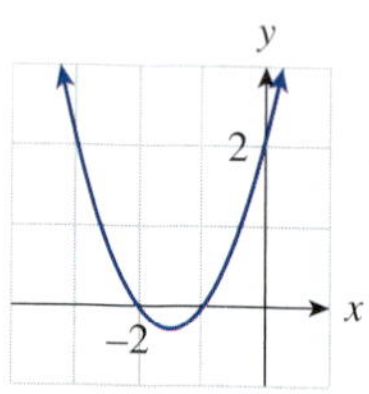

3. Vertex: (2,0), y-intercept: -4, x-intercept: 2

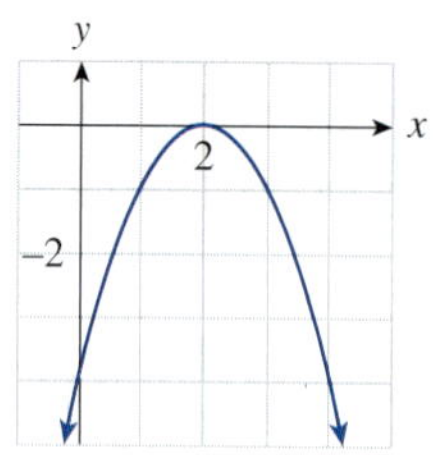

5. Vertex: $(-20, 900)$, y-intercept: 500, x-intercepts: -50, 10

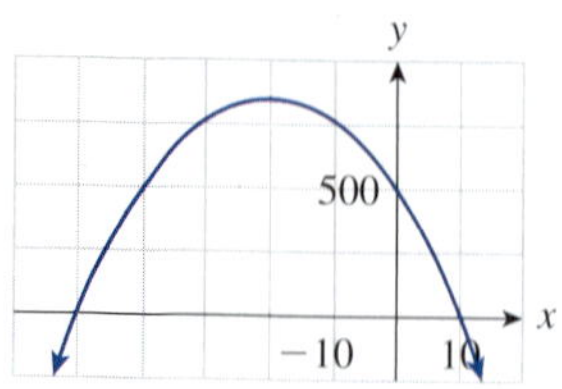

7. Vertex: $(-1/2, -5/4)$, y-intercept: -1, x-intercepts: $-1/2 \pm \sqrt{5}/2$

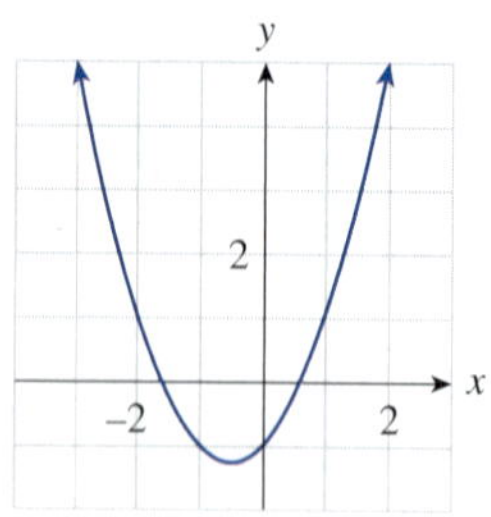

9. Vertex: (0, 1), y-intercept: 1, no x-intercepts

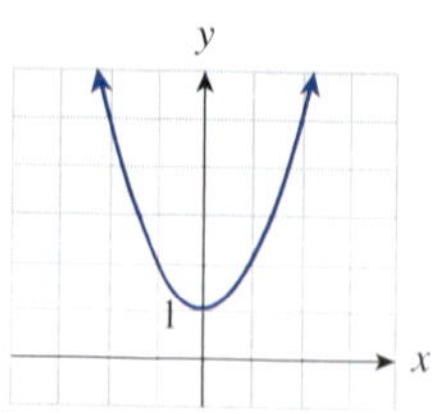

11. $R = -4p^2 + 100p$; Maximum revenue when $p = \$12.50$

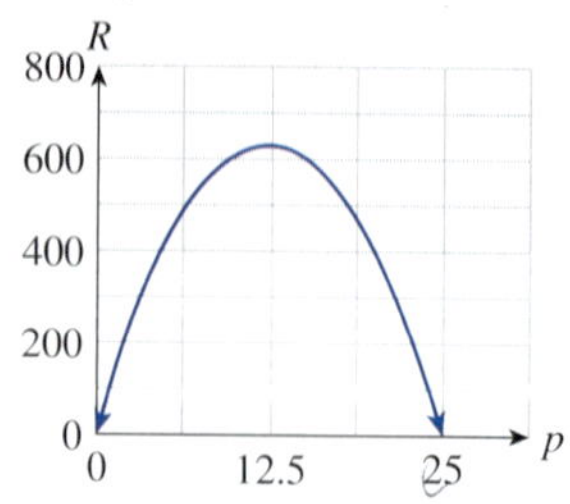

13. $R = -2p^2 + 400p$; Maximum revenue when $p = \$100$

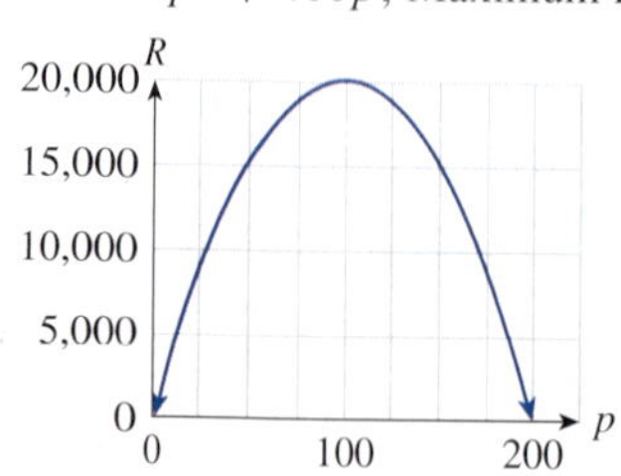

15. $y = -0.7955x^2 + 4.4591x - 1.6000$
17. $y = -1.1667x^2 - 6.1667x - 3.0000$
19. a. Positive because the data suggest a curve that is concave up. **b.** (C) **c.** 1998. Extrapolating in the positive direction leads one to predict more and more steeply rising military expenditure, which may or may not occur; extrapolating in the negative direction predicts continually more an more steeply rising military expenditure as we go back in time, contradicting history. **21.** 2003; about 1.6 million barrels/day **23.** 5,000 pounds. The model is not trustworthy for vehicle weights larger than 5,000 pounds, as it predicts increasing fuel economy with increasing weight, and 5,000 is close to the upper limit of the domain of the function. **25.** Maximum revenue when $p = \$140$, $R = \$9,800$ **27.** Maximum revenue with 70 houses, $R = \$9,800,000$ **29. a.** $q = -4,500p + 41,500$ **b.** \$4.61 for a daily revenue of \$95,680.55 **c.** No
31. a. $q = -560x + 1,400$; $R = -560x^2 + 1,400x$
b. $P = -560x^2 + 1,400x - 30$; $x = \$1.25$; $P = \$845$ per month **33.** $C = -200x + 620$; $P = -400x^2 + 1,400x - 620$; $x = \$1.75$ per log-on; $P = \$605$ per month **35. a.** $q = -10p + 400$ **b.** $R = -10p^2 + 400p$ **c.** $C = -30p + 4,200$ **d.** $P = -10p^2 + 430p - 4,200$; $p = \$21.50$
37. $f(t) = 6.25t^2 - 100t + 1,200$; \$1,425 billion, which is \$25 billion lower than the actual value.
39. a. $S(t) = -391t^2 + 3,278t - 4,844$;
Graph:

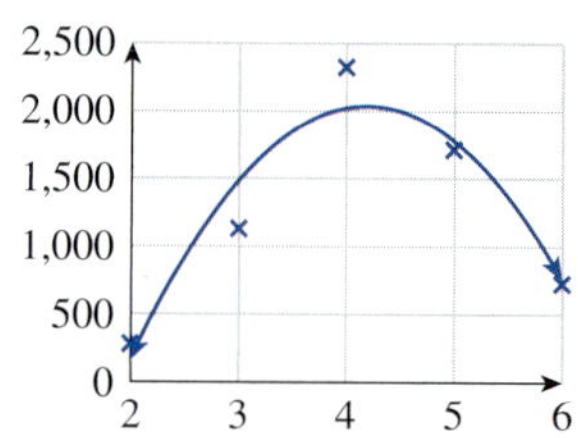

b. $-1,057$ thousand units. This makes no sense, and shows the danger of extrapolating mathematical models beyond their domains. Actual sales in the third quarter of 2008 were 6.9 million. **41.** The graph is a straight line. **43.** (C) **45.** Positive; the x coordinate of the vertex is negative, so $-b/(2a)$ must be negative. Because a is positive (the parabola is concave up), this means that b must also be positive to make $-b/(2a)$ negative. **47.** The x coordinate of the vertex represents the unit price that leads to the maximum revenue, the y coordinate of the vertex gives the maximum possible revenue, the x-intercepts give the unit prices that result in zero revenue, and the y-intercept gives the revenue resulting from zero unit price (which is obviously zero). **49.** Graph the data to see whether the points suggest a curve rather than a straight line. It the curve suggested by the graph is concave up or concave down, then a quadratic model would be a likely candidate. **51.** No; the graph of a quadratic function is a parabola. In the case of a concave-up parabola, the curve would unrealistically predict sales increasing without bound in the future. In the case of a concave-down parabola, the curve would predict "negative" sales from some point on. **53.** If $q = mp + b$ (with $m < 0$), then the revenue is given by $R = pq = mp^2 + bp$. This is the equation of a parabola with $a = m < 0$, and so is concave down. Thus, the vertex is the highest point on the parabola, showing that there is a single highest value for R, namely, the y coordinate of the vertex. **55.** Since $R = pq$, the demand must be given by

$$q = \frac{R}{p} = \frac{-50p^2 + 60p}{p} = -50p + 60.$$

Section 2.2

1. `4^x`

x	−3	−2	−1	0	1	2	3
$f(x)$	$\frac{1}{64}$	$\frac{1}{16}$	$\frac{1}{4}$	1	4	16	64

3. `3^(-x)`

x	−3	−2	−1	0	1	2	3
$f(x)$	27	9	3	1	$\frac{1}{3}$	$\frac{1}{9}$	$\frac{1}{27}$

5. `2*2^x` or `2*(2^x)`

x	−3	−2	−1	0	1	2	3
$f(x)$	$\frac{1}{4}$	$\frac{1}{2}$	1	2	4	8	16

7. `-3*2^(-x)`

x	−3	−2	−1	0	1	2	3
$f(x)$	−24	−12	−6	−3	$-\frac{3}{2}$	$-\frac{3}{4}$	$-\frac{3}{8}$

9. `2^x-1`

x	−3	−2	−1	0	1	2	3
$f(x)$	$-\frac{7}{8}$	$-\frac{3}{4}$	$-\frac{1}{2}$	0	1	3	7

11. `2^(x-1)`

x	−3	−2	−1	0	1	2	3
$f(x)$	$\frac{1}{16}$	$\frac{1}{8}$	$\frac{1}{4}$	$\frac{1}{2}$	1	2	4

13. $y = 3^{-x}$

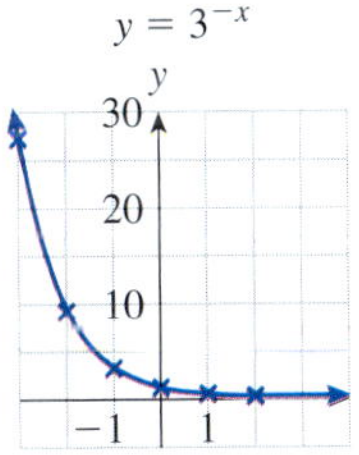

15. $y = 2(2^x)$

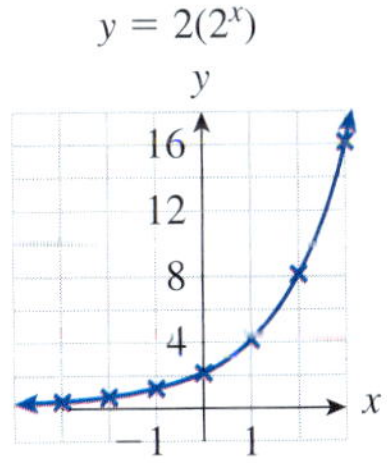

17. $y = -3(2^{-x})$

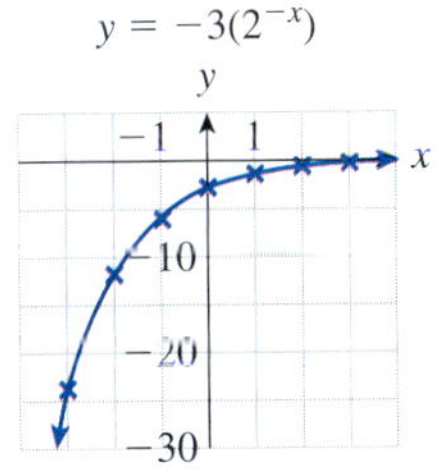

19. Both; $f(x) = 4.5(3^x)$. $g(x) = 2(1/2)^x$, or $2(2^{-x})$
21. Neither **23.** g; $g(x) = 4(0.2)^x$
25. `e^(-2*x)` or `EXP(-2*x)`

x	-3	-2	-1	0	1	2	3
$f(x)$	403.4	54.60	7.389	1	0.1353	0.01832	0.002479

27. `1.01*2.02^(-4*x)`

x	-3	-2	-1	0	1	2	3
$f(x)$	4662	280.0	16.82	1.01	0.06066	0.003643	0.0002188

29. `50*(1+1/3.2)^(2*x)`

x	-3	-2	-1	0	1	2	3
$f(x)$	9.781	16.85	29.02	50	86.13	148.4	255.6

31. `2^(x-1)` *not* `2^x-1` **33.** `2/(1-2^(-4*x))` *not* `2/1-2^-4*x` *and not* `2/1-2^(-4*x)`
35. `(3+x)^(3*x)/(x+1)` or `((3+x)^(3*x))/(x+1)` *not* `(3+x)^(3*x)/x+1` *and not* `(3+x^(3*x))/(x+1)`
37. `2*e^((1+x)/x)` or `2*EXP((1+x)/x)` *not* `2*e^1+x/x` *and not* `2*e^(1+x)/x` *and not* `2*EXP(1+x)/x`

39.

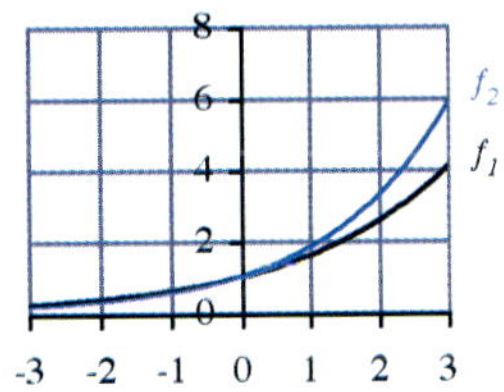

`y1=1.6^x` `y2=1.8^x`

41.

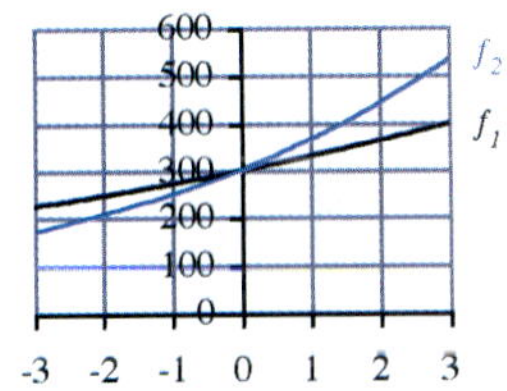

`y1=300*1.1^x`
`y2=300*1.1^(2*x)`

43.

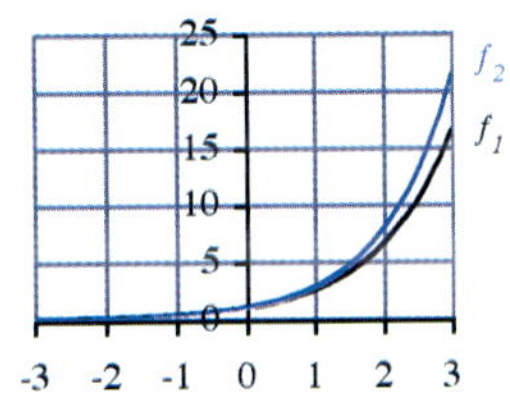

`y1=2.5^(1.02*x)`
`y2=e^(1.02*x)`
or `exp(1.02*x)`

45.

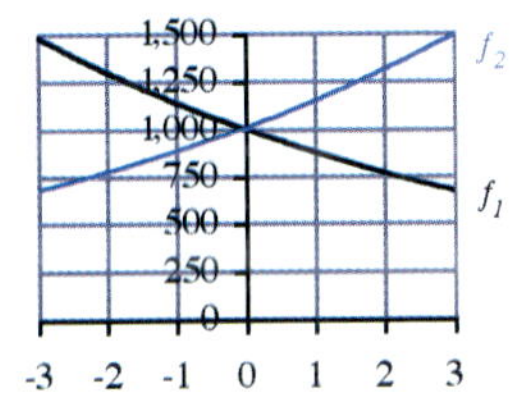

`y1=1000*1.045^(-3*x)`
`y2=1000*1.045^(3*x)`

47. $f(x) = 500(0.5)^x$ **49.** $f(x) = 10(3)^x$
51. $f(x) = 500(0.45)^x$ **53.** $f(x) = -100(1.1)^x$
55. $y = 4(3^x)$ **57.** $y = -1(0.2^x)$ **59.** $y = 2.1213(1.4142^x)$
61. $y = 3.6742(0.9036^x)$ **63.** $f(t) = 5{,}000e^{0.10t}$
65. $f(t) = 1{,}000e^{-0.063t}$ **67.** $y = 1.0442(1.7564)^x$
69. $y = 15.1735(1.4822)^x$ **71.** $f(t) = 300(0.5)^t$; 9.375 mg
73. a. Linear model: $F = 65t - 60$. Exponential model: $F = 97.2(1.20)^t$. The exponential model is more appropriate. **b.** 418 tons, not too far off the projected figure
75. a. $P = 180(1.01091)^t$ million **b.** 5 significant digits **c.** 345 million **77. a.** $y = 50{,}000(1.5^{t/2})$, t = time in years since two years ago **b.** 91,856 tags **79.** $y = 1{,}000(2^{t/3})$; 65,536,000 bacteria after 2 days **81.** $A(t) = 167(1.18)^t$; 1,695 cases **83.** $A(t) = 5{,}000(1 + 0.0494/12)^{12t}$; \$7,061
85. At the beginning of 2016 **87.** 491.82 **89.** $A(t) = 1.3e^{-0.3t}$; 2008: 0.53 million; 2009: 0.39 million

91. a.

Year	1950	2000	2050	2100
$C(t)$ parts per million	561	669	799	953

b. 2010 ($t = 260$)
93. a. $P(t) = 0.339(1.169)^t$.
Graph:

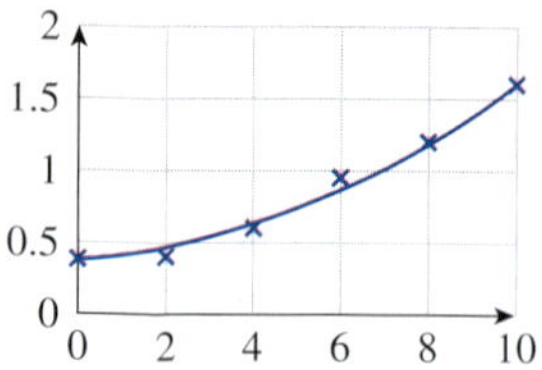

b. \$1.9 million
95. a. $n = 1.127(3.544)^t$
Graph:

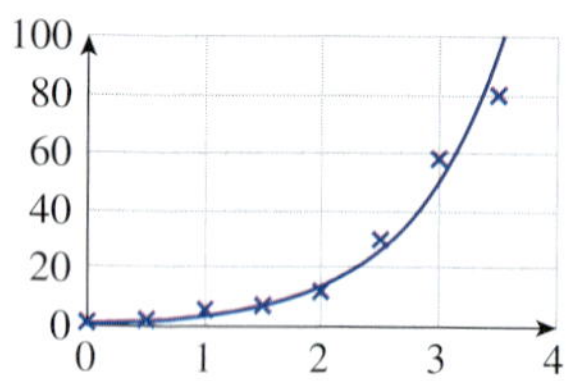

b. 3.544 **c.** 178 million **97.** (B) **99.** Exponential functions of the form $f(x) = Ab^x$ $(b > 0)$ increase rapidly for large values of x. In real-life situations, such as population growth, this model is reliable only for relatively short periods of growth. Eventually, population growth tapers off because of pressures such as limited resources and overcrowding. **101.** The article was published about a year before the "housing bubble" burst in 2006, whereupon house prices started to fall, contrary to the prediction of the graph, as documented in Exercise 90. This shows the danger of using any mathematical model to extrapolate. The blogger was, however, cautious in the choice of words, claiming only to be estimating what the future U.S. median house price "might be." **103.** Linear functions better: cost models where there is a fixed cost and a variable cost; simple interest, where interest is paid on the original amount invested. Exponential models better: compound interest, population growth. (In both of these, the rate of growth depends on the present number of items, rather than on some fixed quantity.) **105.** Take the ratios y_2/y_1 and y_3/y_2. If they are the same, the points fit on an exponential curve. **107.** This reasoning is suspect—the bank need not use its computer resources to update all the accounts every minute, but can instead use the continuous compounding formula to calculate the balance in any account at any time.

Section 2.3

1.

Logarithmic Form	$\log_{10} 10{,}000 = 4$	$\log_4 16 = 2$	$\log_3 27 = 3$	$\log_5 5 = 1$	$\log_7 1 = 0$	$\log_4 \frac{1}{16} = -2$

3.

Exponential Form	$(0.5)^2 = 0.25$	$5^0 = 1$	$10^{-1} = 0.1$	$4^3 = 64$	$2^8 = 256$	$2^{-2} = \frac{1}{4}$

5. 1.4650 **7.** −1.1460 **9.** −0.7324 **11.** 6.2657
13.

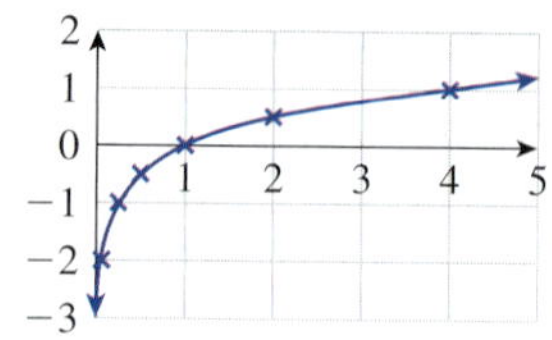

15.

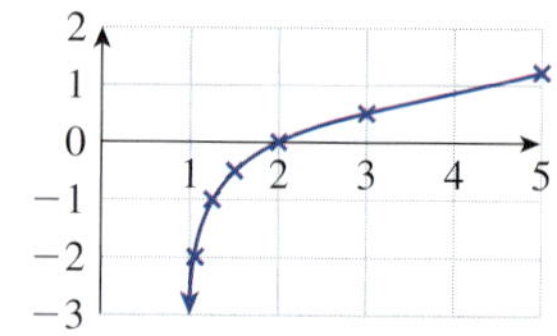

17.

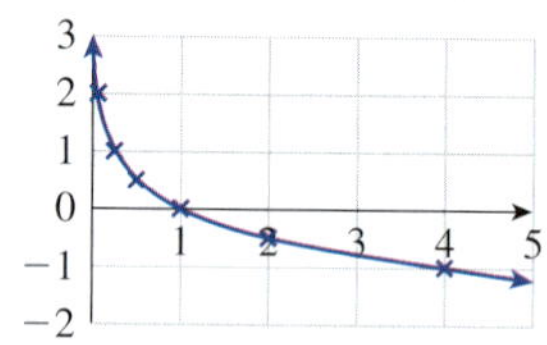

19. $Q = 1000e^{-t \ln 2}$ **21.** $Q = 1000e^{t(\ln 2)/2}$
23. Doubling time = 2 ln 2 **25.** Half-life = (ln 2)/4
27. $f(x) = 4(7.389)^x$ **29.** $f(t) = 2.1e^{0.000\,9995t}$
31. $f(t) = 10e^{-0.01309t}$. **33.** 3.36 years **35.** 11 years
37. 23.1% **39.** 63,000 years old **41.** 11 years
43. 207 months **45.** 12 years **47.** 17.77 years
49. 1,600 years **51. a.** $b = 3^{1/6} \approx 1.20$ **b.** 3.8 months
53. a. $Q(t) = Q_0e^{-0.139t}$ **b.** 3 years **55.** 2,360 million years
57. 3.2 hours **59.** 3.89 days
61. a. $P(t) = 6.591 \ln(t) - 17.69$ **b.** 1 digit **c.** (A)
63. $57.51 \ln t + 30.96$
Graph:

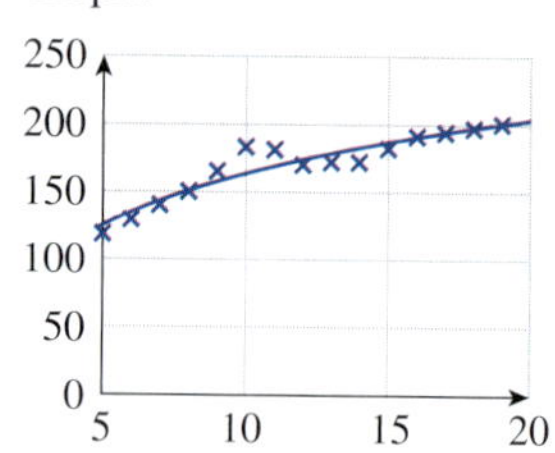

Extrapolating in the positive direction results in a prediction of ever-increasing R&D expenditures by industry. This is reasonable to a point, as expenditures cannot reasonably be expected to increase without bound. Extrapolating in the negative direction eventually leads to negative values, which does not model reality. **65. a.** About 1.259×10^{24} ergs **b.** about 2.24% **c.** $E = 10^{1.5R+11.8}$ **d.** Proof **e.** 1,000 **67. a.** 75 dB, 69 dB, 61 dB **b.** $D = 95 - 20 \log r$ **c.** 57,000 feet
69. Graph:

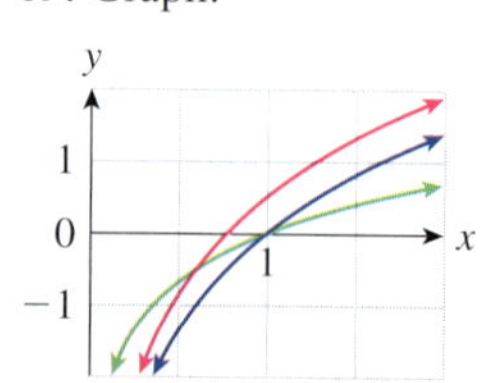

The green curve is $y = \ln x$. The blue curve is $y = 2\ln x$, and the red curve is $y = 2\ln x + 0.5$. Multiplying by A stretches the graph in the y-direction by a factor of A. Adding C moves the graph C units vertically up. **71.** The logarithm of a negative number, were it defined, would be the power to which a base must be raised to give that negative number. But raising a base to a power never results in a negative number, so there can be no such number as the logarithm of a negative number. **73.** Any logarithmic curve $y = \log_b t + C$ will eventually surpass 100%, and hence not be suitable as a long-term predictor of market share. **75.** $\log_4 y$ **77.** 8 **79.** x **81.** Time is increasing logarithmically with population; solving $P = Ab^t$ for t gives $t = \log_b(P/A) = \log_b P - \log_b A$, which is of the form $t = \log_b P + C$. **83.** (Proof)

Section 2.4

1. $N = 7, A = 6, b = 2;$
`7/(1+6*2^-x)`

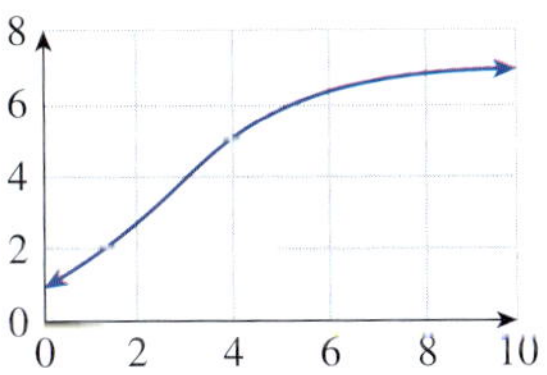

3. $N = 10, A = 4, b = 0.3;$
`10/(1+4*0.3^-x)`

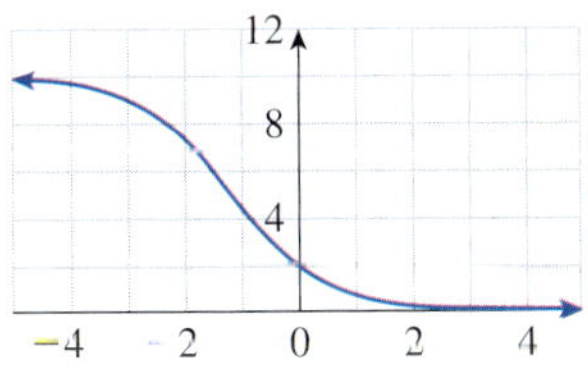

5. $N = 4, A = 7, b = 1.5;$
`4/(1+7*1.5^-x)`

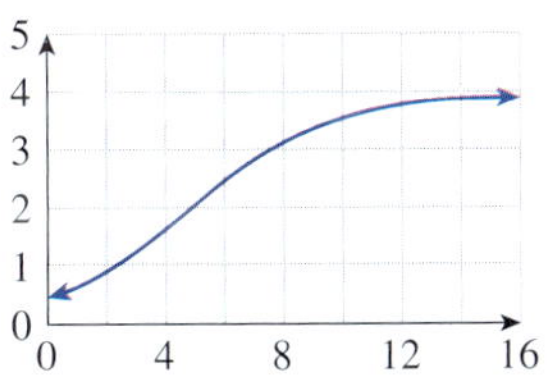

7. $f(x) = \dfrac{200}{1 + 19(2^{-x})}$ **9.** $f(x) = \dfrac{6}{1 + 2^{-x}}$

11. (B) **13.** (B) **15.** (C)

17. $y = \dfrac{7.2}{1 + 2.4(1.05)^{-x}}$

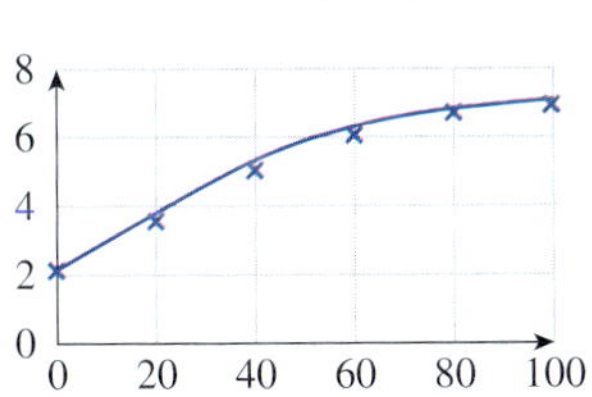

19. $y = \dfrac{97}{1 + 2.2(0.942)^{-x}}$

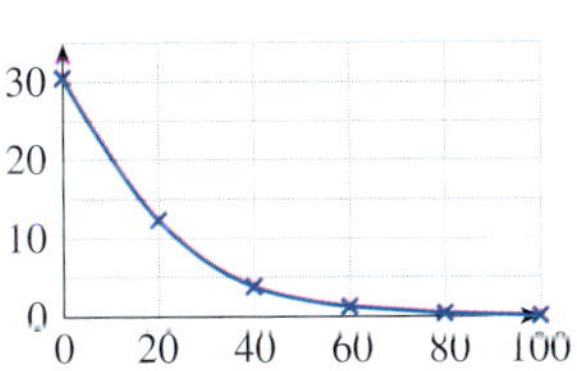

21. a. (A) **b.** 2003 **23. a.** (A) **b.** 20% per year

25. a. 91% **b.** $P(x) \approx 14.33(1.05)^x$ **c.** \$38,000

27. $N(t) = \dfrac{10{,}000}{1 + 9(1.25)^{-t}}$; $N(7) \approx 3{,}463$ cases

29. $N(t) = \dfrac{3{,}000}{1 + 29(2^{1/5})^{-t}}$; $t = 16$ days

31. a. $A(t) = \dfrac{6.3}{1 + 4.8(1.2)^{-t}}$; 6,300 articles

b. 5,200 articles **33. a.** $B(t) = \dfrac{1{,}070}{1 + 0.391(1.10)^{-t}}$;

Leveling off value $\approx$ 1,070

b. $t \approx -9.85$. According to the model, the number of teams was rising fastest about 9.85 years *prior* to 1990; that is, some time during 1980. **c.** The number of men's basketball teams was growing by about 10% per year around 1990.

35. $y = \dfrac{4{,}500}{1 + 1.1466(1.0357)^{-t}}$; 2013

37. Just as diseases are communicated via the spread of a pathogen (such as a virus), new technology is communicated via the spread of information (such as advertising and publicity). Further, just as the spread of a disease is ultimately limited by the number of susceptible individuals, so the spread of a new technology is ultimately limited by the size of the potential market. **39.** It can be used to predict where the sales of a new commodity might level off. **41.** The curve is still a logistic curve, but decreases when $b > 1$ and increases when $b < 1$. **43.** (Proof)

Chapter 2 Review

1.

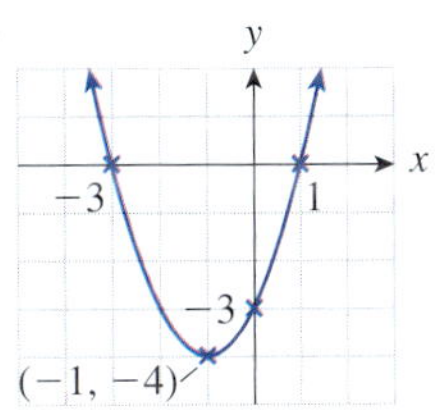

3. f: $f(x) = 5(1/2)^x$, or $5(2^{-x})$

5.

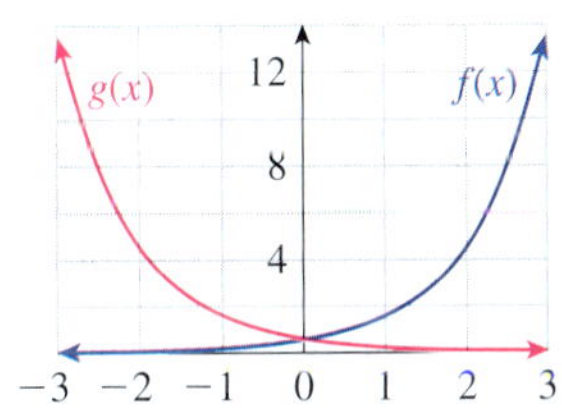

7.

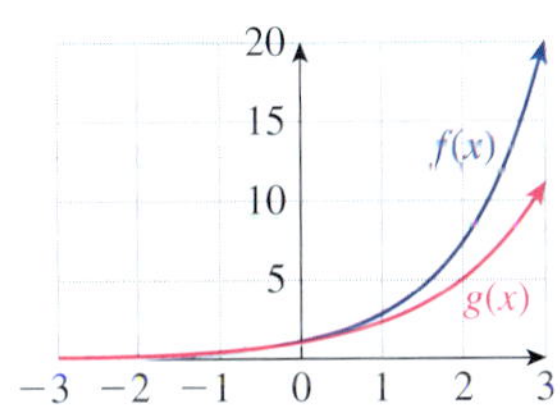

9. \$3,484.85 **11.** \$3,705.48 **13.** \$3,485.50

15. $f(x) = 4.5(9^x)$ **17.** $f(x) = \dfrac{2}{3}3^x$ **19.** $-\dfrac{1}{2}\log_3 4$

21. $\dfrac{1}{3}\log 1.05$

23.

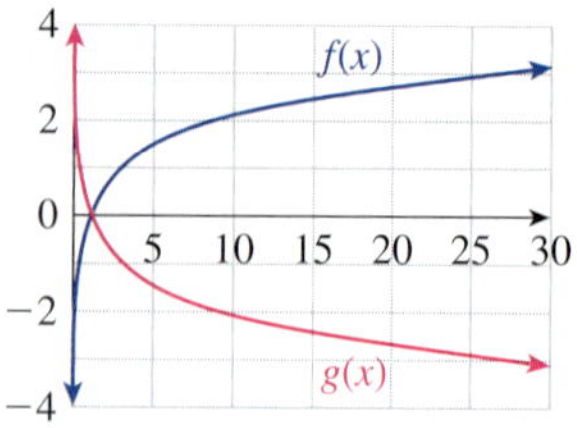

25. $Q = 5e^{-0.00693t}$ **27.** $Q = 2.5e^{0.347t}$ **29.** 10.2 years **31.** 10.8 years **33.** $f(x) = \dfrac{900}{1 + 8(1.5)^{-x}}$ **35.** $f(x) = \dfrac{20}{1 + 3(0.8)^{-x}}$ **37. a.** \$8,500 per month; an average of approximately 2,100 hits per day **b.** \$29,049 per month **c.** The fact that −0.000005, the coefficient of c^2, is negative. **39. a.** $R = -60p^2 + 950p$; $p = \$7.92$ per novel, Monthly revenue = \$3,760.42 **b.** $P = -60p^2 + 1{,}190p - 4{,}700$; $p = \$9.92$ per novel, Monthly profit = \$1,200.42 **41. a.** 10, 34 **b.** About 360,000 pounds **43.** 2008 **45.** 299,000 pounds **47.** $n(t) = 10(0.66^t)$ million pounds of lobster **49.** (C)

Chapter 3

Section 3.1

1. 0 **3.** 4 **5.** Does not exist **7.** 1.5 **9.** 0.5 **11.** Diverges to $+\infty$ **13.** 0 **15.** 1 **17.** 0 **19. a.** −2 **b.** −1 **21. a.** 2 **b.** 1 **c.** 0 **d.** $+\infty$ **23. a.** 0 **b.** 2 **c.** −1 **d.** Does not exist **e.** 2 **f.** $+\infty$ **25. a.** 1 **b.** 1 **c.** 2 **d.** Does not exist **e.** 1 **f.** 2 **27. a.** 1 **b.** $+\infty$ **c.** $+\infty$ **d.** $+\infty$ **e.** not defined **f.** −1 **29. a.** −1 **b.** $+\infty$ **c.** $-\infty$ **d.** Does not exist **e.** 2 **f.** 1 **31.** 210 trillion pesos per year. In the long term, the model predicts that the value of sold goods in Mexico will approach 210 trillion pesos per year. **33.** 7.0; in the long term, the number of research articles in *Physical Review* written by researchers in Europe approaches 7,000 per year. **35.** 470. This suggests that students whose parents earn an exceptionally large income score an average of 470 on the SAT verbal test. **37.** $\lim_{t\to+\infty} p(t) = 15$; the home price index will approach a value 15% above the 2003 level in the long term. **39.** $\lim_{t\to1^-} C(t) = 0.06$, $\lim_{t\to1^+} C(t) = 0.08$, so $\lim_{t\to1} C(t)$ does not exist. **41.** $\lim_{t\to+\infty} I(t) = +\infty$, $\lim_{t\to+\infty}(I(t)/E(t)) \approx 2.5$. In the long term, U.S. imports from China will rise without bound and be 2.5 times U.S. exports to China. In the real world, imports and exports cannot rise without bound. Thus, the given models should not be extrapolated far into the future. **43.** To approximate $\lim_{x\to a} f(x)$ numerically, choose values of x closer and closer to, and on either side of $x = a$, and evaluate $f(x)$ for each of them. The limit (if it exists) is then the number that these values of $f(x)$ approach. A disadvantage of this method is that it may never give the exact value of the limit, but only an approximation. (However, we can make this as accurate as we like.) **45.** Any situation in which there is a sudden change can be modeled by a function in which $\lim_{t\to a^+} f(t)$ is not the same as $\lim_{t\to a^-} f(t)$. One example is the value of a stock market index before and after a crash: $\lim_{t\to a^-} f(t)$ is the value immediately before the crash at time $t = a$, while $\lim_{t\to a^+} f(t)$ is the value immediately after the crash. Another example might be the price of a commodity that is suddenly increased from one level to another. **47.** It is possible for $\lim_{x\to a} f(x)$ to exist even though $f(a)$ is not defined. An example is $\lim_{x\to1} \dfrac{x^2 - 3x + 2}{x - 1}$. **49.** The limit may not be defined and, even if it is, may not equal $f(a)$. See, for example, Exercises 23 and 24. **51.** An example is $f(x) = (x - 1)(x - 2)$.

Section 3.2

1. Continuous on its domain **3.** Continuous on its domain **5.** Discontinuous at $x = 0$ **7.** Discontinuous at $x = -1$ **9.** Continuous on its domain **11.** Discontinuous at $x = -1$ and 0 **13.** (A), (B), (D), (E) **15.** 0 **17.** −1 **19.** No value possible **21.** −1 **23.** Continuous on its domain **25.** Continuous on its domain **27.** Discontinuity at $x = 0$ **29.** Discontinuity at $x = 0$ **31.** Continuous on its domain **33.** Not unless the domain of the function consists of all real numbers. (It is impossible for a function to be continuous at points not in its domain.) For example, $f(x) = 1/x$ is continuous on its domain—the set of nonzero real numbers—but not at $x = 0$. **35.** True. If the graph of a function has a break in its graph at any point a, then it cannot be continuous at the point a. **37.** Answers may vary. $f(x) = 1/[(x - 1)(x - 2)(x - 3)]$ is such a function; it is undefined at $x = 1, 2, 3$ and so its graph consists of three distinct curves. **39.** Answers may vary.

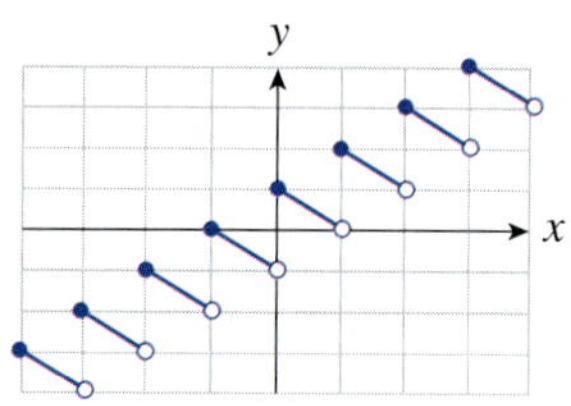

41. Answers may vary. The price of OHaganBooks.com stocks suddenly drops by \$10 as news spreads of a government investigation. Let $f(x)$ = Price of OHaganBooks.com stocks.

Section 3.3

1. $x = 1$ **3.** 2 **5.** Determinate; diverges to $+\infty$ **7.** Determinate; does not exist **9.** Determinate; diverges to $-\infty$ **11.** Determinate; 0 **13.** Indeterminate; $-1/3$ **15.** Indeterminate; 0 **17.** Determinate; 0 **19.** Determinate; −60 **21.** 1 **23.** 2 **25.** 0 **27.** 6 **29.** 4 **31.** 2 **33.** 0 **35.** 0 **37.** 12 **39.** $+\infty$ **41.** Does not exist; left and right (infinite) limits differ. **43.** $-\infty$ **45.** Does not exist **47.** $+\infty$ **49.** $3/2$ **51.** $1/2$ **53.** $+\infty$ **55.** 0 **57.** $3/2$ **59.** $1/2$ **61.** $-\infty$ **63.** 0 **65.** 12 **67.** 0 **69.** $+\infty$ **71.** 0 **73.** Discontinuity at $x = 0$ **75.** Continuous everywhere **77.** Discontinuity at $x = 0$ **79.** Discontinuity at $x = 0$ **81. a.** $\lim_{t\to5^-} N(t) = \lim_{t\to5^+} N(t) = 4.1$; shortly before and after 2000 ($t = 5$) the number of workers employed in manufacturing jobs in Mexico was close to 4.1 million. **b.** N is continuous at $t = 5$; no abrupt changes **83. a.** 0.49, 1.16. Shortly before 1999 annual advertising expenditures were close to \$0.49 billion. Shortly after 1999 annual advertising expenditures were close to \$1.16 billion. **b.** Not continuous; movie advertising expenditures jumped suddenly in 1999. **85.** 1.59; if the trend continued indefinitely, the annual spending on police would be 1.59 times the annual spending on courts in the long run. **87.** 825; in the long term, annual revenues will approach \$825 million. **89.** $\lim_{t\to+\infty} I(t) = +\infty$, $\lim_{t\to+\infty}(I(t)/E(t)) = 2.5$. In the long term, U.S. imports from China will rise without bound and be 2.5 times U.S. exports to China. In the real world, imports and exports cannot rise without bound. Thus, the given models should not be extrapolated far into the future. **91.** $\lim_{t\to+\infty} p(t) = 100$. The percentage of children who learn to speak approaches 100% as their age increases. **93.** To evaluate $\lim_{x\to a} f(x)$ algebraically, first check whether $f(x)$ is a closed-form function. Then check whether $x = a$ is in its domain. If so, the limit is just $f(a)$; that is, it is obtained by substituting $x = a$. If not, then try to first simplify $f(x)$ in such a way as to transform it into a new function such that $x - a$ is in its domain, and then substitute. A disadvantage of this method is that it is sometimes extremely difficult to evaluate limits algebraically, and rather sophisticated methods are often needed. **95.** She is wrong. Closed-form functions are continuous only at points in their domains, and $x = 2$ is not in the domain of the closed-form function $f(x) = 1/(x-2)^2$. **97.** Answers may vary. (1) See Example 2: $\lim_{x\to 2} \frac{x^3-8}{x-2}$, which leads to the indeterminate form 0/0 but the limit is 12. (2) $\lim_{x\to+\infty} \frac{60x}{2x}$, which leads to the indeterminate form ∞/∞, but where the limit exists and equals 30. **99.** The statement may not be true, for instance, if $f(x) = \begin{cases} x+2 & \text{if } x < 0 \\ 2x-1 & \text{if } x \ge 0 \end{cases}$, then $f(0)$ is defined and equals -1, and yet $\lim_{x\to 0} f(x)$ does not exist. The statement can be corrected by requiring that f be a closed-form function: "If f is a closed form function, and $f(a)$ is defined, then $\lim_{x\to a} f(x)$ exists and equals $f(a)$." **101.** Answers may vary, for example $f(x) = \begin{cases} 0 & \text{if } x \text{ is any number other than 1 or 2} \\ 1 & \text{if } x = 1 \text{ or } 2 \end{cases}$

103. Answers may vary.

(1) $\lim_{x\to+\infty} [(x+5) - x] = \lim_{x\to+\infty} 5 = 5$

(2) $\lim_{x\to+\infty} [x^2 - x] = \lim_{x\to+\infty} x(x-1) = +\infty$

(3) $\lim_{x\to+\infty} [(x-5) - x] = \lim_{x\to+\infty} -5 = -5$

Section 3.4

1. −3 **3.** 0.3 **5.** −\$25,000 per month **7.** −200 items per dollar **9.** \$1.33 per month **11.** 0.75 percentage point increase in unemployment per 1 percentage point increase in the deficit **13.** 4 **15.** 2 **17.** 7/3

19.

h	Ave. Rate of Change
1	2
0.1	0.2
0.01	0.02
0.001	0.002
0.0001	0.0002

21.

h	Ave. Rate of Change
1	−0.1667
0.1	−0.2381
0.01	−0.2488
0.001	−0.2499
0.0001	−0.24999

23.

h	Ave. Rate of Change
1	9
0.1	8.1
0.01	8.01
0.001	8.001
0.0001	8.0001

25. a. \$25 billion per year; world military expenditure increased at an average rate of \$25 billion per year during 1994–2006. **b.** \$50 billion per year; world military expenditure increased at an average rate of \$50 billion per year during 1998–2006. **27. a.** −20,000 barrels/year; during 2002–2007, daily oil production by Pemex was decreasing at an average rate of 20,000 barrels of oil per year. **b.** (C) **29. a.** 1.7; the percentage of mortgages classified as subprime was increasing at an average rate of around 1.7 percentage points per year between 2000 and 2006. **b.** 2004–2006 **31. a.** The second and third quarter of 2007. During the second and third quarter of 2007 iPhone sales were increasing at an average rate of 1,022,500 phones per quarter. **b.** Fourth quarter of 2007 and first quarter of 2008. During the fourth quarter of 2007 and first quarter of 2008 iPhone sales were decreasing at an average rate of 797,500 phones per quarter. **33. a.** [3, 5]; −0.25 thousand articles per year. During the period 1993–1995, the number of articles authored by U.S. researchers decreased at an average rate of 250 articles per year. **b.** Percentage rate ≈ −0.1765, Average rate = −0.09 thousand articles/year. Over the period 1993–2003, the number of articles authored by U.S. researchers decreased at an average rate of 90 per year, representing a 17.65% decrease over that period. **35. a.** 12 teams per year **b.** Decreased **37. a.** (C) **b.** (A) **c.** (B) **d.** Approximately −0.0063 (to two significant digits) billion dollars per year (−\$6,300,000 per year). This is much less than the (positive) slope of the regression line, 0.0125 ≈ 0.013 billion dollars per year (\$13,000,000 per year). **39.** Answers may vary. Graph:

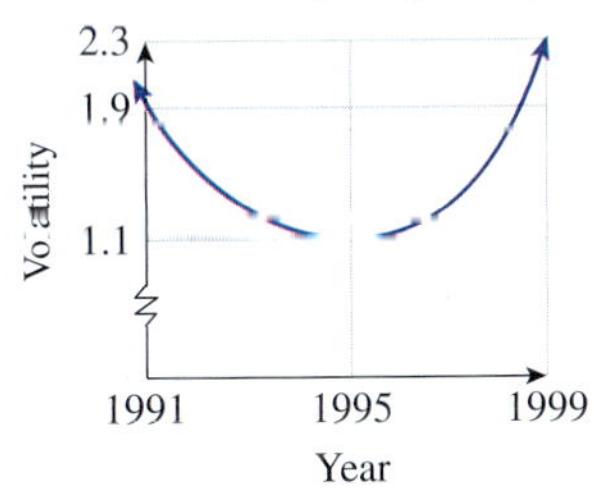

41. The index was increasing at an average rate of 300 points per day. **43. a.** \$0.15 per year **b.** No; according to the model, during that 25-year period the price of oil went down from around \$93 to a low of around \$25 in 1993 before climbing back up. **45. a.** 47.3 new cases per day; the number of SARS cases was growing at an average rate of 47.3 new cases per day over the period March 17 to March 23. **b.** (A) **47. a.** 8.85 manatee deaths per 100,000 boats; 23.05 manatee deaths per 100,000 boats **b.** More boats result in more manatee deaths per additional boat. **49. a.** -0.88, -0.79, -0.69, -0.60, -0.51, -0.42 **b.** For household incomes between \$40,000 and \$40,500, the poverty rate decreases at an average rate of 0.69 percentage points per \$1,000 increase in the median household income. **c.** (B) **d.** (B) **51.** The average rate of change of f over an interval $[a, b]$ can be determined numerically, using a table of values, graphically, by measuring the slope of the corresponding line segment through two points on the graph, or algebraically, using an algebraic formula for the function. Of these, the least precise is the graphical method, because it relies on reading coordinates of points on a graph. **53.** No, the formula for the average rate of a function f over $[a, b]$ depends only on $f(a)$ and $f(b)$, and not on any values of f between a and b. **55.** Answers will vary. Graph:

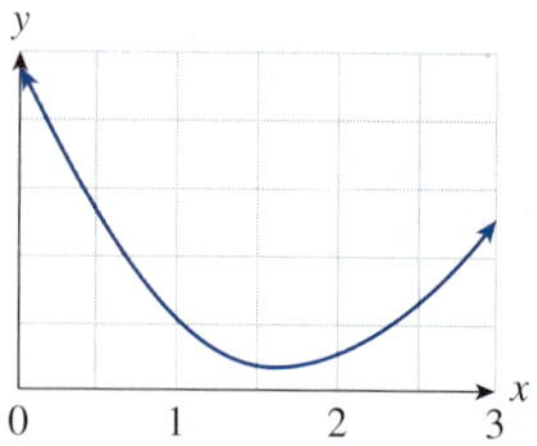

57. 6 units of quantity A per unit of quantity C **59.** (A)
61. Yes. Here is an example:

Year	2000	2001	2002	2003
Revenue (\$ billion)	\$10	\$20	\$30	\$5

63. (A)

Section 3.5

1. 6 **3.** -5.5

5.

h	1	0.1	0.01
Ave. rate	39	39.9	39.99

Instant. Rate = \$40 per day

7.

h	1	0.1	0.01
Ave. Rate	140	66.2	60.602

Instant. Rate = \$60 per day

9.

h	10	1
C_{ave}	4.799	4.7999

11.

h	10	1
C_{ave}	99.91	99.90

$C'(1{,}000) = \$4.8$ per item $\quad C'(100) = \$99.9$ per item

13. 1/2 **15.** 0 **17. a.** R **b.** P **19. a.** P **b.** R **21. a.** Q **b.** P **23. a.** Q **b.** R **c.** P **25. a.** R **b.** Q **c.** P **27. a.** $(1, 0)$ **b.** None **c.** $(-2, 1)$ **29. a.** $(-2, 0.3)$, $(0, 0)$, $(2, -0.3)$ **b.** None **c.** None **31.** $(a, f(a))$; $f'(a)$ **33.** (B) **35. a.** (A) **b.** (C) **c.** (B) **d.** (B) **e.** (C) **37.** -2 **39.** -1.5 **41.** -5 **43.** 16 **45.** 0 **47.** -0.0025

49. a. 3 **b.** $y = 3x + 2$

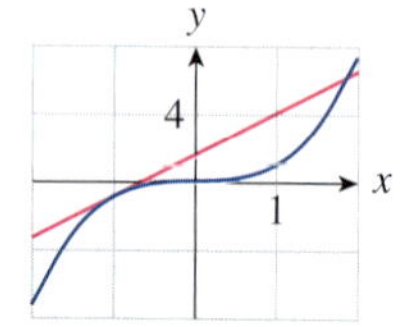

51. a. $\frac{3}{4}$ **b.** $y = \frac{3}{4}x + 1$

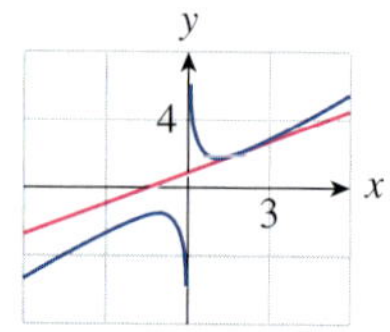

53. a. $\frac{1}{4}$ **b.** $y = \frac{1}{4}x + 1$

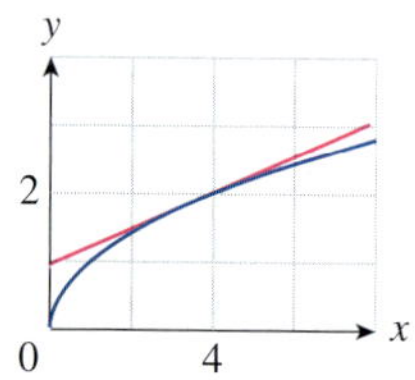

55. 1.000 **57.** 1.000 **59.** (C) **61.** (A) **63.** (F)
65. Increasing for $x < 0$; decreasing for $x > 0$. **67.** Increasing for $x < -1$ and $x > 1$; decreasing for $-1 < x < 1$
69. Increasing for $x > 1$; decreasing for $x < 1$. **71.** Increasing for $x < 0$; decreasing for $x > 0$. **73.** $x = -1.5$, $x = 0$
Graph:

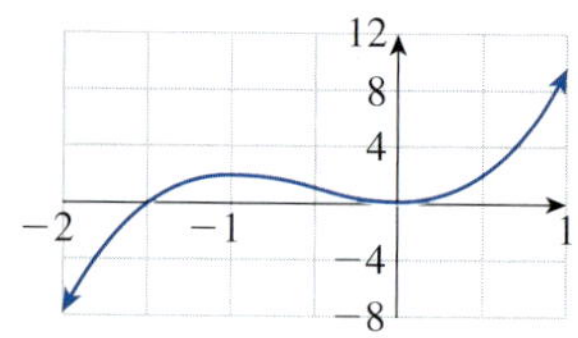

75. Note: Answers depend on the form of technology used.
Excel ($h = 0.1$):

	A	B	C	D	E
1	x	f(x)	f'(x)	xmin	4
2	4	6	-4.545454545	h	0.1
3	4.1	5.545454545	-3.787878788		
4	4.2	5.166666667	-3.205128205		
5	4.3	4.846153846	-2.747252747		
6	4.4	4.571428571	-2.380952381		
7	4.5	4.333333333	-2.083333333		
8	4.6	4.125	-1.838235294		
9	4.7	3.941176471	-1.633986928		
10	4.8	3.777777778	-1.461988304		
11	4.9	3.631578947	-1.315789474		
12	5	3.5			
13					
14					

Graphs:

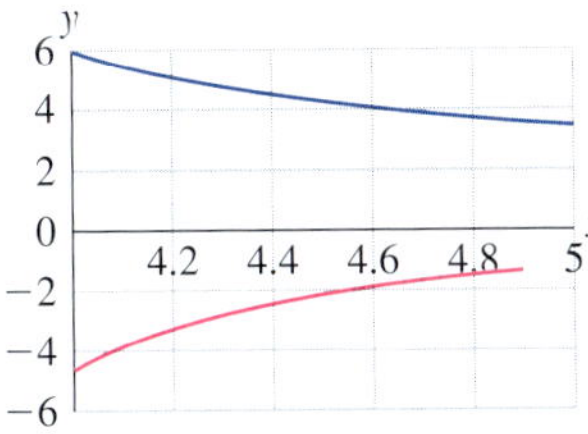

77. $q(100) = 50{,}000$, $q'(100) = -500$. A total of 50,000 pairs of sneakers can be sold at a price of \$100, but the demand is decreasing at a rate of 500 pairs per \$1 increase in the price. **79. a.** -0.05; daily oil imports from Mexico in 2005 were 1.6 million barrels and declining at a rate of 0.05 million barrels (or 50,000 barrels) per year. **b.** Decreasing **81. a.** (B) **b.** (B) **c.** (A) **d.** 1992 **e.** 0.05; in 1996, the total number of state prisoners was increasing at a rate of approximately 50,000 prisoners per year. **83. a.** -96 ft/s **b.** -128 ft/s **85. a.** \$0.60 per year; the price per barrel of crude oil in constant 2008 dollars was growing at an average rate of about 60¢ per year over the 28-year period beginning at the start of 1980. **b.** $-\$12$ per year; the price per barrel of crude oil in constant 2008 dollars was dropping at an instantaneous rate of about \$12 per year at the start of 1980. **c.** The price of oil was decreasing in January 1980, but eventually began to increase (making the average rate of change in part (a) positive). **87. a.** 144.7 new cases per day; the number of SARS cases was growing at a rate of about 144.7 new cases per day on March 27. **b.** (A) **89.** $S(5) \approx 109$, $\left.\frac{dS}{dt}\right|_{t=5} \approx 9.1$. After 5 weeks, sales are 109 pairs of sneakers per week, and sales are increasing at a rate of 9.1 pairs per week each week. **91.** $A(0) = 4.5$ million; $A'(0) = 60{,}000$ subscribers/week **93. a.** 60% of children can speak at the age of 10 months. At the age of 10 months, this percentage is increasing by 18.2 percentage points per month. **b.** As t increases, p approaches 100 percentage points (all children eventually learn to speak), and dp/dt approaches zero because the percentage stops increasing. **95. a.** $A(6) \approx 12.0$; $A'(6) \approx 1.4$; at the start of 2006, about 12% of U.S. mortgages were subprime, and this percentage was increasing at a rate of about 1.4 percentage points per year
b. Graphs:

Graph of A:

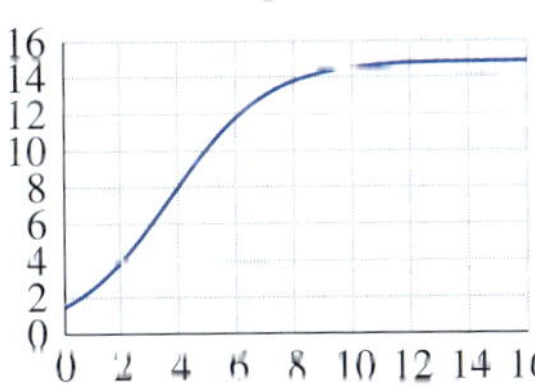

Graph of A':

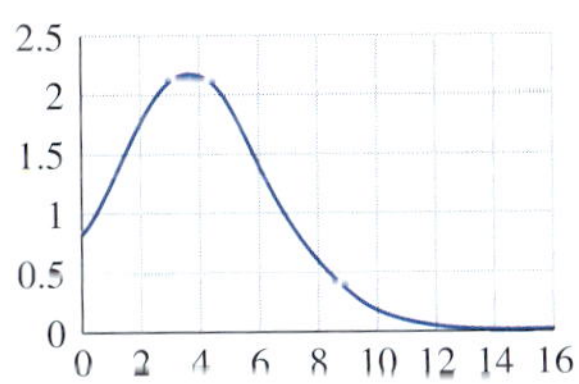

From the graphs, $A(t)$ approaches 15 as t becomes large (in terms of limits, $\lim_{x\to+\infty} A(t) = 15$) and $A'(t)$ approaches 0 as t becomes large (in terms of limits, $\lim_{x\to+\infty} A'(t) = 0$).

Interpretation: If the trend modeled by the function A had continued indefinitely, in the long term 15% of U.S. mortgages would have been subprime, and this percentage would not be changing. **97. a.** (D) **b.** 33 days after the egg was laid **c.** 50 days after the egg was laid. Graph:

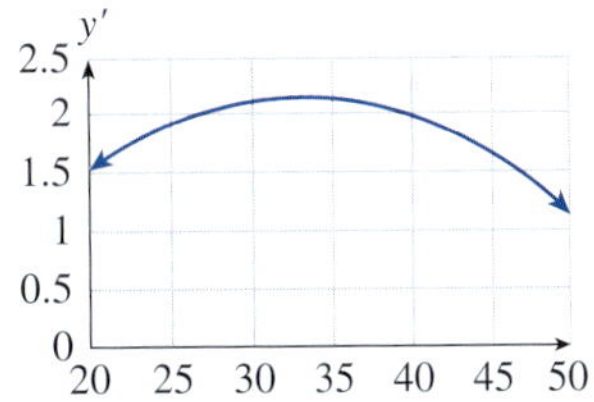

99. $L(.95) = 31.2$ meters and $L'(.95) = -304.2$ meters/warp. Thus, at a speed of warp 0.95, the spaceship has an observed length of 31.2 meters and its length is decreasing at a rate of 304.2 meters per unit warp, or 3.042 meters per increase in speed of 0.01 warp. **101.** The difference quotient is not defined when $h = 0$ because there is no such number as 0/0. **103.** (D) **105.** The derivative is positive and decreasing toward zero. **107.** Company B. Although the company is currently losing money, the derivative is positive, showing that the profit is increasing. Company A, on the other hand, has profits that are declining. **109.** (C) is the only graph in which the instantaneous rate of change on January 1 is greater than the one-month average rate of change. **111.** The tangent to the graph is horizontal at that point, and so the graph is almost horizontal near that point. **113.** Answers may vary.

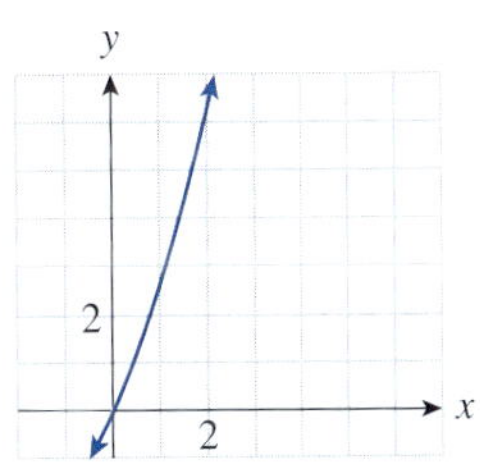

115. If $f(x) = mx + b$, then its average rate of change over any interval $[x, x + h]$ is $\frac{m(x+h) + b - (mx + b)}{h} = m$. Because this does not depend on h, the instantaneous rate is also equal to m. **117.** Increasing because the average rate of change appears to be rising as we get closer to 5 from the left. (See the bottom row.)

119. Answers may vary.

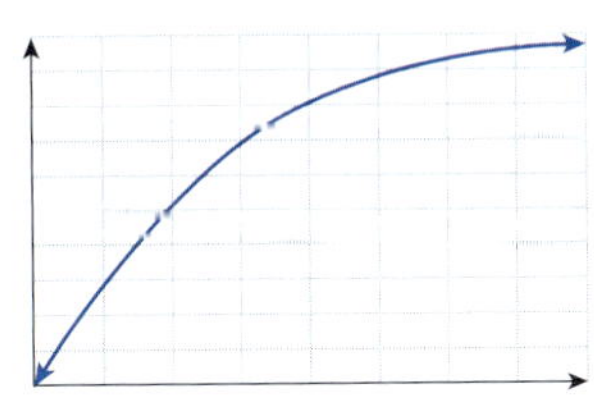

121. Answers may vary.

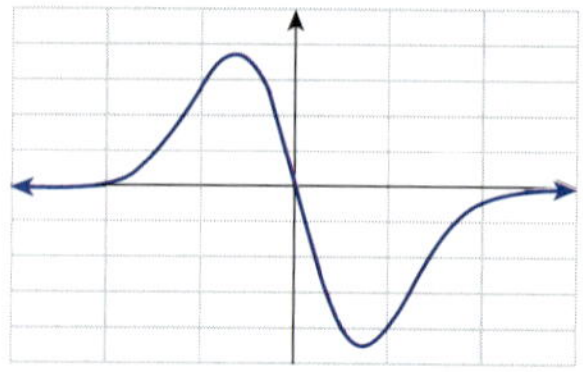

123. (B) **125.** Answers will vary.

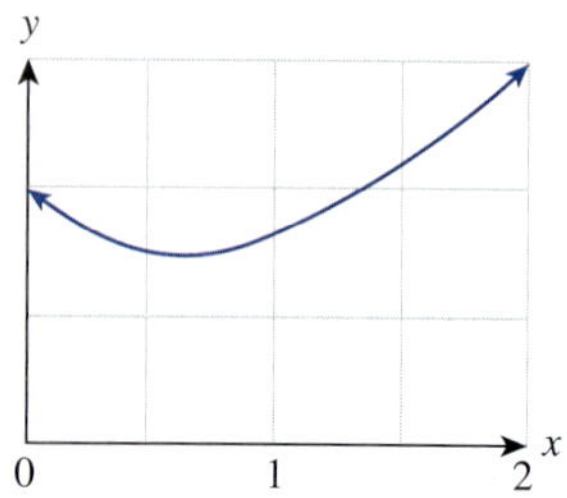

Section 3.6

1. 4 **3.** 3 **5.** 7 **7.** 4 **9.** 14 **11.** 1 **13.** m **15.** $2x$ **17.** 3 **19.** $6x+1$ **21.** $2-2x$ **23.** $3x^2+2$ **25.** $1/x^2$ **27.** m **29.** -1.2 **31.** 30.6 **33.** -7.1 **35.** 4.25 **37.** -0.6 **39.** $y=4x-7$ **41.** $y=-2x-4$ **43.** $y=-3x-1$ **45.** $s'(t)=-32t$; $s'(4)=-128$ ft/s
47. $\frac{dI}{dt}=-0.030t+0.1$, daily oil imports were decreasing at a rate of 0.11 million barrels per year. **49.** $R'(t)=24t+500$; annual U.S. sales of bottled water were increasing by 620 million gallons per year in 2005. **51.** $f'(8)=26.6$ manatee deaths per 100,000 boats. At a level of 800,000 boats, the number of manatee deaths is increasing at a rate of 26.6 manatees per 100,000 additional boats. **53. a.** Yes; $\lim_{t\to 8^-} C(t)=\lim_{t\to 8^+} C(t)=1.24=C(8)$. **b.** No; $\lim_{t\to 8^-} C'(t)=0.08$ while $\lim_{t\to 8^+} C'(t)=0.13$. Until 1998, the cost of a Super Bowl ad was increasing at a rate of \$80,000 per year. Immediately thereafter, it was increasing at a rate of \$130,000 per year. **55.** The algebraic method because it gives the exact value of the derivative. The other two approaches give only approximate values (except in some special cases).
57. The error is in the second line: $f(x+h)$ is *not* equal to $f(x)+h$. For instance, if $f(x)=x^2$, then $f(x+h)=(x+h)^2$, whereas $f(x)+h=x^2+h$. **59.** The error is in the second line: One could only cancel the h if it were a *factor* of both the numerator and denominator; it is not a factor of the numerator. **61.** Because the algebraic computation of $f'(a)$ is exact and not an approximation, it makes no difference whether one uses the balanced difference quotient or the ordinary difference quotient in the algebraic computation.
63. The computation results in a limit that cannot be evaluated.

Chapter 3 Review

1. 5 **3.** Does not exist **5. a.** -1 **b.** 3 **c.** Does not exist **7.** $-4/5$ **9.** -1 **11.** Does not exist **13.** Does not exist **15.** $+\infty$ **17.** Diverges to $-\infty$ **19.** 0 **21.** 2/5 **23.** 1

25.

h	1	0.01	0.001
Ave. Rate of Change	-0.5	-0.9901	-0.9990

Slope ≈ -1

27.

h	1	0.01	0.001
Avg. Rate of Change	6.3891	2.0201	2.0020

Slope ≈ 2

29. (i) P **(ii)** Q **(iii)** R **(iv)** S **31. (i)** Q **(ii)** None **(iii)** None **(iv)** None **33. a.** (B) **b.** (B) **c.** (B) **d.** (A) **e.** (C)
35. $2x+1$ **37.** $2/x^2$

39.

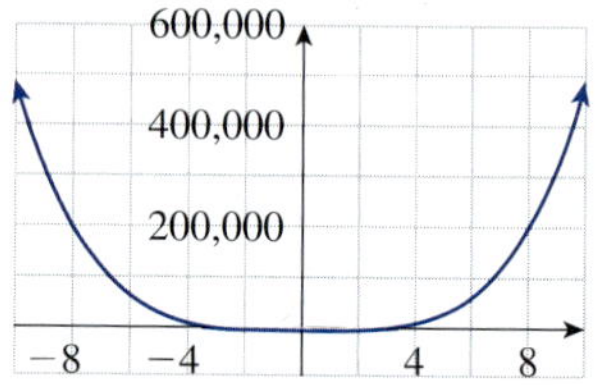

41.

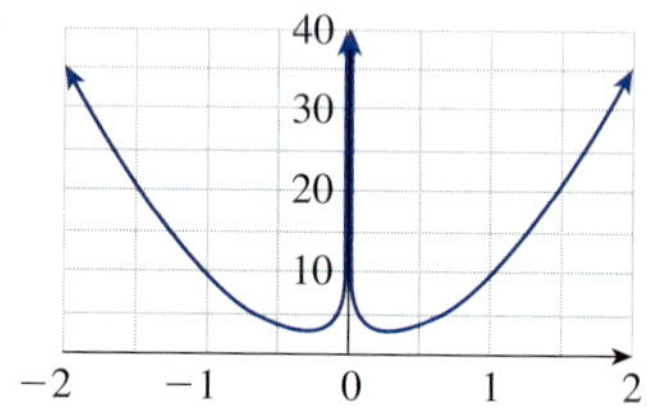

43. a. $P(3)=25$: O'Hagan purchased the stock at \$25. $\lim_{t\to 3^-} P(t)=25$: The value of the stock had been approaching \$25 up the time he bought it. $\lim_{t\to 3^+} P(t)=10$: The value of the stock dropped to \$10 immediately after he bought it.
b. Continuous but not differentiable. Interpretation: the stock price changed continuously but suddenly reversed direction (and started to go up) the instant O'Hagan sold it.
45. a. $\lim_{t\to 3} p(t)\approx 40$; $\lim_{t\to +\infty} p(t)=+\infty$. Close to 2007 ($t=3$), the home price index was about 40. In the long term, the home price index will rise without bound. **b.** 10 (The slope of the linear portion of the curve is 10.) In the long term, the home price index will rise about 10 points per year.
47. a. 500 books per week **b.** [3, 4], [4, 5] **c.** [3, 5]; 650 books per week **49. a.** 3 percentage points per year **b.** 0 percentage points per year **c.** (D) **51. a.** $72t+250$ **b.** 322 books per week **c.** 754 books per week.

Chapter 4

Section 4.1

1. $5x^4$ **3.** $-4x^{-3}$ **5.** $-0.25x^{-0.75}$ **7.** $8x^3+9x^2$
9. $-1-1/x^2$ **11.** $\frac{dy}{dx}=10(0)=0$ (constant multiple and power rule) **13.** $\frac{dy}{dx}=\frac{d}{dx}(x^2)+\frac{d}{dx}(x)$ (sum rule) $=2x+1$ (power rule) **15.** $\frac{dy}{dx}=\frac{d}{dx}(4x^3)+\frac{d}{dx}(2x)-\frac{d}{dx}(1)$ (sum and difference)

17. $f'(x) = 2x - 3$ **19.** $f'(x) = 1 + 0.5x^{-0.5}$

21. $g'(x) = -2x^{-3} + 3x^{-2}$ **23.** $g'(x) = -\frac{1}{x^2} + \frac{2}{x^3}$

25. $h'(x) = -\frac{0.8}{x^{1.4}}$ **27.** $h'(x) = -\frac{2}{x^3} - \frac{6}{x^4}$

29. $r'(x) = -\frac{2}{3x^2} + \frac{0.1}{2x^{1.1}}$ **31.** $r'(x) = \frac{2}{3} - \frac{0.1}{2x^{0.9}} - \frac{4.4}{3x^{2.1}}$

33. $t'(x) = |x|/x - 1/x^2$ **35.** $s'(x) = \frac{1}{2\sqrt{x}} - \frac{1}{2x\sqrt{x}}$

37. $s'(x) = 3x^2$ **39.** $t'(x) = 1 - 4x$ **41.** $2.6x^{0.3} + 1.2x^{-2.2}$
43. $1.2(1 - |x|/x)$ **45.** $3at^2 - 4a$ **47.** $5.15x^{9.3} - 99x^{-2}$

49. $-\frac{2.31}{t^{2.1}} - \frac{0.3}{t^{0.4}}$ **51.** $4\pi r^2$ **53.** 3 **55.** -2 **57.** -5

59. $y = 3x + 2$

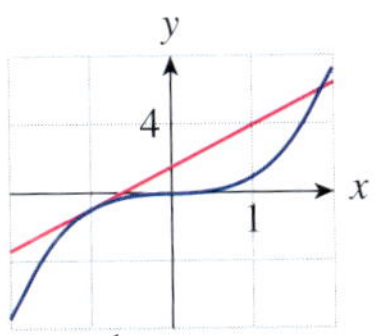

61. $y = \frac{3}{4}x + 1$

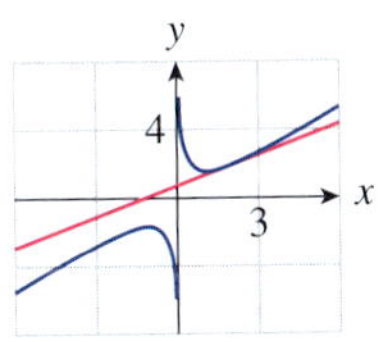

63. $y = \frac{1}{4}x + 1$

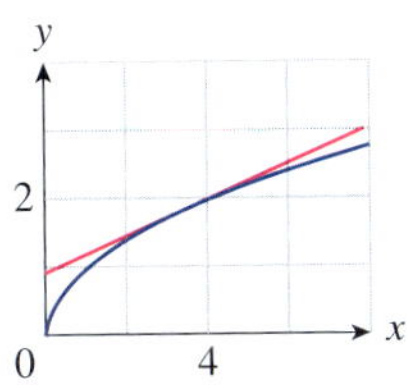

65. $x = -3/4$

67. No such values

69. $x = 1, -1$

73.

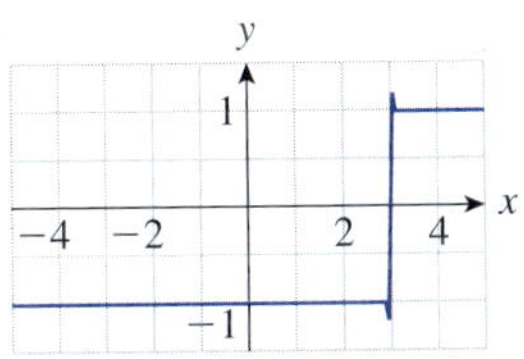

(a) $x = 3$ **(b)** None

75.

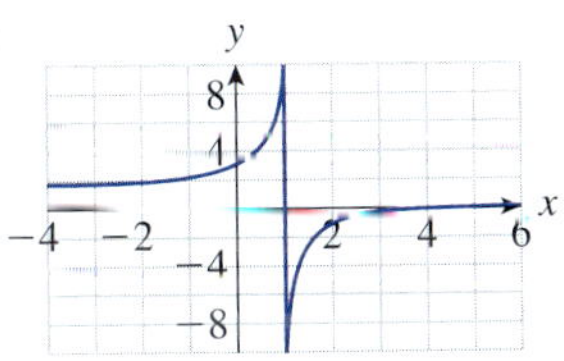

(a) $x = 1$ **(b)** $x = 4.2$

77. a. $f'(1) = 1/3$ **b.** Not differentiable at 0
79. a. Not differentiable at 1 **b.** Not differentiable at 0
81. Yes; 0 **83.** Yes; 12 **85.** No; 3 **87.** Yes; 3/2
89. Yes; diverges to $-\infty$ **91.** Yes; diverges to $-\infty$
93. $P'(t) = 0.9t - 12$; $P'(20) = 6$; the price of a barrel of crude oil was increasing at a rate of \$6 per year in 2000.

95. a. $n'(t) = -1.12t + 14$ **b.** 7 teams/year **c.** Decreases; $n'(t)$ is a linear function with negative slope, so it decreases with increasing t. **97.** 0.55 **99. a.** $s'(t) = -32t$; 0, -32, -64, -96, -128 ft/s **b.** 5 seconds; downward at 160 ft/s
101. a. $S'(t) = -780t + 3{,}300$; dropping at a rate of 600,000 iPhones per quarter. **b.** (B); geometrically: The graph is rising but less steeply, over [2, 4]. Algebraically: The slope is the derivative: $-780t + 3{,}300$, which is positive but decreasing over [2, 4]. **103. a.** $f'(x) = 7.1x - 30.2$ manatees per 100,000 boats. **b.** Increasing; the number of manatees killed per additional 100,000 boats increases as the number of boats increases. **c.** $f'(8) = 26.6$ manatees per 100,000 additional boats. At a level of 800,000 boats, the number of manatee deaths is increasing at a rate of 26.6 manatees per 100,000 additional boats. **105. a.** $c(t) - m(t)$ measures the combined market share of the other three providers (Comcast, Earthlink, and AOL); $c'(t) - m'(t)$ measures the rate of change of the combined market share of the other three providers. **b.** (A) **c.** (A) **d.** 3.72% per year. In 1992, the combined market share of the other three providers was increasing at a rate of about 3.72 percentage points per year. **107.** After graphing the curve $y = 3x^2$, draw the line passing through $(-1, 3)$ with slope -6. **109.** The slope of the tangent line of g is twice the slope of the tangent line of f. **111.** $g'(x) = -f'(x)$ **113.** The left-hand side is not equal to the right-hand side. The *derivative* of the left-hand side is equal to the right-hand side, so your friend should have written $\frac{d}{dx}(3x^4 + 11x^5) = 12x^3 + 55x^4$.
115. The derivative of a constant times a function is the constant times the derivative of the function, so that $f'(x) = (2)(2x) = 4x$. Your enemy mistakenly computed the *derivative* of the constant times the derivative of the function. (The derivative of a product of two functions is not the product of the derivative of the two functions. The rule for taking the derivative of a product is discussed later in the chapter.).
117. For a general function f, the derivative of f is defined to be $f'(x) = \lim_{h \to 0} \frac{f(x+h) - f(x)}{h}$. One then finds by calculation that the derivative of the specific function x^n is nx^{n-1}. In short, nx^{n-1} is the derivative of a specific function: $f(x) = x^n$, it is not the *definition* of the derivative of a general function or even the definition of the derivative of the function $f(x) = x^n$.
119. Answers may vary.

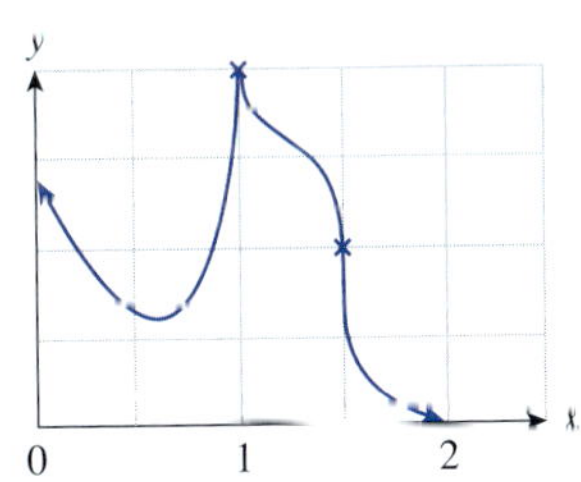

Section 4.2

1. $C'(1{,}000) = \$4.80$ per item **3.** $C'(100) = \$99.90$ per item **5.** $C'(x) = 4$; $R'(x) = 8 - x/500$; $P'(x) = 4 - x/500$; $P'(x) = 0$ when $x = 2{,}000$. Thus, at a production level of 2,000, the profit is stationary (neither increasing nor decreasing) with respect to the production level. This may indicate a maximum profit at a production level of 2,000. **7. a.** (B) **b.** (C) **c.** (C) **9. a.** $C'(x) = 2{,}250 - 0.04x$. The cost is going up at a rate of \$2,249,840 per television commercial. The exact cost of airing the fifth television commercial is $C(5) - C(4) = \$2{,}249{,}820$. **b.** $\overline{C}(x) = 150/x + 2{,}250 - 0.02x$; $\overline{C}(4) = \$2{,}287{,}420$ per television commercial. The average cost of airing the first four television commercials is \$2,287,420. **11. a.** $R'(x) = 0.90$, $P'(x) = 0.80 - 0.002x$ **b.** Revenue: \$450, Profit: \$80, Marginal revenue: \$0.90, Marginal profit: −\$0.20. The total revenue from the sale of 500 copies is \$450. The profit from the production and sale of 500 copies is \$80. Approximate revenue from the sale of the 501st copy is 90¢. Approximate loss from the sale of the 501st copy is 20¢. **c.** $x = 400$. The profit is a maximum when you produce and sell 400 copies. **13.** The profit on the sale of 1,000 DVDs is \$3,000, and is decreasing at a rate of \$3 per additional DVD sold. **15.** $P \approx \$257.07$ and $dP/dx \approx 5.07$. Your current profit is \$257.07 per month, and this would increase at a rate of \$5.07 per additional magazine in sales. **17. a.** \$2.50 per pound **b.** $R(q) = 20{,}000/q^{0.5}$ **c.** $R(400) = \$1{,}000$. This is the monthly revenue that will result from setting the price at \$2.50 per pound. $R'(400) = -\$1.25$ per pound of tuna. Thus, at a demand level of 400 pounds per month, the revenue is decreasing at a rate of \$1.25 per pound. **d.** The fishery should raise the price (to reduce the demand). **19.** $P'(50) = \$350$. This means that, at an employment level of 50 workers, the firm's daily profit will increase at a rate of \$350 per additional worker it hires.
21. a. (B) **b.** (B) **c.** (C)
23. a. $C(x) = 500{,}000 + 1{,}600{,}000x - 100{,}000\sqrt{x}$;

$$C'(x) = 1{,}600{,}000 - \frac{50{,}000}{\sqrt{x}};$$

$$\overline{C}(x) = \frac{500{,}000}{x} + 1{,}600{,}000 - \frac{100{,}000}{\sqrt{x}}$$

b. $C'(3) \approx \$1{,}570{,}000$ per spot, $\bar{C}(3) \approx \$1{,}710{,}000$ per spot. The average cost will decrease as x increases.
25. a. $C'(q) = 200q$; $C'(10) = \$2{,}000$ per one-pound reduction in emissions. **b.** $S'(q) = 500$. Thus $S'(q) = C'(q)$ when $500 = 200q$, or $q = 2.5$ pounds per day reduction. **c.** $N(q) = C(q) - S(q) = 100q^2 - 500q + 4{,}000$. This is a parabola with lowest point (vertex) given by $q = 2.5$. The net cost at this production level is $N(2.5) = \$3{,}375$ per day. The value of q is the same as that for part (b). The net cost to the firm is minimized at the reduction level for which the cost of controlling emissions begins to increase faster than the subsidy. This is why we get the answer by setting these two rates of increase equal to each other. **27.** $M'(10) \approx 0.0002557$ mpg/mph. This means that, at a speed of 10 mph, the fuel economy is increasing at a rate of 0.0002557 miles per gallon per 1-mph increase in speed. $M'(60) = 0$ mpg/mph. This means that, at a speed of 60 mph, the fuel economy is neither increasing nor decreasing with increasing speed. $M'(70) \approx -0.00001799$. This means that, at 70 mph, the fuel economy is decreasing at a rate of 0.00001799 miles per gallon per 1-mph increase in speed. Thus 60 mph is the most fuel-efficient speed for the car.
29. (C) **31.** (D) **33.** (B) **35.** Cost is often measured as a function of the number of items x. Thus, $C(x)$ is the cost of producing (or purchasing, as the case may be) x items.
a. The average cost function $\overline{C}(x)$ is given by $\overline{C}(x) = C(x)/x$. The marginal cost function is the derivative, $C'(x)$, of the cost function. **b.** The average cost $\overline{C}(r)$ is the slope of the line through the origin and the point on the graph where $x = r$. The marginal cost of the rth unit is the slope of the tangent to the graph of the cost function at the point where $x = r$.
c. The average cost function $\overline{C}(x)$ gives the average cost of producing the first x items. The marginal cost function $C'(x)$ is the rate at which cost is changing with respect to the number of items x, or the incremental cost per item, and approximates the cost of producing the $(x + 1)$st item. **37.** The marginal cost
39. Not necessarily. For example, it may be the case that the marginal cost of the 101st item is larger than the average cost of the first 100 items (even though the marginal cost is decreasing). Thus, adding this additional item will *raise* the average cost. **41.** The circumstances described suggest that the average cost function is at a relatively low point at the current production level, and so it would be appropriate to advise the company to maintain current production levels; raising or lowering the production level will result in increasing average costs.

Section 4.3

1. 3 **3.** $3x^2$ **5.** $2x + 3$ **7.** $210x^{1.1}$ **9.** $-2/x^2$
11. $2x/3$ **13.** $3(4x^2 - 1) + 3x(8x) = 36x^2 - 3$
15. $3x^2(1 - x^2) + x^3(-2x) = 3x^2 - 5x^4$
17. $2(2x + 3) + (2x + 3)(2) = 8x + 12$ **19.** $3\sqrt{x}/2$
21. $(x^2 - 1) + 2x(x + 1) = (x + 1)(3x - 1)$
23. $(x^{-0.5} + 4)(x - x^{-1}) + (2x^{0.5} + 4x - 5)(1 + x^{-2})$
25. $8(2x^2 - 4x + 1)(x - 1)$
27. $(1/3.2 - 3.2/x^2)(x^2 + 1) + 2x(x/3.2 + 3.2/x)$
29. $2x(2x + 3)(7x + 2) + 2x^2(7x + 2) + 7x^2(2x + 3)$

31. $5.3(1-x^{2.1})(x^{-2.3}-3.4) - 2.1x^{1.1}(5.3x-1)(x^{-2.3}-3.4) - 2.3x-3.3(5.3x-1)(1-x^{2.1})$

33. $\frac{1}{2\sqrt{x}}\left(\sqrt{x}+\frac{1}{x^2}\right)+(\sqrt{x}+1)\left(\frac{1}{2\sqrt{x}}-\frac{2}{x^3}\right)$ **35.** $\frac{2(3x-1)-3(2x+4)}{(3x-1)^2} = -14/(3x-1)^2$

37. $\frac{(4x+4)(3x-1)-3(2x^2+4x+1)}{(3x-1)^2} = (6x^2-4x-7)/(3x-1)^2$

39. $\frac{(2x-4)(x^2+x+1)-(x^2-4x+1)(2x+1)}{(x^2+x+1)^2} = (5x^2-5)/(x^2+x+1)^2$

41. $\frac{(0.23x^{-0.77}-5.7)(1-x^{-2.9})-2.9x^{-3.9}(x^{0.23}-5.7x)}{(1-x^{-2.9})^2}$ **43.** $\frac{\frac{1}{2}x^{-1/2}(x^{1/2}-1)-\frac{1}{2}x^{-1/2}(x^{1/2}+1)}{(x^{1/2}-1)^2} = \frac{-1}{\sqrt{x}\left(\sqrt{x}-1\right)^2}$

45. $-3/x^4$ **47.** $\frac{[(x+1)+(x+3)](3x-1)-3(x+3)(x+1)}{(3x-1)^2} = (3x^2-2x-13)/(3x-1)^2$

49. $\frac{[(x+1)(x+2)+(x+3)(x+2)+(x+3)(x+1)](3x-1)-3(x+3)(x+1)(x+2)}{(3x-1)^2}$

51. $4x^3-2x$ **53.** 64 **55.** 3 **57.** Difference; $4x^3-12x^2+2x-480$ **59.** Sum; $1+2/(x+1)^2$

61. Product; $\left[\frac{x}{x+1}\right]+(x+2)\frac{1}{(x+1)^2}$

63. Difference; $2x-1-2/(x+1)^2$ **65.** $y=12x-8$ **67.** $y=x/4+1/2$ **69.** $y=-2$ **71.** $q'(5)=1{,}000$ units/month (sales are increasing at a rate of 1,000 units per month); $p'(5)=-\$10$/month (the price of a sound system is dropping at a rate of \$10 per month); $R'(5)=900{,}000$ (revenue is increasing at a rate of \$900,000 per month). **73.** \$242 million; increasing at a rate of \$39 million per year. **75.** Decreasing at a rate of \$1 per day **77.** Decreasing at a rate of approximately \$0.10 per month

79. $M'(x)=\frac{3{,}000(3{,}600x^{-2}-1)}{\left(x+3{,}600x^{-1}\right)^2}$; $M'(10)\approx 0.7670$ mpg/mph.

This means that, at a speed of 10 mph, the fuel economy is increasing at a rate of 0.7670 miles per gallon per one mph increase in speed. $M'(60)=0$ mpg/mph. This means that, at a speed of 60 mph, the fuel economy is neither increasing nor decreasing with increasing speed. $M'(70)\approx -0.0540$. This means that, at 70 mph, the fuel economy is decreasing at a rate of 0.0540 miles per gallon per one mph increase in speed. 60 mph is the most fuel-efficient speed for the car. (In the next chapter we shall discuss how to locate largest values in general.) **81. a.** $P(t)-I(t)$ represents the daily production of oil in Mexico that was not exported to the United States. $I(t)/P(t)$ represents U.S. imports of oil from Mexico as a fraction of the total produced there. **b.** -0.023 per year; at the start of 2008, the fraction of oil produced in Mexico that was imported by the United States was decreasing at a rate of 0.023 (or 2.3 percentage points) per year. **83.** Increasing at a rate of about \$3,420 million per year.

85. $R'(p)=-\frac{5.625}{(1+0.125p)^2}$; $R'(4)=-2.5$ thousand organisms per hour, per 1,000 organisms. This means that the reproduction rate of organisms in a culture containing 4,000 organisms is declining at a rate of 2,500 organisms per hour, per 1,000 additional organisms. **87.** Oxygen consumption is decreasing at a rate of 1,600 milliliters per day. This is due to the fact that the number of eggs is decreasing, because $C'(25)$ is positive. **89.** 20; 33 **91.** $5/4$; $-17/16$ **93.** The analysis is suspect, as it seems to be asserting that the annual increase in revenue, which we can think of as dR/dt, is the product of the annual increases, dp/dt in price, and dq/dt in sales. However, because $R=pq$, the product rule implies that dR/dt is not the product of dp/dt and dq/dt, but is instead

$$\frac{dR}{dt}=\frac{dp}{dt}\cdot q+p\cdot\frac{dq}{dt}.$$

95. Answers will vary. $q=-p+1{,}000$ is one example. **97.** Mine; it is increasing twice as fast as yours. The rate of change of revenue is given by $R'(t)=p'(t)q(t)$ because $q'(t)=0$. Thus, $R'(t)$ does not depend on the selling price $p(t)$. **99.** (A)

Section 4.4

1. $4(2x+1)$ **3.** $(x-1)^{-2}$ **5.** $2(2-x)^{-3}$ **7.** $(2x+1)^{-0.5}$ **9.** $-4(4x-1)^{-2}$ **11.** $-3/(3x-1)^2$ **13.** $4(x^2+2x)^3(2x+2)$ **15.** $-4x(2x^2-2)^{-2}$ **17.** $-5(2x-3)(x^2-3x-1)^{-6}$ **19.** $-6x/(x^2+1)^4$ **21.** $1.5(0.2x-4.2)(0.1x^2-4.2x+9.5)^{0.5}$ **23.** $4(2s-0.5s^{-0.5})(s^2-s^{0.5})^3$ **25.** $-x/\sqrt{1-x^2}$ **27.** $-[(x+1)(x^2-1)]^{-3/2}(3x-1)(x+1)$ **29.** $6.2(3.1x-2)+6.2/(3.1x-2)^3$ **31.** $2[(6.4x-1)^2+(5.4x-2)^3][12.8(6.4x-1)+16.2(5.4x-2)^2]$ **33.** $-2(x^2-3x)^{-3}(2x-3)(1-x^2)^{0.5} -x(x^2-3x)^{-2}(1-x^2)^{-0.5}$

35. $-56(x+2)/(3x-1)^3$ **37.** $3z^2(1-z^2)/(1+z^2)^4$ **39.** $3[(1+2x)^4-(1-x)^2]^2[8(1+2x)^3+2(1-x)]$ **41.** $-0.43(x+1)^{-1.1}[2+(x+1)^{-0.1}]^{3.3}$

43. $-\dfrac{\left(\dfrac{1}{\sqrt{2x+1}}-2x\right)}{\left(\sqrt{2x+1}-x^2\right)^2}$

45. $54(1+2x)^2\left(1+(1+2x)^3\right)^2\left(1+\left(1+(1+2x)^3\right)^3\right)^2$

47. $(100x^{99}-99x^{-2})dx/dt$ **49.** $(-3r^{-4}+0.5r^{-0.5})dr/dt$ **51.** $4\pi r^2 dr/dt$ **53.** $-47/4$ **55.** $1/3$ **57.** $-5/3$ **59.** $1/4$

61. $\left.\dfrac{dP}{dn}\right|_{n=10} = 146{,}454.9$. At an employment level of 10 engineers, Paramount will increase its profit at a rate of \$146,454.90 per additional engineer hired. **63.** $y=35(7+0.2t)^{-0.25}$; -0.11 percentage points per month. **65.** $-\$30$ per additional ruby sold. The revenue is decreasing at a rate of \$30 per additional ruby sold.

67. $\dfrac{dy}{dt}=\dfrac{dy}{dx}\dfrac{dx}{dt}=(1.5)(-2)=-3$ murders per 100,000 residents/yr each year. **69.** $5/6\approx 0.833$; relative to the 2003 levels, home sales were changing at a rate of 0.833 percentage points per percentage point change in price. (Equivalently, home sales in 2008 were dropping at a rate of 0.833 percentage points per percentage point drop in price.) **71.** 12π mi²/h **73.** $\$200{,}000\pi$/week $\approx$ \$628,000/week **75. a.** $q'(4)\approx 333$ units per month **b.** dR/dq = \$800/unit **c.** $dR/dt\approx$ \$267,000 per month **77.** 3% per year **79.** 8% per year **81.** The glob squared, times the derivative of the glob. **83.** The derivative of a quantity cubed is three times the *original quantity* squared, times the derivative of the quantity, not three times the derivative of the quantity squared. Thus, the correct answer is $3(3x^3-x)^2(9x^2-1)$. **85.** First, the derivative of a quantity cubed is three times the *original quantity* squared times the derivative of the quantity, not three times the derivative of the quantity squared. Second, the derivative of a quotient is not the quotient of the derivatives; the quotient rule needs to be used in calculating the derivative of $\dfrac{3x^2-1}{2x-2}$. Thus, the correct result (before simplifying) is

$$3\left(\frac{3x^2-1}{2x-2}\right)^2\left(\frac{6x(2x-2)-(3x^2-1)(2)}{(2x-2)^2}\right).$$

87. Following the calculation thought experiment, pretend that you were evaluating the function at a specific value of x. If the last operation you would perform is addition or subtraction, look at each summand separately. If the last operation is multiplication, use the product rule first; if it is division, use the quotient rule first; if it is any other operation (such as raising a quantity to a power or taking a radical of a quantity) then use the chain rule first. **89.** An example is

$$f(x)=\sqrt{x+\sqrt{x+\sqrt{x+\sqrt{x+\sqrt{x+1}}}}}.$$

Section 4.5

1. $1/(x-1)$ **3.** $1/(x\ln 2)$ **5.** $2x/(x^2+3)$ **7.** e^{x+3} **9.** $-e^{-x}$ **11.** $4^x\ln 4$ **13.** $2^{x^2-1}2x\ln 2$ **15.** $1+\ln x$ **17.** $2x\ln x+(x^2+1)/x$ **19.** $10x(x^2+1)^4\ln x+(x^2+1)^5/x$ **21.** $3/(3x-1)$ **23.** $4x/(2x^2+1)$ **25.** $(2x-0.63x^{-0.7})/(x^2-2.1x^{0.3})$ **27.** $-2/(-2x+1)+1/(x+1)$ **29.** $3/(3x+1)-4/(4x-2)$ **31.** $1/(x+1)+1/(x-3)-2/(2x+9)$ **33.** $5.2/(4x-2)$ **35.** $2/(x+1)-9/(3x-4)-1/(x-9)$

37. $\dfrac{1}{(x+1)\ln 2}$ **39.** $\dfrac{1-1/t^2}{(t+1/t)\ln 3}$ **41.** $\dfrac{2\ln|x|}{x}$

43. $\dfrac{2}{x}-\dfrac{2\ln(x-1)}{x-1}$ **45.** $e^x(1+x)$

47. $1/(x+1)+3e^x(x^3+3x^2)$ **49.** $e^x(\ln|x|+1/x)$ **51.** $2e^{2x+1}$ **53.** $(2x-1)e^{x^2-x+1}$ **55.** $2xe^{2x-1}(1+x)$ **57.** $4(e^{2x-1})^2$ **59.** $2\cdot 3^{2x-4}\ln 3$ **61.** $2\cdot 3^{2x+1}\ln 3+3e^{3x+1}$

63. $\dfrac{2x3^{x^2}[(x^2+1)\ln 3-1]}{(x^2+1)^2}$ **65.** $-4/(e^x-e^{-x})^2$

67. $5e^{5x-3}$ **69.** $-\dfrac{\ln x+1}{(x\ln x)^2}$ **71.** $2(x-1)$ **73.** $\dfrac{1}{x\ln x}$

75. $\dfrac{1}{2x\ln x}$ **77.** $y=(e/\ln 2)(x-1)\approx 3.92(x-1)$

79. $y=x$ **81.** $y=-[1/(2e)](x-1)+e$ **83.** \$163 billion and increasing at a rate of \$5.75 billion per year **85.** \$163 billion and increasing at a rate of \$5.75 billion per year. **87.** $-1{,}653$ years per gram; the age of the specimen is decreasing at a rate of about 1,653 years per additional one gram of carbon 14 present in the sample. (Equivalently, the age of the specimen is increasing at a rate of about 1,653 years per additional one gram less of carbon 14 in the sample.) **89.** Average price: \$1.4 million; increasing at a rate of about \$220,000 per year. **91. a.** $N(15)\approx 1{,}762\approx 1{,}800$ (rounded to 2 significant digits) wiretap orders; $N'(15)\approx 89.87\approx 90$ wiretap orders per year (rounded to 2 significant digits). The constants in the model are specified to 2 significant digits, so we cannot expect the answer to be accurate to more than 2 digits. **b.** In 2005, the number of people whose communications were intercepted was about 180,000 and increasing at a rate of about 9,000 people per year. **c.** (C) **93.** \$451.00 per year **95.** \$446.02 per year **97.** $A(t)=167(1.18)^t$; 280 new cases per day **99. a.** (A) **b.** The verbal SAT increases by approximately 1 point. **c.** $S'(x)$ decreases with increasing x, so that as parental income increases, the effect on SAT scores decreases. **101. a.** -6.25 years/child; when the fertility rate is 2 children per woman, the average age of a population is dropping at a rate of 6.25 years per one-child increase in the fertility rate. **b.** 0.160 **103.** 3,300,000 cases/week; 11,000,000 cases/week; 640,000 cases/week **105.** 2.1 percentage points per year; the rate of change is the slope of the tangent at $t=3$. This is also approximately the average rate of change over [2, 4], which is about $4/2=2$, in approximate agreement with the answer. **107.** 2.1 percentage points per year **109.** 277,000 people per year **111.** 0.000283 g/yr

113. a.

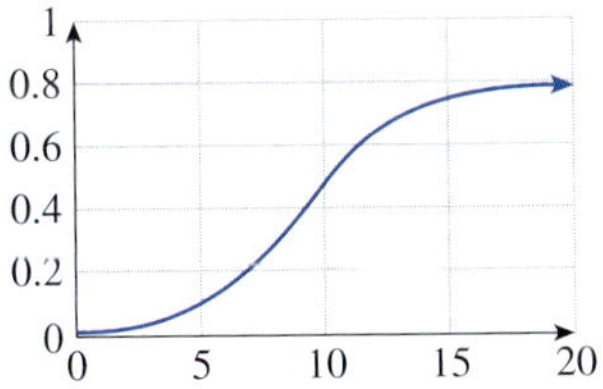

$p'(10) \approx 0.09$, so the percentage of firms using numeric control is increasing at a rate of 9 percentage points per year after 10 years. **b.** 0.80. Thus, in the long run, 80% of all firms will be using numeric control.
c. $p'(t) = 0.3816e^{4.46-0.477t}/(1 + e^{4.46-0.477t})^2$. $p'(10) = 0.0931$.
Graph:

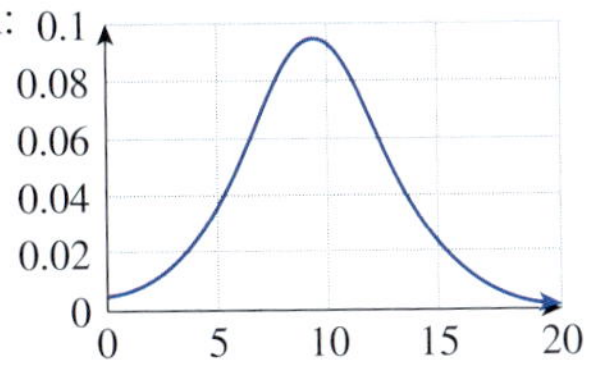

115. $R(t) = 350e^{-0.1t}(39t + 68)$ million dollars; $R(2) \approx$ \$42 billion; $R'(2) \approx$ \$7 billion per year **117.** e raised to the glob, times the derivative of the glob. **119.** 2 raised to the glob, times the derivative of the glob, times the natural logarithm of 2. **121.** The derivative of $\ln|u|$ is not $\frac{1}{|u|}\frac{du}{dx}$; it is $\frac{1}{u}\frac{du}{dx}$. Thus, the correct derivative is $\frac{3}{3x+1}$. **123.** The power rule does not apply when the exponent is not constant. The derivative of 3 raised to a quantity is 3 raised to the quantity, times the derivative of the quantity, times ln 3. Thus, the correct answer is $3^{2x}\, 2 \ln 3$. **125.** No. If $N(t)$ is exponential, so is its derivative. **127.** If $f(x) = e^{kx}$, then the fractional rate of change is $\frac{f'(x)}{f(x)} = \frac{ke^{kx}}{e^{kx}} = k$, the fractional growth rate.
129. If $A(t)$ is growing exponentially, then $A(t) = A_0e^{kt}$ for constants A_0 and k. Its percentage rate of change is then $\frac{A'(t)}{A(t)} = \frac{kA_0e^{kt}}{A_0e^{kt}} = k$, a constant.

Section 4.6

1. $-2/3$ **3.** x **5.** $(y-2)/(3-x)$ **7.** $-y$
9. $-\frac{y}{x(1+\ln x)}$ **11.** $-x/y$ **13.** $-2xy/(x^2-2y)$
15. $-(6+9x^2y)/(9x^3-x^2)$ **17.** $3y/x$
19. $(p+10p^2q)/(2p-q-10pq^2)$
21. $(ye^x - e^y)/(xe^y - e^x)$ **23.** $se^{st}/(2s - te^{st})$
25. $ye^x/(2e^x + y^3e^y)$ **27.** $(y-y^2)/(-1+3y-y^2)$
29. $-y/(x+2y-xye^y-y^2e^y)$ **31. a.** 1 **b.** $y = x - 3$
33. a. -2 **b.** $y = -2x$ **35. a.** -1 **b.** $y = -x + 1$
37. a. $-2{,}000$ **b.** $y = -2{,}000x + 6{,}000$ **39. a.** 0 **b.** $y = 1$ **41. a.** -0.1898 **b.** $y = -0.1898x + 1.4721$
43. $\frac{2x+1}{4x-2}\left[\frac{2}{2x+1} - \frac{4}{4x-2}\right]$

45. $\frac{(3x+1)^2}{4x(2x-1)^3}\left[\frac{6}{3x+1} - \frac{1}{x} - \frac{6}{2x-1}\right]$

47. $(8x-1)^{1/3}(x-1)\left[\frac{8}{3(8x-1)} + \frac{1}{x-1}\right]$

49. $(x^3+x)\sqrt{x^3+2}\left[\frac{3x^2+1}{x^3+x} + \frac{1}{2}\frac{3x^2}{x^3+2}\right]$

51. $x^x(1+\ln x)$ **53.** $-$\$3,000 per worker. The monthly budget to maintain production at the fixed level P is decreasing by approximately \$3,000 per additional worker at an employment level of 100 workers and a monthly operating budget of \$200,000. **55.** -125 T-shirts per dollar; when the price is set at \$5, the demand is dropping by 125 T-shirts per \$1 increase in price. **57.** $\left.\frac{dk}{de}\right|_{e=15} = -0.307$ carpenters per electrician. This means that, for a \$200,000 house whose construction employs 15 electricians, adding one more electrician would cost as much as approximately 0.307 additional carpenters. In other words, one electrician is worth approximately 0.307 carpenters.
59. a. 22.93 hours. (The other root is rejected because it is larger than 30.) **b.** $\frac{dt}{dx} = \frac{4t-20x}{0.4t-4x}$; $\left.\frac{dt}{dx}\right|_{x=3.0} \approx -11.2$ hours per grade point. This means that, for a 3.0 student who scores 80 on the examination, 1 grade point is worth approximately 11.2 hours.
61. $\frac{dr}{dy} = 2\frac{r}{y}$, so $\frac{dr}{dt} = 2\frac{r}{y}\frac{dy}{dt}$ by the chain rule.
63. x, y, y, x **65.** Let $y = f(x)g(x)$. Then $\ln y = \ln f(x) + \ln g(x)$, and

$$\frac{1}{y}\frac{dy}{dx} = \frac{f'(x)}{f(x)} + \frac{g'(x)}{g(x)}, \text{ so } \frac{dy}{dx} = y\left(\frac{f'(x)}{f(x)} + \frac{g'(x)}{g(x)}\right) = f(x)g(x)\left(\frac{f'(x)}{f(x)} + \frac{g'(x)}{g(x)}\right) = f'(x)g(x) + f(x)g'(x).$$

67. Writing $y = f(x)$ specifies y as an explicit function of x. This can be regarded as an equation giving y as an *implicit* function of x. The procedure of finding dy/dx by implicit differentiation is then the same as finding the derivative of y as an explicit function of x: We take d/dx of both sides.
69. Differentiate both sides of the equation $y = f(x)$ with respect to y to get $1 = f'(x)\cdot\frac{dx}{dy}$, giving $\frac{dx}{dy} = \frac{1}{f'(x)} = \frac{1}{dy/dx}$.

Chapter 4 Review

1. $50x^4 + 2x^3 - 1$ **3.** $9x^2 + x^{-2/3}$ **5.** $1 - 2/x^3$
7. $-\frac{4}{3x^2} + \frac{0.2}{x^{1.1}} + \frac{1.1x^{0.1}}{3.2}$ **9.** $e^x(x^2+2x-1)$
11. $20x(x^2-1)^9$ **13.** $e^x(x^2+1)^9(x^2+20x+1)$
15. $3^x[(x-1)\ln 3 - 1]/(x-1)^2$ **17.** $2xe^{x^2-1}$
19. $2x/(x^2-1)$ **21.** $x = 7/6$ **23.** $x = \pm 2$
25. $x = (1-\ln 2)/2$ **27.** None **29.** $\frac{2x-1}{2y}$ **31.** $-y/x$

33. $\dfrac{(2x-1)^4(3x+4)}{(x+1)(3x-1)^3}\left[\dfrac{8}{2x-1}+\dfrac{3}{3x+4}-\dfrac{1}{x+1}-\dfrac{9}{3x-1}\right]$

35. $y = -x/4 + 1/2$ **37.** $y = -3ex - 2e$ **39.** $y = x + 2$ **41. a.** 274 books per week **b.** 636 books per week **c.** The function w begins to decrease more and more rapidly after $t = 14$ Graph:

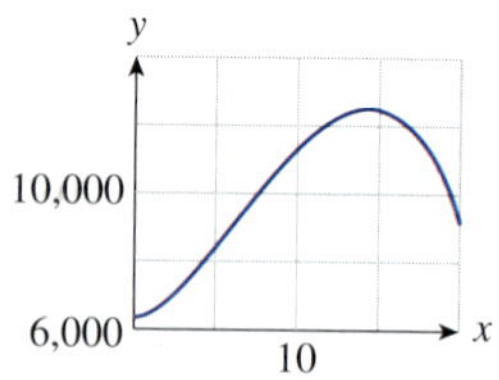

d. Because the data suggest an upward curving parabola, the long-term prediction of sales for a quadratic model would be that sales will increase without bound, in sharp contrast to (c) **43. a.** \$2.88 per book **b.** \$3.715 per book **c.** Approximately −\$0.000104 per book, per additional book sold. **d.** At a sales level of 8,000 books per week, the cost is increasing at a rate of \$2.88 per book (so that the 8,001st book costs approximately \$2.88 to sell), and it costs an average of \$3.715 per book to sell the first 8,000 books. Moreover, the average cost is decreasing at a rate of \$0.000104 per book, per additional book sold. **45. a.** \$3,000 per week (rising) **b.** 300 books per week

47. $R = pq$ gives $R' = p'q + pq'$. Thus, $R'/R = R'/(pq) = (p'q + pq')/pq = p'/p + q'/q$ **49.** \$110 per year

51. a. $s'(t) = \dfrac{2{,}475e^{-0.55(t-4.8)}}{(1+e^{-0.55(t-4.8)})^2}$; 556 books per week

b. 0; In the long term, the rate of increase of weekly sales slows to zero. **53.** 616.8 hits per day per week. **55. a.** −17.24 copies per \$1. The demand for the gift edition of *The Complete Harry Potter* is dropping at a rate of about 17.24 copies per \$1 increase in the price. **b.** \$138 per dollar is positive, so the price should be raised.

Chapter 5

Section 5.1

1. Absolute min: (−3, −1), relative max: (−1, 1), relative min: (1, 0), absolute max: (3, 2) **3.** Absolute min: (3, −1) and (−3, −1), absolute max: (1, 2) **5.** Absolute min: (−3, 0) and (1, 0), absolute max: (−1, 2) and (3, 2) **7.** Relative min: (−1, 1) **9.** Absolute min: (−3, −1), relative max: (−2, 2), relative min: (1, 0), absolute max: (3, 3) **11.** Relative max: (−3, 0), absolute min: (−2, −1), stationary non-extreme point: (1, 1) **13.** Absolute max: (0, 1), absolute min: (2, −3), relative max: (3, −2) **15.** Absolute min: (−4, −16), absolute max: (−2, 16), absolute min: (2, −16), absolute max: (4, 16) **17.** Absolute min: (−2, −10), absolute max: (2, 10) **19.** Absolute min: (−2, −4), relative max: (−1, 1), relative min: (0, 0) **21.** Relative max: (−1, 5), absolute min: (3, −27) **23.** Absolute min: (0, 0) **25.** Absolute maxima at (0, 1) and (2, 1), absolute min at (1, 0) **27.** Relative maximum at (−2, −1/3), relative minimum at (−1, −2/3), absolute maximum at (0, 1) **29.** Relative min: (−2, 5/3), relative max: (0, −1), relative min: (2, 5/3) **31.** Relative max: (0, 0); absolute min: $(1/3, -2\sqrt{3}/9)$ **33.** Relative max: (0, 0), absolute min: (1, −3) **35.** No relative extrema **37.** Absolute min: (1, 1) **39.** Relative max: $(-1, 1 + 1/e)$, absolute min: (0, 1), absolute max: $(1, e - 1)$ **41.** Relative max: (−6, −24), relative min: (−2, −8) **43.** Absolute max $(1/\sqrt{2}, \sqrt{e/2})$, absolute min: $(-1/\sqrt{2}, -\sqrt{e/2})$ **45.** Relative min at (0.15, −0.52) and (2.45, 8.22), relative max at (1.40, 0.29) **47.** Absolute max at (−5, 700), relative max at (3.10, 28.19) and (6, 40), absolute min at (−2.10, −392.69) and relative min at (5, 0). **49.** Stationary minimum at $x = -1$ **51.** Stationary minima at $x = -2$ and $x = 2$, stationary maximum at $x = 0$ **53.** Singular minimum at $x = 0$, stationary non-extreme point at $x = 1$ **55.** Stationary minimum at $x = -2$, singular non-extreme points at $x = -1$ and $x = 1$, stationary maximum at $x = 2$

57. Answers will vary.

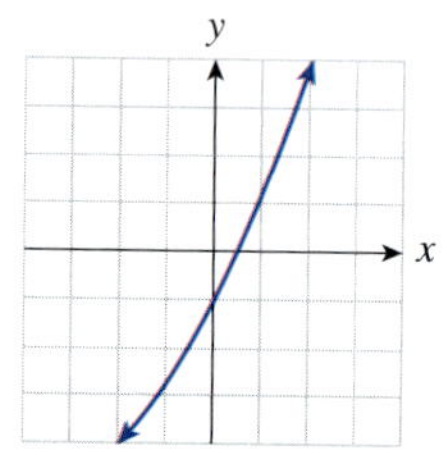

59. Answers will vary.

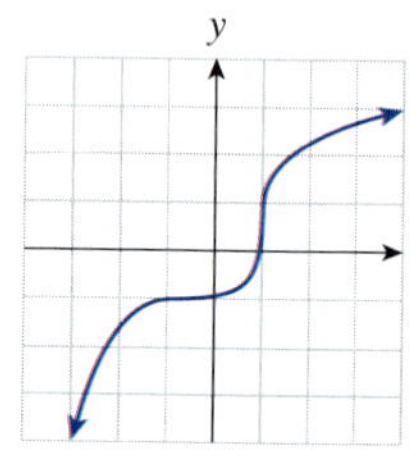

61. Not necessarily; it could be neither a relative maximum nor a relative minimum, as in the graph of $y = x^3$ at the origin. **63.** Answers will vary.

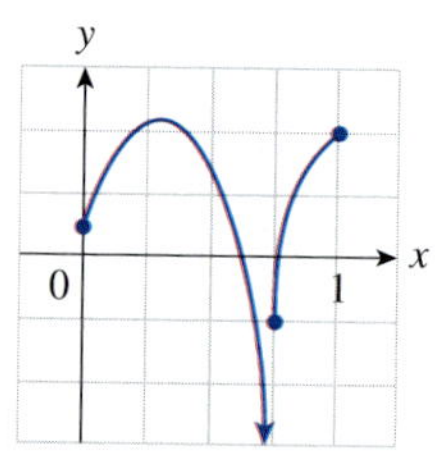

65. The graph oscillates faster and faster above and below zero as it approaches the end-point at 0, so 0 cannot be either a relative minimum or maximum.

Section 5.2

1. $x = y = 5$; $P = 25$ **3.** $x = y = 3$; $S = 6$ **5.** $x = 2$, $y = 4$; $F = 20$ **7.** $x = 20$, $y = 10$, $z = 20$; $P = 4{,}000$ **9.** 5×5 **11.** 1,500 per day for an average cost of \$130 per iPod **13.** $\sqrt{40} \approx 6.32$ pounds of pollutant per day, for an average cost of about \$1,265 per pound **15.** 2.5 lb **17.** 5×10 **19.** 50×10 for an area of 250 sq. ft. **21.** 11×22 **23.** \$10 **25.** \$78 for a quarterly revenue of \$6,084 million, or \$6.084 billion **27.** \$4.61 for a daily revenue of \$95,680.55 **29. a.** \$1.41 per pound **b.** 5,000 pounds **c.** \$7,071.07 per month **31.** 34.5¢ per pound, for an annual (per capita) revenue of \$5.95 **33.** \$98 for an annual profit of \$3,364 million, or \$3.364 billion **35. a.** 656 headsets, for a

profit of \$28,120 **b.** \$143 per headset **37.** Height = Radius of base ≈ 20.48 cm. **39.** Height ≈ 10.84 cm; Radius ≈ 2.71 cm; Height/Radius = 4 **41.** $13\frac{1}{3}$ in × $3\frac{1}{3}$ in × $1\frac{1}{3}$ in for a volume of 1,600/27 ≈ 59 cubic inches **43.** 5 × 5 × 5 cm **45.** $l = w = h \approx 20.67$ in, volume ≈ 8,827 in^3 **47.** $l = 30$ in, $w = 15$ in, $h = 30$ in **49.** $l = 36$ in, $w = h = 18$ in, $V = 11{,}664$ in^3 **51. a.** 1.6 years, or year 2001.6; **b.** R_{max} = \$28,241 million **53.** $t = 2.5$ or midway through 1972; $D(2.5)/S(2.5) \approx 4.09$. The number of new (approved) drugs per \$1 billion of spending on research and development reached a high of around four approved drugs per \$1 billion midway through 1972. **55.** 30 years from now **57.** 55 days **59.** 1,600 copies. At this value of x, average profit equals marginal profit; beyond this the marginal profit is smaller than the average. **61.** 40 laborers and 250 robots **63.** 71 employees **65.** Increasing most rapidly in 1990; increasing least rapidly in 2007 **67.** Maximum when $t = 17$ days. This means that the embryo's oxygen consumption is increasing most rapidly 17 days after the egg is laid.
69. Graph of derivative:

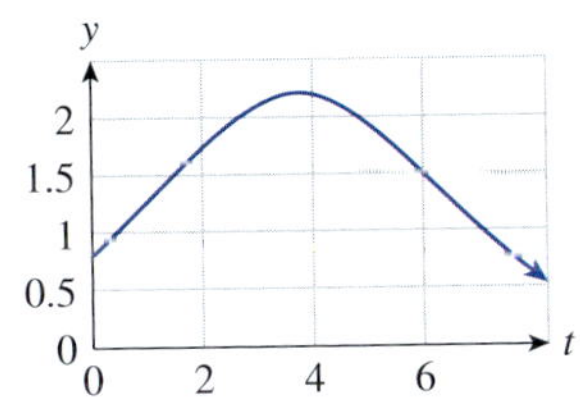

The absolute maximum occurs at approximately (3.7, 2.2) during the year 2003. The percentage of mortgages that were subprime was increasing most rapidly during 2003, when it increased at a rate of around 2.2 percentage points per year. **71.** You should sell them in 17 years' time, when they will be worth approximately \$3,960. **73.** 25 additional trees **75.** (D) **77.** (A); (B) **79.** The problem is uninteresting because the company can accomplish the objective by cutting away the entire sheet of cardboard, resulting in a box with surface area zero. **81.** Not all absolute extrema occur at stationary points; some may occur at an endpoint or singular point of the domain, as in Exercises 29, 30, 65, and 66. **83.** The minimum of dq/dp is the fastest that the demand is dropping in response to increasing price.

Section 5.3

1. 6 **3.** $4/x^3$ **5.** $-0.96x^{-1.6}$ **7.** $e^{-(x-1)}$ **9.** $2/x^3 + 1/x^2$ **11. a.** $a = -32$ ft/s^2 **b.** $a = -32$ ft/s^2 **13. a.** $a = 2/t^3 + 6/t^4$ ft/s^2 **b.** $a = 8$ ft/s^2 **15. a.** $a = -1/(4t^{3/2}) + 2$ ft/s^2 **b.** $a = 63/32$ ft/s^2 **17.** (1, 0) **19.** (1, 0) **21.** None **23.** (−1, 0), (1, 1) **25.** Points of inflection at $x = -1$ and $x = 1$ **27.** One point of inflection, at $x = -2$ **29.** Points of inflection at $x = -2, x = 0, x = 2$ **31.** Points of inflection at $x = -2$ and $x = 2$ **33.** $x = 2$; minimum **35.** Maximum at $x = -2$, minimum at $x = 2$ **37.** Maximum at $t = -1/\sqrt{3}$, minimum at $t = 1/\sqrt{3}$ **39.** Non-extreme stationary point at $x = 0$ minimum at $x = 3$ **41.** Maximum at $x = 0$ **43.** Minimum at $x = -1/\sqrt{2}$; maximum at $x = 1/\sqrt{2}$ **45.** $f'(x) = 8x - 1$; $f''(x) = 8$; $f'''(x) = f^{(4)}(x) = \dots = f^{(n)}(x) = 0$ **47.** $f'(x) = -4x^3 + 6x$; $f''(x) = -12x^2 + 6$; $f'''(x) = -24x$; $f^{(4)}(x) = -24$; $f^{(5)}(x) = f^{(6)}(x) = \dots = f^{(n)}(x) = 0$ **49.** $f'(x) = 8(2x+1)^3$; $f''(x) = 48(2x+1)^2$; $f'''(x) = 192(2x+1)$; $f^{(4)}(x) = 384$; $f^{(5)}(x) = f^{(6)}(x) = \dots = f^{(n)}(x) = 0$ **51.** $f'(x) = -e^{-x}$; $f''(x) = e^{-x}$; $f'''(x) = -e^{-x}$; $f^{(4)}(x) = e^{-x}$; $f^{(n)}(x) = (-1)^n e^{-x}$ **53.** $f'(x) = 3e^{3x-1}$; $f''(x) = 9e^{3x-1}$; $f'''(x) = 27e^{3x-1}$; $f^{(4)}(x) = 81e^{3x-1}$; $f^{(n)}(x) = 3^n e^{3x-1}$ **55.** −3.8 m/s^2 **57.** $6t - 2$ ft/s^2; increasing **59.** Accelerating by 24 million gals/yr^2 **61. a.** 400 ml **b.** 36 ml/day **c.** −1 ml/day^2 **63. a.** 0.6% **b.** Speeding up **c.** Speeding up for $t < 3.33$ (prior to 1/3 of the way through March) and slowing for $t > 3.33$ (after that time) **65. a.** December 2005: −0.202% (deflation rate of 0.202%) February 2006: 0.363% **b.** Speeding up **c.** Decreasing for $t < 4.44$ (prior to mid-November) and increasing for $t > 4.44$ (after that time). **67.** Concave up for $8 < t < 20$, concave down for $0 < t < 8$, point of inflection around $t = 8$. The percentage of articles written by researchers in the United States was decreasing most rapidly at around $t = 8$ (1991). **69. a.** (B) **b.** (B) **c.** (A) **71.** Graphs:

$A(t)$:

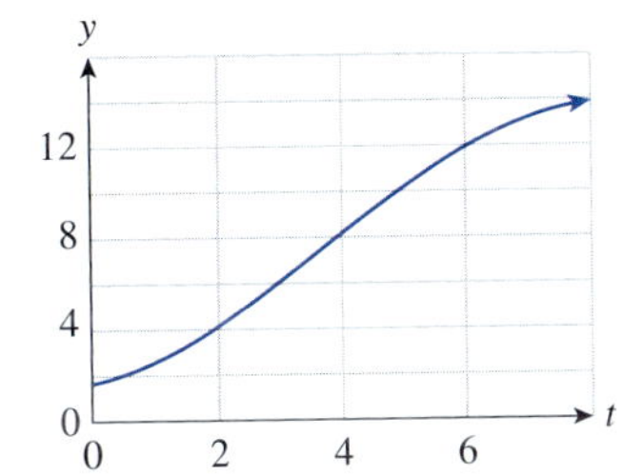

$A'(t)$:

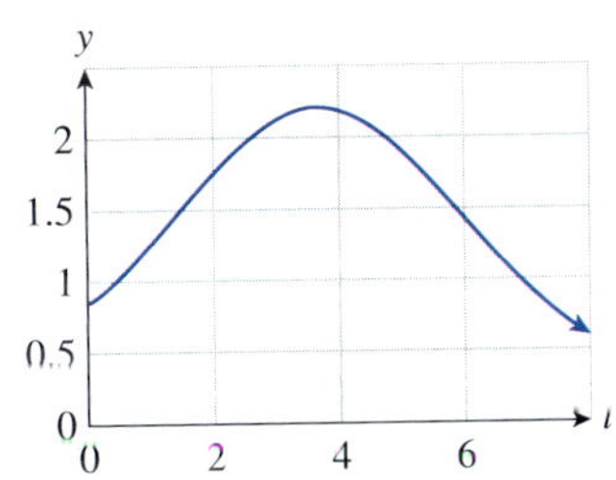

$A''(t)$:

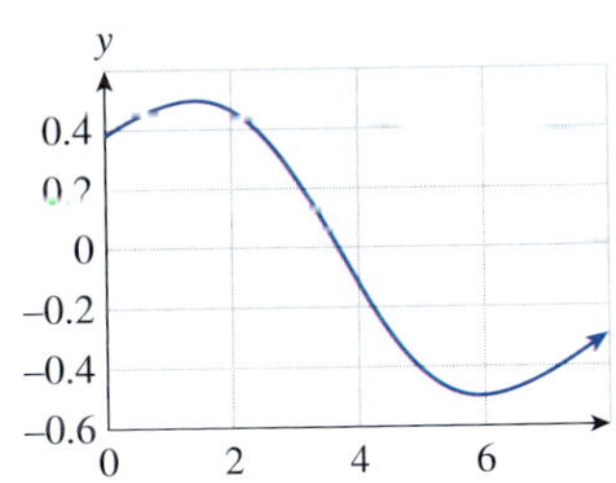

Concave up when $t < 4$; concave down when $t > 4$; point of inflection when $t \approx 4$. The percentage of U.S. mortgages that were subprime was increasing fastest at the beginning of 2004. **73. a.** 2 years into the epidemic **b.** 2 years into the epidemic **75. a.** 2024 **b.** 2026 **c.** 2022; (A) **77. a.** There are no points of inflection in the graph of S. **b.** Because the graph is concave up, the derivative of S is increasing, and so the rate of *decrease* of SAT scores with increasing numbers of prisoners is diminishing. In other words, the apparent effect of more prisoners is diminishing. **79. a.** $\left.\frac{d^2n}{ds^2}\right|_{s=3} = -21.494$. Thus, for a firm with annual sales of \$3 million, the rate at which new patents are produced decreases with increasing firm size. This means that the returns (as measured in the number of new patents per increase of \$1 million in sales) are diminishing as the firm size increases. **b.** $\left.\frac{d^2n}{ds^2}\right|_{s=7} = 13.474$. Thus, for a firm with annual sales of \$7 million, the rate at which new patents are produced increases with increasing firm size by 13.474 new patents per \$1 million increase in annual sales. **c.** There is a point of inflection when $s \approx 5.4587$, so that in a firm with sales of \$5,458,700 per year, the number of new patents produced per additional \$1 million in sales is a minimum.

81. Graphs:

$I(t)/P(t)$:

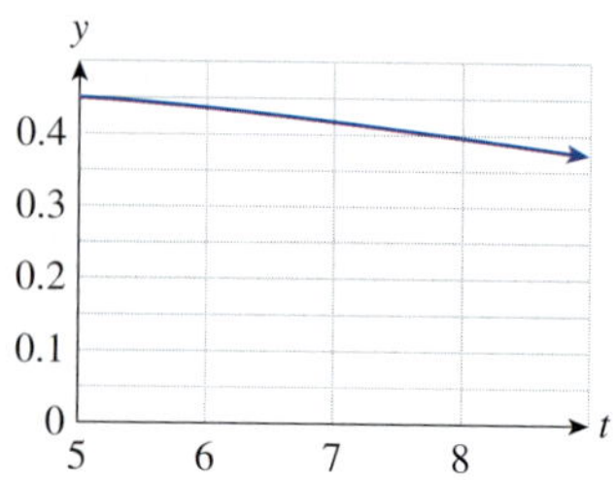

$[I(t)/P(t)]'$:

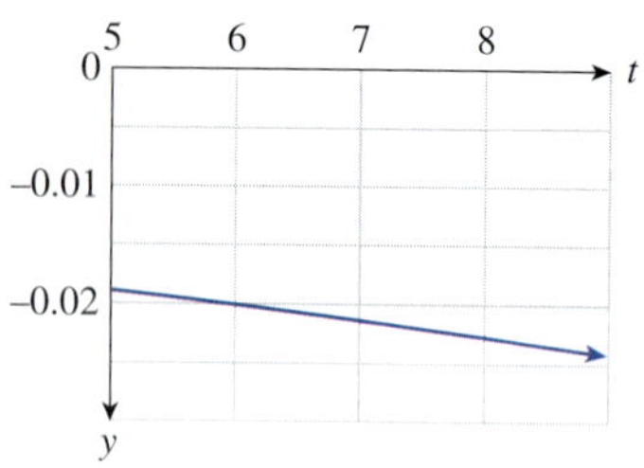

Concave down; (D) **83.** (Proof) **85.** $t \approx 4$; the population of Puerto Rico was increasing fastest in 1954. **87.** About \$570 per year, after about 12 years **89.** Increasing most rapidly in 17.64 years, decreasing most rapidly now (at $t = 0$) **91.** Non-negative **93.** Daily sales were decreasing most rapidly in June 2002.

95.

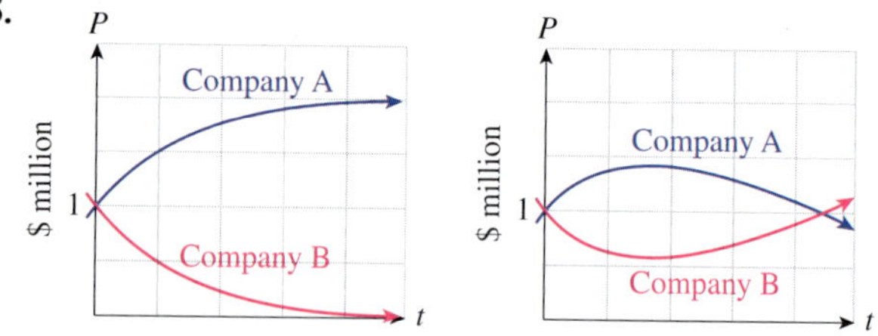

97. At a point of inflection, the graph of a function changes either from concave up to concave down, or vice versa. If it changes from concave up to concave down, then the derivative changes from increasing to decreasing, and hence has a relative maximum. Similarly, if it changes from concave down to concave up, the derivative has a relative minimum.

Section 5.4

1. a. x-intercept: -1; y-intercept: 1 **b.** Absolute min at $(-1, 0)$ **c.** None **d.** None **e.** $y \to +\infty$ as $x \to \pm\infty$
Graph:

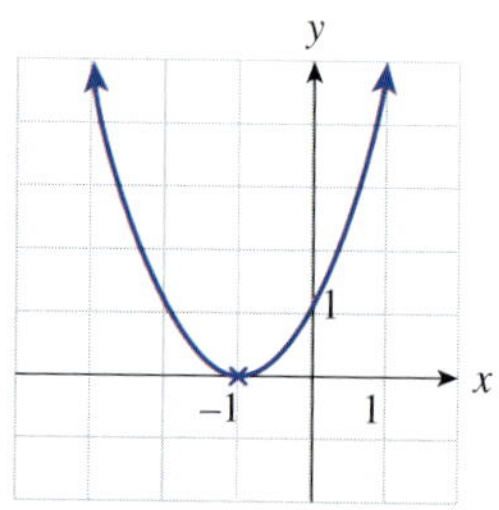

3. a. x-intercepts: $-\sqrt{12}, 0, \sqrt{12}$; y-intercept: 0 **b.** Absolute min at $(-4, -16)$ and $(2, -16)$, absolute max at $(-2, 16)$ and $(4, 16)$ **c.** $(0, 0)$ **d.** None **e.** None
Graph:

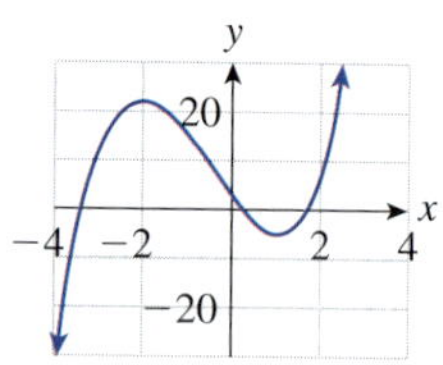

5. a. x-intercepts: $-3.6, 0, 5.1$; y-intercept: 0 **b.** Relative max at $(-2, 44)$, relative min at $(3, -81)$ **c.** $(0.5, -18.5)$ **d.** None **e.** $y \to -\infty$ as $x \to -\infty$; $y \to +\infty$ as $x \to +\infty$
Graph:

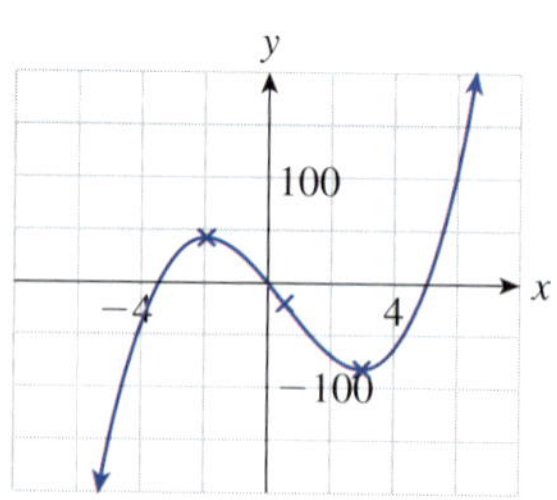

7. a. x-intercepts: 0.1, 1.8; y-intercept: 1 **b.** Relative max at $(-2, 21)$, relative min at $(1, -6)$ **c.** $(-1/2, 15/2)$ **d.** None **e.** $y \to -\infty$ as $x \to -\infty$; $y \to +\infty$ as $x \to +\infty$
Graph:

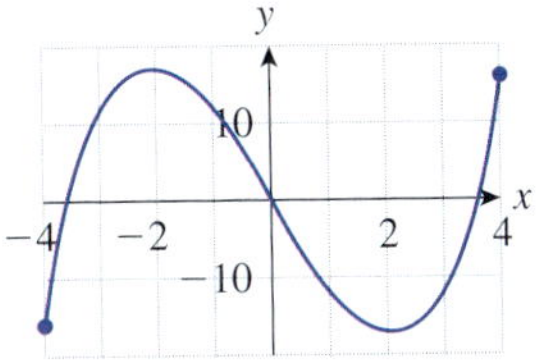

9. a. x-intercepts: -2.9, 4.2; y-intercept: 10 **b.** Relative max at $(-2, 74)$, relative min at $(0, 10)$, absolute max at $(3, 199)$ **c.** $(-1.12, 44.8)$, $(1.79, 117.3)$ **d.** None
e. $y \to -\infty$ as $x \to -\infty$; $y \to -\infty$ as $x \to +\infty$
Graph:

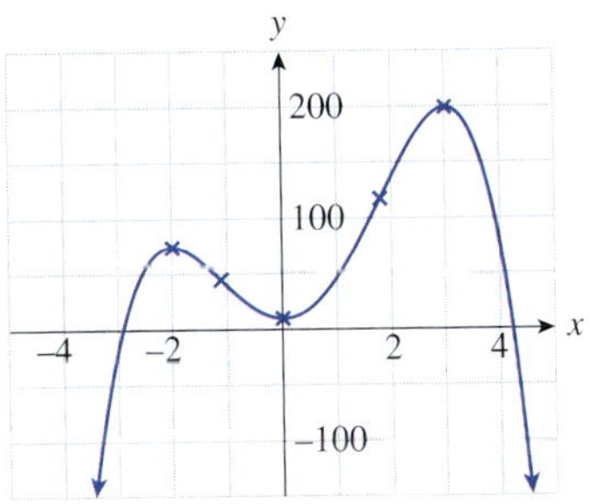

11. a. x-intercepts: None; y-intercept: 0 **b.** Absolute min at $(0, 0)$ **c.** $(1/3, 11/324)$ and $(1, 1/12)$ **d.** None
e. $y \to +\infty$ as $x \to -\infty$; $y \to +\infty$ as $x \to +\infty$
Graph:

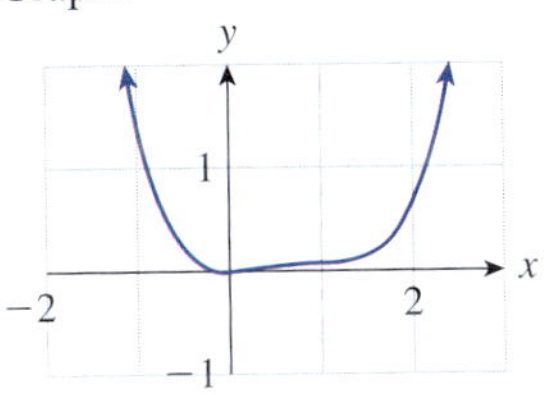

13. a. x-intercepts: None; y-intercept: None **b.** Relative min at $(1, 2)$, relative max at $(-1, -2)$ **c.** None **d.** $y \to -\infty$ as $x \to 0^-$; $y \to +\infty$ as $x \to 0^+$, so there is a vertical asymptote at $x = 0$. **e.** $y \to -\infty$ as $x \to -\infty$; $y \to +\infty$ as $x \to +\infty$
Graph:

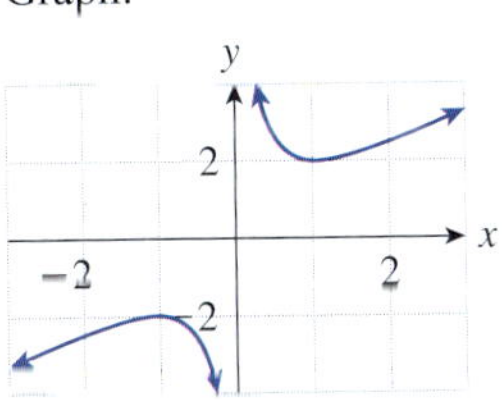

15. a. x-intercept: 0; y-intercept: 0 **b.** None **c.** $(0, 0)$, $(-3, -9/4)$, and $(3, 9/4)$ **d.** None **e.** $y \to -\infty$ as $x \to -\infty$; $y \to +\infty$ as $x \to +\infty$
Graph:

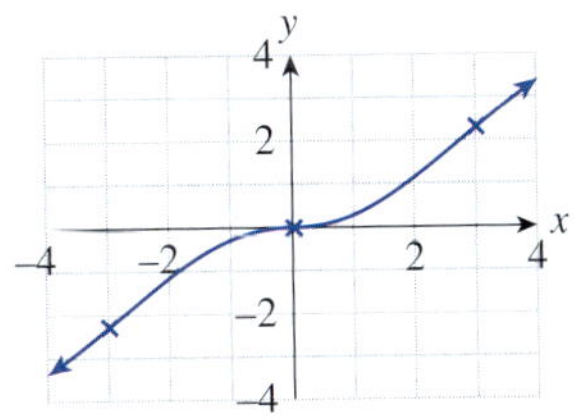

17. a. t-intercepts: None; y-intercept: -1 **b.** Relative min at $(-2, 5/3)$ and $(2, 5/3)$, relative max at $(0, -1)$ **c.** None **d.** $y \to +\infty$ as $t \to -1^-$; $y \to -\infty$ as $t \to -1^+$; $y \to -\infty$ as $t \to 1^-$; $y \to +\infty$ as $t \to 1^+$; so there are vertical asymptotes at $t = \pm 1$. **e.** None
Graph:

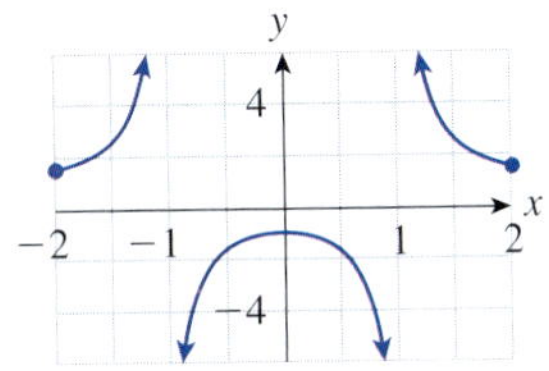

19. a. x-intercepts: -0.6; y-intercept: 1 **b.** Relative maximum at $(-2, -1/3)$, relative minimum at $(-1, -2/3)$ **c.** None **d.** None. **e.** $y \to -\infty$ as $x \to -\infty$; $y \to +\infty$ as $x \to +\infty$
Graph:

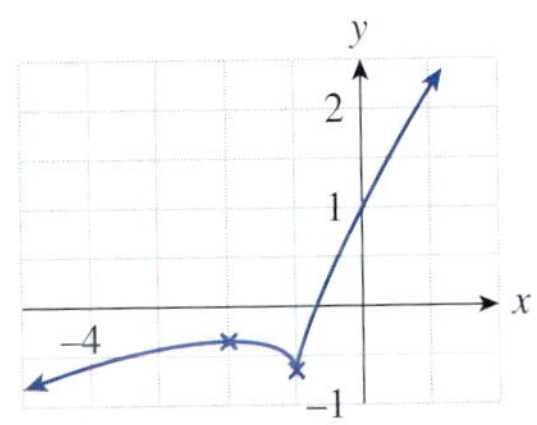

21. a. x-intercepts: None; y-intercept: None **b.** Absolute min at $(1, 1)$ **c.** None **d.** Vertical asymptote at $x = 0$
e. $y \to +\infty$ as $x \to +\infty$
Graph:

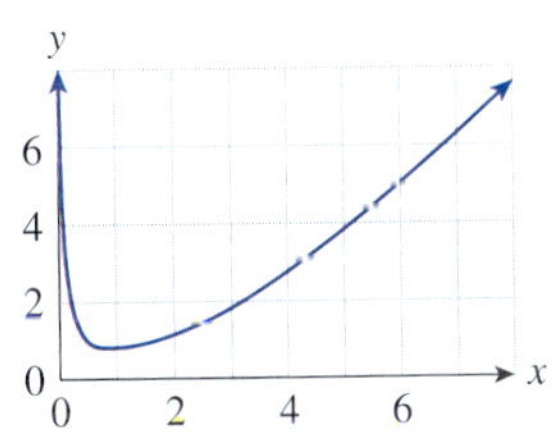

23. a. x-intercepts: ± 0.8; x-intercept: None **b.** None **c.** $(1, 1)$ and $(-1, 1)$ **d.** $y \to -\infty$ as $x \to 0$; vertical asymptote at $x = 0$
e. $y \to +\infty$ as $x \to \pm\infty$

Graph:

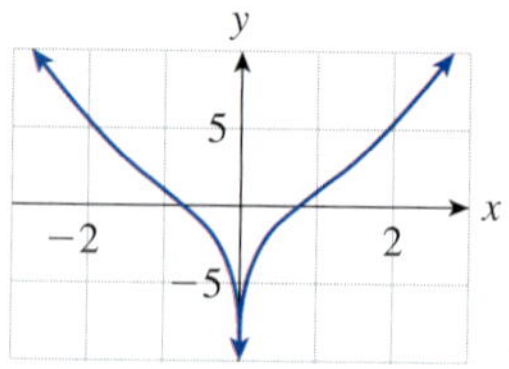

25. a. t-intercepts: None; y-intercept: 1 **b.** Absolute min at (0, 1). Absolute max at $(1, e - 1)$, relative max at $(-1, e^{-1} + 1)$. **c.** None **e.** None
Graph:

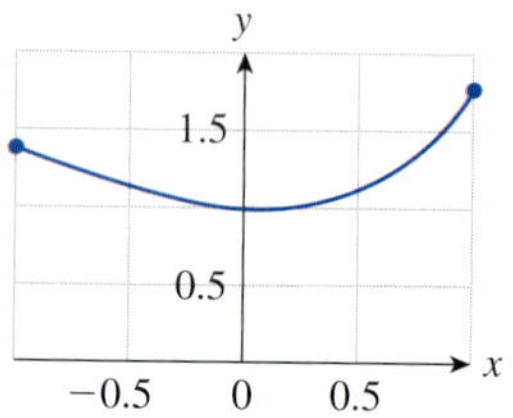

27. Absolute min at (1.40, −1.49); points of inflection: (0.21, 0.61), (0.79, −0.55)
Graph:

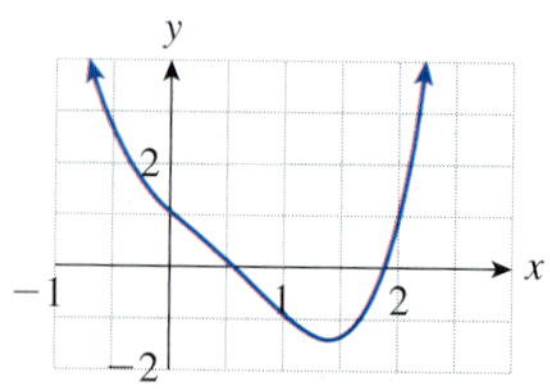

29. $f(x) = e^x - x^3$. Relative min at (−0.46, 0.73), relative max at (0.91, 1.73), absolute min at (3.73, −10.22); points of inflection at (0.20, 1.22) and (2.83, −5.74)
Graph:

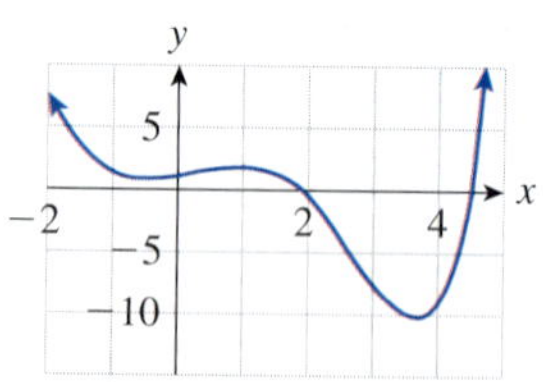

31. y-intercept: 10; t-intercepts: None; the home price index was 10 percentage points at the start of 2004; absolute minimum at (0, 10); absolute maximum at (2, 40); the home price index was lowest in 2004 when it stood at 10 percentage points. The index was at its highest in 2006 at 40 points. Point of inflection at (3.5, 30); the home price index was decreasing most rapidly midway through 2007 when it stood at 30 percentage points. As $t \to +\infty$, $y \to 15$; assuming the trend shown in the graph continues indefinitely, the home price index will approach a value of 15 percentage points in the long term. **33. a.** Intercepts: No t-intercept; y-intercept at $I(0) = 195$. The CPI was never zero during the given period; in July 2005 the CPI was 195. Absolute min at (0, 195), absolute max at (8, 199.3), relative max at (2.9, 198.7), relative min at (6.0, 197.8). The CPI was at a low of 195 in July 2005, rose to 198.7 around October 2005, dipped to 197.8 around January 2006, and then rose to a high of 199.3 in March 2006. There is a point of inflection at (4.4, 198.2). The rate of change of the CPI (inflation) reached a minimum around mid-November 2005 when the CPI was 198.2.
b. The inflation rate was zero at around October 2005 and January 2006. **35.** Extrema: Relative max at (0, 100), absolute min at (1, 99); point of inflection (0.5, 99.5); $s \to +\infty$ as $t \to +\infty$. At time $t = 0$ seconds, the UFO is stationary, 100 feet away from the observer, and begins to move closer. At time $t = 0.5$ seconds, when the UFO is 99.5 feet away, its distance is decreasing most rapidly (it is moving toward the observer most rapidly). It then slows down to a stop at $t = 1$ second when it is at its closest point (99 feet away) and then begins to move further and further away.
Graph:

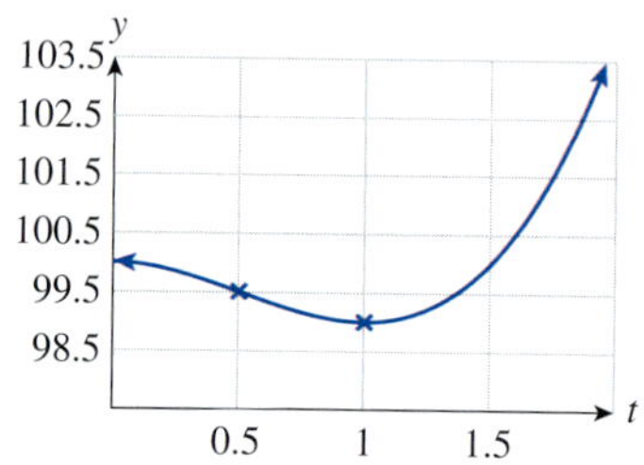

37. Intercepts: None; Absolute minimum at (1,500, 130); No points of inflection; vertical asymptote at $x = 0$. As $x \to +\infty$, $y \to +\infty$. The average cost is never zero, nor is it defined for zero iPods. The average cost is a minimum ($130) when 1,500 iPods are manufactured per day. The average cost becomes extremely large for very small or very large numbers of iPods.
Graph:

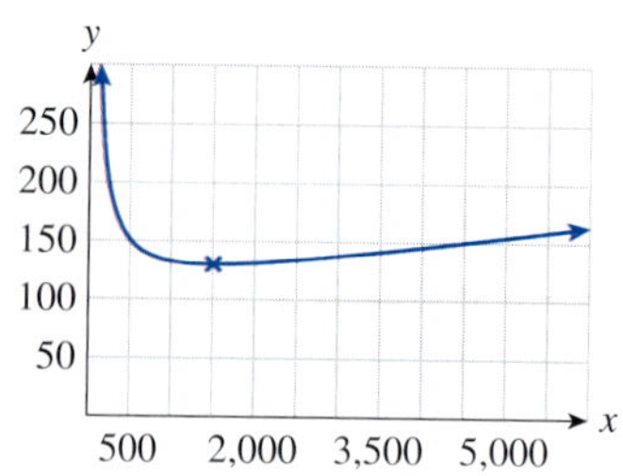

39. Graph of derivative:

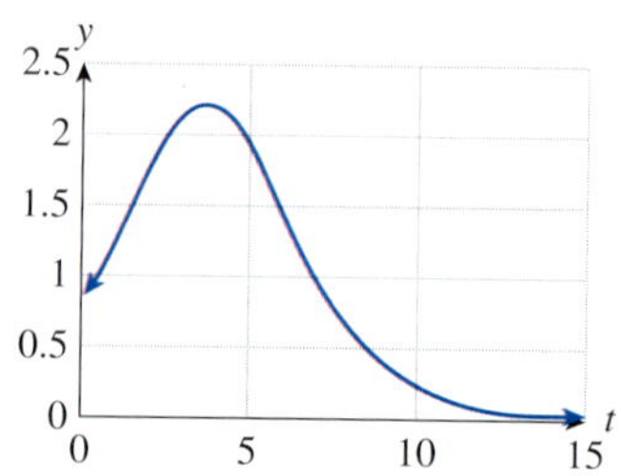

The absolute maximum occurs at approximately (3.7, 2.2); during the year 2003. The percentage of mortgages that were subprime was increasing most rapidly during 2003, when it increased at a rate of around 2.2 percentage points per year. As $t \to +\infty$, $A'(t) \to 0$; In the long term, assuming the trend shown in the model continues, the rate of change of the percentage of mortgages that were subprime approaches zero; that is, the percentage of mortgages that were subprime approaches a constant value. **41.** No; yes. Near a vertical asymptote the value of y increases without bound, and so the graph could not be included between two horizontal lines; hence, no vertical asymptotes are possible. Horizontal asymptotes are possible, as for instance in the graph in Exercise 31. **43.** It too has a vertical asymptote at $x = a$; the magnitude of the derivative increases without bound as $x \to a$. **45.** No. If the leftmost critical point is a relative maximum, the function will decrease from there until it reaches the rightmost critical point, so can't have a relative maximum there. **47.** Between every pair of zeros of $f(x)$ there must be a local extremum, which must be a stationary point of $f(x)$, hence a zero of $f'(x)$.

Section 5.5

1. $P = 10{,}000$; $\frac{dP}{dt} = 1{,}000$ **3.** Let R be the annual revenue of my company, and let q be annual sales. $R = 7{,}000$ and $\frac{dR}{dt} = -700$. Find $\frac{dq}{dt}$. **5.** Let p be the price of a pair of shoes, and let q be the demand for shoes. $\frac{dp}{dt} = 5$. Find $\frac{dq}{dt}$.
7. Let T be the average global temperature, and let q be the number of Bermuda shorts sold per year. $T = 60$ and $\frac{dT}{dt} = 0.1$. Find $\frac{dq}{dt}$. **9. a.** $6/(100\pi) \approx 0.019$ km/sec **b.** $6/(8\sqrt{\pi}) \approx 0.4231$ km/sec **11.** $3/(4\pi) \approx 0.24$ ft/min **13.** 7.5 ft/s **15.** Decreasing at a rate of \$1.66 per player per week **17.** Monthly sales will drop at a rate of 26 T-shirts per month. **19.** Raise the price by 3¢ per week. **21.** Increasing at a rate of \$233.68 million per year **23.** 1 laborer per month **25.** The daily operating budget is dropping at a rate of \$2.40 per year. **27.** The price is decreasing at a rate of approximately 31¢ per pound per month. **29.** $2{,}300/\sqrt{4{,}100} \approx 36$ miles/hour. **31.** 10.7 ft/s **33.** The y coordinate is decreasing at a rate of 16 units per second. **35.** \$1,814 per year **37.** Their prior experience must increase at a rate of approximately 0.97 years every year. **39.** $\frac{2{,}500}{9\pi}\left(\frac{3}{5{,}000}\right)^{2/3} \approx 0.63$ m/sec **41.** $\frac{\sqrt{1+128\pi}}{4\pi} \approx 1.6$ cm/sec **43.** 0.5137 computers per household, and increasing at a rate of 0.0230 computers per household per year. **45.** The average SAT score was 904.71 and decreasing at a rate of 0.11 per year. **47.** Decreasing by 2 percentage points per year **49.** The section is called "related rates" because the goal is to compute the rate of change of a quantity based on a knowledge of the rate of change of a related quantity. **51.** Answers may vary: A rectangular solid has dimensions 2 cm × 5 cm × 10 cm, and each side is expanding at a rate of 3 cm/second. How fast is the volume increasing? **53.** (Proof) **55.** Linear **57.** Let x = my grades and y = your grades. If $dx/dt = 2\,dy/dt$, then $dy/dt = (1/2)\,dx/dt$.

Section 5.6

1. $E = 1.5$; the demand is going down 1.5% per 1% increase in price at that price level; revenue is maximized when $p = \$25$; weekly revenue at that price is \$12,500. **3. a.** $E = 6/7$; the demand is going down 6% per 7% increase in price at that price level; thus, a price increase is in order. **b.** Revenue is maximized when $p = 100/3 \approx \$33.33$ **c.** Demand would be $(100 - 100/3)^2 = (200/3)^2 \approx 4{,}444$ cases per week.
5. a. $E = (4p - 33)/(-2p + 33)$ **b.** 0.54; the demand for $E = mc^2$ T-shirts is going down by about 0.54% per 1% increase in the price. **c.** \$11 per shirt for a daily revenue of \$1,331 **7. a.** $E = 1.81$. Thus, the demand is elastic at the given tuition level, showing that a decrease in tuition will result in an increase in revenue. **b.** They should charge an average of \$2,250 per student, and this will result in an enrollment of about 4,950 students, giving a revenue of about \$11,137,500.
9. a. $E = 51$; the demand is going down 51% per 1% increase in price at that price level; thus, a large price decrease is advised. **b.** Revenue is maximized when $p = ¥0.50$. **c.** Demand would be $100e^{-3/4+1/2} \approx 78$ paint-by-number sets per month.
11. a. $E = -\frac{mp}{mp+b}$ **b.** $p = -\frac{b}{2m}$ **13. a.** $E = r$ **b.** E is independent of p. **c.** If $r = 1$, then the revenue is not affected by the price. If $r > 1$, then the revenue is always elastic, while if $r < 1$, the revenue is always inelastic. This is an unrealistic model because there should always be a price at which the revenue is a maximum. **15. a.** $q = -1{,}500p + 6{,}000$. **b.** \$2 per hamburger, giving a total weekly revenue of \$6,000.
17. $E \approx 0.77$. At a family income level of \$20,000, the fraction of children attending a live theatrical performance is increasing by 0.77% per 1% increase in household income.
19. a. $E \approx 0.46$. The demand for computers is increasing by 0.46% per one percent increase in household income.
b. E decreases as income increases. **c.** Unreliable; it predicts a likelihood greater than 1 at incomes of \$123,000 and above. In a more appropriate model, one would expect the curve to level off at or below 1. **d.** $E \approx 0$ **21.** $\frac{Y}{Q} \cdot \frac{dQ}{dY} = \beta$. An increase in income of x% will result in an increase in demand of βx%. (Note that we do *not* take the negative here, because we expect an increase in income to produce an *increase* in demand.) **23. a.** $q = 1{,}000e^{-0.30p}$ **b.** At $p = \$3$, $E = 0.9$; at $p = \$4$, $E = 1.2$; at $p = \$5$, $E = 1.5$ **c.** $p = \$3.33$
d. $p = \$5.36$. Selling at a lower price would increase demand, but you cannot sell more than 200 pounds anyway. You should charge as much as you can and still be able to sell all 200 pounds. **25.** The price is lowered. **27.** Start with $R = pq$, and differentiate with respect to p to obtain $\frac{dR}{dp} = q + p\frac{dq}{dp}$. For a stationary point, $dR/dp = 0$, and so $q + p\frac{dq}{dp} = 0$.

Rearranging this result gives $p\frac{dq}{dp} = -q$, and hence $-\frac{dq}{dp} \cdot \frac{p}{q} = 1$, or $E = 1$, showing that stationary points of R correspond to points of unit elasticity. **29.** The distinction is best illustrated by an example. Suppose that q is measured in weekly sales and p is the unit price in dollars. Then the quantity $-dq/dp$ measures the drop in weekly sales per \$1 increase in price. The elasticity of demand E, on the other hand, measures the *percentage* drop in sales per *one percent* increase in price. Thus, $-dq/dp$ measures absolute change, while E measures fractional, or percentage, change.

Chapter 5 Review

1. Relative max: $(-1, 5)$, absolute min: $(-2, -3)$ and $(1, -3)$ **3.** Absolute max: $(-1, 5)$, absolute min: $(1, -3)$ **5.** Absolute min: $(1, 0)$ **7.** Absolute min: $(-2, -1/4)$ **9.** Relative max at $x = 1$, point of inflection at $x = -1$ **11.** Relative max at $x = -2$, relative min at $x = 1$, point of inflection at $x = -1$ **13.** One point of inflection, at $x = 0$ **15. a.** $a = 4/t^4 - 2/t^3$ m/sec^2 **b.** 2 m/sec^2 **17.** Relative max: $(-2, 16)$; absolute min: $(2, -16)$; point of inflection: $(0, 0)$; no horizontal or vertical asymptotes

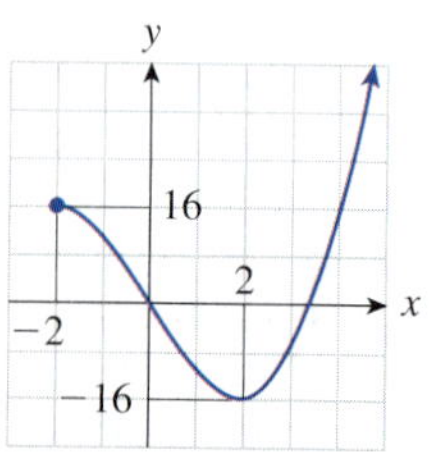

19. Relative min: $(-3, -2/9)$; relative max: $(3, 2/9)$; inflection: $(-3\sqrt{2}, -5\sqrt{2}/36)$, $(3\sqrt{2}, 5\sqrt{2}/36)$; vertical asymptote: $x = 0$; horizontal asymptote: $y = 0$

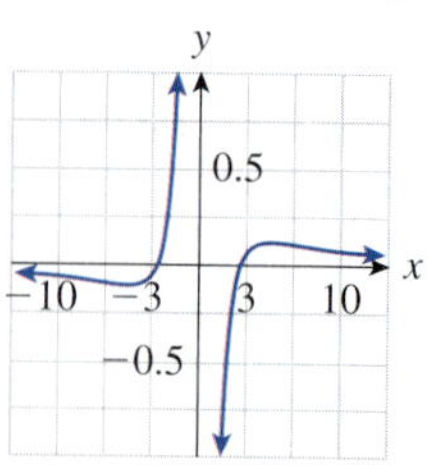

21. Relative max at $(0, 0)$, absolute min at $(1, -2)$, no asymptotes

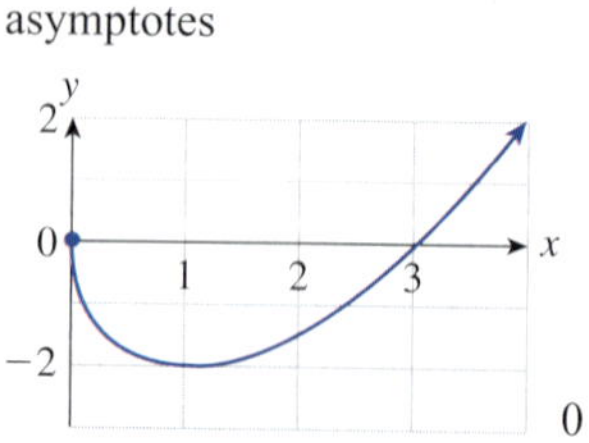

23. \$22.14 per book **25. a.** Profit $= -p^3 + 42p^2 - 288p - 181$ **b.** \$24 per copy; \$3,275 **c.** For maximum revenue, the company should charge \$22.14 per copy. At this price, the cost per book is decreasing with increasing price, while the revenue is not decreasing (its derivative is zero). Thus, the profit is increasing with increasing price, suggesting that the maximum profit will occur at a higher price. **27.** 12 in × 12 in × 6 in, for a volume of 864 in^3 **29. a.** $E = \frac{2p^2 - 33p}{-p^2 + 33p + 9}$ **b.** 0.52, 2.03; when the price is \$20, demand is dropping at a rate of 0.52% per 1% increase in the price; when the price is \$25, demand is dropping at a rate of 2.03% per 1% increase in the price. **c.** \$22.14 per book **31. a.** $E = 2p^2 - p$ **b.** 6; the demand is dropping at a rate of 6% per 1% increase in the price. **c.** \$1.00, for a monthly revenue of \$1,000 **33. a.** week 5 **b.** Point of inflection on the graph of s; maximum on the graph of s', t-intercept in the graph of s''. **c.** 10,500; if weekly sales continue as predicted by the model, they will level off at around 10,500 books per week in the long term. **d.** 0; if weekly sales continue as predicted by the model, the rate of change of sales approaches zero in the long term. **35. a.–d.** $10/\sqrt{2}$ ft/s

Chapter 6

Section 6.1

1. $x^6/6 + C$ **3.** $6x + C$ **5.** $x^2/2 + C$ **7.** $x^3/3 - x^2/2 + C$ **9.** $x + x^2/2 + C$ **11.** $-x^{-4}/4 + C$ **13.** $x^{3.3}/3.3 - x^{-0.3}/0.3 + C$ **15.** $u^3/3 - \ln|u| + C$ **17.** $\frac{4x^{5/4}}{5} + C$ **19.** $3x^5/5 + 2x^{-1} - x^{-4}/4 + 4x + C$ **21.** $2\ln|u| + u^2/8 + C$ **23.** $\ln|x| - \frac{2}{x} + \frac{1}{2x^2} + C$ **25.** $3x^{1.1}/1.1 - x^{5.3}/5.3 - 4.1x + C$ **27.** $\frac{x^{0.9}}{0.3} + \frac{40}{x^{0.1}} + C$ **29.** $2.55t^2 - 1.2\ln|t| - \frac{15}{t^{0.2}} + C$ **31.** $2e^x + 5\ln|x| + x/4 + C$ **33.** $12.2x^{0.5} + x^{1.5}/9 - e^x + C$ **35.** $\frac{2^x}{\ln 2} - \frac{3^x}{\ln 3} + C$ **37.** $\frac{100(1.1^x)}{\ln(1.1)} + C$ **39.** $1/x - 1/x^2 + C$ **41.** $f(x) = x^2/2 + 1$ **43.** $f(x) = e^x - x - 1$ **45.** $\frac{3x|x| - x^2}{2} + C$ **47.** $C(x) = 5x - x^2/20{,}000 + 20{,}000$ **49.** $C(x) = 5x + x^2 + \ln x + 994$ **51. a.** $M(t) = 4t^3 - 10t^2 + 10t + 1$ **b.** 85 million **53.** $I(t) = 42{,}000 + 1200t$; \$48,000 **55.** $S(t) = 4t^3 + 250t^2 + 4700t$; 38,064 million gallons **57. a.** $H'(t) = 3.5t + 65$ billion dollars per year **b.** $H(t) = 1.75t^2 + 65t + 700$ billion dollars **59. a.** $-0.4t + 1$ percentage points per year **b.** $-0.2t^2 + t + 13$; 13.8 **61. a.** $s = t^3/3 + t + C$ **b.** $C = 1$; $s = t^3/3 + t + 1$ **63.** 320 ft/s downward **65. a.** $v(t) = -32t + 16$ **b.** $s(t) = -16t^2 + 16t + 185$; zenith at $t = 0.5$ sec $s = 189$ feet, 4 feet above the top of the tower. **67. a.** $s = 525t + 25t^2$ **b.** 3 hours **c.** The negative solution indicates that, at time $t = -24$, the tail wind would have been large and negative, causing the plane to be moving backward through that position 24 hours prior to departure and arrive at the starting point of the flight at time 0! **69.** (Proof) **71.** $(1{,}280)^{1/2} \approx 35.78$ ft/s **73. a.** 80 ft/s **b.** 60 ft/s **c.** 1.25 seconds **75.** $\sqrt{2} \approx 1.414$ times as fast **77.** The term *indefinite* refers to the arbitrary constant term in the indefinite integral; we do not obtain a definite value for C and hence the integral is "not definite." **79.** Constant; because the derivative of a linear function is constant, linear functions are antiderivatives of constant functions. **81.** No; there are infinitely many antiderivatives of a given function, each pair of them differing by a constant. Knowing the value of the function at a specific point suffices. **83.** They differ by a constant, $G(x) - F(x) =$ Constant **85.** Antiderivative, marginal **87.** $\int f(x)\,dx$ represents the total cost of manufacturing x items. The units of $\int f(x)\,dx$ are the product of the units of $f(x)$ and the units of x. **89.** $\int (f(x) + g(x))\,dx$ is, by definition, an antiderivative of $f(x) + g(x)$. Let $F(x)$ be an antiderivative of $f(x)$ and let $G(x)$ be an antiderivative of $g(x)$. Then, because the derivative of $F(x) + G(x)$ is $f(x) + g(x)$ (by the rule for sums of derivatives), this means that $F(x) + G(x)$ is an antiderivative of $f(x) + g(x)$. In symbols, $\int (f(x) + g(x))\,dx = F(x) + G(x) + C = \int f(x)\,dx + \int g(x)\,dx$, the sum of the indefinite integrals. **91.** Answers will vary. $\int x \cdot 1\,dx = \int x\,dx = x^2/2 + C$, whereas $\int x\,dx \cdot \int 1\,dx = (x^2/2 + D)\cdot(x + E)$, which is not the same as $x^2/2 + C$, no matter what values we choose for the constants C, D, and E. **93.** Derivative; indefinite integral; indefinite integral; derivative

Section 6.2

1. $(3x - 5)^4/12 + C$ **3.** $(3x - 5)^4/12 + C$ **5.** $-e^{-x} + C$ **7.** $-e^{-x} + C$ **9.** $\frac{1}{2}e^{(x+1)^2} + C$ **11.** $(3x + 1)^6/18 + C$ **13.** $(-2x + 2)^{-1}/2 + C$ **15.** $1.6(3x - 4)^{3/2} + C$ **17.** $2e^{(0.6x+2)} + C$ **19.** $(3x^2 + 3)^4/24 + C$ **21.** $(x^2 + 1)^{2.3}/4.6 + C$ **23.** $x + 3e^{3.1x-2} + C$ **25.** $2(3x^2 - 1)^{3/2}/9 + C$ **27.** $-(1/2)e^{-x^2+1} + C$ **29.** $-(1/2)e^{-(x^2+2x)} + C$ **31.** $(x^2 + x + 1)^{-2}/2 + C$ **33.** $(2x^3 + x^6 - 5)^{1/2}/3 + C$ **35.** $(x - 2)^7/7 + (x - 2)^6/3 + C$ **37.** $4[(x + 1)^{5/2}/5 - (x + 1)^{3/2}/3] + C$ **39.** $20 \ln|1 - e^{-0.05x}| + C$ **41.** $3e^{-1/x} + C$ **43.** $(e^x - e^{-x})/2 + C$ **45.** $\ln(e^x + e^{-x}) + C$ **47.** $(e^{2x^2-2x} + e^{x^2})/2 + C$ **53.** $e^{-x} + C$ **55.** $(1/2)e^{2x-1} + C$ **57.** $(2x + 4)^3/6 + C$ **59.** $(1/5)\ln|5x - 1| + C$ **61.** $(1.5x)^4/6 + C$ **63.** $\dfrac{1.5^{3x}}{3\ln(1.5)} + C$ **65.** $\frac{1}{4}(2x + 4)|2x + 4| + C$ **67.** $\dfrac{2^{3x+4} - 2^{-3x+4}}{3\ln 2} + C$ **69.** $f(x) = (x^2 + 1)^4/8 - 1/8$ **71.** $f(x) = (1/2)e^{x^2} - 1$ **73.** $210t + 1{,}240e^{-0.05t} - 1{,}240$ **75.** $825t + 600e^{-0.4t} - 600$; \$7,660 million **77.** $C(x) = 5x - 1/(x + 1) + 995.5$ **79. a.** $N(t) = 35\ln(5 + e^{0.2t}) - 63$ **b.** 80,000 articles **81.** $S(t) = 3{,}600[\ln(3 + e^{0.25t}) - \ln 4]$; 6,310 sets **83. a.** $s = (t^2 + 1)^5/10 + t^2/2 + C$ **b.** $C = 9/10$; $s = (t^2 + 1)^5/10 + t^2/2 + 9/10$ **85.** $S(t) = 4(t - 2{,}000)^3 + 250(t - 2{,}000)^2 + 4{,}700(t - 2{,}000) - 16{,}458$; 39,190 million gallons **87.** None; the substitution $u = x$ simply replaces the letter x throughout by the letter u, and thus does not change the integral at all. For instance, the integral $\int x(3x^2 + 1)\,dx$ becomes $\int u(3u^2 + 1)\,du$ if we substitute $u = x$. **89.** (D); to compute the integral, first break it up into a sum of two integrals: $\displaystyle\int \frac{x}{x^2 - 1}\,dx + \int \frac{3x}{x^2 + 1}\,dx$, and then compute the first using $u = x^2 - 1$ and the second using $u = x^2 + 1$. **91.** The purpose of substitution is to introduce a new variable that is defined in terms of the variable of integration. One cannot say $u = u^2 + 1$, since u is not a new variable. Instead, define $w = u^2 + 1$ (or any other letter different from u). **93.** The integral $\int x(x^2 + 1)\,dx$ can be solved by the substitution $u = x^2 + 1$, because it leads to $\displaystyle\frac{1}{2}\int u\,du = \frac{1}{4}u^2 + C = \frac{(x^2 + 1)^2}{4} + C.$

Section 6.3

1. 30 **3.** 22 **5.** −2 **7.** 0 **9.** 4 **11.** 6 **13.** 0.7456 **15.** 2.3129 **17.** 2.5048 **19.** 1 **21.** 1/2 **23.** 1/4 **25.** 2 **27.** 0 **29.** 6 **31.** 0 **33.** 0.5 **35.** 3.3045, 3.1604, 3.1436 **37.** 0.0275, 0.0258, 0.0256 **39.** \$99.95 **41.** 29 billion gallon **43.** 4,900 million items **45. a.** Left sum: about 46,000 articles, right sum: about 55,000 articles **b.** 50.5; a total of about 50,500 articles in *Physical Review* were written by researchers in Europe in the 16-year period beginning 1983. **47.** 118,750 degrees **49.** Left sum: 20.5; right sum: 21; left sum, as it is the sum of the quarterly net incomes for 2006. **51.** 54,000 students **53.** −84.8; after 4 seconds, the stone is about 84.8 ft below where it started. **55.** 91.2 ft **57.** 9,450; a total of about 9,450 million barrels of oil was produced by **Pemex** from the start of 2001 to the start of 2009. **59.** 18,400; a total of 18,400 wiretaps were authorized by U.S. state and federal courts during the 15-year period starting January 1990. **61.** Yes. The Riemann sum gives an estimated area of 420 square feet. **63. a.** Graph:

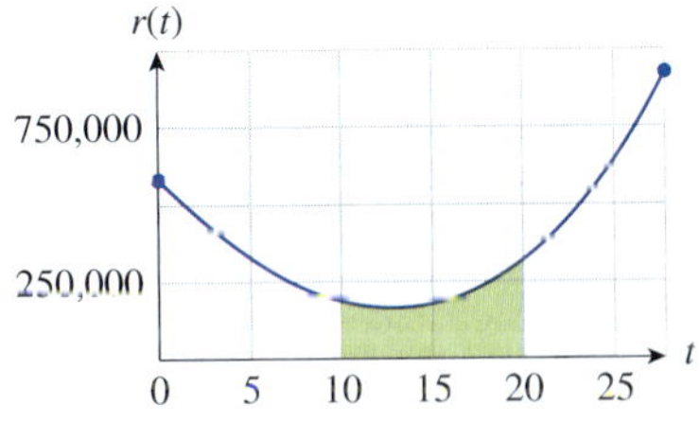

Total expenditure on oil in the United States from 1990 to 2000. **b.** 2,010,000; a total of \$2,010,000 million, or \$2.01 trillion, was spent on oil in the U.S. from 1990 to 2000. **65. a.** 99.4% **b.** 0 (to at least 15 decimal places) **67.** Stays the same **69.** Increases **71.** The area under the curve and above the x axis equals the area above the curve and below the x axis. **73.** Answers will vary. One example: Let $r(t)$ be the rate of

change of net income at time t. If $r(t)$ is negative, then the net income is decreasing, so the change in net income, represented by the definite integral of $r(t)$, is negative.
75. Answers may vary.

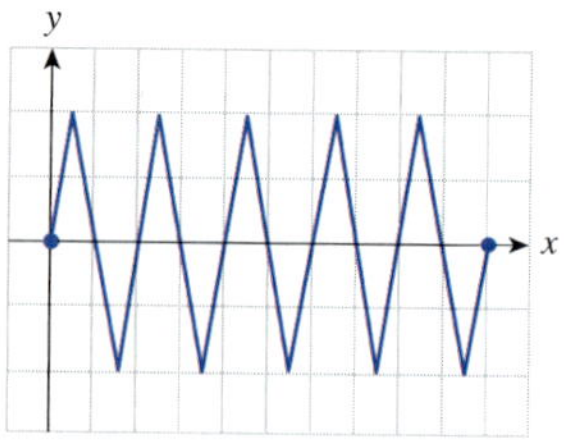

77. The total cost is $c(1) + c(2) + \cdots + c(60)$, which is represented by the Riemann sum approximation of $\int_1^{61} c(t)\,dt$ with $n = 60$. **79.** $[f(x_1) + f(x_2) + \cdots + f(x_n)]\Delta x = \sum_{k=1}^{n} f(x_k)\Delta x$

Section 6.4

1. $14/3$ **3.** 5 **5.** 0 **7.** $40/3$ **9.** -0.9045 **11.** $2(e-1)$ **13.** $2/3$ **15.** $1/\ln 2$ **17.** $4^6 - 1 = 4{,}095$ **19.** $(e^1 - e^{-3})/2$ **21.** $3/(2\ln 2)$ **23.** $50(e^{-1} - e^{-2})$ **25.** $40/3$ **27.** $e^{2.1} - e^{-0.1}$ **29.** 0 **31.** $(5/2)(e^3 - e^2)$ **33.** $(1/3)[\ln 26 - \ln 7]$ **35.** $\dfrac{0.1}{2.2\ln(1.1)}$ **37.** $e - e^{1/2}$ **39.** $2 - \ln 3$ **41.** $-4/21$ **43.** $3^{5/2}/10 - 3^{3/2}/6 + 1/15$ **45.** $1/2$ **47.** $16/3$ **49.** $9/2$ **51.** $56/3$ **53.** $1/2$ **55.** \$783 **57.** 296 miles **59.** 260 ft **61.** 30 billion gallons **63.** 9,451 million barrels **65.** 18,470; a total of 18,470 wiretaps were authorized by U.S. state and federal courts during the 15-year period starting January 1990. **67.** $210t + 1{,}240e^{-0.05t} - 1{,}240$ **69.** 9 gallons **71.** 907 T-shirts **73.** 68 milliliters **75.** 4,200; a total of around 4,200 million items were listed on **eBay** during 2005 and 2006. **77.** Change in cost $= \int_0^x m(t)\,dt = C(x) - C(0)$ by the FTC, so $C(x) = C(0) + \int_0^x m(t)\,dt$. $C(0)$ is the *fixed cost*. **79.** 80,000 articles **81. a., b.** (Proofs) **c.** 2,401 thousand students **85.** They are related by the Fundamental Theorem of Calculus, which (briefly) states that the definite integral of a suitable function can be calculated by evaluating the indefinite integral at the two end-points and subtracting. **87.** Computing its definite integral from a to b. **89.** Calculate definite integrals by using antiderivatives. **91.** An example is $v(t) = t - 5$. **93.** An example is $f(x) = e^{-x}$. **95.** By the FTC, $\int_a^x f(t)\,dt = G(x) - G(a)$ where G is an antiderivative of f. Hence, $F(x) = G(x) - G(a)$. Taking derivatives of both sides, $F'(x) = G'(x) + 0 = f(x)$, as required. The result gives us a formula, in terms of area, for an antiderivative of any continuous function.

Chapter 6 Review

1. $x^3/3 - 5x^2 + 2x + C$ **3.** $4x^3/15 + 4/(5x) + C$ **5.** $-e^{-2x+11}/2 + C$ **7.** $\dfrac{1}{22}(x^2+4)^{11} + C$ **9.** $-\dfrac{5}{2}e^{-2x} + C$ **11.** $(x+2) - \ln|x+2| + C$ or $x - \ln|x+2| + C$ **13.** 1 **15.** 2.75 **17.** -0.24 **19.** $0.7778, 0.7500, 0.7471$ **21.** 0 **23.** $1/4$ **25.** $2 + e - e^{-1}$ **27.** $52/9$ **29.** $[\ln 5 - \ln 2]/8 = \ln(2.5)/8$ **31.** $32/3$ **33.** $(1 - e^{-25})/2$ **35. a.** $N(t) = 196t + t^2/2 - 0.16t^6/6$ **b.** 4 books **37. a.** At a height of $-16t^2 + 100t$ ft **b.** 156.25 ft **c.** After 6.25 seconds **39.** 5,000 copies. **41.** 0 books **43.** 39,200 hits **45.** \$8,200 **47.** About 86,000 books **49.** About 35,800 books

Chapter 7

Section 7.1

1. $2e^x(x-1) + C$ **3.** $-e^{-x}(2+3x) + C$ **5.** $e^{2x}(2x^2 - 2x - 1)/4 + C$ **7.** $-e^{-2x+4}(2x^2 + 2x + 3)/4 + C$ **9.** $2^x[(2-x)/\ln 2 + 1/(\ln 2)^2] + C$ **11.** $-3^{-x}[(x^2-1)/\ln 3 + 2x/(\ln 3)^2 + 2/(\ln 3)^3] + C$ **13.** $-e^{-x}(x^2 + x + 1) + C$ **15.** $\dfrac{1}{7}x(x+2)^7 - \dfrac{1}{56}(x+2)^8 + C$ **17.** $-\dfrac{x}{2(x-2)^2} - \dfrac{1}{2(x-2)} + C$ **19.** $(x^4 \ln x)/4 - x^4/16 + C$ **21.** $(t^3/3 + t)\ln(2t) - t^3/9 - t + C$ **23.** $(3/4)t^{4/3}(\ln t - 3/4) + C$ **25.** $x\log_3 x - x/\ln 3 + C$ **27.** $e^{2x}(x/2 - 1/4) - 4e^{3x}/3 + C$ **29.** $e^x(x^2 - 2x + 2) - e^{x^2}/2 + C$ **31.** e **33.** $38{,}229/286$ **35.** $\dfrac{1}{3}(3x-4)(2x-1)^{3/2} - \dfrac{1}{5}(2x-1)^{5/2} + C$ **37.** $(7/2)\ln 2 - 3/4$ **39.** $1/4$ **41.** $1 - 11e^{-10}$ **43.** $4\ln 2 - 7/4$ **45.** $\dfrac{1}{2}x(x-3)|x-3| - \dfrac{1}{6}(x-3)^2|x-3| + C$ **47.** $2x|x-3| - (x-3)|x-3| + C$ **49.** $-x^2(-x+4)|-x+4| - \dfrac{2}{3}x(-x+4)^2|-x+4| - \dfrac{1}{6}(-x+4)^3|-x+4| + C$ **51.** $\dfrac{1}{2}(x^2 - 2x + 3)(x-4)|x-4| - \dfrac{1}{3}(x-1)(x-4)^2|x-4| + \dfrac{1}{12}(x-4)^3|x-4| + C$ **53.** $28{,}800{,}000(1 - 2e^{-1})$ ft **55.** $5{,}001 + 10x - 1/(x+1) - [\ln(x+1)]/(x+1)$ **57.** \$4,252 billion **59.** 11,919 million gallons **61.** 19 billion square feet **63.** $e^{0.1x}(250[-8x^2 + 70x + 1{,}000] - 2{,}500[-16x + 70] - 400{,}000) + 325{,}000$ **65.** 33,598 **67. a.** $r(t) = -0.075t + 2.75 + [0.025t - 0.25]\dfrac{|t-10|}{t-10}$ **b.** 42.5 million people **69.** Answers will vary. Examples are xe^{x^2} and $e^{x^2} = 1 \cdot e^{x^2}$. **71.** Answers will vary. Examples are Exercises 35 and 36, or more simply, integrals like $\int x(x+1)^5\,dx$. **73.** Substitution **75.** Parts **77.** Substitution **79.** Parts **81.** $n + 1$ times

Section 7.2

1. $16/3$ **3.** 9.75 **5.** 2 **7.** $31/3$ **9.** $8/3$ **11.** 4 **13.** $1/3$ **15.** 1 **17.** 2 **19.** 39 **21.** $e - 3/2$ **23.** $2/3$ **25.** $3/10$

27. 1/20 **29.** 4/15 **31.** 1/3 **33.** 32 **35.** $2\ln 2 - 1$
37. $8\ln 4 + 2e - 16$ **39.** 0.9138 **41.** 0.3222 **43.** 112.5. This represents your total profit for the week, \$112.50. **45.** 608; there were approximately 608,000 housing starts from the start of 2002 to the start of 2006 not for sale purposes.
47. a. Graph: (The upper curve is **Myspace**.)

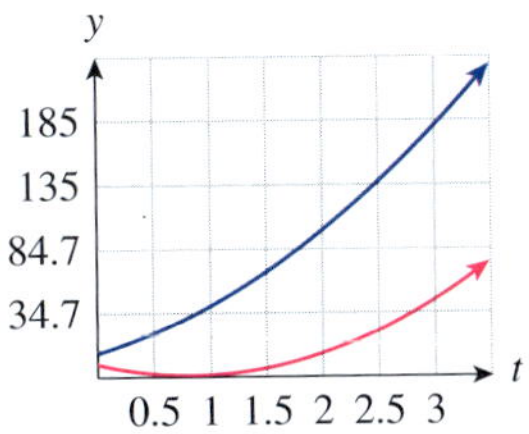

Myspace; about 93 million members **b.** The area between the curves $y = f(t)$ and $y = m(t)$ over [0.5, 2]
49. $16{,}078(e^{0.051t} - 1) - 7{,}333.3(e^{0.06t} - 1)$; the total number of wiretaps authorized by federal courts from the start of 1990 up to time t was about $16{,}078(e^{0.051t} - 1) - 7{,}333.3(e^{0.06t} - 1)$.
51. Wrong: It could mean that the graphs of f and g cross, as shown in the caution at the start of this topic in the textbook.
53. The area between the export and import curves represents the United States's accumulated trade deficit (that is, the total excess of imports over exports) from 1960 to 2007. **55.** (A)
57. The claim is wrong because the area under a curve can only represent income if the curve is a graph of income *per unit time*. The value of a stock price is not income per unit time—the income can be realized only when the stock is sold, and it amounts to the current market price. The total net income (per share) from the given investment would be the stock price on the date of sale minus the purchase price of \$50.

Section 7.3

1. Average = 2

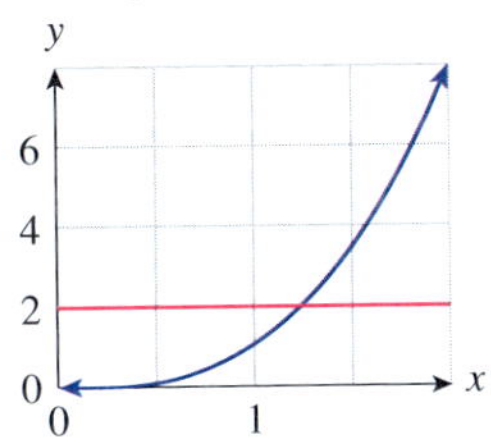

3. Average = 1

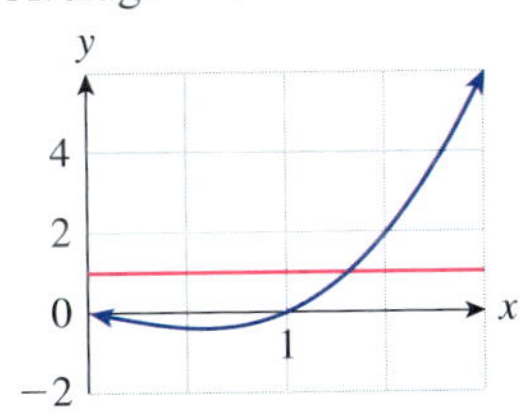

5. Average = $(1 - e^{-2})/2$

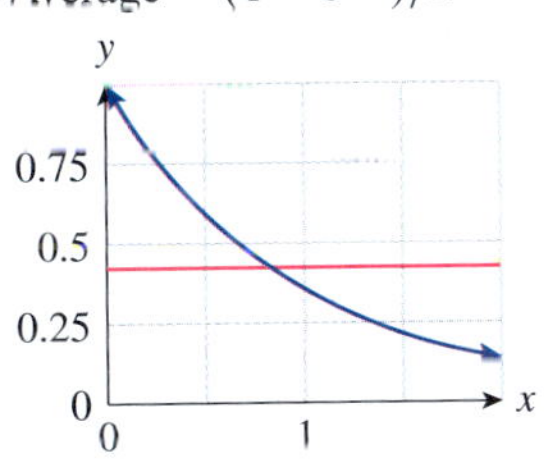

7. Average = 17/8

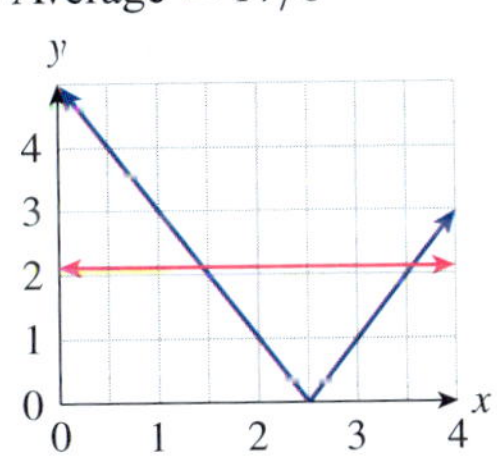

9.

x	0	1	2	3	4	5	6	7
$r(x)$	3	5	10	3	2	5	6	7
$\bar{r}(x)$			6	6	5	10/3	13/3	6

11.

x	0	1	2	3	4	5	6	7
$r(x)$	1	2	6	7	11	15	10	2
$\bar{r}(x)$			3	5	8	11	12	9

13. Moving average:

$$\bar{f}(x) = x^3 - (15/2)x^2 + 25x - 125/4$$

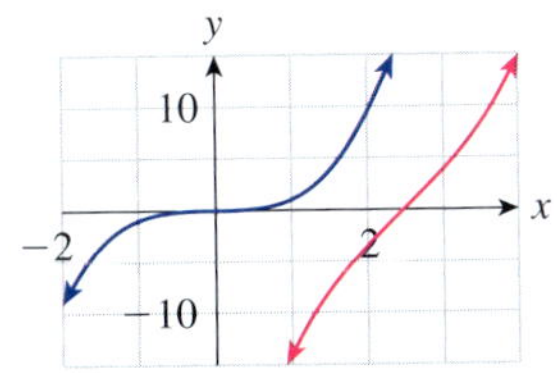

15. Moving average:

$$\bar{f}(x) = (3/25)[x^{5/3} - (x - 5)^{5/3}]$$

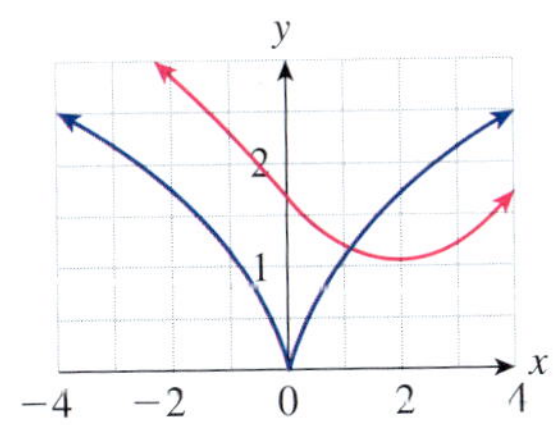

17. $\bar{f}(x) = \dfrac{2}{5}(e^{0.5x} - e^{0.5(x-5)})$

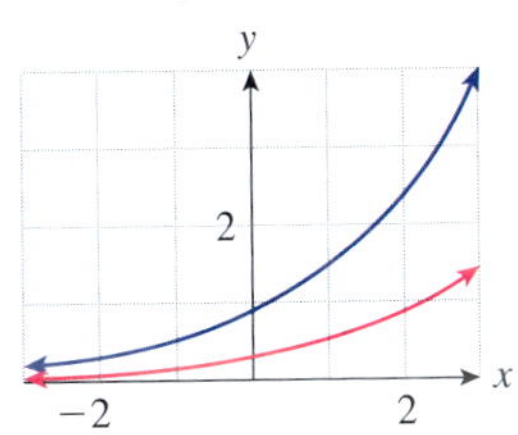

19. $\bar{f}(x) = \dfrac{2}{15}(x^{3/2} - (x - 5)^{3/2})$

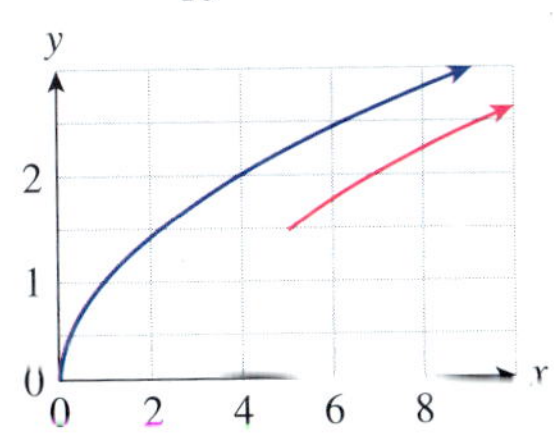

21. $\bar{f}(x) = \dfrac{1}{5}\left[5 - \dfrac{1}{2}|2x - 1| + \dfrac{1}{2}|2x - 11|\right]$

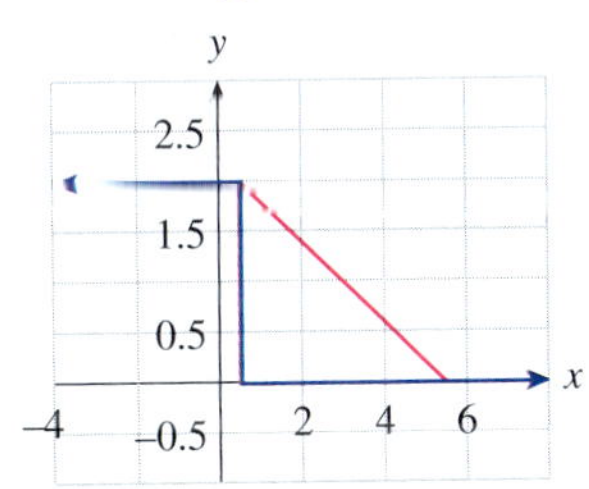

23. $\bar{f}(x) = \frac{1}{5}\left[2x - \frac{1}{2}(x+1)|x+1| + \frac{1}{2}x|x| - 2(x-5) + \frac{1}{2}(x-4)|x-4| - \frac{1}{2}(x-5)|x-5|\right]$

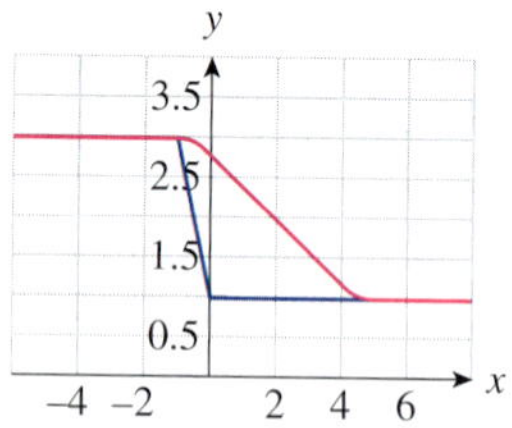

25.

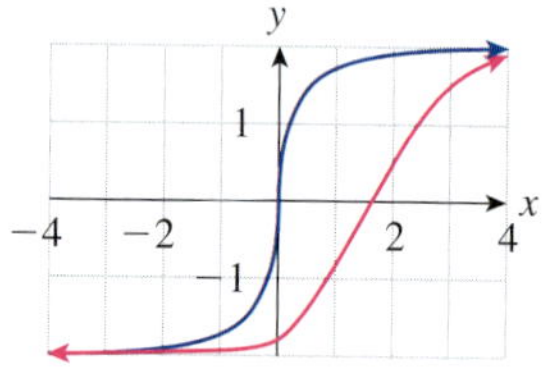

27.

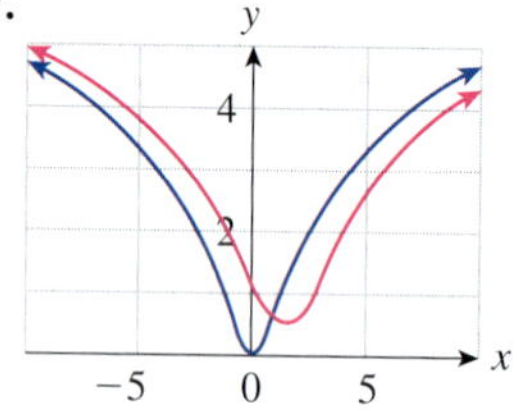

29.

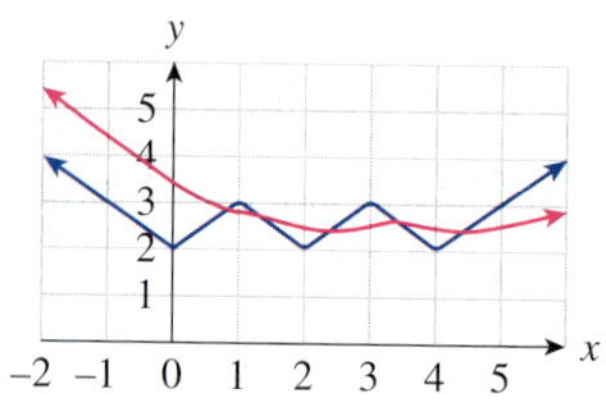

31.

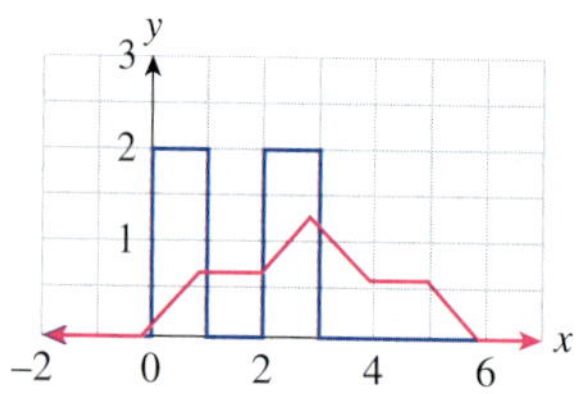

33.

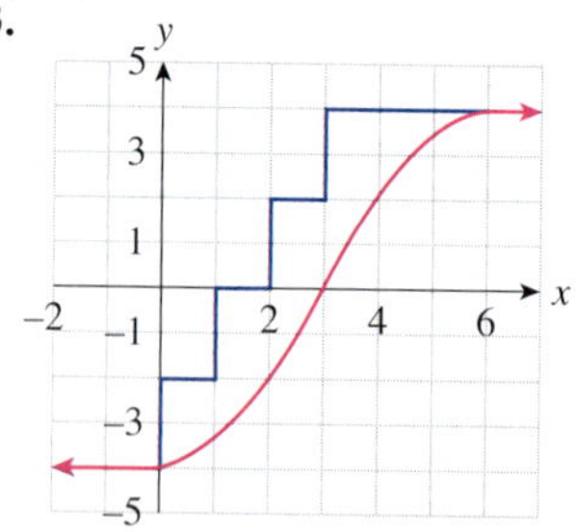

35. $6,718.8 million **37.** $1.955 million **39.** 16 million members per year **41.** 277 tons **43.** $10,410.88 **45.** $1,500 **47.** 537 million items per quarter

49.

Year t	1999	2000	2001	2002	2003	2004	2005	2006	2007	2008
Stock Price	36	44	41	36	38	49	56	71	92	62
Moving Average (rounded)				39	40	41	45	54	67	70

The moving average continued up at a lower rate.

51. a. To obtain the moving averages from January to June, use the fact that the data repeats every 12 months. Graph:

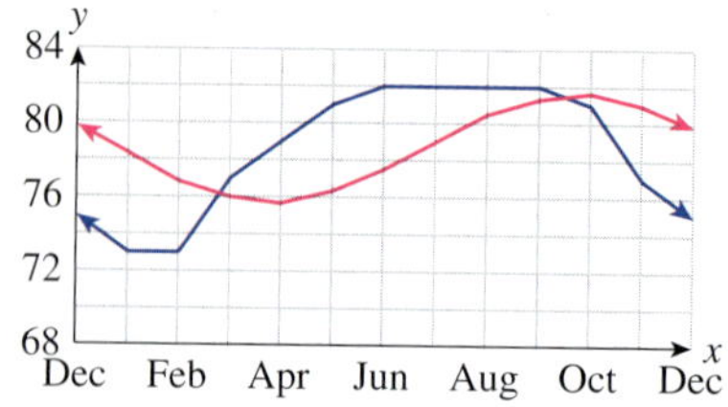

b. The 12-month moving average is constant and equal to the year-long average of approximately 79°.

53. a. Graph:

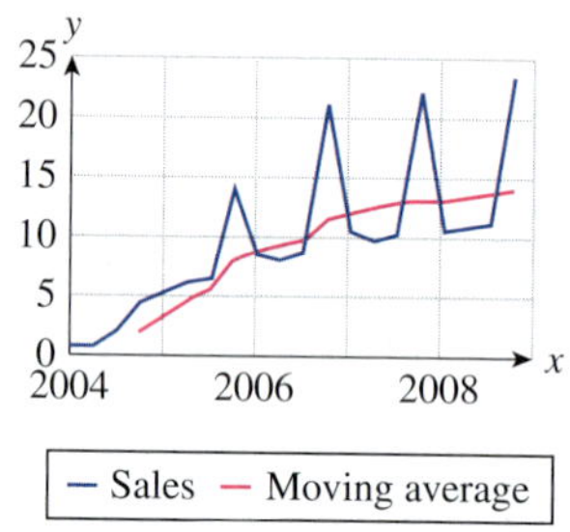

b. Approximately 0.3 million iPods per quarter **55. a.** 7,000 million gallons per year **b.** $\frac{1}{2}[4[t^3 - (t-2)^3] + 250[t^2 - (t-2)^2] + 9{,}400]$ **c.** Quadratic **57. a.** $s = 14.4t + 240$ **b.** $\bar{s}(t) = 14.4t + 211.2$ **c.** The slope of the moving average is the same as the slope of the original function. **59.** $\bar{f}(x) = mx + b - \frac{ma}{2}$ **61.** The moving average "blurs" the effects of short-term oscillations in the price, and shows the longer-term trend of the stock price. **63.** They repeat every six months. **65.** The area above the x-axis equals the area below the x-axis. Example: $y = x$ on $[-1,1]$ **67.** This need not be the case; for instance, the function $f(x) = x^2$ on [0, 1] has average value 1/3, whereas the value midway between the maximum and minimum is 1/2. **69.** (C) A shorter term moving average most closely approximates the original function because it averages the function over a shorter period, and continuous functions change by only a small amount over a small period.

Section 7.4

1. $6.25 **3.** $512 **5.** $119.53 **7.** $900 **9.** $416.67 **11.** $326.27 **13.** $25 **15.** $0.50 **17.** $386.29 **19.** $225 **21.** $25.50 **23.** $12,684.63 **25.** $TV = \$300{,}000$, $FV = \$434{,}465.45$ **27.** $TV = \$350{,}000$, $FV = \$498{,}496.61$ **29.** $TV = \$389{,}232.76$, $FV = \$547{,}547.16$ **31.** $TV = \$100{,}000$, $PV = \$82{,}419.99$ **33.** $TV = \$112{,}500$, $PV = \$92{,}037.48$ **35.** $TV = \$107{,}889.50$, $PV = \$88{,}479.69$ **37.** $\bar{p} = \$5{,}000$, $\bar{q} = 10{,}000$, CS = $25 million, PS = $100 million. The total social gain is $125 million. **39.** $CS = \frac{1}{2m}(b - m\bar{p})^2$ **41.** €220 billion **43.** $1,380 billion

45. €250 billion **47.** \$1,210 billion **49.** \$1,943,162.44 **51.** \$3,086,245.73 **53.** \$58,961.74 **55.** \$1,792,723.35 **57.** Total **59.** She is correct, provided there is a positive rate of return, in which case the future value (which includes interest) is greater than the total value (which does not). **61.** $PV < TV < FV$

Section 7.5

1. Diverges **3.** Converges to $2e$ **5.** Converges to e^2 **7.** Converges to ½ **9.** Converges to $1/108$ **11.** Converges to $3 \times 5^{2/3}$ **13.** Diverges **15.** Diverges **17.** Converges to $\frac{5}{4}(3^{4/5} - 1)$ **19.** Diverges **21.** Converges to 0 **23.** Diverges **25.** Diverges **27.** 0.9, 0.99, 0.999, . . . ; converges to 1. **29.** 7.602, 95.38, 993.1, . . . ; diverges. **31.** 1.368, 1.800, 1.937, 1.980, 1.994, 1.998, 1.999, 2.000, . . . ; converges to 2. **33.** 9.000, 99.00, 999.0, . . . ; diverges to $+\infty$. **35.** 2.79 million homes **37.** 13,600 billion cigarettes **39.** No; you will not sell more than 2,000 of them. **41.** The integral diverges, and so the number of graduates each year will rise without bound. **43. a.** $R(t) = 350e^{-0.1t}(39t + 68)$ million dollars/yr; **b.** \$1,603,000 million **45.** \$20,700 billion **47.** $\int_0^{+\infty} N(t)\,dt$ diverges, indicating that there is no bound to the expected future total online sales of mousse. $\int_{-\infty}^{0} N(t)\,dt$ converges to approximately 2.006, indicating that total online sales of mousse prior to the current year amounted to approximately 2 million gallons. **49.** 1 **51.** 0.1587 **53.** \$70,833 **55. a.** 2.468 meteors on average **b.** The integral diverges. We can interpret this as saying that the number of impacts by meteors smaller than 1 megaton is very large. (This makes sense because, for example, this number includes meteors no larger than a grain of dust.) **57. a.** $\Gamma(1) = 1$; $\Gamma(2) = 1$ **59.** The integral does not converge, so the number given by the FTC is meaningless. **61.** Yes; the integrals converge to 0, and the FTC also gives 0. **63. a.** Not improper. $|x|/x$ is not defined at zero, but $\lim_{x\to 0^-} |x|/x = -1$ and $\lim_{x\to 0^+} |x|/x = 1$. Because these limits are finite, the integral is not improper. **b.** Improper, because $x^{-1/3}$ has infinite left and right limits at 0. **c.** Improper, since $(x - 2)/(x^2 - 4x + 4) = 1/(x - 2)$, which has an infinite left limit at 2. **65.** In all cases, you need to rewrite the improper integral as a limit and use technology to evaluate the integral of which you are taking the limit. Evaluate for several values of the endpoint approaching the limit. In the case of an integral in which one of the limits of integration is infinite, you may have to instruct the calculator or computer to use more subdivisions as you approach $+\infty$.

Section 7.6

1. $y = \frac{x^3}{3} + \frac{2x^{3/2}}{3} + C$ **3.** $\frac{y^2}{2} = \frac{x^2}{2} + C$ **5.** $y = Ae^{x^2/2}$ **7.** $y = -\frac{2}{(x+1)^2 + C}$ **9.** $y = \pm\sqrt{(\ln x)^2 + C}$ **11.** $y = \frac{x^4}{4} - x^2 + 1$ **13.** $y = (x^3 + 8)^{1/3}$ **15.** $y = 2x$ **17.** $y = e^{x^2/2} - 1$ **19.** $y = -\frac{2}{\ln(x^2+1) + 2}$ **21.** With $s(t)$ = monthly sales after t months, $\frac{ds}{dt} = -0.05s$; $s = 1{,}000$ when $t = 0$. Solution: $s = 1{,}000e^{-0.05t}$ quarts per month. **23. a.** $75 + 125e^{-0.05t}$ **b.** 64.4 minutes **25.** $k \approx 0.04274$; $H(t) = 75 + 115e^{-0.04274t}$ degrees Fahrenheit after t minutes. **27.** With $S(t)$ = total sales after t months, $\frac{ds}{dt} = 0.1(100{,}000 - S)$; $S(0) = 0$. Solution: $S = 100{,}000(1 - e^{-0.1t})$ monitors after t months. **29.** $q = 0.6078e^{-0.05p}p^{1.5}$ **31.** $y = e^{-t}(t + 1)$ **33.** $y = e^{t/2}\left[-2te^{-t/2} - 4e^{-t/2} + 5\right]$ **35.** $i = 5e^{-t}(e^t - e)\left[1 + \frac{|t-1|}{t-1}\right]$

Graph:

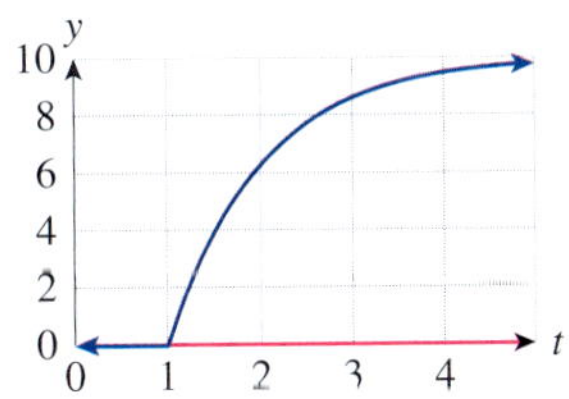

37. a. $\frac{dp}{dt} = k(D(p) - S(p)) = k(20{,}000 - 1{,}000p)$ **b.** $p = 20 - Ae^{-kt}$ **c.** $p = 20 - 10e^{-0.2231t}$ dollars after t months. **41.** $S = \frac{2/1{,}999}{e^{-0.5t} + 1/1{,}999}$

Graph:

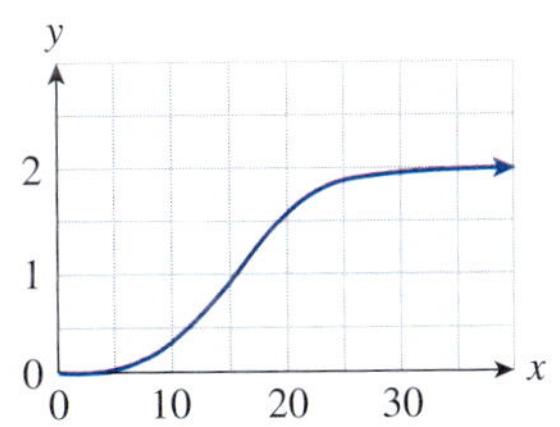

It will take about 27 months to saturate the market.

43. a. $y = be^{Ae^{-at}}$, A = constant **b.** $y = 10e^{-0.69315e^{-t}}$

Graph:

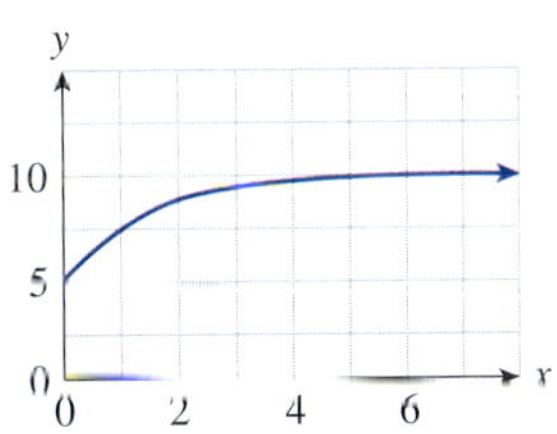

45. A general solution gives all possible solutions to the equation, using at least one arbitrary constant. A particular solution is one specific function that satisfies the equation. We obtain a particular solution by substituting specific values for any arbitrary constants in the general solution. **47.** Example: $y'' = x$ has general solution $y = \frac{1}{6}x^3 + Cx + D$ (integrate twice). **49.** $y' = -4e^{-x} + 3$

Chapter 7 Review

1. $(x^2 - 2x + 4)e^x + C$ **3.** $(1/3)x^3 \ln 2x - x^3/9 + C$ **5.** $\frac{1}{2}x(2x+1)|2x+1| - \frac{1}{12}(2x+1)^2|2x+1| + C$ **7.** $-5x|-x+3| - \frac{5}{2}(-x+3)|-x+3| + C$ **9.** $-e^2 - 39/e^2$ **11.** $\frac{3}{2 \cdot 2^{1/3}} - \frac{1}{2}$ **13.** $\frac{2\sqrt{2}}{3}$ **15.** -1 **17.** $e - 2$ **19.** $3x - 2$ **21.** $\frac{3}{14}[x^{7/3} - (x-2)^{7/3}]$ **23.** \$1,600 **25.** \$2,500 **27.** 1/4 **29.** Diverges **31.** 1 **33.** $y = -\frac{3}{x^3 + C}$ **35.** $y = \sqrt{2 \ln |x| + 1}$ **37.** \$18,200 **39. a.** \$2,260 **b.** $50{,}000e^{0.01t}(1 - e^{-0.04}) \approx 1{,}960.53e^{0.01t}$ **41. a.** $\bar{p} = 20$, $\bar{q} = 40{,}000$ **b.** $CS = \$240{,}000$, $PS = \$30{,}000$ **43.** Approximately \$910,000 **45. a.** \$5,549,000 **b.** Principal: \$5,280,000, interest: \$269,000 **47.** \$51 million **49.** The amount in the account would be given by $y = 10{,}000/(1 - t)$ where t is time in years, so would approach infinity 1 year after the deposit.

Chapter 8

Section 8.1

1. a. 1 **b.** 1 **c.** 2 **d.** $a^2 - a + 5$ **e.** $y^2 + x^2 - y + 1$ **f.** $(x+h)^2 + (y+k)^2 - (x+h) + 1$ **3. a.** 0 **b.** 0.2 **c.** -0.1 **d.** $0.18a + 0.2$ **e.** $0.1x + 0.2y - 0.01xy$ **f.** $0.2(x+h) + 0.1(y+k) - 0.01(x+h)(y+k)$ **5. a.** 1 **b.** e **c.** e **d.** e^{x+y+z} **e.** $e^{x+h+y+k+z+l}$ **7. a.** Does not exist **b.** 0 **c.** 0 **d.** $xyz/(x^2 + y^2 + z^2)$ **e.** $(x+h)(y+k)(z+l)/[(x+h)^2 + (y+k)^2 + (z+l)^2]$ **9. a.** Increases; 2.3 **b.** Decreases; 1.4 **c.** Decreases; 1 unit increase in z **11.** Neither **13.** Linear **15.** Linear **17.** Interaction **19. a.** 107 **b.** -14 **c.** -113

21.

$y \downarrow$ \ $x \rightarrow$	10	20	30	40
10	52	107	162	217
20	94	194	294	394
30	136	281	426	571
40	178	368	558	748

25. 18, 4, 0.0965, 47,040 **27.** 6.9078, 1.5193, 5.4366, 0 **29.** Let z = annual sales of Z (in millions of dollars), x = annual sales of X, and y = annual sales of Y. The model is $z = -2.1x + 0.4y + 16.2$.

31.

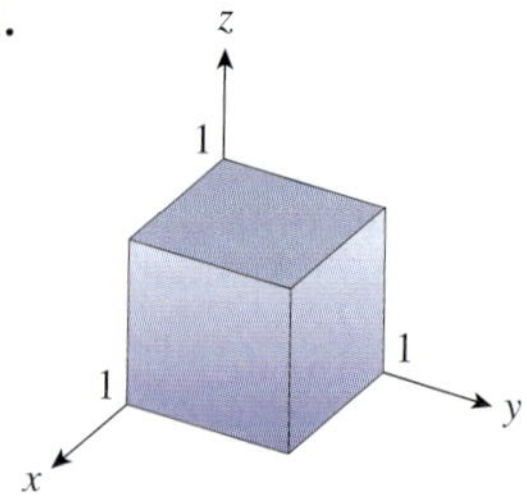

33.

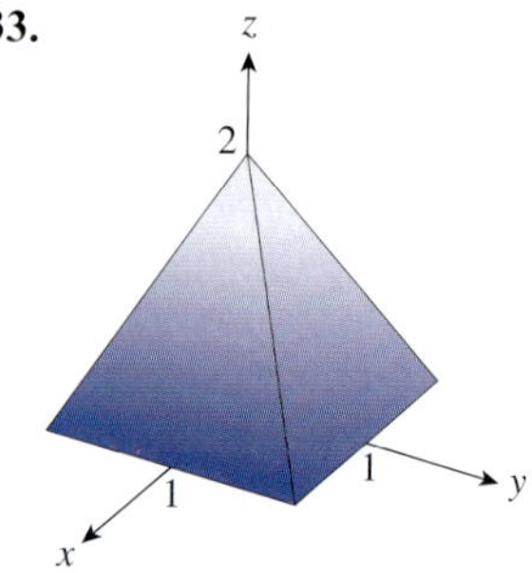

35.

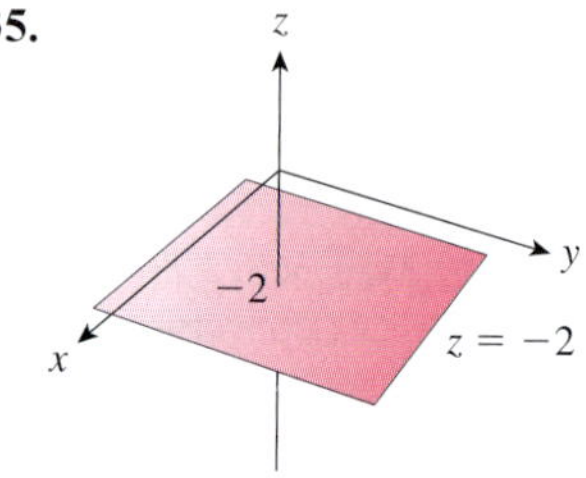

37.

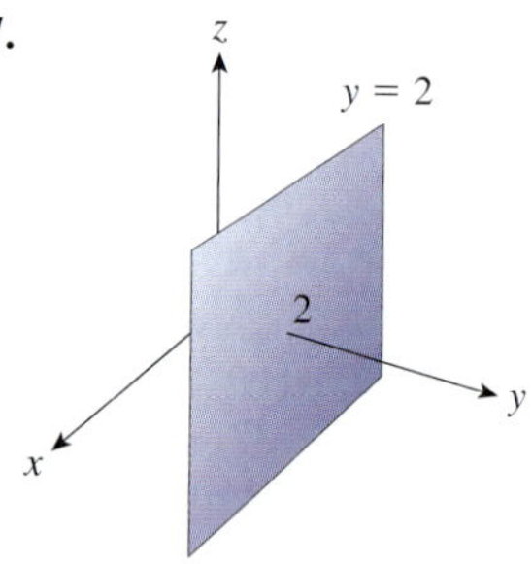

39.

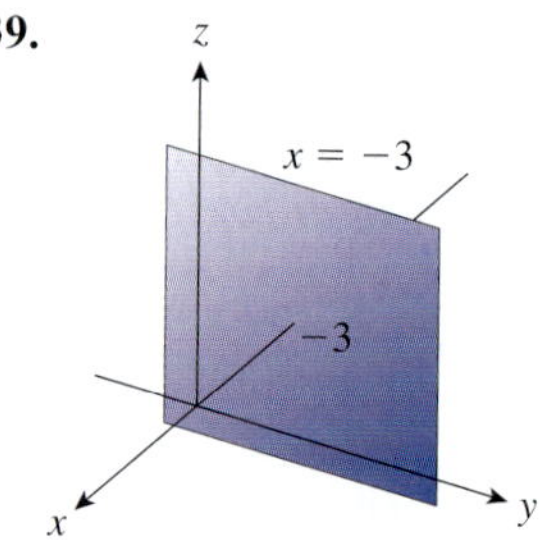

41. (H) **43.** (B) **45.** (F) **47.** (C)

49.

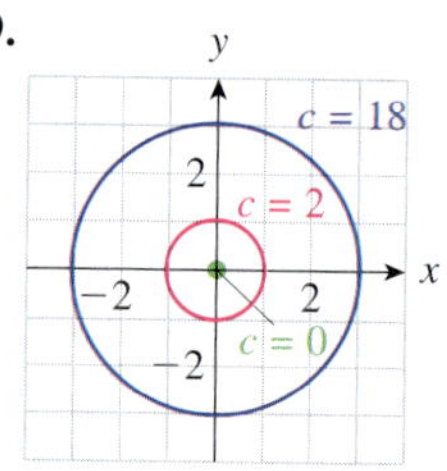

51.

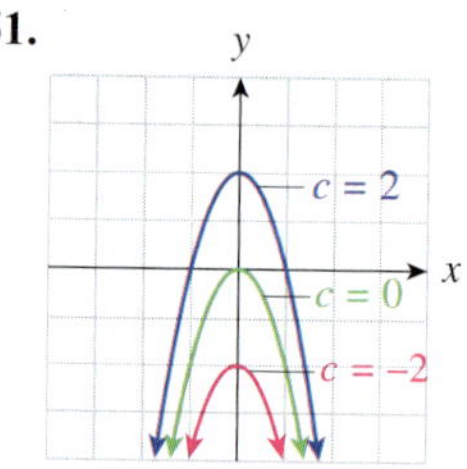

53.

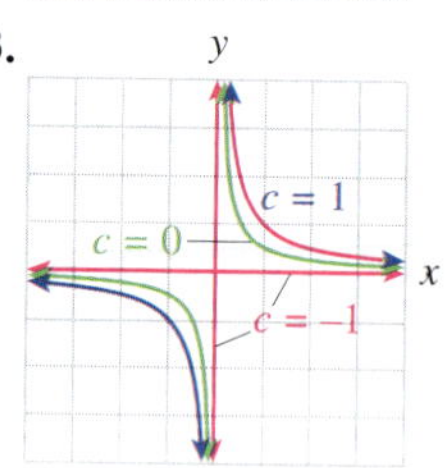

55.

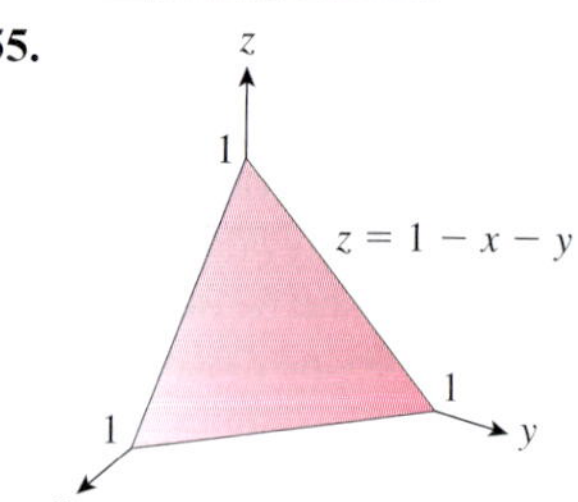

57.

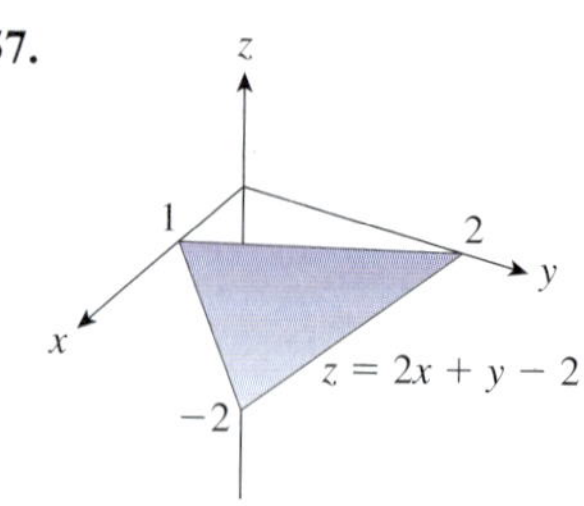

59.

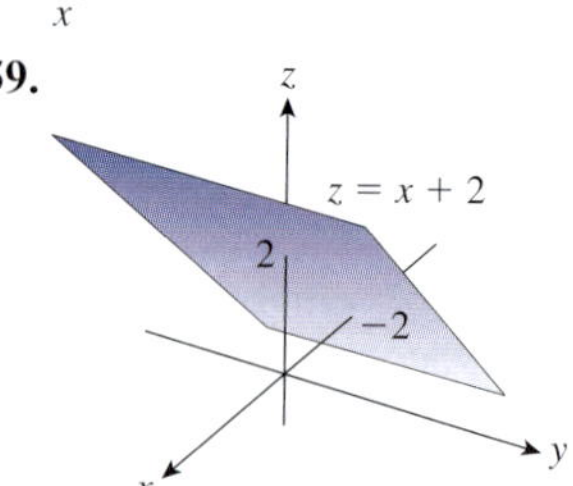

61.

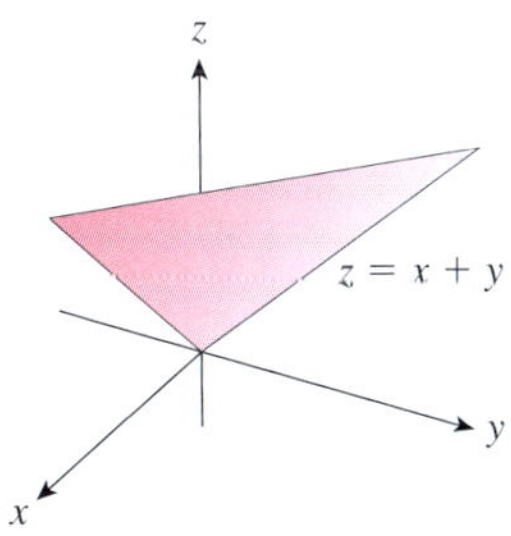

63.

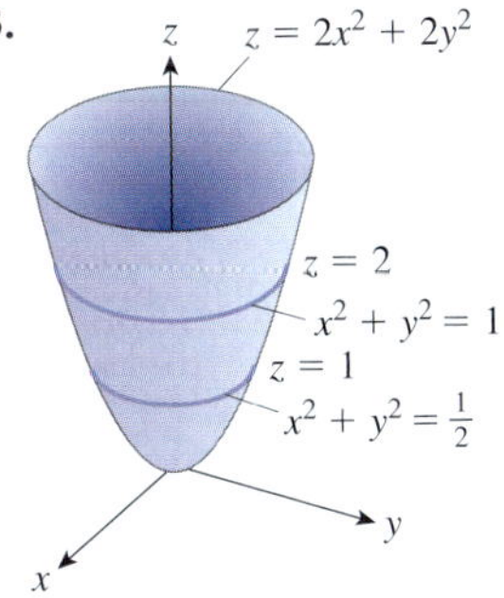

65.

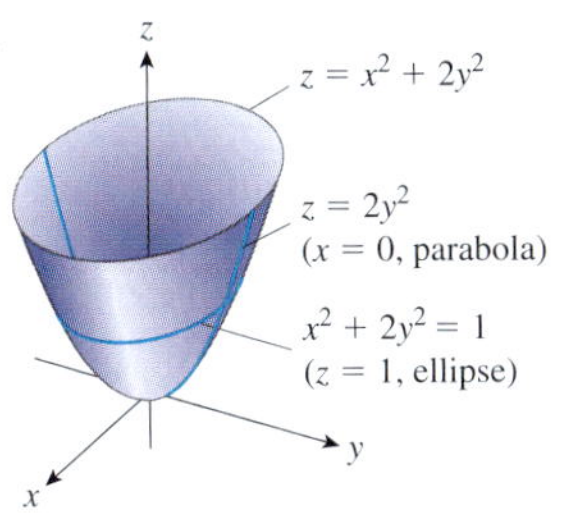

67.

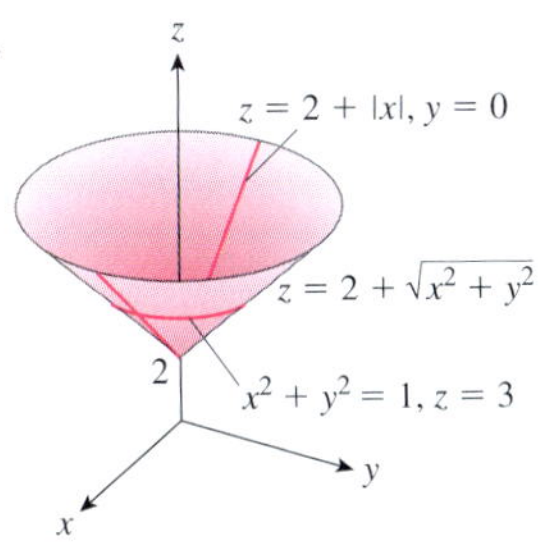

69.

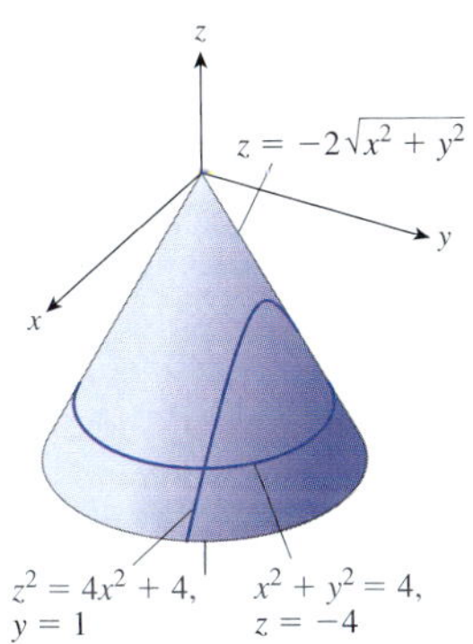

71.

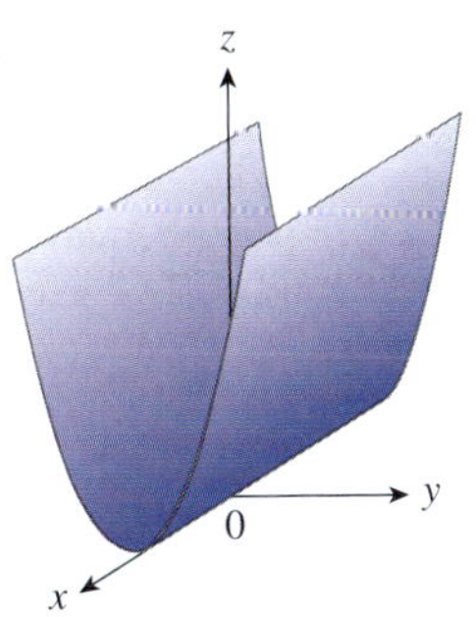

73.

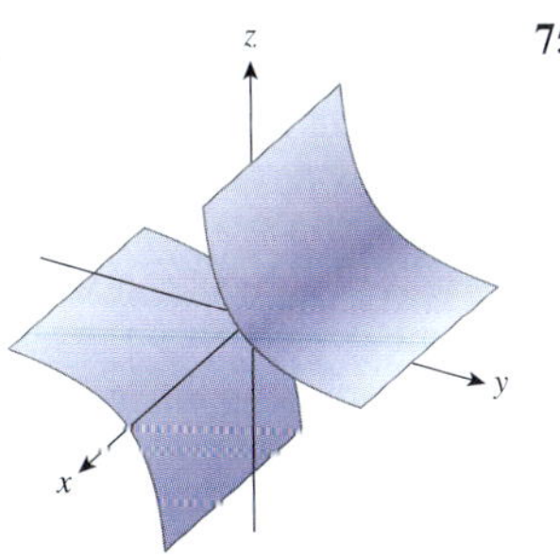

75.

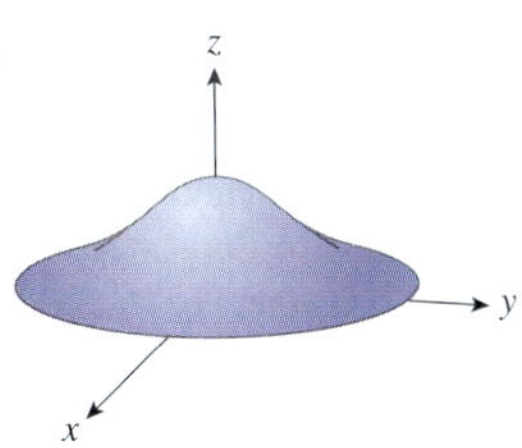

77. a. The marginal cost of cars is \$6,000 per car. The marginal cost of trucks is \$4,000 per truck. **b.** The graph is a plane with x-intercept -40, y-intercept -60, and z-intercept 240,000 **c.** The slice $x = 10$ is the straight line with equation $z = 300{,}000 + 4{,}000y$. It describes the cost function for the manufacture of trucks if car production is held fixed at 10 cars per week. **d.** The level curve $z = 480{,}000$ is the straight line $6{,}000x + 4{,}000y = 240{,}000$. It describes the number of cars and trucks you can manufacture to maintain weekly costs at \$480,000. **79.** $C(x, y) = 10 + 0.03x + 0.04y$ where C is the cost in dollars, x = # video clips sold per month, y = # audio clips sold per month **81. a.** 28% **b.** 21% **c.** Percentage points per year **83.** The graph is a plane with x_1-intercept 0.3, x_2-intercept 33, and x_3-intercept 0.66. The slices by x_1 = constant are straight lines that are parallel to each other. Thus, the rate of change of **General Motors**' share as a function of **Ford**'s share does not depend on **Chrysler**'s share. Specifically, GM's share decreases by 0.02 percentage points per one percentage-point increase in **Ford**'s market share, regardless of **Chrysler**'s share. **85. a.** 189 thousand prisoners **b.** $y = 300$: $N = -0.02x + 51$; $y = 500$: $N = 0.02x + 67$; when there are 300,000 prisoners in local jails, the number in federal prisons decreases by 20 per 1,000 additional prisoners in state prisons. When there are 500,000 prisoners in local jails, the number in federal prisons increases by 20 per 1,000 additional prisoners in state prisons. **87. a.** The slices x = constant and y = constant are straight lines. **b.** No. Even though the slices x = constant and y = constant are straight lines, the level curves are not, and so the surface is not a plane. **c.** The slice $x = 10$ has a slope of 3,800. The slice $x = 20$ has a slope of 3,600. Manufacturing more cars lowers the marginal cost of manufacturing trucks. **89. a.** \$9,980 **b.** $R(z) = 9{,}850 + 0.04z$ **91.** $f(x, t) = 6x - 2$; \$313 billion **93.** $U(11, 10) - U(10, 10) \approx 5.75$. This means that, if your company now has 10 copies of Macro Publish and 10 copies of Turbo Publish, then the purchase of one additional copy of Macro Publish will result in a productivity increase of approximately 5.75 pages per day. **95.** Answers will vary. $(a, b, c) = (3, 1/4, 1/\pi)$; $(a, b, c) = (1/\pi, 3, 1/4)$. **97.** 7,000,000 **99. a.** $100 = K(1{,}000)^a(1{,}000{,}000)^{1-a}$; $10 = K(1{,}000)^a(10{,}000)^{1-a}$ **b.** $\log K - 3a = -4$; $\log K - a = -3$ **c.** $P = 71$ pianos (to the nearest piano) **101. a.** 4×10^{-3} gram per square meter **b.** The total weight of sulfates in the earth's atmosphere **103. a.** The value of N would be doubled. **b.** $N(R, f_p, n_e, f_l, f_i, L) = R\, f_p\, n_e\, f_l\, f_i\, L$, where here L is the average lifetime of an intelligent civilization **c.** Take the logarithm of both sides, since this would yield the linear function $\ln(N) = \ln(R) + \ln(f_p) + \ln(n_e) + \ln(f_l) + \ln(f_i) + \ln(f_c) + \ln(L)$. **105.** They are reciprocals of each other. **107.** For example, $f(x, y) = x^2 + y^2$. **109.** For example, $f(x, y, z) = xyz$ **111.** For example, take $f(x, y) = x + y$. Then setting $y = 3$ gives $f(x, 3) = x + 3$. This can be viewed as a function of the single variable x. Choosing other values for y gives other functions of x. **113.** The slope is independent of the choice of $y = k$. **115.** That CDs cost more than cassettes **117.** Plane **119. (B)** Traveling in the direction B results in the shortest trip to nearby isotherms, and hence the fastest rate of increase in temperature. **121.** Agree: Any slice through a plane is a straight line. **123.** The graph of a function of three or more variables lives in four-dimensional (or higher) space, which makes it difficult to draw and visualize.

Section 8.2

1. $f_x(x, y) = -40$; $f_y(x, y) = 20$; $f_x(1, -1) = -40$; $f_y(1, -1) = 20$ **3.** $f_x(x, y) = 6x + 1$; $f_y(x, y) = -3y^2$; $f_x(1, -1) = 7$; $f_y(1, -1) = -3$ **5.** $f_x(x, y) = -40 + 10y$; $f_y(x, y) = 20 + 10x$; $f_x(1, -1) = -50$; $f_y(1, -1) = 30$

7. $f_x(x, y) = 6xy$; $f_y(x, y) = 3x^2$; $f_x(1, -1) = -6$; $f_y(1, -1) = 3$ **9.** $f_x(x, y) = 2xy^3 - 3x^2y^2 - y$; $f_y(x, y) = 3x^2y^2 - 2x^3y - x$; $f_x(1, -1) = -4$; $f_y(1, -1) = 4$ **11.** $f_x(x, y) = 6y(2xy + 1)^2$; $f_y(x, y) = 6x(2xy + 1)^2$; $f_x(1, -1) = -6$; $f_y(1, -1) = 6$ **13.** $f_x(x, y) = e^{x+y}$; $f_y(x, y) = e^{x+y}$; $f_x(1, -1) = 1$; $f_y(1, -1) = 1$ **15.** $f_x(x, y) = 3x^{-0.4}y^{0.4}$; $f_y(x, y) = 2x^{0.6}y^{-0.6}$; $f_x(1, -1)$ undefined; $f_y(1, -1)$ undefined **17.** $f_x(x, y) = 0.2ye^{0.2xy}$; $f_y(x, y) = 0.2xe^{0.2xy}$; $f_x(1, -1) = -0.2e^{-0.2}$; $f_y(1, -1) = 0.2e^{-0.2}$ **19.** $f_{xx}(x, y) = 0$; $f_{yy}(x, y) = 0$; $f_{xy}(x, y) = f_{yx}(x, y) = 0$; $f_{xx}(1, -1) = 0$; $f_{yy}(1, -1) = 0$; $f_{xy}(1, -1) = f_{yx}(1, -1) = 0$ **21.** $f_{xx}(x, y) = 0$; $f_{yy}(x, y) = 0$; $f_{xy}(x, y) = f_{yx}(x, y) = 10$; $f_{xx}(1, -1) = 0$; $f_{yy}(1, -1) = 0$; $f_{xy}(1, -1) = f_{yx}(1, -1) = 10$ **23.** $f_{xx}(x, y) = 6y$; $f_{yy}(x, y) = 0$; $f_{xy}(x, y) = f_{yx}(x, y) = 6x$; $f_{xx}(1, -1) = -6$; $f_{yy}(1, -1) = 0$; $f_{xy}(1, -1) = f_{yx}(1, -1) = 6$ **25.** $f_{xx}(x, y) = e^{x+y}$; $f_{yy}(x, y) = e^{x+y}$; $f_{xy}(x, y) = f_{yx}(x, y) = e^{x+y}$; $f_{xx}(1, -1) = 1$; $f_{yy}(1, -1) = 1$; $f_{xy}(1, -1) = f_{yx}(1, -1) = 1$ **27.** $f_{xx}(x, y) = -1.2x^{-1.4}y^{0.4}$; $f_{yy}(x, y) = -1.2x^{0.6}y^{-1.6}$; $f_{xy}(x, y) = f_{yx}(x, y) = 1.2x^{-0.4}y^{-0.6}$; $f_{xx}(1, -1)$ undefined; $f_{yy}(1, -1)$ undefined; $f_{xy}(1, -1)$ & $f_{yx}(1, -1)$ undefined **29.** $f_x(x, y, z) = yz$; $f_y(x, y, z) = xz$; $f_z(x, y, z) = xy$; $f_x(0, -1, 1) = -1$; $f_y(0, -1, 1) = 0$; $f_z(0, -1, 1) = 0$ **31.** $f_x(x, y, z) = 4/(x+y+z^2)^2$; $f_y(x, y, z) = 4/(x+y+z^2)^2$; $f_z(x, y, z) = 8z/(x + y + z^2)^2$; $f_x(0, -1, 1)$ undefined; $f_y(0, -1, 1)$ undefined; $f_z(0, -1, 1)$ undefined **33.** $f_x(x, y, z) = e^{yz} + yze^{xz}$; $f_y(x, y, z) = xze^{yz} + e^{xz}$; $f_z(x, y, z) = xy(e^{yz} + e^{xz})$; $f_x(0, -1, 1) = e^{-1} - 1$; $f_y(0, -1, 1) = 1$; $f_z(0, -1, 1) = 0$ **35.** $f_x(x, y, z) = 0.1x^{-0.9}y^{0.4}z^{0.5}$; $f_y(x, y, z) = 0.4x^{0.1}y^{-0.6}z^{0.5}$; $f_z(x, y, z) = 0.5x^{0.1}y^{0.4}z^{-0.5}$; $f_x(0, -1, 1)$ undefined; $f_y(0, -1, 1)$ undefined, $f_z(0, -1, 1)$ undefined **37.** $f_x(x, y, z) = yze^{xyz}$, $f_y(x, y, z) = xze^{xyz}$, $f_z(x, y, z) = xye^{xyz}$; $f_x(0, -1, 1) = -1$; $f_y(0, -1, 1) = f_z(0, -1, 1) = 0$ **39.** $f_x(x, y, z) = 0$; $f_y(x, y, z) = -\dfrac{600z}{y^{0.7}(1 + y^{0.3})^2}$; $f_z(x, y, z) = \dfrac{2{,}000}{1 + y^{0.3}}$; $f_x(0, -1, 1)$ undefined; $f_y(0, -1, 1)$ undefined; $f_z(0, -1, 1)$ undefined **41.** $\partial C/\partial x = 6{,}000$, the marginal cost to manufacture each car is \$6,000. $\partial C/\partial y = 4{,}000$, the marginal cost to manufacture each truck is \$4,000. **43.** $\partial y/\partial t = -0.78$. The number of articles written by researchers in the United States was decreasing at a rate of 0.78 percentage points per year. $\partial y/\partial x = -1.02$. The number of articles written by researchers in the United States was decreasing at a rate of 1.02 percentage points per one percentage point increase in articles written in Europe. **45.** \$5,600 per car **47. a.** $\partial M/\partial c = -3.8$, $\partial M/\partial f = 2.2$. For every 1 point increase in the percentage of Chrysler owners who remain loyal, the percentage of Mazda owners who remain loyal decreases by 3.8 points. For every 1 point increase in the percentage of Ford owners who remain loyal, the percentage of Mazda owners who remain loyal increases by 2.2 points. **b.** 16% **49. a.** \$36,600; \$52,900 **b.** \$270 per year; \$410 per year **c.** Widening **d.** The rate at which the income gap is widening **51.** The marginal cost of cars is $\$6{,}000 + 1{,}000e^{-0.01(x+y)}$ per car. The marginal cost of trucks is $\$4{,}000 + 1{,}000e^{-0.01(x+y)}$ per truck. Both marginal costs decrease as production rises.

53. $\bar{C}(x, y) = \dfrac{200{,}000 + 6{,}000x + 4{,}000y - 100{,}000e^{-0.01(x+y)}}{x + y}$; $\bar{C}_x(50, 50) = -\$2.64$ per car. This means that at a production level of 50 cars and 50 trucks per week, the average cost per vehicle is decreasing by \$2.64 for each additional car manufactured. $\bar{C}_y(50, 50) = -\$22.64$ per truck. This means that at a production level of 50 cars and 50 trucks per week, the average cost per vehicle is decreasing by \$22.64 for each additional truck manufactured. **55.** No; your marginal revenue from the sale of cars is $\$15{,}000 - \dfrac{2{,}500}{\sqrt{x + y}}$ per car and $\$10{,}000 - \dfrac{2{,}500}{\sqrt{x + y}}$ per truck from the sale of trucks. These increase with increasing x and y. In other words, you will earn more revenue per vehicle with increasing sales, and so the rental company will pay more for each additional vehicle it buys. **57.** $P_z(10, 100{,}000, 1{,}000{,}000) \approx 0.0001010$ papers/\$ **59. a.** $U_x(10, 5) = 5.18$, $U_y(10, 5) = 2.09$. This means that if 10 copies of Macro Publish and 5 copies of Turbo Publish are purchased, the company's daily productivity is increasing at a rate of 5.18 pages per day for each additional copy of Macro purchased and by 2.09 pages per day for each additional copy of Turbo purchased. **b.** $\dfrac{U_x(10, 5)}{U_y(10, 5)} \approx 2.48$ is the ratio of the usefulness of one additional copy of Macro to one of Turbo. Thus, with 10 copies of Macro and 5 copies of Turbo, the company can expect approximately 2.48 times the productivity per additional copy of Macro compared to Turbo. **61.** 6×10^9 N/sec **63. a.** $A_P(100, 0.1, 10) = 2.59$; $A_r(100, 0.1, 10) = 2{,}357.95$; $A_t(100, 0.1, 10) = 24.72$. Thus, for a \$100 investment at 10% interest, after 10 years the accumulated amount is increasing at a rate of \$2.59 per \$1 of principal, at a rate of \$2,357.95 per increase of 1 in r (note that this would correspond to an increase in the interest rate of 100%), and at a rate of \$24.72 per year. **b.** $A_P(100, 0.1, t)$ tells you the rate at which the accumulated amount in an account bearing 10% interest with a principal of \$100 is growing per \$1 increase in the principal, t years after the investment. **65. a.** $P_x = Ka\left(\dfrac{y}{x}\right)^b$ and $P_y = Kb\left(\dfrac{x}{y}\right)^a$. They are equal precisely when $\dfrac{a}{b} = \left(\dfrac{x}{y}\right)^b\left(\dfrac{x}{y}\right)^a$. Substituting $b = 1 - a$ now gives $\dfrac{a}{b} = \dfrac{x}{y}$. **b.** The given information implies that $P_x(100, 200) = P_y(100, 200)$. By part (a), this occurs precisely when $a/b = x/y = 100/200 = 1/2$. But $b = 1 - a$, so $a/(1 - a) = 1/2$, giving $a = 1/3$ and $b = 2/3$. **67.** Decreasing at 0.0075 parts of nutrient per part of water/sec **69.** f is increasing at a rate of s units per unit of x, f is increasing at a rate of t units per unit of y, and the value of

$\underline{f}$ is $\underline{r}$ when $x = \underline{a}$ and $y = \underline{b}$ **71.** The marginal cost of building an additional orbicus; zonars per unit. **73.** Answers will vary. One example is $f(x, y) = -2x + 3y$. Others are $f(x, y) = -2x + 3y + 9$ and $f(x, y) = xy - 3x + 2y + 10$. **75. a.** b is the z-intercept of the plane. m is the slope of the intersection of the plane with the xz-plane. n is the slope of the intersection of the plane with the yz-plane. **b.** Write $z = b + rx + sy$. We are told that $\partial z/\partial x = m$, so $r = m$. Similarly, $s = n$. Thus, $z = b + mx + ny$. We are also told that the plane passes through (h, k, l). Substituting gives $l = b + mh + nk$. This gives b as $l - mh - nk$. Substituting in the equation for z therefore gives $z = l - mh - nk + mx + ny = l + m(x - h) + n(y - k)$, as required.

Section 8.3

1. P: relative minimum; Q: none of the above; R: relative maximum **3.** P: saddle point; Q: relative maximum; R: none of the above **5.** Relative minimum **7.** Neither **9.** Saddle point **11.** Relative minimum at $(0, 0, 1)$ **13.** Relative maximum at $(-1/2, 1/2, 3/2)$ **15.** Saddle point at $(0, 0, 0)$ **17.** Minimum at $(4, -3, -10)$ **19.** Maximum at $(-7/4, 1/4, 19/8)$ **21.** Relative maximum at $(0, 0, 0)$, saddle points at $(\pm 4, 2, -16)$ **23.** Relative minimum at $(0, 0, 0)$, saddle points at $(-1, \pm 1, 1)$ **25.** Relative minimum at $(0, 0, 1)$ **27.** Relative minimum at $(-2, \pm 2, -16)$, $(0, 0)$ a critical point that is not a relative extremum **29.** Saddle point at $(0, 0, -1)$ **31.** Relative maximum at $(-1, 0, e)$ **33.** Relative minimum at $(2^{1/3}, 2^{1/3}, 3(2^{2/3}))$ **35.** Relative minimum at $(1, 1, 4)$ and $(-1, -1, 4)$ **37.** Absolute minimum at $(0, 0, 1)$ **39.** None; the relative maximum at $(0, 0, 0)$ is not absolute. (Look at, say, $(10, 10)$.) **41.** Minimum of $1/3$ at $(c, f) = (2/3, 2/3)$. Thus, at least $1/3$ of all Mazda owners would choose another new Mazda, and this lowest loyalty occurs when $2/3$ of Chrysler and Ford owners remain loyal to their brands. **43.** It should remove 2.5 pounds of sulfur and 1 pound of lead per day. **45.** You should charge \$580.81 for the Ultra Mini and \$808.08 for the Big Stack **47.** $l = w = h \approx 20.67$ in, volume $\approx 8{,}827$ cubic inches **49.** 18 in $\times$ 18 in $\times$ 36 in, volume $= 11{,}664$ cubic inches

51.

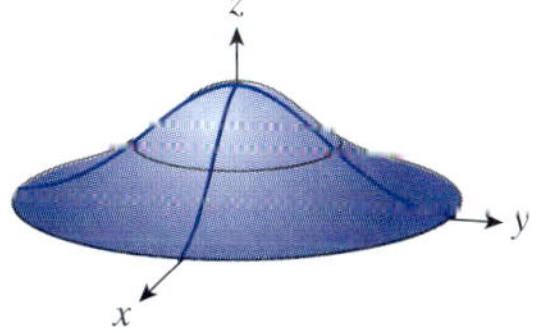

53.

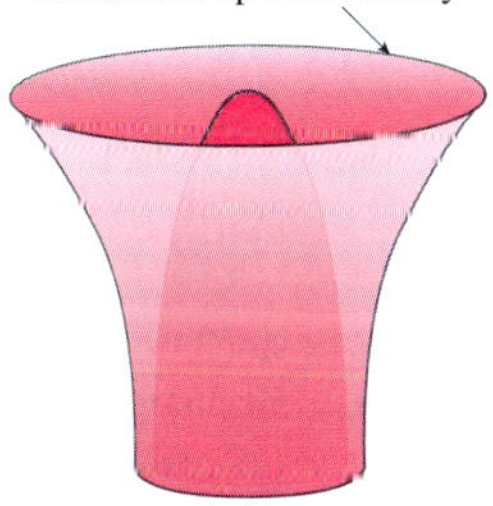

55. H must be positive. **57.** No. In order for there to be a relative maximum at (a, b), *all* vertical planes through (a, b) should yield a curve with a relative maximum at (a, b). It could happen that a slice by another vertical plane through (a, b) (such as $x - a = y - b$) does not yield a curve with a relative maximum at (a, b). [An example is $f(x, y) = x^2 + y^2 - \sqrt{xy}$, at the point $(0, 0)$. Look at the slices through $x = 0$, $y = 0$ and $y = x$.]

59. $\bar{C}_x = \dfrac{\partial}{\partial x}\left(\dfrac{C}{x + y}\right) = \dfrac{(x + y)C_x - C}{(x + y)^2}$. If this is zero, then $(x + y)C_x = C$, or $C_x = \dfrac{C}{x + y} = \bar{C}$. Similarly, if $\bar{C}_y = 0$ then $C_y = \bar{C}$. This is reasonable because if the average cost is decreasing with increasing x, then the average cost is greater than the marginal cost C_x. Similarly, if the average cost is increasing with increasing x, then the average cost is less than the marginal cost C_x. Thus, if the average cost is stationary with increasing x, then the average cost equals the marginal cost C_x. (The situation is similar for the case of increasing y.)

61. The equation of the tangent plane at the point (a, b) is $z = f(a, b) + f_x(a, b)(x - a) + f_y(a, b)(y - b)$. If f has a relative extremum at (a, b), then $f_x(a, b) = 0 = f_y(a, b)$. Substituting these into the equation of the tangent plane gives $z = f(a, b)$, a constant. But the graph of $z = constant$ is a plane parallel to the xy-plane.

Section 8.4

1. 1; $(0, 0, 0)$ **3.** 1.35; $(1/10, 3/10, 1/2)$ **5.** Minimum value $= 6$ at $(1, 2, 1/2)$ **7.** 200; $(20, 10)$ **9.** 16; $(2, 2)$ and $(-2, -2)$ **11.** 20; $(2, 4)$ **13.** 1; $(0, 0, 0)$ **15.** 1.35; $(1/10, 3/10, 1/2)$ **17.** Minimum value $= 6$ at $(1, 2, 1/2)$ **19.** $5 \times 10 - 50$ sq. ft. **21.** \$10 **23.** $(1/\sqrt{3}, 1/\sqrt{3}, 1/\sqrt{3})$, $(-1/\sqrt{3}, -1/\sqrt{3}, 1/\sqrt{3})$, $(1/\sqrt{3}, -1/\sqrt{3}, -1/\sqrt{3})$, $(-1/\sqrt{3}, 1/\sqrt{3}, -1/\sqrt{3})$ **25.** $(0, 1/2, -1/2)$ **27.** $(-5/9, 5/9, 25/9)$ **29.** $l \times w \times h = 1 \times 1 \times 2$ **31.** 18 in $\times$ 18 in $\times$ 36 in, volume $= 11{,}664$ cubic inches **33.** $(2l/h)^{1/3} \times (2l/h)^{1/3} \times 2^{1/3}(h/l)^{2/3}$, where $l =$ cost of lightweight cardboard and $h =$ cost of heavy-duty cardboard per square foot **35.** $1 \times 1 \times 1/2$ **37.** 6 laborers, 10 robots for a productivity of 368 pairs of socks per day

39. Method 1: Solve $g(x, y, z) = 0$ for one of the variables and substitute in $f(x, y, z)$. Then find the maximum value of the resulting function of 2 variables. Advantage (answers may vary): We can use the second derivative test to check whether the resulting critical points are maxima, minima, saddle points, or none of these. Disadvantage (answers may vary): We may not be able to solve $g(x, y, z) = 0$ for one of the variables. Method 2: Use the method of Lagrange multipliers. Advantage (answers may vary): We do not need to solve the constraint equation for one of the variables. Disadvantage (answers may vary): The method does not tell us whether the critical points obtained are maxima, minima, points of inflection, or none of these.

41. If the only constraint is an equality constraint, and if it is impossible to eliminate one of the variables in the objective function by substitution (solving the constraint equation for a variable or some other method). **43.** Answers may vary: Maximize $f(x, y) = 1 - x^2 - y^2$ subject to $x = y$.

45. Yes. There may be relative extrema at points on the boundary of the domain of the function. The partial derivatives of the function need not be 0 at such points. **47.** If the solution were located in the interior of one of the line segments making up the boundary of the domain of f, then the derivative of a certain function would be 0. This function is obtained by substituting the linear equation $C(x, y) = 0$ in the linear objective function. But since the result would again be a linear function, it is either constant, or its derivative is a non-zero constant. In either event, extrema lie on the boundary of that line segment; that is, at one of the corners of the domain.

Section 8.5

1. $-1/2$ **3.** $e^2/2 - 7/2$ **5.** $(e^3 - 1)(e^2 - 1)$ **7.** $7/6$ **9.** $[e^3 - e - e^{-1} + e^{-3}]/2$ **11.** $1/2$ **13.** $(e - 1)/2$ **15.** $45/2$ **17.** $8/3$ **19.** $4/3$ **21.** 0 **23.** $2/3$ **25.** $2/3$ **27.** $2(e - 2)$ **29.** $2/3$

31. $\int_0^1 \int_0^{1-x} f(x, y)\,dy\,dx$

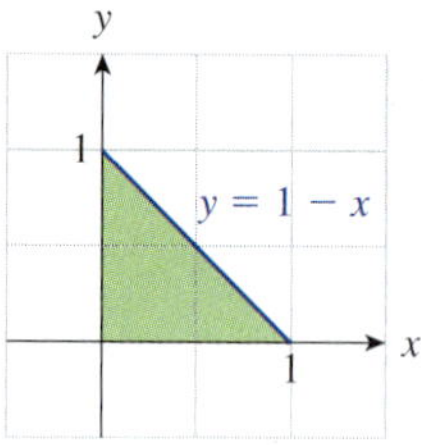

33. $\int_0^1 \int_0^{1-x} f(x, y)\,dy\,dx$

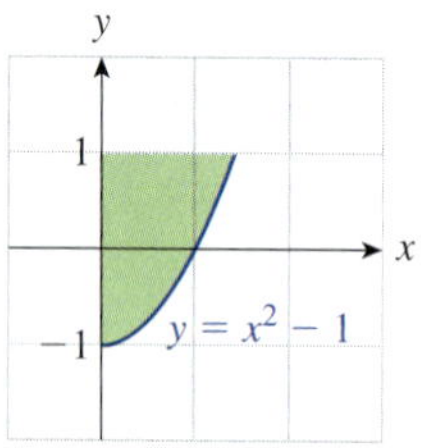

35. $\int_1^4 \int_1^{2/\sqrt{y}} f(x, y)\,dx\,dy$

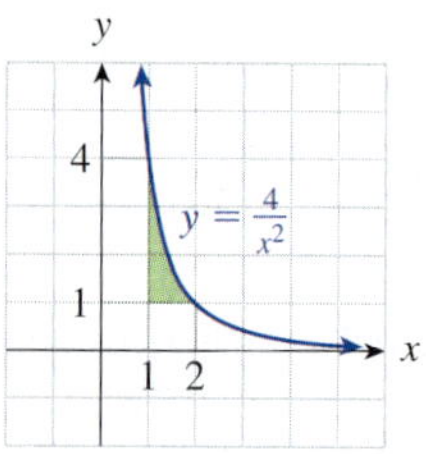

37. $4/3$ **39.** $1/6$ **41.** 162,000 gadgets **43.** \$312,750 **45.** \$17,500 **47.** 8,216 **49.** 1 degree **51.** The area between the curves $y = r(x)$ and $y = s(x)$ and the vertical lines $x = a$ and $x = b$ is given by $\int_a^b \int_{r(x)}^{s(x)} dy\,dx$ assuming that $r(x) \le s(x)$ for $a \le x \le b$. **53.** The first step in calculating an integral of the form $\int_a^b \int_{r(x)}^{s(x)} f(x, y)\,dy\,dx$ is to evaluate the integral $\int_{r(x)}^{s(x)} f(x, y)\,dy$, obtained by holding x constant and integrating with respect to y. **55.** Paintings per picasso per dali **57.** Left-hand side is $\int_a^b \int_c^d f(x)\,g(y)\,dx\,dy = \int_a^b \left(g(y) \int_c^d f(x)\,dx\right) dy$ (because $g(y)$ is treated as a constant in the inner integral) $= \left(\int_c^d f(x)\,dx\right) \int_a^b g(y)\,dy$ (because $\int_c^d f(x)\,dx$ is a constant and can therefore be taken outside the integral).

$$\int_0^1 \int_1^2 ye^x\,dx\,dy = \frac{1}{2}(e^2 - e)$$ no matter how we compute it.

Chapter 8 Review

1. 0; 14/3; 1/2; $\frac{1}{1+z} + z^3$; $\frac{x+h}{y+k+(x+h)(z+l)} + (x+h)^2(y+k)$ **3.** Decreases by 0.32 units; increases by 12.5 units **5.** Reading left to right, starting at the top: 4, 0, 0, 3, 0, 1, 2, 0, 2 **7.** Answers may vary; two examples are $f(x, y) = 3(x - y)/2$ and $f(x, y) = 3(x - y)^3/8$. **9.** $f_x = 2x + y$, $f_y = x$, $f_{yy} = 0$ **11.** 0 **13.** $\frac{\partial f}{\partial x} = \frac{-x^2 + y^2 + z^2}{(x^2 + y^2 + z^2)^2}$, $\frac{\partial f}{\partial y} = \frac{2xy}{(x^2 + y^2 + z^2)^2}$, $\frac{\partial f}{\partial z} = -\frac{2xz}{(x^2 + y^2 + z^2)^2}$, $\left.\frac{\partial f}{\partial x}\right|_{(0,1,0)} = 1$ **15.** Absolute minimum at $(1, 3/2)$ **17.** Saddle point at $(0, 0)$ **19.** Absolute maximum at each point on the circle $x^2 + y^2 = 1$ **21.** $1/27$ at $(1/3, 1/3, 1/3)$ **23.** $(0, 2, \sqrt{2})$ **25.** 4; $(\sqrt{2}, \sqrt{2})$ and $(-\sqrt{2}, -\sqrt{2})$ **27.** Minimum value $= 5$ at $(2, 1, 1/2)$ **29.** 2 **31.** ln 5 **33.** 1 **35. a.** $h(x, y) = 5{,}000 - 0.8x - 0.6y$ hits per day (x = number of new customers at JungleBooks.com, y = number of new customers at FarmerBooks.com) **b.** 250 **c.** $h(x, y, z) = 5{,}000 - 0.8x - 0.6y + 0.0001z$ (z = number of new Internet shoppers) **d.** 1.4 million **37. a.** 2,320 hits per day **b.** $0.08 + 0.00003x$ hits (daily) per dollar spent on television advertising per month; increases with increasing x **c.** \$4,000 per month **39.** (A) **41. a.** About 15,800 additional orders per day **b.** 11 **43.** \$23,050

Chapter 9

Section 9.1

1.

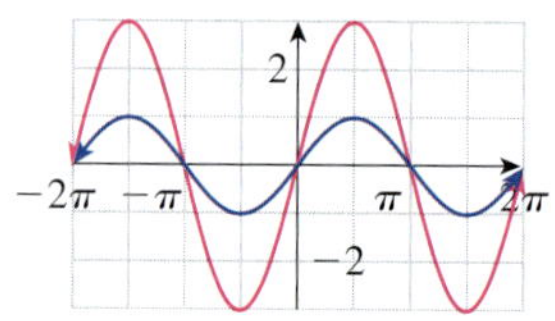

3.

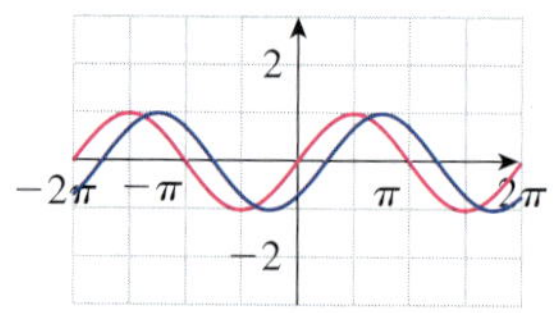

5.

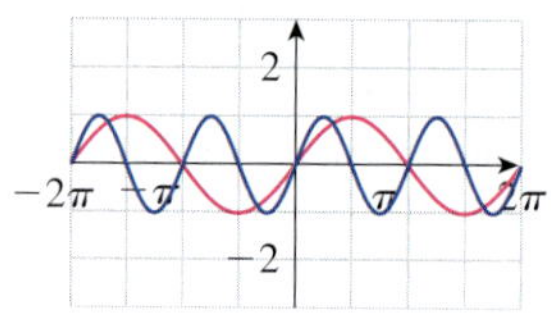

7.

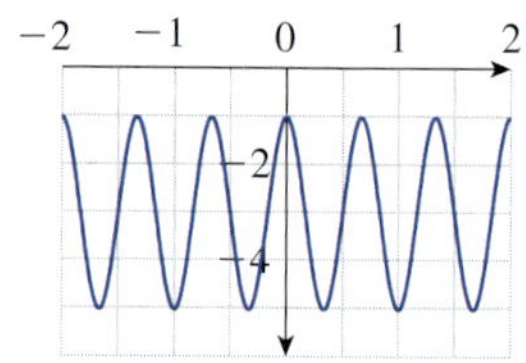

9.

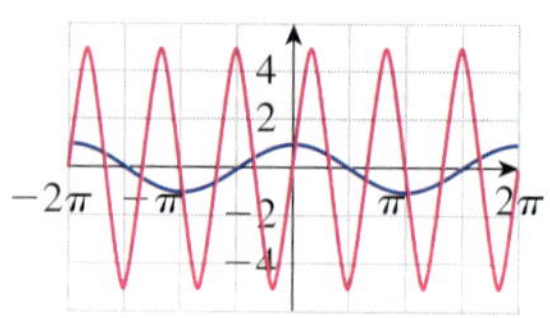

11.

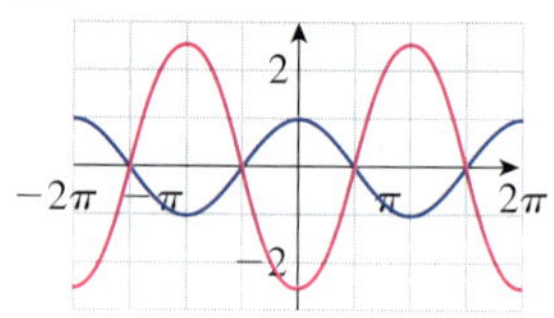

13. $f(x) = \sin(2\pi x) + 1$ **15.** $f(x) = 1.5\sin(4\pi(x - 0.25))$ **17.** $f(x) = 50\sin(\pi(x - 5)/10) - 50$ **19.** $f(x) = \cos(2\pi x)$ **21.** $f(x) = 1.5\cos(4\pi(x - 0.375))$ **23.** $f(x) = 40\cos(\pi(x - 10)/10) + 40$ **25.** $f(t) = 4.2\sin(\pi/2 - 2\pi t) + 3$ **27.** $g(x) = 4 - 1.3\sin[\pi/2 - 2.3(x - 4)]$ **31.** $\sqrt{3}/2$ **37.** $\tan(x + \pi) = \tan(x)$ **39. a.** $2\pi/0.602 \approx 10.4$ years. **b.** Maximum: $58.8 + 57.7 = 116.5 \approx 117$; minimum: $58.8 - 57.7 = 1.1 \approx 1$ **c.** $1.43 + P/4 + P = 1.43 + 13.05 \approx 14.5$ years, or midway through 2011

41. a. Maximum sales occurred when $t \approx 4.5$ (during the first quarter of 1996). Minimum sales occurred when $t \approx 2.2$ (during the third quarter of 1995) and $t \approx 6.8$ (during the third quarter of 1996). **b.** Maximum quarterly revenues were \$0.561 billion; minimum quarterly revenues were \$0.349 billion. **c.** Maximum: $0.455 + 0.106 = 0.561$; minimum: $0.455 - 0.106 = 0.349$ **43.** Amplitude $= 0.106$, vertical offset $= 0.455$, phase shift $= -1.61/1.39 \approx -1.16$, angular frequency $= 1.39$, period $= 4.52$. In 1995 and 1996, quarterly revenue from the sale of computers at **Computer City** fluctuated in cycles of 4.52 quarters about a baseline of \$0.455 billion. Every cycle, quarterly revenue peaked at \$0.561 billion (\$0.106 billion above the baseline) and dipped to a low of \$0.349 billion. Revenue peaked early in the middle of the first quarter of 1996 (at $t = -1.16 + (5/4) \times 4.52 = 4.49$).
45. $P(t) = 7.5\sin[\pi(t - 13)/26] + 12.5$
47. $T(t) = 3.5\cos[\pi(t - 7)/6] + 78.5$
49. $n(t) = 1.25\sin[\pi(t - 1)/2] + 5.75$
51. $n(t) = 1.25\cos[\pi(t - 2)/2] + 5.75$
53. $d(t) = 5\sin(2\pi(t - 1.625)/13.5) + 10$
55. a. $u(t) = 2.5\sin(2\pi(t - 0.75)) + 7.5$
b. $c(t) = 1.04^t[2.5\sin(2\pi(t - 0.75)) + 7.5]$
57. a. $C \approx 50$, $A \approx 8$, $P \approx 12$, $\beta \approx 6$

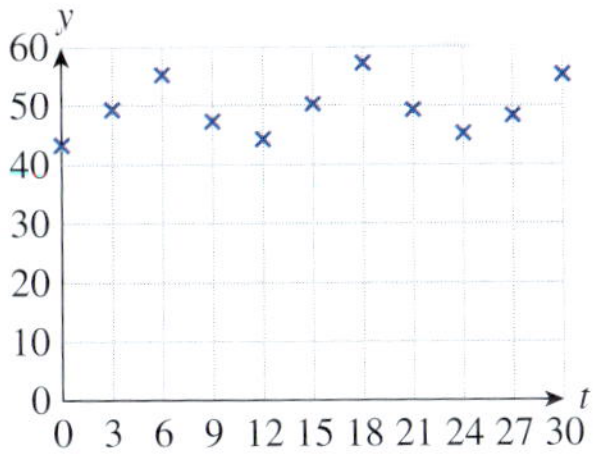

b. $f(t) = 5.882\cos[2\pi(t - 5.696)/12.263] + 49.238$

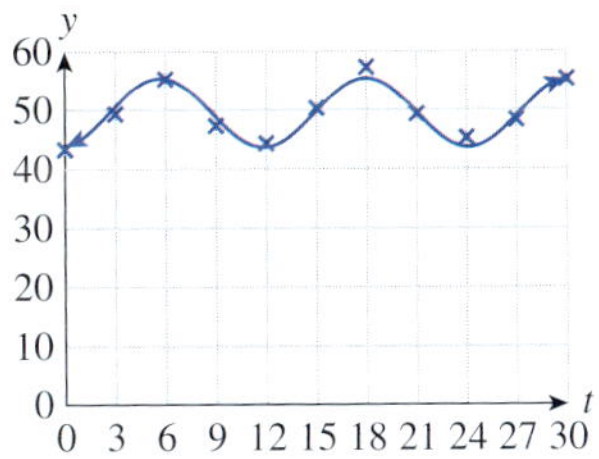

c. 12, 43; 55

59. a

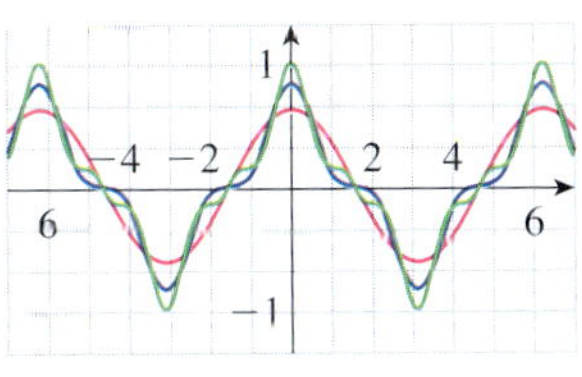

b. $y_{11} = \dfrac{2}{\pi}\cos x + \dfrac{2}{3\pi}\cos 3x + \dfrac{2}{5\pi}\cos 5x + \dfrac{2}{7\pi}\cos 7x + \dfrac{2}{9\pi}\cos 9x + \dfrac{2}{11\pi}\cos 11x$

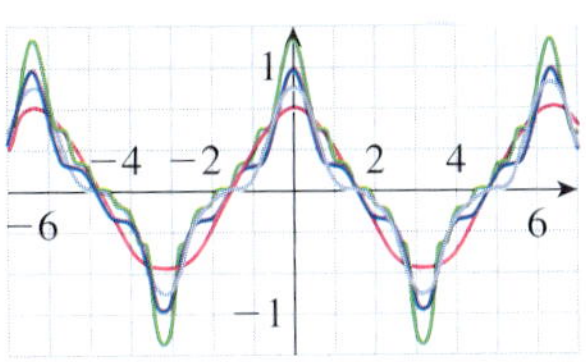

61. The period is approximately 12.6 units

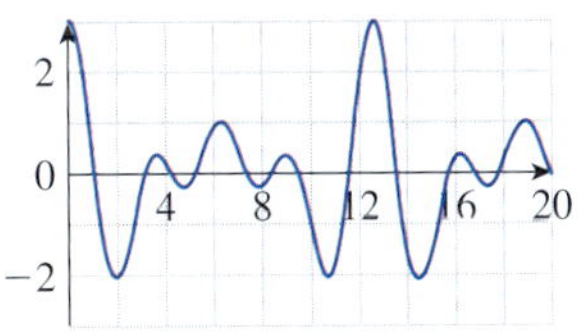

63. Lows: $B - A$; highs: $B + A$. **65.** He is correct. The other trig functions can be obtained from the sine function by first using the formula $\cos x = \sin(x + \pi/2)$ to obtain cosine, and then using the formulas $\tan x = \dfrac{\sin x}{\cos x}$, $\cot x = \dfrac{\cos x}{\sin x}$, $\sec x = \dfrac{1}{\cos x}$, $\csc x = \dfrac{1}{\sin x}$ to obtain the rest **67.** The largest B can be is A. Otherwise, if B is larger than A, the low figure for sales would have the negative value of $A - B$.

Section 9.2

1. $\cos x + \sin x$ **3.** $(\cos x)(\tan x) + (\sin x)(\sec^2 x)$
5. $-2\csc x\cot x - \sec x\tan x + 3$ **7.** $\cos x - x\sin x + 2x$
9. $(2x - 1)\tan x + (x^2 - x + 1)\sec^2 x$
11. $-[\csc^2 x(1 + \sec x) + \cot x\sec x\tan x]/(1 + \sec x)^2$
13. $-2\cos x\sin x$ **15.** $2\sec^2 x\tan x$ **17.** $3\cos(3x - 5)$
19. $2\sin(-2x + 5)$ **21.** $\pi\cos\left[\dfrac{\pi}{5}(x - 4)\right]$
23. $-(2x - 1)\sin(x^2 - x)$
25. $(2.2x^{1.2} + 1.2)\sec(x^{2.2} + 1.2x - 1)\tan(x^{2.2} + 1.2x - 1)$
27. $\sec x\tan x\tan(x^2 - 1) + 2x\sec x\sec^2(x^2 - 1)$
29. $e^x[-\sin(e^x) + \cos x - \sin x]$ **31.** $\sec x$
37. $e^{-2x}[-2\sin(3\pi x) + 3\pi\cos(3\pi x)]$
39. $1.5[\sin(3x)]^{-0.5}\cos(3x)$
41. $\dfrac{x^4 - 3x^2}{(x^2 - 1)^2}\sec\left(\dfrac{x^3}{x^2 - 1}\right)\tan\left(\dfrac{x^3}{x^2 - 1}\right)$
43. $\dfrac{\cot(2x - 1)}{x} - 2\ln|x|\csc^2(2x - 1)$
45. a. Not differentiable at 0 **b.** $f'(1) \approx 0.5403$ **47.** 0
49. 2 **51.** Does not exist **53.** $1/\sec^2 y$
55. $-[1 + y\cos(xy)]/[1 + x\cos(xy)]$
57. $c'(t) = 7\pi\cos[2\pi(t - 0.75)]$; $c'(0.75) \approx \$21.99$ per year $\approx \$0.42$ per week **59.** $N'(6) \approx -32.12$ On January 1, 2003, the number of sunspots was decreasing at a rate of 32.12 sunspots per year. **61.** $c'(t) = 1.035^t \times [\ln(1.035)(0.8\sin(2\pi t) + 10.2) + 1.6\pi\cos(2\pi t)]$; $c'(1) = 1.035[10.2\ln|1.035| + 1.6\pi] \approx \$5.57$ per year, or \$0.11 per week. **63. a.** $d(t) = 5\cos(2\pi t/13.5) + 10$
b. $d'(t) = -(10\pi/13.5)\sin(2\pi t/13.5)$; $d'(7) \approx 0.270$. At noon, the tide was rising at a rate of 0.270 feet per hour.
65. $\dfrac{dV}{dt} = 11{,}000\pi\dfrac{|\sin(100\pi t)|}{\sin(100\pi t)}\cos(100\pi t)$

ANSWERS TO SELECTED EXERCISES

Graph:

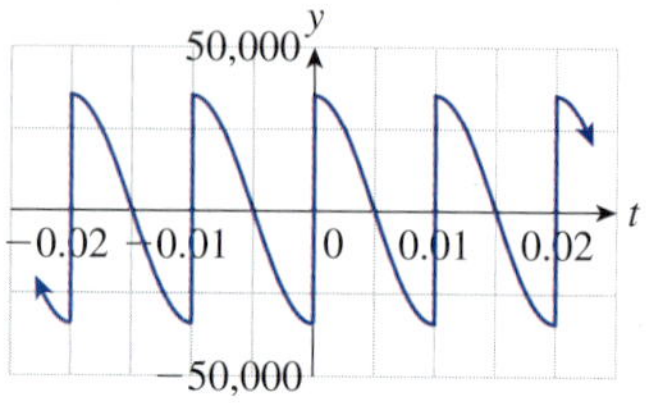

The sudden jumps in the graph are due to the non-differentiability of V at the times 0, ± 0.01, ± 0.02, ... The derivative is negative immediately to the left and positive immediately to the right of these points. **67. a.** 1.2 cm above the rest position **b.** 0 cm/sec; not moving; moving downward at 18.85 cm/sec **c.** 2.5 cycles per second **69. a.** Moving downward at 0.12 cm/sec; moving downward at 18.66 cm/sec **b.** 0.1 sec

Graphs: of p:

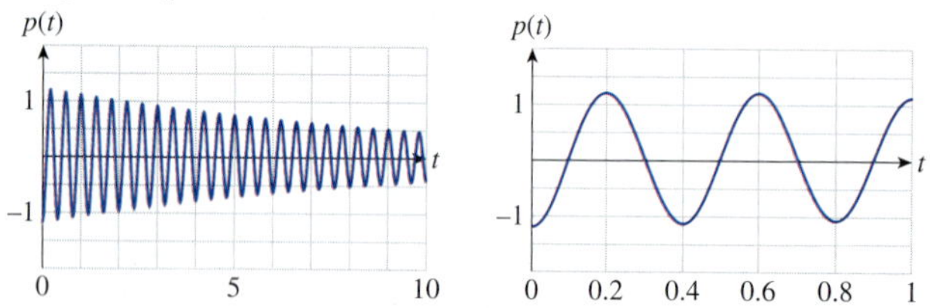

Graphs of p':

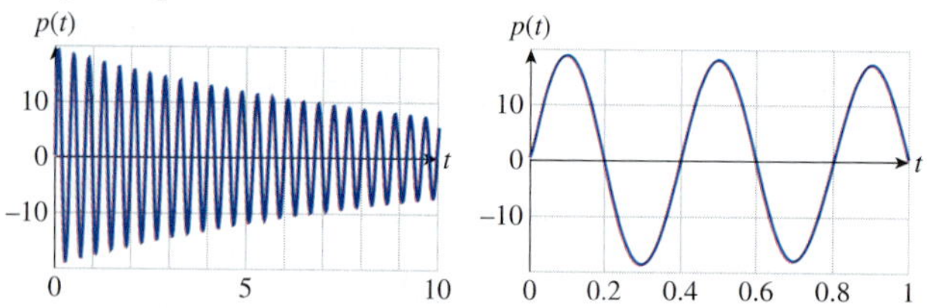

71. a. (III) **b.** Increasing at a rate of 0.157 degrees per thousand years **73.** −6; 6 **75.** Answers will vary. Examples: $f(x) = \sin x$; $f(x) = \cos x$ **77.** Answers will vary. Examples: $f(x) = e^{-x}$; $f(x) = -2e^{-x}$ **79.** The graph of $\cos x$ slopes down over the interval $(0, \pi)$, so that its derivative is negative over that interval. The function $-\sin x$, and not $\sin x$, has this property. **81.** The velocity is $p'(t) = A\omega\cos(\omega t + d)$, which is a maximum when its derivative, $p''(t) = -A\omega^2\sin(\omega t + d)$, is zero. But this occurs when $\sin(\omega t + d) = 0$, so that $p(t)$ is zero as well, meaning that the stock is at yesterday's close. **83.** The derivative of $\sin x$ is $\cos x$. When $x = 0$, this is $\cos(0) = 1$. Thus, the tangent to the graph of $\sin x$ at the point (0, 0) has slope 1, which means it slopes upward at 45°.

Section 9.3

1. $-\cos x - 2\sin x + C$ **3.** $2\sin x + 4.3\cos x - 9.33x + C$
5. $3.4\tan x + (\sin x)/1.3 - 3.2e^x + C$
7. $(7.6/3)\sin(3x - 4) + C$ **9.** $-(1/6)\cos(3x^2 - 4) + C$
11. $-2\cos(x^2 + x) + C$ **13.** $(1/6)\tan(3x^2 + 2x^3) + C$
15. $-(1/6)\ln|\cos(2x^3)| + C$
17. $3\ln|\sec(2x - 4) + \tan(2x - 4)| + C$
19. $(1/2)\sin(e^{2x} + 1) + C$ **21.** −2 **23.** ln(2) **25.** 0 **27.** 1
33. $-\frac{1}{4}\cos(4x) + C$ **35.** $-\sin(-x + 1) + C$
37. $[\cos(-1.1x - 1)]/1.1 + C$ **39.** $-\frac{1}{4}\ln|\sin(-4x)| + C$
41. 0 **43.** 2π **45.** $-x\cos x + \sin x + C$
47. $\left[\frac{x^2}{2} - \frac{1}{4}\right]\sin(2x) + \frac{x}{2}\cos(2x) + C$
49. $-\frac{1}{2}e^{-x}\cos x - \frac{1}{2}e^{-x}\sin x + C$ **51.** $\pi^2 - 4$
53. Average $= 2/\pi$

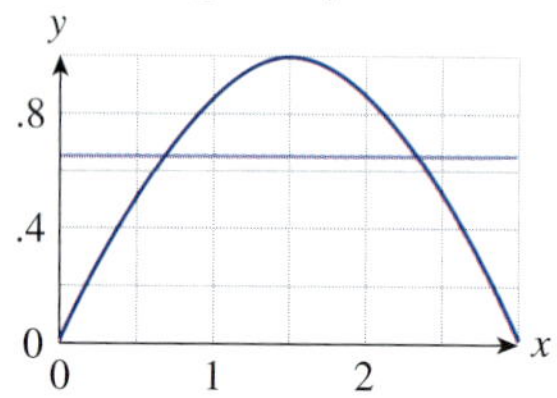

55. Diverges **57.** Converges to 1/2
59. $C(t) = 0.04t + \frac{2.6}{\pi}\cos\left[\frac{\pi}{26}(t - 25)\right] + 1.02$
61. 12 feet **63.** 79 sunspots
65. $P(t) = 7.5\sin[(\pi/26(t - 13)] + 12.5$; 7.7%
67. a. Average voltage over [0, 1/6] is zero; 60 cycles per second.
b.

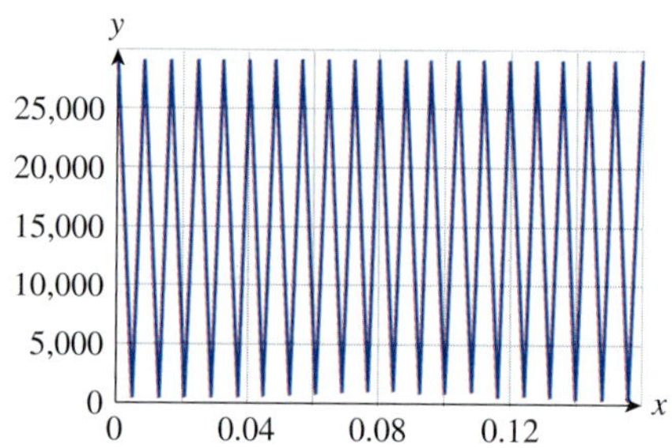

c. 116.673 volts **69.** $50,000 **71.** It is always zero. **73.** 1
75. $s = -\frac{K}{\omega^2}\sin(\omega t - \alpha) + Lt + M$ for constants L and M

Section 9 Review

1. $f(x) = 1 + 2\sin x$
3. $f(x) = 2 + 2\sin[\pi(x - 1)] = 2 + 2\sin[\pi(x + 1)]$
5. $f(x) = 1 + 2\cos(x - \pi/2)$
7. $f(x) = 2 + 2\cos[\pi(x + 1/2)] = 2 + 2\cos[\pi(x - 3/2)]$
9. $-2x\sin(x^2 - 1)$ **11.** $2e^x\sec^2(2e^x - 1)$
13. $4x\sin(x^2)\cos(x^2)$ **15.** $2\sin(2x - 1) + C$
17. $\tan(2x^2 - 1) + C$ **19.** $-\frac{1}{2}\ln|(\cos(x^2 + 1)| + C$
21. 1 **23.** $-x^2\cos x + 2x\sin x + 2\cos x + C$
25. $s(t) = 10{,}500 + 1{,}500\sin[(2\pi/52)t - \pi] = 10{,}500 + 1{,}500\sin(0.12083t - 3.14159)$ **27.** Decreasing at a rate of $3,852 per month **29.** $2,029,700
31. $150t - \frac{100}{\pi}\cos\left[\frac{\pi}{2}(t - 1)\right]$ grams

Index

INDEX

INDEX

INDEX